T0206107

The Standard Model and Beyond

SECOND EDITION

Series in High Energy Physics, Cosmology, and Gravitation

This series of books covers all aspects of theoretical and experimental high energy physics, cosmology and gravitation and the interface between them. In recent years the fields of particle physics and astrophysics have become increasingly interdependent and the aim of this series is to provide a library of books to meet the needs of students and researchers in these fields.

Other recent books in the series:

The Standard Model and Beyond, Second Edition
Paul Langacker

An Introduction to Beam Physics
Martin Berz, Kyoko Makino, and Weishi Wan

Neutrino Physics, Second Edition
K Zuber

Group Theory for the Standard Model of Particle Physics and Beyond
Ken J Barnes

The Standard Model and Beyond
Paul Langacker

Particle and Astroparticle Physics
Utpal Sakar

Joint Evolution of Black Holes and Galaxies
M Colpi, V Gorini, F Haardt, and U Moschella (Eds)

Gravitation: From the Hubble Length to the Planck Length
I Ciufolini, E Coccia, V Gorini, R Peron, and N Vittorio (Eds)

The Galactic Black Hole: Lectures on General Relativity and Astrophysics
H Falcke, and F Hehl (Eds)

The Mathematical Theory of Cosmic Strings: Cosmic Strings in the Wire Approximation
M R Anderson

Geometry and Physics of Branes
U Bruzzo, V Gorini, and U Moschella (Eds)

Modern Cosmology
S Bonometto, V Gorini, and U Moschella (Eds)

Gravitation and Gauge Symmetries
M Blagojevic

Gravitational Waves
I Ciufolini, V Gorini, U Moschella, and P Fré (Eds)

Series in High Energy Physics, Cosmology, and Gravitation

The Standard Model and Beyond

SECOND EDITION

Paul Langacker

Institute for Advanced Study
Princeton, New Jersey, USA

CRC Press
Taylor & Francis Group
6000 Broken Sound Parkway NW, Suite 300
Boca Raton, FL 33487-2742

First issued in paperback 2020

ISBN 13: 978-0-367-57344-7 (pbk)
ISBN 13: 978-1-4987-6321-9 (hbk)

DOI: 10.1201/b22175

Library of Congress Cataloging-in-Publication Data

Names: Langacker, P., author.
Title: The standard model and beyond / Paul Langacker.
Other titles: Series in high energy physics, cosmology, and gravitation.
Description: Second edition. | Boca Raton, FL : CRC Press, Taylor & Francis Group, [2017] | Series: Series in high energy physics, cosmology and gravitation | Includes bibliographical references and index.
Identifiers: LCCN 2016057676| ISBN 9781498763219 (hardback ; alk. paper) | ISBN 1498763219 (hardback ; alk. paper) | ISBN 9781315170626 (e-book) | ISBN 1315170620 (e-book)
Subjects: LCSH: Standard model (Nuclear physics)
Classification: LCC QC794.6.S75 L36 2017 | DDC 539.7/2--dc23
LC record available at https://lccn.loc.gov/2016057676

Visit the Taylor & Francis Web site at
http://www.taylorandfrancis.com

and the CRC Press Web site at
http://www.crcpress.com

Contents

Preface

PREFACE TO SECOND EDITION

Much has (or has *not*) happened in the seven years or so since the publication of the first edition of this book. Most notably, a new spin-0 particle with mass ~125 GeV was discovered at the LHC by the ATLAS and CMS collaborations, which is either the elementary Higgs boson or something that closely resembles it. The Higgs discovery completes the $\gtrsim$ 40 year saga of verifying the standard model (SM). Moreover, its mass is almost maximally interesting: it is near the top of the range predicted by the most popular SM extension, minimal supersymmetry, and near the minimum value consistent with the unextended SM (and then only if the vacuum is metastable)!

However, the notorious problems of the SM are still unresolved. Perhaps the most pressing is the apparently fine-tuned hierarchy between the weak interaction and gravity scales. Extensive searches at the LHC and elsewhere have so far not yielded any compelling evidence for new TeV-scale physics such as supersymmetry, extra space dimensions, or strong coupling that had been proposed to explain or at least stabilize the hierarchy. Equally puzzling is the nature of the dark energy and its incredibly tiny magnitude compared to most theoretical expectations. Similarly, numerous experimental attempts to identify the mysterious dark matter inferred from its gravitational effects have not as yet had any positive results and have excluded much of the parameter space for supersymmetric dark matter. And despite the great experimental success of the SM, it is a very complicated theory, involving several interactions with different properties, and two apparently superfluous heavier copies of the fundamental particles that constitute ordinary matter under ordinary conditions. For these and other reasons, many theorists have started exploring less canonical possibilities, such as a dark matter sector that is at most very weakly coupled to ordinary particles, or, more radically, that the Universe is part of a vast *multiverse* of regions (presumably associated with a superstring landscape of vacua) with different laws of physics.

The existing experimental programs in high energy physics will continue for many years. These include high luminosity running at the LHC; active programs around the world in neutrino, flavor, and dark matter physics; and observational probes of the dark energy. There are also proposed next generation facilities such as new e^+e^- colliders that can serve as Higgs factories, and $\mathcal{O}(100 \text{ TeV})$ hadron colliders. We will most likely find evidence for any multi-TeV scale physics relevant to the hierarchy problem or that is "just there" as a remnant of a more basic underlying theory; hopefully identify the dark matter and energy and shed light on the origin of the baryon asymmetry; perhaps progress toward a fundamental grand unification, superstring, or other theory that no one has yet imagined; and even reconsider such paradigms as naturalness, uniqueness, and minimality.[1]

Like the first edition, this volume is intended to serve as a detailed text and reference on the formalism, technology, phenomenology, and experimental verification of the standard model and its possible extensions. In addition to updating all of the experimental and phenomenological results, it contains expanded discussions of collider, Higgs, neutrino, and dark matter physics, and includes many new problems. The book website at

[1] For more extensive speculations along these lines, see, e.g., (Langacker, 2017).

www.sas.upenn.edu/~pgl/SMB2/ includes various supplemental materials, suggestions for use in a one-semester course, and corrections.

I would like to thank Vernon Barger and Jonathan Heckman for critiquing parts of this new edition, all those who have commented on the first one, and Irmgard for her extreme patience during the preparation of this new version.

Paul Langacker
December 8, 2016

PREFACE TO FIRST EDITION

In the last few decades there has been a tremendous advance in our understanding of the elementary particles and their interactions. We now have a mathematically consistent theory of the strong, electromagnetic, and weak interactions—the standard model—most aspects of which have been successfully tested in detail at colliders, accelerators, and non-accelerator experiments. It also provides a successful framework and has been strongly constrained by many observations in cosmology and astrophysics. The standard model is almost certainly an approximately correct description of Nature down to a distance scale $1/1000th$ the size of the atomic nucleus.

However, nobody believes that the standard model is the ultimate theory: it is too complicated and arbitrary, does not provide an understanding of the patterns of fermion masses and mixings, does not incorporate quantum gravity, and it involves several severe fine-tunings. Furthermore, the origins of electroweak symmetry breaking, whether by the Higgs mechanism or something else, are uncertain. The recent discovery of non-zero neutrino mass can be incorporated, but in more than one way, with different implications for physics at very short distance scales. Finally, the observations of dark matter and energy suggest new particle physics beyond the standard model.

Most current activity is directed toward discovering the new physics which must underlie the standard model. Much of the theoretical effort involves constructing models of possible new physics at the TeV scale, such as supersymmetry or alternative models of spontaneous symmetry breaking. Others are examining the extremely promising ideas of superstring theory, which offer the hope of an ultimate unification of all interactions including gravity. There is a lively debate about the implications of a *landscape* of possible string vacua, and serious efforts are being made to explore the consequences of string theory for the TeV scale. It is likely that a combination of such bottom-up and top-down ideas will be necessary for progress. In any case, new experimental data are urgently needed. At the time of this writing, the particle physics community is eagerly awaiting the results of the Large Hadron Collider (LHC) and is optimistic about a possible future International Linear Collider. Future experiments to elucidate the properties of neutrinos and to explore aspects of flavor, and more detailed probes of the dark energy and dark matter, are also anticipated.

The purpose of this volume is to provide an advanced introduction to the physics and formalism of the standard model and other non-abelian gauge theories, and thus to provide a thorough background for topics such as supersymmetry, string theory, extra dimensions, dynamical symmetry breaking, and cosmology. It is intended to provide the tools for a researcher to understand the structure and phenomenological consequences of the standard model, construct extensions, and to carry out calculations at tree level. Some "old-fashioned" topics which may still be useful are included. This is not a text on field theory, and does not substitute for the excellent texts that already exist. Ideally, the reader will have completed a standard field theory course. Nevertheless, Chapter 2 of this book presents a largely self-contained treatment of the complicated technology needed for tree-level calculations involving spin-0, spin-$\frac{1}{2}$, and spin-1 particles, and should be useful for those who have not studied field theory recently, or whose exposure has been more formal than calculational.[2] It does *not* attempt to deal systematically with the subtleties of renormalization, gauge issues, or higher-order corrections. An introductory-level background in the ideas of particle physics is assumed, with occasional reference to topics such as gluons or supersymmetry before they are formally introduced. Similarly, occasional reference is made to applications to and constraints from astrophysics and cosmology. The necessary background material may be found in the sources listed in the bibliography.

Chapter 1 is a short summary of notations and conventions and of some basic mathematical machinery. Chapter 2 contains a review of calculational techniques in field theory and the status of quantum electrodynamics. Chapters 3 and 4 are concerned with global and local symmetries and the construction of non-abelian gauge theories. Chapter 5 examines the strong interactions and the structure and tests of Quantum Chromodynamics (QCD). Chapters[3] 6 and 7 examine the electroweak interactions and theory, including neutrino masses. Chapter 8 considers the motivations for extending the standard model, and examines supersymmetry, extended gauge groups, and grand unification. There are short appendices on additional topics. The bibliographies list many useful reference books, review articles, research papers, and websites. No attempt has been made to list all relevant original articles, with preference given instead to later articles and books that can be used to track down the original ones. Supplementary materials and corrections are available at `http://www.sns.ias.edu/~pgl/SMB/`. Comments, corrections, and typographical errors can also be sent through that site.

I would like to thank Mirjam Cvetič, Jens Erler, Hye-Sung Lee, Gil Paz, Liantao Wang, and Itay Yavin for reading and commenting on parts of the manuscript, Lisa Fleischer for help in the preparation of the manuscript, and my wife Irmgard for her extreme patience during the writing.

Paul Langacker
July 4, 2009

[2]Most calculations, especially at the tree-level, are now carried out by specialized computer programs, many of which are included in the list of websites, but it is still important to understand the techniques that go into them. Some examples may be found in the notebooks on the book website.

[3]These chapter numbers refer to the first edition.

CHAPTER 1

Notation and Conventions

DOI: 10.1201/b22175-1

In this chapter we briefly survey our notation and conventions.

Conventions

We generally follow the conventions used in (Langacker, 1981). In particular, (μ, ν, ρ, σ) are Lorentz indices; $(i, j, k = 1 \cdots 3)$ are three-vector indices; $(i, j, k = 1 \cdots N)$ are also used to label group generators or elements of the adjoint representation; (a, b, c) run over the elements of a representation, while (α, β, γ) and (r, s, t) refer to the special cases of color and flavor, respectively. (α, β) are also occasionally used for Dirac indices. (m, n) are used as horizontal (family) indices, labeling repeated fermions, scalars, and representations. The summation convention applies to all repeated indices except where indicated. Operators are represented by capital letters (T^i, Q, Y), their eigenvalues by the same symbols or by lower case letters[1] (t^i, q, y), and their matrix representations by (L^i, L_Q, L_Y). In Feynman diagrams, ordinary fermions are represented by solid lines; spin-0 particles by dashed lines; gluons by curly lines; other gauge bosons by wavy lines; and gluinos, neutralinos, and charginos by double lines. Experimental errors are usually quoted as a single number, with statistical, systematic, and theoretical uncertainties combined in quadrature and asymmetric errors symmetrized.

Units and Physical Constants

We take $\hbar = c = 1$, implying that E, p, m, $\frac{1}{x}$, $\frac{1}{t}$ have "energy units," such as electron volts (eV).[2] Related energy units are

$$\begin{aligned} 1 \text{ eV} &= 10^3 \text{ meV} = 10^{-3} \text{ keV} = 10^{-6} \text{ MeV} = 10^{-9} \text{ GeV} \\ &= 10^{-12} \text{ TeV} = 10^{-15} \text{ PeV} = 10^{-18} \text{ EeV}, \end{aligned} \tag{1.1}$$

where the prefixes represent milli, kilo, Mega, Giga, Tera, Peta, and Exa, respectively. One can restore conventional units at the end of a calculation using the values of $\hbar$, c, and $\hbar c$ listed in Table 1.1. We use Heaviside-Lorentz units, in which the fine structure constant is $\alpha = e^2/4\pi$, where $e > 0$ is the charge of the positron.

[1] Or sometimes e_r for the electric charge of the rth quark.

[2] Most likely only dimensionless quantities, such as α or ratios of masses, are fundamental.

TABLE 1.1 Conversions and physical constants.[a]

$\hbar \sim 6.6 \times 10^{-22}$ MeV-s	$c \sim 3.0 \times 10^{10}$ cm/s	$\hbar c \sim 197$ MeV-fm
$\alpha^{-1} \sim 137.04$	$\alpha^{-1}(M_Z^2) \sim 128.9$	$\sin^2 \hat{\theta}_W(M_Z^2) \sim 0.2313$
$\alpha_g(M_Z^2) \sim 0.034$	$\alpha_{g'}(M_Z^2) \sim 0.010$	$\alpha_s(M_Z^2) \sim 0.118$
$G_F \sim 1.17 \times 10^{-5}$ GeV^{-2}	$M_W \sim 80.39$ GeV	$M_Z \sim 91.19$ GeV
$m_e \sim 0.511$ MeV	$m_\mu \sim 105.7$ MeV	$m_\tau \sim 1.78$ GeV
$m_p \sim 938$ MeV	$m_{\pi^\pm} \sim 140$ MeV	$m_{K^\pm} \sim 494$ MeV
$M_P \sim 1.22 \times 10^{19}$ GeV	$M_H \sim 125$ GeV	1 g $\sim 5.6 \times 10^{23}$ GeV
$k \sim 1.16 \times 10^4$ °K/eV	1 barn $= 10^{-24}$ cm^2	1 yr $\sim 3.16 \times 10^7$ s $\sim \pi \times 10^7$ s

[a] For more precise values, see (Patrignani, 2016). The Planck constant is $M_P = G_N^{-1/2}$, where G_N is the gravitational constant.

Operators and Matrices

The commutator and anti-commutator of two operators or matrices are

$$[A, B] = AB - BA, \qquad \{A, B\} = AB + BA. \tag{1.2}$$

The transpose, adjoint, and trace of an $n \times n$ matrix M are

$$\text{transpose: } M^T \quad (M^T_{ab} = M_{ba}), \qquad \text{adjoint: } M^\dagger = M^{T*} \tag{1.3}$$

$$\text{trace : } \mathrm{Tr}\, M = \sum_{a=1}^{n} M_{aa}, \qquad \mathrm{Tr}\,(M_1 M_2) = \mathrm{Tr}\,(M_2 M_1), \qquad \mathrm{Tr}\, M = \mathrm{Tr}\, M^T. \tag{1.4}$$

Vectors, Metric, and Relativity

Three-vectors and unit vectors are denoted by $\vec{x}$ and $\hat{x} = \vec{x}/|\vec{x}|$, respectively. We do not distinguish between upper and lower indices for three-vectors; e.g., the inner (dot) product $\vec{x} \cdot \vec{y}$ may be written as $x^i y^i, x^i y_i$, or $x_i y_i$. The Levi-Civita tensor ϵ_{ijk}, where $i, j, k = 1 \cdots 3$, is totally antisymmetric, with $\epsilon_{123} = 1$. Its contractions are

$$\epsilon_{ijk}\epsilon_{ijk} = 6, \qquad \epsilon_{ijk}\epsilon_{ijm} = 2\delta_{km}, \qquad \epsilon_{ijk}\epsilon_{imn} = \delta_{jm}\delta_{kn} - \delta_{jn}\delta_{km}, \tag{1.5}$$

where the Kronecker delta function is

$$\delta_{ij} = \begin{cases} 1, & i = j \\ 0, & i \neq j \end{cases}. \tag{1.6}$$

ϵ_{ijk} is useful for vector cross products and their identities. For example,

$$(\vec{A} \times \vec{B})_i = \epsilon_{ijk} A_j B_k \tag{1.7}$$

$$(\vec{A} \times \vec{B}) \cdot (\vec{C} \times \vec{D}) = \epsilon_{ijk}\epsilon_{ilm} A_j B_k C_l D_m = (\vec{A} \cdot \vec{C})(\vec{B} \cdot \vec{D}) - (\vec{A} \cdot \vec{D})(\vec{B} \cdot \vec{C}). \tag{1.8}$$

Notations for four-vectors and the metric are given in Table 1.2.

The four-momentum of a particle with mass m is $p^\mu = (E, \vec{p})$ with $p^2 = E^2 - \vec{p}^{\,2} = m^2$.

(The symbol p is occasionally used to represent $|\vec{p}|$ rather than a four-vector, but the meaning should always be clear from the context.) The velocity $\vec{\beta}$ and energy are given by

$$\vec{\beta} = \frac{\vec{p}}{E}, \qquad \gamma \equiv \frac{E}{m} = \frac{1}{\sqrt{1-\beta^2}}. \tag{1.9}$$

Under a Lorentz boost by velocity $\vec{\beta}_L$ (the relativistic addition of $\vec{\beta}_L$ to $\vec{\beta}$, which is equivalent to going to a new Lorentz frame moving with $-\vec{\beta}_L$)

$$p^\mu \to p'^\mu = (E', \vec{p}\,'), \tag{1.10}$$

where

$$E' = \gamma_L(E + \vec{\beta}_L \cdot \vec{p}), \qquad \vec{p}\,' = \vec{p}_\perp + \gamma_L(\vec{p}_\parallel + \vec{\beta}_L E), \tag{1.11}$$

with

$$\vec{p}_\parallel = \hat{\beta}_L \hat{\beta}_L \cdot \vec{p}, \qquad \vec{p}_\perp = \vec{p} - \vec{p}_\parallel, \qquad \gamma_L = \frac{1}{\sqrt{1-\beta_L^2}}. \tag{1.12}$$

TABLE 1.2 Notations and conventions for four-vectors and the metric

Contravariant four-vector	$A^\mu = (A^0, \vec{A}), \qquad x^\mu = (t, \vec{x})$
Covariant four-vector	$A_\mu = g_{\mu\nu} A^\nu = (A^0, -\vec{A}), \qquad x_\mu = (t, -\vec{x})$
Metric	$g_{\mu\nu} = g^{\mu\nu} = \text{diag}(1, -1, -1, -1)$
	$g_\mu^\nu \equiv g^{\nu\sigma} g_{\mu\sigma} = \delta_\mu^\nu = \begin{cases} 1, & \mu = \nu \\ 0, & \mu \neq \nu \end{cases}$
Lorentz invariant	$A \cdot B \equiv A_\mu B^\mu = g_{\mu\nu} A^\mu B^\nu = A^0 B^0 - \vec{A} \cdot \vec{B}$
Derivatives	$\partial_\mu \equiv \frac{\partial}{\partial x^\mu} = \left(\frac{\partial}{\partial t}, \vec{\nabla}\right), \qquad \partial^\mu \equiv \frac{\partial}{\partial x_\mu} = \left(\frac{\partial}{\partial t}, -\vec{\nabla}\right)$
	$\Box \equiv \partial_\mu \partial^\mu = \frac{\partial^2}{\partial t^2} - \vec{\nabla}^2$
	$\partial \cdot A = \partial_\mu A^\mu = \frac{\partial A^0}{\partial t} + \vec{\nabla} \cdot \vec{A}$
	$a \overleftrightarrow{\partial}^\mu b = a\, \partial^\mu b - (\partial^\mu a)\, b$
Antisymmetric tensor	$\epsilon^{\mu\nu\rho\sigma}$, with $\epsilon_{0123} = +1$ and $\epsilon^{0123} = -1$
Contractions	$\epsilon^{\mu\nu\rho\sigma} \epsilon_{\mu\nu\rho\sigma} = -24 \qquad \epsilon^{\mu\nu\rho\sigma} \epsilon_{\mu\nu\rho\tau} = -6 g_\tau^\sigma$
	$\epsilon^{\mu\nu\rho\sigma} \epsilon_{\mu\nu\tau\omega} = -2\,(g_\tau^\rho g_\omega^\sigma - g_\omega^\rho g_\tau^\sigma)$

Translation Invariance

Let P^μ be the momentum operator, $|i\rangle$ and $|f\rangle$ momentum eigenstates,

$$P^\mu |i\rangle = p_i^\mu |i\rangle, \qquad P^\mu |f\rangle = p_f^\mu |f\rangle, \tag{1.13}$$

and let $\mathcal{O}(x)$ be an operator defined at spacetime point x, so that

$$\mathcal{O}(x) = e^{iP\cdot x} \mathcal{O}(0) e^{-iP\cdot x}. \tag{1.14}$$

Then the x dependence of the matrix element $\langle f|\mathcal{O}(x)|i\rangle$ is given by

$$\langle f|\mathcal{O}(x)|i\rangle = e^{i(p_f - p_i)\cdot x} \langle f|\mathcal{O}(0)|i\rangle. \tag{1.15}$$

The combination of Lorentz and translation invariance is *Poincaré invariance*.

The Pauli Matrices

The 2×2 *Pauli matrices* $\vec{\sigma} = (\sigma_1, \sigma_2, \sigma_3)$ (also denoted by $\vec{\tau}$, especially for internal symmetries) are Hermitian, $\sigma_i = \sigma_i^\dagger$, and defined by

$$[\sigma_i, \sigma_j] = 2i\epsilon_{ijk}\sigma_k. \tag{1.16}$$

A convenient representation is

$$\sigma_1 = \begin{pmatrix} 0 & 1 \\ 1 & 0 \end{pmatrix}, \qquad \sigma_2 = \begin{pmatrix} 0 & -i \\ i & 0 \end{pmatrix}, \qquad \sigma_3 = \begin{pmatrix} 1 & 0 \\ 0 & -1 \end{pmatrix}. \tag{1.17}$$

There is no distinction between σ_i and σ^i. Some useful identities include

$$\begin{aligned} \operatorname{Tr}\sigma_i &= \sum_{a=1}^{2} \sigma_{iaa} = 0, \qquad \operatorname{Tr}(\sigma_i\sigma_j) = 2\delta_{ij} \\ \{\sigma_i, \sigma_j\} &= 2\delta_{ij} I \;\Rightarrow\; \sigma_i^2 = I, \qquad \sigma_i\sigma_j = \delta_{ij} I + i\epsilon_{ijk}\sigma_k. \end{aligned} \tag{1.18}$$

The last identity implies

$$(\vec{A}\cdot\vec{\sigma})\,(\vec{B}\cdot\vec{\sigma}) = \vec{A}\cdot\vec{B}\, I + i(\vec{A}\times\vec{B})\cdot\vec{\sigma}, \tag{1.19}$$

where $\vec{A}$ and $\vec{B}$ are any three-vectors (including operators) and $\vec{A}\cdot\vec{\sigma}$ is a 2×2 matrix. Thus, $(\vec{A}\cdot\vec{\sigma})^2 = A^2 I$ for an ordinary real vector $\vec{A}$ with $A \equiv |\vec{A}\,|$, and

$$e^{i\vec{A}\cdot\vec{\sigma}} = (\cos A) I + i(\sin A)\hat{A}\cdot\vec{\sigma}. \tag{1.20}$$

Any 2×2 matrix M can be expressed in terms of $\vec{\sigma}$ and the identity by

$$M = \frac{1}{2}\operatorname{Tr}(M) I + \frac{1}{2}\operatorname{Tr}(M\vec{\sigma})\cdot\vec{\sigma}. \tag{1.21}$$

The $SU(2)$ *Fierz identity* is given in Problem 1.1.

The Delta and Step Functions

The Dirac delta function $\delta(x)$ is defined (for our purposes) by

$$\int_{-\infty}^{+\infty} \delta(x-a) g(x) dx = g(a) \tag{1.22}$$

for sufficiently well-behaved $g(x)$. Useful representations of $\delta(x)$ include

$$\delta(x-a) = \frac{1}{2\pi}\int_{-\infty}^{+\infty} e^{ik(x-a)} dk = \frac{1}{\pi}\lim_{\gamma\to 0}\left[\frac{\gamma}{(x-a)^2+\gamma^2}\right]. \tag{1.23}$$

The derivative of $\delta(x)$ is defined by integration by parts,

$$\int_{-\infty}^{+\infty} \delta'(x-a) g(x) dx \equiv \int_{-\infty}^{+\infty} \frac{d\delta(x-a)}{dx} g(x) dx = -\left.\frac{dg}{dx}\right|_{x=a}. \tag{1.24}$$

Suppose a well-behaved function $f(x)$ has zeroes at x_{0i}. Then

$$\delta(f(x)) = \sum_i \frac{\delta(x - x_{0i})}{|df/dx|_{x_{0i}}}. \tag{1.25}$$

The step function, $\Theta(x)$, is defined by

$$\Theta(x - x') = \begin{cases} 1, & x > x' \\ 0, & x < x' \end{cases}, \tag{1.26}$$

from which $\delta(x) = d\Theta/dx$.

Useful Integrals

$$\begin{aligned} &\text{Gaussian:} \quad \int_{-\infty}^{+\infty} e^{-\alpha x^2 + \beta x} dx = e^{\beta^2/4\alpha} \sqrt{\frac{\pi}{\alpha}} \text{ for } \Re e\, \alpha > 0 \\ &\text{Yukawa:} \quad \int d^3\vec{x}\, \frac{e^{-i\vec{q}\cdot\vec{x}} e^{-\mu|\vec{x}\,|}}{|\vec{x}\,|} = \frac{4\pi}{\mu^2 + |\vec{q}\,|^2}. \end{aligned} \tag{1.27}$$

1.1 PROBLEMS

1.1 Let χ_n, $n = 1 \cdots 4$, be arbitrary Pauli spinors (i.e., two-component complex column vectors). Then the bilinear form $\chi_m^\dagger \sigma_i \chi_n$ is an ordinary number. Prove the Fierz identity

$$(\chi_4^\dagger \vec{\sigma} \chi_3) \cdot (\chi_2^\dagger \vec{\sigma} \chi_1) = 2\eta_F (\chi_4^\dagger \chi_1)(\chi_2^\dagger \chi_3) - (\chi_4^\dagger \chi_3)(\chi_2^\dagger \chi_1),$$

where $\eta_F = +1$. (The identity also holds for anticommuting two-component fields if one sets $\eta_F = -1$.) Hint: expand the 2×2 matrix $\chi_1 \chi_2^\dagger$ in $(\chi_4^\dagger \chi_1)(\chi_2^\dagger \chi_3)$ using (1.21).

1.2 Justify the result (1.25) for $\delta(f(x))$.

1.3 Calculate the surface area $\int d\Omega_n$ of a unit sphere in n-dimensional Euclidean space, so that $\int d^n\vec{k} = \int d\Omega_n \int_0^\infty k^{n-1} dk$. Show that the general formula yields

$$\int d\Omega_1 = 2, \qquad \int d\Omega_2 = 2\pi, \qquad \int d\Omega_3 = 4\pi, \qquad \int d\Omega_4 = 2\pi^2.$$

Hint: Use the Gaussian integral formula (1.27) to integrate $\int d^n\vec{k}\, e^{-\alpha \vec{k}^{\,2}}$ in both Euclidean and spherical coordinates.

1.4 Show that the Lorentz boost in (1.11) can be written as

$$\begin{pmatrix} E' \\ p'_{\|} \end{pmatrix} = \begin{pmatrix} \cosh y_L & \sinh y_L \\ \sinh y_L & \cosh y_L \end{pmatrix} \begin{pmatrix} E \\ p_{\|} \end{pmatrix},$$

where

$$y_L = \frac{1}{2} \ln \frac{1 + \beta_L}{1 - \beta_L}$$

is the *rapidity* of the boost.

CHAPTER 2

Review of Perturbative Field Theory

DOI: 10.1201/b22175-2

Field Theory is the basic language of particle physics (i.e., of point particles). It combines quantum mechanics, relativistic kinematics, and the notion of particle creation and annihilation. The basic framework is remarkably successful and well-tested. In this book we will work mainly with *perturbative* field theory, characterized by weak coupling.

The *Lagrangian* of a field theory contains the interaction vertices. Combined with the *propagators* for virtual or unstable particles one can compute scattering and decay amplitudes using *Feynman diagrams*. Here we will review the rules (but not the derivations) for carrying out field theory calculations of amplitudes and the associated kinematics for processes involving spin-0, spin-$\frac{1}{2}$, and spin-1 particles in four dimensions of space and time, and give examples, mainly at tree level. Much more detail may be found in such field theory texts as (Bjorken and Drell, 1964, 1965; Weinberg, 1995; Peskin and Schroeder, 1995).

2.1 CREATION AND ANNIHILATION OPERATORS

Let $|0\rangle$ represent the *ground state* or *vacuum*, which we define as the no particle state (we are ignoring for now the complications of spontaneous symmetry breaking). The vacuum is normalized $\langle 0|0\rangle = 1$. We will use a *covariant normalization* convention.[1] For a spin-0 particle, define the creation and annihilation operators for a state of momentum $\vec{p}$ as $a^\dagger(\vec{p})$ and $a(\vec{p})$, respectively, i.e.,

$$a(\vec{p})|0\rangle = 0, \qquad a^\dagger(\vec{p})|0\rangle = |\vec{p}\rangle, \tag{2.1}$$

where $|\vec{p}\rangle$ describes a single-particle state with three-momentum $\vec{p}$, energy $E_p = \sqrt{\vec{p}^{\,2} + m^2}$, and velocity $\vec{\beta} = \vec{p}/E_p$. We assume the commutation rules (for Bose-Einstein statistics)

$$[a(\vec{p}), a^\dagger(\vec{p}\,')] = (2\pi)^3 2E_p \delta^3(\vec{p} - \vec{p}\,'), \qquad [a(\vec{p}), a(\vec{p}\,')] = [a^\dagger(\vec{p}), a^\dagger(\vec{p}\,')] = 0, \tag{2.2}$$

which correspond to the state normalization

$$\langle \vec{p}|\vec{p}\,'\rangle = (2\pi)^3 2E_p\, \delta^3(\vec{p} - \vec{p}\,'). \tag{2.3}$$

[1] Some formulas are simpler in the alternative *non-covariant* convention $[a_n(\vec{p}), a_n^\dagger(\vec{p}\,')] = \delta^3(\vec{p}-\vec{p}\,')$, with $a_n(\vec{p}) = a(\vec{p})/\sqrt{(2\pi)^3 2E_p}$. The corresponding single-particle state is $|\vec{p}\rangle_n \equiv a_n^\dagger(\vec{p})\,|0\rangle$, and the integration over physical states is $\int d^3\vec{p}$. Yet another possibility is (periodic) *box normalization* in a volume $V = L^3$, leading to discrete three-momenta with i^{th} component $p_i = n_i \frac{2\pi}{L}, n_i = 0, \pm 1, \pm 2, \cdots$, and commutators $[a_B(\vec{p}), a_B^\dagger(\vec{p}\,')] = \delta_{\vec{p}\,\vec{p}\,'}$.

This is Lorentz invariant because of the E_p factor. This can be seen from the fact that the integration over physical momenta is

$$\frac{d^3\vec{p}}{(2\pi)^3 2E_p} = \frac{d^4p}{(2\pi)^3}\,\delta(p^2 - m^2)\,\Theta(p^0), \tag{2.4}$$

where $\Theta(x)$ is the step function and we have used (1.25). The right-hand side of (2.4) is manifestly invariant. The additional $2(2\pi)^3$ factor is for convenience. The corresponding projection operator onto single-particle states is

$$I_{1p} \equiv \int \frac{d^3\vec{p}}{(2\pi)^3 2E_p} |\vec{p}\rangle\langle\vec{p}|. \tag{2.5}$$

The interpretation of (2.2) is that each momentum $\vec{p}$ of a non-interacting particle can be described by a simple harmonic oscillator. The *number operator* $\mathcal{N}(\vec{p})$, which counts the number of particles with momentum $\vec{p}$ in a state, and the total number operator N, which counts the total number of particles, are given by

$$\mathcal{N}(\vec{p}) \equiv a^\dagger(\vec{p})a(\vec{p}), \qquad N \equiv \int \frac{d^3\vec{p}}{(2\pi)^3 2E_p}\,\mathcal{N}(\vec{p}). \tag{2.6}$$

(2.1)–(2.6) actually hold for any real or complex bosons, provided one adds appropriate labels for particle type and (in the case of spin-1, 2, $\cdots$) for spin. For real (i.e., describing particles that are their own antiparticles) spin-0 fields of particle types a and b, for example,

$$[a_a(\vec{p}), a_b^\dagger(\vec{p}\,')] = \delta_{ab}(2\pi)^3 2E_p\,\delta^3(\vec{p} - \vec{p}\,'), \qquad [a_a(\vec{p}), a_b(\vec{p}\,')] = [a_a^\dagger(\vec{p}), a_b^\dagger(\vec{p}\,')] = 0. \tag{2.7}$$

Similarly, for a complex scalar, describing a spin-0 particle with a distinct antiparticle, it is conventional to use the symbols $a^\dagger$ and $b^\dagger$ for the particle and antiparticle creation operators, respectively. (Which state is called the particle and which the antiparticle is a convention.) For example, for the π^+ state,

$$a^\dagger(\vec{p})|0\rangle = |\pi^+(\vec{p})\rangle, \qquad b^\dagger(\vec{p})|0\rangle = |\pi^-(\vec{p})\rangle, \tag{2.8}$$

with

$$[b(\vec{p}), b^\dagger(\vec{p}\,')] = (2\pi)^3 2E_p\,\delta^3(\vec{p} - \vec{p}\,'), \qquad [a(\vec{p}), b^\dagger(\vec{p}\,')] = [a(\vec{p}), b(\vec{p}\,')] = 0. \tag{2.9}$$

The creation and annihilation operators for fermions are similar, except that they obey the anti-commutation rules appropriate to Fermi-Dirac statistics. The creation operator for a spin-$\frac{1}{2}$ particle is $a^\dagger(\vec{p}, s)$, where s refers to the particle's spin orientation, which may be taken with respect to a fixed z axis or with respect to $\hat{p}$ (*helicity*). Then,

$$\{a(\vec{p}, s), a^\dagger(\vec{p}\,', s')\} \equiv a(\vec{p}, s)a^\dagger(\vec{p}\,', s') + a^\dagger(\vec{p}\,', s')a(\vec{p}, s) = (2\pi)^3 2E_p\,\delta^3(\vec{p} - \vec{p}\,')\delta_{ss'}. \tag{2.10}$$

Similarly, for the antiparticle,

$$\{b(\vec{p}, s), b^\dagger(\vec{p}\,', s')\} = (2\pi)^3 2E_p\,\delta^3(\vec{p} - \vec{p}\,')\delta_{ss'}, \tag{2.11}$$

while

$$\{a, a\} = \{b, b\} = \{a, b\} = \{a, b^\dagger\} = 0 \tag{2.12}$$

for all values of $\vec{p}, \vec{p}\,'$, s, and s'. Fermion and boson operators commute with each other, e.g., $[a_{\text{boson}}, a_{\text{fermion}}] = 0$.

Non-interacting multi-particle states are constructed similarly. For example, the state for two identical bosons is

$$|\vec{p}_1\vec{p}_2\rangle = |\vec{p}_2\vec{p}_1\rangle = a^\dagger(\vec{p}_1)a^\dagger(\vec{p}_2)|0\rangle \tag{2.13}$$

with

$$\begin{aligned}\langle\vec{p}_1\vec{p}_2|\vec{p}_3\vec{p}_4\rangle = (2\pi)^3 2E_1(2\pi)^3 2E_2 \big[&\delta^3(\vec{p}_1-\vec{p}_3)\delta^3(\vec{p}_2-\vec{p}_4)\\ &+\delta^3(\vec{p}_1-\vec{p}_4)\delta^3(\vec{p}_2-\vec{p}_3)\big].\end{aligned} \tag{2.14}$$

Similarly, for two identical fermions,

$$|\vec{p}_1 s_1;\vec{p}_2 s_2\rangle = -|\vec{p}_2 s_2;\vec{p}_1 s_1\rangle = a^\dagger(\vec{p}_1 s_1)a^\dagger(\vec{p}_2 s_2)|0\rangle, \tag{2.15}$$

with

$$\begin{aligned}\langle\vec{p}_1 s_1;\vec{p}_2 s_2|\vec{p}_3 s_3;\vec{p}_4 s_4\rangle = (2\pi)^3 2E_1(2\pi)^3 2E_2 \big[&\delta^3(\vec{p}_1-\vec{p}_3)\delta_{s_1 s_3}\delta^3(\vec{p}_2-\vec{p}_4)\delta_{s_2 s_4}\\ &-\delta^3(\vec{p}_1-\vec{p}_4)\delta_{s_1 s_4}\delta^3(\vec{p}_2-\vec{p}_3)\delta_{s_2 s_3}\big].\end{aligned} \tag{2.16}$$

2.2 LAGRANGIAN FIELD THEORY

Consider a real or complex *field* $\phi(x)$, where $x \equiv (t, \vec{x})$. The (Hermitian) Lagrangian density

$$\mathcal{L}(\phi(x), \partial_\mu\phi(x), \phi^\dagger(x), \partial_\mu\phi^\dagger(x)) \tag{2.17}$$

contains information about the kinetic energy, mass, and interactions of ϕ. We will generally use the simpler notation $\mathcal{L}(\phi, \partial_\mu\phi)$, or just $\mathcal{L}(x)$, with the understanding that for a complex field $\mathcal{L}$ can depend on both ϕ and its Hermitian conjugate $\phi^\dagger$. Equation (2.17) is trivially generalized to the case in which there is more than one field. It is useful to also introduce the Lagrangian $L(t)$ and the *action* $\mathcal{S}$ by integrating $\mathcal{L}$ over space and over space-time, respectively,

$$L(t) = \int d^3\vec{x}\ \mathcal{L}(\phi, \partial_\mu\phi), \qquad \mathcal{S} = \int_{-\infty}^{+\infty} dt\ L(t) = \int d^4x\ \mathcal{L}(\phi, \partial_\mu\phi). \tag{2.18}$$

The *Euler-Lagrange* equations of motion for ϕ are obtained by minimizing the action with respect to $\phi(x)$ and $\phi^\dagger(x)$,

$$\frac{\delta\mathcal{L}}{\delta\phi} - \partial_\mu\frac{\delta\mathcal{L}}{\delta\partial_\mu\phi} = 0, \tag{2.19}$$

and similarly for $\phi^\dagger$. The fields ϕ are interpreted as operators in the *Heisenberg picture*, i.e., they are time-dependent while the states are time independent. Other quantities, such as the conjugate momentum, the Hamiltonian, and the canonical commutation rules, are summarized in Appendix A.

2.3 THE HERMITIAN SCALAR FIELD

A *real* (or more accurately, *Hermitian*) spin-0 (scalar) field satisfies $\phi(x) = \phi^\dagger(x)$. It is suitable for describing a particle such as the π^0 that has no internal quantum numbers and is therefore the same as its antiparticle.

2.3.1 The Lagrangian and Equations of Motion

The Lagrangian density for a Hermitian scalar is

$$\mathcal{L}(\phi, \partial_\mu \phi) = \frac{1}{2}\left[(\partial_\mu \phi)^2 - m^2\phi^2\right] - V_I(\phi), \tag{2.20}$$

where $(\partial_\mu \phi)^2$ is a shorthand for $(\partial_\mu \phi)(\partial^\mu \phi)$. The first two terms correspond, respectively, to canonical kinetic energy and mass (the $\frac{1}{2}$ is special to Hermitian fields), while the last describes interactions.

The interaction *potential* is

$$V_I(\phi) = \kappa \frac{\phi^3}{3!} + \lambda \frac{\phi^4}{4!} + c + d_1 \phi + d_5 \frac{\phi^5}{5!} + \cdots + \text{ non} - \text{perturbative}, \tag{2.21}$$

where the $k!$ factors are for later convenience in cancelling combinatoric factors,[2] and "non-perturbative" allows for the possibility of non-polynomial interactions. The constant c is irrelevant unless gravity is included. A non-zero d_1 (*tadpole*) term will induce a non-zero *vacuum expectation value* (VEV), $\langle 0|\phi|0\rangle \neq 0$, suggesting that one is working in the wrong vacuum. The d_1 term can be eliminated by a redefinition of $\phi \to \phi' = \text{ constant} + \phi$, as will be described in Chapter 3. $\mathcal{L}$ and ϕ have dimensions of 4 and 1, respectively, in mass units, so the coefficient of ϕ^k has the mass dimension $4-k$. The d_k terms with $k \geq 5$ are known as *non-renormalizable* or *higher-dimensional operators.* They lead to new divergences in each order of perturbation theory, with d_k typically of the form $d_k = c_k/\mathcal{M}^{k-4}$, where c_k is dimensionless and $\mathcal{M}$ is a large scale with dimensions of mass. Such terms would be absent in a *renormalizable* theory, but may occur in an *effective* theory at low energy, where they describe the effects of the exchange of heavy particles (or other degrees of freedom) of mass $\mathcal{M}$ that are not explicitly taken into account in the field theory. (In Chapters 7 and 8 we will see that an example of this is the four-fermi operator that is relevant to describing the weak interactions at low energy.) Keeping just the renormalizable terms (and $c = d_1 = 0$), one has

$$V_I(\phi) = \kappa \frac{\phi^3}{3!} + \lambda \frac{\phi^4}{4!}, \tag{2.22}$$

where κ (dimensions of mass) and λ (dimensionless) describe three- and four-point interactions, respectively, as illustrated in Figure 2.1. From the Euler-Langrange equation (2.19), one obtains the field equation

$$\left(\Box + m^2\right)\phi + \frac{\partial V_I}{\partial \phi} = \left(\Box + m^2\right)\phi + \kappa \frac{\phi^2}{2} + \lambda \frac{\phi^3}{6} = 0, \tag{2.23}$$

where $\Box + m^2 = \partial_\mu \partial^\mu + m^2$ is the *Klein-Gordon* operator. The expression for the Hamiltonian density is given in Appendix A.

2.3.2 The Free Hermitian Scalar Field

Let $\phi_0 = \phi_0^\dagger$ be the solution of (2.23) in the *free* (or non-interacting) limit $\kappa = \lambda = 0$, i.e.,

$$\left(\Box + m^2\right)\phi_0(x) = 0. \tag{2.24}$$

Equation (2.24) can be solved exactly, and small values of the interaction parameters κ and λ can then be treated perturbatively (as Feynman diagrams). The general solution is

$$\phi_0(x) = \phi_0^\dagger(x) = \int \frac{d^3\vec{p}}{(2\pi)^3 2E_p}\left[a(\vec{p})e^{-ip\cdot x} + a^\dagger(\vec{p})e^{+ip\cdot x}\right], \tag{2.25}$$

[2] Conventions for such factors may change, depending on the context.

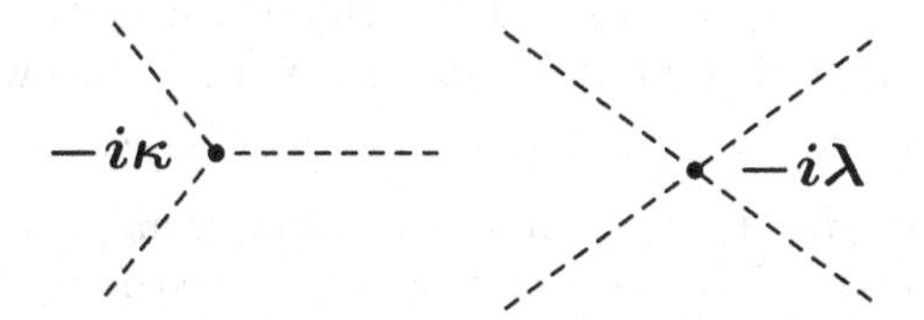

Figure 2.1 Three- and four-point interactions of a Hermitian scalar field ϕ. The factor is the coefficient of $\phi^n/n!$ in $i\mathcal{L}$, as described in Appendix B.

where $x = (t, \vec{x})$ and $p = (E_p, \vec{p})$ with $E_p \equiv \sqrt{\vec{p}^2 + m^2}$, i.e., the four-momentum p in the Fourier transform is an on-shell momentum for a particle of mass m. The canonical commutation rules for ϕ_0 and its conjugate momentum given in Appendix A will be satisfied if the Fourier coefficients $a(\vec{p})$ satisfy the creation-annihilation operator rules in (2.2).

It is useful to define the *Feynman propagator* for ϕ_0,

$$i\Delta_F(x - x') \equiv \langle 0|\mathcal{T}[\phi_0(x), \phi_0(x')]|0\rangle, \tag{2.26}$$

where

$$\mathcal{T}[\phi_0(x), \phi_0(x')] \equiv \Theta(t - t')\phi_0(x)\phi_0(x') + \Theta(t' - t)\phi_0(x')\phi_0(x) \tag{2.27}$$

represents the *time-ordered product* of $\phi_0(x)$ and $\phi_0(x')$. In (2.27) $\Theta(t - t')$ is the step function, defined in (1.26). $\Delta_F(x - x')$ is just the Green's function of the Klein-Gordon operator, i.e.,

$$\left(\Box_x + m^2\right) \Delta_F(x - x') = -\delta^4(x - x'), \tag{2.28}$$

where $\Box_x$ refers to derivatives w.r.t. x. The momentum space propagator is

$$\Delta_F(k) \equiv \int d^4x e^{+ik\cdot x} \Delta_F(x) = \frac{1}{k^2 - m^2 + i\epsilon} = \frac{1}{k_0^2 - \vec{k}^2 - m^2 + i\epsilon}. \tag{2.29}$$

k is an arbitrary four-momentum, i.e., it need not be on shell. The on-shell limit is correctly handled by the $i\epsilon$ factor in the denominator, where ϵ is a small positive quantity that can be taken to 0 at the end of the calculation.

2.3.3 The Feynman Rules

The *Feynman rules* allow a systematic diagrammatic representation of the terms in the perturbative expansion (in κ and λ) of the transition amplitude M_{fi} between an initial state i and a final state f. The derivation is beyond the scope of this book, but can be found in any standard field theory text. (The derivation of a simple example is sketched in Appendix B.) Heuristic derivations may also be found, e.g., in (Bjorken and Drell, 1964; Renton, 1990). For the Hermitian scalar field with the potential (2.22), the rules are

> Draw each connected topologically distinct diagram in momentum space corresponding to initial (final) states i (f), with internal lines corresponding to virtual (intermediate) particles. The internal and external lines are joined at three- and four-point vertices corresponding to the interactions in V_I. Each external and internal line has an associated four-momentum, which is off-shell for the virtual particles. It is convenient to put an arrow on the line to indicate the direction of momentum flow. This

direction is only a convention, and for the Hermitian scalar field (with no internal quantum numbers) there is no restriction on how many arrows flow into or out of a diagram.

There is a factor of $-i\kappa$ at every three-point vertex and a factor $-i\lambda$ at each four-point vertex, as in Figure 2.1. These correspond to the coefficients of $\phi^3/3!$ and $\phi^4/4!$, respectively, in $i\mathcal{L}$, with the $1/3!$ for κ cancelling against 3! ways to associate the three lines with the three fields in ϕ^3, and similarly for the λ term.

There is a factor of $i\Delta_F(k) = \frac{i}{k^2-m^2+i\epsilon}$ for each internal line with four-momentum k.

Four-momentum is conserved at each vertex, implying that the overall four-momentum is conserved (i.e., M_{fi} is only defined for $\Sigma p_i = \Sigma p_f$).

Integrate over each unconstrained internal momentum (there will be one for each internal loop in the diagram), with a factor $\int \frac{d^4k}{(2\pi)^4}$.

There may be additional combinatoric factors[3] associated with the interchange of internal lines for fixed vertices, e.g., a factor of $1/n!$ if n internal lines connect the same pair of vertices, as in Figure 2.2.

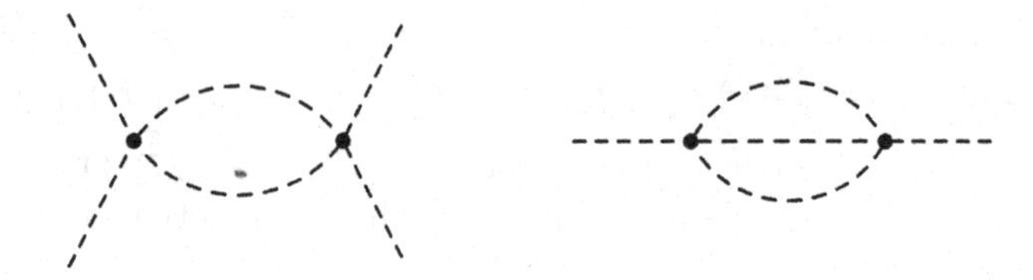

Figure 2.2 Diagrams with additional factors of 1/2! (left) and 1/3! (right), required because not all of the $(4!)^2$ ways of associating the four fields at each vertex with four lines lead to distinct diagrams.

The ordering and arrangement of the external lines in a Feynman diagram is usually irrelevant, although the relative ordering between two diagrams does matter for fermions. In this book we will usually, but not always, place the initial state particles at the bottom and the final particles at the top.

As a simple example, the tree-level diagrams for the $2 \rightarrow 2$ scattering amplitude $M_{fi} = \langle \vec{p}_3\vec{p}_4|M|\vec{p}_1\vec{p}_2\rangle$ are shown in Figure 2.3. Applying the Feynman rules, these diagrams correspond to the expression

$$M_{fi} = -i\lambda + (-i\kappa)^2 \left[\frac{i}{s-m^2} + \frac{i}{t-m^2} + \frac{i}{u-m^2}\right], \tag{2.30}$$

where s, t, and u are the *Mandelstam variables*

$$\begin{aligned} s &\equiv (p_1+p_2)^2 = (p_3+p_4)^2 \\ t &\equiv (p_1-p_3)^2 = (p_4-p_2)^2 \\ u &\equiv (p_1-p_4)^2 = (p_3-p_2)^2. \end{aligned} \tag{2.31}$$

[3]In general, there may be subtleties involving combinatorial factors (or, signs when fermions are involved), especially in higher-order diagrams, which are best resolved by returning to the original derivation.

For the present (equal mass) case, $s = E_{CM}^2 \geq 4m^2$, $t \leq 0$, and $u \leq 0$, where E_{CM} is the total energy in the center of mass. The internal lines are never on-shell ($s, t, u \neq m^2$) for physical (on-shell) external momenta, so one can drop the $+i\epsilon$. It is implicit that the external momenta satisfy the four-momentum conservation $p_1 + p_2 = p_3 + p_4$. The second, third, and fourth diagrams in Figure 2.3 are said to have *s-channel, t-channel,* and *u-channel poles*, respectively.

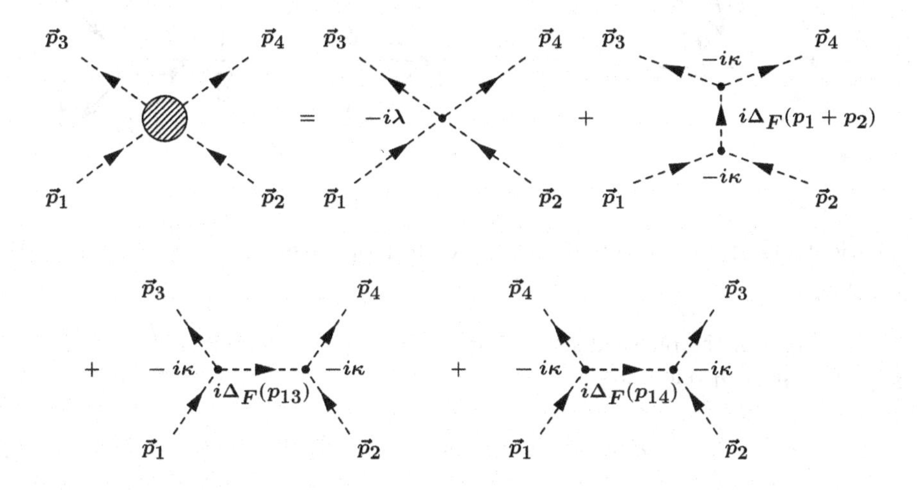

Figure 2.3 Tree level diagrams for $M_{fi} = \langle \vec{p}_3 \vec{p}_4 | M | \vec{p}_1 \vec{p}_2 \rangle$ for a Hermitian scalar field. The arrows label the directions of momentum flow, and $p_{ij} \equiv p_i - p_j$. The arrangement of lines in the last diagram is modified to allow the diagram to be drawn without crossing lines.

2.3.4 Kinematics and the Mandelstam Variables

Let us digress to generalize to the case of a $2 \to 2$ scattering process $1 + 2 \to 3 + 4$, where we allow for the possibility of inelastic scattering with unequal masses for 1, 2, 3, and 4. In the absence of spin, the scattering amplitude can be expressed in terms of the Lorentz invariant Mandelstam variables defined in (2.31). s, t, and u are not independent, but are related by

$$s + t + u = m_1^2 + m_2^2 + m_3^2 + m_4^2. \tag{2.32}$$

Of course, $s = m_1^2 + m_2^2 + 2p_1 \cdot p_2$, etc.

The kinematics is simplest in the *center of mass* (CM) frame, which is more accurately the center of momentum, in which the total three-momentum of the initial and final state vanishes:

$$p_1 = (E_1, \vec{p}_i), \qquad p_2 = (E_2, -\vec{p}_i), \qquad p_3 = (E_3, \vec{p}_f), \qquad p_4 = (E_4, -\vec{p}_f), \tag{2.33}$$

where $\vec{p}_i$ and $\vec{p}_f$ are, respectively, the initial and final three-momenta; the energy and velocity of particle 1 are

$$E_1^2 = \vec{p}_i^{\,2} + m_1^2, \quad \vec{\beta}_1 = \frac{\vec{p}_i}{E_1}, \tag{2.34}$$

and similarly for 2, 3, and 4; and the CM scattering angle θ is related by

$$\vec{p}_i \cdot \vec{p}_f = p_i p_f \cos\theta, \tag{2.35}$$

as shown in Figure 2.4. In the CM frame, $s = (E_1 + E_2)^2 = (E_3 + E_4)^2$ is just the square

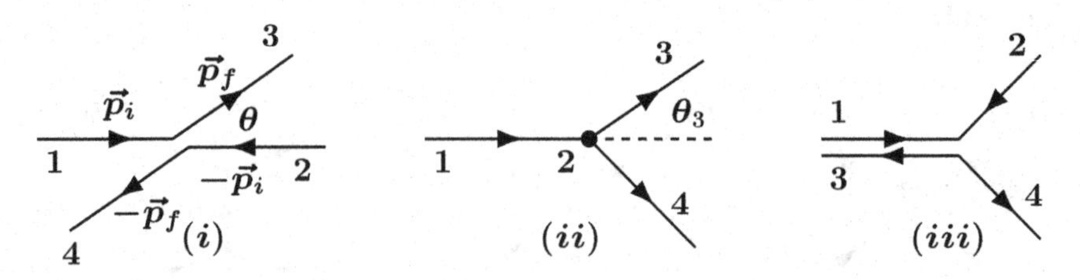

Figure 2.4 Scattering kinematics in (i) the center of mass, (ii) the lab, and (iii) the Breit frames.

of the total energy. In the physical scattering region, $s \geq (m_1 + m_2)^2$ and $s \geq (m_3 + m_4)^2$. Using $p_2 = p_1 + p_2 - p_1$ one finds

$$E_1 = \frac{s + m_1^2 - m_2^2}{2\sqrt{s}} \xrightarrow[m_1=m_2]{} \frac{\sqrt{s}}{2}, \tag{2.36}$$

so that

$$\begin{aligned} p_i = \sqrt{E_1^2 - m_1^2} &= \frac{\left[s - (m_1 + m_2)^2\right]^{1/2} \left[s - (m_1 - m_2)^2\right]^{1/2}}{2\sqrt{s}} \\ &\xrightarrow[m_1=m_2]{} \frac{\left[s - 4m_1^2\right]^{1/2}}{2}, \end{aligned} \tag{2.37}$$

with similar expressions for the other particles. This is sometimes written as

$$p_i = \frac{\lambda^{1/2}(s, m_1^2, m_2^2)}{2\sqrt{s}}, \tag{2.38}$$

where

$$\lambda(x, y, z) \equiv x^2 + y^2 + z^2 - 2xy - 2xz - 2yz. \tag{2.39}$$

The t and u variables, which describe the momentum transfer between particles 1 and 3 and between 1 and 4, respectively, are given in the CM by

$$\begin{aligned} t &= m_1^2 + m_3^2 - 2E_1E_3 + 2p_ip_f\cos\theta \xrightarrow[\substack{m_1=m_3 \\ m_2=m_4}]{} -2p^2(1 - \cos\theta) = -4p^2\sin^2\frac{\theta}{2} \leq 0 \\ u &= m_1^2 + m_4^2 - 2E_1E_4 - 2p_ip_f\cos\theta \xrightarrow[\substack{m_1=m_4 \\ m_2=m_3}]{} -2p^2(1 + \cos\theta) = -4p^2\cos^2\frac{\theta}{2} \leq 0, \end{aligned} \tag{2.40}$$

where the last expressions are for elastic scattering ($m_1 = m_3$, $m_2 = m_4$ or $m_1 = m_4$, $m_2 = m_3$), for which $p_i = p_f \equiv p$. Note that t and u are negative at high energies (e.g., the last expressions in (2.40) hold with $p \sim \sqrt{s}/2$ when the masses can be neglected), but may be positive at low energies if the masses are not all equal.

Fixed target experiments are carried out in the *laboratory* frame, in which 2 is at rest,

$$p_1 = (E_1, \vec{p}_1), \qquad p_2 = (m_2, \vec{0}), \tag{2.41}$$

with

$$E_1 = \frac{s - m_1^2 - m_2^2}{2m_2}. \tag{2.42}$$

A sometimes useful relation between the CM and laboratory variables is

$$p_i = |\vec{p}_1| m_2/\sqrt{s}, \tag{2.43}$$

where p_i is the CM momentum in (2.37). From energy and momentum conservation,

$$E_3 + E_4 = E_1 + m_2, \qquad |\vec{p}_4|^2 = |\vec{p}_1|^2 + |\vec{p}_3|^2 - 2|\vec{p}_1||\vec{p}_3|\cos\theta_3, \tag{2.44}$$

so that E_3 and E_4 can be expressed in terms of s and the laboratory scattering angle θ_3 for particle 3. The relations between the laboratory variables and t and u are straightforward. For example,

$$t = m_3^2 + m_1^2 - 2E_1E_3 + 2|\vec{p}_1||\vec{p}_3|\cos\theta_3, \qquad u = m_2^2 + m_3^2 - 2m_2E_3. \tag{2.45}$$

These formulae become especially simple in the special case $m_1 = m_3 = 0$ and $m_2 = m_4$, e.g.,

$$\frac{p_3}{p_1} = \frac{1}{1 + \frac{p_1}{m_2}(1 - \cos\theta_3)}, \qquad t = -2p_1p_3(1 - \cos\theta_3), \tag{2.46}$$

with $p_1 \equiv |\vec{p}_1| = E_1$ and $p_3 \equiv |\vec{p}_3| = E_3$.

Yet another frame, especially useful for theoretical purposes, is the *Breit* or *brick wall* frame (e.g., Hagedorn, 1964; Renton, 1990), in which the scattered particle 3 simply reverses the direction of 1. For $m_1 = m_3$, the momenta are

$$p_1 = (E_1, \vec{p}), \qquad p_3 = (E_1, -\vec{p}), \tag{2.47}$$

so that

$$t = -4|\vec{p}|^2. \tag{2.48}$$

We will see an example in discussing the simple parton model in Section 5.5.

2.3.5 The Cross Section and Decay Rate Formulae

In this section we sketch the derivation of the relation of the transition amplitude M_{fi} to the *cross section* or *decay rate*. The results apply to any field theory, not just Hermitian scalars.

Two-Body Scattering

First consider the cross section for the $2 \to n$ process $i \to f$, where $|i\rangle = |\vec{p}_1\vec{p}_2\rangle$ and $|f\rangle = |\vec{p}_{f_1} \cdots \vec{p}_{f_n}\rangle$, as shown in Figure 2.5. (Particle-type and spin labels are suppressed.)

As described in Appendix B, the transition matrix element U_{fi} and transition (scattering) amplitude M_{fi} are related by (B.1), so the transition probability is

$$|U_{fi}|^2 = \left|(2\pi)^4\delta^4\Big(\sum_k p_{f_k} - p_1 - p_2\Big)M_{fi}\right|^2. \tag{2.49}$$

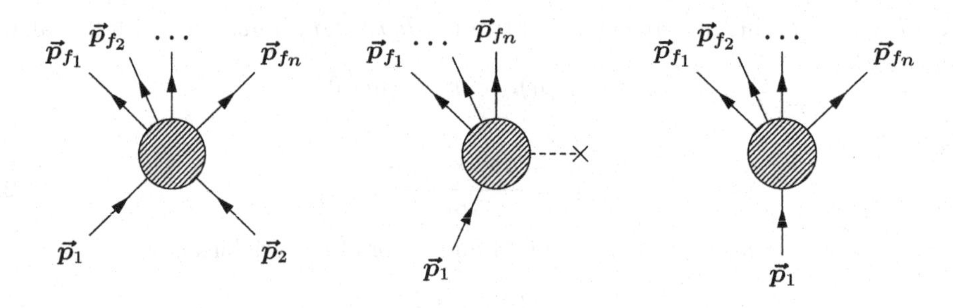

Figure 2.5 The $2 \to n$ scattering process $p_1 p_2 \to p_{f_1} \cdots p_{f_n}$ (left); $1 \to n$ scattering $p_1 \to p_{f_1} \cdots p_{f_n}$ from a potential, represented by a cross (middle); and the $1 \to n$ decay process $p_1 \to p_{f_1} \cdots p_{f_n}$ (right).

To interpret the square of the delta function, it is convenient to temporarily assume a finite volume V for space and a total transition time T, which can be taken to ∞ at the end. Then

$$\begin{aligned}\left|(2\pi)^4\delta^4\Big(\sum_k p_{f_k} - p_1 - p_2\Big)\right|^2 &= \left|\int_{V,T} d^4x\ e^{i(\sum_k p_{f_k} - p_1 - p_2)\cdot x}\right|^2 \\ &= VT\ (2\pi)^4\delta^4\Big(\sum_k p_{f_k} - p_1 - p_2\Big),\end{aligned} \tag{2.50}$$

where we have used one delta function to replace the integrand of the other integral by unity.

A scattering cross section is defined as the transition rate divided by the relative flux. The differential cross section to scatter into $\prod_k d^3\vec{p}_{f_k}$ is therefore

$$\begin{aligned} d\sigma &= \frac{|U_{fi}|^2}{T}\frac{V}{|\vec{\beta}_1 - \vec{\beta}_2|}\left(\frac{1}{2E_1V}\right)\left(\frac{1}{2E_2V}\right)\left[\prod_{k=1}^{n}\frac{d^3\vec{p}_{f_k}}{2E_{f_k}V}\frac{V}{(2\pi)^3}\right] \\ &= \frac{(2\pi)^4\delta^4(\sum_k p_{f_k} - p_1 - p_2)}{4E_1E_2|\vec{\beta}_1 - \vec{\beta}_2|}\prod_{k=1}^{n}\frac{d^3\vec{p}_{f_k}}{(2\pi)^3 2E_{f_k}}\,|M_{fi}|^2. \end{aligned} \tag{2.51}$$

In the first line, $|U_{fi}|^2/T$ is the transition rate, $|\vec{\beta}_1 - \vec{\beta}_2|/V$ is the relative flux, the $1/2EV$ factors correct for the normalization of the covariant states,[4] and the n factors of $V/(2\pi)^3$ represent the density of momentum states of the n final particles. Note that the factors of V and T cancel in the final expression in (2.51), and that all the particles are on-shell ($E^2 = \vec{p}^{\,2} + m^2$). The flux factor $|\vec{\beta}_1 - \vec{\beta}_2|$ is evaluated using $\vec{\beta}_j = \vec{p}_j/E_j$.

It should be intuitively clear that the differential cross section is the same in the lab frame and in *collinear* frames related to the lab by boosts along the $\vec{p}_1$ direction, such as the CM frame, but is *not* the same in arbitrary frames. This can be seen by the fact that the factor $E_1E_2|\vec{\beta}_1 - \vec{\beta}_2|$ in the denominator is equal to the Lorentz invariant quantity $[(p_1 \cdot p_2)^2 - m_1^2 m_2^2]^{1/2}$ in collinear frames. In fact, the cross section formula is often written

[4]The inner product $\langle \vec{p}|\vec{p}\,'\rangle = (2\pi)^3 2E_p\,\delta^3(\vec{p} - \vec{p}\,') = 2E_p \int d^3x e^{-i(\vec{p}-\vec{p}\,')\cdot\vec{x}}$ goes to $2E_pV\delta_{\vec{p}\vec{p}\,'}$ in a finite volume V.

in terms of that quantity, though that is only strictly valid in the collinear frames.[5] The δ^4 function and the scattering amplitude M_{fi} in (2.51) are manifestly Lorentz invariant. The final state phase space factors are also invariant, as can be seen in (2.4). The differential cross section can be integrated over the ranges of momenta of interest. The total cross section for the specific process is obtained by integrating over all final momenta,

$$\sigma = \prod_l S_l \int d\sigma, \tag{2.52}$$

where the statistical factor $S_l \equiv 1/l!$ must be included for any set of l identical final particles to avoid multiple counting of the same final state.

Now, consider the example of $2 \to 2$ scattering in the CM, as illustrated in Figure 2.4(i). For a given CM energy $\sqrt{s} = E_{CM}$, the initial and final momenta p_i, p_f, and the energies E_a, $a = 1 \cdots 4$, are fixed by (2.36) and (2.37). It is convenient to introduce spherical coordinates, with the z axis along $\vec{p}_i$, so that $\vec{p}_f$ has polar angle θ and azimuthal angle φ. For spin-0 particles (or unpolarized initial particles with non-zero spin) the scattering amplitude M_{fi} is independent of φ by rotational invariance.

The differential cross section is given by

$$d\sigma = \frac{(2\pi)^4 \delta^4(p_3 + p_4 - p_1 - p_2)}{4E_1 E_2 |\vec{\beta}_1 - \vec{\beta}_2|} \frac{d^3\vec{p}_3}{(2\pi)^3 2E_3} \frac{d^3\vec{p}_4}{(2\pi)^3 2E_4} |M_{fi}|^2. \tag{2.53}$$

In the CM,

$$E_1 E_2 |\vec{\beta}_1 - \vec{\beta}_2| = E_1 E_2 \left| \vec{p}_i \left(\frac{1}{E_1} + \frac{1}{E_2} \right) \right| = p_i (E_1 + E_2) = p_i \sqrt{s}. \tag{2.54}$$

The implicit $\vec{p}_4$ integral can be done using the four-momentum conservation,

$$\begin{aligned} \delta^4(p_3 + p_4 - p_1 - p_2) \frac{d^3\vec{p}_3}{E_3} \frac{d^3\vec{p}_4}{E_4} &= \delta(E_3 + E_4 - \sqrt{s}) \frac{d^3\vec{p}_3}{E_3 E_4} \\ &= 2\pi \, d\cos\theta \; \delta(E_3 + E_4 - \sqrt{s}) \frac{p_3^2 \, dp_3}{E_3 E_4}, \end{aligned} \tag{2.55}$$

where E_3 and E_4 represent $\sqrt{\vec{p}_3{}^2 + m_3^2}$ and $\sqrt{\vec{p}_3{}^2 + m_4^2}$, respectively. In the last expression the azimuthal angle has been (optionally) integrated over, i.e., $d^3\vec{p}_3 = d\varphi \; d\cos\theta \; p_3^2 \, dp_3 \to 2\pi \, d\cos\theta \; p_3^2 \, dp_3$. The remaining delta function determines p_3. Using $p_3 \, dp_3 = E_3 \, dE_3 = E_4 \, dE_4$, $d(E_3 + E_4)/dE_3 = 1 + E_3/E_4$, and the identity (1.25), the quantity (2.55) is equal to

$$2\pi \, d\cos\theta \; \frac{p_f}{\sqrt{s}}. \tag{2.56}$$

The differential cross section is therefore

$$\frac{d\sigma}{d\cos\theta} = \frac{1}{32\pi s} \frac{p_f}{p_i} |M_{fi}|^2 \xrightarrow[m_2 = m_4]{m_1 = m_3} \frac{|M_{fi}|^2}{32\pi s}. \tag{2.57}$$

The initial momentum p_i is given in (2.37), while p_f is of the same form with $m_1 \to m_3$ and

---

[5]In applications to astrophysics one is usually interested in the thermal average of $\sigma|\vec{\beta}_1 - \vec{\beta}_2|$, so the flux factor cancels (e.g., Kolb and Turner, 1990).

$m_2 \to m_4$. The last expression in (2.57) is for elastic scattering, for which $p_i = p_f$. Closely related quantities are

$$\frac{d\sigma}{dt} = \frac{1}{2p_i p_f} \frac{d\sigma}{d\cos\theta}, \qquad \frac{d\sigma}{d\Omega} = \frac{1}{2\pi} \frac{d\sigma}{d\cos\theta}, \tag{2.58}$$

where t is the Mandelstam invariant given by (2.40), and $d\Omega \equiv d\varphi\, d\cos\theta$ is the solid angle element. The second form is useful if $d\varphi$ is not integrated over.

In our example of the Hermitian scalar field with V_I given by (2.22), the differential cross section at tree level is obtained from (2.30) and (2.57),

$$\frac{d\sigma}{d\cos\theta} = \frac{1}{32\pi s} \left| \lambda + \kappa^2 \left[\frac{1}{s - m^2} + \frac{1}{t - m^2} + \frac{1}{u - m^2} \right] \right|^2, \tag{2.59}$$

where

$$s = E_{CM}^2 = 4(p^2 + m^2), \quad t = -2p^2(1 - \cos\theta), \quad u = -2p^2(1 + \cos\theta), \tag{2.60}$$

and $p \equiv p_i = p_f$. The total cross section is then

$$\sigma = \frac{1}{2} \int_{-1}^{+1} \frac{d\sigma}{d\cos\theta} d\cos\theta = \int_{0}^{+1} \frac{d\sigma}{d\cos\theta} d\cos\theta, \tag{2.61}$$

where the 1/2 is because the final particles are identical.

One can also use (2.53) to calculate the cross section in the lab frame, using

$$E_1 E_2 |\vec{\beta}_1 - \vec{\beta}_2| = E_1 m_2 |\vec{\beta}_1| = p_1 m_2. \tag{2.62}$$

The phase space integral can be carried out explicitly in the lab frame, or can be obtained by Lorentz transforming the CM result. In analogy with the derivation of (2.56)

$$\delta^4(p_3 + p_4 - p_1 - p_2) \frac{d^3\vec{p}_3}{E_3} \frac{d^3\vec{p}_4}{E_4} = \frac{2\pi |\vec{p}_3|\, d\cos\theta_3}{E_4 \left| 1 + \frac{dE_4}{dE_3} \right|}, \tag{2.63}$$

where θ_3 is the laboratory scattering angle of particle 3. dE_4/dE_3 can be calculated using the second equation in (2.44), yielding

$$\frac{d\sigma}{d\cos\theta_3} = \frac{|M_{fi}|^2}{32\pi p_1 m_2} \frac{p_3}{E_1 + m_2 - \frac{p_1 E_3}{p_3} \cos\theta_3}. \tag{2.64}$$

Using (2.46) this implies

$$\frac{d\sigma}{d\cos\theta_3} = \frac{|M_{fi}|^2}{32\pi m_2^2} \left(\frac{p_3}{p_1} \right)^2 \tag{2.65}$$

for the special case $m_1 = m_3 = 0$ and $m_2 = m_4$.

Potential Scattering

Consider the $1 \to n$ scattering of a single particle from a static source (i.e., potential scattering), as illustrated in Figure 2.5. This may be an approximation to scattering from a heavy target particle. In particular, suppose $\mathcal{L}$ contains an interaction term

$$\mathcal{L}_I(x) = \mathcal{L}_p(x)\, \Phi(0, \vec{x}\,), \tag{2.66}$$

where $\Phi(0,\vec{x}\,)$ is the static source and $\mathcal{L}_p$ involves ordinary quantum fields. Then, in analogy with Equation (B.4) from Appendix B, the tree-level transition amplitude for $i \to f$, with $|i\rangle = |\vec{p}_1\rangle$ and $|f\rangle = |\vec{p}_{f_1} \cdots \vec{p}_{f_n}\rangle$, is

$$\begin{aligned} U_{fi} &\simeq \int d^4x \, \langle f|i\mathcal{L}_p(x)|i\rangle \Phi(0,\vec{x}\,) \\ &= \frac{1}{(2\pi)^3} \int d^4x \, d^3\vec{q} \, \langle f|i\mathcal{L}_p(0)|i\rangle \tilde{\Phi}(\vec{q}\,) e^{i(p_f - p_1 - q)\cdot x}, \end{aligned} \tag{2.67}$$

where

$$\tilde{\Phi}(\vec{q}\,) = \int d^3\vec{x}\,' e^{-i\vec{q}\cdot\vec{x}\,'} \Phi(0,\vec{x}\,') \tag{2.68}$$

is the Fourier transform of Φ, $q = (0,\vec{q}\,)$, and we have used translation invariance for the matrix element. Then, carrying out the x integral,

$$U_{fi} = 2\pi\delta\,(E_f - E_1)\, M_{fi}, \tag{2.69}$$

where

$$M_{fi} = \langle f|i\mathcal{L}_p(0)|i\rangle \tilde{\Phi}(\vec{p}_f - \vec{p}_1) \tag{2.70}$$

and $p_f = \sum_k p_{f_k}$. Proceeding as in (2.51), the differential cross section is

$$\begin{aligned} d\sigma &= \frac{|U_{fi}|^2}{T} \frac{V}{|\vec{\beta}_1|} \left(\frac{1}{2E_1 V} \right) \left[\prod_{k=1}^{n} \frac{d^3\vec{p}_{f_k}}{2E_{f_k} V} \frac{V}{(2\pi)^3} \right] \\ &= \frac{2\pi\delta(E_2 - E_1)}{2E_1|\vec{\beta}_1|} \prod_{k=1}^{n} \frac{d^3\vec{p}_{f_k}}{(2\pi)^3 2E_{f_k}} |M_{fi}|^2, \end{aligned} \tag{2.71}$$

so that energy but not 3-momentum is conserved, as expected. For the important special case of elastic scattering, i.e., $n = 1$ with $m_2 = m_1$ and $p_2 \equiv p_{f_1}$,

$$|\vec{p}_{1,2}| = \beta E_{1,2} \equiv p, \text{ with } \beta \equiv |\vec{\beta}_1|. \tag{2.72}$$

But,

$$\delta\,(E_2 - E_1)\, d^3\vec{p}_2 = p_2 E_2 d\cos\theta d\varphi, \tag{2.73}$$

where θ and φ are the polar and azimuthal scattering angles. Therefore,

$$\frac{d\sigma}{d\cos\theta d\varphi} = \frac{1}{16\pi^2}|M_{fi}|^2 \quad \to \quad \frac{d\sigma}{d\cos\theta} = \frac{1}{8\pi}|M_{fi}|^2. \tag{2.74}$$

The last form holds when M_{fi} depends only on

$$|\vec{q}\,|^2 = |\vec{p}_2 - \vec{p}_1|^2 = 2p^2(1 - \cos\theta) = 4p^2 \sin^2\frac{\theta}{2}, \tag{2.75}$$

as in the case of spinless or unpolarized particles and a radially symmetric source. As a simple example, consider

$$\mathcal{L}_I = -\frac{1}{2}\phi^2\Phi(r), \tag{2.76}$$

where $\Phi(r)$ is the spherically symmetric Yukawa potential

$$\Phi(0,\vec{x}\,) = \kappa \frac{e^{-\mu r}}{r} \qquad \longleftrightarrow \qquad \tilde{\Phi}(\vec{q}\,) = \frac{4\pi\kappa}{\mu^2 + |\vec{q}\,|^2}, \tag{2.77}$$

with $r \equiv |\vec{x}\,|$. Φ can be thought of as arising from the exchange of a heavy scalar of mass μ between a nucleus and the scattered particle described by ϕ. (This could be an approximate model for the contribution to pion-nucleon scattering from the exchange of a heavy scalar resonance, such as the σ or the $f_0(980)$.) From (2.70) and (2.74) we obtain $M_{fi} = -i\tilde{\Phi}(\vec{q}\,)$ and

$$\frac{d\sigma}{d\cos\theta} = \frac{2\pi\kappa^2}{(\mu^2 + |\vec{q}\,|^2)^2}. \tag{2.78}$$

Decays

One can also consider the $1 \to n$ decay process with $|i\rangle = |\vec{p}_1\rangle$ and $|f\rangle = |\vec{p}_{f_1} \cdots \vec{p}_{f_n}\rangle$ (Figure 2.5). Similar to (2.49) and (2.51), one has

$$|U_{fi}|^2 = VT(2\pi)^4\delta^4\Big(\sum_k p_{f_k} - p_1\Big)|M_{fi}|^2 \tag{2.79}$$

and differential decay rate

$$\begin{aligned} d\Gamma = d\,\frac{1}{\tau} &= \frac{|U_{fi}|^2}{T}\left(\frac{1}{2E_1 V}\right)\left[\prod_{k=1}^{n}\frac{d^3\vec{p}_{f_k}}{2E_{f_k}V}\frac{V}{(2\pi)^3}\right] \\ &\xrightarrow[\text{rest frame}]{} \frac{(2\pi)^4\delta^4(\sum_k p_{f_k} - p_1)}{2m_1}\prod_{k=1}^{n}\frac{d^3\vec{p}_{f_k}}{(2\pi)^3 2E_{f_k}}|M_{fi}|^2, \end{aligned} \tag{2.80}$$

where Γ is the decay rate of particle 1 and τ is its lifetime, and the second expression is specialized to the rest frame of the decaying particle. The total decay rate is obtained by integrating over the final particle phase space,

$$\Gamma = \prod_l S_l \int d\Gamma, \tag{2.81}$$

where $S_l \equiv 1/l!$ is a statistical factor for l identical particles, analogous to (2.52).

For a 2-body decay, $1 \to 2 + 3$, Equation (2.80) simplifies to

$$\begin{aligned} d\Gamma &= \frac{(2\pi)^4\delta^4(p_3 + p_2 - p_1)}{2m_1}\frac{d^3\vec{p}_2}{(2\pi)^3 2E_2}\frac{d^3\vec{p}_3}{(2\pi)^3 2E_3}|M_{fi}|^2 \\ &= \frac{|M_{fi}|^2}{32\pi^2 m_1}\left[\frac{p_2 E_2 dE_2}{E_2 E_3}\delta(E_2 + E_3 - m_1)\right]d\Omega = \frac{p_2}{32\pi^2 m_1^2}|M_{fi}|^2 d\Omega, \end{aligned} \tag{2.82}$$

where $E_{2,3} \equiv \sqrt{\vec{p}_2^{\;2} + m_{2,3}^2}$ in the second line. The quantity in square brackets evaluates to

$$\frac{p_2}{m_1} = \frac{\left[m_1^2 - (m_2 + m_3)^2\right]^{1/2}\left[m_1^2 - (m_2 - m_3)^2\right]^{1/2}}{2m_1^2}, \tag{2.83}$$

where the expression for p_2 is analogous to (2.37).

As an example, consider three distinct Hermitian scalar fields ϕ_i, $i = 1 \cdots 3$, with masses m_i. If $m_1 > m_2 + m_3$ it is possible for particle 1 to decay into $2 + 3$ as shown in Figure 2.6, provided there is an interaction term to drive the decay. The simplest appropriate

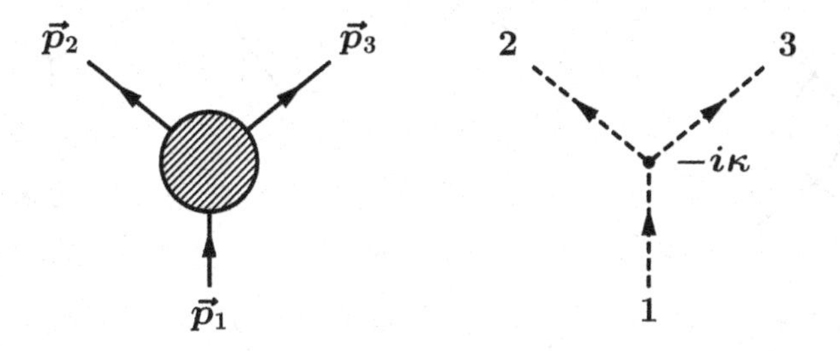

Figure 2.6 The two-body decay of the spin-0 scalar 1 into 2 + 3, and the associated Feynman diagram corresponding to (2.84).

Lagrangian is

$$\mathcal{L} = \sum_{i=1}^{3} \frac{1}{2} \left[(\partial_\mu \phi_i)^2 - m_i^2 \phi_i^2 \right] - \kappa \phi_1 \phi_2 \phi_3, \tag{2.84}$$

where the kinetic energy and mass terms generalize (2.20). No counting factor is needed in the interaction term because the fields are distinct. The tree-level decay amplitude is therefore

$$M_{fi} = -i\kappa, \tag{2.85}$$

i.e., the coefficient of $\phi_1\phi_2\phi_3$ in $i\mathcal{L}$. The decay rate is therefore

$$\Gamma = \frac{\kappa^2 p_2}{32\pi^2 m_1^2} \underbrace{\int d\Omega}_{4\pi} = \frac{\kappa^2 p_2}{8\pi m_1^2}. \tag{2.86}$$

More general techniques for evaluating phase space integrals are discussed in Appendix D and in (Barger and Phillips, 1997).

2.3.6 Loop Effects

In addition to the tree diagrams shown in Figure 2.3 for the potential in (2.22), there are a large number of loop diagrams. A sample of the many one-loop diagrams is shown in Figure 2.7. Of course, the relative strength of these and higher-loop diagrams depends on the magnitudes of λ and κ^2.

The first diagram is shown in more detail in Figure 2.8, with the momenta labeled. The external lines are on-shell, while the two internal lines are not. Four-momentum is conserved at each vertex, so only the single momentum k is integrated over. From the Feynman rules, this diagram contributes

$$M = \frac{(-i\lambda)^2}{2} \int \frac{d^4k}{(2\pi)^4} \frac{i}{k^2 - m^2 + i\epsilon} \frac{i}{(k - p_1 - p_2)^2 - m^2 + i\epsilon} \tag{2.87}$$

to the scattering amplitude, where the $\frac{1}{2}$ is due to the identical particles. The techniques for evaluating such integrals will be illustrated in Appendix E and are developed in detail in field theory texts. Here we just note that the integral is logarithmically divergent for large k (the

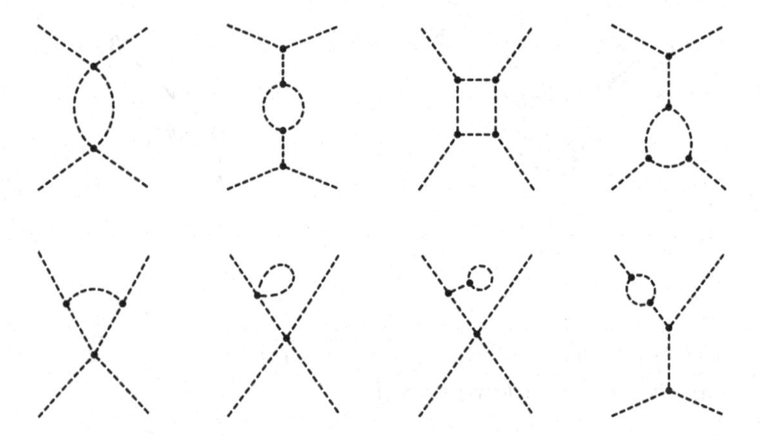

Figure 2.7 Examples of one-loop diagrams contributing to the process $p_1 p_2 \to p_3 p_4$.

integrand goes as d^4k/k^4). Fortunately, the theory is renormalizable,[6] so the divergences can all be absorbed in the observed ("renormalized") values of m, λ, and the "wave functions," leading to finite and calculable quantities. The pure ϕ^3 theory (i.e., $\lambda = 0$, $\kappa \neq 0$) is *super-renormalizable*: the only divergent diagrams are the two one-loop diagrams shown on the right in Figure 2.8. (These may appear as components of larger diagrams.) It should be noted

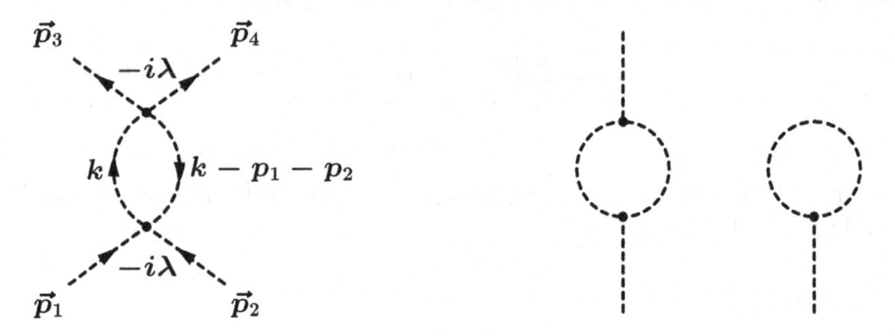

Figure 2.8 Left: A typical one-loop diagram, with the momenta flowing through each line labeled. Right: Divergent (sub)diagrams in the super-renormalizable ϕ^3 theory. The second (*tadpole*) diagram contributes to the d_1 term in (2.21) and must be included in the field redefinition that eliminates it.

that the modern view of divergences is that the integrals are not truly infinite. Rather, they reflect the view that a low energy field theory is an approximate description valid below some energy scale Λ, such as the Planck (gravity) scale $M_P = G_N^{-1/2} \sim 1.2 \times 10^{19}$ GeV, above which there is a more complete theory. A logarithmically divergent integral such as (2.87) is therefore proportional to $\lambda^2 \ln(\Lambda/m)$. As long as this quantity is small, the sensitivity to the new physics scale is not great. In addition, the leading logarithmic contributions

[6]In a renormalizable theory all divergences can be absorbed in a finite number of quantities, which can either be measured or drop out of final expressions for observables. Non-renormalizable theories, on the other hand, encounter new divergences at each order in perturbation theory.

can be summed by the renormalization group equations. This will be discussed further in Section 2.12.2.

2.4 THE COMPLEX SCALAR FIELD

Now let us consider the *complex*, i.e., non-Hermitian, scalar field $\phi \neq \phi^\dagger$, which describes a spin-0 particle that is not the same as its own antiparticle. For example, ϕ_{π^+} annihilates a π^+ or creates a π^-, while $\phi_{\pi^-} = \phi^\dagger_{\pi^+}$ is the corresponding π^- field.

The Lagrangian density for a complex scalar field is

$$\mathcal{L} = \mathcal{L}^\dagger = (\partial_\mu \phi)^\dagger (\partial^\mu \phi) - m^2 \phi^\dagger \phi - V_I(\phi, \phi^\dagger), \tag{2.88}$$

where the (Hermitian) potential V_I is given by[7]

$$V_I(\phi, \phi^\dagger) = \frac{\lambda}{4} (\phi^\dagger \phi)^2 + \text{non-renormalizable.} \tag{2.89}$$

There is no $\frac{1}{2}$ in the kinetic energy and mass terms for a complex field, and the $1/4$ in V_I is for later convenience. Keeping only the quartic term in V_I, the field equation is

$$\frac{\delta \mathcal{L}}{\delta \phi^\dagger} - \partial_\mu \frac{\delta \mathcal{L}}{\delta \partial_\mu \phi^\dagger} = 0 \quad \Rightarrow \quad \left(\Box + m^2\right) \phi + \frac{\lambda}{2} \phi (\phi^\dagger \phi) = 0. \tag{2.90}$$

A complex scalar field ϕ can always be expressed in terms of two real fields $\phi_i = \phi_i^\dagger$, $i = 1, 2$, by[8]

$$\phi = \frac{1}{\sqrt{2}} \left(\phi_1 + i \phi_2\right), \qquad \phi^\dagger = \frac{1}{\sqrt{2}} \left(\phi_1 - i \phi_2\right). \tag{2.91}$$

In terms of these, $\mathcal{L}$ in (2.88) can be written

$$\mathcal{L} = \sum_{i=1}^{2} \frac{1}{2} \left[(\partial_\mu \phi_i)^2 - m^2 \phi_i^2 \right] - \frac{\lambda}{16} (\phi_1^2 + \phi_2^2)^2. \tag{2.92}$$

This is not the most general renormalizable Lagrangian for two real scalars because the masses are the same and the potential is of a special form that leads to a conserved quantum number.

The solution for the free complex field (i.e., $\lambda = 0$) is

$$\begin{aligned} \phi_0(x) &= \int \frac{d^3 \vec{p}}{(2\pi)^3 2E_p} \left[a(\vec{p}) e^{-ip\cdot x} + b^\dagger(\vec{p}) e^{+ip\cdot x} \right] \\ \phi_0^\dagger(x) &= \int \frac{d^3 \vec{p}}{(2\pi)^3 2E_p} \left[b(\vec{p}) e^{-ip\cdot x} + a^\dagger(\vec{p}) e^{+ip\cdot x} \right], \end{aligned} \tag{2.93}$$

where $a^\dagger(\vec{p})$ and $b^\dagger(\vec{p})$ create π^+ and π^- states, respectively, as in (2.8). They satisfy the

[7]There are other possible renormalizable terms in V_I, such as $\sigma_n[(\phi)^n + (\phi^\dagger)^n]$, $n = 1 \cdots 4$, or $\rho_m[(\phi)^m + (\phi^\dagger)^m]\phi^\dagger\phi$, $m = 1, 2$. These would not allow a conserved charge, and would lead to processes such as $\pi^+\pi^+ \to \pi^-\pi^-$ (Problem 2.6). Equivalently, they would not exhibit the $U(1)$ phase symmetry discussed in Section 2.4.1.

[8]It is a convention whether $\phi = (\phi_1 + i\phi_2)/\sqrt{2}$ is identified as ϕ_{π^+}, as is done here, or with ϕ_{π^-}. In later chapters we will often take the opposite choice, especially to be consistent with isospin conventions.

commutation rules in (2.2) and (2.9). a and b can be expressed in terms of the creation operators for the real fields ϕ_{10} and ϕ_{20} by

$$\begin{aligned} a(\vec{p}) &= \frac{1}{\sqrt{2}}[a_1(\vec{p}) + ia_2(\vec{p})], & b(\vec{p}) &= \frac{1}{\sqrt{2}}[a_1(\vec{p}) - ia_2(\vec{p})] \\ a^\dagger(\vec{p}) &= \frac{1}{\sqrt{2}}[a_1^\dagger(\vec{p}) - ia_2^\dagger(\vec{p})], & b^\dagger(\vec{p}) &= \frac{1}{\sqrt{2}}[a_1^\dagger(\vec{p}) + ia_2^\dagger(\vec{p})]. \end{aligned} \tag{2.94}$$

The Feynman propagator for the complex field is

$$\langle 0|\mathcal{T}[\phi_0(x), \phi_0^\dagger(x')]|0\rangle = i\Delta_F(x - x') = i\int \frac{d^4k}{(2\pi)^4} \frac{e^{-ik\cdot(x-x')}}{k^2 - m^2 + i\epsilon}, \tag{2.95}$$

so that an internal line carrying momentum k in a Feynman diagram is associated with the factor $i\Delta_F(k)$ defined in (2.29).

In Feynman diagrams, the arrows represent the direction of flow of positive charge. For example, an external line with an arrow going into the diagram may represent either an initial state π^+ *or* a final state π^-, and conversely for an arrow leaving the diagram. Each vertex is associated with a factor $-i\lambda$ and must involve the same number of entering and exiting lines (charge conservation). It is conventional to label internal momenta so that the arrow also coincides with the direction of momentum flow, while the momenta for external lines are the *physical* momenta, which are entering the diagram for initial particles and exiting for final particles. Thus, the physical momentum flows opposite to the arrow for a π^-, as illustrated in Figure 2.9. We also introduce the symbol $\bar{p}$ for the momentum flowing in the direction of the arrow, i.e., $\bar{p}_{\pi^\pm} = \pm p_{\pi^\pm}$.

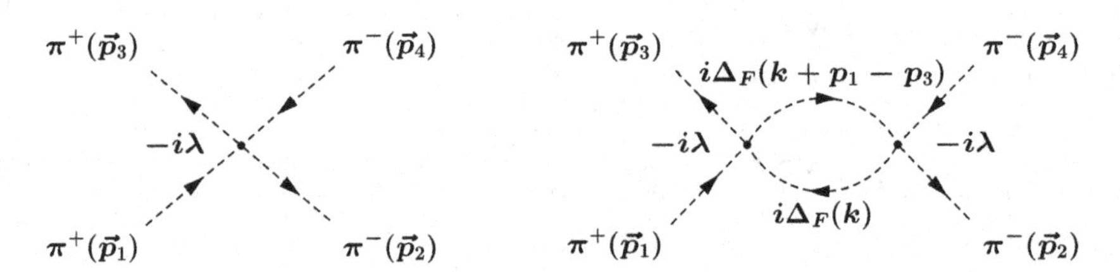

Figure 2.9 Left: Tree-level diagram for $\pi^+(\vec{p}_1)\,\pi^-(\vec{p}_2) \to \pi^+(\vec{p}_3)\,\pi^-(\vec{p}_4)$, where $\vec{p}_{1,2}$ ($\vec{p}_{3,4}$) are the physical initial (final) momenta. Right: An example of a one-loop diagram. $\Delta_F(k)$ is defined in (2.29).

2.4.1 $U(1)$ Phase Symmetry and the Noether Theorem

The Lagrangian density in (2.88) is invariant under the *phase transformations*

$$\phi(x) \ \to \ \phi'(x) \equiv e^{i\beta}\phi(x) \xrightarrow[\beta \text{ small}]{} \phi(x) + \delta\phi(x), \tag{2.96}$$

where $\delta\phi(x) \equiv i\beta\phi(x)$, which means that

$$\mathcal{L}(\phi, \phi^\dagger) = \mathcal{L}(\phi', \phi'^\dagger). \tag{2.97}$$

The invariance holds for all β, i.e., $\mathcal{L}$ has a $U(1)$ *symmetry group.* $U(1)$ is the group of 1×1 unitary matrices, which is a fancy expression for phase factors. The symmetry is *global,*[9] i.e., β is a constant, independent of x.

According to the *Noether theorem,* any continuous symmetry of the action leads to a conserved current and conserved charge. For example, translation invariance of the action under $x_\mu \rightarrow x'_\mu = x_\mu - a_\mu$ leads to energy and momentum conservation, with the four-momentum operator $P^\mu = (H, \vec{P})$ in (A.5) on page 507 conserved (see, e.g., Bjorken and Drell, 1965). In the present case, invariance of $\mathcal{L}$ automatically implies that the action is invariant also. As will be shown below, the Noether theorem then implies that

$$\delta\mathcal{L} = -\beta \partial_\mu J^\mu, \tag{2.98}$$

where

$$\delta\mathcal{L} \equiv \mathcal{L}(\phi', \phi'^\dagger) - \mathcal{L}(\phi, \phi^\dagger) = 0 \tag{2.99}$$

is the change in the Lagrangian density under (2.96) for infinitesimal β. In (2.98), J^μ is the conserved *Noether current*

$$J^\mu = -i\left[(\partial^\mu \phi)^\dagger \phi - \phi^\dagger \partial^\mu \phi\right] = i\phi^\dagger \overleftrightarrow{\partial}^\mu \phi. \tag{2.100}$$

In the last expression, we have introduced the symbol $\overleftrightarrow{\partial}^\mu$, defined by

$$a \overleftrightarrow{\partial}^\mu b \equiv a\, \partial^\mu b - (\partial^\mu a)\, b \tag{2.101}$$

for any a and b.

The *Noether charge*

$$Q \equiv \int d^3\vec{x}\; J^0(x) \tag{2.102}$$

is then conserved, assuming there is no current flow at infinity, because

$$\begin{aligned} \frac{\partial Q}{\partial t} &= \int d^3\vec{x}\; \frac{\partial J^0(x)}{\partial t} = \int d^3\vec{x} \left[\frac{\partial J^0(x)}{\partial t} + \underbrace{\vec{\nabla} \cdot \vec{J}(x)}_{\text{integral } = 0}\right] \\ &= \int d^3\vec{x}\; \partial_\mu J^\mu(x) = 0. \end{aligned} \tag{2.103}$$

This is most easily interpreted by explicitly calculating Q for non-interacting fields ϕ_0, for which one has easily that

$$Q = \int \frac{d^3\vec{p}}{(2\pi)^3 2E_p} \left[a^\dagger(\vec{p}) a(\vec{p}) - b^\dagger(\vec{p}) b(\vec{p})\right] = N_{\pi^+} - N_{\pi^-}. \tag{2.104}$$

That is, the conserved Noether charge of a state is just the total number of particles minus the number of antiparticles, which coincides with electric charge for the specific example. It is straightforward to show that the conservation law continues to hold in every order of perturbation theory provided that the interaction term respects the symmetry (commutes with Q).

---

[9]The derivative terms in $\mathcal{L}$ would not be invariant for $\beta = \beta(x)$. Later, we will discuss how to modify $\mathcal{L}$ to yield *local (gauge)* symmetries under which $\phi(x) \rightarrow e^{i\beta(x)}\phi(x)$.

Derivation of the Noether Current for the Complex Scalar Field

Under the infinitesimal phase transformation (2.96), the change in the Lagrangian density is

$$\delta\mathcal{L} = \frac{\delta\mathcal{L}}{\delta\phi}\delta\phi + \frac{\delta\mathcal{L}}{\delta(\partial_\mu\phi)}\delta(\partial_\mu\phi) + \frac{\delta\mathcal{L}}{\delta\phi^\dagger}\delta\phi^\dagger + \frac{\delta\mathcal{L}}{\delta(\partial_\mu\phi)^\dagger}\delta(\partial_\mu\phi)^\dagger, \tag{2.105}$$

where

$$\begin{aligned} \delta\phi(x) &= i\beta\phi(x), & \delta\phi(x)^\dagger &= -i\beta\phi(x)^\dagger \\ \delta(\partial_\mu\phi) &= i\beta\partial_\mu\phi(x), & \delta(\partial_\mu\phi)^\dagger &= -i\beta\left(\partial_\mu\phi(x)\right)^\dagger . \end{aligned} \tag{2.106}$$

Using the Euler-Lagrange equations (2.19), this is

$$\delta\mathcal{L} = i\beta\left[\partial_\mu\left(\frac{\delta\mathcal{L}}{\delta(\partial_\mu\phi)}\phi\right) - \partial_\mu\left(\frac{\delta\mathcal{L}}{\delta(\partial_\mu\phi)^\dagger}\phi^\dagger\right)\right] = -\beta\partial_\mu J^\mu, \tag{2.107}$$

where the Noether current is

$$J^\mu \equiv -i\left[\frac{\delta\mathcal{L}}{\delta(\partial_\mu\phi)}\phi - \frac{\delta\mathcal{L}}{\delta(\partial_\mu\phi)^\dagger}\phi^\dagger\right] = +i\phi^\dagger\overleftrightarrow{\partial^\mu}\phi. \tag{2.108}$$

Finally, for a symmetry[10], one has in addition that the Lagrangian density is unchanged, i.e., $\delta\mathcal{L} = 0$, so that $\partial_\mu J^\mu = 0$.

2.5 ELECTROMAGNETIC AND VECTOR FIELDS

Define the electromagnetic *vector potential* $A^\mu(x)$ and the *field strength tensor*

$$F^{\mu\nu} = -F^{\nu\mu} = \partial^\mu A^\nu - \partial^\nu A^\mu, \tag{2.109}$$

related to the electric and magnetic fields $\vec{E}$ and $\vec{B}$ by

$$F^{\mu\nu} = \begin{pmatrix} 0 & -E_x & -E_y & -E_z \\ E_x & 0 & -B_z & B_y \\ E_y & B_z & 0 & -B_x \\ E_z & -B_y & B_x & 0 \end{pmatrix}. \tag{2.110}$$

In the absence of sources, one has the free-field Lagrangian density

$$\mathcal{L}_0(A^\mu) = -\frac{1}{4}F_{\mu\nu}F^{\mu\nu} = \frac{1}{2}\left(\vec{E}^2 - \vec{B}^2\right), \tag{2.111}$$

leading to the free-field equation of motion $\Box A^\mu - \partial^\mu(\partial_\nu A^\nu) = 0$. (We have omitted the free-field subscript 0 on F and A for simplicity.)

$\mathcal{L}_0$ is *gauge invariant*, i.e., $F^{\mu\nu} = F'^{\mu\nu} = \partial^\mu A'^\nu - \partial^\nu A'^\mu$, and therefore $\mathcal{L}_0(A^\mu) = \mathcal{L}_0(A'^\mu)$, where

$$A'^\mu = A^\mu - \frac{1}{e}\partial^\mu\beta(x), \tag{2.112}$$

[10]The condition for a symmetry is actually weaker. It is sufficient for the action in (2.18) to be invariant. We will see examples of this for discrete symmetries in Section 2.10 and Chapter 7. For a continuous symmetry, it is sufficient to have $\delta\mathcal{L} = -\beta\partial_\mu a^\mu$, where a^μ transforms as a Lorentz four-vector. In that case, the current $J^\mu - a^\mu$ is conserved.

and $\beta(x)$ is an arbitrary differentiable function of $(t, \vec{x})$. (The $1/e$ factor, where $e > 0$ is the electric charge of the positron, is inserted for convenience.) This is known as a $U(1)$ invariance because there is only one function $\beta(x)$. The $U(1)$ gauge invariance will continue to hold for the full Lagrangian including interactions. It is convenient to work in the *Lorenz gauges*, $\partial_\nu A^\nu = 0$. This still leaves the freedom of making a further gauge transformation provided $\Box\beta = 0$, and one can exploit this freedom to simultaneously choose $A^0 = 0$ and $\vec{\nabla} \cdot \vec{A} = 0$, the *Coulomb* or *radiation gauge.* The radiation gauge does not have manifest Lorentz or gauge invariance, but is attractive in that it doesn't involve any unphysical degrees of freedom. The two non-zero components of A^μ are transverse to the photon momentum and correspond to the two polarization directions. In the radiation gauge, the free-field expression for the vector potential is

$$\vec{A}(x) = \int \frac{d^3\vec{p}}{(2\pi)^3 2E_p} \sum_{\lambda=1,2} \left[\vec{\epsilon}(\vec{p},\lambda)a(\vec{p},\lambda)e^{-ip\cdot x} + \vec{\epsilon}^{\,*}(\vec{p},\lambda)a^\dagger(\vec{p},\lambda)e^{+ip\cdot x}\right], \tag{2.113}$$

where $E_p = |\vec{p}\,|$, a and $a^\dagger$ are bosonic annihilation and creation operators satisfying

$$\left[a(\vec{p},\lambda), a^\dagger(\vec{p}\,',\lambda')\right] = (2\pi)^3 2E_p\, \delta^3(\vec{p} - \vec{p}\,')\delta_{\lambda\lambda'}, \tag{2.114}$$

$\lambda = 1, 2$ refers to the two possible polarization states, and $\vec{\epsilon}(\vec{p},\lambda)$ is the photon *polarization* in the direction of the photon electric field. The radiation gauge condition $\vec{\nabla} \cdot \vec{A} = 0$ implies

$$\vec{p} \cdot \vec{\epsilon}(\vec{p},\lambda) = 0, \tag{2.115}$$

and the normalization is

$$\vec{\epsilon}^{\,*}(\vec{p},\lambda) \cdot \vec{\epsilon}(\vec{p},\lambda') = \delta_{\lambda\lambda'}. \tag{2.116}$$

For a linear polarization basis, the $\vec{\epsilon}(\vec{p},\lambda)$ are real and can be chosen

$$\vec{\epsilon}(\vec{p},1) \times \vec{\epsilon}(\vec{p},2) = \hat{p}. \tag{2.117}$$

Thus, one can choose $\vec{\epsilon}(\vec{p},1) = \hat{p}_\perp$ where $\hat{p}_\perp$ is an arbitrary unit vector orthogonal to $\vec{p}$, and $\vec{\epsilon}(\vec{p},2) = \hat{p} \times \hat{p}_\perp$; e.g., for $\vec{p}$ in the z direction, $\vec{\epsilon}(\vec{p},1) = (1,0,0)$ and $\vec{\epsilon}(\vec{p},2) = (0,1,0)$. One can also use the basis of left- and right-handed circular polarization vectors

$$\vec{\epsilon}_{L,R}(\vec{p}) = \frac{\vec{\epsilon}(\vec{p},1) \mp i\vec{\epsilon}(\vec{p},2)}{\sqrt{2}}, \tag{2.118}$$

which correspond to photon helicities (spin measured with respect to the momentum direction) of ∓ 1, respectively. It is useful to introduce the polarization four-vectors $\epsilon^\mu(\vec{p},\lambda) = (0, \vec{\epsilon}(\vec{p},\lambda))$, where

$$p \cdot \epsilon(\vec{p},\lambda) = 0, \qquad \epsilon^*(\vec{p},\lambda) \cdot \epsilon(\vec{p},\lambda') = -\delta_{\lambda\lambda'}. \tag{2.119}$$

Thus, the free-field expression for A^μ is obtained from (2.113) by replacing $\vec{\epsilon}(\vec{p},\lambda) \to \epsilon^\mu(\vec{p},\lambda)$.

From (2.113) one sees that the amplitude for a physical process is of the form $\epsilon^\mu(\vec{p},\lambda) \cdot M_\mu$ for an initial photon, or $\epsilon^{\mu *}(\vec{p},\lambda) \cdot M_\mu$ for a final photon, where the amplitude M_μ depends on p_μ and the other momenta and spins in the process. (The rules for constructing M_μ will be described in later sections.) Often, the initial photon beam is unpolarized, or a final photon polarization is not measured. In that case, one averages (sums) over the initial (final) polarization directions, yielding a rate proportional to

$$\frac{1}{2}\left[\sum_{\lambda=1,2} \epsilon^\mu(\vec{p},\lambda)\epsilon^{\nu *}(\vec{p},\lambda)\right] M_\mu M_\nu^* \tag{2.120}$$

(or $\frac{1}{2}M_\mu M_\nu^* \to M_\mu^* M_\nu$ for a final photon). This is easily evaluated using

$$\sum_{\lambda=1,2} \epsilon^\mu(\vec{p},\lambda)\epsilon^{\nu *}(\vec{p},\lambda) = -g^{\mu\nu} + \frac{p^\mu p_r^\nu + p^\nu p_r^\mu}{p \cdot p_r}, \tag{2.121}$$

where $p_r \equiv (E_p, -\vec{p})$ so that $p_r \cdot p = 2E_p^2$. One can use gauge invariance to show that $p^\mu M_\mu = 0$, so that the second term in (2.121) does not contribute,[11] i.e., (2.120) is just $-M_\mu M^{\mu *}$.

The Feynman propagator for the free photon field is

$$\langle 0|\mathcal{T}[A^\mu(x), A^\nu(x')]|0\rangle = ig^{\mu\nu} \int \frac{d^4k}{(2\pi)^4} e^{-ik\cdot(x-x')} D_F(k) + \text{ gauge terms}, \tag{2.122}$$

where

$$D_F(k) = -\frac{1}{k^2 + i\epsilon}. \tag{2.123}$$

The additional terms in (2.122) are gauge-dependent (e.g., for the radiation gauge they involve the generalization of the second term in (2.121) to off-shell momenta). A full treatment of gauge issues is beyond the scope of this book (but see the discussion of the R_ξ gauges in Chapter 4). However, the additional terms do not contribute to physical processes by gauge invariance, so in practice one can simply assign a factor $ig^{\mu\nu}D_F(k) = -ig^{\mu\nu}/(k^2 + i\epsilon)$ for an internal photon line carrying momentum k in a Feynman diagram.

2.5.1 Massive Neutral Vector Field

The results for the free electromagnetic field are easily generalized to the case of a massive Hermitian vector (spin-1) boson V^μ. The free-field Lagrangian density is

$$\mathcal{L}_0 = -\frac{1}{4}G_{\mu\nu}G^{\mu\nu} + \frac{1}{2}m^2 V_\mu V^\mu, \text{ with } G^{\mu\nu} = \partial^\mu V^\nu - \partial^\nu V^\mu. \tag{2.124}$$

It should be emphasized that $\mathcal{L}_0$ is *not* gauge invariant under

$$V^\mu \to V^\mu + \partial^\mu \beta, \tag{2.125}$$

because of the mass term, and does not lead to a renormalizable theory. However, it is useful to display the relevant formulae useful for phenomenological tree-level calculations.

The free field is given by

$$V^\mu(x) = \int \frac{d^3\vec{p}}{(2\pi)^3 2E_p} \sum_{\lambda=1}^{3} \left[\epsilon^\mu(\vec{p},\lambda)a(\vec{p},\lambda)e^{-ip\cdot x} + \epsilon^{\mu *}(\vec{p},\lambda)a^\dagger(\vec{p},\lambda)e^{+ip\cdot x}\right], \tag{2.126}$$

with $E_p = \sqrt{\vec{p}^2 + m^2}$. A massive vector has three polarization states. These include the two transverse states $\epsilon(\vec{p},1)$ and $\epsilon(\vec{p},2)$ (or $\epsilon_{L,R}(\vec{p})$) similar to the massless case, and a third *longitudinal* (helicity-0) state

$$\epsilon(\vec{p},3) = \frac{1}{m}(|\vec{p}|, 0, 0, E_p), \tag{2.127}$$

[11]The analogous situation for QCD and other non-abelian theories is trickier; one can only drop the second term if ghost contributions are included (e.g., Peskin and Schroeder, 1995). Another subtlety will be encounted in the example of Compton scattering in Section 2.6.

where we have taken $\vec{p}$ along the z direction for definiteness. Note that the magnitudes of the components of $\epsilon(\vec{p},3)$ become large at high energy, with $\epsilon^\mu(\vec{p},3) \sim p^\mu/m$. The three polarization vectors satisfy (2.119), and the polarization sum is

$$\sum_{\lambda=1}^{3} \epsilon^\mu(\vec{p},\lambda)\epsilon^{\nu *}(\vec{p},\lambda) = -g^{\mu\nu} + \frac{p^\mu p^\nu}{m^2}. \tag{2.128}$$

The momentum space propagator associated with an internal line in a Feynman diagram is

$$iD_V^{\mu\nu}(k) = i\left[\frac{-g^{\mu\nu} + \frac{k^\mu k^\nu}{m^2}}{k^2 - m^2 + i\epsilon}\right]. \tag{2.129}$$

The second term in D_V does *not* drop out of calculations and leads to bad ultraviolent (large k) behavior and thus to non-renormalizability. One also sees that the limit $m \to 0$ is not smooth.

A renormalizable gauge invariant theory of massive spin-1 fields, in which the mass is obtained by the *Higgs mechanism*[12] rather than as an elementary term in $\mathcal{L}$, will be discussed in Chapter 4. In that case, (2.129) will still correspond to the *unitary gauge*, in which the physical degrees of freedom are manifest, but there are other gauges in which the $m \to 0$ limit is smooth.

2.6 ELECTROMAGNETIC INTERACTION OF CHARGED PIONS

We are now ready to combine the results of Sections 2.4 and 2.5. Consider the Lagrangian density

$$\mathcal{L}(\phi, A) = (\partial_\mu \phi)^\dagger \partial^\mu \phi - m^2 \phi^\dagger \phi - \frac{1}{4} F_{\mu\nu}F^{\mu\nu} - V_I(\phi,\phi^\dagger) + \mathcal{L}_{\phi A}(\phi, A) \tag{2.130}$$

for a complex scalar field ϕ and the electromagnetic field A^μ. The scalar self-interaction V_I is defined in (2.89), and $\mathcal{L}_{\phi A}$ describes the electromagnetic interaction. Its form is dictated by the requirement of gauge invariance under (2.112). The only way to do this (without introducing non-renormalizable interactions) is the *minimal electromagnetic substitution*, familiar from classical and quantum mechanics. One replaces

$$p^\mu \to p^\mu - qA^\mu \Longleftrightarrow i\partial^\mu - qA^\mu \tag{2.131}$$

in the Lagrangian density in (2.88), where $q = e > 0$ is the charge of the π^+, and identifies the additional terms with $\mathcal{L}_{\phi A}$. $\mathcal{L}$ will then be invariant under the generalized gauge (or *local*) transformation

$$\begin{aligned} A^\mu \to A'^\mu &= A^\mu - \frac{1}{e}\partial^\mu \beta(x) \\ \phi \to \phi' &= e^{iq\beta(x)/e}\phi = e^{i\beta(x)}\phi, \end{aligned} \tag{2.132}$$

i.e., $\mathcal{L}(\phi, A) = \mathcal{L}(\phi', A')$. Equation (2.132) generalizes the global symmetry of Section 2.4.1, i.e., $\phi \to e^{i\beta}\phi$ where $\beta =$ constant.

[12]It is also possible to construct a $U(1)$ gauge invariant theory for a massive vector without the Higgs or other spontaneous symmetry breaking mechanism by the *Stückelberg mechanism* (Stueckelberg, 1938; Cianfrani and Lecian, 2007).

Using the minimal substitution,

$$\begin{aligned}\mathcal{L} &= \underbrace{[(\partial_\mu + iqA_\mu)\phi]^\dagger}_{(\partial_\mu - iqA_\mu)\phi^\dagger}(\partial^\mu + iqA^\mu)\phi - m^2\phi^\dagger\phi - \frac{1}{4}F_{\mu\nu}F^{\mu\nu} - V_I(\phi,\phi^\dagger) \\ &= [D_\mu\phi]^\dagger[D^\mu\phi] - m^2\phi^\dagger\phi - \frac{1}{4}F_{\mu\nu}F^{\mu\nu} - V_I(\phi,\phi^\dagger),\end{aligned} \tag{2.133}$$

where

$$D^\mu \equiv \partial^\mu + iqA^\mu \tag{2.134}$$

is known as the *gauge covariant derivative*.

The gauge invariance of (2.133) is obvious except for the first term. Under (2.132),

$$\begin{aligned}&D^\mu\phi \to D'^\mu\phi' = \left[\partial^\mu + iqA^\mu - i\frac{q}{e}\partial^\mu\beta(x)\right]\left(e^{i\beta(x)}\phi\right) = e^{i\beta(x)}D^\mu\phi \\ &(D_\mu\phi)^\dagger \to (D_\mu\phi)^\dagger\, e^{-i\beta(x)}.\end{aligned} \tag{2.135}$$

That is, the shift in A^μ compensates for the change in $\partial^\mu\phi$ due to the derivative of β, leaving $\mathcal{L}$ gauge invariant. Thus, gauge invariance for the charged scalar requires the existence of a spin-1 field, dictates the form of the interaction with A^μ, forbids an elementary mass term (i.e., $\frac{1}{2}m_A^2 A_\mu A^\mu$ is not gauge invariant), and restricts the form of the scalar self-interactions. It turns out that the theory is then renormalizable.

From (2.133) one can read off $i\mathcal{L}_{\phi A}$:

$$\begin{aligned}i\mathcal{L}_{\phi A} &= -q(\partial^\mu\phi)^\dagger\phi A_\mu + q(\phi^\dagger\partial^\mu\phi)A_\mu + iq^2A_\mu A^\mu\phi^\dagger\phi \\ &\equiv q(\phi^\dagger\overleftrightarrow{\partial}^\mu\phi)A_\mu + iq^2A_\mu A^\mu\phi^\dagger\phi,\end{aligned} \tag{2.136}$$

where $i\phi^\dagger\overleftrightarrow{\partial}^\mu\phi$ is the Noether current for the free complex scalar field[13] in (2.108). The Feynman vertex rules can be found from (2.136), inserting the free-field expressions for ϕ and $\phi^\dagger$ in (2.93) and recalling that $a(\vec{p})$ and $b(\vec{p})$ are the annihilation operators for π^+ and π^-, respectively. For this purpose, it is convenient to explicitly display

$$\begin{aligned}\partial^\mu\phi_0(x) &= \int\frac{d^3\vec{p}}{(2\pi)^3 2E_p}\left[-ip^\mu a(\vec{p})e^{-ip\cdot x} + ip^\mu b^\dagger(\vec{p})e^{+ip\cdot x}\right] \\ \partial^\mu\phi_0^\dagger(x) &= \int\frac{d^3\vec{p}}{(2\pi)^3 2E_p}\left[-ip^\mu b(\vec{p})e^{-ip\cdot x} + ip^\mu a^\dagger(\vec{p})e^{+ip\cdot x}\right],\end{aligned} \tag{2.137}$$

where $p = (E_p, \vec{p})$ with $E_p \equiv \sqrt{\vec{p}^2 + m^2}$ as usual.

The vertices are displayed in Figure 2.10 for $q = e$. The three-point vertices include a contribution $-iep^\mu_{\pi^+}$ for each π^+ and a $+iep^\mu_{\pi^-}$ for each π^-. These are always the physical momenta, whether initial or final. They can both be written as $-ie\bar{p}^\mu_{\pi^\pm}$, where $\bar{p}_{\pi^\pm} = \pm p_{\pi^\pm}$ is the momentum in the direction of the arrow. There is also a four-point (*seagull*) vertex $2ie^2g^{\mu\nu}$, which is needed for gauge invariance. The Lorentz indices are contracted with $\epsilon_\mu(k,\lambda)$ for an initial photon of momentum k, with $\epsilon^*_\mu(k,\lambda)$ for a final photon, and with $ig_{\mu\nu}D_F(k) = -ig_{\mu\nu}/(k^2 + i\epsilon)$ for an internal virtual photon line. Each internal pion has a propagator $i\Delta_F(k) = i/(k^2 - m^2 + i\epsilon)$, and there is an integral $\int d^4k/(2\pi)^4$ over each unconstrained internal momentum.

---

[13]Interaction terms usually do not modify the form of the Noether currents. Gauge interactions of complex scalars are an exception because the interaction terms involve derivatives. For example, the conserved Noether current for $\mathcal{L}$ in (2.133) is $J^\mu = i\phi^\dagger\overleftrightarrow{\partial}^\mu\phi - 2q\phi^\dagger A^\mu\phi$, which appears in the field equation for A in the familiar Maxwell form $\partial_\mu F^{\mu\nu} = qJ^\nu$.

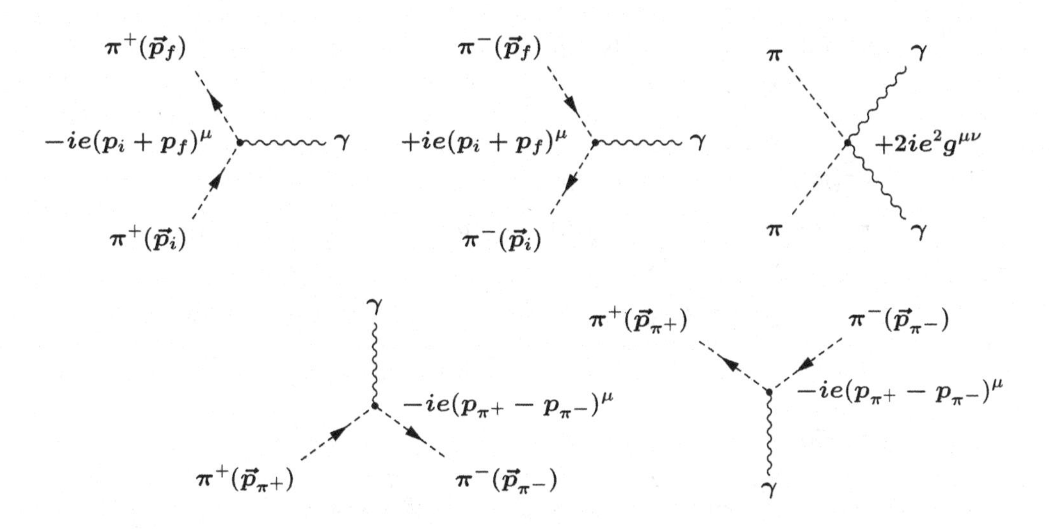

Figure 2.10 Vertices involving charged pions and one or two photons. In the one-photon diagrams the initial pions enter from the bottom and the final leave from the top. Antiparticle (π^-) vertices are obtained from particle ones by twisting the lines and replacing p by $\bar{p} \equiv -p$, where p is the physical four-momentum and $\bar{p}$ follows the direction of the arrow. The wavy lines represent photons.

πK and $\pi\pi$ Scattering

As a first example, consider the electromagnetic scattering $\pi^- K^+ \to \pi^- K^+$, where $K^\pm$ is a complex field with the same electromagnetic couplings (but different mass) as $\pi^\pm$. They are introduced here to avoid identical particle effects. We ignore strong interactions of other non-electromagnetic couplings. There is a single tree-level diagram, as shown in Figure 2.11. The Feynman rules lead to the transition amplitude

$$\begin{aligned} M_{fi} &= ie(p_1+p_3)^\mu \left[\frac{-ig_{\mu\nu}}{(p_1-p_3)^2}\right](-ie)(p_2+p_4)^\nu \\ &= -4\pi i\alpha \frac{(p_1+p_3)\cdot(p_2+p_4)}{(p_1-p_3)^2} = -4\pi i\alpha\left(\frac{s-u}{t}\right), \end{aligned} \tag{2.138}$$

where $\alpha \equiv e^2/4\pi$ is the fine structure constant. The Mandelstam variables s, t, and u are defined in (2.31) and are related to the CM scattering angle in (2.40). From (2.57) the differential cross section in the CM is

$$\begin{aligned} \frac{d\sigma}{d\cos\theta} &= \frac{1}{32\pi s}|M_{fi}|^2 = \frac{\pi\alpha^2}{2s}\left(\frac{s/2p^2+1+\cos\theta}{1-\cos\theta}\right)^2 \\ &\xrightarrow[\sqrt{s}\gg m_K]{} \frac{\pi\alpha^2}{2s}\left(\frac{3+\cos\theta}{1-\cos\theta}\right)^2, \end{aligned} \tag{2.139}$$

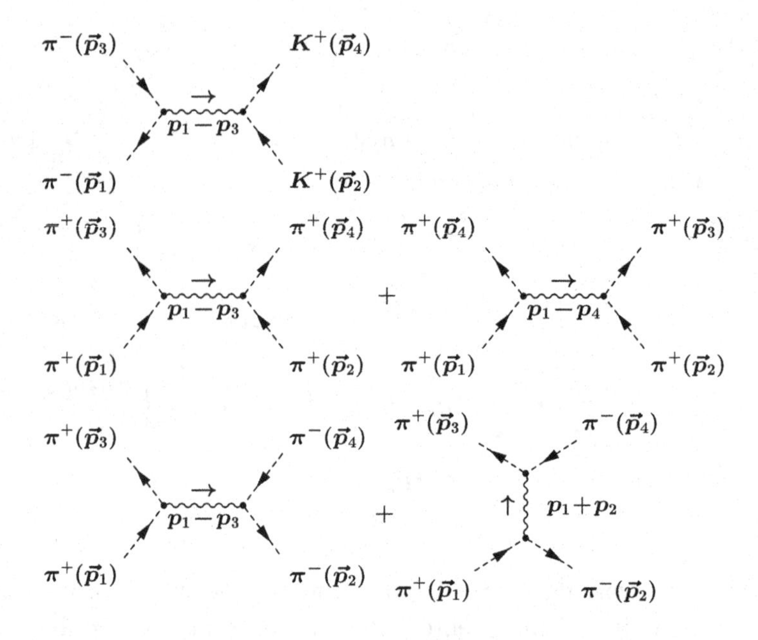

Figure 2.11 Feynman diagrams for $\pi^- K^+ \to \pi^- K^+$ (top), $\pi^+\pi^+ \to \pi^+\pi^+$ (middle), and $\pi^+\pi^- \to \pi^+\pi^-$ (bottom).

where $p \sim (s - m_K^2)/2\sqrt{s}$ since $m_K \gg m_\pi$. The total cross section is[14]

$$\sigma = \int_{-1}^{+1} \frac{d\sigma}{d\cos\theta} d\cos\theta. \tag{2.140}$$

For $\pi^+\pi^+ \to \pi^+\pi^+$ there are two Feynman diagrams due to the different ways of associating the fields in $i\mathcal{L}_{\phi A}$ with the identical external particles, as shown in Figure 2.11. The transition amplitude is

$$\begin{aligned} M_{fi} &= ie^2 \left[\frac{(p_1+p_3)\cdot(p_2+p_4)}{(p_3-p_1)^2} + \frac{(p_1+p_4)\cdot(p_2+p_3)}{(p_4-p_1)^2} \right] \\ &= 4\pi i\alpha \left[\frac{s-u}{t} + \frac{s-t}{u} \right]. \end{aligned} \tag{2.141}$$

The differential cross section is again given by (2.57), but in this case there is an extra factor $\frac{1}{2}$ in the total cross section because the final particles are identical. At high energies, $\sqrt{s} \gg m_\pi$, one finds

$$\frac{d\sigma}{d\cos\theta} = \frac{2\pi\alpha^2}{s}\left(\frac{3+\cos^2\theta}{\sin^2\theta}\right)^2, \qquad \sigma = \frac{1}{2}\int_{-1}^{+1} \frac{d\sigma}{d\cos\theta} d\cos\theta. \tag{2.142}$$

[14]The integral is divergent for forward scattering due to the massless photon propagator pole, i.e., the long range Coulomb force. This divergence is already present for classical Coulomb scattering, and disappears for realistic situations in which screening by other charges or the finite resolution of the detector are taken into account. Similar comments apply to $\pi^+\pi^+$ for both forward and backward scattering and to $\pi^-\pi^+$.

There are similarly two diagrams for $\pi^+\pi^- \to \pi^+\pi^-$, yielding

$$\begin{aligned} M_{fi} &= ie^2\left[\frac{-(p_1+p_3)\cdot(p_2+p_4)}{(p_3-p_1)^2}+\frac{(p_1-p_2)\cdot(p_3-p_4)}{(p_1+p_2)^2}\right] \\ &= 4\pi i\alpha\left[\frac{-s+u}{t}+\frac{u-t}{s}\right] = 4\pi i\alpha\left(\frac{3+\cos^2\theta}{1-\cos\theta}\right). \end{aligned} \tag{2.143}$$

In this case, however, the final particles are not identical. Note that the amplitude for $\pi^+\pi^-$ can be obtained from that for $\pi^+\pi^+$ by the formal substitutions $p_4 \to -p_2$ and $p_2 \to -p_4$, an example of *crossing* symmetry. That is, the amplitude for an outgoing π^+ of momentum p is the same as that for an incoming π^- with momentum $-p$, as is apparent from the Feynman rules. Of course, the physical values of p are different for the two cases, since p^0 $(-p^0)$ must be positive for the first (second) one.

Pion Compton Scattering

Now consider pion Compton scattering, $\gamma(\vec{k}_1,\lambda_1)\,\pi^+(\vec{p}_1) \to \gamma(\vec{k}_2,\lambda_2)\,\pi^+(\vec{p}_2)$, with $s = (k_1+p_1)^2$, $t = (k_2-k_1)^2$, and $u = (p_2-k_1)^2$. The amplitude corresponding to the diagrams in Figure 2.12 is

$$\begin{aligned} M_{fi} = \epsilon^*_{2\mu}\epsilon_{1\nu}\Big[&-ie(k_2+2\,p_2)^\mu\frac{i}{s-m^2}(-ie)(k_1+2\,p_1)^\nu \\ &-ie(2\,p_1-k_2)^\mu\frac{i}{u-m^2}(-ie)(2\,p_2-k_1)^\nu \;+\; 2\,ie^2g^{\mu\nu}\Big], \end{aligned} \tag{2.144}$$

where $\epsilon_2 \equiv \epsilon(\vec{k}_2,\lambda_2)$ and $\epsilon_1 \equiv \epsilon(\vec{k}_1,\lambda_1)$. It is straightforward to show that M_{fi} vanishes

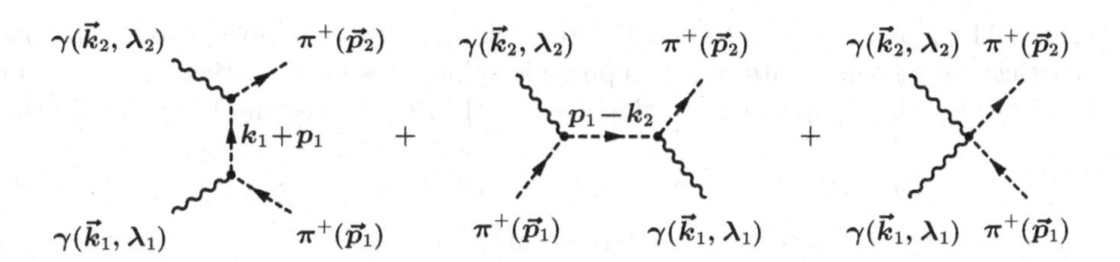

Figure 2.12 Diagrams for pion Compton scattering. The third (seagull) diagram is required by gauge invariance.

if $\epsilon_{1\nu}$ is replaced by $k_{1\nu}$ or $\epsilon^*_{2\mu}$ by $k_{2\mu}$, which is a manifestation of gauge invariance. This would *not* occur without the $2ie^2g^{\mu\nu}$.

Recall that our radiation gauge formalism is not *manifestly* Lorentz invariant. It is convenient to choose $A^0 = 0$ and $\vec{\nabla}\cdot\vec{A} = 0$ to hold in the laboratory frame, in which $p_1 = (m,\ \vec{0})$. Then, $p_1\cdot\epsilon_1 = p_1\cdot\epsilon_2^* = 0$, as well as $k_1\cdot\epsilon_1 = k_2\cdot\epsilon_2^* = 0$, so that M_{fi} takes the extremely simple form

$$M_{fi} = 8\pi i\alpha\epsilon_2^*\cdot\epsilon_1. \tag{2.145}$$

From (2.65) and (2.46),

$$\frac{d\sigma}{d\cos\theta_L} = 2\pi\frac{\alpha^2}{m^2}\left(\frac{k_2}{k_1}\right)^2|\epsilon_1\cdot\epsilon_2^*|^2, \tag{2.146}$$

where θ_L is the laboratory angle of the final photon and

$$\frac{k_2}{k_1} = \frac{1}{1 + \frac{k_1}{m}(1 - \cos\theta_L)}. \tag{2.147}$$

For an unpolarized initial photon and unobserved final polarization one must average (sum) over the polarization states λ_1 (λ_2), yielding

$$\frac{d\bar{\sigma}}{d\cos\theta_L} \equiv \frac{1}{2} \sum_{\lambda_1, \lambda_2} \frac{d\sigma}{d\cos\theta_L}. \tag{2.148}$$

This can be evaluated using (2.121) for both the λ_1 and λ_2 sums,[15] with

$$\begin{aligned} k_1^\mu &= k_1(1, 0, 0, 1), & k_2^\mu &= k_2(1, \sin\theta_L, 0, \cos\theta_L) \\ k_{1r}^\mu &= k_1(1, 0, 0, -1), & k_{2r}^\mu &= k_2(1, -\sin\theta_L, 0, -\cos\theta_L), \end{aligned} \tag{2.149}$$

yielding

$$\frac{1}{2} \sum_{\lambda_1 \lambda_2} |\epsilon_1 \cdot \epsilon_2^*|^2 = \frac{1}{2}(1 + \cos^2\theta_L). \tag{2.150}$$

In this case, however, it is simpler to evaluate the sum using the explicit forms

$$\begin{aligned} &\epsilon_1(1) = (0, 1, 0, 0), \qquad \epsilon_2(1) = (0, \cos\theta_L, 0, -\sin\theta_L) \\ &\epsilon_1(2) = \epsilon_2(2) = (0, 0, 1, 0). \end{aligned} \tag{2.151}$$

for the transverse polarization vectors.

2.7 THE DIRAC FIELD

The Dirac field $\psi_\alpha(x)$, where $\alpha = 1 \cdots 4$ is the spinor index, describes a four-component spin-$\frac{1}{2}$ particle, i.e., ψ annihilates the two possible spin states for a particle and creates the two spin states for the antiparticle. In the absence of interactions, the Lagrangian density is

$$\mathcal{L}_0 = \bar{\psi}(x)_\alpha \left(i \not\partial - m\right)_{\alpha\beta} \psi(x)_\beta = \bar{\psi}(x) \left(i \not\partial - m\right) \psi(x). \tag{2.152}$$

The sum over α and β is written in matrix notation in the second form, in which ψ is a four-component column vector; the *Dirac adjoint*

$$\bar{\psi}(x) \equiv \psi^\dagger(x)\gamma^0 \tag{2.153}$$

is a four-component row vector; and

$$\not\partial \equiv \gamma^\mu \frac{\partial}{\partial x^\mu} = \gamma^\mu \partial_\mu, \tag{2.154}$$

where $\gamma^\mu, \mu = 0 \cdots 3$, are the 4×4 *Dirac* matrices, defined by

$$\{\gamma^\mu, \gamma^\nu\} = 2g^{\mu\nu}. \tag{2.155}$$

(There is an implicit 4×4 identity matrix on the right side of (2.155) and after m in (2.152).) The γ^μ must also satisfy

$$(\gamma^\mu)^\dagger = \gamma^0 \gamma^\mu \gamma^0 = \gamma_\mu, \tag{2.156}$$

[15] Since we have already evaluated M_{fi} in a specific gauge, we *cannot* drop the second term on the right of (2.121).

so that $\mathcal{L}_0 = \mathcal{L}_0^\dagger$. It is useful to define

$$\gamma^5 = \gamma_5 \equiv i\gamma^0\gamma^1\gamma^2\gamma^3, \qquad \sigma^{\mu\nu} \equiv \frac{i}{2}[\gamma^\mu, \gamma^\nu]. \tag{2.157}$$

γ^5 enters for spin and chirality projections, for coupling fermions to pseudoscalars, and for axial vector currents in the weak interactions. From (2.155), $\gamma^5 = \gamma^{5\dagger}$, $\gamma^5\gamma^\mu = -\gamma^\mu\gamma^5$, and $(\gamma^5)^2 = I$. $\sigma^{\mu\nu}$ is useful, e.g., in the description of electric and magnetic dipole moments. An arbitrary 4×4 matrix can be written as a linear combination of $I, \gamma^5, \gamma^\mu, \gamma^\mu\gamma^5$, and $\sigma^{\mu\nu}$. For example, $\sigma^{\mu\nu}\gamma^5$ is given in Problem 2.9.

2.7.1 The Free Dirac Field

From (2.152), one obtains the free-field (Dirac) equation

$$(i\not{\partial} - m)\,\psi(x) = 0. \tag{2.158}$$

The solution to (2.158) is

$$\psi(x) = \int \frac{d^3\vec{p}}{(2\pi)^3 2E_p} \sum_{s=1}^{2} \left[u(\vec{p}, s)\; a(\vec{p}, s)e^{-ip\cdot x} + v(\vec{p}, s)b^\dagger(\vec{p}, s)e^{+ip\cdot x}\right], \tag{2.159}$$

where $a^\dagger(\vec{p}, s)$ and $b^\dagger(\vec{p}, s)$ are the creation operators for e^- and e^+ states, respectively,[16]

$$a^\dagger(\vec{p}, s)|0\rangle = |e^-(\vec{p}, s)\rangle, \qquad b^\dagger(\vec{p}, s)|0\rangle = |e^+(\vec{p}, s)\rangle. \tag{2.160}$$

$\vec{p}$ refers to the physical momentum for both $|e^\mp(\vec{p}, s)\rangle$, and s runs over the two independent spin states. a and b and their adjoints satisfy the anticommutation rules in (2.10)–(2.12), e.g., $\{a(\vec{p}, s), a^\dagger(\vec{p}\,', s')\} = (2\pi)^3 2E_p\, \delta^3(\vec{p} - \vec{p}\,')\delta_{ss'}$.

In (2.159), $u(\vec{p}, s)$ and $v(\vec{p}, s)$ are four-component *Dirac spinors* (complex column vectors). From (2.158) they are the solutions to the momentum space Dirac equation, i.e.,

$$\begin{aligned} (\not{p} - m)u(\vec{p}, s) &\equiv (p_\mu\gamma^\mu - m)u(\vec{p}, s) = 0 \\ (\not{p} + m)v(\vec{p}, s) &= 0. \end{aligned} \tag{2.161}$$

Taking the adjoint of (2.161) and using (2.156),

$$\begin{aligned} [(\not{p} - m)u]^\dagger &= u^\dagger(p_\mu\gamma^{\mu\dagger} - m) = u^\dagger(p_\mu\gamma^0\gamma^\mu\gamma^0 - (\gamma^0)^2 m) \\ &= \bar{u}(\not{p} - m)\gamma^0 = 0, \end{aligned} \tag{2.162}$$

where $\bar{u} \equiv u^\dagger\gamma^0$ and $\bar{v} \equiv v^\dagger\gamma^0$ are the Dirac adjoints. Thus,

$$\bar{u}(\vec{p}, s)(\not{p} - m) = \bar{v}(\vec{p}, s)(\not{p} + m) = 0. \tag{2.163}$$

Before proceeding to the electrodynamics of fermions, let us consider some properties of the Dirac matrices and spinors.

[16]By convention, we are taking $\psi \equiv \psi_{e^-}$ to be the e^- field. The e^+ field $\psi^c \equiv \psi_{e^+}$ is of the same form as (2.159) except a and b are reversed. Charge conjugation and space reflection will be discussed in more detail in Section 2.10.

2.7.2 Dirac Matrices and Spinors

Explicit Forms for the Dirac Matrices

The Dirac matrices are defined by (2.155) and (2.156). For most calculations one does not need their explicit form, but it is occasionally useful to have one. For example, the *Pauli-Dirac* representation is useful for studying the non-relativistic limit of an interaction, while the *chiral* representation is useful at high energy and for *Weyl* or *Majorana* fields, encounted in neutrino physics and supersymmetry. In the Pauli-Dirac representation

$$\begin{aligned} \gamma^0 &= \begin{pmatrix} I & 0 \\ 0 & -I \end{pmatrix} \qquad \gamma^i = \begin{pmatrix} 0 & \sigma^i \\ -\sigma^i & 0 \end{pmatrix} \qquad \gamma^5 = \begin{pmatrix} 0 & I \\ I & 0 \end{pmatrix} \\ \sigma^{0i} &= i \begin{pmatrix} 0 & \sigma^i \\ \sigma^i & 0 \end{pmatrix} \qquad \sigma^{ij} = \epsilon_{ijk} \begin{pmatrix} \sigma^k & 0 \\ 0 & \sigma^k \end{pmatrix}, \end{aligned} \tag{2.164}$$

where I is the 2×2 identity matrix and σ^i are the Pauli matrices in (1.17). Similarly, in the chiral representation

$$\begin{aligned} \gamma^0 &= \begin{pmatrix} 0 & I \\ I & 0 \end{pmatrix} \qquad \gamma^i = \begin{pmatrix} 0 & \sigma^i \\ -\sigma^i & 0 \end{pmatrix} \qquad \gamma^5 = \begin{pmatrix} -I & 0 \\ 0 & I \end{pmatrix} \\ \sigma^{0i} &= i \begin{pmatrix} -\sigma^i & 0 \\ 0 & \sigma^i \end{pmatrix} \qquad \sigma^{ij} = \epsilon_{ijk} \begin{pmatrix} \sigma^k & 0 \\ 0 & \sigma^k \end{pmatrix}. \end{aligned} \tag{2.165}$$

It is sometimes convenient to rewrite the chiral representation matrices as

$$\gamma^\mu = \begin{pmatrix} 0 & \sigma^\mu \\ \bar{\sigma}^\mu & 0 \end{pmatrix}, \tag{2.166}$$

where

$$\sigma^\mu \equiv (I, \vec{\sigma}), \qquad \bar{\sigma}^\mu \equiv (I, -\vec{\sigma}) = \sigma_\mu. \tag{2.167}$$

Traces and Products of Dirac Matrices

Most calculations can be carried out without using the specific forms of the Dirac matrices by using various *trace identities*, where the trace is $\mathrm{Tr}\, A = \sum_\alpha A_{\alpha\alpha}$ for any square matrix A. Note that $\mathrm{Tr}\, A = \mathrm{Tr}\, A^T = (\mathrm{Tr}\, A^\dagger)^*$ and $\mathrm{Tr}\,(AB) = \mathrm{Tr}\,(BA)$. For any representation, (2.155) and (2.157) imply

$$\mathrm{Tr}\, I = 4, \qquad \mathrm{Tr}\, \gamma^\mu = \mathrm{Tr}\, \gamma^5 = 0. \tag{2.168}$$

Define the 4×4 matrix

$$\not{a} = \gamma^\mu a_\mu \tag{2.169}$$

for an arbitrary four-vector a_μ. Then, from (2.155)

$$\not{a} \not{b} = - \not{b} \not{a} + 2a \cdot b I \quad \Rightarrow \quad \not{a} \not{a} = a^2 I. \tag{2.170}$$

One has immediately that

$$\mathrm{Tr}\,(\not{a} \not{b}) = 4\, a \cdot b. \tag{2.171}$$

Other useful trace identities include (see, e.g., Bjorken and Drell, 1964),

$$\begin{aligned} \mathrm{Tr}\,(\not{a} \not{b} \not{c} \not{d}) &= 4(a \cdot b \;\; c \cdot d + a \cdot d \;\; b \cdot c - a \cdot c \;\; b \cdot d) \\ \mathrm{Tr}\,(\gamma^5 \not{a} \not{b}) &= 0 \\ \mathrm{Tr}\,(\gamma^5 \not{a} \not{b} \not{c} \not{d}) &= 4i\epsilon^{\mu\nu\rho\sigma}\, a_\mu b_\nu c_\rho d_\sigma, \end{aligned} \tag{2.172}$$

where $\epsilon^{\mu\nu\rho\sigma}$ is the totally antisymmetric tensor with $\epsilon_{0123} = +1$ and $\epsilon^{0123} = -1$. Also,

$$\begin{aligned}
\mathrm{Tr}\,(\not a_1 \cdots \not a_n) &= \mathrm{Tr}\,(\gamma^5 \not a_1 \cdots \not a_n) = 0 \quad (\text{for } n \text{ odd}) \\
\mathrm{Tr}\,(\not a_1 \cdots \not a_n) &= a_1 \cdot a_2 \mathrm{Tr}\,(\not a_3 \cdots \not a_n) - a_1 \cdot a_3 \mathrm{Tr}\,(\not a_2 \not a_4 \cdots \not a_n) \\
&\quad \cdots + a_1 \cdot a_n \mathrm{Tr}\,(\not a_2 \cdots \not a_{n-1}) \quad (\text{for } n \text{ even}) \\
\mathrm{Tr}\,(\not a_1 \not a_2 \cdots \not a_n) &= \mathrm{Tr}\,(\not a_n \cdots \not a_2 \not a_1).
\end{aligned} \tag{2.173}$$

Related useful identities are

$$\begin{aligned}
\gamma_\mu \gamma^\mu &= 4I, & \gamma_\mu \not a \gamma^\mu &= -2 \not a \\
\gamma_\mu \not a \not b \gamma^\mu &= 4\, a \cdot b\, I, & \gamma_\mu \not a \not b \not c \gamma^\mu &= -2 \not c \not b \not a.
\end{aligned} \tag{2.174}$$

Finally, we record the identities

$$\begin{aligned}
&\mathrm{Tr}\,\left[\gamma_\mu \not a \gamma_\nu \not b (1 + \lambda\gamma^5)\right] \mathrm{Tr}\,\left[\gamma^\mu \not c \gamma^\nu \not d\, (1 \pm \lambda\gamma^5)\right] \\
&= \begin{cases} 64\, a \cdot c\, b \cdot d \text{ for } + \\ 64\, a \cdot d\, b \cdot c \text{ for } - \end{cases},
\end{aligned} \tag{2.175}$$

where $\lambda = \pm 1$. These results can be derived using the identities in (2.172) and Table 1.2, as will be shown in detail in Section 7.2.1. They are extremely useful for calculating four-fermion polarization effects, as well as for weak interaction decay and scattering processes.

Spinor Normalization and Projections

The Dirac spinors satisfy the normalization and orthogonality relations

$$\begin{aligned}
\bar u(\vec p, s) u(\vec p, s') &= -\bar v(\vec p, s) v(p, s') = 2m\, \delta_{ss'} \\
u^\dagger(\vec p, s) u(\vec p, s') &= v^\dagger(\vec p, s) v(\vec p, s') = 2E_p\, \delta_{ss'} \\
\bar u(\vec p, s) v(\vec p, s') &= u^\dagger(\vec p, s) v(-\vec p, s') = \bar v(\vec p, s) u(\vec p, s') = v^\dagger(\vec p, s) u(-\vec p, s') = 0,
\end{aligned} \tag{2.176}$$

which are just numbers (e.g., Bjorken and Drell, 1965). The projections

$$\sum_s u(\vec p, s)\ \bar u(\vec p, s) = \not p + m, \qquad \sum_s v(\vec p, s)\ \bar v(\vec p, s) = \not p - m, \tag{2.177}$$

which are useful in summing over spin orientations in physical rates, are 4×4 matrices. The form of (2.177) follows from the Dirac equation, while the normalization follows by taking the trace and using the first equation in (2.176). (The second equation (2.176) for $s = s'$ then follows by right-multiplying (2.177) by γ^0 and then taking the trace.)

The projections

$$\begin{aligned}
u(\vec p, s)\ \bar u(\vec p, s) &= (\not p + m) \left(\frac{1 + \gamma^5 \not s}{2} \right) \\
v(\vec p, s)\ \bar v(\vec p, s) &= (\not p - m) \left(\frac{1 + \gamma^5 \not s}{2} \right)
\end{aligned} \tag{2.178}$$

are useful when one does *not* want to sum over spins. Here, s^μ is the *spin four-vector*. In the particle rest frame, $p = (m, \vec 0\,)$, it is just a unit vector

$$s = (0, \hat s) \tag{2.179}$$

in the spin direction, so that $s^2 = -1$ and $p \cdot s = 0$. Boosting to an arbitrary frame in which $\vec{p} = \vec{\beta} E_p = \gamma \vec{\beta} m$,

$$s = (\gamma \vec{\beta} \cdot \hat{s},\ \gamma \hat{s}_{\parallel} + \hat{s}_{\perp}), \tag{2.180}$$

where $\hat{s}_{\parallel} = \hat{s} \cdot \hat{\beta}\hat{\beta}$ and $\hat{s}_{\perp} = \hat{s} - \hat{s}_{\parallel}$ are, respectively, the components of $\hat{s}$ parallel and perpendicular to $\hat{\beta}$. Thus, $s = (0, \hat{s})$ for $\hat{s} \cdot \hat{\beta} = 0$, while $s \equiv s_{\pm} = \pm\gamma(\beta, \hat{\beta})$ for $\hat{s} \cdot \hat{\beta} = \pm 1$. The latter are known as the positive and negative helicity states, respectively, or alternatively as right- and left-handed states. The projections in (2.178) simplify greatly for the helicity states in the relativistic limit:

$$\begin{aligned}(\not{p} + m)\left(\frac{1+\gamma^5 \not{s}_{\pm}}{2}\right) &\xrightarrow[m\to 0]{} \frac{1 \pm \gamma^5}{2}\not{p} \equiv P_{R,L}\not{p} \\ (\not{p} - m)\left(\frac{1+\gamma^5 \not{s}_{\pm}}{2}\right) &\xrightarrow[m\to 0]{} \frac{1 \mp \gamma^5}{2}\not{p} = P_{L,R}\not{p},\end{aligned} \tag{2.181}$$

where the chiral projection operators $P_{R,L}$ will be discussed below.

The Propagator

From (2.159) and the spinor sums (2.177), one obtains the Feynman propagator for the free Dirac field[17]

$$\langle 0|T[\psi(x), \bar{\psi}(x')]|0\rangle = i\int \frac{d^4k}{(2\pi)^4} e^{-ik\cdot(x-x')} S_F(k), \tag{2.182}$$

where the momentum space propagator $S_F(k)$, which is a 4×4 matrix, is

$$S_F(k) = \frac{1}{\not{k} - m + i\epsilon} = \frac{\not{k} + m}{k^2 - m^2 + i\epsilon}. \tag{2.183}$$

The last equality follows from (2.170).

Explicit Spinor Forms

Explicit forms for the Dirac spinors are occasionally useful, e.g., for considering non-relativistic limits. In the Pauli-Dirac representation

$$u(\vec{p}, s) = \sqrt{E_p + m}\begin{pmatrix} \phi_s \\ \frac{\vec{\sigma}\cdot\vec{p}}{E_p+m}\phi_s \end{pmatrix}, \tag{2.184}$$

where ϕ_s is a two-component Pauli spinor describing the spin orientation. Thus,

$$\phi_1 = \begin{pmatrix} 1 \\ 0 \end{pmatrix}, \qquad \phi_2 = \begin{pmatrix} 0 \\ 1 \end{pmatrix} \tag{2.185}$$

describe spin orientations in the $\pm\hat{z}$ directions, respectively. $\phi_{\pm}$, defined by

$$h\phi_{\pm} \equiv \frac{1}{2}\vec{\sigma} \cdot \hat{p}\, \phi_{\pm} = \pm\frac{1}{2}\phi_{\pm}, \tag{2.186}$$

[17] The time-ordered product of two fermion fields is defined as in (2.27) except there is a minus sign before the second term.

describe positive (negative) helicity states. Similarly, the v spinors are

$$v(\vec{p},s) = \sqrt{E_p+m}\begin{pmatrix} \frac{\vec{\sigma}\cdot\vec{p}}{E_p+m}\chi_s \\ \chi_s \end{pmatrix}, \tag{2.187}$$

where

$$\chi_s \equiv -i\sigma^2\phi_s^*. \tag{2.188}$$

Thus,

$$\chi_1 = \begin{pmatrix} 0 \\ 1 \end{pmatrix}, \qquad \chi_2 = \begin{pmatrix} -1 \\ 0 \end{pmatrix} \tag{2.189}$$

represent spins in the $\pm\hat{z}$ directions. The positive (negative) helicity spinors, which continue to mean that the physical spin is parallel (antiparallel) to the physical momentum, satisfy

$$\frac{1}{2}\vec{\sigma}\cdot\hat{p}\,\chi_\pm = \mp\frac{1}{2}\chi_\pm. \tag{2.190}$$

The counter-intuitive results in (2.189) and (2.190) are considered in Problem 2.13 and in the discussion of charge conjugation in Section 2.10. Explicit forms for the helicity spinors are given in Table 2.1, using a phase convention for which

$$i\sigma^2\phi_\pm^* = \mp\phi_\mp, \qquad i\sigma^2\chi_\pm^* = \mp\chi_\mp. \tag{2.191}$$

TABLE 2.1 Explicit forms and properties of the helicity spinors[a] corresponding to spherical angles (θ,φ) for $\hat{p}$.

Rotation:	$R(\theta,\varphi) = e^{-i\sigma_3\frac{\varphi}{2}}e^{-i\sigma_2\frac{\theta}{2}}e^{i\sigma_3\frac{\varphi}{2}}$
$\phi_+(\hat{z}) = -\chi_-(\hat{z}) = \begin{pmatrix} 1 \\ 0 \end{pmatrix}$	$\phi_-(\hat{z}) = \chi_+(\hat{z}) = \begin{pmatrix} 0 \\ 1 \end{pmatrix}$
$\phi_+(\hat{p}) = -\chi_-(\hat{p}) = \begin{pmatrix} \cos\frac{\theta}{2} \\ \sin\frac{\theta}{2}e^{i\varphi} \end{pmatrix}$	$\phi_-(\hat{p}) = \chi_+(\hat{p}) = \begin{pmatrix} -\sin\frac{\theta}{2}e^{-i\varphi} \\ \cos\frac{\theta}{2} \end{pmatrix}$
$\phi_+(-\hat{p}) = -\chi_-(-\hat{p}) = \begin{pmatrix} \sin\frac{\theta}{2} \\ -\cos\frac{\theta}{2}e^{i\varphi} \end{pmatrix}$	$\phi_-(-\hat{p}) = \chi_+(-\hat{p}) = \begin{pmatrix} \cos\frac{\theta}{2}e^{-i\varphi} \\ \sin\frac{\theta}{2} \end{pmatrix}$
$\phi_\pm(\hat{p}) = \pm e^{\pm i\varphi}\phi_\mp(-\hat{p})$	$\chi_\pm(\hat{p}) = \pm e^{\mp i\varphi}\chi_\mp(-\hat{p})$
$\chi_+(\hat{p}) = -i\sigma^2\phi_+(\hat{p})^* = \phi_-(\hat{p})$	$\chi_-(\hat{p}) = -i\sigma^2\phi_-(\hat{p})^* = -\phi_+(\hat{p})$
Orthonormality:	$\phi_i^\dagger(\hat{p})\phi_j(\hat{p}) = \delta_{ij} = \chi_i^\dagger(\hat{p})\chi_j(\hat{p})$
Completeness:	$\sum_i \phi_i(\hat{p})\phi_i^\dagger(\hat{p}) = \sum_i \chi_i(\hat{p})\chi_i^\dagger(\hat{p}) = I$

[a]The spinors are constructed using the rotation $\mathfrak{s}_\pm(\hat{p}) = R(\theta,\varphi)\mathfrak{s}_\pm(\hat{z})$, where $\mathfrak{s} = \phi$ or χ. The spherical angles of $-\vec{p}$ are $(\pi-\theta,\ \pi+\varphi)$, so that $-\hat{z} = (\pi,\pi)$. The orthonormality and completeness relations also apply to the fixed spin axis basis.

The explicit u and v spinors in the chiral representation are

$$u(\vec{p},s) = \sqrt{\frac{E_p+m}{2}} \begin{pmatrix} \left(I - \frac{\vec{\sigma}\cdot\vec{p}}{E_p+m}\right)\phi_s \\ \left(I + \frac{\vec{\sigma}\cdot\vec{p}}{E_p+m}\right)\phi_s \end{pmatrix} = \begin{pmatrix} \sqrt{p\cdot\sigma}\phi_s \\ \sqrt{p\cdot\bar{\sigma}}\phi_s \end{pmatrix}$$
$$v(\vec{p},s) = \sqrt{\frac{E_p+m}{2}} \begin{pmatrix} \left(I - \frac{\vec{\sigma}\cdot\vec{p}}{E_p+m}\right)\chi_s \\ -\left(I + \frac{\vec{\sigma}\cdot\vec{p}}{E_p+m}\right)\chi_s \end{pmatrix} = \begin{pmatrix} \sqrt{p\cdot\sigma}\chi_s \\ -\sqrt{p\cdot\bar{\sigma}}\chi_s \end{pmatrix}, \tag{2.192}$$

where σ^μ and $\bar{\sigma}^\mu$ are defined in (2.167). The second form is easily verified using (1.19) on page 4. In the helicity basis, these are

$$u(\pm) = \sqrt{\frac{E+m}{2}} \begin{pmatrix} \lambda_\pm\, \phi_\pm \\ \lambda_\mp\, \phi_\pm \end{pmatrix}, \qquad v(\pm) = \sqrt{\frac{E+m}{2}} \begin{pmatrix} \lambda_\mp\, \chi_\pm \\ -\lambda_\pm\, \chi_\pm \end{pmatrix}, \tag{2.193}$$

where

$$\lambda_\pm(p) \equiv 1 \mp \frac{p}{E+m}. \tag{2.194}$$

The chiral representation is especially useful in the case of massless or relativistic fermions, for which the upper (lower) two components of the positive (negative) helicity u spinors vanish, and oppositely for v,

$$u(\vec{p},+) \xrightarrow[m\to 0]{} \sqrt{2E}\begin{pmatrix} 0 \\ \phi_+ \end{pmatrix}, \qquad u(\vec{p},-) \xrightarrow[m\to 0]{} \sqrt{2E}\begin{pmatrix} \phi_- \\ 0 \end{pmatrix}$$
$$v(\vec{p},+) \xrightarrow[m\to 0]{} \sqrt{2E}\begin{pmatrix} \chi_+ \\ 0 \end{pmatrix}, \qquad v(\vec{p},-) \xrightarrow[m\to 0]{} \sqrt{2E}\begin{pmatrix} 0 \\ -\chi_- \end{pmatrix}. \tag{2.195}$$

Chiral Fields

For a fermion field ψ, one can define left (L)- and right (R)-*chiral projections*

$$\psi_L = P_L\psi = \frac{1-\gamma^5}{2}\psi, \qquad \psi_R = P_R\psi = \frac{1+\gamma^5}{2}\psi, \tag{2.196}$$

where $P_{L,R}^2 = P_{L,R}$, $P_L P_R = P_R P_L = 0$, $P_{L,R}^\dagger = P_{L,R}$, and $P_L + P_R = I$. ψ_L and ψ_R can be viewed as independent degrees of freedom, with $\psi = \psi_L + \psi_R$. For a massless fermion the L- and R-chiral components correspond to particles with negative and positive helicity, respectively, i.e., ψ_L and ψ_R annihilate fermions with helicity $h = \mp\frac{1}{2}$. For antifermions it is just the reverse, ψ_L and ψ_R create antifermion states with $h = \pm\frac{1}{2}$. For mass $m \neq 0$ the chiral states of energy E associated with $\psi_{L,R}$ have admixtures of $\mathcal{O}(m/E)$ of the "wrong" helicity (Problem 2.15). The free-field Lagrangian density in (2.152) can be rewritten in terms of the chiral projections as

$$\mathcal{L} = \bar{\psi}_L i\,\partial\!\!\!/\,\psi_L + \bar{\psi}_R i\,\partial\!\!\!/\,\psi_R - m\left(\bar{\psi}_L\psi_R + \bar{\psi}_R\psi_L\right), \tag{2.197}$$

where $\bar{\psi}_{L,R}$ are defined[18] by

$$\bar{\psi}_L \equiv (\psi_L)^\dagger\gamma^0 = \psi^\dagger P_L\gamma^0 = \bar{\psi}P_R$$
$$\bar{\psi}_R \equiv (\psi_R)^\dagger\gamma^0 = \psi^\dagger P_R\gamma^0 = \bar{\psi}P_L. \tag{2.198}$$

[18] Some authors use the notation $\overline{\psi_L}$ or $\overline{(\psi_L)}$ rather than $\bar{\psi}_L$ to emphasize that the Dirac adjoint operation acts on ψ_L rather than on ψ.

The chiral projections $\psi_{L,R}$ are also known as *Weyl spinors* or *Weyl two-component fields.* They can be described as above in *four-component notation*, i.e., as two-dimensional projections of four-component fields ψ, but it is often convenient to discard the superfluous components and work in *two-component notation.* As the name "chiral" suggests, this is most conveniently displayed in the chiral representation, for which

$$P_L = \begin{pmatrix} I & 0 \\ 0 & 0 \end{pmatrix}, \qquad P_R = \begin{pmatrix} 0 & 0 \\ 0 & I \end{pmatrix}. \tag{2.199}$$

Thus

$$\psi = \begin{pmatrix} \Psi_L \\ \Psi_R \end{pmatrix} \Rightarrow P_L\psi = \begin{pmatrix} \Psi_L \\ 0 \end{pmatrix}, \quad P_R\psi = \begin{pmatrix} 0 \\ \Psi_R \end{pmatrix}, \tag{2.200}$$

where $\Psi_{L,R}$ are the Weyl two-component fields.

The Lagrangian density (2.152) for the free Dirac field can be written in terms of the Weyl fields as

$$\mathcal{L}_0 = \Psi_L^\dagger i\bar{\sigma}^\mu \partial_\mu \Psi_L + \Psi_R^\dagger i\sigma^\mu \partial_\mu \Psi_R - m\left(\Psi_L^\dagger \Psi_R + \Psi_R^\dagger \Psi_L\right), \tag{2.201}$$

where the four-vectors σ^μ and $\bar{\sigma}^\mu$ are the 2×2 matrices defined in (2.167). The Dirac mass term couples the L and R components, while the kinetic energy terms are diagonal. The free-field Dirac equation becomes

$$i\bar{\sigma}^\mu \partial_\mu \Psi_L - m\Psi_R = 0, \qquad i\sigma^\mu \partial_\mu \Psi_R - m\Psi_L = 0. \tag{2.202}$$

Above, we introduced the chiral fields as projections of a four-component Dirac field. Alternatively, one can simply define chiral fields as those satisfying $\psi_L = P_L\psi_L$ or $\psi_R = P_R\psi_R$, i.e., not necessarily as projections of another field ψ, and in fact this was done for the L-chiral neutrinos in the original formulation of the standard model. Equivalently, Weyl fields Ψ_L or Ψ_R can be introduced independently of each other. For example, a single Weyl L field with $\mathcal{L}_0 = \Psi_L^\dagger i\bar{\sigma}^\mu \partial_\mu \Psi_L$ would describe a massless negative helicity particle and a positive helicity antiparticle.

One can also define the chiral projections of the u and v spinors,

$$u_{L,R}(\vec{p}, s) \equiv P_{L,R}u(\vec{p}, s), \qquad v_{L,R}(\vec{p}, s) \equiv P_{L,R}v(\vec{p}, s). \tag{2.203}$$

If one writes

$$u = \begin{pmatrix} \mathfrak{u}_L \\ \mathfrak{u}_R \end{pmatrix} \Rightarrow u_L = \begin{pmatrix} \mathfrak{u}_L \\ 0 \end{pmatrix}, \quad u_R = \begin{pmatrix} 0 \\ \mathfrak{u}_R \end{pmatrix}, \tag{2.204}$$

then $\mathfrak{u}_{L,R}$ satisfy the Dirac equation

$$\begin{aligned} p \cdot \bar{\sigma}\, \mathfrak{u}_L &= (E_p I + \vec{\sigma} \cdot \vec{p})\, \mathfrak{u}_L = m\mathfrak{u}_R \\ p \cdot \sigma\, \mathfrak{u}_R &= (E_p I - \vec{\sigma} \cdot \vec{p})\, \mathfrak{u}_R = m\mathfrak{u}_L. \end{aligned} \tag{2.205}$$

The chiral components of $v = (\mathfrak{v}_L \ \mathfrak{v}_R)^T$ satisfy similar equations, only with $m \to -m$. It follows easily that the solutions for u and v are given by (2.192).

For $m \to 0$ the equations for $\mathfrak{u}_{L,R}$ decouple. The u and v spinors of definite helicity given in (2.195) then coincide with the left- and right-chiral projections,

$$\begin{aligned} P_R u(\vec{p}, +) &= u(\vec{p}, +), \qquad & P_L u(\vec{p}, -) &= u(\vec{p}, -) \\ P_R v(\vec{p}, -) &= v(\vec{p}, -), \qquad & P_L v(\vec{p}, +) &= v(\vec{p}, +), \end{aligned} \tag{2.206}$$

showing the flip between chirality and helicity for the v spinors, consistent with (2.181). The free-field expressions for $\psi_{L,R}$ or $\Psi_{L,R}$ are especially simple in this case. From (2.159) and (2.195)

$$\Psi(x)_{L,R} = \int \frac{d^3\vec{p}}{(2\pi)^3 2E_p} \sqrt{2E_p} \left[\phi_{\mp}(\hat{p})\, a(\vec{p}, \mp) e^{-ip\cdot x} \pm \chi_{\pm}(\hat{p}) b^\dagger(\vec{p}, \pm) e^{+ip\cdot x} \right], \tag{2.207}$$

which is similar to the free Dirac field except there is no sum on spins.

The Weyl two-component formalism is further developed in Section 2.11 and in Chapters 9 and 10.

Bilinear Forms

Consider the bilinear form $\bar{w}_2 M w_1$, where $w_{1,2}$ are any two Dirac u or v spinors. They may even correspond to particles with different masses, which is relevant, e.g., for weak interaction transitions. M is an arbitrary 4×4 matrix. Then,

$$(\bar{w}_2 M w_1)^* = \bar{w}_1 \overline{M} w_2, \tag{2.208}$$

where $\overline{M} \equiv \gamma^0 M^\dagger \gamma^0$ is the *Dirac adjoint* of M. One finds

$$\begin{aligned} \overline{M} &= \quad M \text{ for } M = I, \gamma^\mu, \gamma^\mu\gamma^5, \sigma^{\mu\nu} \\ \overline{M} &= -M \text{ for } M = \gamma^5, \sigma^{\mu\nu}\gamma^5 \\ \overline{M_1 M_2} &= \overline{M_2}\, \overline{M_1} \Rightarrow \overline{\not{a}_1 \not{a}_2 \cdots \not{a}_n} = \not{a}_n \cdots \not{a}_2 \not{a}_1. \end{aligned} \tag{2.209}$$

There is an equivalent relation,

$$\left(\bar{\psi}_2 M \psi_1\right)^\dagger = \bar{\psi}_1 \overline{M} \psi_2, \tag{2.210}$$

for two Dirac fields, which may be the same or different. Then, for example,

$$\begin{aligned} \sum_{s_1,s_2} |\bar{u}_2 M u_1|^2 &= \sum_{s_1,s_2} \bar{u}_2 M u_1 \bar{u}_1 \overline{M} u_2 \\ &= \sum_{s_2} \bar{u}_{2\alpha} \left[M(\not{p}_1 + m_1)\overline{M} \right]_{\alpha\beta} u_{2\beta} \\ &= \text{Tr} \left[M(\not{p}_1 + m_1)\overline{M} \Big(\sum_{s_2} u_2 \bar{u}_2 \Big) \right] \\ &= \text{Tr} \left[M(\not{p}_1 + m_1)\overline{M}(\not{p}_2 + m_2) \right], \end{aligned} \tag{2.211}$$

which allow one to express a physical rate in terms of a trace. (This is sometimes referred to as the Casimir trick.)

For chiral spinors

$$\bar{w}_{L,R} \equiv \left(w_{L,R}\right)^\dagger \gamma^0 = w^\dagger P_{L,R} \gamma^0 = w^\dagger \gamma^0 P_{R,L} = \bar{w} P_{R,L}. \tag{2.212}$$

Therefore, for any two spinors w_1 and w_2,

$$\bar{w}_{1L} \Gamma w_{2L} = 0 = \bar{w}_{1R} \Gamma w_{2R} \tag{2.213}$$

for $\Gamma = I, \gamma^5, \sigma^{\mu\nu}$, or $\sigma^{\mu\nu}\gamma^5$, while

$$\bar{w}_{1L} \Gamma w_{2R} = 0 = \bar{w}_{1R} \Gamma w_{2L} \tag{2.214}$$

for $\Gamma = \gamma^\mu$ or $\gamma^\mu\gamma^5$. Equivalent relations for the chiral fields $\psi_{L,R}$ were used in (2.197). Thus, scalar, pseudoscalar, and tensor transitions reverse the chirality between an initial and final fermion, while vector and axial vector transitions maintain it.

The Fierz Identities

The *Fierz identities* are

$$
\begin{aligned}
\left(\bar{\psi}_{1L}\gamma^\mu\psi_{2L}\right)\left(\bar{\psi}_{3L}\gamma_\mu\psi_{4L}\right) &= -\eta_F\left(\bar{\psi}_{1L}\gamma^\mu\psi_{4L}\right)\left(\bar{\psi}_{3L}\gamma_\mu\psi_{2L}\right)\\
\left(\bar{\psi}_{1R}\gamma^\mu\psi_{2R}\right)\left(\bar{\psi}_{3R}\gamma_\mu\psi_{4R}\right) &= -\eta_F\left(\bar{\psi}_{1R}\gamma^\mu\psi_{4R}\right)\left(\bar{\psi}_{3R}\gamma_\mu\psi_{2R}\right)\\
\left(\bar{\psi}_{1R}\gamma^\mu\psi_{2R}\right)\left(\bar{\psi}_{3L}\gamma_\mu\psi_{4L}\right) &= 2\eta_F\left(\bar{\psi}_{1R}\psi_{4L}\right)\left(\bar{\psi}_{3L}\psi_{2R}\right)\\
\left(\bar{\psi}_{1R}\psi_{2L}\right)\left(\bar{\psi}_{3R}\psi_{4L}\right) &= \frac{\eta_F}{2}\left(\bar{\psi}_{1R}\psi_{4L}\right)\left(\bar{\psi}_{3R}\psi_{2L}\right)+\frac{\eta_F}{8}\left(\bar{\psi}_{1R}\sigma^{\mu\nu}\psi_{4L}\right)\left(\bar{\psi}_{3R}\sigma_{\mu\nu}\psi_{2L}\right),
\end{aligned}
\tag{2.215}
$$

where ψ_{iL} and ψ_{jR} are anticommuting chiral fields and $\eta_F = -1$. There are analogous relations for u and v spinors, but with $\eta_F = +1$, e.g.,

$$
\left(\bar{w}_{1L}\gamma^\mu w_{2L}\right)\left(\bar{w}_{3L}\gamma_\mu w_{4L}\right) = -\left(\bar{w}_{1L}\gamma^\mu w_{4L}\right)\left(\bar{w}_{3L}\gamma_\mu w_{2L}\right). \tag{2.216}
$$

The Fierz identities are easily derived by expressing the 4×4 matrices such as $\psi_{2L}\bar{\psi}_{3L}$ in terms of the complete set $I, \gamma^5, \gamma^\mu, \gamma^\mu\gamma^5$, and $\sigma^{\mu\nu}$. A related useful identity is

$$
\left(\bar{\psi}_{1R}\sigma^{\mu\nu}\psi_{2L}\right)\left(\bar{\psi}_{3L}\sigma_{\mu\nu}\psi_{4R}\right) = 0. \tag{2.217}
$$

The Fierz identities are frequently very useful in computations, and are often used in conjunction with those for charge conjugation (Section 2.10).

2.8 QED FOR ELECTRONS AND POSITRONS

Just as for pions, one can obtain the Lagrangian density for *quantum electrodynamics* (QED), i.e., for the gauge theory of electrons and positrons interacting with photons, by combining the free-field Lagrangians in (2.111) and (2.152) and applying the minimal electromagnetic substitution, $p^\mu \to p^\mu - qA^\mu \Longleftrightarrow i\partial^\mu - qA^\mu$, where $q = -e < 0$ for the electron field. Thus,

$$
\begin{aligned}
\mathcal{L} &= \bar{\psi}(x)\left(i\not{D} - m\right)\psi(x) - \frac{1}{4}F_{\mu\nu}F^{\mu\nu}\\
&= \bar{\psi}(x)\left(i\not{\partial} + e\not{A} - m\right)\psi(x) - \frac{1}{4}F_{\mu\nu}F^{\mu\nu}\\
&= \bar{\psi}(x)\left(i\not{\partial} - m\right)\psi(x) - eA_\mu(x)J^\mu_Q(x) - \frac{1}{4}F_{\mu\nu}F^{\mu\nu},
\end{aligned}
\tag{2.218}
$$

where

$$
\not{D} = \gamma^\mu D_\mu, \qquad D_\mu \equiv \partial_\mu - ieA_\mu. \tag{2.219}
$$

$\mathcal{L}$ is clearly invariant under the gauge transformation

$$
A^\mu \to A^\mu - \frac{1}{e}\partial^\mu\beta(x), \qquad \psi \to e^{iq\beta(x)/e}\psi = e^{-i\beta(x)}\psi, \tag{2.220}
$$

since $D_\mu\psi \to e^{-i\beta(x)}D_\mu\psi$, in analogy with (2.135). In the last form of $\mathcal{L}$

$$
J^\mu_Q(x) \equiv -\bar{\psi}(x)\gamma^\mu\psi(x) \tag{2.221}
$$

is the (conserved) *electromagnetic current*, i.e., the Noether current.

In Feynman diagrams electrons and positrons are represented by solid lines, with an arrow indicating the direction of flow of negative electric charge (or opposite the flow of positive charge). The interaction term in (2.218) implies three-point vertices involving one photon and two charged fermions, with a factor $ie\gamma_\mu$. Charge conservation implies that there is always one arrow entering (i.e., entering e^- *or* exiting e^+) and one leaving (exiting e^- or entering e^+) the vertex. A final e^- has a factor $\bar{u}(\vec{p}_f, s_f)$, while an initial e^- has a factor $u(\vec{p}_i, s_i)$. The corresponding factors for a final or initial e^+ are $v(\vec{p}_f, s_f)$ and $\bar{v}(\vec{p}_i, s_i)$, respectively. These are always the physical momenta and spin. An internal fermion line carrying momentum k corresponds to the propagator $iS_F(k) = i\frac{\not{k}+m}{k^2-m^2+i\epsilon}$. The spinor indices always arrange themselves so that each fermion line running through the diagram is a bilinear form starting with $\bar{u}$ or $\bar{v}$ and ending with u or v. These rules are illustrated in Figure 2.13. As in Section 2.6 there is also a factor ϵ^*_μ (ϵ_μ) for a final (initial) photon, a propagator $ig_{\mu\nu}D_F(k) = -ig_{\mu\nu}/(k^2 + i\epsilon)$ for an internal virtual photon, and an integral $\int d^4k/(2\pi)^4$ over each unconstrained internal momentum.

Since $e^\pm$ are fermions, the overall sign of a diagram depends on the ordering of particles in the states. The overall sign is rarely needed, but there may be crucial *relative* signs between two diagrams due to the anti-commutation rules for the creation and annihilation operators. In particular, there is a relative minus sign between diagrams involving the exchange of two external lines, and a factor of -1 for a closed fermion loop, as illustrated in Figure 2.14. More complicated sign ambiguities are best resolved by going directly to the expression for the transition amplitude.

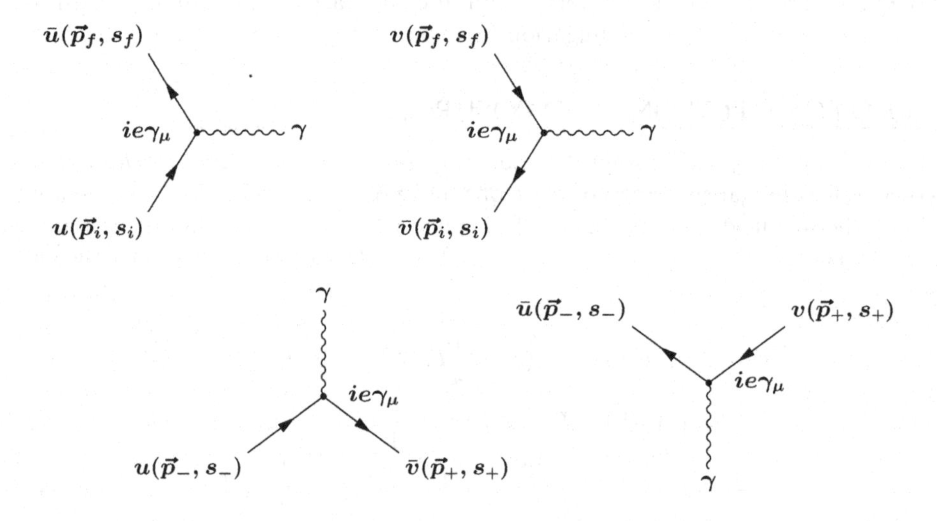

Figure 2.13 Vertices involving the interactions of $e^\pm$ with a photon. The initial fermions enter from the bottom and the final leave from the top. $\pm$ refer to a positron or electron, respectively. Note the crossing symmetry, i.e., up to an overall sign the amplitude for an initial positron can be obtained from that for a final electron except $\bar{u} \to \bar{v}$, always using the physical momentum and spin, and similarly for the relation of a final positron and initial electron.

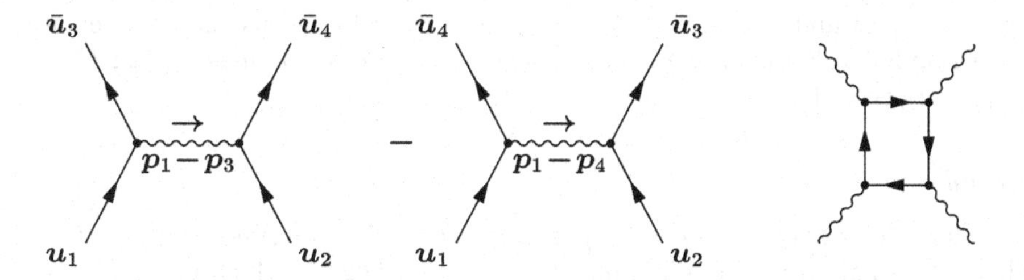

Figure 2.14 Left and center: relative minus signs between diagrams involving exchanged fermion lines, with $u_i \equiv u(\vec{p}_i, s_i)$. Right: a closed loop diagram, with an extra factor of -1.

$e\pi$ Scattering

First consider $e^-(\vec{k}_1)\,\pi^+(\vec{p}_1) \to e^-(\vec{k}_2)\,\pi^+(\vec{p}_2)$. The lowest order diagram, shown in Figure 2.15, yields the transition amplitude

$$M_{fi} = ie\bar{u}(\vec{k}_2, s_2)\gamma_\mu u(\vec{k}_1, s_1)\left(\frac{-ig^{\mu\alpha}}{(k_1-k_2)^2}\right)[-ie(p_2+p_1)_\alpha], \tag{2.222}$$

where the Feynman rules for the pion vertex are taken from Figure 2.10. Using (2.57), the

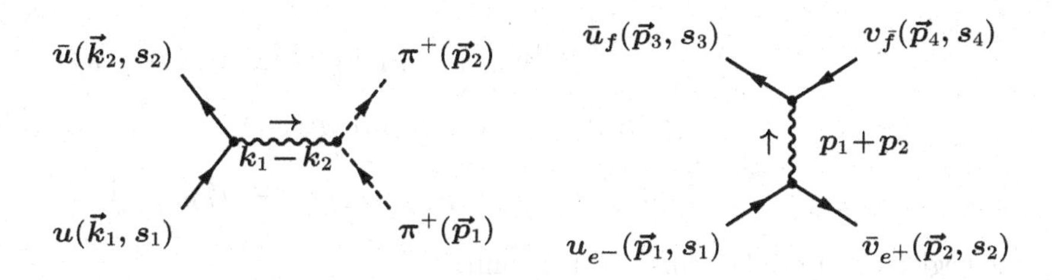

Figure 2.15 Left: lowest order diagram for $e^-(\vec{k}_1)\,\pi^+(\vec{p}_1) \to e^-(\vec{k}_2)\,\pi^+(\vec{p}_2)$. Right: diagram for $e^-(\vec{p}_1)\,e^+(\vec{p}_2) \to f(\vec{p}_3)\,\bar{f}(\vec{p}_4)$, where $f \neq e^\pm$.

unpolarized differential cross section in the CM is

$$\frac{d\bar{\sigma}}{d\cos\theta} \equiv \frac{1}{2}\sum_{s_1,s_2}\frac{d\sigma}{d\cos\theta} = \frac{1}{32\pi s}\left(\frac{1}{2}\sum_{s_1,s_2}|M_{fi}|^2\right) \equiv \frac{1}{32\pi s}|\bar{M}_{fi}|^2, \tag{2.223}$$

where $s = (p_1+k_1)^2$ and we have averaged (summed) over s_1 (s_2). The squared matrix element is evaluated using (2.209), (2.211) and the trace identities in (2.171) and (2.172),

$$\begin{aligned} |\bar{M}_{fi}|^2 &= \frac{1}{2}\frac{e^4}{t^2}(p_2+p_1)_\mu(p_2+p_1)_\nu \sum_{s_1,s_2}(\bar{u}_2\gamma^\mu u_1)(\bar{u}_1\gamma^\nu u_2) \\ &= \frac{1}{2}\frac{e^4}{t^2}(p_2+p_1)_\mu(p_2+p_1)_\nu \operatorname{Tr}[\gamma^\mu(\not{k}_1+m_e)\gamma^\nu(\not{k}_2+m_e)] \\ &= 2\frac{e^4}{t^2}(p_2+p_1)_\mu(p_2+p_1)_\nu\left[k_1^\mu k_2^\nu + k_1^\nu k_2^\mu - g^{\mu\nu}(k_1\cdot k_2 - m_e^2)\right], \end{aligned} \tag{2.224}$$

where $u_i = u(\vec{k}_i, s_i)$ and $t = (k_1 - k_2)^2$. This is easily evaluated using the kinematic relations in Section 2.3.4. For example, ignoring the pion and electron masses, (2.224) becomes $16\pi^2\alpha^2([(s-u)/t]^2 - 1)$.

$e^- e^+$ *Annihilation*

Now, consider $e^-(\vec{p}_1)\, e^+(\vec{p}_2) \to f(\vec{p}_3)\, \bar{f}(\vec{p}_4)$, where $f \neq e^\pm$ is a fermion with electric charge $Q_f e$. There is a single tree-level diagram, shown in Figure 2.15, which implies

$$\begin{aligned} M_{fi} &= (-iQ_f e\; \bar{u}_3\gamma_\mu v_4)\left(\frac{-ig^{\mu\rho}}{s}\right)(+ie\; \bar{v}_2\gamma_\rho u_1) \\ &= \frac{-iQ_f e^2}{s}\bar{u}_3\gamma_\mu v_4\; \bar{v}_2\gamma^\mu u_1, \end{aligned} \tag{2.225}$$

where $s = (p_1 + p_2)^2$; u_3 and v_4 are spinors for a mass m_f; and v_2 and u_1 correspond to mass m_e. Averaging (summing) over $s_{1,2}$ ($s_{3,4}$), the unpolarized cross section is

$$\frac{d\bar{\sigma}}{d\cos\theta} = \frac{1}{32\pi s}\frac{p_f}{p_i}|\bar{M}_{fi}|^2, \tag{2.226}$$

where

$$\begin{aligned} |\bar{M}_{fi}|^2 \equiv \frac{1}{4}\sum_{s_i,\; i=1\cdots 4} |M_{fi}|^2 &= \frac{Q_f^2 e^4}{4s^2}\,\mathrm{Tr}\left[\gamma_\mu(\not{p}_4 - m_f)\gamma_\nu(\not{p}_3 + m_f)\right] \\ &\quad\times \mathrm{Tr}\left[\gamma^\mu(\not{p}_1 + m_e)\gamma^\nu(\not{p}_2 - m_e)\right] \\ &= \frac{4Q_f^2 e^4}{s^2}\left[p_{4\mu}p_{3\nu} + p_{4\nu}p_{3\mu} - g_{\mu\nu}(p_3\cdot p_4 + m_f^2)\right] \\ &\quad\times \left[p_1^\mu p_2^\nu + p_1^\nu p_2^\mu - g^{\mu\nu}(p_1\cdot p_2 + m_e^2)\right] \\ \xrightarrow[m_e\sim 0]{} &\frac{8Q_f^2 e^4}{s^2}\left[p_1\cdot p_4\, p_2\cdot p_3 + p_1\cdot p_3\, p_2\cdot p_4 + m_f^2\, p_1\cdot p_2\right], \end{aligned} \tag{2.227}$$

where m_e is neglected in the last line. In that limit,

$$p_{1,2} = (E, 0, 0, \pm E), \qquad p_{3,4} = (E, \pm p_f\sin\theta, 0, \pm p_f\cos\theta), \tag{2.228}$$

in the CM, where

$$E = p_i = \frac{\sqrt{s}}{2}, \qquad p_f = \frac{\sqrt{s - 4m_f^2}}{2} = \beta_f E. \tag{2.229}$$

Then,

$$\begin{aligned} &p_1\cdot p_4 = p_2\cdot p_3 = E^2(1 + \beta_f\cos\theta) \\ &p_1\cdot p_3 = p_2\cdot p_4 = E^2(1 - \beta_f\cos\theta) \\ &p_1\cdot p_2 = 2E^2, \qquad m_f^2 = (1 - \beta_f^2)E^2, \end{aligned} \tag{2.230}$$

implying

$$\frac{d\bar{\sigma}}{d\cos\theta} = \frac{\pi Q_f^2\alpha^2\beta_f}{2s}\left(2 - \beta_f^2 + \beta_f^2\cos^2\theta\right). \tag{2.231}$$

Integrating over $\cos\theta$, the total cross section is

$$\bar{\sigma} = \frac{4\pi Q_f^2\alpha^2\beta_f}{3s}\left(\frac{3 - \beta_f^2}{2}\right) \xrightarrow[\beta_f\to 1]{} \frac{4\pi Q_f^2\alpha^2}{3s}. \tag{2.232}$$

Bhabha Scattering

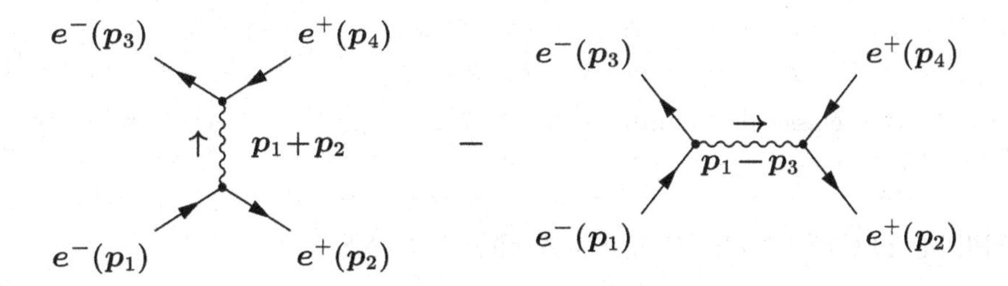

Figure 2.16 Feynman diagrams for $e^-(\vec{p}_1)\,e^+(\vec{p}_2) \to e^-(\vec{p}_3)\,e^+(\vec{p}_4)$. The relative minus sign between the two diagrams is independent of state conventions.

The diagrams for Bhabha ($e^-e^+ \to e^-e^+$) scattering are shown in Figure 2.16. The amplitude is

$$M = ie^2 \left[\frac{\bar{u}_3\gamma_\mu v_4 \bar{v}_2 \gamma^\mu u_1}{(p_1+p_2)^2} - \frac{\bar{u}_3\gamma_\mu u_1 \bar{v}_2 \gamma^\mu v_4}{(p_1-p_3)^2} \right]. \tag{2.233}$$

Ignoring the e^- mass, the unpolarized cross section is then $d\bar{\sigma}/d\cos\theta = |\bar{M}_{fi}|^2/32\pi s$, where

$$\begin{aligned} |\bar{M}_{fi}|^2 &\equiv \frac{1}{4}\sum_{spins}|M|^2 \\ &= \frac{e^4}{4}\left[\frac{\mathrm{Tr}\,(\gamma_\mu\, \not{p}_4 \gamma_\nu\, \not{p}_3)\mathrm{Tr}\,(\gamma^\mu\, \not{p}_1 \gamma^\nu\, \not{p}_2)}{s^2} + \frac{\mathrm{Tr}\,(\gamma_\mu\, \not{p}_1 \gamma_\nu\, \not{p}_3)\mathrm{Tr}\,(\gamma^\mu\, \not{p}_4 \gamma^\nu\, \not{p}_2)}{t^2} \right. \\ &\left. - \frac{\mathrm{Tr}\,(\gamma_\mu\, \not{p}_4 \gamma_\nu\, \not{p}_2 \gamma^\mu\, \not{p}_1 \gamma^\nu\, \not{p}_3)}{st} - \frac{\mathrm{Tr}\,(\gamma_\mu\, \not{p}_1 \gamma_\nu\, \not{p}_2 \gamma^\mu\, \not{p}_4 \gamma^\nu\, \not{p}_3)}{st} \right]. \end{aligned} \tag{2.234}$$

The traces are easily evaluated using the identities in Section 2.7.2, yielding

$$\frac{d\bar{\sigma}}{d\cos\theta} = \frac{\pi\alpha^2}{s}\left[\frac{t^2+u^2}{s^2} + \frac{s^2+u^2}{t^2} + \frac{2u^2}{st} \right]. \tag{2.235}$$

Electron Compton Scattering

The Feynman diagrams for e^- Compton scattering $\gamma(k_1)e^-(p_1) \to \gamma(k_2)e^-(p_2)$ are the same as the first two diagrams in Figure 2.12 except that the pions are replaced by electrons. The corresponding amplitude is

$$M = \epsilon^*_{2\mu}\epsilon_{1\nu}(ie)^2\bar{u}_2 \left[\gamma^\mu \frac{i}{\not{p}_1 + \not{k}_1 - m_e}\gamma^\nu + \gamma^\nu \frac{i}{\not{p}_1 - \not{k}_2 - m_e}\gamma^\mu \right] u_1. \tag{2.236}$$

It is straightforward to show that the differential laboratory cross section averaged (summed) over the initial (final) electron spins is given by the Klein-Nishina formula

$$\frac{d\bar{\sigma}}{d\cos\theta_L} = \frac{\pi\alpha^2}{2m_e^2}\left(\frac{k_2}{k_1}\right)^2 \left[\frac{k_2}{k_1} + \frac{k_1}{k_2} + 4(\epsilon_1\cdot\epsilon_2^*)^2 - 2 \right]. \tag{2.237}$$

k_2/k_1 has the same form as in pion Compton scattering, Equation (2.147), except $m \to m_e$. If one also averages and sums over the initial and final photon spins, one obtains (using (2.150))

$$\frac{d\bar{\sigma}}{d\cos\theta_L} = \frac{\pi\alpha^2}{m_e^2}\left(\frac{k_2}{k_1}\right)^2\left[\frac{k_2}{k_1} + \frac{k_1}{k_2} - \sin^2\theta_L\right], \tag{2.238}$$

which yields the classical Thomson cross section $\sigma_{\text{tot}} = 8\pi\alpha^2/3m_e^2$ in the limit $k_2 \sim k_1 \ll m_e$.

2.9 SPIN EFFECTS AND SPINOR CALCULATIONS

We have so far mainly focused on spin-averaged calculations. However, many experiments involve polarized initial particles or measure the final polarization, e.g., by their decay distributions. The calculation of fermion polarization effects can always be carried out using the standard trace techniques, which yield the absolute squares of amplitudes, provided one uses the spin projections in (2.178) and (2.181). However, it is often simpler to calculate the amplitudes directly using the explicit forms for the γ matrices and Dirac spinors (e.g., Hagiwara and Zeppenfeld, 1986). Further simplifications are achieved because of space reflection invariance, which relates different spin amplitudes. (Even for the weak interactions, which violate reflection invariance, different amplitudes may be related up to a known overall coefficient.) However, such calculations are carried out in a specific Lorentz frame and are therefore not manifestly invariant. Both techniques can be tedious for non-zero masses, so in this section we will illustrate calculations assuming that the masses are negligible. An example of a calculation for a massive fermion is given in Problem 2.26.

The spin-averaged amplitude-squared for $e^-(\vec{p}_1)\, e^+(\vec{p}_2) \to f(\vec{p}_3)\, \bar{f}(\vec{p}_4)$, given in (2.227), reduces to

$$|\bar{M}|^2 = Q_f^2 e^4 \left(1 + \cos^2\theta\right) \tag{2.239}$$

for $m_e = m_f = 0$. Now consider the amplitude for definite spins in the helicity basis, which we denote by $M(h_3 h_4, h_1 h_2)$, where $h_i = \pm\frac{1}{2}$ is the ith particle helicity. For massless particles, by (2.214) the only nonzero amplitudes are for $h_1 = -h_2$ and $h_3 = -h_4$, since vector and axial vector interactions do not reverse chirality. We will see in Section 2.10 that reflection invariance implies that there are only two independent amplitudes, and that

$$M(+-,+-) = M(-+,-+), \qquad M(-+,+-) = M(+-,-+), \tag{2.240}$$

where we have simplified the notation by only writing the sign of h_i. Momentarily keeping the masses, the absolute square for arbitrary spins is

$$\begin{aligned} |M|^2 = &\frac{Q_f^2 e^4}{s^2}\text{Tr}\left[\gamma_\mu\left(\not{p}_4 - m_f\right)\left(\frac{1+\gamma^5\not{s}_4}{2}\right)\gamma_\nu\left(\not{p}_3 + m_f\right)\left(\frac{1+\gamma^5\not{s}_3}{2}\right)\right] \\ &\times \text{Tr}\left[\gamma^\mu\left(\not{p}_1 + m_e\right)\left(\frac{1+\gamma^5\not{s}_1}{2}\right)\gamma^\nu\left(\not{p}_2 - m_e\right)\left(\frac{1+\gamma^5\not{s}_2}{2}\right)\right], \end{aligned} \tag{2.241}$$

where s_i is the spin vector given in (2.180). Taking the masses to zero and using (2.181), this reduces to

$$\begin{aligned} |M(+-,+-)|^2 &= |M(-+,-+)|^2 \\ &= \frac{Q_f^2 e^4}{s^2}\text{Tr}\left(\gamma_\mu\, \not{p}_4 \gamma_\nu\, \not{p}_3 P_L\right)\text{Tr}\left(\gamma^\mu\, \not{p}_1 \gamma^\nu\, \not{p}_2 P_L\right) \\ &= \frac{Q_f^2 e^4}{s^2}\left(p_1\cdot p_4\; p_2\cdot p_3\right) = Q_f^2 e^4 \left(1+\cos\theta\right)^2, \end{aligned} \tag{2.242}$$

where the calculation of the traces and contraction of the Lorentz indices was carried out using (2.175). Similarly,

$$|M(+-,-+)|^2 = |M(-+,+-)|^2 = \frac{Q_f^2 e^4}{s^2}(p_1 \cdot p_3 \, p_2 \cdot p_4) = Q_f^2 e^4 (1-\cos\theta)^2. \quad (2.243)$$

Equations (2.242) and (2.243) reproduce the spin-averaged result in (2.239).

Now let us repeat the calculation using the explicit gamma matrices and spinors in the chiral representation, given in (2.165) and (2.192), noting that

$$\gamma^0\gamma^0 = I, \qquad \gamma^0\gamma^i = -\gamma^0\gamma_i = \begin{pmatrix} -\sigma^i & 0 \\ 0 & +\sigma^i \end{pmatrix}. \quad (2.244)$$

Then [using $s = (\sqrt{2}E)^4$],

$$\begin{aligned} M(+-,+-) &= \frac{-iQ_f e^2}{s}\bar{u}_+(3)\gamma_\mu v_-(4)\,\bar{v}_-(2)\gamma^\mu u_+(1) \\ &= -iQ_f e^2\left[\phi_+(3)^\dagger\chi_-(4)\,\chi_-(2)^\dagger\phi_+(1) - \phi_+(3)^\dagger\sigma^i\chi_-(4)\,\chi_-(2)^\dagger\sigma^i\phi_+(1)\right], \end{aligned} \quad (2.245)$$

and

$$\begin{aligned} M(-+,+-) = &+iQ_f e^2\left[\phi_-(3)^\dagger\chi_+(4)\,\chi_-(2)^\dagger\phi_+(1)\right. \\ &\left. + \phi_-(3)^\dagger\sigma^i\chi_+(4)\,\chi_-^\dagger(2)\sigma^i\phi_+(1)\right]. \end{aligned} \quad (2.246)$$

From (2.228) and Table 2.1 the helicity spinors are

$$\begin{aligned} &\phi_+(1) = \begin{pmatrix} 1 \\ 0 \end{pmatrix}, \quad \chi_-(2) = \begin{pmatrix} 0 \\ 1 \end{pmatrix}, \quad \phi_+(3) = \chi_+(4) = \begin{pmatrix} \cos\frac{\theta}{2} \\ \sin\frac{\theta}{2} \end{pmatrix} \\ &\phi_-(3) = \chi_-(4) = \begin{pmatrix} -\sin\frac{\theta}{2} \\ \cos\frac{\theta}{2} \end{pmatrix}. \end{aligned} \quad (2.247)$$

The first terms in (2.245) and (2.246) vanish. The second terms are most easily evaluated using the $SU(2)$ Fierz identities in Problem 1.1,

$$\phi(3)^\dagger\sigma^i\chi(4)\,\chi(2)^\dagger\sigma^i\phi(1) = 2\phi(3)^\dagger\phi(1)\,\chi(2)^\dagger\chi(4) - \phi(3)^\dagger\chi(4)\,\chi(2)^\dagger\phi(1), \quad (2.248)$$

yielding

$$\begin{aligned} M(+-,+-) &= M(-+,-+) = 2iQ_f e^2\cos^2\frac{\theta}{2} = iQ_f e^2(1+\cos\theta) \\ M(-+,+-) &= M(+-,-+) = -2iQ_f e^2\sin^2\frac{\theta}{2} = -iQ_f e^2(1-\cos\theta). \end{aligned} \quad (2.249)$$

2.10 THE DISCRETE SYMMETRIES P, C, CP, T, AND CPT

Space reflection and charge conjugation are symmetries of the strong and electromagnetic interactions, but are violated by the weak interactions because of their chiral nature. The product CP is also violated in the weak sector, though more feebly, due ultimately to complex phases in the Yukawa interactions between the fermion and Higgs fields. Such phases may be large, but their effects are small because they require mixing between all three fermion families to be observable. Time reversal is also violated, as is expected because of the CPT theorem, which states that any local, Lorentz-invariant, unitary field theory must be invariant under the product CPT (Streater and Wightman, 2000). In this section we introduce the basic formalism for the discrete symmetries. Physical consequences and tests will be described in Chapters 7 and 8. More detailed descriptions of the discrete symmetries may be found in, e.g., (Bjorken and Drell, 1965; Gasiorowicz, 1966; Weinberg, 1995; Sozzi, 2008).

Space Reflection

Under *space reflection*[19] (P) a classical *vector* changes sign, e.g.,

$$\vec{x} \underset{P}{\rightarrow} -\vec{x}, \qquad \vec{p} \underset{P}{\rightarrow} -\vec{p}. \tag{2.250}$$

However, an *axial vector* such as orbital angular momentum $\vec{L} = \vec{x} \times \vec{p}$ is left invariant, and it is reasonable to define spin and total angular momentum to have the same property

$$\vec{J} \underset{P}{\rightarrow} \vec{J}. \tag{2.251}$$

Both vectors and axial vectors transform as vectors under rotations. *Scalars* are rotational scalars that do not change sign under P. Examples are t, E, and $|\vec{p}\,|^2$. *Pseudoscalars*, such as $\vec{J} \cdot \vec{p}$, *do* change sign. In particular, helicity, $h = \vec{S} \cdot \hat{p}$, is a pseudoscalar and reverses under P. Vector and axial[20] four-vectors transform, respectively, as

$$V^\mu \underset{P}{\rightarrow} V_\mu, \qquad A^\mu \underset{P}{\rightarrow} -A_\mu, \tag{2.252}$$

so that $x = (t, \vec{x}\,)$ and $p = (E, \vec{p}\,)$ are vectors. The fermion spin vector s_μ in (2.180) is an axial four-vector. The angular momentum $\vec{J}$ is part of a second rank antisymmetric tensor, $L^i = \frac{1}{2}\epsilon_{ijk} L^{jk}$, where

$$L^{\mu\nu} \equiv x^\mu p^\nu - x^\nu p^\mu \underset{P}{\rightarrow} L_{\mu\nu}. \tag{2.253}$$

A single particle state is assumed to transform similarly,

$$P|\vec{p}\, s\rangle = \eta_P| - \vec{p}\, s\rangle, \qquad P|\vec{p}\, h\rangle = \eta_P| - \vec{p} - h\rangle, \tag{2.254}$$

where (in this section) s represents spin with respect to a fixed axis and h represents helicity.[21] The phase η_P is the intrinsic parity, which depends on the particle type. One must have $\eta_P = \pm 1$ to ensure $P^2 = I$. Invariance of the action in (2.18) requires invariance of the Lagrangian, and therefore that

$$P\mathcal{L}(t, \vec{x}\,) P^{-1} = \mathcal{L}(t, -\vec{x}\,). \tag{2.255}$$

For the example of the free complex scalar,

$$\mathcal{L} = (\partial_\mu \phi)^\dagger (\partial^\mu \phi) - m^2 \phi^\dagger \phi, \tag{2.256}$$

[19] More precisely, an inversion or point reflection.

[20] We use the symbol A_μ both for gauge fields and for axial vectors. The meaning should always be clear from the context.

[21] In the helicity form, there may be additional $(\vec{p}, h)$-dependent phases, depending on the phase conventions for the helicity states.

this can be accomplished for

$$P\phi(t,\vec{x})P^{-1} = \eta_P \phi(t,-\vec{x}). \tag{2.257}$$

The mass term is obviously invariant. For the kinetic term,

$$\partial_\mu \phi(x) \to \eta_P \partial_\mu \phi(x') = \eta_P \frac{\partial}{\partial x^\mu}\phi(x') = \eta_P \frac{\partial}{\partial x'_\mu}\phi(x') = \eta_P \partial'^\mu \phi(x'), \tag{2.258}$$

where $x' \equiv (t,-\vec{x})$. Thus, $|\partial_\mu \phi(x)|^2 \to |\partial'_\mu \phi(x')|^2$. Equation (2.257) follows from the free-field expression in (2.93) provided

$$Pa^\dagger(\vec{p})P^{-1} = \eta_P a^\dagger(-\vec{p}), \qquad Pb^\dagger(\vec{p})P^{-1} = \eta_P b^\dagger(-\vec{p}), \tag{2.259}$$

where $a^\dagger(\vec{p})$ and $b^\dagger(\vec{p})$ are the particle and antiparticle creation operators. The transformation of a Hermitian scalar field is the same as in (2.257) and (for $a^\dagger(\vec{p})$) in (2.259).

For the free Dirac field, with

$$\mathcal{L} = \bar{\psi}(t,\vec{x})\left[i\not{\partial} - m\right]\psi(t,\vec{x}), \tag{2.260}$$

we must choose $P\psi(t,\vec{x})P^{-1}$ in such a way that

$$P\bar{\psi}(t,\vec{x})\psi(t,\vec{x})P^{-1} = \bar{\psi}(t,-\vec{x})\psi(t,-\vec{x}) \tag{2.261}$$

and

$$P\bar{\psi}(t,\vec{x})\gamma^\mu\psi(t,\vec{x})P^{-1} = \bar{\psi}(t,-\vec{x})\gamma_\mu\psi(t,-\vec{x}). \tag{2.262}$$

The lowering of the index on γ^μ compensates for $\partial_\mu = \partial'^\mu$, analogous to (2.258), i.e., $\bar{\psi}(x)\not{\partial}\psi(x) = \bar{\psi}(x')\not{\partial}'\psi(x')$. These conditions can be satisfied if

$$\begin{aligned} &P\psi(t,\vec{x})P^{-1} = \gamma^0\psi(t,-\vec{x}) \\ &\Rightarrow P\bar{\psi}(t,\vec{x})P^{-1} = \left(\gamma^0\psi(t,-\vec{x})\right)^\dagger\gamma^0 = \psi(t,-\vec{x})^\dagger\gamma^0\gamma^0 = \bar{\psi}(t,-\vec{x})\gamma^0, \end{aligned} \tag{2.263}$$

using $\gamma^0\gamma^\mu\gamma^0 = \gamma_\mu$ from (2.156). (We have taken $\eta_{P\psi} = +1$ for simplicity.) For chiral fields,

$$\begin{aligned} P\psi_{L,R}(t,\vec{x})P^{-1} &= \gamma^0\psi_{R,L}(t,-\vec{x}) \\ P\bar{\psi}_{L,R}(t,\vec{x})P^{-1} &= \left(\gamma^0\psi_{R,L}(t,-\vec{x})\right)^\dagger\gamma^0 = \bar{\psi}_{R,L}(t,-\vec{x})\gamma^0. \end{aligned} \tag{2.264}$$

Using the expression (2.159) for the free Dirac field and the relations[22]

$$\gamma^0 u(\vec{p},s) = u(-\vec{p},s), \qquad \gamma^0 v(\vec{p},s) = -v(-\vec{p},s), \tag{2.265}$$

which follow from the explicit spinor forms in Section 2.7.2, we see that (2.263) is equivalent to

$$Pa^\dagger(\vec{p},s)P^{-1} = a^\dagger(-\vec{p},s), \qquad Pb^\dagger(\vec{p},s)P^{-1} = -b^\dagger(-\vec{p},s). \tag{2.266}$$

Thus, a fermion and its antifermion must have opposite intrinsic parity.

An arbitrary fermion bilinear form transforms as

$$P\bar{\psi}_a(t,\vec{x})\Gamma\psi_b(t,\vec{x})P^{-1} = \eta_{Pa}\eta_{Pb}\bar{\psi}_a(t,-\vec{x})\gamma^0\Gamma\gamma^0\psi_b(t,-\vec{x}). \tag{2.267}$$

The values of $\Gamma_p \equiv \gamma^0\Gamma\gamma^0$ are listed in Table 2.2. In particular, $\Gamma = 1, \gamma^5, \gamma^\mu$, and $\gamma^\mu\gamma^5$

[22]For the phase conventions in Table 2.1, the transformations in the helicity basis are $\gamma^0 u(\vec{p},\phi_\pm(\hat{p})) = \pm e^{\pm i\varphi}u(-\vec{p},\phi_\mp(-\hat{p}))$ and $\gamma^0 v(\vec{p},\chi_\pm(\hat{p})) = \mp e^{\mp i\varphi}v(-\vec{p},\chi_\mp(-\hat{p}))$, where φ is the azimuthal angle of $\vec{p}$.

TABLE 2.2 Transformation of the complete set of Dirac matrices and of their related chiral forms under Hermitian conjugation (the Dirac adjoint $\overline{\Gamma}$) in (2.210), space reflection (Γ_p) in (2.267), charge conjugation (Γ_c) in (2.296), and time reversal (Γ_t) in (2.325).

	Γ	$\overline{\Gamma}$ $\gamma^0\Gamma^\dagger\gamma^0$	Γ_p $\gamma^0\Gamma\gamma^0$	Γ_c $\mathcal{C}\Gamma^T\mathcal{C}^{-1}$	Γ_t $\mathcal{T}\Gamma^*\mathcal{T}^{-1}$
Scalar	1	1	1	1	1
Pseudoscalar	γ^5	$-\gamma^5$	$-\gamma^5$	γ^5	γ^5
Vector	γ^μ	γ^μ	γ_μ	$-\gamma^\mu$	γ_μ
Axial vector	$\gamma^\mu\gamma^5$	$\gamma^\mu\gamma^5$	$-\gamma_\mu\gamma^5$	$\gamma^\mu\gamma^5$	$\gamma_\mu\gamma^5$
Tensor	$\sigma^{\mu\nu}$	$\sigma^{\mu\nu}$	$\sigma_{\mu\nu}$	$-\sigma^{\mu\nu}$	$-\sigma_{\mu\nu}$
Pseudotensor[a]	$\sigma^{\mu\nu}\gamma^5$	$-\sigma^{\mu\nu}\gamma^5$	$-\sigma_{\mu\nu}\gamma^5$	$-\sigma^{\mu\nu}\gamma^5$	$-\sigma_{\mu\nu}\gamma^5$
$S \mp P$	$P_{L,R}$	$P_{R,L}$	$P_{R,L}$	$P_{L,R}$	$P_{L,R}$
$V \mp A$	$\gamma^\mu P_{L,R}$	$\gamma^\mu P_{L,R}$	$\gamma_\mu P_{R,L}$	$-\gamma^\mu P_{R,L}$	$\gamma_\mu P_{L,R}$
	$\sigma^{\mu\nu} P_{L,R}$	$\sigma^{\mu\nu} P_{R,L}$	$\sigma_{\mu\nu} P_{R,L}$	$-\sigma^{\mu\nu} P_{L,R}$	$-\sigma_{\mu\nu} P_{L,R}$

[a]The pseudotensor $\sigma^{\mu\nu}\gamma^5$ is not independent (Problem 2.9), but is included for completeness.

transform as a scalar, pseudoscalar, vector, and axial vector, respectively, for $\eta_{Pa}\eta_{Pb} = 1$.

An important example involves the left-chiral ($V - A$) currents $\bar{\psi}_{aL}\gamma^\mu\psi_{bL}$, which transform as

$$P\bar{\psi}_{aL}\gamma^\mu\psi_{bL}P^{-1} = \bar{\psi}_{aR}\gamma_\mu\psi_{bR}. \tag{2.268}$$

We will see in Chapter 7 that the weak charged current (WCC) interactions involve only $V - A$, so (2.268) implies that the WCC is *not* P invariant.

As will be discussed in Chapters 3 and 5, the Yukawa interaction between a neutral pion π^0 and nucleons can be phenomenologically described by

$$\mathcal{L}_{\pi N} = i g_\pi \left(\bar{p}\gamma^5 p - \bar{n}\gamma^5 n\right)\pi^0, \tag{2.269}$$

where g_π is a real coupling, and where p and n, respectively, represent the proton and neutron fields. Since the nucleon term is a pseudoscalar, reflection invariance requires that the π^0 must be a pseudoscalar, i.e., $\eta_\pi = -1$. Of course, this was originally ascertained by experiment (see, e.g., Gasiorowicz, 1966). The pseudoscalar nature of the π^- was determined by the observation of the reaction $\pi^- D \to nn$, where the deuteron D has $J^P = 1^+$ and the π^- was established to be in an S-wave. The neutrons had to be in an odd-parity $^3P_1$ state, since it is the only antisymmetric $J = 1$ state available. Similarly, the π^0 was shown to have odd parity by the angular distribution in $\pi^0 \to e^+e^-\, e^+e^-$ (Abouzaid et al., 2008). The i in (2.269) is required by Hermiticity.

The interaction terms in (2.133) and (2.218) indicate that the electromagnetic field A_Q^μ couples to vector currents, where in this section, we use the symbol A_Q^μ to distinguish it from an axial vector. Reflection invariance of the electromagnetic interactions therefore requires that A_Q^μ transforms as a vector

$$PA_Q^\mu(t,\vec{x})P^{-1} = A_{Q\mu}(t,-\vec{x}), \tag{2.270}$$

which is consistent with the classical expectation $\vec{E} \to -\vec{E}$ and $\vec{B} \to +\vec{B}$. Using the free-field results in Section 2.5, this is equivalent to

$$Pa^\dagger(\vec{p},\lambda_h)P^{-1} = a^\dagger(-\vec{p},-\lambda_h), \tag{2.271}$$

where the polarization label λ_h represents the photon helicity. This can be seen from the free-field expression (2.113), using the relation

$$\epsilon^\mu(\vec{p}, \lambda_h) = \epsilon_\mu(-\vec{p}, -\lambda_h), \tag{2.272}$$

which follows from (2.117) and (2.118) under the convention

$$\epsilon^\mu(\vec{p}, \lambda) = (-1)^{\lambda+1}\epsilon_\mu(-\vec{p}, \lambda),\ \lambda = 1, 2 \tag{2.273}$$

for the linear polarization vectors. All of these results also apply to massive vectors, with three helicity states.

The explicit transformations of the fields in (2.257), (2.263), and (2.270) were justified for free fields. However, the same transformation laws will continue to hold in the presence of the strong and electromagnetic interactions, or any others that respect reflection invariance. This is apparent in the interaction picture (see, e.g., Peskin and Schroeder, 1995, and Appendix B), in which the interacting fields can be formally expressed in terms of the non-interacting ones and the interaction Hamiltonian H_I, and using $[P, H_I] = 0$. To illustrate the importance of this, consider the matrix elements of the vector and axial currents, $\bar{\psi}_a\gamma_\mu\psi_b$ and $\bar{\psi}_a\gamma_\mu\gamma^5\psi_b$, respectively, where $\psi_{a,b}$ represent strongly interacting fields such as nucleon or quark fields, and a can be the same as or different from b.

$$\begin{aligned}
\langle a(\vec{p}_a, s_a)|\bar{\psi}_a\gamma_\mu\psi_b|b(\vec{p}_b, s_b)\rangle &= \bar{u}(\vec{p}_a, s_a)\Gamma^V_\mu u(\vec{p}_b, s_b)\\
\langle a(\vec{p}_a, s_a)|\bar{\psi}_a\gamma_\mu\gamma^5\psi_b|b(\vec{p}_b, s_b)\rangle &= \bar{u}(\vec{p}_a, s_a)\Gamma^A_\mu u(\vec{p}_b, s_b).
\end{aligned} \tag{2.274}$$

The fields are evaluated at $x = 0$ (otherwise, by translation invariance, there would be a factor $e^{i(p_a - p_b)\cdot x}$ on the right). In the absence of strong or electromagnetic interactions, one would have $\Gamma^V_\mu = \gamma_\mu$ and $\Gamma^A_\mu = \gamma_\mu\gamma^5$. However, such corrections can yield more complicated matrix elements. From Lorentz invariance, $\Gamma^{V,A}_\mu$ can involve linear combinations of $\gamma_\mu, \sigma_{\mu\nu}q^\nu$, $q_\mu, \gamma_\mu\gamma_5, \sigma_{\mu\nu}q^\nu\gamma^5$, and $q_\mu\gamma^5$, where $q \equiv p_a - p_b$. Each of these can be multiplied by an arbitrary *form factor* that can depend on the invariant q^2. Other four-vectors, such as those involving $p_{a\mu} + p_{b\mu}$ are not independent (see Problem 2.10). However, reflection invariance of the strong and electromagnetic interactions restricts the possibilities. Assuming the same intrinsic parities for a and b,

$$\begin{aligned}
\langle a(\vec{p}_a, s_a)|\bar{\psi}_a\gamma_\mu\psi_b|b(\vec{p}_b, s_b)\rangle &= \langle a(\vec{p}_a, s_a)|P^{-1}P\bar{\psi}_a\gamma_\mu\psi_b P^{-1}P|b(\vec{p}_b, s_b)\rangle\\
&= \langle a(-\vec{p}_a, s_a)|\bar{\psi}_a\gamma^0\gamma_\mu\gamma^0\psi_b|b(-\vec{p}_b, s_b)\rangle\\
&= \bar{u}(-\vec{p}_a, s_a)\Gamma^{V\mu}(q')u(-\vec{p}_b, s_b)\\
&= \bar{u}(\vec{p}_a, s_a)\gamma^0\Gamma^{V\mu}(q')\gamma^0 u(\vec{p}_b, s_b),
\end{aligned} \tag{2.275}$$

where we have used (2.265). The notation $\Gamma^{V\mu}(q')$ indicates that it is evaluated using $q'_\mu = p'_{a\mu} - p'_{b\mu} = q^\mu$. Reflection invariance therefore requires

$$\Gamma^V_\mu(q) = \gamma^0\Gamma^{V\mu}(q')\gamma^0. \tag{2.276}$$

Comparison with Table 2.2 shows that Γ^V_μ can only contain $\gamma_\mu, \sigma_{\mu\nu}q^\nu$, and q_μ. Similarly, Γ^A_μ can only contain $\gamma_\mu\gamma^5, \sigma_{\mu\nu}q^\nu\gamma^5$, and $q_\mu\gamma^5$. (Higher-order terms in the weak interactions, which violate reflection invariance, could induce the wrong terms. However, one can always treat such effects explicitly in perturbation theory, using P invariance for calculating the matrix elements.)

As another application, consider $M(a_i(\vec{p}_i, h_i))$, the matrix element of a scattering or decay process involving k external particles with particle types a_i, momenta $\vec{p}_i$, and helicities h_i, $i = 1 \cdots k$, some of which are in the initial and some in the final state. If the relevant interactions are reflection invariant, then it follows from the interaction picture expression that the amplitude is the same as the amplitude for the reflection-reversed process,

$$M(a_i(\vec{p}_i, h_i)) = \eta M(a_i(-\vec{p}_i, -h_i)), \tag{2.277}$$

in which $\eta = \pm 1$ is associated with the intrinsic parities and the spinor phase conventions (see Footnote 22). This result is especially useful in certain special cases for which $M(a_i(-\vec{p}_i, -h_i)) = M(a_i(\vec{p}_i, -h_i))$. For example, this follows from ordinary rotational invariance for $2 \to 2$ scattering in the center of mass frame,[23] or for $1 \to 2$ or $1 \to 3$ decays of a spinless particle in its rest frame, because the momenta lie in a plane and can be reversed by a rotation by π around an axis perpendicular to the plane. In such cases,

$$M(a_i(\vec{p}_i, h_i)) = \eta M(a_i(\vec{p}_i, -h_i)), \tag{2.278}$$

which can be very useful in simplifying calculations.

Charge Conjugation

Charge conjugation changes particles into antiparticles, without affecting their momenta or spin, e.g.,

$$C|a(\vec{p}, s)\rangle = \eta_{Ca}|a^c(\vec{p}, s)\rangle, \tag{2.279}$$

where a^c is the antiparticle to a and η_{Ca} is a phase factor. (It is basically a convention as to which is called the particle and which the antiparticle.) For a complex scalar field, this implies

$$C\phi^\dagger C^{-1} = \eta_{C\phi}\phi, \qquad C\phi C^{-1} = \eta^*_{C\phi}\phi^\dagger, \tag{2.280}$$

since $\phi^\dagger$ is the field for the antiparticle to ϕ. Equation (2.280) implies

$$Ca^\dagger(\vec{p})C^{-1} = \eta_{C\phi}b^\dagger(\vec{p}), \qquad Cb^\dagger(\vec{p})C^{-1} = \eta^*_{C\phi}a^\dagger(\vec{p}) \tag{2.281}$$

for the creation operators in (2.93). One can always make a phase transformation $\phi' = \sqrt{\pm\eta_{C\phi}}\phi$ so that the new field has $\eta_{C\phi'} = \pm 1$. For a Hermitian scalar, one must have $\eta_{C\phi} = \pm 1$,

$$C\phi C^{-1} = \pm\phi, \qquad Ca^\dagger(\vec{p})C^{-1} = \pm a^\dagger(\vec{p}). \tag{2.282}$$

If one expresses a complex scalar ϕ with $\eta_{C\phi} = \pm 1$ in terms of two Hermitian scalars $\phi_{1,2}$, as in (2.91), then $\eta_{C\phi_1} = -\eta_{C\phi_2} = \pm 1$. Results similar to (2.280)–(2.282) hold for the transformation of a vector field A_μ. For QED the gauge field couples to a C-odd vector current, so one must choose $A_{Q\mu} \to -A_{Q\mu}$. (C is not conserved for the chiral theories that will be introduced in Chapter 4, which involve both vector and axial currents with opposite C transformations. It is nevertheless useful to define the transformations of the gauge fields in this way.) The intrinsic C phases can be determined experimentally for (C-invariant) strong and electromagnetic processes. For example, π^0 decays to two photons, so $\eta_{C\pi^0} = \eta^2_{C\gamma} = +1$.

For QED, the electron state is transformed into a positron,

$$C|e^-(\vec{p}, s)\rangle = \eta_{Ce}|e^+(\vec{p}, s)\rangle, \tag{2.283}$$

[23] Helicity is not Lorentz invariant in general.

and similarly for other fermions. One requires

$$C\psi C^{-1} = \eta^*_{C\psi}\psi^c, \qquad C\bar{\psi}C^{-1} = \eta_{C\psi}\bar{\psi}^c \equiv \eta_{C\psi}\left(\psi^c\right)^\dagger \gamma_0, \tag{2.284}$$

where ψ is the e^- (or fermion) field, and ψ^c is the e^+ (or antifermion) field. Only the relative phases of different fermion fields in bilinears are usually relevant;[24] in the following we will set $\eta_{C\psi} = 1$ for simplicity. From (2.159), the free fermion fields are given by

$$\begin{aligned} \psi(x) &= \int \frac{d^3\vec{p}}{(2\pi)^3 2E_p} \sum_{s=1}^{2} \left[u(\vec{p},s)\, a(\vec{p},s)e^{-ip\cdot x} + v(\vec{p},s)b^\dagger(\vec{p},s)e^{+ip\cdot x}\right] \\ \psi^c(x) &= \int \frac{d^3\vec{p}}{(2\pi)^3 2E_p} \sum_{s=1}^{2} \left[u(\vec{p},s)\, b(\vec{p},s)e^{-ip\cdot x} + v(\vec{p},s)a^\dagger(\vec{p},s)e^{+ip\cdot x}\right], \end{aligned} \tag{2.285}$$

so one must choose

$$Ca^\dagger(\vec{p},s)C^{-1} = b^\dagger(\vec{p},s), \qquad Cb^\dagger(\vec{p},s)C^{-1} = a^\dagger(\vec{p},s). \tag{2.286}$$

From (2.285) it is apparent that ψ^c is related to the Hermitian conjugate $\psi^\dagger$. The precise relation, which is extremely useful even if a theory is not charge conjugation invariant, is

$$\psi^c = \mathcal{C}\bar{\psi}^T = \mathcal{C}\left(\psi^\dagger\gamma^0\right)^T = \mathcal{C}\gamma^{0T}\psi^{\dagger T}, \qquad \bar{\psi}^c = -\psi^T\mathcal{C}^{-1} \tag{2.287}$$

(in indices, $\psi^c_\alpha = \mathcal{C}_{\alpha\beta}\bar{\psi}_\beta = \left(\mathcal{C}\gamma^{0T}\right)_{\alpha\beta}\psi^\dagger_\beta$). $\mathcal{C}$ is a 4×4 Dirac matrix, which can be determined by the requirement that the free Lagrangian density in (2.152) is C-invariant, i.e.,

$$C\mathcal{L}_0 C^{-1} = \bar{\psi}^c(x)\left(i\,\partial\!\!\!/ - m\right)\psi^c(x) = \mathcal{L}_0. \tag{2.288}$$

Equation (2.288) is satisfied if and only if $\mathcal{C}$ satisfies

$$\mathcal{C}^{-1}\gamma_\mu\mathcal{C} = -\gamma_\mu^T, \tag{2.289}$$

where one must use the anticommuting nature of the fermion field, and the fields are considered to be normal-ordered so that c-numbers can be ignored.[25] Consistency between (2.285) and (2.287) (or between (2.161) and (2.163)) further requires that the u and v spinors are related by

$$v(\vec{p},s) = \mathcal{C}\bar{u}(\vec{p},s)^T = \left(\mathcal{C}\gamma^{0T}\right)u(\vec{p},s)^*, \qquad u(\vec{p},s) = \mathcal{C}\bar{v}(\vec{p},s)^T, \tag{2.290}$$

which implies

$$\bar{u}(\vec{p},s) = -v(\vec{p},s)^T\mathcal{C}^\dagger, \qquad \bar{v}(\vec{p},s) = -u(\vec{p},s)^T\mathcal{C}^\dagger. \tag{2.291}$$

These relations hold both for the case of a fixed spin axis or for helicity spinors. The explicit form of $\mathcal{C}$ depends on the representation used for the γ matrices and is seldom needed. However, $\mathcal{C}$ takes the simple form

$$\mathcal{C} = -\mathcal{C}^{-1} = -\mathcal{C}^\dagger = -\mathcal{C}^T = \pm i\gamma^2\gamma^0 \tag{2.292}$$

in the Pauli-Dirac ($+i\gamma^2\gamma^0$) and chiral ($-i\gamma^2\gamma^0$) representations. The off-diagonal nature of σ^2 accounts for the extra minus sign in (2.190).

---

[24] Majorana masses, which we will encounter in connection with neutrinos in Chapter 9 and supersymmetry in Section 10.2, are an important exception.

[25] From (2.287) and (2.289) we have that ψ and ψ^c have opposite intrinsic parities under space reflection, consistent with the comment following (2.266) and ensuring that C and P commute.

For arbitrary fermion fields $\psi_{a,b}$ (with the same intrinsic phases),

$$\begin{aligned} \mathcal{C}\bar{\psi}_a\gamma^\mu\psi_b\mathcal{C}^{-1} &= \bar{\psi}_a^c\gamma^\mu\psi_b^c = -\psi_a^T\mathcal{C}^{-1}\gamma^\mu\mathcal{C}\bar{\psi}_b^T \\ &= +\psi_a^T\gamma^{\mu T}\bar{\psi}_b^T = -\bar{\psi}_b\gamma^\mu\psi_a = -(\bar{\psi}_a\gamma^\mu\psi_b)^\dagger. \end{aligned} \tag{2.293}$$

In particular, the electromagnetic current

$$J_Q^\mu = -\bar{\psi}_{e^-}\gamma^\mu\psi_{e^-} \tag{2.294}$$

in QED transforms as

$$\mathcal{C}J_Q^\mu\mathcal{C}^{-1} = -\bar{\psi}_{e^+}\gamma^\mu\psi_{e^+} = +\bar{\psi}_{e^-}\gamma^\mu\psi_{e^-} = -J_Q^\mu, \tag{2.295}$$

so that the interaction $-eA_{Q\mu}J_Q^\mu$ in (2.218) is invariant for $A_{Q\mu} \to -A_{Q\mu}$. An arbitrary fermion bilinear transforms as

$$\mathcal{C}\bar{\psi}_a\Gamma\psi_b\mathcal{C}^{-1} = \bar{\psi}_a^c\Gamma\psi_b^c = \bar{\psi}_b\Gamma_c\psi_a, \tag{2.296}$$

where

$$\Gamma_c \equiv \left(\mathcal{C}^{-1}\Gamma\mathcal{C}\right)^T = \mathcal{C}\Gamma^T\mathcal{C}^{-1}. \tag{2.297}$$

The Γ_c are listed in Table 2.2. The spinor identity corresponding to (2.296) differs by a sign, i.e.,

$$\bar{w}_a\Gamma w_b = -\bar{w}_b^c\Gamma_c w_a^c, \tag{2.298}$$

where each w can be either a u or v spinor, and

$$u^c(\vec{p},s) \equiv v(\vec{p},s), \qquad v^c(\vec{p},s) \equiv u(\vec{p},s) \tag{2.299}$$

is the charge conjugate spinor.

The chiral fermion fields transform as

$$\mathcal{C}\psi_L\mathcal{C}^{-1} = \psi_L^c = P_L\psi^c, \qquad \mathcal{C}\psi_R\mathcal{C}^{-1} = \psi_R^c = P_R\psi^c, \tag{2.300}$$

where $\psi_L^c(\psi_R^c)$ is the field that annihilates a left (right)-chiral antiparticle or creates a right (left)-chiral particle. From (2.287) and (2.300)

$$\begin{aligned} \psi_L^c &= P_L\psi^c = \mathcal{C}\bar{\psi}_R^T, & \psi_R^c &= P_R\psi^c = \mathcal{C}\bar{\psi}_L^T \\ \bar{\psi}_L^c &= -\psi_R^T\mathcal{C}^{-1}, & \bar{\psi}_R^c &= -\psi_L^T\mathcal{C}^{-1}, \end{aligned} \tag{2.301}$$

where, e.g., $\bar{\psi}_R^T \equiv \left[(\psi_R)^\dagger\gamma^0\right]^T = \gamma^{0T}(\psi_R)^{\dagger T}$. Thus, $\psi_L^c \sim \psi_R^\dagger$ up to Dirac indices. Similar results hold for the transformations (2.290) and (2.291) of the chiral spinors. The flip of chirality in (2.301) is at first confusing. It is associated with the fact that $\mathcal{C}$ does *not* change the spin or helicity, but the relation between chirality and helicity is reversed for antiparticles, as can be seen in (2.195). For example, in the massless limit ψ_L^c and $\psi_R^\dagger$ both create positive helicity particles or annihilate negative helicity antiparticles. Note that γ^μ and $\gamma^\mu\gamma^5$ transform with opposite signs, so that a left-chiral current is mapped onto a right-chiral one,

$$\mathcal{C}\bar{\psi}_{aL}\gamma^\mu\psi_{bL}\mathcal{C}^{-1} = \bar{\psi}_{aL}^c\gamma^\mu\psi_{bL}^c = -\bar{\psi}_{bR}\gamma^\mu\psi_{aR}. \tag{2.302}$$

Thus, from (2.268) and (2.302) the $(V-A)$ charged current weak interactions violate both space reflection and charge conjugation invariance. Similarly,

$$\begin{aligned} \mathcal{C}\bar{\psi}_{aL}\psi_{bR}\mathcal{C}^{-1} &= \bar{\psi}_{aL}^c\psi_{bR}^c = +\bar{\psi}_{bL}\psi_{aR} \\ \mathcal{C}\bar{\psi}_{aL}\sigma^{\mu\nu}\psi_{bR}\mathcal{C}^{-1} &= \bar{\psi}_{aL}^c\sigma^{\mu\nu}\psi_{bR}^c = -\bar{\psi}_{bL}\sigma^{\mu\nu}\psi_{aR}. \end{aligned} \tag{2.303}$$

The relations (2.296), (2.298), and (2.302) are frequently used in conjunction with the Fierz identities in (2.215). For example, combining (2.215) and (2.302),

$$\begin{aligned}(\bar{\psi}_{aR}\gamma^\mu\psi_{bR})\ (\bar{\psi}_{cL}\gamma_\mu\psi_{dL}) &= -\left(\bar{\psi}^c_{bL}\gamma^\mu\psi^c_{aL}\right)\ (\bar{\psi}_{cL}\gamma_\mu\psi_{dL}) \\ &= -\left(\bar{\psi}^c_{bL}\gamma^\mu\psi_{dL}\right)\ \left(\bar{\psi}_{cL}\gamma_\mu\psi^c_{aL}\right).\end{aligned} \tag{2.304}$$

The transformation laws for boson and fermion fields continue to hold in the presence of the strong and electromagnetic interactions, which are C-invariant. Even including the weak interactions, which violate C, the same relations hold for the unperturbed fields when the weak effects are treated perturbatively, analogous to the discussion of reflection non-invariance. The spinor identities in (2.290) and (2.291) are valid independent of whether C is violated.

The implications of C-invariance and the related concept of *G-parity* for strong transitions and for matrix elements of operators will be discussed in Section 3.2.5 and Appendix G.

CP Transformations

It is straightforward to write Lagrangians that violate P or C, but more difficult to find ones that violate CP invariance, at least for a small number of fields, due to the combination of Hermiticity, Lorentz invariance, and the fact that the overall phases of physical fields or states are not observable in quantum mechanics. In particular, one can always choose the phases of the fields so that any given term in a Lagrangian is manifestly CP invariant. CP violation can therefore only emerge when two or more terms clash and cannot be invariant simultaneously. Note that C and P commute.

To make this more explicit, the P and C transformations of spin-0 fields ϕ and generic spin-1 fields W^μ defined in the previous sections always imply

$$\phi(x) \underset{CP}{\longrightarrow} \eta^*_\phi \phi^\dagger(x'), \qquad W^\mu(x) \underset{CP}{\longrightarrow} \eta^*_W W^\dagger_\mu(x'), \tag{2.305}$$

where $x' \equiv (t, -\vec{x}\,)$, η_ϕ and η_W are products of C and P phases, and the $\dagger$ are omitted for Hermitian fields. Similarly,

$$\begin{aligned}\psi(x) &\underset{CP}{\longrightarrow} \eta^*_\psi \gamma^0 \psi^c(x') = -\eta^*_\psi \mathcal{C}\psi^{\dagger T}(x') \\ \psi_{L,R}(x) &\underset{CP}{\longrightarrow} \eta^*_\psi \gamma^0 \psi^c_{R,L}(x') = -\eta^*_\psi \mathcal{C}\psi^{\dagger T}_{L,R}(x').\end{aligned} \tag{2.306}$$

Thus, CP maps $\psi_{L,R}$ onto its own Hermitian conjugate. In contrast, P and C each maps a chiral fermion field onto a *different* one, $\psi_{L,R} \underset{P}{\rightarrow} \psi_{R,L}$, $\psi_{L,R} \underset{C}{\rightarrow} \psi^\dagger_{R,L}$. The elementary fermion bilinear forms $\bar{\psi}_a(x)\Gamma\psi_b(x)$ for the Γ's in Table 2.2 all satisfy

$$\bar{\psi}_a(x)\Gamma\psi_b(x) \underset{CP}{\longrightarrow} \eta^*_{\Gamma ab}\bar{\psi}_b(x')\bar{\Gamma}\psi_a(x') = \eta^*_{\Gamma ab}\left[\bar{\psi}_a(x')\Gamma\psi_b(x')\right]^\dagger, \tag{2.307}$$

where $\eta^*_{\Gamma ab} = \eta_a\eta^*_b\eta_\Gamma$, with $\eta_\Gamma = +1$ for $\Gamma = 1$ or γ^5, and $\eta_\Gamma = -1$ for $\Gamma = \gamma^\mu, \gamma^\mu\gamma^5, \sigma^{\mu\nu}$, and $\sigma^{\mu\nu}\gamma^5$. It is understood that Lorentz indices in Γ are lowered in the second and third expressions. Equation (2.307) can be easily verified from the transformations in (2.210), (2.267), and (2.296), along with Table 2.2. The upshot of (2.305) and (2.307) is that any Poincaré invariant term $\mathcal{L}_i(x)$ with field content such as

$$|\phi^\dagger_a\phi_b|^2, \qquad \phi^3_a, \qquad \phi^\dagger_a\partial_\mu\phi_b W^\mu, \qquad \bar{\psi}_a\psi_b\phi_c, \qquad \bar{\psi}_a\gamma_\mu\psi_b W^\mu, \tag{2.308}$$

as well as any mass or kinetic energy term, is mapped onto its Hermitian conjugate evaluated at x' up to an overall phase $\mathcal{L}_i(x) \underset{CP}{\longrightarrow} \eta_i^* \mathcal{L}_i^\dagger(x')$. The phase η_i depends on the intrinsic CP phases $\eta_{\phi,W,\psi}$ and on any complex coefficients (which are not complex-conjugated under CP, but are in $\mathcal{L}_i^\dagger$). The intrinsic phases can always be chosen (or the fields redefined) so that $\eta_i = 1$ for any given i, and therefore the Hermitian combination $\mathcal{L}_i(x) + \mathcal{L}_i^\dagger(x) \underset{CP}{\longrightarrow} \mathcal{L}_i^\dagger(x') + \mathcal{L}_i(x')$ (or just $\mathcal{L}_i(x) \underset{CP}{\longrightarrow} \mathcal{L}_i(x')$ if it is Hermitian). CP invariance of L results if this can be done for all of the terms *simultaneously*.

Let us illustrate by an example involving two complex scalars ϕ_a and ϕ_b, with

$$\mathcal{L} = \sum_{r=a,b} \left(|\partial_\mu \phi_r|^2 - m_r^2 |\phi_r|^2 \right) - V(\phi_a, \phi_b). \tag{2.309}$$

For the special case

$$V = g_{abb} \phi_a \phi_b^2 + g_{abb}^* \phi_a^\dagger \phi_b^{\dagger 2}, \tag{2.310}$$

one has

$$V \underset{CP}{\longrightarrow} g_{abb} \eta_a^* \eta_b^{*2} \phi_a^\dagger \phi_b^{\dagger 2} + g_{abb}^* \eta_a \eta_b^2 \phi_a \phi_b^2. \tag{2.311}$$

The model is CP invariant even if $g_{abb} = |g_{abb}| e^{i\alpha}$ is complex because we have the freedom to choose the CP phases $\eta_{a,b}$. For example, $\mathcal{L}(x) \to \mathcal{L}(x')$ for the choice $\eta_a = \exp(2i\alpha)$ and $\eta_b = 1$. This can be cast in a simpler form by a *field redefinition*, in which one rewrites $\mathcal{L}$ in terms of a new field $\hat{\phi}_a \equiv \phi_a \exp(i\alpha)$, so that

$$V = |g_{abb}| \left(\hat{\phi}_a \phi_b^2 + \hat{\phi}_a^\dagger \phi_b^{\dagger 2} \right), \tag{2.312}$$

with the other terms unchanged in form. This is invariant for $\hat{\eta}_a = \eta_b = 1$ (for which the transformations of ϕ_a and $\hat{\phi}_a$ are consistent). On the other hand, the potential

$$V = g_{abb} \phi_a \phi_b^2 + g_{aab} \phi_a^2 \phi_b + g_{aaa} \phi_a^3 + h.c. \tag{2.313}$$

with complex coefficients is *not* CP invariant in general because the two CP phases $\eta_{a,b}$ are not sufficient to make all three terms invariant (except for special cases such as $g_{aaa} = 0$ or all phases $= 0$). Equivalently, one can perform two phase redefinitions of the fields, but that is not usually enough to make three coefficients real. This example illustrates the origin of CP violation in the standard model, in which there are not enough quark field redefinitions to remove all of the phases from both their mass terms and the charged current weak couplings simultaneously.

Time Reversal and CPT

Under a time reversal (T) transformation, classical trajectories are replaced by their time-reversed ones,

$$\vec{x}(t) \underset{T}{\rightarrow} \vec{x}(-t), \qquad \vec{p}(t) \underset{T}{\rightarrow} -\vec{p}(-t), \qquad \vec{J}(t) \underset{T}{\rightarrow} -\vec{J}(-t). \tag{2.314}$$

The electromagnetic current and gauge potential satisfy

$$J_{Q\mu}(x) \underset{T}{\rightarrow} J_Q^\mu(x''), \qquad A_{Q\mu}(x) \underset{T}{\rightarrow} A_Q^\mu(x''), \tag{2.315}$$

where $x'' \equiv (-t, \vec{x})$, so that $\vec{E}(x) \to \vec{E}(x'')$ and $\vec{B}(x) \to -\vec{B}(x'')$. Time reversal is more complicated in quantum mechanics. In analogy with (2.314) one defines a time reversal operator T so that

$$T x_i T^{-1} = x_i, \qquad T p_i T^{-1} = -p_i, \tag{2.316}$$

where x_i and p_i are the position and momentum operators. However, consistency with the canonical commutation rules $[x_i, p_j] = i\delta_{ij}$ (or with the Schrödinger equation) requires that T is *antiunitary*, i.e., $TcT^{-1} = c^*$ for any c-number c. This extra complex conjugation significantly complicates many calculations. The antiunitarity is also required for the consistency of the canonical commutation relations in field theory. A consequence of antiunitarity is that

$$\langle a|\mathcal{O}|b\rangle = \langle Ta|T\mathcal{O}T^{-1}|Tb\rangle^*, \tag{2.317}$$

where $\mathcal{O}$ is an operator and $|Tb\rangle$ and $\langle Ta|$ are time reversed states. For example,

$$T|a(\vec{p}, s)\rangle \equiv |Ta(\vec{p}, s)\rangle = \pm|a(-\vec{p}, -s)\rangle, \tag{2.318}$$

for a single particle state with spin orientation s measured with respect to a fixed axis. For multiparticle states, T also interchanges initial (in) states with final (out) states.

Invariance of the action under time reversal requires $T\mathcal{L}(x)T^{-1} = \mathcal{L}(x'')$. This is ensured for free spin-0 fields for

$$T\phi(x)T^{-1} = \eta_{T\phi}\phi(x''), \tag{2.319}$$

where ϕ can be either Hermitian or complex, and $\eta_{T\phi}$ is a phase. (In the complex case $\phi = (\phi_1 + i\phi_2)/\sqrt{2}$ this requires $\eta_{T1} = -\eta_{T2}$ for the Hermitian components ϕ_1 and ϕ_2.) For an electrically charged complex field, (2.319) implies that (2.315) is satisfied for the contribution

$$J_Q^{\phi\mu}(x) = iq\phi^\dagger(x)\overleftrightarrow{\partial}^\mu\phi(x) \underset{T}{\rightarrow} -iq\phi^\dagger(x'')\overleftrightarrow{\partial}^\mu\phi(x'') = J^\phi_{Q\mu}(x'') \tag{2.320}$$

since $\partial^\mu = -\partial''_\mu$. Similarly,

$$TW^\mu(x)T^{-1} = \eta_{TW}W_\mu(x'') \tag{2.321}$$

for an arbitrary Hermitian or complex spin-1 field.

For a Dirac field, invariance of the free-field action or the requirement (2.315) for J_Q implies

$$T\psi(x)T^{-1} = \eta_{T\psi}\mathcal{T}\psi(x''), \tag{2.322}$$

where $\mathcal{T}$ is a Dirac matrix satisfying

$$\mathcal{T}\gamma^\mu\mathcal{T}^{-1} = \gamma^{\mu T} = \gamma^*_\mu. \tag{2.323}$$

In the Pauli-Dirac and chiral representations

$$\mathcal{T} = \mathcal{T}^\dagger = \mathcal{T}^{-1} = -\mathcal{T}^* = i\gamma^1\gamma^3. \tag{2.324}$$

A fermion bilinear therefore transforms as

$$T\bar{\psi}_a(x)\Gamma\psi_b(x)T^{-1} = \bar{\psi}_a(x'')\Gamma_t\psi_b(x''), \qquad \Gamma_t \equiv \mathcal{T}\Gamma^*\mathcal{T}^{-1}, \tag{2.325}$$

assuming the same T-phases for $\psi_{a,b}$. The values of Γ_t are listed in Table 2.2. Implications of T-invariance for matrix elements will be discussed in Appendix G. The u and v spinors transform as

$$\mathcal{T}u(\vec{p}, s) = u(-\vec{p}, -s)^* e^{i\alpha}, \qquad \mathcal{T}v(\vec{p}, s) = v(-\vec{p}, -s)^* e^{i\beta}, \tag{2.326}$$

where α and β are phases that depend on s. For the conventions in (2.184) and (2.192), $\exp(i\alpha) = \exp(i\beta) = \mp i$ for s in the $\pm\hat{z}$ direction.

The CPT theorem, i.e., $(CPT)\mathcal{L}(x)(CPT)^{-1} = \mathcal{L}(-x)$ for a Hermitian, local, Poincaré-invariant $\mathcal{L}$, can be demonstrated heuristically by an extension of the discussion following

(2.308). There it was argued that CP maps the field part of each term onto its Hermitian conjugate, up intrinsic CP phase factors, but does not complex conjugate the coefficient. CPT *does* conjugate the coefficients, and is such that the product of intrinsic phase factors can always be chosen to be +1. To see this, by combining the transformations for T, P, and C, we find that

$$\begin{aligned} \phi(x) &\underset{CPT}{\longrightarrow} \hat{\eta}_\phi \phi(-x)^\dagger, & V_\mu(x) &\underset{CPT}{\longrightarrow} \hat{\eta}_V V_\mu(-x)^\dagger \\ \psi(x) &\underset{CPT}{\longrightarrow} \mp i\hat{\eta}_\psi \gamma^5 \psi(-x)^{\dagger T}, & \frac{\partial}{\partial x^\mu} &= -\frac{\partial}{\partial(-x^\mu)}, \end{aligned} \tag{2.327}$$

where the upper (lower) sign is for the Pauli-Dirac (chiral) representation, and $\hat{\eta} = \eta_T \eta_P \eta_C^*$. Analogous to (2.307), the fermion bilinear forms transform as

$$\bar{\psi}_a(x)\Gamma\psi_b(x) \underset{CPT}{\longrightarrow} \hat{\eta}_{\Gamma ab}\left[\bar{\psi}_a(-x)\Gamma\psi_b(-x)\right]^\dagger, \tag{2.328}$$

where $\hat{\eta}_{\Gamma ab} = \hat{\eta}_a^* \hat{\eta}_b \hat{\eta}_\Gamma$, with $\hat{\eta}_\Gamma = 1$ for $\Gamma = 1, \gamma^5, \sigma^{\mu\nu}$, or $\sigma^{\mu\nu}\gamma^5$, and $\hat{\eta}_\Gamma = -1$ for $\Gamma = \gamma^\mu$ or $\gamma^\mu\gamma^5$. Therefore, if one chooses the phases so that $\hat{\eta}_\phi = +1$ for all ϕ, $\hat{\eta}_V = -1$ for all V, and $\hat{\eta}_\psi = -1$ for all ψ every Poincaré invariant term in $\mathcal{L}(x)$ will be mapped onto its Hermitian conjugate evaluated at $-x$, leaving the action invariant. (Any universal value for $\hat{\eta}_\psi$ would suffice for a fermion number conserving theory, but $\hat{\eta}_\psi$ = real is required for manifest CPT invariance in the presence of Majorana mass terms.) More rigorous derivations of the CPT theorem may be found in, e.g., (Streater and Wightman, 2000; Weinberg, 1995).

The CPT theorem implies that the mass, intrinsic properties, and total lifetime of a particle and its antiparticle must be equal (see, e.g., Sozzi, 2008). However, CPT does allow partial rate asymmetries (Okubo, 1958) if CP is violated, i.e., the decays rates $\Gamma(a \to b_{1,2})$ for a decay of a into b_1 or b_2 can differ from the corresponding antiparticle decay rates $\Gamma(a^c \to b^c_{1,2})$, provided that the sums of the partial rates are the same. This is important for many models of baryogenesis (Chapter 10). Experimental searches for CPT and Lorentz invariance violation are reviewed in (Kostelecky and Russell, 2011; Liberati, 2013; Patrignani, 2016).

2.11 TWO-COMPONENT NOTATION AND INDEPENDENT FIELDS

Weyl two-component fields (or Weyl spinors) were briefly introduced in (2.200) and the subsequent discussion. The description of the discrete symmetries for fermions is somewhat simpler when reexpressed in the two-component language. First, consider the case of a four-component Dirac field ψ, written in terms of two-component Weyl L and R fields as in (2.200), i.e., $\psi = \begin{pmatrix} \Psi_L \\ \Psi_R \end{pmatrix}$. Using the explicit chiral forms

$$\mathcal{C} = \begin{pmatrix} -i\sigma^2 & 0 \\ 0 & i\sigma^2 \end{pmatrix}, \qquad \mathcal{C}\gamma^0 = \begin{pmatrix} 0 & -i\sigma^2 \\ i\sigma^2 & 0 \end{pmatrix}, \qquad \mathcal{T} = -\begin{pmatrix} \sigma^2 & 0 \\ 0 & \sigma^2 \end{pmatrix}, \tag{2.329}$$

the transformations under P and C in (2.263) and (2.287) become

$$\Psi_{L,R}(x) \underset{P}{\rightarrow} \Psi_{R,L}(x'), \qquad \Psi_{L,R}(x) \underset{C}{\rightarrow} \Psi^c_{L,R}(x), \tag{2.330}$$

where $x' = (t, -\vec{x})$ and we have set $\eta_{P\psi} = \eta_{C\psi} = +1$. The charge conjugate Weyl fields are defined as

$$\Psi^c_R = i\sigma^2 \Psi^*_L, \qquad \Psi^c_L = -i\sigma^2 \Psi^*_R, \tag{2.331}$$

where $*$ is a shorthand for $\dagger T$, i.e.,

$$\Psi^*_{L,R} \equiv \Psi^{\dagger T}_{L,R} \equiv (\Psi_{L,R})^{\dagger T} \tag{2.332}$$

is a column vector with components that are the adjoints of those of $\Psi_{L,R}$. Equations (2.330) and (2.331) require that $\Psi^c_{L,R}(x) \underset{P}{\rightarrow} -\Psi^c_{R,L}(x')$, i.e., the fields and their charge conjugates have the opposite intrinsic parity. Both P and C map one Weyl field onto another, e.g., Ψ_L is mapped unto Ψ_R or Ψ^*_R. On the other hand, CP maps a Weyl field onto its own adjoint:

$$\begin{aligned} \Psi_L(x) &\underset{CP}{\longrightarrow} \Psi^c_R(x') = +i\sigma^2\Psi^*_L(x') \\ \Psi_R(x) &\underset{CP}{\longrightarrow} \Psi^c_L(x') = -i\sigma^2\Psi^*_R(x'), \end{aligned} \tag{2.333}$$

with the $i\sigma^2$ acting like a raising/lowering operator on the helicity indices (cf. Equation (2.191)). Thus, the CP transformation always exists, even in a theory involving a single Weyl field (and independent of whether the Lagrangian is invariant). On the other hand, P and C are only defined when one has both Ψ_L and Ψ_R.

Under time reversal,

$$\Psi_{L,R}(x) \underset{T}{\rightarrow} -\sigma^2\Psi_{L,R}(x''), \tag{2.334}$$

where $x'' = (-t, \vec{x}\,)$ and we have used (2.322) with $\eta_{T\psi} = 1$. Combining (2.333) and (2.334)

$$\Psi_{L,R}(x) \underset{CPT}{\longrightarrow} \mp i\Psi_{L,R}(-x)^*, \tag{2.335}$$

consistent with (2.327).

Fermion mass and kinetic energy terms are expressed in two-component notation in (2.201). Some important fermion bilinear forms become

$$\begin{aligned} \bar{\psi}_a\gamma^\mu\psi_b &= \Psi^\dagger_{aL}\bar{\sigma}^\mu\Psi_{bL} + \Psi^\dagger_{aR}\sigma^\mu\Psi_{bR} \\ \bar{\psi}_a\gamma^\mu\gamma^5\psi_b &= -\Psi^\dagger_{aL}\bar{\sigma}^\mu\Psi_{bL} + \Psi^\dagger_{aR}\sigma^\mu\Psi_{bR} \\ \bar{\psi}_a\psi_b &= \Psi^\dagger_{aL}\Psi_{bR} + \Psi^\dagger_{aR}\Psi_{bL} \\ \bar{\psi}_a\gamma^5\psi_b &= \Psi^\dagger_{aL}\Psi_{bR} - \Psi^\dagger_{aR}\Psi_{bL} \\ \frac{1}{2}\bar{\psi}_a\sigma^{\mu\nu}\psi_b &= \Psi^\dagger_{aL}\,\bar{\mathfrak{s}}^{\mu\nu}\Psi_{bR} + \Psi^\dagger_{aR}\,\mathfrak{s}^{\mu\nu}\Psi_{bL} \\ \frac{1}{2}\bar{\psi}_a\sigma^{\mu\nu}\gamma^5\psi_b &= \Psi^\dagger_{aL}\,\bar{\mathfrak{s}}^{\mu\nu}\Psi_{bR} - \Psi^\dagger_{aR}\,\mathfrak{s}^{\mu\nu}\Psi_{bL}, \end{aligned} \tag{2.336}$$

where σ^μ and $\bar{\sigma}^\mu$ are defined in (2.167), and

$$\mathfrak{s}^{\mu\nu} \equiv \frac{i}{4}\left(\sigma^\mu\bar{\sigma}^\nu - \sigma^\nu\bar{\sigma}^\mu\right), \qquad \bar{\mathfrak{s}}^{\mu\nu} \equiv \frac{i}{4}\left(\bar{\sigma}^\mu\sigma^\nu - \bar{\sigma}^\nu\sigma^\mu\right) \tag{2.337}$$

$$\mathfrak{s}^{0i} = -\bar{\mathfrak{s}}^{0i} = -i\frac{\sigma^i}{2}, \qquad \mathfrak{s}^{ij} = +\bar{\mathfrak{s}}^{ij} = +\frac{1}{2}\epsilon_{ijk}\sigma^k. \tag{2.338}$$

The transformations of the bilinear forms under the discrete symmetries can easily be rewritten from those given in Section 2.10 and Table 2.2. Especially useful are the charge conjugation identities such as

$$\Psi^{c\dagger}_{aL}\Psi^c_{bR} = \Psi^\dagger_{bL}\Psi_{aR}, \qquad \Psi^{c\dagger}_{aL}\bar{\sigma}^\mu\Psi^c_{bL} = -\Psi^\dagger_{bR}\sigma^\mu\Psi_{aR}, \tag{2.339}$$

analogous to (2.302) and (2.303). The Fierz identities for the two-component fields can be read off from (2.215) or derived directly from Problem 1.1. For example,

$$\begin{aligned} \left(\Psi^\dagger_{aL}\bar\sigma^\mu\Psi_{bL}\right)\left(\Psi^\dagger_{cL}\bar\sigma_\mu\Psi_{dL}\right) &= \left(\Psi^\dagger_{aL}\bar\sigma^\mu\Psi_{dL}\right)\left(\Psi^\dagger_{cL}\bar\sigma_\mu\Psi_{bL}\right) \\ \left(\Psi^\dagger_{aR}\sigma^\mu\Psi_{bR}\right)\left(\Psi^\dagger_{cL}\bar\sigma_\mu\Psi_{dL}\right) &= -2\left(\Psi^\dagger_{aR}\Psi_{dL}\right)\left(\Psi^\dagger_{cL}\Psi_{bR}\right). \end{aligned} \tag{2.340}$$

Combining (2.339) and (2.340) one finds identities such as

$$\left(\Psi^\dagger_{aL}\bar\sigma^\mu\Psi_{bL}\right)\left(\Psi^\dagger_{cL}\bar\sigma_\mu\Psi_{dL}\right) = 2\left(\Psi^{c\dagger}_{bR}\Psi_{dL}\right)\left(\Psi^\dagger_{cL}\Psi^c_{aR}\right). \tag{2.341}$$

As a very simple example, let us repeat the calculation of the amplitude for $e^-(\vec p_1)\,e^+(\vec p_2) \to f(\vec p_3)\,\bar f(\vec p_4)$ given in Section 2.9 one more time, using two-component notation. The QED interaction for the e^- is

$$\mathcal{L} = -eA_\mu J^\mu_Q = +eA_\mu\left(\Psi^\dagger_L\bar\sigma^\mu\Psi_L + \Psi^\dagger_R\sigma^\mu\Psi_R\right), \tag{2.342}$$

and similarly for the other fermions, so that

$$\begin{aligned} M(+-,+-) &= \frac{-iQ_f e^2}{s}\langle f_+(3)\bar f_-(4)|\Psi^\dagger_R\sigma^\mu\Psi_R|0\rangle\langle 0|\Psi^\dagger_R\sigma_\mu\Psi_R|e_+(1)e^c_-(2)\rangle \\ M(-+,+-) &= \frac{-iQ_f e^2}{s}\langle f_-(3)\bar f_+(4)|\Psi^\dagger_L\bar\sigma^\mu\Psi_L|0\rangle\langle 0|\Psi^\dagger_R\sigma_\mu\Psi_R|e_+(1)e^c_-(2)\rangle \end{aligned} \tag{2.343}$$

in an obvious notation. Using the free-field expression in (2.207), the amplitudes become

$$\begin{aligned} M(+-,+-) &= -iQ_f e^2\left[\phi_+(3)^\dagger\sigma^\mu(-\chi_-(4))\,(-\chi_-(2))^\dagger\sigma_\mu\phi_+(1)\right] \\ M(-+,+-) &= -iQ_f e^2\left[\phi_-(3)^\dagger\bar\sigma^\mu(+\chi_+(4))\,(-\chi_-(2))^\dagger\sigma_\mu\phi_+(1)\right], \end{aligned} \tag{2.344}$$

which reproduces (2.245) and (2.246).

We will mainly utilize the two-component notation in connection with neutrino masses and for displaying the interaction terms in a supersymmetric theory. Much more extensive expositions, including such topics as Feynman rules and propagators in two-component form are described in the books on supersymmetry and in (Dreiner et al., 2010). A more compact version of the two-component notation will be introduced in Chapter 10.

Independent Fermion Fields

It was emphasized in (2.196) that the left- and right-chiral projections $\psi_{L,R} = P_{L,R}\psi$ or their associated Weyl fields $\Psi_{L,R}$ could be considered as independent degrees of freedom. However, from (2.287) and (2.301) it is clear that the conjugate fields ψ^c and $\psi^c_{L,R}$ are not independent, but are related by $\psi^c \sim \psi^\dagger$ and $\psi^c_{L,R} \sim \psi^\dagger_{R,L}$. In many cases it is convenient to work in terms of the $\psi_{L,R}$ (or $\Psi_{L,R}$) fields, as is conventionally done for QED, quantum chromodynamics, and most treatments of the standard electroweak model. However, it is sometimes easier to work in terms of the L-chiral fields for both the particles and antiparticles, i.e., ψ_L and ψ^c_L (or Ψ_L, Ψ^c_L) instead, regarding the ψ_R and ψ^c_R as dependent. This is typically done when discussing grand unification or supersymmetry, for example. To illustrate this, the QED electromagnetic current J^μ_Q can be written in a number of equivalent ways, including

$$\begin{aligned} J^\mu_Q &= -\bar\psi_e\gamma^\mu\psi_e = +\bar\psi_{e^c}\gamma^\mu\psi_{e^c} \\ &= -\bar\psi_{eL}\gamma^\mu\psi_{eL} - \bar\psi_{eR}\gamma^\mu\psi_{eR} = -\bar\psi_{eL}\gamma^\mu\psi_{eL} + \bar\psi_{e^cL}\gamma^\mu\psi_{e^cL}, \end{aligned} \tag{2.345}$$

where ψ_e and $\psi_{e^c} \equiv \psi_e^c$ are, respectively, the e^- and e^+ fields. In two-component notation

$$J_Q^\mu = -\Psi_{eL}^\dagger \bar{\sigma}^\mu \Psi_{eL} - \Psi_{eR}^\dagger \sigma^\mu \Psi_{eR} = -\Psi_{eL}^\dagger \bar{\sigma}^\mu \Psi_{eL} + \Psi_{e^c L}^\dagger \bar{\sigma}^\mu \Psi_{e^c L}. \tag{2.346}$$

Similarly, a fermion mass term can be reexpressed as

$$\begin{aligned} -\mathcal{L} = m\bar{\psi}\psi &= m\left(\bar{\psi}_L \psi_R + \bar{\psi}_R \psi_L\right) = m\left(\bar{\psi}_L \mathcal{C} \bar{\psi}_L^{cT} + \psi_L^{cT} \mathcal{C} \psi_L\right) \\ &= m\left(\Psi_L^\dagger \Psi_R + \Psi_R^\dagger \Psi_L\right) = m\left(\Psi_L^\dagger i\sigma^2 \Psi_L^{c*} - \Psi_L^{cT} i\sigma^2 \Psi_L\right), \end{aligned} \tag{2.347}$$

where we have used (2.301) and (2.292). The two forms in the second line emphasize two different interpretations of a Dirac mass. The $\Psi_R^\dagger \Psi_L$ expression, for example, can be viewed as annihilating an L-chiral field and creating an R-chiral one, while the equivalent $-\Psi_L^{cT} i\sigma^2 \Psi_L$ form can be interpreted as the annihilation of two distinct L-chiral fields.

2.12 QUANTUM ELECTRODYNAMICS (QED)

Quantum electrodynamics represents the merger of three great ideas of modern physics: classical electrodynamics as synthesized by Maxwell, quantum mechanics, and special relativity. The basic formulation of QED was completed by 1930; it combined the Dirac theory of the electron (with its correct predictions of the lowest order electron magnetic dipole moment and the existence of positrons) with the quantization of the electromagnetic field into individual photons. A workable prescription for handling the divergent integrals by renormalization of the electron mass, charge, and wave function was completed by the early 1950s by Bethe, Feynman, Tomonaga, Schwinger, Dyson and others (Schwinger, 1958). QED therefore became a mathematically consistent and well-defined theory, which was subsequently tested to incredible precision.

In order to consider QED seriously, one must include higher-order (loop) effects in perturbation theory, both because of the precision of many of the tests and because some effects (such as light by light scattering via the interaction of photons with virtual charged particle loops) only occur at loop level. The calculation of higher-order effects is greatly complicated by divergences and the need to renormalize (while maintaining gauge invariance). A systematic study is beyond the scope of this book, and only brief comments are made. The subject is studied in detail in standard texts on field theory. Another complication is that strong interaction effects are very important and cannot be ignored in studying the electromagnetic interactions of strongly interacting particles (*hadrons*), such as protons, neutrons, and pions. They even enter at higher orders (through loops involving virtual hadrons) in the electrodynamics of electrons and muons. Fortunately, a great deal can be said about these strong interaction corrections using symmetry principles.

2.12.1 Higher-Order Effects

Analogous to the Hermitian scalar case in Section 2.3.6, QED has logarithmically divergent diagrams such as those shown in Figure 2.17. To take this into account, let us modify the notation in the QED Lagrangian density in (2.218) to

$$\begin{aligned} \mathcal{L} &= \bar{\psi}(x)\left(i\partial\!\!\!/ + e_0 A\!\!\!/ - m_0\right)\psi(x) - \frac{1}{4}F_{\mu\nu}F^{\mu\nu} \\ &= \bar{\psi}(x)\left(i\partial\!\!\!/ - m_0\right)\psi(x) - e_0 J_Q^\mu A_\mu - \frac{1}{4}F_{\mu\nu}F^{\mu\nu}, \end{aligned} \tag{2.348}$$

where $J_Q^\mu = -\bar{\psi}\gamma^\mu\psi$ is the electromagnetic current operator defined in (2.294) and ψ is the electron field. e_0 and m_0 are the bare positron charge and mass, respectively, i.e., the

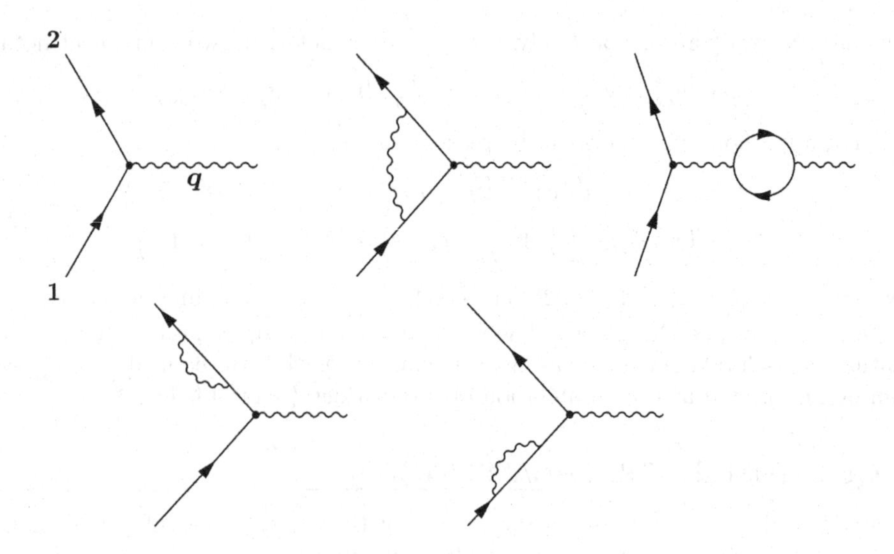

Figure 2.17 Electron-photon vertex (top left), and one-loop corrections, corresponding to the vertex correction (top middle), the vacuum polarization (or photon self-energy) correction (top right), and electron self-energy corrections (bottom).

parameters that appear in $\mathcal{L}$. We redefine e and m as the physical parameters, i.e., the ones actually measured. The renormalizability of QED implies that the divergences enter only in the relations between the bare and physical parameters and in the unobservable *wave function renormalization* of the fields,[26] and that observable quantities are finite to all orders in perturbation theory when expressed in terms of e and m.

The one and higher-loop vertex corrections imply that the lowest order electron-photon vertex $ie_0\gamma^\mu$ is replaced by a function $ie_0\,\Gamma^\mu(p_2,p_1)/Z_1$ that can depend on the external momenta. Z_1 is the (divergent) vertex renormalization constant, defined by the requirement that $\Gamma^\mu = \gamma^\mu$ at the on-shell point $p_2^2 = p_1^2 = m^2$ and $q^2 = (p_2 - p_1)^2 = 0$. Similarly, the electron self-energy and vacuum polarization diagrams modify the electron and photon propagators so that on shell they take the same form as the free ones except they are multiplied by divergent wave function renormalization factors $Z_{2,3}$,

$$S_F(p) \to \frac{Z_2}{\not{p} - m + i\epsilon}, \qquad ig^{\mu\nu}D_F(q) \to -ig^{\mu\nu}\frac{Z_3}{q^2 + i\epsilon}, \tag{2.349}$$

where the position of the pole in $S_F(p)$ defines the physical electron mass. It is convenient to associate a factor of $Z_2^{1/2}$ or $Z_3^{1/2}$ from each electron or photon line with the vertex. The other $Z^{1/2}$ is associated with the vertex or external state at the other end of the line. The overall vertex factor therefore becomes

$$ie_0\gamma^\mu \to ie_0\frac{Z_2 Z_3^{1/2}}{Z_1}\Gamma^\mu(p_2,p_1). \tag{2.350}$$

[26]In addition to the *ultraviolet* divergences (associated with large momenta in the integrals) discussed here, there are also *infrared* divergences associated with low momentum virtual photons. These can be regulated by introducing a fictitious photon mass; they cancel against similar terms associated with the emission of soft real photons at energies below the threshold of the detector.

The renormalization factors $Z_{1,2,3}$ can be combined with e_0 to define the physical charge e. The *Ward-Takahashi identity* ensures that $Z_1 = Z_2$ to all orders, so we can identify

$$e = e_0 \frac{Z_2 Z_3^{1/2}}{Z_1} = e_0 Z_3^{1/2}. \tag{2.351}$$

It can then be shown that $\Gamma^\mu(p_2, p_1)$ is finite to all orders when expressed in terms of e and m, and in particular, $\Gamma^\mu(p_2, p_1) = \gamma^\mu + \mathcal{O}(\alpha)$ where $\alpha = e^2/4\pi \sim 1/137$ is the *fine structure constant.* The charge renormalization depends only on the photon vacuum polarization diagrams and is therefore the same for all particles, e.g., for electrons with charge $-e_0 \to -e$, and for u-quarks with charge $+2e_0/3 \to 2e/3$. Similarly, after removing the $Z_{2,3}$ factors, the electron and photon propagators behave like the free-field ones near the physical $p^2 = m^2$ or $q^2 = 0$ poles, but can have momentum dependent corrections away from the physical masses. These can also be shown to be finite to all orders when expressed in terms of e and m. The vacuum polarization corrections will be discussed in Section 2.12.2 in connection with running couplings.

The relation of such renormalized quantities to physical on-shell amplitudes and matrix elements is considered in field theory texts, but it should be plausible from the above discussion that the on-shell matrix element of $-J_Q^\mu(x)$ between physical electron states is

$$\langle \vec{p}_2\, s_2 | \bar{\psi}(x)\gamma^\mu \psi(x) | \vec{p}_1\, s_1 \rangle = \bar{u}_2 \Gamma^\mu(p_2, p_1) u_1 e^{iq\cdot x}, \tag{2.352}$$

where $q = p_2 - p_1$ and the x dependence follows from translation invariance, Equation (1.15) on page 3. The form of $\Gamma^\mu(p_2, p_1)$ is strongly restricted by symmetry considerations, which continue to hold in the presence of the strong interactions. In particular, Lorentz invariance implies that the r.h.s of (2.352) must be a four-vector, which can only be constructed from p_1^μ, p_2^μ, and the Dirac matrices. The Gordon identities in Problem 2.10 indicate that it is sufficient to consider $p_{1,2}^\mu$ in the combination q^μ only. Furthermore, QED (and the strong interactions) are reflection invariant. Therefore, using (2.276) the most general allowed form is

$$\bar{u}_2 \Gamma^\mu(p_2, p_1) u_1 = \bar{u}_2 \left[\gamma^\mu F_1(q^2) + \frac{i\sigma^{\mu\nu}}{2m} q_\nu F_2(q^2) + q^\mu F_3(q^2) \right] u_1, \tag{2.353}$$

where the *form factors* $F_{1,2,3}(q^2)$ are Lorentz invariant functions of q^2. Charge conservation, $\partial_\mu J_Q^\mu = 0$, which can be derived to all orders from the equations of motion or from the Noether theorem of Section 3.2.2, requires that $q^2 F_3(q^2) = 0$. One does not expect a δ function to develop to any order, so $F_3(q^2)$ must vanish. Also, the Hermiticity of J_Q^μ implies

$$\Gamma^\mu(p_2, p_1) = \overline{\Gamma}^\mu(p_1, p_2), \tag{2.354}$$

where $\overline{\Gamma}^\mu$ is the Dirac adjoint. From Table 2.2, $F_{1,2}$ must therefore be real. Finally, we have normalized so that $F_1(0) = 1$.

We have seen that $eF_1(0)$ is just the physical electric charge. To interpret $F_2(0)$, consider the interaction of an electron with a static classical gauge potential $A_\mu(x)$ (cf., (2.66) and Problem 2.24). The matrix element of the interaction Hamiltonian

$$H = -e_0 \int d^3\vec{x}\; \bar{\psi}(x)\gamma^\mu \psi(x)\; A_\mu(x) \tag{2.355}$$

between one-electron states is

$$\langle \vec{p}_2\, s_2 | H | \vec{p}_1\, s_1 \rangle = -e\bar{u}_2 \Gamma^\mu u_1 \int A_\mu(\vec{x}) e^{-i\vec{q}\cdot\vec{x}} d^3\vec{x} \equiv -e\bar{u}_2 \Gamma^\mu u_1\; \tilde{A}_\mu(\vec{q}). \tag{2.356}$$

Expanding the fermion bilinear in (2.356) to linear order in the non-relativistic limit $|\vec{p}_i| \ll m$, this reduces (Problem 2.30) to

$$\phi^\dagger_{s_2}\left(-2me\tilde{A}_0(\vec{q}) + e(\vec{p}_1 + \vec{p}_2)\cdot\tilde{\vec{A}}(\vec{q}) + e\left[1 + F_2(0)\right]\vec{\sigma}\cdot\tilde{\vec{B}}(\vec{q})\right)\phi_{s_1}, \tag{2.357}$$

where $\phi_{s_{1,2}}$ are the two-component Pauli spinors and $\tilde{\vec{B}}$ is the Fourier transform of the magnetic field $\vec{B} = \vec{\nabla}\times\vec{A}$. The conventional non-relativistic interaction Hamiltonian for an e^- in an external field is

$$H_I = -eA_0(\vec{x}) + \frac{e}{2m}\left(\vec{p}\cdot\vec{A}(\vec{x}) + \vec{A}(\vec{x})\cdot\vec{p}\right) - \vec{\mu}_e\cdot\vec{B}(\vec{x}) + \mathcal{O}(e^2), \tag{2.358}$$

where $\vec{\mu}_e = -g\mu_B\vec{S}$ is the electron magnetic dipole moment operator, $\mu_B = e/2m$ is the Bohr magneton, $\vec{S} = \vec{\sigma}/2$ is the spin operator, and g is the electron g-factor, which is not predicted in the non-relativistic theory. The three terms correspond to the Coulomb interaction, and the orbital and spin magnetic moment interactions, respectively. The momentum space matrix element of H_I (corrected to agree with our covariant state normalization) is

$$\langle\vec{p}_2\, s_2|H|\vec{p}_1\, s_1\rangle = 2m\int d^3\vec{x}\, e^{-i\vec{q}\cdot\vec{x}}\phi^\dagger_{s_2}H_I\phi_{s_1}, \tag{2.359}$$

with the $\vec{p}$ operators in H_I replaced by the $\vec{p}_{2,1}$ eigenvalues. This coincides with (2.357) provided that one identifies the g-factor as $g = 2\left[1 + F_2(0)\right]$, where the 2 is the relativistic Dirac contribution (from $\gamma^\mu F_1(0)$), and $F_2(0) = (g-2)/2$ is the anomalous QED term, usually denoted by a_e. The leading contribution, from the vertex diagram in Figure 2.17, is $F_2(0) = \alpha/2\pi$, as first calculated by Schwinger. The calculation is sketched in Appendix E as an illustration of the techniques for calculating Feynman loop integrals. The relation of field theory matrix elements to non-relativistic potentials is further discussed in Problem 2.33.

2.12.2 The Running Coupling

We saw in (2.351) that the renormalization of the electric charge is due to the vacuum polarization diagram of Figure 2.17, with the divergent parts of the electron self-energy and vertex diagrams cancelling by the Ward-Takahashi identity. Now, let us consider the vacuum polarization as a function of q^2. The one-loop vacuum polarization diagram in Figure 2.18, can be added to the tree level photon propagator to yield (see, e.g., Peskin and Schroeder, 1995)

$$\frac{-ig_{\mu\nu}}{q^2}e_0^2 \to \frac{-ig_{\mu\nu}}{q^2}e_0^2\left[1 - \frac{e_0^2}{12\pi^2}\ln\frac{\Lambda^2}{m^2} + e_0^2\Pi(q^2)\right], \tag{2.360}$$

where we have also included a factor of e_0^2 from the outside vertices. Λ is an *ultraviolet cutoff* of the divergent momentum integral. In practice, the cutoff must be introduced in a way that respects gauge invariance, such as in the Pauli-Villars or dimensional regularization schemes. $\Pi(q^2)$ is a finite function that vanishes at $q^2 = 0$, so we can identify the photon wave function renormalization factor as

$$Z_3 = \left[1 - \frac{e_0^2}{12\pi^2}\ln\frac{\Lambda^2}{m^2}\right], \tag{2.361}$$

which diverges for $\Lambda \to \infty$. The r.h.s. of (2.360) can be rewritten

$$\frac{-ig_{\mu\nu}}{q^2}e^2\left[1 + \frac{e_0^2}{Z_3}\Pi(q^2)\right] \sim \frac{-ig_{\mu\nu}}{q^2}\frac{e^2}{1 - e^2\Pi(q^2)}, \tag{2.362}$$

where the corrections to the last form are higher order in e_0^2 or e^2. One can show that this form holds and is finite to all orders, provided $\Pi(q^2)$ is expressed in terms of e^2 and m^2. The $e^{2n}\Pi^n$ term in the expansion $(1-e^2\Pi)^{-1} = 1 + e^2\Pi + e^4\Pi^2 + \cdots$ corresponds to the diagram with n separate vacuum polarization bubbles along the line.

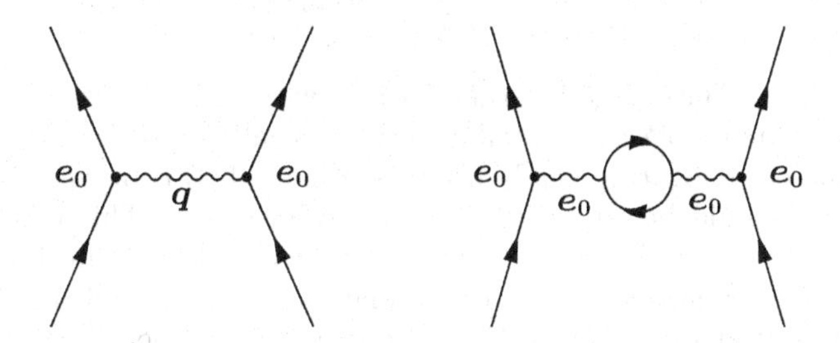

Figure 2.18 One photon exchange and one-loop vacuum polarization "bubble."

This illustrates how the divergences disappear from the expressions for physical observables when they are written in terms of the renormalized quantities. However, one may still have a nagging doubt about the underlying divergences. In fact, the modern view is that the "divergent" momentum integrals are actually cut off physically at the scale at which the theory is replaced by a more complete one, and that Λ should be associated with that scale and not taken to infinity. One can then interpret the renormalizations as finite quantities describing, e.g., the difference between a coupling as measured at a scale much smaller than Λ and the value it would have at Λ. A logarithmically "divergent" term in a weak coupling theory such as QED is then actually a small effect, e.g., $\frac{e_0^2}{12\pi^2}\ln\frac{\Lambda^2}{m^2} \sim 0.08$ for $e_0^2 \sim e^2$ and $\Lambda \sim M_P \sim 10^{19}$ GeV (the Planck scale). Most of the divergences in renormalizable theories are of this logarithmic nature,[27] and results are therefore insensitive to the details of the new physics above Λ. Non-renormalizable theories, on the other hand, typically encounter new divergences of order Λ^{2n} at n-loop level, and are very sensitive.

The function $\Pi(q^2)$ in (2.362) is

$$\Pi(q^2) = \frac{1}{2\pi^2}\int_0^1 dz\; z(1-z)\ln\left[\frac{m^2 - q^2 z(1-z)}{m^2}\right], \tag{2.363}$$

which vanishes as $\Pi(q^2) \to -q^2/60\pi^2 m^2$ for $q^2 \to 0$. It is well behaved for $q^2 \leq 0$ (i.e., t-channel exchange, as in Figure 2.18), but has a branch point at $q^2 = 4m^2$ associated with the e^+e^- threshold in s-channel processes. In the limit $Q^2 \equiv -q^2 \gg m^2$ (positive for t-channel exchange),

$$\Pi(q^2) \to \frac{1}{12\pi^2}\ln\frac{Q^2}{m^2}. \tag{2.364}$$

The replacement in (2.362) is universal for all photon exchanges. It is useful to introduce a *running* or effective fine structure constant $\alpha_{eff}(Q^2) \sim e^2(Q^2)/4\pi = \alpha/(1-4\pi\alpha\Pi(-Q^2))$. Such considerations are most useful for large Q^2, so we will actually define the running quantity as

$$\alpha\left(Q^2\right) = \frac{\alpha(Q_0^2)}{1 - \frac{\alpha(Q_0^2)}{3\pi}\ln\frac{Q^2}{Q_0^2}}, \tag{2.365}$$

[27]The quadratically divergent (but renormalizable) corrections to scalar self-energies, such as the Higgs mass-square in the standard model, are a critical exception. This will be discussed in Chapter 10.

where Q_0^2 is an arbitrary reference scale, such as m^2. This coincides with α_{eff} for $Q^2 \gg Q_0^2 \sim m^2$, and approaches the bare coupling $\alpha_0 = e_0^2/4\pi$ for $Q^2 \to \Lambda^2$ (up to higher-order corrections). Equation (2.365) is equivalent to

$$\frac{1}{\alpha(Q^2)} = \frac{1}{\alpha(Q_0^2)} - b' \ln \frac{Q^2}{Q_0^2}, \tag{2.366}$$

where $b' \equiv 4\pi b = 1/3\pi$. Thus, $\alpha(Q^2)^{-1}$ runs linearly with $\ln Q^2$ at large Q^2. (Higher-order corrections to the vacuum polarization bubble lead to small nonlinear effects). The running $\alpha(Q^2)$ therefore increases logarithmically from its value $\sim 1/137$ for small Q^2, so the QED interaction strength should be larger for high energy processes. This was motivated here for spacelike momentum transfers $q^2 < 0$, but continues to hold even in the timelike region, where higher-order corrections are minimized if one uses $\alpha(|q^2|)$. Equivalently, the effective Coulomb interaction scales as $\alpha(Q \sim 1/r)/r$, and therefore the effective α increases at smaller separations. There is a simple interpretation of this: an electron charge in a dielectric medium is screened for larger separations, leading to an interaction strength falling faster than $1/r$. At small separations, however, the screening is less effective and a test charge feels the full electric charge. The same mechanism applies here, except the dielectric is really due to the quantum fluctuations of the vacuum, as represented by the virtual e^+e^- loop. We will see in Chapter 5 that the gluon self-interactions in quantum chromodynamics have the opposite effect of antiscreening (for which there is no simple classical analog), leading to a *decrease* in the strong coupling at large momenta or short distance (*asymptotic freedom*).

The running of $\alpha(Q^2)$ is described by the *renormalization group equation* (RGE)

$$\frac{d\alpha(Q^2)}{d\ln Q^2} = \beta(q^2) = b'\alpha^2(Q^2) + \mathcal{O}(\alpha^3), \tag{2.367}$$

where $b' = 1/3\pi$ is due to the one-loop diagram in Figure 2.18, and the other terms are higher-loop contributions to the vacuum polarization bubbles. Equation (2.366) is the solution in one-loop approximation. The running in (2.366) is valid for $Q^2 \gg m^2$; there is little effect for $Q^2 \lesssim m^2$. However, for $Q^2 \gg m_\mu^2$, where $m_\mu \sim 200\, m_e$ is the muon mass and $m_e \equiv m$ is the e^- mass, one should also include the effect of the muon loop in the photon propagator, i.e., $b' \to 2/3\pi$, and

$$\frac{1}{\alpha(Q^2)} = \frac{1}{\alpha(m_\mu^2)} - \frac{2}{3\pi} \ln \frac{Q^2}{m_\mu^2} \tag{2.368}$$

for $Q^2 > m_\mu^2$. Thus, b' changes discontinuously and $1/\alpha$ has a kink near the particle threshold. (The exact form of the threshold, including constant terms, whether it occurs at m_μ or $2m_\mu$, etc, depends on the details of the renormalization scheme.) Quark loops also contribute to the vacuum polarization and the running. If one could ignore the strong interactions, then a quark of flavor r (e.g., u, d, s) would contribute a term $3q_r^2/3\pi$ to b' for $Q^2 > m_r^2$, where q_r is the quark electric charge in units of e, m_r is its mass, and the 3 is because there are 3 quark colors. Thus,

$$b' = \frac{1}{3\pi} \sum_{m_r < Q} C_r q_r^2, \tag{2.369}$$

where the sum includes both quarks and charged leptons, with $C_r = 1$ (leptons) and $C_r = 3$ (quarks). Unfortunately, this is a poor approximation for the strongly interacting particles, which cannot really be treated as free quarks at low energies. Multiple gluon exchanges between the quarks in the vacuum polarization diagram, or hadronic bound state effects, invalidate the quark part of (2.369). A more reliable approximation can be obtained by replacing the perturbative loop by a dispersion relation

$$\frac{3q_r^2}{3\pi}\ln\frac{Q^2}{Q_0^2} \rightarrow \int_{4m_\pi^2}^{\infty} F(Q^2,s)R(s)ds \tag{2.370}$$

in the $1/\alpha$ equation. $R(s) \equiv \sigma_{e^+e^-}(s)/(4\pi\alpha^2/3s)$ is the ratio of the total cross section for $e^+e^- \rightarrow$ hadrons at CM energy $\sqrt{s}$ divided by the cross section for $e^+e^- \rightarrow \mu^+\mu^-$ and $F(Q^2,s)$ is a known function (see, e.g., Eidelman and Jegerlehner, 1995). The low-energy part of $R(s)$ can be taken from experiment, while the high-energy part is predicted by QCD. One therefore finds that $1/\alpha$ decreases from ~ 137 for $Q \sim 0$ to ~ 129 at the the Z mass. The extrapolation can be done quite reliably, but the small ($\mathcal{O}(0.02\%)$) uncertainty is still the largest theoretical uncertainty in the precision electroweak program. Closely related hadronic uncertainties are also significant in the interpretation of the measured anomalous magnetic moment of the muon.

The running $\alpha(Q^2)$ effect was sketched here in an *on-shell* renormalization scheme, i.e., e was defined in terms of the electron-photon vertex on shell, and m as the location of the pole in the propagator. In the minimal subtraction schemes ('t Hooft, 1973; Bardeen et al., 1978) one defines renormalized couplings and masses at an arbitrary renormalization scale μ, at which the poles in dimensional regularization are subtracted. Both the masses and charges run as a function of μ. Higher-order corrections to physical processes involving a single large scale Q are usually minimized by evaluating the couplings and masses at $\mu = Q$.

2.12.3 Tests of QED

Quantum electrodynamics is the most successful theory in physics when judged in terms of the theoretical and experimental precision of its tests. A detailed review is given in (Kinoshita, 1990). The classical atomic tests of QED, such as the Lamb shift, atomic hyperfine splittings, muonium (μ^+e^- bound states), and positronium (e^+e^- bound states) are reviewed in (Karshenboim, 2005). More recent results and the experimental values of α and other physical constants are surveyed in (Mohr et al., 2012). Measurements of α and of possible deviations from QED are listed in Table 2.3.

TABLE 2.3 Most precise determinations of the fine structure constant $\alpha = e^2/4\pi$ and other QED quantities.[a]

Experiment	Quantity	Value	Precision
$a_e = (g_e - 2)/2$	α^{-1}	137.035 999 157(33)	2.5×10^{-10}
$h/m(Rb)$	α^{-1}	137.035 999 049(90)	6.6×10^{-10}
Solar wind	m_γ	$< 10^{-18}$ eV	–
CMB	Q_γ	$< 10^{-35}$	–
$e^- \not\rightarrow \nu\gamma$	τ_{e^-}	$> 6.6 \times 10^{28}$ yr	–
Neutrality of SF_6	$\lvert Q_p + Q_{e^-}\rvert$, Q_n	$< 10^{-21}$	–

[a] Q_i is the electric charge of particle i in units of e. Detailed descriptions, caveats, and references are given in (Mohr et al., 2012; Aoyama et al., 2015; Patrignani, 2016).

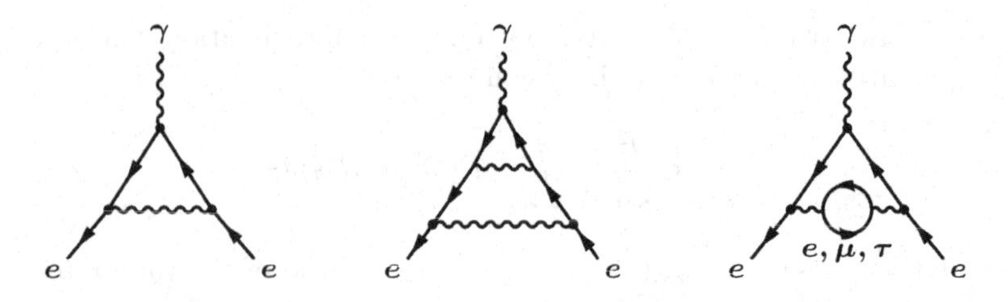

Figure 2.19 One-loop and typical two-loop diagrams contributing to the anomalous magnetic moment of the electron.

The most precise measurement of α is from $a_e = 1.159\ 652\ 180\ 73(28) \times 10^{-3}$, the anomalous magnetic moment of the electron, which was measured at Harvard University using electrons confined in a Penning trap (Hanneke et al., 2008). α is extracted using the theoretical (standard model) prediction (Aoyama et al., 2015)

$$\begin{aligned} a_e^{SM} &= \frac{\alpha}{2\pi} - 0.328\ 478\ 444\ 00 \left(\frac{\alpha}{\pi}\right)^2 + 1.181\ 234\ 017 \left(\frac{\alpha}{\pi}\right)^3 \\ &- 1.912\ 05(84) \left(\frac{\alpha}{\pi}\right)^4 + 8.73(34) \left(\frac{\alpha}{\pi}\right)^5 + a_e^{EW} + a_e^{had}, \end{aligned} \tag{2.371}$$

where the first five terms are the pure-QED contributions, involving photons, electrons, muons, and taus. A few representative diagrams are shown in Figure 2.19. The one-loop term is calculated in Appendix E. The two- and three-loop contributions have been calculated fully analytically, while the four-loop (891 diagrams) and five-loop (12,672 diagrams) terms require numerical integration of the Feynman parametric integrals, resulting in the quoted error in the coefficient.[28] The electroweak and hadronic contributions involving W, Z, Higgs, and strongly interacting particles are estimated to be $\sim 0.02973(52) \times 10^{-12}$ and $\sim 1.706(20) \times 10^{-12}$, respectively (for reviews of the theory, see Czarnecki and Marciano, 2001; Mohr et al., 2012; Aoyama et al., 2015).

a_e yields the single most accurate determination of α, but to test QED it is necessary to compare the values obtained in two or more types of experiment. The second most precise involves the measurement of the recoil velocity of ^{87}Rb atoms after emitting or absorbing a photon (Bouchendira et al., 2011). This yields $h/m(Rb)$, which can be combined with measured mass ratios and the Rydberg constant to determine α. It can be seen in Table 2.3 that the values of α^{-1} obtained by the two methods agree within 1.1σ. Equivalently, one can use the ^{87}Rb value as input to predict a_e, yielding

$$a_e^{exp} - a_e^{SM} = -0.91(0.82) \times 10^{-12}. \tag{2.372}$$

A number of other, somewhat less precise, values of α^{-1} obtained from measurements of a_e at the University of Washington using electrons and positrons (which are the same by CPT), the quantum Hall effect, the p and ^3He gyromagnetic ratios using the Josephson effect, $h/m(Cs)$, and other quantities are also in agreement (Mohr et al., 2012).

The impressive agreement in (2.372), which involves a heroic calculation of a_e to tenth order, validates not only QED but the entire formalism of gauge invariance and renormalization theory. However, one caveat is in order. In interpreting very precise measurements,

[28]The extracted value of α has changed significantly from the first edition of this book due to the correction of an error in the calculation of the coefficient of $(\alpha/\pi)^4$ in (2.371).

especially in extracting parameters from them, one must always be concerned that possible effects from (unknown) physics beyond the standard model might lead to an error. This possibility can never be totally eliminated. However, in the case of a_e the effects of physics at scale $\mathcal{M}$ are usually of $\mathcal{O}[(m_e/\mathcal{M})^2]$, which is $< 10^{-12}$ for $\mathcal{M} \sim 1$ TeV. Including realistic couplings this is usually very small compared to (2.372). (For further discussion and possible loopholes, see Giudice et al., 2012).

Other basic predictions of gauge invariance (assuming it is not spontaneously broken – see Chapter 4), are that the photon mass m_γ and its charge Q_γ (in units of e) should vanish. The current upper bounds, based on astrophysical effects (the survival of the solar magnetic field and the isotropy of the cosmic microwave radiation (CMB)), are listed in Table 2.3. If QED *were* spontaneously broken one would expect electric charge nonconservation, which would allow the electron to decay, e.g., into $\nu\gamma$ (Problem 2.31). The experimental limit (from the Borexino experiment (Agostini et al., 2015a)) is $\tau_e > 6.6 \times 10^{28}$ yr.

The charge assignments of the fundamental fermions are technically arbitrary in QED, but it is usually assumed that atoms are electrically neutral, so that $Q_p = -Q_e$ and $Q_n = 0$. (See the discussion in Section 10.1.) Experiments on bulk matter (Bressi et al., 2011) indicate that this holds to high precision (assuming $Q_n = Q_p + Q_e$).

QED has also been tested at high energies, especially at the e^+e^- colliders PEP (at SLAC), PETRA (DESY), and TRISTAN (KEK), which operated below the Z pole where the s-channel photon diagram dominates (Wu, 1984; Kiesling, 1988). While not as precise as the low energy tests, these results all confirmed the QED predictions. QED was also tested indirectly in the Z pole experiments at LEP (CERN) and SLC (SLAC), and above the Z pole at LEP 2, where it was an important ingredient for calibrations and entered interferences and radiative corrections. Finally, the running of α with Q^2 has been confirmed experimentally (for a review, see Mele, 2006).

Despite all of the successes of QED, there are two discrepancies, possibly due to new physics beyond the standard model: the magnetic moment of the muon and the proton charge radius.

The Anomalous Magnetic Moment of the Muon

The anomalous magnetic moment of the muon a_μ has been measured to high precision in the Brookhaven 821 experiment (Bennett et al., 2006), in which the precession of the μ magnetic moment relative to its momentum in a storage ring was monitored using the parity-violating correlation between the momentum of the decay $e^\pm$ and the $\mu^\pm$ spin direction (Section 7.2). The value obtained was $a_\mu^{exp} = 116\ 592\ 091(54)(33) \times 10^{-11}$, where the two errors are, respectively, statistical and systematic. a_μ is especially important because it is expected to be more sensitive to new physics than most of the other probes, with deviations typically of $\mathcal{O}[(m_\mu/\mathcal{M})^2]$, i.e., $\mathcal{O}[(m_\mu/m_e)^2]$ larger than those for a_e. The SM expectation is [for reviews, see (Miller et al., 2007; Jegerlehner and Nyffeler, 2009; Mohr et al., 2012) and the Höcker and Marciano article in (Patrignani, 2016)] $a_\mu^{SM} = a_\mu^{QED} + a_\mu^{EW} + a_\mu^{had}$. The pure QED part, from diagrams analogous to Figure 2.19, has been calculated to five loops (three analytically) (Aoyama et al., 2012),

$$\begin{aligned} a_\mu^{QED} =& \frac{\alpha}{2\pi} + 0.765\ 857\ 425(17)\left(\frac{\alpha}{\pi}\right)^2 + 24.050\ 509\ 96(32)\left(\frac{\alpha}{\pi}\right)^3 \\ &+ 130.879\ 6(6\,3)\left(\frac{\alpha}{\pi}\right)^4 + 753.3(1.0)\left(\frac{\alpha}{\pi}\right)^5 \\ =& 116\ 584\ 718.95(0.08) \times 10^{-11}, \end{aligned} \tag{2.373}$$

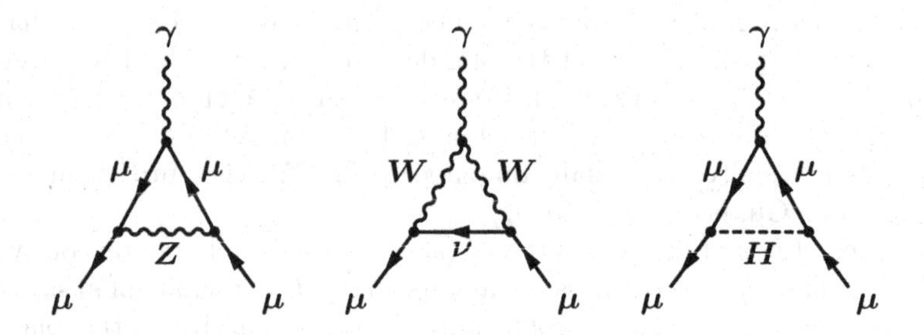

Figure 2.20 One-loop electroweak contributions to $a_\mu = (g_\mu - 2)/2$.

where the uncertainties in the coefficients are from the lepton mass ratios and the numerical integrations. The total in (2.373) uses α from $h/m(Rb)$ in Table 2.3 to avoid theoretical correlations between a_e and a_μ. However, the value obtained from $\alpha(a_e)$ differs only by $\mathcal{O}(10^{-12})$, which is negligible compared to the other uncertainties.

There are additional electroweak and hadronic contributions to a_μ (Patrignani, 2016). The electroweak diagrams include the contributions of the W, Z and Higgs (H) bosons, shown in Figure 2.20. The first two diagrams dominate, yielding

$$a_\mu^{EW} \sim \frac{G_F m_\mu^2}{8\sqrt{2}\pi^2}\left(\frac{5}{3} + \frac{1}{3}(1 - 4\sin^2\theta_W)\right) \sim 194.8 \times 10^{-11} \tag{2.374}$$

at one loop, where G_F and $\sin^2\theta_W$ are, respectively, the Fermi constant and weak angle. Including the two-loop diagrams, which are enhanced by large logarithms,

$$a_\mu^{EW} = 153.6(1.0) \times 10^{-11}. \tag{2.375}$$

An original goal of the BNL 821 experiment was to achieve sensitivity to a_μ^{EW}. However, there is a significant contribution and uncertainty from the hadronic contributions a_μ^{had}, such as shown in Figure 2.21. In particular, the hadronic vacuum polarization diagram

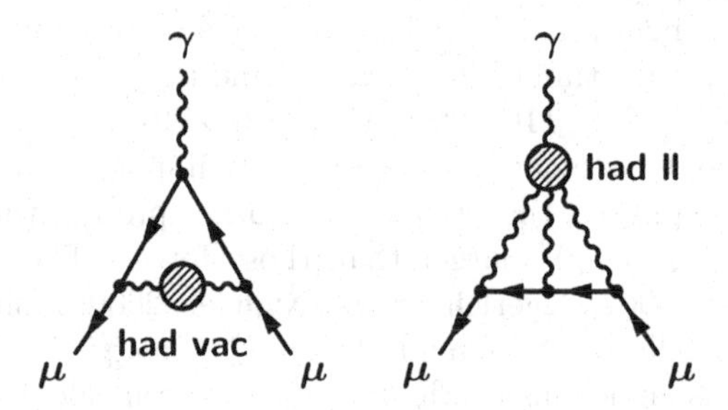

Figure 2.21 Two-loop hadronic vacuum polarization (left) and hadronic light by light (right) diagrams. The shaded blobs represent hadronic states.

yields a large contribution that cannot be reliably calculated in perturbation theory. It is closely related to the hadronic contributions to the running of α mentioned in Section 2.12.2. It can be estimated from the cross section for $e^+e^- \to$ hadrons using a dispersion relation similar to (2.370), but with a slightly different weighting function. The integral is dominated by the experimentally determined low energy region (such as the ρ resonance), yielding (e.g., Davier et al., 2011) $a_\mu^{had} \sim 6\,923(42) \times 10^{-11}$. Although this is quite precise, the uncertainty is comparable to the experimental uncertainty in a_μ. The situation is confused by the fact that one can also obtain a_μ^{had} from hadronic τ decays, which are related by isospin. Applying isospin breaking corrections leads to a value $7\,015(47) \times 10^{-11}$, which differs by $\sim 1.8\sigma$. There are also small but nonnegligible hadronic light by light diagrams (e.g., Prades et al., 2009) and three-loop vacuum polarization diagrams. The former cannot be directly related to experimental data and relies on model calculations. The sum is around $7(26) \times 10^{-11}$. Using the e^+e^- value for a_μ^{had}, one finds the standard model expectation[29]

$$a_\mu^{SM} = 116\,591\,803(49) \times 10^{-11} \;\Rightarrow\; \Delta a_\mu = a_\mu^{exp} - a_\mu^{SM} = 288(80) \times 10^{-11}, \qquad (2.376)$$

a 3.6σ discrepancy. Using the τ decay value of a_μ^{had} reduces the discrepancy to 2.4σ.

If one accepts the value of a_μ^{SM} in (2.376) then there is a strong suggestion of new physics contributions to a_μ. For example, in the supersymmetric extension of the standard model the diagrams in Figure 2.22 would give an additional contribution that would be significant

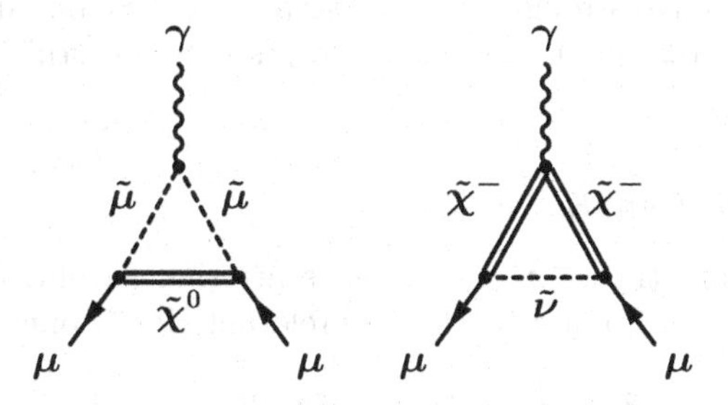

Figure 2.22 New contributions to a_μ in the supersymmetric extension of the SM. $\tilde{\mu}$ and $\tilde{\nu}$ are the spin-0 superpartners of the μ and ν, while $\tilde{\chi}^{\pm,0}$ are spin-$\frac{1}{2}$ superpartners of the electroweak gauge bosons and Higgs fields.

for relatively low masses for the supersymmetric partners and/or relatively large $\tan\beta$ (the ratio of the expectation values of the two Higgs doublets in the theory). The central value of the discrepancy in (2.376) would be accounted for (Czarnecki and Marciano, 2001) if

$$m_{SUSY} \sim 67\sqrt{\tan\beta}\ \text{GeV}, \qquad (2.377)$$

where m_{SUSY} is the typical mass of the new particles in Figure 2.22. (The Higgsino mass parameter μ, introduced in (10.148), must also be positive.)

Another possibility is that a_μ could be enhanced by vertex corrections involving a *dark photon* (e.g., Pospelov, 2009), a hypothetical light (e.g., 10–100 MeV) gauge boson with very weak couplings to ordinary matter. These have been suggested in connection with dark matter and can be searched for in many decay and low-energy processes (Essig et al., 2013; Alexander et al., 2016).

New experiments at Fermilab and J-PARC (e.g., Gorringe and Hertzog, 2015) are expected to reduce the uncertainty in a_μ^{exp} by at least a factor of 2.

---

[29] All of these estimates follow (Patrignani, 2016). However, other analyses (e.g., Hagiwara et al., 2011) yield similar results. The hadronic uncertainties are reviewed in (Benayoun et al., 2014).

The Proton Charge Radius

The proton charge radius R_p (defined precisely in (2.399)) can be measured in hydrogen/deuterium spectroscopy, which also determines the Rydberg, and independently in *ep* scattering, yielding the combined average 0.8775(51) fm (Mohr et al., 2012). Recently, a very accurate measurement of the Lamb shift in muonic hydrogen ($\mu^- p$) has been performed at the PSI in Switzerland (Pohl et al., 2010). Muonic atoms are very sensitive to R_p because of the large μ mass and consequent small extent of its wave function. Surprisingly, R_p was determined to be 0.84087(30) fm, some 7σ lower than the e^- value. So far, no one has found a plausible theoretical or experimental explanation for this discrepancy in the context of QED or the standard model. The muonic Lamb shift and a_μ suggest the possibility of some new physics that especially affects the μ, such as the exchange of a dark photon or other light particle with weak couplings. However, the R_p anomaly is much larger than the a_μ one, and it is nontrivial (but not impossible) to account for both simultaneously while satisfying other constraints. All of these matters are reviewed in (Pohl et al., 2013; Carlson, 2015).

2.12.4 The Role of the Strong Interactions

Many QED processes involve strongly interacting particles. Strong interaction effects complicate the calculations, but in simple enough cases one can still say a great deal using symmetry principles.

The Pion Electromagnetic Form Factor

Let us start by revisiting the $e^-\pi^+ \to e^-\pi^+$ scattering amplitude in Figure 2.15 and Equation (2.222). From (2.133) and (2.218) the relevant Hamiltonian interaction density is

$$\mathcal{H}_I = -\mathcal{L}_I = eJ_Q^\mu A_\mu - e^2 A^2 \phi^\dagger \phi \text{ with } J_Q^\mu = -\bar{\psi}\gamma^\mu \psi + i\phi^\dagger \overleftrightarrow{\partial}^\mu \phi, \tag{2.378}$$

where ψ and ϕ are, respectively, the e^- and π^+ fields. The amplitude is schematically

$$\begin{aligned} M_{fi} &\sim \langle e^-\pi^+ | T\big[-i\int \mathcal{H}_I(x) d^4x, -i\int \mathcal{H}_I(x') d^4x'\big] | e^-\pi^+\rangle \\ &\sim (-ie)^2 \langle e^-(\vec{k}_2, s_2)| J_Q^\mu | e^-(\vec{k}_1, s_1)\rangle \langle 0 | T(A_\mu, A_\nu)|0\rangle \langle \pi^+(\vec{p}_2)| J_Q^\nu | \pi^+(\vec{p}_1)\rangle \\ &\sim e^2 \bar{u}(\vec{k}_2, s_2)\gamma^\mu u(\vec{k}_1, s_1)\left(\frac{-ig_{\mu\nu}}{q^2}\right)\langle \pi^+(\vec{p}_2)| J_Q^\nu | \pi^+(\vec{p}_1)\rangle, \end{aligned} \tag{2.379}$$

where $q = p_2 - p_1$ and the details of the Fourier transforms and momentum-conserving δ functions are not displayed. In the absence of strong interactions, one can replace J_Q^ν in the last expression by $i\phi^\dagger \overleftrightarrow{\partial}^\mu \phi$ where ϕ and $\phi^\dagger$ are free fields, yielding

$$\langle \pi^+(\vec{p}_2)| J_Q^\nu(x) | \pi^+(\vec{p}_1)\rangle = (p_2 + p_1)^\nu e^{iq\cdot x}. \tag{2.380}$$

In the presence of the strong interactions J_Q^μ is still defined by the expression in (2.378). However, the fields are no longer free, but rather are Heisenberg or interaction picture fields w.r.t. the strong interactions. For example, consider the interaction terms

$$\mathcal{L}_{\pi N} = i\sqrt{2} g_\pi \left(\bar{\psi}_p \gamma^5 \psi_n \phi_{\pi^+} + \bar{\psi}_n \gamma^5 \psi_p \phi_{\pi^+}^\dagger\right) - 4\lambda \left(\phi_{\pi^+}^\dagger \phi_{\pi^+}\right)^2, \tag{2.381}$$

where ψ_p, ψ_n, and ϕ_{π^+} are, respectively, the proton, neutron, and π^+ fields. $\mathcal{L}_{\pi N}$ corresponds to the interaction vertices shown in Figure 2.23, and represents a subset of those in

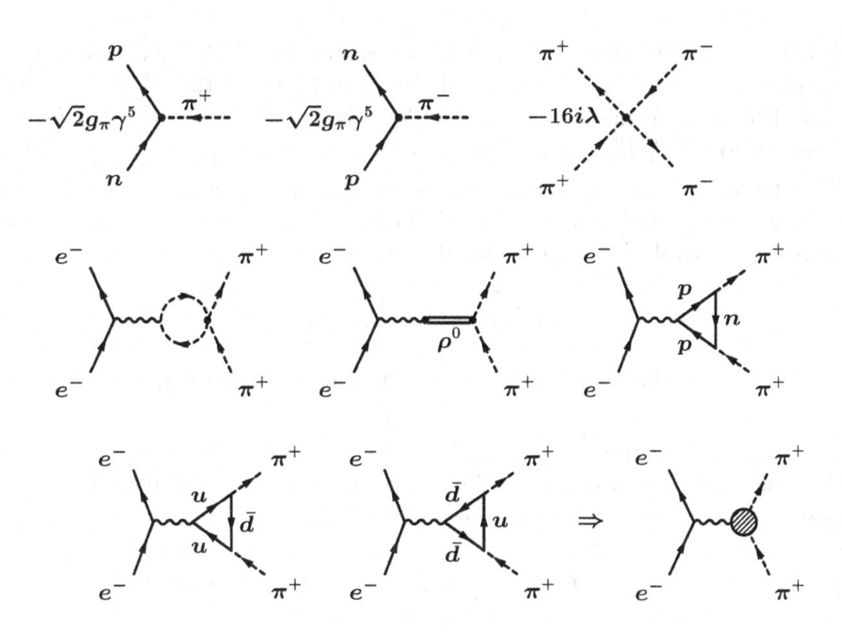

Figure 2.23 Top: interaction vertices described by (2.381). Middle: representative strong interaction diagrams contributing to $\langle \pi^+ | J_Q^\mu | \pi^+ \rangle$ involving nucleons, pions, and the ρ^0 resonance. Bottom left: diagrams in the quark model description in which the π^+ is a $u\bar{d}$ bound state. Bottom right: parametrization of the strong interaction effects by a form factor, represented by a shaded circle.

the isospin conserving model described in Section 3.2.3. (The π^- field is $\phi_{\pi^-} = \phi_{\pi^+}^\dagger$, while the π^0 and such states as the ρ resonance can easily be added.) An alternative and more fundamental description is to regard the π^+ as a bound state of a u (charge $+2/3$) and $\bar{d}$ (charge $+1/3$) quark and antiquark. In practice, one cannot treat the strong interactions perturbatively using either description because the couplings g_π and λ, or the vertex relating the bound state π^+ to the quarks, are too large. However, one can parametrize our uncertainty by writing matrix elements such as $\langle \pi^+ | J_Q^\nu | \pi^+ \rangle$ in a way that reflects the symmetries of the theory and that involves one or more form factors, analogous to those introduced for higher-order QED effects in Section 2.12.1, for the residual unknowns. These form factors can be measured experimentally, computed using non-perturbative lattice techniques, or estimated in other theoretical models.

Consider $\langle \pi^+(p_2) | J_Q^\mu(x) | \pi^+(p_1) \rangle$. By an argument similar to the one in Section 2.12.1, the most general form is

$$\langle \pi^+(p_2) | J_Q^\mu(x) | \pi^+(p_1) \rangle = \left[(p_2 + p_1)^\mu f_+^Q(q^2) + (p_2 - p_1)^\mu f_-^Q(q^2) \right] e^{iq\cdot x}. \qquad (2.382)$$

The x dependence follows from translation invariance, while the only available Lorentz four-vectors are $(p_2 \pm p_1)^\mu$. The form factors $f_\pm^Q$ can depend on q^2, which is the only Lorentz invariant available since the pions are assumed to be on shell (they could also depend on $p_{1,2}^2$ if we extended the discussion to off-shell pions). The electric charge, in units of e, is $f_+^Q(0)$. In the absence of strong interactions we would have $f_+^Q(0) = 1$ and $f_-^Q(q^2) = 0$. Since the strong interactions conserve electric charge, we expect that $f_+^Q(0) = 1$ to be maintained, i.e., that the electric charge is not renormalized. If this were not the case, the π^+ or proton electric charges would differ from those of the e^+ due to the strong interactions. The nonrenormalization will be demonstrated in Section 7.2.4. Furthermore, current conservation,[30] $\partial_\mu J^\mu_Q = 0$, requires $f^Q_-(q^2) = 0$, analogous to $F_3 = 0$ in (2.353).

$f^Q_+(q^2)$ is known as the *electromagnetic form factor of the pion.* Although $f^Q_+(0) = 1$, the strong interactions (and higher-order QED effects) are expected to induce a nontrivial q^2 dependence. For small q^2 one can expand

$$f^Q_+(q^2) \sim 1 + \frac{1}{6} R^2_\pi q^2, \tag{2.383}$$

where R_π is known as the *pion charge radius.* The term is motivated by non-relativistic quantum mechanics. Suppose the π^+ were a bound state of a non-relativistic constituent with charge $+e$ bound in a potential with wave function $\psi(\vec{x})$ (this roughly but not exactly mimics the effects of $u\bar{d}$ constituents). Then the non-relativistic analog of $f^Q_+(q^2)$, which plays a similar role in pion scattering from a static field, is

$$F(\vec{q}) = \int d^3\vec{x}\, e^{i\vec{q}\cdot\vec{x}} |\psi(\vec{x})|^2. \tag{2.384}$$

It is obvious in the non-relativistic case that $F(0) = 1$ by the normalization of the wave function, i.e., charge is not renormalized by the bound state effects. For small $|\vec{q}|^2$, and assuming a spherically symmetric wave function (which is reasonable since the pion spin is 0), it is straightforward to show that

$$F(\vec{q}) \sim 1 - \frac{1}{6} R^2_\pi |\vec{q}|^2 \text{ with } R^2_\pi \equiv \int d^3\vec{x}\, |\vec{x}|^2 |\psi(\vec{x})|^2, \tag{2.385}$$

so R_π is the RMS charge radius. The non-relativistic approximation is not really valid, but the motivation for the terminology remains.

$f^Q_+(q^2)$ can be measured in the spacelike region $q^2 < 0$ by scattering pions from an atomic target, or in the reaction $e^- p \to e^- \pi^+ n$, in which the e^- scatters from a virtual π^+ emitted by the proton, working in a kinematic region where the exchanged pion is as close as possible to being on shell. f^Q_+ can be approximated for small $|q^2|$ by

$$f^Q_+(q^2) \sim \frac{1}{1 - \frac{R^2_\pi q^2}{6}}, \tag{2.386}$$

with $\sqrt{R^2_\pi} = 0.672 \pm 0.008$ fm (Patrignani, 2016). It can be measured in the timelike region for $q^2 > 4m^2_\pi$ by $e^+e^- \to \pi^+\pi^-$. For large q^2 it falls off rapidly as $1/q^2$, while for lower q^2 it is dominated by resonances, especially the ρ^0, which is a spin-1 resonance with $m_\rho \sim 770$ MeV and a width $\Gamma_\rho \sim 150$ MeV. Near the ρ the form factor assumes a classic Breit-Wigner resonance form (Appendix F), which to first approximation is

$$\left| f^Q_+(q^2) \right|^2 \sim \frac{m^4_\rho}{\left(q^2 - m^2_\rho\right)^2 + m^2_\rho \Gamma^2_\rho}. \tag{2.387}$$

Corrections to this formula are given in (Gounaris and Sakurai, 1968) and plots in (Ambrosino et al., 2011).

[30] As mentioned in Section 2.6, the conserved current is actually $i\phi^\dagger \overleftrightarrow{\partial^\mu} \phi - 2e\phi^\dagger A^\mu \phi$. This is irrelevant here as we are working to lowest order in e.

The Proton and Neutron Form Factors

Now consider the elastic scattering process $e^- p \to e^- p$. Similar to (2.378) and (2.379), the electromagnetic interaction is

$$\mathcal{L}_I = -eJ^\mu_Q A_\mu \quad \text{with} \quad J^\mu_Q = -\bar{\psi}_e \gamma^\mu \psi_e + \bar{\psi}_p \gamma^\mu \psi_p, \tag{2.388}$$

where ψ_e and ψ_p represent the electron and proton fields, respectively, and the lowest order amplitude can be written

$$M_{fi} = ie\,\bar{u}(\vec{k}_2)\gamma^\mu u(\vec{k}_1)\left(\frac{-ig_{\mu\nu}}{q^2}\right)(-ie)\langle p(\vec{p}_2)|J^\nu_Q|p(\vec{p}_1)\rangle, \tag{2.389}$$

where $q = p_2 - p_1$ and the spin labels are suppressed. Let us first treat the proton as a pointlike elementary particle. Then

$$\langle p(\vec{p}_2)|J^\nu_Q|p(\vec{p}_1)\rangle = \bar{u}(\vec{p}_2)\gamma^\nu u(\vec{p}_1). \tag{2.390}$$

The cross section in the lab frame, i.e., the proton rest frame, is

$$\frac{d\bar{\sigma}}{d\Omega_L} = \frac{\alpha^2}{4k_1^2 \sin^4\frac{\theta_L}{2}}\frac{k_2}{k_1}\left[\cos^2\frac{\theta_L}{2} - \frac{q^2}{2m_p^2}\sin^2\frac{\theta_L}{2}\right], \tag{2.391}$$

where we have neglected the electron mass, the proton mass is m_p, θ_L is the angle of the scattered electron, $d\Omega_L = d\varphi d\cos\theta_L \to 2\pi d\cos\theta_L$, the final electron energy is given by (2.46),

$$\frac{k_2}{k_1} = \frac{1}{1 + \frac{k_1}{m_p}(1-\cos\theta_L)} = \frac{1}{1 + \frac{2k_1}{m_p}\sin^2\frac{\theta_L}{2}}, \tag{2.392}$$

and $q^2 = -4k_1k_2\sin^2\frac{\theta_L}{2}$.

When we turn on the strong interactions, we must include the full matrix element of J^ν_Q in (2.389). This is denoted by the shaded circle in Figure 2.24, which also contains some representative strong interaction corrections to the vertex, both in the approximation of treating the virtual states as physical hadrons, or in the quark model, for which the proton is a *uud* bound state. Just as for the pion, these effects cannot be summed perturbatively. They must be expressed in terms of form factors that parametrize the matrix element in a way that takes into account the symmetries of the theory, and that can be determined by experiment, by model calculations, or by lattice or other techniques. After making use of the Gordon identities in Problem 2.10 the most general Lorentz and translation invariant matrix element is

$$\langle p(\vec{p}_2)|J^\mu_Q(x)|p(\vec{p}_1)\rangle = \bar{u}(\vec{p}_2)\Gamma^\mu_Q(q)u(\vec{p}_1)e^{iq\cdot x}, \tag{2.393}$$

where

$$\begin{aligned}\Gamma^\mu_Q(q) = \Big[&\gamma^\mu F^p_1(q^2) + \frac{i\sigma^{\mu\nu}}{2m_p}q_\nu F^p_2(q^2) + q^\mu F^p_3(q^2)\\ &+ \gamma^\mu\gamma^5 g^p_1(q^2) + \frac{i\sigma^{\mu\nu}\gamma^5}{2m_p}q_\nu g^p_2(q^2) + q^\mu\gamma^5 g^p_3(q^2)\Big],\end{aligned} \tag{2.394}$$

where the form factors F^p_i and g^p_i depend on q^2. We have so far not imposed additional symmetries such as reflection invariance, so the discussion would be valid even when weak and higher-order electromagnetic effects are included. Similar to the discussion in Section

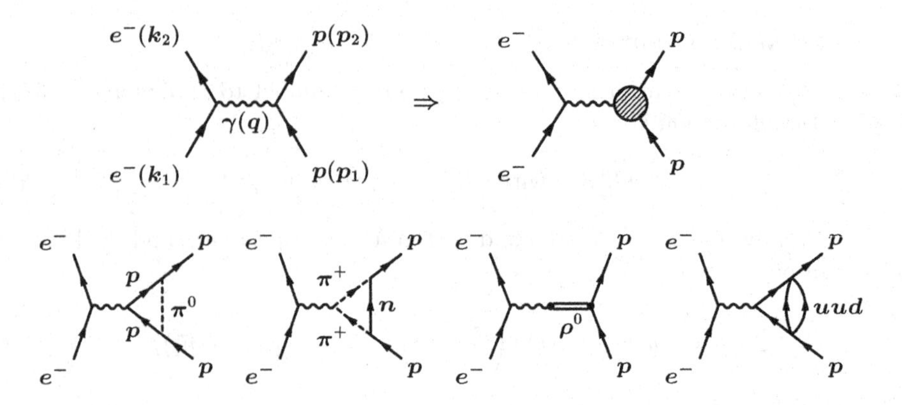

Figure 2.24 Top: tree-level amplitude for $e^-p \to e^-p$ for a point proton and with proton form factors (shaded circle). Bottom: representative strong interaction corrections in terms of virtual hadrons (three left diagrams) or in terms of quark constituents (right).

2.12.1, the proton electric charge is $eF_1^p(0)$, so the nonrenormalization of charge (Section 7.2.4) implies $F_1^p(0) = 1$. $\kappa_p \equiv F_2^p(0)$ is the anomalous magnetic moment of the proton, i.e.,

$$\vec{\mu}_p = +g_p \mu_N \vec{S}_p, \qquad g_p = 2(1+\kappa_p), \tag{2.395}$$

where $\mu_N = e/2m_p$ is the nuclear magneton. The experimental value $\kappa_p \sim 1.79$ is very much larger than the electromagnetic one-loop contribution of $\alpha/2\pi$, indicating the importance and nonperturbative nature of the strong interaction corrections. The value, and also the ratio of the proton and neutron magnetic moments, can be approximately understood in the non-relativistic quark model (see Problem 3.22). Electromagnetic current conservation further implies $q_\mu \bar{u}(\vec{p}_2)\Gamma_Q^\mu(q)u(\vec{p}_1) = 0$, and therefore

$$F_3^p(q^2) = 0, \qquad 2m_p g_1^p(q^2) + q^2 g_3^p(q^2) = 0. \tag{2.396}$$

From the discussion in Section 2.10 and Appendix G, $g_i^p(q^2) \neq 0$ requires the violation of space reflection invariance since J_Q^μ transforms as a vector. $g_{1,3}^p \neq 0$ would require the violation of charge conjugation invariance as well. Since the weak interactions violate P and C they are expected to generate a nonzero $g_{1,3}^p$ subject to (2.396), known as an *anapole moment*, at some level, e.g., due to W or Z exchange across the photon vertex or a parity-violating scalar correction to the pion-nucleon coupling. The anapole moment has been observed in parity violating transitions in the ^{133}Cs atom (Haxton and Wieman, 2001; Haxton et al., 2002) at the expected order of magnitude, though the calculations are difficult and model comparisons with other forms of hadronic parity violation are not in perfect agreement. A nonzero value for $g_2^p \neq 0$ would require both space reflection and time reversal violation, and would correspond to an *electric dipole moment* (EDM) for the proton (see Problem 2.30). P is violated by the weak interactions, but T violation is much weaker, especially for the light fermions, so EDMs are predicted to be extremely small in the standard model, as discussed in Section 8.6. However, many types of physics beyond the standard model predict EDMs much larger than in the SM, so experimental searches for the neutron, electron, atomic, or other EDMs are extremely important.

Let us now restrict our attention to strong interaction corrections to Γ^μ_Q, so that

$$\Gamma^\mu_Q(q) = \left[\gamma^\mu F_1^p(q^2) + \frac{i\sigma^{\mu\nu}}{2m_p} q_\nu F_2^p(q^2)\right], \tag{2.397}$$

with $F_1^p(0) = 1$ and $F_2^p(0) = \kappa_p$, analogous to the higher-order QED corrections to the e^- vertex in (2.353). For $q^2 < 0$ it is conventional to define $Q^2 = -q^2 > 0$, and to define the *Sachs*, or *electric* and *magnetic*,[31] form factors

$$G_E(q^2) = F_1^p(q^2) - \tau F_2^p(q^2), \qquad G_M(q^2) = F_1^p(q^2) + F_2^p(q^2), \tag{2.398}$$

where $\tau \equiv Q^2/4m_p^2 \equiv -q^2/4m_p^2$, so that $G_E(0) = 1$ and $G_M(0) = 1 + \kappa_p$. The proton charge radius R_p is defined by

$$R_p^2 = 6 \left. \frac{dG_E}{dq^2} \right|_{q^2=0}, \tag{2.399}$$

analogous to the pion charge radius in (2.383).

One can then generalize (2.391) to the *Rosenbluth cross section* formula for elastic $e^-p \to e^-p$ scattering

$$\frac{d\bar{\sigma}}{d\Omega_L} = \frac{\alpha^2}{4k_1^2 \sin^4 \frac{\theta_L}{2}} \frac{k_2}{k_1} \left[\frac{G_E^2 + \tau G_M^2}{1+\tau} \cos^2 \frac{\theta_L}{2} + 2\tau G_M^2 \sin^2 \frac{\theta_L}{2} \right]. \tag{2.400}$$

The neutron also has electromagnetic properties due to strong interaction and bound state effects, just as a neutral atom has electromagnetic moments due to its internal charge distribution. This is most obvious in the quark picture, where it is a *udd* bound state, but can also be seen in the virtual hadron language, e.g., induced by the diagrams analogous to Figure 2.24 in which the neutron turns into a virtual $p\pi^-$. The neutron electromagnetic matrix element is

$$\langle n(\vec{p}_2)|J^\mu_Q(x)|n(\vec{p}_1)\rangle = \bar{u}(\vec{p}_2)\left[\gamma^\mu F_1^n(q^2) + \frac{i\sigma^{\mu\nu}}{2m_n} q_\nu F_2^n(q^2)\right] u(\vec{p}_1), \tag{2.401}$$

where the neutron total charge is $F_1^n(0) = 0$ and $F_2^n(0) = \kappa_n \equiv g_n/2 \sim -1.91$ is the neutron anomalous magnetic moment. $G^n_{E,M}$ and the charge radius R_n are defined as in (2.398) and (2.399). It is sometimes useful to define the *isovector* and *isoscalar* form factors

$$F^V_{1,2} = \frac{1}{2}\left(F^p_{1,2} - F^n_{1,2}\right), \qquad F^S_{1,2} = \frac{1}{2}\left(F^p_{1,2} + F^n_{1,2}\right), \tag{2.402}$$

and similarly for $G^{V,S}_{E,M}$.

One can extract the form factors $G^p_{E,M}(Q^2)$ or $F^p_{1,2}(Q^2)$ by measuring the cross section for $e^-p \to e^-p$ while varying k_1 and θ_L for fixed Q^2, and also by scattering from polarized protons (for reviews, see Hyde-Wright and de Jager, 2004; Arrington et al., 2007; Pacetti et al., 2014; Punjabi et al., 2015). They are described by an approximate *dipole* form

$$G_E^p(Q^2) \sim \frac{G_M^p(Q^2)}{1+\kappa_p} \sim \frac{1}{\left[1 + Q^2/Q_0^2\right]^2}, \tag{2.403}$$

where $Q_0^2 \sim 0.71$ GeV$^2 \sim 18.2$ fm^{-2}, which holds to $\sim$10% for $Q^2 \lesssim 10$ GeV^2. The neutron

[31]The terms are motivated by the form of the matrix element in the Breit frame (Problem 2.32), which generalizes the discussion of the non-relativistic limit of the e^- vertex at the end of Section 2.12.1.

form factors are obtained from scattering from nuclear targets, such as deuterium (D) or polarized ^3He, and are approximated by $G_M^n(Q^2)/\kappa_n \sim G_E^p(Q^2)$ for $Q^2 \lesssim$ few GeV^2, with $|G_E^n(Q^2)|$ much smaller (a few %).

The approximate $1/Q^2$ and $1/Q^4$ behaviors of the pion and proton form factors for large $1/Q^2$ are consistent with the expectation (see, e.g., Lepage and Brodsky, 1980) of $(Q^2)^{-(n-1)}$ for a bound state of n constituent quarks based on dimension counting rules derivable from QCD. The more rapid falloff for larger n may be thought of as due to the greater difficulty for the system to hold together in the scattering.

2.13 OPERATOR DIMENSIONS AND CLASSIFICATION

The classification of interaction terms for Hermitian scalars in Section 2.3.1 can be generalized to an arbitrary field theory. In particular, in four space-time dimensions the Lagrangian density has mass dimension of four (so that the action is dimensionless), boson and fermion fields have dimensions of 1 and 3/2, respectively, and ordinary or covariant derivatives have dimension 1. Thus, an operator $\mathcal{O}^k$ consisting of the product of n_B boson fields, n_F fermion fields, and n_D derivatives will have dimension $k = n_B + 3n_F/2 + n_D$, and the coefficient of $\mathcal{O}^k$ in $\mathcal{L}$ must be of the form $c_k/\mathcal{M}^{k-4}$, where c_k is dimensionless and $\mathcal{M}$ is a mass.

Operators with $k < 4$ are super-renormalizable, those with $k = 4$ are renormalizable,[32] and those with $k > 4$ are non-renormalizable (higher-dimensional). Examples include

$$\begin{aligned} \text{Super-renormalizable:} &\quad \phi^\dagger\phi, \quad \bar{\psi}\psi, \quad \phi^3 \\ \text{Renormalizable:} &\quad (\partial^\mu\phi)^\dagger\partial_\mu\phi, \quad \bar{\psi}i\not{D}\psi, \quad F_{\mu\nu}F^{\mu\nu}, \quad \bar{\psi}\psi\phi, \quad (\phi^\dagger\phi)^2 \\ \text{Non-renormalizable:} &\quad \bar{\psi}\psi\,\phi^\dagger\phi, \quad (\bar{\psi}\psi)^2, \quad (D^\mu\phi)^\dagger(D_\mu\phi)(\phi^\dagger\phi), \quad (\phi^\dagger\phi)^3. \end{aligned} \tag{2.404}$$

Non-renormalizable operators often emerge as a low-energy effective theory approximation to a more fundamental field theory, superstring theory, etc., in which heavy fields or other degrees of freedom of mass of $\mathcal{O}(\mathcal{M})$ are integrated out (Section 8.2.3).

2.14 MASS AND KINETIC MIXING

In general, the Lagrangian density for a theory with more than one field of the same type (i.e., Hermitian or complex scalar, massive vector, or fermion) may include non-canonical kinetic energy terms and off-diagonal mass terms. For example, the Lagrangian density in (2.20) can be generalized for n Hermitian fields $\hat{\phi}_a, a = 1 \cdots n$ to

$$\mathcal{L}(\hat{\phi}) = \frac{1}{2}\left[\partial_\mu\hat{\phi}^T\mathcal{K}\partial^\mu\hat{\phi} - \hat{\phi}^T\hat{m}^2\hat{\phi}\right] - \hat{V}_I(\hat{\phi}), \tag{2.405}$$

where $\hat{\phi} = (\hat{\phi}_1\hat{\phi}_2\cdots\hat{\phi}_n)^T$ is an n-component column vector, $\mathcal{K}$ is a real symmetric $n \times n$ matrix with positive eigenvalues, and $\hat{m}^2$ is a real symmetric matrix.[33] A non-canonical kinetic energy term (i.e., $\mathcal{K} \neq 1$), which may arise if the theory descends from a more fundamental underlying theory, must be put into canonical form prior to quantization to maintain the classical relation $E^2 = \vec{p}^{\,2} + m^2$. This can be accomplished by defining new fields $\phi^0 \equiv S^{1/2}O_\mathcal{K}^T\hat{\phi}$, where $O_\mathcal{K}$ is the orthogonal matrix that diagonalizes $\mathcal{K}$ and S is the

[32] The term renormalizable is sometimes extended to include $k < 4$. Super-renormalizable, renormalizable, and non-renormalizable operators are also referred to as relevant, marginal, and irrelevant, respectively, motivated by their low energy behavior.

[33] The $\hat{m}^2$ eigenvalues must be non-negative to avoid spontaneous symmetry breaking.

(non-orthogonal) diagonal matrix consisting of the eigenvalues of $\mathcal{K}$. Then

$$\partial_\mu \hat{\phi}^T \mathcal{K} \partial^\mu \hat{\phi} = \partial_\mu \phi^{0T} S^{-1/2} O_\mathcal{K}^T \mathcal{K} O_\mathcal{K} S^{-1/2} \partial^\mu \phi^0 = \partial_\mu \phi^{0T} \partial^\mu \phi^0. \tag{2.406}$$

The mass and interaction terms must then be rewritten in terms of these rotated and rescaled fields,

$$\mathcal{L}(\phi^0) = \frac{1}{2}\left[\partial_\mu \phi^{0T} \partial^\mu \phi^0 - \phi^{0T} m^2 \phi^0\right] - V_I^0(\phi^0), \tag{2.407}$$

where $m^2 = S^{-1/2} O_\mathcal{K}^T \hat{m}^2 O_\mathcal{K} S^{-1/2}$ and $V_I^0(\phi^0) = \hat{V}_I(O_\mathcal{K} S^{-1/2} \phi^0)$. We will see examples of the effects of this *kinetic mixing* in Chapter 10.

The standard way to interpret (2.407) is to perform a second orthogonal transformation (or a first one if $\mathcal{K}$ was equal to I initially) to diagonalize m^2, i.e., $\phi \equiv O_m^T \phi^0$, where $O_m^T m^2 O_m = m_D^2$ and m_D^2 is the diagonal matrix $\text{diag}(m_1^2\ m_2^2 \cdots m_n^2)$. Then

$$\mathcal{L}(\phi) = \frac{1}{2}\sum_{a=1}^{n}\left[(\partial_\mu \phi_a)^2 - m_a^2 \phi_a^2\right] - V_I(\phi), \tag{2.408}$$

where ϕ_a and m_a are, respectively, the ath *mass eigenstate field* and *mass eigenvalue*, and $V_I(\phi) = V_I^0(O_m \phi)$. The free-field propagator in (2.29) generalizes trivially in this mass eigenstate basis to

$$\Delta_{Fab}(k) = \frac{\delta_{ab}}{k^2 - m_a^2 + i\epsilon} \tag{2.409}$$

in an obvious notation, so that one can calculate scattering and decay processes in a straightforward way, with the index a (possibly) changing at the vertices but not along internal lines.

The same considerations apply to massive vectors fields. For complex scalars the only difference is that the analogs of $\mathcal{K}$ and $\hat{m}^2$ must be Hermitian, and those of $O_\mathcal{K}$ and O_m must be unitary (with T replaced by $\dagger$). Such unitary transformations also apply to Dirac fields in the special case of Hermitian kinetic and mass matrices. Extensions involving γ^5's (i.e., non-Hermitian matrices) and Majorana fermions are discussed in Problem 3.32 and in Chapter 8.

In some cases, however, it is easier to treat the small elements (usually the off-diagonal ones) of m^2 by *mass insertions*, in which they are ignored in the calculation of the propagators but are instead treated as if they were interaction terms. We will illustrate this for the case of two Hermitian scalars $\phi_{1,2}^0$ interacting with a single fermion ψ,

$$\mathcal{L} = \mathcal{L}_\psi + \mathcal{L}_{KE_{\phi^0}} - \frac{1}{2} m_1^2 \phi_1^{02} - \frac{1}{2} m_2^2 \phi_2^{02} - m_{12}^2 \phi_1^0 \phi_2^0 + \bar{\psi}\left[h_1 \phi_1^0 + h_2 \phi_2^0\right]\psi, \tag{2.410}$$

where $\mathcal{L}_\psi$ is the free-field density for ψ and $\mathcal{L}_{KE_{\phi^0}}$ is the canonical kinetic energy for $\phi_{1,2}^0$.

The mass eigenstates are

$$\begin{pmatrix} \phi_a \\ \phi_b \end{pmatrix} = O^T \begin{pmatrix} \phi_1^0 \\ \phi_2^0 \end{pmatrix}, \qquad O \equiv \begin{pmatrix} \cos\theta & \sin\theta \\ -\sin\theta & \cos\theta \end{pmatrix}, \tag{2.411}$$

where the expressions for θ and the mass eigenvalues $m_{a,b}^2$ are elementary and will not be displayed. The interactions of $\phi_{a,b}$ are given by $\bar{\psi}\left[h_a \phi_a + h_b \phi_b\right]\psi$, where

$$\begin{pmatrix} h_a \\ h_b \end{pmatrix} = O^T \begin{pmatrix} h_1 \\ h_2 \end{pmatrix}. \tag{2.412}$$

The amplitude for $\psi_1\psi_2 \to \psi_3\psi_4$ is then given in terms of the mass eigenstates by

$$M = i^3 \bar{u}_4 u_2 \bar{u}_3 u_1 \left[\frac{h_a^2}{t - m_a^2} + \frac{h_b^2}{t - m_b^2} \right] + (3 \leftrightarrow 4). \tag{2.413}$$

Now, let us go to the limit in which m_{12}^2 is a perturbation, i.e., $|m_{12}^2| \ll m_1^2,\, m_2^2,\, |m_1^2 - m_2^2|$. Then,

$$m_a^2 \sim m_1^2, \qquad m_b^2 \sim m_2^2, \qquad \theta \sim \frac{m_{12}^2}{m_2^2 - m_1^2}, \tag{2.414}$$

and the bracketed quantity in (2.413) reduces to

$$\frac{h_1^2}{t - m_1^2} + \frac{h_2^2}{t - m_2^2} + 2\theta h_1 h_2 \left[\frac{1}{t - m_2^2} - \frac{1}{t - m_1^2} \right]. \tag{2.415}$$

Alternatively, one can treat the m_{12}^2 term as a perturbation, and work in terms of the propagators for $\phi_{1,2}^0$ exchange. Then

$$M = i^3 \bar{u}_4 u_2 \bar{u}_3 u_1 \left[\frac{h_1^2}{t - m_1^2} + \frac{h_2^2}{t - m_2^2} + \frac{2 m_{12}^2 h_1 h_2}{(t - m_1^2)(t - m_2^2)} \right] + (3 \leftrightarrow 4), \tag{2.416}$$

where the m_{12}^2 term is from the mass insertion diagrams in Figure 2.25. Expressing m_{12}^2 in terms of θ from (2.414) one recovers the result in (2.413) and (2.415), but (at least in more complicated examples) with considerably less effort. In addition to the computational simplicity, mass insertions allow one to keep track of small effects such as symmetry breaking systematically.

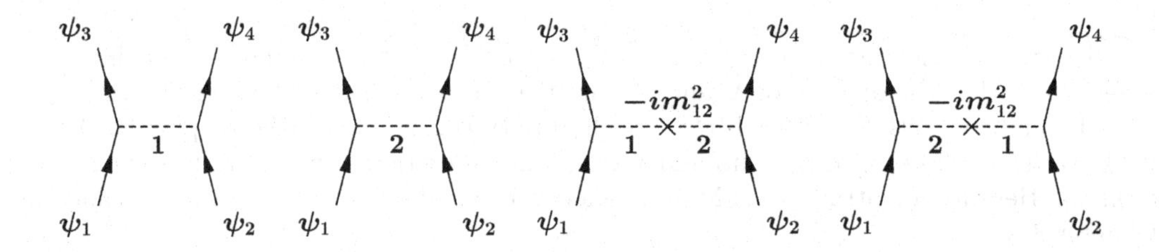

Figure 2.25 t-channel diagrams for $\psi_1\psi_2 \to \psi_3\psi_4$, treating the m_{12}^2 term in (2.410) as a perturbative mass insertion. There are additional u channel diagrams with $(3 \leftrightarrow 4)$.

2.15 PROBLEMS

2.1 From (A.7), the Hamiltonian for a free complex scalar field is

$$H = \int d^3\vec{x} \left[\left(\frac{\partial \phi^\dagger}{\partial t} \right) \left(\frac{\partial \phi}{\partial t} \right) + (\vec{\nabla}\phi^\dagger) \cdot (\vec{\nabla}\phi) + m^2 \phi^\dagger \phi \right].$$

Show that

$$H = \int \frac{d^3\vec{p}}{(2\pi)^3 2E_p} E_p \left[a^\dagger(\vec{p}) a(\vec{p}) + b^\dagger(\vec{p}) b(\vec{p}) \right].$$

Assume that H is normal ordered.

2.2 The CM differential and total cross sections for the elastic scattering of a Hermitian scalar of mass m with potential (2.22) is given in (2.59) and (2.61). Choose $m = 0.5$ GeV, $\lambda = 0.3$, and $\kappa = 1.2$ GeV. Plot the CM differential cross section $d\sigma/d\cos\theta$ in units of $\text{fm}^2 = 10^{-26}\ \text{cm}^2$ as a function of $\cos\theta$ for $s = 5m^2$, $6m^2$, and $7m^2$, and plot the total cross section as a function of $\sqrt{s}$ for $2m < \sqrt{s} < 3m$. Use any convenient plotting program.

2.3 Derive (2.64) for the special case $m_1 = m_3 = 0$ and $m_2 = m_4$ by Lorentz transforming (2.57). Hint: use the fact that σ, s, and t are invariant.

2.4 Consider the process $\pi^+(\vec{p}_1)\pi^-(\vec{p}_2) \to \pi^+(\vec{p}_3)\pi^-(\vec{p}_4)$, with $p_1 \neq p_3$ and $p_2 \neq p_4$, in the theory of a complex scalar field with Lagrangian density given in (2.88) with

$$V_I = \frac{\lambda}{4}(\phi^\dagger\phi)^2.$$

As shown in Appendix B, the tree-level amplitude M_{fi} is given by

$$(2\pi)^4\delta^4(p_3 + p_4 - p_1 - p_2)\, M_{fi} = \int d^4x\, \langle \vec{p}_3\vec{p}_4| - iV_I(\phi_0(x), \phi_0^\dagger(x))|\vec{p}_1\vec{p}_2\rangle.$$

Calculate this explicitly using the free-field expression for ϕ_0, and show that $M_{fi} = -i\lambda$.

2.5 Consider the Lagrangian density in (2.84) for three non-identical Hermitian fields. Calculate the lowest order differential cross section in the center of mass for $\phi_1(\vec{p}_1)\phi_2(\vec{p}_2) \to \phi_1(\vec{p}_3)\phi_2(\vec{p}_4)$ as a function of s and the CM scattering angle θ for the special case $m_1 = m_2 \neq m_3$.

2.6 Suppose that the interaction potential for a complex scalar field were

$$V_I = \sigma_4\left(\phi^4 + \phi^{\dagger 4}\right)$$

(rather than $\lambda(\phi^\dagger\phi)^2/4$), which is *not* $U(1)$ invariant. Show that charge is not conserved and calculate the lowest order amplitude for $\pi^+(\vec{p}_1)\,\pi^+(\vec{p}_2) \to \pi^-(\vec{p}_3)\,\pi^-(\vec{p}_4)$.

2.7 Consider charged pion QED with a *massive* photon, i.e., add the term $\frac{1}{2}M_A^2 A_\mu A^\mu$ to the Lagrangian density in (2.133) (with $(V_I = 0)$. Assume $M_A > 2m$. (The photon mass term is not gauge invariant, but the model still makes sense at tree level.)
(a) Calculate the decay rate for $\gamma \to \pi^+\pi^-$ in the photon rest frame at tree level for an unpolarized massive photon.
(b) Calculate the π^+ angular distribution $d\Gamma/d\cos\theta$ for a polarized photon, where θ is the angle between the photon polarization direction in the rest frame and the π^+ direction.
(c) Show that one recovers the result in (a) when $d\Gamma/d\cos\theta$ is integrated over $\cos\theta$.

2.8 Consider $\pi^+(\vec{p}_1)\pi^-(\vec{p}_2) \to \pi^+(\vec{p}_3)\pi^-(\vec{p}_4)$ in *massive scalar electrodynamics*, i.e., with

$$V_I = g\phi^\dagger\phi A,$$

where A, the analog of the electromagnetic field, is a Hermitian spin-0 field with mass $\mu \neq 0$. The analog of the charge is g, which now has dimensions of mass.
(a) Find expressions for the differential and total cross sections in the CM to lowest non-trivial order, in terms of s, m, μ, g, and $\cos\theta$.
(b) Define the dimensionless variable $x = s/m^2 \geq 4$, and specialize to the values $m = g = 1$ GeV, $\mu = 0.5$ GeV. Plot $d\sigma/d\cos\theta$ vs $\cos\theta$ in units of 1 $\text{fm}^2 = 10^{-26}\ \text{cm}^2$ for $x = 4, 4.2$, and 4.4. Use any plotting program.
(c) Plot σ in units of fm^2 versus x for the same parameter values and the range $4 \leq x \leq 5$.

2.9 Prove directly from the defining relations that

$$\sigma^{\mu\nu}\gamma^5 = -\frac{i}{2}\epsilon^{\mu\nu\rho\sigma}\sigma_{\rho\sigma}.$$

2.10 Prove the *Gordon decomposition* formulas

$$\begin{aligned}
2m\,(\bar{u}_2\gamma^\mu u_1) &= \bar{u}_2(p_2+p_1)^\mu u_1 + i\bar{u}_2\sigma^{\mu\nu}(p_2-p_1)_\nu u_1\\
2m\,(\bar{u}_2\gamma^\mu\gamma^5 u_1) &= \bar{u}_2(p_2-p_1)^\mu\gamma^5 u_1 + i\bar{u}_2\sigma^{\mu\nu}(p_2+p_1)_\nu\gamma^5 u_1\\
0 &= \bar{u}_2(p_2-p_1)^\mu u_1 + i\bar{u}_2\sigma^{\mu\nu}(p_2+p_1)_\nu u_1\\
0 &= \bar{u}_2(p_2+p_1)^\mu\gamma^5 u_1 + i\bar{u}_2\sigma^{\mu\nu}(p_2-p_1)_\nu\gamma^5 u_1,
\end{aligned}$$

where $u_{1,2}$ are two Dirac u spinors for a particle of mass m.

2.11 Prove the identity

$$\gamma^\mu\gamma^\nu\gamma^\rho = \gamma^\mu g^{\nu\rho} + \gamma^\rho g^{\mu\nu} - \gamma^\nu g^{\mu\rho} + i\epsilon^{\sigma\mu\nu\rho}\gamma_\sigma\gamma^5.$$

2.12 Show by explicit construction that the Pauli-Dirac and chiral representations are related by a unitary transformation, i.e., that there exists a unitary matrix U such that

$$U\gamma^\mu_{PD}U^\dagger = \gamma^\mu_{ch}, \qquad U u_{PD}(\vec{p},s) = u_{ch}(\vec{p},s), \qquad U v_{PD}(\vec{p},s) = -v_{ch}(\vec{p},s).$$

(The extra sign in the v-spinor transformation is due to a sign convention.)

2.13 The angular momentum operator for a Dirac field can be written as

$$J^i = \int d^3\vec{x}\,\psi^\dagger(x)\frac{1}{4}\epsilon^{ijk}\sigma_{jk}\psi(x) + \text{orbital},$$

where normal ordering is implied. Show, in the free field limit, that J^3 has the expected behavior

$$J^3|\psi(\vec{p},s_{1,2})\rangle = \pm\frac{1}{2}|\psi(\vec{p},s_{1,2})\rangle, \qquad J^3|\psi^c(\vec{p},s_{1,2})\rangle = \pm\frac{1}{2}|\psi^c(\vec{p},s_{1,2})\rangle,$$

where ψ and ψ^c represent particle and antiparticle states, $\vec{p}$ is in the $\hat{z}$ direction (so that the orbital terms do not enter), and $s = s_1$ or s_2 represent spins in the $\pm\hat{z}$ direction.

2.14 Derive the results in (2.181) for the helicity projections in the massless limit.

2.15 Weak charged current transitions involve the chiral spinors $u_L(\vec{p},s)$ and $v_L(\vec{p},s)$ defined in (2.203). Show that in the relativistic limit such transitions mainly involve negative helicity particles or positive helicity antiparticles, and estimate the suppression factor for transitions involving the "wrong" helicity.

2.16 Show in two ways that $|\bar{u}(\vec{p}_2,+)u(\vec{p}_1,-)|^2 = 2p_1\cdot p_2$, where $u(\vec{p},\pm)$ are the helicity spinors for a massless fermion: (a) directly from the form of the spinors in the chiral representation, (b) using trace techniques.

2.17 Prove the Fierz identity in (2.216).

2.18 Suppose a fermion ψ of mass m interacts with a Hermitian scalar ϕ of mass μ with

$$\mathcal{L}_I = h\bar{\psi}\psi\phi,$$

where h is small.
(a) Calculate the spin-averaged differential cross section for $\psi(\vec{p}_1)\psi(\vec{p}_2) \to \psi(\vec{p}_3)\psi(\vec{p}_4)$ in the CM in terms of the invariants s, t, and u.
(b) Specialize to $m = \mu = 0$. Show that the scattering is isotropic in that limit and calculate the total cross section.

2.19 Consider $e^-(\vec{k}_1)\,\pi^+(\vec{p}_1) \to e^-(\vec{k}_2)\,\pi^+(\vec{p}_2)$ elastic scattering. Show that the spin-averaged differential cross section in the pion rest frame is

$$\frac{d\bar{\sigma}}{d\cos\theta_L} = \frac{\pi\alpha^2\cos^2\frac{\theta_L}{2}}{2k_1^2\sin^4\frac{\theta_L}{2}\left[1+\frac{2k_1}{m_\pi}\sin^2\frac{\theta_L}{2}\right]},$$

where θ_L is the electron scattering angle and we have neglected the electron mass. Hint: use (2.224).

2.20 Calculate the CM differential cross section for the process $e^-(\vec{p}_1)\mu^+(\vec{k}_1) \to e^-(\vec{p}_2)\mu^+(\vec{k}_2)$ in terms of $s = E^2_{CM}$, the CM scattering angle θ, and the muon mass m_μ. Neglect the electron mass.

2.21 Verify the expressions for Bhabha scattering in (2.234) and (2.235). Rewrite the final result in terms of s and $\cos\theta$.

2.22 Calculate the differential cross section for unpolarized Møller scattering, $e^-e^- \to e^-e^-$, both in terms of the invariants and θ.

2.23 Calculate the spin-average differential cross section $d\bar{\sigma}/d\cos\theta$ in the center of mass for $e^-(p_1)e^+(p_2) \to \pi^-(p_3)\pi^+(p_4)$, and the total cross section $\bar{\sigma}$. Neglect the electron mass but not the pion mass. Ignore strong interaction effects. The angular distribution should be proportional to $\sin^2\theta$. Interpret this result.

2.24 (a) Consider the Mott scattering process in which an electron of momentum $p = \beta E$ scatters from a static Coulomb potential of charge Ze,

$$A^\mu(x) = \frac{Ze}{4\pi|\vec{x}\,|}(1,0,0,0).$$

Show that the unpolarized (Mott) cross section for scattering angle θ is

$$\frac{d\bar{\sigma}}{d\cos\theta} = \frac{(Z\alpha)^2\pi(1-\beta^2\sin^2\frac{\theta}{2})}{2\beta^2p^2\sin^4\frac{\theta}{2}} \xrightarrow[\beta\ll 1]{} \frac{Z^2\alpha^2\pi}{2\beta^2p^2\sin^4\frac{\theta}{2}}.$$

The last formula is the Rutherford cross section.
(b) Suppose the Coulomb potential for an electron in a nuclear field transformed as a scalar rather than as the time component of a four-vector, i.e.,

$$\mathcal{H}_I = -e\bar{\psi}(x)\psi(x)\phi(x), \qquad \phi(x) = \frac{Ze}{4\pi|\vec{x}\,|}.$$

Calculate the unpolarized differential cross section, and compare it with the Mott formula.

2.25 The interaction of the Z (a massive neutral vector boson in the electroweak theory) with a fermion f is

$$\mathcal{L} = -G\bar{\psi}(x)\gamma^\mu \left(g_V - g_A\gamma^5\right)\psi(x)\, Z_\mu(x),$$

where G, g_V, and g_A are real constants. Calculate the width for $Z \to f\bar{f}$. Let M_Z and m be the Z and f masses, and set $G = 1$.

2.26 The Λ is a heavy spin-$\frac{1}{2}$ hyperon that decays into $p\pi^-$ via the non-leptonic weak interactions. The decay interaction can be modeled by

$$\mathcal{L}_I = \bar{\psi}_p(g_S - g_P\gamma^5)\psi_\Lambda\phi_{\pi^+} + h.c.,$$

where g_S and g_P are complex constants that lead, respectively, to S and P-wave final states.
(a) Calculate the width Γ and the differential width $d\Gamma/d\cos\theta$ in the Λ rest frame for a polarized Λ, where θ is the angle between $\hat{s}_\Lambda$ and the proton momentum $\vec{p}_p$. Use trace techniques.
(b) Show that $d\Gamma/d\cos\theta$ is not reflection invariant for $\Re e\,(g_P g_S^*) \neq 0$, i.e., that it is not invariant under $\vec{p}_p \to -\vec{p}_p$, $\hat{s}_\Lambda \to \hat{s}_\Lambda$.
(c) Repeat (a), but use explicit expressions for the Λ and p spinors in the Pauli-Dirac representation. Justify the claim that g_S and g_P generate S and P-wave amplitudes.

2.27 A vector resonance V_μ of mass M_V and width Γ_V couples to massless fermions a and b with the interaction in (F.12) of Appendix F. Calculate the total spin-averaged cross section for $a\bar{a} \to b\bar{b}$. Assume that the propagator in (2.129) is modified to the Breit-Wigner form

$$iD_V^{\mu\nu}(k) = i\left[\frac{-g^{\mu\nu} + \frac{k^\mu k^\nu}{M_V^2}}{k^2 - M_V^2 + iM_V\Gamma_V}\right],$$

and express the result in a form similar to (F.11).

2.28 Consider the interaction

$$\mathcal{L}_I = g\left(\bar{\psi}_{aL}\psi_{bR}\phi + \bar{\psi}_{bR}\psi_{aL}\phi^\dagger\right)$$

between distinct fermions ψ_a and ψ_b, where ϕ is a complex scalar and g is real. Show that the Lagrangian violates P and C, but is CP invariant.

2.29 Consider $e^-(\vec{k}_1)\,\pi^+(\vec{p}_1) \to e^-(\vec{k}_2)\,\pi^+(\vec{p}_2)$ scattering, as in Figure 2.15. Use the two-component formalism of Section 2.11 to calculate the amplitudes $M(-,-)$ and $M(+,+)$ for $m_e = 0$ and $m_\pi \neq 0$. Express your results in terms of α, β_π, and the CM scattering angle θ. (The two amplitudes should be equal up to a possible sign by (2.278).)

2.30 Consider the non-relativistic limit of the matrix element (2.356) for an e^- in a static external field.
(a) Compute the limit to linear order in the momenta. Hint: use the explicit forms for the spinors in the Pauli-Dirac representation. It simplifies the calculation to rewrite $\bar{u}_2\Gamma^\mu u_1$ using the Gordon decomposition.
(b) Suppose that $\Gamma^\mu(p_2, p_1)$ in (2.352) contained a term $\frac{i\sigma^{\mu\nu}}{2m}q_\nu\gamma^5 G_2(q^2)$. This violates P and T but in principle could be generated by a new interaction. Show how $G_2(0)$ is related to the electric dipole moment $\vec{d}_e$ of the electron, which is defined by the non-relativistic interaction $H_{EDM} = -\vec{d}_e \cdot \vec{E}(\vec{x})$, where $\vec{E}$ is an external electric field. Note that the Hermiticity condition (2.354) requires that G_2 is pure imaginary.

2.31 Suppose there is a small electric charge-violating coupling between the electron and a massless left-chiral neutrino ν_L, with

$$\mathcal{L}_{e\nu} = -\delta e A_\mu \bar{\psi}_{\nu_L} \gamma^\mu \psi_e + h.c.$$

Calculate the lifetime for $e^- \to \nu_L \gamma$, and find the value of δ corresponding to the limit in Table 2.3.

2.32 Consider the proton matrix element $\bar{u}(\vec{p}_2)\Gamma^\mu_Q(q)u(\vec{p}_1)$ of the electromagnetic current in (2.393), with Γ^μ_Q given by (2.397). Calculate this explicitly in the Breit frame, in which $q^0 = 0$, i.e.,

$$q = (0, 0, 0, \sqrt{Q^2}), \qquad p_1 = (E, 0, 0, -\sqrt{Q^2}/2), \qquad p_2 = (E, 0, 0, +\sqrt{Q^2}/2).$$

Express the time and space components in terms of the electric and magnetic form factors defined in (2.398) and interpret the results.

2.33 Let $V(\vec{x}_1 - \vec{x}_2)$ be the potential between two non-identical spin-$\frac{1}{2}$ particles in non-relativistic quantum mechanics (NRQM). (V may also depend on their spin and momentum operators). One shows in time-dependent perturbation theory that the transition amplitude U_{fi} from $|i\rangle = |\vec{p}_1 s_1, \vec{p}_2 s_2\rangle$ to $|f\rangle = |\vec{p}_3 s_3, \vec{p}_4 s_4\rangle$ with $m_1 = m_3, m_2 = m_4$ is

$$U_{fi} = -i(2\pi)^4 \delta^4(p_3 + p_4 - p_1 - p_2)\, \phi_3^\dagger \phi_4^\dagger \left(\int d^3\vec{r}\, e^{-i\vec{q}\cdot\vec{r}} V(\vec{r}) \right) \phi_1 \phi_2,$$

where ϕ_i is a two-component Pauli spinor, and V contains appropriate spin matrices. Note that these states are in our covariant normalization convention, which has an extra factor $\sqrt{2\pi)^3 2E_i} \sim \sqrt{2\pi)^3 2m_i}$ for the ith external particle compared to the usual conventions of NRQM. The corresponding formula in field theory is

$$U_{fi} = (2\pi)^4 \delta^4(p_3 + p_4 - p_1 - p_2)\, M,$$

where M is the scattering amplitude with the phase convention of Appendix B. Comparing these results, we can read off the equivalent non-relativistic potential corresponding to a given scattering amplitude. Specifically, for

$$\vec{p}_1 = -\vec{p}_2 = \vec{p} - \frac{\vec{q}}{2}, \qquad \vec{p}_3 = -\vec{p}_4 = \vec{p} + \frac{\vec{q}}{2},$$

the non-relativistic limit $\vec{p} \to 0, |\vec{q}|^2 \ll m_i^2$ yields

$$M \to -i(2m_1)(2m_2)\phi_3^\dagger \phi_4^\dagger \tilde{V}(\vec{q})\phi_1\phi_2 \equiv -i(2m_1)(2m_2)\phi_3^\dagger \phi_4^\dagger \int d^3\vec{r}\, e^{-i\vec{q}\cdot\vec{r}} V(\vec{r})\phi_1\phi_2.$$

Non-leading terms in $\vec{p}$ can be interpreted as the non-relativistic momentum operator.
(a) Calculate the potential corresponding to the effective four-fermi Hamiltonian density $\mathcal{H}_I = \lambda \bar{\psi}_1 \psi_1 \, \bar{\psi}_2 \psi_2$.
(b) Consider the interaction

$$\mathcal{L}_I = \left[g_1 \bar{\psi}_1 \psi_1 + g_2 \bar{\psi}_2 \psi_2 \right] \phi$$

between two fermions and a Hermitian scalar of mass m_ϕ. Calculate the potential between

ψ_1 and ψ_2 generated by t-channel ϕ exchange, and show that it is attractive for $g_1 g_2 > 0$.
(c) Calculate the potential generated by

$$\mathcal{L}_I = \left[g_1 \bar{\psi}_1 \gamma^\mu \psi_1 + g_2 \bar{\psi}_2 \gamma^\mu \psi_2 \right] V_\mu,$$

where V_μ is a spin-1 particle of mass M_V. Show that it is repulsive for $g_1 g_2 > 0$.
(d) Repeat parts (b) and (c) for the case $g_1 g_2 > 0$, but for the potential between antiparticle $\bar{1}$ and particle 2 and interpret the results. Hint: it is slightly easier to use the charge conjugation formalism of Section 2.10.
(e) Consider the interaction in (2.269) of a π^0 with protons and neutrons. Calculate the tree-level amplitude for $p(\vec{p}_1)\, n(\vec{p}_2) \to p(\vec{p}_3)\, n(\vec{p}_4)$ by t-channel π^0 exchange, and show that it leads to the non-relativistic potential

$$V(\vec{r}) = -\frac{g_\pi^2 m_\pi^3}{16\pi m_p^2} \left[\frac{1}{3} \vec{\sigma}_p \cdot \vec{\sigma}_n \frac{e^{-x}}{x} + \frac{S}{3} \left(\frac{1}{x} + \frac{3}{x^2} + \frac{3}{x^3} \right) e^{-x} \right],$$

where $m_p \sim m_n$, $\vec{\sigma}_p (\vec{\sigma}_n)$ are the Pauli matrices acting on the $p\,(n)$ spin, $x = m_\pi r$, and S is the tensor operator

$$S = 3\, \vec{\sigma}_p \cdot \hat{r}\, \vec{\sigma}_n \cdot \hat{r} - \vec{\sigma}_p \cdot \vec{\sigma}_n.$$

2.34 Suppose that a new lepton-flavor violating interaction leads to the effective interaction

$$\mathcal{L}_{eff} = -i \frac{e}{2} F^{\mu\nu} \bar{e} \sigma_{\mu\nu} \left[A + B\gamma^5 \right] \mu + h.c.,$$

where A and B are, respectively, magnetic and electric dipole transition moments. Calculate the decay rate for $\mu \to e\gamma$, neglecting m_e.

CHAPTER 3

Lie Groups, Lie Algebras, and Symmetries

DOI: 10.1201/b22175-3

Lie groups and algebras are used to describe continuous global and gauge symmetries in classical and quantum mechanics and in field theory. A familiar example is the description of rotational invariance in quantum mechanics. In particle physics Lie groups are useful not only for space-time symmetries such as translations, rotations, and Lorentz transformations, but also for internal symmetries such as isospin. It is assumed that the reader is familiar with such basic notions as irreducible representations (IRREPs), direct products, and irreducible tensor operators. Excellent introductions more detailed than the treatment here include (Georgi, 1999; Yndurain, 2007; Gilmore, 2005; Ramond, 2010; Barnes, 2010). Finite discrete groups are treated in detail in (Ramond, 2010), and more briefly in Sections 2.10 and 3.2.5.

3.1 BASIC CONCEPTS

3.1.1 Groups and Representations

A *group* G is a set of elements $g_1, g_2 \cdots$ that has

An associative multiplication law, under which $g_1 g_2 = g_3$ for each $g_{1,2} \in G$, with $g_3 \in G$ (closure) and $(g_1 g_2) g_3 = g_1 (g_2 g_3)$ (associative).

An identity element $I \in G$ with $Ig = gI = g$ for all $g \in G$.

A unique inverse element g^{-1} for each $g \in G$, such that $gg^{-1} = g^{-1}g = I$.

The elements may be discrete (with either a finite or countably infinite number) or may depend on a continuous parameter. An *abelian* (commutative) group is a special case, defined by $g_1 g_2 = g_2 g_1$ for all $g_{1,2} \in G$. Otherwise, G is *non-abelian*. A subset of the elements that itself forms a group under the same multiplication law is a *subgroup* of G.

(a) The set of integers n with the operation of ordinary addition is an example of an abelian group with a countable number of elements. The identity element is 0 (i.e., $0 + n = n + 0 = n$) and the inverse is $-n$.

(b) The set of rational numbers r other than 0 under ordinary multiplication forms another countable abelian group. The identity element is 1 and the inverse is $1/r$.

(c) The cyclic group Z_n consists of the nth roots of unity, i.e., $G = \{1, \omega_n, \omega_n^2 \cdots \omega_n^{n-1}\}$ where $\omega_n = e^{2\pi i/n}$. Z_n is abelian and finite.

(d) The quaternion group is finite and non-abelian. It consists of the eight 2×2 matrices $\{\pm I, \pm i\sigma^i\}$, where I is the 2×2 identity and σ^i, $i = 1, 2, 3$, are the Pauli matrices. Multiplication is defined by the ordinary matrix product in (1.18) on page 4, so, e.g., $(i\sigma^i)^{-1} = -i\sigma^i$. The subset $\{\pm I, \pm i\sigma^3\}$ forms an abelian subgroup.

(e) The symmetric group S_n is the group of permutations of n objects. It has $n!$ elements and is non-abelian for $n > 2$. The alternating group A_n is the subgroup of even permutations. It is non-abelian for $n > 3$ and has $n!/2$ elements.

(f) The set of non-singular $m \times m$ matrices A forms a non-abelian continuous group under ordinary matrix multiplication. The identity is the $m \times m$ identity matrix and the inverse is the matrix inverse A^{-1}.

A *Lie group* G is a continuous group for which the multiplication law involves differentiable functions of the parameters that label the group elements. Most of the Lie groups of interest in particle physics are *compact*, which means that the parameters form a compact manifold (the Lorentz group, described in Section 10.2.2, is a notable exception). A Lie group and its multiplication law can be defined, at least for elements close to the identity (*infinitesimal transformations*), in terms of its associated *Lie algebra*, which consists of N *generators* (operators) T^i, $i = 1, 2 \cdots N$, and their commutation rules

$$[T^i, T^j] = ic_{ijk} T^k, \tag{3.1}$$

where a summation on k is implied and the $c_{ijk} = -c_{jik}$ are the *structure constants* of G. Without loss of generality, one can choose the T^i to be Hermitian, in which case the structure constants are real. If all of the $c_{ijk} = 0$, then G is abelian; otherwise, it is non-abelian. An element of a compact G can be represented as a formal power series involving the generators, by the unitary operators

$$U_G(\vec{\beta}) = \exp[-i \sum_{i=1}^{N} \beta^i T^i] \equiv e^{-i\vec{\beta}\cdot\vec{T}} \equiv \sum_{k=0}^{\infty} \frac{(-i\vec{\beta}\cdot\vec{T})^k}{k!}, \tag{3.2}$$

where $\beta^1 \cdots \beta^N$ are N continuous real parameters and the T^i are Hermitian. In particular, the identity element is $U_G(0) = I$, and the inverse of $U_G(\vec{\beta})$ is

$$U_G(\vec{\beta})^{-1} = U_G(-\vec{\beta}) = e^{i\vec{\beta}\cdot\vec{T}} = U_G(\vec{\beta})^\dagger. \tag{3.3}$$

For small $|\vec{\beta}|$ it is sufficient to keep just the linear term in (3.2),

$$U_G(\vec{\beta}) \simeq I - i\vec{\beta}\cdot\vec{T} + O(\beta_i\beta_j), \tag{3.4}$$

i.e., the generators of the Lie algebra describe the group elements close to the identity. The Lie algebra also defines the group multiplication law for arbitrary $\vec{\beta}$. That is,

$$U_G(\vec{\alpha})U_G(\vec{\beta}) = e^{-i\vec{\alpha}\cdot\vec{T}} e^{-i\vec{\beta}\cdot\vec{T}} \equiv U_G(\vec{\gamma}). \tag{3.5}$$

$\vec{\gamma}(\vec{\alpha}, \vec{\beta})$ can in principle be expressed in terms of the Lie algebra (the *Baker-Campbell-Hausdorff* construction), although there is no closed form expression in general. However, for small $|\vec{\alpha}|$ and $|\vec{\beta}|$

$$\vec{\gamma}(\vec{\alpha}, \vec{\beta}) \cdot \vec{T} = (\vec{\alpha} + \vec{\beta}) \cdot \vec{T} - \frac{i}{2}[\vec{\alpha}\cdot\vec{T}, \vec{\beta}\cdot\vec{T}] + \text{h.o.t.} \tag{3.6}$$

Now consider a set of $n \times n$ dimensional matrices L^i, $i = 1, 2 \cdots N$. If the L^i satisfy the same algebra as the generators of a Lie algebra,

$$[L^i, L^j] = ic_{ijk}L^k, \tag{3.7}$$

then the L^i (sometimes written L^i_n) are said to form a *representation* of the algebra, and are Hermitian for the choice of Hermitian T^i. That is, we are considering the T^i to be abstract operators, while the L^i are a specific matrix realization. Similarly, the $n \times n$ matrices $e^{-i\vec{\beta}\cdot\vec{L}}$ form a representation of the group elements $U_G(\vec{\beta})$ and have the same multiplication law. For the compact groups one can choose Hermitian generators normalized such that

$$\mathrm{Tr}\,(L^iL^j) = T(L)\delta_{ij}. \tag{3.8}$$

The *Dynkin index* $T(L) > 0$ depends on which representation is being considered and on an overall normalization convention, but is independent of i and j. With these conventions, one can show that c_{ijk} is totally antisymmetric in all three indices (Problem 3.2).

3.1.2 Examples of Lie Groups

The simplest example of a Lie group is the abelian $G = U(1)$, with a single generator T, so that

$$U_G(\beta) = e^{-i\beta T}, \qquad U_G(\alpha)U_G(\beta) = e^{-i(\alpha+\beta)T}. \tag{3.9}$$

There is an $n = 1$ dimensional representation, with $L = 1$ and group elements $U_G(\beta) \to e^{-i\beta}$. $U(1)$ is named for this representation, i.e., the 1×1 dimensional unitary matrices (phase factors).

$G = SU(2)$ is a non-abelian group with $N = 3$ generators and $c_{ijk} = \epsilon_{ijk}$. $SU(2)$ is named for its *defining* representation, the 2×2 unitary matrices ($U(2)$) with the extra constraint that they are *special*, i.e., their determinant is unity ($SU(2)$). The generators of the defining representation are $L^i = \frac{\tau^i}{2}$, where $\tau^i \equiv \sigma^i$ are the Pauli matrices in (1.17), so their Dynkin index is $\frac{1}{2}$ by (1.18). The L^i are Hermitian, so the group representation elements

$$U(\vec{\beta}) \equiv e^{-i\vec{\beta}\cdot\frac{\vec{\tau}}{2}} = \cos\frac{\beta}{2}I - i\sin\frac{\beta}{2}\hat{\beta}\cdot\vec{\tau} \tag{3.10}$$

are unitary 2×2 matrices. Furthermore, $\mathrm{Tr}\,\tau^i = 0$, so they are special,

$$\det\left[e^{-i\vec{\beta}\cdot\frac{\vec{\tau}}{2}}\right] = e^{-i\mathrm{Tr}(\vec{\beta}\cdot\frac{\vec{\tau}}{2})} = e^{i0} = 1. \tag{3.11}$$

The *adjoint* representation of $SU(2)$ is the 3×3 representation constructed from the structure constants, $(L^i_{adj})_{jk} = -i\epsilon_{ijk}$. There are additional representations for $n = 4, 5 \cdots \infty$. $SU(2)$ is useful in nature for describing rotational invariance, the approximate isospin invariance of the strong interactions, and the weak isospin gauge symmetry of the electroweak interactions.

The group $SU(3)$ plays two major roles in the standard model: as a gauge symmetry associated with color for the strong interactions (QCD), and as an approximate global flavor symmetry of the strong interactions (the *eightfold way*). $SU(3)$ can be defined in terms of its defining representation, the 3×3 unitary matrices with determinant one. There are $N = 8$ generators, with (Hermitian) matrices in the defining relation given by $L^i_3 = \lambda^i/2$, where the Gell-Mann matrices $\lambda^i, i = 1 \cdots 8$, are given in Table 3.1. The structure constants $c_{ijk} = f_{ijk}$, $i, j, k = 1 \cdots 8$, are listed in Table 3.2. The Gell-Mann matrices satisfy

$$\mathrm{Tr}\left(\frac{\lambda^i}{2}\frac{\lambda^j}{2}\right) = \frac{1}{2}\,\delta^{ij}, \qquad \mathrm{Tr}\,\lambda^i = 0 \to \det\left(e^{-i\frac{\vec{\beta}\cdot\vec{\lambda}}{2}}\right) = 1. \tag{3.12}$$

There are two diagonal matrices, λ^3 and λ^8, i.e., $SU(3)$ has *rank* two. The λ^i satisfy the commutation (Lie algebra) and anticommutation rules.[1]

$$[\lambda^i, \lambda^j] = 2if_{ijk}\lambda^k, \qquad \{\lambda^i, \lambda^j\} = \frac{4}{3}\delta^{ij} I + 2d_{ijk}\lambda^k. \tag{3.13}$$

The d_{ijk} are symmetric in all 3 indices, with the nonzero ones listed in Table 3.2. $SU(3)$ has many other representations, such as a second inequivalent 3-dimensional *conjugate* representation $L^i_{3^*} = -\lambda^{i*}/2 = -\lambda^{iT}/2$, and the 8 dimensional adjoint $(L^i_{adj})_{jk} = -if_{ijk}$. There are several $SU(2)$ and $U(1)$ subgroups of $SU(3)$, such as the $SU(2)$ associated with $L^{1,2,3}$.

TABLE 3.1 The Gell-Mann matrices.[a]

$$\lambda^i = \begin{pmatrix} & \tau^i & 0 \\ & & 0 \\ 0 & 0 & 0 \end{pmatrix} \qquad \lambda^4 = \begin{pmatrix} 0 & 0 & 1 \\ 0 & 0 & 0 \\ 1 & 0 & 0 \end{pmatrix} \qquad \lambda^5 = \begin{pmatrix} 0 & 0 & -i \\ 0 & 0 & 0 \\ i & 0 & 0 \end{pmatrix}$$

$$\lambda^6 = \begin{pmatrix} 0 & 0 & 0 \\ 0 & 0 & 1 \\ 0 & 1 & 0 \end{pmatrix} \qquad \lambda^7 = \begin{pmatrix} 0 & 0 & 0 \\ 0 & 0 & -i \\ 0 & i & 0 \end{pmatrix} \qquad \lambda^8 = \frac{1}{\sqrt{3}}\begin{pmatrix} 1 & 0 & 0 \\ 0 & 1 & 0 \\ 0 & 0 & -2 \end{pmatrix}$$

[a] $i = 1, 2, 3$ in the first entry.

TABLE 3.2 The nonzero (totally antisymmetric) structure constants f_{ijk} for $SU(3)$, and the nonzero (totally symmetric) d_{ijk} defined by the anticommutators of the Gell-Mann matrices.

$f_{123} = 1$	$f_{345} = \frac{1}{2}$	$d_{118} = \frac{1}{\sqrt{3}}$	$d_{338} = \frac{1}{\sqrt{3}}$	$d_{558} = -\frac{1}{2\sqrt{3}}$
$f_{147} = \frac{1}{2}$	$f_{367} = -\frac{1}{2}$	$d_{146} = \frac{1}{2}$	$d_{344} = \frac{1}{2}$	$d_{668} = -\frac{1}{2\sqrt{3}}$
$f_{156} = -\frac{1}{2}$	$f_{458} = \frac{\sqrt{3}}{2}$	$d_{157} = \frac{1}{2}$	$d_{355} = \frac{1}{2}$	$d_{778} = -\frac{1}{2\sqrt{3}}$
$f_{246} = \frac{1}{2}$	$f_{678} = \frac{\sqrt{3}}{2}$	$d_{228} = \frac{1}{\sqrt{3}}$	$d_{366} = -\frac{1}{2}$	$d_{888} = -\frac{1}{\sqrt{3}}$
$f_{257} = \frac{1}{2}$		$d_{247} = -\frac{1}{2}$	$d_{377} = -\frac{1}{2}$	
		$d_{256} = \frac{1}{2}$	$d_{448} = -\frac{1}{2\sqrt{3}}$	

3.1.3 More on Representations and Groups

The *rank* of a Lie group is the number of generators that are simultaneously diagonalizable. The diagonal generators correspond to conserved quantum numbers if they commute with the Hamiltonian. $U(1), SU(2)$, and $SU(3)$ have rank 1, 1, and 2, respectively.

Two $n \times n$ representations L^i and L'^i of G are *equivalent* if all N of them are simultaneously related by a similarity transformation, i.e., if there exists an $n \times n$ unitary matrix U such that

$$L'^i = UL^iU^\dagger \quad \text{for} \quad i = 1 \cdots N. \tag{3.14}$$

Otherwise they are inequivalent.

[1] It is sometimes convenient to define $\lambda^0 \equiv \sqrt{2/3}\, I$. Then, $\mathrm{Tr}\,(\lambda^i\lambda^j) = 2\delta^{ij}$ and $\{\lambda^i, \lambda^j\} = 2d_{ijk}\lambda^k$, for $i, j, k = 0, 1 \cdots 8$, with $d_{0jk} = \sqrt{2/3}\delta_{jk}$.

A representation L^i is *reducible* if it is equivalent to a representation

$$L'^i = \begin{pmatrix} L'^i_A & 0 & 0 & 0 \\ 0 & L'^i_B & 0 & 0 \\ 0 & 0 & L'^i_C & 0 \\ 0 & 0 & 0 & \ddots \end{pmatrix} \tag{3.15}$$

in which each element is simultaneously block diagonal (with the same block dimensions). Otherwise, it is *irreducible* (an IRREP). States transforming according to a reducible representation separate into sectors not related by the symmetry, while all of the states in an IRREP are related. Simple Lie groups have an infinite number of IRREPs, and they frequently have inequivalent IRREPs of the same dimension.

A *fundamental representation* is, roughly speaking, a representation from which the others can be generated by direct products, in analogy to the way that any angular momentum j in quantum mechanics may be generated by combining $2j$ angular momenta $\frac{1}{2}$. The defining representations (m) of $SU(m)$, such as the 2 of $SU(2)$ in the example, are fundamentals.

The *adjoint* or *regular representation* of a Lie group is the $N \times N$ dimensional representation constructed from the structure constants,

$$\left(L^i_{adj}\right)_{jk} = -ic_{ijk}. \tag{3.16}$$

It is straightforward to show that L^i_{adj} satisfy (3.7) (Problem 3.3). The adjoint is essential for defining the self-interactions of the gauge fields in a non-abelian gauge theory.

If L^i_n is an n dimensional representation of a Lie algebra, then the *conjugate* $L^i_{n^*} \equiv -L^{i*}_n = -L^{iT}_n$ is also a representation. L_n is *real*[2] if it is equivalent to $L^i_{n^*}$, i.e., if there exists a unitary U such that $-L^{i*}_n = UL^i_nU^\dagger$ for $i = 1 \cdots N$. Otherwise, it is *complex*. The adjoint representation L^i_{adj} is always real, with $U = I$. The 2 of $SU(2)$ is real, i.e.,

$$L_{2^*} = -\frac{\tau^{i*}}{2} = \tau^2 \frac{\tau^i}{2} \tau^2, \tag{3.17}$$

so that $U = \tau^2$. The higher-dimensional $SU(2)$ representations are also real. On the other hand, the m of $SU(m)$ for $m > 2$ is *not* equivalent to the m^*, which is also a fundamental represention. For example, $L_{3^*} = -\frac{\lambda^{i*}}{2}$ in $SU(3)$ is not equivalent to $L_3 = \frac{\lambda^i}{2}$. This is important for the Higgs Yukawa couplings in extensions of the electroweak $SU(2)$ group to higher symmetries.

The Simple Lie Groups

Two groups G_1 and G_2 commute if $[g_i, \hat{g}_j] = 0$ for all $g_i \in G_1$, $\hat{g}_j \in G_2$. Then, one can define the *direct product group* $G = G_1 \times G_2$ with elements $g_i\hat{g}_j$, or direct products of more than two groups, such as the standard model group $SU(3) \times SU(2) \times U(1)$. A *simple* group is (non-rigorously) a non-abelian group such as $SU(3)$ that is *not* a direct product.[3] A *semi-simple* group is basically a direct product of simple groups, i.e., a Lie group with no $U(1)$ factors, such as $SU(3) \times SU(2)$.

[2]Mathematics books typically work in terms of iL, motivating the term "real".

[3]More precisely, a subgroup H of a Lie group G is an *invariant subgroup* if $ghg^{-1} \in H$ for all $g \in G, h \in H$. G is simple if it contains no invariant subgroups (other than the identity and G itself), and semi-simple if it contains no abelian invariant subgroups. Compact semi-simple Lie groups are either simple or the direct product of two or more simple groups.

Cartan has given a classification of the simple Lie algebras. The classification as well as the IRREPs and their properties are elegantly derived from *Dynkin diagrams* (Slansky, 1981), but here we only give the results. There are four countably infinite series of *classical* Lie algebras, and five *exceptional* algebras, as listed in Table 3.3. The four series correspond to simple matrix conditions for the *defining* representations of the associated groups:

$SU(m), m = 2 \cdots \infty$, correspond to the $m \times m$ complex unitary matrices $U_G = e^{-i\vec{\beta}\cdot\vec{L}}$ with unit determinant,

$$U_G U_G^\dagger = I, \qquad \det U_G = 1, \tag{3.18}$$

which implies that $\vec{\beta} \cdot \vec{L}$ are the traceless Hermitian matrices.[4] U_G leaves invariant the inner product of two m-dimensional complex vectors, i.e., $y^\dagger x = y'^\dagger x'$, where $x' = U_G^\dagger x$ and similarly for y'.

$SO(m)$ are the $m \times m$ real orthogonal matrices O_G with unit determinant[5] (i.e., rotations in an m-dimensional real space)

$$O_G O_G^T = I, \qquad \det O_G = 1, \tag{3.19}$$

so that $O_G = e^{-i\vec{\beta}\cdot\vec{L}}$ with $i\vec{\beta} \cdot \vec{L}$ real and antisymmetric. The inner product $y^T x$ of two real vectors is left invariant under an O_G transformation.

$Sp(2m)$ are the real $2m \times 2m$ symplectic matrices M, defined by

$$M^T S M = S, \tag{3.20}$$

where S is the skew symmetric matrix $S = \begin{pmatrix} 0_m & I_m \\ -I_m & 0_m \end{pmatrix}$, where 0_m and I_m are, respectively, the $m \times m$ zero and identity matrices. They therefore leave invariant the quadratic form $y^T S x$, where x and y are $2m$-dimensional real vectors.

The defining representations of $SU(m)$ and $Sp(2m)$ are also fundamental and can be used to generate the higher-dimensional IRREPS as direct products. For $SO(m)$ one can derive higher tensor representations (including the adjoint) from the defining or *vector* (m). However, there are additional double-valued fundamental *spinor* representations, similar to the familiar 2 of $SO(3) \sim SU(2)$ (see, e.g., Li, 1974; Slansky, 1981). All of the IRREPS can be generated as direct products of the fundamental spinor. $SU(m)$, $SO(m)$, and some of the exceptional groups have found considerable application in physics. Recently, $Sp(2m)$ has emerged in connection with string theory.

Casimir Invariants

A *Casimir invariant* is a function $f(T)$ of the group generators T^i that commutes with them, $[f(T), T^i] = 0$. By *Schur's lemma* the corresponding function $f(L)$ of an $n \times n$ IRREP L is a multiple of the identity. The coefficient may depend on the representation and may be used to label it. The simplest example is the *quadratic Casimir*

$$\vec{L}^2 = \sum_{i=1}^{N} L^i L^i \equiv C_2(L) I. \tag{3.21}$$

[4]The $U(m)$ group (which is not simple) is related by $U(m) = SU(m) \times U(1)$, where $U(1)$ is defined in (3.9) with T the $m \times m$ identity matrix.

[5]The orthogonal group $O(m)$ consists of the transformations O_G and RO_G, where $O_G \in SO(m)$ and R, which represents a reflection in an odd number of dimensions, is a diagonal matrix with elements ± 1 and $\det R = -1$.

TABLE 3.3 The Cartan classification of simple Lie algebras.[a]

Cartan label	Classical group	N	Range
A_ℓ	$SU(\ell+1)$	$\ell(\ell+2)$	$\ell \geq 1$
B_ℓ	$SO(2\ell+1)$	$\ell(2\ell+1)$	$\ell \geq 2$
C_ℓ	$Sp(2\ell)$	$\ell(2\ell+1)$	$\ell \geq 3$
D_ℓ	$SO(2\ell)$	$\ell(2\ell-1)$	$\ell \geq 4$
G_2		14	
F_4		52	
E_6		78	
E_7		133	
E_8		248	

[a]The groups $SO(6) \sim SU(4)$, $SO(4) \sim SU(2) \times SU(2)$, $SO(3) \sim SU(2) \sim Sp(2)$, $Sp(4) \sim SO(5)$, and $SO(2) \sim U(1)$ have the same Lie algebras but may differ for non-infinitesimal transformations. The subscript in the first column is the rank.

A familiar example is $\vec{J}^2 = j(j+1)I$ for the angular momentum j representation of the rotation group. $C_2(L)$ is related to the Dynkin index $T(L)$ defined in (3.8) by

$$T(L)N = C_2(L)n. \tag{3.22}$$

The quadratic Casimir of the adjoint $T(L_{adj}) = C_2(L_{adj})$ is also written as $C_2(G)$. From (3.16),

$$c_{ikl}c_{jkl} = C_2(G)\delta_{ij}. \tag{3.23}$$

The quadratic Casimirs and Dynkin indices for the defining and adjoint representations of the classical Lie algebras are given in Table 3.4 (see also van Ritbergen et al., 1999). Other useful identities (which can be used to construct other invariants) are

$$\begin{aligned} L^i L^j L^i &= \left[C_2(L) - \frac{1}{2}C_2(G)\right] L^j \\ c_{ijk}L^i L^j &= \frac{i}{2}C_2(G)L^k, \end{aligned} \tag{3.24}$$

from which $\mathrm{Tr}\, L^i = 0$ for the generators of a simple Lie group.

TABLE 3.4 Quadratic Casimirs and Dynkin indices for the defining representation L_n and adjoint representation of the classical Lie algebras.

G	N	$C_2(G)$	n	$T(L_n)$	$C_2(L_n)$
$SU(m)$	m^2-1	m	m	$\frac{1}{2}$	$\frac{m^2-1}{2m}$
$SO(m)$	$\frac{m(m-1)}{2}$	$2(m-2)$	m	2	$m-1$
$Sp(2m)$	$m(2m+1)$	$m+1$	$2m$	$\frac{1}{2}$	$\frac{2m+1}{4}$

More on $SU(m)$

Properties of the $SU(m)$ IRREPs and their direct products can be found systematically from the *Young tableaux* (e.g., Cheng and Li, 1984; Patrignani, 2016) or the more general

Dynkin methods. However, many aspects of $SU(m)$ are simple enough to "do it yourself." For example, the fundamental $L^i_m \equiv \lambda^i/2$ with $\text{Tr}(\lambda^i\lambda^j) = 2\delta^{ij}$ can be written as an obvious generalization of the 3×3 matrices in Table 3.1, and the structure constants can be calculated from them. An important property of $SU(m)$ (that does not generalize to the other simple groups) is that the L^i_m along with the identity form a complete set that can be used to write any $m\times m$ matrix,

$$M = \frac{1}{m}\text{Tr}(M)I + \frac{1}{2}\sum_{i=1}^{m^2-1}\text{Tr}(M\lambda^i)\lambda^i, \tag{3.25}$$

where the coefficients are generally complex. In particular, the anticommutator $\{\lambda^i,\lambda^j\}$ is a linear combination

$$\{\lambda^i,\lambda^j\} = \frac{4}{m}\delta^{ij}I + 2d_{ijk}\lambda^k. \tag{3.26}$$

where the totally symmetric d_{ijk} generalize those in Table 3.2. This allows one to generalize the $SU(2)$ Fierz identity in Problem 1.1 on page 5 to $SU(m)$,

$$(\chi_4^\dagger\vec{\lambda}\chi_3)\cdot(\chi_2^\dagger\vec{\lambda}\chi_1) = 2\eta_F(\chi_4^\dagger\chi_1)(\chi_2^\dagger\chi_3) - \frac{2}{m}(\chi_4^\dagger\chi_3)(\chi_2^\dagger\chi_1), \tag{3.27}$$

where $\eta_F = +1$ if the χ_i are m-dimensional complex vectors or complex scalar fields and $\eta_F = -1$ for anticommuting fermion fields.

$SU(m)$ tensor methods, discussed in Section 3.2.3, are especially useful for constructing $SU(m)$ singlets from direct products.

3.2 GLOBAL SYMMETRIES IN FIELD THEORY

3.2.1 Transformation of Fields and States

In field theory, groups consist of symmetry operations that leave the equations of motion unchanged in form. These may be discrete symmetries, such as P, C, T discussed in Section 2.10, or discrete internal symmetries, which will be considered in Section 3.2.5. Here we are more concerned with continuous groups. One important class is the *space-time symmetries*, such as space rotations, Lorentz boosts, and translations. Another, considered in this section, is *internal symmetries*, involving the interchanges of fields with similar properties, changes in their phase, etc. To formalize this, let $\Phi_a(x), a = 1, 2\cdots n$, be n fields (which may be spin-0, $\frac{1}{2}$, 1, etc.) related by a symmetry. Furthermore, consider a Lie group G of operators $U_G(\vec{\beta}) = e^{-i\vec{\beta}\cdot\vec{T}}$. Then, define a set of n transformed fields

$$\begin{aligned}\Phi'_a &= e^{-i\vec{\beta}\cdot\vec{T}}\Phi_a(x)e^{+i\vec{\beta}\cdot\vec{T}} \\ &= \Phi_a - i[\vec{\beta}\cdot\vec{T},\Phi_a] + \frac{(-i)^2}{2!}[\vec{\beta}\cdot\vec{T},[\vec{\beta}\cdot\vec{T},\Phi_a]] + \cdots,\end{aligned} \tag{3.28}$$

where the last form follows from the operator identity in Problem 3.5. Thus, the transformation of the fields is determined by their commutators with the group generators T^i. Since we assume that the Φ_a are transformed into each other, one must have that

$$[T^i,\Phi_a(x)] = -L^i_{ab}\Phi_b(x), \tag{3.29}$$

where L^i_{ab} are the components of an $n\times n$ matrix L^i, which is easily shown to form an $n\times n$ dimensional representation of the Lie algebra of G. From (3.28) and (3.29) one has that

$$\Phi'_a = (e^{+i\vec{\beta}\cdot\vec{L}})_{ab}\Phi_b \equiv U(\vec{\beta})_{ab}\Phi_b \xrightarrow[|\vec{\beta}|\text{ small}]{} \Phi_a + i\vec{\beta}\cdot\vec{L}_{ab}\Phi_b, \tag{3.30}$$

which defines how the n fields corresponding to representation L are transformed into each other. One usually considers the case that L is irreducible.

There is a frequently useful *matrix notation* for fields A^i transforming under the adjoint representation[6] (3.16), in which the A^i are reexpressed in terms of the elements of an $n \times n$ matrix

$$\mathcal{A} \equiv \sum_{i=1}^{N} A^i L^i \longleftrightarrow A^i = \frac{\text{Tr}\,(\mathcal{A}L^i)}{T(L)}, \tag{3.31}$$

where L^i is an arbitrary non-trivial IRREP of dimension n (usually taken to be the fundamental or defining). It is then easy to show (Problem 3.7) that the transformation of $A^i \to A'^i \equiv (e^{i\vec{\beta}\cdot\vec{L}_{adj}})_{ij} A^j$ can be expressed in terms of representation L by

$$\mathcal{A} \to \mathcal{A}' \equiv \sum_{i=1}^{N} A'^i L^i = e^{+i\vec{\beta}\cdot\vec{L}} \mathcal{A} e^{-i\vec{\beta}\cdot\vec{L}}, \qquad A'^i = \frac{\text{Tr}\,(\mathcal{A}'L^i)}{T(L)}. \tag{3.32}$$

As described in Chapter 2, if Φ_a corresponds to a particle, then the antiparticle field is given by or closely related to $\Phi_a^\dagger$. There are two possibilities for the transformations of non-Hermitian fields. One is that the fields for the particle and antiparticle are in the same IRREP, such as the pions

$$\Phi = \begin{pmatrix} \pi^+ \\ \pi^0 \\ \pi^- \end{pmatrix}, \tag{3.33}$$

which transform as a triplet under $SU(2)$ isospin. This requires that the representation is real, such as the adjoint in this example. Alternatively, the particle and antiparticle fields can be in different IRREPs, such as the kaons under isospin

$$\Phi = \begin{pmatrix} K^+ \\ K^0 \end{pmatrix} \quad \Phi^\dagger = \begin{pmatrix} K^- \\ \bar{K}^0 \end{pmatrix}. \tag{3.34}$$

Then, if Φ transforms under the n representation L_n^i, $\Phi^\dagger$ transforms under the conjugate representation $L_{n^*}^i = -L_n^{i*} = -L_n^{iT}$, which follow by taking the adjoint of (3.29) and using that T^i is Hermitian. Of course, L_n^i may be real, as in the $SU(2)$ example or for the adjoint of $SU(3)$.

From the expressions (2.93) or (2.159) on pages 23 and 35 one sees that for free fields the single particle state corresponding to Φ_a may be constructed by

$$|a\rangle = a^\dagger|0\rangle \sim \Phi_a^\dagger|0\rangle, \tag{3.35}$$

where in the second expression it is understood that a Fourier transformation and appropriate projections of Dirac spinors, etc., are to be performed. Thus, the states $|a\rangle$ and $\Phi_a^\dagger|0\rangle$ transform the same way under G. This continues to hold for interacting fields as long as the T^i commute with the Hamiltonian H. Then, the action of the generator on the state is

$$T^i|a\rangle \sim T^i\Phi_a^\dagger|0\rangle = \Phi_a^\dagger T^i|0\rangle + L_{ab}^{iT}\Phi_b^\dagger|0\rangle. \tag{3.36}$$

Assume for now that the ground state is invariant, i.e., $T^i|0\rangle = 0$. Then,

$$T^i|a\rangle = |b\rangle L_{ba}^i = |b\rangle\langle b|T^i|a\rangle, \tag{3.37}$$

so that the representation matrix $L_{ba}^i = \langle b|T^i|a\rangle$ is just the matrix element of T^i in the n-dimensional space of particles.

[6]The adjoint representation is real, so there is no distinction between upper and lower indices, i.e., $A^i = A_i$.

3.2.2 Invariance (Symmetry) and the Noether Theorem

The Lagrangian density $\mathcal{L}$ is *invariant* or *symmetric* under a group of transformations $U_G(\vec{\beta})$ if they commute, i.e., if

$$\mathcal{L}' \equiv U_G(\vec{\beta})\mathcal{L}U_G(\vec{\beta})^{-1} = \mathcal{L} \tag{3.38}$$

for all $\vec{\beta}$. (A similar definition applies for invariance under discrete transformations.) Since

$$\mathcal{L}' = \mathcal{L} - i\left[\vec{\beta}\cdot\vec{T}, \mathcal{L}\right] \tag{3.39}$$

for small $|\vec{\beta}|$, invariance holds if and only if

$$\left[T^i, \mathcal{L}\right] = 0 \tag{3.40}$$

for all i.

The first part of (3.38) defines the transformation of $\mathcal{L}$ whether or not there is an exact symmetry. Since $\mathcal{L}$ is a function of $(\Phi_a, \partial_\mu\Phi_a)$, and (for a non-Hermitian field) of $(\Phi_a^\dagger, \partial_\mu\Phi_a^\dagger)$ one has that

$$\mathcal{L}' = U_G(\vec{\beta})\mathcal{L}U_G(\vec{\beta})^{-1} = \mathcal{L}(\Phi_a', \partial_\mu\Phi_a', \Phi_a'^\dagger, \partial_\mu\Phi_a'^\dagger), \tag{3.41}$$

where Φ_a' is given in (3.30), with an analogous expression for $\partial_\mu\Phi_a'$ (since we are considering global transformations, $\vec{\beta} =$ constant). The expressions for $\Phi_a'^\dagger$ and $\partial_\mu\Phi_a'^\dagger$ are similar except that $L^i \to -L^{i*}$. It is frequently useful to consider *explicit symmetry breaking*, i.e.,

$$\delta\mathcal{L} \equiv \mathcal{L}' - \mathcal{L} \neq 0 \text{ (but small)}. \tag{3.42}$$

One can also have *spontaneous symmetry breaking*

$$\mathcal{L}' = \mathcal{L} \quad \text{but } T^i|0\rangle \neq 0, \tag{3.43}$$

i.e., the Lagrangian is invariant but the ground state breaks the symmetry (cf., the breaking of rotational invariance in a ferromagnet). Both of these cases will be considered extensively below, but for now consider an exact symmetry,

$$[T^i, \mathcal{L}] = 0 \text{ and } T^i|0\rangle = 0. \tag{3.44}$$

This implies degenerate multiplets of particles and definite relations between their interactions. It also implies conserved currents and charges according to the *Noether theorem*, which generalizes the result for a single complex scalar field discussed in Section 2.4.1. The Noether theorem for internal symmetries is

$$\partial_\mu J^{i\mu} = 0, \qquad \frac{d}{dt}Q^i = 0, \tag{3.45}$$

where the Noether current and charge are

$$J_\mu^i \equiv -i\frac{\delta\mathcal{L}}{\delta\partial^\mu\Phi_a}L_{ab}^i\Phi_b - i\frac{\delta\mathcal{L}}{\delta\partial^\mu\Phi_a^\dagger}(-L_{ab}^{i*})\Phi_b^\dagger, \tag{3.46}$$

$$Q^i = \int d^3\vec{x}J_0^i(t,\vec{x}). \tag{3.47}$$

Normal ordering on the fields is implied. The Noether theorem can be derived from the

Euler-Lagrange equations, in analogy with the derivation for the $U(1)$ case in Section 2.4.1. One can use the canonical commutation rules to show that

$$[Q^i, Q^j] = ic_{ijk}Q^k, \qquad \left[Q^i, \Phi_a(x)\right] = -L^i_{ab}\Phi_b(x). \tag{3.48}$$

That is, one can identify $Q^i = T^i$ as a concrete construction of the generators in terms of the fields.

The Noether currents are also useful for explicitly broken symmetries. The Noether charges are no longer time independent, but one can use the canonical commutation rules to show that the commutation rules in (3.48) still hold provided the charges and fields are evaluated at equal times. For example, suppose $\mathcal{L} = \mathcal{L}_0 + \mathcal{L}_1$, where only $\mathcal{L}_0$ is invariant,

$$\left[T^i, \mathcal{L}_0\right] = 0 \text{ (all } i), \qquad \left[T^i, \mathcal{L}_1\right] \neq 0 \text{ (some } i). \tag{3.49}$$

Then, the change in $\mathcal{L}$ is related to the divergence of $\vec{J}$,

$$\delta\mathcal{L} = \mathcal{L}' - \mathcal{L} = \left[-i\vec{\beta}\cdot\vec{T}, \mathcal{L}_1\right] = -\vec{\beta}\cdot\partial^\mu\vec{J}_\mu. \tag{3.50}$$

This immediately implies

$$\partial^\mu J^i_\mu = i\left[T^i, \mathcal{L}\right] \Rightarrow \int d^3\vec{x}\, \partial^\mu J^i_\mu = -i\left[T^i, H\right], \tag{3.51}$$

where the last step assumes that any symmetry breaking is in the mass and interaction terms (i.e., that kinetic energy terms are canonical). The (non-conserved) T^i and $\partial^\mu J^i_\mu$ are evaluated at the same t. Equation (3.51) implies that

$$\langle a(p_a)|T^i|b(p_b)\rangle = i(2\pi)^3\delta^3(\vec{p}_a - \vec{p}_b)\frac{\langle a(p_a)|\partial^\mu J^i_\mu|b(p_b)\rangle}{E_b - E_a}. \tag{3.52}$$

This relation is useful when states a and b are not related by the symmetry and are not degenerate in the symmetry limit. One then has that the leakage of $T^i|b\rangle$ into $|a\rangle$ is proportional to the symmetry breaking.

The Complex Scalar

As a first example, consider an IRREP relating n complex scalars $\phi_a, a = 1\cdots n$, transforming as $\left[T^i, \phi_a\right] = -L^i_{ab}\phi_b$ under some Lie algebra that will be determined from the symmetries of the Lagrangian density,

$$\mathcal{L} = \sum_{a=1}^{n} (\partial_\mu\phi_a)^\dagger(\partial^\mu\phi_a) + \text{ non-derivative terms.} \tag{3.53}$$

The Noether currents are

$$J^i_\mu = i\phi^\dagger_a L^i_{ab}\overleftrightarrow{\partial}_\mu \phi_b, \tag{3.54}$$

where $f\overleftrightarrow{\partial}_\mu g \equiv f(\partial_\mu g) - (\partial_\mu f)g$. The derivative (kinetic energy) terms are invariant under the group $U(n) = SU(n) \times U(1)$. Under an $SU(n)$ transformation

$$\phi \to e^{i\vec{\beta}\cdot\vec{L}}\phi, \qquad \phi^\dagger \to \phi^\dagger e^{-i\vec{\beta}\cdot\vec{L}}, \tag{3.55}$$

where ϕ is the n-component column vector $(\phi_1\ \phi_2 \cdots \phi_n)^T$, $\phi^\dagger$ is the row vector $(\phi_1^\dagger\ \phi_2^\dagger \cdots \phi_n^\dagger)$, and L^i is the fundamental representation matrix L_n^i of $SU(n)$. The $SU(n)$ invariance is obvious with this matrix notation

$$\mathcal{L}_{KE} \equiv (\partial_\mu \phi)^\dagger (\partial^\mu \phi) \to (\partial_\mu \phi)^\dagger\, e^{-i\vec{\beta}\cdot\vec{L}} e^{+i\vec{\beta}\cdot\vec{L}}\, (\partial^\mu \phi) = (\partial_\mu \phi)^\dagger\, (\partial^\mu \phi)\,. \tag{3.56}$$

$\mathcal{L}_{KE}$ is also invariant under $U(1)$ transformations, $\phi \to e^{i\beta I}\phi$. $U(n)$ is the maximal possible symmetry group of the system; depending on the mass and interaction terms the symmetry may be smaller.

Including mass terms

$$\mathcal{L} = (\partial_\mu \phi)^\dagger\, (\partial^\mu \phi) - \phi^\dagger \mu^2 \phi, \tag{3.57}$$

where $\phi^\dagger \mu^2 \phi = \phi_a^\dagger \mu_{ab}^2 \phi_b$ and μ^2 is an $n \times n$ matrix with elements μ_{ab}^2. The Hermiticity of $\mathcal{L}$ requires that μ^2 is Hermitian. The eigenvectors and eigenvalues of μ^2 correspond to states of definite mass and to their mass-squares, respectively. For now, however, let us assume that μ^2 is already diagonal, $\mu^2 = \text{diag}\left(\mu_1^2\, \mu_2^2 \cdots \mu_n^2\right)$. Under a group transformation,

$$\phi^\dagger \mu^2 \phi \to \phi^\dagger e^{-i\vec{\beta}\cdot\vec{L}} \mu^2 e^{+i\vec{\beta}\cdot\vec{L}} \phi, \tag{3.58}$$

so the requirement for invariance is that

$$e^{-i\vec{\beta}\cdot\vec{L}} \mu^2 e^{+i\vec{\beta}\cdot\vec{L}} = \mu^2 \tag{3.59}$$

for all $\vec{\beta}$, which is equivalent to

$$\left[\vec{\beta}\cdot\vec{L}, \mu^2\right] = 0. \tag{3.60}$$

Equation (3.60) determines what subgroup of $U(n)$ survives. For example, if all of the masses are the same, $\mu^2 = \mu_1^2 I$, then $\mathcal{L}$ is invariant under $U(n)$. If all the masses μ_a^2 are different, then the symmetry is reduced to

$$U(1)^n = U(1)_1 \times U(1)_2 \times \cdots \times U(1)_n, \tag{3.61}$$

where only ϕ_a transforms nontrivially under $U(1)_a$, i.e.,

$$\phi_a \to e^{i\beta_a}\phi_a, \qquad \phi_b \to \phi_b \text{ for } b \neq a. \tag{3.62}$$

For the intermediate case of

$$\mu^2 = \text{diag}\left(\underbrace{\mu_1^2 \cdots \mu_1^2}_{n_1}\ \underbrace{\mu_2^2 \cdots \mu_2^2}_{n_2} \right) \tag{3.63}$$

with n_1 fields of mass μ_1 and n_2 of mass μ_2, the symmetry group is $U(n_1) \times U(n_2)$, with the first (second) set of fields transforming under $U(n_1)$ ($U(n_2)$).

One can also add quartic interaction terms,

$$\mathcal{L} = (\partial_\mu \phi)^\dagger\, (\partial^\mu \phi) - \phi^\dagger \mu^2 \phi - \sum_{abcd} \lambda_{abcd} \phi_a^\dagger \phi_b\, \phi_c^\dagger \phi_d, \tag{3.64}$$

where $\lambda_{abcd}^* = \lambda_{badc}$ (from Hermiticity), and Bose symmetry projects out the parts of λ_{abcd} that are symmetric in ac and in bd. In general,[7] this reduces the symmetry to $U(1)$, but there

[7]Even the $U(1)$ would be broken in the presence of terms like $\phi^4 + \phi^{\dagger 4}$ or $(\phi + \phi^\dagger)\phi^\dagger\phi$.

could be a higher symmetry for specific λ's. For example, full $U(n)$ invariance is restored for

$$\begin{aligned} \lambda_{abcd} &= \lambda\delta_{ab}\delta_{cd}, \qquad \mu^2 = \mu_1^2 I \\ \Rightarrow \mathcal{L} &= (\partial_\mu\phi)^\dagger (\partial^\mu\phi) - \mu_1^2\phi^\dagger\phi - \lambda\left(\phi^\dagger\phi\right)^2. \end{aligned} \tag{3.65}$$

The Hermitian Scalar

Hermitian scalars, $\phi_a = \phi_a^\dagger$, transform as

$$\left[T^i, \phi_a\right] = -L^i_{ab}\phi_b. \tag{3.66}$$

Consistency requires that $L^i = -L^{i*}$, i.e., the representation is real. The Lagrangian density

$$\mathcal{L} = \frac{1}{2}\left(\partial_\mu\phi_a\right)\left(\partial^\mu\phi_a\right) + \text{ non-derivative terms} \tag{3.67}$$

implies Noether currents

$$J^i_\mu = -i\left(\partial_\mu\phi_a\right) L^i_{ab}\phi_b \tag{3.68}$$

(the second term in (3.46) is absent for Hermitian fields). The special case

$$\mathcal{L} = \frac{1}{2}\left[\left(\partial_\mu\phi_a\right)^2 - \mu^2\phi_a\phi_a\right] - \lambda\left(\phi_a\phi_a\right)^2 \tag{3.69}$$

is $O(n)$ invariant (Problem 3.8). However,

$$\begin{aligned} \mathcal{L} =& \frac{1}{2}\left(\partial_\mu\phi_a\right)^2 - \frac{1}{2}\phi_a\mu^2_{ab}\phi_b \\ &- \kappa_{abc}\phi_a\phi_b\phi_c - \lambda_{abcd}\phi_a\phi_b\phi_c\phi_d \end{aligned} \tag{3.70}$$

has no symmetries at all in general.

Complex Scalar in a Hermitian Basis

A complex scalar field ϕ can always be written in terms of two Hermitian scalars as in (2.91) on page 23, $\phi = (\phi_R + i\phi_I)/\sqrt{2}$, where $\phi_{R,I}$ are Hermitian. For complex fields that are in the same representation as their adjoints, such as in (3.33), it is almost always simpler to rewrite the theory in terms of Hermitian fields. In the case such as (3.34) that the complex fields and their adjoints transform separately, it is still sometimes useful to go to a Hermitian basis (especially for formal manipulations), although the complex basis is usually simpler for explicit calculations. The relation between the bases is straightforward, but can be confusing.

Suppose ϕ is an n-component complex field transforming with representation matrices L^i_ϕ. It is then convenient to introduce the complex $2n$-component vector $\Phi = \begin{pmatrix} \phi \\ \phi^\dagger \end{pmatrix}$, with half of the components redundant. Φ transforms under the reducible representation

$$L^i_\Phi = \begin{pmatrix} L^i_\phi & 0 \\ 0 & -L^{i*}_\phi \end{pmatrix}. \tag{3.71}$$

One can introduce $2n$ Hermitian fields ϕ_{aR}, ϕ_{aI}, by $\phi_a = \frac{1}{\sqrt{2}}(\phi_{aR} + i\phi_{aI})$, and the $2n$-component real vector $\phi_h = \begin{pmatrix} \phi_R \\ \phi_I \end{pmatrix}$. The two bases are related by the unitary transformation

$$\Phi = A\phi_h, \qquad A = \frac{1}{\sqrt{2}} \begin{pmatrix} I & iI \\ I & -iI \end{pmatrix}, \tag{3.72}$$

where I is the $n \times n$ identity. Hence, the representation matrices for the symmetry generators in the Hermitian basis are

$$L^i_{\phi_h} = A^\dagger L^i_\Phi A = \frac{1}{2} \begin{pmatrix} L^i_\phi - L^{i*}_\phi & i(L^i_\phi + L^{i*}_\phi) \\ -i(L^i_\phi + L^{i*}_\phi) & L^i_\phi - L^{i*}_\phi \end{pmatrix}, \tag{3.73}$$

which are manifestly imaginary and antisymmetric for Hermitian L^i_ϕ.

Examples: (a) Consider the $U(1)$ group acting on a single complex $\phi \to \exp(+i\beta)\phi$, so that $L_\phi = 1$. A and the representation matrices are therefore

$$A = \frac{1}{\sqrt{2}} \begin{pmatrix} 1 & i \\ 1 & -i \end{pmatrix}, \qquad L_\Phi = \begin{pmatrix} 1 & 0 \\ 0 & -1 \end{pmatrix}, \qquad L_{\phi_h} = \begin{pmatrix} 0 & i \\ -i & 0 \end{pmatrix}, \tag{3.74}$$

while a finite $U(1)$ transformation is just a rotation

$$\begin{pmatrix} \phi_R \\ \phi_I \end{pmatrix} \to e^{-i\beta\tau^2} \begin{pmatrix} \phi_R \\ \phi_I \end{pmatrix} = \begin{pmatrix} \cos\beta & -\sin\beta \\ \sin\beta & \cos\beta \end{pmatrix} \begin{pmatrix} \phi_R \\ \phi_I \end{pmatrix}. \tag{3.75}$$

(b) Consider two complex fields $\phi = \begin{pmatrix} \phi^+ \\ \phi^0 \end{pmatrix}$ transforming as a doublet under $SU(2)$, $L^i = \tau^i/2$. (The superscripts look ahead to applications to the Higgs.) In the (reducible) 4-dimensional Hermitian basis $\phi_h = (\phi_1\ \phi_3\ \phi_2\ \phi_4)^T$, where $\phi^+ = (\phi_1 + i\phi_2)/\sqrt{2}$ and $\phi^0 = (\phi_3 + i\phi_4)/\sqrt{2}$, the representation matrices are

$$L^{1,3}_{\phi_h} = \frac{i}{2} \begin{pmatrix} 0 & \tau^{1,3} \\ -\tau^{1,3} & 0 \end{pmatrix}, \quad L^2_{\phi_h} = \frac{1}{2} \begin{pmatrix} \tau^2 & 0 \\ 0 & \tau^2 \end{pmatrix}. \tag{3.76}$$

Fermions

Now consider the case of n fermions ψ_a, with

$$\mathcal{L} = \bar{\psi}_a i \partial\!\!\!/ \psi_a - \bar{\psi}_a m_{ab} \psi_b = \bar{\psi} i \partial\!\!\!/ \psi - \bar{\psi} m \psi. \tag{3.77}$$

In the second form, ψ is the n-component column vector $(\psi_1\ \psi_2 \cdots \psi_n)^T$ and $m = m^\dagger$ is an $n \times n$ Hermitian matrix[8] that can be taken to be diagonal. ψ and $\psi^\dagger$ transform as

$$[T^i, \psi_a] = -L^i_{ab}\psi_b, \qquad [T^i, \psi^\dagger_a] = +L^{i*}_{ab}\psi^\dagger_b \tag{3.78}$$

under a symmetry transformation, and the corresponding Noether current is

$$J^i_\mu = \bar{\psi}_a \gamma_\mu L^i_{ab} \psi_b. \tag{3.79}$$

[8]Generalized fermion mass terms involving γ^5 or, equivalently, non-Hermitian matrices, are considered in Problem 3.32 and Chapter 8.

As in the scalar case, the symmetry group is determined by $\mathcal{L}$. One has

$$\mathcal{L} \to \mathcal{L}' = \bar{\psi} i \not{\partial} e^{-i\vec{\beta}\cdot\vec{L}} e^{+i\vec{\beta}\cdot\vec{L}} \psi - \bar{\psi} e^{-i\vec{\beta}\cdot\vec{L}} m e^{+i\vec{\beta}\cdot\vec{L}} \psi. \tag{3.80}$$

The kinetic term is invariant under $U(n) = SU(n) \times U(1)$, but the invariance condition for the mass term is

$$e^{-i\vec{\beta}\cdot\vec{L}} m e^{+i\vec{\beta}\cdot\vec{L}} = m \leftrightarrow \left[\vec{\beta}\cdot\vec{L}, m\right] = 0. \tag{3.81}$$

The full $U(n)$ is maintained for n degenerate masses, $m = m_1 I$, while the symmetry is reduced to $U(1)^n$ for n distinct masses.

3.2.3 Isospin and $SU(3)$ Symmetries

$SU(2)$ Isospin

Isospin is an approximate symmetry of the strong interactions, broken by ~1%. The breaking is ultimately due to the $u - d$ quark mass differences. This is usually viewed as intrinsic to the strong interactions when discussing QCD, though the masses are actually associated with the electroweak sector. There is a comparable breaking from electromagnetism. We first describe a simple model of isospin in terms of the nucleons and pions. Introduce the nucleon (proton and neutron) and pion fields

$$\psi = \begin{pmatrix} \psi_p \\ \psi_n \end{pmatrix}, \qquad \pi = \begin{pmatrix} \pi_1 \\ \pi_2 \\ \pi_3 \end{pmatrix}, \qquad \pi^\pm = \frac{\pi_1 \mp i\pi_2}{\sqrt{2}}, \qquad \pi^0 = \pi_3, \tag{3.82}$$

where $\pi_i = \pi_i^\dagger$ and $\pi^\pm$ annihilate the states $|\pi^\pm\rangle$. ψ and π transform as a doublet (fundamental) and triplet (adjoint), respectively,[9] under $SU(2)$,

$$L_\psi^i = \frac{\tau^i}{2}, \qquad \left(L_\pi^i\right)_{jk} = -i\epsilon_{ijk}. \tag{3.83}$$

The diagonal generator is T^3 (L_π^3 is diagonal in the $\pi^{\pm,0}$ basis). Consider the Lagrangian density

$$\begin{aligned}\mathcal{L}_0 &= \bar{\psi}\left(i\not{\partial} - mI + ig_\pi\gamma^5\vec{\pi}\cdot\vec{\tau}\right)\psi + \frac{1}{2}\left[(\partial_\mu\vec{\pi})^2 - \mu^2\vec{\pi}^{\,2}\right] - \lambda\left(\vec{\pi}^{\,2}\right)^2 \\ &= \bar{\psi}\left(i\not{\partial} - mI + i\sqrt{2}g_\pi\gamma^5\Pi\right)\psi + \frac{1}{2}\mathrm{Tr}\left[(\partial_\mu\Pi)^2 - \mu^2\Pi^2\right] - \lambda\left(\mathrm{Tr}\,\Pi^2\right)^2,\end{aligned} \tag{3.84}$$

with $\vec{\pi}^{\,2} \equiv \sum_{i=1}^3 \pi_i^2$ and $\Pi \equiv \frac{\vec{\pi}\cdot\vec{\tau}}{\sqrt{2}} = \begin{pmatrix} \frac{\pi^0}{\sqrt{2}} & \pi^+ \\ \pi^- & \frac{-\pi^0}{\sqrt{2}} \end{pmatrix}$. The g_π term is the *Yukawa interaction* between the pion and nucleon.[10] The second (matrix) form in (3.84) makes it especially easy to read off the Feynman rules for the $\bar{p}p\pi^0$, $\bar{n}n\pi^0$, $\bar{n}p\pi^-$, and $\bar{p}n\pi^+$ vertices, namely, $-g_\pi\gamma^5$, $+g_\pi\gamma^5$, $-\sqrt{2}g_\pi\gamma^5$, and $-\sqrt{2}g_\pi\gamma^5$, respectively.

We should emphasize that experimentally g_π is very large: the experimental πN coupling

[9] The convention for the $\pi^\pm$ fields in (3.82) is common in particle physics, and convenient because $\pi^+ = (\pi^-)^\dagger$. However, the Condon-Shortley phase conventions usually employed for states in rotational multiplets in quantum mechanics (and in standard Clebsch-Gordan coefficient tables) would instead require the convention $\pi^\pm = \mp(\pi_1 \mp i\pi_2)/\sqrt{2}$ (Problem 3.12).

[10] We use the term Yukawa interaction in a generalized sense, i.e., for any 3-point interaction between a spin-0 and spin-$\frac{1}{2}$ particles.

G_π observed from low energy πN and NN interactions, which may differ from g_π by strong interaction corrections, is $\sim 13.05(8)$ (e.g., Gorringe and Fearing, 2004). Therefore, (3.84) should be viewed as a model to illustrate symmetry considerations and not as a serious perturbative field theory for the strong interactions. $\mathcal{L}_0$ is $SU(2)$ and reflection invariant (the γ^5 is because the pions are pseudoscalar); using (3.32),

$$\psi \to e^{+i\vec{\beta}\cdot\frac{\vec{\tau}}{2}}\psi, \qquad \bar{\psi} \to \bar{\psi}e^{-i\vec{\beta}\cdot\frac{\vec{\tau}}{2}}, \qquad \Pi \to e^{i\vec{\beta}\cdot\frac{\vec{\tau}}{2}}\,\Pi e^{-i\vec{\beta}\cdot\frac{\vec{\tau}}{2}}, \tag{3.85}$$

from which both $\bar{\psi}\gamma^5\Pi\psi$ and $\mathrm{Tr}\,\Pi^2$ are invariant under $SU(2)$. The invariance of the pion self-interaction is further discussed in Problem 3.15.

Now, introduce $SU(2)$ breaking by $\mathcal{L} = \mathcal{L}_0 + \mathcal{L}_1$, where

$$\mathcal{L}_1 = -\epsilon\bar{\psi}\tau_3\psi = -\epsilon\left(\bar{\psi}_p\psi_p - \bar{\psi}_n\psi_n\right). \tag{3.86}$$

$\mathcal{L}_1$ thus represents a splitting between the proton and neutron masses,

$$m_p = m + \epsilon \quad m_n = m - \epsilon, \tag{3.87}$$

which parametrizes contributions both from the quark masses and from electromagnetism.[11] It is straightforward to show that

$$\left[T^3, \mathcal{L}_1\right] = 0, \qquad \left[T^{1,2}, \mathcal{L}_1\right] \neq 0. \tag{3.88}$$

Thus, $SU(2)$ is broken to $U(1)_{T^3}$, corresponding to a conserved charge T^3 with values $t^3_{\pi^\pm} = \pm 1, t^3_{\pi^0} = 0$, and $t^3_p = -t^3_n = 1/2$. Actually, $\mathcal{L}$ is invariant under $U(1)_{T^3} \times U(1)_B$, where the second $U(1)$ corresponds to a conserved fermion (or in this case, baryon) number $B_{p,n} = 1, B_\pi = 0$. The conservation of B and T^3 is equivalent to that of B and electric charge Q, where $Q = T^3 + B/2$ when restricted to the fields in this example. $\mathcal{L}_1$ transforms as the $T = 1, T^3 = 0$ component of an irreducible tensor operator. One can therefore use the Wigner-Eckart theorem for relations between its matrix elements, in exact analogy to broken rotational invariance in quantum mechanics.

SU(3) Symmetry

$SU(3)$ is an approximate global symmetry of the strong interactions that extends the $SU(2)$ isospin subgroup. It is valid at the $\sim 25\%$ level for masses, but works better for relations between couplings and matrix elements. $SU(3)$ was proposed independently by M. Gell-Mann and Y. Ne'eman in the early 1960s to account for the fact that the low lying mesons and baryons could be associated in octets (the eightfold way). As described in Section 3.1.2, $SU(3)$ has eight generators. The fundamental representation matrices $L^i_3 = \lambda^i/2$ and structure constants are listed in Tables 3.1 and 3.2. $SU(3)$ has rank 2, so two generators, T^3 and T^8, can be simultaneously diagonalized with the Hamiltonian. Empirically, these correspond to the strong interaction quantum numbers by the *Gell-Mann-Nishijima relation*

$$Q = I_3 + \frac{\mathcal{Y}}{2}, \qquad I_3 = T^3, \qquad \mathcal{Y} = \frac{2}{\sqrt{3}}T^8, \tag{3.89}$$

where Q is electric charge, I_3 is the third component of isospin, $\mathcal{Y} = B + S$ is strong hypercharge, B is baryon number, and S is strangeness ($S = 0$ for the pions and nucleons).

[11]For simplicity we ignore isospin breaking terms in the interactions or from $\mu^2_{\pi^\pm} - \mu^2_{\pi^0}$. The latter are easily shown to come only from electromagnetism to leading order, and not from the quark mass differences.

The low-dimensional representations of $SU(3)$ are the $1, 3, 3^*, 6, 6^*, 8, 10, 10^*$, and 27, where the 1, 8 (adjoint), and 27 are real and the others complex. The observed light hadrons can be assigned to the 1, 8, 10, and 10^*. It is convenient to display the states on *weight diagrams*, with the axes corresponding to I_3 and $\mathcal{Y}$. Then, the other generators of the Lie algebra describe transitions from one state to another. For example, the lowest lying $J^P = \frac{1}{2}^+$ *baryons*[12] and $J^P = 0^-$ *mesons* (J is the spin and P is the intrinsic parity) both transform under the adjoint (octet) representation, as shown in Figure 3.1. In the absence of $SU(3)$ breaking, the states in each octet would be degenerate. Each consists of two isospin doublets with $\mathcal{Y} = \pm 1$, and one isotriplet and one isosinglet with $\mathcal{Y} = 0$. The baryon fields, which annihilate states with $B = 1$ and strangeness $S = \mathcal{Y} - B$, are given by

$$\begin{aligned}
\Sigma^{\pm} &= \frac{1}{\sqrt{2}} (\psi_1 \mp i\psi_2), & \Sigma^0 &= \psi_3, \qquad \Lambda = \psi_8 \\
p &= \frac{1}{\sqrt{2}} (\psi_4 - i\psi_5), & n &= \frac{1}{\sqrt{2}} (\psi_6 - i\psi_7) \\
\Xi^0 &= \frac{1}{\sqrt{2}} (\psi_6 + i\psi_7), & \Xi^- &= \frac{1}{\sqrt{2}} (\psi_4 + i\psi_5),
\end{aligned} \tag{3.90}$$

while the pseudoscalar meson ($B = 0$) fields are

$$\begin{aligned}
\pi^{\pm} &= \frac{1}{\sqrt{2}} (\phi_1 \mp i\phi_2), & \pi^0 &= \phi_3, \qquad \eta = \phi_8 \\
K^{\pm} &= \frac{1}{\sqrt{2}} (\phi_4 \mp i\phi_5), & K^0\,(\bar{K}^0) &= \frac{1}{\sqrt{2}} (\phi_6 \mp i\phi_7).
\end{aligned} \tag{3.91}$$

The η', which also has $I_3 = \mathcal{Y} = 0$, is an $SU(3)$ singlet. The π, K, η, η' system is referred to as a *nonet* $= 8 + 1$.

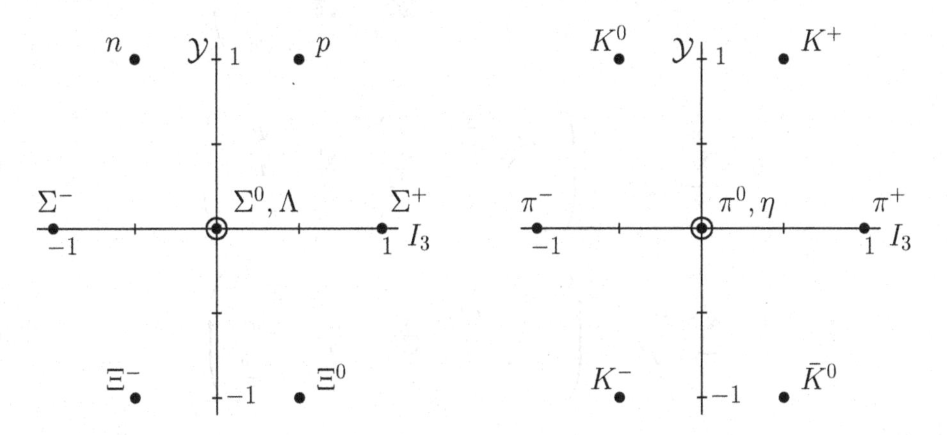

Figure 3.1 Weight diagrams for the $J^P = \frac{1}{2}^+$ baryon and $J^P = 0^-$meson octets. The anti-baryons are in a separate octet, while the mesons and their antiparticles are in the same octet.

Analogous to (3.82) one can define 8-component vectors ψ and ϕ, so that $\psi \rightarrow$

[12]A baryon (meson) is a hadron with half-integer (integer) spin. A *hyperon* is a baryon with nonzero strangeness (but no heavy quantum numbers such as charm).

$\exp(i\vec{\beta}\cdot\vec{L}_{adj})\psi$, $\phi\to\exp(i\vec{\beta}\cdot\vec{L}_{adj})\phi$ under an $SU(3)$ transformation. The extension of (3.84) to the baryon and pseudoscalar meson octets is

$$\begin{aligned}\mathcal{L}_0 =&\bar{\psi}_i\,(i\,\partial\!\!\!/ - m)\,\psi_i + \frac{1}{2}\left[(\partial_\mu\phi_j)^2 - \mu^2\phi_j^2\right] + ig_f\, f_{ijk}\,\bar{\psi}_i i\gamma^5\psi_j\phi_k \\ &+ g_d\, d_{ijk}\,\bar{\psi}_i i\gamma^5\psi_j\phi_k - \alpha(\phi_j^2)^2 - \beta\, d_{ijk}\, d_{imn}\,\phi_j\phi_k\phi_m\phi_n,\end{aligned} \tag{3.92}$$

where $\phi_j^2\equiv\phi_j\phi_j$ and f_{ijk} and d_{ijk} are the coefficients given in Table 3.2. There are two independent $SU(3)$-invariant meson-baryon interactions, known as the "F" and "D" couplings, which are, respectively, antisymmetric and symmetric in the $SU(3)$ indices. This can be understood as follows. An invariant coupling is just a singlet component of the direct product of the fields in the interaction term. For $SU(2)$ the meson-baryon interaction in (3.84) involves the direct product of two doublets and one triplet, $(2\times 2)\times 3 = (1+3)\times 3 = 3+[1+3+5]$, where the IRREPs are labeled by their dimensionality $2I+1$. The singlet only occurs once in the decomposition, so there is only one invariant. For $SU(3)$, however, there are two independent ways to form an invariant from $8\times 8\times 8$,

$$\begin{aligned}8\times 8 &= 1+8+8+10+10^*+27\\ 8\times(8\times 8) &= 8\times 1 + \underbrace{(8\times 8)}_{1+\cdots} + \underbrace{(8\times 8)}_{1+\cdots} + 8\times[10+10^*+27].\end{aligned} \tag{3.93}$$

Similarly, there are now two invariant meson self-interaction terms, associated with the singlet and symmetric octet components of 8×8.

$\mathcal{L}_0$ can be written in matrix notation, using

$$M\equiv\sum_{i=1}^{8}\frac{\lambda^i\phi_i}{\sqrt{2}} = \begin{pmatrix} \frac{\pi^0}{\sqrt{2}}+\frac{\eta}{\sqrt{6}} & \pi^+ & K^+ \\ \pi^- & \frac{-\pi^0}{\sqrt{2}}+\frac{\eta}{\sqrt{6}} & K^0 \\ K^- & \bar{K}^0 & \frac{-2\eta}{\sqrt{6}} \end{pmatrix}, \tag{3.94}$$

as well as

$$B\equiv\sum_{i=1}^{8}\frac{\lambda^i\psi_i}{\sqrt{2}} = \begin{pmatrix} \frac{\Sigma^0}{\sqrt{2}}+\frac{\Lambda^0}{\sqrt{6}} & \Sigma^+ & p \\ \Sigma^- & \frac{-\Sigma^0}{\sqrt{2}}+\frac{\Lambda^0}{\sqrt{6}} & n \\ \Xi^- & \Xi^0 & \frac{-2\Lambda^0}{\sqrt{6}} \end{pmatrix} \tag{3.95}$$

and

$$\bar{B}\equiv\sum_{i=1}^{8}\frac{\lambda^i\bar{\psi}_i}{\sqrt{2}} = \begin{pmatrix} \frac{\bar{\Sigma}^0}{\sqrt{2}}+\frac{\bar{\Lambda}^0}{\sqrt{6}} & \bar{\Sigma}^- & \bar{\Xi}^- \\ \bar{\Sigma}^+ & \frac{-\bar{\Sigma}^0}{\sqrt{2}}+\frac{\bar{\Lambda}^0}{\sqrt{6}} & \bar{\Xi}^0 \\ \bar{p} & \bar{n} & \frac{-2\bar{\Lambda}^0}{\sqrt{6}} \end{pmatrix}. \tag{3.96}$$

Therefore,

$$\begin{aligned}\mathcal{L}_0 =&\text{Tr}\left[\bar{B}\,(i\,\partial\!\!\!/ - m)\,B\right] + \frac{1}{2}\text{Tr}\left[(\partial_\mu M)^2 - \mu^2 M^2\right] + \frac{g_f}{\sqrt{2}}\text{Tr}\left(\bar{B}i\gamma^5\,[B,M]\right)\\ &+\frac{g_d}{\sqrt{2}}\text{Tr}\left(\bar{B}i\gamma^5\,\{B,M\}\right) - (\alpha-\frac{2}{3}\beta)(\text{Tr}\,M^2)^2 - 2\beta\,\text{Tr}\,M^4,\end{aligned} \tag{3.97}$$

which is invariant from (3.32).

$SU(3)$ *Breaking*

The degeneracy of the $SU(3)$ multiplets is only good to around 25%, though the predictions for couplings and amplitudes are typically much better. The *Gell-Mann-Okubo* (GMO) ansatz is that the breaking can be described by an operator that transforms as an octet, i.e.,

$$\mathcal{L} = \mathcal{L}_0 + \epsilon_8 \mathcal{L}_8, \tag{3.98}$$

where $\mathcal{L}_0$ is a singlet (i.e., invariant), ϵ_8 is a small coefficient, and $\mathcal{L}_8$ transforms as the 8th component of an octet of operators, $\mathcal{L}_i, i = 1 \cdots 8$:

$$U_G \mathcal{L}_0 U_G^{-1} = \mathcal{L}_0, \qquad U_G \mathcal{L}_8 U_G^{-1} = \left(e^{i\vec{\beta}\cdot\vec{L}_{adj}}\right)_{8j} \mathcal{L}_j \tag{3.99}$$

or equivalently,

$$\left[T^i, \mathcal{L}_0\right] = 0, \qquad \left[T^i, \mathcal{L}_8\right] = -\left(L^i_{adj}\right)_{8j} \mathcal{L}_j = i f_{i8j} \mathcal{L}_j. \tag{3.100}$$

When first postulated, the actual form of $\mathcal{L}_8$ was not known, only its transformation properties. The power of such an ansatz is that it allows matrix elements of $\mathcal{L}_8$ to be related by $SU(3)$ in terms of one or more parameters that can be measured (or calculated in a more detailed theory). To illustrate this, recall the Wigner-Eckart theorem for $SU(2)$, which relates the matrix elements of an irreducible tensor operator[13] T^k_q, which carries angular momentum (or isospin) k and z- component q, between states α_1 and α_2 with angular momenta and z components $j_{1,2}$ and $m_{1,2}$:

$$\langle \alpha_2\, j_2\, m_2 | T^k_q | \alpha_1\, j_1\, m_1 \rangle = \langle \alpha_2\, j_2\, \| T^k \|\, \alpha_1\, j_1 \rangle \langle j_2\, m_2 \mid k\, q\, j_1\, m_1 \rangle. \tag{3.101}$$

The double-barred quantity is the reduced matrix element, which depends on the operator and states, but is independent of m_1, m_2, and q, while the second quantity is a Clebsch-Gordan (CG) coefficient, which leads to selection rules, $m_2 = q + m_1$ and $|k - j_1| \leq j_2 \leq k + j_1$, and relations between the nonzero matrix elements. [An excellent table of CG coefficients can be found in the *Review of Particle Properties* (Patrignani, 2016, or their website).] Similarly, for $SU(3)$ the matrix element of an octet operator between two octets can be written in terms of two quantities that depend on the dynamics and the f and d symbols, which are analogs of the CG coefficients.[14] Thus,

$$-\epsilon_8 \langle i | \mathcal{L}_8 | j \rangle = \alpha\, i f_{i8j} + \beta d_{i8j}, \tag{3.102}$$

where α and β are proportional to ϵ_8. The symmetry-breaking term in the Hamiltonian is $-\epsilon_8 \int d^3\vec{x}\, \mathcal{L}_8(x)$, implying that the shift in the mass m_{B_r} of baryon B_r due to the $SU(3)$-breaking is

$$\Delta m_{B_r} \sim \frac{\langle B_r | -\epsilon_8 \mathcal{L}_8 | B_r \rangle}{2 m_{B_r}}, \tag{3.103}$$

where the $2m_{B_r}$ in the denominator is from our covariant normalization convention. Therefore,

$$m_{B_r} = m_{B_0} + c^r_f m_\alpha + c^r_d m_\beta, \tag{3.104}$$

where m_{B_0} is a common ($SU(3)$-invariant) mass, $c^r_{f,d}$ are the "CG" coefficients obtained from if and d by taking the appropriate linear combinations of indices, $m_\alpha \sim \alpha/(2m_{B_0})$,

[13] $\left[T^3, T^k_q\right] = q T^k_q$, $\left[T^1 \pm i T^2, T^k_q\right] = \sqrt{(k \mp q)(k \pm q + 1)}\, T^k_{q\pm 1}$.

[14] One can generalize the CG coefficients to arbitrary $SU(3)$ representations using *isoscalar factors* (de Swart, 1963; Patrignani, 2016).

and similarly for m_β. Ignoring isospin breaking, there are four masses, $M_N \equiv (M_p + M_n)/2$, M_Σ, M_Ξ, and M_Λ, which can be expressed in terms of three parameters, $m_{B_{0,\alpha,\beta}}$. The latter cannot be predicted a priori, but there is one linear relation, the *Gell-Mann-Okubo relation* (Problem 3.20)

$$\frac{M_\Xi + M_N}{2} = \frac{3M_\Lambda + M_\Sigma}{4} + \mathcal{O}\left(\epsilon_8^2\right), \tag{3.105}$$

where the result holds to leading order in ϵ_8. Experimentally, the GMO relation works extremely well. The individual masses in GeV,

$$M_N \sim 939, \qquad M_\Xi \sim 1318, \qquad M_\Sigma \sim 1193, \qquad M_\Lambda \sim 1116, \tag{3.106}$$

indicate $SU(3)$ breaking at the 20% level, but the left- and right-hand sides of (3.105) are 1129 and 1135, respectively, equal to better than 1%. Similar formulae apply to other low-lying hadronic states, such as the pseudoscalar ($J^P = 0^-$) mass-squares $\mu_\pi^2, \mu_K^2, \mu_\eta^2$ and the lowest vector ($J^P = 1^-$) meson mass-squares, $\mu_\rho^2, \mu_{K^*}^2, \mu_\phi^2$. (One must include the effects of octet-singlet mixing between the η and η' and between the ϕ and ω. See Section 5.8.3.)

The situation is simpler for the $J^P = \frac{3}{2}^+$ states, which transform as a 10 (decuplet). There is only one invariant of the form $10 \times 8 \times 10^*$, so there is only one reduced matrix element in $\langle 10|\mathcal{L}_8|10\rangle$, leading to a linear spacing in the masses as one goes to lower $\mathcal{Y}$. This works very well, as can be seen in Figure 3.2. In fact, the $\mathcal{Y} = -2, S = -3$ baryon Ω^- was not known at the time $SU(3)$ was proposed. The prediction of its existence and mass from the GMO ansatz was a great triumph for $SU(3)$.

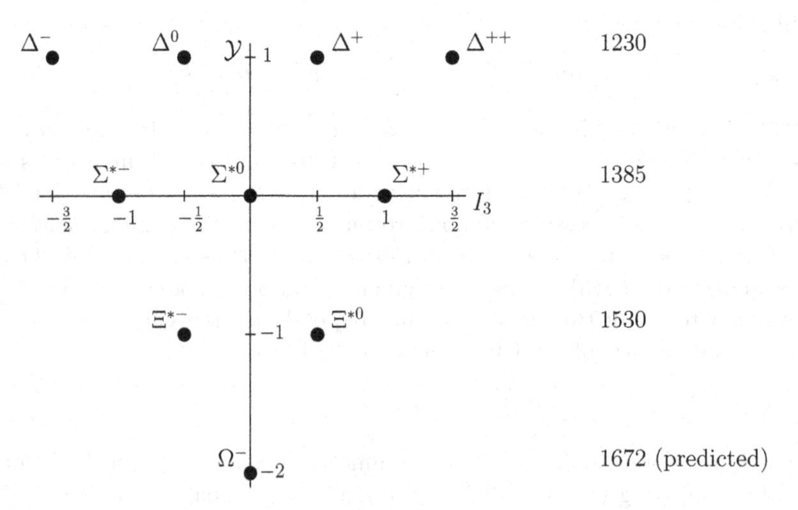

Figure 3.2 Weight diagram for the $J^P = \frac{3}{2}^+$ decuplet.

When $SU(3)$ was proposed it was generally believed that isospin was an exact symmetry of the strong interactions, and that the observed small (1%) breaking is due to electromagnetic and weak interactions. We now understand that there is a small breaking component intrinsic to the strong interactions as well (in part because estimates of the electromagnetic contribution to the proton-neutron mass difference predict a heavier proton). This can be incorporated by writing

$$\mathcal{L} = \mathcal{L}_0 + \epsilon_3 \mathcal{L}_3 + \epsilon_8 \mathcal{L}_8, \tag{3.107}$$

where $\epsilon_3 = \mathcal{O}(1\%)$ and $\epsilon_8 = \mathcal{O}(25\%)$. (A simple model for $\mathcal{L}_3$ was given in (3.86), but we do not assume that specific form here.) Additional predictive power comes from the assumption that $\mathcal{L}_3$ and $\mathcal{L}_8$ are both operators from the same octet. $SU(3)$ is very successful, and is extremely useful for relating various strong and electroweak matrix elements. It is even more successful when extended to chiral symmetry. The existence of heavier hadrons associated with the charm (c) and heavier quarks suggests the possibility of considering $SU(4)$ or higher. However, $SU(4)$ is so badly broken that such an extension is not very useful.

The Quark Model and $SU(3)$

The above description of $SU(3)$ in terms of baryons and mesons becomes simpler when reexpressed in terms of the three lightest quarks and their fields, u, d, and s, which transform under isospin[15] as a doublet (u, d) and singlet (s). Under $SU(3)$ they transform as a fundamental (triplet) while the antiquarks (and the $\bar{q}$ fields) transform as 3^*

$$q = \begin{pmatrix} u \\ d \\ s \end{pmatrix} \equiv \begin{pmatrix} q_1 \\ q_2 \\ q_3 \end{pmatrix} \to e^{i\frac{\vec{\beta}\cdot\vec{\lambda}}{2}} q, \qquad \bar{q} = (\bar{u}\ \bar{d}\ \bar{s}) \to \bar{q} e^{-i\frac{\vec{\beta}\cdot\vec{\lambda}}{2}}, \tag{3.108}$$

which is equivalent to

$$\left[T^i, q_a\right] = \frac{-\lambda^i_{ab}}{2} q_b, \qquad \left[T^i, q_a^\dagger\right] = \frac{+\lambda^{i*}_{ab}}{2} q_b^\dagger. \tag{3.109}$$

The quark electric charges and strong hypercharges are given by

$$Q = \frac{\lambda^3}{2} + \frac{\lambda^8}{2\sqrt{3}} = \frac{1}{3}\begin{pmatrix} 2 & & \\ & -1 & \\ & & -1 \end{pmatrix}, \qquad \mathcal{Y} = \frac{\lambda^8}{\sqrt{3}} = \frac{1}{3}\begin{pmatrix} 1 & & \\ & 1 & \\ & & -2 \end{pmatrix}. \tag{3.110}$$

The weight diagrams are shown in Figure 3.3.

A quark and antiquark can combine to form an octet and singlet, i.e., $3 \times 3^* = 8 + 1$. In an obvious matrix notation,

$$\begin{aligned} q \times \bar{q} &= \begin{pmatrix} u\bar{u} & u\bar{d} & u\bar{s} \\ d\bar{u} & d\bar{d} & d\bar{s} \\ s\bar{u} & s\bar{d} & s\bar{s} \end{pmatrix} = \underbrace{M}_{\text{octet}} + \underbrace{\frac{\eta'}{\sqrt{3}} I}_{\text{singlet}} \\ &= \begin{pmatrix} \frac{2u\bar{u}-d\bar{d}-s\bar{s}}{3} & u\bar{d} & u\bar{s} \\ d\bar{u} & \frac{2d\bar{d}-u\bar{u}-s\bar{s}}{3} & d\bar{s} \\ s\bar{u} & s\bar{d} & \frac{2s\bar{s}-u\bar{u}-d\bar{d}}{3} \end{pmatrix} + \frac{1}{3}\left(u\bar{u} + d\bar{d} + s\bar{s}\right)\begin{pmatrix} 1 & 0 & 0 \\ 0 & 1 & 0 \\ 0 & 0 & 1 \end{pmatrix}, \end{aligned} \tag{3.111}$$

where M is the pseudoscalar octet (3.94) expressed in terms of quarks, and $\eta' = (u\bar{u} + d\bar{d} + s\bar{s})/\sqrt{3}$ is the singlet.

The $J^P = \frac{1}{2}^+$ baryons also transform as an octet, but are constructed from three quarks qqq:

$$\begin{aligned} 3 \times 3 &= 3^* + 6 \\ 3 \times 3 \times 3 &= 3 \times 3^* + 3 \times 6 = 1 + 8 + 8 + 10. \end{aligned} \tag{3.112}$$

[15]In addition to isospin, $SU(3)$ has two other (non-commuting) $SU(2)$ subgroups, U-spin and V-spin. In quark language, these are associated with the $d \leftrightarrow s$ and $u \leftrightarrow s$ transitions, respectively. These are more badly broken than isospin, but are occasionally useful for specific applications, e.g., Problem 3.21 and (Commins and Bucksbaum, 1983).

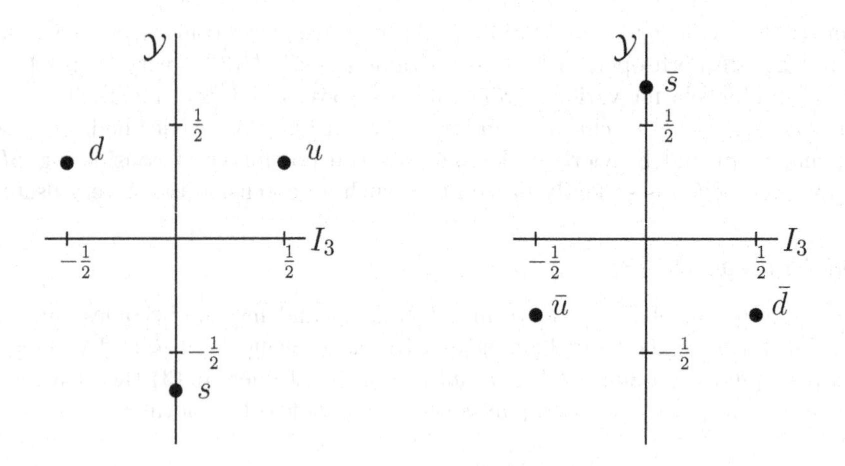

Figure 3.3 Weight diagrams for the 3 and 3* of $SU(3)$.

Thus, both the octet baryons and the $J^P = \frac{3}{2}^+$ decuplet can be constructed from three quarks. Of course, one must also include space, spin,[16] and color indices in the construction (see Problem 3.22); in fact, one of the reasons for the prediction of the color quantum number is that otherwise the Ω^- would be a totally symmetric ($J = 3/2$) composite of three s quarks, in violation of the spin and statistics theorem.

The free quark Lagrangian density is

$$\begin{aligned}\mathcal{L}_{\text{quark}} &= \sum_{a=u,d,s} \bar{q}_a \left(i \not{\partial} - m_a\right) q_a \\ &= \sum \bar{q}_a \left(i \not{\partial} - m_0\right) q_a - m_3 \left(\bar{u}u - \bar{d}d\right) - m_8 \left(\bar{u}u + \bar{d}d - 2\bar{s}s\right)/\sqrt{3} \\ &= \underbrace{\bar{q} \left(i \not{\partial} - m_0\right) I q}_{\mathcal{L}_0} - \underbrace{m_3 \bar{q} \lambda^3 q}_{-\epsilon_3 \mathcal{L}_3} - \underbrace{m_8 \bar{q} \lambda^8 q}_{-\epsilon_8 \mathcal{L}_8}.\end{aligned} \tag{3.113}$$

$\mathcal{L}_{\text{quark}}$ would be invariant for degenerate masses, but the $m_{3,8}$ terms automatically transform as components of an octet, so the GMO ansatz for $SU(3)$ breaking is automatically realized in the quark model. The $u, d,$ and s masses are related by

$$m_u = m_0 + m_3 + \frac{m_8}{\sqrt{3}}, \qquad m_d = m_0 - m_3 + \frac{m_8}{\sqrt{3}}, \qquad m_s = m_0 - 2m_8/\sqrt{3}. \tag{3.114}$$

They will be further discussed following the extension of $SU(3)$ to a chiral symmetry.

SU(m) Tensor Notation

There is a powerful *tensor* notation for $SU(m)$ (de Swart, 1963), which generalizes the matrix notation introduced for the adjoint in (3.31). Denote the components of a field ψ transforming as a fundamental m with a lower index, ψ_a, and those of an antifundamental

[16] For the non-relativistic quark model it is sometimes useful to invoke a mixed $SU(6)$ spin-flavor symmetry, which is quite successful in describing the baryon spectrum and properties such as magnetic moments (e.g., Hey and Kelly, 1983; Georgi, 1999; Capstick and Roberts, 2000; Close et al., 2007; Donoghue et al., 2014; Patrignani, 2016).

(m^*) field χ by an upper index, χ^a,

$$[T^i, \psi_a] = -L_a^{i\,b}\psi_b, \qquad [T^i, \chi^a] = +\chi^b L_b^{i\,a}, \tag{3.115}$$

where $L_a^{i\,b} \equiv (L_m^i)_{ab} = -(L_{m^*}^i)_{ba}$. Of course, the conjugate $\psi^\dagger$ is antifundamental, $(\psi_a)^\dagger = (\psi^\dagger)^a$. Higher-dimensional representations can be formed as products of fundamentals, such as ψ_{ab}, which can be regarded either as an elementary field or as a product of two fundamentals $\psi_{1a}\psi_{2b}$. IRREPs can be formed by symmetrizing and antisymmetrizing, e.g.,

$$\underbrace{\psi_{ab}}_{m\times m} = \frac{1}{2}\underbrace{(\psi_{ab} - \psi_{ba})}_{m(m-1)/2} + \frac{1}{2}\underbrace{(\psi_{ab} + \psi_{ba})}_{m(m+1)/2}. \tag{3.116}$$

Mixed tensors, such as η_a^b, can again be either an elementary field or a product $\psi_a\chi^b$. The η_a^b can be decomposed into an adjoint and singlet,

$$\underbrace{\eta_a^b}_{m\times m^*} = \underbrace{\eta_a^b - \frac{\delta_a^b}{m}\eta_c^c}_{m^2-1} + \underbrace{\frac{\delta_a^b}{m}\eta_c^c}_{1}. \tag{3.117}$$

The transformation of a general mixed tensor is

$$\left[T^i, P_{a_1\cdots a_k}^{b_1\cdots b_l}\right] = -L_{a_1}^{i\,c} P_{c\cdots a_k}^{b_1\cdots b_l} \cdots - L_{a_k}^{i\,c} P_{a_1\cdots c}^{b_1\cdots b_l} + P_{a_1\cdots a_k}^{c\cdots b_l} L_c^{i\,b_1} \cdots + P_{a_1\cdots a_k}^{b_1\cdots c} L_c^{i\,b_l}. \tag{3.118}$$

It is straightforward to prove that the antisymmetric product of $m-1$ fundamentals is an m^*, i.e.,

$$\underbrace{\epsilon^{\alpha_1\cdots\alpha_m}\psi_{\alpha_2\cdots\alpha_m}}_{m^{m-1}} \equiv \underbrace{\eta^{\alpha_1}}_{m^*}, \tag{3.119}$$

where $\epsilon^{\alpha_1\cdots\alpha_m} = \epsilon_{\alpha_1\cdots\alpha_m}$ is the totally antisymmetric tensor in m indices. With this machinery, it is simple to construct the $SU(m)$ invariant operators by contracting upper and lower indices of the operators and the ϵ tensor. For example,

$$\psi_a\eta^a, \qquad \psi_{ab}\eta^{ab}, \qquad \epsilon^{\alpha_1\cdots\alpha_m}\psi_{\alpha_1\cdots\alpha_m}, \qquad \epsilon_{\alpha_1\cdots\alpha_m}\chi^{\alpha_1\cdots\alpha_m}, \tag{3.120}$$

are all invariant (Problem 3.24). The first two operators are also invariant under the extension to $U(m) = SU(m) \times U(1)$, with $\psi_a \to e^{i\beta}\psi_a$, $\chi^a \to \chi^a e^{-i\beta}$, and $P_{a_1\cdots a_k}^{b_1\cdots b_l} \to e^{i(k-l)\beta}P_{a_1\cdots a_k}^{b_1\cdots b_l}$. The contractions involving the ϵ tensor are *not* $U(1)$ invariant. Using tensor notation, the πN Lagrangian density in (3.84) becomes

$$\mathcal{L}_0 = \bar\psi^a\left(i\partial\!\!\!/ - m\right)\psi_a + i\sqrt{2}g_\pi\bar\psi^a\gamma^5\pi_a^{\ b}\psi_b + \frac{1}{2}\left[(\partial_\mu\pi)^2 - \mu^2\pi^2\right] - \lambda\left(\pi^2\right)^2, \tag{3.121}$$

where $\pi_a^{\ b} \equiv \Pi_{ab}$ and $\pi^2 \equiv \pi_a^{\ b}\pi_b^{\ a}$.

3.2.4 Chiral Symmetries

We saw in Section 2.7.2 that one can view the left (L) and right (R) chiral projections $\psi_{L,R} = P_{L,R}\psi$ of a fermion field ψ as independent degrees of freedom, with $\psi = \psi_L + \psi_R$. The significance of the chiral projections is that they can have different transformation properties under a global or local symmetry group G. Suppose

$$\left[T^i, \psi_{aL}\right] = -L^i_{Lab}\psi_{bL}, \qquad \left[T^i, \psi_{aR}\right] = -L^i_{Rab}\psi_{bR}. \tag{3.122}$$

If $L_L^i \neq L_R^i$ the transformation is chiral; otherwise it is non-chiral. For example, the weak interactions are associated with a chiral gauge symmetry, implying parity violation. The strong interactions obey a non-chiral gauge symmetry, but have approximate chiral global symmetries. Even for a chiral symmetry, the fermion representation matrices may be reducible, with some of the fermions chiral (i.e., their L and R components transform differently), and others non-chiral or *vector* (i.e., they transform the same way). The fermion Noether current for a chiral symmetry is

$$J_\mu^i = \bar{\psi}_{aL}\gamma_\mu \left(L_L^i\right)_{ab}\psi_{bL} + \bar{\psi}_{aR}\gamma_\mu \left(L_R^i\right)_{ab}\psi_{bR}, \tag{3.123}$$

which reduces to (3.79) for the non-chiral case $L_L^i = L_R^i = L^i$. In matrix notation with $\psi \equiv (\psi_1 \cdots \psi_n)^T$,

$$J_\mu^i = \bar{\psi}\gamma_\mu \left[L_L^i P_L + L_R^i P_R\right]\psi = \frac{1}{2}\bar{\psi}\gamma_\mu \left[(L_L^i + L_R^i) - (L_L^i - L_R^i)\gamma^5\right]\psi, \tag{3.124}$$

where $L_L^i P_L + L_R^i P_R$ contains both Dirac and internal indices. Under a finite transformation,

$$\psi_L \to e^{i\vec{\beta}\cdot\vec{L}_L}\psi_L, \qquad \psi_R \to e^{i\vec{\beta}\cdot\vec{L}_R}\psi_R, \qquad \psi \to e^{i\vec{\beta}\cdot\left[\vec{L}_L P_L + \vec{L}_R P_R\right]}\psi. \tag{3.125}$$

As an example, consider the free fermion Lagrangian density

$$\begin{aligned}\mathcal{L} &= \bar{\psi}_a i \not{\partial}\psi_a - \bar{\psi}_a m_{ab}\psi_b \\ &= \bar{\psi}_{aL} i \not{\partial}\psi_{aL} + \bar{\psi}_{aR} i \not{\partial}\psi_{aR} - \bar{\psi}_{aL} m_{ab}\psi_{bR} - \bar{\psi}_{aR} m_{ab}\psi_{bL},\end{aligned} \tag{3.126}$$

for $a, b = 1 \cdots n$. m is an $n \times n$ mass matrix which we assume to be Hermitian (one can generalize to a non-Hermitian mass matrix, or, equivalently, one involving $P_{L,R}$, as in Problem 3.32). The kinetic part of $\mathcal{L}$ is invariant under the chiral flavor symmetry $U(n)_L \times U(n)_R = SU(n)_L \times SU(n)_R \times U(1)_V \times U(1)_A$. The subscripts L and R have no group theoretical significance. Rather, they indicate that the L-chiral fermions transform as an n under $SU(n)_L$, but as a singlet under $SU(n)_R$, and the reverse for the R-chiral fermions,

$$\begin{aligned} SU(n)_L: &\quad \underbrace{\psi_{aL} \to \left(e^{i\vec{\beta}_L\cdot\vec{L}_n}\right)_{ab}\psi_{bL}}_{n}, \qquad \underbrace{\psi_{aR} \to \psi_{aR}}_{1} \\ SU(n)_R: &\quad \underbrace{\psi_{aL} \to \psi_{aL}}_{1}, \qquad \underbrace{\psi_{aR} \to \left(e^{i\vec{\beta}_R\cdot\vec{L}_n}\right)_{ab}\psi_{bR}}_{n}, \end{aligned} \tag{3.127}$$

where the $\vec{\beta}_L$ and $\vec{\beta}_R$ are independent parameters. The non-chiral (vector) $U(1)_V$ generator is the sum of the $U(1)_L$ and $U(1)_R$ ones. It corresponds to fermion number, i.e., $\psi_{aL,R} \to e^{i\beta}\psi_{aL,R}$. The *axial* $U(1)_A$ generator is the difference of the $U(1)_{L,R}$ ones. In QCD the $U(1)_A$ generator is broken by non-perturbative effects, as will be discussed in Chapter 5.

The mass term $\bar{\psi}_L m \psi_R + h.c.$ is only invariant for

$$e^{-i\vec{\beta}_L\cdot\vec{L}_n} m e^{i\vec{\beta}_R\cdot\vec{L}_n} = m. \tag{3.128}$$

First consider the case that m is diagonal with degenerate entries, $m = m_1 I$:

(a) For $m_1 \neq 0$ the mass term breaks the chiral symmetry to the subgroup with $\vec{\beta}_L = \vec{\beta}_R$. This corresponds to the non-chiral $SU(n)$ with $L_L^i = L_R^i = L_n^i$.

(b) For $m = 0$ the full $SU(n)_L \times SU(n)_R \times U(1)_V$ chiral symmetry is preserved, i.e., chiral symmetries involve vanishing fermion masses.

Of course, one can have hybrid situations, e.g., in which m is diagonal with blocks of degenerate eigenvalues. Each block with k equal but nonzero masses will correspond to a non-chiral $SU(k) \times U(1)_V$ subgroup of the original symmetry, while a block with l massless fermions will have a chiral $SU(l)_L \times SU(l)_R \times U(1)_V$ symmetry.

It turns out that the quark masses are $m_u \lesssim m_d \sim$ a few MeV, which is much less than typical hadronic mass scales. Thus, the strong interactions have an approximate $SU(2)_L \times SU(2)_R$ chiral flavor symmetry, explicity broken by $m_{u,d}$. This is usefully extended to $SU(3)_L \times SU(3)_R$, though the latter is broken by the much larger $m_s = \mathcal{O}(100)$ MeV. The precise meaning of these *current* (Lagrangian) quark masses, and of the *constituent* quark masses $\sim M_{p,n}/3 \sim 300$ MeV, will be discussed in Chapter 5.

3.2.5 Discrete Symmetries

In addition to the continuous symmetries, many theories have *discrete* symmetries characterized by a group with discrete elements. As an example, consider a single Hermitian field ϕ, with

$$\mathcal{L} = \frac{1}{2}(\partial_\mu \phi)^2 - \frac{1}{2}\mu^2\phi^2 + a\phi - \frac{\kappa\phi^3}{3} - \frac{\lambda\phi^4}{4}. \tag{3.129}$$

For general couplings $\mathcal{L}$ has no internal symmetries. However, for the special case $a = \kappa = 0$, $\mathcal{L}$ in invariant under the discrete two-element group $Z_2 = \{I, R\}$, where $\phi \to -\phi$ under R. The Z_2 symmetry has the obvious consequence that the number of particles is conserved modulo 2 in any reaction.

As another example, the Lagrangian density for two Hermitian fields

$$\mathcal{L} = \frac{1}{2}(\partial_\mu \phi_1)^2 + \frac{1}{2}(\partial_\mu \phi_2)^2 - \frac{1}{2}\mu^2(\phi_1^2 + \phi_2^2) - \frac{\lambda_{11}\phi_1^4}{4} - \frac{\lambda_{22}\phi_2^4}{4} - \frac{\lambda_{12}\phi_1^2\phi_2^2}{2} \tag{3.130}$$

is $O(2)$ invariant for $\lambda_{11} = \lambda_{22} = \lambda_{12}$. For $\lambda_{11} = \lambda_{22} \neq \lambda_{12}$ the symmetry is broken to a discrete subgroup consisting of rotations by $\{0, \pi/2, \pi, 3\pi/2\}$ and of reflections times the same rotations, i.e., to the interchange between ϕ_1 and ϕ_2 as well as possible sign changes for one or both fields. For $\lambda_{11} \neq \lambda_{22}$ only the sign changes survive. Discrete symmetries are not associated with conserved charges and often (as in the Z_2 and the $\lambda_{11} \neq \lambda_{22}$ examples above) do not imply that particles fall into degenerate multiplets. Their major consequence is usually to relate couplings or to remove certain couplings that would otherwise be possible, and they are often introduced ad hoc in model building for that purpose. They may also come about as a low energy remnant of a continuous symmetry in an underlying theory that is valid at high energies.

A more realistic example is G-parity, which is an approximate discrete symmetry of the strong interactions:

$$G = Ce^{i\pi T^2}, \tag{3.131}$$

where C is the charge conjugation operator and T^2 is the second generator of the $SU(2)$ isospin group. The mesons consisting of u and d quarks and antiquarks are eigenstates of G-parity. G-invariance, which follows from QCD if isospin breaking is neglected, leads to important constraints on their interactions. For example, the pions have $G = -1$ while the η has $G = +1$. G-parity therefore forbids the decays $\eta \to 3\pi$ as well as any transitions between an odd number of pions by the isospin-conserving strong interactions. ($\eta \to 3\pi$

does proceed much more slowly through isospin breaking effects, especially the $u-d$ quark mass difference.) Consequences for nucleon matrix elements are considered in Appendix G.

The G parity assignments can be justified experimentally. For the pions, $C\pi^0 C^{-1} = \pi^0$ since $\pi^0 \to 2\gamma$ is observed, and $C\pi^+ C^{-1} = +\pi^-$ in our phase conventions. Similarly, the observed $\eta \to 2\gamma$ decay implies $C\eta C^{-1} = +\eta$. The G-parity then follows because the π and η are, respectively, an isotriplet and isosinglet. The assignments also follow from the quark model. It is shown in Problem 3.27 that any color-singlet meson $|q_1\bar{q}_2\rangle$ with $q_{1,2} = u$ or d has $G = (-1)^{L+S+I}$, where L, S, and I are, respectively, the total orbital angular momentum, spin, and isospin. The π and η have $L = S = 0$ so $G_\eta = -G_\pi = 1$. Similarly, the vector mesons ρ^0 and ω (Section 5.8.3) are mainly $\frac{1}{\sqrt{2}}|u\bar{u} \mp d\bar{d}\rangle$, respectively, in an $L=0, S=1$ state, so $G_\rho = -G_\omega = 1$ and the decays $\rho \to 2\pi$ and $\omega \to 3\pi$ are allowed.

3.3 SYMMETRY BREAKING AND REALIZATION

Symmetries of the equations of motion may be broken *explicitly* by small terms in the equations themselves, or *spontaneously* in the solutions. Simple quantum-mechanical analogs of many of the possibilities, including no breaking, explicit breaking, spontaneous breaking, and combined explicit-spontaneous breaking for discrete and continuous symmetries, are described in Appendix H. More extensive discussions than the one here can be found in, e.g., (Lee, 1972; Pagels, 1975; Coleman, 1985).

3.3.1 A Single Hermitian Scalar

Consider the Lagrangian density

$$\mathcal{L} = \frac{1}{2}\left(\partial_\mu \phi\right)^2 - V(\phi) \tag{3.132}$$

for a single Hermitian scalar, where the potential is

$$V(\phi) = \frac{\mu^2\phi^2}{2} + \frac{\lambda\phi^4}{4}. \tag{3.133}$$

$\mathcal{L}$ has no continuous internal symmetries, but does have a discrete Z_2 symmetry under $\phi \to -\phi$ (this would not be the case for the more general potential in (3.129)). The equation of motion for ϕ is

$$\left(\frac{\partial^2}{\partial t^2} - \vec{\nabla}^2\right)\phi = -\frac{\partial V}{\partial \phi} = -\left[\mu^2 + \lambda\phi^2\right]\phi. \tag{3.134}$$

First consider the solutions to (3.134) for a *classical* field ϕ_{class}. The lowest energy classical solution can be interpreted as the *vacuum expectation value* (VEV) or the ground state value of ϕ, $\langle 0|\phi|0\rangle \equiv \langle\phi\rangle$. This can be thought of as a coherent state or Bose condensation effect (cf., classical solutions to Maxwell's equations). From the expression for the Hamiltonian density in (A.8), the lowest energy solution is for $\phi(t,\vec{x}) =$ *constant*,[17] with a value $\langle\phi\rangle$ that minimizes the potential

$$\left.\frac{\partial V}{\partial\phi}\right|_{\langle\phi\rangle} = 0, \qquad \left.\frac{\partial^2 V}{\partial\phi^2}\right|_{\langle\phi\rangle} > 0, \tag{3.135}$$

[17]Any x dependence of the ground state would also violate translation invariance, while any VEV for higher-spin fields would violate Lorentz invariance. However, as we will see in the next section, higher energy classical solutions are merely excitations on the ground state and *can* involve x dependence or higher spin bosons.

where the first condition guarantees an extremum, and the second that the solution is stable (a minimum). One must choose $\lambda > 0$ so that V is bounded from below. However, the sign of μ^2 is arbitrary. The shape of $V(\phi)$ is shown in Figure 3.4 for both signs. $\mu^2 > 0$ is the

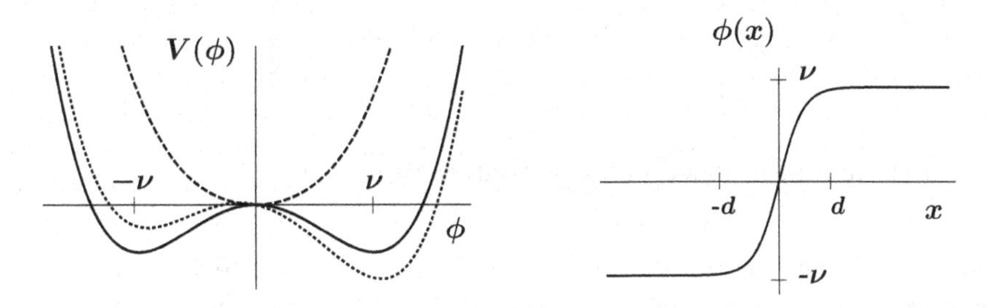

Figure 3.4 Left: potential in (3.133) for $\mu^2 > 0$ (dashed), $\mu^2 < 0$ (solid), or $\mu^2 < 0$ with an explicit symmetry breaking term $-a\phi$ (dotted). Right: domain wall solution of the classical field equation for ϕ.

familiar case. From (3.134) or Figure 3.4 the minimum occurs for $\langle\phi\rangle = 0$, i.e., the vacuum is just "empty space." One can then quantize ϕ as in Chapter 2, μ is the mass of the scalar particle, and the Z_2 symmetry is unbroken.

For $\mu^2 < 0$ there are three extrema, at $\phi = 0$ and at $\pm\nu \equiv \pm\sqrt{-\mu^2/\lambda}$. The extremum $\phi = 0$ is a maximum; it is unstable, since for a small initial perturbation $\phi > 0$

$$\frac{d^2\phi}{dt^2} \sim -\mu^2\phi \geq 0. \tag{3.136}$$

The other two are minima, as can be seen by rewriting

$$V(\phi) = \frac{\lambda}{4}\left(\phi^2 - \nu^2\right)^2 - \frac{\lambda\nu^4}{4}. \tag{3.137}$$

Thus, there are two possible degenerate ground states, with $\langle\phi\rangle = \pm\nu$. One can quantize around whichever of them is (randomly) chosen, and the Z_2 symmetry ($\phi \to -\phi$) is *spontaneously* broken. For the + solution, one can write

$$\phi = \nu + \phi', \tag{3.138}$$

where ϕ' is a normal quantum field with $\langle\phi'\rangle = 0$. With this substitution,

$$\mathcal{L}(\phi) = \mathcal{L}(\nu + \phi') = \frac{1}{2}\left(\partial_\mu\phi'\right)^2 - V(\phi') \tag{3.139}$$

with

$$V(\phi') = \frac{-\mu^4}{4\lambda} - \mu^2\phi'^{\,2} + \lambda\nu\phi'^{\,3} + \frac{\lambda}{4}\phi'^{\,4}. \tag{3.140}$$

The first term is a constant, which is irrelevant until gravity is considered (then it becomes a contribution to the cosmological constant). The second term shows that the ϕ' field corresponds to a particle with (zeroth-order) mass-squared $\mu^2_{\phi'} = -2\mu^2 > 0$. The third is a cubic self-interaction for ϕ' induced by the spontaneous symmetry breaking. It is due to the quartic ϕ interaction, with one of the legs replaced by $\langle\phi\rangle = \nu$, as seen in Figure 3.5.

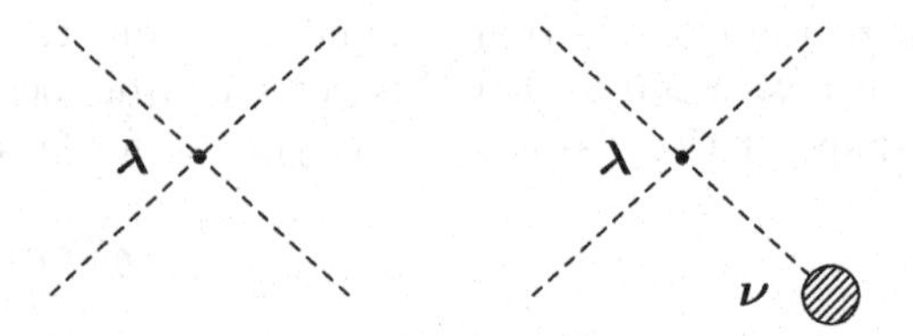

Figure 3.5 Quartic and induced cubic self-interactions for ϕ'.

The induced cubic interaction is a manifestation that the discrete Z_2 symmetry of $\mathcal{L}$ is spontaneously broken by the ground state (classical) solution, and has the effect that particle number modulo 2 is no longer conserved. The last term is a quartic self-interaction that is unaffected by the symmetry breaking.

For the intermediate case $\mu^2 = 0$ it is not sufficient to consider the theory classically. One must extend the discussion to consider the *effective potential*, which reduces to the potential at tree-level in an expansion in the number of loops. The model considered here is difficult to study; the apparent minimum is at nonzero ν but occurs outside of the range of validity of the expansion. However, in more realistic theories, such as the standard model, the symmetry *is* spontaneously broken for $\mu^2 = 0$ (Coleman and Weinberg, 1973). The effective potential is extremely useful for incorporating higher-order effects in the study of symmetry breaking and for considering field theory at finite temperature (see, e.g., Dolan and Jackiw, 1974; Weinberg, 1995; Quiros, 1999).

One can perturb the potential in (3.133) by linear or cubic terms that explicitly break the Z_2 symmetry, as in (3.129). For definiteness, we will consider the linear *tadpole* operator $-a\phi$ with $a > 0$,

$$V(\phi) = \frac{\mu^2\phi^2}{2} - a\phi + \frac{\lambda\phi^4}{4}. \tag{3.141}$$

An immediate consequence is that the Z_2 symmetry must be violated in the ground state as well, i.e., $\langle\phi\rangle \neq 0$. For $\mu^2 > 0$ the VEV is induced by the explicit breaking, $\nu = \langle\phi\rangle = a/\mu^2 + \mathcal{O}(a^3)$. The most important consequence is the presence of a small cubic term for the shifted field $\phi' = \phi - \nu$,

$$V(\phi') = \frac{\mu^2}{2}\phi'^{\,2} + \lambda\nu\phi'^{\,3} + \frac{\lambda}{4}\phi'^{\,4} + \mathcal{O}(a^2), \tag{3.142}$$

similar to the (larger) one in (3.140). For $\mu^2 < 0$ the potential is shifted, as seen in the dotted curve in Figure 3.4. The deepest (*global* or *true*) minimum is at

$$\nu = \nu_0 + \frac{a}{2\nu_0^2} + \mathcal{O}(a^2), \tag{3.143}$$

where $\nu_0 = \sqrt{-\mu^2/\lambda}$ is the unperturbed minimum. For small enough a there is another metastable *local* or *false* minimum for $\nu < 0$. Such metastable local minima are frequently encountered in field theories, and in some cases it makes sense to quantize around them rather than around the global minimum (e.g., Kusenko et al., 1996; Intriligator et al., 2006; Degrassi et al., 2012; Camargo-Molina et al., 2014). The relevant issues for a realistic theory with a metastable vacuum are: (a) Is the lifetime of the metastable vacuum due to tunneling (Linde, 1983) long compared to the 10^{10} year age of the observed Universe? (b) Which vacuum would have been occupied initially as the Universe cooled?

3.3.2 A Digression on Topological Defects

There are also more energetic classical solutions to (3.134). For example, the static *domain wall* solution is an infinite wall perpendicular, e.g., to the x direction, with $\phi(x)$ varying from $\phi(-\infty) = -\nu$ to $\phi(+\infty) = +\nu$. This is illustrated in Figure 3.4, where the center of the wall is at $x = 0$ and the wall is parallel to the y and z directions. Energy is stored in the wall in the transition region near $x = 0$. The thickness d of the transition region can be estimated by minimizing the sum of the kinetic energy density $\sim (\nu/d)^2 = |\mu|^2/\lambda d^2$ and the potential energy density (with respect to the minimum) $\sim \mu^4/\lambda$, leading to $d \sim 1/|\mu|$. Thus, the energy density per unit area is $\sim d\mu^4/\lambda \sim |\mu|^3/\lambda$. Since the wall is infinite in extent, it would take infinite energy to tunnel to one of the ground states $\phi(x) = \pm\nu$, so the wall is stable. (Such objects are known as *topological defects*.)

This simple model illustrates a generic difficulty with spontaneously broken discrete symmetries. Such walls would presumably have formed in the early Universe as it cooled from a temperature T much larger than $|\mu|$ because causally disconnected regions would have fallen randomly into either of the two minima, somewhat like the formation of ferromagnetic domains. Both walls and anti-walls (making transitions from $+\nu$ to $-\nu$) would have formed. Most would presumably have been annihilated, but one would expect at least one to survive in a volume $V \sim R^3$ of the size of our observable Universe, contributing to the energy density and anisotropy of our Universe. To get an idea of the magnitude, let us assume the average energy per unit volume due to a single domain wall is bounded by the observed average energy density ρ_{tot},

$$\frac{\text{energy}}{\text{volume}} \sim \frac{|\mu|^3}{\lambda R} < \rho_{tot} \sim \left(3 \times 10^{-3}\text{eV}\right)^4. \tag{3.144}$$

Of course, this underestimates the constraint since the observed energy density is extremely isotropic, unlike a domain wall. Using $R \sim 1.4 \times 10^{10}$ yr $\sim 4 \times 10^{17}$ sec, this yields $|\mu|/\lambda^{1/3} < 30$ MeV. Discrete symmetries spontaneously broken at a larger scale are therefore cosmologically dangerous.[18]

Spontaneous symmetry breaking in particle physics can lead to other defects of possible cosmological relevance, which may involve classical values for gauge fields as well as scalars. These include *monopoles*, which occur when non-abelian symmetries are broken down to a subgroup containing a $U(1)$ factor, and *cosmic strings*, associated with $U(1)$ symmetries (see Problem 4.1). In another class are *textures*, which are not topologically stable but may be long-lived. These matters are discussed in much more detail in (Coleman, 1985; Vilenkin, 1985; Kolb and Turner, 1990; Brandenberger, 2013).

3.3.3 A Complex Scalar: Explicit and Spontaneous Symmetry Breaking

Consider a complex scalar ϕ with

$$\mathcal{L}_0 = (\partial_\mu \phi)^\dagger \partial^\mu \phi - V(\phi), \qquad V(\phi) = \mu^2 \phi^\dagger \phi + \lambda \left(\phi^\dagger \phi\right)^2, \tag{3.145}$$

where $\lambda > 0$ so that V is bounded from below. As discussed in Section 2.4.1, $\mathcal{L}_0$ is invariant under the $U(1)$ phase transformations $\phi \to e^{i\beta}\phi$, with a conserved charge corresponding to particle number. It is convenient to go to a Hermitian basis, $\phi = (\phi_1 + i\phi_2)/\sqrt{2}$, as in (2.91),

[18] They can be avoided if the reheating temperature of the Universe after a period of inflation is smaller than μ. Also, the addition of a small explicit symmetry breaking term $-\kappa\phi^3$ or $-a\phi$ to (3.133) would eliminate the problem; the energy difference would lead to an attractive force between domain walls and anti-walls and therefore more rapid annihilation.

so that

$$\mathcal{L}_0 = \frac{1}{2}\left[(\partial_\mu\phi_1)^2 + (\partial_\mu\phi_2)^2\right] - V(\phi_1, \phi_2), \quad V = \frac{\mu^2}{2}\left(\phi_1^2 + \phi_2^2\right) + \frac{\lambda}{4}\left(\phi_1^2 + \phi_2^2\right)^2. \quad (3.146)$$

As shown in (3.75), the $U(1)$ transformation takes the form of a rotation[19] in this basis

$$\begin{pmatrix} \phi_1 \\ \phi_2 \end{pmatrix} \to \begin{pmatrix} \cos\beta & -\sin\beta \\ \sin\beta & \cos\beta \end{pmatrix}\begin{pmatrix} \phi_1 \\ \phi_2 \end{pmatrix}. \quad (3.147)$$

(In fact, $U(1)$ and $SO(2)$ are equivalent.) Again, the vacuum corresponds to a classical solution of the field equations

$$\left(\frac{\partial^2}{\partial t^2} - \vec{\nabla}^2\right)\phi_a = -\left[\mu^2 + \lambda\left(\phi_1^2 + \phi_2^2\right)\right]\phi_a \quad (3.148)$$

with minimum energy. This implies a constant $\nu_a \equiv \langle\phi_a\rangle$, with

$$\left.\frac{\partial V}{\partial\phi_a}\right|_{\nu_1,\nu_2} = 0. \quad (3.149)$$

The minimum condition requires that the eigenvalues $m_{1,2}^2$ of

$$\left.\begin{pmatrix} \frac{\partial^2 V}{\partial\phi_1^2} & \frac{\partial^2 V}{\partial\phi_1\partial\phi_2} \\ \frac{\partial^2 V}{\partial\phi_1\partial\phi_2} & \frac{\partial^2 V}{\partial\phi_2^2} \end{pmatrix}\right|_{\nu_1,\nu_2} \quad (3.150)$$

must be non-negative. (These are interpreted as the mass-squares of the physical mass eigenstate particles when one expands around the minimum.)

For $\mu^2 > 0$, the minimum is at $\nu_1 = \nu_2 = 0$, as seen in Figure 3.6. One can quantize around this point, obtaining degenerate $\phi_{1,2}$ with mass-square μ^2 and the relations between the quartic couplings unbroken (or, equivalently, degenerate ϕ and $\phi^\dagger$, with a conserved particle number). Thus, there is an unbroken symmetry in both the equations of motion and the ground state. This is known as the *Wigner-Weyl* realization of the symmetry.

There are two ways in which a Lagrangian symmetry can be broken. One is to add small explicit breaking terms, as we did for $SU(3)$ in (3.113). For example, with $\mu^2 > 0$ and

$$\mathcal{L} = \mathcal{L}_0 - \frac{\epsilon}{2}\phi_2^2 \quad (3.151)$$

the $U(1) \sim SO(2)$ symmetry would be broken, with nondegenerate masses

$$m_1^2 = \mu^2, \qquad m_2^2 = \mu^2 + \epsilon, \quad (3.152)$$

corresponding to the Hermitian mass eigenstates ϕ_1 and ϕ_2. In this example, the quartic relations are not modified at tree-level, though there would be finite corrections induced at loop-level.[20] However, there would no longer be a conserved particle number (Problem 3.29).

The other possibility is *spontaneous symmetry breaking* (SSB), also known as the *Nambu-Goldstone* realization of the symmetry (Nambu, 1960; Nambu and Jona-Lasinio, 1961; Goldstone, 1961). For $\mu^2 < 0$ the *Mexican hat* (or *wine bottle*) potential in (3.146) has its minima

[19]The full symmetry of $\mathcal{L}_0$ is $O(2)$. The extra $\phi_2 \to -\phi_2$ reflection symmetry corresponds to $\phi \leftrightarrow \phi^\dagger$ in the complex basis.

[20]They are finite because the symmetry breaking is *soft*, i.e., the coefficient has a positive power in mass units.

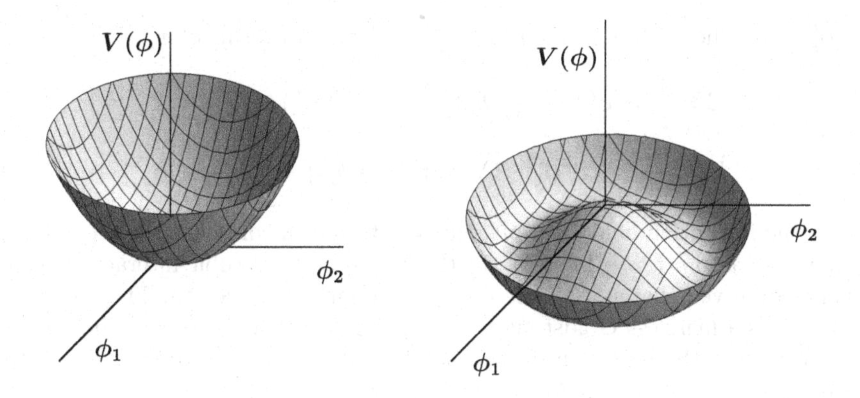

Figure 3.6 Left: the potential in (3.146) for $\mu^2 > 0$. Right: potential for $\mu^2 < 0$.

away from the origin, as seen in Figure 3.6. The rotational symmetry leads to a circle of degenerate minima at

$$\phi_1^2 + \phi_2^2 = \nu^2 \equiv \frac{-\mu^2}{\lambda} > 0, \tag{3.153}$$

as illustrated in Figure 3.7. The true ground state must pick a specific point on this circle, spontaneously breaking the rotational symmetry in much the same way that the spins in a ferromagnetic domain line up in a specific direction. Since the classical field carries a $U(1)$ charge, the symmetry is broken and the charge is not conserved.

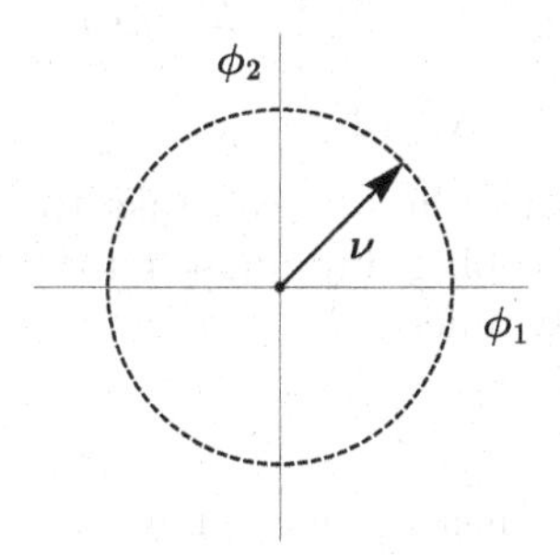

Figure 3.7 Top view of $V(\phi)$ for $\mu^2 < 0$. The origin is a local maximum (unstable). The points on the dashed circle with radius ν are degenerate minima.

Another consequence of the symmetry breaking is the existence of a massless *Nambu-Goldstone boson*. The Nambu-Goldstone theorem states that for every spontaneously broken generator of a continuous global symmetry there exists a massless spin-0 particle, the *Goldstone boson*.

Let us demonstrate these statements in this example. Without loss of generality, one can choose axes in the $\phi_{1,2}$ plane so that the ground state is at $\nu_1 = \nu$ and $\nu_2 = 0$. (Equivalently, for any choice of ground state one can make an $SO(2)$ rotation so that this holds for the rotated fields.) Then, define

$$\phi_1 = \nu + \phi_1', \qquad \phi_2 = \phi_2', \tag{3.154}$$

where $\langle \phi_i' \rangle = 0$. One can quantize ϕ_i' in the usual way. In terms of ϕ_i',

$$\mathcal{L} = \frac{1}{2}\left(\partial_\mu \phi_1'\right)^2 + \frac{1}{2}\left(\partial_\mu \phi_2'\right)^2 - V\left(\phi_1', \phi_2'\right)$$
$$V = \frac{-\mu^4}{4\lambda} - \mu^2 \phi_1'^2 + \lambda\nu\phi_1'\left(\phi_1'^2 + \phi_2'^2\right) + \frac{\lambda}{4}\left(\phi_1'^2 + \phi_2'^2\right)^2. \tag{3.155}$$

Similar to the Hermitian field example, the first term is a constant, the second implies that ϕ_1' has mass-squared $m_1^2 = -2\mu^2 > 0$, the third is an induced cubic interaction which implies there is no conserved charge, and the last is the quartic interaction. There is no mass term for ϕ_2', i.e., it is a massless Goldstone boson. The interpretation is that the potential is flat as one moves from the minimum in the ϕ_2' direction, so that excitations from the ground state are massless "rolling" modes rather than massive oscillations.

It is possible to combine explicit and spontaneous symmetry breaking. Suppose there is a small explicit breaking term as in (3.151), with $\mu^2 < 0$ and $0 < \epsilon \ll |\mu|^2$. This tilts the potential in Figure 3.6 so that there is a minimum at (3.154) (which is unique except for $\phi_1 \to -\nu$), with

$$m_1^2 = -2\mu^2, \qquad m_2^2 = \epsilon \ll m_1^2. \tag{3.156}$$

Thus, the Goldstone boson acquires a small mass from the explicit breaking (it is known as a *pseudo-Goldstone boson*).

3.3.4 Spontaneously Broken Chiral Symmetry

Let us extend the discussion of spontaneous symmetry breaking to a chiral symmetry. Consider a single chiral fermion $\psi = \psi_L + \psi_R$, and a complex scalar ϕ, with

$$\mathcal{L} = \bar{\psi}_L i \not\partial \psi_L + \bar{\psi}_R i \not\partial \psi_R - h\bar{\psi}_L\psi_R\phi - h^*\bar{\psi}_R\psi_L\phi^\dagger + \left(\partial_\mu\phi\right)^\dagger \partial^\mu\phi - V(\phi)$$
$$\text{with} \quad V(\phi) = \mu^2\phi^\dagger\phi + \lambda\left(\phi^\dagger\phi\right)^2, \tag{3.157}$$

where $\lambda > 0$. Without loss of generality we can take the Yukawa coupling h to be real and positive (if necessary by absorbing any phase into $\psi_L \to e^{i(\arg h)}\psi_L$, which is *not* a symmetry). $\mathcal{L}$ has a chiral $U(1)$ symmetry[21]

$$\phi \to e^{i\beta}\phi, \qquad \psi_L \to \psi_L, \qquad \psi_R \to e^{-i\beta}\psi_R, \tag{3.158}$$

which forbids elementary fermion mass terms. For $\mu^2 < 0$ the symmetry is spontaneously broken. The scalar part of $\mathcal{L}$ can be rewritten in terms of the shifted fields defined in (3.154), just as in (3.155). The Yukawa terms become

$$\mathcal{L}_{Yuk} = -\frac{h\nu}{\sqrt{2}}\bar{\psi}\psi\left(1 + \frac{\phi_1'}{\nu}\right) - \frac{h}{\sqrt{2}} i\bar{\psi}\gamma^5\psi\phi_2', \tag{3.159}$$

where we have used $\bar{\psi}_L\psi_R + \bar{\psi}_R\psi_L = \bar{\psi}\psi$ and $\bar{\psi}_L\psi_R - \bar{\psi}_R\psi_L = \bar{\psi}\gamma^5\psi$. Thus, ψ acquires an effective mass

$$m_\psi = \frac{h\nu}{\sqrt{2}} \tag{3.160}$$

from the spontaneous breaking of the chiral symmetry. The Hermitian scalars $\phi_{1,2}'$ couple as a scalar and pseudoscalar, respectively, to ψ, with a Yukawa coupling $h/\sqrt{2} = m_\psi/\nu$.

[21] The individual chiral transformations of the ψ_L and ψ_R are somewhat arbitrary. All that matters is that $\bar{\psi}_L\psi_R \to e^{-i\beta}\bar{\psi}_L\psi_R$. Changing the prescription is equivalent to adding terms proportional to the unbroken fermion number generator to the chiral charges.

3.3.5 Field Redefinition

In field redefinition one introduces new fields as functions of the original ones. We saw simple examples in Section 2.10 involving the change in the overall phases of complex scalar or fermion fields. In some cases, such phase rotations correspond to $U(1)$ symmetries. However, they are also useful in the absence of a symmetry for removing or changing the phases of the constants appearing in $\mathcal{L}$. Field redefinitions were also encountered in Section 2.14 where they were used to put kinetic energy and mass terms in canonical form. Yet another example was seen in the model of a spontaneously broken chiral $U(1)$ symmetry described in Section 3.3.4, which can be described in terms of the shifted fields $\phi'_{1,2}$ with interactions given in (3.155) and (3.159).

It is sometimes convenient to rewrite the theory in terms of new fields that are more complicated functions of the original ones. Such field redefinitions can lead to theories that look very different from the original ones, though for a wide class of cases they are equivalent in the sense that they have the same on-shell S (transition) matrices (Haag, 1958; Coleman et al., 1969; Callan et al., 1969). In the case of SSB, instead of the rectangular Hermitian basis in (3.146), one can define the fields in a polar basis, i.e.,

$$\phi = \frac{(\nu + \eta)}{\sqrt{2}} e^{i\xi/\nu}, \tag{3.161}$$

where η and ξ are Hermitian fields. η is invariant under a $U(1)$ transformation, while ξ is *shifted*

$$\eta \to \eta, \qquad \xi \to \xi + \beta\nu. \tag{3.162}$$

The ϕ part of $\mathcal{L}$ becomes

$$\begin{aligned} \mathcal{L}_\phi &= \frac{1}{2}(\partial_\mu \eta)^2 + \frac{1}{2}(\partial_\mu \xi)^2 \left[1 + \frac{\eta}{\nu}\right]^2 - V(\eta) \\ V(\eta) &= \frac{-\mu^4}{4\lambda} - \mu^2\eta^2 + \lambda\nu\eta^3 + \frac{\lambda}{4}\eta^4, \end{aligned} \tag{3.163}$$

i.e., ξ drops out of the potential and is therefore massless, illustrating that the Goldstone boson is just the phase of ϕ. The self-interaction terms in V for η are the same as those for ϕ'_1 in (3.155), and $m_\eta^2 = -2\mu^2$. Both η and ξ have canonical kinetic energy terms (the $1/\nu$ normalization of the phase in (3.161) was chosen for that purpose), but there are now derivative couplings of $(\partial_\mu\xi)^2$ to η and η^2 with coupling strength $1/\nu$, which enter through the ϕ kinetic energy terms. These replace the ϕ'_2 interactions in (3.155). The Yukawa interactions in (3.159) become

$$\mathcal{L}_{Yuk} = -\frac{h\nu}{\sqrt{2}}\bar{\psi}_L\psi_R \left(1 + \frac{\eta}{\nu}\right) e^{i\xi/\nu} + h.c., \tag{3.164}$$

i.e., there are interactions with ξ to all orders. However, this too can be simplified by a redefinition

$$\psi'_R = e^{i\xi/\nu}\psi_R. \tag{3.165}$$

ξ disappears from $\mathcal{L}_{Yuk}$, but reemerges as a derivative coupling to the right-chiral fermion that enters from $\bar{\psi}_R i \not\partial \psi_R$. Altogether, $\mathcal{L} = \mathcal{L}_f + \mathcal{L}_\phi$, where

$$\mathcal{L}_f = \bar{\psi} i \not\partial \psi + \frac{\partial_\mu \xi}{\nu}\bar{\psi}_R\gamma^\mu\psi_R - \frac{h\nu}{\sqrt{2}}\bar{\psi}\psi\left(1 + \frac{\eta}{\nu}\right). \tag{3.166}$$

The prime on ψ_R has been dropped. It is straightforward to show that the derivative couplings of $\partial_\mu\xi/\nu$ in (3.163) and (3.166) are to the Noether current associated with the $U(1)$.

The reformulation of the model in the polar basis is not manifestly renormalizable, due to the derivative couplings and/or the higher-order couplings in (3.164), and for considering high energy or higher-order calculations the original rectangular basis is more useful. However, the new variables define an *effective theory* that is very useful at low energies. The physical particle content and nature of the Goldstone boson are transparent. Also, all of the ξ interactions involve derivatives, which translate into factors of the ξ four-momentum p_ξ in physical amplitudes. Therefore, such amplitudes must vanish for $p_\xi \to 0$. This finds application in the *soft pion theorems* in the more realistic case of QCD. Even for $p_\xi \neq 0$, the formulation is useful for small external momenta compared with ν, where the interactions are small. A similar field redefinition is useful for displaying the physical particle content in spontaneously broken gauge theories (the Higgs mechanism). Calculations in the various formulations are compared in Problem 3.30.

3.3.6 The Nambu-Goldstone Theorem

As we saw in an example in Section 3.3.3, there are two possible realizations of an exact continuous symmetry of the Lagrangian (the *Goldstone alternative*). In the Wigner-Weyl realization, the ground state also respects the symmetry. Particles fall into degenerate multiplets with relations between their couplings, and for a chiral symmetry the chiral fermions are massless. In the Nambu-Goldstone realization, the symmetry is not respected by the vacuum, massless Nambu-Goldstone bosons appear, and chiral fermions acquire effective masses. The converse statement is Coleman's theorem (Coleman, 1966), i.e., that a continuous symmetry of the vacuum is also a symmetry of the Lagrangian. The various possibilities for the realization and breaking of continuous symmetries are summarized in Table 3.5.

TABLE 3.5 Possibilities for the realization and breaking of a continuous symmetry.

Exact Lagrangian Symmetry ($[U_G, \mathcal{L}] = 0$)	
Wigner-Weyl: $U_G\|0\rangle = \|0\rangle$ Exact symmetry	*Nambu-Goldstone*: $U_G\|0\rangle \neq \|0\rangle$ Spontaneous symmetry breaking
Degenerate multiplets Conserved charges Relations between interactions Chiral: massless fermions Gauge: massless gauge bosons	Chiral: fermions acquire mass Global: Goldstone bosons Gauge: gauge bosons acquire mass by Higgs or dynamical mechanism
Explicit Breaking ($[U_G, \mathcal{L}] \neq 0$) (global only)	
Multiplet splitting, etc. Chiral: fermions acquire mass	Multiplet splitting, etc. Goldstone bosons acquire mass

Let us extend the proof of the Nambu-Goldstone theorem at tree level to an arbitrary scalar sector, which can always be represented by n Hermitian scalars ϕ_a, $a = 1 \cdots n$. One can write the n fields as a column vector $\phi = (\phi_1\ \phi_2 \cdots \phi_n)^T$, and their possible VEVs by a column vector $\nu \equiv \langle\phi\rangle = (\nu_1\ \nu_2 \cdots \nu_n)^T$. Some or all of the ν_a may be zero. The Lagrangian density consists of the kinetic energy terms for ϕ, a potential $V(\phi)$, and possible terms involving fermions. The most general renormalizable potential is the quartic polynomial

$$V(\phi) = V_0 - \sigma_a \phi_a + \frac{1}{2}\mu^2_{ab}\phi_a \phi_b + \kappa_{abc}\phi_a \phi_b \phi_c + \lambda_{abcd}\phi_a \phi_b \phi_c \phi_d. \tag{3.167}$$

The coefficients σ_a, μ^2_{ab}, κ_{abc}, and λ_{abcd} are real and symmetric in the indices. ν is determined by minimizing the potential, i.e., $(\partial V/\partial \phi_a)|_\nu = 0$, as in (3.149). Define the shifted fields $\phi'_a = \phi_a - \nu_a$ with $\langle \phi'_a \rangle = 0$ and rewrite V in terms of ϕ',

$$V(\phi') = \hat{V}_0 + \frac{1}{2}\hat{\mu}^2_{ab}\phi'_a\,\phi'_b + \hat{\kappa}_{abc}\phi'_a\,\phi'_b\,\phi'_c + \hat{\lambda}_{abcd}\phi'_a\,\phi'_b\,\phi'_c\,\phi'_d, \tag{3.168}$$

where

$$\begin{aligned} \hat{\mu}^2_{ab} &= \left(\frac{\partial^2 V}{\partial\phi_a\partial\phi_b}\right)\Big|_\nu, \qquad \hat{\kappa}_{abc} = \frac{1}{3!}\left(\frac{\partial^3 V}{\partial\phi_a\partial\phi_b\partial\phi_c}\right)\Big|_\nu \\ \hat{\lambda}_{abcd} &= \frac{1}{4!}\left(\frac{\partial^4 V}{\partial\phi_a\partial\phi_b\partial\phi_c\partial\phi_d}\right)\Big|_\nu = \lambda_{abcd}. \end{aligned} \tag{3.169}$$

There is no linear term in $V(\phi')$ since we are at a minimum. The eigenvalues of the mass-squared matrix $\hat{\mu}^2_{ab}$ are the physical mass-squares, and the eigenvectors are in the directions of the mass eigenstate fields. The eigenvalues are guaranteed to be non-negative by the definition of a minimum.

Now consider the role of a continuous internal symmetry, with generator representation matrices L^i. Under an infinitesimal transformation (and returning to the original fields)

$$V(\phi) \to V(\phi) + \delta V(\phi), \tag{3.170}$$

where

$$\delta V(\phi) = \frac{\partial V}{\partial \phi_a}\delta\phi_a = i\frac{\partial V}{\partial\phi_a}\left(\vec{\beta}\cdot\vec{L}\right)_{ab}\phi_b. \tag{3.171}$$

Invariance under the transformation requires $\delta V(\phi) = 0$, and therefore

$$\frac{\partial V}{\partial\phi_a}L^i_{ab}\phi_b = 0. \tag{3.172}$$

Differentiating (3.172) with respect to ϕ_c and evaluating at $\phi = \nu$, one has

$$\hat{\mu}^2_{ab}\left(L^i\nu\right)_b = 0, \quad i = 1\cdots N. \tag{3.173}$$

Let us label the generators so that

$$L^i\nu = 0, \quad i = 1\cdots M, \qquad L^i\nu \neq 0, \quad i = M+1\cdots N. \tag{3.174}$$

The subgroup $G' \in G$ generated by $T^1\cdots T^M$ leaves the vacuum invariant, $T^i|0\rangle = 0$, so G' is unbroken.[22] However, symmetries associated with the remaining $N-M$ generators are spontaneously broken, $T^i|0\rangle \neq 0$, as can easily be seen from (3.29). ($M = 0$ [G completely broken], and $M = N$ [no breaking] are special cases.) From (3.173) and (3.174) the scalar mass-square matrix has $N - M$ linearly independent eigenvectors $L^i\nu$ with eigenvalue zero. Thus, there are $N - M$ massless Goldstone bosons, one for each spontaneously broken generator. μ^2_{ab} also has $p = n - (N - M)$ (generally) non-zero eigenvalues, corresponding to p (generally) massive scalar particles.

The Nambu-Goldstone theorem holds quite generally.[23] It can be proved to all orders

[22]The generators in (3.174) may be linear combinations of the original ones. The precise meaning is that there are M linearly independent generators for which $L^i\nu = 0$ and $N - M$ for which $L^i\nu \neq 0$.

[23]A way around this, for the axial $U(1)$ generator, will be described in Section 5.8.3.

using the effective potential. It even holds non-perturbatively (Goldstone et al., 1962; Weinberg, 1995), as will be shown in an example in Section 5.8. Goldstone bosons do not appear to exist in nature. However, pseudo-Goldstone bosons, which acquire a small mass from explicit breaking, are relevant to the chiral flavor symmetries of the strong interactions.[24] Furthermore, as will be seen in Chapter 4, for a gauge symmetry the Goldstone bosons are reinterpreted as the longitudinal modes of massive gauge bosons (the Higgs mechanism).

3.3.7 Boundedness of the Potential

To have a sensible theory there should be a lowest energy state, i.e., that the vacuum is stable, or at least a local (metastable) minimum. Absolute stability requires that the full effective potential, including loop effects, running couplings, and possible non-renormalizable contributions, is bounded from below within the range of the validity of the theory. However, let us consider the simpler question of whether the renormalizable tree-level potential V in (3.167) is bounded from below for all values of the fields. V is well-behaved for finite ϕ, so all we have to do is make sure that it is bounded from below as $\phi \to \infty$ for all orientations of the n-dimensional vector ϕ. It suffices to consider the case $\phi(x) =$ constant, and since we are concerned with the large ϕ behavior we can ignore spontaneous symmetry breaking. An arbitrary (constant) ϕ can be written[25]

$$\phi_a = r e_a, \tag{3.175}$$

where $r = (\sum_a \phi_a^2)^{1/2}$ is a radius vector and $\hat{e}$ is an n-dimensional unit vector with components e_a. One requires

$$\lim_{r \to \infty} V(r, \hat{e}) = c(\hat{e}), \tag{3.176}$$

where the limit $c(\hat{e})$ is either a finite number or $+\infty$ for all orientations $\hat{e}$. If $V \to -\infty$ for any $\hat{e}$ there is no well-defined ground state and the theory is unstable.

In most cases the quartic terms in V determine the asymptotic behavior. If $C(\hat{e}) \equiv \lambda_{abcd}\, e_a e_b e_c e_d > 0$ for *all* unit vectors $\hat{e}$, then the theory is bounded, while if $C(\hat{e}) < 0$ for *any* $\hat{e}$ the potential is unbounded. If $C(\hat{e}) = 0$ for some directions $\hat{e}$, then one must investigate the other terms in V along those directions.

It sometimes happens that there are some directions for which $V(r, \hat{e})$ is independent of r, i.e., the quadratic, cubic, and quartic terms vanish so that $V(r, \hat{e}) = V_0$. This is especially common in supersymmetric theories (which have $V_0 = 0$). If such *flat* directions correspond to the minimum of V the ground state is not uniquely defined at the renormalizable tree-level. In some cases, loop corrections to the effective potential, higher-dimensional (non-renormalizable) terms, or soft supersymmetry breaking terms lift the flatness to yield a unique minimum.

As an example, consider the potential

$$V(\phi_1, \phi_2) = \frac{1}{2}\mu_1^2 \phi_1^2 + \frac{1}{2}\mu_2^2 \phi_2^2 + \lambda_2 \phi_2^4, \tag{3.177}$$

with $\lambda_2 > 0$. One has

$$\phi = \begin{pmatrix} \phi_1 \\ \phi_2 \end{pmatrix} = r\hat{e}, \tag{3.178}$$

[24] Possible pseudo-Goldstone bosons associated with the breaking of a symmetry by anomalies (Section 4.5) are known as *axions*. Applications to the strong CP problem and to dark matter are mentioned in Chapter 10.

[25] This form is also often useful for finding the minimum of V.

where $r = (\phi_1^2 + \phi_2^2)^{1/2}$ and $\hat{e} = (\cos\theta \ \sin\theta)^T$ is a two-dimensional unit vector. The quartic coefficient is thus $C(\hat{e}) = \lambda_2 \sin^4\theta$, which is positive for all directions except $\sin\theta = 0$. In that case ($\phi_1 = \pm r$, $\phi_2 = 0$) the potential is bounded below, flat, or unbounded for $\mu_1^2 > 0, 0$, or < 0, respectively.

3.3.8 Example: Two Complex Scalars

As a major example of some of these considerations, let

$$\phi_I = \frac{\phi_1 + i\phi_2}{\sqrt{2}}, \qquad \phi_{II} = \frac{\phi_3 + i\phi_4}{\sqrt{2}} \tag{3.179}$$

be two complex scalars with Hermitian components ϕ_i, $i = 1 \cdots 4$, and Lagrangian density

$$\begin{aligned} \mathcal{L} &= |\partial_\mu \phi_I|^2 + |\partial_\mu \phi_{II}|^2 - V(\phi_I, \phi_{II}) \\ V &= \mu_I^2 |\phi_I|^2 + \mu_{II}^2 |\phi_{II}|^2 + \lambda_I |\phi_I|^4 + \lambda_{II} |\phi_{II}|^4 + \lambda_{III} |\phi_I|^2 |\phi_{II}|^2 \\ &\quad - A\phi_I \phi_{II} - A^* \phi_I^\dagger \phi_{II}^\dagger, \end{aligned} \tag{3.180}$$

where, e.g., $|\phi_I|^2 \equiv \phi_I^\dagger \phi_I$ and $|\phi_I|^4 \equiv (\phi_I^\dagger \phi_I)^2$. For V to be bounded below it is sufficient to require that the quartic terms are positive for all values of ϕ_J, $J = I, II$, which holds provided

$$\lambda_{I,II} > 0, \qquad \lambda_{III} > -2\lambda_I^{1/2} \lambda_{II}^{1/2}. \tag{3.181}$$

(The limiting case $\lambda_I = \lambda_{II} = -\lambda_{III}/2$, which occurs in supersymmetry, is considered in Problem 3.35.) μ_J^2 are real, and A is an arbitrary complex number. $\mathcal{L}$ has an internal $U(1)$ symmetry, $\phi_J \to e^{iq_J\beta}\phi_J$ with $q_I = -q_{II} = 1$ (which is promoted to $U(1) \times U(1)$ for $A = 0$). W.l.o.g. one can redefine the phases of $\phi_{I,II}$ to make A real and non-negative. Then, if the parameters are such that the ϕ_J both acquire VEVs, the minimum will occur for $\langle\phi_I\rangle\langle\phi_{II}\rangle$ real and positive. We will assume that the individual VEVs are real and positive, i.e., $\nu_{1,3} = \langle\phi_{1,3}\rangle > 0$ and $\nu_{2,4} = \langle\phi_{2,4}\rangle = 0$. The $U(1)$ symmetry implies that there are degenerate minima that differ from this by $U(1)$ transformations.

In the Hermitian basis with A real and non-negative, the last term in V is

$$V_A = -A(\phi_1\phi_3 - \phi_2\phi_4), \tag{3.182}$$

while the other terms depend only on $\phi_1^2 + \phi_2^2$ and $\phi_3^2 + \phi_4^2$. The nontrivial extremum conditions for $\nu_{2,4} = 0$ are

$$\begin{aligned} \left.\frac{\partial V}{\partial \phi_1}\right|_\nu &= \left[\mu_I^2 + \lambda_I \nu_1^2 + \frac{1}{2}\lambda_{III}\nu_3^2\right]\nu_1 - A\nu_3 = 0 \\ \left.\frac{\partial V}{\partial \phi_3}\right|_\nu &= \left[\mu_{II}^2 + \lambda_{II} \nu_3^2 + \frac{1}{2}\lambda_{III}\nu_1^2\right]\nu_3 - A\nu_1 = 0. \end{aligned} \tag{3.183}$$

These have the trivial solution $\nu_1 = \nu_3 = 0$, and possible nontrivial solutions with $\nu_1 \neq 0$, $\nu_3 \neq 0$. The mass-square matrix for $\phi_{1,3}$ (there is no mixing with $\phi_{2,4}$ for either case) corresponding to the trivial solution is

$$\mu_{Re}^2 \equiv \left.\begin{pmatrix} \frac{\partial^2 V}{\partial \phi_1^2} & \frac{\partial^2 V}{\partial \phi_1 \partial \phi_3} \\ \frac{\partial^2 V}{\partial \phi_1 \partial \phi_3} & \frac{\partial^2 V}{\partial \phi_3^2} \end{pmatrix}\right|_{\nu_{1,3}=0} = \begin{pmatrix} \mu_I^2 & -A \\ -A & \mu_{II}^2 \end{pmatrix}. \tag{3.184}$$

This is a minimum provided the two eigenvalues $\mu^2_{a,b}$ are both non-negative. The trace and determinant are preserved by the unitary transformation which diagonalizes μ^2_{Re},

$$\mu_a^2 + \mu_b^2 = \mu_I^2 + \mu_{II}^2, \qquad \mu_a^2\mu_b^2 = \mu_I^2\mu_{II}^2 - A^2, \tag{3.185}$$

so the origin is a minimum provided both of these are positive, in which case there is no spontaneous symmetry breaking.[26] Otherwise, the origin is a saddlepoint with one negative eigenvalue (for $\mu_I^2\mu_{II}^2 - A^2 < 0$), or a maximum with two negative eigenvalues (for $\mu_I^2 + \mu_{II}^2 < 0$ and $\mu_I^2\mu_{II}^2 - A^2 > 0$). In these cases, there will be SSB. $\nu_{1,3}$ can be found numerically from (3.183), and the mass-square matrix for $\phi_{1,3}$ can be written

$$\mu^2_{Re} \equiv \left.\begin{pmatrix} \frac{\partial^2 V}{\partial \phi_1^2} & \frac{\partial^2 V}{\partial \phi_1 \partial \phi_3} \\ \frac{\partial^2 V}{\partial \phi_1 \partial \phi_3} & \frac{\partial^2 V}{\partial \phi_3^2} \end{pmatrix}\right|_{\nu_{1,3}} = \begin{pmatrix} 2\lambda_I \nu_1^2 + A\nu_3/\nu_1 & \lambda_{III}\nu_1\nu_3 - A \\ \lambda_{III}\nu_1\nu_3 - A & 2\lambda_{II}\nu_3^2 + A\nu_1/\nu_3 \end{pmatrix}, \tag{3.186}$$

where $\mu^2_{I,II}$ have been eliminated using (3.183).

More interesting is the mass-square matrix μ^2_{Im} for $\phi_{2,4}$. For $\nu_{1,3} = 0$, μ^2_{Im} is the same as (3.184) except $A \to -A$. This has the same eigenvalues as μ^2_{Re}, and the symmetry is unbroken. For the SSB case, one can again use (3.183) to obtain

$$\mu^2_{Im} \equiv \left.\begin{pmatrix} \frac{\partial^2 V}{\partial \phi_2^2} & \frac{\partial^2 V}{\partial \phi_2 \partial \phi_4} \\ \frac{\partial^2 V}{\partial \phi_2 \partial \phi_4} & \frac{\partial^2 V}{\partial \phi_4^2} \end{pmatrix}\right|_{\substack{\nu_{1,3}\neq 0 \\ \nu_{2,4}=0}} = A \begin{pmatrix} \frac{\nu_3}{\nu_1} & 1 \\ 1 & \frac{\nu_1}{\nu_3} \end{pmatrix}. \tag{3.187}$$

This has a zero eigenvalue, corresponding to the Goldstone boson of the broken $U(1)$. The unnormalized eigenvector is $(\nu_1 \ -\nu_3)^T$. This is in the $L\nu$ direction, as expected from (3.174), since in the Hermitian basis,

$$L = \begin{pmatrix} -\tau^2 & 0 \\ 0 & \tau^2 \end{pmatrix}, \qquad \nu = \begin{pmatrix} \nu_1 \\ 0 \\ \nu_3 \\ 0 \end{pmatrix}, \tag{3.188}$$

where the 0 in L is the 2×2 zero matrix, in analogy with (3.74) or (3.147). The other eigenvalue is $A(\frac{\nu_3}{\nu_1} + \frac{\nu_1}{\nu_3})$, with (unnormalized) eigenvector $(\nu_3 \ \ \nu_1)^T$. For $A \to 0$ this state also becomes massless, i.e., both generators of the enhanced $U(1) \times U(1)$ are broken and there are two Goldstone bosons.

An alternative derivation of these results utilizes a polar basis, analogous to (3.161), i.e.,

$$\phi_J = \frac{(\nu_J + \eta_J)}{\sqrt{2}} e^{i\xi_J/\nu_J}, \tag{3.189}$$

where $\nu_I = \nu_1$, $\nu_{II} = \nu_3$, and where η_J and ξ_J are Hermitian. The normalization of the exponents is chosen so that the kinetic energy terms

$$\mathcal{L}_{KE} = \sum_J |\partial_\mu \phi_J|^2 = \frac{1}{2}\sum_J \left((\partial_\mu \eta_J)^2 + (\partial_\mu \xi_J)^2 \left[1 + \frac{\eta_K}{\nu_K}\right]^2 \right) \tag{3.190}$$

are canonical for real fields, although there are new three- and four-point interactions involving derivatives that we will not be concerned with here. Under a $U(1)$ transformation the ξ_J are shifted while the η_J are unchanged,

$$\xi_{I,II} \to \xi_{I,II} \pm \nu_{I,II}\beta. \tag{3.191}$$

[26]There are no other minima provided (3.181) is satisfied.

The ξ_J only enter the V_A part of the potential

$$V_A = -A(\eta_I + \nu_I)(\eta_{II} + \nu_{II}) \cos\left(\frac{\xi_I}{\nu_I} + \frac{\xi_{II}}{\nu_{II}}\right), \tag{3.192}$$

which has minima for $\frac{\xi_I}{\nu_I} + \frac{\xi_{II}}{\nu_{II}} = 0$. We will consider the minimum at $\xi_J = 0$, with the other values related by $U(1)$ transformations. The calculation of the VEVs and the mass matrix for the η_J proceeds as before. The mass matrix for the ξ_J is obtained by expanding V_A to second order,

$$V_A = \frac{A\nu_1\nu_3}{2}\left(\frac{\xi_I}{\nu_I} + \frac{\xi_{II}}{\nu_{II}}\right)^2 = \frac{1}{2}A\left(\frac{\nu_3}{\nu_1} + \frac{\nu_1}{\nu_3}\right)\frac{(\nu_3\xi_I + \nu_1\xi_{II})^2}{\nu_1^2 + \nu_3^2}. \tag{3.193}$$

This reproduces our previous result, with one Goldstone boson and one massive state. Approximating $\phi_J \sim \frac{(\eta_J + \nu_J + i\xi_J)}{\sqrt{2}}$, the eigenvectors and eigenvalues are the same as found previously.

3.4 PROBLEMS

3.1 Verify the Lie group multiplication rule in (3.6).

3.2 Show that with the normalization (3.8) the structure constants c_{ijk} are antisymmetric in all three indices.

3.3 Prove the *Jacobi identity*

$$[L^i, [L^j, L^k]] + [L^j, [L^k, L^i]] + [L^k, [L^i, L^j]] = 0$$

for an arbitrary representation and use it to prove that the adjoint representation matrices in (3.16) satisfy the commutation rules in (3.7).

3.4 Use (3.10) to find an explicit expression for $\vec{\gamma}(\vec{\alpha}, \vec{\beta})$ defined by the multiplication in (3.5) for the group $SU(2)$.

3.5 Prove the formal power series identity

$$e^A B e^{-A} = B + [A, B] + \frac{1}{2!}[A, [A, B]] + \frac{1}{3!}[A, [A, [A, B]]] + \cdots,$$

where A and B are any two operators or square matrices of the same dimension.

3.6 Prove that $\det[e^{\alpha A}] \equiv \sum_k (\alpha A)^k / k! = e^{\alpha \mathrm{Tr} A}$ for any Hermitian matrix A. This implies that the condition for a unitary or orthogonal matrix to be special (unit determinant) is that the generators be traceless.

3.7 Prove the transformation law (3.32) for a field transforming according to the adjoint representation.

3.8 Show that

$$\mathcal{L} = \frac{1}{2}\left[(\partial_\mu \phi_a)^2 - \mu^2 \phi_a^2\right] - \lambda\left(\phi_a^2\right)^2,$$

with $\phi_a = \phi_a^\dagger$, is invariant under infinitesimal $SO(m)$ transformations (i.e., $\left[T^i, \mathcal{L}\right] = 0$), with ϕ_a, $a = 1 \cdots m$ transforming as the vector representation.

3.9 Derive the Noether currents in (3.54) and (3.79) for complex scalars and fermions, respectively. In each case, use the canonical commutation rules in Appendix A to show that the associated Noether charges satisfy the Lie algebra commutation rules. Assume there are no interaction terms involving field derivatives.

3.10 In the free-field limit the isospin generators for the pion-nucleon fields in (3.82) can be written as

$$T^i = \int \frac{d^3\vec{p}}{(2\pi)^3 2E_p} \sum_s \left[b_c^\dagger(\vec{p},s) \left(\frac{\tau^i_{cb}}{2}\right) b_b(\vec{p},s) + d_c^\dagger(\vec{p},s) \left(-\frac{\tau^i_{bc}}{2}\right) d_b(\vec{p},s) \right]$$
$$+ \int \frac{d^3\vec{p}}{(2\pi)^3 2E_p} a_j^\dagger(\vec{p}) \left(-i\epsilon_{ijk}\right) a_k(\vec{p}),$$

where b_a and d_a are, respectively, the annihilation operators for nucleon $a = p$ or n and for its antiparticle, and a_i is the annihilation operator for π^i. Show that the T^i satisfy the $SU(2)$ Lie algebra and that they have the right commutation rules with the fields.

3.11 Show that the pion-nucleon Lagrangian density in (3.84) is $SU(2)$ invariant by proving that $\left[T^i, \mathcal{L}_0\right] = 0$. Use the representation matrices in (3.83).

3.12 Show that the Condon-Shortley convention (analogous to that used for rotations)

$$(L^1 \pm iL^2)|I\, m\rangle = \sqrt{(I \mp m)(I \pm m + 1)}|I\, m \pm 1\rangle$$

for the states in an isospin I multiplet with I_3 eigenvalue m would require $\pi^\pm = \mp\left(\pi_1 \mp i\pi_2\right)/\sqrt{2}$ instead of (3.82).

3.13 The pion and nucleon transform as isospin triplets and doublets, respectively. The approximate isospin invariance of the strong interactions implies that all of the amplitudes for $\pi_i N_a \to \pi_j N_b$ can be expressed in terms of two amplitudes M_I, corresponding to total isospin $I = 1/2$ and $3/2$. Write the amplitudes for $\pi^+ p \to \pi^+ p$, $\pi^- p \to \pi^- p$, $\pi^- p \to \pi^0 n$, and $\pi^0 n \to \pi^0 n$ in terms of the M_I. (Note that at low energy the $M_{3/2}$ amplitude, which is dominated by the Δ resonance, strongly dominates. Such calculations can be extended to $SU(3)$ by means of isoscalar factors (de Swart, 1963; Patrignani, 2016).)

3.14 It is sometimes useful to introduce a *spurion*, which is a fictitious or composite field with the right quantum numbers to parametrize the effects of symmetry breaking, and which can therefore be used as a bookkeeping device. For example, it is known that the effective operator for strangeness-changing nonleptonic weak transitions, which could in principle involve both $\Delta I = \frac{1}{2}$ and $\Delta I = \frac{3}{2}$ components, is strongly dominated by $\Delta I = \frac{1}{2}$. Therefore, the transitions $\Sigma^+ \to \pi^0 p$, $\Sigma^+ \to \pi^+ n$, and $\Sigma^- \to \pi^- n$ should be related in the same way as the isospin-conserving amplitudes for $\Sigma^+ S \to \pi^0 p$, $\Sigma^+ S \to \pi^+ n$, and $\Sigma^- S \to \pi^- n$, where S is a spin-0 spurion carrying $I = 1/2$ and $I_3 = -1/2$. Show that the amplitudes (using the Condon-Shortley convention for π^+ and Σ^+) should be related by

$$M(\Sigma^+ \to \pi^+ n) - M(\Sigma^- \to \pi^- n) = -\sqrt{2} M(\Sigma^+ \to \pi^0 p).$$

This relation, which holds separately for the S and P wave amplitudes (Problem 2.26), works reasonably well. (This simple example could be handled easily using the Wigner-Eckart theorem, but the spurion formalism is more convenient in more complicated cases.)

3.15 Generalize the pion self-interaction term in (3.84) to the case of four distinct types of pions, $\pi_{Ai} \equiv A_i, \pi_{Bi} \equiv B_i, \pi_{Ci} \equiv C_i, \pi_{Di} \equiv D_i$, each transforming as an adjoint under $SU(2)$. In general, there are cubic and quartic interactions such as

$$\mathcal{L}_I = -c_{ijkl} A_i B_j C_k D_l - d_{ijk} A_i B_j C_k.$$

(There are also other possible terms, such as A^2BC or ABD, which we will not consider.) However, $\mathcal{L}_I$ is only $SU(2)$ invariant for the combinations

$$\begin{aligned}\mathcal{L}_I = &- c_\alpha \mathrm{Tr}\,(AB)\,\mathrm{Tr}\,(CD) - c_\beta \mathrm{Tr}\,(AC)\,\mathrm{Tr}\,(BD)\\ &- c_\gamma \mathrm{Tr}\,(AD)\,\mathrm{Tr}\,(BC) - d\,\mathrm{Tr}\,(ABC),\end{aligned}$$

where $A = \vec{A} \cdot \vec{\tau}/\sqrt{2}$, etc., so that $\mathrm{Tr}\,(AB) = \vec{A} \cdot \vec{B}$. (If A, B, and C are all pseudoscalars, then reflection invariance would require $d = 0$.)
(a) Show explicitly that invariants like $\mathrm{Tr}\,(ABCD)$ are not independent.
(b) The existence of only 3 quartic invariants can be justified as follows: A, B, C, and D are all isospin-1 operators. Thus, the product AB is a direct sum of operators $\mathcal{O}_{AB}^{I_{AB}}$ with isospin $I_{AB} = 0, 1$, or 2, and similarly for CD. However, the only overall singlets involve the pairing with $I_{AB} = I_{CD}$. Applying similar reasoning to the cubic term,

$$\mathcal{L}_I = -\sum_{I=0}^{2} c_I \mathcal{O}_{AB}^I \mathcal{O}_{CD}^I - d_1 \mathcal{O}_{AB}^1 C.$$

Construct these invariants explicitly, and relate the coefficients c_I and d_1 to $c_{\alpha,\beta,\gamma}$ and d.
(c) Show why there is only a single invariant in (3.84).

3.16 The effective (non-renormalizable) interaction of a hypothetical doublet of mesons $\mathcal{A}$ with the nucleons is assumed to take the isospin-invariant form

$$\mathcal{L}_{AN} = \lambda(\bar{\psi}\vec{\tau}\psi) \cdot (\mathcal{A}^\dagger \vec{\tau} \mathcal{A}),$$

where

$$\psi = \begin{pmatrix} \psi_p \\ \psi_n \end{pmatrix}, \qquad \mathcal{A} = \begin{pmatrix} \mathcal{A}^+ \\ \mathcal{A}^0 \end{pmatrix},$$

and τ^i are the Pauli matrices. Calculate the tree-level spin-averaged CM cross sections for $\mathcal{A}^+ p \to \mathcal{A}^+ p$, $\mathcal{A}^0 p \to \mathcal{A}^0 p$, and $\mathcal{A}^+ n \to \mathcal{A}^0 p$. Assume that $m_p = m_n$ and $m_{\mathcal{A}^+} = m_{\mathcal{A}^0}$.

3.17 By comparing the $SU(2)$ and $SU(3)$ expressions in (3.84) and (3.92), show that $g_\pi = \frac{1}{2}(g_f + g_d)$.

3.18 Let A_j and B_k transform as $SU(3)$ octets. Prove that $F_i \equiv i f_{ijk} A_j B_k$ and $D_i \equiv d_{ijk} A_j B_k$ also transform as octets. Note that the result for F_i generalizes to any simple Lie algebra, while that for D_i generalizes to $SU(m)$.

3.19 The interactions of pseudoscalar mesons with hyperons can be described by

$$\mathcal{L} = g_{\pi^+ pn}\, \pi^+ \bar{p} i\gamma^5 n + g_{K^+ p\Lambda}\, K^+ \bar{p} i\gamma^5 \Lambda + g_{K^+ n\Sigma^-}\, K^+ \bar{n}\, i\gamma^5 \Sigma^- + \cdots + h.c., \qquad (3.194)$$

where the dots refer to other mesons and hyperons. Show that $g_{K^+ n\Sigma^-}$ can be expressed in terms of $g_{\pi^+ pn}$ and $g_{K^+ p\Lambda}$ in the $SU(3)$ limit, and find the expression.

3.20 Starting from the ansatz in (3.98), derive the Gell-Mann-Okubo mass relation (3.105) for baryons.

3.21 From (3.89), the electromagnetic current can be written as $J_{Q\mu} = J^3_\mu + \frac{1}{\sqrt{3}} J^8_\mu$, where J^3_μ and J^8_μ are elements of an $SU(3)$ octet of currents. $J_{Q\mu}$ is a singlet under the U-spin subgroup of $SU(3)$, which has generators $U^3 = -\frac{1}{2}T^3 + \sqrt{\frac{3}{4}}T^8$, $U^\pm = T^6 \pm iT^7$. (a) Use U-spin to show that in the $SU(3)$ limit the hyperon magnetic moments are related by

$$\mu_{\Sigma^+} = \mu_p, \qquad \mu_n = \mu_{\Xi^0} = -\frac{1}{2}\mu_{\Sigma^0} + \frac{3}{2}\mu_\Lambda$$

$$\mu_{\Sigma^-} = \mu_{\Xi^-}, \qquad \mu_{\Sigma\Lambda} = \frac{\sqrt{3}}{2}\left(\mu_{\Sigma^0} - \mu_\Lambda\right),$$

where $\mu_{\Sigma\Lambda}$ is the transition moment observed in $\Sigma^0 \to \Lambda\gamma$.
(b) Use isospin to show that $\mu_{\Sigma^0} = \frac{1}{2}\left(\mu_{\Sigma^+} + \mu_{\Sigma^-}\right)$.
(c) Use $SU(3)$ to show that $\mu_{\Sigma^0} = -\mu_\Lambda$, allowing all of the magnetic moments to be expressed in terms of μ_p and μ_n. Hint: use the analog of (3.102).

3.22 In the non-relativistic quark model the baryon octet is approximated as three-quark states that are totally symmetric in flavor, space, and spin but antisymmetric in color. The total and internal orbital angular momenta are zero.
(a) Show that the flavor-spin part of the proton wave function is

$$|p+\rangle = \frac{1}{3\sqrt{2}}\Big[|uud\rangle \times \Big(2|++-\rangle - |+-+\rangle - |-++\rangle\Big) + \big\{|123\rangle \to |231\rangle\big\} + \big\{(|123\rangle \to |312\rangle\big\}\Big],$$

where $\pm$ denote spin projections of $\pm\frac{1}{2}$, and the second and third terms are cyclic permutations on the three quarks. The neutron wave function is similar except $u \leftrightarrow d$.
(b) From (2.395) the proton magnetic moment is $\vec{\mu}_p = g_p \mu_N \vec{S}_p = (1+\kappa_p)\mu_N \vec{\sigma}_p$, where $\kappa_p \sim 1.79$ is the anomalous magnetic moment. Similarly, the neutron moment is $\vec{\mu}_n = g_n \mu_N \vec{S}_n = \kappa_n \mu_N \vec{\sigma}_n$, with $\kappa_n \sim -1.91$. These can be estimated in the quark model by assuming that the quarks are pointlike, i.e., that their magnetic moment operator is $\vec{\mu} = Q\frac{e}{2M}\vec{\sigma}$, where Q has eigenvalues $e_u = \frac{2}{3}$ and $e_d = -\frac{1}{3}$, and $M \sim m_p/3$ is the constituent quark mass. The nucleon moments can then be calculated by taking the expectation value of $\sum_{a=1,2,3} \vec{\mu}_a$, where $\vec{\mu}_a$ acts on the ath quark. Calculate the ratio g_p/g_n and compare with the experimental value.
(c) Calculate the individual values of $g_p/2$ and $g_n/2$ and compare with experiment.
See (Georgi, 1999) for a more detailed discussion of the magnetic moments of the baryon octet and the relation to $SU(6)$.

3.23 Draw the Feynman diagrams and write down the amplitudes for the processes $\gamma p \to \pi^0 p$ and $\gamma p \to K^+\Sigma^0$ in the $SU(3)$ limit, using the meson-baryon Lagrangian density in (3.97). Ignore strong interaction effects in the couplings of the photon. (It is *not* necessary to actually calculate $|M|^2$.)

3.24 Prove that the operator $\epsilon^{\alpha_1\cdots\alpha_m}\psi_{\alpha_1\cdots\alpha_m}$ in (3.120) is invariant under $SU(m)$ but not under $U(m)$.

3.25 Using (3.116), the fields

$$\psi^A_{ab} \equiv \frac{1}{\sqrt{2}}(\psi_{ab} - \psi_{ba}), \qquad \psi^S_{ab} \equiv c_{ab}(\psi_{ab} + \psi_{ba}) \quad \text{[no sum]}$$

transform, respectively, as antisymmetric (A) and symmetric (S) products of two $SU(m)$ fundamentals. Here, $c_{ab} = 1/\sqrt{2}$ for $a \neq b$ and $c_{aa} = \frac{1}{2}$. The coefficients are chosen so that the corresponding states are properly normalized.
(a) Calculate the representation matrices $\left(L_A^i\right)_{ab;cd}$ (in terms of the fundamental $L^i \equiv L_m^i$) and the Dynkin index $T(L_A)$. Show that $T(L_A)$ has the expected value for $m = 2$ and 3.
(b) Calculate $\left(L_S^i\right)_{ab;cd}$ and $T(L_S)$. Show that the L_S^i for $m = 2$ are of the correct form. Hint: be careful to avoid double counting in sums over the elements of $\psi^{A,S}$.

3.26 Let ϕ_{ab}, $a, b = 1 \cdots n$ be a complex scalar transforming as (n, n^*) under $SU(n)_L \times SU(n)_R$, i.e.,

$$[T_L^i, \phi_{ab}] = -(L_n^i)_{ac}\phi_{cb}, \qquad [T_R^i, \phi_{ab}] = +\phi_{ac}(L_n^i)_{cb},$$

where T_L^i and T_R^i are, respectively, the generators of $SU(n)_L$ and $SU(n)_R$.
(a) Calculate a finite transformation of ϕ.
(b) Show that the Yukawa interaction

$$\mathcal{L}_Y = h\bar{\psi}_{aL}\phi_{ab}\psi_{bR} + h^*\bar{\psi}_{bR}(\phi_{ab})^\dagger\psi_{aL} \equiv h\bar{\psi}_L\phi\psi_R + h^*\bar{\psi}_R\phi^\dagger\psi_L$$

and the mass term

$$\mathcal{L}_m = -\mu^2(\phi_{ab})^\dagger\phi_{ab} \equiv -\mu^2\mathrm{Tr}\,(\phi^\dagger\phi)$$

are invariant under $U(n)_L \times U(n)_R$. In the second forms, ϕ represents an $n \times n$ matrix with components ϕ_{ab}, and $(\phi^\dagger)_{ba} \equiv (\phi_{ab})^\dagger$.
(c) Construct the most general $U(n)_L \times U(n)_R$ invariant renormalizable Lagrangian density for ψ and ϕ.
(d) Display the $U(n)_L \times U(n)_R$ Noether currents for the Lagrangian density in (c). Hint: both ϕ and $\phi^\dagger$ must be included in the sum over fields in (3.46).

3.27 Prove that the color-singlet states $|u\bar{d}\rangle$, $|d\bar{u}\rangle$, and $|u\bar{u} \pm d\bar{d}\rangle$ are eigenstates of G parity with eigenvalue $(-1)^{L+S+I}$, where L, S, and I are, respectively, the total orbital angular momentum, spin, and isospin.

3.28 Find the exact domain wall solution to (3.134) corresponding to Figure 3.4 and use this to justify that the wall thickness is $\mathcal{O}(|\mu|^{-1})$. Use the Hamiltonian density in (A.7) to compute the energy density (with respect to the ground state) at a distance x from the wall, and compute the energy per unit area in the wall. Hint: look for a solution of the form $\phi(x) = a\tanh(bx)$.

3.29 The $O(2)$ invariance of (3.146) is explicitly broken by the mass term in (3.151).
(a) Identify any unbroken symmetries and describe their consequences.
(b) Show that charge is not conserved and draw a Feynman diagram that demonstrates this, e.g., by allowing the reaction $\phi^+\phi^- \to \phi^+\phi^+\phi^+\phi^-$, where $\phi^+ = (\phi^-)^\dagger$ is the particle created by $\phi^\dagger$. Treat the symmetry breaking term as a mass insertion on an internal line. (There can also be mixing associated with the external legs, leading to processes such as $\phi^+\phi^- \to \phi^+\phi^+$. However, the degeneracy of the states for $\epsilon \to 0$ makes this case more complicated to treat. It is very much like $K^0 - \bar{K}^0$ mixing treated in Chapter 8.) Note that this is a spin-0 analog of a Majorana neutrino mass term.

3.30 Consider the spontaneously broken theory of a complex scalar field in Section 3.3.3, which is described after symmetry breaking by the Lagrangian density in (3.155) for a rectangular basis, or by (3.163) in a polar basis.

(a) Show that the on-shell amplitudes for $\phi_1'(p_1)\phi_2'(p_2) \to \phi_1'(p_3)\phi_2'(p_4)$ and $\eta(p_1)\xi(p_2) \to \eta(p_3)\xi(p_4)$ are the same at tree-level.
(b) Now consider pseudoscalar-fermion scattering $\phi_2'(p_1)\psi(p_2) \to \phi_2'(p_3)\psi(p_4)$. The Yukawa interactions are given for three parametrizations in (3.159), (3.164), and (3.166), and the corresponding scalar interactions in (3.155) and (3.163). Show that the tree-level amplitudes are the same in all three parametrizations.

3.31 In theories involving scalars and fermions it is sometimes the case that each mass eigenstate scalar field couples either as a scalar or a pseudoscalar to fermions, i.e., that there is a conserved parity. The SM and the MSSM at tree level are examples of this. As a simpler example, consider a single Dirac fermion ψ and a single complex scalar ϕ, with

$$\begin{aligned}\mathcal{L} =& \bar{\psi} i \not{\partial} \psi + (\partial^\mu \phi)^\dagger (\partial_\mu \phi) - m \bar{\psi}_L \psi_R - m^* \bar{\psi}_R \psi_L \\ &+ h \bar{\psi}_L \phi \psi_R + h^* \bar{\psi}_R \phi^\dagger \psi_L - \mu^2 \phi^\dagger \phi - \kappa \phi \phi - \kappa^* \phi^\dagger \phi^\dagger,\end{aligned}$$

where μ^2 is real, while m, h, and κ are complex. We ignore ϕ self-interactions for simplicity, and choose $\mu^2 > 2|\kappa| \geq 0$ to ensure that ϕ does not acquire a VEV. Consider the cases (a) $m = \kappa = 0$, (b) $m = 0, \kappa \neq 0$, (c) and $m \neq 0, \kappa \neq 0$, always with $h \neq 0$, $\mu^2 \neq 0$. In each case, find the spectrum, interpret it in terms of the symmetries of $\mathcal{L}$, and determine whether P, C, and CP are conserved. Hint: identify which of the phases in

$$\psi_L \to e^{i\beta_L} \psi_L, \qquad \psi_R \to e^{i\beta_R} \psi_R, \qquad \phi \to e^{i\alpha} \phi$$

are symmetries and which are field redefinitions.

3.32 The most general fermion number conserving Lagrangian density for a free Dirac field is

$$\mathcal{L} = \bar{\psi} i \not{\partial} [a + b\gamma^5] \psi - \bar{\psi} [c + i\gamma^5 d] \psi,$$

where Hermiticity requires a, b, c, and d to be real. The canonical form in (2.152), which corresponds to $a = 1$, $c = m$, and $b = d = 0$, is needed to maintain the classical relation $E^2 = \vec{p}^{\,2} + m^2$. However, the more general form is acceptable (and may emerge from an underlying theory) provided that $\mathcal{L}$ can be put into canonical form by a field redefinition prior to quantization.
(a) Show that for $a > |b| \geq 0$ (and all of the parameters real), one can find a multiplicative redefinition, $\psi' \equiv \mathcal{F}\psi$, with

$$\mathcal{F}(a, b, c, d) = \mathcal{F}_I(a, b, c, d) + \mathcal{F}_5(a, b, c, d)\gamma^5,$$

such that $\mathcal{L}(\psi')$ is canonical, i.e., $\mathcal{L} = \bar{\psi}'(i\not{\partial} - m)\psi'$, with $m \geq 0$. Construct $\mathcal{F}$ explicitly, and find $m(a, b, c, d)$. Hint: write ψ in terms of its chiral components, $\psi = \psi_L + \psi_R = P_L\psi + P_R\psi$, where it will be seen that d corresponds to a complex mass.
(b) Specialize your results to the "wrong sign mass" case, $a = 1$, $c = -|m|$, $b = d = 0$, and interpret the result.

3.33 The potential corresponding to (3.130) can be written for $\lambda_{11} = \lambda_{22}$ as

$$V(\phi_1, \phi_2) = \frac{1}{2}\mu^2(\phi_1^2 + \phi_2^2) + \frac{\lambda}{4}\left(\phi_1^2 + \phi_2^2\right)^2 + \frac{\lambda' \phi_1^2 \phi_2^2}{2},$$

which is clearly $O(2)$ invariant for $\lambda' = 0$, but only invariant under sign changes and $\phi_{1,2}$ interchange for $\lambda' \neq 0$. Assume that $\mu^2 < 0$ and $\lambda > 0$. Find the minimum, the physical particle masses, and the unbroken symmetries for the three cases $\lambda' > 0$, $\lambda' = 0$, and $\lambda' < 0$. In the last case impose the necessary boundedness condition.

3.34 Consider the potential

$$V(\phi_1, \phi_2, \phi_3) = \frac{\mu^2}{2}(\phi_1^2 + \phi_2^2 + \phi_3^2) + \mu_{12}^2 \phi_1\phi_2 + \mu_{13}^2 \phi_1\phi_3 + \mu_{23}^2 \phi_2\phi_3 + \lambda_{12} (\phi_1^2 - \phi_2^2)^2 + \lambda_3 \phi_3^4.$$

Find the conditions for boundedness (assume $\lambda_{12} \neq 0$, $\lambda_3 \neq 0$), and the special cases for which there are flat directions.

3.35 Specialize the example in Section 3.3.8 to the case $\lambda_I = \lambda_{II} = -\lambda_{III}/2 \equiv \lambda > 0$, for which the boundedness condition (3.181) does *not* hold. This is a simplified version of the two Higgs doublets in the MSSM, in which the $U(1)$ is a gauge symmetry.
(a) Find the condition on $\mu_{I,II}^2$, A, and λ for the existence of a minimum of V. Assume that A is real and non-negative.
(b) Show that $A > 0$ is a necessary condition to have both $\nu_1 \neq 0$ and $\nu_3 \neq 0$.
(c) For $A > 0$, what is the condition that the origin ($\nu_{1,3} = 0$) is not a minimum? If that is the case, is the origin a maximum or a saddlepoint?
(d) Find explicit expressions for $\nu^2 \equiv \nu_1^2 + \nu_3^2$ and $\gamma \equiv \tan^{-1}(\nu_3/\nu_1)$ in terms of $\mu_{I,II}^2$, A, and λ.
(e) Calculate the nonzero eigenvalue μ_A^2 of μ_{Im}^2 in terms of A and γ.
(f) Express the mass-square matrix μ_{Re}^2 in terms of μ_A^2, γ and $M_Z^2 \equiv 2\lambda\nu^2$ (motivated by the MSSM prototype).
(g) Prove the sum rule $\mu_a^2 + \mu_b^2 = \mu_A^2 + M_Z^2$, where $\mu_{a,b}^2$ are the μ_{Re}^2 eigenvalues.
(h) Prove that $\mu_a^2 \leq M_Z^2 \cos^2 2\gamma$ and $\mu_b^2 > \mu_A^2 + M_Z^2 \sin^2 2\gamma$, where the labelling is such that $\mu_a^2 < \mu_b^2$.
(i) Show that these inequalities are saturated for large μ_A^2/M_Z^2, and interpret this limit.

3.36 This problem involves a chiral fermion and complex scalar, with an internal global symmetry that may be both spontaneously and explicitly broken. Consider the Lagrangian density

$$\begin{aligned}\mathcal{L} &= \bar{\psi} i \not{\partial} \psi + \partial_\mu \phi^\dagger \partial^\mu \phi - h[\bar{\psi}_L \phi \psi_R + \bar{\psi}_R \phi^\dagger \psi_L] \\ &\quad - \mu^2 \phi^\dagger \phi - \lambda (\phi^\dagger \phi)^2 + a(\phi + \phi^\dagger),\end{aligned}$$

where $\psi = \psi_L + \psi_R$ is a fermion, ϕ is a complex scalar, $\lambda > 0$, $h > 0$, and $a \geq 0$.
(a) Show that $\mathcal{L}$ has a global chiral symmetry $U(1) \times U(1)$ for the case $a = 0$.
(b) Calculate the spectrum of the model (i.e., the masses) for the cases (i) $\mu^2 > 0, a = 0$ and (ii) $\mu^2 < 0, a = 0$.
(c) Calculate the spectrum for the cases (i) $\mu^2 > 0, a > 0$ and (ii) $\mu^2 < 0, a > 0$. In each case assume that $\sqrt{\lambda} a/|\mu|^3 \ll 1$, and keep only the leading nonzero term in that parameter.
(d) Interpret the spectrum in each of the above cases in terms of the symmetries and symmetry breaking.
(e) Now add the Majorana mass term $-\frac{m_L}{2}(\bar{\psi}_L \psi_R^c + \bar{\psi}_R^c \psi_L)$ to $\mathcal{L}$, where ψ_R^c defined in (2.301) is essentially the CP conjugate of ψ_L. (An analogous term could be added for ψ_R.) Show that this violates the $U(1)$ fermion number symmetry. Calculate the divergence of the corresponding Noether current.

CHAPTER 4

Gauge Theories

DOI: 10.1201/b22175-4

In Chapter 3 we considered continuous *global* symmetries, parametrized by real constants β^i. In *local* or *gauge* symmetries[1] the β^i are promoted to arbitrary differentiable functions $\beta^i(x)$ of space and time. Gauge invariance is sometimes motivated on esthetic grounds. For example, why should the phase of the electron field on the Earth be correlated with its phase on Mars? This is not entirely compelling because the standard model does involve global symmetries (though they may derive from gauge symmetries in an underlying theory). In any case, we will take the pragmatic view that gauge invariance is a powerful tool for constructing well-behaved field theories and are the unique renormalizable field theories for spin-1 particles. In particular, each generator of a gauge invariant theory must correspond to an (apparently) massless spin-1 *gauge boson*, which mediates an (apparently) long-range force, and the diagonal generators of an unbroken gauge theory correspond to conserved charges (Weyl, 1929). The gauge interactions are uniquely prescribed once one specifies the gauge group, the representations of the matter fields, and a gauge coupling constant g for each group factor. This approach is opposite to historical development: Maxwell's equations of classical electrodynamics were first derived from observation and consistency, and then it was noticed that they were invariant under gauge transformations, i.e., that the vector and scalar potential involved redundant degrees of freedom. In this chapter, we outline the construction of gauge invariant Lagrangian densities. More detailed treatments include (Abers and Lee, 1973; Weinberg, 1973c,d, 1995; Peskin and Schroeder, 1995).

Let Φ_a, $a = 1 \cdots n$, represent n spin-0 or $\frac{1}{2}$ fields that transform as

$$\Phi_a(x) \to \left(e^{i\vec{\beta}(x)\cdot\vec{L}}\right)_{ab} \Phi_b(x) \equiv \left[U\big(\vec{\beta}(x)\big)\right]_{ab} \Phi_b(x) \tag{4.1}$$

under a gauge transformation (generalizing (3.30)). L^i, $i = 1 \cdots N$, are $n \times n$ representation matrices of the Lie algebra of the gauge group G. Equation (4.1) is equivalent to

$$\Phi(x) \to \Phi'(x) \equiv e^{i\vec{\beta}(x)\cdot\vec{L}}\Phi(x) = U\big(\vec{\beta}(x)\big)\Phi(x), \tag{4.2}$$

where $\Phi(x)$ is an n component column vector with components $\Phi_a(x)$. If a theory involves n_ψ fermion fields ψ_a in a column vector ψ and n_ϕ real or complex scalars ϕ_c in a vector ϕ, then

$$\begin{aligned} \psi(x) &\to e^{i\vec{\beta}(x)\cdot\vec{L}_\psi}\psi(x) \equiv U_\psi\big(\vec{\beta}(x)\big)\psi(x) \\ \phi(x) &\to e^{i\vec{\beta}(x)\cdot\vec{L}_\phi}\phi(x) \equiv U_\phi\big(\vec{\beta}(x)\big)\phi(x), \end{aligned} \tag{4.3}$$

[1]Gauge transformations are often referred to as redundancies rather than symmetries because they refer to unobservable degrees of freedom rather than relating different systems.

where L_ψ and L_ϕ are respectively the fermion and scalar representation matrices. In the case of a chiral symmetry,

$$L^i_\psi = L^i_L P_L + L^i_R P_R \text{ with } L^i_L \neq L^i_R, \tag{4.4}$$

so that

$$\begin{aligned}\psi_L(x) &\to e^{i\vec{\beta}(x)\cdot\vec{L}_L}\psi_L(x) \equiv U_L(\vec{\beta}(x))\psi_L(x)\\ \psi_R(x) &\to e^{i\vec{\beta}(x)\cdot\vec{L}_R}\psi_R(x) \equiv U_R(\vec{\beta}(x))\psi_R(x).\end{aligned} \tag{4.5}$$

The transformations of the necessary gauge fields will be detailed below.

4.1 THE ABELIAN CASE

As a first example, we take $G = U(1)$. We already considered the electromagnetic interactions of charged pions and electrons in Sections 2.6 and 2.8, but repeat the key results. Under a (non-chiral) gauge transformation

$$\psi \to e^{-i\beta(x)}\psi, \qquad \phi_{\pi^\pm} \to e^{\pm i\beta(x)}\phi_{\pi^\pm}, \qquad A_\mu \to A_\mu - \frac{1}{e}\partial_\mu\beta, \tag{4.6}$$

where ψ is the electron field, ϕ_{π^+} and $\phi_{\pi^-} = \phi^\dagger_{\pi^+}$ are respectively the π^+ and π^- fields, A is the photon (gauge) field γ, and $e > 0$. The gauge covariant derivatives transform as

$$\begin{aligned}D^\mu\psi &\equiv (\partial^\mu - ieA^\mu)\,\psi \to e^{-i\beta(x)}D^\mu\psi\\ D^\mu\phi_{\pi^\pm} &\equiv (\partial^\mu \pm ieA^\mu)\,\phi_{\pi^\pm} \to e^{\pm i\beta(x)}D^\mu\phi_{\pi^\pm}.\end{aligned} \tag{4.7}$$

Thus,

$$\mathcal{L} = \bar{\psi}\,(i\not{D} - m)\,\psi + (D_\mu\phi_{\pi^+})^\dagger D^\mu\phi_{\pi^+} - \frac{1}{4}F_{\mu\nu}F^{\mu\nu} - \mu^2\phi^\dagger_{\pi^+}\phi_{\pi^+} - \lambda\left(\phi^\dagger_{\pi^+}\phi_{\pi^+}\right)^2, \tag{4.8}$$

where $F_{\mu\nu} = \partial_\mu A_\nu - \partial_\nu A_\mu$ is gauge invariant. However, an explicit photon mass term $\frac{M_A^2}{2}A_\mu A^\mu$ is *not* gauge invariant, so the γ must be massless. The Feynman rules for the gauge vertices are given in Figures 2.10 and 2.13. They are unique except for e and the charge assignments. However, the mass parameters are arbitrary. Only one pion self-interaction is consistent with $U(1)$, but the coefficient λ is arbitrary.

These considerations are easily generalized to an arbitrary non-chiral $U(1)$, with n_ψ fermions ψ_a and n_ϕ complex scalars ϕ_b, with charges q_a and q'_b in units of the gauge coupling g. (The simple QED example had $q_{e^-} = -1$, $q'_{\pi^+} = +1$, and $g = e$.) The fields transform as

$$\psi_a \to e^{iq_a\beta(x)}\psi_a, \qquad \phi_b \to e^{iq'_b\beta(x)}\phi_b, \qquad A_\mu \to A_\mu - \frac{1}{g}\partial_\mu\beta(x), \tag{4.9}$$

where A is the gauge boson. One can think of the charges as the elements of the (reducible) diagonal fermion and scalar representation matrices $L_\psi = \text{diag}\,(q_1\,q_2\cdots q_{n_\psi})$ and $L_\phi = \text{diag}\left(q'_1\,q'_2\cdots q'_{n_\phi}\right)$. The gauge covariant derivatives are

$$\begin{aligned}D^\mu\psi_a &= (\partial^\mu + igq_aA^\mu)\,\psi_a \to e^{iq_a\beta(x)}D^\mu\psi_a\\ D^\mu\phi_b &= (\partial^\mu + igq'_bA^\mu)\,\phi_b \to e^{iq'_b\beta(x)}D^\mu\phi_b,\end{aligned} \tag{4.10}$$

or in column vector notation,

$$D^\mu \psi = (\partial^\mu I + igA^\mu L_\psi)\,\psi \to e^{i\beta(x)L_\psi} D^\mu \psi$$
$$D^\mu \phi = (\partial^\mu I + igA^\mu L_\phi)\,\phi \to e^{i\beta(x)L_\phi} D^\mu \phi, \tag{4.11}$$

where I is the $n_\psi \times n_\psi$ or $n_\phi \times n_\phi$ identity. The kinetic terms

$$\mathcal{L}_0 = \bar{\psi} i \not{D} \psi \;+\; \underbrace{(D_\mu \phi)^\dagger}_{\phi^\dagger\left(\overleftarrow{\partial}_\mu I - igA_\mu L_\phi\right)} D^\mu \phi - \frac{1}{4} F_{\mu\nu} F^{\mu\nu} \tag{4.12}$$

are gauge invariant. These lead to vertices with the massless gauge boson similar to those in Figures 2.10 and 2.13, except $e \to g q'_b$ for ϕ_b and $-e \to g q_a$ for ψ_a. (There are no off-diagonal transitions such as $\psi_1 \to \psi_2$.)

It is clear that only gq_a and gq'_b are physical. One can always rescale g provided q_a, q'_b are rescaled accordingly. For example, this freedom is used in QED to set the electron charge to -1. In a pure $U(1)$ theory, the relative values of the charges are arbitrary. However, if the $U(1)$ is a subgroup of a simple group, then the relative charges are fixed by the higher symmetry.

It is straightforward to extend these considerations to a chiral $U(1)$, i.e., in which the L and R charges $q_{aL,R}$ of ψ_a are different (this is *not* the case for QED). Define $L_\psi = L_L P_L + L_R P_R$, where $L_{L,R}$ are the diagonal charge matrices of $\psi_{L,R}$ and $P_{L,R}$ are the chiral projections in (2.196) on page 40. Then, the fermion term in (4.12) becomes

$$\bar{\psi} i \not{D} \psi = \bar{\psi}\,(i\not{\partial} I - g\not{A} L_\psi)\,\psi = \bar{\psi}\,(i\not{\partial} I - g\not{A}[L_L P_L + L_R P_R])\,\psi, \tag{4.13}$$

and the ψ_a vertex in Figure 2.13 is

$$-ig\gamma_\mu \left[\left(\frac{q_{aL} + q_{aR}}{2} \right) - \left(\frac{q_{aL} - q_{aR}}{2} \right) \gamma^5 \right]. \tag{4.14}$$

One can add mass and additional non-derivative interaction terms to $\mathcal{L}_0$ provided they are invariant under the global $U(1)$. For example,[2]

$$\mathcal{L} = \mathcal{L}_0 - \bar{\psi}_a m_{ab} \psi_b - \phi_c^\dagger \mu_{cd}^2 \phi_d$$
$$+ \left[\bar{\psi}_a \Gamma^c_{ab} \psi_b \phi_c + \bar{\psi}_a i \hat{\Gamma}^c_{ab} \gamma^5 \psi_b \phi_c + h.c. \right] + \lambda_{abcd} \phi_a^\dagger \phi_b^\dagger \phi_c \phi_d \tag{4.15}$$

would be invariant provided $q_a = q_b$ for nonzero m_{ab}, $q'_c = q'_d$ for nonzero μ^2_{cd}, $q_a = q_b + q'_c$ for nonzero Γ^c_{ab} or $\hat{\Gamma}^c_{ab}$, and $q'_a + q'_b = q'_c + q'_d$ for nonzero λ_{abcd}.

4.2 NON-ABELIAN GAUGE THEORIES

Now consider a non-abelian gauge symmetry (Yang and Mills, 1954), in which the spin-0 and $\frac{1}{2}$ fields transform according to (4.3), where G is a simple group. Any mass terms, Yukawa interactions, or non-derivative scalar self-interactions that are invariant under the corresponding global symmetry are automatically gauge invariant as well. However, the fermion or scalar kinetic energy terms contain derivatives and are therefore not gauge invariant,

$$\partial^\mu \Phi \to \partial^\mu \left[U(\vec{\beta}(x))\,\Phi \right] = U \partial^\mu \Phi + [\partial^\mu U]\,\Phi. \tag{4.16}$$

[2]In (4.15) the matrix μ^2 must be Hermitian, and there are constraints on the λ_{abcd} from Hermiticity and Bose statistics. m_{ab} is assumed Hermitian, but non-Hermitian fermion mass matrices can be introduced by rewriting in terms of $\psi_{L,R}$ (or allowing γ^5 terms). Other types of cubic or quartic scalar self-interactions could be allowed if they are globally invariant, such as $\phi_a \phi_b \phi_c + h.c.$ for $q'_a + q'_b + q'_c = 0$.

One must therefore introduce a gauge covariant derivative

$$\partial^\mu \to D^\mu \equiv \partial^\mu + ig\vec{A}^\mu \cdot \vec{L}, \tag{4.17}$$

where A^i_μ, $i = 1 \cdots N$, are real vector gauge fields (one per generator), g is an arbitrary (real) gauge coupling, $\vec{A}^\mu \cdot \vec{L} = A^{i\mu} L^i$, and L^i are the representation matrices for Φ, i.e., $L^i = L^i_\phi, L^i_\psi$, or $L^i_L P_L + L^i_R P_R$. Of course, $\partial^\mu \equiv \partial^\mu I$, where I is the $n \times n$ identity matrix. In terms of components,

$$(D^\mu \Phi)_a = \left[\partial^\mu \delta_{ab} + ig\vec{A}^\mu \cdot \vec{L}_{ab}\right] \Phi_b. \tag{4.18}$$

These gauge covariant derivatives specify the interactions of the spin-0 or spin-$\frac{1}{2}$ fields with the gauge bosons in terms of a single gauge coupling g (or one per group factor for a non-simple group), once the group and representations are specified. The Feynman rules for the gauge interactions are shown in Figure 4.1. For example, the fermion kinetic energy term

$$\mathcal{L}_{KE_\psi} = \bar{\psi} i \not{D} \psi \tag{4.19}$$

implies that the amplitude for ψ_b to absorb or emit gauge boson A^i and turn into ψ_a is just

$$-ig\gamma^\mu \left[L^i_{Lab} \frac{1-\gamma^5}{2} + L^i_{Rab} \frac{1+\gamma^5}{2}\right] \xrightarrow[L_L = L_R = L]{} -ig\gamma^\mu L^i_{ab}. \tag{4.20}$$

This must be sandwiched between appropriate u or v spinors and contracted with a gauge polarization vector, as described in Chapter 2. Unlike the abelian case, the non-diagonal generators imply transitions between the different members of the fermion (or scalar) IRREP that are related by the symmetry.

Similarly, the kinetic energy term for a complex scalar representation becomes

$$(D^\mu \phi)^\dagger D_\mu \phi = (\partial^\mu \phi)^\dagger \partial_\mu \phi - igA^i_\mu \left(\phi^\dagger \overleftrightarrow{\partial}^\mu L^i \phi\right) + g^2 A^{i\mu} A^j_\mu \phi^\dagger L^i L^j \phi. \tag{4.21}$$

The second term on the right yields the three-point vertex in Figure 4.1 since $\partial^\mu \phi \to -i\bar{p}^\mu \phi$ ($\partial^\mu \phi^\dagger \to +i\bar{p}^\mu \phi^\dagger$), where $\bar{p}$ is the momentum entering (exiting) the vertex for ϕ ($\phi^\dagger$), i.e., $\bar{p}$ always flows in the direction of the arrow. The four-point vertex from $i\mathcal{L}_{KE_\phi}$ follows from the last term, taking into account that there are two ways to contract $A^{i\mu} A^j_\mu$ with the external gauge fields. The vertices for Hermitian scalars are considered in Problem 4.6.

We must still determine the transformation of A^i_μ. The first requirement is

$$D^\mu \Phi \to U D^\mu \Phi. \tag{4.22}$$

This implies that the fermion and scalar kinetic energy terms

$$\begin{aligned}\mathcal{L}_{KE_{\psi,\phi}} &= \bar{\psi} i \not{D}_\psi \psi + \left(D^\mu_\phi \phi\right)^\dagger D_{\phi\mu} \phi \\ &\to \bar{\psi} i U^\dagger_\psi U_\psi \not{D}_\psi \psi + \underbrace{\left(D^\mu_\phi \phi\right)^\dagger}_{\phi^\dagger \left(\overleftarrow{\partial}^\mu I - ig\vec{A}^\mu \cdot \vec{L}_\phi\right)} U^\dagger_\phi U_\phi D_{\phi\mu} \phi\end{aligned} \tag{4.23}$$

are gauge invariant, since $U^\dagger U = I$. The necessary transformation is given by

$$\vec{A}_\mu \cdot \vec{L} \to \vec{A}'_\mu \cdot \vec{L} \equiv U \vec{A}_\mu \cdot \vec{L} U^{-1} + \frac{i}{g} (\partial_\mu U) U^{-1}, \tag{4.24}$$

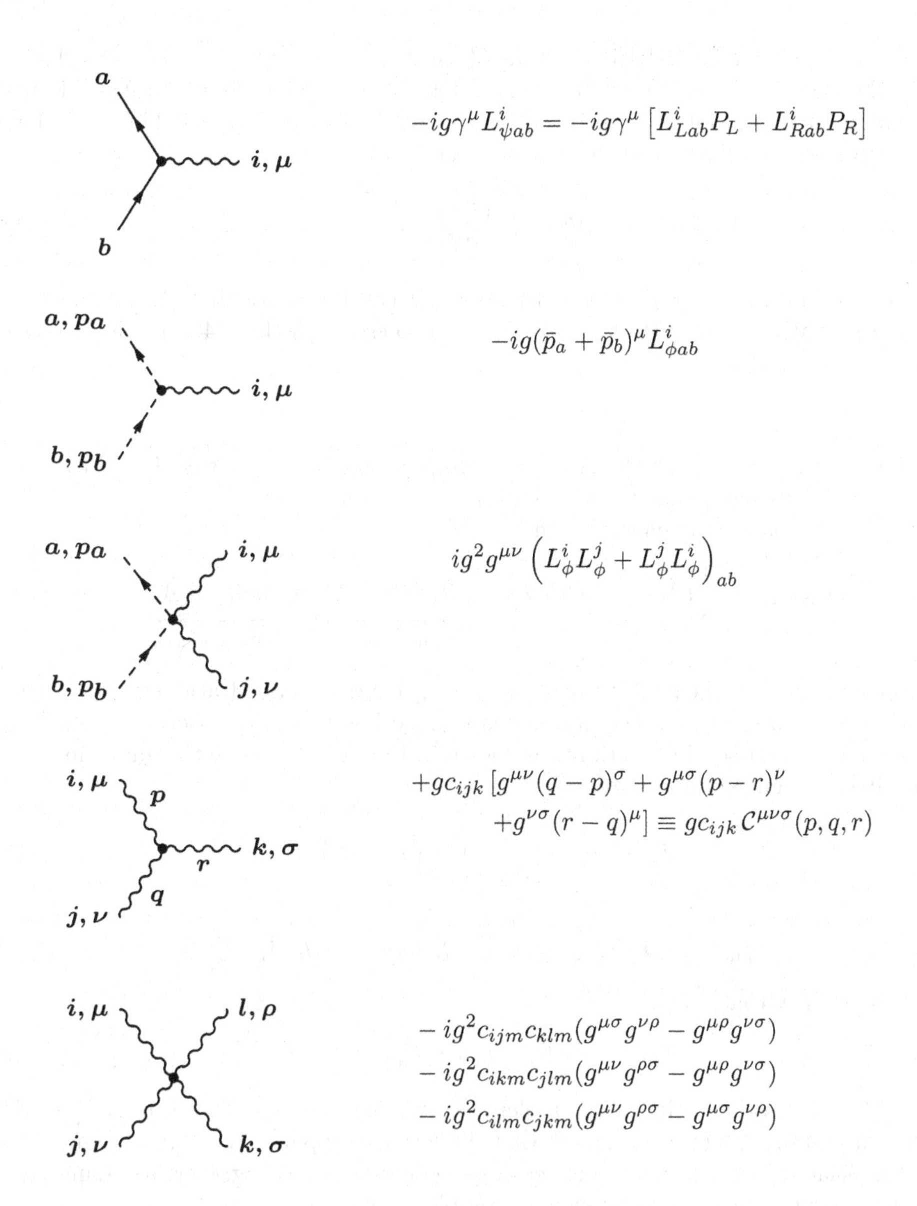

Figure 4.1 Feynman rules for the interactions of gauge bosons with fermions, scalars, and gauge self-interactions. The scalar vertices apply to both complex or Hermitian scalars. $\bar{p}_a$ and $\bar{p}_b$ in the second vertex are the momenta flowing in the direction of the arrows, which in some cases are the negative of the physical momenta. For example, $\bar{p}_a = -p_a$ if the a is twisted downward to represent an incident antiparticle or Hermitian scalar, while $\bar{p}_b = -p_b$ for an outgoing antiparticle or Hermitian scalar, as in Figure 2.10. Recall that $L^i_{\phi ab} = -L^i_{\phi ba}$ for a Hermitian scalar. In the triple gauge vertex the momenta p, q, and r all flow into the vertex and satisfy $p+q+r=0$. The function $\mathcal{C}^{\mu\nu\sigma}(p,q,r)$ is totally antisymmetric if the indices and corresponding momenta are both interchanged.

where L^i is any nontrivial IRREP, such as L^i_ψ or L^i_ϕ, and $U \equiv U(x) = e^{i\vec{\beta}(x)\cdot\vec{L}}$. The first term on the right is the expected transformation for an adjoint representation. It would be present even for a global transformation, as in (3.32) on page 97. The second term generalizes (3.32) to a gauge transformation. One has

$$A'^i_\mu \equiv \frac{\text{Tr}\,(L^i \vec{A}'_\mu \cdot \vec{L})}{T(L)}, \tag{4.25}$$

where the Dynkin index $T(L)$ is defined in (3.8). It can be shown that A'^i_μ is independent of the representation L. It is then straightforward to establish that (4.22) holds. For small $\vec{\beta}(x)$, (4.24) reduces to

$$A'^i_\mu = A^i_\mu - c_{ijk}\beta^j A^k_\mu - \frac{1}{g}\partial_\mu \beta^i, \tag{4.26}$$

where the first term represents the global transformation of an adjoint field, and the second is reminiscent of abelian gauge transformations.

The gauge boson kinetic energy term is

$$\mathcal{L}_{KE_A} = -\frac{1}{4}F^i_{\mu\nu}F^{i\mu\nu} \text{ with } F^i_{\mu\nu} = \underbrace{\partial_\mu A^i_\nu - \partial_\nu A^i_\mu}_{\text{kinetic energy}} - \underbrace{g c_{ijk} A^j_\mu A^k_\nu}_{\text{self-interactions}}. \tag{4.27}$$

Unlike the abelian case, the field strength tensor $F^i_{\mu\nu}$ contains a quadratic term proportional to the structure constant, i.e., the non-abelian gauge bosons are themselves charged. This leads to 3 and 4 point self-interactions, as shown in Figure 4.1. To see the gauge invariance of $\mathcal{L}_{KE_A}$ it is convenient to rewrite

$$\mathcal{L}_{KE_A} = -\frac{1}{4T(L)}\text{Tr}\left[\left(\vec{F}_{\mu\nu}\cdot\vec{L}\right)^2\right] \tag{4.28}$$

where

$$\vec{F}_{\mu\nu}\cdot\vec{L} = \partial_\mu \vec{A}_\nu \cdot \vec{L} - \partial_\nu \vec{A}_\mu \cdot \vec{L} + ig\left[\vec{A}_\mu \cdot \vec{L}, \vec{A}_\nu \cdot \vec{L}\right]. \tag{4.29}$$

One can then show that

$$\vec{F}_{\mu\nu}\cdot\vec{L} \to \vec{F}'_{\mu\nu}\cdot\vec{L} = U\vec{F}_{\mu\nu}\cdot\vec{L}U^{-1}, \tag{4.30}$$

so that (4.28) is invariant. However, an elementary gauge boson mass term $\frac{1}{2}M^2_{ij}A^{i\mu}A^j_\nu$ is not invariant, establishing the statement that there is one (apparently) massless gauge boson and the associated long range force for each gauge generator. Altogether, the Langrangian density for a gauge theory of fermions and complex scalars is

$$\mathcal{L} = -\frac{1}{4}F^i_{\mu\nu}F^{i\mu\nu} + \bar{\psi}i\not{D}_\psi\psi + \left(D^\mu_\phi\phi\right)^\dagger D_{\phi\mu}\phi + \cdots \tag{4.31}$$

Scalar or fermion mass terms, Yukawa couplings, and non-derivative scalar self-interactions consistent with the global symmetry can be included.

Direct Product Groups

The formalism can easily be extended to the case in which the gauge group G is a direct product of two or more simple or abelian factors, e.g., $G = G_1 \times G_2$. The gauge covariant derivatives in (4.18) for scalar or fermion fields become

$$D^\mu\Phi = \left[\partial^\mu + ig_1\vec{A}^\mu_1\cdot\vec{L}_1 + ig_2\vec{A}^\mu_2\cdot\vec{L}_2\right]\Phi, \tag{4.32}$$

where g_m, $\vec{A}_m$, and $\vec{L}_m$, $m = 1, 2$, are, respectively, the gauge coupling, gauge bosons, and representation matrices for G_m. Similarly, the gauge boson kinetic energy terms become a sum over those for the group factors

$$\mathcal{L}_{KE_A} = -\frac{1}{4}F^i_{1\mu\nu}F^{i\mu\nu}_1 - \frac{1}{4}F^i_{2\mu\nu}F^{i\mu\nu}_2. \tag{4.33}$$

The field strength tensors are unchanged,

$$F^i_{m\mu\nu} = \partial_\mu A^i_{m\nu} - \partial_\nu A^i_{m\mu} - g_m\, c_{mijk} A^j_{m\mu} A^k_{m\nu}. \tag{4.34}$$

where c_{mijk} are the structure constants for G_m. Thus, the fermion and scalar 3-point vertices in Figure 4.1 apply separately to G_1 and G_2. The same is true for the gauge self-interactions, which do not connect the gauge bosons of different group factors because they commute. However, scalar multiplets that transform under both groups lead to mixed seagull diagrams. The Feynman rule for the mixed $\phi^\dagger_a \phi_b A^{i\mu}_1 A^{j\nu}_2$ vertex is

$$2ig_1g_2g^{\mu\nu}\left(L^i_{1\phi}L^j_{2\phi}\right)_{ab}. \tag{4.35}$$

4.3 THE HIGGS MECHANISM

We have seen that gauge theories do not allow elementary mass terms for gauge bosons, because they would break the gauge invariance and spoil the renormalizability. This appears problematic for the weak interactions, which are short-ranged and require massive mediators.

Another potential problem, for theories with spontaneous symmetry breaking, is that they imply massless Goldstone bosons, but there are no known exactly-massless Goldstone bosons associated with the elementary particle interactions. Fortunately, when G is a gauge symmetry the two problems of the unwanted Goldstone bosons and the unwanted massless gauge bosons can cure each other when the symmetry is spontaneously broken (Anderson, 1963). A particularly simple implementation is the Higgs mechanism, involving elementary spin-0 fields (Higgs, 1964, 1966; Englert and Brout, 1964; Guralnik et al., 1964). Instead of existing as a massless spin-0 particle, the degree of freedom carried by the Goldstone boson manifests itself as the longitudinal spin component of a gauge boson, which has in the process acquired a mass. (One says that the Goldstone boson has been "eaten.") Remarkably, SSB via the Higgs mechanism preserves the renormalizability of the theory ('t Hooft, 1971a; 't Hooft and Veltman, 1972; Lee and Zinn-Justin, 1972, 1973).

To illustrate this, consider a $U(1)$ gauge theory with a single complex scalar field ϕ, with

$$\mathcal{L} = -\frac{1}{4}F_{\mu\nu}F^{\mu\nu} + [(\partial_\mu + igA_\mu)\phi]^\dagger (\partial^\mu + igA^\mu)\phi - \mu^2\phi^\dagger\phi - \lambda(\phi^\dagger\phi)^2 \tag{4.36}$$

as in (2.133) or (4.8). We studied the analogous problem of a global $U(1)$ symmetry in Section 3.3.3. It was found that for $\mu^2 < 0$, ϕ acquired a VEV,

$$\langle 0|\phi|0\rangle = \frac{\nu}{\sqrt{2}}, \qquad \nu = \sqrt{\frac{-\mu^2}{\lambda}}. \tag{4.37}$$

Expanding $\phi \equiv (\nu + \sigma + i\chi)/\sqrt{2}$ around $\nu/\sqrt{2}$, there was one massive physical scalar η and one massless Goldstone boson χ.

The minimization of the potential for a gauge symmetry is the same as for the global case, and (4.37) continues to hold. Rewriting (4.36) in terms of the new fields,

$$\begin{aligned}\mathcal{L} = &-\frac{1}{4}F_{\mu\nu}F^{\mu\nu} + \frac{1}{2}(\partial_\mu\chi)^2 + \frac{g^2\nu^2}{2}A_\mu A^\mu + g\nu A_\mu\partial^\mu\chi \\ &+\frac{g^2}{2}A_\mu A^\mu\left[2\nu\sigma + \sigma^2 + \chi^2\right] + gA^\mu\left(\sigma\overleftrightarrow{\partial}_\mu\chi\right) \\ &+\frac{1}{2}(\partial_\mu\sigma)^2 - \frac{\mu_\sigma^2}{2}\sigma^2 - \lambda\nu\sigma(\sigma^2+\chi^2) - \frac{\lambda}{4}(\sigma^2+\chi^2)^2,\end{aligned} \tag{4.38}$$

where $\mu_\sigma^2 = -2\mu^2$ and the constant term in (3.155) is not displayed. We see that in addition to the kinetic energy terms for the gauge fields and for σ and χ there are two new quadratic terms. The third term in (4.38) has the form of a mass term for the gauge field with mass $g\nu$. This mass can be interpreted as arising from the interaction of the gauge field with the condensate of Higgs fields. The fourth term in (4.38) is proportional to the gauge field times the derivative of the Goldstone boson field χ. To interpret it, we can recombine the χ and A terms as

$$\frac{g^2\nu^2}{2}\left(A'_\mu\right)^2, \qquad A'_\mu \equiv A_\mu + \frac{\partial_\mu\chi}{g\nu}, \tag{4.39}$$

suggesting that $\partial_\mu\chi/g\nu$ is the longitudinal component of a now massive vector field A'_μ. However, χ still enters the cubic and quartic terms in a way that is hard to interpret.

To see what is going on it is useful to use the Kibble reparametrization (Kibble, 1967), which makes the physical particle content manifest and was already introduced for the global case in Section 3.3.5. While working at the level of the classical field theory, one can define new fields η, ξ related to σ and χ by a non-linear field redefinition:

$$\phi = \frac{\nu+\sigma+i\chi}{\sqrt{2}} = \exp(i\xi/\nu)\left(\frac{\nu+\eta}{\sqrt{2}}\right). \tag{4.40}$$

For small σ and χ one has approximately

$$\exp(i\xi/\nu)\left(\frac{\nu+\eta}{\sqrt{2}}\right) \sim \frac{\nu+\eta+i\xi}{\sqrt{2}}, \tag{4.41}$$

so that $\eta\sim\sigma$ and $\xi\sim\chi$. The fields η and ξ therefore represent the massive and Goldstone scalars, respectively. In terms of the new fields,

$$\mathcal{L} = -\frac{1}{4}F'_{\mu\nu}F^{\mu\nu\prime} + \frac{g^2\nu^2}{2}A'_\mu A'^\mu\left[1+\frac{\eta}{\nu}\right]^2 + \frac{1}{2}(\partial_\mu\eta)^2 - \frac{\mu_\eta^2}{2}\eta^2 - \lambda\nu\eta^3 - \frac{\lambda}{4}\eta^4, \tag{4.42}$$

where

$$A'_\mu \equiv A_\mu + \frac{\partial_\mu\xi}{g\nu}, \qquad F'_{\mu\nu} = \partial_\mu A'_\nu - \partial_\nu A'_\mu, \tag{4.43}$$

which is clearly the Lagrangian density for a massive vector field, including 3- and 4-point gauge and self interactions for the massive scalar η. The Goldstone boson has disappeared from the theory.

One can interpret (4.42) as the result of choosing a special gauge. Before quantizing one can make a $U(1)$ gauge transformation[3] as in (4.6), choosing $\beta(x) = -\xi(x)/\nu$. Then

$$\phi(x)\to\phi'(x) = e^{-i\xi(x)/\nu}\phi = \left(\frac{\nu+\eta}{\sqrt{2}}\right), \qquad A_\mu(x)\to A'_\mu = A_\mu + \frac{\partial_\mu\xi}{g\nu}. \tag{4.44}$$

[3] $\xi\to\xi' = \xi + \beta(x)\nu$, which is known as a shift transformation.

Equation (4.42) then follows from (4.36) by gauge invariance, $\mathcal{L}(A, \phi) = \mathcal{L}(A', \phi')$. This *unitary gauge* is useful because it makes the physical particle content of the theory manifest. However, it is not so useful for explicit calculations, especially in higher orders where it is rather singular.

The number of degrees of freedom was not changed by the Higgs mechanism. Before SSB there were two massless gauge degrees of freedom and two Hermitian scalars, while afterwards there are three gauge degrees of freedom and one massive real scalar.

Now let us consider the non-abelian case with an arbitrary scalar sector. Just as in the general discussion of the Nambu-Goldstone theorem, it is more convenient to choose a Hermitian basis for the scalars for formal manipulations, though a complex basis may be simpler for concrete calculations. We assume there are n Hermitian scalar fields ϕ_a, which can be arranged in a column vector $\phi = (\phi_1 \cdots \phi_n)^T$ with VEV $\nu = \langle 0|\phi|0\rangle$. As discussed in Section 3.3.6, we assume that M of the generators are not broken, $L^i\nu = 0$, $i = 1 \cdots M$, while the remaining $N-M$ are broken, i.e., $L^i\nu \neq 0$, $i = M+1 \cdots N$. Then, for a global symmetry we expect that there will be $N-M$ massless Goldstone bosons and $p = n-(N-M)$ massive physical scalar particles. According to (3.173) the Goldstone bosons are linear combinations of the original Hermitian fields, corresponding to the directions of the massless eigenvectors $iL^i\nu$, $i = M+1 \cdots N$. These span an $N-M$ dimensional vector space, but need not be orthogonal. (The i is because the L^i are imaginary). We can therefore label the n scalars so that

$$\phi = \begin{pmatrix} \nu_1 + \sigma_1 \\ \vdots \\ \nu_p + \sigma_p \\ \chi_{p+1} \\ \vdots \\ \chi_n \end{pmatrix} = e^{\left[i\sum_{i=M+1}^{N}\xi^i L^i\right]} \begin{pmatrix} \nu_1 + \eta_1 \\ \vdots \\ \nu_p + \eta_p \\ 0 \\ \vdots \\ 0 \end{pmatrix} \equiv e^{i\vec{\xi}\cdot\vec{L}}(\nu + \eta), \tag{4.45}$$

in analogy with (4.40). The $n - p = N - M$ fields χ_i in the first form are associated with the subspace spanned by the $iL^i\nu$. There are no VEV's in those directions: we are working in a Hermitian basis, which implies $L^i = -L^{iT} = -L^{i*}$ and therefore that $\langle\nu|L^i\nu\rangle = 0$. The p fields σ_i will be associated with the massive scalars. Not all of the ν_i, $i = 1 \cdots p$, are necessarily non-zero. The second form is a polar reparametrization. Only the broken generators are included in the exponent, and the ξ are the Goldstone boson fields. Unlike (4.40) we have not attempted to normalize the ξ^i by factors of ν, as it is not needed for a gauge symmetry in unitary gauge.[4]

The Goldstone fields may then be removed by going to the unitary gauge, i.e., the Kibble transformation, just as for the $U(1)$ example. Gauge invariance ensures that $\mathcal{L}$ will be unchanged in form under the gauge transformations in (4.2) and (4.24). In particular, choose $\beta^i(x) = -\xi^i(x)$, $i = M+1 \cdots N$, and $\beta^i(x) = 0$ otherwise. Then, $\mathcal{L}(A, \phi) = \mathcal{L}(A', \phi')$, where $\phi' = \nu + \eta$. A' is given generally by (4.24), but for small ξ^i is

$$A'^i_\mu = A^i_\mu + c_{ijk}\xi^j A^k_\mu + \frac{1}{g}\partial_\mu \xi^i, \tag{4.46}$$

which contains the longitudinal term $\partial_\mu \xi^i$. The relevant term for our present considerations is the gauge covariant kinetic term for the Hermitian scalars. This becomes, in terms of the

[4]For a global symmetry the ξ fields would have a non-canonical kinetic energy matrix, and field redefinitions similar to those in Section 2.14 to restore the correct diagonal form $\frac{1}{2}\sum_i(\partial_\mu\xi)^2$ would be required.

new variables,

$$\begin{aligned}\frac{1}{2}D_\mu\phi D^\mu\phi &\to \frac{1}{2}(\nu+\eta)^T(\overleftarrow{\partial}_\mu - ig\vec{A}^\mu\cdot\vec{L})(\overrightarrow{\partial}_\mu + ig\vec{A}_\mu\cdot\vec{L})(\nu+\eta)\\ &= \frac{1}{2}M^2_{ij}A^{i\mu}A^j_\mu + \frac{1}{2}D_\mu\eta D^\mu\eta + g^2\left(\nu^T L^i L^j\eta\right)A^{i\mu}A^j_\mu,\end{aligned} \tag{4.47}$$

where the primes have been dropped for notational simplicity. The Goldstone fields ξ have disappeared from the theory, reemerging as the longitudinal gauge field components, and the gauge bosons corresponding to the broken generators have acquired mass, described by the first term in the second line. The second term is the normal gauge covariant kinetic energy for the physical η fields, which include the three and four point gauge interactions, and the last is an induced cubic gauge interaction.

The gauge boson mass matrix in (4.47) is

$$\begin{aligned}M^2_{ij} &= M^2_{ji} = g^2\nu^T L^i L^j\nu\\ &= g^2\langle\nu|L^iL^j|\nu\rangle = g^2\langle L^i\nu|L^j\nu\rangle = g^2\sum_{a=p+1}^{n}(L^i\nu)^*_a(L^j\nu)_a,\end{aligned} \tag{4.48}$$

where the inner product is defined by $\langle x|y\rangle = \sum x^*_a y_a$. The induced cubic term in (4.47) may be similarly rewritten as

$$\begin{aligned}g^2\left(\nu^T L^i L^j\eta\right)A^{i\mu}A^j_\mu &= g^2\langle L^i\nu|L^j\eta\rangle\,A^{i\mu}A^j_\mu = g^2\langle\nu|L^iL^j\eta\rangle\,A^{i\mu}A^j_\mu\\ &= g^2\sum_{a=p+1}^{n}(L^i\nu)^*_a(L^j\eta)_a\,A^{i\mu}A^j_\mu.\end{aligned} \tag{4.49}$$

It is easy to prove, using the explicit form in (4.48) and the fact that the representation matrices L^i are antisymmetric and purely imaginary, that M^2 is real, symmetric, and has non-negative eigenvalues. From (3.174), M^2 must have the block-diagonal form

$$M^2 = \begin{pmatrix} 0 & 0 \\ 0 & \mathcal{M}^2 \end{pmatrix}, \tag{4.50}$$

where the upper diagonal block is $M\times M$ dimensional and $\mathcal{M}^2$ is $(N-M)\times(N-M)$ dimensional. There are therefore M massless gauge bosons $A^1\cdots A^M$, corresponding to the unbroken generators, and $N-M$ massive gauge bosons, corresponding to the $N-M$ non-zero eigenvalues of $\mathcal{M}^2$. The $N-M$ Goldstone bosons have been eaten to become the longitudinal modes of $N-M$ massive gauge bosons $A^{M+1}\cdots A^N$.

4.4 THE R_ξ GAUGES

The unitary gauge for a spontaneously broken theory, introduced in Section 4.3, is extremely useful for identifying the physical states of the theory, but it is not very convenient for explicit calculations, especially when higher-order loop corrections are involved, because it is very singular. It is useful to work instead in a new class of gauges called the R_ξ gauges (Fujikawa et al., 1972; Weinberg, 1973c,d; Lee and Zinn-Justin, 1973), which are less singular, though the particle content is less obvious. They are therefore better behaved in higher-order calculations and for proving the renormalizability of spontaneously broken gauge theories.

Consider a gauge theory with n Hermitian scalars $\phi_a, a = 1 \cdots n$, arranged in a column vector ϕ, and m fermion fields represented by a column vector ψ. The most general renormalizable Lagrangian density with a conserved fermion number is then

$$\mathcal{L} = -\frac{1}{4}F^i_{\mu\nu}F^{i\mu\nu} + \frac{1}{2}(D^\mu\phi)(D_\mu\phi) - V(\phi) + \bar{\psi}(i\not{D} - m_0)\psi + \bar{\psi}\Gamma^a\psi\phi_a, \tag{4.51}$$

where $F^i_{\mu\nu}, i = 1 \cdots N$, are the field strength tensors for the gauge group G, D_μ and $\not{D}$ are the covariant derivatives for the scalars and fermions, and V is the scalar potential containing gauge-invariant terms up to order ϕ^4. There may be present a fermion bare mass term described by the $m \times m$ matrix $m_0 = m_{0L}P_L + m_{0R}P_R$, with $m_{0L} = m^\dagger_{0R}$, if it is allowed by the symmetries of the theory. Similarly, there will in general be Yukawa interactions between the fermions and scalars described by $m \times m$ matrices $\Gamma^a = \Gamma^a_L P_L + \Gamma^a_R P_R$, with $\Gamma^a_L = \Gamma^{a\dagger}_R$.

Just as in Section 4.3, we define a column vector $\nu = \langle 0|\phi|0\rangle$ of VEVs, where $\nu_a = 0$ and some $(L^i\nu)_a \neq 0$ for $a = p+1 \cdots n$, where $n - p = N - M$ is the number of broken generators. In an arbitrary gauge one can write $\phi = \nu + \phi'$, where ϕ' is the shifted field with $\langle 0|\phi'|0\rangle = 0$,

$$\phi = \nu + \phi' = \begin{pmatrix} \nu_1 + \phi'_1 \\ \vdots \\ \nu_n + \phi'_n \end{pmatrix}. \tag{4.52}$$

We saw in (4.45) that ϕ'_a, $a = 1 \cdots p$, represent physical scalars, while $\phi'_a, a = p+1 \cdots n$, are associated with the Goldstone degrees of freedom. The latter disappear in the unitary gauge, which can be defined by the condition

$$\langle L^i\nu|\phi'\rangle = 0, \quad i = 1 \cdots N. \tag{4.53}$$

In a general gauge, ϕ' will include the unphysical Goldstone bosons. Similar to (4.38) we can rewrite (4.51) in terms of the shifted fields

$$\begin{aligned}\mathcal{L} = &-\frac{1}{4}F^i_{\mu\nu}F^{i\mu\nu} + \frac{1}{2}(D^\mu\phi')(D_\mu\phi') \\ &+ \frac{M^2_{ij}}{2}A^{\mu i}A^j_\mu - ig\langle\nu|L^i\partial^\mu\phi'\rangle A^i_\mu + g^2\langle\nu|L^iL^j\phi'\rangle A^i_\mu A^{j\mu} \\ &- V(\nu+\phi') + \bar{\psi}(i\not{D} - m_0 + \Gamma^a\nu_a)\psi + \bar{\psi}\Gamma^a\psi\phi'_a,\end{aligned} \tag{4.54}$$

where $L \equiv L_\phi = -L^T$ are the representation matrices for the Hermitian scalars. The terms in the first line represent the gauge and Higgs kinetic energies and gauge interactions, just as in the non-spontaneously broken case (Figure 4.1). Both the physical and the Goldstone boson states are included. The first term in the second line is the induced gauge boson mass term. It is of the same form as in the unitary gauge, with M^2 given in (4.48). The next will be cancelled by terms added to $\mathcal{L}$ to fix the gauge, and the third is the induced cubic interaction, leading to the $\phi'_b A^i_\mu A^j_\nu$ vertex

$$ig^2 g_{\mu\nu}\nu_a(L^iL^j + L^jL^i)_{ab} \tag{4.55}$$

shown in Figure 4.2. In the last line, V becomes the scalar potential for ϕ', including mass terms for the physical states and scalar self-interactions. The $n \times n$ dimensional mass-squared matrix $\hat{\mu}^2$ is given by (3.169), just as for the case of a global symmetry. It has p (generally) nonzero eigenvalues corresponding to the physical scalars, and $n - p$ zero eigenvalues corresponding to the Goldstone bosons. The next term in (4.54) includes the fermion mass matrix

$$m = m_0 - \Gamma^a \nu_a \equiv m_L P_L + m_R P_R, \qquad m_L = m_R^\dagger, \tag{4.56}$$

which may in general have both bare (m_0) and spontaneously-generated pieces. m may involve γ^5's unless $m_L = m_R$. Techniques for "diagonalizing" m to obtain the fermion mass eigenvalues and eigenvectors are described in the standard model context in Section 8.2.2. The Yukawa interactions involving the shifted fields are given by the last term in (4.54).

The R_ξ gauges are defined by the condition

$$\mathcal{F}^i\left(A(x), \phi(x), \xi\right) = \alpha^i(x), \tag{4.57}$$

where

$$\mathcal{F}^i \equiv \sqrt{\xi}\left[\partial^\mu A^i_\mu + \frac{ig}{\xi}\langle \nu | L^i \phi' \rangle\right], \tag{4.58}$$

ξ is a real parameter which runs from 0 to ∞ and specifies the gauge, and $\alpha^i(x)$ is a number that depends on x only. α^i is averaged with an exponential weighting factor in the quantization procedure, so its precise value is unimportant. The limit $\xi \to 0$ corresponds to the unitary gauge.

The quantization procedure in the R_ξ gauges introduces additional terms into the effective Lagrangian density that modify the structure of the vector and scalar propagators. The Goldstone bosons have not been removed from the Lagrangian, and occur in internal lines in Feynman diagrams. They do not occur as external states (except in connection with the equivalence theorem, introduced in Section 8.5.1). The quantization also introduces terms that can be represented by a set of N *ghost* fields η_i, $i = 1, \cdots, N$. These are fictitious particles that do not correspond to physical states but do circulate in internal loops. They are needed to ensure unitarity and renormalizability, and can be treated like complex scalar fields except that they obey Fermi statistics. The ghost vertices are given by the effective interaction

$$\mathcal{L}_{ghost} = g(\partial^\mu \eta_i^\dagger) c_{ijk}\, \eta_j A^k_\mu + \frac{g^2}{\xi} \eta_i^\dagger \eta_j \langle \nu | L^i L^j \phi' \rangle \tag{4.59}$$

and are shown in Figure 4.2. There is a factor of -1 for each closed ghost loop. $\eta_i^\dagger$ and η_i are to be viewed as independent anticommuting c-numbers, not necessarily related by Hermitian or complex conjugation. The notation emphasizes that there is an effective propagator $i\mathcal{D}_G(x - x') = \langle 0 | \mathcal{T}[\eta(x), \eta^\dagger(x')] | 0 \rangle$. Clear discussions and derivations of the formulae may be found in (Weinberg, 1995; Pokorski, 2000).

The momentum space propagator for the gauge bosons in an arbitrary R_ξ gauge is

$$iD_V^{\mu\nu}(k) = -i\left[g^{\mu\nu} - \frac{k^\mu k^\nu (1 - 1/\xi)}{k^2 - M^2/\xi}\right] \frac{1}{k^2 - M^2}, \tag{4.60}$$

which is an $N \times N$ matrix in the space of gauge indices.[5] For practical calculations it is often convenient to rewrite this (within the subspace of broken generators) as

$$iD_V^{\mu\nu}(k) = -i\left(g^{\mu\nu} - \frac{k^\mu k^\nu}{M^2}\right)\frac{1}{k^2 - M^2} - \frac{i}{M^2}\frac{k^\mu k^\nu}{k^2 - M^2/\xi}. \tag{4.61}$$

The ξ dependent piece, which will ultimately cancel other ξ dependent terms in the Higgs

[5]There is an implicit $+i\epsilon$ in each propagator, e.g., $1/(k^2 - M^2) \to 1/(k^2 - M^2 + i\epsilon)$.

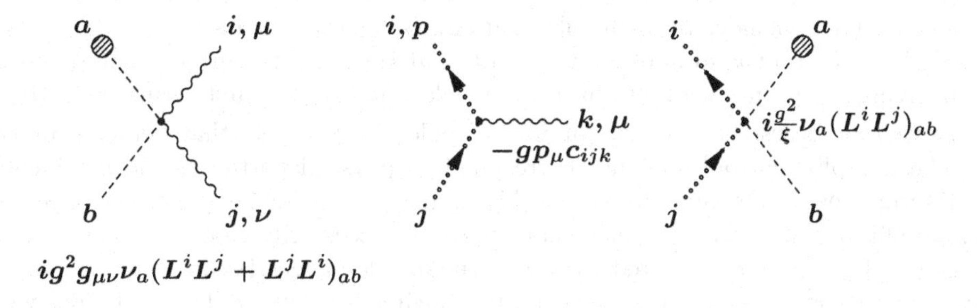

Figure 4.2 Left: Induced three-point vertex between one scalar and two gauge fields, derived from a four-point interaction with one external scalar replaced by its vacuum expectation value. Vertices for ghost interactions with vector (middle) and scalar (right) fields. The dotted lines represent ghost propagators. Ghost-ghost-vector vertices involve one outgoing and one incoming momentum, with the outgoing one appearing in the vertex factor. (Some authors place a dot near the outgoing line to indicate which momentum appears.)

and ghost propagators, is singled out. The ghost propagator, which is also an $N \times N$ matrix, is

$$i\mathcal{D}_G(k) = \frac{-i}{k^2 - M^2/\xi}. \tag{4.62}$$

The Higgs propagator is an $n \times n$ matrix in the space of scalar field indices. It can be written as

$$i\Delta_\phi(k) = (I - \mathcal{P})\frac{i}{k^2 - \mu^2} + \mathcal{P}\frac{i}{k^2 - M^2/\xi}, \tag{4.63}$$

where $\mathcal{P}$ is defined as the projection operator onto the $N-M$ dimensional space of Goldstone fields spanned by the vectors $iL^i\nu$, $i = M+1 \cdots N$. It is given explicitly by

$$\mathcal{P} = g^2 |L^i \nu\rangle \left(\frac{1}{M^2}\right)_{ij} \langle L^j \nu| \Rightarrow \mathcal{P}_{ab} = g^2 (L^i\nu)_a \left(\frac{1}{M^2}\right)_{ij} (L^j\nu)^*_b, \tag{4.64}$$

where

$$\frac{1}{M^2} \equiv \begin{pmatrix} 0 & 0 \\ 0 & \frac{1}{\mathcal{M}^2} \end{pmatrix} \tag{4.65}$$

is the inverse of the vector mass-squared matrix in the subspace of massive vectors (and zero otherwise). The scalar propagator in (4.63) decomposes into two pieces. The first, proportional to $I - \mathcal{P}$, represents the propagation of real physical scalars with mass-squared matrix μ^2. The second term is the Goldstone boson contribution, which is present in an arbitrary ξ gauge. It is shorthand for

$$\left[\mathcal{P}\frac{1}{k^2 - M^2/\xi}\right]_{ab} \equiv g^2 (L^i v)_a \left[\frac{1}{M^2}\frac{1}{k^2 - M^2/\xi}\right]_{ij} (L^j v)^*_b. \tag{4.66}$$

Using $\mathcal{P}\mu^2 = 0$, (4.63) can be rewritten

$$i\Delta_\phi(k)_{ab} = \left[\frac{i}{k^2 - \mu^2}\right]_{ab} + \frac{ig^2}{\xi k^2}(L^i\nu)_a \left(\frac{1}{k^2 - M^2/\xi}\right)_{ij} (L^j\nu)^*_b, \tag{4.67}$$

where the first term includes both the physical and Goldstone scalars.

We see that the propagators of gauge, ghost, and Higgs fields can be decomposed into pieces involving the propagation of physical particles, namely the first terms in (4.61) and (4.63), as well as pieces that involve unphysical poles at $k^2 = \frac{M^2}{\xi}$ that depend upon the gauge. These unphysical poles do not correspond to physical particles, and are included only in internal lines in Feynman diagrams. They cancel in the final expressions for physical on-shell amplitudes, i.e., S matrix elements, which are necessarily gauge invariant.

R_ξ gauges for $\xi \neq 0$ are referred to as renormalizable gauges because the propagators are well-behaved for $k^2 \to \infty$, falling as $1/k^2$. In principle it is best to do calculations with ξ arbitrary so that (a) the Feynman diagrams are well-behaved and manifestly renormalizable, and (b) so that the cancellation of the ξ dependence in the physical S matrix elements can be used as a check on the correctness of the calculation. In practice, however, it is messy to carry out calculations for an arbitrary ξ, and therefore people often make use of particular gauges for calculations or formal arguments. The $\xi = 1$ gauge is known as the 't Hooft-Feynman gauge. The gauge propagator takes the particularly simple form

$$iD_V^{\mu\nu} = \frac{-ig^{\mu\nu}}{k^2 - M^2}, \tag{4.68}$$

so the $\xi = 1$ gauge is often convenient for carrying out concrete calculations. Another useful gauge is $\xi \to \infty$, known as the renormalizable or Landau gauge. The propagators again take a relatively simple form in this limit

$$iD_V^{\mu\nu}(k) = \frac{-i\left(g^{\mu\nu} - \frac{k^\mu k^\nu}{k^2}\right)}{k^2 - M^2}, \qquad i\mathcal{D}_G(k) = \frac{-i}{k^2}, \qquad i\Delta_\phi(k) = \frac{i}{k^2 - \mu^2}, \tag{4.69}$$

but this gauge still contains ghosts and unphysical scalar fields.

The above gauges are best for calculations of higher orders. However, the $\xi \to 0$ or unitary gauge is particularly simple for displaying the physical particle content of the theory. The propagators become

$$iD_V^{\mu\nu}(k) = \frac{-i\left(g^{\mu\nu} - \frac{k^\mu k^\nu}{M^2}\right)}{k^2 - M^2}, \qquad i\mathcal{D}_G(k) = 0, \qquad i\Delta_\phi(k) = (I - \mathcal{P})\frac{i}{k^2 - \mu^2}. \tag{4.70}$$

That is, there are no ghost fields and only the physical, non-Goldstone scalar fields survive. If one is only interested in calculating at tree level, the unitary gauge is very convenient. However, the gauge boson propagator is badly behaved as $k \to \infty$; it approaches a constant rather than falling like $1/k^2$. It therefore induces severe ultraviolet divergences in higher-order calculations that must be handled very carefully.

The ghost fields do not entirely disappear from the theory in the unitary gauge (Weinberg, 1973c,d). There is an effective multiscalar interaction

$$\mathcal{L}_{\mathcal{J}} = -i\delta^4(0)\text{Tr } \ln(I + \mathcal{J}), \tag{4.71}$$

where

$$\mathcal{J}_{ij} = g^2 \left(\frac{1}{M^2}\right)_{ik} \langle \nu | L^k L^j | \phi' \rangle. \tag{4.72}$$

The trace and matrices in (4.71) and (4.72) are restricted to the $N - M$ dimensional subspace of broken generators of G. $\mathcal{L}_{\mathcal{J}}$ is a remnant of the ghost loops that survives as $\xi \to 0$ because of the factors ξ^{-1} in the ghost-ghost-scalar vertices, which cancel the zeroes in the ghost propagators. $\mathcal{L}_{\mathcal{J}}$ is necessary to cancel divergences due to gauge boson loops in multiscalar interactions.

Complex Scalars

It is straightforward to reexpress the results for the interaction vertices and propagators in a complex scalar basis, either by "starting from scratch" or by using the formal results in (3.71)–(3.73). Writing $\phi = v + \phi'$, where the fields and VEVs are now complex, the induced cubic and ghost-Higgs vertices in (4.54) and (4.59) are replaced by

$$\begin{aligned} \mathcal{L}_{AA\phi'} &= g^2 \left[\langle v|L^i L^j \phi'\rangle + \langle \phi'|L^j L^i v\rangle \right] A^i_\mu A^{j\mu} \\ \mathcal{L}_{\eta\eta\phi'} &= \frac{g^2}{\xi} \eta_i^\dagger \eta_j \left[\langle v|L^i L^j \phi'\rangle + \langle \phi'|L^j L^i v\rangle \right], \end{aligned} \tag{4.73}$$

respectively, while the gauge boson mass matrix in (4.48) becomes

$$M^2_{ij} = g^2 \langle v|L^i L^j + L^j L^i|v\rangle. \tag{4.74}$$

The projection operator $\mathcal{P}$ onto the Goldstone subspace is usually obvious, but is best worked out in the Hermitian basis for complicated cases.

Let us illustrate the R_ξ constructions for an $SU(2)$ gauge theory involving a single complex scalar $SU(2)$ doublet. (This is a simplified version of the standard $SU(2) \times U(1)$ model with the $U(1)$ gauge coupling, and therefore electromagnetism, turned off.) The Hermitian $SU(2)$ gauge fields A^i_μ, $i = 1 \cdots 3$, will sometimes be re-expressed as

$$A^\pm_\mu = \frac{A^1_\mu \mp i A^2_\mu}{\sqrt{2}}, \qquad A^0_\mu = A^3_\mu, \tag{4.75}$$

where $A^+ = (A^-)^\dagger$, and the labels $\pm$ and 0 are suggestive of the electric charges that will emerge when the model is promoted to $SU(2) \times U(1)$. The gauge self-interactions for the A fields are derived in Problem 4.8. Similarly, the scalar doublet can be written

$$\phi = \begin{pmatrix} \phi^+ \\ \phi^0 \end{pmatrix}, \qquad \phi^\dagger = \begin{pmatrix} \phi^- \\ \phi^{0*} \end{pmatrix}, \tag{4.76}$$

where the conjugate fields are defined as $\phi^- \equiv (\phi^+)^\dagger$ and $\phi^{0*} \equiv (\phi^0)^\dagger$. The most general renormalizable gauge invariant potential for ϕ,

$$V(\phi) = +\mu^2 \phi^\dagger \phi + \lambda (\phi^\dagger \phi)^2, \tag{4.77}$$

is identical to the Higgs potential in the standard model, and will be described in more detail in Section 8.2. Here we note that $\lambda > 0$ is needed for vacuum stability. For $\mu^2 > 0$ there is no SSB, i.e., $v \equiv \langle 0|\phi|0\rangle = 0$, while $v \neq 0$ for $\mu^2 < 0$.

In the unbroken case, $v = 0$, the complex scalar fields ϕ^+ and ϕ^0 are degenerate with mass μ, and their self-interaction terms are

$$\mathcal{L}_I = -V_I = -\lambda(\phi^\dagger \phi)^2 = -\lambda(\phi^- \phi^+ + \phi^{0*}\phi^0)^2. \tag{4.78}$$

The scalar gauge interactions[6] can be obtained by writing the general form in (4.21) in terms of components, using $L^i = \tau^i/2$ and $A^i \tau^i = A^0 \tau^3 + \sqrt{2} A^+ \tau^+ + \sqrt{2} A^- \tau^-$:

$$\begin{aligned} \mathcal{L}_\phi = &-i\frac{g}{2} \left(\phi^- \overleftrightarrow{\partial^\mu} \phi^+ - \phi^{0*} \overleftrightarrow{\partial^\mu} \phi^0 \right) A^0_\mu - i\frac{g}{\sqrt{2}} \left(\phi^{0*} \overleftrightarrow{\partial^\mu} \phi^+ \right) A^-_\mu \\ &- i\frac{g}{\sqrt{2}} \left(\phi^- \overleftrightarrow{\partial^\mu} \phi^0 \right) A^+_\mu + \frac{g^2}{4} (\phi^- \phi^+ + \phi^{0*}\phi^0) \vec{A}^\mu \cdot \vec{A}_\mu, \end{aligned} \tag{4.79}$$

[6]The quartic term is the product of the $SU(2)$ singlets $\vec{A}^\mu \cdot \vec{A}_\mu$ and $\phi^-\phi^+ + \phi^{0*}\phi^0 = \phi^\dagger \phi$. This is because the product of two doublets transforms as $0+1$ under $SU(2)$ while that for two triplets is $0+1+2$. However, the 1 is antisymmetric for the triplets and vanishes in this case, so only the singlet components can enter.

where $\vec{A}^{\mu} \cdot \vec{A}_{\mu} = 2A^{+\mu}A^{-}_{\mu} + A^{0\mu}A^{0}_{\mu}$. The diagonal quantum number associated with T^3 is conserved.

For $\mu^2 < 0$ the $SU(2)$ symmetry is completely broken. Similar to the $U(1)$ example in (3.153) on page 119 the minimum occurs for $v = \nu/\sqrt{2}$, where $|\nu|^2 = -\mu^2/\lambda$. Without loss of generality, we can choose ν to be real and in the ϕ^0 direction: other orientations of the minimum differ by $SU(2)$ transformations. Thus, we can write

$$v = \frac{1}{\sqrt{2}} \begin{pmatrix} 0 \\ \nu \end{pmatrix}, \qquad \phi = \begin{pmatrix} w^+ \\ \frac{\nu + H + iz}{\sqrt{2}} \end{pmatrix}, \tag{4.80}$$

where H and z are the Hermitian components of $\phi^{0\prime}$ and $w^+ \equiv \phi^+$. Rewriting the potential in the new variables,

$$V(\phi) = -\mu^2 H^2 + \lambda \nu H \left[H^2 + z^2 + 2w^+w^-\right] + \frac{\lambda}{4}\left[H^2 + z^2 + 2w^+w^-\right]^2, \tag{4.81}$$

where $\nu = \sqrt{-\mu^2/\lambda}$ and we have dropped the additive constant. We therefore recognize that H is a physical scalar field with mass $M_H = \sqrt{-2\mu^2} = \sqrt{2\lambda}\nu$, while z, w^+, and $w^- = (w^+)^\dagger$ are the Goldstone bosons that should disappear in the unitary gauge. (The notation is chosen to coincide with the analogous standard model case.) The second term on the right is an induced cubic interaction.

In the notation following (3.71) on page 101, ϕ can be written in a Hermitian basis as

$$\phi_h = \begin{pmatrix} w^1 \\ \nu + H \\ w^2 \\ z \end{pmatrix} \equiv \nu_h + \phi_h', \qquad \nu_h = \begin{pmatrix} 0 \\ \nu \\ 0 \\ 0 \end{pmatrix}, \tag{4.82}$$

where $w^\pm = (w^1 \pm iw^2)/\sqrt{2}$ and the ordering of the components differs slightly from the convention in (4.52). Then, using the $SU(2)$ representation matrices in (3.76),

$$L_h^1 \nu_h = -\frac{i}{2}\begin{pmatrix} 0 \\ 0 \\ \nu \\ 0 \end{pmatrix}, \qquad L_h^2 \nu_h = -\frac{i}{2}\begin{pmatrix} \nu \\ 0 \\ 0 \\ 0 \end{pmatrix}, \qquad L_h^3 \nu_h = +\frac{i}{2}\begin{pmatrix} 0 \\ 0 \\ 0 \\ \nu \end{pmatrix}, \tag{4.83}$$

establishing that all three generators are broken and that the vectors $iL_h^i \nu_h$ span the Goldstone subspace.

Substituting the representation in (4.80) for ϕ, the scalar gauge interactions become

$$\begin{aligned} \mathcal{L}_\phi =& \frac{1}{2} M_A^2 \, \vec{A}^\mu \cdot \vec{A}_\mu + \frac{g}{2}\left(-iw^- \overleftrightarrow{\partial}^\mu w^+ + z \overleftrightarrow{\partial}^\mu H\right) A^0_\mu \\ & - i\frac{g}{2}\left([H - iz] \overleftrightarrow{\partial}^\mu w^+\right) A^-_\mu - i\frac{g}{2}\left(w^- \overleftrightarrow{\partial}^\mu [H + iz]\right) A^+_\mu \\ & + \frac{g^2}{4}\nu H \vec{A}^\mu \cdot \vec{A}_\mu + \frac{g^2}{4}\left(w^- w^+ + \frac{H^2 + z^2}{2}\right) \vec{A}^\mu \cdot \vec{A}_\mu, \end{aligned} \tag{4.84}$$

where we have omitted the irrelevant $\phi' - A$ mixing terms. The three gauge bosons A^i (or $A^\pm, A^0$) have acquired a common mass

$$M_A = \frac{g\nu}{2} \tag{4.85}$$

by the Higgs mechanism. They are degenerate because the Lagrangian has an unbroken $O(3)$ global symmetry after the breaking, as is discussed in Section 8.2.2. The derivative cubic terms always connect H to a Goldstone field (or two Goldstone fields to each other), so they disappear in the unitary gauge. (There are important analogs involving physical Higgs fields in models with extended Higgs sectors, such as in the *minimal supersymmetric extension of the standard model (MSSM).*) The induced cubic term $\nu H\vec{A}^2$ involves only the physical scalar field. The projection operator $\mathcal{P}$ in (4.63) projects onto $w^\pm$ and z, while $I-\mathcal{P}$ projects onto H.

As a simple example, let us consider the amplitude for $HH \to A^+A^-$ in an R_ξ gauge, with the tree-level diagrams shown in Figure 4.3. Using the gauge and scalar propagators in (4.61) and (4.63), and the interaction vertices in (4.81) and (4.84), the amplitude is

$$\begin{aligned} M = &\left(\frac{ig^2\nu}{2}\right)\epsilon^*_{3\mu}\, iD^{\mu\nu}_{A^+}(p_3-p_1)\left(\frac{ig^2\nu}{2}\right)\epsilon^*_{4\nu} + (p_1 \leftrightarrow p_2) \\ &+\left[i\left(\frac{-ig}{2}\right)\epsilon_3^{\mu *}[-i(p_3-p_1)_\mu + ip_{1\mu}]\, iD_{w^+}(p_3-p_1)\right] \\ &\quad\times\left[i\left(\frac{+ig}{2}\right)\epsilon_4^{\nu *}[+i(p_2-p_4)_\nu + ip_{2\nu}]\right] + (p_1\leftrightarrow p_2) \\ &+\left(\frac{ig^2\nu}{2}\right)\epsilon_3^*\cdot\epsilon_4^*\, iD_H(p_1+p_2)(-i\lambda\nu)3!, \end{aligned} \tag{4.86}$$

where the 3! in the last term is combinatoric. The propagators are

$$\begin{aligned} iD^{\mu\nu}_{A^+}(k) &= -i\left(g^{\mu\nu} - \frac{k^\mu k^\nu}{M_A^2}\right)\frac{1}{k^2-M_A^2} - \frac{i}{M_A^2}\frac{k^\mu k^\nu}{k^2 - M^2/\xi} \\ iD_{w^+}(k) &= \frac{i}{k^2 - M_A^2/\xi}, \qquad iD_H(k) = \frac{i}{k^2-M_H^2}. \end{aligned} \tag{4.87}$$

Using (4.85) as well as $p_3\cdot\epsilon_3^* = p_4\cdot\epsilon_4^* = 0$, the w^+ exchange term cancels the ξ-dependent part of the A^+ term, leaving the A^+ and H exchange diagrams evaluated in unitary gauge. This illustrates the general result that physical on-shell amplitudes are independent of ξ.

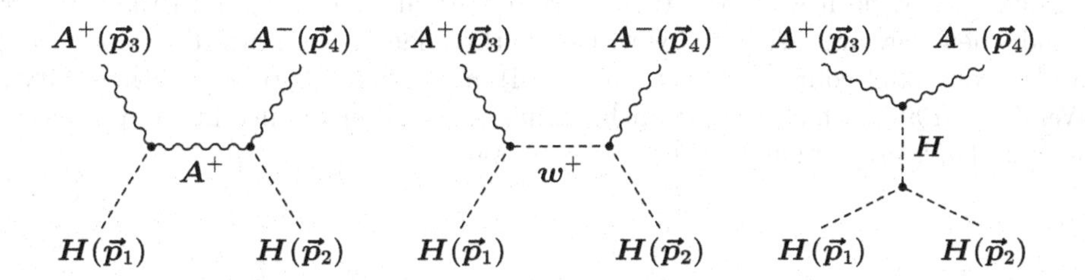

Figure 4.3 Tree-level diagrams for $HH \to A^+A^-$ in an R_ξ gauge. There are two additional u-channel diagrams obtained from the first two by $p_1 \leftrightarrow p_2$,

4.5 ANOMALIES

Anomalies refer to quantum effects that break the symmetries associated with the classical equations of motion. In particular, they may occur when the diverences in a theory cannot be regularized in a way that is consistent with the original symmetries. The Adler-Bell-Jackiw anomalies (Adler, 1969; Adler and Bardeen, 1969; Bell and Jackiw, 1969; Bardeen, 1969; Adler, 2004) are singularities associated with the fermion triangle diagram contributions to the vertex of three currents, as shown in Figure 4.4. If one or three of the vertices involve

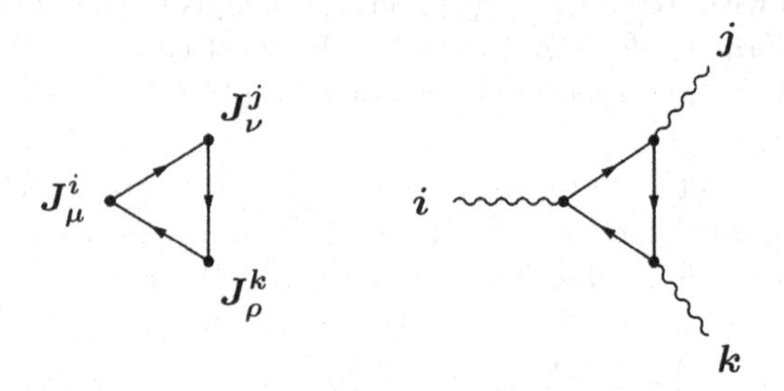

Figure 4.4 Left: The anomalous triangle diagram for the $J^i_\mu J^j_\nu J^k_\rho$ vertex. Right: The analogous diagram for the triple gauge boson vertex.

an axial vector (γ^5) coupling, as can occur for a chiral symmetry, the diagram diverges linearly, leading to an anomalous divergence of the currents in perturbation theory that is not revealed by formal manipulation of the field equations. If one or more of the currents is associated with a global symmetry of the theory, the anomalous divergence does not cause any particular problems, and it can even be useful (Adler, 1969; 't Hooft, 1976a), as we will see in Sections 5.2 and 5.8.3. If the currents are all associated with gauge symmetries, however, then the diagram contributes to the triple gauge vertex and cannot be regularized in a way consistent with gauge invariance, implying that the renormalizability of the theory is lost.

The anomaly coefficient A_{ijk} in the vertex of currents i, j, and k is (Georgi and Glashow, 1972)

$$A_{ijk} = 2\text{Tr}\, L^i_L\{L^j_L, L^k_L\} - 2\text{Tr}\, L^i_R\{L^j_R, L^k_R\}, \tag{4.88}$$

independent of the fermion masses. We require that each A_{ijk} must vanish for a renormalizable gauge theory, both when the three currents are all associated with the same group factor and when they are associated with two or three factors in a direct product group.

Another constraint comes from the (universal) interaction of fermions with gravity (see, e.g., Weinberg, 1995), which leads to a breaking of gauge invariance in the presence of a gravitational field, proportional to the trace anomaly

$$T_i = \text{Tr}\, L^i_L - \text{Tr}\, L^i_R. \tag{4.89}$$

We therefore require for this to vanish as well.

A_{ijk} and T_i vanish for chiral theories ($L^i_L = L^i_R$), and also if both L^i_L and L^i_R are real (equivalent to $-L^{iT}_{L,R}$ in a Hermitian basis) since $\text{Tr}\, M = \text{Tr}\, M^T$. The only simple Lie algebras in Table 3.3 that admit complex representations are $SU(m)$ for $m \geq 3$ (which includes $SO(6) \sim SU(4)$), $SO(4m+2)$ for $m \geq 2$, and E_6, so anomalies can occur only for gauge groups that include these factors[7] or $U(1)$'s.

There are no anomalies for pure QED or QCD, since they are non-chiral. The full SM is

[7]The anomalies associated with three $SO(4m+2)$ for $m \geq 2$ or three E_6 factors vanish even for complex representations (Georgi and Glashow, 1972; Okubo, 1977).

based on $SU(3) \times SU(2) \times U(1)$, with $SU(2) \times U(1)$ chiral, but as will be seen in Section 8.1 the anomalies cancel between quarks and leptons or (in non-trivial ways) between L and R.

We discussed in Section 2.10 that instead of working in terms of the L and R-chiral particle fields $\psi_{L,R}$, one could just as well express the theory in terms of the L-chiral particle and antiparticle fields ψ_L and ψ_L^c, where ψ_R and ψ_L^c are related by (2.301). The $2n$ fields $(\psi_L \; \psi_L^c)^T$ transform under the reducible $2n \times 2n$-dimensional representation matrices

$$\mathcal{L}^i = \begin{pmatrix} L_L^i & 0 \\ 0 & L_L^{ci} \end{pmatrix} = \begin{pmatrix} L_L^i & 0 \\ 0 & -L_R^{iT} \end{pmatrix}, \tag{4.90}$$

where the ψ_L^c transform as $-L_R^{iT}$ since they are basically the adjoints of the ψ_R. The anomaly conditions in this basis are

$$A_{ijk} = 2\mathrm{Tr}\,\mathcal{L}^i\{\mathcal{L}^j, \mathcal{L}^k\} = 0, \qquad T_i = \mathrm{Tr}\,\mathcal{L}^i = 0, \tag{4.91}$$

which are equivalent to (4.88) and (4.89).

4.6 PROBLEMS

4.1 The spontaneous breaking of a global or local $U(1)$ symmetry allows one-dimensional cosmic string classical solutions, analogous to the two-dimensional domain walls considered in Section 3.3.2. Consider the complex scalar ϕ in a $U(1)$ gauge theory, with

$$\mathcal{L} = -\frac{1}{4} F_{\mu\nu} F^{\mu\nu} + (D_\mu \phi)^\dagger D^\mu \phi - V(\phi), \qquad V(\phi) = \mu^2 \phi^\dagger \phi + \lambda \left(\phi^\dagger \phi\right)^2,$$

where $D^\mu \phi = (\partial^\mu + igA^\mu)\phi$, $F_{\mu\nu}$ is the field strength tensor, $\lambda > 0$, and $\mu^2 < 0$. It is convenient to rewrite

$$V(\phi) = \lambda \Big(\phi^\dagger \phi - \frac{1}{2}\nu^2\Big)^2, \qquad \nu = \sqrt{\frac{-\mu^2}{\lambda}},$$

where we have dropped the irrelevant constant $-\lambda \nu^4/4$. It was shown (Nielsen and Olesen, 1973) that there is a classical solution for $\phi(x)$ and $A^\mu(x)$ corresponding to a vortex or string running along the z axis from $-\infty$ to $+\infty$. In cylindrical coordinates (r, θ, z) the solution exhibits

$$\phi \xrightarrow[r\to\infty]{} \frac{\nu}{\sqrt{2}} e^{-in\theta},$$

so that $V \to 0$ for $r \to \infty$. Single-valuedness requires that n is an integer. Find the asymptotic expression for the vector potential A^μ for the string solution, for which $F^{\mu\nu} \to 0$, $D^\mu \phi \to 0$, and calculate the magnetic flux $\int \vec{B} \cdot d\vec{S} = \int (\vec{\nabla} \times \vec{A}) \cdot d\vec{S}$ for the solution.

4.2 As mentioned in Section 3.2.5 it often occurs that the spontaneous breaking of a continuous symmetry leaves a discrete subgroup unbroken. If the original symmetry was local, the remaining subgroup is known as a *discrete gauge symmetry* (e.g., Ibáñez and Ross, 1992). As a simple example, consider a model involving three complex scalars ϕ_i, $i = 1 \cdots 3$, with a $U(1)$ gauge symmetry. Suppose the potential is of the form

$$V(\phi_1, \phi_2, \phi_3) = \sum_{i=1}^{3} V_i(\phi_i^\dagger \phi_i) + V_{cubic},$$

where $V_i(\phi_i^\dagger \phi_i)$ are quadratic functions of $\phi_i^\dagger \phi_i$ (i.e., quadratic and quartic in ϕ_i and $\phi_i^\dagger$) and

$$V_{cubic} = \sigma_1 \phi_1 \phi_2 \phi_3 + \sigma_2 \phi_1^2 \phi_2^\dagger + h.c.$$

(a) Find the $U(1)$ charges of the ϕ_i for which the theory is $U(1)$ invariant.
(b) Show that $\sigma_{1,2}$ can be taken to be real w.l.o.g.
(c) Suppose the V_i are such that the minimum of V occurs for $\langle 0|\phi_3|0\rangle \neq 0$ but $\langle 0|\phi_{1,2}|0\rangle = 0$. Show that a discrete Z_3 symmetry remains unbroken.

4.3 Let $\Phi^i, i = 1 \cdots m^2 - 1$, be m Hermitian scalars transforming according to the adjoint representation of $SU(m)$. Define the $m \times m$ matrix $\Phi = \sum_i \Phi^i L^i$ as in (3.31), where L^i are the fundamental representation matrices L^i_m. Show that the gauge covariant derivative is

$$D_\mu \Phi = \partial_\mu \Phi + ig \left[\vec{A}_\mu \cdot \vec{L}, \Phi \right].$$

4.4 From (3.19) the defining (vector) representation of $SO(m)$ consists of the $m(m-1)/2$ antisymmetric imaginary $m \times m$ matrices. These are conveniently labeled by two indices, with

$$\left(L^{ij}\right)_{ab} = -i(\delta^i_a \delta^j_b - \delta^i_b \delta^j_a),$$

where i, j, a, and b all range from 1 to m. Clearly, $L^{ij} = -L^{ji}$ and $L^{ii} = 0$. One could restrict the indices so that $i < j$, but it is convenient not to do so, provided one is careful about double counting. The trace is

$$\text{Tr}\left(L^{ij} L^{kl}\right) = 2(\delta^{ik}\delta^{jl} - \delta^{il}\delta^{jk}).$$

From the vector representation, one has the Lie algebra

$$\left[T^{ij}, T^{kl}\right] = i\left(-\delta^{jk} T^{il} - \delta^{il} T^{jk} + \delta^{ik} T^{jl} + \delta^{jl} T^{ik}\right),$$

where the generators T^{ij} have the same labelling convention as L^{ij}. (It is instructive to specialize these relations to $m = 3$.)
(a) Calculate the gauge covariant derivative $(D^\mu \Phi)_a$ for a scalar or fermion field Φ_a, $a = 1 \cdots m$, transforming as a vector under $SO(m)$, labeling the gauge bosons by $A^{ij}_\mu = -A^{ji}_\mu$ and the gauge coupling as g.
(b) Calculate the field strength tensor $F^{ij}_{\mu\nu}$ for A^{ij}_μ.

4.5 The gauge transformation for a non-abelian gauge field is given in (4.24).
(a) Verify that (4.24) reduces to (4.26) for small $|\beta^i|$.
(b) Verify the transformation (4.22) for a matter field.
(c) Use (4.24) to prove (4.30) for all $\vec{\beta}$ (i.e., *not* just small $|\beta^i|$). This implies that $\mathcal{L}_{KE_A}$ is invariant.
(d) Rederive (4.30) by the simpler method of first proving $ig\vec{F}_{\mu\nu} \cdot \vec{L} = [D_\mu, D_\nu]$, with D_μ from (4.17), and then showing that $U D_\mu U^{-1} = D'_\mu \equiv \partial_\mu + ig\vec{A}'_\mu \cdot \vec{L}$.

4.6 Derive the gauge vertices in Figure 4.1 for Hermitian scalars.

4.7 Derive the triple gauge vertex rule shown in Figure 4.1.

4.8 Specialize the 3- and 4-point gauge vertices in Figure 4.1 to the case of $SU(2)$. Define the fields $A^\pm$ and A^0, as in (4.75). Show that the vertices for $A^+_\mu(p)A^0_\nu(q)A^-_\sigma(r)$, $A^+_\mu A^0_\nu A^0_\sigma A^-_\rho$, and $A^+_\mu A^+_\nu A^-_\sigma A^-_\rho$ are, respectively,

$$ig\,\mathcal{C}_{\mu\nu\sigma}(p,q,r), \qquad -ig^2\mathcal{Q}_{\mu\rho\nu\sigma}, \qquad ig^2\mathcal{Q}_{\mu\nu\rho\sigma},$$

and that the others vanish. All of the particles and momenta flow into the vertices. $\mathcal{C}_{\mu\nu\sigma}(p,q,r)$ is defined in Figure 4.1, while

$$\mathcal{Q}_{\mu\nu\rho\sigma} \equiv 2g_{\mu\nu}g_{\rho\sigma} - g_{\mu\rho}g_{\nu\sigma} - g_{\mu\sigma}g_{\nu\rho}.$$

Note that $\mathcal{Q}$ is symmetric in $\mu \leftrightarrow \nu$ or $\rho \leftrightarrow \sigma$, that $\mathcal{Q}_{\mu\nu\rho\sigma} = \mathcal{Q}_{\rho\sigma\mu\nu}$, and that

$$\mathcal{Q}_{\mu\nu\rho\sigma} + \mathcal{Q}_{\mu\sigma\nu\rho} + \mathcal{Q}_{\mu\rho\nu\sigma} = 0.$$

4.9 Prove that the gauge boson mass matrix in (4.48) is real, symmetric, has non-negative eigenvalues, and that $N - M$ eigenvalues are non-zero. Hint: define an eigenvalue and normalized eigenvector of M^2 as λ and w, i.e., $M^2 w = \lambda w$. Use the fact that $\lambda = w^T M^2 w = w_i M^2_{ij} w_j$. Recall that the vectors $L^i \nu$ span an $N - M$ dimensional space.

4.10 Derive the effective multiscalar interaction in (4.71).

4.11 Extend the $SU(2)$ model on pages 149-151 (with $\mu^2 < 0$) by the addition of a non-chiral fermion doublet $\psi = \begin{pmatrix} \psi^+ \\ \psi^0 \end{pmatrix}$ with mass m_ψ.
(a) Find the interaction vertices for $\psi^+ \to \psi^0 A^+$, $\psi^+ \to \psi^+ A^0$, and $A^0 \to A^+ A^-$.
(b) Write the tree-level amplitude for $\psi^+(p_1)\psi^-(p_2) \to A^+(p_3)A^-(p_4)$ in the R_ξ gauge. Show that it is independent of ξ.
(c) Show that the individual diagrams (for the longitudinal polarizations) grow $\propto s$ for $s \gg M_A^2, m_\psi^2$, but that the leading term cancels in the full amplitude. (The implications for unitarity and renormalizability will be discussed in Chapter 7.1.)

4.12 (a) Calculate the anomaly coefficient for a left-chiral fundamental representation of $SU(m)$ in terms of the d_{ijk} defined in (3.26).
(b) Consider a field ψ_{ab} that transforms as the (reducible) direct product of two $SU(m)$ fundamentals. It follows from (3.118) that the corresponding representation matrices are $\left(L^i_D\right)_{ab;cd} = \left(L^i_{ac}\delta_{bd} + \delta_{ac}L^i_{bd}\right)$, which can conveniently be written in direct product notation as $L^i_D = L^i \otimes I + I \otimes L^i$, where I is the $m \times m$ identity. Calculate the anomaly coefficient for a left-chiral field transforming as ψ_{ab}. Hint: $\mathrm{Tr}\, A \otimes B = \mathrm{Tr}\, A\, \mathrm{Tr}\, B$, and $(A \otimes B)(C \otimes D) = AC \otimes BC$.

CHAPTER 5

The Strong Interactions and QCD

DOI: 10.1201/b22175-5

The modern theory of the strong interactions is *quantum chromodynamics* (QCD). The basic ingredient is that each of the six *flavors* or types of quark, u, d, s, c, b, and t, has an additional quantum number *color*, which takes the values $\alpha = 1, 2, 3$, or red (R), green (G), blue (B), and that there is an unbroken non-chiral $SU(3)$ gauge symmetry acting on the color index. Thus, there are 8 massless gauge bosons (*gluons*), G^i, and a strong gauge coupling g_s and strong fine structure constant $\alpha_s = g_s^2/4\pi$. In this chapter we survey the properties of the strong interactions and QCD, especially the symmetry aspects at short and long distances. For more detailed treatments, see, e.g., (Brock et al., 1995; Pich, 1999; Ellis et al., 2003; Sterman, 2004; Kronfeld and Quigg, 2010; Salam, 2010a; Skands, 2013; Altarelli, 2013; Trócsányi, 2015; Patrignani, 2016). For the historical development, see, e.g., (Gross, 2005; Leutwyler, 2014; Fritzsch and Gell-Mann, 2015).

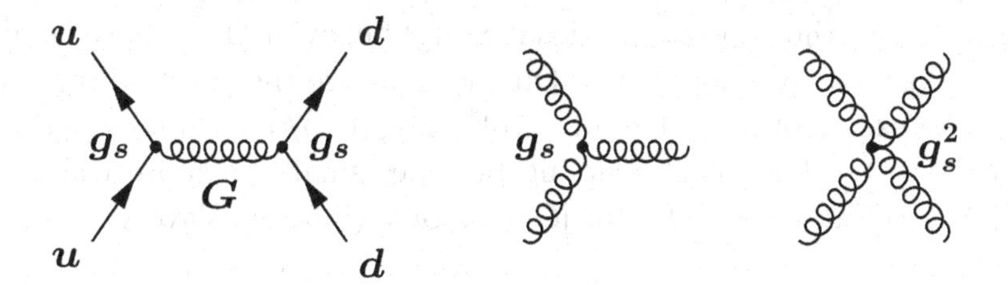

Figure 5.1 QCD interactions.

The quark-quark interaction diagram via one-gluon exchange and the gluon three- and four-point self interactions are shown schematically in Figure 5.1. As discussed in Chapter 2, the dominant higher-order vacuum polarization diagrams in the gluon propagator, shown in Figure 5.2, can be absorbed into an effective (logarithmically) running coupling $g_s(\mu^2)$ that depends on the renormalization scale. Higher-order corrections are minimized if one chooses this scale comparable to the momentum Q carried by the gluon. The quark-loop contributions screen the color charge at long distance, making the effective force weaker. However, the gluon loops anti-screen. The latter dominate for six flavors, so the strong force becomes stronger at long distances or low momentum (*infrared slavery*), and weaker at short distances or high momentum (*asymptotic freedom*). This is in contrast to QED, which has only the charged particle loops, so that it becomes stronger at short distances.

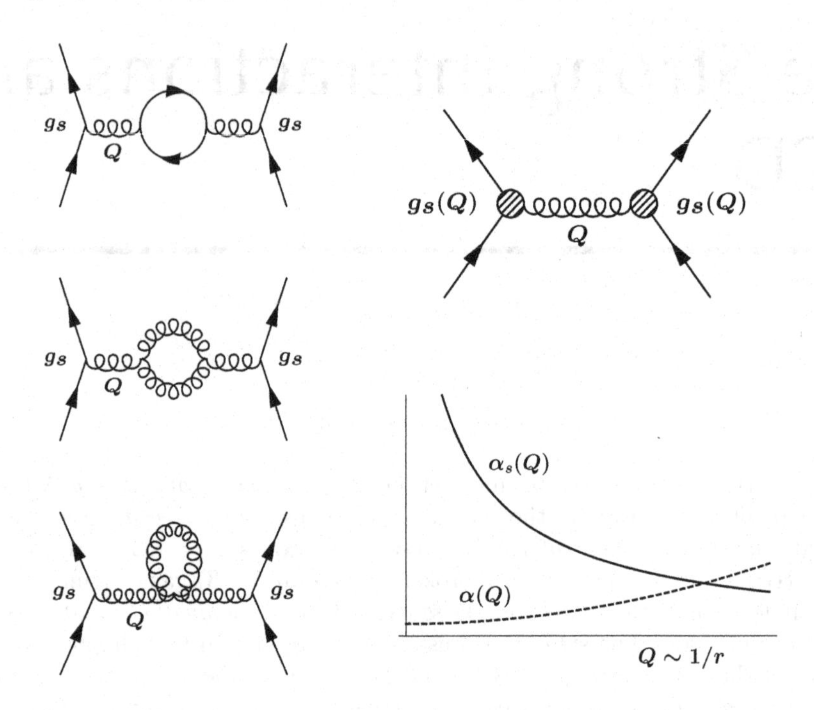

Figure 5.2 The running couplings in QCD. The upper left (quark-loop) diagram leads to a decrease in the effective $g_s(Q^2)$ at small momentum Q (i.e., long distance $r \sim 1/Q$) due to screening of the charge, analogous to QED. The lower left (gluon-loop) diagrams have the opposite (anti-screening) behavior. They have no analog in QED. Ghost diagrams are not shown. On the right are the effective running QCD coupling, and a sketch of the values of $\alpha_s(Q^2) = g_s^2(Q^2)/4\pi$ and the QED coupling $\alpha(Q^2) = e^2(Q^2)/4\pi$. The QCD coupling becomes small (asymptotically free) for $|Q| \gg 1$ GeV, and large (of $\mathcal{O}(1)$) for $|Q| \lesssim 1$ GeV (infrared slavery).

The Long Distance Regime

The long distance regime, relevant for momenta $|Q| \lesssim 1$ GeV, is characterized by strong coupling, $g_s = \mathcal{O}(1)$. It is therefore non-perturbative. Isolated quarks and gluons have not been observed; presumably they are *confined* and cannot emerge as free particles due to the strong coupling and gluon-self couplings [infrared slavery (Fritzsch et al., 1973)] that only allow color-singlet asymptotic states (*hadrons*). These include $q\bar{q}$ mesons M, and qqq baryons B, in the color-singlet states[1]

$$|M\rangle \sim \frac{1}{\sqrt{3}}\delta^{\alpha\beta}|q_{i\alpha}\bar{q}_{j\beta}\rangle, \qquad |B\rangle \sim \frac{1}{\sqrt{6}}\epsilon^{\alpha\beta\gamma}|q_{i\alpha}q_{j\beta}q_{k\gamma}\rangle. \tag{5.1}$$

[1]Other possible color-singlets will be mentioned in Section 5.9.

In (5.1) α, β, and γ are the color indices, and i, j, and k are collectively the flavor, spin, and space indices. Their strong interactions are approximately described by phenomenological models, such as the Yukawa model of pion exchange between nucleons. Within the underlying QCD theory this is interpreted as an approximation to higher-order effects, as illustrated in Figure 5.3. Even though perturbation theory is not useful in this regime, one can still utilize the approximate global flavor symmetries described in Chapter 3, i.e., $SU(2)$ isospin symmetry relating u and d at the few % level, and the extension to $SU(3)$ relating u, d, and s at the 25% level, and their chiral extensions.

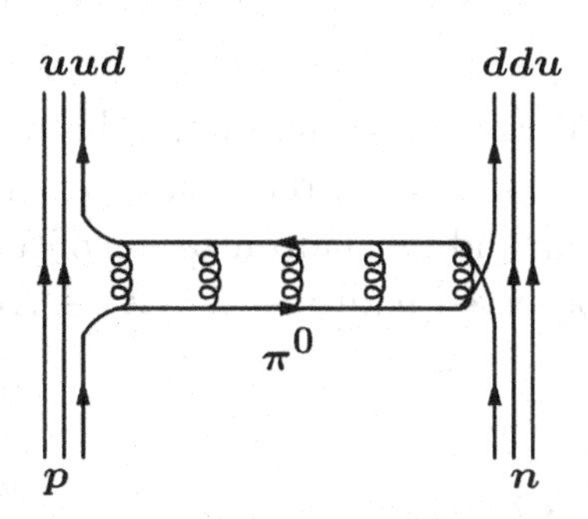

Figure 5.3 Effective Yukawa interaction, due to quark-antiquark exchange with the $q\bar{q}$ pair interacting via gluon exchange.

The Short Distance Regime

The short distance regime, $|Q| \gg 1$ GeV, has weak coupling, $g_s \ll 1$ (asymptotic freedom). It can therefore be described in terms of point-like quarks and gluons with interactions that can be calculated in perturbation theory. For example, in a *deep inelastic scattering* process, $e^- p \to e^- X$, the final e^- is observed but the final hadronic states, represented by X, are not observed and are summed over. (Such processes are known as *inclusive*, as opposed to *exclusive* ones such as $e^- p \to e^- p + 3\pi$ or $e^- p \to e^- \Delta^+$ involving a definite final state.) For $Q^2 \equiv -q^2 \gg 1$ GeV^2 the process is described to first approximation by the photon interacting with a point-like quark, as shown in Figure 5.4. Other parts of the diagram, involving the distribution of quarks within the proton, and the process of the remaining quarks in the proton and the scattered quark turning into hadrons (*hadronization*), are non-perturbative effects. The concept of describing inclusive sums over final states of hadrons in terms of perturbative calculations involving quarks and gluons is known as *quark-hadron duality* (e.g., Melnitchouk et al., 2005).

5.1 THE QCD LAGRANGIAN

The quark fields are denoted $q_{r\alpha}$, where $\alpha = 1, 2, 3$ or R,G,B is the gauged color and $r = u, d, s, c, b, t$ is the (ungauged) flavor index. An alternate notation is $u_\alpha = q_{u\alpha}$, etc. The Dirac indices are suppressed. The quarks transform under the fundamental (3) representation of $SU(3)$, and are non-chiral, $L^i_L = L^i_R = \frac{\lambda^i}{2}$. The Dirac adjoint field transforms as a 3^* and is written $\bar{q}^\alpha_r$. There are 8 Hermitian gauge fields (gluons), $G^i = G^{i\dagger}, i = 1 \cdots 8$. The QCD

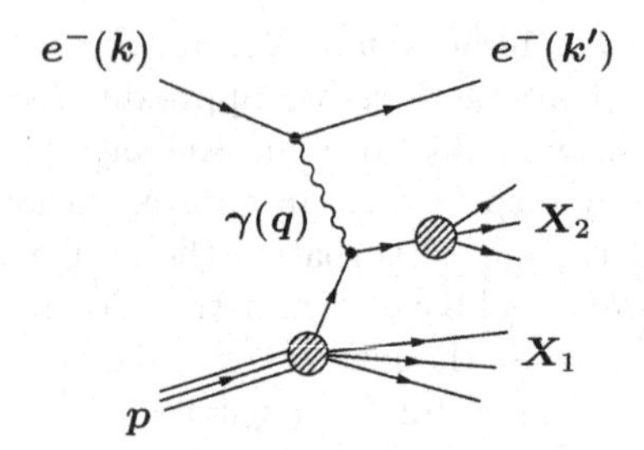

Figure 5.4 Deep inelastic scattering. The interaction between the lepton e^- and the quark can be described perturbatively by one photon exchange. However, the distribution of quarks in the proton, and the hadronization of the remaining and scattered quarks (the blobs) into unobserved hadrons $X = X_1 + X_2$ are non-perturbative.

Lagrangian density is

$$\mathcal{L}_{QCD} = -\frac{1}{4} G^i_{\mu\nu} G^{i\mu\nu} + \sum_r \bar{q}^{\alpha}_r i \not{D}^{\beta}_{\alpha} q_{r\beta} - \sum_r m_r \bar{q}^{\alpha}_r q_{r\alpha} + \frac{\theta_{QCD}}{32\pi^2} g_s^2 G^i_{\mu\nu} \tilde{G}^{i\mu\nu}, \tag{5.2}$$

where the field strength tensor is

$$G^i_{\mu\nu} = \partial_\mu G^i_\nu - \partial_\nu G^i_\mu - g_s f_{ijk} G^j_\mu G^k_\nu. \tag{5.3}$$

The quark gauge covariant derivative is

$$D^{\mu\beta}_{\alpha} \equiv (D^\mu)_{\alpha\beta} = \partial^\mu \delta^\beta_\alpha + \frac{i g_s}{\sqrt{2}} G^{\mu\beta}_{\alpha}, \tag{5.4}$$

where

$$G^\beta_\alpha = (G^\alpha_\beta)^\dagger = \sum_{i=1}^{8} G^i \frac{\lambda^i_{\alpha\beta}}{\sqrt{2}}, \tag{5.5}$$

with $G^\alpha_\alpha = 0$, represents the gluon field in tensor or matrix notation. The m_r in (5.2) are the *current* quark masses. They are actually generated by spontaneous symmetry breaking in the full standard model (including the chiral electroweak part), but can be considered as bare masses when considering QCD alone. Without loss of generality (for QCD) they can be taken to be real, nonnegative, and diagonal in flavor.

The interaction terms in (5.3) and (5.4) are the same for all six flavors, so they are invariant under a global chiral $U(6) \times U(6)$ flavor symmetry. However, m_c, m_b, and m_t are large compared to the typical scale of the strong interactions, $\Lambda \sim (200 - 300)$ MeV, so the symmetry is badly broken except at very high energy. More useful is an approximate $SU(3)$ flavor symmetry in the limit $m_u \sim m_d \sim m_s$, which is valid at the 25% level, and the even better (1%) $SU(2)$ isospin symmetry, which holds in the limit $m_d \sim m_u$.[2] These symmetries are enhanced to become chiral for $m = 0$; e.g., for $m_u = m_d = 0$ the continuous symmetries of $\mathcal{L}_{QCD}$ become

$$\underbrace{SU(3)_{\text{color}}}_{\text{gauge}} \times \underbrace{SU(2) \times SU(2)}_{\text{global}} \times \underbrace{U(1) \times U(1)}_{\text{global } B, B_A}, \tag{5.6}$$

[2]In fact, m_u, m_d, and m_s are not really degenerate compared to each other. However, m_u and m_d are both very small compared to Λ, so they are "degenerate" in the sense that $m_u/\Lambda \sim m_d/\Lambda \sim 0$. Similarly, $SU(3)$ holds approximately because $m_s \sim 100$ MeV ($\gg m_{u,d}$) is smaller than Λ though non-negligible.

while the $SU(2) \times SU(2)$ becomes $SU(3) \times SU(3)$ if $m_s \to 0$ as well. The $U(1)$ factors are separate L and R chiral baryon numbers. The sum (difference) of the generators corresponds to baryon ("axial baryon") number $B\,(B_A)$. However, B_A is *not* a good symmetry of the strong interactions, and its unsuccessful prediction in the quark model is known as the *axial* $U(1)_A$ *problem*. Its resolution by non-perturbative effects in QCD will be commented on in Section 5.8.3.

Another difficulty is the strong CP problem, which refers to the last term in (5.2). The strong interactions are observed to be reflection invariant (i.e., parity P is conserved), as well as invariant under charge conjugation (C), time reversal (T), and the products CP and CPT. The first three terms in (5.2) respect these symmetries. However, it is possible to add the final *strong* CP term to $\mathcal{L}_{QCD}$, where θ_{QCD} is a dimensionless constant, and

$$\tilde{G}^i_{\mu\nu} \equiv \frac{1}{2}\epsilon_{\mu\nu\rho\sigma}G^{i\rho\sigma} \tag{5.7}$$

is the dual field strength tensor. (G and $\tilde{G}$ are related by exchanging the analogs of the electric and magnetic fields.) The strong CP term is gauge invariant and does not spoil the renormalizability of QCD. However, for $\theta_{QCD} \neq 0$ it violates P, T, and CP symmetries, and stringent experimental limits on the electric dipole moment of the neutron require $|\theta_{QCD}| \lesssim 10^{-11} - 10^{-10}$ (e.g., Kim and Carosi, 2010). For pure QCD it is possible to simply impose these symmetries, i.e., to take $\theta_{QCD} = 0$. However, as will be discussed in Chapter 10, this becomes problematic in the context of the full standard model, which has other sources of CP violation.

5.2 EVIDENCE FOR QCD

QCD is the unique renormalizable field theory consistent with the observations that existed by ca. 1970, and since that time there has not been any serious competing theory.[3] Nevertheless, it is interesting to review some of the evidence for the ingredients of QCD.

Spin-$\frac{1}{2}$ Quarks and Spin-1 Gluons

The first evidence for spin-$\frac{1}{2}$ quarks was the success of the constituent quark model, which successfully classified a large number of hadrons in terms of three flavors of quarks. The spin-$\frac{1}{2}$ nature is essential for this classification, as is evident from the construction of the nucleons out of three quarks or of the spin-1 ρ from a $q\bar{q}$ pair in an S-wave. The very simple description of the approximate flavor symmetries of the strong interactions and their breaking in the quark model, i.e., (3.113) on page 110, also provided evidence.

By 1970, dynamical evidence emerged from the deep-inelastic scattering process $e^-p \to e^-X$, with $Q^2 = -q^2 \gg 1$ GeV^2, shown in Figure 5.4 and to be described in Section 5.5. The rate observed at the Stanford Linear Accelerator Center (SLAC) at large Q^2 was much larger than would be expected for a "big fuzzy" proton, calling for the existence of point-like *parton* constituents of the proton, reminiscent of the discovery of the atomic nucleus in the Rutherford experiment. In principle, the partons could be spin-0, spin-$\frac{1}{2}$, or higher. However, the angular distribution of the scattered electron established that they are spin-$\frac{1}{2}$ (the Callan-Gross relation), consistent with being quarks. This was later confirmed in

[3] In fact, prior to the development of QCD many physicists seriously considered abandoning the ideas of field theory or of any fundamental dynamical equations for the strong interactions, in favor of the *bootstrap*, which postulated that there was a unique S-matrix consistent with the ideas of unitarity, analyticity, crossing, etc. (See, e.g., Eden et al., 1966; Collins, 1977).

the distributions observed in deep inelastic $\mu^\pm N$ and $\overset{(-)}{\nu} N$ scattering and in their relative strengths.

Additional evidence emerged a few years later in the study of $e^+e^- \to$ hadrons. The observed cross section falls as $1/s$, where s is the square of the CM energy. This is consistent with the behavior expected for the production of point-like quarks (up to higher-order QCD corrections), as in (2.232) on page 46, and not with the more rapid falloff expected without such constituents. These later hadronize into collimated clusters of hadrons known as *jets*. The spin-$\frac{1}{2}$ nature was established by the observed $1+\cos^2\theta$ angular distribution of (2.231) for the jets, as opposed to the $\sin^2\theta$ distribution predicted for spin-0 (Problem 2.23). On the other hand, the positive evidence for quarks was apparently contradicted by the non-observation of isolated quarks. The resolution of that conflict had to await the development of QCD and the notion of infrared slavery.

The first direct evidence for spin-1 gluons came from the observation of distinct 3-jet events from $e^+e^- \to q\bar{q}G$ and of the planar broadening of events in which the third jet could not be resolved, by the TASSO and other collaborations at PETRA (DESY) in 1979 (see, e.g., Wu, 1984; Bethke, 2007). Another type of compelling evidence is indirect, i.e., the observed asymptotic freedom of the strong interactions requires a non-abelian gauge theory. Other advantages of color octet gluons were especially emphasized in (Fritzsch et al., 1973).

Evidence for Color

The observed hadrons are all color singlets. Nevertheless, with the benefit of hindsight, the color quantum number was already needed in the original quark model. That is because the quark assignments for the baryons and hyperons (baryons involving an s quark) were totally symmetric in the flavor, spin, and space indices. In particular, the Ω^-, which was successfully predicted by the $SU(3)$ model (Section 3.2.3), was interpreted as

$$|\Omega^-\rangle = |s^\uparrow s^\uparrow s^\uparrow\rangle, \tag{5.8}$$

where the arrows all represent spin-up with respect to a reference axis. The Ω^- is therefore symmetric in flavor (all s quarks), spin (all spins in the same direction), and in space indices (the orbital angular momenta are zero). However, this violates the spin and statistics theorem, which follows from the union of relativity and quantum mechanics (see, e.g., Streater and Wightman, 2000), and which requires that all physical spin-$\frac{1}{2}$ states should be antisymmetric. This fundamental difficulty with the quark model is easily resolved by the introduction of the color quantum number, under the assumption that the Ω^- and other baryon/hyperon states are color singlets, since the projection of $3\times 3\times 3$ onto the singlet is totally antisymmetric, as in (5.1),

$$|\Omega^-\rangle \propto \epsilon^{\alpha\beta\gamma}|s^\uparrow_\alpha s^\uparrow_\beta s^\uparrow_\gamma\rangle. \tag{5.9}$$

Thus, the color quantum number was needed even before the development of QCD,[4], where it played the additional role of a gauge quantum number.

There are other tests based on counting the number of colors that contribute to an amplitude or rate. The leading diagrams for $e^+e^- \to$ hadrons are shown in Figure 5.5. At short enough distance, one may regard the process as first producing quark $q_{r\alpha}$ and its antiquark, which may be computed perturbatively, followed by hadronization, in which the

[4]An apparently alternative resolution of the statistics problem, parastatistics (Greenberg, 2008), is in fact equivalent to the existence of the color quantum number.

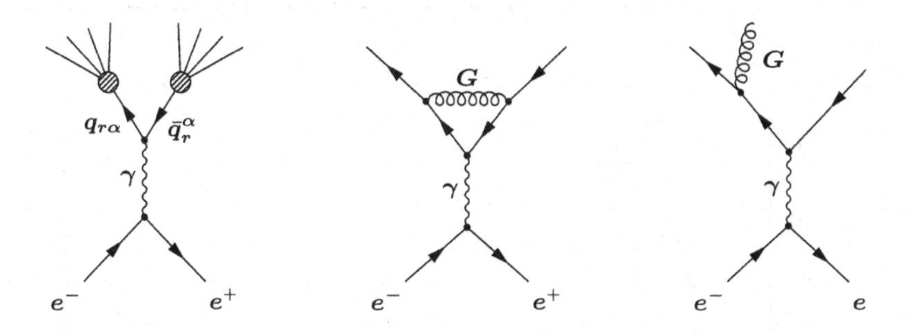

Figure 5.5 $e^+e^- \to q_{r\alpha}\bar{q}_r^\alpha$. Left: the blobs represent the quark hadronization. Right: higher-order QCD corrections.

quarks turn into jets of hadrons such as pions (low momentum [*soft*] gluons or quarks may be exchanged between the jets to ensure color neutrality). There are also higher-order QCD corrections, which can be calculated perturbatively. One usually expresses the theoretical and experimental result in terms of the ratio

$$R(s) = \frac{\sigma(e^+e^- \to \text{ hadrons})}{\sigma(e^+e^- \to \mu^+\mu^-)} \tag{5.10}$$

at CM energy $\sqrt{s}$, where the denominator is the lowest order theoretical expression in (2.232) with a running coupling, $\sigma(e^+e^- \to \mu^+\mu^-) = 4\pi\alpha^2(s)/3s$. $R(s)$ is convenient because the largest energy dependence cancels in the ratio, and experimentally because the luminosity also cancels. $R(s)$ counts the number of quark colors and flavors, weighted by the quark electric charge-squared e_r^2, where $e_r = \frac{2}{3}$ for $[u,\ c,\ t]$ and $-\frac{1}{3}$ for $[d,\ s,\ b]$. The lowest-order prediction (i.e., ignoring α_s) is $R = N_c \sum_r e_r^2$, where N_c is the number of colors and only the n_q quarks lighter than $\sqrt{s}/2$ should be included in the sum. For QCD ($N_c = 3$) this is

$$R = \frac{5}{3} \text{ for } [u,d]\,,\ \frac{6}{3} \text{ for } [uds]\,,\ \frac{10}{3} \text{ for } [udsc]\,,\ \frac{11}{3} \text{ for } [udscb]\,. \tag{5.11}$$

The higher-order QCD corrections have been computed to 4 loops (Baikov et al., 2012). Neglecting quark masses,[5] one predicts

$$R = N_c \sum_r e_r^2 \left[1 + \frac{\alpha_s}{\pi} + c_2^{n_q}\left(\frac{\alpha_s}{\pi}\right)^2 + c_3^{n_q}\left(\frac{\alpha_s}{\pi}\right)^3 + c_4^{n_q}\left(\frac{\alpha_s}{\pi}\right)^4 + \cdots\right] \tag{5.12}$$

for n_q quark flavors. The higher-order terms use the value of the running α_s at s. For $n_q = 5$, for example, $c_2^5 = 1.40902$, $c_3^5 = -12.80$, and $c_4^5 = -80.434$. References to quark mass and Z exchange corrections are given in (Patrignani, 2016). The QCD ($N_c = 3$) prediction is in excellent agreement with the experimental result in Figure 5.6. R also excludes an alternative quark model involving integer electric charges (Nambu and Han, 1974), at least

[5]Similar to QED, there are infrared singularities associated with both virtual and real gluons as their energy approaches zero, as well as *mass* (or *collinear*) *singularities* that occur in the limit of a massless quark due to gluons radiated parallel to the quark. These can be shown to cancel under realistic conditions using dimensional regularization or by introducing a fictitious gluon mass. (The latter is only possible for diagrams not involving the non-abelian gauge vertices, since it would violate gauge invariance). See (Field, 1989; Salam, 2010a) for detailed discussions.

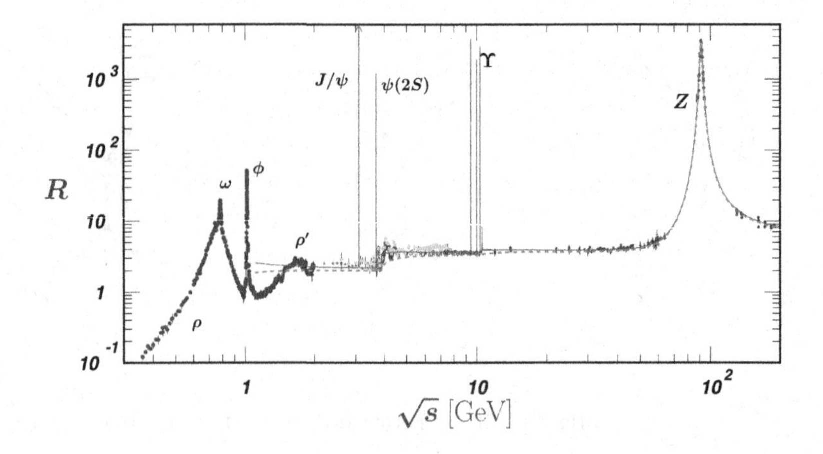

Figure 5.6 Experimental data on $R(s)$ compared with the lowest order (dashes) and three-loop (solid) QCD predictions. There are steps at the s, c, and b thresholds, given approximately by the locations of the $\phi\,(s\bar{s})$, $J/\psi\,(c\bar{c})$, and $\Upsilon\,(b\bar{b})$ resonances. The perturbative prediction works extremely well above a few GeV, provided one includes the threshold resonances. At high energies, Z boson exchange strongly dominates over one photon exchange. Plot courtesy of the Particle Data Group (Patrignani, 2016).

under the assumption that the quarks can be treated as pointlike (Problem 5.1). Further tests of QCD in e^+e^- annihilation are reviewed in (Kluth, 2006).

Other evidence for color and QCD includes the ratio of nonleptonic and leptonic decays of the W, since $W^- \to q\bar{q}$, where $q\bar{q} = d\bar{u}$, $s\bar{c}$ counts the number of colors, while the leptonic decays into $\ell^-\bar{\nu}_\ell$, $\ell = e, \mu, \tau$, do not. In particular, the branching ratio

$$B\left(W^- \to e^-\nu^-\right) = \frac{\Gamma\left(W^- \to e^-\nu^-\right)}{\Gamma\left(W^- \to q\bar{q}\right) + \Gamma\left(W^- \to \ell^-\bar{\nu}\right)} \simeq \frac{1}{3 + 2N_c} \xrightarrow[N_c=3]{} 11\% \tag{5.13}$$

(up to small corrections from QCD, fermion mass, quark mixing, etc.), in agreement with the experimental value $\sim 10.7\%$.

Another probe is the *Drell-Yan* process in which

$$p\overset{(-)}{p} \to \ell^+\ell^- + \text{hadrons}, \quad \ell^- = e^-, \mu^-, \tau^- \tag{5.14}$$

at large $\left(p_{\ell^+} + p_{\ell^-}\right)^2 \gg 1\ \text{GeV}^2$. This is dominated by the $q\bar{q}$ annihilation through a γ or Z, as shown in Figure 5.7, and can be thought of as the inverse to $e^+e^- \to q\bar{q}$. The $q_{r\alpha}$ and $\bar{q}_r^\beta$ can each be in one of three color states, but only the combination in which $\alpha = \beta$ can contribute, leading to a cross section $1/N_c$ compared to what would be expected without color, in agreement with observation. To see this, consider $p\bar{p}$ scattering in the approximation of considering the valence quarks only. Using the baryon wave function in (5.1),

$$\sigma_{\text{color}} = 3 \times 2 \times 2 \times \left(\frac{1}{\sqrt{6}}\right)^2 \times \left(\frac{1}{\sqrt{6}}\right)^2 \sigma_{\text{no-color}} = \frac{1}{3}\sigma_{\text{no-color}}, \tag{5.15}$$

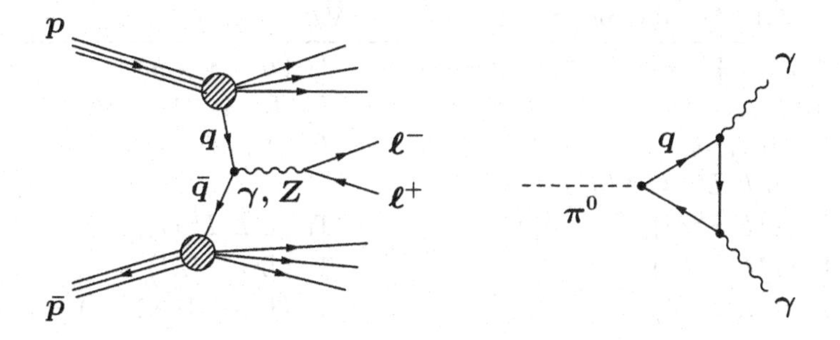

Figure 5.7 Left: Drell-Yan process, $\bar{p}p \to \gamma, Z \to \ell^-\ell^+$. Right: The $\pi^0 \to 2\gamma$ decay.

where the 3 represents the three colors that can annihilate, and 2^2 represents the color assignments of the remaining quarks, all of which add incoherently.[6]

One can use $SU(2) \times SU(2)$ chiral symmetry to show that the chiral anomaly associated with the triangle diagram in Figure 5.7 (with π^0 coupling to the axial isospin generator $T_R^3 - T_L^3$) dominates the $\pi^0 \to 2\gamma$ decay amplitude in the $m_\pi \to 0$ limit (Adler, 1969; Donoghue et al., 2014). This again counts the number of quark colors, so that

$$\Gamma\left(\pi^0 \to 2\gamma\right) \sim \left(\frac{N_c}{3}\right)^2 \frac{\alpha^2 m_{\pi^0}^3}{32\pi^3 f_\pi^2} = 7.75(2) \left(\frac{N_c}{3}\right)^2 \text{ eV}, \tag{5.16}$$

where $f_\pi = 130.5(1)$ MeV is the pion decay constant, which is associated with the spontaneous breaking of the chiral symmetry and is measured in $\pi^+ \to \mu^+\nu$ decay. The prediction for $N_c = 3$ is increased to ~ 8.10 eV by chiral-breaking corrections (Bernstein and Holstein, 2013), consistent with the experimental value, 7.6(3) eV.

5.3 SIMPLE QCD PROCESSES

In this section we sketch the derivation of some simple QCD processes at tree level. Some of the calculations are similar to the QED calculations in Chapter 2, except that one has to properly take the color factors into account. Others involve the gluon self-interactions, which have no QED analog. Processes involving non-abelian vertices are extremely tedious to carry out by hand, and are best handled by specialized computer algebra programs (see the list of websites in the bibliography and the example notebooks on the book website). However, we will illustrate one relatively simple example. Higher-order calculations involve all of the subtleties of gauges and ghost loops (the latter may even appear in tree-level calculations involving external gluons), which are treated in standard field theory texts.

Color Identities

The calculations are greatly simplified by the use of certain color identities listed in Table 5.1 for the fundamental representation matrices $L_3^i = \lambda^i/2$ (denoted in this section by L^i) and the $SU(3)$ structure constants f_{ijk} defined in Tables 3.1 and 3.2. They follow easily from or are special cases of the identities given in Sections 3.1.2 and 3.1.3 and in the Problems in Chapter 3.

[6] An equivalent derivation is to simply average over the N_c colors of the interacting q and $\bar{q}$, so that $\sigma \propto (1/N_c)^2 \sum_{\alpha,\beta=1}^{N_c} \delta^{\alpha\beta} = 1/N_c$.

TABLE 5.1 $SU(3)$ color identities[a] for $L^i \equiv \lambda^i/2$ and f_{ijk}.

$L^iL^i = \frac{4}{3}I$	$\mathrm{Tr}\left(L^iL^j\right) = \frac{1}{2}\delta^{ij}$
$f_{ijk}f_{ijm} = 3\delta_{km}$	$f_{ijk}f_{ijk} = 24$
$f_{ijm}f_{klm}f_{ijn}f_{kln} = 72$	$f_{ijm}f_{klm}f_{ikn}f_{jln} = 36$
$\mathrm{Tr}\left(L^iL^jL^k\right) = \frac{1}{4}\left(d_{ijk} + if_{ijk}\right)$	
$\mathrm{Tr}\left(L^iL^jL^k\right) if_{ijm} = -\frac{3}{4}\delta^{km}$	$\mathrm{Tr}\left(L^iL^jL^k\right) if_{ijk} = -6$
$\mathrm{Tr}\left(L^iL^jL^iL^k\right) = -\frac{1}{12}\delta_{jk}$	$\mathrm{Tr}\left(L^iL^jL^iL^j\right) = -\frac{2}{3}$
$\mathrm{Tr}\left(L^iL^jL^jL^i\right) = \frac{16}{3}$	$\mathrm{Tr}\left(L^iL^j\right)\mathrm{Tr}\left(L^iL^j\right) = 2$
$f_{ilm}d_{jmk} - f_{imk}d_{jlm} + f_{ijm}d_{mlk} = 0$	
$f_{ilm}f_{jmk} - f_{imk}f_{jlm} + f_{ijm}f_{mlk} = 0$ (Jacobi identity)	

[a] $L^iL^i \equiv \sum_i L^iL^i$, I is the 3×3 identity matrix, and the indices run from 1 to 8. There is no distinction between upper and lower indices.

qq and $q\bar{q}$ Scattering

qq and $q\bar{q}$ scattering via gluon exchange are very much like the simple QED processes described in Section 2.8. Here we will neglect the quark masses for simplicity, but it is straightforward to include them (as would be necessary for the production of a heavy quark, such as $u\bar{u} \to t\bar{t}$). The process $q_{r\beta}\bar{q}_{r\alpha} \to q_{s\gamma}\bar{q}_{s\delta}$, where α, β, γ, and δ are color indices and $r \neq s$ are flavor indices, proceeds through an s channel gluon, as shown in the first diagram in Figure 5.8. The only differences compared with the QED process $e^-e^+ \to f\bar{f}$ in (2.225) on page 46 and Figure 2.15 are that $-e^2Q_f \to g_s^2$ and that there are color factors on the vertices and gluon propagator, as shown in Figure 4.1, yielding

$$\begin{aligned} M_{fi} &= \left(-ig_s\, \bar{u}_3\gamma_\mu L^k_{\gamma\delta}v_4\right)\left(\frac{-ig^{\mu\rho}\delta^{ik}}{s}\right)\left(-ig_s\, \bar{v}_2\gamma_\rho L^i_{\alpha\beta}u_1\right) \\ &= \frac{ig_s^2}{s}L^i_{\alpha\beta}L^i_{\gamma\delta}\left(\bar{u}_3\gamma_\mu v_4\, \bar{v}_2\gamma^\mu u_1\right). \end{aligned} \tag{5.17}$$

The calculation of the spin-average cross section proceeds as in $e^-e^+ \to f\bar{f}$, Equation (2.227), except it is now convenient to do a color average as well, in which one averages (sums) over initial (final) quark colors $\alpha = 1\cdots 3$. Similarly, in processes involving external gluons one averages (sums) over initial (final) gluon color indices $i = 1\cdots 8$. Neglecting the

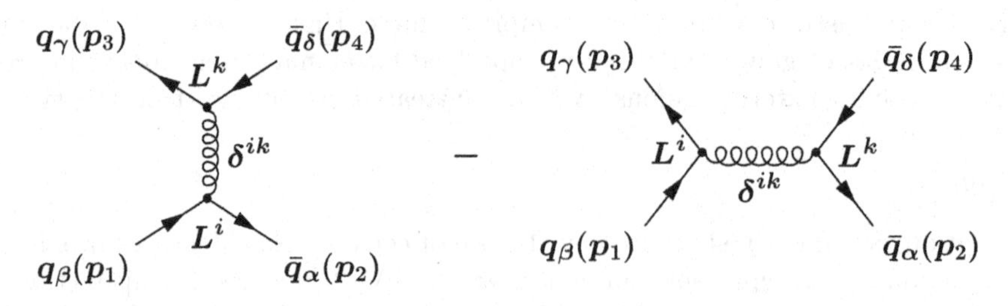

Figure 5.8 Diagrams for $q_{r\beta}\bar{q}_{r\alpha} \to q_{s\gamma}\bar{q}_{s\delta}$, where r and s are flavor indices. Only the diagram on the left contributes for $r \neq s$.

quark masses, the expression $2Q_f^2 e^4(t^2+u^2)/s^2$, derivable from (2.227), is replaced by

$$\begin{aligned} |\bar{M}_{fi}|^2 &\equiv \left(\frac{1}{3}\right)^2 \sum_{\alpha,\beta,\gamma,\delta=1}^{3} L^i_{\alpha\beta} L^i_{\gamma\delta} L^{j*}_{\alpha\beta} L^{j*}_{\gamma\delta} (2g_s^4) \left(\frac{t^2+u^2}{s^2}\right) \\ &= \frac{1}{9} \text{Tr}\left(L^i L^j\right) \text{Tr}\left(L^i L^j\right) (2g_s^4) \left(\frac{t^2+u^2}{s^2}\right) = \frac{4}{9} g_s^4 \left(\frac{t^2+u^2}{s^2}\right). \end{aligned} \tag{5.18}$$

For annihilation into the same flavor, $q_{r\beta}\bar{q}_{r\alpha} \to q_{r\gamma}\bar{q}_{r\delta}$, the calculation is similar to Bhabha scattering in (2.233)–(2.235). The first two terms are obtained by the replacement $e^4 \to 2g_s^4/9$, just as in (5.18). However, the third (interference) term now has the color factor

$$\frac{1}{9} L^i_{\alpha\beta} L^i_{\gamma\delta} L^{j*}_{\gamma\beta} L^{j*}_{\alpha\delta} = \frac{1}{9} \text{Tr}\left(L^i L^j L^i L^j\right) = -\frac{2}{27}, \tag{5.19}$$

so that

$$|\bar{M}_{fi}|^2 = \frac{4}{9} g_s^4 \left[\frac{t^2+u^2}{s^2} + \frac{s^2+u^2}{t^2} - \frac{2}{3}\frac{u^2}{st}\right]. \tag{5.20}$$

The spin and color-averaged squared matrix elements for a number of $2 \to 2$ QCD processes, neglecting masses, are listed in Table 5.2. More extensive listings, including mass effects and extensions to supersymmetry, may be found in (Patrignani, 2016).

TABLE 5.2 Spin and color-averaged squared amplitudes $|\bar{M}|^2/g_s^4$ for various QCD subprocesses,[a] characterized by kinematic subprocess invariants s, t, and u.

	$\|\bar{M}\|^2/g_s^4$	90°
$\begin{pmatrix} q_r q_s \to q_r q_s \\ q_r \bar{q}_s \to q_r \bar{q}_s \end{pmatrix}$	$\frac{4}{9}\left(\frac{s^2+u^2}{t^2}\right)$	2.2
$q_r \bar{q}_r \to q_s \bar{q}_s$	$\frac{4}{9}\left(\frac{t^2+u^2}{s^2}\right)$	0.2
$q\bar{q} \to q\bar{q}$	$\frac{4}{9}\left(\frac{t^2+u^2}{s^2} + \frac{s^2+u^2}{t^2} - \frac{2}{3}\frac{u^2}{st}\right)$	2.6
$qq \to qq$	$\frac{4}{9}\left(\frac{s^2+u^2}{t^2} + \frac{s^2+t^2}{u^2} - \frac{2}{3}\frac{s^2}{tu}\right)$	3.3
$\begin{pmatrix} GG \to q\bar{q} \\ q\bar{q} \to GG \end{pmatrix}$	$\left[\frac{1}{6}\left(\frac{t^2+u^2}{tu}\right) - \frac{3}{8}\left(\frac{t^2+u^2}{s^2}\right)\right] \times \begin{pmatrix} 1 \\ 64/9 \end{pmatrix}$	$\begin{pmatrix} 0.1 \\ 1.0 \end{pmatrix}$
$qG \to qG$	$\left(\frac{s^2+u^2}{t^2}\right) - \frac{4}{9}\left(\frac{s^2+u^2}{su}\right)$	6.1
$GG \to GG$	$\frac{9}{4}\left(\frac{s^2+u^2}{t^2} + \frac{s^2+t^2}{u^2} + \frac{t^2+u^2}{s^2} + 3\right)$	30.4
$q\bar{q} \to q_0\bar{q}_0$	$\frac{4}{9}\frac{tu}{s^2}$	0.1
$GG \to q_0\bar{q}_0$	$\frac{3}{32}\left(\frac{u-t}{s}\right)^2 + \frac{7}{96}$	0.07

[a] $r \neq s$ when flavor indices are given. The last two processes involve the production of a hypothetical spin-0 color-triplet q_0, such as one encounters in supersymmetry. Masses are neglected. The last column is the numerical value for CM scattering angle 90°, where $t = u = -s/2$ (neglecting masses). Expanded from (Combridge et al., 1977; Barger and Phillips, 1997).

$GG \to q_0\bar{q}_0$

To illustrate a non-trivial non-abelian vertex, consider the process $GG \to q_0\bar{q}_0$, where q_0 is a hypothetical spin-0 color triplet, such as a scalar quark in supersymmetry. There are four tree-level diagrams, as shown in Figure 5.9 Using the vertex factors from Figure 4.1, the

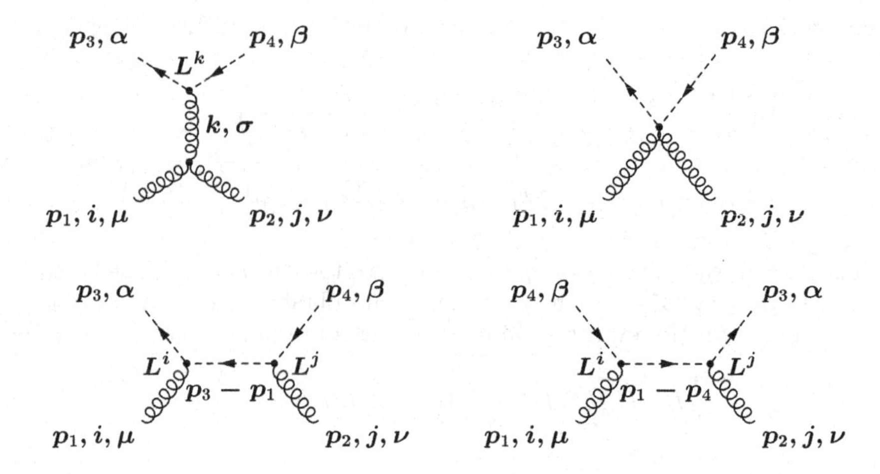

Figure 5.9 Diagrams for $G^i_\mu(p_1)G^j_\nu(p_2) \to q_{0\alpha}(p_3)\bar{q}_{0\beta}(p_4)$, where q_0 is a hypothetical spin-0 color triplet.

amplitude in Feynman gauge is

$$\begin{aligned} M = \epsilon_{1\mu}\epsilon_{2\nu} \Bigg[& -ig_s L^k_{\alpha\beta}(p_3 - p_4)_\sigma \left(\frac{-i}{s}\right) g_s f_{ijk} \\ & \times \left[g^{\mu\nu}(p_2 - p_1)^\sigma + g^{\mu\sigma}(2p_1 + p_2)^\nu - g^{\nu\sigma}(p_1 + 2p_2)^\mu\right] \\ & + ig_s^2 g^{\mu\nu} \left\{L^i, L^j\right\}_{\alpha\beta} \\ & + (-ig_s)^2 \left(L^i L^j\right)_{\alpha\beta} \left(\frac{i}{t - m_0^2}\right) (2p_3 - p_1)^\mu (p_3 - p_1 - p_4)^\nu \\ & + (-ig_s)^2 \left(L^j L^i\right)_{\alpha\beta} \left(\frac{i}{u - m_0^2}\right) (p_1 - 2p_4)^\mu (p_1 + p_3 - p_4)^\nu \Bigg], \end{aligned} \tag{5.21}$$

where m_0 is the q_0 mass, and $\epsilon_{1\mu} \equiv \epsilon_\mu(\vec{p}_1, \lambda_1)$ and $\epsilon_{2\nu} \equiv \epsilon_\nu(\vec{p}_2, \lambda_2)$ are the gluon polarization vectors. The straightforward way to calculate $|\bar{M}|^2$ would be to first take the absolute square and then use (2.121) on page 28 for the gluon polarization sums. However, this would be extremely tedious. The calculation would be simplified if the second term in (2.121) did not contribute, but this requires the calculation to be done in a gauge invariant way (cf., the discussion of Compton scattering in Section 2.8). In particular, for a non-abelian theory one must include the negative contribution of fictitious ghost pairs (Sterman, 1993, Section 8.5; Peskin and Schroeder, 1995, Section 17.4). Although this is straightforward, it can be avoided by introducing explicit expressions for the polarization vectors, just as we did in (2.151).

For the $q_0\bar{q}_0$ final state it is simpler to calculate the amplitudes rather than their absolute squares, analogous to the fermion helicity calculations in Section 2.9. The four-momenta in the CM frame are

$$p_{1,2} = k(1, 0, 0, \pm 1), \qquad p_{3,4} = E_f(1, \pm\beta_f \sin\theta, 0, \pm\beta_f \cos\theta), \tag{5.22}$$

where $k = \sqrt{s}/2$, $k_f = \sqrt{s - 4m_0^2}/2 = \beta_f E_f = \beta_f k$, and θ is the CM scattering angle. The gluon polarization vectors are

$$\epsilon_{1,2}(1) = (0, \pm 1, 0, 0), \qquad \epsilon_{1,2}(2) = (0, 0, 1, 0) \tag{5.23}$$

in a linear basis, where $\epsilon_n(\lambda_n) \equiv \epsilon_n(\vec{p}_n, \lambda_n), n = 1, 2$. (The sign for $\epsilon_2(1)$ was chosen to be consistent with the space reflection convention in (2.273) on page 53.) Thus,

$$\begin{aligned}
\epsilon_1(\lambda_1) \cdot \epsilon_2(\lambda_2) &= (-1)^{\lambda_1+1}\delta^{\lambda_1\lambda_2} \\
p_1 \cdot \epsilon_n(\lambda) &= p_2 \cdot \epsilon_n(\lambda) = 0 \\
p_3 \cdot \epsilon_1(1) &= -p_4 \cdot \epsilon_1(1) = -p_3 \cdot \epsilon_2(1) = p_4 \cdot \epsilon_2(1) = -k_f \sin\theta \\
p_3 \cdot \epsilon_n(2) &= p_4 \cdot \epsilon_n(2) = 0,
\end{aligned} \tag{5.24}$$

which greatly simplify the calculation. Denoting the amplitude by $M(\lambda_1, \lambda_2)$, one has $M(1,2) = M(2,1) = 0$, and only the $g^{\mu\nu}$ terms contribute to $M(2,2)$,

$$M(2,2) = -ig_s^2 \left[-if_{ijk}L^k_{\alpha\beta} \left(\frac{u-t}{s} \right) + \{L^i, L^j\}_{\alpha\beta} \right]. \tag{5.25}$$

t and u are related to θ by

$$t - m_0^2 = -2k^2(1 - \beta_f \cos\theta), \qquad u - m_0^2 = -2k^2(1 + \beta_f \cos\theta), \tag{5.26}$$

so that

$$\left(\frac{u-t}{s} \right) = -\frac{k_f}{k} \cos\theta. \tag{5.27}$$

$M(1,1)$ has the same $g^{\mu\nu}$ terms as $M(2,2)$, as well as contributions from the t and u-channel pole terms. These are easily calculated, yielding

$$\begin{aligned}
M(1,1) &= M(2,2) \left[-1 + \frac{2k_f^2 s \sin^2\theta}{(t-m_0^2)(u-m_0^2)} \right] \\
&= M(2,2) \left[1 + 2m_0^2 \left(\frac{1}{t-m_0^2} + \frac{1}{u-m_0^2} \right) \right] \\
&\equiv M(2,2) \left[1 + X \right],
\end{aligned} \tag{5.28}$$

where (5.26) was used to obtain the second form. The spin and color-average amplitude squared is

$$|\bar{M}|^2 = \frac{1}{4} \frac{1}{64} \sum_{i,j} \sum_{\alpha,\beta} \sum_{\lambda_1,\lambda_2} |M(\lambda_1, \lambda_2)|^2, \tag{5.29}$$

where the factors of $1/4$ and $1/64$ are, respectively, due to the averages over the gluon spins and colors. There is no interference between the two terms in $M(2,2)$ because they are, respectively, antisymmetric and symmetric in i and j. From Table 5.1, the color sums for the squares of these terms are 12 and $28/3$, respectively, so that

$$|\bar{M}|^2 = \frac{g_s^4}{64} \left[3 \left(\frac{u-t}{s} \right)^2 + \frac{7}{3} \right] \left[1 + (1+X)^2 \right], \tag{5.30}$$

and

$$\frac{d\bar{\sigma}}{d\cos\theta} = \frac{1}{32\pi s} \frac{k_f}{k} |\bar{M}|^2, \qquad \frac{d\bar{\sigma}}{dt} = \frac{1}{16\pi s^2} |\bar{M}|^2. \tag{5.31}$$

One can obtain the amplitudes for left and right circularly polarized gluons using (2.118),

$$\begin{aligned}
M(L,L) = M(R,R) &= \frac{1}{2}[M(1,1) - M(2,2)] = -M(2,2) \left[1 - \frac{k_f^2 s \sin^2\theta}{(t-m_0^2)(u-m_0^2)} \right] \\
M(L,R) = M(R,L) &= \frac{1}{2}[M(1,1) + M(2,2)] = M(2,2) \left[\frac{k_f^2 s \sin^2\theta}{(t-m_0^2)(u-m_0^2)} \right].
\end{aligned} \tag{5.32}$$

$M(L,R)$ and $M(R,L)$ vanish in the forward and backward directions because of angular momentum conservation, i.e., the gluon spins are in the same direction and cannot be compensated by orbital angular momentum.[7] $M(L,L)=M(R,R)$ and $M(L,R)=M(R,L)$ follow from reflection invariance.

5.4 THE RUNNING COUPLING IN NON-ABELIAN THEORIES

It was already described in the introduction and (for QED) in Section 2.12.2 that many of the higher-order corrections to QCD (and other field theories) can be absorbed into an effective $g_s(\mu^2)$ or $\alpha_s(\mu^2)=g_s(\mu^2)^2/4\pi$, where μ is the renormalization scale. Higher-order corrections are usually minimized when μ is taken to be a typical momentum scale of the process, such as $\mu^2=Q^2\equiv|q|^2$, where q might be the four-momentum carried by an exchanged gluon. The gluon self-interactions in QCD imply asymptotic freedom, i.e., that $\alpha_s(\mu^2)$ becomes small at large $\mu\gg\mathcal{O}(1\text{ GeV})$ (short distance), so that one can treat the quarks and gluons as weakly coupled, and processes such as deep inelastic scattering can be calculated in perturbation theory. For small μ (large distance), $\alpha_s(\mu^2)$ becomes large and perturbation theory no longer holds. The strong coupling and gluon self-interactions presumably lead to the confinement of quarks, gluons, and any colored states, so that only color singlet hadron states can emerge. In the real world, both regimes may be relevant to different aspects of a process, e.g., a quark may scatter in the short distance regime, but its initial distribution in the proton and its subsequent hadronization are low momentum processes. Fortunately, for many processes the short and long distance effects can be *factorized*, with the former calculable. The latter must be taken from experiment, although their logarithmnic Q^2 dependence is predicted by QCD.

The deep inelastic scattering experiments at SLAC could be understood in the *simple parton model* of point-like constituents of the nucleon, which had been developed somewhat earlier to understand hadronic processes (see, e.g., Field and Feynman, 1977). However, it was quickly understood that the interactions of the partons had to be asymptotically free, so that they could appear point-like at short distances and still be confined in the nucleon. The breakthrough came in 1973, when it was shown that non-abelian gauge theories could be asymptotically free (if there are not too many matter fields), and furthermore that they are the unique asymptotically free renormalizable theories in four dimensions (Gross and Wilczek, 1973; Politzer, 1973). Combined with the evidence for three colors, QCD emerged as the unique candidate theory.

5.4.1 The RGE Equations for an Arbitrary Gauge Theory

The running of the effective gauge coupling g in a gauge theory with a single group factor is described by the *renormalization group equation* (RGE)

$$\frac{dg^2}{d\ln Q^2}\equiv 4\pi\beta(g^2)=\underbrace{bg^4}_{\text{1 loop}}+\underbrace{O(g^6)}_{\text{2 loop}}+\cdots, \tag{5.33}$$

where the coefficient of the one-loop term in the β *function* is

$$b=-\frac{1}{(4\pi)^2}\left[\frac{11}{3}C_2(G)-\frac{4}{3}T_F-\frac{1}{3}T_{\phi_c}-\frac{1}{6}T_{\phi_h}\right], \tag{5.34}$$

[7]The $\sin^2\theta$ factor is cancelled by the singularities from the t- and u-channel poles for $m_0^2\to 0$, as can be seen from (5.28) and (5.32).

where $C_2(G)$ is the quadratic Casimir defined after (3.22), with $C_2(SU(m)) = m$ and $C_2(U(1)) = 0$. T_F, T_{ϕ_c}, and T_{ϕ_h} are, respectively, the fermion, complex scalar, and Hermitian scalar Dynkin indices

$$\begin{aligned} T_F \delta_{ij} &= \frac{1}{2}\text{Tr}\left(L_L^i L_L^j\right) + \frac{1}{2}\text{Tr}\left(L_R^i L_R^j\right) \xrightarrow[L_L = L_R]{} \text{Tr}\left(L_\psi^i L_\psi^j\right) \\ T_{\phi_c}\delta_{ij} &= \text{Tr}\left(L_{\phi_c}^i L_{\phi_c}^j\right), \qquad T_{\phi_h}\delta_{ij} = \text{Tr}\left(L_{\phi_h}^i L_{\phi_h}^j\right). \end{aligned} \tag{5.35}$$

$T(L_m) = \frac{1}{2}$ in $SU(m)$ for a fundamental representation; for $U(1)$ a set of fields with $U(1)$ charges q_a yields $T = \sum_a q_a^2$ (or $\frac{1}{2}\sum_a (q_{aL}^2 + q_{aR}^2)$ for chiral fermions). One must sum over all of the IRREPs in which the particle masses are smaller than Q, so that there are discontinuities in the slope at the particle thresholds. Equation (5.33) is easily solved analytically in one-loop approximation,

$$\frac{1}{\alpha(Q^2)} = \frac{1}{\alpha(M^2)} - 4\pi b \ln\frac{Q^2}{M^2} + \mathcal{O}\left[\alpha(Q^2)\right], \tag{5.36}$$

where $\alpha \equiv g^2/4\pi$ and M is an arbitrary reference scale. Thus, $1/\alpha(Q^2)$ varies linearly with $\ln Q^2$. Asymptotic freedom[8] occurs for $b < 0$. Since $C_2(G)$ and the Dynkin indices are nonnegative, asymptotic freedom always occurs in a pure non-abelian gauge theory, but not in $U(1)$. In the presence of fermion and scalar fields, asymptotic freedom may or may not hold depending on their contribution relative to the gauge terms.

The two-loop contributions to β are also known (see, e.g., Martin and Vaughn, 1994; Luo et al., 2003). [They are known to four loops for pure QCD (van Ritbergen et al., 1997).] However, at this order one must also include diagrams involving other interactions, such as Yukawa interactions or other gauge factors, which lead to a coupling between their RGEs. More careful treatment of thresholds and of the renormalization scheme are also needed at this order.

For QCD, one has $C_2 = 3$, $T_F = \frac{n_q}{2}$, and $T_{\phi_{c,h}} = 0$, where n_q is the number of quark flavors lighter than Q, e.g., $n_q = 3$ for $m_s < Q < m_c$ (perturbative results are not expected to be valid below m_s), $n_q = 4$ for $m_c < Q < m_b$, and $n_q = 5$ for $m_b < Q < m_t$. Thus, at one loop $b = -(33 - 2n_q)/48\pi^2$ and

$$\frac{1}{\alpha_s(Q^2)} = \frac{1}{\alpha_s(M^2)} + \frac{33 - 2n_q}{12\pi}\ln\frac{Q^2}{M^2} \equiv \frac{1}{\alpha_s(M^2)} + b_2^{n_q}\ln\frac{Q^2}{M^2}. \tag{5.37}$$

This can be rewritten

$$\alpha_s(Q^2) = \frac{1}{b_2^{n_q}\ln\frac{Q^2}{\Lambda^2}}, \tag{5.38}$$

where the scale[9] Λ is defined by

$$\Lambda = M\exp\left(\frac{-1}{2b_2^{n_q}\alpha_s(M^2)}\right) \tag{5.39}$$

(see Problem 5.7 for a generalization to two loops). In addition to $\alpha_s(Q^2)$ becoming small for large Q, it also goes to ∞ at the scale $\Lambda_{QCD} \equiv \Lambda$, at least in one-loop approximation.

---

[8]Other renormalizable interactions in four dimensions also satisfy RGE for their running couplings, but these are never asymptotically free (Coleman and Gross, 1973; Zee, 1973).

[9]One can define $\Lambda^{(n_q)}$ as the value relevant to the region with n_q light flavors, with the discontinuities at the thresholds fixed so that α_s is continuous.

Of course, the one-loop approximation and perturbation theory break down in this limit, but one may nevertheless loosely interpret Λ as the scale at which α_s becomes large. This treatment is easily extended to higher orders, inclusion of quark thresholds, etc. α_s has been determined in many ways at different scales (d'Enterria and Skands, 2015; Deur et al., 2016; and the QCD review in Patrignani, 2016), often from the corrections to simple parton model results. These include hadronic τ decays, Υ spectroscopy and decays, e^+e^- annihilation event shapes above and below the Z, deep inelastic scattering, jet and $t\bar{t}$ cross sections at the LHC, and the width for $Z \to$ hadrons. Other determinations are made by comparing perturbative calculations of quantities such as current correlation functions or the static energy between color sources at close distance with the corresponding lattice evaluations. The running predicted by QCD is clearly confirmed, as illustrated in Figure 5.10. It is

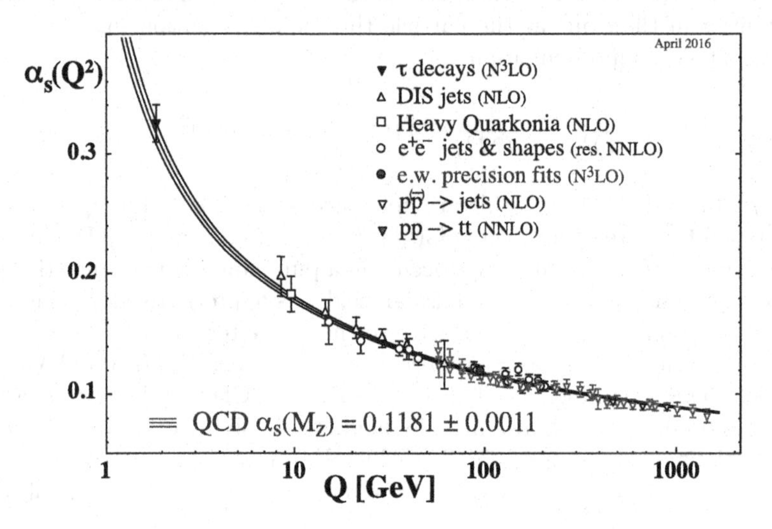

Figure 5.10 Running of the QCD coupling as a function of the scale $\mu \sim Q$. The data points are various experimental determinations and the band is the best fit QCD prediction. Plot courtesy of the Particle Data Group (Patrignani, 2016).

convenient to quote the value of α_s at the mass of the Z because that is the scale at which the electroweak couplings are best determined. There is a direct measurement of $\alpha_s(M_Z^2)$ from the hadronic Z width, and measurements at other scales can be extrapolated to M_Z using the higher-order QCD running and the value of Λ obtained in a global fit. The QCD review in (Patrignani, 2016) gives the precise average $\alpha_s = 0.1181 \pm 0.0011$ from an analysis based on the quantities with small and manageable theoretical uncertainties.

The observed running corresponds to $\Lambda \sim (100 - 400)$ MeV, depending on the exact definition. This sets the scale at which the strong interactions become strong and determines the approximate scale of such strong interaction quantities as the nucleon and ρ masses. In fact, the physical hadrons not involving the c or b quark receive relatively little contribution from the bare quark masses (this is especially true of the non-strange ones), and have masses dominated by the dynamical or constitutent quark mass $M_{dyn} \sim \Lambda$, which (along with the pion decay constant $f_\pi(\Lambda)$) is associated with the spontaneous breaking of the $SU(2) \times SU(2)$ or $SU(3) \times SU(3)$ chiral symmetry. The one exception is the pseudoscalar octet, which are pseudo-Goldstone bosons with masses generated by the explicit chiral breaking from the bare quark masses (see Sections 3.3.3 and 5.8).

If one ignores the bare masses, there are no dimensionful parameters in the QCD Lagrangian. However, Λ emerges by *dimensional transmutation*, i.e., it is the scale at which the dimensionless coupling becomes large. This implies that only dimensionless ratios like m_p/Λ or m_ρ/Λ are meaningful, at least in the $m_r = 0$ limit. One could choose to simply define units so that $\Lambda = 1$, analogous to $c = \hbar = 1$. However, it is more convenient to keep the "historical units" for masses, or to scale everything in terms of easily measured quantities such as m_p. Of course, when one considers other interactions, new scales, such as the Z mass of the Planck scale, also become relevant.

5.5 DEEP INELASTIC SCATTERING

Let us consider deep inelastic lepton nucleon scattering (DIS) in more detail. In this chapter we focus on $e^-p \to e^-X$, as shown in Figure 5.11, where the momentum transfer $Q^2 \equiv -q^2 \gg 1$ GeV^2. This limit is in the short distance regime, so to first approximation one can describe the underlying process as the scattering of the virtual photon with a free quark, as in Figure 5.4 (the simple parton model or SPM). Only the final electron is observed in the scattering, and the unobserved hadrons X are summed over. These typically involve complicated multihadron states. Other deep inelastic reactions substitute $e^+, \mu^\pm$, and $\overset{(-)}{\nu}_\mu$ for the e^-, with neutrino scattering proceeding by either charged or neutral current processes through the exchange of $W^\pm$ or Z. [The HERA $e^\pm p$ scatterings involved all of these processes (Diaconu et al., 2010; Abramowicz et al., 2015).] The initial proton may be replaced by a nuclear target, which either gives a weighted sum of p and n cross sections, or, for bubble chamber experiments, sometimes allows one to identify whether the scattering was from a p or n. We will concentrate on unpolarized scattering, but some experiments have involved polarized $\ell^\pm$ and/or hadrons.

It is difficult to overestimate the importance of deep inelastic scattering. DIS via γ or $W^\pm$ exchange was especially important for establishing the existence of point-like quarks, while the test of the higher-order corrections to the SPM helped establish QCD. Similarly, the neutral current processes probed in $\overset{(-)}{\nu}_\mu N \to \overset{(-)}{\nu}_\mu X$ were extremely important early tests of the standard electroweak theory. More detailed descriptions may be found in (Renton, 1990; Barger and Phillips, 1997; Ellis et al., 2003; De Roeck and Thorne, 2011; Perez and Rizvi, 2013; Blumlein, 2013; Patrignani, 2016).

5.5.1 Deep Inelastic Kinematics

The kinematics of DIS are indicated in Figure 5.11. Since the final hadrons are not directly observed, the independent variables for the unpolarized case are the four-momenta p, k, and k' of the proton and of the initial and final electron. Except for the HERA $e^\pm p$ collider, all of the experiments have been performed in the proton or nuclear rest frame (the lab frame), where the observables are the energies $E_{k,k'}$ of the initial and final electrons, and the electron scattering angle θ.

There are a number of useful related kinematic variables, which we express both in Lorentz invariant form and in terms of the lab frame observables. The total CM energy squared is

$$s = (k+p)^2 \underset{\text{lab}}{\longrightarrow} m_e^2 + M^2 + 2E_k M \sim M^2 + 2kM, \tag{5.40}$$

where M is the nucleon mass, and $k \sim E_k$ is the magnitude of the e^- three-momentum. In the remainder, we neglect the electron mass m_e. The momentum transfer-square and

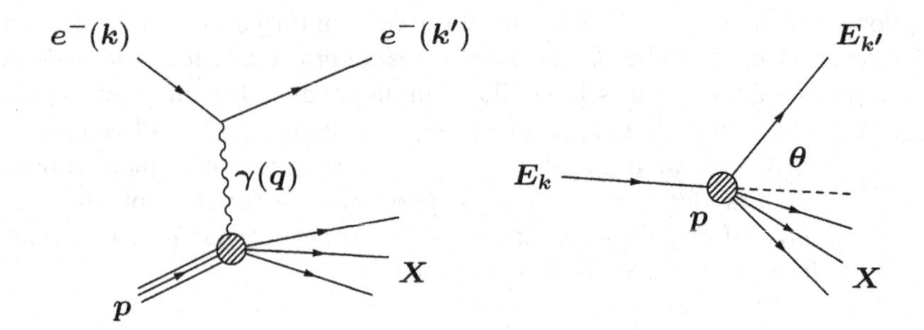

Figure 5.11 Left: deep inelastic scattering $e^- p \to e^- X$, where X represents unobserved hadrons. $q = k - k'$ is the four-momentum of the virtual photon. Right: kinematics in the proton rest frame. E_k and $E_{k'}$ are the energies of the initial and final electrons, and θ is the laboratory scattering angle.

energy transfer to the hadrons are

$$
\begin{aligned}
Q^2 &\equiv -q^2 = -(k-k')^2 \underset{\text{lab}}{\longrightarrow} 2kk'(1-\cos\theta) \\
\nu &\equiv \frac{p\cdot q}{M} \underset{\text{lab}}{\longrightarrow} E_k - E_{k'} \sim k - k'.
\end{aligned}
\tag{5.41}
$$

The invariant mass-square of the unobserved final hadrons is

$$
\begin{aligned}
W^2 &\equiv P_X^2 = (p+q)^2 = M^2 + 2M\nu - Q^2 \\
&\underset{\text{lab}}{\longrightarrow} M^2 + 2M(k-k') - 2kk'(1-\cos\theta).
\end{aligned}
\tag{5.42}
$$

Elastic scattering $X = p$ is a special case with $W^2 = M^2$ and $Q^2 = 2M\nu$. The next hadronic threshold is for a nucleon plus one pion, i.e., $X = p{+}\pi^0$ or $n{+}\pi^+$, corresponding to $W^2 \geq (M{+}m_\pi)^2$. Still larger W^2 can involve more complicated many-particle states, which are summed over. The three independent kinematic variables are (s, Q^2, ν), or equivalently (k, k', θ) (only two are independent in the special case of elastic scattering). However, the hadronic part of the process can only depend on the two Lorentz invariants Q^2 and ν. For a fixed initial energy k, the variables Q^2 and ν (and therefore W^2) can be varied and determined from k' and θ.

It is convenient to define dimensionless variables

$$
x \equiv \frac{Q^2}{2M\nu}, \qquad y \equiv \frac{\nu}{E_k} = \frac{E_k - E_{k'}}{E_k} \sim \frac{k-k'}{k}. \tag{5.43}
$$

x is defined kinematically. However, in the SPM x will be interpreted as the fraction of the proton momentum carried by the scattered parton.[10] y is the fraction of the e^- energy in the lab frame that is transferred to the hadrons. Their ranges are

$$
0 \leq x \leq 1, \qquad 0 \leq y \leq \frac{1}{1+\frac{xM}{2k}} \underset{k\gg M}{\longrightarrow} 1. \tag{5.44}
$$

The relation of x, W^2, and θ to ν and Q^2 is shown in Figure 5.12.

[10] Early papers often used the variable $\omega \equiv 1/x$.

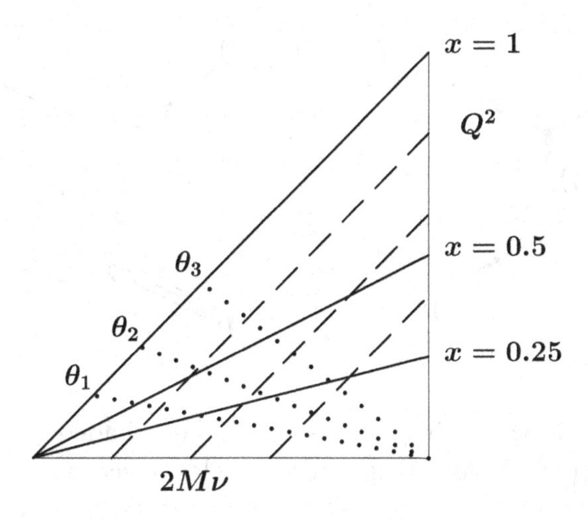

Figure 5.12 Kinematic variables for deep inelastic scattering for fixed initial energy $E_k \sim k \gg m_e$. The horizontal and vertical axes are, respectively, $2M\nu = 2My/k$ and $Q^2 = 2Mxy/k$, each running from 0 to $2Mk$. The sloping solid lines are for fixed $0 \le x \equiv Q^2/2M\nu \le 1$. The dashed lines are for fixed hadronic invariant mass-square $W^2 = M^2 + 2M\nu - Q^2$, where $W^2 = M^2$ along the $x = 1$ line, increasing as one moves down and to the right. The dotted lines are for fixed laboratory scattering angle θ, with $\theta_3 > \theta_2 > \theta_1$.

5.5.2 The Cross Section and Structure Functions

First consider elastic scattering $e^- p \to e^- p$ for a hypothetical point-like proton. The spin-averaged cross section from the first diagram in Figure 5.13 (with a point vertex) is

$$d\bar{\sigma} = \frac{(2\pi)^4 \delta^4(k' + p' - k - p)}{4Mk} \frac{d^3 k'}{(2\pi)^3 2E_{k'}} \frac{d^3 p'}{(2\pi)^3 2E_{p'}} \frac{e^4}{q^4} L_e^{\mu\nu} L_{p\mu\nu}, \tag{5.45}$$

where $q = k - k' = p' - p$, and the traces are collected in the leptonic and hadronic tensors

$$\begin{aligned} L_e^{\mu\nu} &= \frac{1}{2} \text{Tr}\,[\gamma^\mu (\not{k} + m_e) \gamma^\nu (\not{k}' + m_e)] = 2\left[k^\mu k'^\nu + k'^\mu k^\nu + g^{\mu\nu} \frac{q^2}{2}\right] \\ L_{p\mu\nu} &= \frac{1}{2} \text{Tr}\,[\gamma_\mu (\not{p} + M) \gamma_\nu (\not{p}' + M)] = 2\left[p_\mu p'_\nu + p'_\mu p_\nu + g_{\mu\nu} \frac{q^2}{2}\right]. \end{aligned} \tag{5.46}$$

The cross section can be rewritten

$$\frac{d\bar{\sigma}}{dk' d\Omega} = \frac{\alpha^2}{q^4} \frac{k'}{k} L_e^{\mu\nu} W_{\mu\nu}, \tag{5.47}$$

where $d\Omega = d\cos\theta d\varphi$, and

$$W_{\mu\nu} \equiv \frac{1}{4\pi M} \frac{1}{2} \sum_{ss'} \int \frac{d^3 p' (2\pi)^4 \delta^4(p' - p - q)}{(2\pi)^3 2E_{p'}} \langle ps | J_{Q\mu}^\dagger(0) | p's' \rangle \langle p's' | J_{Q\nu}(0) | ps \rangle. \tag{5.48}$$

The current in (5.48) is the proton part of the electromagnetic current

$$J_Q^\mu = \bar{\psi}_p \gamma^\mu \psi_p = J_Q^{\mu\dagger}, \tag{5.49}$$

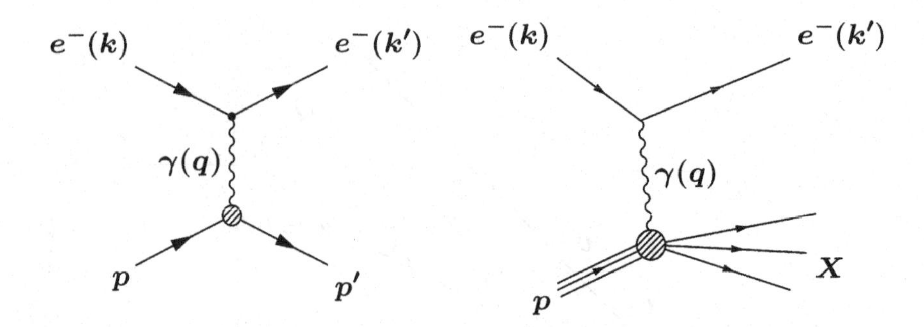

Figure 5.13 Left: elastic scattering $e^- p \to e^- p$. The shaded circle represents the effects of the strong interactions. Right: deep inelastic scattering $e^- p \to e^- X$.

with

$$\langle p's'|J_Q^\nu(0)|ps\rangle = \bar{u}(p',s')\gamma^\nu u(p,s). \tag{5.50}$$

Of course,

$$\langle ps|J_Q^{\mu\dagger}|p's'\rangle = \langle p's'|J_Q^\mu|ps\rangle^*. \tag{5.51}$$

J_Q^μ is actually Hermitian, but it is useful to write (5.48) and (5.50) in a more general form for a later extension to the weak interactions. The tensor $W_{\mu\nu}$ contains all of the information about the hadrons.

The cross section expression in (5.47) can be immediately extended to elastic scattering from a physical (strongly-interacting) proton, provided one replaces (5.50) by

$$\langle p's'|J_Q^\nu(0)|ps\rangle \to \bar{u}(p',s')\Gamma_Q^\nu u(p,s), \tag{5.52}$$

where $\Gamma_Q^\nu(p',p)$ is the vertex function that includes strong corrections. As discussed in Section 2.12.4, the combination of Lorentz covariance, electromagnetic current conservation, and the observed reflection invariance of the strong interactions restricts $\Gamma_Q^\nu(p',p)$ to the form

$$\Gamma_Q^\nu = \gamma^\nu F_1^p(q^2) + \frac{i\sigma^{\nu\rho}q_\rho}{2M} F_2^p(q^2), \tag{5.53}$$

where $F_{1,2}^p(q^2)$ are form factors that can depend on q^2 with $F_1^p(0) = 1$ and $F_2^p(0) = \kappa_p$, where $\kappa_p \sim 1.79$ is the anomalous magnetic moment of the proton. For elastic scattering k' and $\cos\theta$ are not independent, but are related by (2.392) on page 77, i.e.,

$$\frac{k'}{k} = \frac{1}{1 + \frac{k}{M}(1 - \cos\theta)}, \tag{5.54}$$

(enforced by an energy-conserving delta function in $W_{\mu\nu}$), with the correspondence of notation $(k_1, k_2) \to (k, k')$, $\theta_L \to \theta$, and $m_p \to M$. From (5.47) and (5.53) one obtains the Rosenbluth cross section formula for elastic scattering given in (2.400).

Expression (5.47) continues to hold for the inelastic case provided the hadronic tensor is redefined as

$$W_{\mu\nu} = \frac{1}{4\pi M}\left(\frac{1}{2}\sum_s\right)\left[\sum_N \int \prod_{n=1}^{N} \frac{d^3p'_n}{(2\pi)^3 2E_{p'_n}} \sum_{s'_n} (2\pi)^4 \delta^4(p+q-\sum_n p'_n)\right.$$
$$\left. \times \langle ps|J^\dagger_{Q\mu}(0)|X_N\rangle\langle X_N|J_{Q\nu}(0)|ps\rangle \right], \tag{5.55}$$

where X_N is an N particle state that may contain both fermions and bosons. In the inelastic case k' and $\cos\theta$ are independent variables related to the invariant mass W^2 by (5.42). For unpolarized protons the tensor $W_{\mu\nu}$ can only depend on the four-vectors p and q (it can also depend on the spin vector s in the polarized case). The only tensors one can construct are

$$\begin{array}{lllll} \text{symmetric:} & p_\mu p_\nu, & q_\mu q_\nu, & p_\mu q_\nu + q_\mu p_\nu, & g_{\mu\nu} \\ \text{antisymmetric:} & p_\mu q_\nu - q_\mu p_\nu, & \epsilon_{\mu\nu\rho\sigma} p^\rho q^\sigma, & & \end{array} \tag{5.56}$$

each of which can be multiplied by a function of the Lorentz invariants Q^2 and ν. One can use the reflection invariance of the strong interactions to show that the $\epsilon_{\mu\nu\rho\sigma}$ term is absent. In any case, the leptonic tensor $L_e^{\mu\nu}$ is symmetric,[11] so one can keep just the symmetric part of $W_{\mu\nu}$. Furthermore, the electromagnetic current is conserved, $\partial^\mu J_{Q\mu} = 0$, which implies

$$q^\mu W_{\mu\nu} = 0, \qquad q^\nu W_{\mu\nu} = 0. \tag{5.57}$$

Thus, only two linear combinations survive, and the most general form is

$$W_{\mu\nu} = \left[-g_{\mu\nu} + \frac{q_\mu q_\nu}{q^2}\right] W_1\left(Q^2,\nu\right) + \frac{1}{M^2}\left[p_\mu - \frac{p\cdot q}{q^2}q_\mu\right]\left[p_\nu - \frac{p\cdot q}{q^2}q_\nu\right] W_2\left(Q^2,\nu\right), \tag{5.58}$$

with $Q^2 = -q^2 > 0$. The real Lorentz invariant functions $W_{1,2}(Q^2,\nu)$ are known as the proton *structure functions.* They generalize the form factors $F_{1,2}(q^2)$ of elastic scattering, and contain all of the information about the strong interactions effects. One can combine (5.47) and (5.58) to obtain

$$\frac{d\bar\sigma}{dk'd\Omega} = \frac{\alpha^2}{4k^2\sin^4\frac{\theta}{2}}\left[W_2\left(Q^2,\nu\right)\cos^2\frac{\theta}{2} + 2W_1\left(Q^2,\nu\right)\sin^2\frac{\theta}{2}\right] \tag{5.59}$$

for the deep inelastic cross section in the proton rest frame. The solid angle element is usually integrated over the azimuthal angle, $d\Omega = d\varphi d\cos\theta \to 2\pi d\cos\theta$, and the kinematic variables can be rewritten

$$dk'd\cos\theta = \frac{1}{2kk'}dQ^2 d\nu = \frac{Mky}{k'}dxdy \tag{5.60}$$

to yield

$$\frac{d\bar\sigma}{dxdy} = 2Mk^2 y\frac{d\bar\sigma}{dQ^2 d\nu} = \frac{\pi M\alpha^2 y}{2kk'\sin^4\frac{\theta}{2}}\left[W_2\left(Q^2,\nu\right)\cos^2\frac{\theta}{2} + 2W_1\left(Q^2,\nu\right)\sin^2\frac{\theta}{2}\right]. \tag{5.61}$$

[11]Both the leptonic and hadronic tensors have $\epsilon_{\mu\nu\rho\sigma}$ terms if there is polarization, or for parity-violating weak processes such as $\nu N \to \nu X$ and $\nu N \to \mu X$ due to the interference between the vector and axial currents.

W_1 and W_2 can be determined separately from the data by varying θ, k, and k' for fixed Q^2 and ν.

An alternative notation is to use the variables x and Q^2 and to define

$$F_1\left(x, Q^2\right) = M W_1\left(Q^2, \nu\right), \qquad F_2\left(x, Q^2\right) = \nu W_2\left(Q^2, \nu\right). \tag{5.62}$$

This is useful because the simple parton model (QCD) predicts that F_i is independent of (slowly varying with) Q^2 for Q^2 large. Then,

$$\begin{aligned} \frac{d^2\bar{\sigma}}{dx dy} &= \frac{8\pi\alpha^2 M k}{Q^4}\left[F_2\left(x, Q^2\right)\left(1 - y - \frac{Mxy}{2k}\right) + 2xF_1\left(x, Q^2\right)\frac{y^2}{2}\right] \\ &= \frac{8\pi\alpha^2 M k}{Q^4}\left[2xF_1\left(\frac{1 + (1-y)^2}{2}\right) + \left(F_2 - 2xF_1\right)(1-y) - \frac{MxyF_2}{2k}\right]. \end{aligned} \tag{5.63}$$

In the second form, the last term vanishes for $k \gg M$, while the middle term vanishes in the simple quark parton model.

The deep inelastic limit is defined as $Q^2, \nu \to \infty$ with $x = Q^2/2M\nu$ fixed. If the proton were an extended fuzzy object, one would expect $F_i(x, Q^2) \to 0$ in this limit. However, the MIT-SLAC experiments (Friedman and Kendall, 1972; Mishra and Sciulli, 1989) circa 1970 showed instead that

$$F_i(x, Q^2) \xrightarrow[Q^2 \to \infty]{} F_i(x) \neq 0, \tag{5.64}$$

a property known as Bjorken *scaling* (Bjorken, 1969). This scaling can be understood in the Feynman parton model (e.g., Field and Feynman, 1977; Drell et al., 1969), in which the proton is made up of hard point-like parton constituents. The observed y distribution showed that the partons have spin-$\frac{1}{2}$, consistent with QCD and asymptotic freedom. Subsequent experiments established that the scaling is only approximate, and in fact the slow (logarithmic) variation of the F_i with Q^2 is what one expects from the higher-order corrections in QCD.[12]

5.5.3 The Simple Quark Parton Model (SPM)

The cross section for elastic scattering $e^- p \to e^- p$ from a point proton is given in (2.391) on page 77, with the kinematic constraint for k' given in (5.54) (see the subsequent comment on notation). It is convenient to rewrite (2.391) as

$$\frac{d\bar{\sigma}}{dQ^2} = \frac{\alpha^2}{4k^2 \sin^4\frac{\theta}{2}} \frac{\pi}{kk'}\left[\cos^2\frac{\theta}{2} + \frac{Q^2}{2M^2}\sin^2\frac{\theta}{2}\right], \tag{5.65}$$

where we have used $dQ^2 = 2k'^2 d\cos\theta$, which follows from (5.41) and (5.54). For elastic scattering, $Q^2 = 2M\nu$ from (5.42), so (5.65) can also be written as

$$\frac{d\bar{\sigma}}{dQ^2\, d\nu} = \frac{\alpha^2}{4k^2 \sin^4\frac{\theta}{2}} \frac{\pi}{kk'}\left[\cos^2\frac{\theta}{2} + \frac{Q^2}{2M^2}\sin^2\frac{\theta}{2}\right]\delta\left(\nu - \frac{Q^2}{2M}\right). \tag{5.66}$$

[12] Scaling also breaks down at low Q^2, where strong coupling effects are important. In particular, hadronic resonances in the γp channel for fixed W^2 are important. The scaling behavior of the structure functions smoothly interpolates these resonances (Bloom and Gilman, 1971; Melnitchouk et al., 2005), as can be understood from *finite energy sum rules* (FESR), derivable from analyticity.

This is a special case of the general formula (5.61) provided we identify the structure functions as

$$\begin{aligned} W_1(Q^2,\nu) &= \frac{Q^2}{4M^2}\delta\left(\nu - \frac{Q^2}{2M}\right) = \frac{Q^2}{4M^2\nu}\,\delta\left(1 - \frac{Q^2}{2M\nu}\right) \\ W_2(Q^2,\nu) &= \delta\left(\nu - \frac{Q^2}{2M}\right) = \frac{1}{\nu}\,\delta\left(1 - \frac{Q^2}{2M\nu}\right). \end{aligned} \tag{5.67}$$

Now assume that the proton is a bound state of point-like quarks. Consider the contribution to the structure functions from an individual quark q_i with electric charge e_i, which carries four-momentum $x_i p$, where p is the proton momentum. It is plausible that (5.67) applies for that quark, provided one replaces $M \to x_i M$ and multiplies by e_i^2, i.e.,

$$W_1(Q^2,\nu) = \frac{e_i^2 Q^2}{4x_i M^2\nu}\,\delta\left(x_i - \frac{Q^2}{2M\nu}\right), \qquad W_2(Q^2,\nu) = \frac{e_i^2 x_i}{\nu}\,\delta\left(x_i - \frac{Q^2}{2M\nu}\right). \tag{5.68}$$

(This result can be better derived in the *infinite momentum frame*, where the proton and quark masses and the transverse momentum of the proton relative to the electron direction are negligible.) The contribution of that quark to the structure functions is

$$\begin{aligned} F_1^i(x,Q^2) &= MW_1^i(Q^2,\nu) = \frac{1}{2}e_i^2\delta\left(x_i - \frac{Q^2}{2M\nu}\right) \\ F_2^i(x,Q^2) &= \nu W_2^i(Q^2,\nu) = e_i^2 x_i\delta\left(x_i - \frac{Q^2}{2M\nu}\right). \end{aligned} \tag{5.69}$$

To find $F_{1,2}(x,Q^2)$ one must sum over the quark types and integrate over their possible momenta. This is done in the cross section (i.e., in $F_{1,2}$) because the different i and x_i lead to different (incoherent) final states. Introduce the *parton distribution function (PDF)* $q_i(x_i)$ as the probability density (the absolute square of the momentum space wave function) for finding quark q_i in the proton with momentum fraction x_i. Then from (5.69) the predicted structure functions are

$$\begin{aligned} F_2\left(x,Q^2\right) &= 2xF_1\left(x,Q^2\right) \\ &= \sum_i \int_0^1 dx_i q_i(x_i) e_i^2 x_i \delta\left(x_i - \frac{Q^2}{2M\nu}\right) = \sum_i e_i^2\, x q_i(x). \end{aligned} \tag{5.70}$$

The simple parton model therefore predicts that the F_i are independent of Q^2 for large Q^2 (Bjorken scaling). The quantity $xq_i(x)$ is interpreted as the momentum distribution for parton q_i.

The predicted relation $F_2 = 2xF_1$, the *Callan-Gross* relation (Callan and Gross, 1969), is a signature of spin-$\frac{1}{2}$ constituents. Scattering from spin-0 constituents would lead to $F_1 = 0, F_2 \neq 0$ and therefore a different angular distribution. An interpretation of this result is that one can show (see, e.g., Renton, 1990) that

$$R(x,Q^2) \equiv \frac{\sigma_L}{\sigma_T} \sim \frac{F_2 - 2xF_1}{2xF_1}, \tag{5.71}$$

where σ_L and σ_T are, respectively, the total cross sections for $\gamma^*_{L,T}p$, where $\gamma^*_{L,T}$ is a virtual photon with momentum q, and L and T refer to longitudinal (photon helicity 0) and transverse (photon helicities ± 1) polarizations. In the Breit frame (discussed in Section 2.3.4), the virtual photon has four-momentum $q = (0,0,0,-Q)$, while the incident [final] parton has momentum $\frac{1}{2}(Q,0,0,Q)$ $[\frac{1}{2}(Q,0,0,-Q)]$, as in Figure 2.4. For spin-0 partons one would have $\sigma_T = 0$ $(R = \infty)$ by angular momentum conservation, since there is no orbital angular momentum along the direction of the photon and parton momenta. For spin-$\frac{1}{2}$ partons, on the other hand, $\sigma_L = R = 0$ using the fact that helicity is conserved for vector transitions of massless spin-$\frac{1}{2}$ particles, e.g., (2.214) on page 42.

Even to the extent that the SPM is valid, a real proton is expected to consist of not only the three *valence* quarks uud of the quark model, but also a *sea* of $q\bar{q}$ pairs and of gluons produced by soft (low momentum) processes such as illustrated in Figure 5.14. The

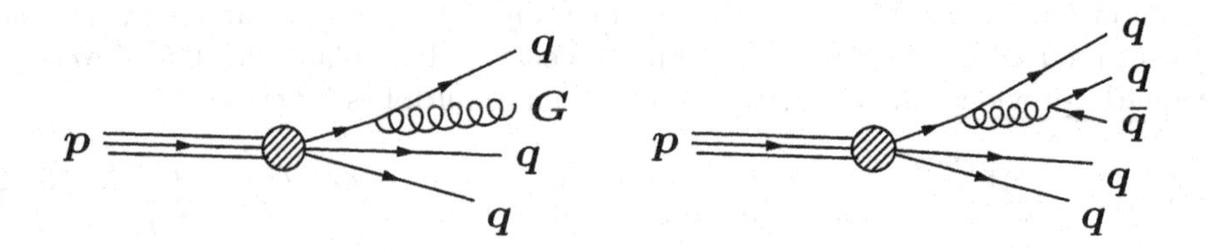

Figure 5.14 Higher-order soft processes that produce a sea of gluons and $q\bar{q}$ pairs.

anitiquarks contribute to the F_i with the same formula, except $q_i(x) \to \bar{q}_i(x)$,

$$F_2\left(x, Q^2\right) = 2xF_1\left(x, Q^2\right) = \sum_i e_i^2\, x(q_i(x) + \bar{q}_i(x)). \tag{5.72}$$

The gluons do not contribute directly to electromagnetic processes (they do contribute to purely hadronic short distance processes). It is convenient to write

$$q_i(x) = q_{Vi}(x) + q_{Si}(x), \qquad \bar{q}_i(x) = \bar{q}_{Si}(x), \tag{5.73}$$

where q_{Vi} and q_{Si} represent the valence and sea quarks, respectively. The quark distribution functions are determined by long distance effects, and at present there is no way to reliably calculate them. However, there are a number of plausible constraints and consistency conditions. One expects

$$q_{Si}(x) \sim \bar{q}_{Si}(x), \tag{5.74}$$

as is suggested by the diagrams in Figure 5.14. However, this is not a rigorous result and there could be small deviations. A more precise result is

$$\int_0^1 dx\, [q_{Si}(x) - \bar{q}_{Si}(x)] = 0. \tag{5.75}$$

This follows from the meaning of valence and sea quarks, and should hold to the extent that the SPM is valid. Similarly, the proton should have two valence u quarks and one valence d,

$$\int_0^1 u_V(x)dx = 2, \qquad \int_0^1 d_V(x)dx = 1, \tag{5.76}$$

which implies

$$\begin{aligned} &\int_0^1 [u(x) - \bar{u}(x)]\, dx = 2, \qquad \int_0^1 \left[d(x) - \bar{d}(x)\right] dx = 1 \\ &\int_0^1 [s(x) - \bar{s}(x)]\, dx = 0, \qquad \int_0^1 [c(x) - \bar{c}(x)]\, dx = 0. \end{aligned} \tag{5.77}$$

These predictions are difficult to measure, but existing data is consistent. Because of quark mass effects one expects $\bar{b} < \bar{c} < \bar{s} < \bar{u}, \bar{d}$, and for the kinematic region relevant to the MIT-SLAC experiments it is reasonable to neglect the $b, \bar{b}, c$, and $\bar{c}$. Since the u and d quark masses are so small compared to Λ it is a reasonable first approximation to expect $\bar{u}(x) \sim \bar{d}(x)$. This is not a rigorous consequence of isospin since the proton is not an isosinglet, but should hold approximately because of the isospin invariance of the gluon coupling. Another constraint is that

$$\sum_i \int_0^1 x \left[q_i(x) + \bar{q}_i(x)\right] dx + \int_0^1 xG(x)dx = 1, \tag{5.78}$$

where $G(x)$ is the gluon probability distribution, which can be probed in hadronic processes. Equation (5.78) states that the total momenta of all of the constituents must add up to the proton momentum.

From (5.72), one predicts

$$F_2(x) \sim \frac{4}{9}x\left[u(x) + \bar{u}(x)\right] + \frac{1}{9}x\left[d(x) + \bar{d}(x)\right] + \frac{1}{9}x\left[s(x) + \bar{s}(x)\right]. \tag{5.79}$$

One cannot distinguish the various distribution functions using $e^- p \to e^- X$ data alone, but separation between the quarks and antiquarks and between flavors can be accomplished by considering scattering from neutrons and other deep inelastic processes, such as

$$\overset{(-)}{\nu}_\mu p \to \mu^\mp X \quad \text{and} \quad e^\pm p \to \overset{(-)}{\nu}_e X, \tag{5.80}$$

as will be described in Chapter 8. One finds that the momentum fractions of the proton constituents are of order

$$\int xu(x)dx \sim 0.30, \qquad \int xd(x)dx \sim 0.15, \qquad \int xG(x)dx \sim 0.50. \tag{5.81}$$

The momenta carried by the $\bar{u} + \bar{d}$ is about 10% that of the u and d, while the $s + \bar{s}$ momentum fraction is about half of that. These rough estimates apply either to SPM analyses of relatively low Q^2 data (e.g., Field and Feynman, 1977), or as the values at a low initial Q^2 (e.g., 5 GeV^2) in the more sophisticated QCD improved model described below. The actual momentum distributions obtained in a recent analysis are shown in Figure 5.15.

The quark distributions and structure functions defined above referred to the proton, and it is sometimes useful to denote them as W_i^p, F_i^p, q_i^p to emphasize that fact. One can define analogous quantities for a neutron target, such as q_i^n. The SPM prediction is

$$F_2^n\left(x, Q^2\right) = 2xF_1^n\left(x, Q^2\right) = \sum_i e_i^2\, x\left(q_i^n(x) + \bar{q}_i^n(x)\right). \tag{5.82}$$

The neutron distribution functions are not independent, but are related to those of the proton by isospin, i.e.,

$$u^n(x) = d^p(x), \qquad d^n(x) = u^p(x), \qquad s^n(x) = s^p(x) \tag{5.83}$$

up to possible small isospin-breaking corrections. Similar results hold for the antiquarks. It is also useful to define isospin-averaged distribution functions

$$\begin{aligned} u^S(x) &= \frac{1}{2}\left[u^p(x) + u^n(x)\right] = \frac{1}{2}\left[u^p(x) + d^p(x)\right] \\ d^S(x) &= \frac{1}{2}\left[d^p(x) + d^n(x)\right] = \frac{1}{2}\left[d^p(x) + u^p(x)\right] = u^S(x), \end{aligned} \tag{5.84}$$

analogous to (2.402). These are the distribution functions per nucleon for nuclei with equal numbers of protons and neutrons such as ^{12}C or (up to a correction) ^{56}Fe. For such targets,

$$F_2^S\left(x,Q^2\right) = 2xF_1^S\left(x,Q^2\right) = \sum_i e_i^2\, x\,\left[q_i^S(x) + \bar{q}_i^S(x)\right]. \tag{5.85}$$

Most experiments are done on heavier nuclei because higher statistics can be achieved, even though information on the isospin structure is lost.

The simple quark parton model was important because it approximately described the observed properties of deep inelastic scattering at moderate Q^2. The approximate scaling established the existence of point-like (i.e., asymptotically free) constituents of the nucleon, and the relation $F_2(x) \sim 2xF_1(x)$ indicated that the constituents have spin-$\frac{1}{2}$. The measured distribution functions $q(x)$ give more detailed information about the distribution of the quarks within the nucleon. Further implications of deep inelastic scattering processes involving $W^\pm$ and Z exchange will be described in Chapter 8. We now turn, however, to the corrections to the SPM as predicted by QCD.

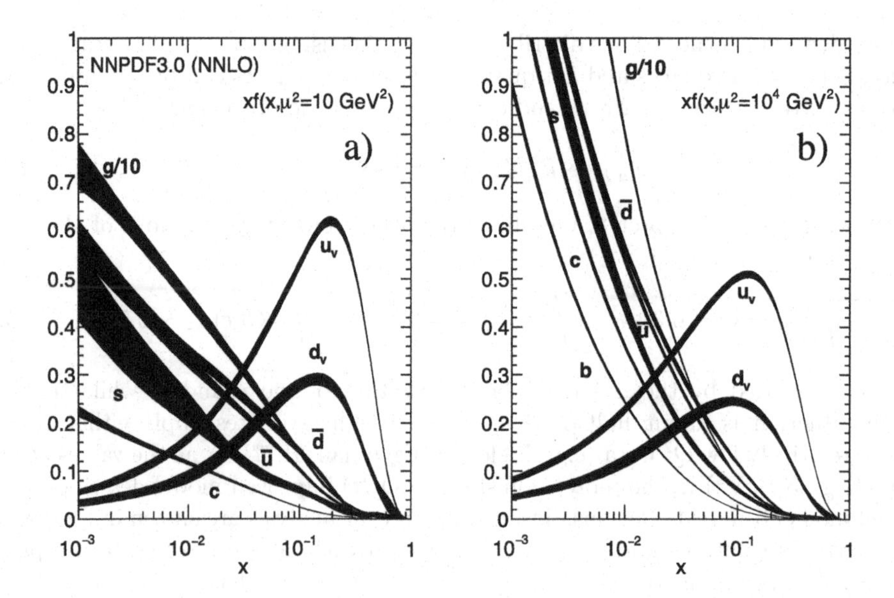

Figure 5.15 Momentum distributions $xf(x)$ of the gluon (g), valence quarks (u_v, d_v), and sea quarks ($\bar{u}, \bar{d}, s, c, b$) from a recent global analysis (Ball et al., 2015), at μ^2 (i.e., Q^2) $= 10$ GeV^2 (left), and $\mu^2 = 10^4$ GeV^2 (right). The heavy quarks become more important for large Q^2 and small x, where mass effects are less important. Plot courtesy of the Particle Data Group (Patrignani, 2016).

5.5.4 Corrections to the Simple Parton Model

The scattering from non-interacting point-like quarks in the SPM leads to Q^2-independent structure functions $F_{1,2}(x,Q^2) \to F_{1,2}(x)$. However, the asymptotic freedom of QCD predicts that $\alpha_s(Q^2)$ is small but nonzero at the observed scales, so one expects higher-order

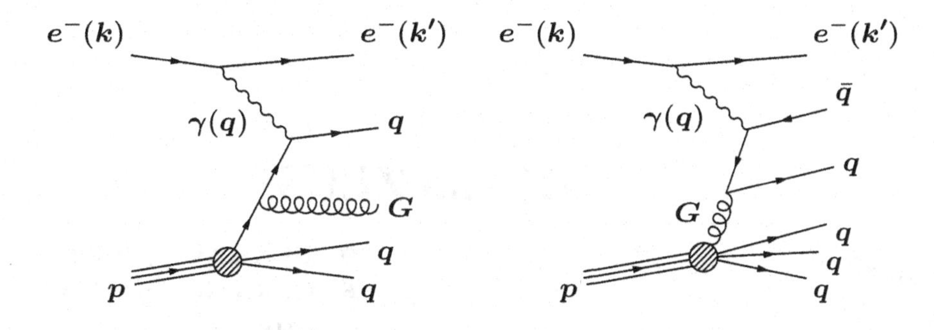

Figure 5.16 Lowest order QCD corrections to the simple parton model.

corrections to scaling and the SPM from diagrams such as those in Figure 5.16. These lead to logarithmic Q^2 effects that can be interpreted as an effective Q^2 dependence of the quark and gluon distribution functions, in what is known as the *QCD-improved parton model.*

A heuristic picture is that at moderate Q^2 (e.g., 10 GeV^2) there are relatively few gluons or sea quarks, at least in the region $x \gtrsim 0.2$. At higher Q^2, on the other hand, the virtual photon or other probe can resolve more $q\bar{q}$ pairs and gluons associated with the splittings in Figures 5.14 and 5.16. One therefore expects a somewhat reduced momentum distribution for the valence quarks, and enhanced sea quark and gluon distributions at low x. This indeed is the case, as can be seen in Figure 5.15. The most dramatic effect is for $x \lesssim 0.05$, which has especially been studied at the HERA *ep* collider at DESY. The observed dependence of $F_2(x, Q^2)$ for fixed x and varying Q^2 is shown in Figure 5.17.

More precisely, the QCD corrections are dominated by gluon emissions and other corrections involving small transverse momenta k_T. These cannot be calculated completely in perturbation theory because of collinear singularities (e.g., Ellis et al., 2003; Salam, 2010a). Rather, those with k_T smaller than some *factorization scale* μ_F are absorbed into the distribution functions,[13] which therefore depend on μ_F in a calculable way, while those with larger k_T are treated explicitly. μ_F is usually taken to be Q for deep inelastic scattering.

The distribution functions $f_i(x, Q^2)$ where $f_i = q_i, \bar{q}_i, G$ therefore depend both on x and Q^2 in a complicated way. Fortunately, one can parametrize them and test QCD by a two-step process. First, $f_i(x, Q_0^2)$ can be *measured* at some convenient reference scale Q_0^2, thus determining the long distance effects that cannot be calculated perturbatively or by other presently available techniques.[14] Then, the $\ln Q^2$ evolution of the $f_i(x, Q^2)$ to larger Q^2 values can be predicted from QCD and compared with the experimental data. In practice, one generally assumes (physically motivated) analytic expressions for the $f_i(x, Q_0^2)$ in terms of several unknown parameters, and then determines those parameters as well as $\alpha_s(Q^2)$ (i.e., Λ) by a fit to the data at all Q^2. The evolution actually depends on all of the distribution functions, including the gluon's. Therefore, modern fits often make use of all relevant data from the various deep inelastic and other short distance processes.

It is not sufficient to consider only the lowest order diagrams in Figure 5.16. These scale as $\alpha_s(Q^2) \ln Q^2$, which is not small. It is possible to sum the leading order (LO) contributions of $\mathcal{O}[(\alpha_s \ln Q^2)^n]$ for all n by integrating the Dokshitzer-Gribov-Lipatov-Altarelli-

[13]This is an infrared analog of renormalization theory for ultraviolet divergences.

[14]In principle, the $f_i(x, Q_0^2)$ could be calculated by lattice techniques, but this would be difficult.

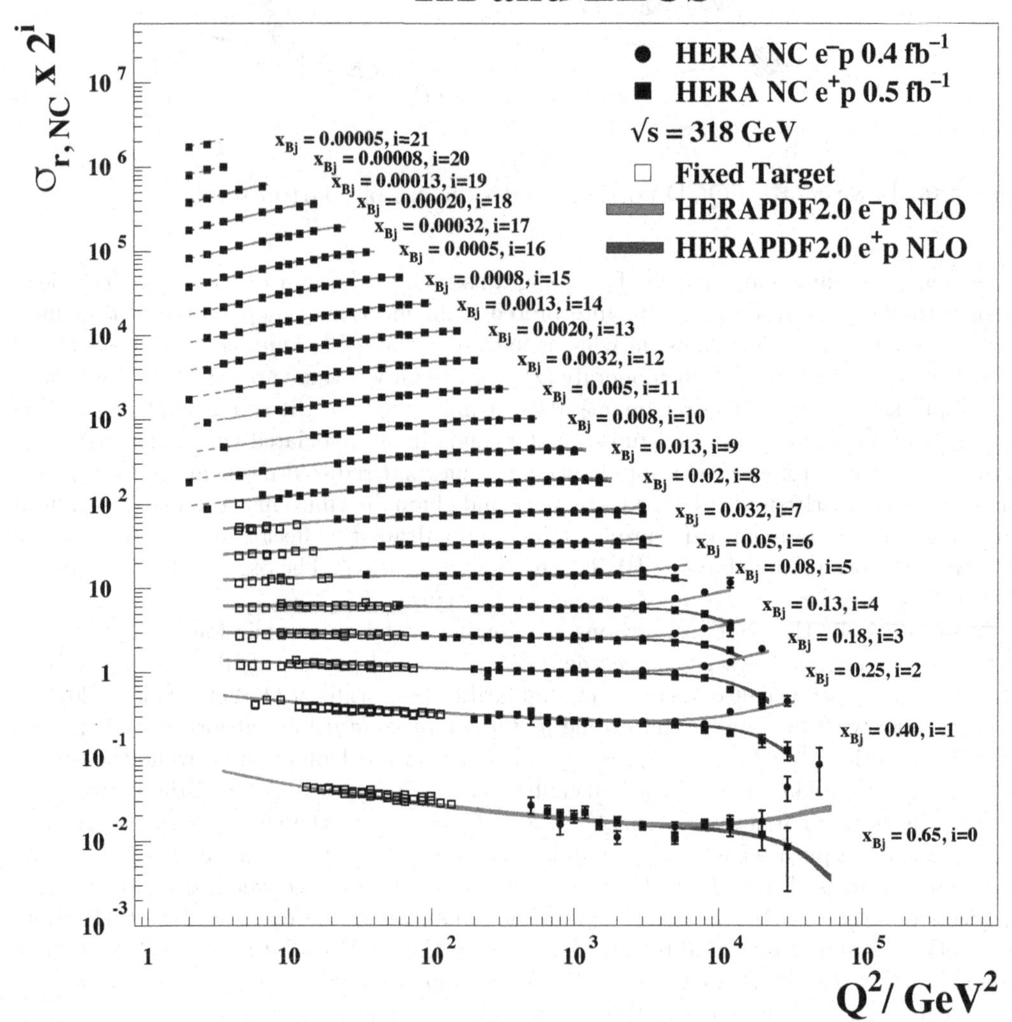

Figure 5.17 Reduced cross sections for $e^{\pm}p \to e^{\pm}X$ as a function of Q^2 for various fixed x as measured at HERA. These are essentially $F_2(x, Q^2)$ except at the highest Q^2, where the effects of Z exchange (which distinguish e^- [upper curve] from e^+ [lower]) become significant. The lines are the result of a QCD fit. Reprinted from (Abramowicz et al., 2015), with kind permission of *The European Physical Journal* (EPJ).

Parisi (DGLAP), or just Altarelli-Parisi, equations (see, e.g., Altarelli and Parisi, 1977)

$$\begin{aligned} \frac{dG(x,Q^2)}{d\ln Q^2} &= \frac{\alpha_s(Q^2)}{2\pi}\int_x^1 \frac{dw}{w}\left[\sum_j P_{Gq}\left(\frac{x}{w}\right)q_j\left(w,Q^2\right)+P_{GG}\left(\frac{x}{w}\right)G\left(w,Q^2\right)\right] \\ \frac{dq_i(x,Q^2)}{d\ln Q^2} &= \frac{\alpha_s(Q^2)}{2\pi}\int_x^1 \frac{dw}{w}\left[P_{qq}\left(\frac{x}{w}\right)q_i\left(w,Q^2\right)+P_{qG}\left(\frac{x}{w}\right)G\left(w,Q^2\right)\right], \end{aligned} \tag{5.86}$$

where q_i includes both q_i and $\bar{q}_i$. In (5.86), the $P_{f_j f_i}(z)$ are the *splitting functions* that describe the probability for parton f_i to emit parton f_j carrying a fraction $z \leq 1$ of the f_i momentum. Their explicit forms are given in, e.g., (Ellis et al., 2003). The splitting functions are perturbative and can be expanded in α_s. The zeroth order term (one factor α_s is already extracted in (5.86)) yields the LO approximation. The next to leading order (NLO) terms, of order $\mathcal{O}[\alpha_s^n(\ln Q^2)^{n-1}]$, involve the $\mathcal{O}(\alpha_s)$ corrections to the splitting functions. The splitting functions have been calculated to NNLO (for the original references, see Patrignani, 2016), which are used in modern analyses.[15]

It is possible to integrate the DGLAP equations numerically. One can also consider the moments (Mellin transformations) of the distribution functions,

$$f_i(N,Q^2) = \int_0^1 x^{N-1} f_i(x,Q^2)dx. \tag{5.87}$$

The moments of the *flavor non-singlet* (ns) distribution functions (such as $q_i - \bar{q}_i$ or $q_i - q_j$) satisfy simple first order equations, which are easily integrated. In LO the non-singlet moments are predicted to scale as

$$f_{ns}(N,Q^2) = f_{ns}(N,Q_0^2)\left(\frac{\alpha_s(Q_0^2)}{\alpha_s(Q^2)}\right)^{\gamma_N}, \tag{5.88}$$

where the *anomalous dimension* γ_N is a known function of the number of light flavors, calculable from P_{qq} and α_s. Similarly, the moments of the G and singlet (i.e., $\sum_i(q_i+\bar{q}_i)$) distribution functions satisfy coupled differential equations. After solving, the distribution functions can be recovered from the inverse Mellin transformation. The non-singlet moment evolutions in principle provide very clean tests of QCD and determinations of α_s, since they are independent of the gluon distribution, but in practice the tests are limited by uncertainties in the distribution functions.

There are many complications to the description of deep inelastic scattering. For very small x, as studied at HERA, there are important $\ln(1/x)$ contributions to the splitting functions, which must be summed. Also, at moderate Q^2 values there are mass and *higher twist* effects, due to diagrams such as gluon exchange between the scattered and unscattered quarks, which are of order $(1/Q^2)^n$, $n \geq 1$. These are difficult to calculate but can be parametrized. Another complication is that $R(x,Q^2)$, defined in (5.71), does not really vanish, due to QCD corrections and effects neglected in the SPM, such as quark mass and transverse momentum within the proton. R is difficult to measure precisely, but is consistent with the expected values $\lesssim 0.1$. Nuclear effects for heavy targets can also be important (e.g., Armesto, 2006).

Detailed global analyses of the *parton distribution functions* (PDFs) at LO, NLO, and NNLO have been carried out by a number of groups, including ABM (Alekhin

[15] Other quantities, such as $\alpha_s(\mu^2)$ and the parton-parton cross sections for short-distance hadronic processes must be calculated to the same order.

et al., 2014), JR (Jimenez-Delgado and Reya, 2014), NNPDF (Ball et al., 2015), MSTW/MMHT (Harland-Lang et al., 2015), HERAPDF (Abramowicz et al., 2015), and CTEQ/CT (Dulat et al., 2016), with the PDFs available from `lhapdf.hepforge.org`. Impressive agreement between the data and the QCD predictions was obtained.

5.6 OTHER SHORT DISTANCE PROCESSES

The formalism for deep inelastic scattering is easily extended to allow for polarized beams and/or targets (e.g., Patrignani, 2016). Experiments measuring elastic and inelastic electron or muon asymmetries with polarized beams have been carried out at SLAC, Jefferson Lab, Bates, and Mainz; experiments involving both polarized leptons and polarized nucleons have been done at DESY, CERN, SLAC and Jefferson Lab; and those with polarized pp scattering at RHIC. These involve not only photon but also Z exchange, and allow tests of the electroweak theory and detailed studies of the nucleon, such as its spin distribution (Burkardt et al., 2010; Aidala et al., 2013; Leader and Lorcé, 2014). Surprisingly, only about 1/3 of the nucleon spin appears to be due to the quark spins, with the remainder presumably from gluons and orbital angular momenta. They also yield information on the strange quark content and matrix elements of the proton (Kaplan and Manohar, 1988; Armstrong and McKeown, 2012), which are relevant to the spin question, the interpretation of dark matter experiments (Section 10.1.2), precision experiments, and the σ term (Problem 5.13).

One can also introduce structure functions for the *photon*. These are useful when a photon turns into a virtual $q\bar{q}$ pair, and a subsequent interaction is more associated with these hadronic constitutents than with the direct photon couplings.[16] (One says that the parton content of the photon has been *resolved*, and the structure function describes the parton distributions.) The photon structure functions can be measured in the deep-inelastic scattering of, e.g., an electron on a quasi-real (i.e., nearly on shell) photon emitted by a positron in e^-e^+ scattering. This can occur by the exchange of a second (highly virtual) photon, similar to the right-hand diagram in Figure 5.16 with the gluon replaced by the quasi-real photon and the proton (and unscattered quarks) replaced by a positron. The overall cross section can be given in terms of the photon structure functions and the effective flux of these quasi-real photons (in the *equivalent photon approximation*). The photon structure has also been studied in e^-p scattering at HERA, where in approppropriate kinematic regimes the constitutents of a quasi-real photon emitted by the e^- scatter from the proton constituents, similar to the short distance hadron-hadron interactions to be considered in Chapter 6. The formalism, measurements, and QCD evolution is reviewed in Nisius (2000).

PDFs can be thought of as probability distributions in momentum space and expressed as diagonal matrix elements of quark or gluon bilinear operators. They can be extended to *generalized parton distributions*. These are off-diagonal matrix elements[17] that carry more information than the PDFs, i.e., the full three-dimensional momentum and spin distributions of the partons within the nucleon. They can be probed experimentally in a variety of exclusive processes (e.g., Diehl, 2003; Belitsky and Radyushkin, 2005).

Many processes involving short-distance scatterings of the constituent quarks and gluons are observed at hadron colliders. These are essential for testing QCD and are both the

[16]Similar considerations apply in other regimes. For example, the electromagnetic form factors of hadrons are dominated in the timelike region by vector mesons with the quantum numbers of the photon, i.e., the ρ, ω, and ϕ, as in Equation (2.387) on page 76. The *vector meson dominance* (VMD) model (e.g., Schildknecht, 2006) further assumed that photoproduction (γN scattering) and other interactions of the photon with hadrons are dominated by their vector meson components.

[17]They are analogous to the density matrix in quantum mechanics, which carries more information than the modulus of the wave function.

backgrounds and possible signals in searches for new physics. These will be discussed in some detail in Chapter 6.

Other powerful techniques allow separation of perturbative and non-perturbative effects, e.g., in the decays of mesons involving a single heavy quark, such as the $B^+ = u\bar{b}$. In particular, heavy quark effective theory (HQET) (Isgur and Wise, 1989; Bigi et al., 1997; Neubert, 1994) and soft collinear effective theory (SCET) (Neubert, 2005; Becher et al., 2015) are, respectively, based on systematic expansions in Λ_{QCD}/m_Q and Λ_{QCD}/E, where m_Q and E are large quark masses and large energies of decay products. Non-relativistic QCD (NRQCD) (Bodwin et al., 1995) is useful for describing states with two heavy quarks, such as upsilonium ($b\bar{b}$).

5.7 THE STRONG INTERACTIONS AT LONG DISTANCES

At low energies or long distances $\alpha_s \to \mathcal{O}(1)$ so one cannot use perturbative techniques. Quarks and gluons are confined, and hadrons become the basic degrees of freedom. Nevertheless, there are a number of tools available for understanding the strong interactions. One class involves phenomenological models, which typically are useful in limited domains of kinematics and parameters. These include S-matrix theory and models, such as dispersion relations, the Veneziano (dual resonance) model, Regge theory and the pomeron[18] (Collins, 1977; Donnachie and Landshoff, 2013); the MIT bag model (Chodos et al., 1974); and one boson exchange potentials in nuclear physics (Epelbaum et al., 2009). Another approach, which we touched on in Chapter 3 and will elaborate on in the next section, involves the chiral flavor symmetries of the strong interactions, including related techniques such as current algebra (Adler and Dashen, 1968; D'Alfaro et al., 1973; Coleman, 1985; Weinberg, 2009); chiral perturbation theory (e.g., Pich, 1995; Bernard, 2008; Cirigliano et al., 2012; Bijnens and Ecker, 2014); QCD sum rules (Colangelo and Khodjamirian, 2000); the $1/N_c$ expansion ('t Hooft, 1974); the Skyrme model (Skyrme, 1962; Adkins et al., 1983; Zahed and Brown, 1986); and the OZI (Zweig) rule (Lipkin, 1984).

Lattice QCD (e.g., Gupta, 1997; DeGrand and DeTar, 2006; Kronfeld, 2012; Patrignani, 2016) is a non-perturbative definition of QCD. It is based on approximating space and time by a discrete lattice of points, allowing the QCD equations to be solved numerically, typically on a supercomputer. Most recent calculations have been able to include dynamical fermions (i.e., in loops), a significant improvement over earlier (*quenched approximation*) calculations. Lattice QCD has had considerable success in computing the hadron (Davies et al., 2004; Dürr et al., 2008) and gluonium spectra, demonstrating quark confinement, calculating form factors necessary for the weak interactions (Aoki et al., 2017), and in computing finite temperature effects (e.g., DeTar and Heller, 2009).

Confinement

It is believed that under ordinary conditions color is confined, preventing the production of isolated quarks and gluons. First consider QED. A well-separated e^- and e^+ act as sources of a classical electromagnetic field. The photons have no self-interactions, and the non-linear effects due to QED are small. The electromagnetic field therefore spreads out in space, leading to the familiar $V(r) \sim \alpha/r$ potential between the charged particles. Given sufficient energy they can get arbitrarily far apart, so there is no confinement. The qualitative picture

[18]Pomeron exchange can be roughly thought of as the exchange of two gluons in a color-singlet state. See, e.g., (Donnachie et al., 2002; Domokos et al., 2009) for a more detailed discussion of the connection to QCD and string theory.

for QCD, which is supported by lattice calculations, is very different. When a quark and antiquark get far enough apart the classical gluon field they produce forms into a long narrow flux tube due to the strong coupling and gluon self-interactions. The energy stored in such a flux tube is proportional to its length, so that there is an effective potential between the q and $\bar{q}$ which grows linearly with their separation, $V(r) \sim \kappa r$. It would take infinite energy to separate them, so they are confined. Actually, once the q and $\bar{q}$ are far enough apart so that $\kappa r \gtrsim 2m_\pi$ the tube can create a $\bar{q}q$ pair and break, leading to two mesons rather than isolated quarks. This is analogous to cutting a bar magnet, which creates two smaller bar magnets rather than a monopole-antimonopole pair.

The linear $q\bar{q}$ potential can be combined with a one-gluon exchange short-distance term,

$$V(r) = -\frac{4}{3}\frac{\alpha_s}{r} + \kappa r, \tag{5.89}$$

in a non-relativistic model[19] for heavy meson bound states, such as charmonium ($c\bar{c}$) and bottomonium ($b\bar{b}$) (e.g., Eichten et al., 2008). The 4/3 is appropriate for color-singlets, and α_s is evaluated at the $q\bar{q}$ mass scale. Fits to the spectrum and decays such as $J/\psi \to e^+e^-$ are reasonably successful and obtain $\kappa \sim 1$ GeV/fm ~ 0.2 GeV^2. See Problems 5.9–5.11.

5.8 THE SYMMETRIES OF QCD

QCD contains a number of *accidental* global symmetries in addition to the gauged color $SU(3)_c$ (Weinberg, 1973a,b), where the subscript c identifies $SU(3)_c$ as the group associated with color. They are referred to as accidental because no terms consistent with gauge invariance and renormalizability can be written to violate them without introducing extra fields. For example, for $\theta_{QCD} = 0$ the Lagrangian density in (5.2) is automatically invariant under the discrete space reflection (P), charge conjugation (C), and time reversal (T) transformations.[20]

5.8.1 Continuous Flavor Symmetries

The gauge interaction terms in $\mathcal{L}_{QCD}$ for n_q quark flavors are invariant under an $SU(n_q)_L \times SU(n_q)_R \times U(1)_B \times U(1)_A$ global chiral flavor symmetry that commutes with $SU(3)_c$. The quark mass terms break the chiral symmetries, and the non-chiral subgroup if they are not degenerate. $m_{c,b,t}$ are all large compared to the QCD scale Λ so the flavor symmetry associated with them is too badly broken to be relevant. Therefore, we will focus on the approximate $SU(3)_L \times SU(3)_R \times U(1)_B \times U(1)_A$ symmetry of the u, d, and s quarks, and, since $m_s \lesssim \Lambda$, on its $SU(2)_L \times SU(2)_R \times U(1)_B \times U(1)_A$ subgroup, extending the discussion in Chapter 3. The relevant quark part of the Lagrangian density is

$$\mathcal{L}_q = \bar{q}\,(i\not{D} - m)\,q = \bar{q}_L i\not{D} q_L + \bar{q}_R i\not{D} q_R - (\bar{q}_L m q_R + \bar{q}_R m q_L)\,, \tag{5.90}$$

where $q = (u\ d\ s)^T$ represent the quark fields, $m = \text{diag}(m_u\ m_d\ m_s)$ is the diagonal quark mass matrix, and the color indices are suppressed. m can be written as

$$m = m_0 I + m_3\lambda^3 + m_8\lambda^8, \tag{5.91}$$

[19] Other potentials, such as $V(r) = \lambda \ln r$, which roughly interpolates between the terms in (5.89), can also be used.

[20] Parity violation in strong processes can be induced by weak interaction perturbations ($W^\pm$ exchange). In a general field theory these could lead to relatively large $\mathcal{O}(\alpha)$ effects in addition to the safer $\mathcal{O}(\alpha/M_W^2)$ ones. However, for QCD and similar theories, these only affect quark masses and can be absorbed into their renormalization (Weinberg, 1973b). They could in principle be problematic for scalar-mediated theories, including supersymmetric QCD, but there they are suppressed by large superpartner masses.

using the notation of (3.113) and (3.114) on page 110.

The two $U(1)$ factors have generators

$$B = \frac{1}{3}\int J_0^B(x)d^3\vec{x}, \qquad B_A = \frac{1}{3}\int J_0^A(x)d^3\vec{x}, \tag{5.92}$$

where $J^{B,A}$ are the associated Noether currents

$$J_\mu^B = \sum_{r\alpha} \bar{q}_r^\alpha \gamma_\mu q_{r\alpha}, \qquad J_\mu^A = \sum_{r\alpha} \bar{q}_r^\alpha \gamma_\mu \gamma^5 q_{r\alpha}, \tag{5.93}$$

with r and α the flavor and color indices, respectively. B is just baryon number ($\frac{1}{3}$ for quarks and $-\frac{1}{3}$ for antiquarks), while B_A is the analogous axial baryon number. They are related to the chiral $U(1)$ generators $B_{L,R}$ (i.e., $B_L = \pm\frac{1}{3}$ for $q_L(\bar{q}_L)$ and 0 for $q_R(\bar{q}_R)$, and the reverse for B_R) by $B(B_A) = B_R \pm B_L$. The baryon number is an accidental symmetry that is conserved to all orders in perturbation theory in the standard model, though there may be a small non-perturbative breaking in the electroweak sector due to vacuum tunneling (*instanton*) effects (e.g., 't Hooft, 1976b; Schäfer and Shuryak, 1998). Many extensions of the SM predict B violation at some very small level, which would lead to proton decay. However, there is no sign of the axial baryon number symmetry in nature, and in fact the pseudoscalar spectrum implies that it must be absent, as will be discussed in Section 5.8.3. It is now believed to be broken by non-perturbative effects associated with the triangle anomaly in the B_A-gluon-gluon vertex and with instantons.

Consider $SU(3)_L \times SU(3)_R$ or its $SU(2)_L \times SU(2)_R$ subgroup. Define the generators of $SU(3)_{L,R}$ as $F^i_{L,R}$, respectively. Their Lie algebra is

$$\left[F_L^i, F_L^j\right] = i f_{ijk} F_L^k, \qquad \left[F_R^i, F_R^j\right] = i f_{ijk} F_R^k, \qquad \left[F_L^i, F_R^j\right] = 0, \tag{5.94}$$

and their commutators with the chiral quark fields are

$$\begin{aligned} \left[F_L^i, q_{aL}\right] &= -\frac{\lambda^i_{ab}}{2} q_{bL}, \qquad & \left[F_L^i, q_{aR}\right] &= 0 \\ \left[F_R^i, q_{aL}\right] &= 0, \qquad & \left[F_R^i, q_{aR}\right] &= -\frac{\lambda^i_{ab}}{2} q_{bR}, \end{aligned} \tag{5.95}$$

i.e., q_L transforms as $(3,1)$, a triplet under $SU(3)_L$ and a singlet under $SU(3)_R$. Similarly, q_R transforms as $(1,3)$, and the antiquarks as $(3^*,1)$ and $(1,3^*)$. $F^i_{L,R}$ are associated with the Noether currents

$$J^i_{L\mu} = \bar{q}_L \gamma_\mu \frac{\lambda^i}{2} q_L \equiv \bar{q}^\alpha_{aL} \gamma_\mu \frac{\lambda^i_{ab}}{2} q_{bL\alpha}, \qquad J^i_{R\mu} = \bar{q}_R \gamma_\mu \frac{\lambda^i}{2} q_R. \tag{5.96}$$

It is useful to also define the vector and axial vector generators

$$F^i = F_R^i + F_L^i = \int V_0^i(x)d^3\vec{x}, \qquad F^{i5} = F_R^i - F_L^i = \int A_0^i(x)d^3\vec{x} \tag{5.97}$$

and their associated currents

$$V^i_\mu = J^i_{R\mu} + J^i_{L\mu} = \bar{q}\gamma_\mu \frac{\lambda^i}{2} q, \qquad A^i_\mu = J^i_{R\mu} - J^i_{L\mu} = \bar{q}\gamma_\mu\gamma^5 \frac{\lambda^i}{2} q. \tag{5.98}$$

The V^i_μ are associated with the weak and electromagnetic currents, while the A^i_μ enter the weak interactions. They do not commute, but rather satisfy

$$[F^i, F^j] = i f_{ijk} F^k, \qquad [F^i, F^{j5}] = i f_{ijk} F^{k5}, \qquad [F^{i5}, F^{j5}] = i f_{ijk} F^k. \tag{5.99}$$

The vector F^i generate the *diagonal* $SU(3)$ subgroup of $SU(3) \times SU(3)$.

The vector and axial currents V^i_μ and A^i_μ are conserved for massless quarks. Turning on $m = \text{diag}(m_u\ m_d\ m_s) \neq 0$, one obtains from (3.51) on page 99 (or directly from the equations of motion) that

$$\partial^\mu V^i_\mu = -i\bar{q}\left[\frac{\lambda^i}{2}, m\right] q, \qquad \partial^\mu A^i_\mu = i\bar{q}\left\{\frac{\lambda^i}{2}, m\right\}\gamma^5 q. \tag{5.100}$$

As expected, the axial symmetries are broken by the quark masses, even if they are degenerate, while the vector symmetries are only broken by mass splittings.

5.8.2 The $(3^*, 3) + (3, 3^*)$ Model

The $(3^*, 3) + (3, 3^*)$ model of $SU(3)_L \times SU(3)_R$ symmetry realization and breaking (Gell-Mann et al., 1968; Glashow and Weinberg, 1968) was inferred from the data prior to QCD or even the general acceptance of the quark model (Pagels, 1975), but emerges naturally in that framework. The basic idea is the assumption that the strong interactions have an $SU(3)_L \times SU(3)_R$ invariant part as well as a small breaking term. The observed hadron spectrum does not exhibit the full $SU(3)_L \times SU(3)_R$ symmetry,[21] so the symmetries associated with the axial generators are assumed to be spontaneously broken. The explicit breaking term transforms as a singlet and octet under ordinary $SU(3)$, and as $(3^*, 3) + (3, 3^*)$ under $SU(3)_L \times SU(3)_R$. That is manifestly the case for the quark mass term in (5.90) and (5.91), which constitutes a concrete realization of the $(3^*, 3) + (3, 3^*)$ model.

The qualitative picture, which is supported by lattice calculations, is therefore that in the limit $m_u = m_d = m_s = 0$ the symmetries associated with the 8 axial generators F^{i5} are spontaneously broken. According to the Nambu-Goldstone theorem of Section 3.3.6 this implies the existence of 8 massless pseudoscalar Goldstone bosons, which are identified with the pseudoscalar octet $\pi^{\pm,0}, K^\pm, K^0, \bar{K}^0; \eta$. The $SU(3)$-singlet η' is not a Goldstone boson because of the non-perturbative breaking of the associated B_A symmetry, as will be discussed in the next section. Unlike the examples in Chapter 3 the SSB of the F^{i5} is not due to the VEV of an elementary scalar, but rather is associated with the VEV or *vacuum condensate* of a composite $\bar{q}q$ operator, i.e., in the massless limit

$$\langle 0|\bar{u}u|0\rangle = \langle 0|\bar{d}d|0\rangle = \langle 0|\bar{s}s|0\rangle \equiv \nu_0 = \mathcal{O}(\Lambda^3) \tag{5.101}$$

due to the non-perturbative long-distance dynamics. Since the chiral symmetry is broken, the quarks must acquire a *dynamical* or *constituent* common mass $M_{dyn} \sim M_p/3 \sim 300$ MeV proportional to Λ from the strong dynamics, i.e., from the interaction with the quarks and gluons in the condensate.

To better justify these statements, let us define the composite pseudoscalar and scalar field operators

$$\begin{aligned}
\pi^i &\equiv -i\bar{q}\frac{\lambda^i}{2}\gamma^5 q = -i\bar{q}_L\frac{\lambda^i}{2}q_R + i\bar{q}_R\frac{\lambda^i}{2}q_L \\
\sigma^i &\equiv \bar{q}\frac{\lambda^i}{2}q = \bar{q}_L\frac{\lambda^i}{2}q_R + \bar{q}_R\frac{\lambda^i}{2}q_L,
\end{aligned} \tag{5.102}$$

for $i = 0 \cdots 8$, with $\lambda^0 = \sqrt{\frac{2}{3}}I$ and $d_{0jk} = \sqrt{\frac{2}{3}}\delta_{jk}$. σ^0 and π^0 are $SU(3)$ singlets and the

[21] One would expect massless chiral fermions, and each meson isomultiplet would need to have at least one degenerate partner with the opposite parity, contrary to observations.

others are octets. In particular $\sigma^0 = (\bar{u}u + \bar{d}d + \bar{s}s)/\sqrt{6}$. It is straightforward to show from (5.95) that

$$\left[F^{i5}, \pi^j\right] = i d_{ijk} \sigma^k, \qquad \left[F^{i5}, \sigma^j\right] = -i d_{ijk} \pi^k. \tag{5.103}$$

Now assume the $\bar{q}q$ condensate in (5.101), which implies that $\langle 0|\sigma^0|0\rangle = \sqrt{\frac{3}{2}}\nu_0$. Taking the expectation value of the first equation in (5.103),

$$\langle 0| \left[F^{i5}, \pi^j\right] |0\rangle = i\delta_{ij}\nu_0. \tag{5.104}$$

This implies that $F^{i5}|0\rangle \neq 0$, i.e., the F^{i5} symmetry is spontaneously broken. It also implies that there must be some angular momentum-0 state in the Hilbert space with a non-zero inner product with $F^{i5}|0\rangle$, which we assume to be a single particle state that we denote $|\pi^i(\vec{q})\rangle$ under the anticipation that it is a bound state with the quantum numbers of the π^i field. Using (5.97),

$$\langle 0|F^{i5}|\pi^k(\vec{q})\rangle = \int d^3\vec{x}\, \langle 0|A_0^i(x)|\pi^k(\vec{q})\rangle. \tag{5.105}$$

But $\langle 0|A_\mu^i(x)|\pi^k(\vec{q})\rangle$ must be of the form

$$\langle 0|A_\mu^i(x)|\pi^k(\vec{q})\rangle = i\frac{f_\pi}{\sqrt{2}} q_\mu \delta_{ik} e^{-iq\cdot x}. \tag{5.106}$$

The q_μ follows from Lorentz invariance because q is the only four-vector available, the δ_{ik} is from the unbroken $SU(3)$, and the $\exp(-iq\cdot x)$ follows from translation invariance. f_π, known as the *pseudoscalar decay constant*, is illustrated diagramatically in Figure 5.18. It

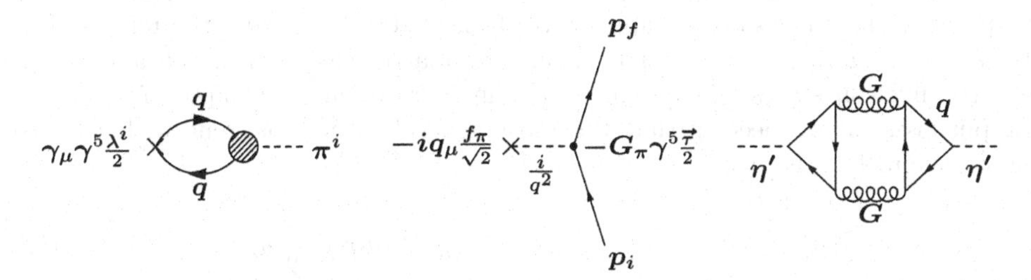

Figure 5.18 Left: the pseudoscalar decay constant. Middle: pion pole contribution to the induced pseudoscalar form factor in the chiral limit. The cross indicates the axial current and $q = p_f - p_i$. Right: two gluon intermediate state coupling to the η'.

is determined from the strong dynamics that lead to the bound state, and is expected to be of $\mathcal{O}(\Lambda)$. We will see in Chapter 7 that $f_\pi \sim 130$ MeV can be measured from pion decay. Inserting (5.106) into (5.105)

$$\langle 0|F^{i5}|\pi^k(\vec{q})\rangle = (2\pi)^3 q_0 i\delta_{ik}\delta^3(\vec{q})\frac{f_\pi}{\sqrt{2}}, \tag{5.107}$$

where we can take $t = 0$ since F^{i5} generates a symmetry of the equations of motion. One can insert a complete set of states between F^{i5} and π^j in (5.104). Assuming that only the single particle states enter, or at least that they dominate, we find

$$\begin{aligned}\int \frac{d^3\vec{q}}{(2\pi)^3 2q_0} \left\{ \langle 0|F^{i5}|\pi^k(\vec{q})\rangle\langle\pi^k(\vec{q})|\pi^j|0\rangle - \langle 0|\pi^j|\pi^k(\vec{q})\rangle\langle\pi^k(\vec{q})|F^{i5}|0\rangle \right\} \\ = i\delta_{ij}\frac{f_\pi}{\sqrt{2}} Z^{1/2} = i\delta_{ij}\nu_0,\end{aligned} \tag{5.108}$$

where $Z^{1/2}$, which has dimensions of mass^2, is associated with the bound state wave function of the π^k by

$$\langle 0|\pi^j|\pi^k(\vec{q})\rangle \equiv Z^{1/2}\delta_{jk}. \tag{5.109}$$

Defining $\nu \equiv \nu_0/Z^{1/2}$, which has dimensions of mass, one has that

$$\frac{f_\pi}{\sqrt{2}} = \nu, \tag{5.110}$$

which is nonzero if the symmetry is spontaneously broken. The Nambu-Goldstone theorem then follows from $\partial^\mu A^i_\mu = 0$ and (5.106)

$$\langle 0|\partial^\mu A^i_\mu|\pi^k\rangle = 0 = \frac{f_\pi}{\sqrt{2}} q^2 \delta_{ik} e^{-iq\cdot x} \Rightarrow q^2 f_\pi = m_\pi^2 f_\pi = 0. \tag{5.111}$$

This summarizes the Goldstone alternative: either $f_\pi = 0$ (no SSB) *or* $m_\pi^2 = 0$ (there exists a massless spin-0 Goldstone boson).

To justify the generation of a dynamical mass $M_{dyn} \neq 0$ let us first derive the *Goldberger-Treiman* (GT) relation. To simplify the notation, we work with the $SU(2)_L \times SU(2)_R$ subgroup. Since the quarks do not appear as physical states, consider the nucleon isospin doublet $\psi = (\psi_p \ \psi_n)^T$, as in Section 3.2.3. The matrix elements of the axial currents between the nucleon states will be considered in more detail in Chapter 7, but for our present purposes all we need is that the most general form consistent with the strong interaction symmetries is

$$\langle \psi_f|A^j_\mu(x)|\psi_i\rangle = \bar{u}_f\left[\gamma_\mu\gamma^5 g_1(q^2) + q_\mu\gamma^5 g_3(q^2)\right]\frac{\tau^j}{2} u_i\, e^{iq\cdot x}, \tag{5.112}$$

where $q = p_f - p_i$ and $g_{1,3}$ are form factors. g_1 is a slowly varying function, with experimental value $g_1(0) \sim 1.27$ from β decay, while the *induced pseudoscalar* g_3 has a pole at $q^2 = 0$ from the massless Goldstone boson, as indicated in Figure 5.18. Using (5.106) it is given by

$$g_3(q^2) \sim -\frac{f_\pi}{\sqrt{2}}\frac{1}{q^2} G_\pi, \tag{5.113}$$

where $G_\pi \sim 13.1$ is the effective pion-nucleon coupling defined following (3.84) on page 103.

Using $\partial^\mu A^i_\mu$ and evaluating the divergence of (5.112) at $q^2 = 0$ one obtains the GT relation

$$g_1(0) M_p = \frac{f_\pi G_\pi}{\sqrt{2}}. \tag{5.114}$$

(5.114) should hold in the chiral limit, and experimentally it is satisfied to better than 2% (Gorringe and Fearing, 2004). Our present purpose, however, is the observation that even in the chiral limit one has that $M_p \propto f_\pi \neq 0$ is induced by the chiral breaking, and that presumably the quarks also acquire dynamical masses $M_{dyn} \sim M_p/3$.

The picture is perturbed when non-zero *current* quark masses m_q are turned on in

$\mathcal{L}_{QCD}$. For $m_u = m_d = m_s \neq 0$ the axial generators F^{i5} are explicitly broken and the pseudoscalars acquire small mass-squares $m_{PS}^2 \propto m_q$, i.e., they become pseudo-Goldstone bosons, analogous to the example in Section 3.3.3. Also, the constituent quark masses are shifted to effective values

$$M_q = M_{dyn} + m_q. \tag{5.115}$$

The pseudoscalar masses may be obtained by taking the matrix element of

$$\partial^\mu A_\mu^i = im\bar{q}\lambda^i\gamma^5 q = -2m\pi^i \tag{5.116}$$

from (5.100) between the vacuum and one Goldstone boson state. Using (5.106) and (5.109),

$$\frac{f_\pi}{\sqrt{2}}m_\pi^2 = -2mZ^{1/2} \equiv \epsilon, \tag{5.117}$$

where the explicit symmetry breaking parameter ϵ has dimensions of mass^3. Finally, it is convenient to define a renormalized field $\hat{\pi}^i \equiv \pi^i/Z^{1/2}$ so that

$$\langle 0|\hat{\pi}^j|\pi^k(\vec{q})\rangle = \delta_{jk}. \tag{5.118}$$

In terms of $\hat{\pi}^i$ (5.116) becomes

$$\partial^\mu A_\mu^i = \frac{f_\pi}{\sqrt{2}}m_\pi^2\hat{\pi}^i, \tag{5.119}$$

which is known as the *partially conserved axial current* (PCAC) relation (Gell-Mann and Levy, 1960). It again displays the Goldstone alternative in the limit $\partial^\mu A_\mu^i = 0$, and can be used as the basis for *soft pion* theorems, which can be used to calculate amplitudes for physical pions at low energy (Adler and Dashen, 1968).

Breaking the degeneracy of the quark masses leads to breaking of the vector generators F^i. For $m_s > m_u = m_d$ the vector $SU(3)$ is broken to $SU(2)$ (isospin), while $m_d \neq m_u$ breaks the isospin symmetry, as described in Section 3.2.3. The diagonal generators F^3 and F^8, which are related to electric charge and strong hypercharge by (3.89), are unbroken. The ratios of the current masses cannot be determined directly because of confinement and because of M_{dyn}. However, they can be inferred from the observed $SU(3)_L \times SU(3)_R$ symmetry breaking effects (Weinberg, 1977; Langacker and Pagels, 1979; Gasser and Leutwyler, 1982). To lowest order in symmetry breaking,

$$\begin{aligned}\frac{m_u}{m_d} &= \frac{2m_{\pi^0}^2 - m_{\pi^+}^2 + m_{K^+}^2 - m_{K^0}^2}{m_{K^0}^2 - m_{K^+}^2 + m_{\pi^+}^2}\\ \frac{m_s}{m_d} &= \frac{m_{K^0}^2 + m_{K^+}^2 - m_{\pi^+}^2}{m_{K^0}^2 - m_{K^+}^2 + m_{\pi^+}^2},\end{aligned} \tag{5.120}$$

or ignoring isospin breaking,

$$\frac{m_K^2}{m_\pi^2} \simeq \frac{m_s}{m_d + m_u}. \tag{5.121}$$

In practice, higher-order corrections are needed, and additional information on isospin breaking is obtained from the proton-neutron and $K^+ - K^0$ mass differences (which also have electromagnetic components that must be estimated), $\rho - \omega$ mixing, the $\eta \to 3\pi$ and $\psi' \to \psi\pi^0$ decays, and isospin breaking effects in nuclear physics (e.g., Gasser and Leutwyler, 1982). Typical estimates of the ratios are

$$\frac{m_s}{m_d} \sim 17 - 22, \qquad \frac{m_u}{m_d} \sim 0.4 - 0.6. \tag{5.122}$$

The larger value of m_d compared to m_u explains why $m_n > m_p$, since the electromagnetic effect would make the proton heavier (the two contributions are of the same order of magnitude). Similar statements apply to $m^2_{K^+} - m^2_{K^0} < 0$, while $m^2_{\pi^+} - m^2_{\pi^0} > 0$ is almost entirely due to electromagnetism (both $m^2_{\pi^+}$ and $m^2_{\pi^0}$ are proportional to $m_u + m_d$ to leading order).

The absolute scales are more difficult, since they depend on strong interaction matrix elements of the operators in (3.113), and also on the renormalization scheme and scale (see the review on Quark Masses in Patrignani, 2016). Most estimates are in the range

$$m_u \sim 2 - 3 \text{ MeV}, \qquad m_d \sim 4 - 5 \text{ MeV}, \qquad m_s \sim 90 - 100 \text{ MeV}. \tag{5.123}$$

These are actually running masses, evaluated at a scale $\mu \sim 2$ GeV. The u and d masses are much smaller than other strong interaction scales, such as the nucleon mass or Λ, i.e., both $SU(2)$ and $SU(2)_L \times SU(2)_R$ hold at the $\lesssim$ few % level. The near exactness of the chiral symmetry is most evident from the fact that m^2_π is so tiny compared to other hadronic mass-squares. The success of the vector isospin symmetry is because m_u and m_d are both extremely small even though their ratio is not close to unity, and also because α is small. m_s is small compared to most hadronic masses, but not negligible. That is why the $SU(3)$ breaking effects are typically of $\mathcal{O}(25)\%$. The one counterexample is m^2_K/m^2_π, which is large because of $m_s \gg m_{u,d}$ and the special role of the K and π as pseudo-Goldstone bosons.

We also record typical values for the heavy quark masses

$$\begin{aligned} m_c(m_c) &\sim 1.2 - 1.3 \text{ GeV} \Rightarrow m_c^{pole} \sim 1.6 - 1.7 \text{ GeV} \\ m_b(m_b) &\sim 4.1 - 4.2 \text{ GeV} \Rightarrow m_b^{pole} \sim 4.8 - 5.0 \text{ GeV} \\ m_t(m_t) &\sim 163.7 \pm 0.7 \text{ GeV} \Rightarrow m_t^{pole} \sim 173.3 \pm 0.8 \text{ GeV}. \end{aligned} \tag{5.124}$$

The first values are the running ($\overline{MS}$) masses, while the second are the propagator *pole* masses, which correspond most closely to kinematic masses. They are related by

$$m^{pole} \sim m(m)\left[1 + \frac{4}{3}\frac{\alpha_s(m^2)}{\pi}\right]. \tag{5.125}$$

The relation is actually known through $\mathcal{O}([\alpha_s(m^2)/\pi]^3)$(Melnikov and Ritbergen, 2000). The series converges very slowly, especially for m_b and m_c, and it is best to work in terms of the running masses to minimize the associated uncertainties.

5.8.3 The Axial $U(1)$ Problem

One aspect of QCD and other quark-gluon models that is not present in the general $(3^*, 3) + (3, 3^*)$ model is that the symmetry is $U(3)_L \times U(3)_R$ rather than just $SU(3)_L \times SU(3)_R \times U(1)_B$ as the quark masses go to zero. Assuming that the dynamics of the extra axial $U(1)$ generator $F^{05} = B_A$ are similar to those of the F^{i5}, one might expect that it is spontaneously broken, $F^{05}|0\rangle \neq 0$, so that there would be a ninth pseudo-Goldstone boson in addition to the π's, K's, and η. Although the $\eta'(958)$ has the appropriate quantum numbers, it turns out that either m_η, $m_{\eta'}$, or both are just too large to be consistent with the π and K masses, independent of the assumptions on the spontaneous breaking. Even if the F^{05} symmetry is not spontaneously broken, the axial $U(1)$ from $U(2)_L \times U(2)_R$ will be, with m_η predicted to be too light when the quark masses are turned on. To be more explicit, using standard current algebra techniques and ignoring $SU(2)$ breaking for simplicity (i.e., $m_3 = 0$), one can show (Weinberg, 1975) that the two mass-squared eigenvalues of the mixed $\eta - \eta'$ system are

$$\frac{3m^2_\pi}{1 + 2z^2}, \qquad \frac{4}{3}m^2_K\left(1 + \frac{1}{2z^2}\right), \tag{5.126}$$

up to corrections smaller by $\mathcal{O}(m_\pi^2/m_K^2)$, where $z = f_0/f_\pi$, with f_0 the decay constant for F^{05} defined analogously to f_π in (5.106). For $z = 1$ the two mass eigenvalues are predicted to be $m_\eta = m_\pi \sim 140$ MeV and $m_{\eta'} = \sqrt{2}m_K \sim 700$ MeV, respectively, compared with the observed η and η' masses of 548 and 958 MeV. For $z \to 0$ (no spontaneous F^{05} breaking) one state becomes very heavy (not a Goldstone boson), but the lighter eigenvalue is $\sqrt{3}m_\pi \sim 243$ MeV. The latter value is the upper limit on the lighter mass. This conflict with observation was known as the $U(1)_A$ problem or the η problem. There were also difficulties associated with the η decay.

To resolve the $U(1)_A$ problem there needs to be a contribution to the explicit breaking that is independent of the quark masses. The $SU(3)$-singlet η' is distinguished from the octet states π, K, and η by the third diagram in Figure 5.18. The two-gluon intermediate state is a singlet under flavor $SU(3)$ and can therefore couple to the η' but not to the octet states in the $SU(3)$ limit. This is not by itself sufficient to solve the problem, however. Another necessary ingredient is the triangle anomaly in the B_A-gluon-gluon vertex, which violates the symmetry and leads to a divergence (Adler, 1969; Bell and Jackiw, 1969)

$$\partial^\mu A^0_\mu = 2i \sum_{r=1}^{n_q} m_r \bar{q}_r \gamma^5 q_r + n_q \frac{g_s^2}{16\pi^2} G^{i\mu\nu} \tilde{G}^i_{\mu\nu}, \tag{5.127}$$

where $A^0_\mu = \sum_{r=1}^{n_q} \bar{q}_r \gamma_\mu \gamma^5 q_r$. We have included the explicit breaking from the quark masses, $n_q = 2$ or 3 is the number of flavors being considered, and $\tilde{G}$ is the gluon dual field strength tensor defined in (5.7). The $G\tilde{G}$ term (which incidentally has the same structure as the θ_{QCD} term in (5.2)) appears to do the job, but there is a further complication in that it can be written as a four-divergence,

$$G^{i\mu\nu} \tilde{G}^i_{\mu\nu} = \partial^\mu \left[\epsilon_{\mu\nu\rho\sigma} G^{i\nu} \left(G^{i\rho\sigma} + \frac{g_s}{3} f_{ijk} G^{j\rho} G^{k\sigma} \right) \right] \equiv \partial^\mu K_\mu. \tag{5.128}$$

The current

$$\tilde{A}^0_\mu \equiv A^0_\mu - n_q \frac{g_s^2}{16\pi^2} K_\mu \tag{5.129}$$

is therefore conserved for $m_r \to 0$, apparently reintroducing the problem. However, K_μ is not gauge invariant, and it was shown ('t Hooft, 1976b, 1986; Schäfer and Shuryak, 1998; Vicari and Panagopoulos, 2009) that there are non-perturbative vacuum tunneling gauge field configurations (instantons) that fall off sufficiently slowly that they contribute to surface terms when the divergence is integrated over space-time. The $U(1)_A$ anomaly and instantons therefore generate an η' mass that survives when the quark masses vanish, solving the $U(1)_A$ problem and evading the Nambu-Goldstone theorem.

There is another consequence of these considerations: since the current $\tilde{A}^0_\mu$ is formally conserved in the chiral limit, one can use the associated charge $\tilde{F}^{05}$ to generate phase rotations on the quark fields. For example, under the transformation $U(\beta) \equiv \exp(-i\beta\tilde{F}^{05})$ the quarks transform as

$$q \to U(\beta) q U^{-1}(\beta) = e^{i\beta\gamma^5} q \iff q_{L,R} \to e^{\mp i\beta} q_{L,R} \tag{5.130}$$

(cf. (3.28) on page 96). However, it can be shown (Callan et al., 1976; Jackiw and Rebbi, 1976) that because of the gauge noninvariance of the K component, the θ_{QCD} term in (5.2) is not chiral invariant, and that θ_{QCD} is shifted to $\theta_{QCD} - 2n_q\beta$ under the same transformation. An equivalent formulation is that there are a continua of possible vacua of QCD, labeled by θ_{QCD}. However, the vacua are not invariant,

$$U(\beta)|\theta_{QCD}\rangle = |\theta_{QCD} - 2n_q\beta\rangle. \tag{5.131}$$

The implications for the strong CP problem will be mentioned in Section 10.1.

The Pseudoscalar and Vector Nonets

In Section 3.2.3 we described the Gell-Mann-Okubo formula for the nucleon masses in the presence of $SU(3)$ breaking. The GMO formula (3.105) on page 108 followed from the assumption that the explicit breaking term in $\mathcal{L}$ transformed as the 8th component of an $SU(3)$ octet, as in the quark model. Similar considerations apply to the nonets of mesons, such as the pseudoscalar ($\pi[138]$, $K[496]$, $\eta[548]$, $\eta'[958]$) and the vector ($\rho[770]$, $K^*[892]$, $\phi[1020]$, $\omega[782]$). In each case, the first state refers to the isospin triplet, the second to the two isospin doublets, and the last two to the isospin singlets, which may be mixtures of the isosinglet in the octet and an $SU(3)$ singlet. The numbers in brackets are the masses in MeV. (We ignore isospin breaking.) The vectors are interpreted as the lowest lying $L = 0, S = 1$ $q\bar{q}$ states in the $SU(3)$ quark model. The pseudoscalars have the dual (and sometimes conflicting) interpretation as pseudo-Goldstone bosons, or as the $L = 0, S = 0$ states in the quark model.

The derivation of the mass formula differs in several ways from that in (3.105). First, as is clear from chiral perturbation theory or the form of the free Lagrangian, the analogs of (3.102) and (3.104) should apply to the meson mass-squares. Second, the meson octets are self-conjugate, so the two isodoublets are related and degenerate by charge conjugation. The apparent loss of a parameter is compensated by the fact that Bose symmetry implies that only the symmetric βd_{i8j} term in (3.102) is present. Finally, the matrix elements in (3.102) only involve the states within the original octet. In principle, $SU(3)$ breaking can mix the octet states with single particle states from outside the octet, or even with multiparticle states. That is not a significant effect for the baryons, but it is for the meson nonets, which involve an $SU(3)$-singlet that can mix significantly. The result is that the analog of (3.105) for the pseudoscalar and vector octets is

$$4m_K^2 = 3m_{\eta_8}^2 + m_\pi^2, \qquad 4m_{K^*}^2 = 3m_{\phi_8}^2 + m_\rho^2, \tag{5.132}$$

where $m_{\eta_8}^2$ and $m_{\phi_8}^2$ are, respectively, m_{P88}^2 and m_{V88}^2, the 88 entries of the pseudoscalar and vector octet mass-squared matrices. However, the $SU(3)$ breaking can also induce $m_{P08}^2 = m_{P80}^2$ or $m_{V08}^2 = m_{V80}^2$ mixing terms between the octet and singlet, so the $\eta - \eta'$ and $\phi - \omega$ systems are really described by 2×2 mass matrices[22] such as

$$-\mathcal{L}_{\eta\eta'} = \frac{1}{2} \begin{pmatrix} \eta_8 & \eta_0 \end{pmatrix} \begin{pmatrix} m_{P88}^2 & m_{P80}^2 \\ m_{P08}^2 & m_{P00}^2 \end{pmatrix} \begin{pmatrix} \eta_8 \\ \eta_0 \end{pmatrix}, \tag{5.133}$$

where m_{P00}^2 is the singlet mass-square in the absence of mixing. η_8 and η_0 are the original octet and singlet states, given in the quark model by

$$\eta_8 = \frac{1}{\sqrt{6}} \left(u\bar{u} + d\bar{d} - 2s\bar{s} \right), \qquad \eta_0 = \frac{1}{\sqrt{3}} \left(u\bar{u} + d\bar{d} + s\bar{s} \right). \tag{5.134}$$

Within this simple two-state approximation, the mass eigenstates are

$$\eta = \eta_8 \cos\theta_P - \eta_0 \sin\theta_P, \qquad \eta' = \eta_8 \sin\theta_P + \eta_0 \cos\theta_P, \tag{5.135}$$

where θ_P is the *pseudoscalar mixing angle*. One finds

$$m_{\eta_8}^2 = \cos^2\theta_P m_\eta^2 + \sin^2\theta_P m_{\eta'}^2. \tag{5.136}$$

[22] Isospin-breaking (ϵ_3) effects would extend the discussion to include $\pi^0 - \eta - \eta'$ and $\rho^0 - \phi - \omega$ mixing. Also, mixing with excited octet or singlet $q\bar{q}$ states or with gluonium is possible. We will ignore these complications.

From the GMO formula in (5.132) as well as confirming (and sign) information from decays involving the η and η' one obtains (Patrignani, 2016) $\theta_P \sim -11.5°$. The rather small value for θ_P is easily understood in terms of our discussion of the $U(1)_A$ problem. In the chiral limit the octet mass-squares all go to zero, while the $m_{\eta_0} \sim m_{\eta'}$ induced by the anomaly/instanton effects is large compared to typical $SU(3)$ mass splittings of $\lesssim$ few hundred MeV. Therefore, one expects $|m^2_{P08}|, m^2_{P88} \ll m^2_{P00}$ and relatively little mixing.

Similar formulas apply to $\phi - \omega$ mixing. In this case, however, the mixing angle θ_V defined by

$$\phi = \phi_8 \cos\theta_V - \phi_0 \sin\theta_V, \qquad \omega = \phi_8 \sin\theta_V + \phi_0 \cos\theta_V, \tag{5.137}$$

turns out to be large, $\theta_V \sim 38.7°$. This is very close to the *ideal* or *magic mixing angle* $\theta_I = 35.3°$ $(\tan\theta_I = 1/\sqrt{2})$ for which ϕ would be purely $s\bar{s}$,

$$\phi \sim -s\bar{s}, \qquad \omega \sim \frac{1}{\sqrt{2}}(u\bar{u} + d\bar{d}), \qquad m^2_{\phi_8} = \frac{2}{3}m^2_\phi + \frac{1}{3}m^2_\omega. \tag{5.138}$$

This, and similar results for other nonets with sufficient data, is again easy to understand. The vector mesons are well described by the non-relativistic quark model. In the chiral limit there is no major distinction between the octet and singlet, and they should have comparable masses, perhaps dominated by $2M_{dyn}$. This is not exact. For example, the ϕ_0 could acquire additional mass from gluon intermediate states, analogous to the two-gluon diagram in Figure 5.18. However, by charge conjugation and gauge invariance at least three gluons are required, so the contribution is suppressed by $\mathcal{O}(\alpha_s^3)$. In any case, as long as the the unperturbed singlet and octet mass-squared difference is small compared to the $SU(3)$-breaking effects one expects the heavier mass eigenstate to be mainly $s\bar{s}$ since $m_s \gg m_{u,d}$. (The precise connection to the 2×2 mass matrix in (5.133) is explored in Problem 5.15.) This ideal mixing picture is further supported by the expectation

$$m^2_\rho \sim m^2_\omega, \qquad m^2_\phi - m^2_{K^*} \sim m^2_{K^*} - m^2_\rho, \tag{5.139}$$

which is obvious from the quark interpretation and agrees reasonably well with observations. For the vector mesons, unlike the pseudoscalars, the approximate picture, including the near ideal mixing, still holds if one assumes the perturbations are linear in the masses.

5.8.4 The Linear σ Model

Some of the formal arguments presented above are more transparent when presented in terms of an effective theory, the *linear σ model,* involving the composite fields defined in (5.102). We will illustrate this for the $SU(2)_L \times SU(2)_R$ limit, with fields

$$\pi^i = -i\bar{q}\frac{\tau^i}{2}\gamma^5 q, \qquad \sigma = \frac{1}{2}\bar{q}q = \frac{1}{2}(\bar{u}u + \bar{d}d), \tag{5.140}$$

where $i = 1, 2, 3$ and the normalization of σ is chosen for convenience. (Neither the σ nor the η field in the next Section corresponds closely to any physical meson.) We assume that the strong QCD dynamics generates the effective interactions in

$$\begin{aligned} \mathcal{L}_\sigma = {} & \bar{\psi} i \partial\!\!\!/ \psi + \frac{1}{2}(\partial_\mu \sigma)^2 + \frac{1}{2}(\partial_\mu \vec{\pi})^2 - g_\pi \bar{\psi}\left[\sigma - i\vec{\pi}\cdot\vec{\tau}\gamma^5\right]\psi \\ & - \frac{\mu^2}{2}\left(\sigma^2 + \vec{\pi}^{\,2}\right) - \frac{\lambda}{4}\left(\sigma^2 + \vec{\pi}^{\,2}\right)^2 + a\sigma, \end{aligned} \tag{5.141}$$

where $\psi = (\psi_p \ \psi_n)^T$ is the nucleon doublet. $\mathcal{L}_\sigma$ is $SU(2)_L \times SU(2)_R$ invariant except for the linear $a\sigma$ term, which mimics the current quark masses and explicitly breaks the symmetry.

W.l.o.g., we can take $a \geq 0$. We see explicitly from (5.141) that if $SU(2)_L \times SU(2)_R$ were unbroken ($a = 0$ and no spontaneous breaking) the nucleons would be massless and the pions would have a degenerate parity-even partner (the σ). The phase in which the axial generators are spontaneously broken is more realistic.

To see the $SU(2)_L \times SU(2)_R$ invariance of $\mathcal{L}_\sigma$ explicitly, define a 2×2 matrix

$$M \equiv \frac{\sigma + i\vec{\pi} \cdot \vec{\tau}}{\sqrt{2}} = \begin{pmatrix} \frac{\sigma + i\pi^0}{\sqrt{2}} & i\pi^+ \\ i\pi^- & \frac{\sigma - i\pi^0}{\sqrt{2}} \end{pmatrix}, \tag{5.142}$$

which is reminiscent of our treatment of the adjoint representation of a simple group in Section 3.2. Then

$$\begin{aligned} \mathcal{L}_\sigma &= \bar{\psi}_L i \not{\partial} \psi_L + \bar{\psi}_R i \not{\partial} \psi_R + \frac{1}{2} \text{Tr} \left[(\partial^\mu M)^\dagger (\partial_\mu M) \right] \\ &- \sqrt{2} g_\pi \left(\bar{\psi}_R M \psi_L + \bar{\psi}_L M^\dagger \psi_R \right) - \frac{\mu^2}{2} \text{Tr} \left(M M^\dagger \right) - \frac{\lambda}{4} \left[\text{Tr} \left(M M^\dagger \right) \right]^2 + a\sigma. \end{aligned} \tag{5.143}$$

$\psi_{L,R}$ transform the same way as $q_{L,R}$, i.e., as $(2, 1)$ and $(1, 2)$,

$$\begin{aligned} SU(2)_L: &\qquad \psi_L \to e^{i\vec{\beta}_L \cdot \frac{\vec{\tau}}{2}} \psi_L, &\qquad \psi_R \to \psi_R \\ SU(2)_R: &\qquad \psi_L \to \psi_L, &\qquad \psi_R \to e^{i\frac{\vec{\beta}_R \cdot \vec{\tau}}{2}} \psi_R. \end{aligned} \tag{5.144}$$

It can be shown (Problem 5.16) that M transforms as $(2^*, 2)$, i.e.,

$$M \to e^{i\vec{\beta}_R \cdot \frac{\vec{\tau}}{2}} M e^{-i\vec{\beta}_L \cdot \frac{\vec{\tau}}{2}}, \tag{5.145}$$

so that $\mathcal{L}_\sigma$ is invariant except for the $a\sigma$ term. The infinitestimal form of (5.145) is

$$\vec{\pi} \to \vec{\pi} - \vec{\alpha} \times \vec{\pi} + \vec{\beta}\sigma, \qquad \sigma \to \sigma - \vec{\beta} \cdot \vec{\pi}, \tag{5.146}$$

where

$$\vec{\beta} \equiv \frac{\vec{\beta}_R - \vec{\beta}_L}{2}, \qquad \vec{\alpha} \equiv \frac{\vec{\beta}_R + \vec{\beta}_L}{2} \tag{5.147}$$

are associated, respectively, with the axial and vector generators. The meson terms in (5.141) and (5.143) are invariant under $SO(4)$ for $a = 0$, with $(\sigma, \vec{\pi})$ transforming as a vector. This is not a new symmetry, but just reflects the equivalence $SO(4) \sim SU(2) \times SU(2)$. Of course, the vector $SU(2)$ symmetry is preserved even for $a \neq 0$.

The minimization of the scalar potential

$$V(\sigma, \vec{\pi}) = \frac{\mu^2}{2} \left(\sigma^2 + \vec{\pi}^2 \right) + \frac{\lambda}{4} \left(\sigma^2 + \vec{\pi}^2 \right)^2 - a\sigma \tag{5.148}$$

is straightforward, i.e.,

$$\begin{aligned} \frac{dV}{d\sigma} &= \left[\mu^2 + \lambda \left(\sigma^2 + \vec{\pi}^2 \right) \right] \sigma - a = 0 \\ \frac{dV}{d\pi^i} &= \left[\mu^2 + \lambda \left(\sigma^2 + \vec{\pi}^2 \right) \right] \pi^i = 0 \end{aligned} \tag{5.149}$$

at the minimum. In the chiral limit, $a = 0$, one has

$$\langle 0|\pi^i|0\rangle = 0, \qquad \langle 0|\sigma|0\rangle \equiv \nu = \begin{cases} 0, & \mu^2 > 0 \\ \sqrt{\frac{-\mu^2}{\lambda}}, & \mu^2 < 0, \end{cases} \tag{5.150}$$

where we assume that any nonzero VEV is in the σ direction to ensure a smooth $a \to 0$ limit (Dashen, 1971). For $a \neq 0$, one has that $\langle 0|\sigma|0\rangle \equiv \nu \neq 0$ and $\langle 0|\pi^i|0\rangle = 0$. The precise value of ν can be obtained by solving the cubic equation in (5.149). However, we have no need for an explicit expression. For both $a = 0$ and $a > 0$ one can expand $V(\nu + \sigma', \vec{\pi})$ to obtain

$$m_\pi^2 = \mu^2 + \lambda\nu^2 \Rightarrow m_\pi^2 \nu = a. \tag{5.151}$$

Again, this displays the Goldstone alternative, i.e., either $m_\pi^2 \to 0$ or $\nu \to 0$ in the symmetry limit $a \to 0$. One also has that a dynamical nucleon mass

$$m_\psi = g_\pi \nu \tag{5.152}$$

is generated for $\nu \neq 0$. The σ' mass is $\sqrt{-2\mu^2}$ for $a = 0$. However, the σ' is not a realistic aspect of the model, so the exact value is not important. From (3.46) the axial current is

$$A_\mu^i = \bar{\psi}\gamma_\mu\gamma^5 \frac{\tau^i}{2}\psi + (\partial_\mu\sigma)\,\pi^i - (\partial_\mu\pi^i)\,\sigma. \tag{5.153}$$

From (5.106) this suggests that the pion decay constant is related to ν by (5.110), at least up to renormalization effects. Then, from the equations of motion or from (3.50) on page 99 one recovers the PCAC relation

$$\partial^\mu A_\mu^i = a\pi^i = \nu m_\pi^2 \pi^i = \frac{f_\pi}{\sqrt{2}} m_\pi^2 \pi^i. \tag{5.154}$$

The phase diagram for the model as a function of μ^2 and a is further discussed in (Lee, 1972).

5.8.5 The Nonlinear σ Model

The PCAC relation leads to various soft pion theorems for the behavior of amplitudes involving low energy pions (Adler, 1965b; Adler and Dashen, 1968; D'Alfaro et al., 1973). However, these can be made more explicit, and the relation of the pseudo-Goldstone boson pion fields to the SSB more obvious, by redefining the fields (Weinberg, 1967b). There are various possible reparametrizations (see Donoghue et al., 2014, for a very nice discussion). One closely related to the discussion in Section 3.3.5 is to define fields η and $\vec{\xi}$ such that

$$M = U(\vec{\xi})^{1/2}\left(\frac{\nu+\eta}{\sqrt{2}}\right) I U(\vec{\xi})^{1/2} = \left(\frac{\nu+\eta}{\sqrt{2}}\right) I U(\vec{\xi}), \tag{5.155}$$

where $\vec{\xi}$ is interpreted as the pion field and

$$U(\vec{\xi}) \equiv e^{i\vec{\xi}\cdot\vec{\tau}/\nu}. \tag{5.156}$$

Then, $\vec{\xi}$ disappears from the meson self-interaction terms, yielding

$$\begin{aligned}\mathcal{L}_\sigma &= \bar{\psi}_L i \not\partial \psi_L + \bar{\psi}_R i \not\partial \psi_R + \frac{1}{2}(\partial_\mu \eta)^2 + \frac{(\nu+\eta)^2}{4} \text{Tr}\left[(\partial^\mu U)^\dagger (\partial_\mu U)\right] \\ &- \sqrt{2} g_\pi \left(\frac{\nu+\eta}{\sqrt{2}}\right)\left(\bar{\psi}_R U \psi_L + \bar{\psi}_L U^\dagger \psi_R\right) - \frac{\mu^2}{2}(\nu+\eta)^2 - \frac{\lambda}{4}(\nu+\eta)^4,\end{aligned} \tag{5.157}$$

where we have taken $a = 0$. However, $\vec{\xi}$ reemerges as derivative interactions from the $|\partial_\mu U|^2$ term, which can be expanded as a power series in ξ/ν to all orders. The leading terms are

$$\frac{(\nu+\eta)^2}{4}\text{Tr}\left[(\partial^\mu U)^\dagger (\partial_\mu U)\right] = \frac{1}{2}\left(\partial_\mu \vec{\xi}\right)^2 + \frac{1}{6\nu^2}\left[\left(\vec{\xi}\cdot\partial_\mu\vec{\xi}\right)^2 - |\vec{\xi}|^2\left(\partial_\mu\vec{\xi}\right)^2\right]. \tag{5.158}$$

Similarly, one can eliminate the (apparently) non-derivative pion-nucleon interaction by defining new fermion fields

$$\psi'_L = U^{1/2}\psi_L, \qquad \psi'_R = U^{-1/2}\psi_R, \tag{5.159}$$

so that the nucleon terms in $\mathcal{L}_\sigma$ become

$$\begin{aligned}\mathcal{L}_\psi &= \bar{\psi}'_L i \not\partial \psi'_L + \bar{\psi}'_R i \not\partial \psi'_R - \sqrt{2} g_\pi \left(\frac{\nu+\eta}{\sqrt{2}}\right)\left(\bar{\psi}'_R\psi'_L + \bar{\psi}'_L\psi'_R\right) \\ &+ i\bar{\psi}'_L U^{1/2}\left(\not\partial U^{-1/2}\right)\psi'_L + i\bar{\psi}'_R U^{-1/2}\left(\not\partial U^{1/2}\right)\psi'_R.\end{aligned} \tag{5.160}$$

$U^{1/2}$ can again be expanded in a power series in ξ^i/ν. The leading pion-nucleon interactions are (dropping the primes)

$$\mathcal{L}_{\xi\psi} = -\frac{1}{\nu}\bar{\psi}\gamma^\mu\gamma^5\frac{\vec{\tau}}{2}\cdot\left(\partial_\mu\vec{\xi}\right)\psi - \frac{1}{2\nu^2}\bar{\psi}\gamma^\mu\frac{\vec{\tau}}{2}\cdot\left(\vec{\xi}\times\partial_\mu\vec{\xi}\right)\psi. \tag{5.161}$$

By comparing the matrix elements between physical nucleon states of the terms linear in $\vec{\pi}$ and $\vec{\xi}$ in (5.141) and (5.161) one immediately recovers the Goldberger-Treiman relation (5.114).

The σ model, especially as formulated in (5.157), is useful for computing processes involving low energy pions. The η field decouples for large mass and can be ignored, yielding the *nonlinear σ model.* This and equivalent representations can be used as a basis for chiral perturbation theory (Pich, 1995; Bernard, 2008), which summarizes the implications of spontaneously broken chiral symmetry and the associated soft pion theorems for low energy hadronic processes.

5.9 OTHER TOPICS

QCD and the strong interactions are an enormous subject, and we have only been able to scratch the surface here. An incomplete list of other topics that were omitted or only touched on includes the connection to nuclear forces (Epelbaum et al., 2009; Machleidt and Entem, 2011); modern highly-efficient methods for the calculation of scattering amplitudes (e.g., Parke and Taylor, 1986; Witten, 2004; Elvang and Huang, 2015; Dixon, 2014); the connection with string theory (e.g., Domokos et al., 2009); the application of ideas derived from or motivated by string theory, such as the AdS/CFT correspondence (Maldacena, 1998; Klebanov, 2000); and computer-based computational methods for hadronic processes. Some of the packages used for the calculation of matrix elements, event generators, and parton distribution functions are listed in the website section of the bibliography.

The Hadron Spectrum

The spectra, decays, and other properties of ordinary $q\bar{q}$ mesons and qqq baryons and their interpretation in QCD are discussed in, e.g., (De Rujula et al., 1975; Godfrey and Isgur, 1985; Rosner, 2007; Eichten et al., 2008; Klempt and Richard, 2010; Patrignani, 2016).

The $q\bar{q}$ and qqq states in (5.1) are not the only possible color-singlet hadrons (Jaffe, 2005). For example, there could exist *gluonium* or *glueballs* (bound states of gluons), which are hard to observe because of complications from mixing with ordinary $q\bar{q}$ states (Klempt and Zaitsev, 2007; Ochs, 2013).

Exotic quark combinations are also possible, such as *tetraquark* ($qq\bar{q}\bar{q}$) mesons and *pentaquark* ($qqqq\bar{q}$) baryons. In recent years a number of "XYZ" tetraquark states involving

$c\bar{c}$ or $b\bar{b}$, but which are not consistent with ordinary charmonium or bottomonium, have been reported (Olsen, 2015). These could be *molecular*, i.e., consisting of pairs of loosely bound ordinary mesons, i.e., $(q\bar{q})(q\bar{q})$, with each $(q\bar{q})$ a color singlet. They could instead be diquark-antidiquark bound states, with the diquark (qq) in a 3^* (or a 6) of color, and the converse for the $(\bar{q}\bar{q})$ antidiquark. They could also be admixtures of these and could contain, e.g., a $c\bar{c}$ component. The LHCb collaboration(Aaij et al., 2015) has reported two $uudc\bar{c}$ pentaquark states in the $\Lambda_b^0 \to J/\psi\, K^- p$ decay, appearing as resonances in the $J/\psi\, p$ channel. Possible interpretations again include molecular bound states of color singlets, such as $(udc)(u\bar{c})$, or diquark-diquark-antiquark combinations like $(ud)(uc)\bar{c}$. The experimental and theoretical situation for tetraquarks and pentaquarks is reviewed in (Chen et al., 2016).

Other possible exotic hadrons include *dibaryons* ($6q$, such as $uuddss$) and *hybrids* ($q\bar{q}G$). The deuteron can be considered a molecular dibaryon, but no other such exotics have been established.

High Temperature and Density

Techniques such as lattice field theory and the AdS/CFT correspondence have been very useful in extending the study of the strong interactions into new domains, such as high temperature and/or density (Das, 1997; Kogut and Stephanov, 2004; DeGrand and DeTar, 2006; DeTar and Heller, 2009; Ohnishi, 2012; Satz, 2012). Under such conditions there may be a phase transition to a plasma consisting of quarks and gluons, such as presumably existed above some critical temperature T_C of order Λ in the early universe (Kolb and Turner, 1990). The chiral symmetry, which is spontaneously broken at $T = 0$, may also be restored at high T, perhaps at or close to T_C. These extreme conditions are also probed in heavy ion collisions in the RHIC collider at Brookhaven and the ALICE experiment at the LHC (Armesto et al., 2008; Florkowski, 2014), and may be relevant in extreme astrophysical environments such as the cores of neutron stars (Alford et al., 2008; Brambilla et al., 2014).

5.10 PROBLEMS

5.1 The Han-Nambu model (Nambu and Han, 1974) was an alternative to the color-quark model involving integer-charged quarks. Like QCD, it assumed that the u, d, and s quarks each transformed as triplets under the ordinary global $SU(3)$ flavor symmetry, and also as triplets under a second $SU(3)_c$, which we will refer to here as color for simplicity. The $SU(3)_c$ might or might not be gauged. (Nambu and Han actually assumed a $(3, 3^*)$ assignment rather than $(3, 3)$.) Like QCD, it was assumed that the ordinary baryons and mesons were $SU(3)$ singlets. However, they allowed a more general electric charge generator than (3.89), i.e., $Q = T^3 + T^8/\sqrt{3} + aT_c^3 + bT_c^8$, where T^i (T_c^i) are the $SU(3)$ ($SU(3)_c$) generators, with b and c chosen so that the electric charges are integer rather than fractional, i.e.,

$$
Q_{HN} = \begin{matrix} \\ u \\ d \\ s \end{matrix} \begin{matrix} \begin{matrix} R & G & B \end{matrix} \\ \begin{pmatrix} 0 & 1 & 1 \\ -1 & 0 & 0 \\ -1 & 0 & 0 \end{pmatrix} \end{matrix}, \quad \text{vs.} \quad Q_{QCD} = \begin{matrix} \\ u \\ d \\ s \end{matrix} \begin{matrix} \begin{matrix} R & G & B \end{matrix} \\ \begin{pmatrix} \frac{2}{3} & \frac{2}{3} & \frac{2}{3} \\ -\frac{1}{3} & -\frac{1}{3} & -\frac{1}{3} \\ -\frac{1}{3} & -\frac{1}{3} & -\frac{1}{3} \end{pmatrix} \end{matrix}.
$$

(a) Show that the ordinary baryons and mesons have the correct electric charges.
(b) Find b and c.
(c) How are the tests (R, Drell-Yan, $\pi^0 \to 2\gamma$) for the number of colors in Section 5.2 modified?

5.2 Calculate the color- and spin-averaged differential cross section $d\bar{\sigma}/d\cos\theta$ in the CM for $cd \to cd$ in QCD at tree level, where c and d are the charm and down quarks, respectively. Keep m_c but neglect m_d. Express your result in terms of g_s, s, m_c, and $\cos\theta$.
Hint: verify that your result agrees with Table 5.2 in the appropriate limit.

5.3 Suppose that the QCD gauge symmetry is extended to the chiral color group $SU(3)_L \times SU(3)_R$, where the two factors have gauge couplings $g_{L,R}$ and gauge bosons $G^i_{L,R}$, $i = 1 \cdots 8$, and the chiral quarks q_L and q_R transform as $(3,1)$ and $(1,3)$, respectively. Thus, the quark gauge interactions are

$$\mathcal{L}_q = -g_L\, \bar{q}_L\, \vec{\not{G}}_L \cdot \vec{L}\, q_L - g_R\, \bar{q}_R\, \vec{\not{G}}_R \cdot \vec{L}\, q_R,$$

where $L^i = \lambda^i/2$ acts on the color indices and the flavor indices are suppressed. (Do not confuse this with the chiral *flavor* symmetries.) Introduce in addition a complex scalar field ϕ_{ab}, $a, b = 1 \cdots 3$, transforming as $(3, 3^*)$. ϕ can be represented as a 3×3 matrix, as in Problem 3.26.
(a) Write the renormalizable gauge, Yukawa, and potential terms in $\mathcal{L}$ involving ϕ.
(b) Suppose the potential is minimized for $\langle \phi \rangle = v_\phi\, I$, where I is the 3×3 identity matrix. Show that the gauge symmetry is broken to the diagonal (vector) $SU(3)$, i.e., to QCD, and show how the QCD gauge coupling, gauge bosons, and generators are related to those of $SU(3)_L \times SU(3)_R$.
(c) Find the mass of the other gauge bosons (the *axigluons*) and their coupling to the quarks. Show that the interaction is purely axial in the special case $g_L = g_R$.
(d) Show that the original $SU(3)_L \times SU(3)_R$ theory has triangle anomalies. Suggest a simple way to avoid the anomalies, and show how the interactions are modified. (Ignore the electroweak gauge symmetries.)

5.4 Calculate the spin and color-averaged CM cross sections for $q\bar{q} \to Q\bar{Q}$ and $q\bar{q} \to q_0\bar{q}_0$, where Q is a heavy quark with mass m_Q and q_0 is a spin-0 color triplet with mass m_0. Neglect the mass of the initial quarks.

5.5 Suppose the heavy q_0 in Problem 5.4 were stable. Then, $q_0\bar{q}_0$ pairs produced in the early universe could annihilate into quarks and gluons. The relic density of scalar quarks that do not annihilate depends on the thermal average of $\bar{\sigma}(q_0\bar{q}_0 \to q\bar{q})\beta_{rel}$ and $\bar{\sigma}(q_o\bar{q}_0 \to GG)\beta_{rel}$, where $\beta_{rel} = |\vec{\beta}_{q_0} - \vec{\beta}_{\bar{q}_0}|$ (Kolb and Turner, 1990). Calculate these quantities in the CM (*not* the thermal average), assuming that m_0 is much heavier even than the top quark. Show that near threshold, $s \gtrsim 4m_0^2$, the rate into quarks is P-wave suppressed (i.e, is proportional to β^2, where β is the q_0 velocity), while the rate into gluons is not.

5.6 Use a computer algebra program to calculate $|\bar{M}|^2/g_s^4$ for:
(a) $q_r\bar{q}_r \to q_s\bar{q}_s$, $r \neq s$, keeping the masses m_r and m_s.
(b) $GG \to q\bar{q}$ keeping the quark mass m.
(c) $GG \to GG$ (i.e., verify the result in Table 5.2).
(d) Plot the total cross sections in nb (~ 2.6 TeV^{-2}) for $q\bar{q} \to t\bar{t}$ and $GG \to t\bar{t}$ vs $\sqrt{s}$ for $2m_t < \sqrt{s} < 4$ TeV. Take $\alpha_s \sim 0.12$ and $m_t \sim 173$ GeV. Ignore m_q.

5.7 The two-loop RGE for the QCD coupling for n_q quark flavors is

$$\frac{d\alpha_s}{d\ln Q^2} = -b_2\alpha_s^2 - b_3\alpha_s^3,$$

where

$$b_2 = \frac{1}{4\pi}\left[11 - \frac{2n_q}{3}\right], \qquad b_3 = \frac{1}{8\pi^2}\left[51 - \frac{19n_q}{3}\right].$$

(The one-loop coefficient b_2 is $-4\pi b$, where b is defined in (5.34), because the equation is written for α_s rather than for g_s^2.) Show that the solution can be written for large Q^2 as

$$\alpha_s(Q^2) = \frac{1}{b_2 \ln \frac{Q^2}{\Lambda^2}} \left[1 - \frac{b_3}{b_2^2} \frac{\ln \ln(\frac{Q^2}{\Lambda^2})}{\ln \frac{Q^2}{\Lambda^2}} \right] + \mathcal{O}\left(\frac{1}{\ln^3 \frac{Q^2}{\Lambda^2}} \right).$$

5.8 In QCD the strong fine structure constant $\alpha_s(Q^2)$ is predicted to run according to the renormalization group equation in (5.37) (obtained from (5.34)). The measured value at $Q = M_Z \sim 91.2$ GeV is $\alpha_s(M_Z^2) \sim 0.12$.
(a) Calculate the predicted value of $\alpha_s(M_P^2)$, where $M_P \sim 1.2 \times 10^{19}$ GeV is the Planck scale. Assume $n_q = 6$ and neglect all fermion masses, including m_t.
(b) In supersymmetric QCD one adds to the particle content 8 *gluinos*, which are Weyl fermions transforming according to the adjoint representation of $SU(3)$, and $2n_q$ *scalar quarks* (or *squarks*), each of which is a complex spin-0 particle transforming according to the fundamental representation. Calculate how (5.37) is modified in supersymmetric QCD and the new predicted value at M_P (still assuming the value 0.12 at M_Z). Neglect the gluino and squark masses.

5.9 Show that the perturbative (short-distance) part of the non-relativistic potential (defined in Problem 2.33) between a quark and antiquark in a color-singlet state is $V(r) = -\frac{4}{3}\frac{\alpha_s(1/r)}{r}$.

5.10 The well-established charmonium (ψ) and bottomonium (Υ) $L = 0$ spin-triplet states are respectively $\psi(nS), n = 1, 2$ and $\Upsilon(nS), n = 1 \ldots 4$, where $\psi(1S) \equiv J/\psi$, with masses $m_{\psi(nS)} = (3.097, 3.686)$ GeV and $m_{\Upsilon(nS)} = (9.490, 10.023, 10.355, 10.579)$ GeV. Show that these masses can be reproduced to $\lesssim 1\%$ in the nonrelativistic model using the potential in (5.89). Allow m_c and m_b to vary in the vicinity of $m_{\psi(1S)}/2$ and $m_{\Upsilon(1S)}/2$, but fix $\alpha_s = 0.24$ for charmonium, $\alpha_s = 0.18$ for bottomonium, and $\kappa = 0.2$ GeV^2 for both. (More complete studies vary κ and α_s, include relativistic and spin-dependent corrections, and include information from decay rates, as in Problem 5.11).
Hint: solve the radial Schrödinger equation numerically, and iterate the values of m_c and m_b until the mass of the $1S$ state is reproduced.

5.11 The $J/\psi(3100)$ is a narrow charmonium ($c\bar{c}$) resonance with orbital angular momentum $L = 0$ and spin-1, which was predicted (Appelquist and Politzer, 1975) before the observation of charm. Assuming that the J/ψ can be considered a non-relativistic bound state, show that the lifetime for $J/\psi \to e^+e^-$ by one photon is

$$\Gamma \sim \frac{4}{3}\pi N_c e_c^2 \alpha^2 \frac{|\psi(0)|^2}{m_c^2},$$

where we have taken $m_{J/\psi} \sim 2m_c$ and neglected the electron mass. $N_c = 3$ is the number of colors, $e_c = 2/3$ is the electric charge of the c quark, and $\psi(0)$ is the wave function at the origin.

5.12 Derive the QCD equations of motion for q and $\bar{q}$ for three quark flavors using $\mathcal{L}_q$ in (5.90), and use them to prove (5.100).

5.13 The *pion-nucleon σ term* $\sigma_{\pi N}$ is defined as

$$\sigma_{\pi N} = \frac{\hat{m}}{2M_N}\langle N|\bar{u}u + \bar{d}d|N\rangle,$$

where $\hat{m} \equiv (m_u + m_d)/2$. It can be thought of as the contribution of the chiral $SU(2)$-breaking quark masses to the nucleon mass, and it can be measured independently in πN scattering (e.g., Cheng and Li, 1984). One way to estimate $\sigma_{\pi N}$ (Cheng, 1976) is to write the matrix element as

$$\langle N|\bar{u}u + \bar{d}d - 2\bar{s}s|N\rangle + 2\langle N|\bar{s}s|N\rangle,$$

and then invoke the OZI rule (based on simple quark model ideas) to neglect the second term. The first matrix element can be related to the shift in the nucleon mass due to $SU(3)$-breaking, as in (3.103). Show that

$$\sigma_{\pi N} \sim \frac{3}{2}\frac{\hat{m}}{m_s - \hat{m}}\left(M_\Lambda + M_\Sigma - 2M_N\right) \sim 25 \text{ MeV}$$

in the OZI approximation. Note that estimates from πN scattering and from lattice QCD (Kronfeld, 2012) are typically higher than this value, suggesting that the OZI approximation is not very good in this case.

5.14 Derive (5.128).

5.15 (a) Derive the meson GMO formula in (5.132).
(b) Express the elements of the mass matrix in (5.133) in terms of the mixing angle and eigenvalues.
(c) Show how m^2_{88}, m^2_{00}, and m^2_{08} must be related to obtain ideal mixing.
(d) Show that ideal mixing and the assumption that the octet and singlet are degenerate in the $SU(3)$ limit lead to (5.139).

5.16 Show that (5.103), (5.145), and (5.146) are all equivalent for infinitestimal $\vec{\beta}_R = -\vec{\beta}_L = \vec{\beta}$. (The transformation of M under $\vec{\alpha}$ follows from the ordinary $SU(2)$ considerations in Chapter 3.)

5.17 (a) Consider the linear σ model in Section 5.8.4. It is claimed that the most general $SU(2)_L \times SU(2)_R$ invariant renormalizable Lagrangian density is given by (5.143) with $a = 0$. Why is there no need for a term

$$\text{Tr}\left(M^\dagger M\ M^\dagger M\right)?$$

(b) Suppose we have two sets of four real fields $M_1 = (\sigma_1 + i\vec{\pi}_1 \cdot \vec{\tau})/\sqrt{2}$ and $M_2 = (\sigma_2 + i\vec{\pi}_2 \cdot \vec{\tau})/\sqrt{2}$, which transform as

$$M_i \to e^{i\vec{\beta}_R\cdot\vec{\tau}/2} M_i e^{-i\vec{\beta}_L\cdot\vec{\tau}/2}.$$

Write the renormalizable $SU(2)_L \times SU(2)_R$ invariant interactions and mass terms that involve both M_1 and M_2. Include any quadratic terms that mix the two.
(c) Find quadratic terms that, when added to the Lagrangian, would break the symmetry to the diagonal $SU(2)$ group with generators $F^i = F^i_L + F^i_R$.
(d) Find quadratic terms that would break the symmetry from $SU(2)_L \times SU(2)_R$ to $SU(2)_L$.

CHAPTER 6

Collider Physics

DOI: 10.1201/b22175-6

Most of our knowledge of the elementary particles and their interactions comes from experiments performed at accelerators and colliders. In *fixed target experiments* primary particle beams (e.g., e^-, p, ion) from an accelerator collide with a fixed macroscopic target. They have the advantage of high relative fluxes because of the target density. They can also generate secondary beams of particles or antiparticles produced directly in the primary collision, by subsequent decays, or by radiation. These may be unstable or neutral, and include mesons, hyperons, e^+, $\mu^\pm$, ν, $\bar{\nu}$, and γ. *Colliders* have the complementary advantage of allowing higher center of mass energies. For example, in the CERN and SLAC e^+e^- experiments at the Z-pole each beam had an energy of $M_Z/2 \sim 45$ GeV, while the production of a Z utilizing an e^+ beam scattering from atomic electrons would require $E_{e^+} \sim M_Z^2/2m_e \sim 8 \times 10^6$ GeV!

The history and machine physics of accelerators and colliders is extensively described in (Panofsky and Breidenbach, 1999; Sessler and Wilson, 2014; Wiedemann, 2015; Patrignani, 2016), and particle detectors in (Kleinknecht, 1998; Green, 2000; Grupen and Shwartz, 2008). For introductions, see, e.g., (Mann, 2010; Tully, 2011).

Here we will mainly be concerned with the particle physics implications of recent high energy e^+e^-, $e^\pm p$, $p\bar{p}$, and pp colliders.[1] e^+e^- results, especially the Z-pole experiments at LEP and the SLC, will be discussed in Chapter 8. The $e^\pm p$ collider HERA at DESY extended neutral current deep inelastic scattering ($e^\pm p \to e^\pm X$ by γ and Z exchange) into new kinematic regions (Figure 6.1), and also studied the charged current ($e^\pm p \to \nu(\bar{\nu})X$ by W exchange) case.

This chapter considers physics at the Tevatron ($p\bar{p}$ at $\sqrt{s} \sim 2$ TeV) and LHC (pp with $\sqrt{s}$ up to 13-14 TeV), especially the short-distance processes that can be viewed as scatterings of the constituent quarks and gluons. These are essential for testing QCD and determining α_s; probing other aspects of the standard model in the production of electroweak gauge bosons, heavy quarks, and the Higgs boson; and for estimating both the signal and the background in searches for new physics beyond the standard model (BSM).

[1]There are a number of proposals for future higher-energy e^+e^- linear or circular colliders to study the Higgs couplings, search for new physics, and perhaps redo the Z-pole measurements at higher precision. The circular collider proposals include $\sqrt{s} = \mathcal{O}(100$ TeV) pp collisions at later stages. There are even more exotic possibilities, such as a $\mu^+\mu^-$ collider or colliders involving one or two high-energy photon beams produced by backscattering laser light from high-energy electrons.

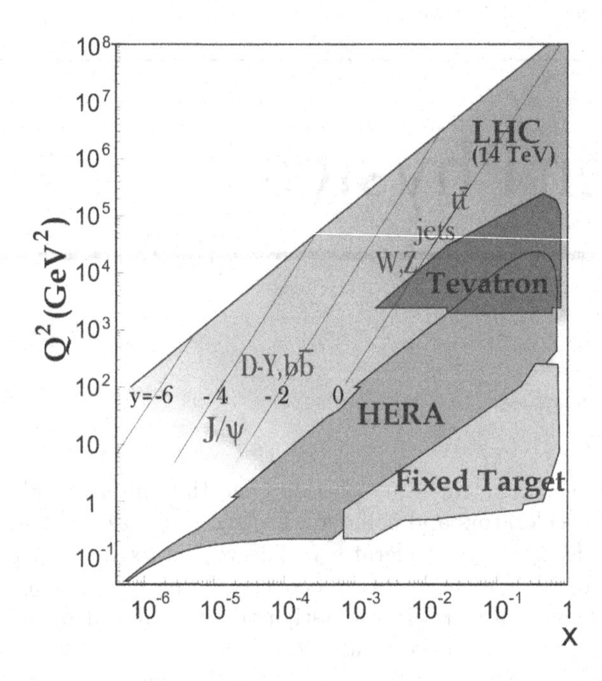

Figure 6.1 Kinematic regions probed by deep-inelastic, pp, and $p\bar{p}$ scattering experiments. Plot courtesy of the Particle Data Group (Patrignani, 2016).

6.1 BASIC CONCEPTS

6.1.1 The Cross Section and Luminosity

The cross section formula (2.51) on page 16 is useful theoretically, but cross sections are actually measured by scattering two beams of particles in a collider or by scattering a beam from a fixed target. In both cases, the reaction rate is $\sigma\mathfrak{L}$, where $\mathfrak{L}$, the instantaneous *luminosity*, is the product of the number of incident particles per unit time and the number of scatterers (in the target or second beam) per unit area (Patrignani, 2016). For example, $\mathfrak{L} = (fN)(nd)$ for a beam consisting of bunches of N particles impinging with frequency f on a fixed target of number density n and thickness d. For a collider in which bunches with $N_1(N_2)$ particles in the first (second) beam collide head-on, $\mathfrak{L} = fN_1N_2/\mathcal{A}$ where f is the crossing frequency and $\mathcal{A}$ is an effective area that characterizes the transverse size and shape of the bunches. An example is given in Problem 6.1. The total number of events is σL, where $L \equiv \int \mathfrak{L}\, dt$ is the total (integrated) luminosity.

Cross sections have units of area, and are conveniently expressed in units of energy^{-2} for theoretical calculations. For experimental considerations, however, it is conventional to work in more practical units based on the barn (b), where $1 \text{ b} \equiv 10^{-24} \text{ cm}^2 = 100 \text{ fm}^2$ is typical of nuclear cross sections. Related units are

$$1 \text{ b} = 10^3 \text{ mb} = 10^6 \ \mu\text{b} = 10^9 \text{ nb} = 10^{12} \text{ pb} = 10^{15} \text{ fb} = 10^{18} \text{ ab}, \tag{6.1}$$

where the prefixes represent milli, micro, nano, pico, femto, and atto, respectively. The cross sections of some typical processes are around 90 nb for $e^+e^- \to \mu^+\mu^-$ at $\sqrt{s} = 1$ GeV (from 2.232), 140 fb for a 20 GeV neutrino scattering from a proton, and 10^8 nb for the total pp

cross section at LHC energies (e.g., Pancheri and Srivastava, 2017). However, most of the pp cross section involves soft (long-distance) processes. The hard (short-distance) events we are mainly concerned with have much smaller σ, e.g., of $\mathcal{O}(1000 \text{ nb})$ for inclusive jet production ($pp \to$ jet $+ X$) at the LHC, 50 nb for inclusive Z production, and 20–50 pb for $pp \to$ Higgs $+X$, as shown in Figure 6.2.

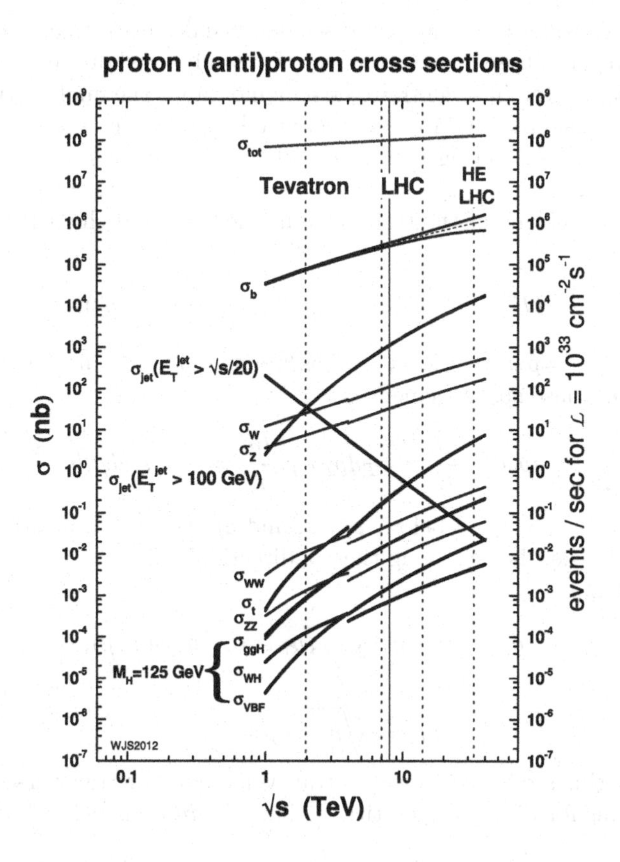

Figure 6.2 pp and $p\bar{p}$ cross sections for SM processes as a function of energy, with LHC energies of 7, 8, and 14 TeV indicated. HE LHC refers to a possible 33 TeV high-energy upgrade. The discontinuities are due to the use of $p\bar{p}$ (pp) at lower (higher) energies. Plot courtesy of W.J. Stirling, private communication.

The corresponding units for $\mathfrak{L}$ are, e.g., "inverse barns per second" (b^{-1}/s):

$$\begin{aligned} 1\ \text{b}^{-1}/\text{s} &= 10^{-3}\ \text{mb}^{-1}/\text{s} = 10^{-6}\ \mu\text{b}^{-1}/\text{s} = 10^{-9}\ \text{nb}^{-1}/\text{s} \\ &= 10^{-12}\ \text{pb}^{-1}/\text{s} = 10^{-15}\ \text{fb}^{-1}/\text{s} = 10^{-18}\ \text{ab}^{-1}/\text{s}. \end{aligned} \tag{6.2}$$

A common rule of thumb for estimating the total number of events in a year's running is to multiply $\sigma\mathfrak{L}$ by 10^7 s, since there are close to $\pi \times 10^7$ s in a year, but experiments typically operate for only about 1/3 of that time.

6.1.2 Collider Kinematics

Here we collect a few definitions and results relevant to kinematics at colliders. For more detail, see (Barger and Phillips, 1997; Han, 2005; Patrignani, 2016).

Single Particle Variables

In Chapter 2 and in considering decay processes one usually uses spherical coordinates to describe the final particle phase space. However, for studying short distance processes in QCD one is usually interested in large transverse momenta for the final particles. Moreover, in a hadron-hadron collision the CM of the incident hadrons does not in general coincide with the CM of the hard scattering subprocess of interest. For these reasons, it is often convenient to work in cylindrical coordinates, with the z axis along the beam direction. p_T and φ represent the magnitude of the transverse momentum and the azimuthal angle of a produced particle, i.e.,

$$p^\mu = (E, p_x, p_y, p_z) = (E, p_T \cos\varphi, p_T \sin\varphi, p_z), \tag{6.3}$$

with $E = \sqrt{m^2 + p^2}$, $p_T = p \sin\theta$, and $p_z = p \cos\theta$, where $p \equiv |\vec{p}|$ and θ is the polar angle. The Lorentz-invariant phase space element is

$$\frac{d^3\vec{p}}{E} = dp_x dp_y \frac{dp_z}{E} = p_T dp_T \, d\varphi \, \frac{dp_z}{E} = p_T dp_T \, d\varphi \, dy, \tag{6.4}$$

where the last form will be defined below. p_T, φ, and dp_z/E are all invariant under longitudinal Lorentz boosts (i.e., boosts along the $\pm z$ direction).

p^μ can be rewritten as

$$p^\mu = (m_T \cosh y, p_T \cos\varphi, p_T \sin\varphi, m_T \sinh y), \tag{6.5}$$

where

$$m_T = \sqrt{m^2 + p_T^2}. \tag{6.6}$$

This quantity is sometimes referred to as the transverse mass or transverse energy, but we will reserve these terms for different quantitites. The *rapidity* y is defined as

$$y = \frac{1}{2} \ln \frac{E + p_z}{E - p_z} = \ln \frac{E + p_z}{m_T} = \tanh^{-1} \frac{p_z}{E}. \tag{6.7}$$

Equation (6.5) follows immediately by taking the cosh and sinh of (6.7). The rapidity is especially useful for expressing the effects of longitudinal boosts. From (6.5), (6.7), and the expression in Problem 1.4, the momentum of the particle as seen in another Lorentz frame moving with velocity $\beta_0 \hat{z}$ is of the same form, except

$$y \to y - y_0 \quad \text{where} \quad y_0 = \frac{1}{2} \ln \frac{1 + \beta_0}{1 - \beta_0} = \tanh^{-1} \beta_0. \tag{6.8}$$

dy is therefore invariant under longitudinal boosts, with $dy = dp_z/E$ from (6.5), leading to (6.4). There is no longitudinal momentum in the frame with $\beta_0 = p_z/E$, where $p^\mu = (m_T, p_T \cos\varphi, p_T \sin\varphi, 0)$.

The measurement of y requires a measurement of both p_z and E, so it is often easier to work with the closely related *pseudorapidity*

$$\eta \equiv \frac{1}{2} \ln \frac{1 + \cos\theta}{1 - \cos\theta} = -\ln\left(\tan\frac{\theta}{2}\right), \tag{6.9}$$

which is measured directly from the direction of the particle, with the range $0 \leq \theta \leq \pi$ corresponding to $+\infty \geq \eta \geq -\infty$. Equation (6.9) is equivalent to

$$\cos\theta = \tanh\eta, \qquad \sin\theta = \frac{1}{\cosh\eta}. \tag{6.10}$$

Rapidity is identical to pseudorapidity for $m = 0$ and reduces to it for $p \gg m$ except for very small angles ($\theta \lesssim m/p$). p_T, φ, and η are therefore useful variables for describing the momenta of the final particles in a high p_T event.[2]

The angular separation between two tracks of momentum p_1 and p_2 is conveniently expressed as the (longitudinal boost-invariant) distance

$$\Delta R = \sqrt{(\Delta y)^2 + (\Delta\varphi)^2} = \sqrt{(y_2 - y_1)^2 + (\varphi_2 - \varphi_1)^2} \tag{6.11}$$

between them in the y-φ plane.[3] (ΔR is often defined in terms of the distance in the $\eta - \varphi$ plane, i.e., as $\sqrt{(\Delta\eta)^2 + (\Delta\varphi)^2}$, rather than (6.11).) Jets can be defined operationally as consisting of the cone of particles separated from the center of the jet by a distance ΔR smaller than some reference value (e.g., 0.7). (Other jet definitions are mentioned in Section 6.2.)

Events are often displayed on plots that indicate the the total transverse energy

$$E_T \equiv E\sin\theta = E/\cosh\eta \tag{6.12}$$

deposited in the electromagnetic and/or hadron calorimeters in cells in the $\eta - \varphi$ (or $y - \varphi$) plane. If the masses of the particles in the jet can be ignored then E_T is just the sum of their m_T. An example of a dijet event from CMS is shown in Figure 6.3.

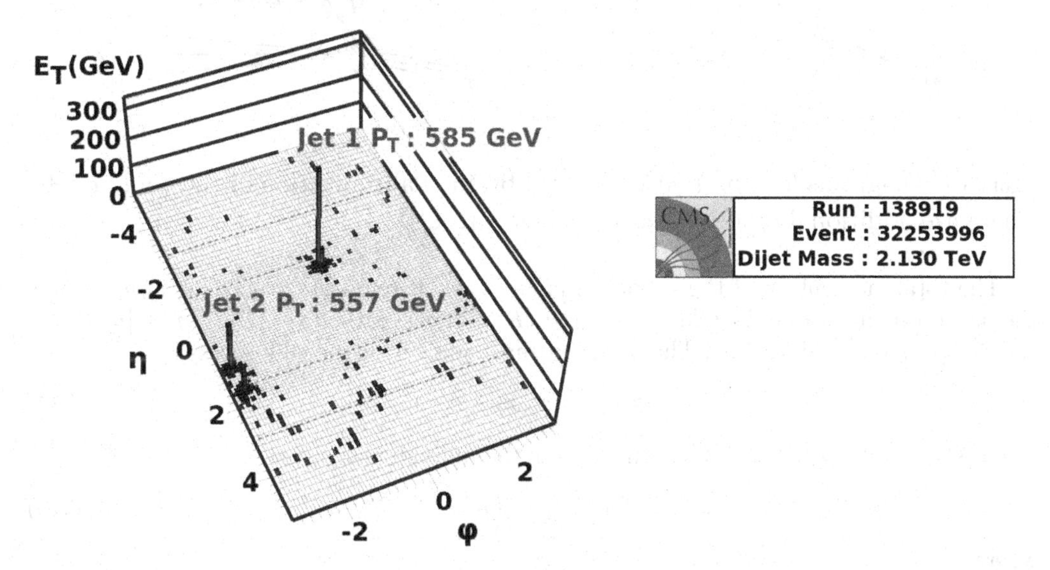

Figure 6.3 An example of an $\eta - \varphi$ plot, showing a CMS dijet event from the 7 TeV run of the LHC. Figure courtesy of CERN.

[2]The various components of the ATLAS and CMS detectors at the LHC have coverage up to $|\eta| \sim 2.5-5$.
[3]For e^+e^- the distance can be defined as the angle between two tracks.

Parton-Parton Scattering

Consider the process $H_A H_B \to F + X$ at a hadron collider, where $H_{A,B}$ refer to the two hadrons (e.g., pp or $p\bar{p}$), F is specific partonic final state, and X are additional unobserved particles. In the short distance regime (e.g., large invariant mass $\hat{s} = p_F^2$, which typically leads to large momenta $\vec{p}_T$ of the final particles), the cross section can be factorized into long distance terms that describe the constituent parton distributions in the initial hadrons and a short distance term describing the hard scattering of the partons to produce F (Drell and Yan, 1971), as illustrated in Figure 6.4. That is,

$$\sigma\left(H_A H_B \to F + X\right) = \sum_{ij} \int dx_A dx_B f_i^A(x_A, \mu_F^2) f_j^B(x_B, \mu_F^2) \sigma_{ij \to F}(x_A, x_B), \qquad (6.13)$$

where i and j label the parton types, such as q, $\bar{q}$, and G; f_i^A and f_j^B are their parton distribution functions, which are the same ones that are relevant to deep inelastic scattering; $x_{A,B}$ are their momentum fractions; and μ_F is the factorization scale, which can be taken to be, e.g., $\sqrt{\hat{s}}$ or p_T. $\sigma_{ij \to F}$ are the parton-parton cross sections, such as $q\bar{q} \to q\bar{q}$ or $GG \to q\bar{q}$. The μ_F^2 evolution of the PDFs and the hadronization of the q and $\bar{q}$ into F will be considered in Section 6.2.

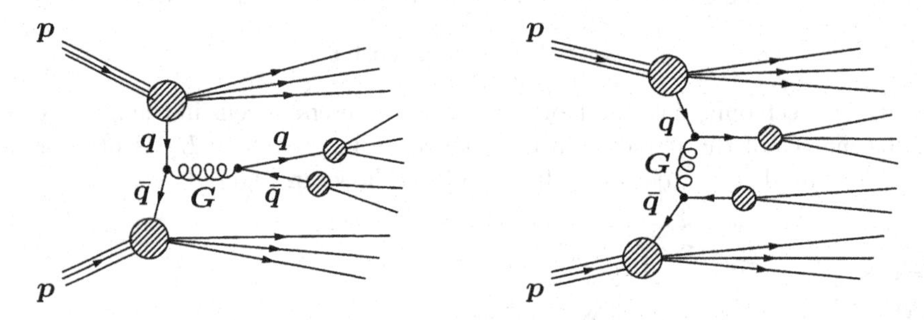

Figure 6.4 Diagrams for $pp \to q\bar{q}X$ followed by the hadronization of the q and $\bar{q}$ jets. There are additional diagrams corresponding to $GG \to q\bar{q}$.

The total momentum of the subprocess $ij \to F$ is therefore $p_F = x_A p_A + x_B p_B$, where the hadron momenta in the CM are $p_{A,B} = (E_{A,B}, \pm\vec{p})$, with $E_A \sim E_B \sim p \equiv |\vec{p}|$ in the (relevant) high energy regime. The invariant mass-square for the subprocess is

$$p_F^2 \equiv \hat{s} \sim x_A x_B s \equiv \tau s, \qquad (6.14)$$

where $s = (p_A + p_B)^2 \sim 4p^2$. One can express the individual $x_{A,B}$ as

$$x_A = \sqrt{\tau} e^{+y}, \qquad x_B = \sqrt{\tau} e^{-y}, \qquad (6.15)$$

where

$$y = \frac{1}{2} \ln \frac{x_A}{x_B} \qquad (6.16)$$

is the rapidity of the CM of the subsystem F (not to be confused with the single-particle rapidity in (6.7)). This interpretation follows from $p_F = (x_A + x_B, 0, 0, x_A - x_B)p$ and the definition (6.5). The velocity of the system in the hadron CM is

$$\beta = \tanh y = \frac{x_A - x_B}{x_A + x_B}. \qquad (6.17)$$

$\sigma_{ij\to F}$ should be invariant under longitudinal boosts, i.e., it depends only $\hat{s}$ and not on x_A or x_B separately.

It is sometimes useful to trade the variables $x_{A,B}$ for $\tau = \hat{s}/s$ and y. In practical applications, the cross section expression (6.13) may involve the factor $\int dx_A dx_B \Theta(\hat{s}-\hat{s}_0)$, where $\hat{s}_0$ is a physical or experimental threshold in the invariant mass of the system. But $\Theta(\hat{s}-\hat{s}_0) = \Theta(\tau - \tau_0) = \Theta(x_A x_B - \tau_0)$, where $\tau_0 = \hat{s}_0/s$. Thus,

$$\int_0^1 dx_A \int_0^1 dx_B \Theta(\hat{s}-\hat{s}_0) = \int_{\tau_0}^1 dx_A \int_{\tau_0/x_A}^1 dx_B = \int_{\tau_0}^1 d\tau \int_{\frac{1}{2}\ln\tau}^{-\frac{1}{2}\ln\tau} dy. \tag{6.18}$$

The relation of the kinematic variables is shown in Figure 6.5.

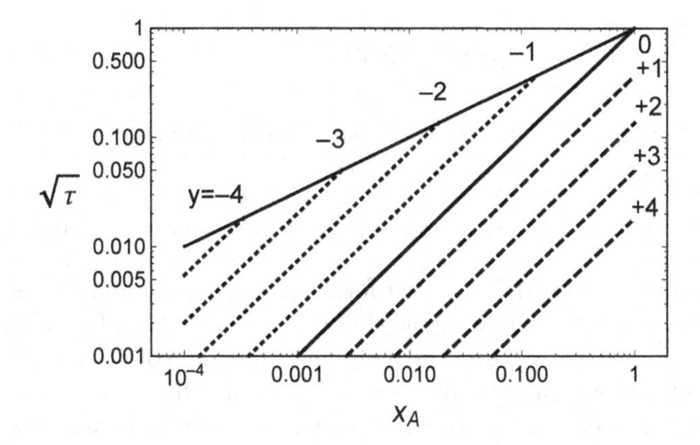

Figure 6.5 The fractional invariant mass $\sqrt{\tau} = \sqrt{\hat{s}/s}$ of a hard subprocess as a function of x_A for various rapidities y. The $x_B - \sqrt{\tau}$ plot is obtained by replacing $y \to -y$. The rapidity range for a given τ is from $\frac{1}{2}\ln\tau$ to $-\frac{1}{2}\ln\tau$, which follows from the requirement that $\tau \le x_{A,B} \le 1$.

Transverse Variables and Missing Momenta

The high p_T particles relevant to a short distance process often involve one or more weakly interacting neutral particles that are stable or long-lived and therefore do not interact or decay in the detector. These could be neutrinos produced in weak decays of other particles, or they could be new "beyond the standard model" particles, such as the lightest supersymmetric partner in R-parity conserving versions of supersymmetry. In principle one can determine the sum of the four momenta of the missing particles by subtracting the total four momentum of the observed particles from the initial momentum $(\sqrt{s}, \vec{0})$. In practice this is impossible at a hadron collider because many undetected beam fragments exit the detector close to the beam direction.[4] This problem can be evaded by working in terms of the transverse components of the momenta, $\vec{p}_T$. In a high p_T process it is a reasonable first approximation to ignore the transverse momenta of the scattering partons, and the transverse momenta of the beam fragments is small. Therefore, the total transverse momentum

[4]In the real world there are other complications ignored here, including large numbers of low energy particles, resolution and threshold effects, inefficiencies in particle identification, cracks in the detector, etc.

of the subprocess can be approximated by zero, and the (missing) transverse momentum of the invisible high p_T particles is given by

$$\vec{p}_T^{\,miss} = -\sum_i \vec{p}_{iT}, \tag{6.19}$$

where the sum is over the observed particles. The magnitude of $\vec{p}_T^{\,miss}$ is the *missing transverse energy* (MET) E_T^{miss} (or $\not{E}_T$). It is sometimes also useful to consider the scalar sum $H_T \equiv \sum_i |\vec{p}_{iT}|$, as well as other variables, especially in searches for new physics such as supersymmetry.

As a simple example, consider $p\bar{p} \to W^- + X$, where $W^- \to e^-\bar{\nu}_e$ and all of the high p_T particles in X are measured. In that case, one can determine the transverse momentum of the unobserved $\bar{\nu}_e$

$$\vec{p}_{\bar{\nu}_e T} = -\vec{p}_{eT} - \sum_{i\in X} \vec{p}_{iT}. \tag{6.20}$$

Since $p_{\bar{\nu}_e z}$ is not known, one cannot directly reconstruct the mass of the W. However, one can determine its *transverse mass* M_{WT}, defined by

$$M_{WT}^2 \equiv (m_{\bar{\nu}_e T} + m_{eT})^2 - (\vec{p}_{\bar{\nu}_e T} + \vec{p}_{eT})^2 \sim 2p_{\bar{\nu}_e T}p_{eT}(1 - \cos\varphi_{e\nu}), \tag{6.21}$$

where the m_T are defined in (6.6). In the last form $\varphi_{e\nu} = \varphi_e - \varphi_{\bar{\nu}_e}$ and we have neglected the lepton masses. It is straightforward to show that $0 \leq M_{WT} \leq M_W$. From a single event one cannot determine M_W. However, from the endpoint and shape of the M_{WT} distribution for a large number of events one can obtain an excellent value. Of course, backgrounds and the finite W width must be taken into account. For more complicated examples, e.g., involving two missing particles, see (Barger and Phillips, 1997; Han, 2005).

6.1.3 Soft Processes in Hadron-Hadron Scattering

Most of the events in high-energy pp, $p\bar{p}$, and other hadron-hadron reactions are soft, characterized by low momentum transfers and (typically) low multiplicity. They are associated with long distance effects and are best described by phenomenological models such as Regge theory. However, they are of great practical relevance in collider experiments, in which the particles from the short-distance processes are only a tiny fraction of those that are actually observed. At LHC energies around 25% of the scatterings are elastic ($pp \to pp$). The remaining inelastic processes can be characterized as singly-diffractive, doubly diffractive, or non-diffractive. In single diffraction one proton remains intact, while the other is gently excited into a small number of particles with low transverse momenta and with rapidities not too different from the original proton. In double diffraction each proton is gently excited. Both types of diffraction (as well as elastic scattering) are therefore characterized by *rapidity gaps*, i.e., regions in the $y - \phi$ plane with little or no activity. They can be approximately described by pomeron exchange.[5] Most of the inelastic events are non-diffractive, in which the produced particles are distributed in y and ϕ with no large rapidity gaps.

It is difficult in practice to measure the total number of inelastic events in a completely "unbiased" way since many of the diffractive scatterings leave little or no trace in the detector. One instead refers to *minimum bias events*. The definition is somewhat vague, but basically refers to all inelastic events observed in the detector with minimal and reasonably

[5]Also possible is central diffractive dissociation, in which both protons remain intact but additional particles are produced with y intermediate between those of the protons, i.e., two rapidity gaps. These may be thought of as due to the "annnihilation" of two pomerons emitted by the protons.

unbiased trigger requirements defined by the experimenters. (Singly diffractive events are often excluded.). Most minimum bias events have low multiplicities and low transverse momenta. Their detailed characterization is important, however, because they occur in the same bunch crossings as the rare hard events, complicating the observations. (Other complications include *pile-up*, i.e., multiple hard interactions involving different protons in the same bunch crossing, and *underlying events*. The latter are associated with the beam remnants, i.e., the partons not involved in the hard scattering, and include their additional low-p_T scatterings (multi-parton interactions) and soft initial state radiation.)

6.2 HADRON-HADRON SCATTERING AT SHORT DISTANCES

Many different types of hard scattering processes $pp \to F + X$ or $p\bar{p} \to F + X$ can occur at hadron colliders. These include pure QCD processes such as $F =$ light quarks or gluons, which subsequently hadronize into jets, as well as standard model processes like $F = \gamma$, Z, or W (Drell-Yan), with the W or Z observed through their subsequently decays to leptons or jets. Heavy quarks, Higgs bosons, and new physics particles are also possible, as are combinations of any of the above. Although the underlying short-distance scattering may appear simple, the full description of the event may be quite complicated. Let us examine the various parts one by one.

The Parton Distribution Functions

The basic formula is given by (6.13), which displays the separation of the cross section into short and long distance terms. This separation can be shown to hold to all orders in perturbation theory, and is expressed in the *factorization theorems* (Collins and Soper, 1987). The PDFs $f_i^{A,B}(x_{A,B}, \mu_F^2)$ are the same as those that enter deep inelastic scattering, provided that one uses the same factorization prescription. Their dependence on the factorization scale μ_F is controlled by the DGLAP equations. These are known to NNLO, as described in Section 5.5.4.

It is sometimes convenient to rewrite (6.13) using (6.18), i.e.,

$$\sigma\,(H_A H_B \to F + X) = \sum_{ij} \int \frac{d\hat{s}}{\hat{s}}\, dy\, \frac{d\mathcal{L}_{ij}}{d\hat{s}\, dy}\, \hat{s}\, \sigma_{ij\to F}(\hat{s}) = \sum_{ij} \int \frac{d\hat{s}}{\hat{s}}\, \frac{d\mathcal{L}_{ij}}{d\hat{s}}\, \hat{s}\, \sigma_{ij\to F}(\hat{s}), \quad (6.22)$$

where

$$\frac{d\mathcal{L}_{ij}}{d\hat{s}\, dy} \equiv \frac{1}{s} f_i^A(x) f_j^B\left(\frac{\tau}{x}\right), \qquad \frac{d\mathcal{L}_{ij}}{d\hat{s}} = \frac{1}{s}\int_\tau^1 \frac{dx}{x}\, f_i^A(x) f_j^B\left(\frac{\tau}{x}\right), \quad (6.23)$$

with $x \equiv \sqrt{\tau} e^y$ and $\tau \equiv \hat{s}/s$. We have suppressed μ_F and the labels A and B on $d\mathcal{L}_{ij}$, and used that $\sigma_{ij\to F}$ is independent of the F rapidity. $\frac{d\mathcal{L}_{ij}}{d\hat{s}} = \frac{\tau}{\hat{s}}\frac{d\mathcal{L}_{ij}}{d\tau}$, which has units of length^2, is known as the *parton luminosity*. It expresses the effectiveness of the partons for hard processes at $\hat{s}$ in a way that is independent of F. We have multiplied and divided by $\hat{s}$ because the logarithmic integral $\int d\hat{s}/\hat{s}$ is better behaved for numerical computations, and because the dimensionless $\hat{s}\,\sigma(\hat{s})$ has a weaker energy dependence. Plots of $d\mathcal{L}_{ij}/d\hat{s}$ (evaluated at $\mu_F^2 = \hat{s}$) and of their ratios at different s (useful for evaluating the relative event rates at different collider energies) may be found in (e.g., Quigg, 2011) or at the websites in the bibliography. Examples are given in Figure 6.6.

In general $d\mathcal{L}_{ij}/d\hat{s}\, dy \neq d\mathcal{L}_{ji}/d\hat{s}\, dy$ for $i \neq j$, and the ij and ji terms must be included separately[6] in the sums in (6.13) and (6.22). For the special case $A = B$ the $d\mathcal{L}/d\hat{s}\,dy$ for ij and ji are related by $y \leftrightarrow -y$ (Problem 6.3). For example, $d\mathcal{L}_{q\bar{q}}/d\hat{s}\,dy$ will be larger in the $y > 0$ direction for pp because there are more q than $\bar{q}$ at large x, while $d\mathcal{L}_{\bar{q}q}/d\hat{s}\,dy$ will be larger for $y < 0$. The integrated luminosities are equal in this case, i.e., $d\mathcal{L}_{ij}/d\hat{s} = d\mathcal{L}_{ji}/d\hat{s}$.

The PDFs for q and $\bar{q}$ are reversed for $\bar{p}$, so that, for example, $\mathcal{L}^{pp}_{q_r q_s} = \mathcal{L}^{p\bar{p}}_{q_r \bar{q}_s}$ (for the same τ and s).

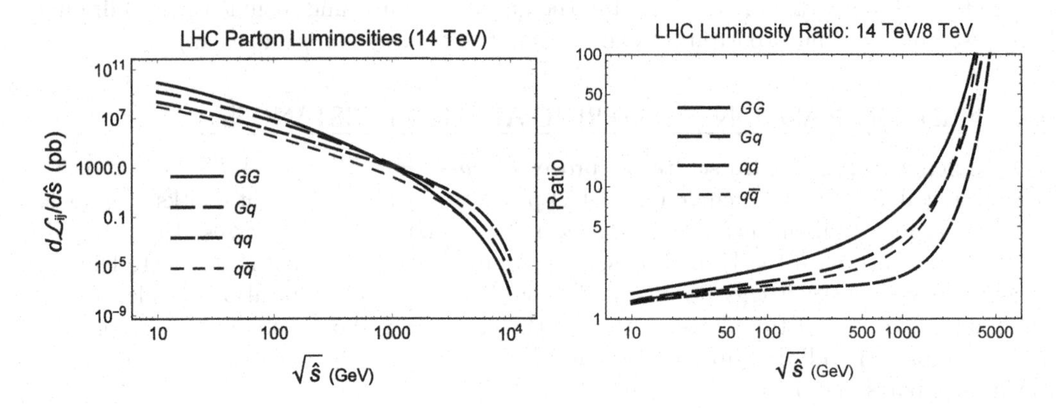

Figure 6.6 Left: LHC parton luminosities at $\sqrt{s} = 14$ TeV as a function of $\sqrt{\hat{s}}$, computed using the MMHT2014 NNLO PDFs (Harland-Lang et al., 2015). Gq refers to $Gu + Gd$, qq to $uu + ud + du + dd$, and $q\bar{q}$ to $u\bar{u} + d\bar{d}$. Right: ratios of the parton luminosities at 14 and 8 TeV.

The Hard Scattering

The hadron-hadron cross sections in (6.13) and (6.22) involve incoherent sums over the color-averaged/summed hard partonic cross sections $\sigma_{ij\to F}(\hat{s})$. The initial parton spins are also averaged (unless one generalizes to polarized hadron beams) and are often summed over final spins as well.

Examples of QCD and standard model parton-level subprocesses include

$$\begin{array}{lllll} qq \to qq, & q\bar{q} \to q\bar{q}, & q\bar{q} \leftrightarrow GG, & Gq \to Gq, & GG \to GG, \\ q\bar{q} \to W, Z, \gamma \to f\bar{f}, & q\bar{q} \to (W, Z, \gamma)G, & q\bar{q} \to WW, & GG \to H, & \end{array} \tag{6.24}$$

where W and Z are the electroweak gauge bosons and H is the Higgs. The leading-order amplitude-squared expressions for QCD processes (for massless partons) are listed in Table 5.2 (one must replace s, t, and u by the invariants $\hat{s}, \hat{t}$, and $\hat{u}$ relevant to the subprocess). More extensive listings, including mass effects and extensions to beyond the standard model processes may be found in, e.g., (Barger and Phillips, 1997; Patrignani, 2016). In practice, $\sigma_{ij\to F}$ must be calculated in perturbation theory to the same order as the PDFs. Many processes are known to NLO[7] and some to NNLO.

The hard inclusive cross sections actually depend not only on the kinematic variables

[6]Some authors define $s\,d\mathcal{L}_{ij}/d\hat{s}\,dy$ as $f_i^A(x)f_j^B(\tau/x) + f_j^A(x)f_i^B(\tau/x)$ for $i \neq j$. We will instead denote this sum by $s\,d\mathcal{L}_{ij+ji}/d\hat{s}\,dy$.

[7]The NLO expressions are sometimes approximated from the LO by a multiplicative K *factor*.

of the observed particles in F, but also on the factorization scale μ_F (e.g., through radiation from an intial parton) and on the renormalization scale μ introduced in Section 5.4. These scales would not appear in physical observables if one could calculate to all orders in perturbation theory, but do enter when one truncates at a finite order, introducing a theoretical uncertainty. One typically chooses central values of μ_F and μ near $\sqrt{\hat{s}}$ or a transverse momentum p_T to minimize some of the logarithmic corrections, and estimates the uncertainties by independently varying them over some range, e.g., by an overall factor of two, with $1/2 < \mu_F/\mu < 2$. Of course, the scale dependence is reduced by going to higher orders. Large logarithmic corrections arise if there is a significant difference in the magnitudes of relevant kinematic quantities.

As an example of (6.22), consider a massive color-singlet vector resonance V_μ of mass M_V and width Γ_V coupling to fermion ψ_a as $\mathcal{L}_a = -g_a \bar{\psi}_a \gamma^\mu \psi_a V_\mu$. It is straightforward to show (Problem 2.27) that the spin and color-averaged Drell-Yan cross section for $a\bar{a} \to b\bar{b}$ via V is

$$\bar{\sigma}_{a\bar{a}\to b\bar{b}}(\hat{s}) = \frac{1}{C_a^2} \frac{12\pi(\hat{s}/M_V^2)\bar{\Gamma}_{a\bar{a}}\bar{\Gamma}_{b\bar{b}}}{(\hat{s}-M_V^2)^2 + M_V^2\Gamma_V^2}, \tag{6.25}$$

where C_a is the color factor (3 for quarks, 1 for leptons),

$$\bar{\Gamma}_{a\bar{a}} = \frac{C_a g_a^2 M_V}{12\pi} \tag{6.26}$$

is the width for $V \to a\bar{a}$, and similarly for $\bar{\Gamma}_{b\bar{b}}$, and we have neglected the fermion masses. When Γ_V is small compared to M_V and to the energy resolution of the detector it is convenient to employ the narrow width approximation in (F.3) on page 523, so that

$$\bar{\sigma}_{a\bar{a}\to V}(\hat{s}) \to \frac{12\pi^2}{C_a^2}\delta(\hat{s}-M_V^2)\frac{\bar{\Gamma}_{a\bar{a}}}{M_V}, \qquad \bar{\sigma}_{a\bar{a}\to b\bar{b}}(\hat{s}) \to \bar{\sigma}_{a\bar{a}\to V}(\hat{s}) B_{b\bar{b}}, \tag{6.27}$$

where $B_{b\bar{b}} = \bar{\Gamma}_{b\bar{b}}/\Gamma_V$ is the branching ratio into $b\bar{b}$. Therefore,

$$\sigma\left(H_A H_B \to V + X\right) = \frac{4\pi^2}{3} \sum_r \left.\frac{d\mathcal{L}_{q_r\bar{q}_r+\bar{q}_r q_r}}{d\hat{s}}\right|_{\hat{s}=M_V^2} \frac{\bar{\Gamma}_{q_r\bar{q}_r}}{M_V}. \tag{6.28}$$

The formulae are easily extended to inclusive and/or differential cross sections. For example, for a two-body final state one can replace $\sigma_{ij}(\hat{s}) \to \int \frac{d\sigma_{ij}(\hat{s},\hat{z})}{d\hat{z}} d\hat{z}$ or $\int \frac{d\sigma_{ij}(\hat{s},\hat{\Omega})}{d\hat{\Omega}} d\hat{\Omega}$, where $\hat{z} = \cos\hat{\theta}$ and $d\hat{\Omega} = d\hat{z}d\hat{\phi}$, with $\hat{\theta}$ and $\hat{\phi}$ the polar and azimuthal scattering angles of one of the final particles in the hard-scattering CM. The hadronic differential cross section for $H_A H_B \to F + X$ for a two-body F is then

$$\frac{d\sigma}{d\hat{s}\,dy\,d\hat{z}} = \sum_{ij} \frac{d\mathcal{L}_{ij}}{d\hat{s}\,dy} \frac{d\sigma_{ij}(\hat{s},\hat{z})}{d\hat{z}}, \qquad \frac{d\sigma}{d\hat{s}\,d\hat{z}} = \sum_{ij} \frac{d\mathcal{L}_{ij}}{d\hat{s}} \frac{d\sigma_{ij}(\hat{s},\hat{z})}{d\hat{z}}. \tag{6.29}$$

One can rewrite the cross section in terms of the (y-invariant) single-particle transverse momentum p_T introduced in (6.3), $p_T = p\sin\theta = \hat{p}\sin\hat{\theta}$, where $\hat{p}$ is the momentum in the parton frame, e.g., $\hat{p} = \sqrt{\hat{s}}/2$ if the masses can be ignored. Then, for example,

$$\frac{d\sigma}{d\hat{s}\,dp_T} = \sum_{ij} \frac{d\mathcal{L}_{ij}}{d\hat{s}} \frac{p_T}{\hat{p}^2} \frac{1}{\sqrt{1-(p_T/\hat{p})^2}} \frac{d\sigma_{ij}(\hat{s},\hat{z})}{d\hat{z}}, \tag{6.30}$$

where $d\sigma_{ij}(\hat{s},\hat{z})/d\hat{z}$ is evaluated at $\hat{z} = \pm\sqrt{1-(p_T/\hat{p})^2}$ (which may be summed). For fixed

$\hat{s}$ there is an enhancement (the *Jacobian peak*) at the maximum value $p_T/\hat{p} = 1$, but the integral over p_T^2 is finite.

Equation (6.13) has been successfully tested and applied in many processes at the Tevatron and the LHC, including Drell-Yan, heavy quark production, high $\vec{p}_T$ jet production, and $W+$ jets (Campbell et al., 2007; Wobisch et al., 2011; Patrignani, 2016).

Hadron Fragmentation

Equation (6.13) or (6.22) can adequately describe the hard scattering process. However, the final quarks and gluons eventually fragment (hadronize) into mesons and baryons, many of which are clustered in jets. For a totally inclusive measurement of the hard process one can equivalently consider either the scattering into the partons or into the final hadrons (quark-hadron duality). However, for more detailed studies, such as the angular, p_T, or number distribution of jets, one must first identify them. One may also want to study the details of the jets themselves, e.g., to determine whether they are more likely associated with quarks or gluons, or with the decays of heavy standard model (e.g., W, Z, t, H) or BSM particles. Adequate modeling of both signals and backgrounds may also require knowledge of the distributions of the individual hadrons.

The hadronization can be parametrized by a *fragmentation function* $D_i^h(E_h/E_i, \mu^2)$, which describes the probability for parton f_i to emit hadron h carrying fraction E_h/E_i of the parton energy; μ is the relevant renormalization or energy scale (Ellis et al., 2003; Buckley et al., 2011; Metz and Vossen, 2016; Patrignani, 2016). One can model the fragmentation as a two step process. First, the initial partons branch or fragment into a number of lower energy ones in a *parton shower*, which can be described by perturbative DGLAP-like equations. The initial hard parton is typically very energetic and far off-shell, with a virtual mass (*virtuality*) comparable to that of the final jet. In each branching, however, the final partons have lower energy and lower virtuality than the parent.

The parton-shower algorithms are efficient at describing branchings that are at relatively low angle or low energy with respect to the initial parton.[8] Harder emissions are usually dealt with by including them as extra contributions to the hard-scattering process, which can be calculated to fixed order, and which can themselves subsequently shower. Of course, care is required to merge the parton-shower and hard scattering and to avoid double counting.

Eventually the partons from the showers are of low enough energy, e.g., $\mathcal{O}(1 \text{ GeV})$, that perturbation theory is no longer valid, and one must resort to a phenomenological long-distance model, which can be tuned with experimental data, to describe the final formation of hadrons. One popular model involves string fragmentation, in which a classical color flux tube is formed between a q and $\bar{q}$, and which can absorb gluons. When enough energy is stored in the tube it creates a $q\bar{q}$ pair and breaks, forming two color singlet objects. Another popular model is cluster hadronization. Following the parton branching, including non-perturbative branching of the gluons into $q\bar{q}$ pairs, hadrons emerge from neighboring $q\bar{q}$ or other color-singlet clusters.

Jet Definitions

One can roughly think of jets as being in one to one correspondence with the partons produced in the hard scattering. In practice, however, it is non-trivial to actually identify the jets, especially in a hadron collider environment, because it is not always certain which

[8] In practice, there are infrared singularities associated with very soft or collinear emissions that can be efficiently dealt with by a resummation technique known as *Sudakov form factors*.

hadrons are associated with a particular jet. This can occur when jets partially overlap, or because of confusion due to multiple scatterings in the same bunch crossing (pile-up) or from hadrons from the underlying event. In fact, there must be some cross-talk between jets or with the underlying event to allow the resulting hadrons to be color neutral.

There are a number of (imperfect) algorithms for identifying the jets in a hard scattering, each of which has its advantantages and disadvantages, and which may be employed depending on the details of the analysis. One important criterion is that an algorithm should be *infrared and collinear safe*, i.e., insensitive to the emission of soft or collinear partons. The major classes are *cone algorithms* and *sequential recombination algorithms*.

The older cone algorithms (Sterman and Weinberg, 1977), extensively used at the Tevatron, define a jet as those hadrons lying within some distance ΔR in (6.11) from the jet center, where in practice the jet center is defined by some iterative process. Not all cone algorithms are infrared and collinear safe, especially at higher orders, so care had to be taken in their use.

Most LHC, e^+e^- and ep analyses utilize sequential recombination algorithms, in which each pair of high p_T particles i and j is characterized by a "distance" d_{ij}, and each particle also has a distance d_{iB} from the incident beams. Jets are identified in a multistep process:

Find the minimum of d_{ij} and d_{iB} for all i, and j.

If the minimum is for pair d_{ij} then i and j are combined into a pseudo-particle. If it is a d_{iB} then define i as a jet and remove it from further consideration.

Iterate this process until all of the initial particles have been merged into jets.

The most common sequential recombination algorithms at hadron colliders define

$$d_{ij} = \min(p_{Ti}^{2n}, p_{Tj}^{2n}) \frac{\Delta R_{ij}^2}{R^2}, \qquad d_{iB} = p_{Ti}^{2n}, \tag{6.31}$$

where ΔR_{ij} is the distance in the $y - \varphi$ or $\eta - \varphi$ plane defined in (6.11), n is an integer, and R is a conveniently chosen constant. $n = +1$, 0, and -1 are known, respectively, as the k_T, *Cambridge-Aachen* (C/A), and *anti-k_T* algorithms, all of which are infrared/collinear safe. The k_T algorithm combines the soft particles first, and models in reverse the likely sequence in which the particles were emitted in a parton shower, while the C/A algorithm is based entirely on the angular separation. However, both lead to irregularly-shaped regions in the $y - \varphi$ plane that are awkward to deal with. The anti-k_T algorithm merges the hardest particles first. Its physical interpretation is less clear, but it leads to more regularly-shaped jet boundaries and is most often employed at the LHC. For specific purposes, however, it is sometimes useful to utilize other definitions.

For a much more detailed discussion of these and related issues, see (Moretti et al., 1998; Ellis et al., 2008; Salam, 2010b; Sapeta, 2016; Patrignani, 2016).

Jet Characteristics and Boosted Decays

Jets can emerge not only from pure QCD processes, but also from the decays of heavy standard model particles, i.e., top quarks, which are too short-lived to directly hadronize, electroweak gauge bosons ($V = W^\pm, Z$), or the Higgs boson H. Heavy BSM particles usually also lead to jets, either produced directly or through intermediate t, V, or H decays. It is therefore important to characterize the jets to determine their likely origin. This is a vast subject and we can only mention a few aspects here.

Jets may be characterized by their *substructure*, referring to the details of the branchings in the parton shower, and by various *jet shape* variables, describing properties such as their mass, angular distributions from the direction of the jet, and number of subjets.

Ordinary QCD jets resulting from light quarks and gluons tend to shower asymmetrically, with most of the energy in each step going to one of the daughters (e.g., because the other is a soft gluon). The angles tend to be small, and the virtuality of the harder parton decreases gradually. Because of different color and splitting factors gluon jets tend to be broader and have more soft radiation than light quark jets.[9] Bottom quark jets, which are extremely common at the LHC, can be efficiently tagged by criteria involving their mass, lifetime, and B decay products (e.g., Voutilainen, 2015).

The energy at the LHC is sufficiently large that the $t, W^\pm, Z$, and H may be produced with p_T (or energy) much larger than their mass ($\mathcal{O}(100-200\text{ GeV})$). This is true in parts of the phase space for standard model production processes, and is even more likely if they result from the decays of very heavy (e.g., $m >$ TeV) BSM particles. The leptonic or hadronic decay products of such highly boosted particles will be highly collimated. Consider, for example, a $t\bar{t}$ pair, each decaying hadronically, $t \to bq\bar{q}'$. If the $t\bar{t}$ pair is produced at rest, one will typically observe six jets, distributed more or less isotropically. If they are sufficiently boosted, on the other hand, the jets will be collimated into two groups, perhaps even appearing to be only two *fat* jets, i.e., the *event shapes* will be very different. Similarly, for a W^- decaying to $\mu^-\bar{\nu}_\mu$ one cannot in general reconstruct the longitudinal momentum of the invisible $\bar{\nu}_\mu$. However, if the W^- is highly boosted one has the additional constraint that the $\bar{\nu}_\mu$ momentum is nearly parallel to that of the μ^-.

Highly boosted decays and detailed studies of the substructure of the resulting jets often allow methods to identify the decaying particle, increase the signal to background ratio, and reduce combinatoric backgrounds. We illustrate by describing a seminal strategy to search for the decay of the Higgs boson into $b\bar{b}$ that inspired a great deal of effort on jet substructure (even though the recipe was not followed in detail by the initial ATLAS or CMS analyses). For $M_H \sim 125$ GeV, $b\bar{b}$ is expected to be the dominant decay mode, with a branching ratio of nearly 60% (Section 8.5). However, the $H \to b\bar{b}$ rate is swamped by QCD and other backgrounds at the LHC, even when the Higgs is produced in association with a Z or $W^\pm$. To ameliorate this difficulty it was suggested in (Butterworth et al., 2008) to concentrate on highly boosted events in which the V and the H are nearly back to back, each with $p_T \gtrsim 200$ GeV. Even though the signal would be greatly reduced, the background would be even more so. In that region the jets from $V + H \to V + b\bar{b}$ or from backgrounds such as $V + G \to V + b\bar{b}$ would appear to be a single fat jet. Relevant events could be identified, e.g., by applying the C/A algorithm with a large $R \sim 1.2$ to capture most of the radiation and tagging the Z or W by its leptonic decays.

The next step is to determine whether the event is more likely due to a Higgs, by undoing (or *unwinding*) the jet recombination sequence one step at a time, starting from the last. Each such step involves the branching of a parent jet into two subjets. If there is a significant *mass drop* (i.e., virtuality drop) between the parent and the more massive subjet, the splitting is not too asymmetric, and the subjets are b-tagged, then one considers the event as a Higgs candidate with its decay into the two quark jets occurring at that branching. If the first two criteria are not satisfied, one discards the less massive subjet and considers the subsequent branching of the more massive one. If one reaches the end of the chain without finding a candidate branching, the event is considered to be background.

The large R value virtually guarantees that the fat jet will be contaminated by radiation

[9]For example, the probability of an ordinary quark or gluon radiating a soft gluon of energy E at a small angle ϑ is $\sim \frac{2\alpha_s C}{\pi}\frac{dE}{E}\frac{d\vartheta}{\vartheta}$, where $C = 4/3$ (quark) or $C = 3$ gluon is the quadratic Casimir of the emitter (e.g., Salam, 2010a).

from the underlying event, decreasing the resolution on the Higgs mass. Some of this is discarded by the unwinding process. However, the event can be further *groomed* (cleaned up) by an additional *filtering* step, in which the jet-finding algorithm is applied again, but this time with a much smaller R (e.g., $\min(0.3, R_{b\bar{b}}/2)$, where $R_{b\bar{b}}$ is the distance between the quark subjets). Keeping only the three hardest jets (to allow for the radiation of a hard gluon from one of the quarks) and applying b tags to the two hardest results in a sample of fairly clean candidate events with a significantly enhanced signal to background (i.e., $S/\sqrt{B}$) ratio.

Many other jet substructure, jet event, and jet grooming (e.g., "pruning" and "trimming") techniques relevant to boosted objects, with applications to Higgs, electroweak boson, top, and BSM physics, are described in detail in (e.g., Abdesselam et al., 2011; Altheimer et al., 2012; Shelton, 2013).

Vector Boson Fusion

We have mainly been concerned with the hard scattering of quarks and gluons. However, the electroweak vector bosons γ, $W^\pm$, and Z can also be radiated from the incident p or $\bar{p}$ (or from the $e^\pm$ in $e^\pm e^-$ or ep colliders), and in some cases can be thought of as additional partons.

First consider e^-e^+. The dominant process at low energies is s-channel annihilation through a photon. However, the cross section falls as $1/s$, as in (2.232) on page 46. There are also *two-photon* diagrams for $e^-e^+ \to e^-e^+F$, where F can be $\ell^-\ell^+$, π^0, $\pi^+\pi^-$, $\cdots$, as shown in Figure 6.7. We encountered such diagrams in considering deep inelastic scattering from a quasi-real (nearly on-shell) photon in Section 5.6, but here we are more interested in the case that both photons are quasi-real. Because of collinear singularities these cross sections grow in energy as a power of $\ln s$ and eventually dominate at high energy despite the extra power of α in the amplitude. To a good approximation the processes can be described by the equivalent photon approximation, that is, by a integral of the cross section $\sigma_{\gamma\gamma\to F}(k_1, k_2)$ for two real transversely-polarized photons of energies $k_{1,2}$ to scatter into F, weighted by the probabilities for the beam particles to radiate them (cf. Equation 6.13). See (Brodsky et al., 1971; Nisius, 2000) and the Cross-Section article in (Patrignani, 2016) for detailed discussions.

Similarly, one can consider the radiation of $V = W^\pm$ or Z from an incident quark (or lepton) at a hadron (or lepton) collider with energy $\sqrt{s} \gg M_V$, with V scattering from a constituent of the other beam particle. At high enough energies this can be approximated by the *effective W (Z) approximation* (Kane et al., 1984; Dawson, 1985; Chanowitz and Gaillard, 1985), i.e., as a flux ("parton distribution function") of nearly on shell transverse (V_T) or longitudinal (V_L) vector bosons times a cross section. In particular, *vector boson fusion* (VBF) describes the process $VV' \to F$ illustrated in Figure 6.7. The W^+W^- or $ZZ \to H$ processes were important in the Higgs discovery, while $W_LW_L \to W_LW_L$ will be described in Section 8.5. The detailed kinematics implies that VBF will typically be accompanied by energetic forward jets associated with the beam remnants, while the color-singlet nature of the V means that there should be a rapidity gap, i.e., relatively little hadronic activity in the central region other than from the F (e.g., Han, 2005).

Collider Observables

Many observables for testing the standard model and searching for new physics are possible at hadron and other colliders, some of which have already been touched on or will be mentioned in subsequent chapters. These include:

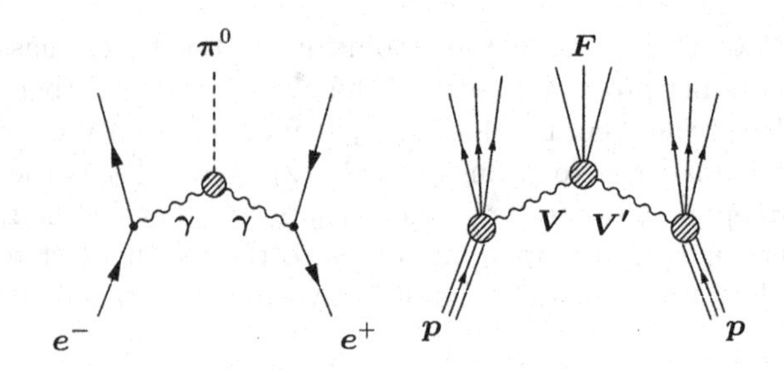

Figure 6.7 Left: two-photon diagram for $e^-e^+ \to e^-e^+\pi^0$. Similar diagrams could produce other final states, such as $\ell^-\ell^+$ or $\pi^+\pi^-$. Right: vector boson fusion (VBF) diagram for $pp \to F+$ forward jets, where V and V' can be $W^\pm$ or Z. Similar diagrams are possible for e^-e^+ at sufficiently high energy.

Jet observables, such as single (inclusive) jet production as a function of p_T and y or η, including boosted jets, jet shape, substructure, composition, and flavor tagging; dijet production as a function of the dijet mass and various angular separations between them; distributions in the numbers and event shapes (Banfi et al., 2010) of multiple jets; forward jets; rapidity gaps; and production in association with other particles.

Production of single leptons (presumably in association with missing $\vec{p}_{\nu T}$), lepton pairs, multiple leptons from BSM particle decays, or collimated *lepton jets* from boosted decays (Arkani-Hamed and Weiner, 2008). Lepton pairs can be same sign (SS), opposite sign (OS), same flavor (SF), or opposite flavor (OF). For example, the Drell-Yan process $q\bar{q} \to \gamma, Z \to \ell^+\ell^-$ should produce OSSF dileptons.

Preferential production of third family particles (t, b, τ), especially in BSM theories associated with electroweak symmetry breaking.

Resonances (bumps in the mass distributions of, e.g., dilepton or diquark pairs). These may be due to the production and decay of new BSM particles such as a new $U(1)'$ gauge boson (Section 10.3.1.), and may be described by a Breit-Wigner distribution or in the narrow width approximation (Appendix F).

Angular distributions, forward-backward asymmetries, and charge asymmetries, e.g., in dilepton or dijet distributions.

Spin correlations, polarizations, and possible T-violating observables.

Associated productions of BSM particles, due to new conserved quantities such as R-parity in supersymmetry (Section 10.2).

Cascade decays, i.e., multistep decay processes of very heavy BSM particles, often associated with kinematic edges and missing energy, such as can occur in supersymmetry.

Missing transverse energy, e.g., due to unobserved heavy stable or quasi-stable particles.

Stable (on collider time scales) strongly interacting particles that either pass out of the detector or stop, possibly decaying much later. Examples include R-hadrons (Kraan et al., 2007; Kang et al., 2008), which are gluinos or heavy exotic quarks bound into hadrons with ordinary quarks, and stabilized by R-parity.

Displaced vertices due the the decays of long-lived particles, such as b quarks or neutral particles in *hidden valley* models (Strassler and Zurek, 2007).

These and other signatures, and their motivations from BSM physics, are reviewed in (Alves et al., 2012).

6.3 PROBLEMS

6.1 (a) Consider an idealized colliding beam experiment in which bunches of particles moving in the z direction collide head-on with frequency f. Show that the instantaneous luminosity defined on page 206 is given by

$$\mathfrak{L} = f d_1 d_2 \int n(x,y)^2 dx\, dy \equiv \frac{f N_1 N_2}{\mathcal{A}},$$

where $d_{1,2}$ are the lengths of the bunches, and $n(x,y)$ is the number density of each bunch as a function of the transverse directions, which is assumed to be the same for each beam and independent of z within the bunch. $N_{1,2}$ are the total number of particles in each bunch, and the effective transverse area $\mathcal{A}$ is defined by the second form.
(b) Suppose $n(x,y)$ is a constant n over a transverse area A and zero outside. Calculate $\mathcal{A}$.
(c) Show that $\mathcal{A} = 4\pi\sigma_x\sigma_y$ for the more realistic Gaussian density profile

$$n(x,y) \propto e^{-\frac{1}{2}[(x/\sigma_x)^2 + (y/\sigma_y)^2]},$$

where $\sigma_{x,y}$ are the RMS radii.

6.2 There are a variety of Fortran, C++, and Mathematica packages available at `www.hep.ucl.ac.uk/mmht/` or other websites listed in the bibliography for generating parton distribution functions obtained from global fits to the data at LO, NLO, and NNLO. Download one of these packages and use it to (qualitatively) reproduce the central values in Figure 5.15. The same package can be used for subsequent problems.

6.3 Plot $d\mathcal{L}_{ij}/d\hat{s}\, dy$ in pb for $ij = \bar{u}u, u\bar{u}, uu, Gu$, and GG as a function of y at the LHC (14 TeV) for $\sqrt{\hat{s}} = 1$ TeV.

6.4 (a) Calculate the cross sections for $p\bar{p} \to t\bar{t}$ at $\sqrt{s} = 2$ TeV (Tevatron) and for $pp \to t\bar{t}$ at 8 and 14 TeV (LHC). Give the individual contributions of $q\bar{q}$ (summed over $q = u, d, s, c$) and GG. The relevant parton level cross sections were calculated in Problem 5.6, can be found in (Patrignani, 2016), or can be inferred from (2.232) for $q\bar{q}$. Use LO PDFs and neglect the running of $\alpha_s \sim 0.1$.
(b) Plot the contributions of $q\bar{q}$, GG, and their sum to $d\sigma(p\bar{p} \to t\bar{t})/dy$ (in units of pb) and to the t-quark p_T distribution $d\sigma(p\bar{p} \to t\bar{t})/dp_T$ (in pb/GeV), both at 2 TeV.

6.5 Suppose there exists a spin-0 resonance R with mass $M_R = 1$ TeV and width $\Gamma_R \ll M_R$, which couples to gluons, photons, $u\bar{u}$, and $d\bar{d}$, with partial (spin and color summed) decay widths $\bar{\Gamma}_{GG}$, $\bar{\Gamma}_{\gamma\gamma}$, $\bar{\Gamma}_{u\bar{u}}$, and $\bar{\Gamma}_{d\bar{d}}$, respectively. Show that the cross section for $pp \to R \to \gamma\gamma$ in the narrow width approximation (Appendix F) is of the form

$$\sigma = \left[\lambda_{GG} B_{GG} + \lambda_{u\bar{u}} B_{u\bar{u}} + \lambda_{d\bar{d}} B_{d\bar{d}}\right] B_{\gamma\gamma},$$

where $B_{ij} \equiv \bar{\Gamma}_{ij}/\Gamma_R$ is the branching ratio into ij. Ignore any contribution of $\gamma\gamma$ to the production. Give explicit expressions for the λ_{ij} in terms of the parton luminosities, M_R, and Γ_R, and calculate their numerical values in fb at LO for the (14 TeV) LHC, assuming $\Gamma_R = 10^{-4} M_R$. Be careful of the color and identical particle factors.

6.6 Perhaps the best known event shape variable is *thrust*,

$$\tau \equiv \max_{\hat{n}_\tau} \frac{\sum_i |\vec{p}_i \cdot \hat{n}_\tau|}{\sum_i |\vec{p}_i|},$$

where the sum is over all of the particles in the event, and the thrust axis $\hat{n}_\tau$ is chosen to maximize the sum. (At hadron colliders one can replace $\vec{p}_i$ by $\vec{p}_{iT}$.) Calculate τ for (a) two back-to-back particles (or highly-collimated jets); (b) many particles, distributed isotropically; (c) three particles in a plane with equal energies and separated by 120°.

CHAPTER 7

The Weak Interactions

DOI: 10.1201/b22175-7

In this chapter we discuss the charged current weak interactions. After a short overview of the developments that led to the standard model, we describe the Fermi theory, which in its modern form still gives an excellent description of a wide variety of weak decay and scattering processes at tree level. (In some cases, radiative corrections, which require the full structure of the standard model, are required.) Some representative processes are calculated and described. Much more detailed treatments may be found in, e.g., (Commins and Bucksbaum, 1983; Renton, 1990; Langacker, 1995; Patrignani, 2016).

7.1 ORIGINS OF THE WEAK INTERACTIONS

We first give a brief history of some of the highlights in the history of weak interactions and the development of the standard electroweak model (for a professional history, see Pais, 1986). The story begins with the discovery of radioactivity by Henri Becquerel in 1896. One of the three types of radioactive decays that were identified was nuclear β decay, in which, apparently, $(N, Z) \to (N-1, Z+1)e^-$ (or in modern terms, subsequent to the discovery of the neutron, $n \to pe^-$). By 1914, experiments by Chadwick established that the e^- spectrum was continuous, suggesting that β decay violated the conservation of energy. As we now understand it, momentum and angular momentum would also have been violated. In 1930 Wolfgang Pauli speculated that the missing energy was carried off by a hypothetical weakly coupled neutral particle, dubbed the neutrino by Enrico Fermi. The neutrino (or anti-neutrino) would be difficult to detect because of its very weak interactions, but not impossible. Its existence was confirmed in 1956 when the electron antineutrino, $\bar{\nu}_e$, was directly observed near a reactor by its rescattering to produce a positron via inverse β decay, $\bar{\nu}_e p \to e^+ n$ (Cowan et al., 1956). The second neutrino type, associated with the muon, was discovered by a group headed by Lederman, Schwartz, and Steinberger at Brookhaven in 1962, who observed its rescattering to produce a muon (Danby et al., 1962). Following the observation of the $\tau^\pm$ in 1975 (Perl et al., 1975), the existence of the third, tau-type, neutrino, ν_τ, was unambiguously inferred from the τ lifetime and decay properties (see, e.g., Langacker, 1989b). However, it was not observed directly until 2000 when the DONUT collaboration at Fermilab observed its rescattering to produce $\tau^\pm$ (Kodama et al., 2001).

In 1934 Fermi proposed a theory of β decay, $n \to pe^-\bar{\nu}_e$, which loosely resembles QED but involves a zero range (non-renormalizable) four-fermion interaction and non-diagonal *charged currents*, as shown in Figure 7.1. The Fermi Hamiltonian density is

$$\mathcal{H} = G_F J_\mu^\dagger J^\mu, \tag{7.1}$$

where

$$J_\mu^\dagger = \bar{p}\gamma_\mu n + \bar{\nu}_e \gamma_\mu e \tag{7.2}$$

is the charge-raising vector current, which describes the transitions $n \to p$ and $e^- \to \nu_e$. The charge-lowering current is

$$J_\mu = \bar{n}\gamma_\mu p + \bar{e}\gamma_\mu \nu_e, \tag{7.3}$$

which describes $p \to n$ and $\nu_e \to e^-$ (or the creation of an $e^-\bar{\nu}_e$ pair). The coefficient

$$G_F \sim 1.17 \times 10^{-5}\ \text{GeV}^{-2} \sim 1.02 \times 10^{-5} m_p^{-2} \tag{7.4}$$

is the *Fermi constant*, which describes the strength of the interaction. It has dimensions of mass^{-2}, characteristic of a non-renormalizable theory. In amplitudes it is typically multiplied by Δ^2, where $\Delta = \mathcal{O}(\text{MeV})$ is a characteristic energy release, so that indeed the interaction is very weak.

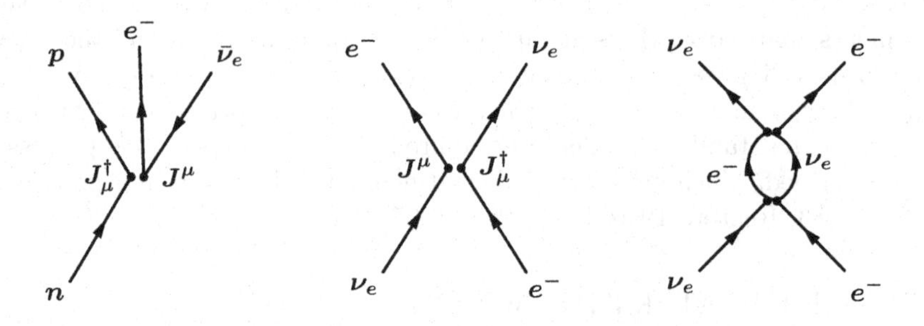

Figure 7.1 Left: four-fermion interaction leading to β decay. The vertices are at the same spacetime point but are displaced for clarity. Middle: four-fermion interaction for $\nu_e e^- \to \nu_e e^-$. Right: a higher-order correction to $\nu_e e^- \to \nu_e e^-$.

The original Fermi theory described β decay and related processes. It has been successfully modified over the years to incorporate new observations,[1] including parity violation (Lee and Yang, 1956; Wu et al., 1957) and the *V − A theory* (Feynman and Gell-Mann, 1958; Sudarshan and Marshak, 1958); μ and τ decays; strangeness changing decays [*Cabibbo mixing* (Cabibbo, 1963)]; the quark model; heavy quarks and mixing [*the Cabibbo-Kobayashi-Maskawa (CKM) matrix*, which allows the incorporation of *CP* violation (Kobayashi and Maskawa, 1973)]; and neutrino mass and mixing. In its modified form it still gives excellent tree level descriptions[2] of a wide variety of charged current mediated decays and scattering processes, including:

Nuclear/neutron β and inverse β decay ($n \to pe^-\bar{\nu}_e; \quad \nu_e n \to e^- p; \quad e^- p \to \nu_e n$)

μ, τ decays ($\mu^- \to e^-\bar{\nu}_e\nu_\mu; \quad \tau^- \to \mu^-\bar{\nu}_\mu\nu_\tau,\ \nu_\tau\pi^-, \cdots$)

π, K decays ($\pi^+ \to \mu^+\nu_\mu,\ \pi^0 e^+\nu_e; \quad K^+ \to \mu^+\nu_\mu,\ \pi^0 e^+\nu_e,\ \pi^+\pi^0$)

hyperon decays ($\Lambda \to p\pi^-; \quad \Sigma^- \to n\pi^-; \quad \Sigma^+ \to \Lambda e^+\nu_e$)

[1] See (Langacker, 1981; Commins and Bucksbaum, 1983; Pais, 1986) for a more complete list of references.

[2] Some of the processes are now well enough measured that one needs to apply higher-order electroweak (e.g., Sirlin and Ferroglia, 2013) and QCD (Buras, 2011) corrections, which often require the full renormalizable standard model to make sense.

heavy quark decays ($c \to s e^+ \nu_e$; $b \to c\mu^- \bar{\nu}_\mu$, $c\pi^-$)

ν scattering ($\nu_\mu e^- \to \mu^- \nu_e$; $\underbrace{\nu_\mu n \to \mu^- p}_{\text{"elastic"}}$; $\underbrace{\nu_\mu N \to \mu^- X}_{\text{deep-inelastic}}$)

However, the Fermi theory violates unitarity at high energy, reflecting its non-renormalizability. For example, the cross section for $\nu_e e^- \to e^- \nu_e$ grows with energy when computed at tree level. In the modern $V - A$ version the middle diagram in Figure 7.1 yields

$$\sigma(\nu_e e^- \to e^- \nu_e) \to \frac{G_F^2 s}{\pi}, \qquad s \equiv E_{CM}^2. \tag{7.5}$$

This is a purely S-wave process, so S-wave unitarity (Appendix C) requires $\sigma < \frac{16\pi}{s}$. Unitarity therefore appears to fail for

$$\frac{E_{CM}}{2} \geq \sqrt{\frac{\pi}{G_F}} \sim 500 \text{ GeV}. \tag{7.6}$$

This is not by itself so serious. In non-relativistic potential scattering, for example, the Born approximation is not unitary, but unitarity is restored by higher-order terms. For the Fermi theory, however, higher-order contributions are divergent, again due to the non-renormalizability. For example, the diagram on the right in Figure 7.1 involves the integral

$$\int d^4k \left(\frac{\not{k}}{k^2}\right)\left(\frac{\not{k}}{k^2}\right), \tag{7.7}$$

which diverges quadratically for $k \to \infty$ (we have neglected the e^- mass and the external momenta, which is valid for large enough k). Although $\nu_e e^- \to e^- \nu_e$ has only been measured at low energies, there is clearly a theoretical inconsistency; the Fermi theory cannot be the full story.

In the *intermediate vector boson theory* (IVB) the four-fermion interaction was eliminated (Yukawa, 1935; Schwinger, 1957). Instead, it was assumed that the process was mediated by a spin-1 particle,[3] analogous to the photon in QED. However, the intermediate bosons $W^\pm$ were assumed to be very massive (compared to the energies of the experiments) and electrically charged, as indicated in Figure 7.2. The coupling to fermions is given by

$$\mathcal{L} = g W_\mu^+ J^{\mu\dagger} + g W_\mu^- J^\mu, \tag{7.8}$$

where g is the coupling strength. For $M_W^2 \gg Q^2 \equiv -q^2$, where q is the momentum transfer, the denominator $q^2 - M_W^2$ of the W propagator can be replaced by $-M_W^2$, and one has effectively a four-fermion interaction. We will see that this reproduces the Fermi theory at tree level with the identification

$$\frac{G_F}{\sqrt{2}} \sim \frac{g^2}{8M_W^2} \text{ for } M_W \gg Q \tag{7.9}$$

(the factors of 2 and $\sqrt{2}$ will become apparent). However, the full propagator leads to a better behaved amplitude for $\nu_e e^- \to e^- \nu_e$ at high energies.

Unfortunately, the difficulties reemerge when we consider processes involving an external W. As was briefly discussed in Section 2.5.1, a vector boson with an elementary mass term

[3]Yukawa actually suggested (in the same paper as the meson theory) that the proposed charged *spin-0 meson* could also have a weak coupling to leptons, leading to β decay.

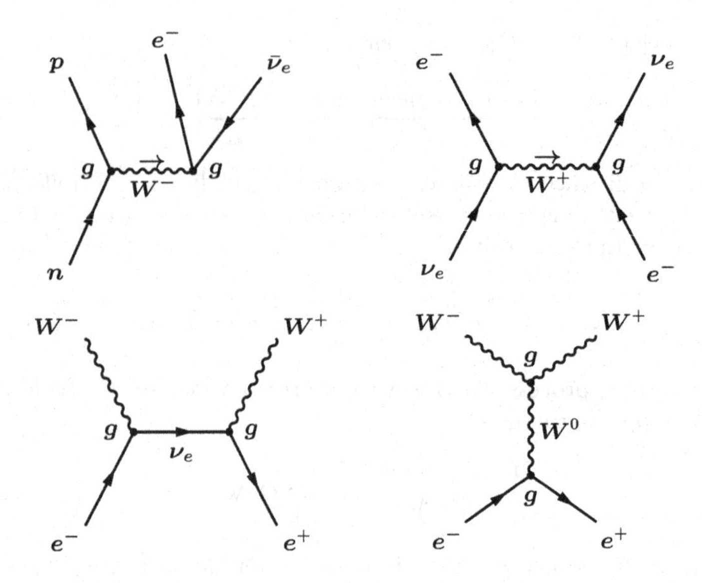

Figure 7.2 Top: intermediate vector bosons mediating β decay and $\nu_e e^- \to e^- \nu_e$. Bottom left: diagram for $e^+e^- \to W^+W^-$ in the IVB theory. Bottom right: additional diagram in the $SU(2)$ theory.

leads to a non-renormalizable theory, and amplitudes involving the longitudinal degree of freedom are badly behaved at high energy due to the polarization vectors $\epsilon_\mu(\vec{k}, 3) \sim k_\mu/M_W$. In particular, the amplitude for $e^+e^- \to W^+W^-$ violates unitarity for $\sqrt{s} \gtrsim 500$ GeV due to the diagram on the lower left in Figure 7.2.

It is possible to resolve these problems by adding more particles in such a way that the bad high energy behaviors cancel. For example, one may add a massive, electrically neutral W^0 boson, so that there is an additional diagram for $e^+e^- \to W^+W^-$, as shown on the bottom right in Figure 7.2. Choosing the neutral current J^0 that describes the W^0-fermion coupling and the triple-vector $W^0W^+W^-$ coupling so that not only $e^+e^- \to W^+W^-$ but also related amplitudes like $\nu_e e^- \to W^0W^-$ are well behaved, one obtains $[J, J^\dagger] \propto J^0$ and 3- and 4- point vector vertices that are equivalent to an $SU(2)$ gauge theory! Thus, one can regard the gauge invariance as a necessary consequence of well-behaved high energy amplitudes.

The $SU(2)$ model just described has no room for electromagnetism and is not realistic. It was extended by Sheldon Glashow in 1961 (Glashow, 1961) to an $SU(2) \times U(1)$ model, with the γ and the prediction of a second neutral boson (the Z) as well as the $W^\pm$. The gauge sector of the Glashow model is in fact the standard model, but at the time there was no satisfactory mechanism for generating masses for the $W^\pm$ and Z. These had to be put in by hand, resulting in difficulties for the behavior of cross sections such as $W^+W^- \to W^+W^-$ at high energy. This was remedied by Steven Weinberg in 1967 (Weinberg, 1967a) and independently by Abdus Salam (Salam, 1968), who invoked the Higgs mechanism to generate their masses and those of the chiral fermions by spontaneous symmetry breaking. Weinberg speculated that the SSB would preserve the renormalizability of the theory. This was proved by 't Hooft and Veltman and others in 1971 ('t Hooft, 1971a,b; 't Hooft and Veltman, 1972; Lee and Zinn-Justin, 1972).

The original Weinberg model considered leptons only, in part because the quark model was not well-established. However, quarks would not have led to a satisfactory extension to hadrons, because only the u, d, and s quarks were known at the time. One could describe the dominant $d \leftrightarrow u$ transitions by assuming they transformed as an $SU(2)$ doublet. The s would have to be an $SU(2)$ singlet because it had no charge-2/3 partner. However, somewhat weaker $s \leftrightarrow u$ charged current transitions were observed experimentally, so there would have to be $d - s$ mixing. That in turn led to strangeness changing neutral currents, i.e., $d - s$ transitions mediated by the Z, which had not been observed experimentally. In particular, both the tree-level and loop diagrams in Figure 7.3 would lead to $K^0 \leftrightarrow \bar{K}^0$ mixing much larger than what was observed. This difficulty could be remedied if there existed a fourth, charge-2/3 *charm (c) quark*. The c and s could transform as an $SU(2)$ doublet. With the d and s transforming the same way, the off-diagonal Z vertex disappeared, and the box diagram was strongly suppressed by the cancellation of the largest parts of the c and u exchange amplitudes [the *GIM mechanism*, (Glashow et al., 1970)]. An early estimate (Gaillard and Lee, 1974b) of the necessary mass, $m_c \sim 1.5$ GeV, was very close to the actual value (see (5.124) on page 194).

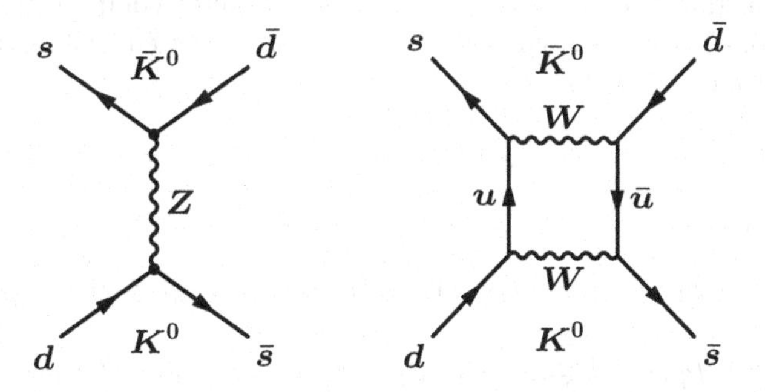

Figure 7.3 Left: $K^0 \leftrightarrow \bar{K}^0$ mixing induced by strangeness changing neutral current vertices in the 3 flavor $SU(2) \times U(1)$ model. Right: box diagram contribution.

However, physicists at the time were reluctant to entertain the possibility of additional quarks (after all, the quark model had been invented to simplify the hadron spectrum). In 1974, however, the J/ψ resonance was discovered simultaneously at Brookhaven and SLAC, and it was soon tentatively identified as a $c\bar{c}$ bound state. The existence of charm was subsequently confirmed by the identification of singly-charmed mesons and in neutrino scattering (see, e.g., Gaillard et al., 1975; Rosner, 1999).

The (flavor diagonal) *weak neutral current processes* mediated by the Z were discovered at CERN and Fermilab in 1973, and extensively probed experimentally in the 1970s and 80s. When combined with the observation of charm, the basic ingredients of the standard model were in place. QCD was established during the 1970s as well. The W and Z were discovered with the expected masses at CERN in 1983, and high precision tests of the Z interactions at the level of loop corrections were carried out at the LEP and SLC colliders at CERN and SLAC, respectively, from 1989–2000. The loop corrections allowed the prediction of the top quark mass, which was confirmed by its direct discovery at Fermilab in 1995. Since $\sim$ 1995 precise tests of the unitarity of the CKM matrix and verification that it describes the observed CP violation were carried out in B physics experiments, especially at Cornell, SLAC, KEK, and later at the Tevatron and the LHC. Finally, hints of nonzero neutrino mass from solar and atmospheric neutrino experiments were confirmed in 1998, and the neutrino sector intensively studied since that time. The Higgs boson, the final ingredient of the standard model (extended to include neutrino mass), was discovered at CERN in 2012.

7.2 THE FERMI THEORY OF CHARGED CURRENT WEAK INTERACTIONS

Let us now describe the Fermi theory of charged current interactions in more detail, mainly in the $V - A$ form that was developed prior to the standard model (the extension to include charm and the third family is straightforward). More detailed discussions of the experiments in β, μ, π, K, and hyperon decays that led to its development, including analyses in a more general framework allowing general S, P, T, V, and A interactions, may be found in (Commins and Bucksbaum, 1983; Renton, 1990; Langacker, 1995; Severijns et al., 2006). The Hamiltonian density is

$$\mathcal{H} = \frac{G_F}{\sqrt{2}} J_\mu^\dagger J^\mu, \tag{7.10}$$

where the Fermi constant G_F is given in (7.4). The $\sqrt{2}$ factor compared to (7.1) is due to the modification from V to $V - A$ currents. The charge raising current is given by the sum of leptonic and hadronic currents,

$$J_\mu^\dagger = J_\mu^{\ell\dagger} + J_\mu^{h\dagger}. \tag{7.11}$$

The leptonic current is

$$J_\mu^{\ell\dagger} = \bar{\nu}_e \gamma_\mu \left(1 - \gamma^5\right) e + \bar{\nu}_\mu \gamma_\mu \left(1 - \gamma^5\right) \mu = 2\left(\bar{\nu}_{eL} \gamma_\mu e_L + \bar{\nu}_{\mu L} \gamma_\mu \mu_L\right), \tag{7.12}$$

where of course $\psi_L = P_L \psi = \frac{1-\gamma^5}{2} \psi$. The $V_\mu - A_\mu$ (i.e., $\gamma_\mu(1 - \gamma^5)$) form corresponds to the maximal amount of parity and charge conjugation violation. One can add $2\nu_{\tau L} \gamma_\mu \tau_L$ to (7.12). The leptonic charge lowering current is

$$J_\mu^\ell = \bar{e} \gamma_\mu \left(1 - \gamma^5\right) \nu_e + \bar{\mu} \gamma_\mu \left(1 - \gamma^5\right) \nu_\mu = 2\left(\bar{e}_L \gamma_\mu \nu_{eL} + \bar{\mu}_L \gamma_\mu \nu_{\mu L}\right). \tag{7.13}$$

The simple extension of (7.2) and (7.13) for the hadronic current would be

$$J_\mu^{h\dagger} \sim \bar{p} \gamma_\mu \left(1 - \gamma^5\right) n \cos\theta_c, \qquad J_\mu^h \sim \bar{n} \gamma_\mu \left(1 - \gamma^5\right) p \cos\theta_c, \tag{7.14}$$

where the $\cos\theta_c$ *Cabibbo angle* factor will be discussed below. This would describe β decay. However, there are other observed hadronic transitions, including hyperon decays such as $\Sigma^- \to n$, and meson decays involving $\pi^+ \to \pi^0, K^+ \to \pi^0$, or $(K^+, \pi^+) \to$ vacuum. These could be described by adding additional terms to (7.14). However, a much simpler and more universal form is obtained by writing it in terms of quark fields. The form prior to the discovery of the c, b, and t quarks was

$$J_\mu^{h\dagger} = \bar{u} \gamma_\mu \left(1 - \gamma^5\right) d' = 2\bar{u}_L \gamma_\mu d'_L, \qquad J_\mu^h = \bar{d}' \gamma_\mu \left(1 - \gamma^5\right) u = 2\bar{d}'_L \gamma_\mu u_L, \tag{7.15}$$

where a sum over color is implied and d' is a rotation of the d and s quark fields

$$d' = d\cos\theta_c + s\sin\theta_c. \tag{7.16}$$

The angle θ_c is the Cabibbo angle, with value $\sin\theta_c \sim 0.23$. It describes the mismatch between the *weak eigenstate* d', i.e., the field that couples to the u quark in $J_\mu^{h\dagger}$, and the *mass eigenstates* d and s. $\tan\theta_c$ measures the ratio between strangeness changing, $\Delta S = 1$, and strangeness conserving, $\Delta S = 0$, transition amplitudes, such as $\Sigma^- \to n e^- \bar{\nu}_e$ and $n \to p e^- \bar{\nu}_e$, respectively. The fact that the hadronic and leptonic currents have the same strength, up to the rotation, reflects *Cabibbo universality.* That is, the squared amplitudes for $\mu^- \to \nu_\mu e^- \bar{\nu}_e$, $n \to p e^- \bar{\nu}_e$, and $\Sigma^- \to n e^- \bar{\nu}_e$ are all different, but they are universal in the sense that if one could ignore strong interaction and mass effects one would have

$$|M(\mu^- \to \nu_\mu e^- \bar{\nu}_e)|^2 = |M(n \to p e^- \bar{\nu}_e)|^2 + |M(\Sigma^- \to n e^- \bar{\nu}_e)|^2, \tag{7.17}$$

since they scale as $1 : \cos^2\theta_c : \sin^2\theta_c \sim 1 : 0.95 : 0.05$, in agreement with the observed strengths. This result emerges naturally in the quark theory, where it is seen to result from the simple rotation in (7.16), but is more mysterious otherwise. Universality generalizes easily to the 4 and 6 quark cases (CKM universality), and in the standard model is seen to be the result of a universal gauge interaction combined with family mixing.

Equation (7.15) incorporates the empirical $\Delta S = \Delta Q$ rule for strangeness changing transitions. For example, in the observed $\Sigma^- \to n$ transition both the strangeness and the hadronic electric charge increase by one unit. However, the $\Delta S = -\Delta Q$ transition $\Sigma^+ \to n e^+ \nu_e$ is *not* observed in nature. Similarly, despite many experimental searches *strangeness changing neutral currents* (or *flavor changing neutral currents (FCNC)*, as they are now called) have never been observed, with the exception of processes that are consistent with being of higher order in the electroweak interactions. Thus, there are no $s \to d$ transitions, which could lead, e.g., to $\Sigma^0 \to n e^+ e^-$. This is illustrated by the weight diagrams in Figure 7.4.

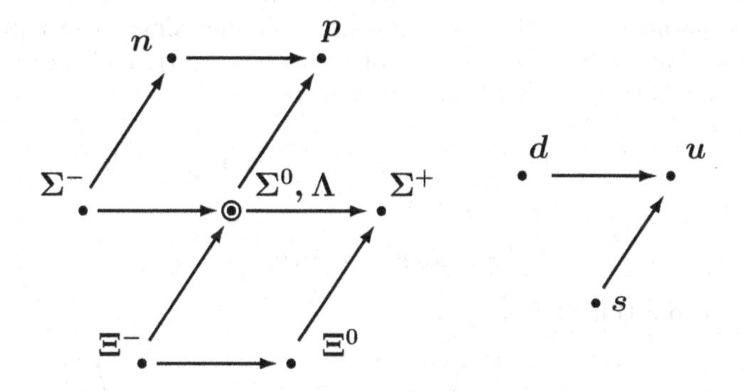

Figure 7.4 Weight diagrams for the baryon octet and for the quarks. The hadronic charge raising current $J^{h\dagger}$ mediates transitions to the right ($\Delta S = 0$) or diagonally up and to the right ($\Delta S = \Delta Q$), but not diagonally up and left (FCNC) or to non-adjacent states ($\Delta S = -\Delta Q$).

Using (7.10), (7.12), and (7.15), the four-fermion Hamiltonian is $H = \int d^3\vec{x}\,\mathcal{H}$, where

$$\mathcal{H} = \frac{G_F}{\sqrt{2}} \left(J_\mu^{\ell\dagger} + J_\mu^{h\dagger}\right)\left(J^{\ell\mu} + J^{h\mu}\right) \equiv \mathcal{H}_l + \mathcal{H}_{sl} + \mathcal{H}_{nl}. \tag{7.18}$$

In (7.18) the *leptonic Hamiltonian* density is

$$\mathcal{H}_l \equiv \frac{G_F}{\sqrt{2}} \left[\left(\sum_m \bar{\nu}_m \gamma_\mu \left(1 - \gamma^5\right) e_m \right) \left(\sum_n \bar{e}_n \gamma^\mu \left(1 - \gamma^5\right) \nu_n \right) \right], \tag{7.19}$$

where $e_m = e$ or μ. It is responsible for such purely leptonic decays and processes as muon decay ($\mu^- \to \nu_\mu e^- \bar{\nu}_e$) and inverse muon decay ($\nu_\mu e^- \to \mu^- \nu_e$). The semi-leptonic density is

$$\begin{aligned}\mathcal{H}_{sl} =& \frac{G_F}{\sqrt{2}} \left[\left(\sum_m \bar{\nu}_m \gamma_\mu \left(1-\gamma^5\right) e_m \right) \left(\bar{d}' \gamma^\mu \left(1-\gamma^5\right) u \right) \right. \\ &+ \left. \left(\bar{u} \gamma_\mu \left(1-\gamma^5\right) d' \right) \left(\sum_m \bar{e}_m \gamma^\mu \left(1-\gamma^5\right) \nu_m \right) \right] .\end{aligned} \tag{7.20}$$

It is responsible for decays of hadrons into final states that include leptons, such as $n \to pe^- \bar{\nu}_e$, $\Sigma^- \to ne^- \bar{\nu}_e$, $K^+ \to \mu^+ \nu_\mu$, and $K^+ \to \pi^0 e^+ \nu_e$, as well as for neutrino scattering and inverse β decay, $\nu_e n \leftrightarrow e^- p$. In the extension to three lepton families it also allows semi-hadronic τ decays, such as $\tau^- \to \nu_\tau \pi^-$. The non-leptonic term

$$\mathcal{H}_{nl} = \frac{G_F}{\sqrt{2}} \left(\bar{u} \gamma_\mu \left(1-\gamma^5\right) d' \right) \left(\bar{d}' \gamma^\mu \left(1-\gamma^5\right) u \right) \tag{7.21}$$

drives decays not involving leptons, e.g., $K^+ \to \pi^+ \pi^0$, $K^+ \to \pi^+ \pi^- \pi^0$, $\Sigma^+ \to p\pi^0$, and $\Lambda^0 \to n\pi^0$ (Commins and Bucksbaum, 1983). It can also generate a parity-violating perturbation in the NN interaction (Ramsey-Musolf and Page, 2006; Haxton and Holstein, 2013; de Vries et al., 2014), leading to effects in polarized pp scattering, nuclear transitions, and an electromagnetic anapole ($\gamma^\mu \gamma^5$) moment (Haxton and Wieman, 2001). It is difficult to calculate matrix elements of $\mathcal{H}_{nl}$ reliably because two hadronic currents are involved.[4]

There is an obvious asymmetry between the quarks and leptons in the 3 quark case. That was remedied subsequently with the discovery of the charm quark, which could partner with the s as an additional term in the hadronic current

$$J_\mu^{h\dagger} = \bar{u} \gamma_\mu \left(1-\gamma^5\right) d' + \bar{c} \gamma_\mu \left(1-\gamma^5\right) s' = 2\bar{u}_L \gamma_\mu d'_L + 2\bar{c}_L \gamma_\mu s'_L, \tag{7.22}$$

where

$$s' = s \cos\theta_c - d \sin\theta_c. \tag{7.23}$$

One can also extend to a third family. Then,

$$J_\mu^\dagger = (\bar{\nu}_e \bar{\nu}_\mu \bar{\nu}_\tau) \gamma_\mu (1-\gamma^5) \begin{pmatrix} e \\ \mu \\ \tau \end{pmatrix} + (\bar{u}\, \bar{c}\, \bar{t}) \gamma_\mu (1-\gamma^5) V_{CKM} \begin{pmatrix} d \\ s \\ b \end{pmatrix}, \tag{7.24}$$

where V_{CKM} is the 3×3 unitary CKM quark mixing matrix. Empirically, the terms in the CKM matrix that mix the third family with the first two are small and can usually be ignored when considering c, s, or d decays. (They are of course critical for b decays.) These extensions will be mentioned in Section 7.2.7 and (along with a further extension to include neutrino mass) described in the context of the full standard model in Chapters 8 and 9.

[4]For example, the $\Delta S = \pm 1$ part can be decomposed into operators with total isospin 1/2 and 3/2. Empirically, the kaon and hyperon decay rates involving the isospin $\Delta I = 1/2$ piece are much larger than those driven by the $\Delta I = 3/2$ term, i.e., the $\Delta I = \frac{1}{2}$ *rule*, or the *octet rule* from the $SU(3)$ perspective (see, e.g., Commins and Bucksbaum, 1983; Donoghue et al., 2014). Especially puzzling are the $K \to 2\pi$ decays, where the ratio of the relevant amplitudes is around 20. Short-distance QCD corrections can account for a factor of ~ 3 (Gaillard and Lee, 1974a; Altarelli and Maiani, 1974), but the rest must be due to long distance non-perturbative effects. There has been some recent progress understanding this from analytic $1/N_c$ (Buras et al., 2014) and lattice calculations (Boyle et al., 2013), and from the AdS/CFT correspondence (Maldacena, 1998; Hambye et al., 2007).

The charged current Fermi Hamiltonian is of the $V - A$ type, i.e., each term is of the form

$$\begin{aligned} H &= \int d^3\vec{x}\,\mathcal{H}(t,\vec{x}) \\ &= \int d^3\vec{x}\,\frac{G_F}{\sqrt{2}}\left[\left(V_{\mu 12}(t,\vec{x})) - A_{\mu 12}(t,\vec{x})\right)\left(V^{\mu}_{34}(t,\vec{x}) - A^{\mu}_{34}(t,\vec{x})\right) + h.c.\right], \end{aligned} \tag{7.25}$$

where

$$V_{\mu ab}(t,\vec{x}) = \bar{\psi}_a(t,\vec{x})\gamma_\mu\psi_b(t,\vec{x}), \quad A_{\mu ab}(t,\vec{x}) = \bar{\psi}_a(t,\vec{x})\gamma_\mu\gamma^5\psi_b(t,\vec{x}) \tag{7.26}$$

are, respectively, vector and axial currents. We saw in Section 2.10 that under space reflection

$$\begin{aligned} V_{\mu ab}(t,\vec{x}) &\to PV_{\mu ab}(t,\vec{x})P^{-1} = V^{\mu}_{ab}(t,-\vec{x}) \\ A_{\mu ab}(t,\vec{x}) &\to PA_{\mu ab}(t,\vec{x})P^{-1} = -A^{\mu}_{ab}(t,-\vec{x}). \end{aligned} \tag{7.27}$$

Changing the sign of the dummy integration variable and using $A^\mu B_\mu = A_\mu B^\mu$,

$$\begin{aligned} H &\to PHP^{-1} \\ &= \int d^3\vec{x}\,\frac{G_F}{\sqrt{2}}\left[\left(V_{\mu 12}(t,\vec{x}) + A_{\mu 12}(t,\vec{x})\right)\left(V^{\mu}_{34}(t,\vec{x}) + A^{\mu}_{34}(t,\vec{x})\right) + h.c.\right], \end{aligned} \tag{7.28}$$

i.e., the interaction changes to $V + A$ and the VA interference terms change sign. Parity is violated maximally, as is manifested by the fact that left-chiral fields (corresponding to left-handed particles and right-handed antiparticles in the massless limit) participate in weak transitions but right-chiral fields do not. It was not initially suspected that parity was not conserved, and in fact most physicists took its conservation for granted.[5] However, in the mid 1950s, the decays $K^+ \to \pi^+\pi^0$ and $K^+ \to \pi^+\pi^-\pi^0$ were both observed. Since the K^+ has spin-0 and the pion has negative intrinsic parity, the first mode would require an even intrinsic parity $\eta_{K^+} = +1$ if parity were conserved, while the second would imply $\eta_{K^+} = -1$.[6] This led Lee and Yang to reexamine the question of whether parity was conserved in the weak interactions (Lee and Yang, 1956), and soon thereafter C. S. Wu et al. established parity violation by observing an asymmetry in the direction of the emitted electron w.r.t. the nuclear spin direction in the β decay of polarized ^{60}Co (Wu et al., 1957).

Similarly, under charge conjugation

$$V_{\mu ab} \to CV_{\mu ab}C^{-1} = -V_{\mu ba} = -V^\dagger_{\mu ab}, \qquad A_{\mu ab} \to CA_{\mu ab}C^{-1} = +A_{\mu ba} = +A^\dagger_{\mu ab}. \tag{7.29}$$

Thus

$$H \to CHC^{-1} = \int d^3\vec{x}\,\frac{G_F}{\sqrt{2}}\left[\left(V_{\mu 12} + A_{\mu 12}\right)^\dagger\left(V^{\mu}_{34} + A^{\mu}_{34}\right)^\dagger + h.c.\right]. \tag{7.30}$$

Charge conjugation therefore changes $V - A$ to $V + A$, just like space reflection. This implies that only left-handed particles and right-handed antiparticles are involved in weak charged

[5]Except for Dirac, who "did not believe in it" (Pais, 1986, p.25).

[6]At first, it was thought that the new modes represented the decays of two distinct particles, $\theta \to \pi^+\pi^0$ and $\tau \to \pi^+\pi^-\pi^0$, but experiments indicated that they must have the same mass, spin (= 0), and lifetime, but opposite parity. This unexpected state of affairs was known as the $\tau - \theta$ puzzle (Commins and Bucksbaum, 1983).

current transitions. Even though C and P are each violated maximally, the product CP restores H to the $V-A$ form and CP is conserved,

$$(CP)H(CP)^{-1} = H. \tag{7.31}$$

For example, the charged current weak interactions of the e_L^- are the same as those of the e_R^+, while e_R^- and e_L^+ do not enter the charged current. Actually, there is observed to be a small amount of CP violation in nature, much weaker than the normal weak interactions. This cannot be accommodated in the Fermi theory, or even its extension to 4 quarks. However, as we will describe in Section 8.6, it can occur in the three family theory.

7.2.1 μ Decay

The muon ($\mu^\pm$) is a heavy version of the $e^\pm$, with mass $m_\mu \sim 105.7$ MeV and lifetime $\tau_\mu = 2.20 \times 10^{-6}$ s. The μ^- decays nearly 100% of the time into $e^-\nu_\mu\bar{\nu}_e$ via the weak charged current, including a small electromagnetic radiative correction leading to $e^-\nu_\mu\bar{\nu}_e\gamma$. (The properties of the μ^+ are analogous by CP invariance.) The diagram for $\mu^- \to e^-\nu_\mu\bar{\nu}_e$ is shown in Figure 7.5. The Lagrangian density[7]

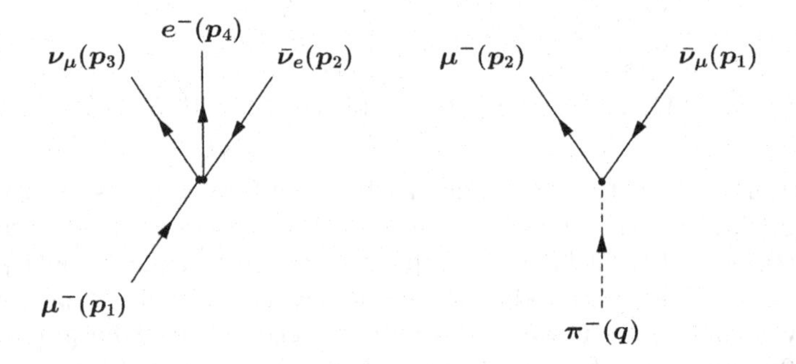

Figure 7.5 Left: Feynman diagram for $\mu^- \to e^-\nu_\mu\bar{\nu}_e$. Right: diagram for $\pi^- \to \mu^-\bar{\nu}_\mu$.

$$-\mathcal{L} = \mathcal{H} = \frac{G_F}{\sqrt{2}}\, \bar{\nu}_\mu \gamma_\mu \left(1-\gamma^5\right) \mu \;\; \bar{e}\gamma^\mu \left(1-\gamma^5\right) \nu_e \tag{7.32}$$

corresponds to a matrix element

$$M = -\frac{iG_F}{\sqrt{2}}\, \bar{u}_3 \gamma_\mu \left(1-\gamma^5\right) u_1 \, \bar{u}_4 \gamma^\mu \left(1-\gamma^5\right) v_2. \tag{7.33}$$

The differential decay rate is

$$d\Gamma = \frac{(2\pi)^4 \delta^4 \left(p_2+p_3+p_4-p_1\right)}{2m_\mu} \frac{d^3p_2 d^3p_3 d^3p_4 |M|^2}{(2\pi)^3\, 2E_2\, (2\pi)^3\, 2E_3\, (2\pi)^3\, 2E_4}. \tag{7.34}$$

[7]It is sometimes convenient to use the Fierz identity in (2.215) on page 43 to rewrite this in the *charge retention form* $-\mathcal{L} = \frac{G_F}{\sqrt{2}}\, \bar{e}\gamma_\mu \left(1-\gamma^5\right) \mu \; \bar{\nu}_\mu \gamma^\mu \left(1-\gamma^5\right) \nu_e$.

$|M|^2$ can be calculated using the standard trace techniques of Chapter 2. Neglecting m_ν,

$$\begin{aligned}|M|^2 =&\frac{G_F^2}{2}\mathrm{Tr}\left[\gamma_\mu\left(1-\gamma^5\right)\left(\not{p}_1+m_\mu\right)\left(\frac{1+\gamma^5\not{s}_\mu}{2}\right)\gamma_\nu\left(1-\gamma^5\right)\not{p}_3\right]\\ &\times\mathrm{Tr}\left[\gamma^\mu\left(1-\gamma^5\right)\not{p}_2\gamma^\nu\left(1-\gamma^5\right)\left(\not{p}_4+m_e\right)\left(\frac{1+\gamma^5\not{s}_e}{2}\right)\right];\end{aligned} \tag{7.35}$$

where we have kept the μ^- and e^- spin projections because many experiments have measured them. For now, however, let us just calculate the total decay rate, obtained by averaging (summing) over the μ^- (e^-) spins (for $m_\nu = 0$ only the left (right)-handed $\nu_\mu(\bar{\nu}_e)$ is produced),

$$|M|^2 \to |\bar{M}|^2 \equiv \frac{1}{2}\sum_{s_\mu, s_e}|M|^2. \tag{7.36}$$

Neglecting the electron mass as well ($m_e/m_\mu \sim 1/200$),

$$\begin{aligned}|\bar{M}|^2 =&\frac{1}{2}\frac{G_F^2}{2}\mathrm{Tr}\left[\gamma_\mu\left(1-\gamma^5\right)\left(\not{p}_1+m_\mu\right)\gamma_\nu\left(1-\gamma^5\right)\not{p}_3\right]\\ &\times\mathrm{Tr}\left[\gamma^\mu\left(1-\gamma^5\right)\not{p}_2\gamma^\nu\left(1-\gamma^5\right)\not{p}_4\right],\end{aligned} \tag{7.37}$$

where the m_μ term doesn't contribute because it involves the trace of an odd number of γ matrices. Despite the formidable appearance, (7.37) is straightforward to evaluate using the identities in (2.172) on page 36. One finds

$$\begin{aligned}|\bar{M}|^2 =&G_F^2\mathrm{Tr}\left[\gamma_\mu\not{p}_1\gamma_\nu\not{p}_3\left(1+\gamma^5\right)\right]\mathrm{Tr}\left[\gamma^\mu\not{p}_2\gamma^\nu\not{p}_4\left(1+\gamma^5\right)\right]\\ =&16G_F^2\left[p_{1\mu}p_{3\nu}+p_{1\nu}p_{3\mu}-g_{\mu\nu}p_1\cdot p_3+i\epsilon_{\mu\tau\nu\omega}p_1^\tau p_3^\omega\right]\\ &\times\left[p_2^\mu p_4^\nu+p_2^\nu p_4^\mu-g^{\mu\nu}p_2\cdot p_4+i\epsilon^{\mu\rho\nu\sigma}p_{2\rho}p_{4\sigma}\right]\\ =&16G_F^2\left[2p_1\cdot p_2p_3\cdot p_4+2p_1\cdot p_4p_2\cdot p_3-\epsilon_{\mu\nu\tau\omega}\epsilon^{\mu\nu\rho\sigma}p_1^\tau p_3^\omega p_{2\rho}p_{4\sigma}\right],\end{aligned} \tag{7.38}$$

where the contraction of the two tensors has been simplified because the first three terms in each are symmetric in μ and ν, and the last is antisymmetric. But

$$\epsilon_{\mu\nu\tau\omega}\epsilon^{\mu\nu\rho\sigma} = -2\left(g_\tau^\rho g_\omega^\sigma - g_\omega^\rho g_\tau^\sigma\right) \tag{7.39}$$

from Table 1.2, so the last term is just

$$2\,p_1\cdot p_2p_3\cdot p_4 - 2\,p_1\cdot p_4p_2\cdot p_3, \tag{7.40}$$

and therefore

$$|\bar{M}|^2 = 64G_F^2p_1\cdot p_2\,p_3\cdot p_4. \tag{7.41}$$

Incidentally, this calculation justifies the identities in (2.175) on page 37, noting that the sign of the last term in (7.38) is reversed for the $1-\lambda\gamma^5$ case.

Assuming that the neutrinos are not detected, we can integrate over their momenta to obtain the unpolarized differential decay rate

$$d\bar{\Gamma} = \frac{4G_F^2}{(2\pi)^5}\frac{d^3p_4p_1^\rho p_4^\sigma}{E_4m_\mu}\int\frac{d^3\vec{p}_2d^3\vec{p}_3}{E_2E_3}\delta^4\left(p_2+p_3-q\right)p_{2\rho}p_{3\sigma}, \tag{7.42}$$

where $q \equiv p_1 - p_4$. To evaluate the integral, we note that

$$I_{\rho\sigma} \equiv \int\frac{d^3\vec{p}_2d^3\vec{p}_3}{E_2E_3}\delta^4\left(p_2+p_3-q\right)p_{2\rho}p_{3\sigma} \tag{7.43}$$

is a second-rank tensor, which can only depend on q. It must be of the form

$$I_{\rho\sigma} = Aq^2 g_{\rho\sigma} + Bq_\rho q_\sigma, \tag{7.44}$$

since $g_{\rho\sigma}$ and $q_\rho q_\sigma$ are the only tensors available. We have extracted a q^2 in the first term so that A is dimensionless. A and B can be obtained by evaluating the much simpler integrals (Problem 7.1)

$$g^{\rho\sigma} I_{\rho\sigma} = 4Aq^2 + Bq^2 = \pi q^2, \qquad q^\rho q^\sigma I_{\rho\sigma} = (A+B)\, q^4 = \frac{\pi}{2} q^4. \tag{7.45}$$

Therefore,

$$A = \frac{\pi}{6}, \quad B = \frac{\pi}{3} \quad \Rightarrow \quad I_{\rho\sigma} = \frac{\pi}{6}\left[q^2 g_{\rho\sigma} + 2q_\rho q_\sigma\right], \tag{7.46}$$

and

$$d\bar{\Gamma} = \frac{2\pi G_F^2}{3(2\pi)^5} \frac{d^3\vec{p}_4}{E_4 m_\mu}\left[(p_1 - p_4)^2\, p_1 \cdot p_4 + 2 p_1 \cdot (p_1 - p_4)\, p_4 \cdot (p_1 - p_4)\right]. \tag{7.47}$$

In the muon rest frame, neglecting the other masses,

$$p_1 = \left(m_\mu, \vec{0}\right), \qquad p_1 \cdot p_4 = m_\mu E_4, \qquad p_1^2 = m_\mu^2, \qquad p_4^2 = 0, \tag{7.48}$$

yielding

$$\begin{aligned} d\bar{\Gamma} &= \frac{2\pi G_F^2}{3(2\pi)^5} \frac{d^3\vec{p}_4}{E_4 m_\mu}\left[\left(m_\mu^2 - 2m_\mu E_4\right) m_\mu E_4 + 2m_\mu \left(m_\mu - E_4\right) m_\mu E_4\right] \\ &= \frac{2\pi G_F^2}{3(2\pi)^5} 2\pi d\cos\theta dE_4 \left[3m_\mu^2 - 4m_\mu E_4\right] E_4^2, \end{aligned} \tag{7.49}$$

where θ is the polar angle of the e^- and $0 \le E_4 \le \frac{m_\mu}{2}$ is its energy. ($E_4 = 0$ corresponds to two back-to-back neutrinos carrying all of the energy, while $E_4 = m_\mu/2$ corresponds to the two neutrinos being in the same direction, so that $p_2 \cdot p_3 = 0$). Integrating over $\cos\theta$ and defining the dimensionless e^- energy $\epsilon = 2E_4/m_\mu$, we obtain the final result for the muon lifetime,

$$\tau^{-1} = \bar{\Gamma} = \frac{G_F^2 m_\mu^5}{192\pi^3} \int_0^1 d\epsilon\, 2\,\epsilon^2 \left[3 - 2\epsilon\right] = \frac{G_F^2 m_\mu^5}{192\pi^3}, \tag{7.50}$$

which can be used to obtain the Fermi constant $G_F = 1.17 \times 10^{-5}$ GeV^{-2} from the observed lifetime τ_μ.

The muon lifetime is known extremely well, $\tau_\mu = 2.1969803(22) \times 10^{-6}$ s, from the MuLan experiment at PSI (Tishchenko et al., 2013), and G_F is a critical parameter for the precision electroweak tests (i.e., for the prediction of the W and Z masses in the standard model), and also for tests of CKM universality. To extract a precise value one must include electron mass and two-loop electromagnetic corrections[8] (some one-loop diagrams are shown in Figure 7.6) to the lifetime formula, yielding (Patrignani, 2016)

$$\tau_\mu^{-1} = \underbrace{\frac{G_F^2 m_\mu^5}{192\pi^3} F\left(\frac{m_e^2}{m_\mu^2}\right)}_{\text{Fermi,with } m_e \neq 0} \times \underbrace{\left[1 + \left(\frac{25}{4} - \pi^2\right)\frac{\hat{\alpha}(m_\mu^2)}{2\pi} + C_2 \frac{\hat{\alpha}^2(m_\mu^2)}{\pi^2}\right]}_{\text{radiative corrections}}, \tag{7.51}$$

[8] The QED radiative corrections to Fermi theory are finite to all orders in α and leading order in G_F (for which there are no $\ln M_W$ effects), provided the usual QED renormalizations are applied (Kinoshita and Sirlin, 1959; Berman and Sirlin, 1962; Sirlin and Ferroglia, 2013). On the other hand, the QED corrections to β decay are logarithmically divergent (they become finite in the full SM). The difference, which has nothing to do with the strong interactions or the neutron mass, is explored in Problem 7.4.

where

$$\begin{aligned}
F(x) &= 1 - 8x + 8x^3 - x^4 - 12x^2 \ln x \\
\hat{\alpha}(m_\mu^2)^{-1} &= \alpha^{-1} - \frac{2}{3\pi} \ln\left(\frac{m_\mu}{m_e}\right) + \mathcal{O}(\alpha) = 135.901 \\
C_2 &= \frac{156815}{5184} - \frac{518}{81}\pi^2 - \frac{895}{36}\zeta(3) + \frac{67}{720}\pi^4 + \frac{53}{6}\pi^2 \ln(2).
\end{aligned} \tag{7.52}$$

$\hat{\alpha}(m_\mu^2)$ is the running QED coupling in the $\overline{MS}$ scheme and $\zeta(3) \sim 1.202$ is the Riemann Zeta function. We have omittted small mixed $m_e - \hat{\alpha}$ and hadronic terms, and a W-propagator correction in the extension to the SM is incorporated in the relation of G_F to $M_{W,Z}$. Using (7.51), one obtains $G_F = 1.1663787(6) \times 10^{-5}$ GeV^{-2}.

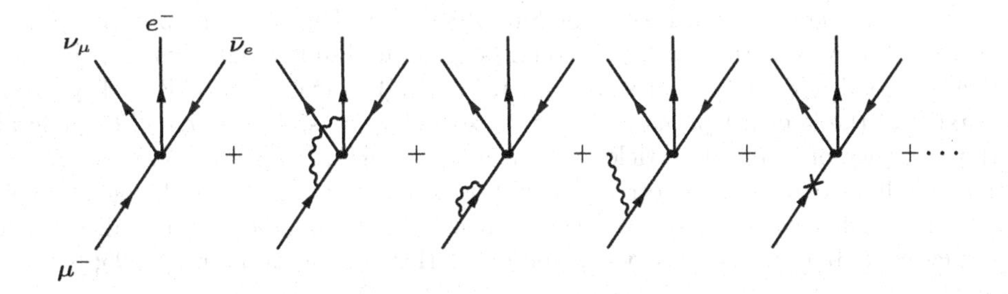

Figure 7.6 Representative low-order diagrams contributing to μ decay in the Fermi theory, including one-loop and initial-state radiation diagrams. The cross represents a mass renormalization counterterm.

The expression in (7.50) can be generalized to allow for a polarized muon and measurement of the electron polarization:

$$d\Gamma = \frac{G_F^2 m_\mu^5}{192\pi^3}\left[2\epsilon^2(3-2\epsilon)\right]\left[1 + \left(\frac{1-2\epsilon}{3-2\epsilon}\right)\cos\theta\right]\left[\frac{1 - \hat{p}_e \cdot \hat{s}_e}{2}\right]\frac{d\epsilon\, d\cos\theta}{2}, \tag{7.53}$$

where $\cos\theta = \hat{p}_e \cdot \hat{s}_\mu$ is the cosine of the angle between the electron momentum and the muon spin direction, and $\hat{p}_e \cdot \hat{s}_e/2$ is the electron helicity. These formulae can of course be extended to include higher-order corrections, the electron mass, and the W propagator (Commins and Bucksbaum, 1983). The $2\epsilon^2(3-2\epsilon)$ factor yields an e^- energy spectrum characteristic of $V-A$ (see Section 7.2.7 and Problem 7.2). The $\hat{p}_e \cdot \hat{s}_\mu$ and $\hat{p}_e \cdot \hat{s}_e$ asymmetries reflect the parity nonconservation. The $\hat{p}_e \cdot \hat{s}_e$ term implies that the e^- helicity is $-\frac{1}{2}$ in the limit $m_e = 0$, another characteristic of $V - A$. The $\cos\theta$ term describes a correlation between the μ^- polarization direction and the electron momentum for a given e^- energy (the coefficient of $\cos\theta$ reverses sign for μ^+ decay). This can be used to determine the polarization direction of a sample of muons, and was used, for example, to determine the muon spin precession in the Brookhaven experiment measuring the muon anomalous magnetic moment (Section 2.12.3). Similarly, the polarization of τ's produced in $e^+e^- \to \tau^+\tau^-$ in the Z-pole experiments at LEP was determined from the angular distributions of the decay products in the analogous leptonic τ decays, such as $\tau^- \to \mu^-\nu_\tau\bar{\nu}_\mu$.

One can also carry out an analysis allowing a more general four-fermion interaction including S, P, and T, as well as V and A interaction terms. There are various parametrizations (see the reviews by Fetscher and Gerber in Langacker, 1995; Patrignani, 2016), such as

$$
\begin{aligned}
H = \frac{4G_F}{\sqrt{2}} & \left(g^V_{LL}\, \bar{e}_L \gamma_\rho \nu_{eL} \bar{\nu}_{\mu L} \gamma^\rho \mu_L \;+\; g^V_{RR}\, \bar{e}_R \gamma_\rho \nu_{eR} \bar{\nu}_{\mu R} \gamma^\rho \mu_R \right.\\
& + g^V_{LR}\, \bar{e}_L \gamma_\rho \nu_{eL} \bar{\nu}_{\mu R} \gamma^\rho \mu_R \;+\; g^V_{RL}\, \bar{e}_R \gamma_\rho \nu_{eR} \bar{\nu}_{\mu L} \gamma^\rho \mu_L \\
& + g^S_{LL}\, \bar{e}_L \nu_{eR} \bar{\nu}_{\mu R} \mu_L \;+\; g^S_{RR}\, \bar{e}_R \nu_{eL} \bar{\nu}_{\mu L} \mu_R \\
& + g^S_{LR}\, \bar{e}_L \nu_{eR} \bar{\nu}_{\mu L} \mu_R \;+\; g^S_{RL}\, \bar{e}_R \nu_{eL} \bar{\nu}_{\mu R} \mu_L \\
& \left. + g^T_{LR}\, \bar{e}_L t_{\rho\sigma} \nu_{eR} \bar{\nu}_{\mu L} t^{\rho\sigma} \mu_R \;+\; g^T_{RL}\, \bar{e}_R t_{\rho\sigma} \nu_{eL} \bar{\nu}_{\mu R} t^{\rho\sigma} \mu_L \right) + h.c.,
\end{aligned}
\tag{7.54}
$$

where $t^{\rho\sigma} = \sigma^{\rho\sigma}/\sqrt{2}$. Equation (7.54) is the most general form assuming lepton-number and lepton-family conservation, but is actually applicable in the more general case in which arbitrary neutrinos and antineutrinos are emitted, if the neutrinos are not observed and their masses are negligible (Langacker and London, 1989). The Fermi theory predicts $g^V_{LL} = 1$ and others $= 0$. However, the other g^V's could be generated in left-right symmetric theories involving a second W_R that couples to $V + A$, i.e., by W_R exchange or $W - W_R$ mixing, or by mixing of the known leptons with heavy exotic fermions. The g^S could be generated from the exchange of a spin-0 particle, such as a non-standard Higgs boson or by R-parity violating couplings in supersymmetry. g^T could be associated with the exchange of a spin-2 particle. The coefficients must be relatively real if T holds. A more detailed discussion of new physics contributions to μ decay is given in (P. Herczeg, in Langacker, 1995; Kuno and Okada, 2001).

There have been many precise experiments on muon decay and inverse muon decay, $\nu_\mu e^- \to \mu^- \nu_e$, including measurements of the electron spectrum and helicity, and of correlations involving the muon spin, at PSI, TRIUMF, and elsewhere. The data are sufficient to establish $|g^V_{LL}| > 0.960$ with (usually stringent) upper limits on the other couplings (Gagliardi et al., 2005; Hillairet et al., 2012; Patrignani, 2016). This excludes the possibility of alternatives to $V - A$ as the dominant contributor to muon decay and inverse decay. One can also search for small deviations from $V - A$, e.g., due to small admixtures of effects involving the types of new physics mentioned above. It is convenient to generalize (7.53) to a general four-fermi interaction,

$$
\begin{aligned}
d\Gamma_{\mu^\mp} = \frac{G_F^2 m_\mu^5}{192\pi^3} \frac{D}{16}\, \epsilon^2 & \left(12(1-\epsilon) + \frac{4}{3}\rho(8\epsilon - 6) \right.\\
& \left. \mp P_\mu \xi \, \cos\theta \left[4(1-\epsilon) + \frac{4}{3}\delta(8\epsilon - 6) \right] \right) d\epsilon \frac{d\cos\theta}{2},
\end{aligned}
\tag{7.55}
$$

where additional terms involving m_e, the $e^\mp$ helicity, and radiative corrections are not displayed. The *Michel spectral parameters* (Michel, 1950; Kinoshita and Sirlin, 1957) ρ, ξ, δ, and η (which appears in the m_e/m_μ corrections) and the overall normalization D are functions of $g^{V,S,T}_{ab}$, while P_μ is the μ polarization from $\pi^\mp \to \mu^\mp \overset{(-)}{\nu}_\mu$. The experimental values in Table 7.1 are in impressive agreements with the predictions of the $V - A$ theory,

$$
\rho = \delta = \frac{3}{4}, \qquad \xi = 1, \qquad \eta = 0, \qquad D = 16, \qquad P_\mu = 1. \tag{7.56}
$$

Very precise measurements of ρ, δ, and $P_\mu \xi$ from the TWIST collaboration at TRIUMF (Hillairet et al., 2012), incorporated in Table 7.1, can also be used to set limits (Bayes et al., 2011) on a possible W_R mass and on a $W - W_R$ mixing angle $|\zeta_{LR}|$ in an extension of the SM to $SU(2)_L \times SU(2)_R \times U(1)$ (Section 10.3). Other tests of the $V - A$ theory of

TABLE 7.1 Experimental values (Patrignani, 2016) of the Michel parameters and their expectations in the $V - A$ or standard models.[a]

Parameter	Experimental Value	$V - A$
ρ	0.74979 ± 0.00026	$\frac{3}{4}$
η	0.057 ± 0.034	0
δ	0.75047 ± 0.00034	$\frac{3}{4}$
$P_\mu\ \xi$	$1.0009^{+0.0016}_{-0.0007}$	1
$P_\mu\ \xi\delta/\rho$	$1.0018^{+0.0016}_{-0.0007}$	1
P_{e^+}	1.00 ± 0.04	1

[a] P_{e^+} is the e^+ longitudinal polarization (twice the helicity).

charged current leptonic interactions are provided by leptonic τ decays, $\tau \to \ell\nu\bar{\nu}$, and by inverse muon decay.

7.2.2 $\nu_e e^- \to \nu_e e^-$

In the Fermi theory the processes $\nu_e e^- \to \nu_e e^-$ and $\bar{\nu}_e e^- \to \bar{\nu}_e e^-$ proceed via the leptonic weak charged current interaction in (7.19), by the diagrams in Figure 7.7. The Lagrangian density is

$$\begin{aligned} -\mathcal{L} = \mathcal{H} &= \frac{G_F}{\sqrt{2}} \bar{\nu}_e \gamma_\mu \left(1 - \gamma^5\right) e \ \bar{e}\gamma^\mu \left(1 - \gamma^5\right) \nu_e \\ &= \frac{G_F}{\sqrt{2}} \bar{\nu}_e \gamma_\mu \left(1 - \gamma^5\right) \nu_e \ \bar{e}\gamma^\mu \left(1 - \gamma^5\right) e \\ &\Rightarrow \quad \frac{G_F}{\sqrt{2}} \bar{\nu}_e \gamma_\mu \left(1 - \gamma^5\right) \nu_e \ \bar{e}\gamma^\mu \left(g_V - g_A \gamma^5\right) e. \end{aligned} \tag{7.57}$$

The second (charge retention) form is obtained using the Fierz identity in (2.215) on page 43. The last form looks ahead to the full standard model, in which there is an additional neutral current contribution. The Fermi theory predicts $g_V = g_A = 1$. The spin-averaged squared

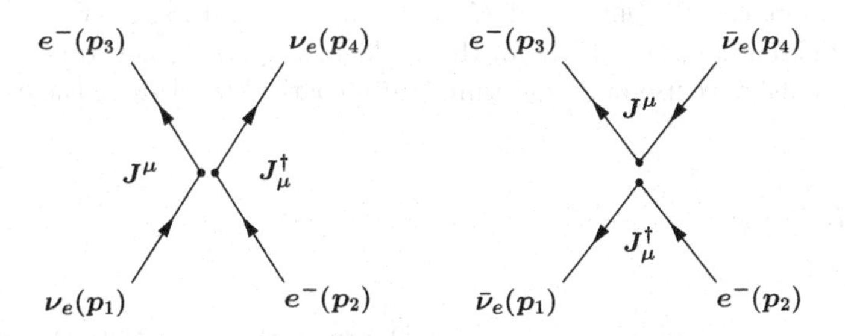

Figure 7.7 Diagrams for $\nu_e e^- \to \nu_e e^-$ and $\bar{\nu}_e e^- \to \bar{\nu}_e e^-$ in the Fermi theory. There are additional weak neutral current contributions in the full standard model.

amplitude can be calculated in the same way as for μ decay. One can use (2.175) to obtain

$$\frac{1}{2}\sum_{s_1s_2s_3s_4}|M|^2 = 16G_F^2\big[(g_V+g_A)^2 p_1\cdot p_2\, p_3\cdot p_4 + (g_V-g_A)^2 p_1\cdot p_3\, p_2\cdot p_4 - (g_V^2-g_A^2)m_e^2\, p_1\cdot p_4\big]. \tag{7.58}$$

(The incident ν_e's all have helicity $-1/2$, so only the e^- spin is averaged. It is convenient to include sums over the ν_e spins, even though the contributions of the $h=+1/2$ states vanish.) Let us define the invariant

$$y = \frac{p_1\cdot p_4}{p_1\cdot p_2} = \begin{cases} \frac{(s-m_e^2)(1+\cos\theta_{13})}{2s}, & \text{CM} \\ \frac{T_e}{E_\nu}, & \text{lab} \end{cases} \tag{7.59}$$

where $s=(p_1+p_2)^2=m_e^2+2m_eE_\nu$, E_ν is the incident ν energy in the electron rest frame (i.e., the lab frame), and T_e is the kinetic energy of the final electron in the lab. y, the fractional energy transfer to the final electron in the lab, runs from 0 to $(1+m_e/2E_\nu)^{-1}$. Then,

$$\begin{aligned} \frac{d\sigma_{\nu e}}{dy} &= \frac{G_F^2 m_e E_\nu}{2\pi}\left[(g_V+g_A)^2+(g_V-g_A)^2(1-y)^2-(g_V^2-g_A^2)\frac{m_e}{E_\nu}y\right] \\ \sigma_{\nu e} &= \int_0^1 \frac{d\sigma}{dy}dy = \frac{G_F^2 m_e E_\nu}{2\pi}\left[(g_V+g_A)^2+\frac{1}{3}(g_V-g_A)^2\right], \end{aligned} \tag{7.60}$$

where we have dropped the last $(m_e y/E_\nu)$ term in the total cross section for simplicity. (The limits on y and the (invariant) cross section formula are most easily derived in the CM and then expressed in lab variables.) Similarly, for $\bar\nu_e e^- \to \bar\nu_e e^-$,

$$\begin{aligned} \frac{d\sigma_{\bar\nu e}}{dy} &= \frac{G_F^2 m_e E_\nu}{2\pi}\left[(g_V+g_A)^2(1-y)^2+(g_V-g_A)^2-(g_V^2-g_A^2)\frac{m_e}{E_\nu}y\right] \\ \sigma_{\bar\nu e} &= \frac{G_F^2 m_e E_\nu}{2\pi}\left[\frac{1}{3}(g_V+g_A)^2+(g_V-g_A)^2\right]. \end{aligned} \tag{7.61}$$

νe scattering was not actually observed until after the standard model was developed, and since then most of the experimental studies have involved ν_μ or $\bar\nu_\mu$ beams, since they are produced much more easily from π and K decays at accelerators. As will be described in Chapter 8, the different y distributions for the analogous $\nu_\mu e$ $(\bar\nu_\mu e)$ neutral current processes in the standard model were useful for determining the parity structure of the neutral current.

7.2.3 π and K Decays

The $\pi_{\ell 2}$ Decays: $\pi\to\mu\nu$, $e\nu$

The pions, $\pi^\pm = u\bar d\ (d\bar u)$ and $\pi^0=(u\bar u - d\bar d)/\sqrt{2}$, are much lighter than the other hadrons, with $m_{\pi^\pm}\sim 139.6$ MeV, which is somewhat larger than $m_{\pi^0}\sim 135.0$ MeV due to electromagnetic corrections. As described in Section 5.8, the small value of m_π can be understood because of their role as pseudo-Goldstone bosons of an approximate global $SU(2)_L\times SU(2)_R$ flavor symmetry of the strong interactions. The $\pi^\pm$ decay via the weak charged current, mainly into $\mu^\pm \overset{(-)}{\nu}_\mu$, denoted $\pi_{\mu 2}$, with a relatively long lifetime of $\tau_{\pi^\pm}\sim 2.6\times 10^{-8}$ s. The π^0 decay is electromagnetic and mainly into 2γ, with a much shorter lifetime $\tau_{\pi^0}\sim 8.4\times 10^{-17}$ s.

The role of the chiral anomaly and the number of colors in the π^0 lifetime was briefly described in Section 5.2. The branching ratios for the principal decay modes of the pions and charged kaons are listed in Table 7.2. Neutral K decays will be considered in Section 8.6.

TABLE 7.2 Branching ratios for the principal decay modes of the pions and charged kaons.[a]

$\pi^+ \to \mu^+\nu_\mu$	$(\pi_{\mu 2})$	99.99%	
$\to e^+\nu_e$	(π_{e2})	1.23×10^{-4}	Universality and V, A tests
$\to e^+\nu_e\pi^0$	(π_{e3})	1.036×10^{-8}	π beta decay
$\to \mu^+\nu_\mu\gamma$		2×10^{-4}	
$\pi^0 \to 2\gamma$		98.8%	Electromagnetic
$\to e^+e^-\gamma$		1.2%	
$\to e^+e^-e^+e^-$		3×10^{-5}	
$K^+ \to \mu^+\nu_\mu$	$(K_{\mu 2})$	63.6%	f_K/f_π
$\to e^+\nu_e$	(K_{e2})	1.6×10^{-5}	
$\to \mu^+\nu_\mu\pi^0$	$(K_{\mu 3})$	3.4%	Universality test
$\to e^+\nu_e\pi^0$	(K_{e3})	5.1%	Universality test
$\to \pi^+\pi^0$	$(K_{2\pi})$	20.7%	Nonleptonic
$\to \pi^+\pi^+\pi^-$	$(K_{3\pi})$	5.6%	
$\to \pi^+\pi^0\pi^0$	$(K_{3\pi})$	1.8%	

[a]The branching ratios of the conjugate π^- and K^- decays are the same by CP invariance.

The Feynman diagram for the semi-leptonic decay process $\pi^- \to \mu^-\bar{\nu}_\mu$ is shown in Figure 7.5. The matrix element is

$$M = -i\frac{G_F}{\sqrt{2}}\,\bar{u}_2\gamma_\mu\left(1-\gamma^5\right)v_1\,\langle 0|J^{h\mu\dagger}(0)|\pi^-(q)\rangle, \tag{7.62}$$

where $p_1 = p_{\bar{\nu}_\mu}$, $p_2 = p_\mu$, $q = p_{\pi^-} = p_1 + p_2$, and the hadronic current is

$$J^{h\dagger}_\mu = V^{h\dagger}_\mu - A^{h\dagger}_\mu = \bar{u}\gamma_\mu\left(1-\gamma^5\right)\left[d\cos\theta_c + s\sin\theta_c\right]. \tag{7.63}$$

The first $(\cos\theta_c)$ term is relevant to π decay. We saw in Section 2.10 that the pion is a pseudoscalar, $P|\pi^i(\vec{p})\rangle = -|\pi^i(-\vec{p})\rangle$. Reflection invariance[9] then implies that only the axial current contributes to $\langle 0|J^{h\mu\dagger}|\pi^-(q)\rangle$. To see this, recall that $PV^{\mu\dagger}P^{-1} = V^\dagger_\mu$ for the vector current, and that the vacuum is invariant, $P|0\rangle = 0$. By Lorentz invariance, the matrix element must be of the form

$$\langle 0|V^{\mu\dagger}|\pi^-(q)\rangle = iaq^\mu, \tag{7.64}$$

where a is a constant, since q is the only four-vector available. (The current is evaluated at $(t,\vec{x}) = 0$.) Therefore, by an argument similar to (2.275) on page 53,

$$iaq^\mu = \langle 0|P^{-1}PV^{\mu\dagger}P^{-1}P|\pi^-(q)\rangle = -\langle 0|V^\dagger_\mu|\pi^-(q')\rangle = -iaq'_\mu, \tag{7.65}$$

where $q = (E_q, \vec{q})$ and $q' = (E_q, -\vec{q})$, i.e., $q'_\mu = q^\mu$. Thus, $a = 0$.

[9]The weak interaction parity violation is explicit in the $V-A$ form of the current. The matrix elements of each part of the current are (reflection invariant) strong interaction quantities, up to negligible higher-order weak effects.

However, $PA^{h\dagger}_{\mu}P^{-1} = -A^{h\mu\dagger}$, so the axial matrix element can be non-vanishing. By Lorentz and translation invariance,

$$\langle 0|A^{h\mu\dagger}(x)|\pi^-(q)\rangle = i\cos\theta_c f_\pi q^\mu e^{-iq\cdot x}, \tag{7.66}$$

where the $\cos\theta_c$ is extracted for convenience and f_π is the pion decay constant introduced in (5.106) on page 191. (We saw in Section 5.8 that f_π is related to the spontaneous breaking of $SU(2)_L \times SU(2)_R$.) Thus,

$$M = \frac{-f_\pi G_F \cos\theta_c}{\sqrt{2}} \bar{u}_2 \left(\not{p}_1 + \not{p}_2\right)\left(1-\gamma^5\right) v_1 \sim \frac{-f_\pi m_\mu G_F \cos\theta_c}{\sqrt{2}} \bar{u}_2 \left(1-\gamma^5\right) v_1, \tag{7.67}$$

where we have neglected m_ν. Summing over the final spins

$$|\bar{M}|^2 = \sum_{s_1 s_2} |M|^2 = \frac{f_\pi^2 m_\mu^2}{2} G_F^2 \cos^2\theta_c \underbrace{\mathrm{Tr}\left[(1-\gamma_5)\not{p}_1(1+\gamma_5)(\not{p}_2 + m_\mu)\right]}_{8p_1\cdot p_2 = 4\left(m_\pi^2 - m_\mu^2\right)}, \tag{7.68}$$

so that

$$\bar{\Gamma} = \tau^{-1} = \frac{p_1}{8\pi m_\pi^2}|\bar{M}|^2 = \frac{G_F^2 \cos^2\theta_c f_\pi^2 m_\pi}{8\pi} m_\mu^2 \left(1 - \frac{m_\mu^2}{m_\pi^2}\right)^2. \tag{7.69}$$

Using G_F from μ decay and $\cos\theta_c \sim 0.9742$ from β decay, the observed lifetime yields $f_\pi = (130.50 \pm 0.14)$ MeV $\sim 0.93 m_{\pi^-}$, where the uncertainty is dominated by higher-order effects (Patrignani, 2016). This is in remarkable agreement with the lattice QCD prediction (Rosner et al., 2015) 130.2 ± 1.7 MeV.

The rate for $\pi^\pm \to e^\pm \nu_e(\bar{\nu}_e)$ is of the same form as (7.69) except $m_\mu \to m_e$. Therefore, one predicts

$$\begin{aligned} R_\pi \equiv \frac{\Gamma(\pi \to e\nu + e\nu\gamma)}{\Gamma(\pi \to \mu\nu + \mu\nu\gamma)} &= \frac{m_e^2}{m_\mu^2}\left(\frac{m_\pi^2 - m_e^2}{m_\pi^2 - m_\mu^2}\right)^2 (1 + \mathcal{O}(\alpha)) \\ &= 1.28 \times 10^{-4}\,(1 + \mathcal{O}(\alpha)) = 1.2352(2) \times 10^{-4}, \end{aligned} \tag{7.70}$$

where $\mathcal{O}(\alpha)$ is a radiative correction (Bryman et al., 2011). This is in perfect agreement with the value $R_\pi = (1.234 \pm 0.003) \times 10^{-4}$ from the PIENU experiment at TRIUMF (Aguilar-Arevalo et al., 2015), and supports the universality of the e and μ charged current interactions. Neutrino beams at accelerators are obtained mainly by the decays of $\pi^\pm$. The small value of R_π is the reason that such beams are mainly ν_μ and $\bar{\nu}_\mu$, with little ν_e or $\bar{\nu}_e$ (some ν_e are produced, e.g., from K or secondary $\mu^\pm$ decays, but these are mainly considered as backgrounds).

The suppression of the π_{e2} mode $\pi \to e\nu$ is a consequence and test of the V, A structure of the charged current. In particular, $V-A$ is just the left-chiral projection $2P_L$. Since chirality and helicity are the same for a massless fermion, weak charged currents imply helicity $h = -1/2$ for e^-, μ^-, ν_e, and ν_μ, up to corrections of $\mathcal{O}(m/E)$ in amplitude. Similarly, $h = +1/2$ for $e^+, \mu^+, \bar{\nu}_e$, and $\bar{\nu}_\mu$. Since the neutrino masses are negligible in this context, the $\bar{\nu}$ in $\pi^- \to \ell^-\bar{\nu}_\ell$ must have $h = +1/2$. But the pion has spin-0, so angular momentum in the rest frame requires $h = +1/2$ for the charged lepton as well. This is the *wrong* helicity, leading to the m_e^2/m_μ^2 factor in R_π. Thus, the electron mode is strongly suppressed, despite the partial compensation from the larger phase space and from the trace. Similar conclusions would actually hold for any combination of V and A, which always lead to opposite helicities in the massless limit. Things would be different for a scalar-pseudoscalar interaction $M \sim \bar{u}_1 \left(1 + a\gamma^5\right) v_2$, which leads to equal ℓ^- and $\bar{\nu}_\ell$ helicities for massless fermions. A pure S, P interaction would predict

$$R_\pi = \left(\frac{m_\pi^2 - m_e^2}{m_\pi^2 - m_\mu^2}\right)^2 = 5.49, \tag{7.71}$$

which is clearly excluded. The value of R_π can also set limits on small S, P perturbations on the dominant $V - A$ and to violations of universality.

The Weak Currents and $SU(3)_L \times SU(3)_R$ Chiral Symmetry

In (5.97) and (5.98) on page 189 we defined an octet of vector and axial generators of $SU(3)_L \times SU(3)_R$ and their associated Noether currents, i.e., $V_\mu^i = \bar{q}\gamma_\mu \frac{\lambda^i}{2} q$ and $A_\mu^i = \bar{q}\gamma_\mu\gamma^5 \frac{\lambda^i}{2} q$. The hadronic weak charged currents are simply related to some of these, e.g.,

$$\bar{u}\gamma_\mu\gamma^5 d = A_\mu^1 + iA_\mu^2, \qquad \bar{u}\gamma_\mu\gamma^5 s = A_\mu^4 + iA_\mu^5, \tag{7.72}$$

and similarly for the vector currents. Defining

$$J_\mu^i = V_\mu^i - A_\mu^i, \tag{7.73}$$

the hadronic weak charged currents are

$$\begin{aligned} J_\mu^{h\dagger} &= \left(J_\mu^1 + iJ_\mu^2\right)\cos\theta_c + \left(J_\mu^4 + iJ_\mu^5\right)\sin\theta_c \\ J_\mu^h &= \left(J_\mu^{h\dagger}\right)^\dagger = \left(J_\mu^1 - iJ_\mu^2\right)\cos\theta_c + \left(J_\mu^4 - iJ_\mu^5\right)\sin\theta_c. \end{aligned} \tag{7.74}$$

Similarly, the hadronic electromagnetic current is

$$\begin{aligned} J_Q^\mu &= \frac{2}{3}\bar{u}\gamma^\mu u - \frac{1}{3}\bar{d}\gamma^\mu d - \frac{1}{3}\bar{s}\gamma^\mu s \\ &= \frac{1}{2}(\bar{u}\gamma^\mu u - \bar{d}\gamma^\mu d) + \frac{1}{6}(\bar{u}\gamma^\mu u + \bar{d}\gamma^\mu d - 2\bar{s}\gamma^\mu s) \\ &= V^{\mu 3} + \frac{1}{\sqrt{3}}V^{\mu 8}. \end{aligned} \tag{7.75}$$

These relations are useful because they allow us to use $SU(3)$ and $SU(3)_L \times SU(3)_R$ to constrain and relate their matrix elements. In particular, in the $SU(2)$ and $SU(3)$ limits one can relate the form factors of the weak interaction vector currents to those measured in electromagnetic transitions, using (7.75). This all seems obvious in the quark model, but was inferred earlier and was known as the *conserved vector current (CVC) hypothesis.* Similarly, the axial currents and the pseudoscalar mesons in (3.91) both transform as $SU(3)$ octets, so in the $SU(3)$ limit one must have

$$\langle 0|A_\mu^i(0)|\pi^j(q)\rangle = +i\frac{f_\pi}{\sqrt{2}}\delta^{ij}q^\mu, \; i,j = 1\cdots 8, \tag{7.76}$$

where the $\sqrt{2}$ is inserted to coincide with the definition in (7.66) of the pion decay constant,[10]

$$\langle 0|\bar{u}\gamma_\mu\gamma_5 d|\pi^-\rangle = \frac{1}{\sqrt{2}}\langle 0|A_\mu^1 + iA_\mu^2|\pi^1 - i\pi^2\rangle = +if_\pi q^\mu. \tag{7.77}$$

Recall that these relations motivated the partially conserved axial current (PCAC) hypothesis in (5.119) on page 193. Of course, $SU(3)$ is broken by $\sim 25\%$, so one expects that the values f_π and f_K actually measured in π and K decays may differ by up to this amount.

[10]Some authors define f_π without the $\sqrt{2}$ in (7.76), so that it takes the value ~ 92 MeV.

The $K_{\ell 2}$ Decays

The $K^\pm = u\bar{s}(s\bar{u})$ are pseudoscalar mesons with $m_{K^\pm} = 493.7$ MeV and a typical weak lifetime of $\tau_{K^\pm} = 1.24 \times 10^{-8}$ s. The neutral kaons $K^0 = d\bar{s}$ and $\bar{K}^0 = s\bar{d}$ are somewhat heavier (497.6 MeV) because the $m_d > m_u$ quark contribution to the splitting is larger than the electromagnetic one. The neutral K decays have special properties due to $K^0 - \bar{K}^0$ mixing and CP violation, and will be described in Section 8.6. The branching ratios for the principal $K^\pm$ decays are given in Table 7.2. The dominant mode is $K_{\mu 2}$, i.e., $K^- \to \mu^- \bar{\nu}_\mu$. The calculation is identical to $\pi_{\mu 2}$. The hadronic matrix element is

$$\langle 0|J^{h\mu\dagger}(x)|K^-(q)\rangle = -i \sin\theta_c f_K q^\mu e^{-iq\cdot x}, \tag{7.78}$$

where the kaon decay constant f_K may differ from f_π due to $SU(3)$-breaking. Again, only the axial part of $J^{h\mu\dagger}$ contributes due to the reflection invariance of the strong interactions. The rate is

$$\bar{\Gamma}_{K^-\to\mu^-\bar{\nu}_\mu} = \frac{G_F^2 \sin^2\theta_c f_K^2 m_K}{8\pi} m_\mu^2 \left(1 - \frac{m_\mu^2}{m_K^2}\right)^2. \tag{7.79}$$

The Cabibbo angle factor $\sin\theta_c \sim 0.2253$ is known independently from $K_{\ell 3}$, β, and hyperon decays, allowing one to determine f_K from the K lifetime and branching ratio, i.e., $\bar{\Gamma}_{K^-\to\mu^-\bar{\nu}_\mu} = B(K^- \to \mu^-\bar{\nu}_\mu)/\tau_{K^-}$. The result is $f_K = (155.7 \pm 0.5)$ MeV $\sim 1.19 f_\pi$, to be compared with the lattice QCD result 155.6 ± 0.4 MeV (Rosner et al., 2015).

The $\pi_{\ell 3}$ and $K_{\ell 3}$ Decays

The $\pi_{\ell 3}$ and $K_{\ell 3}$ decays are the three-body semi-leptonic decays, $\pi^+ \to \pi^0 e^+ \nu_e$ (pion beta decay), $K^+ \to \pi^0 \ell^+ \nu_\ell$, and $K_{L,S} \to \pi^\pm \ell^\mp \nu_\ell(\bar{\nu}_\ell)$, and their CP conjugates, where $\ell = \mu$ or e. $K_{L,S}$ refer to the mass eigenstate neutral kaons $\sim (K^0 \pm \bar{K}^0)/\sqrt{2}$ (Section 8.6).

The matrix element for $\pi^+ \to \pi^0 e^+ \nu_e$ is

$$M = -i\frac{G_F}{\sqrt{2}}\, \bar{u}_\nu \gamma_\mu (1-\gamma^5) v_e \,\langle \pi^0(p')|J^{h\mu}(0)|\pi^+(p)\rangle, \tag{7.80}$$

where the relevant part of $J^{h\mu}$ is the first term in (7.74). It is straightforward to show (Problem 7.9) that only the vector current contributes due to reflection invariance. Furthermore, by Lorentz invariance, the matrix element must be of the form

$$\begin{aligned}\langle \pi^0(p')|J^{h\mu}(x)|\pi^+(p)\rangle &= \langle \pi^0(p')|V^{1\mu}(x) - iV^{2\mu}(x)|\pi^+(p)\rangle \cos\theta_c \\ &= \left[f_+(q^2)(p'+p)^\mu + f_-(q^2)(p'-p)^\mu\right] \cos\theta_c e^{i(p'-p)\cdot x},\end{aligned} \tag{7.81}$$

where $f_\pm$ are form factors analogous to the electromagnetic matrix element in Section 2.12.4. Similarly, the matrix element for $K^+ \to \pi^0 \ell^+ \nu_\ell$ is

$$M = -i\frac{G_F}{\sqrt{2}}\, \bar{u}_\nu \gamma_\mu (1-\gamma^5) v_\ell \,\langle \pi^0(p')|J^{h\mu}(0)|K^+(p)\rangle, \tag{7.82}$$

where

$$\begin{aligned}\langle \pi^0(p')|J^{h\mu}(x)|K^+(p)\rangle &= \langle \pi^0(p')|V^{4\mu}(x) - iV^{5\mu}(x)|K^+(p)\rangle \sin\theta_c \\ &= \left[f_+^K(q^2)(p'+p)^\mu + f_-^K(q^2)(p'-p)^\mu\right] \sin\theta_c e^{i(p'-p)\cdot x}.\end{aligned} \tag{7.83}$$

Time reversal invariance implies that f_+ and f_- are relatively real, as are $f_\pm^K$.

It might seem difficult to say anything, given the existence of two form factors factors for each process. However, in the $SU(2)$ or $SU(3)$ symmetry limits one can say a great deal. That is because the V^i_μ are the Noether currents associated with the $SU(3)$ generators F^i. In particular, they are conserved, $\partial^\mu V^i_\mu = 0$, implying $f_-(q^2) = 0$. Furthermore, the pseudoscalar fields transform as an $SU(3)$ octet, i.e.,

$$\left[F^i, \pi^j\right] = -\left(L^i_{adj}\right)_{jk} \pi^k = i f_{ijk} \pi^k, \quad i,j,k = 1 \cdots 8, \tag{7.84}$$

using (3.16) on page 93. This implies that

$$F^i|\pi^j\rangle = \int d^3\vec{x}\, V^i_0(x)|\pi^j\rangle = i f_{ijk}|\pi^k\rangle, \tag{7.85}$$

since $F^i|0\rangle = 0$ and since $F^i|\pi^j\rangle$ should not contain states outside of the octet (this argument will be made more precise in Section 7.2.4). From the general form analogous to (7.81) or (7.83) this implies

$$\langle \pi^k(p')|V^i_\mu(x)|\pi^j(p)\rangle = i f_{ijk} f(q^2)(p'+p)_\mu e^{i(p'-p)\cdot x}, \tag{7.86}$$

with $f(0) = 1$. For pion β decay the q^2 dependence can be ignored since $m_e^2 \leq q^2 \leq (m_{\pi^+} - m_{\pi^0})^2 \ll m_\pi^2$. Therefore,

$$\begin{aligned}\langle \pi^0|V^1_\mu - iV^2_\mu|\pi^+\rangle &= \frac{1}{\sqrt{2}}\langle \pi^3|V^1_\mu - iV^2_\mu|\pi^1 + i\pi^2\rangle \\ &= \frac{i}{\sqrt{2}}\left[i\epsilon_{123} - i\epsilon_{213}\right](p'+p)_\mu = -\sqrt{2}(p'+p)_\mu,\end{aligned} \tag{7.87}$$

i.e., the prediction of CVC is $f_+(q^2) \sim f_+(0) \sim -\sqrt{2}$. (See the comment on the π^+ phase convention in Problem 3.12.)

Pion Beta Decay

From (7.80) and (7.87), the matrix element for pion beta decay is

$$M = iG_F \cos\theta_c\, (p+p')_\mu\, \bar{u}_\nu \gamma^\mu \left(1-\gamma^5\right) v_e. \tag{7.88}$$

A good first approximation to the rate is obtained by neglecting m_e and working to leading order in

$$\Delta \equiv m_{\pi^+} - m_{\pi^0} \sim 4.6 \text{ MeV}\,. \tag{7.89}$$

Since Δ is much smaller than the π^+ mass, the π^0 is produced nearly at rest and the 3-momentum $\vec{p}_{\pi^0} = -\vec{p}_e - \vec{p}_\nu$ is very small. Therefore,

$$M \sim 2iG_F \cos\theta_c m_\pi \bar{u}_\nu \gamma^0 \left(1-\gamma^5\right) v_e. \tag{7.90}$$

The leptonic part can be evaluated by writing out the Dirac spinors explicitly, most conveniently in the chiral basis using (2.195) on page 40,

$$\begin{aligned}\bar{u}_\nu \gamma^0 \left(1-\gamma^5\right) v_e &= 2u^\dagger_\nu(-)v_e(+) \\ &= -4\sqrt{E_e E_\nu}\; \phi_-\left(\vec{p}_\nu\right)^\dagger \chi_+\left(\vec{p}_e\right) = -4\sqrt{E_e E_\nu}\cos\frac{\theta_{e\nu}}{2},\end{aligned} \tag{7.91}$$

where in the massless limit only the $+(-)$ helicity $e^+(\nu_e)$ is produced. We have used the explicit helicity spinors from Table 2.1, and $\theta_{e\nu}$ is the angle between $\vec{p}_e$ and $\vec{p}_\nu$. Therefore,

$$|M|^2 = 32G_F^2 \cos^2\theta_c m_\pi^2 E_e E_\nu \left(1+\cos\theta_{e\nu}\right), \tag{7.92}$$

which could also have been obtained using standard trace techniques.

The decay rate is

$$d\Gamma = \frac{1}{16(2\pi)^5 m_\pi} \int \frac{d^3\vec{p}_{\pi^0}}{E_{\pi^0}} \frac{d^3\vec{p}_\nu}{E_\nu} \frac{d^3\vec{p}_e}{E_e} \delta^4\left(p_{\pi^+} - p_{\pi^0} - p_\nu - p_e\right)|M|^2. \tag{7.93}$$

We will evaluate the phase space integral directly, though the more powerful techniques in Appendix D are more useful in the general case in which Δ and m_e are not small. One can use the δ function to do the $\vec{p}_{\pi^0}$ integral, and approximate

$$E_{\pi^0} = \sqrt{(\vec{p}_\nu + \vec{p}_e)^2 + m_{\pi^0}^2} \sim m_{\pi^0}, \tag{7.94}$$

as was already assumed in (7.90). The $\vec{p}_e$ and $\vec{p}_\nu$ integrals are therefore unconstrained except for an energy-conserving $\delta(\Delta - E_\nu - E_e)$ factor. The $\cos\theta_{e\nu}$ term integrates to 0 in the overall rate. Therefore,

$$\begin{aligned}
\Gamma &= \frac{1}{16(2\pi)^5 m_\pi^2} \int \frac{d^3\vec{p}_\nu}{E_\nu} \frac{d^3\vec{p}_e}{E_e} \delta\left(\Delta - E_\nu - E_e\right)|M|^2 \\
&= \frac{G_F^2 \cos^2\theta_c}{\pi^3} \int E_e^2 dE_e\, E_\nu^2 dE_\nu\, \delta\left(\Delta - E_\nu - E_e\right) \\
&= \frac{G_F^2 \cos^2\theta_c}{\pi^3} \Delta^5 \int_0^1 dx\, x^2 \left(1-x\right)^2 = \frac{G_F^2 \cos^2\theta_c \Delta^5}{30\pi^3}.
\end{aligned} \tag{7.95}$$

Including the leading corrections (see (Sirlin, 1978) and Problem 7.7), one predicts

$$\Gamma \sim \frac{G_F^2 \cos^2\theta_c}{30\pi^3} \Delta^5 \left[1 - \frac{3}{2}\frac{\Delta}{m_\pi} - \frac{5m_e^2}{\Delta^2} + \text{radiative} + \cdots\right] \sim 0.399(1)\ \text{s}^{-1}, \tag{7.96}$$

in excellent agreement with the experimental value $0.3980(23)$ s^{-1} by the PIBETA collaboration at PSI (Pocanic et al., 2004).

$K_{\ell 3}$ Decays

In addition to pion beta decay, $\pi^+ \to \pi^0 e^+ \nu_e$, the $K_{\ell 3}$ decays $K^+ \to \pi^0 \ell^+ \nu_\ell$ and $K_{L,S} \to \pi^\mp \ell^\pm \nu_\ell(\bar{\nu}_\ell)$ are very important as tests of CVC and especially as a measurement of $\sin\theta_c$. In the $SU(3)$ limit the values of $f_+(t)$ at $t \equiv q^2 = 0$ are predicted from (7.86) to be

$$f_+^{\pi^+\to\pi^0} = -\sqrt{2}, \qquad f_+^{K^+\to\pi^0} = -\frac{1}{\sqrt{2}}, \qquad f_+^{\bar{K}^0\to\pi^+} = -1, \qquad f_+^{K^0\to\pi^-} = 1. \tag{7.97}$$

The $f_+^{\pi^+\to\pi^0}$ prediction follows from $SU(2)$, while the others require $SU(3)$. However, the relative values for K^+ and K^0 follow from isospin, and the relation between K^0 and $\bar{K}^0$ follows from the charge conjugation invariance of the strong interactions. Since isospin is a good symmetry at the percent level, one expects the $\pi^+ \to \pi^0$ prediction to be quite reliable. However, one might expect a large deviation of the $SU(3)$ result for $f^{K\to\pi}$, perhaps as large as the $f_K/f_\pi \sim 1.19$ we found for the axial matrix elements from $K_{\ell 2}/\pi_{\ell 2}$. In fact, the relation of the vector currents to the generators leads to the *Ademollo-Gatto theorem*, which states that the deviations of the $f_+(0)$ due to $SU(2)$ and $SU(3)$ breaking are actually of second order, and should therefore be small. We will derive the Ademollo-Gatto theorem in the next section, but first will consider the $K_{\ell 3}$ decays in more detail.

The $K_{\ell 3}$ decay rate is (Leutwyler and Roos, 1984; Antonelli et al., 2010a; Cirigliano et al., 2012; Patrignani, 2016)

$$\Gamma = \frac{G_F^2}{192\pi^3} m_K^5 |V_{us} f_+(0)|^2 I_K^\ell, \tag{7.98}$$

where

$$I_K^\ell = \frac{1}{m_K^8} \int_{m_\ell^2}^{(m_K - m_\pi)^2} dt\, \lambda^{3/2}(m_K^2, m_\pi^2, t) \left(1 + \frac{m_\ell^2}{2t}\right) \left(1 - \frac{m_\ell^2}{t}\right)^2 F(t). \tag{7.99}$$

λ is the kinematic function defined in (2.39) on page 14 and

$$F(t) \equiv \bar{f}_+(t)^2 + \frac{3m_\ell^2 (m_K^2 - m_\pi^2)^2}{(2t + m_\ell^2)\lambda} \bar{f}_0(t)^2, \tag{7.100}$$

with

$$\bar{f}_+(t) \equiv \frac{f_+(t)}{f_+(0)}, \qquad \bar{f}_0(t) \equiv \bar{f}_+(t) + \frac{t}{m_K^2 - m_\pi^2} \frac{f_-(t)}{f_+(0)}. \tag{7.101}$$

There are also radiative correction factors that we do not display. A special case of (7.98) is derived in Problem 7.8. A complication in the $K_{\ell 3}$ decays is that t is not necessarily small, so the t dependence of f_+ must be included. Fortunately, this can be directly measured, e.g., by the pion energy distribution. Even though the symmetry breaking effects in $f_+(0)$ are second order, they must still be estimated by chiral perturbation theory or lattice techniques and taken into account. Finally, the symmetry breaking may also induce an f_- form factor. This leads to a contribution proportional to m_ℓ in the amplitude, which is non-negligible for $K_{\mu 3}$. Despite these complications the $K_{\ell 3}$ system has been carefully studied, because it leads to the most precise determination of $\sin\theta_c$ (or its generalization to the CKM element V_{us} in the three-family case.)

7.2.4 Nonrenormalization of Charge and the Ademollo-Gatto Theorem

Consider the question of whether electric charge is renormalized by the strong interactions[11] for the example of the pion. As discussed in Section 2.12.4, the matrix element of the electromagnetic current is

$$\langle \pi^+(p')|J_Q^\mu(x)|\pi^+(p)\rangle = \left[(p'+p)^\mu f_+^Q(q^2) + (p'-p)^\mu f_-^Q(q^2)\right] e^{iq\cdot x}, \tag{7.102}$$

where $q = p' - p$. The electric charge, in units of e, is $f_+^Q(0)$, so the nonrenormalization requires $f_+^Q(0) = 1$ in the presence of the strong interactions.

To see how this comes about, recall that the $U(1)_Q$ generator is

$$Q(t) = \int d^3\vec{x}\, J_Q^0(t, \vec{x}), \tag{7.103}$$

[11] The gauge coupling e *is* renormalized, and may depend on the scale, but that effect is universal for all charged particles. See (D'Alfaro et al., 1973) for an extension of this argument to include higher-order electromagnetic effects. Higher-order electromagnetic effects also contribute a small q^2 dependence to f_+.

that current conservation $\partial_\mu J^\mu_Q = 0$ implies $\frac{dQ}{dt} = 0$, and that the transformations of the charged pion fields, which can be viewed either as elementary or as composite as in (5.102) on page 190, are

$$\left[Q, \pi^\pm\right] = \mp\pi^\pm. \tag{7.104}$$

All of these statements are obviously true for free fields, but continue to hold to all orders in the strong interactions since they are $U(1)_Q$ invariant, i.e., $[Q, H] = 0$. Taking the divergence of (7.102), current conservation immediately implies $q^2 f^Q_-(q^2) = 0$, and therefore $f^Q_-(q^2) = 0$ since strong interaction effects are not expected to produce a delta function. The nonrenormalization can be established by taking the matrix element of (7.104) between a single pion state and the vacuum:

$$\begin{aligned}\langle\pi^+(p)|\left[Q, \pi^-\right]|0\rangle &= \int \frac{d^3\vec{k}}{(2\pi)^3 2E_k} \langle\pi^+(p)|Q|\pi^+(k)\rangle\langle\pi^+(k)|\pi^-|0\rangle \\ &+ \sum_{n\neq\pi^+} \langle\pi^+(p)|Q|n\rangle\langle n|\pi^-|0\rangle = +\langle\pi^+(p)|\pi^-|0\rangle.\end{aligned} \tag{7.105}$$

We have assumed that $U(1)_Q$ is not spontaneously broken, i.e., $Q|0\rangle = 0$. The last $\sum_n$ term is a sum or integral over all single and multi-particle states other than the π^+. But,

$$\begin{aligned}\langle\pi^+(p)|Q|\pi^+(k)\rangle &= \int d^3\vec{x}\, e^{i(p-k)\cdot x} \left[(p+k)^0 f^Q_+(q^2) + (p-k)^0 f^Q_-(q^2)\right] \\ &= (2\pi)^3\delta^3(\vec{p}-\vec{k}) 2E_p f^Q_+(0),\end{aligned} \tag{7.106}$$

so the first term in (7.105) is $f^Q_+(0)\langle\pi^+(p)|\pi^-|0\rangle$. Similarly,

$$\langle\pi^+(p)|Q|n\rangle = (2\pi)^3\delta^3(\vec{p}-\vec{p}_n)\langle\pi^+|J^0_Q(0)|n\rangle e^{i(E_p-E_n)t}. \tag{7.107}$$

Since $dQ/dt = 0$, only (positively charged) states with $E_p = E_n$ and $\vec{p} = \vec{p}_n$, and therefore $m_\pi = m_n$, can have nonzero matrix elements. (This can also be seen from (3.52) on page 99.) There are no such states, so the last term in (7.105) vanishes and $f^Q_+(0) = 1$. We note that the strong interactions *will*, however, introduce a strong q^2 dependence to f^Q_+, and also induce a magnetic form factor for nucleon matrix elements. These effects can be taken from experiment.

Exactly the same reasoning can be used to establish the nonrenormalization of the $SU(2)$ and $SU(3)$ generators in the symmetry limit, as expressed in (7.85) and (7.86), by taking the matrix element of (7.84) between $\langle\pi^k|$ and $|0\rangle$. The electric charge generator is a special case.

Now let us turn on $SU(3)$ breaking and establish that the deviations of the $K_{\ell 3}$ form factors $f^K_+(0)$ from the predictions in (7.97) are of second order. This is most easily seen from the commutator relation

$$\left[F^4 + iF^5, F^4 - iF^5\right] = F^3 + \sqrt{3}F^8 = I_3 + \frac{3}{2}\mathcal{Y}. \tag{7.108}$$

This is expected to hold even in the presence of $SU(3)$ breaking mass and interaction terms, as discussed in Section 3.2.2. I_3 and $\mathcal{Y}$ are, respectively, the unbroken third component of isospin and strong hypercharge, defined in (3.89) on page 104. Taking the matrix element,

$$\begin{aligned}
\langle\pi^+(p')|I_3+\frac{3}{2}\mathcal{Y}|\pi^+(p)\rangle &= (2\pi)^3 2E_p\delta^3(\vec{p}\,'-\vec{p}) \\
&= \langle\pi^+(p')|\left[F^4+iF^5,F^4-iF^5\right]|\pi^+(p)\rangle \\
&= \int\frac{d^3\vec{k}}{(2\pi)^3 2E_k}\langle\pi^+(p')|F^4+iF^5|\bar{K}^0(k)\rangle\langle\bar{K}^0(k)|F^4-iF^5|\pi^+(p)\rangle \\
&+\sum_{n\neq\bar{K}^0}\langle\pi^+|F^4+iF^5|n\rangle\langle n|F^4-iF^5|\pi^+\rangle \\
&-\sum_n\langle\pi^+|F^4-iF^5|n\rangle\langle n|F^4+iF^5|\pi^+\rangle.
\end{aligned}\tag{7.109}$$

The $\bar{K}^0$ term can be evaluated using the analog of (7.83), but with $f_\pm^{\bar{K}^0\to\pi^+}$. It is a Lorentz invariant, and is conveniently evaluated in the $|\vec{p}|\to\infty$ frame, to obtain

$$\left(f_+^{\bar{K}^0\to\pi^+}(0)\right)^2(2\pi)^3 2E_p\delta^3(\vec{p}\,'-\vec{p}).\tag{7.110}$$

The remaining terms, involving the leakage out of the pseudoscalar octet, are of second order in $SU(3)$ breaking, as is apparent from (3.52). Since $f_+^{\bar{K}^0\to\pi^+}(0)\to -1$ in the symmetry limit, we obtain the final result

$$f_+^{\bar{K}^0\to\pi^+}(0) = -1+\mathcal{O}(\epsilon_8^2),\tag{7.111}$$

where ϵ_8^2 is the $SU(3)$ breaking parameter defined in (3.107) on page 108. This is the Ademollo-Gatto theorem (Ademollo and Gatto, 1964; Behrends and Sirlin, 1960; Fubini and Furlan, 1965; Marshak et al., 1969; D'Alfaro et al., 1973), which implies that $SU(3)$ is fairly reliable for $f_+^K(0)$. Similarly, deviations from the $SU(2)$ prediction in (7.87) are expected to be second order in isospin breaking, and therefore very small. Similar results hold for the vector (but not axial vector) form factors in nucleon and hyperon β decay.

One can extend these considerations to chiral $SU(3)_L\times SU(3)_R$. In the chiral limit the pseudoscalars become massless and deviations from the symmetry predictions are sometimes dominated by calculable non-analytic terms generated by infrared singularities (Pagels, 1975). In the case of the $f_+(0)$, the leading corrections to the chiral limit are of order ϵ_8^2/ϵ_0 (Langacker and Pagels, 1973), where ϵ_0 represents the explicit chiral breaking, such as the common quark mass term in (5.91) on page 188. This is of first order in chiral $SU(3)$ but second order in ordinary $SU(3)$, and leads to a 2% reduction in the prediction for $f_+^K(0)$. More precise estimates from chiral perturbation theory and lattice techniques are surveyed in the reviews cited in the $K_{\ell 3}$ section.

7.2.5 β Decay

β decay refers both to free neutron decay and to nuclear beta decay. These and closely related processes are listed in Table 7.3. The amplitude for the free neutron decay corresponding to Figure 7.1 is

$$M=-i\frac{G_F}{\sqrt{2}}\,\bar{u}_e\gamma_\mu\left(1-\gamma^5\right)v_\nu\,\langle p|\,\bar{u}\gamma^\mu\left(1-\gamma^5\right)d\,|n\rangle\cos\theta_c.\tag{7.112}$$

TABLE 7.3 β decay and related processes.[a]

$n \to p e^- \bar{\nu}_e$	Neutron decay
$(N, Z) \to (N-1, Z+1) e^- \bar{\nu}_e$	Nuclear (heavy nuclei, e.g., in reactor)
$(N, Z) \to (N+1, Z-1) e^+ \nu_e$	Nuclear (light nuclei, e.g., in Sun)
$e^- p \to \nu_e n$	Electron capture (atomic electron)
$\mu^- p \to \nu_\mu n$	μ capture (muonic atom)
$\nu_e n \to e^- p$	Inverse β decay
$\bar{\nu}_e p \leftrightarrow e^+ n$	

[a] N and Z are the numbers of neutrons and protons in a nucleus. All of these processes are driven by the $\cos\theta_c$ part of $\mathcal{H}_{sl}$ in (7.20).

The Vector and Axial Form Factors

We have seen in Chapter 2 that the matrix element of the vector current is restricted by Lorentz and translation invariance, and by the reflection invariance of the strong interactions, to be of the form

$$\langle p|\bar{u}\gamma^\mu d(x)|n\rangle = \bar{u}_p \left[\gamma^\mu F_1(q^2) + \frac{i\sigma^{\mu\nu}}{2m} q_\nu F_2(q^2) + q^\mu F_3(q^2)\right] u_n e^{iq\cdot x}, \tag{7.113}$$

where $q = p_p - p_n$ and we have used the Gordon decomposition in Problem 2.10 to eliminate possible $\bar{u}_p(p_p^\mu + p_n^\mu)u_n$ and $\bar{u}_p \sigma^{\mu\nu}(p_{p\nu} + p_{n\nu})u_n$ terms. Time reversal invariance implies that $F_{1,2,3}$ are relatively real (Appendix G). We also have that the hadronic vector current is $\bar{u}\gamma^\mu d = V^{1\mu} + iV^{2\mu}$, where $V^{1,2}$ are the Noether currents related to the isospin generators. Equivalently, they are in the same isomultiplet as the isovector part V^3 of the electromagnetic current, $J_Q = V^3 + \frac{1}{\sqrt{3}}V^8$ (CVC). The matrix elements are related by the $SU(2)$ relation $\langle a|V^i|b\rangle \propto \tau^i_{ab}/2$, where $a, b = p$ or n and $i = 1, 2, 3$. The isovector part of J_Q can be separated from the isoscalar V^8 by considering the difference between the corresponding proton and neutron form factors,

$$\begin{aligned} F_1(0) &= 2F_1^V(0) = F_1^p(0) - F_1^n(0) = 1 \\ F_2(0) &= 2F_2^V(0) = \kappa_p - \kappa_n = 1.79 + 1.91 \sim 3.70, \end{aligned} \tag{7.114}$$

where $F_1^p(0)(F_1^n(0)) = 1(0)$ is the proton (neutron) electric charge, $\kappa_{p,n}$ are their measured anomalous magnetic moments, and $F_{1,2}^V$ are the isovector form factors defined in Section 2.12.4. The $F_2(0)$ effect, known as *weak magnetism*, has been observed in the $e^\pm$ spectra in ${}^{12}B \to {}^{12}C e^- \bar{\nu}_e$ and ${}^{12}N \to {}^{12}C e^+ \nu_e$, in agreement with the CVC prediction. CVC also predicts that $F_3(q^2) = 0$, since $\partial^\mu V_\mu^i = 0$. (This also follows from G-parity.) However, its contribution would be proportional to m_e and small even if it were present. For most processes q^2 is small enough that its effects can be ignored.

The vector current form factors are therefore under control for neutron decay. In fact, by the Ademollo-Gatto theorem, corrections are of second order in isospin breaking and therefore tiny, i.e., $(m_d - m_u)^2$, $Z\alpha^2$, and $Z\alpha(m_d - m_u)$, where we have included a Z for the case of nuclear decay.

The axial matrix element is more difficult, because the axial generators of $SU(2)_L \times SU(2)_R$ are spontaneously broken. The most general form consistent with P, Lorentz, and translation invariance is

$$\begin{aligned}\langle p|\bar{u}\gamma^\mu\gamma^5 d(x)|n\rangle &= \langle p|A^{1\mu}(x)+iA^{2\mu}(x)|n\rangle \\ &= \bar{u}_p\left[\gamma^\mu\gamma^5 g_1(q^2)+\frac{i\sigma^{\mu\nu}\gamma^5}{2m}q_\nu g_2(q^2)+q^\mu\gamma^5 g_3(q^2)\right]u_n e^{iq\cdot x},\end{aligned} \tag{7.115}$$

where we have again used the Gordon decomposition. Time reversal requires that g_1, g_2, and g_3 are relatively real,[12] and g_2 is required to vanish by G-parity (Appendix G). As mentioned following (5.112) on page 192, the induced pseudoscalar form factor g_3 is present because of the chiral symmetry breaking, which generates a Goldstone boson pole. However, the effect of this term is usually negligible[13] because it generates a contribution proportional to m_ℓ. $g_1(0) = 1.272(2)$ is measured in the rates and various parity-violating asymmetries in neutron and nuclear β decay. Unlike the vector form factor, there is no nonrenormalization theorem to prevent $g_1(0)$ from being changed from its pointlike value of 1, even in the chiral limit. The proof of nonrenormalization for the vector generators F^i relied on the fact that $F^i|a\rangle$, where a is a single particle state such as a pion or nucleon, can only have a nonzero matrix element with another state with the same invariant mass as a, and there were no such states with the necessary quantum numbers. For the axial generators, however, the pions become massless Goldstone bosons in the symmetry limit, so one can have nonzero matrix elements, e.g., $\langle\pi N|F^{i5}|N\rangle$.

To understand this better, we outline the *Adler-Weisberger relation* (Adler, 1965a; Weisberger, 1965). Recall that the commutators of the axial $SU(2)$ generators are

$$\left[F^{i5}, F^{j5}\right] = i\epsilon_{ijk}F^k, \quad i,j,k = 1,2,3. \tag{7.116}$$

Just as we did for the Ademollo-Gatto theorem, take the matrix element of (7.116) between single particle states. In this case, we will use protons and ignore isospin breaking. Then,

$$\begin{aligned}\frac{1}{2}\sum_{s_1,s_2}\delta_{s_2 s_1}\langle p(p_2,s_2)|2F^3|p(p_1,s_1)\rangle &= (2\pi)^3\delta^3(\vec{p}_2-\vec{p}_1)2E_1 \\ =\frac{1}{2}\sum_{s_1,s_2}\delta_{s_2 s_1}\langle p(p_2,s_2)|&\left[F^{15}+iF^{25}, F^{15}-iF^{25}\right]|p(p_1,s_1)\rangle.\end{aligned} \tag{7.117}$$

Now, insert a complete set of intermediate states between the generators. The single neutron contribution I_n is readily evaluated using (7.115):

$$\begin{aligned}I_n &= \frac{(2\pi)^3}{2}\frac{\delta^3(\vec{p}_2-\vec{p}_1)}{2E_1}|g_1(0)|^2\mathrm{Tr}\left[\gamma_0\gamma^5(\not{p}_1+M_n)\gamma_0\gamma^5(\not{p}_1+M_n)\right] \\ &= (2\pi)^3\delta^3(\vec{p}_2-\vec{p}_1)2E_1\left[|g_1(0)|^2\left(1-\frac{M_n^2}{E_1^2}\right)\right] \\ &\to (2\pi)^3\delta^3(\vec{p}_2-\vec{p}_1)2E_1|g_1(0)|^2,\end{aligned} \tag{7.118}$$

where the last form is valid in the $|\vec{p}|\to\infty$ frame. The remainder of the sum is

$$\begin{aligned}I_{m\neq n} = \frac{1}{2}\sum_{s_1 s_2}\delta_{s_1 s_2}\sum_{m\neq n}\Big\{&\langle p|F^{15}+iF^{25}|m\rangle\langle m|F^{15}-iF^{25}|p\rangle \\ -&\langle p|F^{15}-iF^{25}|m\rangle\langle m|F^{15}+iF^{25}|p\rangle\Big\}.\end{aligned} \tag{7.119}$$

---

[12] g_2 would have to be imaginary for a Hermitian current such as $\bar{u}\gamma^\mu\gamma^5 u$, so it would have to vanish if time reversal holds. The same is true for the g_2 in the matrix element of the electromagnetic current J_Q^μ (where it corresponds to an intrinsic electric dipole moment), even if parity were violated.

[13] It has been measured in μ capture (e.g., Gorringe and Hertzog, 2015).

The matrix elements in (7.119) can be related to matrix elements of the $\pi^\pm$ field using (3.52) and (5.119), which in turn are proportional to the strong interaction amplitudes for $\pi^\pm p \to m$, to obtain the Adler-Weisberger formula

$$|g_1(0)|^2 = 1 + \frac{2f_\pi^2}{\pi} \int_{M_n+m_\pi}^{\infty} \frac{w dw}{w^2 - M_n^2} \left[\sigma_{\pi^+p}(w) - \sigma_{\pi^-p}(w)\right], \tag{7.120}$$

where $\sigma_{\pi^\pm p}(w)$ is the total cross section for $\pi^\pm p$ scattering (with a massless pion) at CM energy w. Using the observed cross sections (and including corrections for the off-shell π), (7.120) yields $g_1(0) \sim 1.25(2)$ (e.g., Beane and Klco, 2016), close to the experimental value ~ 1.27. This agreement was influential in establishing current algebra, and as we have seen relies not only on the Fermi theory but also on our understanding of chiral $SU(2)_L \times SU(2)_R$ as an approximate spontaneously broken global symmetry of the strong interactions.

Neutron and Nuclear Decay

To a good approximation, the hadronic matrix element for neutron decay is therefore

$$\langle p|J_\mu^{h\dagger}|n\rangle = \bar{u}_p \gamma_\mu \left(g_V - g_A \gamma^5\right) u_n, \tag{7.121}$$

where

$$g_V = f_1(0) \cos\theta_c \sim \cos\theta_c, \qquad g_A = g_1(0) \cos\theta_c. \tag{7.122}$$

It is then straightforward to work out the neutron lifetime and various decay distributions (Problem 7.10). The total lifetime is

$$\tau_n^{-1} = \Gamma_n = \frac{G_F^2}{2\pi^3} \left(|g_V|^2 + 3|g_A|^2\right) \int_{m_e}^{\Delta} p_e E_e \left(\Delta - E_e\right)^2 dE_e, \tag{7.123}$$

where $\Delta = m_n - m_p \sim 1.29$ MeV. The integral,[14] obtained by a calculation similar to the one in (7.95) for pion beta decay, is most easily done numerically, yielding $1.64 m_e^5$ for the observed value of Δ/m_e. From the observed lifetime $\sim 880.2(1.0)$ s (Wietfeldt and Greene, 2011; Patrignani, 2016) one can obtain $g_V^2 + 3g_A^2$. The ratio $\lambda \equiv g_A/g_V \sim 1.2723(23)$ can be extracted from various asymmetries. For example, for a polarized neutron, the angular distribution of the e^- w.r.t. the neutron spin direction is

$$\frac{d\Gamma}{d\cos\theta_e} \propto 1 - 2\beta_e \left(\frac{\lambda^2 - \lambda}{1 + 3\lambda^2}\right) \cos\theta_e, \tag{7.124}$$

where $\cos\theta_e = \hat{p}_e \cdot \hat{s}_n$ and $\beta_e = |\vec{p}_e|/E_e$. From g_V and g_A one can determine $g_1(0)$ (actually $g_1(0)/f_1(0)$) and $\cos\theta_c$. Similarly, the polarization of the produced electron is just

$$P_{e^-} = \frac{\Gamma(h_e = +\frac{1}{2}) - \Gamma(h_e = -\frac{1}{2})}{\Gamma(h_e = +\frac{1}{2}) + \Gamma(h_e = -\frac{1}{2})} = -\beta_e, \tag{7.125}$$

which approaches -1 as expected for $m_e \to 0$.

For nuclear β decay the nuclear matrix element, either $\langle N-1, Z+1|J_\mu^{h\dagger}(x)|N, Z\rangle$ or $\langle N+1, Z-1|J_\mu^{h}(x)|N, Z\rangle$, must be determined, from theory or by relating to other measured matrix elements. Transitions involving V_μ^h are known as Fermi transitions. Especially important are the superallowed transitions between 0^+ states in the same isomultiplet (Hardy and

[14] This is known as the Fermi integral. A more detailed treatment includes a factor $F(Z, E_e)$ that corrects for the distortion of the $e^\pm$ wave function by the Coulomb field of the nucleus and atomic electrons.

Towner, 2015). These are pure Fermi transitions, with the relevant form factors determined by the isospin generators. Hence, there is little theoretical uncertainty except for second order isospin violation (the Ademollo-Gatto theorem). The superallowed transitions yield the cleanest determinations of $\cos\theta_c$ (i.e., the element V_{ud} of the quark mixing matrix in the full SM). The axial currents lead to Gamow-Teller transitions, such as those with $\Delta J = \pm 1$. Mixed transitions, such as neutron decay, involve both. There has been an enormous amount of experimental and theoretical work on nuclear and neutron β decay. The formalism has been developed for arbitrary admixtures of S, P, T, V, and A interactions (Jackson et al., 1957). Many different types of asymmetries, polarizations, and correlations have been measured or searched for. These have established the basic $V-A$ nature of the interactions; tested the C, P, T, and G-parity properties; searched for perturbations on $V-A$ from such types of new physics as extended gauge groups involving $V+A$ couplings, leptoquark interactions, and mixing between the ordinary neutrinos or other fermions with heavy exotic states; tested the radiative corrections, which require the full standard model; extracted the form factors; and determined V_{ud} for universality tests. Detailed accounts may be found in (Commins and Bucksbaum, 1983; Renton, 1990; Deutsch and Quin in Langacker, 1995; Herczeg, 2001). Recent reviews include (Severijns et al., 2006; Abele, 2008; Dubbers and Schmidt, 2011; Bhattacharya et al., 2012; Cirigliano et al., 2013; Vos et al., 2015).

7.2.6 Hyperon Decays

There have been extensive studies of the semi-leptonic hyperon decays. These include the $\Delta S = 0$ decays $n \to p e^- \bar{\nu}_e$, $\Sigma^\pm \to \Lambda e^\pm \nu_e(\bar{\nu}_e)$, $\Sigma^- \to \Sigma^0 e^- \bar{\nu}_e$, and $\Xi^- \to \Xi^0 e^- \bar{\nu}_e$, as well as the $\Delta S = \Delta Q = 1$ decays $\Lambda \to p \ell^- \bar{\nu}_\ell$, $\Sigma^- \to n \ell^- \bar{\nu}_\ell$, $\Xi^- \to \Lambda \ell^- \bar{\nu}_\ell$, $\Xi^- \to \Sigma^0 \ell^- \bar{\nu}_\ell$, and $\Xi^0 \to \Sigma^+ \ell^- \bar{\nu}_\ell$, where $\ell = e$ or μ.

In the $SU(3)$ limit the hyperons transform as an octet, as shown in (3.90) on page 105 and Figure 3.1. Since the hadronic weak currents also transform as octets, one can use $SU(3)$ symmetry to related the matrix elements. For an arbitrary $SU(3)$ octet of operators $O^j, j = 1 \cdots 8$, one has that

$$\langle B_k | O^i | B_j \rangle = i f_{ijk} O_F + d_{ijk} O_D, \tag{7.126}$$

where the f and d coefficients are defined in Table 3.2, and $O_{F,D}$ are two reduced matrix elements that are independent of i, j, and k. This is the analog of the Wigner-Eckart theorem for $SU(2)$ and has already been used in (3.102) in the derivation of the Gell-Mann-Okubo formula. The hadronic vector currents are associated with the $SU(3)$ generators, so only the f_{ijk} term is non-vanishing, i.e.,

$$\langle B_k | V^i_\mu(0) | B_j \rangle = i f_{ijk} \bar{u}_k \gamma_\mu u_j f_1(q^2), \tag{7.127}$$

with $q = p_k - p_j$ and $f_1(0) = 1$, exactly analogous to (7.86). The Ademollo -Gatto theorem applies to the hyperons as well as the pseudoscalars, so one expects the predictions to be quite accurate, $f_1(0) = 1 + \mathcal{O}(\epsilon_8^2)$, where the coefficient of the correction can depend on i, j, k. One can also include weak magnetism corrections to (7.127), which are also related by $SU(3)$. The predictions for the vector and axial form factors are summarized in Table 7.4.

The axial current matrix elements can have both f and d type contributions. It is conventional to define the coefficients F and D by

$$\langle B_k | A^i_\mu(0) | B_j \rangle = \left[i f_{ijk} F(q^2) + d_{ijk} D(q^2) \right] \bar{u}_k \gamma_\mu \gamma^5 u_j, \tag{7.128}$$

where $F \equiv F(0)$ and $D \equiv D(0)$. The corrections are first order in $SU(3)$ breaking, so the

TABLE 7.4 $SU(3)$ predictions for the values of the vector and axial form factors at $q^2 = 0$.

	Decay	Vector	Axial
$\Delta S = 0$	$n \to pe^- \bar{\nu}_e$	1	$F + D$
	$\Sigma^\pm \to \Lambda e^\pm \nu_e (\bar{\nu}_e)$	0	$\sqrt{\frac{2}{3}} D$
	$\Sigma^- \to \Sigma^0 e^- \bar{\nu}_e$	$\sqrt{2}$	$\sqrt{2} F$
	$\Xi^- \to \Xi^0 e^- \bar{\nu}_e$	1	$F - D$
$\Delta S = 1$	$\Lambda \to p\ell^- \bar{\nu}_\ell$	$-\sqrt{\frac{3}{2}}$	$-\sqrt{\frac{3}{2}}(F + \frac{D}{3})$
	$\Sigma^- \to n\ell^- \bar{\nu}_\ell$	-1	$-(F - D)$
	$\Xi^- \to \Lambda \ell^- \bar{\nu}_\ell$	$\sqrt{\frac{3}{2}}$	$\sqrt{\frac{3}{2}}(F - \frac{D}{3})$
	$\Xi^- \to \Sigma^0 \ell^- \bar{\nu}_\ell$	$\frac{1}{\sqrt{2}}$	$\sqrt{2}(F + D)$
	$\Xi^0 \to \Sigma^+ \ell^- \bar{\nu}_\ell$	1	$F + D$

predictions are less reliable than for the vector currents. F and D and the q^2 dependence must be taken from experiment. One combination is obtained from the $g_1(0)$ from β decay in (7.115),

$$\langle p|\bar{u}\gamma_\mu \gamma^5 d|n\rangle = \langle \frac{1}{\sqrt{2}}(B_6 + iB_7)|A^1_\mu + iA^2_\mu|\frac{1}{\sqrt{2}}(B_4 + iB_5)\rangle, \tag{7.129}$$

yielding $F + D = g_1(0)$. There have been extensive studies of the decay rates, the ℓ energy spectrum, correlation between the $\hat{p}_\ell$ and the initial hyperon polarization, the $\ell - \nu$ angular correlations, and the final baryon polarization, both to test the Fermi theory and to obtain an accurate value for $\sin\theta_c$. The data agree quite well with the theoretical expectations, and yield

$$F = 0.462(11), \qquad D = 0.808(6), \qquad V_{us}\,(\sim \sin\theta_c) = 0.226(5) \tag{7.130}$$

from an overall fit (Gaillard and Sauvage, 1984; Cabibbo et al., 2003; Mateu and Pich, 2005).

7.2.7 Heavy Quark and Lepton Decays

The $\tau^\pm$ lepton, with $m_\tau \sim 1.777$ GeV and lifetime $\sim 2.9 \times 10^{-13}$ s, was the first member of the third family of quarks and leptons to be discovered. It is similar to the electron and muon except for its large mass, $m_\tau/m_\mu \sim 17$ and $m_\tau/m_e \sim 3500$. It can therefore undergo both leptonic decays, $\tau^- \to \nu_\tau \ell^- \bar{\nu}_\ell$, $\ell = e$ or μ, and *semi-hadronic* (also referred to as hadronic) decays, $\tau \to \nu_\tau d\bar{u}$ or $\nu_\tau s\bar{u}$, with the quarks emerging as mesons such as π^- or $\pi^-\pi^0$. The τ decays have been very useful in testing the Fermi theory (and SM) couplings, extracting the QCD coupling at the τ scale, and searching for new physics (see, e.g., Davier et al., 2006; Pich, 2014; Patrignani, 2016).

The leptonic decays are very similar to μ decay, with branching ratios $\sim 18\%$ each for the e and μ channels, consistent with the simple parton model expectation that the branching ratios for the e, μ, and $q\bar{q}$ decays should be in the ratio 1, 1, 3 (because of the 3 quark colors). They have been extensively studied (Patrignani, 2016), to establish the $V - A$ nature of the τ current and to verify weak universality, i.e., that the strength of the interaction is the same as the one for muon decay, as is expected in the SM extension of the Fermi theory. For example, the Michel ρ parameter, analogous to the Michel parameter for μ decay in (7.55), is measured to be 0.745(8), consistent with $V - A$ value of 3/4. Similarly (Pich, 2014),

$$\frac{G^2_{\tau\to e}}{G^2_F} = 1.002(3), \qquad \frac{G^2_{\tau\to\mu}}{G^2_F} = 1.006(3), \qquad \frac{G^2_{\tau\to\pi}}{G^2_F} = 0.992(5), \tag{7.131}$$

where, e.g., $G_{\tau\to e}$ is the coefficient of the four-fermi interaction measured for $\tau^- \to \nu_\tau e^- \bar{\nu}_e$. The values in (7.131) are in reasonable agreement with universality.

The semi-hadronic decays $\tau^- \to h^- \nu_\tau$ include $h = \pi^-(\sim 11\%)$, $\pi^-\pi^0(\sim 26\%)$, and $\pi^- + n\pi^0(n \leq 4)$, as well as modes involving 3 or 5 charged particles and modes involving kaons (the latter are suppressed by $\sin^2\theta_c$ and by phase space). The $\tau^- \to \pi^-\nu_\tau$ decay rate can be predicted using f_π extracted from $\pi \to \mu\nu$ (Equation 7.66), leading to the effective coupling quoted in (7.131). The semi-hadronic decays can also be used to determine the strong coupling $\alpha_s(m_\tau^2)$, the CKM matrix element $V_{us} \sim \sin\theta_c$, and were historically useful for setting a direct upper bound $\sim$ 18 MeV on the ν_τ mass in modes involving 3 or 5 charged pions (for which little phase space is available).

There have also been extensive searches for rare decays, such as $\tau \to \ell\gamma$, which are forbidden in the Fermi theory and standard model up to negligibly small neutrino mass effects, but could be generated by new physics (Section 8.6.6).

In the four-quark model, the c decays via the hadronic weak current

$$J^h_\mu = \bar{s}\gamma_\mu(1-\gamma^5)c\cos\theta_c - \bar{d}\gamma_\mu(1-\gamma^5)c\sin\theta_c + u \text{ terms} \tag{7.132}$$

in (7.23), i.e., to s and d with relative strengths $\cos\theta_c$ and $\sin\theta_c$. In the 3-family extension, the c decays are little affected because the mixing between the third family and the first two is small. However, these small effects allow b quark decays via

$$J^{h\dagger}_\mu = \bar{u}\gamma_\mu(1-\gamma^5)b\,V_{ub} + \bar{c}\gamma_\mu(1-\gamma^5)b\,V_{cb} + \cdots, \tag{7.133}$$

where V_{ub} and V_{cb} are the relevant elements of the CKM matrix in (7.24). Empirically $|V_{ub}| \ll |V_{cb}| \ll |V_{us}| \sim \sin\theta_c$. The small values for $|V_{ub}|$ and $|V_{cb}|$ lead to longer lifetimes for the b-flavored hadrons than would be naively expected.

Measurements of two-body leptonic decays $D^+ = c\bar{d} \to \mu^+\nu_\mu$, $D_s^+ = c\bar{s} \to \mu^+\nu_\mu$, $D_s^+ \to \tau^+\nu_\tau$, and $B^- \to \tau^-\bar{\nu}_\tau$ have been made by BaBar, Belle, BES, and CLEO, allowing extraction of the decay constants (analogous to f_π and f_K). The results are in reasonable agreement with the theoretical expectations, mainly based on lattice calculations (Rosner et al., 2015), although there are some $\lesssim 2\sigma$ tensions.

The c quark pole mass, at $\sim$ 1.6–1.7 GeV (Equation 5.124), is marginally large enough to attempt to use use simple parton model (SPM) ideas to describe its other decays, especially for the semi-leptonic modes. In the *spectator model* one ignores the hadronic aspects of the initial charmed (or b-flavored) hadron and simply considers the decay of a free quark, treating the light quarks or antiquarks as irrelevant spectators. For example, one would expect branching ratios for $c \to (s \text{ or } d)e^+\nu_e \sim 20\%$ by naive counting arguments, compared to the observed 16% for D^+ decays. The D^+ lifetime $\sim 1.04\times10^{-12}$ s is also close to the naive free quark estimate of $(m_\mu/m_c)^5\tau_\mu/5 \sim 6\times10^{-13}$ s (obtained from scaling the μ lifetime, obtained for $m_c \sim 1.6$ GeV and neglecting the final hadron masses). However, the spectator model is less successful for other charmed hadrons. For example, the $D^0 = c\bar{u}$ meson would have the same lifetime and semi-leptonic branching ratio, whereas experimentally $B(D^0 \to e^+ + X)$ is only $\sim 6.5\%$. (The lifetime, 4.1×10^{-13} s, is actually closer to the naive scaling estimate.) The charm mass scale is clearly too low for the SPM to be reliable, especially for the non-leptonic modes, and there must be important contributions to the decay amplitudes in addition to the simple c-quark decay diagram. More detailed descriptions employ various phenomenological and non-perturbative QCD techniques (e.g., Artuso et al., 2008b; Ryd and Petrov, 2012; Butler et al., 2013). For example, it is convenient to classify the contributions in terms of *topological diagrams* (Chau, 1983), as shown in Figure 7.8. It is also useful to distinguish between Cabibbo favored decays like $c \to s u \bar{d}$ (no powers of $\sin\theta_c$ in amplitude); Cabibbo suppressed, like $c \to d u \bar{d}$ or $c \to s u \bar{s}$ (one power); and doubly suppressed, like $c \to d u \bar{s}$ (two powers). $SU(3)$ and factorization assumptions are often employed as well.

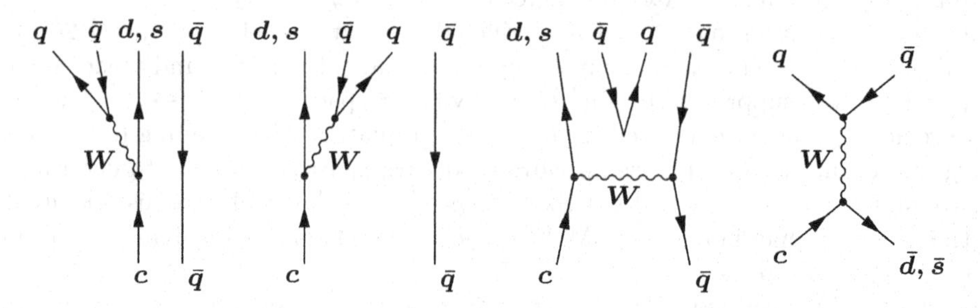

Figure 7.8 Tree-level topological diagrams for D decays, written in terms of an intermediate vector boson W. Gluons may be attached in all possible ways. The first (spectator) diagram represents free quark decay, the second (internal emission) diagram has a reduced color factor (analogous to (5.15)), the third is the exchange diagram, and the last is the annihilation diagram. There are additional one-loop (penguin) diagrams (Section 8.6).

The SPM estimates for the decay of the b quark are expected to be more reliable because of its larger pole mass of $4.8 - 5.0$ GeV. For example, the $B^+ = \bar{b}u$ semi-leptonic branching ratio into $\ell^+\nu_\ell + X$ of 10.99(28)% (this is the *average* of $e^+\nu_e$ and $\mu^+\nu_\mu$) agrees with the naive SPM expectation of 1/9, based on possible decays into three lepton and two quark families. The $B^0 = \bar{b}d \to \ell^+\nu_\ell + X$ branching ratio is similar (10.33(28)%), and the B^+ and B^0 lifetimes agree to within 7%.

One can use the semi-leptonic lifetimes to determine V_{cb}, either using the *inclusive* decays (summed over all charmed hadronic final states), with appropriate QCD corrections and quark masses (Benson et al., 2003), or the complementary *exclusive* measurements of specific decays such as $\bar{B} \to D\ell\bar{\nu}_\ell$ or $\bar{B} \to D^*\ell\bar{\nu}_\ell$. The latter require measurements of the shapes of form factors and HQET and lattice calculations of their normalization (see the review of V_{cb} and V_{ub} in Patrignani, 2016). The results, which are only marginally consistent, are $|V_{cb}| = 0.0422(8)$ (inclusive) and $|V_{cb}| = 0.0392(7)$ (exclusive), with a combined value

$$|V_{cb}| = 0.0405(15), \qquad |V_{ub}| = 0.00409(39). \tag{7.134}$$

The uncertainty has been increased because of the discrepancy, using the PDG procedure (Patrignani, 2016). This small value corresponds to the relatively long lifetime, $\tau_{B^+} \sim 1.6 \times 10^{-12}$ s, comparable to τ_{D^+} despite the much larger B mass, determined by the *displaced vertex* for the decay ~ 0.5 mm from the production vertex. The much rarer $b \to u$ transitions can also be studied by the inclusive semi-leptonic decays (exploiting the higher energy leptons that are not kinematically allowed in $b \to c\ell\bar{\nu}_\ell$), or in the exclusive $\bar{B} \to \pi\ell\bar{\nu}_\ell$ decays (with similar form factor complications as in the charm modes). There is again tension in the two methods, with $|V_{ub}| = 0.00449(23)$ (inclusive) and $|V_{ub}| = 0.00372(19)$ (exclusive), yielding the value for $|V_{ub}|$ in (7.134), again with an increased uncertainty.

B decays are reviewed in (Artuso et al., 2009; Butler et al., 2013), while the charmonium ($c\bar{c}$) and bottomonium ($b\bar{b}$) systems are described in (Eichten et al., 2008; Brambilla et al., 2011; Patrignani et al., 2013). The $K^0 - \bar{K}^0$, $D^0 - \bar{D}^0$, and $B^0 - \bar{B}^0$ systems exhibit or are predicted to have interesting (second order weak) mixing and CP violation effects that will be described along with other aspects of CP violation in Section 8.6.

The t quark, with pole mass $\sim$ 173 GeV, is much heavier than the $W^\pm$ mass of $\sim$ 80 GeV (Chapter 8). It is therefore predicted in the SM to decay directly to bW^+ (or much more rarely into sW^+ or dW^+) with the W on shell (Problem 8.15). Because it is so heavy, the t decays without hadronizing. Its production and decays are excellent probes of QCD, the electroweak theory, and V_{tb} (Déliot et al., 2014; Kröninger et al., 2015; Boos et al., 2015, and Chapter 8). The top couples strongly to the Higgs in the standard model, and its mass is critical to the issue of vacuum stability and to the Higgs mass prediction in supersymmetry (Section 8.5). It also plays an important role in many alternative models of spontaneous symmetry breaking and other types of BSM physics (e.g., Atwood et al., 2001; Zhang and Willenbrock, 2011; Boos et al., 2015).

Lepton Energy Distributions

We conclude this section with a comment on lepton energy distributions. In the four-fermi $V-A$ theory, neglecting the final fermion masses, strong interaction effects, etc., the energy distribution of each final fermion i is either $2\epsilon_i^2(3-2\epsilon_i)$ or $12\epsilon_i^2\,(1-\epsilon_i)$, where $\epsilon_i = 2E_i/M$ and M is the mass of the decaying particle. The first (second) distribution holds for the e^- ($\bar{\nu}_e$) in $\mu^- \to e^-\nu_\mu\bar{\nu}_e$, as is seen in (7.50) and Problem 7.2. The difference is due to the sign of the last term in (7.38), which in turn is associated with the relative $V-A$ or $V+A$ character of the currents (see (2.175) on page 37). In particular, the relevant product of currents for $b \to c\,e^-\bar{\nu}_e$ is

$$\bar{c}\gamma_\mu(1-\gamma^5)b\;\bar{e}\gamma^\mu(1-\gamma^5)\nu_e. \tag{7.135}$$

The b and e^- fields are related the same as for μ decay in (7.32), so the e^- energy distribution will be of the $2\epsilon^2(3-2\epsilon)$ type. For $c \to s\,e^+\nu_e$, on the other hand, the currents are

$$\bar{s}\gamma_\mu(1-\gamma^5)c\;\bar{\nu}_e\gamma^\mu(1-\gamma^5)e = -\bar{s}\gamma_\mu(1-\gamma^5)c\;\bar{e}^c\gamma^\mu(1+\gamma^5)\nu_e^c, \tag{7.136}$$

where we have used the charge conjugation identity (2.302) on page 56. The positron current is therefore $V+A$. This will give the other sign in (7.38) and a positron spectrum $12\epsilon^2\,(1-\epsilon)$.

These considerations are important for probing possible extensions or modifications of the $V-A$ theory or the standard model. For example, if the $b \to c$ transition had been $V+A$, then the e^- spectrum type would have been reversed.

7.3 PROBLEMS

7.1 Calculate the neutrino phase space coefficients A and B defined in (7.44) for μ decay.

7.2 (a) Suppose that the muon weak current were $V+A$ rather than $V-A$, so that the interaction for μ decay becomes

$$\mathcal{H} = \frac{G_F}{\sqrt{2}}\,\bar{\nu}_\mu\gamma_\mu\left(1+\gamma^5\right)\mu\;\bar{e}\gamma^\mu\left(1-\gamma^5\right)\nu_e$$

rather than (7.32). Calculate the electron energy distribution, and show that it corresponds to $\rho = 0$, where ρ is the Michel parameter defined in (7.55). Hint: the results for each part

can be written down without additional trace calculations.
(b) Calculate ρ for the (parity-conserving) vector interaction

$$\mathcal{H} = \sqrt{2} G_F \, \bar{\nu}_\mu \gamma_\mu \mu \;\, \bar{e} \gamma^\mu \nu_e,$$

where the $\sqrt{2}$ is inserted to obtain the same overall rate.
(c) Return to the Fermi $V - A$ theory for unpolarized $\mu^- \to e^- \nu_\mu \bar{\nu}_e$. Give the energy distributions for the ν_μ and for the $\bar{\nu}_e$. In each case the momenta of the other two particles are integrated over. Give a simple helicity argument why one of them vanishes at the endpoint, $\epsilon = 1$, but not the other.

7.3 Verify (7.58), (7.60), and the kinematic limits on y for elastic νe scattering.

7.4 Discuss the difference between the QED corrections to μ and β decay in the Fermi theory. Hint: Use an effective interaction

$$-\mathcal{L}_\beta = \frac{G_F}{\sqrt{2}} \, \bar{p} \gamma_\mu \left(1 - \gamma^5\right) n \;\, \bar{e} \gamma^\mu \left(1 - \gamma^5\right) \nu_e$$

for β decay, and write the four-fermi interactions in each case in charge retention form.

7.5 The formula (7.53) for polarized μ^- decay implies that the angular distribution is $1 - \cos\theta$ for a massless e^- of the maximum allowed energy $m_\mu/2$, where θ is the angle between the e^- direction and the muon spin. The analogous formula for μ^+ decay is $1+\cos\theta$. Interpret these results in terms of angular momentum conservation.

7.6 Use the spin projection in (2.178) on page 37 to verify the $\cos\theta = \hat{p}_e \cdot \hat{s}_\mu$ dependence of the expression (7.53) for polarized μ decay. Hint: show how the calculation leading to (7.49) is modified by the spin projection.

7.7 Use the formalism developed in Appendix D to verify the $3\Delta/2m_\pi$ correction in (7.96) to the pion beta decay rate. Neglect m_e.

7.8 Verify the $K_{\ell 3}$ decay rate in (7.98) in the special case $m_\pi \sim m_\ell \sim 0$, $f_+(t) \sim f_+(0)$, and $f_-(t) \sim 0$.

7.9 Use reflection invariance to show that $\langle \pi^i(\vec{p}_2) | A^j_\mu | \pi^k(\vec{p}_1) \rangle$, the matrix element of the axial current between single pion states, is zero.

7.10 Calculate (a) the neutron lifetime, (b) the e^- polarization, and (c) the electron asymmetry with respect to the neutron spin direction in terms of g_V and g_A defined in (7.121). Work in the approximation that the proton momentum is negligible compared to its mass. Hint: write out the amplitude explicitly using the Pauli-Dirac representation for the spinors and γ matrices. Use the helicity basis for the leptons.
(d) Repeat the lifetime calculation, this time using trace techniques, summing and averaging over all spins, and calculate the $e\nu$ angular correlation coefficient $\alpha_{e\nu}$. The latter is defined by the angular distribution $1+\alpha_{e\nu}\beta_e \hat{p}_e \cdot \hat{p}_\nu$, where $\vec{p}_\nu$ can be measured using $\vec{p}_\nu \equiv -(\vec{p}_p + \vec{p}_e)$.

7.11 Calculate the predicted rate and the angular distribution of the π^- with respect to the τ spin direction for $\tau^- \to \pi^- \nu_\tau$ in the Fermi theory. Show, using the measured lifetime $\tau_\tau = 2.9 \times 10^{-13}$ s, that it leads to a branching ratio in agreement with the observed value of 11%. Neglect m_π.

CHAPTER 8

The Standard Electroweak Theory

DOI: 10.1201/b22175-8

In this chapter we describe the standard $SU(3) \times SU(2) \times U(1)$ model in detail, including the structure of the Lagrangian density, the spontaneous symmetry breaking mechanism, and the rewriting of the Lagrangian after SSB. The electroweak gauge interactions, the properties of the W and Z, the Higgs, and CP violation are described, while neutrino mass and mixing is considered in Chapter 9. More detailed treatments may be found in the books and reviews listed in the bibliography, while the historical development is described in (Weinberg, 2004) and (Quigg, 2015).

8.1 THE STANDARD MODEL LAGRANGIAN

The standard model is based on the gauge group $G = SU(3) \times SU(2) \times U(1)$. The $SU(3)$ (QCD) factor has gauge coupling g_s and 8 gauge bosons (gluons) $G^i, i = 1 \cdots 8$. It is non-chiral, and acts on the color indices of the L- and R-chiral quarks $q_{r\alpha}$, where $\alpha = 1, 2, 3$ refers to color and r to flavor. QCD was described in detail in Chapter 5. However, the bare (current) masses introduced in (5.2) on page 160 are not allowed in the context of the full standard model, but must be generated by the Higgs mechanism. QCD itself is not spontaneously broken, and the gluons remain massless.

In contrast to QCD, the electroweak $SU(2) \times U(1)$ factor is chiral. The $SU(2)$ group has gauge coupling g, gauge bosons $W^i, i = 1, 2, 3$, and acts only on flavor indices of the L-chiral fermions. It leads to the charged current interactions of the Fermi theory, and also includes a neutral boson W^0 associated with a phase symmetry. The abelian $U(1)$ factor has gauge coupling g' and gauge boson B. It is also chiral, acting on both L and R fermions but with different charges. After spontaneous symmetry breaking (SSB), $SU(2) \times U(1)$ is broken to a single unbroken $U(1)_Q$, incorporating QED with the photon a linear combination of W^0 and B. The orthogonal combination (Z), as well as the $W^\pm$, acquire masses. G is sometimes written as $SU(3)_c \times SU(2)_L \times U(1)_Y$. The subscripts have no group-theoretic significance, but refer to the physical application, i.e., c refers to color, L to the left-chiral nature of the $SU(2)$ coupling, and Y to the *weak hypercharge* quantum number.

The standard model Langrangian density is

$$\mathcal{L} = \mathcal{L}_{gauge} + \mathcal{L}_f + \mathcal{L}_\phi + \mathcal{L}_{Yuk}, \tag{8.1}$$

which refer, respectively, to the gauge, fermion, Higgs, and Yukawa sectors of the theory.

There are in addition ghost and gauge-fixing terms that enter into the quantization. The gauge terms are

$$\mathcal{L}_{gauge} = -\frac{1}{4}G^i_{\mu\nu}G^{\mu\nu i} - \frac{1}{4}W^i_{\mu\nu}W^{\mu\nu i} - \frac{1}{4}B_{\mu\nu}B^{\mu\nu}, \tag{8.2}$$

where the field strength tensors for $SU(3)$, $SU(2)$, and $U(1)$ are, respectively,

$$\begin{aligned} G^i_{\mu\nu} &= \partial_\mu G^i_\nu - \partial_\nu G^i_\mu - g_s f_{ijk} G^j_\mu G^k_\nu, \quad i,j,k = 1\cdots 8 \\ W^i_{\mu\nu} &= \partial_\mu W^i_\nu - \partial_\nu W^i_\mu - g\epsilon_{ijk} W^j_\mu W^k_\nu, \quad i,j,k = 1\cdots 3 \\ B_{\mu\nu} &= \partial_\mu B_\nu - \partial_\nu B_\mu. \end{aligned} \tag{8.3}$$

These include the gauge boson kinetic energy terms as well as the three- and four-point self-interactions for the G^i and W^i, as shown for an arbitrary non-abelian theory in Figure 4.1. The abelian $U(1)$ gauge boson has no self-interactions. We ignore the possibility of $F\tilde{F}$ terms such as the θ_{QCD} one in (5.2).

The fermion part of the standard model involves $F = 3$ families of quarks and leptons. Each family consists of

$$\begin{aligned} L-\text{doublets}: \quad & q^0_{mL} = \begin{pmatrix} u^0_m \\ d^0_m \end{pmatrix}_L, \qquad \ell^0_{mL} = \begin{pmatrix} \nu^0_m \\ e^0_m \end{pmatrix}_L \\ R-\text{singlets}: \quad & u^0_{mR}, \; d^0_{mR}, \; e^0_{mR}, \; \nu^0_{mR}, \end{aligned} \tag{8.4}$$

in which the L-chiral fields are $SU(2)$ doublets and the R fields are singlets, leading to parity breaking in $SU(2)$. The superscripts ^{0} refer to the fact that these fields are *weak eigenstates*, i.e., they have definite gauge transformation properties, with the elements of each doublet transforming into each other under $SU(2)$, and $m = 1, 2, 3$ labels the family. After spontaneous symmetry breaking, these will become mixtures of mass eigenstate fields. The u^0 and d^0 are quarks, which will (after SSB) be identified as having electric charges $2/3$ and $-1/3$, respectively. There are altogether $2F = 6$ quark flavors (u^0 and d^0 for each family). Each carries a color index $u^0_{mL,R\alpha}$ or $d^0_{mL,R\alpha}$ not displayed in (8.4), so there are really 3 quark doublets per family. The $SU(2)$ and $SU(3)$ commute, so the QCD interactions do not change the flavor, and vice versa. ν^0 and e^0 are the leptons. They are color singlets and will have electric charges 0 and -1. We have tentatively included $SU(2)$-singlet right-handed neutrinos ν^0_{mR} in (8.4), because they are required in many models for neutrino mass. However, they are not necessary for the consistency of the theory or for some models of neutrino mass, and it is not certain whether they exist or are part of the low-energy theory. All of these fields except the ν^0_{mR} carry weak hypercharge Y, which is defined by

$$Y = Q - T^3_L, \tag{8.5}$$

where T^3_L is the third generator of $SU(2)_L$ and Q is the electric charge.[1] Weak hypercharge should not be confused with the strong hypercharge $\mathcal{Y}$ defined in (3.89) on page 104. $U(1)_Y$ commutes with $SU(3)_c$ and $SU(2)_L$, so it has the same value for all members of $SU(3) \times SU(2)$ multiplets. The Y eigenvalue is $y = q - t^3_L = \frac{1}{6}, \frac{2}{3}$, and $-\frac{1}{3}$ for q^0_{mL}, u^0_{mR}, and d^0_{mR}, respectively. For the leptons, $y = -\frac{1}{2}, 0$, and -1 for ℓ^0_{mL}, ν^0_{mR}, and e^0_{mR}.

The representations can be summarized by the symbol $\{n_3, n_2, y\}_\psi$ for fermion ψ, where n_3 and n_2 are the $SU(3)$ and $SU(2)$ representations and y is its hypercharge. Thus, the 16

[1] Some authors define $Q - T^3_L = Y/2$.

fields (15 without ν^0_{mR}) of each family transform as

$$
\begin{aligned}
\{3,2,\tfrac{1}{6}\}_{q^0_{mL}} \quad &\Rightarrow \quad q_{u^0_{mL}} = +1/2 + 1/6 = 2/3 \\
& \quad q_{d^0_{mL}} = -1/2 + 1/6 = -1/3 \\
\{1,2,-\tfrac{1}{2}\}_{\ell^0_{mL}} \quad &\Rightarrow \quad q_{\nu^0_{mL}} = +1/2 - 1/2 = 0 \\
& \quad q_{e^0_{mL}} = -1/2 - 1/2 = -1 \\
\{3,1,\tfrac{2}{3}\}_{u^0_{mR}} \quad &\Rightarrow \quad q_{u^0_{mR}} = 2/3 \\
\{3,1,-\tfrac{1}{3}\}_{d^0_{mR}} \quad &\Rightarrow \quad q_{d^0_{mR}} = -1/3 \\
\{1,1,-1\}_{e^0_{mR}} \quad &\Rightarrow \quad q_{e^0_{mR}} = -1 \\
\{1,1,0\}_{\nu^0_{mR}} \quad &\Rightarrow \quad q_{\nu^0_{mR}} = 0,
\end{aligned}
\tag{8.6}
$$

where $q = t^3_L + y$ is the electric charge eigenvalue.

The fermion representation is highly reducible, since each weak eigenstate family transforms only into itself. The existence of three families is empirical. $\mathcal{L}_f$ is actually invariant under a global $U(3)^6$ family symmetry, in which the 3 families of $q^0_{mL}, \ell^0_{mL}, u^0_{mR}, d^0_{mR}, e^0_{mR}$, and ν^0_{mR} transform into each other. However, there is no evidence that these symmetries are gauged. In any case most of the generators are broken by the Yukawa interactions. (The unbroken vector generators are baryon and lepton number.)

The SM is anomaly free (Section 4.5) for the assumed fermion content. There are no $SU(3)^3$ anomalies because the quark assignment is non-chiral, and no $SU(2)^3$ anomalies because the representations are real (Section 3.1.3). The $SU(2)^2Y$ and Y^3 anomalies cancel between the quarks and leptons in each family, by what appears to be an accident. The $SU(3)^2Y$ and Y anomalies cancel between the L and R fields, ultimately because the hypercharge assignments are made in such a way that $U(1)_Q$ will be non-chiral.

The $SU(2)_L$ and $U(1)_Y$ representations are chiral, so no fermion mass terms are allowed.[2] $\mathcal{L}_f$ therefore consists entirely of gauge-covariant kinetic energy terms,

$$
\begin{aligned}
\mathcal{L}_f = \sum_{m=1}^{F} & \left(\bar{q}^0_{mL} i \not{D} q^0_{mL} + \bar{\ell}^0_{mL} i \not{D} \ell^0_{mL} \right. \\
& \left. + \bar{u}^0_{mR} i \not{D} u^0_{mR} + \bar{d}^0_{mR} i \not{D} d^0_{mR} + \bar{e}^0_{mR} i \not{D} e^0_{mR} + \bar{\nu}^0_{mR} i \not{D} \nu^0_{mR} \right),
\end{aligned}
\tag{8.7}
$$

where we have allowed for an arbitrary number F of fermion families. The first term in (8.7) is

$$
\begin{aligned}
\bar{q}^0_{mL} i \not{D} q^0_{mL} = i & \sum_{\alpha,\beta=1}^{3} \left(\begin{array}{cc} \bar{u}^{0\alpha}_{mL} & \bar{d}^{0\alpha}_{mL} \end{array} \right) \gamma^\mu \\
& \times \left[\left(\partial_\mu I + \frac{ig}{2} \vec{\tau} \cdot \vec{W}_\mu + \frac{ig'}{6} I B_\mu \right) \delta_{\alpha\beta} + \frac{ig_s}{2} \vec{\lambda}_{\alpha\beta} \cdot \vec{G}_\mu I \right] \left(\begin{array}{c} u^0_{mL\beta} \\ d^0_{mL\beta} \end{array} \right),
\end{aligned}
\tag{8.8}
$$

where the I is the 2×2 $SU(2)$ identity matrix. It is clear from (8.8) that the $SU(3)_c$ and $SU(2)_L \times U(1)_Y$ groups commute. We will simplify the notation by suppressing the color

[2]The possibility of Majorana mass terms for the ν^0_{mR} will be considered in Chapter 9.

indices on the quark fields. Then, the fermion gauge covariant derivatives become

$$D_\mu q^0_{mL} = \left(\partial_\mu + \frac{ig}{2}\vec{\tau}\cdot\vec{W}_\mu + \frac{ig'}{6}B_\mu\right) q^0_{mL}, \qquad D_\mu u^0_{mR} = \left(\partial_\mu + \frac{2ig'}{3}B_\mu\right) u^0_{mR}$$
$$D_\mu \ell^0_{mL} = \left(\partial_\mu + \frac{ig}{2}\vec{\tau}\cdot\vec{W}_\mu - \frac{ig'}{2}B_\mu\right) \ell^0_{mL}, \qquad D_\mu d^0_{mR} = \left(\partial_\mu - \frac{ig'}{3}B_\mu\right) d^0_{mR}$$
$$D_\mu e^0_{mR} = \left(\partial_\mu - ig'B_\mu\right) e^0_{mR}, \qquad D_\mu \nu^0_{mR} = \partial_\mu \nu^0_{mR}, \tag{8.9}$$

where it is understood that there are also gluon couplings for the q^0_{mL}, u^0_{mR}, and d^0_{mR}. The fermion gauge interactions can be read off from (8.7).

The Higgs part of $\mathcal{L}$ is

$$\mathcal{L}_\phi = (D^\mu\phi)^\dagger D_\mu\phi - V(\phi), \tag{8.10}$$

where $\phi = \begin{pmatrix}\phi^+\\ \phi^0\end{pmatrix}$ is a complex Higgs scalar, transforming as $\{1, 2, \frac{1}{2}\}_\phi$. Its adjoint $\{1, 2^*, -\frac{1}{2}\}_{\phi^\dagger}$ is $\phi^\dagger = \begin{pmatrix}\phi^-\\ \phi^{0\dagger}\end{pmatrix}$. The gauge covariant derivative is

$$D_\mu\phi = \left(\partial_\mu + \frac{ig}{2}\vec{\tau}\cdot\vec{W}_\mu + \frac{ig'}{2}B_\mu\right)\phi. \tag{8.11}$$

The square of the covariant derivative leads to three- and four-point interactions between the gauge and Higgs fields. $V(\phi)$ is the Higgs potential. The combination of $SU(2)\times U(1)$ invariance and renormalizability restricts V to the form

$$V(\phi) = +\mu^2\phi^\dagger\phi + \lambda(\phi^\dagger\phi)^2, \tag{8.12}$$

where $\phi^\dagger\phi = \phi^-\phi^+ + \phi^{0\dagger}\phi^0$. For $\mu^2 < 0$ there will be spontaneous symmetry breaking, and the VEV of $\langle 0|\phi^0|0\rangle$ will generate the W and Z masses. The λ term describes a quartic self-interaction between the Higgs fields. Vacuum stability requires $\lambda > 0$.

The last term in (8.1) represents the Yukawa couplings between the Higgs doublet and the fermions, which are needed to generate fermion masses by the spontaneous breaking of the chiral gauge symmetries. For F fermion families it takes the form

$$\begin{aligned}\mathcal{L}_{Yuk} = -\sum_{m,n=1}^{F} \Big[&\Gamma^u_{mn}\bar{q}^0_{mL}\tilde{\phi}u^0_{nR} + \Gamma^d_{mn}\bar{q}^0_{mL}\phi d^0_{nR} \\ &+ \Gamma^e_{mn}\bar{\ell}^0_{mL}\phi e^0_{nR} + \Gamma^\nu_{mn}\bar{\ell}^0_{mL}\tilde{\phi}\nu^0_{nR}\Big] + h.c.,\end{aligned} \tag{8.13}$$

where

$$\phi = \begin{pmatrix}\phi^+\\ \phi^0\end{pmatrix}, \qquad \tilde{\phi} \equiv i\tau^2\phi^\dagger = \begin{pmatrix}\phi^{0\dagger}\\ -\phi^-\end{pmatrix} \tag{8.14}$$

are, respectively, the Higgs doublet and a conjugate form that will be discussed below. $\Gamma^u, \Gamma^d, \Gamma^e$, and Γ^ν are completely arbitrary $F\times F$ matrices, which ultimately determine the fermion masses and mixings. They do not have to be Hermitian, symmetric, diagonal, or real[3] (the hermiticity of $\mathcal{L}_{Yuk}$ is ensured by the addition of the $h.c.$ to the displayed terms).

[3]One can even generalize to non-square matrices of dimension $F_L\times F_R$, where $F_L \neq F_R$ are the numbers of L- and R-chiral fermions of a given type ψ. This would guarantee the existence of one or more massless 2-component fermions ψ_L or ψ_R unless there are also Majorana mass terms. Neither option is experimentally viable except for the neutrinos.

They are the most arbitrary aspect of the SM, introduce most of the free parameters, and break most of the $U(F)^6$ family symmetries of the rest of $\mathcal{L}$. The Γ^ν term would of course be absent if there are no ν^0_{mR}. For example, the mn term of Γ^d yields

$$\Gamma^d_{mn}\bar{q}^{\,0}_{mL}\phi d^0_{nR} = \Gamma^d_{mn}\left[\bar{u}^0_{mL}\phi^+ d^0_{nR} + \bar{d}^{\,0}_{mL}\phi^0 d^0_{nR}\right] \tag{8.15}$$

and its Hermitian conjugate

$$\Gamma^{d\dagger}_{nm}\bar{d}^{\,0}_{nR}\phi^\dagger q^0_{mL} = \Gamma^{d*}_{mn}\left[\bar{d}^{\,0}_{nR}\phi^- u^0_{mL} + \bar{d}^{\,0}_{nR}\phi^{0\dagger} d^0_{mL}\right], \tag{8.16}$$

which lead to the interaction vertices in Figure 8.1. Electric charge is conserved at the

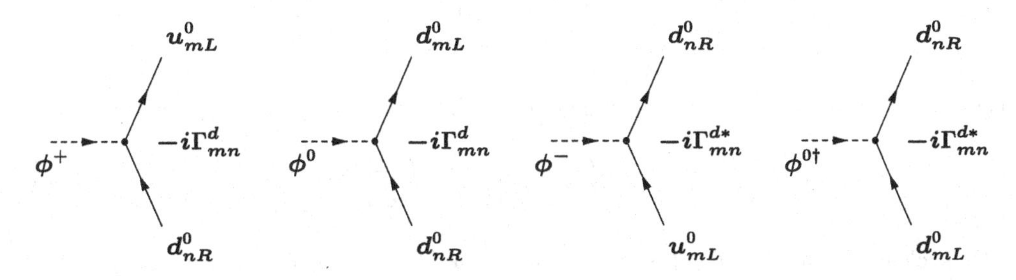

Figure 8.1 Yukawa couplings associated with $\Gamma^d_{mn}\bar{q}^{\,0}_{mL}\phi d^0_{nR} + h.c.$

vertices (guaranteed since Q is embedded in G), while chirality is flipped, which is characteristic of a Yukawa vertex. The family is changed for $m \neq n$. After SSB, the VEV $\langle 0|\phi^0|0\rangle$ will generate effective mass terms for the d^0 and other fermions.

It is straightforward to write the Yukawa couplings involving $\Gamma^{d,e}$ in (8.13). However, the Γ^u and Γ^ν terms require the existence of a complex scalar Higgs doublet $\Phi = \begin{pmatrix} \Phi^0 \\ \Phi^- \end{pmatrix}$ transforming as $\{1, 2, -\frac{1}{2}\}_\Phi$ so that one can write invariant terms $\bar{q}^{\,0}_{mL}\Phi u^0_{nR} + h.c.$ and $\bar{\ell}^{\,0}_{mL}\Phi\nu^0_{nR} + h.c.$ Neither ϕ nor $\phi^\dagger$ have the needed quantum numbers. However, in the SM one can avoid the need to introduce a second Higgs doublet Φ by utilizing the "tilde trick," namely that $\tilde{\phi}$ defined in (8.14) indeed transforms as $\{1, 2, -\frac{1}{2}\}_{\tilde{\phi}}$. That is because the 2^* representation of $SU(2)$ is equivalent to the 2, as shown in (3.17) on page 93. Thus, $\tilde{\phi}$ can play the role of Φ, with the identification $\phi^{0\dagger} = \Phi^0$ and $-\phi^- = \Phi^-$,

$$\Gamma^u_{mn}\bar{q}^{\,0}_{mL}\tilde{\phi} u^0_{nR} = \Gamma^u_{mn}\left[\bar{u}^0_{mL}\phi^{0\dagger} u^0_{nR} - \bar{d}^{\,0}_{mL}\phi^- u^0_{nR}\right]. \tag{8.17}$$

The equivalence of the fundamental and its conjugate does not generalize to higher unitary groups. Furthermore, in supersymmetric extensions of the standard model the supersymmetry forbids the use of a single Higgs doublet in both ways in $\mathcal{L}$, and one must add a second Higgs doublet. Similar statements apply to many theories with an additional $U(1)'$ gauge factor, i.e., a heavy Z' boson, or to the $SO(10)$ grand unified theory. One *can* get by with a single (but larger) Higgs multiplet in the extensions of the SM to the non-supersymmetric versions of $SU(2)_L \times SU(2)_R \times U(1)$ and $SU(5)$.

8.2 SPONTANEOUS SYMMETRY BREAKING

The SM as written is not realistic because bare mass terms are not allowed for the electroweak gauge bosons or for the fermions. However, as described in Sections 3.3 and 4.3

effective masses may be generated by spontaneous symmetry breaking, and in fact the Higgs doublet ϕ was introduced for that purpose. If its neutral component ϕ^0 acquires a nonzero VEV the $SU(2)_L \times U(1)_Y$ electroweak gauge symmetry will be broken to electromagnetism, $U(1)_Q$, generating masses for the chiral fermions. By the Higgs mechanism the Goldstone bosons will be absorbed to become the longitudinal components of the massive $W^\pm$ and Z.

8.2.1 The Higgs Mechanism

Similar to the case of a single complex scalar in Section 3.3.3 it is convenient to rewrite ϕ in a Hermitian basis as

$$\phi = \begin{pmatrix} \phi^+ \\ \phi^0 \end{pmatrix} = \begin{pmatrix} \frac{1}{\sqrt{2}}(\phi_1 + i\phi_2) \\ \frac{1}{\sqrt{2}}(\phi_3 + i\phi_4) \end{pmatrix}, \tag{8.18}$$

where $\phi_i = \phi_i^\dagger$ represents four Hermitian fields. In this new basis the Higgs potential becomes

$$V(\phi) = \frac{1}{2}\mu^2 \left(\sum_{i=1}^{4} \phi_i^2 \right) + \frac{1}{4}\lambda \left(\sum_{i=1}^{4} \phi_i^2 \right)^2 . \tag{8.19}$$

$V(\phi)$ is clearly $O(4) \sim SU(2) \times SU(2)$ invariant. This is an example of an *accidental symmetry*; the most general potential consistent with the $SU(2) \times U(1)$ gauge invariance and renormalizability exhibits a higher symmetry. The extra (global) generators are explicitly broken by the Yukawa and gauge interactions.

Without loss of generality we can choose the axis in this four-dimensional space so that $\langle 0|\phi_i|0\rangle = 0, \quad i = 1, 2, 4,$ and $\langle 0|\phi_3|0\rangle = \nu \geq 0$. (Other directions can be transformed into this form by an $SU(2) \times U(1)$ rotation.) Thus,

$$\begin{aligned} \phi \to \langle 0|\phi|0\rangle \equiv v &= \frac{1}{\sqrt{2}} \begin{pmatrix} 0 \\ \nu \end{pmatrix} \\ V(\phi) \to V(v) &= \frac{1}{2}\mu^2\nu^2 + \frac{1}{4}\lambda\nu^4, \end{aligned} \tag{8.20}$$

which must be minimized with respect to ν, analogous to the case of a single complex scalar in (3.146) on page 118 or a Hermitian scalar in (3.133). Two important cases are illustrated by the dashed and solid curves in Figure 3.4. For $\mu^2 > 0$ the minimum occurs at $\nu = 0$ and $SU(2) \times U(1)$ is unbroken at the minimum. On the other hand, for $\mu^2 < 0$ the $\nu = 0$ symmetric point is unstable, and the minimum occurs for $\nu \neq 0$, breaking the $SU(2) \times U(1)$ symmetry. The point is found by requiring

$$V'(\nu) = \nu(\mu^2 + \lambda\nu^2) = 0, \tag{8.21}$$

which has the solution

$$\nu = \sqrt{\frac{-\mu^2}{\lambda}} \tag{8.22}$$

at the minimum. The dividing point $\mu^2 = 0$ cannot be treated classically. It is necessary to consider the one-loop corrections to the effective potential, in which case it is found that the symmetry is again spontaneously broken (Coleman and Weinberg, 1973).

We are interested in the case $\mu^2 < 0$, for which the Higgs doublet is replaced, in first approximation, by its classical value v in (8.20). The generators corresponding to L^1, L^2, and $L^3 - Y$ are spontaneously broken,

$$L^i v = \frac{\tau^i}{2} \frac{1}{\sqrt{2}} \begin{pmatrix} 0 \\ \nu \end{pmatrix} \neq 0, \; i = 1, 2, 3, \qquad Yv = \frac{I}{2} \frac{1}{\sqrt{2}} \begin{pmatrix} 0 \\ \nu \end{pmatrix} \neq 0. \tag{8.23}$$

However, the vacuum carries no electric charge ($Qv \equiv (L^3 + Y)v = 0$), so the $U(1)_Q$ of electromagnetism is not broken,[4] and $SU(2)_L \times U(1)_Y \to U(1)_Q$.

We therefore expect the photon A, associated with the unbroken generator $Q \equiv L^3 + Y$, as well as the eight gluons, to remain massless, while $W^\pm = (W^1 \mp iW^2)/\sqrt{2}$ and Z, associated with $T^3 - Y$, become massive. To see this, one must quantize around the classical vacuum, i.e., write $\phi = v + \phi'$, where ϕ' are quantum fields with zero vacuum expectation value. To display the physical particle content it is useful to rewrite the four Hermitian components of ϕ' in terms of a new set of variables using the Kibble transformation, as discussed for a general gauge theory in Section 4.3.

$$\phi = \frac{1}{\sqrt{2}} e^{i(\sum_{i=1}^{3} \xi^i L'^i)} \begin{pmatrix} 0 \\ \nu + H \end{pmatrix}, \tag{8.24}$$

where the L'^i are the three broken generators L^1, L^2, and $L^3 - Y$, and H is a Hermitian scalar field, the physical *Higgs boson.*

If we had been dealing with a spontaneously broken global symmetry the three Hermitian fields ξ^i would be the massless pseudoscalar Goldstone bosons. These would have no potential and would only appear as derivatives. In a gauge theory they disappear from the physical spectrum, as we saw in Section 4.3. It is useful to quantize in the unitary gauge,

$$\phi \to \phi' = e^{-i\sum \xi^i L'^i}\phi = \frac{1}{\sqrt{2}} \begin{pmatrix} 0 \\ \nu + H \end{pmatrix}, \qquad \tilde{\phi} \to \frac{1}{\sqrt{2}} \begin{pmatrix} \nu + H \\ 0 \end{pmatrix}, \tag{8.25}$$

along with the corresponding transformations on the other fields. The unitary gauge is simplest for displaying the particle content of the theory, because the Goldstone bosons disappear and only the physical degrees of freedom remain. As described in Section 4.4 it is better to use other gauges for calculating higher-order contributions to physical processes because the unitary gauge is highly singular, and delicate cancellations between diagrams are required.

8.2.2 The Lagrangian in Unitary Gauge after SSB

Let us rewrite the Lagrangian in (8.1) after spontaneous symmetry breaking in the unitary gauge.

The Gauge and Higgs Sectors

The Higgs covariant kinetic energy term takes the simple form

$$(D_\mu\phi)^\dagger D^\mu\phi = \frac{1}{2}(0\ \nu)\left[\frac{g}{2}\tau^i W^i_\mu + \frac{g'}{2}B_\mu\right]^2 \begin{pmatrix} 0 \\ \nu \end{pmatrix} + H \text{ terms.} \tag{8.26}$$

We will return to the kinetic energy and gauge interaction terms of the physical H field later, but for now we concentrate on the part depending only on ν. Equation (8.26) can be rewritten using

$$\tau^i W^i = \tau^3 W^3 + \sqrt{2}\tau^+ W^+ + \sqrt{2}\tau^- W^-, \tag{8.27}$$

where

$$W^\pm = \frac{W^1 \mp iW^2}{\sqrt{2}}, \qquad \tau^\pm = \frac{\tau^1 \pm i\tau^2}{2}, \tag{8.28}$$

[4]It is automatic that a $U(1)$ remains unbroken in the minimal theory with a single Higgs doublet. However, in extended versions of the SM, e.g., with two or more Higgs doublets, all four generators might be broken, depending on the scalar potential. See Problems (8.3) and (8.4).

to obtain

$$\frac{g^2\nu^2}{4}W^{+\mu}W^-_\mu + \frac{1}{2}(g^2+g'^2)\frac{\nu^2}{4}\left[\frac{-g'B_\mu + gW^3_\mu}{\sqrt{g^2+g'^2}}\right]^2$$
$$\equiv M^2_W W^{+\mu}W^-_\mu + \frac{M^2_Z}{2}Z^\mu Z_\mu, \tag{8.29}$$

where $W^\pm$ are the complex charged gauge bosons that will mediate the charged current interactions, and

$$Z \equiv \frac{-g'B + gW^3}{\sqrt{g^2+g'^2}} = -\sin\theta_W B + \cos\theta_W W^3 \tag{8.30}$$

is a massive Hermitian vector boson that will mediate the new neutral current interaction predicted by $SU(2)\times U(1)$. In the second form, θ_W is the weak angle, defined by

$$\tan\theta_W \equiv \frac{g'}{g} \quad \Rightarrow \quad \sin\theta_W = \frac{g'}{g_Z}, \quad \cos\theta_W = \frac{g}{g_Z}, \tag{8.31}$$

where[5]

$$g_Z \equiv \sqrt{g^2+g'^2}. \tag{8.32}$$

The combination of B and W^3 orthogonal to Z is the photon (γ), with field

$$A = \cos\theta_W B + \sin\theta_W W^3, \tag{8.33}$$

which remains massless. The (tree-level) masses are predicted to be

$$M_W = \frac{g\nu}{2}, \qquad M_Z = \frac{g_Z\nu}{2} = \frac{M_W}{\cos\theta_W}, \qquad M_A = 0, \tag{8.34}$$

implying the relation

$$\sin^2\theta_W = 1 - \frac{M^2_W}{M^2_Z}. \tag{8.35}$$

One can think of the generation of masses as due to the fact that the W and Z interact constantly with the condensate of scalar fields and therefore acquire masses, as in Figure 8.2, in analogy with a photon propagating through a plasma. Each Goldstone boson has disappeared from the theory but has re-emerged as the longitudinal mode of a massive vector. The number of field degrees of freedom is unchanged. Before SSB, there are 4 massless electroweak gauge bosons, each with 2 helicities, and 4 Hermitian scalars, ϕ_i, for a total of 12. After SSB, there are 3 massive bosons, each with 3 helicities, the photon with 2, and one Hermitian scalar H, again totalling 12. One sees from (8.34) that in the limit $g' \to 0$ one would have $M_W = M_Z$. That is because the global $O(4)$ symmetry of (8.19) is broken to $O(3) \sim SU(2)$ by SSB. This global *custodial* symmetry is respected by the $SU(2)$ gauge interactions in (8.26) for $g' = 0$, so that $M_{W^\pm} = M_{W^3} = M_Z$. On a deeper level, the custodial $SO(3)$ ensures that the coefficient ν^2 is the same for the $W^\pm$ and Z mass terms in (8.29), even for $g' \neq 0$, implying $M_W = M_Z\cos\theta_W$. Since this relation is well satisfied experimentally, alternative models of spontaneous symmetry breaking often involve a custodial $SU(2)$ global symmetry to maintain it.

[5]There is a confusing variety of notations for g, g', $\sqrt{\frac{5}{3}}g'$, and g_Z in the literature (even in the author's own papers). Similarly, there is no uniformity as to whether symbols such as ν represent $\langle\phi^0\rangle$ or $\sqrt{2}\langle\phi^0\rangle$.

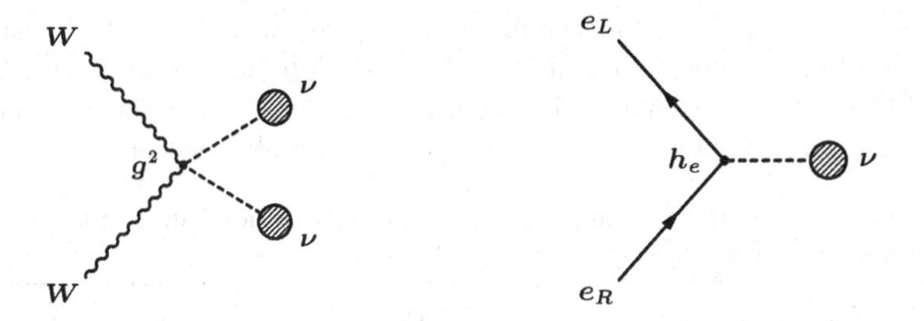

Figure 8.2 Effective masses for the W and electron generated by the VEV ν of the neutral Higgs field.

It will be seen below that $G_F/\sqrt{2} \sim g^2/8M_W^2$, where $G_F = 1.1663787(6) \times 10^{-5}$ GeV^{-2} is the Fermi constant determined by the muon lifetime. The weak scale ν is therefore

$$\nu = 2M_W/g \simeq (\sqrt{2}G_F)^{-1/2} \simeq 246 \text{ GeV}. \tag{8.36}$$

Similarly, we will see that $g = e/\sin\theta_W$, where e is the electric charge of the positron. Hence, to lowest order

$$M_W = M_Z \cos\theta_W \sim \frac{(\pi\alpha/\sqrt{2}G_F)^{1/2}}{\sin\theta_W}, \tag{8.37}$$

where $\alpha \sim 1/137.036$ is the fine structure constant. We will see that $\sin^2\theta_W$ can be measured from the predicted neutral current scattering to have a value ~ 0.23, so one expects $M_W \sim 78$ GeV, and $M_Z \sim 89$ GeV. (These predictions are increased by ~ 2 GeV by loop corrections, including the running of α.) The W and Z were discovered at CERN by the UA1 (Arnison et al., 1986) and UA2 (Ansari et al., 1987) collaborations in 1983, in the reactions $\bar{p}p \to W + X \to \ell\nu + X$ and $\bar{p}p \to Z + X \to \ell^+\ell^- + X$, where X represents unobserved hadrons. Subsequent measurements of their masses and other properties have been in excellent agreement with the standard model expectations, including the higher-order corrections (Patrignani, 2016). The current values are

$$M_W = 80.385 \pm 0.015 \text{ GeV}, \qquad M_Z = 91.1876 \pm 0.0021 \text{ GeV}. \tag{8.38}$$

The full Higgs part of $\mathcal{L}$ is

$$\begin{aligned}\mathcal{L}_\phi =&(D^\mu\phi)^\dagger D_\mu\phi - V(\phi)\\ =&M_W^2 W^{\mu+}W_\mu^- \left(1+\frac{H}{\nu}\right)^2 + \frac{1}{2}M_Z^2 Z^\mu Z_\mu \left(1+\frac{H}{\nu}\right)^2 + \frac{1}{2}\left(\partial_\mu H\right)^2 - V(\phi).\end{aligned} \tag{8.39}$$

The second line in (8.39) includes the mass terms, and also describes the ZZH^2, $W^+W^-H^2$ and the induced ZZH and W^+W^-H interactions, as shown in Table 8.1 and Figure 8.3. It also contains the canonical kinetic energy term and potential for the H. Similar to (3.140) on page 115, the Higgs potential (8.12) in unitary gauge becomes

$$V(\phi) \to -\frac{\mu^4}{4\lambda} - \mu^2 H^2 + \lambda\nu H^3 + \frac{\lambda}{4}H^4. \tag{8.40}$$

The first term is a constant, $\langle 0|V(\nu)|0\rangle = -\mu^4/4\lambda$. It reflects the fact that V was defined so that $V(0) = 0$, and therefore $V < 0$ at the minimum. Such a constant term is irrelevant to physics in the absence of gravity, but will be seen in Section 10.1 to be one of the most serious problems of the SM when gravity is incorporated because it acts like a cosmological constant much larger (and of opposite sign) than is allowed by observations.

TABLE 8.1 Feynman rules ($\sim i\mathcal{L}$) for the gauge and Higgs interactions after SSB, taking combinatoric factors into account.[a]

$W^+_\mu W^-_\nu H$:	$\frac{1}{2} i g_{\mu\nu} g^2 \nu = 2 i g_{\mu\nu} \frac{M_W^2}{\nu}$	$W^+_\mu W^-_\nu H^2$:	$\frac{1}{2} i g_{\mu\nu} g^2 = 2 i g_{\mu\nu} \frac{M_W^2}{\nu^2}$
$Z_\mu Z_\nu H$:	$\frac{1}{2} i g_{\mu\nu} g_Z^2 \nu = 2 i g_{\mu\nu} \frac{M_Z^2}{\nu}$	$Z_\mu Z_\nu H^2$:	$\frac{1}{2} i g_{\mu\nu} g_Z^2 = 2 i g_{\mu\nu} \frac{M_Z^2}{\nu^2}$
H^3:	$-6 i \lambda \nu = -3 i \frac{M_H^2}{\nu}$	H^4:	$-6 i \lambda = -3 i \frac{M_H^2}{\nu^2}$
$H \bar{f} f$:	$-i h_f = -i \frac{m_f}{\nu}$		
$W^+_\mu(p) \gamma_\nu(q) W^-_\sigma(r)$:	$ie\, \mathcal{C}_{\mu\nu\sigma}(p,q,r)$		
$W^+_\mu(p) Z_\nu(q) W^-_\sigma(r)$:	$ig \cos\theta_W \mathcal{C}_{\mu\nu\sigma}(p,q,r)$		
$W^+_\mu W^+_\nu W^-_\sigma W^-_\rho$:	$ig^2 \mathcal{Q}_{\mu\nu\rho\sigma}$		
$W^+_\mu Z_\nu \gamma_\sigma W^-_\rho$:	$-ieg \cos\theta_W \mathcal{Q}_{\mu\rho\nu\sigma}$		
$W^+_\mu Z_\nu Z_\sigma W^-_\rho$:	$-ig^2 \cos^2\theta_W \mathcal{Q}_{\mu\rho\nu\sigma}$		
$W^+_\mu \gamma_\nu \gamma_\sigma W^-_\rho$:	$-ie^2 \mathcal{Q}_{\mu\rho\nu\sigma}$		
$\mathcal{C}_{\mu\nu\sigma}(p,q,r) \equiv g_{\mu\nu}(q-p)_\sigma + g_{\mu\sigma}(p-r)_\nu + g_{\nu\sigma}(r-q)_\mu$			
$\mathcal{Q}_{\mu\nu\rho\sigma} \equiv 2 g_{\mu\nu} g_{\rho\sigma} - g_{\mu\rho} g_{\nu\sigma} - g_{\mu\sigma} g_{\nu\rho}$			

[a]The momenta and quantum numbers flow into the vertex. Note the dependence on M/ν or M^2/ν.

The second term in V represents a (tree-level) mass

$$M_H = \sqrt{-2\mu^2} = \sqrt{2\lambda}\,\nu \tag{8.41}$$

for the Higgs boson. The weak scale is given in (8.36), but the quartic Higgs coupling λ is unknown. A priori, λ could be anywhere in the range $0 < \lambda < \infty$, so there is no theoretical prediction for M_H. The Higgs was discovered at the LHC in 2012, with mass $M_H \sim 125$ GeV, completing the experimental verification of the standard model. The Higgs discovery, properties, and their implications, as well as the theoretical upper (from unitarity and perturbativity) and lower (from vacuum stability) limits on M_H, and indirect constraints from precision experiments, will be described in Section 8.5.

The last two terms in V are, respectively, the induced cubic and the quartic self-interactions of the Higgs.

The $SU(2)$ gauge kinetic energy terms in (8.2) lead to 3- and 4-point gauge self-interactions for the W's,

$$\begin{aligned} \mathcal{L}_{W3} = &- ig(\partial_\rho W^3_\nu) W^+_\mu W^-_\sigma \left[g^{\rho\mu} g^{\nu\sigma} - g^{\rho\sigma} g^{\nu\mu}\right] \\ &- ig(\partial_\rho W^+_\mu) W^3_\nu W^-_\sigma \left[g^{\rho\sigma} g^{\mu\nu} - g^{\rho\nu} g^{\mu\sigma}\right] \\ &- ig(\partial_\rho W^-_\sigma) W^3_\nu W^+_\mu \left[g^{\rho\nu} g^{\mu\sigma} - g^{\rho\mu} g^{\nu\sigma}\right], \end{aligned} \tag{8.42}$$

and

$$\mathcal{L}_{W4} = \frac{g^2}{4} \left[W^+_\mu W^+_\nu W^-_\sigma W^-_\rho \mathcal{Q}^{\mu\nu\rho\sigma} - 2 W^+_\mu W^3_\nu W^3_\sigma W^-_\rho \mathcal{Q}^{\mu\rho\nu\sigma}\right], \tag{8.43}$$

where $\mathcal{Q}_{\mu\nu\rho\sigma}$ is defined in Table 8.1. These carry over to the W, Z, and γ self-interactions

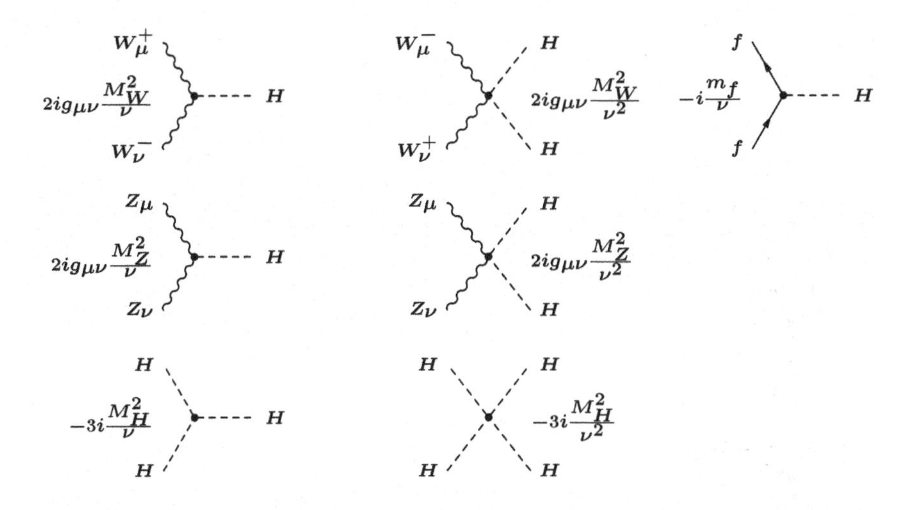

Figure 8.3 Higgs interaction vertices in the standard model.

provided we replace W^3 by $\cos\theta_W Z + \sin\theta_W A$ using (8.30) and (8.33) (the B has no self-interactions). The resulting vertices follow from the matrix element of $i\mathcal{L}$ after including identical particle factors and using $g = e/\sin\theta_W$. They are listed in Table 8.1 and shown in Figure 8.4. (They also follow from Problem 4.8.)

The Yukawa Sector

The fermions acquire masses by the SSB, as illustrated in Figure 8.2. Inserting the unitary gauge expressions for ϕ and $\tilde{\phi}$ in (8.25) into (8.13) yields (for F families)

$$\begin{aligned} -\mathcal{L}_{Yuk} &= \sum_{m,n=1}^{F} \bar{u}^0_{mL} \Gamma^u_{mn} \left(\frac{\nu + H}{\sqrt{2}} \right) u^0_{nR} + (d, e, \nu) \text{ terms} + h.c. \\ &= \bar{u}^0_L \left(M^u + h^u H \right) u^0_R + (d, e, \nu) \text{ terms} + h.c., \end{aligned} \tag{8.44}$$

where in the second form $u^0_L = \left(u^0_{1L} u^0_{2L} \cdots u^0_{FL} \right)^T$ is an F-component column vector, with a similar definition for u^0_R. M^u is an $F \times F$ fermion mass matrix,

$$M^u_{mn} = \Gamma^u_{mn} \frac{\nu}{\sqrt{2}}, \tag{8.45}$$

induced by spontaneous symmetry breaking, and

$$h^u = \frac{\Gamma^u}{\sqrt{2}} = \frac{M^u}{\nu} = \frac{g M^u}{2 M_W} \tag{8.46}$$

is the Yukawa coupling matrix. We have already emphasized that Γ^u, and therefore M^u and h^u, need not be diagonal, Hermitian, or symmetric. To identify the physical particle content it is necessary to diagonalize M by separate unitary transformations A_L and A_R

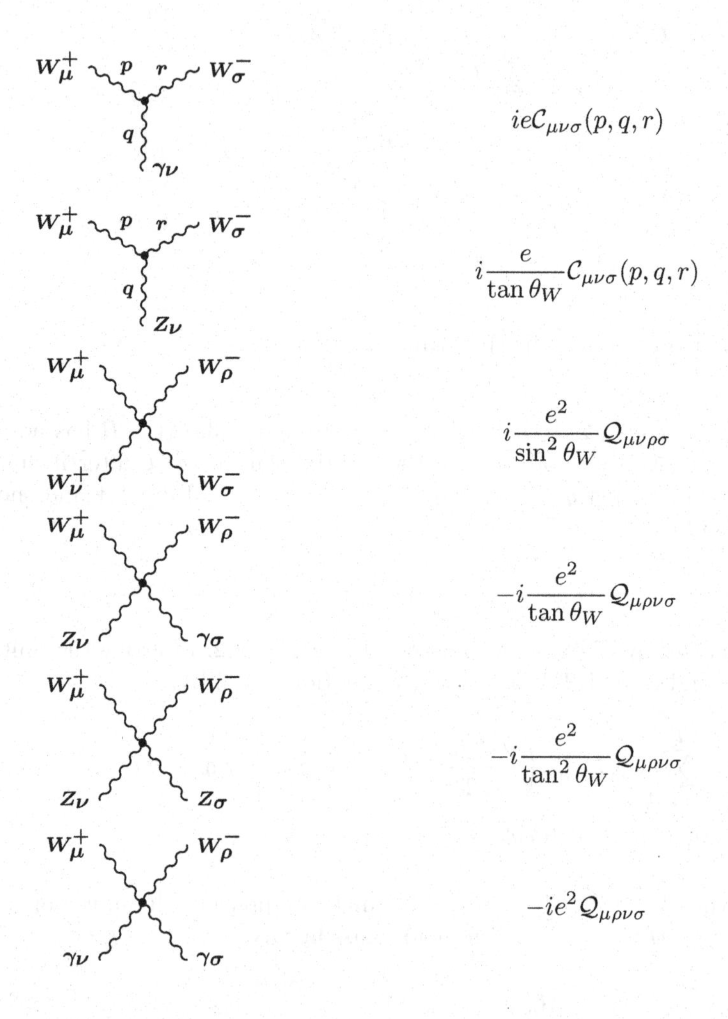

Figure 8.4 The three- and four-point self-interactions of gauge bosons in the standard electroweak model. The momenta and charges flow into the vertices.

on the left- and right-handed fermion fields.[6] Then (for $F = 3$),

$$A_L^{u\dagger} M^u A_R^u = M_D^u = \begin{pmatrix} m_u & 0 & 0 \\ 0 & m_c & 0 \\ 0 & 0 & m_t \end{pmatrix} \tag{8.47}$$

is a diagonal matrix with real non-negative eigenvalues equal to the physical masses of the charge $\frac{2}{3}$ quarks. Similarly, we denote the eigenvalues of the Yukawa matrix h^u by h_u, h_c, and h_t. One also diagonalizes the down quark, charged lepton, and neutrino mass matrices by

$$A_L^{d\dagger} M^d A_R^d = M_D^d, \qquad A_L^{e\dagger} M^e A_R^e = M_D^e, \qquad A_L^{\nu\dagger} M^\nu A_R^\nu = M_D^\nu. \tag{8.48}$$

In terms of these unitary matrices we can define mass eigenstate fields $u_L = A_L^{u\dagger} u_L^0 = (u_L \; c_L \; t_L)^T$, with analogous definitions for $u_R = A_R^{u\dagger} u_R^0$, $d_{L,R} = A_{L,R}^{d\dagger} d_{L,R}^0$, $e_{L,R} = A_{L,R}^{e\dagger} e_{L,R}^0$, and $\nu_{L,R} = A_{L,R}^{\nu\dagger} \nu_{L,R}^0$, so that, e.g.,

$$\bar{u}_L^0 M^u u_R^0 = \bar{u}_L M_D^u u_R. \tag{8.49}$$

For the quarks, the mass eigenvalues in $M_D^{u,d}$ are the masses that appeared as bare or current masses in the QCD Lagrangian density (5.2) on page 160, even though in the context of the full SM they are really spontaneously generated. They are not to be confused with the dynamical masses of $\mathcal{O}(300 \text{ MeV})$ generated by the spontaneous breaking of the global $SU(3)_L \times SU(3)_R$ chiral symmetry of QCD, as described in Section 5.8.

So far we have only allowed for ordinary Dirac mass terms of the form $\bar{\nu}_{mL}^0 \nu_{nR}^0$ for the neutrinos, which can be generated by the ordinary Higgs mechanism. Another possibility are lepton number violating *Majorana masses*, which (for ν_L^0) require an extended Higgs sector or higher-dimensional operators. It is not clear yet whether Nature utilizes Dirac masses, Majorana masses, or both. These issues will be discussed in Chapter 9. What is known is that the neutrino mass eigenvalues are tiny compared to the other masses, $\lesssim \mathcal{O}(0.1 \text{ eV})$, and most experiments are insensitive to them. In describing such processes, one can ignore Γ^ν, and the ν_R effectively decouple. Since $M^\nu \sim 0$ the three mass eigenstates are effectively degenerate with eigenvalues 0, and the eigenstates are arbitrary. That is, there is nothing to distinguish them except their weak interactions,[7] so we can simply define ν_e, ν_μ, ν_τ as the weak interaction partners of the e, μ, and τ, which is equivalent to choosing $A_L^\nu \equiv A_L^e$ so that $\nu_L = A_L^{e\dagger} \nu_L^0$. Of course, this is not appropriate for physical processes, such as oscillation experiments, that *are* sensitive to the masses or mass differences.

The unitary matrices $A_{L,R}$ can be constructed by noting that $MM^\dagger$ and $M^\dagger M$ are Hermitian. From (8.47) and its conjugate one has

$$\hat{A}_L^{u\dagger} M^u M^{u\dagger} \hat{A}_L^u = \hat{A}_R^{u\dagger} M^{u\dagger} M^u \hat{A}_R^u = M_D^{u2} = \begin{pmatrix} m_{u_1}^2 & 0 & 0 & 0 \\ 0 & m_{u_2}^2 & 0 & 0 \\ 0 & 0 & m_{u_3}^2 & 0 \\ 0 & 0 & 0 & \ddots \end{pmatrix}. \tag{8.50}$$

$\hat{A}_{L,R}^u$ and the eigenvalues can be constructed by elementary techniques. The Hermiticity of $MM^\dagger$ and $M^\dagger M$ guarantees that the $m_{u_r}^2$ are real and that the eigenvectors are orthogonal,

---

[6]The special case of a Hermitian (symmetric) M can be diagonalized by $A_L = A_R$ ($A_L = A_R^*$). However, additional phases in A_R, analogous to (8.51), may be required to ensure positive (and real) eigenvalues.

[7]As discussed in Section 8.6.6, this implies that the lepton flavors L_e, L_μ and L_τ are separately conserved for $m_\nu = 0$.

while their special form[8] implies that $m^2_{u_r} \geq 0$. However, the $\hat{A}^u_{L,R}$ are not unique, as is indicated by the hats. We have implicitly assumed in writing (8.50) that the eigenstates are ordered in the same way for $MM^\dagger$ and $M^\dagger M$. Even then, $A^u_{L,R}$ are only determined up to phases. That is, if $\hat{A}^u_{L,R}$ are solutions to (8.50), then so are

$$A^u_L \equiv \hat{A}^u_L K^u_L, \qquad A^u_R \equiv \hat{A}^u_R K^u_R, \tag{8.51}$$

where

$$K^u_{L,R} = \begin{pmatrix} e^{i\phi^u_{1L,R}} & 0 & 0 & 0 \\ 0 & e^{i\phi^u_{2L,R}} & 0 & 0 \\ 0 & 0 & e^{i\phi^u_{3L,R}} & 0 \\ 0 & 0 & 0 & \ddots \end{pmatrix} \tag{8.52}$$

are arbitrary diagonal phase matrices that correspond to the unobservable phases of the mass eigenstate $u_{rL,R}$ fields. (If there are degenerate eigenvalues, then there is additional freedom in $K^u_{L,R}$ associated with arbitrary unitary transformations in the space of degenerate eigenvectors.) For a particular choice of transformations, $\hat{A}^{u\dagger}_L M^u \hat{A}^u_R$ will be diagonal (provided that the eigenvectors are ordered the same way), but there is no guarantee that the elements will be real or positive. The usual prescription is to choose K^u_L for convenience, e.g., to remove unobservable phases from the CKM matrix. Then one can choose the phases in K^u_R so that the m^u_r are real and non-negative. The A^u_R are not observable in the SM, i.e., they don't enter the Lagrangian, though in principle they could be observable in some extensions of the SM involving new interactions of the R fields.

Finally, the fermion terms in $\mathcal{L}$ must be rewritten in terms of the mass eigenstate fields. For the u quarks, the kinetic energy and Yukawa terms are

$$\begin{aligned} \mathcal{L}_u &= \bar{u}^0_L i \not\partial u^0_L + \bar{u}^0_R i \not\partial u^0_R - \left[\bar{u}^0_L M^u u^0_R \left(1 + \frac{H}{\nu}\right) + h.c. \right] \\ &= \bar{u}_L i A^{u\dagger}_L A^u_L \not\partial u_L + \bar{u}_R i A^{u\dagger}_R A^u_R \not\partial u_R - \left[\bar{u}_L A^{u\dagger}_L M^u A^u_R u_R \left(1 + \frac{H}{\nu}\right) + h.c. \right] \\ &= \bar{u}_L i \not\partial u_L + \bar{u}_R i \not\partial u_R - \left[\bar{u}_L M^u_D u_R \left(1 + \frac{H}{\nu}\right) + h.c. \right] \\ &= \sum_{r=1}^{F} \bar{u}_r \left[i \not\partial - m_{u_r} \left(1 + \frac{H}{\nu}\right) \right] u_r, \end{aligned} \tag{8.53}$$

where $u_r \equiv u_{rL} + u_{rR}$. The kinetic energy terms remain in their canonical form since $A^u_{L,R}$ are unitary, and the mass terms are of canonical form since M^u_D is diagonal. The coupling of the physical Higgs boson H to u_r is $-ih_{u_r} = -im_{u_r}/\nu$, i.e., for a given ν the Higgs coupling is proportional to mass, just as for the gauge and self-couplings in Table 8.1.

All of these considerations apply to the d quarks and leptons as well, so

$$\mathcal{L}_\psi = \sum_r \bar{\psi}_r \left[i \not\partial - m_r \left(1 + \frac{H}{\nu}\right) \right] \psi_r, \tag{8.54}$$

where the sum runs over all of the fermions $\psi = u, d, e$, and ν (with the usual ν caveats). The coupling of H to ψ_r is $-im_r/\nu = -igm_r/2M_W$, which is very small except for the top quark. The decay $H \to b\bar{b}$ dominates since $M_H \ll 2M_W$. The branching ratios for

[8]That is, $m^2_{u_r} = \langle M^u x_r | M^u x_r \rangle \geq 0$, where x_r is the normalized eigenvector of $M^{u\dagger} M^u$.

$H \to \bar{c}c$ and $\tau^+\tau^-$ are smaller and the other $f\bar{f}$ rates are negligible, making the Higgs difficult to produce and difficult to observe. The Higgs Yukawa couplings are scalar and are flavor-diagonal in the minimal model: there is just one Yukawa matrix for each type of fermion, so the mass and Yukawa matrices are diagonalized by the same transformations. In generalizations in which more than one Higgs doublet couples to each type of fermion there will in general be flavor-changing Yukawa interactions involving the physical neutral Higgs fields (Glashow and Weinberg, 1977). There are stringent limits on such couplings; for example, the $K_L - K_S$ mass difference implies $h_{ds}/M_H < 10^{-6}\,\text{GeV}^{-1}$, where h_{ds} is the $\bar{d}s$ Yukawa coupling (Gaillard and Lee, 1974b; Langacker, 1991; Nir, 2015). One must also be careful that $U(1)_Q$ is not broken in such extensions.[9]

The Weak Charged Current (WCC)

We now rewrite the fermion gauge interactions in terms of the mass eigenstate fields. These can be read off from (8.7), using the expressions for the gauge covariant derivatives in (8.9); the expressions in (8.27), (8.30), and (8.33) for the gauge boson mass eigenstates; and the unitary transformations defined following (8.48), to obtain the fermion part of $\mathcal{L}$,

$$\begin{aligned} \mathcal{L}_f + \mathcal{L}_{Yuk} &= \mathcal{L}_\psi - \frac{g}{2\sqrt{2}}\left(J_W^\mu W_\mu^- + J_W^{\mu\dagger} W_\mu^+\right) \\ &\quad - \frac{gg'}{\sqrt{g^2+g'^2}} J_Q^\mu A_\mu - \frac{\sqrt{g^2+g'^2}}{2} J_Z^\mu Z_\mu \\ &\equiv \mathcal{L}_\psi + \mathcal{L}_W + \mathcal{L}_Q + \mathcal{L}_Z, \end{aligned} \tag{8.55}$$

where $\mathcal{L}_\psi$ is given in (8.54) and the other terms are the fermion gauge interactions. The $W^\pm$ terms are the charged current interaction, with charge raising and lowering currents

$$\begin{aligned} J_W^{\mu\dagger} &= \sum_{m=1}^{F} \left[\bar{\nu}_m^0 \gamma^\mu (1-\gamma^5) e_m^0 + \bar{u}_m^0 \gamma^\mu (1-\gamma^5) d_m^0\right] \\ J_W^{\mu} &= \sum_{m=1}^{F} \left[\bar{e}_m^0 \gamma^\mu (1-\gamma^5) \nu_m^0 + \bar{d}_m^0 \gamma^\mu (1-\gamma^5) u_m^0\right] \end{aligned} \tag{8.56}$$

in the weak eigenstate basis. We saw in Chapter 7 that these violate P and C maximally, but are CP conserving. The fermion gauge vertices are shown in Figure 8.5.

The amplitude for a t-channel four-fermion interaction in the SM is second order, as shown in Figure 8.6. It is

$$M \sim \frac{(i\mathcal{L}_W)^2}{2} = \left(\frac{-ig}{2\sqrt{2}}\right)^2 J_W^{\nu\dagger}\left(g_{\nu\mu} - \frac{q_\nu q_\mu}{M_W^2}\right)\left(\frac{-i}{q^2 - M_W^2}\right) J_W^\mu. \tag{8.57}$$

For small momentum transfer, $|q^\mu| \ll M_W$, this leads to the effective Hamiltonian density

$$\mathcal{H}_{eff} = iM = \frac{G_F}{\sqrt{2}} J_{W\mu}^\dagger J_W^\mu, \tag{8.58}$$

[9] The minimal supersymmetric extension of the SM does *not* involve flavor-changing neutral Higgs couplings to fermions or electric charge violation due to the Higgs fields even though there are two Higgs doublets. There are, however, potential problems associated with possible charge and color violating VEVs of scalar quark fields in some regions of parameter space.

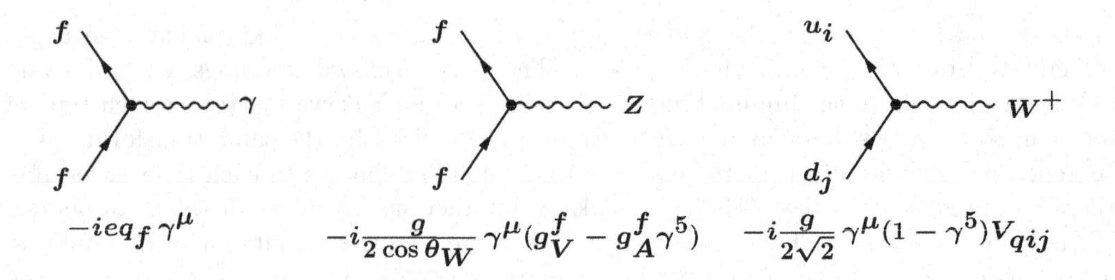

Figure 8.5 The fermion gauge interaction vertices in the standard electroweak model. $g_V^f \equiv t_{fL}^3 - 2\sin^2\theta_W q_f$ and $g_A^f \equiv t_{fL}^3$, where $t_{uL}^3 = t_{\nu L}^3 = +\frac{1}{2}$ and $t_{dL}^3 = t_{eL}^3 = -\frac{1}{2}$. The $\bar{d}_j u_i W^-$ vertex is the same as for $\bar{u}_i d_j W^+$ except $V_{qij} \to \left(V_q^\dagger\right)_{ji} = V_{qij}^*$. The lepton-$W^\pm$ vertices are obtained by $u_i \to \nu_i$, $d_j \to e_j^-$, and $V_q \to V_\ell$.

which reproduces the effective Fermi theory provided we identify the Fermi constant as

$$\frac{G_F}{\sqrt{2}} = \frac{g^2}{8M_W^2} = \frac{1}{2\nu^2}. \tag{8.59}$$

(Of course, the correspondence carries over to decay processes and s and u channel exchanges.) (8.59) allows us to determine the weak scale $\nu \sim 246$ GeV, as in (8.36). The phenomenology of the weak charged current interactions was discussed in Chapter 7.

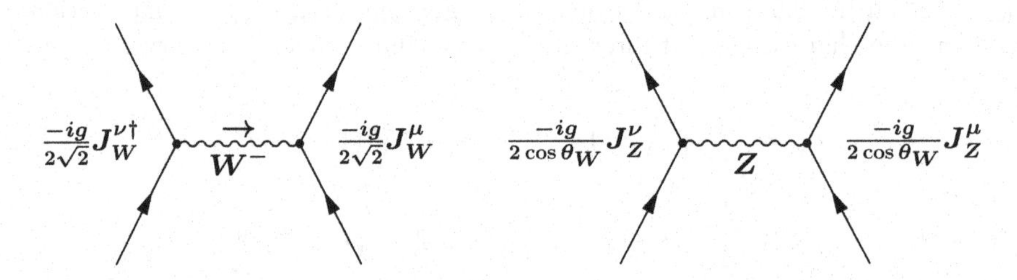

Figure 8.6 Charged and neutral current four-fermion t-channel exchange processes in the standard model. The W^- or Z carries four-momentum q.

Rewriting the currents in terms of mass eigenstates

$$J_W^{\mu\dagger} = 2\bar{\nu}_L\gamma^\mu \underbrace{A_L^{\nu\dagger}A_L^e}_{V_\ell} e_L + 2\bar{u}_L\gamma^\mu \underbrace{A_L^{u\dagger}A_L^d}_{V_q} d_L, \qquad J_W^\mu = 2\bar{e}_L\gamma^\mu V_\ell^\dagger \nu_L + 2\bar{d}_L\gamma^\mu V_q^\dagger u_L, \tag{8.60}$$

where u_L, d_L, e_L, and ν_L are the F-component column vectors defined following (8.48). The $F \times F$ unitary quark mixing matrix

$$V_q \equiv A_L^{u\dagger}A_L^d \tag{8.61}$$

describes the mismatch between the unitary transformations relating the weak and mass eigenstates for the up and down-type quarks. It is ultimately due to the mismatch between the gauge and Yukawa interactions. V_ℓ is the analogous leptonic mixing matrix. It is critical for describing neutrino oscillations and other processes sensitive to neutrino masses, and will be described in Chapter 9. As commented above (8.50), however, for processes for which the neutrino masses are negligible we can effectively set $V_\ell = I$ (more precisely, V_ℓ will only enter such processes in the combination $V_\ell^\dagger V_\ell = I$, so it can be ignored).

An arbitrary complex $F \times F$ matrix involves $2F^2$ real parameters. However, V_q is unitary, which imposes F^2 constraints, $(V_q^\dagger V_q)_{mn} = \delta_{mn}$, so it can be described by F^2 real parameters. Not all of these are observable, however. Recall that the F fields u_L each have an arbitrary unobservable phase, as do the d_L fields, described by the matrices $K_L^{u,d}$ in (8.52). One can use the freedom in choosing $K_L^{u,d}$ to remove $2F-1$ phase differences from V_q, so that altogether there are

$$F^2 - (2F-1) = (F-1)^2 = \underbrace{\frac{F(F-1)}{2}}_{\text{angles}} + \underbrace{\frac{(F-1)(F-2)}{2}}_{\text{phases}} \tag{8.62}$$

observable parameters. These consist of $F(F-1)/2$ rotation angles (the number of angles in an $O(F)$ rotation), with the remainder being observable CP-violating phases. To see that such phases are CP-violating, one has that under CP

$$\begin{aligned} W^{\pm\mu} &\leftrightarrow -W^{\mp}_\mu \\ \bar{u}_{mL}\gamma_\mu V_{qmn} d_{nL} &\rightarrow -\bar{d}_{nL}\gamma^\mu V_{qmn} u_{mL} = -\bar{d}_{nL}\gamma^\mu V^{\dagger *}_{qnm} u_{mL}, \end{aligned} \tag{8.63}$$

where it is understood that $(t, \vec{x}\,) \rightarrow (t, -\vec{x}\,)$ as well. (The transformation of $W^{\pm\mu}$ in (8.63) was defined so that the Lagrangian would be invariant in the absence of phases.) The hadronic part of the charged current interaction therefore transforms into

$$\begin{aligned} \mathcal{L}^h_W(t,\vec{x}\,) &= -\frac{g}{\sqrt{2}}\left[\bar{d}_L\gamma^\mu V_q^\dagger u_L W^-_\mu + \bar{u}_L\gamma^\mu V_q d_L W^+_\mu\right] \\ &\rightarrow -\frac{g}{\sqrt{2}}\left[\bar{u}_L\gamma^\mu V_q^* d_L W^+_\mu + \bar{d}_L\gamma^\mu V_q^{\dagger *} u_L W^-_\mu\right], \end{aligned} \tag{8.64}$$

which differs from $\mathcal{L}^h_W(t, -\vec{x}\,)$ if there are observable phases[10] in V_q. One must have $F \geq 3$ families to obtain CP breaking in the weak charged current interactions, and all three must be involved in a given process to lead to an observable effect. These issues will be described in detail in Section 8.6.

For $F=2$ families, there is one angle and no phase, so V_q is just the Cabibbo rotation

$$V_{Cabibbo} = \begin{pmatrix} \cos\theta_c & \sin\theta_c \\ -\sin\theta_c & \cos\theta_c \end{pmatrix}, \tag{8.65}$$

where $\sin\theta_c \sim 0.23$, as in (7.22) on page 230. Thus, in the four quark theory, the amplitudes for $\ell_L \rightarrow \nu_{\ell L}$, $(d_L \rightarrow u_L$ or $s_L \rightarrow c_L)$, and $(d_L \rightarrow c_L$ or $s_L \rightarrow u_L)$ are proportional to $1, \cos\theta_c$, and $\sin\theta_c$, respectively. The Cabibbo rotation in (8.65) is an excellent approximation to transitions amongst the first two families, even in the full three family case, because the elements of V_q connecting the first two families to the third are very small.

For $F=3$ one has $V_q = V_{CKM}$, which involves three mixing angles and one observable CP-violating phase,

$$V_{CKM} = \begin{pmatrix} V_{ud} & V_{us} & V_{ub} \\ V_{cd} & V_{cs} & V_{cb} \\ V_{td} & V_{ts} & V_{tb} \end{pmatrix} \sim \begin{pmatrix} 1 & \lambda & \lambda^3 \\ \lambda & 1 & \lambda^2 \\ \lambda^3 & \lambda^2 & 1 \end{pmatrix}, \tag{8.66}$$

[10]There are no such phases in the weak basis, so the charged current vertices are CP-conserving while the quark mass matrix (and therefore the propagators) are CP-violating. The situation is reversed in the mass eigenstate basis, but the two descriptions are equivalent for physical amplitudes.

where $\lambda = \sin\theta_c$. The second form is an easy to remember approximation to the observed magnitude of each element, which displays a suggestive but not well understood hierarchical structure. These are order of magnitude only; each element may be multiplied by a phase and a coefficient of $\mathcal{O}(1)$. There have been extensive studies of K, D, and B decays as well as other decays and scattering processes to determine the elements of V_{CKM}, to test its unitarity (which could appear to be violated in the presence of new physics), and to test whether the observed CP violation can be described by the phase in V_{CKM}. These tests and parametrizations of V_{CKM} will be described in Section 8.6. Here we just note that V_{CKM} can most likely account for the CP violation observed in particle processes, but an additional source of CP breaking is required to account for *baryogenesis*, i.e., the origin of the baryon (matter-antimatter) asymmetry of the universe.

QED

The second gauge interaction term in (8.55) is the QED interaction

$$\mathcal{L}_Q = -\frac{gg'}{\sqrt{g^2+g'^2}} J_Q^\mu A_\mu, \tag{8.67}$$

where A_μ in (8.33) is the massless photon field and the (Hermitian) electromagnetic current

$$\begin{aligned} J_Q^\mu &= \sum_r q_r \bar{\psi}_r^0 \gamma^\mu \psi_r^0 = \sum_{m=1}^F \left[\frac{2}{3}\bar{u}_m^0\gamma^\mu u_m^0 - \frac{1}{3}\bar{d}_m^0\gamma^\mu d_m^0 - \bar{e}_m^0\gamma^\mu e_m^0\right] \\ &\equiv \frac{2}{3}\bar{u}^0\gamma^\mu u^0 - \frac{1}{3}\bar{d}^0\gamma^\mu d^0 - \bar{e}^0\gamma^\mu e^0 \end{aligned} \tag{8.68}$$

is the sum over all fermion vector currents weighted by their charges q_r. J_Q^μ is purely vector since we chose the weak hypercharge assignments $y = q - t_L^3$ for both the left and right chiral fields to yield the *same* q even though their t_L^3 differ. Comparing with (2.218) on page 43, we identify the positron electric charge with

$$e = \frac{gg'}{\sqrt{g^2+g'^2}} = g\sin\theta_W \quad \Rightarrow \quad \frac{1}{e^2} = \frac{1}{g^2} + \frac{1}{g'^2}. \tag{8.69}$$

J_Q^μ takes the same form in the mass eigenstate basis,

$$J_Q^\mu = \sum_{m=1}^F \left[\frac{2}{3}\bar{u}_m\gamma^\mu u_m - \frac{1}{3}\bar{d}_m\gamma^\mu d_m - \bar{e}_m\gamma^\mu e_m\right] \equiv \frac{2}{3}\bar{u}\gamma^\mu u - \frac{1}{3}\bar{d}\gamma^\mu d - \bar{e}\gamma^\mu e, \tag{8.70}$$

since, e.g.,

$$\bar{u}^0\gamma^\mu u^0 = \bar{u}_L^0\gamma^\mu u_L^0 + \bar{u}_R^0\gamma^\mu u_R^0 = \bar{u}_L\gamma^\mu A_L^{u\dagger}A_L^u u_L + \bar{u}_R\gamma^\mu A_R^{u\dagger}A_R^u u_R = \bar{u}\gamma^\mu u. \tag{8.71}$$

This is ultimately due to the fact that only fields of the same charge and chirality, such as the u_{mL}^0, are able to mix with each other. J_Q^μ is therefore flavor diagonal and family universal, as well as P, C, and CP preserving. The standard model incorporates QED and all of its successes, as described in Section 2.12.3.

The Weak Neutral Current (WNC)

The Fermi theory of charged current weak interactions and QED were well established before the development of the standard model and were incorporated into it. However, the weak neutral current (WNC) interaction (along with the W and Z bosons) was a new ingredient predicted by the $SU(2) \times U(1)$ unification.

The weak neutral current interaction is described by the last term in (8.55),

$$\mathcal{L}_Z = -\frac{\sqrt{g^2 + g'^2}}{2} J_Z^\mu Z_\mu = -\frac{g}{2\cos\theta_W} J_Z^\mu Z_\mu = -\frac{g_Z}{2} J_Z^\mu Z_\mu, \tag{8.72}$$

where Z_μ is the massive neutral boson defined in (8.30). The (Hermitian) weak neutral current is

$$\begin{aligned} J_Z^\mu &= \sum_r \bar{\psi}_r^0 \gamma^\mu \left[t_{rL}^3 (1 - \gamma^5) - 2 q_r \sin^2\theta_W \right] \psi_r^0 \\ &= \sum_r t_{rL}^3 \bar{\psi}_r^0 \gamma^\mu (1 - \gamma^5) \psi_r^0 - 2 \sin^2\theta_W \, J_Q^\mu, \end{aligned} \tag{8.73}$$

summed over all fermions. Specializing to the SM quantum numbers,

$$\begin{aligned} J_Z^\mu &= \bar{u}_L^0 \gamma^\mu u_L^0 - \bar{d}_L^0 \gamma^\mu d_L^0 + \bar{\nu}_L^0 \gamma^\mu \nu_L^0 - \bar{e}_L^0 \gamma^\mu e_L^0 - 2\sin^2\theta_W \, J_Q^\mu \\ &= \bar{u}_L \gamma^\mu u_L - \bar{d}_L \gamma^\mu d_L + \bar{\nu}_L \gamma^\mu \nu_L - \bar{e}_L \gamma^\mu e_L - 2\sin^2\theta_W \, J_Q^\mu, \end{aligned} \tag{8.74}$$

where the $t_L^3 = \pm\frac{1}{2}$ have been absorbed into the P_L, e.g., $\frac{1}{2}\bar{u}\gamma^\mu(1 - \gamma^5)u = \bar{u}_L \gamma^\mu u_L$. The neutral current has two contributions. The first only involves the left-chiral fields and is purely $V - A$. The second is proportional to the electromagnetic current with coefficient $\sin^2\theta_W$ and is purely vector. P and C are therefore violated in the neutral current interaction, though not maximally. There are no phases, so CP is conserved. J_Z^μ is sometimes written in terms of its chiral or its V, A components as

$$J_Z^\mu = 2\sum_r \bar{\psi}_r \gamma^\mu \left[\epsilon_L(r) P_L + \epsilon_R(r) P_R\right] \psi_r = \sum_r \bar{\psi}_r \gamma^\mu \left[g_V^r - g_A^r \gamma^5\right] \psi_r, \tag{8.75}$$

where $g_{V,A}^r = \epsilon_L(r) \pm \epsilon_R(r)$. This form is especially convenient for generalizing to alternative or extended gauge theories, and also for incorporating radiative corrections. In the SM

$$\begin{aligned} \epsilon_L(r) &= t_{rL}^3 - \sin^2\theta_W \, q_r, \qquad & \epsilon_R(r) &= -\sin^2\theta_W \, q_r \\ g_V^r &= t_{rL}^3 - 2\sin^2\theta_W \, q_r, \qquad & g_A^r &= t_{rL}^3. \end{aligned} \tag{8.76}$$

Like the electromagnetic current J_Z^μ is flavor-diagonal and has the same form in the weak and mass bases in the standard model; all fermions that have the same electric charge and chirality and therefore can mix with each other have the same $SU(2) \times U(1)$ assignments, so the form is not affected by the unitary transformations that relate the bases. It was for this reason that the GIM mechanism (Glashow et al., 1970) was introduced into the model, along with its prediction of the charm quark. Without it the d and s quarks would *not* have had the same $SU(2) \times U(1)$ assignments, and flavor-changing neutral currents would have resulted. To see this, suppose there were only one left-chiral quark doublet, $(u_L^0 \; d_L^0)^T$, with s_L^0 and the right-handed quarks being $SU(2)_L$ singlets. Then, in terms of the mass eigenstates $u_L = u_L^0$ and

$$\begin{pmatrix} d_L \\ s_L \end{pmatrix} = \begin{pmatrix} \cos\theta_c & -\sin\theta_c \\ \sin\theta_c & \cos\theta_c \end{pmatrix} \begin{pmatrix} d_L^0 \\ s_L^0 \end{pmatrix}, \tag{8.77}$$

the hadronic part of J_Z^μ would be

$$\begin{aligned} J_Z^{h\mu} =& \bar{u}_L^0 \gamma^\mu u_L^0 - \bar{d}_L^0 \gamma^\mu d_L^0 - 2\sin^2\theta_W \, J_Q^{h\mu} \\ =& \bar{u}_L \gamma^\mu u_L - \bar{d}_L \gamma^\mu d_L \cos^2\theta_c - \bar{s}_L \gamma^\mu s_L \sin^2\theta_c \\ & - \left(\bar{d}_L \gamma^\mu s_L + \bar{s}_L \gamma^\mu d_L\right) \cos\theta_c \sin\theta_c - 2\sin^2\theta_W \, J_Q^{h\mu}, \end{aligned} \tag{8.78}$$

leading to unacceptable strangeness changing neutral current transitions (as well as large box diagram effects), as shown in Figure 7.3. The absence of such effects is also a restriction on extensions of the standard model involving exotic fermions (Langacker and London, 1988b).

A typical four-fermion process mediated by the Z in the t channel is shown in Figure 8.6. In the limit that the momentum transfer is small compared to M_Z one can neglect the q-dependent terms in the propagator, and the interaction reduces to an effective four-fermi interaction[11]

$$\mathcal{H}_{eff}^{NC} = \frac{G_F}{\sqrt{2}} J_Z^\mu J_{Z\mu}. \tag{8.79}$$

The coefficient is the same as in the charged current case because

$$\frac{G_F}{\sqrt{2}} = \frac{g^2}{8M_W^2} = \frac{g^2 + g'^2}{8M_Z^2}. \tag{8.80}$$

That is, the difference in Z couplings compensates the difference in masses in the propagator. However, unlike the charged current, J_Z^μ is Hermitian, so there is an extra combinatoric factor of 2 when one takes a matrix element of $\mathcal{H}_{eff}^{NC}$.

The weak neutral current was discovered at CERN in 1973 by the Gargamelle bubble chamber collaboration (Hasert et al., 1973) and by HPW at Fermilab (Benvenuti et al., 1974) shortly thereafter, and since that time Z exchange and $\gamma - Z$ interference processes have been extensively studied in many interactions, including $\nu e \to \nu e$, $\nu N \to \nu N$, $\nu N \to \nu X$; polarized e^--hadron and μ-hadron scattering; atomic parity violation; and in e^+e^- and Z-pole reactions. Along with the properties of the W and Z they have been the primary quantitative test of the unification part of the standard electroweak model.

8.2.3 Effective Theories

Let us digress briefly on *effective theories*, of which the four-fermion WCC and WNC interactions derived from the SM are an example. An effective field theory is a description of physics at a given energy scale in terms of the degrees of freedom that can actually appear as physical states at that energy. In particular, it is often convenient to *integrate out* the fields corresponding to particles too heavy to produce, so that they do not appear explicitly in the theory but whose effects are described by non-renormalizable operators.[12] This can be done easily in the path integral formalism, but the results can also be obtained by examining Feynman diagrams (as we have done for WCC and WNC interactions in (8.58) and (8.79)), or by solving the Euler-Lagrange equation for the heavy field, neglecting the kinetic terms. To illustrate the latter method, let us rederive the fermion WNC interaction in (8.79). If we ignore the Higgs and self-interaction terms, the terms in $\mathcal{L}$ involving the Z are

$$\mathcal{L}_Z = -\frac{1}{4} Z_{\mu\nu} Z^{\mu\nu} + \frac{M_Z^2}{2} Z_\mu Z^\mu - \frac{g}{2\cos\theta_W} Z_\mu J_Z^\mu. \tag{8.81}$$

[11] Care must be taken with factors of 2 from t_{rL}^3, $P_{L,R}$, the coefficient in (8.72), etc., in deriving (8.79). Also, an additional factor of 2 will arise in the off-diagonal terms in the square of J_Z or in taking matrix elements of the diagonal terms. Finally, some authors absorb the $\frac{1}{2}$ in (8.72) into the definition of J_Z^μ, but fortunately the 2 is explicitly removed from the r.h.s. of (8.75) so that the $\epsilon_{L,R}(r)$ and $g_{V,A}^r$ are the same.

[12] According to the *decoupling theorem* (Appelquist and Carazzone, 1975) heavy particles do not enter the effective low energy theory except through renormalized parameters and non-renormalizable operators. However, care must be taken in the application of this result when the heavy particles violate a symmetry and therefore require strong coupling, as we will see in Section 8.3.6 in connection with the oblique parameters.

The Euler-Lagrange equation for the Z is therefore

$$\partial_\nu \left(\frac{\delta \mathcal{L}}{\delta \partial_\nu Z_\mu} \right) - \frac{\delta \mathcal{L}}{\delta Z_\mu} = 0 \Rightarrow \left(\Box + M_Z^2 \right) Z_\mu - \partial_\mu \partial_\nu Z^\nu = \frac{g}{2 \cos \theta_W} J_{Z\mu}. \tag{8.82}$$

At low energies compared to M_Z the derivatives will ultimately be replaced by factors of E, where E is a typical external energy for the process being considered. Therefore, (8.82) can be solved, yielding

$$Z_\mu = \frac{g J_{Z\mu}}{2 M_Z^2 \cos \theta_W} + \mathcal{O}\left(\frac{E^2}{M_Z^2} \right). \tag{8.83}$$

This can be reinserted into $\mathcal{L}_Z$, to yield the effective four-fermion operator

$$\mathcal{L}_{eff}^{NC} = -\mathcal{H}_{eff}^{NC} = -\frac{g^2}{8 M_Z^2 \cos^2 \theta_W} J_\mu^Z J_{Z\mu} = -\frac{G_F}{\sqrt{2}} J_\mu^Z J_{Z\mu}. \tag{8.84}$$

More generally, in an effective theory (e.g., Weinberg, 1995; Pich, 1998; Burgess, 2007; Willenbrock and Zhang, 2014; Ellis et al., 2015) one writes the most general Lagrangian density $\mathcal{L}_{eff}$ for the low energy fields that is consistent with the symmetries, including higher-dimensional (non-renormalizable) terms such as those in (8.84). Powers of derivatives ∂_μ / M, where M is the heavy particle scale, can also be included, so $\mathcal{L}_{eff}$ is a systematic expansion in powers of energy. The coefficients of the terms in $\mathcal{L}_{eff}$ can be obtained from experiment, or can be computed in terms of the underlying theory (as in (8.84)). One can treat the low energy effective theory like any other if one uses a mass-independent renormalization scheme (such as $\overline{MS}$), with a finite number of counterterms for any given power of energy. Major applications of effective theories are to describe the low energy limit of a known theory, focusing on the important aspects and symmetries (as in the example above), or to parametrize the observable effects of still unknown new physics associated with a higher scale M (such as the use of the four-fermi interaction to describe the WCC before the SM was developed). Another major example is chiral perturbation theory, touched on in Section 5.8, in which the low energy strong interactions are expressed in terms of an effective theory of mesons and baryons incorporating the spontaneously broken $SU(3) \times SU(3)$ flavor symmetry, the soft pion theorems, etc. The heavy quark and soft collinear effective theories were mentioned in Section 5.6.

8.2.4 The R_ξ Gauges

It is straightforward to write the interaction vertices and propagators in an arbitrary R_ξ gauge, as was discussed for a general non-abelian theory in Section 4.4. This is useful both for higher-order calculations and for applications of the equivalence theorem (Section 8.5). Following SSB, instead of the Kibble representation in (8.24) the Higgs doublet can be rewritten in terms of shifted fields

$$\phi = \begin{pmatrix} w^+ \\ \frac{\nu + H + iz}{\sqrt{2}} \end{pmatrix}, \tag{8.85}$$

just as in the $SU(2)$ example in Section 4.4. H is the physical Higgs scalar, while z, w^+, and $w^- = w^{+\dagger}$ are the Goldstone boson fields that disappear in the unitary gauge. We also define the ghost fields

$$\begin{aligned} \eta_Z &= \cos \theta_W \, \eta_3 - \sin \theta_W \, \eta_Y, & \eta_\gamma &= \sin \theta_W \, \eta_3 + \cos \theta_W \, \eta_Y \\ \eta^\pm &= \frac{\eta_1 \mp i \eta_2}{\sqrt{2}}, & \eta_C &= \cos 2\theta_W \, \eta_Z + \sin 2\theta_W \, \eta_\gamma, \end{aligned} \tag{8.86}$$

where η_i and η_Y are associated with $SU(2)_L$ and $U(1)_Y$, respectively. The anti-ghost fields are $\eta_Z^\dagger$, $\eta_\gamma^\dagger$, and

$$\eta^{\pm\dagger} = \frac{\eta_1^\dagger \pm i\eta_2^\dagger}{\sqrt{2}}. \tag{8.87}$$

The propagators for the mass eigenstate fields are listed in Table 8.2.

TABLE 8.2 Propagators for the standard model mass eigenstate fields in the R_ξ gauges.[a]

$iD_\gamma^{\mu\nu}(k)$:	$\frac{-i}{k^2}\left[g^{\mu\nu} - \frac{k^\mu k^\nu(1-1/\xi)}{k^2}\right]$		
$iD_{W^\pm}^{\mu\nu}(k)$:	$\frac{-i}{k^2-M_W^2}\left[g^{\mu\nu} - \frac{k^\mu k^\nu(1-1/\xi)}{k^2-M_W^2/\xi}\right]$		
$iD_Z^{\mu\nu}(k)$:	$\frac{-i}{k^2-M_Z^2}\left[g^{\mu\nu} - \frac{k^\mu k^\nu(1-1/\xi)}{k^2-M_Z^2/\xi}\right]$		
$i\Delta_H(k)$:	$\frac{i}{k^2-M_H^2}$	$i\Delta_z(k)$:	$\frac{i}{k^2-M_Z^2/\xi}$
$i\Delta_{w^\pm}(k)$:	$\frac{i}{k^2-M_W^2/\xi}$	$i\mathcal{D}_{\eta_\gamma}(k)$:	$\frac{-i}{k^2}$
$i\mathcal{D}_{\eta^\pm}(k)$:	$\frac{-i}{k^2-M_W^2/\xi}$	$i\mathcal{D}_{\eta_Z}(k)$:	$\frac{-i}{k^2-M_Z^2/\xi}$
$iS_{f\neq\nu}(k)$:	$\frac{i}{\not{k}-m_f} = i\frac{\not{k}+m_f}{k^2-m_f^2}$	$iS_\nu(k)$:	$\frac{i}{\not{k}}P_L = i\frac{\not{k}}{k^2}P_L$

[a] $\zeta = 0, 1$, and ∞ correspond, respectively, to the unitary, 't Hooft-Feynman, and Landau (renormalizable) gauges. A $+i\epsilon$ is implicit in each denominator. Neutrino masses are neglected. (The propagators for massive Dirac and Majorana neutrinos are considered in Section 9.1.4.)

The Higgs potential in (8.12) takes the form in (4.81), which is repeated here for convenience

$$V(\phi) = -\mu^2 H^2 + \lambda\nu H\left[H^2 + z^2 + 2w^+w^-\right] + \frac{\lambda}{4}\left[H^2 + z^2 + 2w^+w^-\right]^2. \tag{8.88}$$

The gauge interactions of the Higgs fields are

$$\begin{aligned}
\mathcal{L}_\phi = &- i\frac{g_Z}{2}\left(w^- \overleftrightarrow{\partial}^\mu w^+\right) C_\mu + \frac{g_Z}{2}\left(z \overleftrightarrow{\partial}^\mu H\right) Z_\mu \\
&- i\frac{g}{2}\left([H - iz]\overleftrightarrow{\partial}^\mu w^+\right) W_\mu^- - i\frac{g}{2}\left(w^- \overleftrightarrow{\partial}^\mu [H + iz]\right) W_\mu^+ \\
&+ \nu H\left(\frac{g^2}{2}W^{+\mu}W_\mu^- + \frac{g_Z^2}{4}Z_\mu^2\right) + \nu\frac{gg'}{2}\left(w^+W^{-\mu} + w^-W^{+\mu}\right) B_\mu \\
&+ w^+w^-\left(\frac{g^2}{2}W^{+\mu}W_\mu^- + \frac{g_Z^2}{4}C_\mu^2\right) + \frac{H^2 + z^2}{2}\left(\frac{g^2}{2}W^{+\mu}W_\mu^- + \frac{g_Z^2}{4}Z_\mu^2\right) \\
&+ \frac{gg'}{2}\left(w^+(H - iz)W^{-\mu} + w^-(H + iz)W^{+\mu}\right)B_\mu,
\end{aligned} \tag{8.89}$$

where

$$C_\mu \equiv \frac{gW_\mu^3 + g'B_\mu}{\sqrt{g^2 + g'^2}} = \cos 2\theta_W\, Z_\mu + \sin 2\theta_W\, A_\mu. \tag{8.90}$$

Equation (8.89) is a generalization of the $SU(2)$ example in (4.84) and of the unitary gauge expression in (8.39). (The gauge boson mass term is the same as (8.26).)

One can similarly extend the discussion of the Higgs Yukawa couplings in Section 8.2.2 to include the Goldstone bosons. Analogous to (8.53) the quark mass and Yukawa terms are

$$
\begin{aligned}
-\mathcal{L}_q &= \bar{u}M_D^u\left[1+\frac{H}{\nu}-i\gamma^5\frac{z}{\nu}\right]u+\bar{d}M_D^d\left[1+\frac{H}{\nu}+i\gamma^5\frac{z}{\nu}\right]d \\
&+\frac{\sqrt{2}}{\nu}\bar{d}\left[-V_q^\dagger M_D^u P_R+M_D^d V_q^\dagger P_L\right]u\,w^-+\frac{\sqrt{2}}{\nu}\bar{u}\left[-M_D^u V_q P_L+V_q M_D^d P_R\right]d\,w^+,
\end{aligned}
\tag{8.91}
$$

where u and d are column vectors of mass eigenstate fields, $M_D^{u,d}$ are the diagonal matrices of mass eigenvalues, and $V_q = A_L^{u\dagger}A_L^d$ is the quark mixing matrix. A similar result applies to the leptons, with $u \to \nu$, $d \to e$, and $V_q \to V_\ell$, provided the neutrino masses can be ignored or they are Dirac.

The ghost interactions with the scalars and gauge fields are

$$
\begin{aligned}
\mathcal{L}_{ghost} =& -ig\Big[\left(\partial^\mu\eta^{-\dagger}\right)\eta^- - \left(\partial^\mu\eta^{+\dagger}\right)\eta^+\Big]W_\mu^3 \\
& - ig\Big[\left(\partial^\mu\eta^{+\dagger}\right)W_\mu^+ - \left(\partial^\mu\eta^{-\dagger}\right)W_\mu^-\Big]\eta_3 - ig\left(\partial^\mu\eta_3^\dagger\right)\left[\eta^+W_\mu^- - \eta^-W_\mu^+\right] \\
& + \left[\frac{gM_W}{2\xi}\left(\eta^{+\dagger}\eta^+ + \eta^{-\dagger}\eta^-\right) + \frac{g_Z M_Z}{2\xi}\eta_Z^\dagger\eta_Z\right]H + \frac{gM_W}{2\xi}\left(\eta^{+\dagger}\eta^+ - \eta^{-\dagger}\eta^-\right)iz \\
& + \frac{gM_Z}{2\xi}\Big[\left(\eta^{+\dagger}\eta_C - \eta_Z^\dagger\eta^-\right)w^+ + \left(\eta^{-\dagger}\eta_C - \eta_Z^\dagger\eta^+\right)w^-\Big].
\end{aligned}
\tag{8.92}
$$

8.3 THE Z, THE W, AND THE WEAK NEUTRAL CURRENT

After the discovery of the weak neutral current in 1973 there were generations of weak neutral current experiments, eventually at the precision of a few % and in a few cases $\sim 0.5\%$. The motivation was in part to determine whether its properties agreed with the predictions of the $SU(2) \times U(1)$ model, and whether the latter could be distinguished experimentally from a number of alternative theories. This goal was largely achieved by the *model-independent analyses* of the data, which allowed arbitrary V and A interactions (but generally assumed family universality and $V - A$ couplings for neutrinos). These were consistent with the SM but eliminated competing gauge theories predicting four-fermi interactions very different from the SM predictions. Other possibilities involving purely S, P, and T interactions were excluded by the observation of weak-electromagnetic interference in processes involving charged particles. More complex gauge theory competitors with similar four-fermi interactions but different gauge bosons were largely ruled out by the later discovery of the Z and W. As later generations of more precise results (including the Z pole, Tevatron, LHC, and other high energy experiments) became available, the emphasis turned more towards precision tests of the SM at the loop level, leading to predictions for the top quark and eventually the Higgs masses, precise measurements of couplings for comparison with unification predictions, and testing the underlying structure of renormalizable gauge theories. They also allowed searches for and constraints on small deviations from the SM predictions that could be attributed to new physics beyond the standard model (BSM).

In this section we describe some of these tests and their implications. More detailed discussions may be found in, e.g., (Kim et al., 1981; Amaldi et al., 1987; Costa et al., 1988; Langacker et al., 1992; Langacker, 1995; Erler and Su, 2013; Patrignani, 2016). For a historical perspective, see (Langacker, 1993).

8.3.1 Purely Weak Processes

$\nu e^- \to \nu e^-$ Elastic Scattering

The neutral current reactions $\nu_\mu e^- \to \nu_\mu e^-$ and $\bar{\nu}_\mu e^- \to \bar{\nu}_\mu e^-$ do not occur in the Fermi theory, but are predicted to proceed by t-channel Z exchange, as shown in Figure 8.7. They were observed and studied in a number of experiments at CERN, Fermilab, and Brookhaven, most precisely by the CHARM II collaboration at CERN (for a review, see J. Panman in Langacker, 1995). The momentum transfers are tiny compared to the Z mass, so the four-fermion effective interaction in (8.79) applies. From (8.74) the effective Lagrangian density is

$$-\mathcal{L}^{\nu e} = \frac{G_F}{\sqrt{2}} \bar{\nu}_\mu \gamma^\mu (1-\gamma^5)\nu_\mu \, \bar{e}\gamma_\mu (g_V^{\nu e} - g_A^{\nu e}\gamma^5)e. \tag{8.93}$$

In the SM one expects,

$$g_V^{\nu e} \sim -\frac{1}{2} + 2\sin^2\theta_W, \qquad g_A^{\nu e} \sim -\frac{1}{2}. \tag{8.94}$$

$g_V^{\nu e}$ is predicted to be small, since $\sin^2\theta_W \sim 0.23$. Equation (8.93) is valid in any gauge theory, provided that any right-chiral neutrino ν_R does not have significant gauge interactions and that any ν mass is negligible. We use the symbols $g_{V,A}^{\nu e}$ to represent the coefficients in the four-fermi interaction. These could differ from the coefficients $g_{V,A}^{e}$ relevant to the Zee vertex in more complicated theories involving multiple gauge bosons. Differences can also be generated when one includes higher-order radiative corrections in the SM. In practice, however, the radiative corrections to elastic νe scattering (Sarantakos et al., 1983) are small compared to the experimental precision except for top quark effects. Radiative corrections will be discussed in Section 8.3.4.

We saw in Section 7.2.2 that there is a charged current contribution to $\nu_e e^- \to \nu_e e^-$ scattering, which can be put in the same form as (8.93) by a Fierz transformation. The results in (7.60) and (7.61) (page 238) can therefore be applied to elastic $\nu_\mu e^-$ and $\bar{\nu}_\mu e^-$ scattering provided we substitute $g_{V,A} \to g_{V,A}^{\nu e}$. (The $m_e y/E_\nu$ terms are not important at accelerator energies.) Elastic $\nu_e e^-$ and $\bar{\nu}_e e^-$ scattering have both charged and neutral current contributions, as shown in Figure 8.7. Formulae (7.60) and (7.61) still apply to those reactions provided we take $g_{V,A} = g_{V,A}^{\nu e} + 1$.

One can determine $g_V^{\nu e}$ and $g_A^{\nu e}$ separately by measuring either the energy (y) distributions or the total cross sections for both $\nu_\mu e$ and $\bar{\nu}_\mu e$, up to a four-fold ambiguity associated with the overall sign and with the interchange of $g_V^{\nu e}$ and $g_A^{\nu e}$. Of course, the neutrino beams are not monochromatic, so the ν_μ and $\bar{\nu}_\mu$ spectra must be modeled. As shown in Figure 8.8, one of the four solutions is consistent with the SM for $\sin^2\theta_W \sim 0.23$. If one assumes the validity of the SM then $\sin^2\theta_W$ can be best determined from the ratio of the $\nu_\mu e$ and $\bar{\nu}_\mu e$ cross sections because many of the systematic uncertainties cancel. The precise value of the extracted $\sin^2\theta_W$ depends on the renormalization scheme, as will be discussed in Section 8.3.4. The value of $\sin^2\theta_W$ from $\overset{(-)}{\nu}_\mu e$ elastic scattering, in the on-shell scheme defined by $\sin^2\theta_W = 1 - M_W^2/M_Z^2$, is 0.2230(77) (Patrignani, 2016). The extracted values of $g_{V,A}^{\nu e}$ for the SM-like solution are compared with the SM expectations, obtained using the SM parameters from the global best fit, in Table 8.3. The agreement is excellent. Limits on small deviations from the SM value can be used to limit certain types of possible new physics, such as additional heavy Z' gauge bosons or mixing with exotic heavy fermions (see, e.g., Amaldi et al., 1987).

There were also measurements of $\bar{\nu}_e e$ at the Savannah River reactor and more recently by the TEXONO collaboration at the Kuo-Sheng reactor in Taiwan (Deniz et al., 2010),

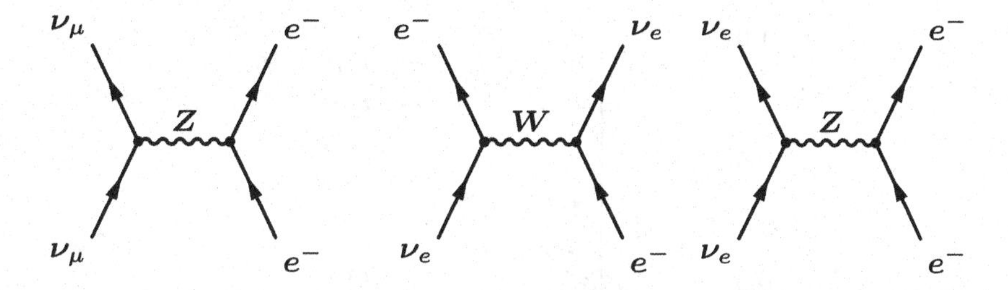

Figure 8.7 Left: diagram for $\nu_\mu e^- \to \nu_\mu e^-$ in the $SU(2) \times U(1)$ model. Middle and right: diagrams for $\nu_e e^- \to \nu_e e^-$.

and of $\nu_e e$ by LSND at Los Alamos. These are not as precise as the $\nu_\mu e$ measurements, but because of the charged current contribution they can help resolve the four-fold ambiguity. As is seen in Figure 8.8 two of the four are excluded. The fourth solution, obtained for $g_V^{\nu e} \leftrightarrow g_A^{\nu e}$, is excluded by $e^+e^- \to \mu^+\mu^-$ data under the plausible assumption that the weak neutral current is dominated by the exchange of a single Z boson. These experiments also demonstrated interference between the charged current and neutral current contributions to $\overset{(-)}{\nu}_e\, e$ scattering. LSND obtained $-1.01(18)$ for $I = 2\sigma^I/\sigma^{WCC}$, where σ^I and σ^{WCC} are, respectively, the interference and WCC contributions to the cross section, compared with the SM expectation of -1.09, while TEXONO found an interference term of $-0.92(34)$ compared to the expected -1. (Other observations of WCC-WNC interference are described in their papers.) This is significant because it provides a confirmation of the conclusion from WNC-electromagnetic interference (Section 8.3.2) that the WNC is really vector and axial.

TABLE 8.3 Values of the model-independent neutral current parameters compared to the SM expectations for the global best fit.[a]

Quantity	Experiment	SM Expectation
$g_V^{\nu e}$	-0.040 ± 0.015	-0.040
$g_A^{\nu e}$	-0.507 ± 0.014	-0.506
g_L^2	0.3005 ± 0.0028	0.3034
g_R^2	0.0329 ± 0.0030	0.0302
θ_L	2.50 ± 0.035	2.46
θ_R	$4.56^{+0.42}_{-0.27}$	5.18
$C_{1u} + 2C_{1d}$	0.489 ± 0.005	0.495
$2C_{1u} - C_{1d}$	-0.708 ± 0.016	-0.719
$2C_{2u} - C_{2d}$	-0.144 ± 0.068	-0.095
g_{AV}^{ee}	0.0190 ± 0.0027	0.0225

[a]From the Electroweak Review in (Patrignani, 2016), where one can also find the full SM expressions including radiative corrections and the correlations on the experimental uncertainties.

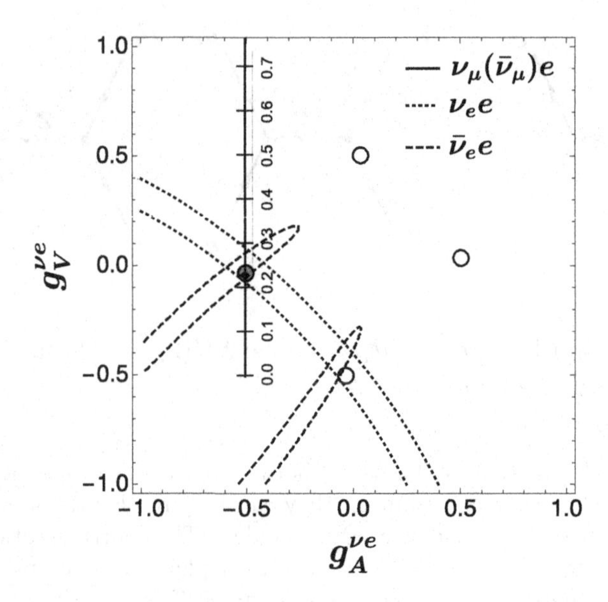

Figure 8.8 Allowed regions in $g_V^{\nu e}$ and $g_A^{\nu e}$ from $\overset{(-)}{\nu}_\mu e^-$, $\nu_e e^-$, and reactor $\bar{\nu}_e e^-$ elastic scattering, compared to the predictions of the SM as a function of $\sin^2\theta_W$ in the $\overline{MS}$ scheme. The SM best fit (shaded area) is almost identical to the experimental region. The $\nu_\mu e$ regions are at 90% c.l., and the bands are at 1σ. Plot courtesy of the Particle Data Group (Patrignani, 2016).

Deep Inelastic Neutrino Scattering

There have been many experiments measuring neutrino hadron scattering, including elastic $\nu_\mu p \to \nu_\mu p$ scattering, inelastic exclusive processes such as $\nu_\mu p \to \nu_\mu \pi N$ (dominated by the Δ resonance), and deep inelastic $\nu_\mu N \to \nu_\mu X$ scattering, as well as their $\bar{\nu}_\mu$ and charged current analogs (for reviews, see F. Perrier in Langacker, 1995; Conrad et al., 1998; Formaggio and Zeller, 2012; Patrignani, 2016). Deep inelastic $e^\mp p \to e^\mp X$ (involving both γ and Z exchange) and $e^\mp p \to \nu_e(\bar{\nu}_e)X$ (via $W^\pm$) have also been studied at the HERA *ep* collider at DESY (Abramowicz et al., 2015, 2016). The WCC experiments have been especially important as tests of QCD and probes of parton distribution functions, providing information complementary to charged lepton scattering. They also constrain the CKM matrix elements $|V_{cd}|$ and $|V_{cs}|$. The neutral current experiments have mainly functioned as tests of the WNC couplings and determinations of parameters such as $\sin^2\theta_W$.

Here we will focus on the deep inelastic neutrino scattering, for which many experiments have been performed, especially at CERN and Fermilab. The neutrinos are mainly produced by $\pi_{\ell 2}$ and $K_{\ell 2}$, where the π and K emerge from proton collisions on a target. These are almost entirely ν_μ and $\bar{\nu}_\mu$, though enough $\nu_e(\bar{\nu}_e)$ are produced, mainly in K_{e3} decays, that these must be taken into account as background. More π^+ and K^+ are produced than π^- and K^-, so ν_μ beams are more intense than $\bar{\nu}_\mu$. The typical neutrino energies at the CERN and Fermilab experiments are in the 10's-100's GeV range with a broad spectrum. The spectrum becomes somewhat narrower in energy in the narrow band beams (NBB), in which the energy of the parent hadron is selected. These actually lead to two peaks in energy, associated with the pion (lower energy) and kaon decays. Even in the NBB there is considerable uncertainty in the incident neutrino energy. E_ν can be determined in WCC events such as $\nu_\mu N \to \mu^- X$ by measuring the outgoing μ^- and hadron momenta, but this is not possible in WNC current events such as $\nu_\mu N \to \nu_\mu X$ because of the unobserved final neutrino. One therefore typically concentrates on the total WNC cross sections. Care is still required to determine or model the incident energy spectrum and average over it appropriately.

We will consider the charged and neutral current processes

$$\begin{aligned} &\text{WCC:} \quad \nu_\mu N \to \mu^- X, \qquad \bar{\nu}_\mu N \to \mu^+ X \\ &\text{WNC:} \quad \nu_\mu N \to \nu_\mu X, \qquad \bar{\nu}_\mu N \to \bar{\nu}_\mu X, \end{aligned} \tag{8.95}$$

where N can be a proton, neutron, or nucleus. The nuclear targets, especially heavy ones, yield more events. They are also easier to interpret theoretically, especially if they are isoscalar (i.e., equal numbers of p and n), such as ^{12}C, or close to isoscalar, such as ^{56}Fe. The p and n-target experiments, performed, e.g., by observing the relevant tracks in a deuterium or neon bubble chamber such as BEBC at CERN, have lower statistics but provide information on the isospin structure of the WNC or of the parton distributions.

Deep Inelastic Charged Current Scattering

The kinematics for deep inelastic neutrino scattering are the same as those for e^- scattering, as described in Section 5.5. The calculation of the cross section is also similar. In the relevant regime $Q^2 \ll M_W^2$ the WCC cross section is

$$\frac{d\bar{\sigma}_{\nu,\bar{\nu}}}{dk' d\Omega} = \frac{G_F^2}{32\pi^2} \frac{k'}{k} L_{\nu,\bar{\nu}}^{\mu\nu} W_{\mu\nu}^{\nu,\bar{\nu}}, \tag{8.96}$$

where

$$\begin{aligned} L_\nu^{\mu\nu} &= \text{Tr}\left[\gamma^\mu(1-\gamma^5)\, \not{k}\, \gamma^\nu (1-\gamma^5)\, \not{k}'\right] \\ &= 8\left[k^\mu k'^\nu + k'^\mu k^\nu + g^{\mu\nu}\frac{q^2}{2} + i\epsilon^{\mu\rho\nu\sigma} k'_\rho k_\sigma\right] \end{aligned} \tag{8.97}$$

is the leptonic tensor, analogous to the one for e^- scattering in (5.46) on page 175 with m_e neglected. (Note that the symbol ν enters several ways: as a Lorentz index, to indicate a neutrino, and as the invariant $p \cdot q/M$. The meaning should be clear from the context.) The antisymmetric $\epsilon^{\mu\rho\nu\sigma}$ is due to the parity-violating vector-axial interference. The tensor $L_{\bar{\nu}}^{\mu\nu}$ for $\bar{\nu}_\mu N \to \mu^+ X$ is obtained by interchanging k and k', so that the $\epsilon^{\mu\rho\nu\sigma}$ term changes sign. Similarly, the hadronic tensor $W_{\mu\nu}^{\nu,\bar{\nu}}$ is defined as in (5.55), except J_Q is replaced by the hadronic charge-raising current $J_W^{h\dagger}$. $W_{\mu\nu}^{\nu,\bar{\nu}}$ can be written in terms of the structure functions $W_{1,2,3}^{\nu,\bar{\nu}}$ in a form similar to (5.58):

$$\begin{aligned} W_{\mu\nu}^{\nu,\bar{\nu}} &= \left[-g_{\mu\nu} + \frac{q_\mu q_\nu}{q^2}\right] W_1^{\nu,\bar{\nu}}\left(Q^2,\nu\right) \\ &+ \frac{1}{M^2}\left[p_\mu - \frac{p\cdot q}{q^2} q_\mu\right]\left[p_\nu - \frac{p\cdot q}{q^2} q_\nu\right] W_2^{\nu,\bar{\nu}}\left(Q^2,\nu\right) + i\epsilon_{\mu\rho\nu\sigma}\frac{p^\rho q^\sigma}{2M^2} W_3^{\nu,\bar{\nu}}\left(Q^2,\nu\right), \end{aligned} \tag{8.98}$$

where the final term is again due to vector-axial interference.[13] The superscripts on the structure functions $W_{1,2}$ indicate that those for e^-, ν, and $\bar{\nu}$ are all different, while W_3^ν can

[13]Strictly speaking, there are additional possible terms in the hadronic tensor since J_W^h is not exactly conserved. However, they lead to effects proportional to the lepton masses and are therefore negligible in the deep inelastic regime.

differ from $W_3^{\bar{\nu}}$. They also depend on the target (p, n, or N). However, to the extent that one can ignore the contributions of the second and third families and of Cabibbo or CKM mixing, one has from isospin that

$$W_{in}^{\nu} \sim W_{ip}^{\bar{\nu}}, \qquad W_{ip}^{\nu} \sim W_{in}^{\bar{\nu}}, \qquad W_{iN}^{\nu} = \frac{1}{2}\left[W_{ip}^{\nu} + W_{in}^{\nu}\right] \sim W_{iN}^{\bar{\nu}}, \tag{8.99}$$

where N is an isoscalar target. In analogy with e^- DIS, we introduce new structure functions

$$F_1^{\nu,\bar{\nu}}\left(x, Q^2\right) = MW_1^{\nu,\bar{\nu}}\left(Q^2, \nu\right), \qquad F_{2,3}^{\nu,\bar{\nu}}\left(x, Q^2\right) = \nu W_{2,3}^{\nu,\bar{\nu}}\left(Q^2, \nu\right), \tag{8.100}$$

anticipating that the quark model or QCD will imply an approximate scaling behavior $F_i^{\nu,\bar{\nu}}\left(x, Q^2\right) \sim F_i^{\nu,\bar{\nu}}(x)$, up to higher-order corrections. In that case, the differential cross section is

$$\frac{d^2\bar{\sigma}_{cc}^{\nu,\bar{\nu}}}{dxdy} = \frac{G_F^2 ME_\nu}{\pi}\left[xy^2 F_1^{\nu,\bar{\nu}}(x) + (1-y)F_2^{\nu,\bar{\nu}}(x) \pm xy(1-\frac{y}{2})F_3^{\nu,\bar{\nu}}(x)\right], \tag{8.101}$$

where the $\pm$ is for $\nu(\bar{\nu})$, and the total spin-averaged cross section

$$\bar{\sigma}_{cc}^{\nu,\bar{\nu}} = \int_0^1 dx \int_0^1 dy \frac{d^2\bar{\sigma}_{cc}^{\nu,\bar{\nu}}}{dxdy} \tag{8.102}$$

is predicted to scale with the neutrino energy in the lab frame, E_ν.

Let us now consider the simple quark parton model approximation, as indicated in Figure 8.9, and initially ignore the heavy families and quark mixing. Then, similarly to (5.70) for e^- scattering, one has

$$\begin{aligned} F_2^{\nu}(x) &= 2xF_1^{\nu}(x) = 2x\left[d(x) + \bar{u}(x)\right], \qquad F_3^{\nu}(x) = 2\left[d(x) - \bar{u}(x)\right] \\ F_2^{\bar{\nu}}(x) &= 2xF_1^{\bar{\nu}}(x) = 2x\left[u(x) + \bar{d}(x)\right], \qquad F_3^{\bar{\nu}}(x) = 2\left[u(x) - \bar{d}(x)\right], \end{aligned} \tag{8.103}$$

where $F_2 = 2xF_1$ is just the Callan-Gross relation, i.e., that the partons have spin-$\frac{1}{2}$, and the $-$ sign for the antiquarks in F_3 is due to $V-A$. Comparing with the analogous relations for e^- scattering in (5.70), we see that ν DIS allows one to separate q from $\bar{q}$ by measuring F_2 and F_3 separately. Equation (8.103) is easily extended to include three families and CKM mixing. However, for the energies and precision of the existing experiments it suffices to consider just the first two families, since the b and t content of the nucleon is small. Using the Cabibbo approximation to the mixing,

$$\begin{aligned} F_{2,3}^{\nu}(x) =& 2\kappa(x)\big[d(x)\left(\cos^2\theta_c + \sin^2\theta_c\xi_c\right) \\ &+ s(x)\left(\sin^2\theta_c + \cos^2\theta_c\xi_c\right) \pm \bar{u}(x) \pm \bar{c}(x)\big], \end{aligned} \tag{8.104}$$

where the terms correspond, respectively, to $d \to u, d \to c, s \to u, s \to c, \bar{u} \to (\bar{d} + \bar{s})$, and $\bar{c} \to (\bar{d} + \bar{s})$. The $\pm$ signs are for $F_2(F_3)$, while $\kappa(x) = x$ for F_2 and 1 for F_3. $\xi_c(Q^2)$ is a kinematic suppression factor associated with the non-zero c quark mass (we ignore $m_{u,d,s}$). One expects $\xi_c \to 1$ for $Q^2 \gg m_c^2$, while the small deviations from unity for finite Q^2 can be extracted from the c production data or estimated theoretically. Similarly,

$$\begin{aligned} F_{2,3}^{\bar{\nu}}(x) =& 2\kappa(x)\big[u(x) + c(x) \pm \bar{d}(x)\left(\cos^2\theta_c + \sin^2\theta_c\xi_c\right) \\ &\pm \bar{s}(x)\left(\sin^2\theta_c + \cos^2\theta_c\xi_c\right)\big]. \end{aligned} \tag{8.105}$$

To include the full CKM matrix, one can replace

$$\begin{aligned} \left(\cos^2\theta_c + \sin^2\theta_c\xi_c\right) &\to \left(|V_{ud}|^2 + |V_{cd}|^2\xi_c\right) \\ \left(\sin^2\theta_c + \cos^2\theta_c\xi_c\right) &\to \left(|V_{us}|^2 + |V_{cs}|^2\xi_c\right). \end{aligned} \tag{8.106}$$

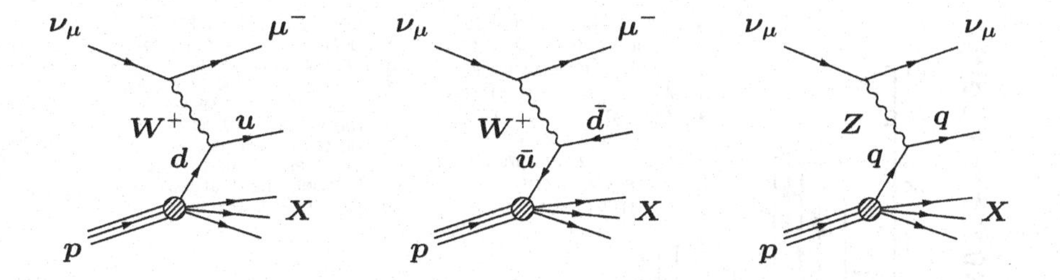

Figure 8.9 Typical diagrams for neutrino deep inelastic scattering in the quark parton model. The initial and final lepton momenta are k and k', and $q = k - k'$.

(In principle, the coefficient of $u(x)$ in $F^{\bar{\nu}}_{2,3}$ is changed to $|V_{ud}|^2 + |V_{us}|^2$, etc., but in practice the difference is negligible.)

An interesting test of the quark model and charge assignments is the $5/18^{th}$ rule, which concerns the ratio of F_2 in e^- DIS from an isoscalar target N (Equation 5.79), to F^{ν}_{2N}. In the simplified model in which one ignores the s and c quarks in the nucleon and sets $\xi_c \sim 1$ one predicts

$$\frac{F_{2N}}{F^{\nu}_{2N}} = \frac{\frac{1}{2}\left(F_{2p} + F_{2n}\right)}{\frac{1}{2}\left(F^{\nu}_{2p} + F^{\nu}_{2n}\right)} = \frac{\left(\frac{4}{9} + \frac{1}{9}\right)\frac{1}{2}x\left(u^p + d^p + \bar{u}^p + \bar{d}^p\right)}{x\left(d^p + \bar{u}^p + u^p + \bar{d}^p\right)} = \frac{5}{18}, \tag{8.107}$$

where q^p is the quark distribution function in the proton and we have used the isospin relation (5.83). Equation (8.107) agrees well with the data, but not with the expectation ~ 0.5 of an early competing model with integer charged quarks (Nambu and Han, 1974, and Problem 5.1).

The parton model expressions for the WCC cross sections are

$$\begin{aligned}\frac{d^2\bar{\sigma}^{\nu}_{cc}}{dxdy} &= \frac{2G_F^2 M E_\nu}{\pi}\left\{xd\underbrace{\left[|V_{ud}|^2 + |V_{cd}|^2\xi_c\right]}_{\lambda_d} + xs\underbrace{\left[|V_{us}|^2 + |V_{cs}|^2\xi_c\right]}_{\lambda_s} + x\left(\bar{u} + \bar{c}\right)(1-y)^2\right\} \\ &\sim \frac{2G_F^2 M E_\nu}{\pi}\left[x\left(d + s\right) + x\left(\bar{u} + \bar{c}\right)(1-y)^2\right]\end{aligned} \tag{8.108}$$

and

$$\begin{aligned}\frac{d^2\bar{\sigma}^{\bar{\nu}}_{cc}}{dxdy} &= \frac{2G_F^2 M E_{\bar{\nu}}}{\pi}\left[x\left(u + c\right)(1-y)^2 + x\bar{d}\lambda_d + x\bar{s}\lambda_s\right] \\ &\sim \frac{2G_F^2 M E_{\bar{\nu}}}{\pi}\left[x\left(u + c\right)(1-y)^2 + x\bar{d} + x\bar{s}\right],\end{aligned} \tag{8.109}$$

where the target labels are not displayed. The second form is valid at sufficiently high energy that the c-quark threshold effects are negligible, and ignores the small CKM mixing with the third family. One sees explicitly that the total cross sections are predicted to grow linearly with energy to the extent that one can ignore the Q^2 dependence of the quark distribution functions, W propagator effects, ξ_c, etc. One sees in Figure 8.10 that this is the case. The ν_μ and $\bar{\nu}_\mu$ data can be combined with e^- DIS to separate the various quark distributions. Of course, QCD predicts logarithmic violation of scaling, which is tested in detail (for a review, see Conrad et al., 1998). The observed structure functions are also an input to the global QCD fits to parton distribution functions described in Section 5.5.

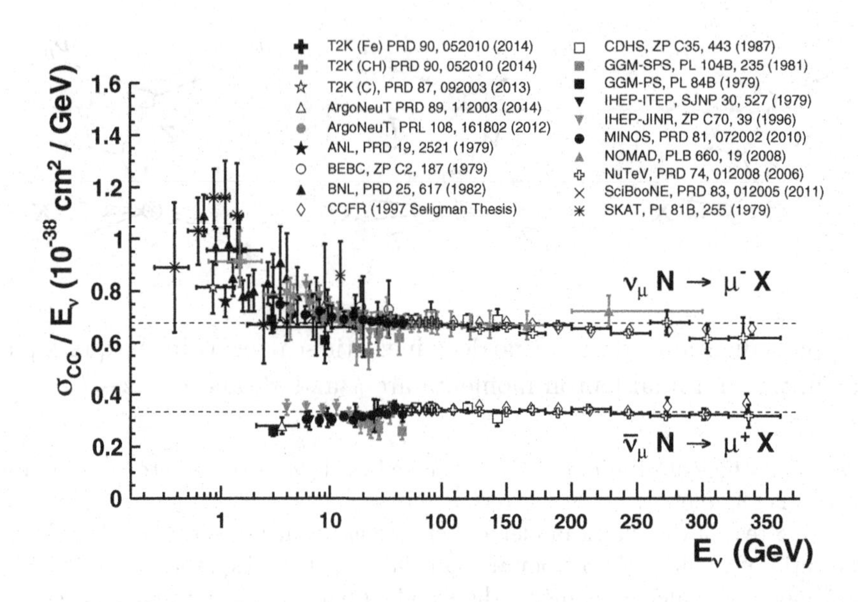

Figure 8.10 Total charged current neutrino cross sections $\bar{\sigma}_{cc}^{\nu,\bar{\nu}}/E_\nu$ vs E_ν. The simple parton model or QCD predict that these should be $\sim$ constant at high energy. Plot courtesy of the Particle Data Group (Patrignani, 2016).

Neutrino induced opposite sign dimuon production is associated with the production of charm (c) quarks from d or s, via $\nu(\bar{\nu})N \to \mu^{\mp}X + c(\bar{c})$, with the subsequent decays $c \to (s,d)\mu^+\nu_\mu$ or $\bar{c} \to (\bar{s},\bar{d})\mu^-\bar{\nu}_\mu$. By measuring these dimuons for both ν_μ and $\bar{\nu}_\mu$ and in different E_ν ranges, one can extract $|V_{cd}|^2$, $S|V_{cs}|^2$, and ξ_c separately from the experimental data, where S is the momentum carried by s quarks, $S = \int xs(x)dx$.

Deep Inelastic Neutral Current Scattering

The neutral current processes $\nu_\mu(\bar{\nu}_\mu)N \to \nu_\mu(\bar{\nu}_\mu)X$ have been especially useful in testing the WNC predictions of the standard $SU(2) \times U(1)$ model. They were extremely important in establishing the correctness of those predictions to leading order and in limiting small deviations. Prior to the Z-pole experiments at LEP and SLC they provided the most precise measurements of $\sin^2\theta_W$.

From (8.75) and (8.79) the effective four-fermi interaction for WNC ν-hadron scattering in the SM is

$$\begin{aligned} -L^{\nu h} =& \frac{G_F}{\sqrt{2}}\, \bar{\nu}\, \gamma^\mu\, (1-\gamma^5)\nu \\ &\times \sum_r \left[\epsilon_L^{\nu h}(r)\, \bar{q}_r\, \gamma_\mu(1-\gamma^5)q_r + \epsilon_R^{\nu h}(r)\, \bar{q}_r\, \gamma_\mu(1+\gamma^5)q_r\right], \end{aligned} \tag{8.110}$$

where[14] from (8.76)

$$\begin{aligned}
\epsilon_L^{\nu h}(u) &= +\frac{1}{2} - \frac{2}{3}\sin^2\theta_W, & \epsilon_R^{\nu h}(u) &= -\frac{2}{3}\sin^2\theta_W \\
\epsilon_L^{\nu h}(d) &= -\frac{1}{2} + \frac{1}{3}\sin^2\theta_W, & \epsilon_R^{\nu h}(d) &= +\frac{1}{3}\sin^2\theta_W.
\end{aligned} \tag{8.111}$$

Just as for νe scattering, (8.110) will continue to hold in an arbitrary gauge theory with neutrinos coupling to $V-A$, or can be extended to include radiative corrections to the SM tree-level predictions (Marciano and Sirlin, 1980; Sirlin and Marciano, 1981), for appropriate values of the $\epsilon_{L,R}^{\nu h}$. We therefore allow for arbitrary $\epsilon_{L,R}^{\nu h}(u,d)$, but will assume family universality, i.e., that $\epsilon_{L,R}^{\nu h}(d) = \epsilon_{L,R}^{\nu h}(s)$ and $\epsilon_{L,R}^{\nu h}(u) = \epsilon_{L,R}^{\nu h}(c)$.

It is then straightforward to show that in the simple parton model,

$$\begin{aligned}
\frac{d^2\bar\sigma_{nc}^{\nu}}{dx dy} = \frac{2G_F^2 M E_\nu}{\pi} & \left[|\epsilon_L^{\nu h}(u)|^2 + |\epsilon_R^{\nu h}(u)|^2(1-y)^2\right](xu + xc\xi_c) \\
& + \left[|\epsilon_L^{\nu h}(d)|^2 + |\epsilon_R^{\nu h}(d)|^2(1-y)^2\right](xd + xs) \\
& + \left[|\epsilon_R^{\nu h}(u)|^2 + |\epsilon_L^{\nu h}(u)|^2(1-y)^2\right](x\bar u + x\bar c\xi_c) \\
& + \left[|\epsilon_R^{\nu h}(d)|^2 + |\epsilon_L^{\nu h}(d)|^2(1-y)^2\right](x\bar d + x\bar s),
\end{aligned} \tag{8.112}$$

while $\epsilon_L^{\nu h}(r) \leftrightarrow \epsilon_R^{\nu h}(r)$ for $\bar\nu$.

Now, consider WCC and WNC scattering from an isoscalar target, and ignore the s and c content of the nucleon and take $\xi_c \sim 1$. (One must correct for these in the actual analysis.) Then, from (8.108), (8.109), and (8.112), the total cross sections are

$$\begin{aligned}
\bar\sigma_{cc}^{\nu} &= \frac{2G_F^2 M E_\nu}{\pi}\int_0^1 \left[xq + \frac{1}{3}x\bar q\right]dx, \qquad \bar\sigma_{cc}^{\bar\nu} = \frac{2G_F^2 M E_{\bar\nu}}{\pi}\int_0^1 \left[\frac{1}{3}xq + x\bar q\right]dx \\
\bar\sigma_{nc}^{\nu} &= \frac{2G_F^2 M E_\nu}{\pi}\int_0^1 \left(g_L^2\left[xq + \frac{1}{3}x\bar q\right] + g_R^2\left[\frac{1}{3}xq + x\bar q\right]\right)dx \\
\bar\sigma_{nc}^{\bar\nu} &= \frac{2G_F^2 M E_{\bar\nu}}{\pi}\int_0^1 \left(g_L^2\left[\frac{1}{3}xq + x\bar q\right] + g_R^2\left[xq + \frac{1}{3}x\bar q\right]\right)dx,
\end{aligned} \tag{8.113}$$

where $q \equiv (u^p + d^p)/2$ and $\bar q \equiv (\bar u^p + \bar d^p)/2$. The effective L and R couplings $g_{L,R}^2$ are

$$\begin{aligned}
g_L^2 &\equiv |\epsilon_L^{\nu h}(u)|^2 + |\epsilon_L^{\nu h}(d)|^2 = \frac{1}{2} - \sin^2\theta_W + \frac{5}{9}\sin^4\theta_W \\
g_R^2 &\equiv |\epsilon_R^{\nu h}(u)|^2 + |\epsilon_R^{\nu h}(d)|^2 = \frac{5}{9}\sin^4\theta_W,
\end{aligned} \tag{8.114}$$

where the expressions on the right are those for the tree-level standard model.

It is useful to consider the ratios of WNC and WCC cross sections, because many of the uncertainties associated with the strong interactions, neutrino fluxes, and systematics cancel in the ratios. Under the same approximations as (8.113) one finds

$$R_\nu \equiv \frac{\bar\sigma_{nc}^{\nu}}{\bar\sigma_{cc}^{\nu}} \sim g_L^2 + g_R^2 r, \qquad R_{\bar\nu} \equiv \frac{\bar\sigma_{nc}^{\bar\nu}}{\bar\sigma_{cc}^{\bar\nu}} \sim g_L^2 + \frac{g_R^2}{r}, \tag{8.115}$$

[14]Similar to $g_{V,A}^{\nu e}$, we use the symbols $\epsilon_{L,R}^{\nu h}(r)$ to indicate that they are the coefficients in the four-fermi interaction, and can differ from the Zqq vertex factors in (8.75) by higher-order corrections or new physics.

where

$$r \equiv \frac{\bar{\sigma}^{\bar{\nu}}_{cc}}{\bar{\sigma}^{\nu}_{cc}} = \frac{\frac{1}{3}+\epsilon}{1+\frac{\epsilon}{3}}, \qquad \epsilon \equiv \frac{\int_0^1 x\bar{q}(x)dx}{\int_0^1 xq(x)dx} \tag{8.116}$$

can be measured directly. ϵ is the ratio of the part of the nucleon's momentum carried by antiquarks to the part carried by quarks. One would have $r = \frac{1}{3}$ for $\bar{q}/q = 0$, but the observed value $r \sim 0.44$ corresponds to $\epsilon \sim 0.125$. It is remarkable that within the stated approximations the details of the quark distributions cancel out, except for r, which can be measured directly. Although this was shown here within the simple parton model, the result essentially follows from isospin symmetry alone, except for a term involving Y^2 that is weighted by a small coefficient of $\sin^4\theta_W$ (Llewellyn Smith, 1983).

The cross section ratios have been measured to 1% or better by the CDHS and CHARM collaborations at CERN and by CCFR at Fermilab during the 1980s and 1990s. Although (8.115) gives a good first approximation, in practice one must correct for nonisoscalar target effects ($N_n \neq N_p$ in ^{56}Fe); $s(x)$ and $c(x)$; the c threshold, ξ_c; third family mixing; the W and Z propagators; radiative corrections; experimental cuts; and the QCD evolution of the quark distributions. Since these are relatively small corrections for the ^{12}C and ^{56}Fe experiments, it suffices to use the QCD improved parton model described in Section 5.5 to estimate them. The same estimates can be used for the n and p target data, for which the isospin argument does not apply: even though the theoretical uncertainties are larger for such targets, the model is adequate since the experiments are less precise.

The isoscalar target experiments yielded the most precise determinations of $\sin^2\theta_W$ prior to the Z-pole era, $\sin^2\theta_W = 0.233(3)(5)$, where the first error is experimental. The second error is theoretical and is largely due to the uncertainties in the charm quark threshold effect ξ_c. Even though c production represents less than 10% of the WCC cross section and ξ_c can be determined to $\sim 15\%$ from dimuon data (page 286), the remaining uncertainty dominated the error. The theoretical uncertainties can be minimized by using the Paschos-Wolfenstein ratio (Paschos and Wolfenstein, 1973),

$$R^- \equiv \frac{\bar{\sigma}^{\nu}_{nc} - \bar{\sigma}^{\bar{\nu}}_{nc}}{\bar{\sigma}^{\nu}_{cc} - \bar{\sigma}^{\bar{\nu}}_{cc}} \sim g_L^2 - g_R^2 \sim \frac{1}{2} - \sin^2\theta_W, \tag{8.117}$$

for an isoscalar target, which follows from isospin. (There is still a small residual uncertainty from the charm threshold, but it is suppressed by $\sin^2\theta_c$.) Utilizing R^- requires a high intensity and high energy $\bar{\nu}_\mu$ beam, which became possible in 1996 using the sign-selected beam at Fermilab.

The NuTeV collaboration (Zeller et al., 2002) utilized the sign-selected beam and a version of (8.117) to obtain a more precise value $\sin^2\theta_W = 0.2277(16)$, which is insensitive to the top quark and Higgs masses and has little uncertainty from the c threshold. However, this value is $\sim 3\sigma$ above the current value of $\sin^2\theta_W = 0.2234(1)$ from the global fit to all data (Patrignani, 2016). The discrepancy increased by about 1σ due to remeasurements of the K_{e3} branching ratio (Section 8.6.1), which affects the $\overset{(-)}{\nu}_e$ contamination of the $\overset{(-)}{\nu}_\mu$ beams, but this was roughly compensated by NuTeV's subsequent measurement of an asymmetry in $\int_0^1 dx\, x[s(x) - \bar{s}(x)]$ (Mason et al., 2007). The origin of the discrepancy is not understood. It could be an indication of some sort of new physics, such as contributions from a Z' (Davidson et al., 2002). However, it could also be associated with additional subtle QCD effects, such as larger than expected isospin breaking, nuclear shadowing, or NLO or electroweak radiative corrections. It is hard to draw any conclusions until there is a full global analysis of these issues.

The neutrino-hadron data can be used in a global analysis to determine $\epsilon^{\nu h}_{L,R}(u,d)$. The results are shown in Figure 8.11 and Table 8.3 (where $\theta_{L,R} \equiv \tan^{-1}[\epsilon^{\nu h}_{L,R}(u)/\epsilon^{\nu h}_{L,R}(d)]$). It is seen that the deep inelastic isoscalar data determine g_L^2 and g_R^2 quite well. The isospin structure, which depends on DIS from p and n targets and other reactions, is well-determined for the L couplings, but only poorly for the $\epsilon_R^{\nu h}$. Nevertheless, the ν-hadron data is consistent with the SM for $\sin^2\theta_W \sim 0.23$.

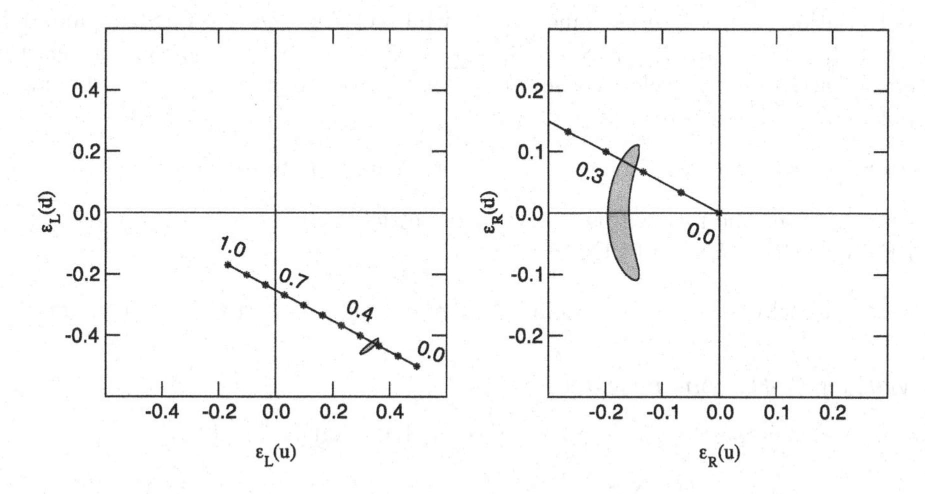

Figure 8.11 90% c.l. allowed regions in $\epsilon^{\nu h}_{L,R}(u)$ vs $\epsilon^{\nu h}_{L,R}(d)$ from $\overset{(-)}{\nu}_\mu N$ elastic, exclusive inelastic, and deep inelastic scattering, compared to the predictions of the SM as a function of $\sin^2\theta_W$, updated from (Amaldi et al., 1987).

8.3.2 Weak-Electromagnetic Interference

Following the discovery of the weak neutral current in the early 1970s, it took some years to establish that it was uniquely consistent with the predictions of the $SU(2)\times U(1)$ model. As described in Section 8.3.1, νe and ν-hadron scattering eventually zeroed in on the SM predictions when analyzed in a general V, A framework. However, it was conceivable that the WNC could be due to S, P, and T couplings, e.g., due to the exchange of spin-0 or 2 particles. These all involve a helicity flip of an initial left-handed neutrino (or right-handed $\bar{\nu}$) produced in a charged current decay, into a neutrino of the opposite helicity (Equations 2.213 and 2.214 on page 42). However, it is not feasible to measure the final helicity. In fact, there is a *confusion theorem* that the final electron or hadron distributions in $\nu_\mu e^-$ elastic scattering or deep inelastic $\nu_\mu N$ scattering predicted by any V, A theory can be duplicated by some combination of S, P, and T (Kayser et al., 1974).

The easiest way to resolve this ambiguity was to observe the interference between WNC amplitudes and those involving other interactions known to be V and/or A in character, which could not occur for S, P, or T at high energies where fermion masses can be ignored. In fact, some interference between WNC and WCC amplitudes was eventually observed, such as in $\nu_e e^-$ and $\bar{\nu}_e e^-$ elastic scattering. However, the more precise tests (and also the earliest chronologically) involved interference between Z exchange and electromagnetism in eq, e^+e^-, and (more recently) in e^-e^- interactions. At low energies, such as in atoms, the WNC is only a tiny perturbation on the Coulomb or other electromagnetic effects. However, QED is purely vector and parity conserving, so observation of VA interference is a clear sign of the WNC. (These are usually parity violating, but in some cases can be parity conserving if they involve the product of two axial currents from the WNC.) At higher energies, the γ and Z exchange amplitudes may be comparable and Z propagator effects may be observable as well. There have by now been many observations of WNC-electromagnetic interference, all of which are in agreement with the predictions of the standard model. They include:

Polarization (or charge) asymmetries in deep inelastic $eD \to eX$ (SLAC and Jefferson Lab), $\mu C \to \mu X$ (CERN), and $e^{\pm}p \to e^{\pm}X$ (DESY); in low energy elastic or quasi-elastic polarized electron scattering at Bates, Mainz, and Jefferson Lab; and in polarized e^-e^- Møller scattering (SLAC).

Atomic parity violation in cesium (Boulder, Paris), thallium, and other atoms.

Cross sections and forward-backward asymmetries in $e^+e^- \to \ell\bar{\ell}, q\bar{q}, c\bar{c}$ and $b\bar{b}$ at PEP, PETRA, TRISTAN, and LEP 2.

Forward-backward asymmetries in $\overset{(-)}{p}p \to e^+e^-$ and $\mu^+\mu^-$ at the Tevatron and LHC.

Parity Violating $\ell^{\pm}$-Hadron Interactions

The parity-violating part of the effective WNC eq Lagrangian density is

$$-\mathcal{L}^{eq} = -\frac{G_F}{\sqrt{2}} \sum_{i=u,d} \left[C_{1i}\, \bar{e}\gamma_\mu\gamma^5 e\, \bar{q}_i\gamma^\mu q_i + C_{2i}\bar{e}\,\gamma_\mu e\, \bar{q}_i\gamma^\mu\gamma^5 q_i \right], \tag{8.118}$$

where the tree-level expressions in the SM are

$$\begin{aligned} C_{1u} &= -\frac{1}{2} + \frac{4}{3}\sin^2\theta_W, \qquad & C_{2u} &= -\frac{1}{2} + 2\sin^2\theta_W \\ C_{1d} &= \frac{1}{2} - \frac{2}{3}\sin^2\theta_W, \qquad & C_{2d} &= \frac{1}{2} - 2\sin^2\theta_W. \end{aligned} \tag{8.119}$$

Early attempts to measure the effects of WNC-electromagnetic interference in atoms were unsuccessful, leading to considerable confusion and doubts concerning the correctness of the $SU(2) \times U(1)$ model. However, those results turned out to be erroneous. The first significant interference observation was in the SLAC measurement (Prescott et al., 1979) of the parity-violating polarization asymmetry $A_{PV} = (\sigma_R - \sigma_L)/(\sigma_R + \sigma_L)$ for deep inelastic e^- scattering on deuterium, where $\sigma_{R,L}$ refer, respectively, to the cross section for a right- or left-handed e^-. One expects

$$\frac{A_{PV}}{Q^2} = a_1 + a_2\, \frac{1-(1-y)^2}{1+(1-y)^2}, \tag{8.120}$$

where Q^2 and y are the usual kinematic variables for DIS, and

$$\begin{aligned} a_1 &= \frac{3G_F}{5\sqrt{2}\pi\alpha}\left(C_{1u} - \frac{1}{2}C_{1d}\right) \underset{\text{SM}}{\longrightarrow} \frac{3G_F}{5\sqrt{2}\pi\alpha}\left(-\frac{3}{4} + \frac{5}{3}\sin^2\theta_W\right) \\ a_2 &= \frac{3G_F}{5\sqrt{2}\pi\alpha}\left(C_{2u} - \frac{1}{2}C_{2d}\right) \to \frac{9G_F}{5\sqrt{2}\pi\alpha}\left(\sin^2\theta_W - \frac{1}{4}\right). \end{aligned} \tag{8.121}$$

They observed an asymmetry, consistent with the SM expectations for $\sin^2\theta_W = 0.224(20)$, agreeing with the neutrino experiments and resolving the confusion. Subsequent measurements of polarization asymmetries in $\mu^{\pm}$ DIS confirmed the result.

More recently, there have been a number of low energy parity violating electron scattering (PVES), i.e., polarization asymmetry, experiments at Jefferson Lab and elsewhere. These include a new eD DIS experiment, PVDIS (Wang et al., 2014), improving on the SLAC results, and an elastic ep asymmetry experiment Q_{weak} (Androic et al., 2013) that yielded the first measurement of the weak charge of the proton, $Q_W^p \equiv -2(2C_{1u} + C_{1d})$. These have constrained $C_{1u,d}$ to a small region consistent with the SM, as can be seen in Table 8.3 and Figure 8.12. These results and future experiments are reviewed in (Kumar et al., 2013; Erler et al., 2014), and the implications for new physics in (Erler et al., 2003; Cirigliano and Ramsey-Musolf, 2013).

Atomic Parity Violation

The parity-violating WNC can induce mixing between S and P wave states in atoms, leading to such effects as the rotation of the polarization plane of linearly polarized light as it passes through an atomic vapor, or differences in transition rates induced by left- and right-circularly polarized photons. Observations have been made in cesium, bismuth, lead, thallium, and ytterbium (for reviews, see Ginges and Flambaum, 2004; Roberts et al., 2015; Patrignani, 2016). The most precise have been measurements of $6S \to 7S$ transitions in cesium in Paris and Boulder, with the most recent Boulder results at the 0.4% level. Furthermore, cesium has a single valence electron outside a tightly bound core, allowing accurate calculation of the matrix elements needed to interpret the results.

The effective interaction in (8.118) leads to a non-relativistic potential for the e^- in an atom Z, N:

$$V(\vec{r}_e) \sim \frac{G_F}{4\sqrt{2}m_e} Q_W \left\{\delta^3(\vec{r}_e), \vec{\sigma}_e \cdot \vec{p}_e\right\}, \tag{8.122}$$

where $\vec{p}_e$ is the electron momentum operator and Q_W is the *weak charge*

$$Q_W = -2\left[C_{1u}\,(2Z+N) + C_{1d}(Z+2N)\right] \sim Z(1-4\sin^2\theta_W) - N \tag{8.123}$$

(see Problem 8.11). In (8.123) we have kept only the $C_{1u,d}$ terms. This involves an axial e^- current and vector hadronic current, with the latter adding coherently over the nucleons in a heavy atom because it is spin independent. In fact, the effects of this term scale as Z^3. One factor is obvious from the coherence effect in Q_W, while the others involve the electron wave function and momentum near the nucleus. There are no such enhancements for the $C_{2u,d}$ (nucleon spin-dependent) terms, which require unpaired nucleons,[15] so the $C_{1u,d}$ strongly dominate.

As mentioned, Q_W has been measured in cesium at the 0.4% level. Neither the theoretical expression in (8.122) nor the treatment of the atom as hydrogen-like are adequate for such precision. In practice, one must correct for the finite nuclear size, treat the electron relativistically using the Dirac equation, carry out a many-body Hartree-Fock calculation of the wave functions, and include both electroweak and QED radiative corrections (Ginges and Flambaum, 2004; Roberts et al., 2015; Patrignani, 2016). The latter are difficult due to the large nuclear charge, but careful treatments have now yielded a precision of $\sim 0.5\%$, allowing a determination $Q_W(Cs) = -72.62(43)$, in reasonable agreement with the SM expectation $-73.25(2)$ (Patrignani, 2016). The corresponding weak angle (in the $\overline{MS}$ scheme near $\mu = 0$) is $\sin^2\theta_W = 0.2358(20)$.

Experiments in hydrogen or deuterium would be extremely clean theoretically, but the

[15]Nevertheless, the small spin-dependent effects can be separated by measuring different hyperfine transitions. They have been observed in cesium, but are dominated not by the WNC but by the larger electromagnetic anapole moments mentioned in Section 2.12.4.

effects are much smaller. To date no atomic measurements have been completed, though $Q_W^p = 0.064(12)$ has been measured in PVES (Androic et al., 2013), to be compared with the SM value 0.0708(3). Other future possibilities include measurements of ratios of effects in different isotopes of the same atom (for which many theoretical uncertainties cancel), or on other atoms or ions with significantly enhanced parity-violating effects.

The results from polarized lepton asymmetries and atomic parity violation can be combined in a global analysis. The combinations $C_{1u} \pm C_{1d}$ and $2C_{2u} - C_{2d}$ are now very well determined and in agreement with the SM, as can be seen in Table 8.3 and Figure 8.12.

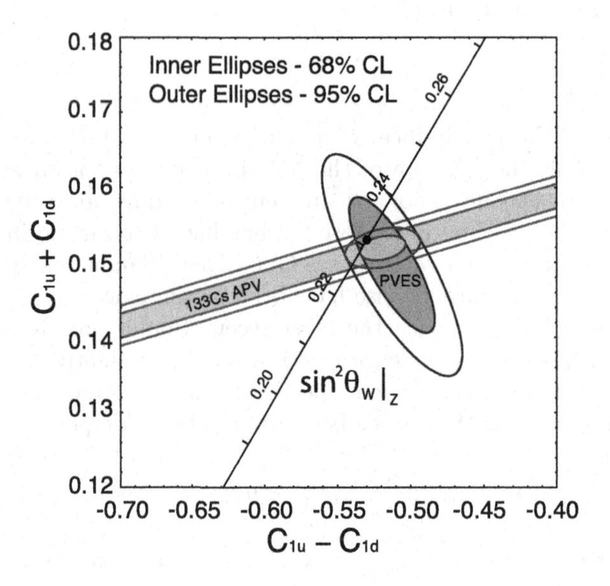

Figure 8.12 Allowed regions in $C_{1u} - C_{1d}$ vs $C_{1u} + C_{1d}$ from atomic parity violation (APV), parity-violating electron scattering (PVES), and the global combined analysis at 68% and 95% c.l. The SM prediction as a function of $\sin^2\theta_W$ in the $\overline{MS}$ scheme and the global best fit prediction (Patrignani, 2016) are indicated. Plot reproduced with permission from (Androic et al., 2013).

Polarized e^-e^- scattering

The parity-violating ee interaction

$$-\mathcal{L}^{ee} = -\frac{G_F}{\sqrt{2}} g_{AV}^{ee}\, \bar{e}\gamma_\mu\gamma^5 e\, \bar{e}\gamma^\mu e, \tag{8.124}$$

where $g_{AV}^{ee} = \frac{1}{2} - 2\sin^2\theta_W$ ($\sim -Q_W^e/2$) at tree level in the SM, leads to a polarization asymmetry A_{PV} in Møller ($e^-e^- \to e^-e^-$) scattering of electrons on an atomic target. A_{PV} has been measured by the E158 collaboration at SLAC at low $Q^2 \sim 0.026$ GeV^2 using a nearly 90% polarized beam (Anthony et al., 2005). The process is dominated by t and u-channel γ exchange, with A_{PV} generated by interference with Z exchange. Their measured asymmetry corresponds (Erler and Su, 2013) to $g_{AV}^{ee} = 0.0190(27)$, compared with the SM value 0.0225. Moreover, it was the first clear confirmation of the SM prediction (Czarnecki and Marciano, 2000) for the running of the weak angle, to be discussed in Section

8.3.4. A more precise measurement of A_{PV} is underway at Jefferson Lab by the MOLLER collaboration.

$e^-e^+ \to f\bar{f}$ *Below the Z*

There have been many measurements of $e^-e^+ \to f\bar{f}$, where $f = e, \mu, \tau, b, c$, or q (i.e., an unidentified quark flavor), below the Z pole at SLAC (SPEAR, PEP), DESY (DORIS, PETRA) and KEK (TRISTAN), as well as measurements at or near the Z-pole (LEP, SLC), and above it (LEP 2). The PEP and PETRA measurements (for reviews, see (Wu, 1984; Kiesling, 1988) and D. Haidt in (Langacker, 1995)) were able to observe effects of the s-channel Z exchange as a perturbation on the dominant one photon contribution, but were at low enough energy that they could be treated as a four-fermi interaction. TRISTAN was at a higher energy, where the virtual Z propagator effects were non-negligible (e.g., Mori et al., 1989). The principal observables relevant to the Z were the total cross sections and the forward-backward asymmetries

$$\sigma = \sigma_F + \sigma_B, \qquad A_{FB} \equiv \frac{\sigma_F - \sigma_B}{\sigma_F + \sigma_B}, \tag{8.125}$$

where

$$\sigma_F = \int_0^1 \frac{d\sigma}{d\cos\theta} d\cos\theta, \qquad \sigma_B = \int_{-1}^0 \frac{d\sigma}{d\cos\theta} d\cos\theta, \tag{8.126}$$

and θ is the angle between the e^- and f in the center of mass. It is difficult to present the results in a model independent form because of the Z propagator. Instead, we will write the tree-level formulae (for arbitrary s) assuming that only the s-channel photon and Z exchanges are relevant, but allow arbitrary Z couplings. (t-channel contributions must be included for Bhabha scattering, $e^-e^+ \to e^-e^+$).

Polarizations and forward-backward asymmetries for massless fermions are easily calculated using straightforward generalizations of the helicity amplitudes in (2.249) on page 49. Suppose the amplitude for $e^-(\vec{p}_1)\, e^+(\vec{p}_2) \to f(\vec{p}_3)\, \bar{f}(\vec{p}_4)$ is

$$\begin{aligned} M = &- i\epsilon_{LL}\, \bar{u}_3\gamma_\mu P_L v_4\; \bar{v}_2\gamma^\mu P_L u_1 - i\epsilon_{RR}\, \bar{u}_3\gamma_\mu P_R v_4\; \bar{v}_2\gamma^\mu P_R u_1 \\ &- i\epsilon_{LR}\, \bar{u}_3\gamma_\mu P_L v_4\; \bar{v}_2\gamma^\mu P_R u_1 - i\epsilon_{RL}\, \bar{u}_3\gamma_\mu P_R v_4\; \bar{v}_2\gamma^\mu P_L u_1, \end{aligned} \tag{8.127}$$

where the first (second) subscript on the ϵ is the chirality of the $f(e^-)$. In the massless limit the four terms in (8.127) do not interfere with each other, and coincide, respectively, with the helicity amplitudes $M(-+,-+)$, $M(+-,+-)$, $M(-+,+-)$, and $M(+-,-+)$ defined below (2.239). Reflection invariance only holds in the special case $\epsilon_{LL} = \epsilon_{RR}$ and $\epsilon_{LR} = \epsilon_{RL}$. Nevertheless, we can use (2.249) to obtain

$$\begin{aligned} M(-+,-+) &= i\epsilon_{LL}\, s\, (1+\cos\theta), \qquad & M(+-,+-) &= i\epsilon_{RR}\, s\, (1+\cos\theta) \\ M(-+,+-) &= i\epsilon_{LR}\, s\, (1-\cos\theta), \qquad & M(+-,-+) &= i\epsilon_{RL}\, s\, (1-\cos\theta), \end{aligned} \tag{8.128}$$

from which any cross section or polarization effect can be calculated. For example, the spin-average cross section and forward-backward asymmetry are

$$\bar{\sigma} = \int_{-1}^{1} \frac{|\bar{M}|^2}{32\pi s} d\cos\theta = \frac{s}{48\pi}\left(|\epsilon_{LL}|^2 + |\epsilon_{RR}|^2 + |\epsilon_{LR}|^2 + |\epsilon_{RL}|^2\right) \tag{8.129}$$

and

$$A_{FB} = \frac{3}{4}\left(\frac{|\epsilon_{LL}|^2 + |\epsilon_{RR}|^2 - |\epsilon_{LR}|^2 - |\epsilon_{RL}|^2}{|\epsilon_{LL}|^2 + |\epsilon_{RR}|^2 + |\epsilon_{LR}|^2 + |\epsilon_{RL}|^2}\right). \tag{8.130}$$

Assuming that only the s-channel γ and Z exchange are relevant, the coefficients in (8.127) are given by

$$\epsilon_{AB} = \frac{Q_f e^2}{s} - 4\sqrt{2} G_F M_Z^2 D(s) \epsilon_A(f) \epsilon_B(e), \tag{8.131}$$

where $A, B = L$ or R. The $\epsilon_{L,R}(r)$ are the chiral couplings of the Z to fermion r. They are defined in (8.75) and their SM values given in (8.76). $D(s) \equiv \left(s - M_Z^2 + i M_Z \Gamma_Z\right)^{-1}$ is the Breit-Wigner form for the propagator of an unstable Z of width Γ_Z, as discussed in Appendix F. It is introduced here so that the expression is valid below, near, and above the Z pole. Equation (8.131) can be rewritten by factoring out the pure QED piece

$$\epsilon_{AB} = \frac{4\pi Q_f \alpha}{s} \left[1 + \frac{4\chi_0}{Q_f} \left(\cos\delta_R + i \sin\delta_R\right) \epsilon_A(f) \epsilon_B(e) \right], \tag{8.132}$$

where

$$\chi_0 = \frac{G_F}{2\sqrt{2}\pi\alpha} \frac{s M_Z^2}{\left[(M_Z^2 - s)^2 + M_Z^2 \Gamma_Z^2\right]^{1/2}}, \qquad \tan\delta_R = \frac{M_Z \Gamma_Z}{M_Z^2 - s}. \tag{8.133}$$

$\cos\delta_R \to 1$ at low energies for $\Gamma_Z / M_Z \ll 1$, while $\cos\delta_R \to -1$ for $s \gg M_Z^2$. At $s = M_Z^2$ we have that $\cos\delta_R \to 0$ and the interference term vanishes. Substituting into (8.129) and (8.130), we find

$$\bar{\sigma} = \bar{\sigma}_0 F_1, \qquad A_{FB} = 3F_2 / 4F_1, \tag{8.134}$$

where $\bar{\sigma}_0 = 4\pi Q_f^2 \alpha^2 / 3s$ is the QED cross section in (2.232), and

$$\begin{aligned} F_1 &= 1 + \frac{2\chi_0}{Q_f} g_V^f g_V^e \cos\delta_R + \frac{\chi_0^2}{Q_f^2} \left(g_V^{f2} + g_A^{f2}\right) \left(g_V^{e2} + g_A^{e2}\right) \\ F_2 &= + \frac{2\chi_0}{Q_f} g_A^f g_A^e \cos\delta_R + \frac{4\chi_0^2}{Q_f^2} g_A^f g_A^e g_V^f g_V^e. \end{aligned} \tag{8.135}$$

We have used that $\epsilon_L(r) \pm \epsilon_R(r) = g_{V,A}^r$ and therefore $\epsilon_L^2(r) - \epsilon_R^2(r) = g_V^r g_A^r$. For $s \ll M_Z^2$, $\chi_0 = \mathcal{O}(s/M_Z^2) \ll 1$ and $\delta_R \sim \Gamma_Z / M_Z \ll 1$, so only the first term in F_2 is relevant. This involves only the axial couplings, which in the SM are independent of $\sin^2\theta_W$. The leading contribution to A_{FB} is therefore an absolute prediction. For $f = \mu^-$ the SM couplings are

$$g_A^\mu = g_A^e = -\frac{1}{2}, \qquad g_V^\mu = g_V^e = -\frac{1}{2} + 2\sin^2\theta_W. \tag{8.136}$$

The observed asymmetry, shown in Figure 8.13, agrees with the SM prediction.

Asymmetries for $f = \tau^-$ and $f = b$ also agreed well, and were important in establishing the canonical doublet assignments of the τ_L^- and b_L and singlet assignments of τ_R^- and b_R (i.e., that the third family is *sequential*), prior to the direct discovery of the ν_τ and t. This was quite important, because the third family could well have been different from the first two. As an example, the JADE collaboration at PETRA obtained the first significant measurement of the $b\bar{b}$ asymmetry, at $\sqrt{s} = 35$ GeV (Bartel et al., 1984), from which one can extract[16] $g_A^b = -0.54 \pm 0.15$. In the context of $SU(2) \times U(1)$, but allowing arbitrary assignment of the b_L and b_R to $SU(2)$ multiplets, one expects $g_A^b = t_{bL}^3 - t_{bR}^3$, where $t_{bL,R}^3$ are their T^3 eigenvalues. The JADE value is consistent with the SM assignment

[16] Subsequent LEP experiments allowed a much more precise determination of both $g_{V,A}^b$ (Schaile and Zerwas, 1992).

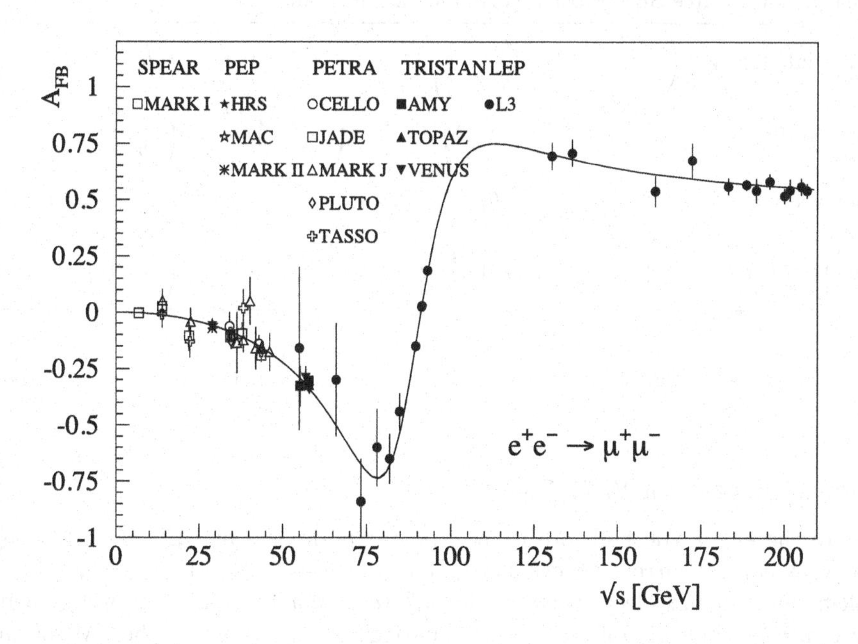

Figure 8.13 Experimental results on the forward-back asymmetry in $e^+e^- \rightarrow \mu^+\mu^-$ as a function of CM energy, compared with the SM prediction. Updated plot reprinted with permission from (Mnich, 1996).

$t^3_{bL} = -1/2$, $t^3_{bR} = 0$, but not with several alternative models for the third generation, such as the anomaly-free models listed in Table 8.4. These include a *mirror* family (Maalampi and Roos, 1990), involving L-singlets and R-doublets, and non-chiral (vector) models (e.g., Frampton et al., 2000). The four models in Table 8.4 predict $g^b_A = -1/2, +1/2, 0$, and 0, respectively, so only the sequential model was viable. In particular, the "topless" singlet vector model was excluded, strongly suggesting that the the top quark had to exist (barring extremely exotic alternatives). A similar result was also obtained by the absence of flavor changing neutral current decays $B \rightarrow \ell^+\ell^- X$ by the CLEO collaboration at CESR, which would have been generated by the violation of the GIM mechanism if the $b_{L,R}$ were singlets that decayed by $b_L - d_L$ or $b_L - s_L$ mixing (Kane and Peskin, 1982). Similarly, the τ couplings were obtained by observations of the τ lifetime, decay distributions, FB asymmetry, and absence of flavor changing neutral current decays $\tau \rightarrow \ell\ell\bar{\ell}$, requiring the existence of the ν_τ. In fact, the canonical assignments for all of the known fermions could be extracted from the data (Langacker, 1989b).

A forward-backward asymmetry does not by itself require parity violation, and in fact a non-trivial asymmetry is generated by higher-order QED corrections that must be taken into account in the analysis. As an example, the effective interaction in (8.127) would be reflection invariant for $\epsilon_{LL} = \epsilon_{RR}$ and $\epsilon_{LR} = \epsilon_{RL} = 0$, though that would be difficult to achieve in a simple gauge theory.

The $\gamma - Z$ interference effects in $e^-e^+ \rightarrow f\bar{f}$ have also been observed above the Z-pole at LEP 2 (Schael et al., 2013). The results are in agreement with the SM. Forward-backward asymmetries in the inverse processes, $\bar{p}p, pp \rightarrow \ell^+\ell^-$, $\ell = e$ or μ, will be mentioned in Section 8.3.5.

TABLE 8.4 Anomaly-free $SU(2)$ assignments for the third family.[a]

Sequential family	$\begin{pmatrix}\nu_\tau\\ \tau^-\end{pmatrix}_L$	$\begin{pmatrix}t\\ b\end{pmatrix}_L$	τ_R^-	t_R	b_R	
Mirror family	τ_L^-	t_L	b_L	$\begin{pmatrix}\nu_\tau\\ \tau^-\end{pmatrix}_R$	$\begin{pmatrix}t\\ b\end{pmatrix}_R$	
Singlet vector	τ_L^-	b_L	τ_R^-	b_R		(topless)
Doublet vector	$\begin{pmatrix}\nu_\tau\\ \tau^-\end{pmatrix}_L$	$\begin{pmatrix}t\\ b\end{pmatrix}_L$	$\begin{pmatrix}\nu_\tau\\ \tau^-\end{pmatrix}_R$	$\begin{pmatrix}t\\ b\end{pmatrix}_R$		

[a]Singlets $\nu_{\tau R}$ or (mirror) $\nu_{\tau L}$ could be added. The leptons could be interchanged in the two vector models, e.g., vector doublet quarks and vector singlet τ. More complicated assignments are also possible.

8.3.3 Implications of the WNC Experiments

Even before the era of the high precision Z-pole experiments at LEP and SLC began in 1989, the weak neutral current experiments and observation of the W and Z had done much to establish the standard electroweak model. *Global analyses* of the data were essential to the program, especially in constraining alternatives or extensions of the SM, because no one type of experiment was sensitive to all of the possible parameters. They also allowed a unified theoretical treatment of similar experiments. The caveat is that one must be careful with the treatment of systematic and theoretical uncertainties and in correlations between uncertainties.

Major results included:

Model-independent fits showed that the four-fermion interactions relevant to νq, νe, and eq were for the most part uniquely determined. They were consistent with the $SU(2) \times U(1)$ SM to first approximation, eliminating many competing gauge theories and alternatives involving spin-0 or 2 exchange, and limiting small deviations. Measurements of M_W and M_Z agreed with the expectations of the $SU(2) \times U(1)$ gauge group and canonical Higgs mechanism, eliminating more complicated alternative models with the same four-fermi interactions.

The combination of WCC and WNC results allowed the $SU(2) \times U(1)$ representations for all of the known fermions to be uniquely determined, i.e., that all of the f_L transformed as $SU(2)$ doublets and the f_R as singlets. That required that the t and ν_τ had to exist, as partners of the b and τ^-.

The precision of the deep inelastic neutrino data was high enough to require the application of QCD-evolved structure functions.

Both WNC and WCC processes were measured precisely enough to require QED and electroweak radiative corrections in their interpretation, and attention had to be paid to the definition of the renormalized $\sin^2\theta_W$, as will be discussed in Section 8.3.4.

$\sin^2\theta_W = 0.230 \pm 0.007$ (using the on-shell definition), and $m_t < 200$ GeV from the electroweak radiative corrections.

The $SU(3)\times SU(2)\times U(1)$ couplings were determined well enough to exclude the gauge coupling constant unification predictions of the simplest form of non-supersymmetric grand unified theories (GUTs). However, they were consistent with the predictions of the simplest supersymmetric GUTs, as will be discussed in Section 10.2.6.

Significant limits could be placed on many types of new physics that could perturb the SM predictions, including a heavy Z', new fermions with exotic quantum numbers, exotic Higgs representations, leptoquarks, and many types of new four-fermi operators generated by other types of underlying new physics. In many cases both WNC and WCC constraints were critical.

8.3.4 Precision Tests of the Standard Model

The precision electroweak program entered a new phase in the late 1980s with the advent of the Z-pole experiments at LEP (CERN) and SLC (SLAC), which eventually allowed tests of the SM at the 0.1% level. With such precision, care was needed in the application of QED, electroweak, QCD, and mixed radiative corrections, and in the definition of the renormalized weak angle $\sin^2\theta_W$. The radiative corrections were sensitive to the top quark and Higgs masses and to the QCD coupling α_s, and therefore allowed these quantities to be constrained or predicted by the data. Here, we briefly survey some of the relevant issues (which were also necessary, though less critical, for the WNC experiments described in Sections 8.3.1 and 8.3.2). Most of the numerical results are from the *Electroweak* review by Erler and Freitas in (Patrignani, 2016).

Input Parameters

The basic input parameters relevant to WNC, WCC, and Z/W properties in the SM are: (a) the $SU(2)\times U(1)$ gauge couplings g and g'; (b) the weak scale, $\nu=\sqrt{2}\langle 0|\varphi^0|0\rangle\sim 246$ GeV, that characterizes the SSB; (c) the Higgs mass M_H defined in (8.41) (or, equivalently, the quartic coupling λ), which enters the radiative corrections; (c) the heavy fermion masses m_t, m_b, etc., which are needed in radiative corrections and phase space factors; and (d) the strong fine structure constant α_s, which enters radiative corrections. Especially important are g, g', and ν, which are needed for weak amplitudes and mass formulae at the tree level. However, it is convenient to trade them for other quantities that are precisely known and more directly measured, and to express the theoretical expressions for other observables in terms of them. A conventional choice is:

The fine structure constant, $\alpha=1/137.035999139(31)$ (Mohr et al., 2012), determined from the anomalous magnetic moment of the electron and other low energy QED tests, as described in Section 2.12.3. It is related by $e=g\sin\theta_W$ (8.69). However, the running α must be extrapolated to $Q^2=M_Z^2$, as described in Section 2.12.2. With the conventional (on-shell) QED renormalization scheme,

$$\alpha^{-1}(M_Z^2)\equiv\frac{1-\Delta\alpha(M_Z^2)}{\alpha}=128.927(17), \tag{8.137}$$

where the uncertainty is mainly from the hadronic contribution $\Delta\alpha_{had}^{(5)}=0.02764(13)$ to $\Delta\alpha$ (for $\alpha_s(M_Z^2)=0.118(2)$). In the $\overline{MS}$ scheme one expects $\hat{\alpha}^{-1}(M_Z^2)\sim 127.950(17)$. Even these tiny uncertainties dominate the theoretical uncertainty in the precision program.

The Fermi constant, $G_F=1/\sqrt{2}\nu^2=1.1663787(6)\times 10^{-5}$ GeV, defined in terms of the muon lifetime using (7.51).

The third parameter was traditionally $\sin^2\theta_W$, defined at tree level by $g'^2/(g^2+g'^2)$ and determined by WNC processes or properties of the Z. However, following the ultra-precise determination of $M_Z = 91.1876(21)$ GeV at LEP, it has become more common to view M_Z as the third input parameter. At tree level $M_Z = e\nu/2\sin\theta_W\cos\theta_W$ by (8.34).

Definitions of the Renormalized $\sin^2\theta_W$

As discussed in Section 2.12.1, higher-order corrections to amplitudes and other observables are usually divergent. In renormalizable theories the divergences can all be absorbed into renormalized (physical) charges, masses, and wave function renormalizations. To actually carry out the renormalization one must first regularize the divergent integrals in a way that preserves the symmetries, such as Pauli-Villars, lattice, or dimensional regularization, or to avoid the divergences in the BPHZ scheme (see, e.g., Collins, 1986). It is also necessary to define the renormalized quantities. Two popular definitions are the *on-shell* and the *modified minimal subtraction* ($\overline{MS}$) schemes. The on-shell scheme was illustrated in Section 2.12.1, where the renormalized electron mass m was defined as the position of the propagator pole, which coincides with the kinematic meaning of mass. The renormalized electric charge is defined in terms of the electron-photon vertex when all of the external particles are on shell, $p_2^2 = p_1^2 = m^2$ and $q^2 = (p_2 - p_1)^2 = 0$. An alternative is the $\overline{MS}$ scheme (see, e.g., Peskin and Schroeder, 1995). One can define the renormalized charge $\hat{e}(\mu)$ at scale μ as $e_0 - \delta e$, where e_0 is the bare charge and the *counterterm* δe is chosen to absorb the $(n-4)^{-1}$ poles that arise in the dimensional regularization of higher-order corrections, as well as some associated constants, $\ln 4\pi - \gamma_E$, where $\gamma_E \sim 0.5772$ is the Euler-Mascheroni constant. Renormalized masses are defined similarly. The $\overline{MS}$ schemes are simple computationally and minimize some higher-order effects, and they lead automatically to running (μ-dependent) couplings and masses. However, they are not as directly related to what is actually measured as the on-shell quantities. Of course, one can write theoretical expressions for all observables in terms of either set of renormalized quantities, and there are α-dependent translations between them.

Now, let us consider the renormalization of $\sin^2\theta_W$ in the $SU(2)\times U(1)$ theory (for reviews, see W. Hollik in Langacker, 1995; Sirlin and Ferroglia, 2013; Patrignani, 2016). In Section 8.2 we encountered a number of expressions for $\sin^2\theta_W$, all of which are equivalent at tree-level. These included

$$\begin{aligned} \sin^2\theta_W &= 1 - \frac{M_W^2}{M_Z^2} \text{ (on-shell)}, & \sin^2\theta_W \cos^2\theta_W &= \frac{\pi\alpha}{\sqrt{2}G_F M_Z^2} \ (Z-\text{mass}) \\ \sin^2\theta_W &= \frac{g'^2}{g^2+g'^2} \ (\overline{MS}), & g_V^\ell &= -\frac{1}{2} + 2\sin^2\theta_W \text{ (effective)}, \end{aligned} \tag{8.138}$$

where g_V^ℓ is the vector $Z\ell\ell$ coupling defined in (8.75). Each of the four expressions in (8.138) can be the basis of the definition of the renormalized $\sin^2\theta_W$, which will be denoted s_W^2, $s_{M_Z}^2$, $\hat{s}_Z^2$, and $\bar{s}_\ell^2$, respectively, and in each case all observables can be expressed in terms of the renormalized $\sin^2\theta_W$ as well as m_t, M_H, α_s, etc. The four schemes are related by calculable corrections of $\mathcal{O}(\alpha)$, which can, however, depend on m_t and M_H. Each has its advantages and disadvantages.

The *on-shell* scheme (Sirlin, 1980) defines $s_W^2 \equiv 1 - M_W^2/M_Z^2$ to all orders in perturbation theory, where $M_{W,Z}$ are the physical (pole masses). It is therefore based more on the spontaneous symmetry breaking (SSB) mechanism than on the WNC vertices. It is the most commonly used scheme, is simple conceptually, and is used in the program ZFITTER (Arbuzov et al., 2006). However, observables[17] involving the $Z\bar{f}f$ vertex receive somewhat artificial m_t and M_H dependence, and the dependence of M_Z on m_t is also somewhat enhanced, so the experimental value $s_W^2 = 0.22336(10)$ has a larger uncertainty than the other schemes. Other drawbacks are that the mixed QCD-EW radiative corrections are large in the scheme, and it becomes awkward in the presence of any "beyond the standard model" physics that affects M_Z or M_W.

The Z-mass scheme is an alternative on-shell scheme (Novikov et al., 1993), which essentially uses the tree-level relation between M_Z and $s_{M_Z}^2$, except the value of the running $\alpha(M_Z^2)$ is used. It is simple conceptually, and the value $s_{M_Z}^2 = 0.23105(5)$ extracted from the measured Z mass has the smallest uncertainty because there is no m_t or M_H dependence in the relation. (The uncertainty is mainly from $\alpha(M_Z^2)$.) However, the m_t and M_H dependence and their resultant uncertainties reemerge when other observables are expressed in terms of $s_{M_Z}^2$. The other difficulties of the on-shell scheme are shared.

The other schemes considered here are based on the coupling constants and therefore the Z vertices more than on the SSB mechanism. The $\overline{MS}$ scheme defines a running

$$\hat{s}_Z^2(\mu^2) \equiv \frac{\hat{g}'^2}{\hat{g}^2 + \hat{g}'^2}, \tag{8.139}$$

where $\hat{g}$ and $\hat{g}'$ are defined by modified minimal subtraction (Marciano and Sirlin, 1981). The scheme, which is used in the radiative correction program GAPP (Erler, 1999), is simple theoretically, is convenient for comparison of running couplings with grand unification theories, and is closely connected with the scheme used in most QCD calculations, thereby minimizing the uncertainties in the mixed QCD-EW radiative corrections. Other advantages are that the value of $\hat{s}_Z$ is usually insensitive to new physics, the Z asymmetries are almost independent of m_t and M_H for fixed $\hat{s}_Z$ (except for $Z \to b\bar{b}$, which involves special vertex corrections involving the t), and the sensitivity of the value obtained from M_Z is reduced compared to the on-shell value, leading to a smaller uncertainty in the experimental value, $\hat{s}_Z^2 \equiv \hat{s}_Z^2(M_Z^2) = 0.23129(5)$. The running of $\hat{s}_Z^2(\mu^2)$ can be tested by determinations based on low energy WNC experiments, as shown in Figure 8.14. The agreement is generally good, though the deviation of the NuTeV deep inelastic measurement (Section 8.3.1) is evident. It is sometimes useful to define $\hat{s}_0^2 \equiv \hat{s}_Z^2(0)$, predicted to be 0.23865(8), for comparison with the present and future (more precise) low energy experiments (Kumar et al., 2013; Erler et al., 2014). The drawback of $\hat{s}_Z^2$ is that it is a "theorist's" definition. All observables can be expressed in terms of it, but there is no simple defining relation. Just as in the on-shell scheme, the experimental value of $\hat{s}_Z^2$ is generally obtained by a global fit to all of the data.

The effective scheme, which has been used very successfully in expressing the results of the Z-pole experiments, defines an effective $\bar{s}_f^2$ for each type of fermion f in such a way that the formulae for the $Z \to f\bar{f}$ partial widths and asymmetries take their tree level form when written in terms of effective axial and vector couplings

$$\bar{g}_A^f = \sqrt{\bar{\rho}_f}\, t_{fL}^3, \qquad \bar{g}_V^f = \sqrt{\bar{\rho}_f}\left(t_{fL}^3 - 2\bar{s}_f^2 q_f\right), \tag{8.140}$$

where $\bar{\rho}_f$ is common to the vector and axial couplings. This form is very simple conceptually and has the advantage that the $Z \to f\bar{f}$ asymmetries are almost independent of m_t and M_H (except for $f = b$). The most precise is the charged lepton effective coupling, $\bar{s}_\ell^2 = 0.23152(5)$, which is closely related to $\hat{s}_Z^2$. Drawbacks are that it is a phenomenological definition, which

[17] It should be emhasized that even in the on-shell scheme the most precise experimental value of s_W^2 is obtained from a global fit of s_W^2 and the other parameters to all of the precision observables, and not just the value obtained from the defining relation.

in detail depends on how all of the radiative corrections to the processes observed near the Z-pole have been applied. Also, $\bar{s}_f^2$ is different for each f, and the scheme is not very useful for describing non-Z-pole observables.

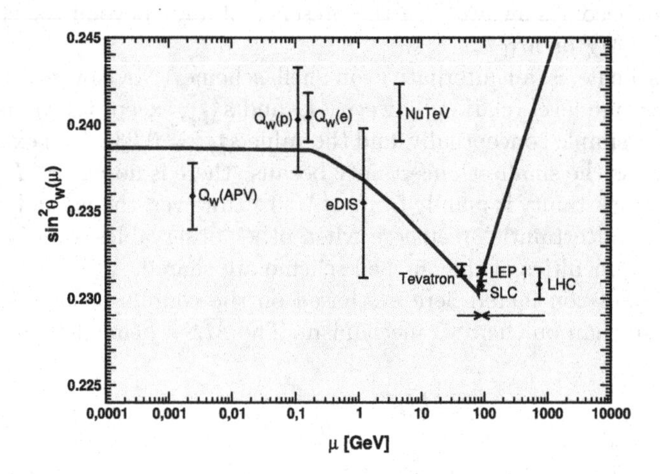

Figure 8.14 Running $\hat{s}_Z^2(\mu^2)$ measured at various scales, compared with the predictions of the SM, updated from (Czarnecki and Marciano, 2000; Erler and Ramsey-Musolf, 2005). eDIS refers to deep inelastic eD scattering. The Tevatron and LHC points are at $\mu = M_Z$ but are displaced for clarity. The discontinuities in slope, which are due to particle thresholds, can be smoothed out in a varient scheme defined in (Czarnecki and Marciano, 2000). Plot courtesy of the Particle Data Group (Patrignani, 2016).

Radiative Corrections

There are several classes of radiative corrections that must be applied to electroweak quantities. The first are the reduced QED corrections consisting of diagrams in which real and virtual photons are attached to charged particles in all possible ways, but not including vacuum polarization (self-energy) diagrams, as illustrated in Figure 8.15. For Z exchange processes this set is often finite and gauge invariant. However, these diagrams depend on the

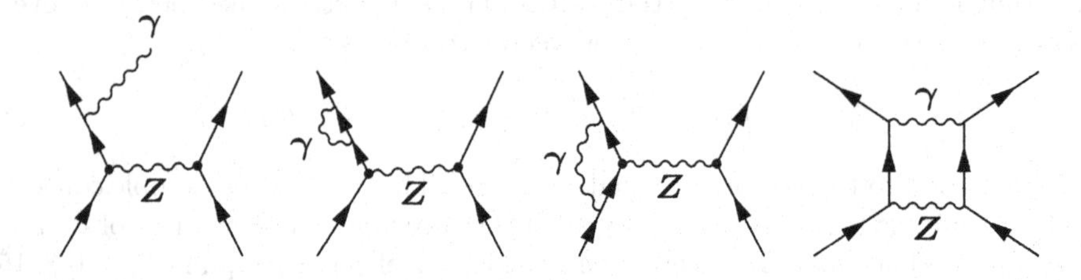

Figure 8.15 Typical reduced QED corrections to Z exchange. Vacuum polarization diagrams are not included.

energies, acceptances, and cuts, and need to be calculated and applied for each individual experiment.

Gauge self-energy (*oblique*) diagrams for the $\gamma\gamma$, γZ, ZZ, and WW propagators involving fermion, gauge, and Higgs loops, as illustrated in Figure 8.16, are the most important. They lead to large m_t and M_H dependence of M_W, M_Z, partial widths, and four-fermi amplitudes, and to the major differences between the renormalization schemes. The dominant effects are quadratic in m_t and logarithmic in M_H, i.e., of $\mathcal{O}(\alpha m_t^2)$ and $\mathcal{O}(\alpha \ln M_H)$. One-loop self-energy diagrams sensitive to the top and Higgs masses are shown in Figure 8.17. These were significant in constraining the values of m_t and M_H from the precision data.

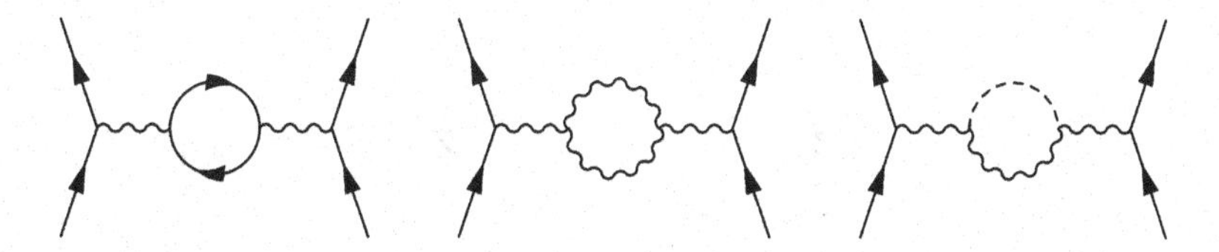

Figure 8.16 Typical one-loop gauge self-energy diagrams, also known as vacuum polarization or oblique diagrams.

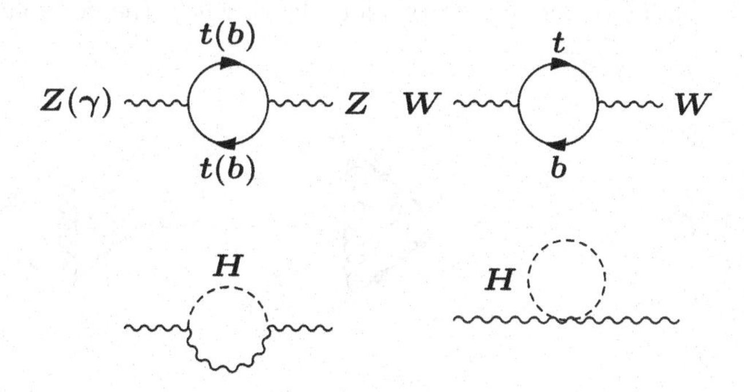

Figure 8.17 Oblique (self-energy) diagrams involving the top and the Higgs.

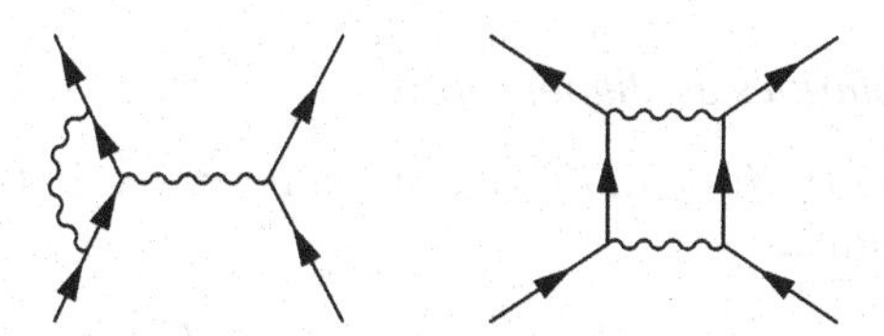

Figure 8.18 Electroweak vertex and box diagrams, involving W and Z bosons.

Electroweak vertex and box diagrams include those in which a W or Z is exchanged across a vertex, or box diagrams involving WW, WZ, or ZZ, as illustrated in Figure 8.18. These are usually small. However, they are needed for gauge invariance and are not always negligible, e.g., in effective four-fermi operators. They are usually absorbed into the parameters in the effective interactions. Especially important are corrections to the $Zb\bar{b}$ vertex involving the t quark, shown in Figure 8.19. Like the oblique corrections, these scale as αm_t^2 for large m_t and therefore make a significant contribution to $Z \to b\bar{b}$ decay, especially the width. The $Zb\bar{b}$ vertex corrections were especially useful for distinguishing the effects of m_t from the Higgs mass and from possible new physics contributions to the oblique corrections.

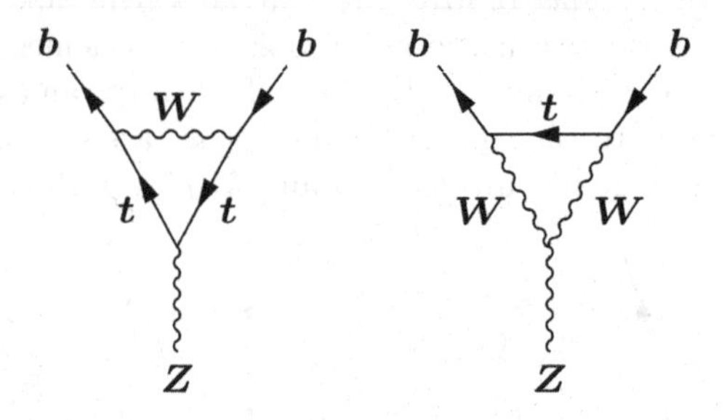

Figure 8.19 One-loop contributions to the $Zb\bar{b}$ vertex involving the t.

QCD and mixed QCD-electroweak diagrams involve gluon exchange between quarks in electroweak processes, as illustrated in Figure 8.20. They affect the hadronic W and Z decay widths at one-loop level, and can therefore be used to determine $\alpha_s(M_Z^2)$ from the Z decays. They also affect the oblique corrections at two-loop level, where the dominant effects are of order $\alpha\alpha_s m_t^2$.

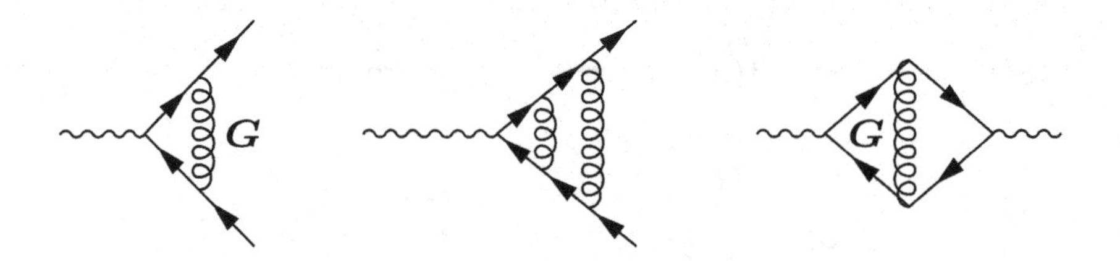

Figure 8.20 QCD corrections to the $Wq\bar{q}$ or $Zq\bar{q}$ vertices and mixed corrections to an electroweak self-energy diagram.

M_W, M_Z, and WNC Amplitudes at Higher Order.

The tree-level expressions for $M_{W,Z}$ in (8.34) can be rewritten in terms of the independent variables α, G_F, and $\sin^2\theta_W$ as

$$M_W = \frac{A_0}{\sin\theta_W}, \qquad M_Z = \frac{M_W}{\cos\theta_W}, \tag{8.141}$$

where $A_0 = (\pi\alpha/\sqrt{2}G_F)^{1/2}$. All of the quantities in (8.141), i.e., $M_{W,Z}$, α, G_F, and $\sin^2\theta_W$, are unrenormalized, though we do not indicate that explicitly for notational simplicity. Including radiative corrections, (8.141) is modified in the $\overline{MS}$ renormalization scheme to

$$M_W = \frac{A_0}{\hat{s}_Z(1-\Delta\hat{r}_W)^{1/2}}, \qquad M_Z = \frac{M_W}{\hat{\rho}^{1/2}\hat{c}_Z}, \tag{8.142}$$

where $\hat{s}_Z^2$ is the $\overline{MS}$ weak angle, $\hat{c}_Z^2 = 1 - \hat{s}_Z^2$, $A_0 = (\pi\alpha/\sqrt{2}G_F)^{1/2} = 37.28039(1)$ GeV using the physical values for α and G_F, and $\Delta\hat{r}_W$ collects the radiative corrections to M_W. (For comparison with Z-pole physics it is convenient to define $\Delta\hat{r}_W$ so that the running $\hat{s}_Z^2$ is evaluated at M_Z.) The largest contribution to $\Delta\hat{r}_W$ is due to the running of α^{-1} to the value $\hat{\alpha}^{-1}(M_Z^2) \sim 128$ at the electroweak scale, and yields

$$\Delta\hat{r}_W \sim \Delta r_0 \equiv 1 - \frac{\alpha}{\hat{\alpha}(M_Z^2)} \sim 0.06630(13). \tag{8.143}$$

Including smaller effects, one expects $\Delta\hat{r}_W \sim 0.06952(13)$ in the SM. The large $m_t - m_b$ mass difference breaks $SU(2)$ symmetry and leads, via the oblique corrections from the diagrams in Figure 8.17, to an increase that depends quadratically on m_t in the ratio M_W/M_Z compared to the bare ratio. The quadratic dependence does not enter $\Delta\hat{r}_W$ because the shift in M_W has been absorbed into the observed strength of the WCC, i.e., into the measured G_F. Rather, it is shifted into the parameter $\hat{\rho}$ in (8.142). There are other contributions to $\hat{\rho}$, but the largest is from m_t,

$$\hat{\rho} \sim 1 + \rho_t, \tag{8.144}$$

where the ubiquitous quantity ρ_t is (Veltman, 1977; Chanowitz et al., 1978)

$$\rho_t = \frac{3G_F m_t^2}{8\sqrt{2}\pi^2} \sim 0.00940 \left(\frac{m_t}{173.34 \text{ GeV}}\right)^2. \tag{8.145}$$

In the on-shell scheme (8.141) becomes

$$M_W = \frac{A_0}{s_W(1 - \Delta r)^{1/2}}, \qquad M_Z = \frac{M_W}{c_W}, \tag{8.146}$$

where $s_W^2 \equiv 1 - M_W^2/M_Z^2$ is the definition of the renormalized angle, $c_W^2 = 1 - s_W^2$, and A_0 is the same as in the on-shell scheme. Δr collects all of the radiative corrections relating α, $\alpha(M_Z)$, G_F, M_W, and M_Z. The largest contributions to Δr are

$$\Delta r \sim \Delta r_0 - \frac{\rho_t}{t_W^2}, \tag{8.147}$$

where $t_W = s_W/c_W$. Δr_0 is due to the running of α^{-1} and is the same as in the $\overline{MS}$ scheme. The ρ_t dependence of Δr is enhanced by the t_W^{-2} in the on-shell scheme, due to the fact that the physical m_t effect of increasing M_W/M_Z forces a downward shift in s_W^2 which must be compensated by Δr. There are additional contributions to Δr from bosonic loops, including those which depend logarithmically on M_H. Altogether, one expects $\Delta r \sim 0.03648(31)$ in the SM.

M_W has been measured separately at the Tevatron and at LEP 2 at the 0.04% level using event shapes and also the behavior of the $e^-e^+ \to W^-W^+$ cross section near threshold. The results agree and can be combined to obtain $M_W = 80.385(15)$ GeV, in agreement with the SM expectation of 80.361(6) GeV using the global best fit parameters. This is about 3.7% higher than the value of ~ 78 GeV that one would expect (for the same $\hat{s}_Z$) at tree-level, therefore confirming the radiative corrections, especially the running α. The value of M_Z will be discussed in Section 8.3.5.

The tree-level amplitude $\mathcal{A}_i^0$ for a WNC process i is of the form

$$\mathcal{A}_i^0 = \frac{G_F^0}{\sqrt{2}} F_i(\sin^2\theta_W), \tag{8.148}$$

where F_i is a function of i, such as defined in (8.79), and G_F^0 and $\sin^2\theta_W$ are the unrenormalized quantities. The effect of higher-order corrections is to replace $\mathcal{A}_i^0$ by the amplitude

$$\mathcal{A}_i = \rho_R^i \frac{G_F}{\sqrt{2}} F_i(\kappa_R^i s_R^2) + \lambda_R^i, \tag{8.149}$$

where G_F is the renormalized Fermi constant defined by (7.51) on page 234, F_i takes the same functional form as in (8.148), and s_R^2 is the renormalized quantity in scheme R. The three parameters ρ_R^i, κ_R^i, and λ_R^i depend on the process and on the renormalization scheme, and one expects $\rho_R^i - 1, \kappa_R^i - 1$, and λ_R^i to be of $\mathcal{O}(\alpha)$. They are defined operationally, i.e., by the way that they modify the tree-amplitude (for an example, see Marciano and Sirlin, 1980). Oblique corrections directly affect ρ_R^i, but may also enter κ_R^i through their effects on s_W^2 in the on-shell scheme. Vertex corrections typically affect ρ_R^i and κ_R^i, while box diagrams may yield contributions to λ_R^i. The λ_R^i are usually small enough to be neglected.

In the $\overline{MS}$ scheme one may write

$$\hat{\rho}^i = \hat{\rho}\hat{\rho}_{rad}^i, \tag{8.150}$$

where $\hat{\rho}$ is a universal part due to the shift of M_Z^2 in (8.142). $\hat{\rho}$ contains the quadratic m_t dependence in (8.144) and the dominant M_H dependence. $\hat{\rho}_{rad}^i$ and $\hat{\kappa}^i$ are close to unity and (like $\Delta\hat{r}_W$) depend only weakly on m_t and M_H (with the exception of processes involving the $Zb\bar{b}$ vertex). In the on-shell scheme both ρ^i and κ^i have universal terms containing the quadratic m_t and the dominant M_H dependence, and much weaker process-dependent contributions. The leading m_t-dependent term is

$$\rho^i \sim 1 + \rho_t, \qquad \kappa^i \sim 1 + \rho_t/t_W^2, \tag{8.151}$$

where (similar to Δr) the κ^i dependence is present to cancel the m_t dependence of the vertices introduced by the definitions of s_W^2.

The W Decay Width

Consider the coupling

$$\mathcal{L} = -\bar{f}_1 \gamma^\mu \left(g_V - g_A \gamma^5\right) f_2 V_\mu + h.c., \tag{8.152}$$

of a massive vector V to fermions $f_{1,2}$. (If V is Hermitian and $f_1 = f_2$ then $g_{V,A}$ are real and there is no $+h.c.$) The partial width for $V \to f_1\bar{f}_2$ is easily shown to be (Problem 2.25)

$$\Gamma(V \to f_1\bar{f}_2) = \frac{M_V}{12\pi}\left(|g_V|^2 + |g_A|^2\right) \tag{8.153}$$

at tree-level and neglecting the fermion masses. Applying this result to leptonic W decays using (8.55) and (8.60), this implies

$$\Gamma(W^+ \to e^+\nu_e) = \frac{g^2 M_W}{48\pi} = \frac{\alpha M_W}{12\sin^2\theta_W}, \tag{8.154}$$

and similarly for $\mu^+\nu_\mu$ and $\tau^+\nu_\tau$. It is convenient to rewrite (8.154) as

$$\Gamma(W^+ \to e^+\nu_e) = \frac{G_F M_W^3}{6\sqrt{2}\pi} \approx 226.27(5)\text{ MeV} \tag{8.155}$$

using $g^2 M_W / 4\sqrt{2} = G_F M_W^3$. This form absorbs the dominant radiative corrections, including the running of α. The numerical values for the partial widths include the small residual electroweak corrections and fermion mass effect. Similarly, the hadronic partial widths are

$$\Gamma(W^+ \to u_i \bar{d}_j) = \frac{C G_F M_W^3}{6\sqrt{2}\pi} |V_{ij}|^2 \approx 705.1(3)\, |V_{ij}|^2 \text{ MeV}, \tag{8.156}$$

where V_{ij} is the CKM matrix element, and

$$C = 3\left(1 + \frac{\alpha_s(M_W^2)}{\pi} + 1.409\frac{\alpha_s^2}{\pi^2} - 12.77\frac{\alpha_s^3}{\pi^3} - 80.0\frac{\alpha_s^4}{\pi^4}\right), \tag{8.157}$$

includes the color counting factor and the four-loop QCD vertex correction for massless quarks. Again, there are small QED and electroweak corrections and one can include fermion mass effects. Adding the contributions, one predicts $\Gamma_W \sim 2.0888(7)$ GeV, in agreement with the experimental world average $\Gamma_W = 2.085(42)$ GeV, which is based largely on decay distributions at LEP (Schael et al., 2013) and the Tevatron (Bandurin et al., 2015).

8.3.5 The Z-Pole and Above

As can be seen in Figure 8.21 the cross section for e^+e^- annihilation is greatly enhanced near the Z-pole. This allowed high statistics studies of the properties of the Z at LEP (CERN) and SLC (SLAC) in $e^-e^+ \to Z \to \ell^-\ell^+$, $q\bar{q}$, and $\nu\bar{\nu}$ (for reviews, see the articles by D. Schaile and by A. Blondel in Langacker, 1995; Grunewald, 1999; Schael et al., 2006a). The four experiments ALEPH, DELPHI, L3, and OPAL at LEP collected some 1.7×10^7 events at or near the Z-pole during the period 1989–1995. The SLD collaboration at the SLC observed some 6×10^5 events during 1992-1998, with the lower statistics compensated by a highly polarized e^- beam with $P_{e^-} \gtrsim 75\%$.

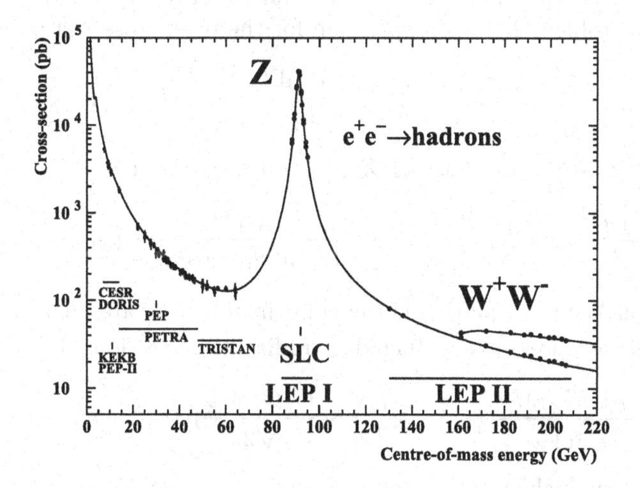

Figure 8.21 Cross section for $e^+e^- \to$ hadrons as a function of CM energy. At low energies the cross section is dominated by 1 photon exchange in the s channel, while at high energies the Z dominates. The approximate energy ranges of various e^+e^- colliders are also shown. Reprinted with permission from (Schael et al., 2006a).

The basic Z-pole observables relevant to the precision program are:

The lineshape variables M_Z, Γ_Z, and σ_{peak} (Appendix F).

The branching ratios for Z to decay into e^-e^+, $\mu^-\mu^+$, or $\tau^-\tau^+$; into $q\bar{q}$, $c\bar{c}$, or $b\bar{b}$; or into invisible channels such as $\nu\bar{\nu}$ (allowing a determination of the number $N_\nu = 2.992 \pm 0.007$ of neutrinos lighter than $M_Z/2$).

Various asymmetries, including forward-backward (FB), hadronic FB charge, polarization (LR), mixed FB-LR, and the polarization of produced τ's.

The branching ratios and FB asymmetries could be measured separately for e, μ, and τ, allowing tests of lepton family universality.

LEP and SLC simultaneously carried out other programs, most notably studies and tests of QCD, and heavy quark physics.

The Z Lineshape

One of the most important observables is the Z lineshape, i.e., the cross section for $e^-e^+ \to f\bar{f}$, where $f = e, \mu, \tau, b, c$, or hadrons, as a function $s = E_{CM}^2$ near the Z-pole. The expected Breit-Wigner form[18] (slightly more sophisticated than the one in (F.2)), is

$$\sigma_f(s) \sim \sigma_f \frac{s\Gamma_Z^2}{(s - M_Z^2)^2 + \frac{s^2\Gamma_Z^2}{M_Z^2}}, \tag{8.158}$$

where σ_f is the peak cross section and Γ_Z is the total Z width. By measuring the cross section at a number of energies near the peak one can determine M_Z, Γ_Z, and σ_f, as illustrated in Figure F.1. For example, the cross section σ_{had} for $e^-e^+ \to$ hadrons is shown in Figure 8.21. The results for the Z are generally expressed in a model-independent way in terms of the partial widths $\Gamma(f\bar{f})$ for $Z \to f\bar{f}$ and in terms of effective couplings to be defined below. From Problem 2.27 the expression for the peak cross section is

$$\sigma_f = \frac{12\pi}{M_Z^2} \frac{\Gamma(e^-e^+)\Gamma(f\bar{f})}{\Gamma_Z^2}. \tag{8.159}$$

Using (8.72) and (8.153), the width for $Z \to f\bar{f}$ at tree-level is

$$\Gamma(f\bar{f}) = \frac{C_f(g^2 + g'^2)M_Z}{48\pi}\left(g_V^{f2} + g_A^{f2}\right) = \frac{C_f \alpha M_Z}{12\sin^2\theta_W \cos^2\theta_W}\left(g_V^{f2} + g_A^{f2}\right), \tag{8.160}$$

where $C_f = 1$ (leptons) and 3 (quarks) is the color factor, $g_{V,A}^f$ are defined in (8.75), and we have neglected the fermion masses. Including radiative corrections, $\Gamma(f\bar{f})$ becomes

$$\Gamma(f\bar{f}) = \frac{C_f\hat{\alpha}(M_Z^2)M_Z}{12\hat{s}_W^2\hat{c}_W^2}\left(\bar{g}_V^{f2} + \bar{g}_A^{f2}\right) \sim \frac{C_f G_F M_Z^3}{6\sqrt{2}\pi}\hat{\rho}\left(\bar{g}_V^{f2} + \bar{g}_A^{f2}\right), \tag{8.161}$$

where for quarks C_f now includes QCD corrections similar to (8.157) evaluated at M_Z^2 and $\bar{g}_{V,A}^f$ are effective vertex factors

$$\bar{g}_A^f = \sqrt{\hat{\rho}_f}t_{fL}^3, \qquad \bar{g}_V^f = \sqrt{\hat{\rho}_f}\left(t_{fL}^3 - 2\hat{\kappa}_f q_f \hat{s}_Z^2\right), \tag{8.162}$$

[18]There are non-negligible corrections from reduced QED diagrams, including initial state photon radiation and s-channel photon exchange, as well as $\gamma - Z$ interference. These are taken into account, usually assuming the SM expressions, in determining the Z parameters.

which are the same as the tree-level expressions in (8.76) except for the electroweak radiative correction factors $\hat{\rho}_f$ and $\hat{\kappa}_f$. Note the similarity of this parametrization to the one in (8.149) for WNC amplitudes. The second form in (8.161) is obtained using (8.142), and contains the quadratic m_t dependence of (8.144) as well as the largest M_H dependence in the universal factor $\hat{\rho}$. Many types of corrections enter into $\hat{\rho}_f$ and $\hat{\kappa}_f$, but the overall corrections are small and only weakly sensitive to m_t and M_H. (The $\hat{\rho}_f$ are analogous to the $\hat{\rho}^i_{rad}$ of (8.150).) For example, one expects $\hat{\rho}_\ell \sim 0.9980$ and $\hat{\kappa}_\ell \sim 1.0010$. The one exception is $f = b$ because of the t contributions to the vertex in Figure 8.19. For the known m_t one has the rather large corrections $\hat{\rho}_b \sim 0.9868$, $\hat{\kappa}_b \sim 1.0065$. In practice, (8.161) must be extended to include fermion mass effects (including those within the radiative corrections), two-loop QED corrections, two-loop mixed QED-QCD corrections, etc. The partial width can also be written in the on-shell scheme as

$$\Gamma(f\bar{f}) \sim \frac{C_f G_F M_Z^3}{6\sqrt{2}\pi}\left(\bar{g}_V^{f2} + \bar{g}_A^{f2}\right) \tag{8.163}$$

with

$$\bar{g}_A^f = \sqrt{\rho_f}\, t^3_{fL}, \qquad \bar{g}_V^f = \sqrt{\rho_f}\left(t^3_{fL} - 2\kappa_f q_f s_W^2\right). \tag{8.164}$$

We have used the same notation for $\bar{g}^f_{V,A}$ in both schemes to avoid a proliferation of symbols, but they differ in detail, e.g., there is no analog of $\hat{\rho}$ in (8.163), with the effects now absorbed into the ρ_f. The dominant m_t dependence is $\rho_f \sim 1 + \rho_t, \kappa_f \sim 1 + \rho_t/t_W^2$ (cf. (8.151)).

It is convenient to define an effective angle $\bar{s}_f^2$

$$\bar{s}_f^2 = \kappa_f s_W^2 = \hat{\kappa}_f \hat{s}_Z^2. \tag{8.165}$$

$\hat{\kappa}_{f\neq b}$ is insensitive to m_t and M_H, so one has $\bar{s}_\ell^2 \sim \hat{s}_Z^2 + 0.00023$. The effective couplings can then be written as in (8.140), i.e., they take the form of tree-level expressions up to an overall factor of $\sqrt{\bar{\rho}_f}$, where $\bar{\rho}_f$ depends on the renormalization scheme. The Z-pole analyses often extract the $\bar{g}^f_{V,A}$ from the data in a model-independent way.

Using (8.161) or (8.163) and additional corrections, the SM predictions are

$$\Gamma(f\bar{f}) \sim \begin{cases} 299.91 \pm 0.19 \text{ MeV } (u\bar{u}), & 167.17 \pm 0.02 \text{ MeV } (\nu\bar{\nu}) \\ 382.80 \pm 0.14 \text{ MeV } (d\bar{d}), & 83.97 \pm 0.01 \text{ MeV } (e^+e^-) \\ 375.69 \mp 0.17 \text{ MeV } (b\bar{b}) & \end{cases} \tag{8.166}$$

for $\alpha_s = 0.1182(16)$ and the observed m_t and M_H. The small partial width into e^+e^- is due to the small value of $\bar{g}_V^e$, which would vanish for $\bar{s}_e^2 = 1/4$. The total width is predicted to be $\Gamma_Z \sim 2.4943(8)$ GeV, compared with the experimental value 2.4952(23).

The results of the four LEP experiments have been carefully combined by the LEP Electroweak Working Group (LEPEWWG) (Schael et al., 2006a, LEPEWWG website), which took into account common systematics, corrected the data for non-Z effects, applied radiative corrections, and carried out SM and model-independent analyses of the Z-pole and other data, in collaboration with SLC, Tevatron and other LEP working groups. For the Z lineshape, results were typically presented in terms of a conventional set of observables: M_Z, Γ_Z, σ_{had}, R_ℓ, R_b, and R_c, where

$$\sigma_{had} \equiv \frac{12\pi}{M_Z^2}\,\frac{\Gamma(e^+e^-)\Gamma(Z \to hadrons)}{\Gamma_Z^2} \tag{8.167}$$

is the peak cross section into hadrons, and

$$R_{q_i} \equiv \frac{\Gamma(q_i\bar{q}_i)}{\Gamma(had)},\quad q_i = (b, c), \qquad R_{\ell_i} \equiv \frac{\Gamma(had)}{\Gamma(\ell_i\bar{\ell}_i)},\quad \ell_i = (e, \mu, \tau) \tag{8.168}$$

are ratios of partial widths to the total width into hadrons, in which many of the radiative corrections and uncertainties cancel. This set had the benefit of being weakly correlated, though the precision still required careful treatment of the full error correlation matrices.

The principal Z-pole observations from LEP and SLC are listed in Table 8.5, along with the standard model expectations using the global best fit values for the input parameters. The precision of the M_Z determination, to ~0.0023%, is unprecedented in this high energy regime. The energy calibration was done using a resonant depolarization technique, and corrections had to be made for the tidal effects of the Sun and Moon, the water table and the water level in Lake Geneva, and leakage currents produced by nearby trains! Many of the other quantities are measured to the 0.1–1% level, and the agreement with the predicted values from the fit is generally quite good, though there are one or two exceptions.

TABLE 8.5 Principal Z-pole observables, their experimental values, theoretical predictions using the SM parameters from the global best fit, and pull (difference from the prediction divided by the uncertainty).[a]

Quantity	Value	Standard Model	Pull
M_Z [GeV]	91.1876 ± 0.0021	91.1880 ± 0.0020	−0.2
Γ_Z [GeV]	2.4952 ± 0.0023	2.4943 ± 0.0008	0.4
$\Gamma(had)$ [GeV]	1.7444 ± 0.0020	1.7420 ± 0.0008	—
$\Gamma(inv)$ [MeV]	499.0 ± 1.5	501.66 ± 0.05	—
$\Gamma(\ell^+\ell^-)$ [MeV]	83.984 ± 0.086	83.995 ± 0.010	—
σ_{had} [nb]	41.541 ± 0.037	41.484 ± 0.008	1.5
R_e	20.804 ± 0.050	20.734 ± 0.010	1.4
R_μ	20.785 ± 0.033		1.6
R_τ	20.764 ± 0.045	20.779 ± 0.010	−0.3
R_b	0.21629 ± 0.00066	0.21579 ± 0.00003	0.8
R_c	0.1721 ± 0.0030	0.17221 ± 0.00003	0.0
$A_{FB}^{0,e}$	0.0145 ± 0.0025	0.01622 ± 0.00009	−0.7
$A_{FB}^{0,\mu}$	0.0169 ± 0.0013		0.5
$A_{FB}^{0,\tau}$	0.0188 ± 0.0017		1.5
$A_{FB}^{0,b}$	0.0992 ± 0.0016	0.1031 ± 0.0003	−2.4
$A_{FB}^{0,c}$	0.0707 ± 0.0035	0.0736 ± 0.0002	−0.8
$A_{FB}^{0,s}$	0.098 ± 0.011	0.1032 ± 0.0003	−0.5
$\bar{s}_\ell^2(A_{FB}^{0,q})$ (LEP)	0.2324 ± 0.0012	0.23152 ± 0.00005	0.7
$\bar{s}_\ell^2(A_{FB}^{0,\ell})$ (Tevatron)	0.23185 ± 0.00035		0.9
$\bar{s}_\ell^2(A_{FB}^{0,\ell})$ (LHC)	0.23105 ± 0.00087		−0.5
A_e (*hadronic*)	0.15138 ± 0.00216	0.1470 ± 0.0004	2.0
(*leptonic*)	0.1544 ± 0.0060		1.2
(P_τ)	0.1498 ± 0.0049		0.6
A_μ	0.142 ± 0.015		−0.3
A_τ (SLD)	0.136 ± 0.015		−0.7
(P_τ)	0.1439 ± 0.0043		−0.7
A_b	0.923 ± 0.020	0.9347	−0.6
A_c	0.670 ± 0.027	0.6678 ± 0.0002	0.1
A_s	0.895 ± 0.091	0.9356	−0.4

[a] From the Electroweak review in (Patrignani, 2016). $\Gamma(had)$, $\Gamma(inv)$, and $\Gamma(\ell^+\ell^-)$ are not independent.

The individual lepton channels could be measured separately, allowing the lepton universality prediction $R_e = R_\mu \sim R_\tau$ (up to a small τ mass correction) to be tested. The data are in agreement with universality within the uncertainties, and can therefore be combined to form an average lepton ratio R_ℓ. The LEP values of R_ℓ are shown in Figure 8.22 and compared with the predictions of the SM as a function of the Higgs mass M_H, obtained using the then known values for M_Z, m_t, and α_s.

Other quantities can be derived from the conventional set, such as the hadronic and leptonic widths, $\Gamma(had)$ and $\Gamma(\ell\bar{\ell})$. Especially important is the *invisible* Z width determined in principle by subtracting the hadronic and charged lepton widths from the total (lineshape) one,

$$\Gamma(inv) = \Gamma_Z - \Gamma(had) - \sum_i \Gamma(\ell_i \bar{\ell}_i). \tag{8.169}$$

In the SM $\Gamma(inv)$ is due to the (unobserved) neutrinos. It can be predicted from the other observables provided one knows the number N_ν of neutrino flavors, implying that the lineshape observables are not all independent for a given N_ν. The hadronic cross section is shown as a function of $\sqrt{s}$ in Figure 8.23 compared with the SM prediction for 2, 3, and 4 flavors of light neutrinos. It is seen that the data are only consistent with $N_\nu = 3$. One can quantify this by allowing $N_\nu \equiv 3 + \Delta N_\nu$ to be a free parameter, with

$$\Gamma(inv) = (3 + \Delta N_\nu)\Gamma(\nu\bar{\nu}), \tag{8.170}$$

where $\Gamma(\nu\bar{\nu})$ is the theoretical width in (8.166) for a single neutrino flavor. One obtains[19] $N_\nu = 2.992 \pm 0.007$ or $\Delta N_\nu = -0.008(7)$. Thus, a fourth family with a light neutrino is clearly excluded. (The exclusion is valid for m_{ν_4} almost as large as $M_Z/2$. However, it does not apply to right-handed (*sterile*) neutrinos, which do not couple directly to the Z.) The result for $\Gamma(inv)$ is also important for constraining other types of possible extensions to the SM, because it would receive contributions from any decays into unobserved particles. This is conveniently parametrized by (8.170), even when ΔN_ν has nothing to do with extra neutrinos. For example, ΔN_ν would be $\frac{1}{2}$ for a single flavor of light scalar neutrino ($Z \to \tilde{\nu}_i \tilde{\nu}_i^c$) in supersymmetry, while $\Delta N_\nu = 2$ would be expected in *triplet Majoron* neutrino mass models with spontaneous L violation, from the decay of the Z into a pseudo-Goldstone boson (Majoron) and a light scalar (Chapter 9).

Z-Pole Asymmetries

Most of the Z-pole asymmetries[20] can be expressed in terms of the quantities

$$A_f \equiv \frac{2\bar{g}_V^f \bar{g}_A^f}{\bar{g}_V^{f2} + \bar{g}_A^{f2}}, \tag{8.171}$$

where $\bar{g}_{V,A}^f$ are the effective couplings defined in (8.140). The asymmetries are easily calculated (ignoring the fermion masses) using (8.127) and (8.128) and the effective coupling

$$\mathcal{L}_Z = -\frac{\hat{e}}{2\hat{s}_Z \hat{c}_Z} \sum_f \bar{f}\gamma^\mu \left(\bar{g}_V^f - \bar{g}_A^f \gamma^5\right) f Z_\mu. \tag{8.172}$$

---

[19]This value was obtained from the global fit including all correlations. The definition used by the LEP collaborations, $N_\nu = \frac{\Gamma(inv)}{\Gamma(\ell\bar{\ell})} \left(\frac{\Gamma(\ell\bar{\ell})}{\Gamma(\nu\bar{\nu})}\right)_{SM}$, where $\Gamma(\ell\bar{\ell})$ is measured and the last factor is the theoretical ratio in the SM, yields $N_\nu = 2.984(8)$.

[20]The *Born asymmetries* A^0 considered here are the idealized asymmetries obtained after subtracting reduced QED and off-pole effects from the data, and dividing by the beam polarization P_{e^-} for the SLC results.

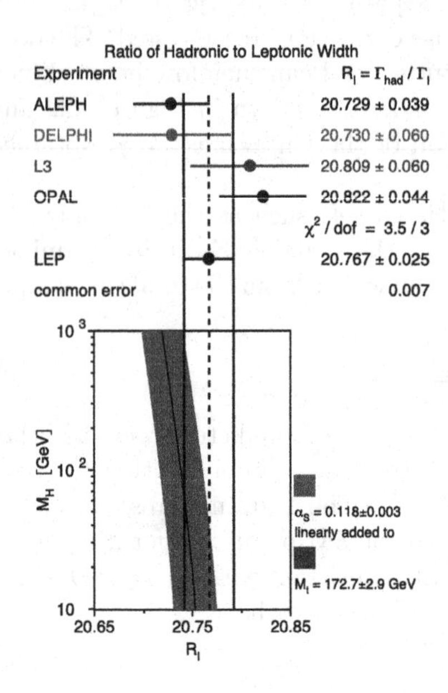

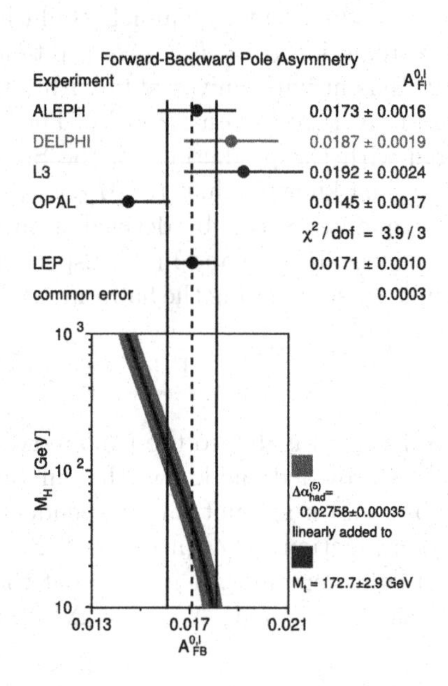

Figure 8.22 R_ℓ and $A_{FB}^{0,\ell}$ from the four LEP experiments, compared with the SM predictions as a function of M_H using the observed M_Z. Both results are consistent with $M_H = 125$ GeV. Reprinted with permission from (Schael et al., 2006a, with LP05/EPS05 updates).

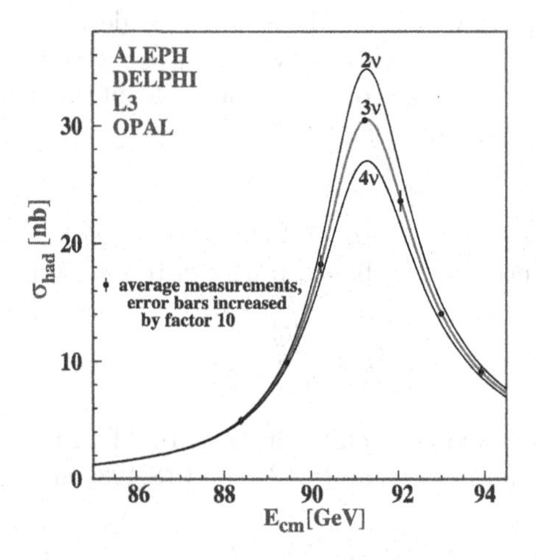

Figure 8.23 Lineshape for $Z \to$ hadrons, compared with the expectation for 2, 3, and 4 ordinary neutrinos lighter than $\sim M_Z/2$. The asymmetry is due to the explicit factor of s in the numerator of the Breit-Wigner formula (cf. Equation F.7) and to radiative corrections. Reprinted with permission from (Schael et al., 2006a).

For example, the forward-backward asymmetry A^{0f}_{FB} for $e^-e^+ \to f\bar{f}$ at the Z-pole, defined by (8.125) just as in the low-energy case, is

$$A^{0f}_{FB} = \frac{3}{4} A_e A_f. \tag{8.173}$$

The LEP experiments measured A^{0f}_{FB} for $f = e, \mu, \tau, b, c$, and (less precisely) s. These and other asymmetries are listed in Table 8.5. The individual measurements of A^{0e}_{FB}, $A^{0\mu}_{FB}$, and $A^{0\tau}_{FB}$ are in reasonable agreement with lepton universality, allowing them to be combined into an average $A^{0\ell}_{FB}$. The LEP experiments also measured a forward-backward asymmetry between positive and negative charge in hadronic Z events. This is more complicated to interpret because it depends on the fragmentation of the final quarks, but nevertheless leads to a measurements of the effective leptonic angle that is denoted $\bar{s}^2_\ell(A^{0,q}_{FB})$ in Table 8.5.

Forward-backward asymmetries in $\bar{p}p \to \ell^+\ell^-$, $\ell = e$ or μ, have been measured by CDF and D0 at the Tevatron, and in $pp \to \ell^+\ell^-$ by ATLAS, CMS, and LHCb at the LHC.[21] These (Drell-Yan) processes are mediated by γ and Z exchange, and for $\ell^+\ell^-$ masses in the vicinity of M_Z they lead to high precision determinations of $\sin^2\theta_W$ comparable to those of the Z-pole experiments (e.g., Patrignani, 2016), as shown in Table 8.5.

As discussed following (7.53), the polarization of a $\tau^\mp$ in $e^-e^+ \to \tau^-\tau^+$ can be measured from the angular distribution of the τ decay products in the τ rest frame. The τ^- polarization, averaged over the τ^- CM angle θ, is

$$P^0_\tau \equiv \frac{\sigma(h_{\tau^-} = +\frac{1}{2}) - \sigma(h_{\tau^-} = -\frac{1}{2})}{\sigma(h_{\tau^-} = +\frac{1}{2}) + \sigma(h_{\tau^-} = -\frac{1}{2})} = -A_\tau, \tag{8.174}$$

which allows the extraction of A_τ. Similarly, the polarization of τ^- produced at a definite angle θ is

$$P^0_\tau(z) = -\frac{A_\tau + A_e \frac{2z}{1+z^2}}{1 + A_\tau A_e \frac{2z}{1+z^2}}, \tag{8.175}$$

where $z = \cos\theta$, allowing A_e to be determined from the angular distribution.

The SLC had a longitudinally polarized e^- beam, with a reversible polarization of $\gtrsim$ 75%. This allowed them to measure the polarization asymmetry

$$A^0_{LR} \equiv \frac{\sigma(h_{e^-} = -\frac{1}{2}) - \sigma(h_{e^-} = +\frac{1}{2})}{\sigma(h_{e^-} = -\frac{1}{2}) + \sigma(h_{e^-} = +\frac{1}{2})} = +A_e, \tag{8.176}$$

which projects out the electron couplings. The measurements were done separately for hadronic and leptonic final states. They also measured mixed forward-backward polarization asymmetries for various final states,

$$A^{0FB}_{LR} = \frac{\sigma^f_{LF} - \sigma^f_{LB} - \sigma^f_{RF} + \sigma^f_{RB}}{\sigma^f_{LF} + \sigma^f_{LB} + \sigma^f_{RF} + \sigma^f_{RB}} = \frac{3}{4} A_f \tag{8.177}$$

in an obvious notation, allowing them to determine the final fermion couplings.

The LEP and SLD lineshape and asymmetry results can be combined to determine the effective couplings $\bar{g}^f_{V,A}$. The results are displayed in Figure 8.24.

[21] Forward-backward asymmetries are possible in pp reactions because of the difference between the PDFs for quarks and antiquarks at nonzero rapidity.

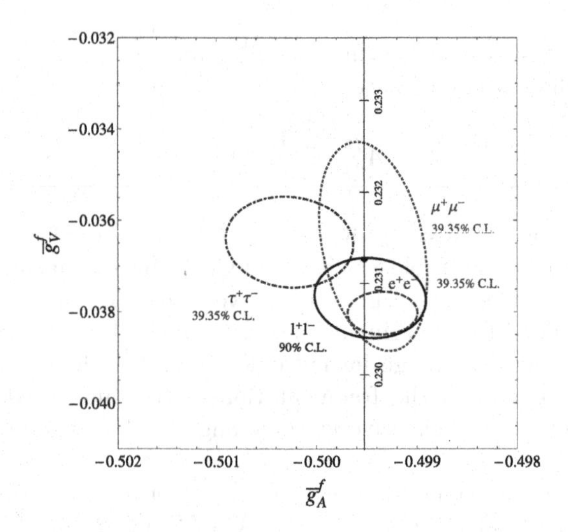

Figure 8.24 Allowed regions in $\bar{g}_V^f$ vs $\bar{g}_A^f$ at 1σ (39.4% c.l.) for $f = e, \mu, \tau$, compared with the SM prediction as a function of $\hat{s}_Z^2$. The 90% c.l. region for $\bar{g}_{V,A}^\ell$ assuming lepton universality, and the global best fit value $\hat{s}_Z^2 = 0.23129$ are also indicated. Plot courtesy of the Particle Data Group (Patrignani, 2016).

A_f involves the ratios of effective couplings, so ρ_f cancels. A_f and the asymmetries are therefore insensitive to m_t and M_H for a fixed $\bar{s}_\ell^2$ or $\hat{s}_Z^2$, allowing the weak angle to be extracted with little ambiguity. (Similar statements apply to M_W and to R_c and R_ℓ in (8.168), although some m_t dependence enters the ratios through the $Zb\bar{b}$ vertices.) However, M_Z and the other lineshape variables do depend on them through $\hat{\rho}$, so the m_t and M_H constraints come mainly from comparing M_Z with the asymmetries and other observables, or, equivalently, by expressing the other quantities in terms of M_Z rather than $\bar{s}_\ell^2$. As an example, the observed $A_{FB}^{0,\ell}$ as well as R_ℓ are compared to the SM expectation as a function of M_H in Figure 8.22, where the SM predictions utilize the observed M_Z. These observables favored relatively low values for the Higgs mass (prior to its discovery), but with large uncertainty due to the logarithmic dependence.

There is some tension between the leptonic and hadronic asymmetries, as can be seen in Figure 8.25. The most precise determinations of $\bar{s}_\ell^2$ are from the polarization asymmetry A_{LR} and from the b FB asymmetry A_{FB}^{0b}. A_{LR} is proportional to $\bar{g}_V^\ell \sim -\frac{1}{2} + 2\bar{s}_\ell^2$, and is therefore extremely sensitive to $\bar{s}_\ell^2$. On the other hand, the leptonic FB asymmetry $A_{FB}^{0\ell}$ is proportional to $\bar{g}_V^{\ell 2}$ and is therefore less sensitive since $\bar{g}_V^\ell$ is small. $A_{FB}^{0b} = \frac{3}{4} A_e A_b$ does not have this difficulty, and is much more sensitive to the e vertex than the b vertex assuming the SM. As can be seen from Table 8.5 and Figure 8.25, there is a possible discrepancy of A_{FB}^{0b} from the value predicted from the best fit, and the corresponding $\bar{s}_\ell^2$ is about 2.4σ higher than the average. Similarly, the $\bar{s}_\ell^2$ from A_{LR} is $\sim 2\sigma$ lower. This could well be due to fluctuations, but the other (less precise) hadronic and leptonic asymmetries tend to follow the same pattern. The hadronic asymmetries favor a larger Higgs mass, while the leptonic ones (and M_W) favor a small M_H, so the average value relies on a delicate balance. As can be seen in Figure 8.25 (which predated the Higgs observation), the average was consistent (with large uncertainties) with the value $M_H \sim 125$ GeV that was subsequently observed. Also, the recent precise Tevatron value $\bar{s}_\ell^2 = 0.23185(35)$ fits comfortably in between.

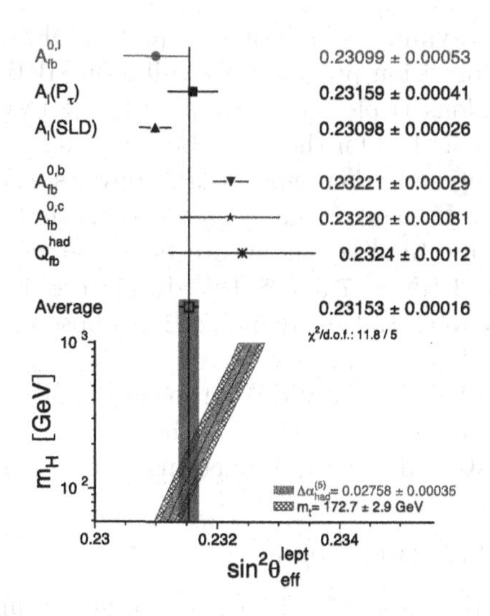

Figure 8.25 $\bar{s}^2_\ell$ from various leptonic and hadronic asymmetries, compared with the SM predictions as a function of M_H using the observed M_Z. Reprinted with permission from (Schael et al., 2006a).

It is possible that there is some kind of new physics that affects A^{0b}_{FB} and A_{LR}, and that their combined implication for M_H was fortuitous. For example, there could be a tree-level effect that couples preferentially to the third family and accounts for the A^{0b}_{FB} value (it would be hard for a new physics effect that only enters at loop level to lead to the needed 4% shift). Possibilities would include a flavor off-diagonal coupling of a heavy Z' gauge boson or the mixing of the b with a heavy exotic fermion. However, there would have to be a compensation between the L and R couplings to avoid a large change in R_b.

Precision Tests at High Energy

The second phase of LEP, LEP 2, ran at CERN from 1996-2000, with energies gradually increasing from ~ 140 to ~ 209 GeV (Schael et al., 2013). The principal electroweak results were precise measurements of the W mass, as well as its width and branching ratios; a measurement of $e^+e^- \to W^+W^-$, ZZ, and single W or Z, as a function of center of mass (CM) energy, which tests the cancellations between diagrams that is characteristic of a renormalizable gauge field theory, or, equivalently, probes the triple gauge vertices; limits on anomalous quartic gauge vertices; measurements of various cross sections and asymmetries for $e^+e^- \to f\bar{f}$ for $f = e^-, \mu^-, \tau^-, q, b$ and c in reasonable agreement with SM predictions; and a stringent lower limit of 114.4 GeV on the Higgs mass. LEP 2 also studied heavy quark properties, tested QCD, and searched for supersymmetric and other exotic particles.

The Tevatron $\bar{p}p$ collider at Fermilab ran from 1983–2011, with an ultimate CM energy of 1.96 TeV (Bandurin et al., 2015). The CDF and D0 collaborations there discovered the top quark in 1995, with a mass consistent with the predictions from the precision electroweak and B/K physics observations; measured the t mass, decay distributions, $t\bar{t}$ cross sections and distributions, and single t production (relevant for the CKM element V_{tb}); and carried out an extensive program of electroweak physics. This included measurements of the W

mass, width, and $W^{\pm}$ charge asymmetry (relevant to quark PDFs); $\ell^+\ell^-$ forward-backward asymmetries, as part of the precision program; and diboson $(WW, ZZ, ZW, Z\gamma, W\gamma)$ production, to constrain anomalous triple gauge couplings and test QCD predictions. The Tevatron collaborations also searched for the Higgs boson (and later constrained its properties), observed $B_s - \bar{B}_s$ mixing and other aspects of B physics; carried out extensive QCD tests; and searched for heavy W' and Z' gauge bosons, exotic fermions, supersymmetry, and other types of new physics. Similar studies are being carried out at higher energy at the LHC (pp), which ran at CERN at 7 and 8 TeV during the first run (2010–2013), with results that included the discovery and study of the Higgs boson. The LHC restarted at 13 TeV in 2015, with an eventual goal of 14 TeV.

The HERA $e^{\pm}p$ collider at DESY observed W propagator and Z exchange effects, probed electroweak couplings, searched for leptoquark and other exotic interactions, and carried out a major program of QCD tests and structure functions studies (e.g., Diaconu et al., 2010).

8.3.6 Implications of the Precision Program

The principal Z-pole observables are listed in Table 8.5, and a number of important non-Z-pole observables in Table 8.6. The direct measurement of m_t is important for controlling the radiative corrections in the analysis more precisely than could be done from the indirect constraints alone, and in allowing sensitivity to the Higgs mass. M_W provides information on radiative corrections independent of the Z lineshape and asymmetries. The low energy WNC measurements play a critical role in constraining many types of new physics even though they are less precise than the Z-pole observations, because the latter are essentially blind to any new physics that doesn't directly affect the Z or its couplings to fermions (such as heavy particle exchanges, new box-diagrams, or four-fermi operators).

TABLE 8.6 Principal non-Z-pole observables, as of 11/15.[a]

Quantity	Value	Standard Model	Pull
m_t	173.34 ± 0.81	173.76 ± 0.76	-0.5
M_W (Tevatron)	80.387 ± 0.016	80.361 ± 0.006	1.6
M_W (LEP 2)	80.376 ± 0.033		0.4
Γ_W (Tevatron)	2.046 ± 0.049	2.089 ± 0.001	-0.9
Γ_W (LEP 2)	2.195 ± 0.083		1.3
M_H	125.09 ± 0.24	125.11 ± 0.24	0.0
$g_V^{\nu e}$	-0.040 ± 0.015	-0.0397 ± 0.0002	0.0
$g_A^{\nu e}$	-0.507 ± 0.014	-0.5064	0.0
Q_W^e (Møller)	-0.0403 ± 0.0053	-0.0473 ± 0.0003	1.3
Q_W^p	0.064 ± 0.012	0.0708 ± 0.0003	-0.6
$Q_W(Cs)$	-72.62 ± 0.43	-73.25 ± 0.02	1.5
$Q_W(Tl)$	-116.4 ± 3.6	-116.91 ± 0.02	0.1
$\hat{s}_Z^2$ (eDIS)	0.2299 ± 0.0043	0.23129 ± 0.00005	-0.3

[a]From (Patrignani, 2016). Masses are in GeV. Other non-Z-pole observables are included in the fits but not listed.

The Standard Model Fit

As of November, 2015, the result of the global fit to the precision data yielded[22]

$$\begin{aligned} &M_H = 125.11(24)\ \text{GeV}, && m_t = 173.76(76)\ \text{GeV} && \\ &\alpha_s(M_Z^2) = 0.1182(16), && \hat{\alpha}(M_Z^2)^{-1} = 127.936(17), && \Delta\alpha_{had}^{(5)} = 0.02774(13) \\ &\hat{s}_Z^2 = 0.23129(5), && \bar{s}_\ell^2 = 0.23152(5), && s_W^2 = 0.22336(10), \end{aligned} \tag{8.178}$$

with an overall χ^2/df of 53.6/42. The three values of the weak angle s^2 refer, respectively, to the $\overline{MS}$, effective Z-lepton vertex, and on-shell values, defined after Equation (8.138). The latter has a larger uncertainty because of a stronger dependence on the top mass. $\Delta\alpha_{had}^{(5)}$ is defined after (8.137).

The value of $\alpha_s(M_Z^2)$ in (8.178), based on Z-pole data and hadronic τ decays, agrees well with the other low-energy determinations mentioned in Section 5.4. The Z-pole results alone yield $\alpha_s(M_Z^2) = 0.1203 \pm 0.0028$, within 1σ of the global world average 0.1181(13). The Z-pole result has negligible theoretical uncertainty if one assumes the exact validity of the standard model, and is also insensitive to oblique (propagator) new physics. However, it is sensitive to non-universal new physics, such as those which affect the $Zb\bar{b}$ vertex.

The precision data alone yield[23] $m_t = 176.7 \pm 2.1$ GeV from loop corrections, in impressive agreement with the direct collider value 173.34 ± 0.81 GeV. The fit actually uses the $\overline{MS}$ mass $\hat{m}_t(\hat{m}_t)$, which is ~ 10 GeV lower, and is converted to the pole mass at the end.

The electroweak precision observables depend logarithmically on the Higgs mass through the oblique diagrams in Figure 8.17. This M_H dependence is correlated with the stronger quadratic m_t dependence, but following the direct measurement of m_t it became possible to derive nontrivial restrictions on M_H. For example, the 2011 Particle Data Group fit predicted 68 GeV $\leq M_H \leq$ 155 GeV at 90% c.l. from the indirect data, while the current data (excluding the direct Higgs observations) yields 66 GeV $\leq M_H \leq$ 134 GeV. This is consistent with the direct measurement of ~ 125 GeV (but see the caveats mentioned in the discussion of Figure 8.25). The individual constraints can be seen in Figure 8.26.

The M_H prediction is fairly robust to many types of new physics, with some exceptions. In particular, a larger M_H would have been allowed for negative new physics contributions to S or positive contributions to T, where S and T are the oblique parameters to be discussed below (see Peskin and Wells, 2001, for some specific examples), or for some Z' models with $Z - Z'$ mixing (e.g., Langacker, 2009). The predicted value would decrease if new physics accounted for the value of $A_{FB}^{(0b)}$ (Chanowitz, 2002). Following the direct observation one can view M_H as a constraint on these extensions.

---

[22] The fit results are from the Particle Data Group (Patrignani, 2016), which uses the fully $\overline{MS}$ program GAPP (Erler, 1999) for the radiative corrections. The results are generally in good agreement with those of other groups, such as those of the LEPEWWG (Schael et al., 2006a, and website) or the Gfitter group (Flacher et al., 2009, and website), which use the on-shell scheme. However, the PDG fits use a more complete set of low energy data.

[23] Prior to LEP and the SLC the WNC current experiments and early W and Z masses could set an upper limit on m_t of $\mathcal{O}(200$ GeV) (see, e.g., Amaldi et al., 1987). The more precise Z-pole experiments led to a prediction for m_t that was correlated with M_H, which was consistent (within the rather large uncertainties) with the value later observed directly by CDF and D0. For example, the first measurement of M_Z by MARK II at the SLC already implied $m_t = 140^{+43}_{-52}$ GeV for $M_H = 100$ GeV, with the central value changing to 128 (165) GeV for $M_H = 10$ (1000) GeV (Langacker, 1989a). By the late 1980s observations of $B - \bar{B}$ oscillations and improved analyses of CP violation in K decays made possible by measurements of the relevant CKM matrix elements allowed independent lower limits on m_t to be set, e.g., $m_t > 130$ GeV (Buras, 1993).

The predicted M_H range and observed value are suggestive of (but certainly do not prove) the possibility that the SM is extended to a supersymmetric version at the TeV scale, since such theories predict a light SM-like Higgs for much of their parameter space. However, $M_H = 125$ GeV is on the upper end of the range allowed in the minimal version, the MSSM. The theoretical constraints on M_H in both the standard model and its supersymmetric extensions will be discussed in Sections 8.5 and 10.2.5.

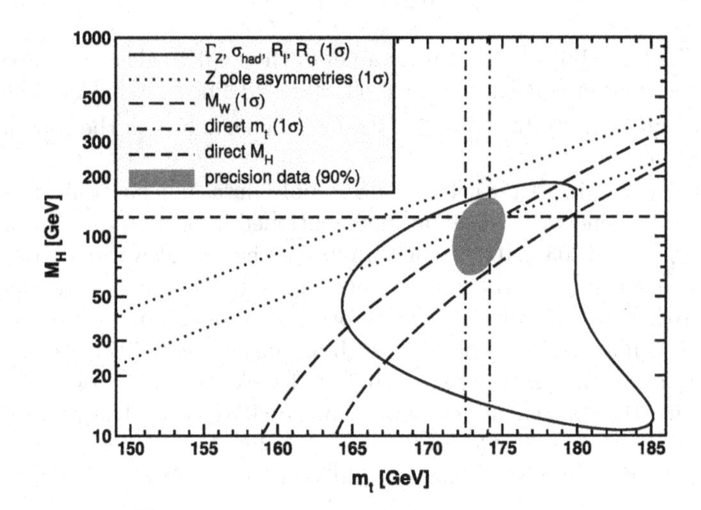

Figure 8.26 1σ ($\Delta\chi^2 = 1$) allowed regions in M_H vs m_t and the 90% c.l. global fit region from precision data, compared with the direct measurements from the LHC and Tevatron. Plot courtesy of the Particle Data Group (Patrignani, 2016).

Beyond the Standard Model

New physics modifies the SM predictions for the precision electroweak observables in several ways. For example, new heavy W' or Z' gauge bosons (Section 10.3); exotic Higgs particles (Accomando et al., 2006); leptoquark bosons (Barbier et al., 2005; Doršner et al., 2016), such as occur in supersymmetric models with R-parity violation; or supersymmetric particles in box diagrams (Section 10.2) can lead to new four-fermi operators, which are especially important below or above the Z-pole (Cho et al., 1998; Cheung, 2001; Han and Skiba, 2005; Schael et al., 2013). The most important effects for the Z-pole experiments are: (a) new physics that affects the $Zf\bar{f}$ vertices, and (b) new physics that modifies the masses and propagators of the W and Z. The first class includes such tree-level effects as the mixing of the ordinary with heavy exotic fermions (Langacker and London, 1988b) and Z-Z' mixing, as well as actual vertex corrections from new gauge bosons, leptoquarks, superpartners, etc. Their implications are model dependent and hard to parametrize in general, since they affect each type of chiral fermion in a different way. For this reason the precision constraints are usually studied for specific models or types of models. The second class, the *oblique* corrections, are described below.

Oblique Parameters

 oblique corrections are universal in that they are independent of the fermions in the cess, and can be described by a small number of parameters. They are especially impor-t for describing new particles that only or mainly affect the precision observables through uum polarization loops, such as new fermion or scalar multiplets with little or no mixing h or direct coupling to the ordinary particles. They can also be used to parametrize the cts of the t quark and Higgs on the radiative corrections (although for the top one must include the $Zb\bar{b}$ vertex diagrams in Figure 8.19). However, we will follow the Particle a Group (Patrignani, 2016) and focus on the new physics contributions to the oblique ameters, treating the m_t and M_H effects separately.

The ρ_0 parameter

$$\rho_0 \equiv \frac{M_W^2}{M_Z^2 \hat{\rho} \hat{c}_Z^2} \tag{8.179}$$

ends the $\overline{MS}$ parameter $\hat{\rho}$ in (8.142). The SM expression for M_W in (8.142) is unmodified. is associated with new physics that breaks the $SU(2)$ vector generators and therefore ates the custodial symmetry relation between M_W and M_Z in (8.141) (and therefore ween the WNC and WCC amplitudes). We assume that the new physics that leads to $\neq 1$ is a small perturbation that does not significantly affect other radiative corrections.[24] n ρ_0 can be viewed as an extra parameter that also multiplies G_F in the expressions for effective WNC interaction in (8.79), the quantity χ_0 in (8.133), and the partial Z-widths 8.163).

$\rho_0 \neq 1$ can be generated by non-degenerate multiplets of heavy particles, analogous to effect of the non-degenerate t and b quarks in (8.144). Non-degenerate $SU(2)$ doublets calars and/or fermions with masses m_{n1}, m_{n2}, such as a fourth family (*sequential*), a ror family (involving right-chiral doublets and left-chiral singlets), vector pairs of quark epton doublets (with an extra factor of 2), heavy Higgs doublets not involved in symmetry aking, or unmixed squark or slepton doublets in supersymmetry, yield (Veltman, 1977; nowitz et al., 1978)

$$\rho_0 - 1 = \frac{3G_F}{8\sqrt{2}\pi^2} \sum_n \frac{C_i}{3} \Delta m_n^2 \tag{8.180}$$

ere

$$\Delta m_n^2 \equiv m_{n1}^2 + m_{n2}^2 - \frac{4m_{n1}^2 m_{n2}^2}{m_{n1}^2 - m_{n2}^2} \ln \frac{m_{n1}}{m_{n2}} \geq (m_{n1} - m_{n2})^2, \tag{8.181}$$

h $C_n = 1(3)$ for color singlets (triplets). There is enough data to determine ρ_0 and the er parameters, with the global fit yielding

$$\rho_0 = 1.00037 \pm 0.00023, \tag{8.182}$$

.6σ from unity, with little change in the other parameters. The ρ_0 value corresponds to 95% c.l. limit

$$\sum_n \frac{C_n}{3} \Delta m_n^2 \leq (49 \text{ GeV})^2. \tag{8.183}$$

h non-degenerate multiplets lead to $\rho_0 \geq 1$. However, additional Higgs doublets that ticipate in SSB or heavy lepton multiplets involving Majorana neutrinos can yield $\rho_0 < 1$ references, see the Electroweak article in Patrignani, 2016). Tree-level effects, such as let or higher-dimensional Higgs representations with non-zero VEVs can contribute to ρ_0

One must still define a fourth renormalized parameter. A convenient set is α, G_F, M_W, and M_Z.

ith either sign (Problem 8.1). Mixing of the Z with a heavy Z' would reduce M_Z, increasing), while $W - W'$ mixing would have the opposite effect. However, such mixings could also ave other important effects on the WNC, WCC, and precision observables (Section 10.3).

Heavy chiral fermions break the axial $SU(2)$ generators as well as the vector ones, so their fects are not adequately parametrized by ρ_0 alone. These can be taken into account by a eneralization to the S, T, and U parameters (Peskin and Takeuchi, 1990, 1992). Related arametrizations are given in (Kennedy and Langacker, 1990; Marciano and Rosner, 1990; olden and Randall, 1991; Altarelli and Barbieri, 1991). Define the vacuum polarization inctions

$$i\Pi_{ij}^{\mu\nu}(q) = i\left[g^{\mu\nu}\Pi_{ij}(q^2) - q^\mu q^\nu \Delta_{ij}(q^2)\right] \tag{8.184}$$

s the one-particle irreducible propagator corrections shown in Figure 8.27. The second term irrelevant for the existing precision experiments because the q^μ or q^ν always yield light pton masses when contracted into the external fermion lines. We are concerned with the

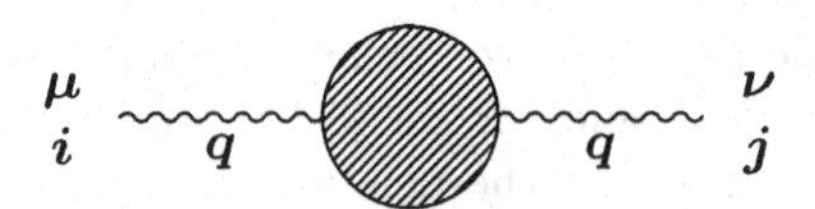

gure 8.27 Vacuum polarization bubbles for the gauge boson propagators. Examples re shown in Figure 8.17.

ew physics contributions $\Pi_{ij}^{new}(q^2)$ to the vacuum polarizations. As long as the new physics much heavier than M_Z it is sufficient to assume that the $\Pi_{ij}^{new}(q^2)$ are approximately linear om $q^2 = 0$ to M_Z^2. The effects on the precision observables can then be expressed in terms f the three parameters

$$\begin{aligned} \hat{\alpha}(M_Z^2)T &\equiv \frac{\Pi_{WW}^{new}(0)}{M_W^2} - \frac{\Pi_{ZZ}^{new}(0)}{M_Z^2} \\ \frac{\hat{\alpha}(M_Z^2)}{4\hat{s}_Z^2\hat{c}_Z^2}S &\equiv \frac{\Pi_{ZZ}^{new}(M_Z^2) - \Pi_{ZZ}^{new}(0)}{M_Z^2} \\ \frac{\hat{\alpha}(M_Z^2)}{4\hat{s}_Z^2}(S+U) &\equiv \frac{\Pi_{WW}^{new}(M_W^2) - \Pi_{WW}^{new}(0)}{M_W^2}. \end{aligned} \tag{8.185}$$

quation (8.185) assumes an $\overline{MS}$ renormalization scheme as defined in (Marciano and osner, 1990). The more general expressions are given in (Patrignani, 2016), along with dditional details and references. These quantities have a factor of α removed, so new hysics contributions are expected to be of $\mathcal{O}(1)$. The T parameter is equivalent to ρ_0,

$$\rho_0 = \frac{1}{1-\alpha T} \sim 1 + \alpha T, \tag{8.186}$$

hile S and U are associated with the axial currents and can be generated even by degenrate heavy chiral multiplets. The contribution of a degenerate multiplet to S is

$$S = \frac{C}{3\pi}\sum_r \left[t_{rL}^3 - t_{rR}^3\right]^2, \tag{8.187}$$

where t^3_{rL} (t^3_{rR}) is the T^3 eigenvalue of ψ_{rL} (ψ_{rR}). A heavy degenerate fourth family or mirror family would contribute $2/3\pi$ to S. Similarly, in technicolor models (Weinberg, 1979; Susskind, 1979) with scaled up QCD-like dynamics, one expects $S \sim 0.45$ for an $SU(2)$ doublet of technifermions with $N_{TC} = 4$ technicolors, and $S \sim 1.62$ for a full technigeneration (Peskin and Takeuchi, 1992; Golden and Randall, 1991). Most types of new physics yield $U = 0$.

The effects of T on the precision observables are the same as ρ_0 (using (8.186)). S and U mainly modify the SM expressions for the Z and W masses

$$\begin{aligned} M_Z^2 &= M_{Z0}^2 \, \frac{1 - \alpha T}{1 - G_F M_{Z0}^2 S/2\sqrt{2}\pi} \\ M_W^2 &= M_{W0}^2 \, \frac{1}{1 - G_F M_{W0}^2 (S+U)/2\sqrt{2}\pi}, \end{aligned} \tag{8.188}$$

where M_{Z0} and M_{W0} are the SM expressions in the $\overline{MS}$ scheme (see Patrignani, 2016, for the modifications to the widths, for which wave function renormalizations must be included, and to the WNC amplitudes). The global fit yields

$$\begin{aligned} &S = 0.05(10), \qquad T = 0.08(12), \qquad U = 0.02(10) \\ &S = 0.07(8), \qquad T = 0.10(7), \end{aligned} \tag{8.189}$$

where U is fixed to be zero in the second line, consistent with the SM values $S = T = U = 0$.

A heavy degenerate fourth or mirror family, which corresponds to $T = 0$, is strongly excluded. This complements the invisible Z width constraint below (8.170) on a fourth neutrino with mass $< M_Z/2$. A heavy fourth family with large mass splittings may be marginally allowed by the precision data due to a compensation between the effects of $S > 0$ and $T > 0$, though at the expense of large Yukawa couplings that may lead to Landau poles at low scales. However, in combination with direct search limits from the LHC and the properties of the Higgs,[25] a perturbative fourth or mirror family is now excluded by more than 5σ (Eberhardt et al., 2012). Similarly, the QCD-like technicolor models are excluded, but versions that do not resemble QCD in their dynamics may be allowed. Supersymmetric extensions of the SM usually give very small contributions to S and T (or to the precision observables). The implications for other types of new physics are reviewed in the *Electroweak* and *Extra Dimensions* articles in (Patrignani, 2016).

The approximation that the vacuum polarizations are linear in q^2 breaks down for new physics that is not much heavier than M_Z. The oblique parameter formalism can be extended to accomodate this by the inclusion of more parameters (Maksymyk et al., 1994). The formalism can also be extended to include effects relevant to LEP 2 (Barbieri et al., 2004). The oblique parameters can also be redefined in terms of higher-dimensional operators (Barbieri et al., 2004; Han, 2008; Skiba, 2011).

Summary

The precision Z-pole, LEP 2, WNC, Tevatron, and LHC experiments have successfully tested the SM at the 0.1% level, including electroweak loops, thus confirming the gauge principle, $SU(2) \times U(1)$ group, representations, and the basic structure of renormalizable field theory. $\sin^2\theta_W$ was precisely determined, m_t was roughly predicted from its indirect loop effects prior to the direct discovery at the Tevatron, and the indirect value of α_s,

[25] The production rate via gluon fusion would be increased by a factor ~ 9 because of the heavy quarks in the loop. See Section 8.5 and Problem 8.21.

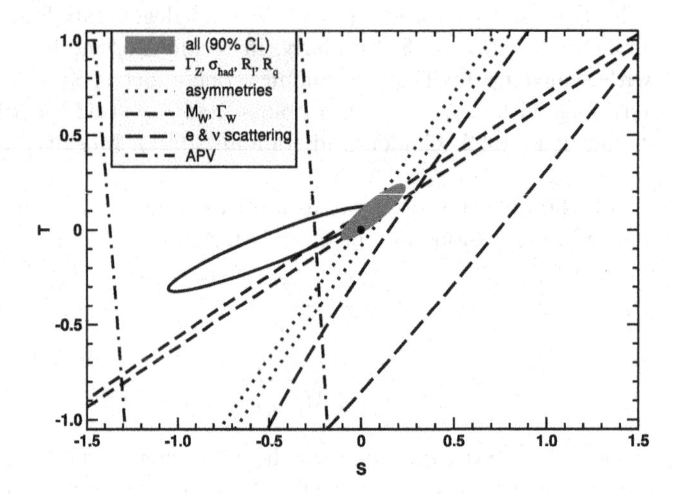

Figure 8.28 $\Delta\chi^2 = 1$ allowed regions in the S-T plane for individual types of measurements and the 90% c.l. global fit region. The S and T shown here parametrize the effects of new physics only. Plot courtesy of the Particle Data Group (Patrignani, 2016).

extracted mainly from the Z-lineshape, agreed with more direct QCD determinations. The precision data also predicted the Higgs mass, though with large uncertainty, consistent with the value subsequently observed at the LHC and suggestive of supersymmetry. The agreement of the data with the SM imposes a severe constraint on possible new physics at the TeV scale, and points towards decoupling theories (such as most versions of supersymmetry and unification), which typically lead to 0.1% effects. On the other hand, generic versions of new TeV-scale dynamics or compositeness (e.g., dynamical symmetry breaking, composite Higgs, or composite fermion models) usually imply deviations of several %, and often large flavor changing neutral currents, although decoupling (or fine-tuned) versions may still be viable. Finally, the precisely measured gauge couplings were consistent with the unification expected in the simplest form of grand unification if the SM is extended to the MSSM (Section 10.4). Most of these results could be made even more precise by a high intensity Z-pole option for a possible future e^+e^- collider (e.g., Bicer et al., 2014).

8.4 GAUGE SELF-INTERACTIONS

The predicted gauge self-interactions of the SM, listed in Table 8.1, are essential probes of the structure and consistency of a spontaneously-broken non-abelian gauge theory. Even tiny deviations in their form or value would destroy the delicate cancellations needed for renormalizability, and would signal the need either for compensating new physics (e.g., from mixing with other gauge bosons or new particles in loops), or of a more fundamental breakdown of the gauge principle, e.g., from some forms of compositeness. Experimental tests of the self-interactions are therefore important, although in practice many types of new physics that could modify them would show up sooner in the Z-pole and WNC experiments (e.g., De Rujula et al., 1992; Burgess et al., 1994).

In the SM, the triple gauge couplings (TGC) for WWZ and $WW\gamma$ are

$$\begin{aligned} W^+_\mu(p)\gamma_\nu(q)W^-_\sigma(r): &\quad ie\,\mathcal{C}_{\mu\nu\sigma}(p,q,r) \\ W^+_\mu(p)Z_\nu(q)W^-_\sigma(r): &\quad ie\cot\theta_W\mathcal{C}_{\mu\nu\sigma}(p,q,r), \end{aligned} \tag{8.190}$$

where the charges and momenta flow into the vertex and

$$\mathcal{C}_{\mu\nu\sigma}(p,q,r) \equiv g_{\mu\nu}(q-p)_\sigma + g_{\mu\sigma}(p-r)_\nu + g_{\nu\sigma}(r-q)_\mu. \tag{8.191}$$

These have been constrained by measuring the total cross section and various decay distributions for $e^-e^+ \to W^-W^+$ at LEP 2, and by observing $\bar{p}p$ $(pp) \to W^+W^-, WZ$, and $W\gamma$ at the Tevatron (LHC). Possible anomalies in the predicted quartic vertices in Table 8.1, and the neutral cubic vertices for ZZZ, $ZZ\gamma$, and $Z\gamma\gamma$, which are absent in the SM, have also been constrained.

The three tree-level diagrams for $e^-e^+ \to W^-W^+$ are shown in Figure 8.29. The cross section from any one or two of these rises rapidly with center of mass energy, but gauge invariance relates the gauge three-point vertices to their couplings to the fermions in such a way that at high energies there is a cancellation. It is another manifestation of the cancellation in a gauge theory that brings higher-order loop integrals under control, leading to a renormalizable theory. It is seen in Figure 8.30 that the expected cancellations do occur.

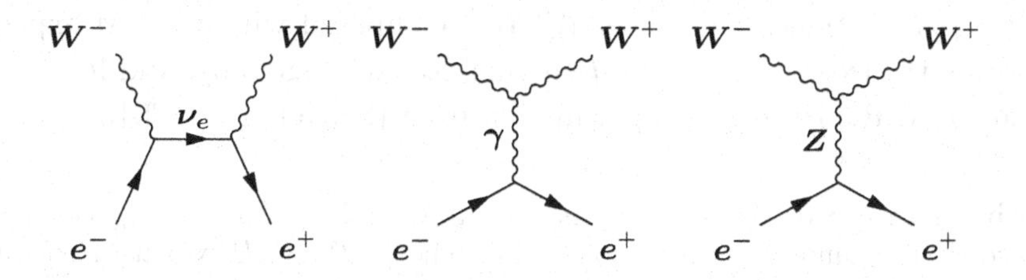

Figure 8.29 Diagrams for $e^-e^+ \to W^-W^+$ in the $SU(2) \times U(1)$ model.

More detailed information can be obtained from the angular distributions. It is useful to parametrize the TGCs in a form more general than the SM predictions. Neglecting terms that are irrelevant for light fermions, there are only seven possible Lorentz invariant form factors for the effective VW^-W^+ vertex, where $V{=}\gamma$ or Z (e.g., Hagiwara et al., 1993; Gounaris et al., 1996; Schael et al., 2013):

$$\begin{aligned} i\mathcal{L}^{VWW}_{eff} =& g_{VWW}\Big[g_1^V V^\mu\left(W^-_{\mu\nu}W^{+\nu} - W^+_{\mu\nu}W^{-\nu}\right) + \kappa_V\, W^+_\mu W^-_\nu V^{\mu\nu} + \frac{\lambda_V}{m_W^2}\, V^{\mu\nu}W^{+\rho}_\nu W^-_{\rho\mu} \\ &+ ig_5^V\,\epsilon_{\mu\nu\rho\sigma}\Big((\partial^\rho W^{-\mu})W^{+\nu} - W^{-\mu}(\partial^\rho W^{+\nu})\Big)V^\sigma + ig_4^V W^+_\mu W^-_\nu(\partial^\mu V^\nu + \partial^\nu V^\mu) \\ &- \frac{\tilde{\kappa}_V}{2} W^-_\mu W^+_\nu \epsilon^{\mu\nu\rho\sigma}V_{\rho\sigma} - \frac{\tilde{\lambda}_V}{2m_W^2} W^-_{\rho\mu}W^{+\mu}{}_\nu\epsilon^{\nu\rho\tau\omega}V_{\tau\omega}\Big], \end{aligned} \tag{8.192}$$

where $F_{\mu\nu} \equiv \partial_\mu F_\nu - \partial_\nu F_\mu$ for $F = W^\pm, Z$, or γ is the non-interacting field strength. The form factors must be real (up to rescattering effects, which are small for a weakly coupled theory), and to first approximation one can neglect any momentum dependence and treat them as constants. The coefficients g_{VWW} are arbitrary normalizations, and can be set to the conventional values $g_{\gamma WW} = e$ and $g_{ZWW} = e\cot\theta_W$. The SM expectations are then

$$g_1^\gamma = g_1^Z = \kappa_\gamma = \kappa_Z = 1, \qquad \lambda_V = g_{4,5}^V = \tilde{\kappa}_V = \tilde{\lambda}_V = 0. \tag{8.193}$$

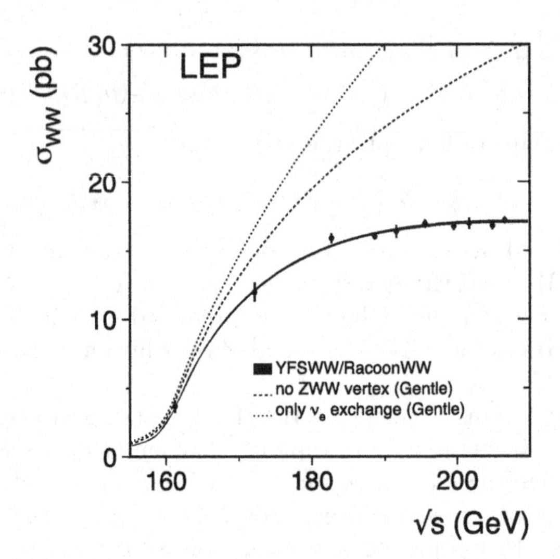

Figure 8.30 Cross section for $e^-e^+ \to W^-W^+$ compared with the SM expectation. Also shown is the expectation from t channel ν_e exchange only, and for the ν_e and γ diagrams only. Reprinted with permission from (Schael et al., 2013).

Going beyond the SM, the g_1^V, κ_V, and λ_V are C and P invariant, g_5^V violates C and P but preserves CP, and the others violate CP. The C, P, or CP violating couplings are relatively weakly constrained at present and are usually neglected. It is reasonable to impose $U(1)_Q$ invariance, since it is believed to be unbroken, in which case $g_1^\gamma = 1$ and $g_4^\gamma = g_5^\gamma = 0$. In fact, g_1^γ, κ_γ, and λ_γ can be interpreted in terms of the $W^\pm$ electromagnetic multipoles

$$q_W = e g_1^\gamma, \quad \mu_W = e(1+\kappa_\gamma+\lambda_\gamma)/2M_W, \quad Q_W = -e(\kappa_\gamma - \lambda_\gamma)/M_W^2, \tag{8.194}$$

which are, respectively, the W^+ charge, magnetic moment, and quadrupole moment, predicted in the SM to assume the values e, e/M_W, and $-e/M_W^2$. The stronger assumption of a global (custodial) $SU(2) \times U(1)$ symmetry in the limit $g' = 0$, yields two more constraints

$$\kappa_Z = g_1^Z - \tan^2\theta_W(\kappa_\gamma - 1), \qquad \lambda_Z = \lambda_\gamma. \tag{8.195}$$

Making all of these assumptions, one is left with three real parameters, g_1^Z, κ_γ, and λ_γ, which are used in most analyses. (In fact, the data is usually analyzed allowing only one to vary from the SM value at a time.) The combined results of the four LEP 2 experiments are (Schael et al., 2013)

$$g_1^Z - 1 = -0.016(19), \quad \kappa_\gamma - 1 = -0.018(42), \quad \lambda_\gamma = -0.022(19), \tag{8.196}$$

in agreement with the SM expectations of zero. The ATLAS and CMS results from Run I were comparable or stronger.

W Helicity Measurements

The polarizations of produced $W^\pm$, which can be measured by the angular distributions of their decay products in the W rest frame, probe the couplings of the W in the production process. In particular, the fraction of longitudinally polarized W's may differ from the SM prediction in models with admixtures of right-handed currents or in alternative models of spontaneous symmetry breaking (e.g., Chen et al., 2005). Polarization studies have been carried out at LEP 2 using $e^-e^+ \to W^-W^+$ and at the Tevatron and LHC using $t \to Wb$. In $t \to bW^+$ (with t at rest) one expects the fraction F_0 of longitudinally polarized W^+ to approach unity for $m_t \gg M_W$, due to the large numerical value of the components of the longitudinal polarization vector $\epsilon(\vec{p}_W, 3) \sim p_W/M_W$ for $E_W \gg M_W$ in Equation (2.127) on page 28, while for the actual masses one expects $F_0 = m_t^2/(m_t^2 + 2M_W^2) \sim 0.70$. (See (Kane et al., 1992) and Problem 8.15.) Similarly, the fraction F_+ of W^+ with helicity $+1$ should vanish for small m_b due to the $V - A$ coupling, so that $F_- \sim 0.30$. A recent CMS observation (Khachatryan et al., 2016a) at 8 TeV yielded $F_0 = 0.681(26)$, $F_- = 0.323(16)$, and $F_+ = -0.004(15)$, to be compared with the full SM prediction, including NNLO QCD corrections (Czarnecki et al., 2010), $F_0 = 0.687(5)$, $F_- = 0.311(5)$, and $F_+ = 0.0017(1)$.

8.5 THE HIGGS

The spontaneous symmetry breaking sector has always been the most uncertain part of the standard model. The simplest possibility is that the SSB is accomplished by the VEV of an elementary Higgs doublet, as described in Section 8.2.1, or by a more complicated Higgs sector involving two or more Higgs doublets (as in the supersymmetric extension of the SM) or other representations. It is also possible that the SSB is associated with a dynamical mechanism not involving elementary scalar fields, as will be touched on at the end of this Section and in Chapter 10. The observation of a Higgs-like boson[26] with mass ~ 125 GeV at the LHC in 2012 strongly supported the notion of an elementary Higgs doublet. However, the initial measurements of the Higgs couplings were not sufficiently precise to exclude some of the alternatives, especially limiting cases involving a boson very similar to the SM Higgs. In this section, we describe some of the theoretical constraints on the SM Higgs and some of its predicted properties for an arbitrary mass. We then discuss the searches for the Higgs, its discovery and study at the LHC, and the implications. For more detailed discussions, see (Gunion et al., 1990; Carena and Haber, 2003; Gomez-Bock et al., 2007; Djouadi, 2008a; de Florian et al., 2016); the Higgs Boson review by Carena, Grojean, Kado, and Sharma in (Patrignani, 2016); and the sites devoted to Higgs production, decay, and experimental constraints listed in the websites section.

In the standard model there remains one physical Higgs particle H after spontaneous symmetry breaking, with a potential given by (8.40) on page 265, and gauge, self-interaction, and Yukawa couplings to fermions given in Table 8.1. The mass is $M_H = \sqrt{2\lambda}\,\nu$, where λ is the quartic self-coupling and $\nu \sim 246$ GeV is $\sqrt{2}\langle 0|\phi^0|0\rangle$. The couplings to a particle of mass M are always proportional to M/ν, M^2/ν, or $(M/\nu)^2$. Thus, the couplings to fermions are small except for the top quark, making the H hard to produce or detect in that way. More optimistic are gauge couplings, such as VVH where $V = W^\pm$ or Z, which are proportional to $g^2\nu \sim M_V^2/\nu$.

8.5.1 Theoretical Constraints

The Higgs mass and quartic coupling are related by

$$M_H^2 = 2\lambda\nu^2, \qquad \lambda = \frac{g^2 M_H^2}{8M_W^2} = \frac{G_F M_H^2}{\sqrt{2}}, \tag{8.197}$$

[26]That is, the SM Higgs boson or something closely resembling it. In the following it will also be referred to as the "Higgs boson" or the "Higgs".

where $\nu \sim 246$ GeV and $G_F \sim 1.2 \times 10^{-5}\text{GeV}^2$ are known. The only constraint we have imposed so far on λ is (tree-level) vacuum stability, $\lambda > 0$, which would allow any M_H from 0 to ∞. However, there are a number of more stringent theoretical constraints that lead to nontrivial upper and lower bounds. In this section we consider issues related to the Higgs and gauge boson self-interactions. Other theoretical issues, involving the Higgs contribution to the vacuum energy and the higher-order corrections to the Higgs mass, are discussed in Section 10.1.

Renormalization Group Constraints

The renormalization group equations for the running gauge couplings are given for an arbitrary gauge theory in (5.33) on page 170. For the SM the β function one-loop coefficients in (5.34) are

$$\begin{aligned} b_{g_s} &= -\frac{1}{16\pi^2}\left[11 - \frac{4F}{3}\right] \xrightarrow[F=3,n_H=1]{} \frac{1}{16\pi^2}(-7) \\ b_g &= -\frac{1}{16\pi^2}\left[\frac{22}{3} - \frac{4F}{3} - \frac{n_H}{6}\right] \xrightarrow[F=3,n_H=1]{} \frac{1}{16\pi^2}\left(-\frac{19}{6}\right) \\ b_{g'} &= +\frac{1}{16\pi^2}\left[+\frac{20F}{9} + \frac{n_H}{6}\right] \xrightarrow[F=3,n_H=1]{} \frac{1}{16\pi^2}\left(+\frac{41}{6}\right) \end{aligned} \tag{8.198}$$

for $SU(3)$, $SU(2)$, and $U(1)$, respectively.[27] These are valid at momenta much larger than m_t and ν, and assume the existence of F fermion families and n_H Higgs doublets. For $F = 3$ and $n_H = 1$, both $SU(3)$ and $SU(2)$ are asymptotically free.

λ and the Yukawa couplings also run. At one-loop (see, e.g., Cheng et al., 1974; Gunion et al., 1990)

$$\begin{aligned} \frac{d\lambda(Q^2)}{d\ln Q^2} &= \frac{1}{32\pi^2}\left[24\lambda^2 + 24\lambda h_t^2 - 24h_t^4 - 3\lambda\left(3g^2 + g'^2\right) + \frac{3}{8}\left(2g^4 + (g^2 + g'^2)^2\right)\right] \\ \frac{dh_t(Q^2)}{d\ln Q^2} &= \frac{1}{32\pi^2}\left[9h_t^3 - h_t\left(8g_s^2 + \frac{9}{4}g^2 + \frac{17}{12}g'^2\right)\right], \end{aligned} \tag{8.199}$$

where it is understood that the couplings on the r.h.s. are the running couplings at Q^2. The quantity in (8.197) is the low energy value $\lambda(\nu^2)$, while $h_t(\nu^2) \sim m_t/\nu$ is the t-quark Yukawa coupling defined in (8.46) on page 267 or in (8.54); we have neglected all of the other Yukawas. The running of h_t is due to vertex, t quark, and Higgs self-energy diagrams (for a scalar coupling there is no analog of the Ward-Takahashi identity described in Section 2.12.1). Typical one-loop diagrams contributing to the running of λ are shown in Figure 8.31.

The first bound that we consider is the *triviality* upper limit on M_H (Cabibbo et al., 1979). From (8.197), $\lambda(\nu^2)$ is larger than unity for $M_H \gtrsim 350$ GeV, while $h_t(\nu^2) \sim 0.7$ for $m_t \sim 173$ GeV and the electroweak gauge coupling terms at low energy are small (Table 1.1). For large M_H one can therefore approximate the λ equation by the first term on the r.h.s., which is immediately solved to obtain

$$\lambda(Q^2) = \frac{\lambda(\nu^2)}{1 - \frac{3\lambda(\nu^2)}{4\pi^2}\ln\frac{Q^2}{\nu^2}}. \tag{8.200}$$

---

[27] In considering grand unified theories it is conventional to introduce the *GUT-normalized* $U(1)$ gauge coupling $g_1 = \sqrt{\frac{5}{3}}g'$, which has the coefficient $b_{g_1} = \frac{3}{5}b_{g'} \rightarrow \frac{1}{16\pi^2}\left(+\frac{41}{10}\right)$.

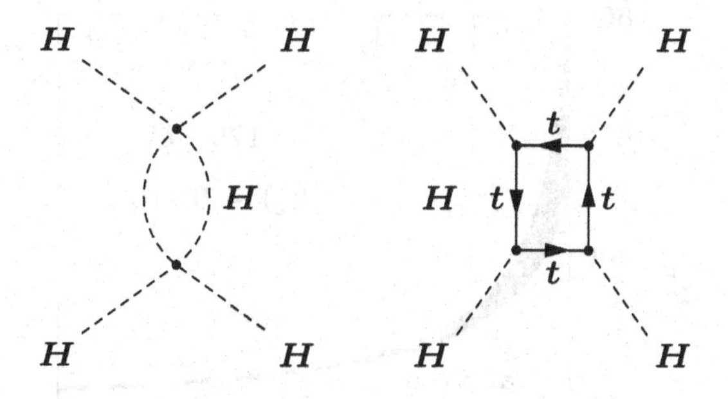

Figure 8.31 Typical diagrams contributing to the running of λ.

This diverges at the *Landau pole*

$$Q_{LP} = \nu e^{2\pi^2/3\lambda(\nu^2)}. \tag{8.201}$$

Presumably, it does not make sense for λ to diverge within the domain of validity of the theory.[28] It suffices that Q_{LP} is larger than the scale Λ at which new physics sets in (and above which the RGE in (8.199) no longer apply). Requiring $Q_{LP} > \Lambda$ leads to the triviality limit

$$M_H < \left(\frac{2\sqrt{2}\pi^2}{3G_F \ln(\Lambda/\nu)} \right)^{1/2} \sim \begin{cases} \mathcal{O}(140)\ \text{GeV},\ \Lambda \sim M_P \\ \mathcal{O}(650)\ \text{GeV},\ \Lambda \sim 1500\ \text{GeV} \end{cases}, \tag{8.202}$$

where $M_P = G_N^{-1/2} \sim 1.2 \times 10^{19}$ GeV is the Planck scale. This limit is somewhat fuzzy, because the one-loop approximation is not valid when λ is large, and in fact perturbation theory breaks down. However, it provides a reasonable estimate (as a function of the new physics scale) of how large a value of M_H is consistent with having a weakly coupled field theory. In particular, the observation of a SM-type Higgs at a mass scale much larger than 200 GeV would have strongly suggested that new physics sets in at a rather low scale. A more detailed evaluation including two-loop effects and the neglected terms in (8.199) is shown in Figure 8.32. One finds that $M_H \lesssim 180$ GeV for $\Lambda \sim M_P$, while $M_H \lesssim 700$ GeV for $\Lambda < 2M_H$. (It would not make much sense to consider an elementary Higgs field for a lower Λ.) The latter upper bound can be justified by non-perturbative lattice calculations (Hasenfratz et al., 1987; Kuti et al., 1988; Luscher and Weisz, 1989), which suggest an absolute upper limit of $650 - 700$ GeV.

There is also a lower limit on the Higgs mass from vacuum stability (e.g., Cabibbo et al., 1979; Degrassi et al., 2012; Espinosa, 2016). To see this, consider the small λ limit of the first equation in (8.199). If we keep only the (dominant) h_t^4 term,[29] and also treat h_t as a constant, then one finds

$$\lambda(Q^2) \sim \lambda(\nu^2) - \frac{3h_t^4}{4\pi^2} \ln \frac{Q^2}{\nu^2}, \tag{8.203}$$

which goes negative at Q^2 for which

$$\lambda(\nu^2) = \frac{3h_t^4}{4\pi^2} \ln \frac{Q^2}{\nu^2}, \qquad M_H^2 = \frac{3h_t^4}{\sqrt{2}\pi^2 G_F} \ln \frac{Q}{\nu}. \tag{8.204}$$

[28]In the pure $\lambda H^4/4$ theory the only way for the one-loop RGE solution to remain finite for $Q^2 \to \infty$ would be to take $\lambda(\nu^2) = 0$, justifying the term triviality.

[29]Lower limits relevant to a much lighter t quark are reviewed in (Gunion et al., 1990).

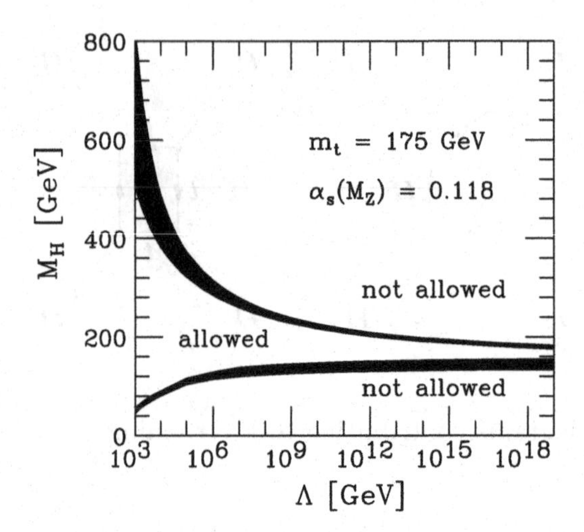

Figure 8.32 Theoretical limits on the SM Higgs mass as a function of the scale Λ at which new physics enters. The upper limit is from the absence of a Landau pole below Λ, while the lower limit is from vacuum stability. Reprinted with permission from (Hambye and Riesselmann, 1997).

A negative λ within the domain of validity of the theory would suggest an unstable vacuum, so we have a lower limit on $\lambda(\nu^2)$ and M_H^2 coinciding with the terms on the r.h.s. in (8.204) for $Q = \Lambda$. In particular, $M_H \gtrsim 85$ GeV for $\Lambda = 1500$ GeV. For large Λ it is not valid to neglect the running of h_t. One needs to integrate the coupled equations, include two-loop effects, and include loop contributions to the effective potential in addition to those in $\lambda(Q^2)$. Typical results are shown as a function of Λ in Figure 8.32. Combining the triviality and vacuum stability limits, M_H is constrained to the rather limited range $130-180$ GeV for $\Lambda = M_P$, with a somewhat larger range for smaller Λ. The lower limit is weakened somewhat if one allows a sufficiently long-lived (i.e., longer than $\sim$13.8 Gy, the age of the Universe) metastable vacuum (Espinosa and Quiros, 1995; Isidori et al., 2001). Early estimates allowed a limit in that case of around 115 GeV for $\Lambda = M_P$ (but see the comments in Section 8.5.3).

These limits do not apply in the minimal supersymmetric extension of the SM (the MSSM), because λ is not an independent parameter and there are additional contributions to the RGE. There is a complementary *upper* limit of $\sim$ 135 GeV on the lightest Higgs scalar in the MSSM (increasing to $\sim$ 150 GeV in singlet extensions of the MSSM), which is close to the SM vacuum stability lower bound for most values of Λ. The observation of a Higgs much heavier or lighter than 135 GeV would have helped distinguish between the SM and supersymmetry, but the observed 125 GeV is inconclusive and somewhat challenging for both.

Tree Unitarity and the Equivalence Theorem

Another theoretical upper limit is based on the (tree-level) unitarity for $W^+W^- \to W^+W^-$ scattering and related processes such as ZZ, ZH, and HH scattering (Lee et al., 1977) (cf. the discussions of unitarity breakdown in high energy $\nu_e e \to \nu_e e$ and $e^+e^- \to W^+W^-$ scattering in Section 7.1). The potential difficulty involves the longitudinal polarization states for the W or Z, which dominate in high energy processes since the polarization vector $\epsilon_\mu(\vec{k},3) \sim k_\mu/M_{W,Z}$ grows with k at high energy. The amplitude for $W_L^+W_L^- \to W_L^+W_L^-$ (the subscript indicates longitudinal) is approximately

$$M = -i\sqrt{2}G_F M_H^2 \left(\frac{s}{s - M_H^2} + \frac{t}{t - M_H^2} \right) \tag{8.205}$$

for $s, M_H^2 \gg M_W^2$, with the relevant diagrams shown in Figure 8.33. If one takes $M_H \to \infty$ this grows linearly with s, leading to violation of S-wave unitarity (Appendix C) at $\sqrt{s} \sim (16\pi\sqrt{2}/G_F)^{1/2} \gtrsim 2.4$ TeV, analogous to the problem in (7.6) on page 225 for $\nu_e e \to \nu_e e$ scattering in the Fermi theory. This illustrates why the existence of the Higgs (or some alternative form of spontaneous symmetry breaking) is essential to the consistency of the theory, since taking $M_H \to \infty$ is equivalent to removing it from the theory. However, there are problems even for finite M_H. It is straightforward to show (Lee et al., 1977; Gunion et al., 1990) that the S-wave amplitude obtained from (8.205) grows with M_H^2 for $s \gg M_H^2$, i.e.,

$$a_0 = -\frac{G_F M_H^2}{4\pi\sqrt{2}}, \tag{8.206}$$

where a_0 is the S-wave projection of the amplitude $T = -iM$, as defined in (C.5) and (C.7). The unitarity condition $|a_0| < 1$ then leads to an upper limit on M_H. This can be strengthened somewhat by considering the coupled channel analysis and a more precise unitarity constraint, leading to the unitarity bound

$$M_H \leq \left(\frac{4\pi\sqrt{2}}{3G_F} \right)^{1/2} \sim 700 \text{ GeV}, \tag{8.207}$$

which is comparable to the lattice version of the triviality bound.

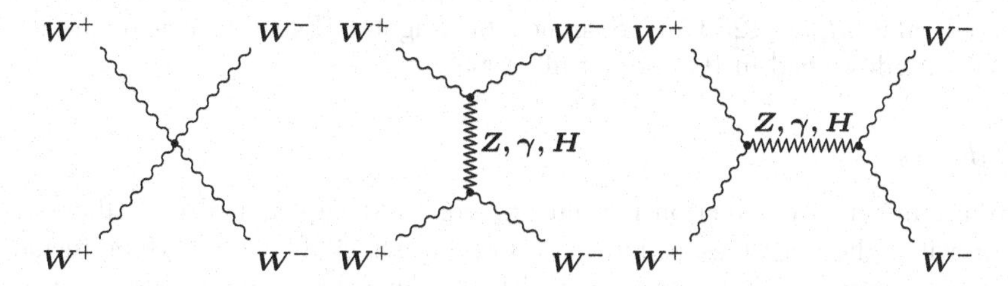

Figure 8.33 Tree-level diagrams for $W^+W^- \to W^+W^-$. The zigzag lines can represent a Z, γ or H.

Of course, (8.207) is based on the tree-level amplitude, and unitarity only applies rigorously to the full amplitude. One should therefore interpret (8.207) as the condition for a weakly coupled (perturbative) Higgs and gauge self-interaction sector, for which higher-order corrections to the amplitude are not expected to be important. Conversely, a violation of (8.207), or the nonobservation of a Higgs below this scale, would have suggested that spontaneous symmetry breaking is associated with a strongly coupled Higgs sector or some strong coupling alternative to the elementary Higgs mechanism. This would presumably be manifested by enhanced WW cross sections at high energy (Chanowitz and Gaillard, 1985) and effects such as WW (bound state) resonances.

In Section 8.2.1 we applied the Kibble transformation to the Higgs doublet following SSB to go to the unitary gauge, in which it is manifest that the Goldstone bosons are eaten to become the longitudinal degrees of freedom of the $W^\pm$ and Z. We also saw above that amplitudes involving the W and Z at high energy are dominated by their longitudinal components. It is therefore not surprising that such high energy amplitudes can be calculated more easily in terms of the original Goldstone degrees of freedom using the *equivalence theorem* (Lee et al., 1977). (More rigorous discussions in more general gauges and including higher-order effects are given in (Chanowitz and Gaillard, 1985; Chanowitz et al., 1987).) We will illustrate this with a simple example, using the expressions (8.85) on page 277 for the Higgs doublet ϕ and (8.88) for the Higgs potential $V(\phi)$ in an R_ξ gauge, in which H is the physical Higgs field, and $w^\pm$ and z are the Goldstone degrees of freedom that disappear in the unitary gauge. (The gauge and Yukawa interactions for ϕ are also given in Section 8.2.4.) The equivalence theorem states that amplitudes involving high energy longitudinal Z's and W's can be obtained by the much simpler calculation of the corresponding amplitudes involving z and $w^\pm$. As a simple example, the $w^+w^- \to w^+w^-$ amplitude is given at tree-level by the four-point $w^+w^-w^+w^-$ vertex derived from (8.88), and by H exchange in the s and t channels. One finds

$$M = -4i\lambda + (-2i\lambda\nu)^2 \left[\frac{i}{s - M_H^2} + \frac{i}{t - M_H^2}\right] = -2i\lambda \left[\frac{s}{s - M_H^2} + \frac{t}{t - M_H^2}\right], \tag{8.208}$$

which reproduces (8.205) after using (8.197).

8.5.2 Higgs Properties, Searches, and Discovery

The Higgs interaction vertices are displayed in Figure 8.3 on page 267. They are always proportional to the mass of a fermion or to mass-squared for a boson, making the Higgs difficult to produce or detect in processes involving light particles. Here we consider searches for a Higgs boson anywhere in the mass range $M_H \gtrsim \mathcal{O}(10 \text{ GeV})$, and then specialize to the observed value $M_H \sim 125$ GeV. Searches for a lighter Higgs in nuclear physics, meson decays, etc., are described in (Gunion et al., 1990).

Higgs Production

There are a number of production mechanisms for the Higgs at hadron colliders, some in association with other particles, as indicated in Figure 8.34. The gluon-gluon fusion (GGF) mechanism $GG \to H$, which proceeds via a virtual t-quark loop, has the largest cross section at both the Tevatron and the LHC. Observation of gluon-gluon fusion allows an indirect constraint on the $t\bar{t}H$ coupling, even for $M_H \ll 2m_t$. There are other modes with lower rates than gluon-gluon fusion, but with the advantage of allowing tagging on the associated particles. For example, the associated production of $W^\pm H$ or ZH (*Higgstrahlung*) is important. The ZH mode can be tagged by $Z \to \ell^-\ell^+$ with $\ell = e$ or μ, allowing a constraint on the ZH rate even for otherwise difficult modes such as $H \to b\bar{b}$ or for (non-standard) Higgs decays into unobserved particles. $WH \to \nu\ell H$ and $ZH \to \nu\bar{\nu}H$ have even larger rates. Vector boson fusion (VBF), i.e., $W^+W^- \to H$ or $ZZ \to H$ with the W or Z radiated from a q or $\bar{q}$, has the second highest rate at the LHC and is associated with two hard jets. Associated $t\bar{t}H$ and $b\bar{b}H$ production is also important at the LHC. QCD (and electroweak) radiative corrections can be large. This is especially true for gluon fusion, where they are of $\mathcal{O}(100\%)$ at the LHC. They are described in the reviews and summarized in (Heinemeyer et al., 2013; de Florian et al., 2016, and Problem 8.20). The various production cross secıs, including higher-order corrections, are shown for pp collisions as functions of M_H and in Figure 8.35.

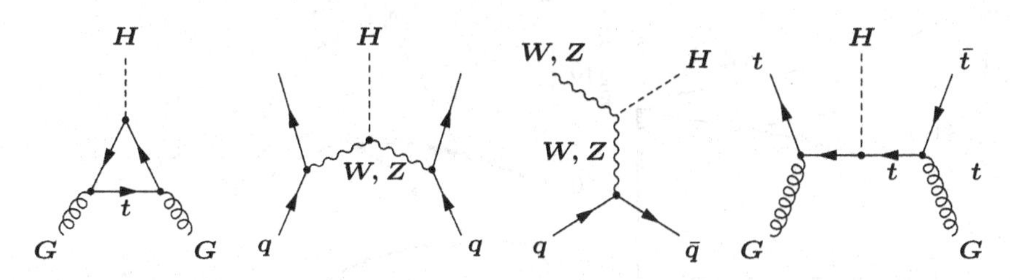

re 8.34 Representative Higgs production diagrams at a hadron collider, including on-gluon fusion (left), WW or ZZ fusion (second left), and typical diagrams for ociated production of WH, ZH, or $t\bar{t}H$.

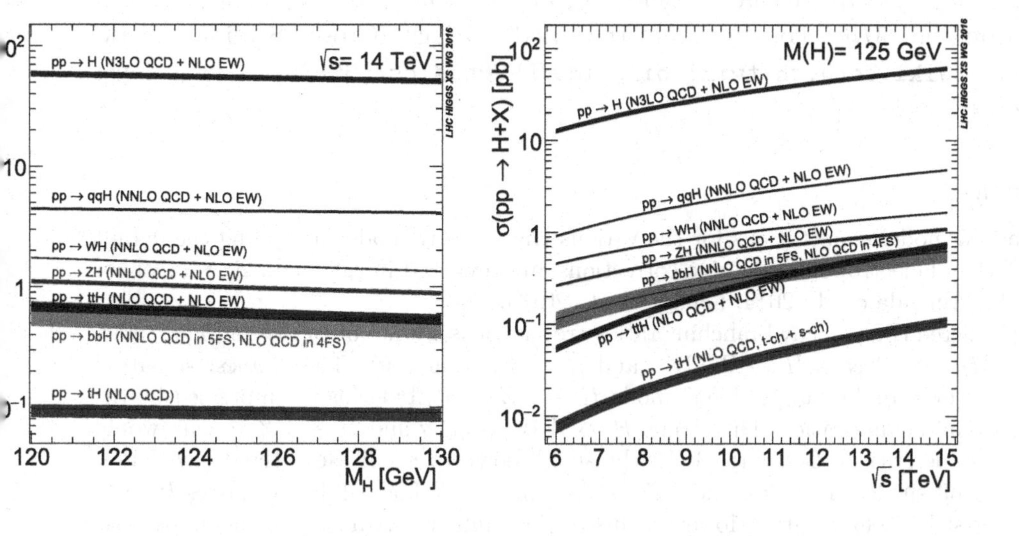

re 8.35 Higgs production cross sections at the LHC as a function of M_H $\sqrt{s} = 14$ TeV (left), and as a function of $\sqrt{s}$ for $M_H = 125$ GeV ght). Figures courtesy of the LHC Higgs Cross Section Working Group, iki.cern.ch/twiki/bin/view/LHCPhysics/LHCHXSWG (de Florian et al., 2016).

The dominant production mechanism in e^-e^+ at lower energies (e.g., LEP) is through ociated Z production (Higgstrahlung), $e^-e^+ \to Z \to ZH$, analogous to the third diagram Figure 8.34. At higher energies ($\sqrt{s} \gtrsim 450$ GeV for $M_H = 125$ GeV), the WW fusion cess $e^-e^+ \to \nu_e W^- W^+ \bar{\nu}_e \to \nu_e \bar{\nu}_e H$ (analogous to the second diagram in Figure 8.34) ninates because the cross section scales as $\ln(s/M_H^2)$ rather than $1/s$ (see page 219). fusion, $e^-e^+ \to e^-e^+H$, is cleaner than WW fusion, but the cross section is an order nagnitude smaller. Smaller still is the cross section for $e^-e^+ \to t\bar{t}H$. However, this is ninated by the radiation of the Higgs from the t or $\bar{t}$ produced in $e^-e^+ \to t\bar{t}$ and could w the determination of the $t\bar{t}H$ coupling in a future high energy collider. The predicted $^+$ cross sections for $M_H = 125$ GeV are shown in Figure 8.36.

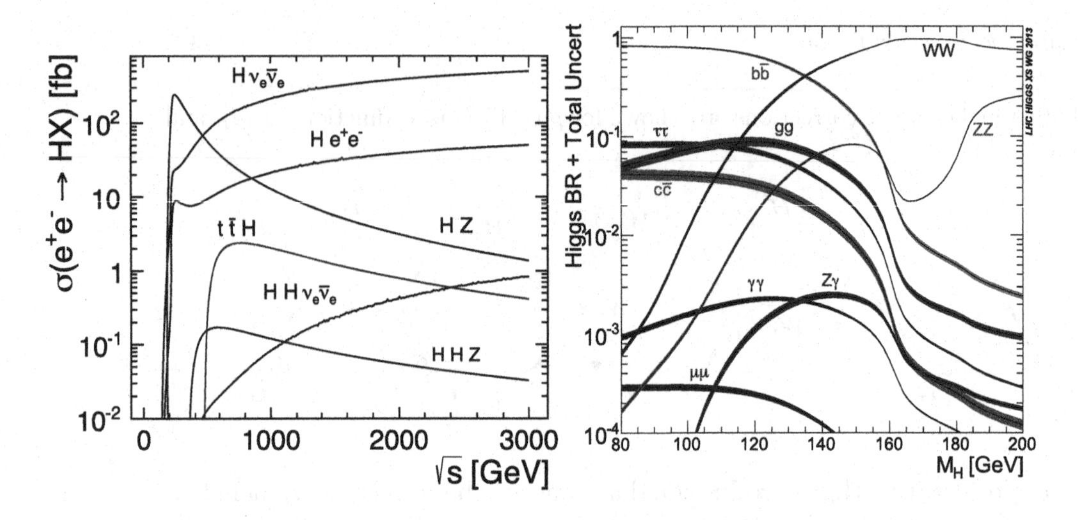

Figure 8.36 Left: SM cross sections for $e^-e^+ \to HX$ ($X = Z, \nu\bar{\nu}, e^-e^+, t\bar{t}, \cdots$) for $M_H = 125$ GeV as a function of $\sqrt{s}$, courtesy of CERN (Lebrun et al., 2012). Right: branching ratios for SM Higgs decays as a function of M_H, courtesy of the LHC Higgs Cross Section Working Group, `twiki.cern.ch/twiki/bin/view/LHCPhysics/LHCHXSWG` (Heinemeyer et al., 2013).

Higgs Decays

The standard model expectations for the various Higgs decay modes, including the details of the QCD and electroweak radiative corrections, are reviewed in (Djouadi, 2008a; Denner et al., 2011; Almeida et al., 2014; Lepage et al., 2014).

The predicted Higgs decay branching ratios are shown as a function of M_H in Figure 8.36. For large M_H, the decays $H \to W^+W^-$ and $H \to ZZ$ dominate. The cleanest signature in this case would be for the "golden" mode $H \to ZZ \to 4\ell$, but this is suppressed by the low leptonic branching ratios. The chains $H \to WW \to q\bar{q}\ell\nu$ and $H \to ZZ \to q\bar{q}\ell\bar{\ell}$ would therefore also be critical. Below the WW threshold, down to ~ 135 GeV, the decay WW^*, where W^* is off-shell, dominates, and ZZ^* is also important. For still lower masses $H \to b\bar{b}$ has the largest branching ratio. However, this mode is difficult at hadron colliders because of the very large QCD background, especially at the LHC. Thus, a Higgs produced by gluon fusion would be unobservable in $b\bar{b}$. The most promising light Higgs channels at the LHC are $H \to \gamma\gamma$, which proceeds at one-loop, and $H \to ZZ^* \to 4\ell$. These have very small branching ratios, but have clean signatures and lower backgrounds than the other modes, and in fact these turned out to be the discovery channels. Other modes with significant branching ratios for a light Higgs include GG, $\tau^+\tau^-$, and $c\bar{c}$. Associated productions of the Higgs with W, Z, or $t\bar{t}$, as well as vector boson fusion with hard forward quark jets, provide powerful additional handles, e.g., allowing the observation of $H \to b\bar{b}$ at hadron colliders.

The partial Higgs decay widths into fermions are

$$\Gamma(H \to f\bar{f}) = C_f \frac{G_F m_f^2}{4\sqrt{2}\pi} \beta_f^3 M_H, \tag{8.209}$$

where $\beta_f = (1 - 4m_f^2/M_H^2)^{1/2}$ is the fermion velocity and $C_f = 1$ (leptons) or 3 (quarks) is the color factor. The dominant QCD corrections for quarks are included by evaluating the running masses at M_H^2, e.g., $m_b[(125 \text{ GeV})^2] \sim 3$ GeV, reducing the width from the leading order expression. There are additional (smaller) corrections, e.g., from vertex diagrams. These fermionic widths are very small (except $H \to t\bar{t}$ for a very heavy H). For example, $\Gamma(b\bar{b})/M_H \sim 1.9 \times 10^{-5}$, so that $\Gamma(b\bar{b}) \sim 2.4$ MeV for $M_H = 125$ GeV. The partial widths for on-shell W^+W^- or ZZ are

$$\Gamma(H \to VV^\dagger) = \delta_V \frac{G_F}{16\sqrt{2}\pi}(1 - x_V)^{1/2}\left(1 - x_V + \frac{3}{4}x_V^2\right) M_H^3, \tag{8.210}$$

where $\delta_W = 2$, $\delta_Z = 1$, and $x_V = 4M_V^2/M_H^2$. The partial width grows as M_H^3. This is because for $M_H \gg M_V$ the decay is dominated by the longitudinal vector states, and their polarization vectors are of order M_H/M_V (see Problem 8.25). Asymptotically,

$$\Gamma(H \to W^+W^- + ZZ)/M_H \sim \frac{1}{2}\left(\frac{M_H}{1 \text{ TeV}}\right)^2, \tag{8.211}$$

so a heavy Higgs is expected to be very broad, while a light one is very narrow. The partial widths for virtual decays, $\Gamma(H \to VV^*)$ are given, e.g., in (Gunion et al., 1990; Djouadi, 2008a).

The decays $H \to \gamma\gamma$, $Z\gamma$, and GG occur at one-loop. At leading order

$$\Gamma(H \to GG) \sim \frac{G_F \alpha_s^2 M_H^3}{36\sqrt{2}\pi^3} \tag{8.212}$$

for $M_H \ll 2m_t$, from the top-loop diagram in Figure 8.37, with α_s evaluated at M_H. There is a very large QCD correction. At the next order (Spira et al., 1995) the rate is enhanced by $1 + \delta\alpha_s/\pi$, where $\delta \sim 95/4 - 7N_f/6$, i.e., an increase of $\sim 60\%$ for $N_f = 5$. The GG decay is of course difficult to observe at a hadron collider but can be probed indirectly by measuring the GG fusion rate. Similarly, the $H \to \gamma\gamma$ decay is dominated by the W and top loops in Figure 8.37, which partially cancel. For $M_H \ll 2M_W$, the SM prediction is

$$\Gamma(H \to \gamma\gamma) \sim \frac{G_F \alpha^2 M_H^3}{128\sqrt{2}\pi^3}\left|-7 + \frac{4}{3}C_t q_t^2\right|^2, \tag{8.213}$$

where the first term is from the W, while the $4C_t q_t^2/3 = 16/9$ is from the top. See the reviews for corrections for finite M_H, higher order corrections, and $\Gamma(H \to Z\gamma)$.

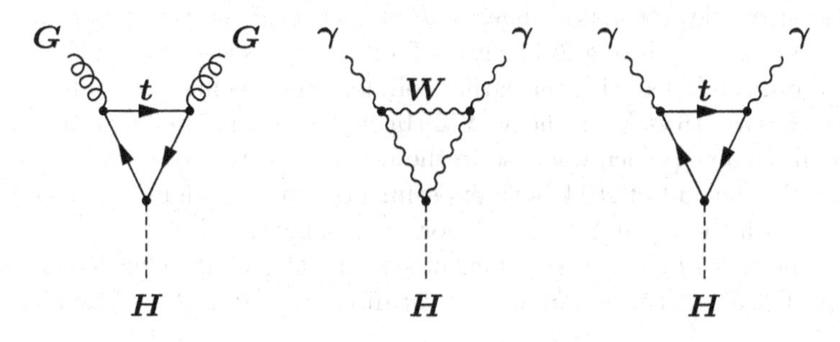

Figure 8.37 Diagrams for the loop-induced decays $H \to GG$ and $H \to \gamma\gamma$.

Precision Electroweak Constraints

As described in Section 8.3.4, the precision electroweak data depend logarithmically on M_H due to the loop contributions to the gauge self-energies, shown in Figure 8.17 on page 301. These favor $66 < M_H < 134$ GeV at 90% cl, with a central value $M_H = 96^{+22}_{-19}$ GeV (Patrignani, 2016), consistent with 125 GeV at 1.3σ. The allowed region in the m_t-M_H plane is shown in Figure 8.26 on page 316.

Direct Searches at Colliders[30]

The Higgs was searched for directly at LEP, through the Higgstrahlung process $e^-e^+ \to Z \to ZH$, which probes the ZZH vertex in Figure 8.3. At LEP 1, the s-channel Z was on-shell and the final one virtual, while at LEP 2, the s-channel Z was virtual. Various combinations of final states were searched for, including $H \to b\bar{b}$ or $\tau^-\tau^+$ and $Z \to q\bar{q}, \ell^-\ell^+$, or $\nu\bar{\nu}$. The LEP experiments were able to exclude a SM Higgs with mass below 114.4 GeV at 95% cl (Barate et al., 2003; Patrignani, 2016). LEP 2 also obtained limits on possible Higgs-like states with masses below 114.4 GeV but with reduced rates compared to the SM expectation, as would be expected in some extended models.

The CDF and D0 experiments at the Tevatron could probe to higher values of the SM Higgs mass once they had acquired sufficient integrated luminosity. Initially, they were mainly sensitive to $H \to W^+W^-$ in the mass range 160–170 GeV near or just above threshold. With increased luminosity their sensitivity increased and they could search for lower Higgs masses via associated VH production ($V = W$ or Z), with $H \to b\bar{b}$ (the Tevatron $b\bar{b}$ background is not so large as that at the LHC). By the time the Tevatron ceased running in 2012 the collaborations had searched for the SM Higgs in the range 90–200 GeV in the decay modes $b\bar{b}$, W^+W^-, ZZ, $\tau^-\tau^+$, and $\gamma\gamma$ at $\sqrt{s} = 1.96$ TeV with integrated luminosities $L = \int \mathcal{L}dt$ up to 10 fb^{-1}, taking into account all of the production mechanisms in Figure 8.34. Their combined analysis (Aaltonen et al., 2013) excluded a SM Higgs in the mass ranges 90–109 and 149–182 GeV, and observed a broad excess of events (mainly $Vb\bar{b}$) from 115–140 GeV. Taking resolution into account this was consistent with the LHC observations and with a SM Higgs, with a local significance[31] of 3.0σ for $M_H = 125$ GeV.

The Higgs Discovery and Properties

Because of the higher energy (and therefore cross sections), and eventually the larger integrated luminosity, the LHC experiments ATLAS and CMS were sensitive to a SM or MSSM Higgs over the entire relevant mass range. In *Run 1* each of the two experiments obtained $L \sim 5$ fb^{-1} at $\sqrt{s} = 7$ TeV during 2011 and ~ 20 fb^{-1} at $\sqrt{s} = 8$ TeV in 2012.

The 7 TeV data, combined with the earlier collider experiments, was sufficient to exclude a light SM Higgs below 116 GeV or a heavy SM Higgs between 127 and 600 GeV, leaving only a relatively small window (which was also in the range expected from precision electroweak measurements). By the end of 2011 both experiments reported significant excesses around 120–126 GeV in both the $\gamma\gamma$ and 4ℓ, $\ell = e^\pm$ or $\mu^\pm$ channels.

By the summer of 2012 there were sufficient statistics for CMS (Chatrchyan et al., 2012) and ATLAS (Aad et al., 2012) to announce $> 5\sigma$ discoveries of a Higgs-like particle of mass

[30] The history is described in, e.g., (Bernardi and Herndon, 2014; Dittmaier and Schumacher, 2013).

[31] The local significance is for a definite M_H, while the lower global significance takes into account the look-elsewhere effect, i.e., that a statistical fluctuation of the background could have occurred anywhere in a given mass range.

$\sim$(125–126) GeV. The most significant channels were again $\gamma\gamma$ and 4ℓ (from $H \to ZZ^* \to 4\ell$), for which there is the best mass resolution. The final combined ATLAS-CMS analysis of the Run 1 data yielded $M_H = 125.09 \pm 0.24$ GeV (Aad et al., 2015a).

The Higgs production and decay properties were extensively studied by ATLAS and CMS in Run 1, as reviewed in, e.g., (Murray and Sharma, 2015; Aad et al., 2016; Patrignani, 2016). They established not only the $\gamma\gamma$ and ZZ^* decays, but also $H \to WW^*$, which is observed as a broad enhancement in 2ℓ (with the neutrinos unobserved), and $H \to \tau^+\tau^-$. There is also some evidence (at about 2σ) for $H \to b\bar{b}$ (mainly from associated ZH and WH production), as well as upper limits on rare decay modes such as $\mu^-\mu^+$ and $Z\gamma$.

The various production mechanisms can be separated experimentally, e.g., by tagging on the associated forward jets, W, Z, or t. Most of the combinations $\sigma_i B_f$ have been determined, where σ_i represent the cross sections for gluon-gluon fusion, vector boson fusion, or associated WH, ZH, or $t\bar{t}H$ production, and $B_f = \Gamma_f/\Gamma_H$ is the branching ratio into $f = \gamma\gamma$, ZZ^*, WW^*, $\tau^+\tau^-$, or $b\bar{b}$. The results for the ratios of $\sigma_i B_f$ to the SM expectations are shown in Figure 8.38. They are consistent with unity, though with large uncertainties. A global analysis over all i and f yields a a rate of 1.09(11) relative to the SM.

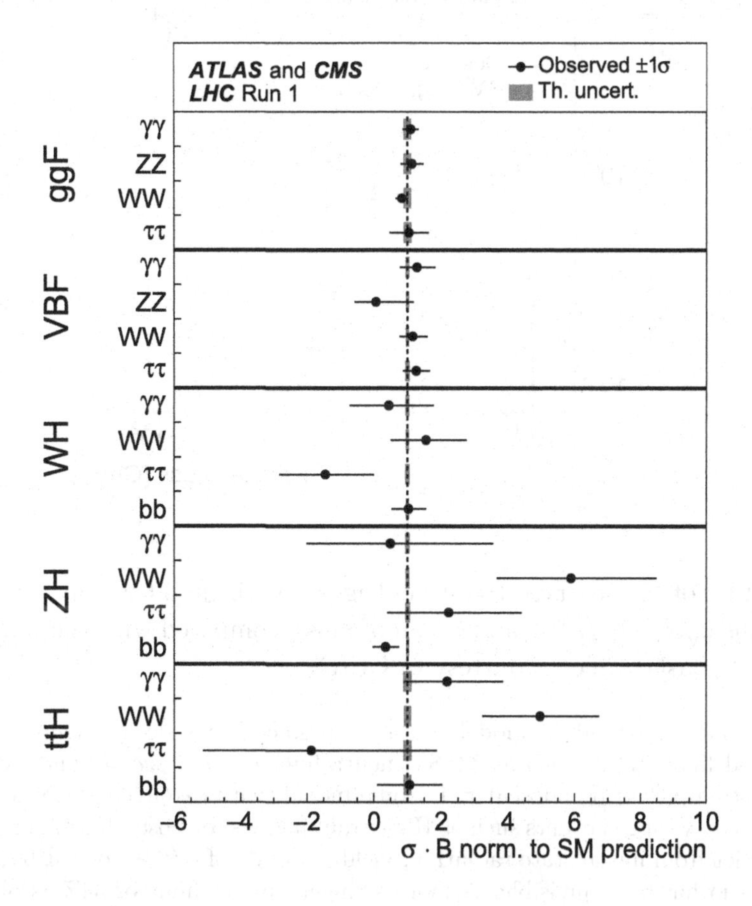

Figure 8.38 Signal strengths μ (ratio of observed rate to the expectation for a SM Higgs) for various production processes and subsequent decay modes for the combined ATLAS and CMS analysis of Run 1 data, from (Aad et al., 2016).

The basic observables $\sigma_i B_f$ at a hadron collider do not allow a separation between the cross sections and branching ratios without additional information, such as a theoretical calculation of the cross section.[32] However, the relative branching ratios into two different final states, the ratios of production cross sections, or the ratios of events at different energies can be determined. All agree with the SM within uncertainties.

An alternative way of analyzing the data is to assume the SM production and decay mechanisms, but to allow the elementary WWH, ZZH, $t\bar{t}H$, $b\bar{b}H$, and $\tau^+\tau^-H$ couplings to vary by factors κ compared to the SM expectations. All of the κ's are consistent with unity within the (10–20)% uncertainties except for $b\bar{b}H$, for which $\kappa_b \sim 0.67^{+0.22}_{-0.27}$ (Aad et al., 2016).[33] These results are displayed in Figure 8.39, in which the observed couplings are seen to agree with the SM prediction that they should be proportional to a power of mass, i.e., to $h_f \equiv m_f/\nu$ for fermion f, or to $h_V \equiv 2M_V^2/\nu$ for $V = W, Z$.

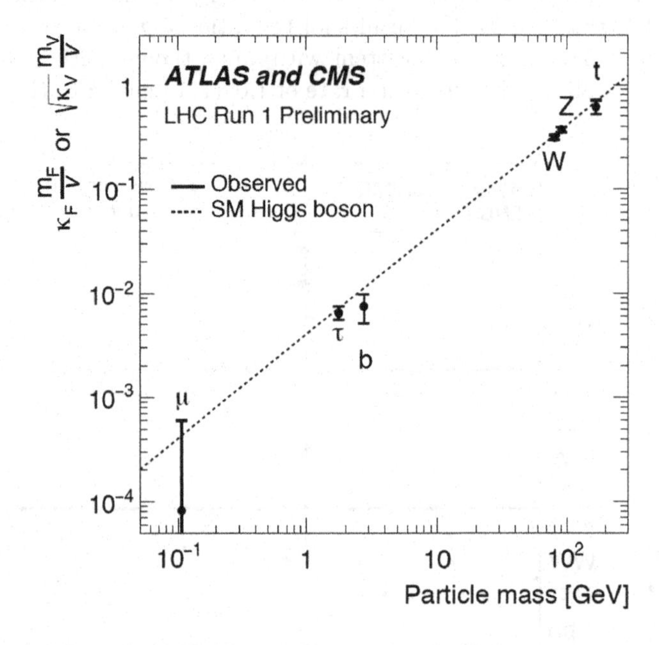

Figure 8.39 Plot of the observed tree-level Higgs couplings to fermions ($\kappa_f m_f/\nu$) and gauge bosons ($\sqrt{\kappa_V} M_V/\nu$) as a function of mass, compared with the SM prediction $\kappa_f = \kappa_V = 1$ (dashed line). Courtesy of CERN.

Some beyond the standard model scenarios predict Higgs decay modes into invisible or unobserved final states, such as MSSM neutralinos or *dark sector* candidates for dark matter, or into a pair of pseudoscalars in some models with extended Higgs sectors. These are constrained by tagged events such as Higgstrahlung, vector boson fusion, or gluon fusion with a monojet from initial state radiation, yielding limits of ~(60–70)% at 95% c.l. on the branching ratio for $H \to$ invisible. A more stringent upper limit of 34% is obtained in a

[32] In principle the cross sections could be separated by tagging some inclusive production mode, such as associated WH or ZH, with W or $Z \to$ jets, $Z \to 2\ell$, or $Z \to b\bar{b}$.

[33] There are additional solutions with different signs due to interference effects.

eralization of the global analysis with arbitrary couplings, allowing for invisible decays new particles in the loops.

A direct determination of the total width Γ_H would be extremely desirable, both for ting branching ratios to absolute partial widths and for further constraining unobserved invisible decay modes. CMS obtained an upper limit of 3.4 GeV at 95% c.l. on Γ_H n the 4ℓ lineshape. This is very much larger than the intrinsic width of $\sim$ 4.1 MeV ected in the SM, and is presumably due almost entirely to the experimental resolution. IS and ATLAS subsequently obtained much more stringent limits $\Gamma_H \lesssim$ 22 MeV at 6 c.l. indirectly by comparing the $H \to ZZ^*$ and WW^* rates on-shell and off-shell, king use of the Breit-Wigner energy dependence in (F.2) on page 523, and assuming extra decay modes are accessible off shell. This is still considerably larger than the SM ectation, however. A precise determination of Γ_H will probably require a measurement he $e^-e^+ \to ZH \to ZZZ^*$ rate at a future e^-e^+ collider (to obtain the ZZH vertex and refore $\Gamma(H \to ZZ^*)$), combined with $B(H \to ZZ^*)$.

Another critical aspect is to determine the spin, parity, and charge conjugation quan- n numbers J^{PC}. These are predicted to be 0^{++} for the SM Higgs, i.e., the couplings to nions and vectors are proportional to $H\bar{\psi}\psi$ and $HV_\mu V^\mu$ (or $HV_{\mu\nu}V^{\mu\nu}$ at loop level), pectively. However, dynamical alternatives to the elementary Higgs mechanism often in- ve pseudoscalars, $J^{PC} = 0^{-+}$, which could be light (Eichten et al., 2012). A pseudoscalar vould couple like $P\bar{\psi}\gamma^5\psi$ and $PV_{\mu\nu}\tilde{V}^{\mu\nu}$, where $\tilde{V}^{\mu\nu} = \frac{1}{2}\epsilon^{\mu\nu\rho\sigma}V_{\rho\sigma}$. If a light pseudoscalar nehow had the appropriate coupling strengths it could mimic the SM Higgs. In principle, Higgs-like particle could also have spin higher than 0.

The fact that the H decays to $\gamma\gamma$ implies that it has $\eta_C = +1$ (or that it is not a eigenstate). It also implies[34] that $J = 0, 1$, or 2 under the plausible assumption of an vave decay. CMS and ATLAS have studied the various possibilities using the observed ributions in the $\gamma\gamma$, ZZ^*, and WW^* decays. All data are consistent with the 0^{++} ignment, and many alternative models involving $J^P = 0^-, 1^\pm$, and 2^+ are excluded at ter than 99% c.l.

.3 Implications of the Higgs Discovery

e various observations described in the previous section leave little doubt that the ob- ved state is either the SM Higgs, or something very similar. Assuming that it really is SM Higgs, its relatively low mass implies some tension with the the vacuum stability siderations discussed in Section 8.5.1. A recent NNLO calculation of the effective po- tial and running couplings (Degrassi et al., 2012; Buttazzo et al., 2013) shows that for observed parameters the quartic coupling $\lambda(Q^2)$ goes negative for Q around $10^{10} - 10^{12}$ V. See Figure 8.40 (left). However, the decrease of λ slows for larger Q^2 due to cancella- ns, implying a metastable vacuum (with the lifetime longer than the observed age of the iverse.[35]) It is remarkable that the observed parameters imply that the SM vacuum lies he narrow strip between instability and absolute stability. In particular, the theoretically iguing possibility $\lambda(M_P^2) = 0$ is excluded at better than 99% c.l. (Buttazzo et al., 2013).

According to the Landau-Yang theorem (Landau, 1948; Yang, 1950), which follows from rotational riance and Bose statistics, a spin-1 particle cannot decay into two identical massless vectors. It can be led for off-shell, massive, or non-identical vectors, for processes that are mistaken for two photons, or off-shell or interference effects for the decaying particle. It is therefore useful to exclude this possibility erimentally.

However, the decay rate could be greatly increased in the presence of Planck-scale higher-dimensional rators (e.g., Lalak et al., 2014; Branchina et al., 2015) or if they are nucleated by small black holes (Burda ., 2016).

It should be reiterated that these considerations only apply if there is no new physics below the Planck scale M_P, or at least below the scale at which λ vanishes.

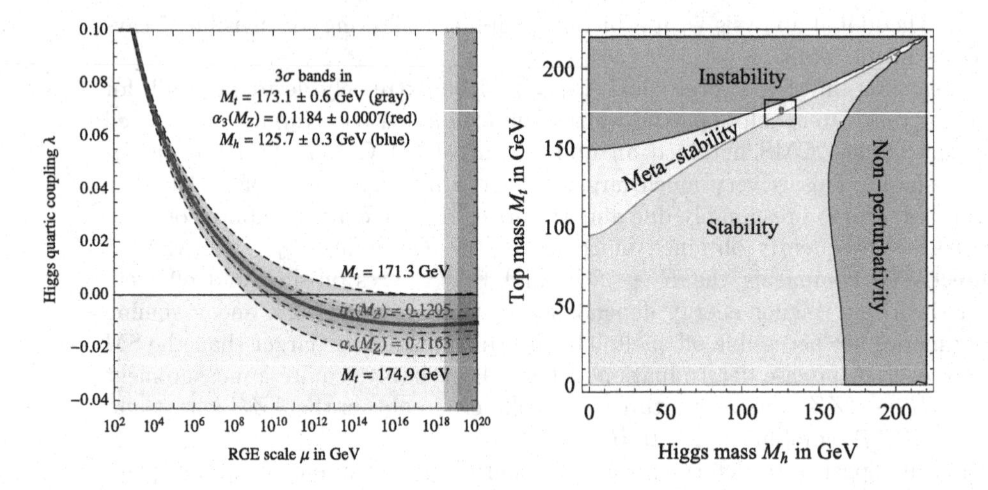

Figure 8.40 Left: the running of the quartic Higgs self-coupling for $M_H = 125$ GeV. The vacuum remains metastable for $\lambda(M_P^2) \gtrsim -0.05$. Right: phase diagram for the standard model as a function of M_H and m_t, assuming that there is no new physics up to the Planck scale. From (Degrassi et al., 2012).

Extended Electroweak Symmetry Breaking Sectors

The observed Higgs-like particle is consistent with the SM Higgs. However, the measurements of its couplings still leave considerable room for deviations. These could be due either to extended sectors involving elementary Higgs fields or to the possibility that the Higgs sector is an effective theory describing underlying strong dynamics at a higher scale. Some of these extensions are motivated by the Higgs/hierarchy problem described in Section 10.1. This basically states that the Higgs mass-squared involves quadratically-divergent corrections of $\mathcal{O}(\lambda, g^2, h^2)\Lambda^2$ (see (10.2) on page 427), where Λ is the new physics scale. This implies that for $\Lambda \gg$ TeV there must be fine-tuned cancellations between the bare mass-squared and the corrections.

Many theories beyond the SM have extended Higgs sectors,[36] involving additional Higgs doublets (such as are required in supersymmetry) and/or Higgs fields transforming under different $SU(2)$ representations, such as singlets or triplets (for reviews, see Gunion et al., 1990; Djouadi, 2008b; Accomando et al., 2006; Barger et al., 2009; Maniatis, 2010; Ellwanger et al., 2010; Branco et al., 2012; Patrignani, 2016). Such additional multiplets allow new physical spin-0 particles, including extra electrically-neutral states, singly-charged states for additional doublets or triplets, and even doubly-charged states for $SU(2)$ triplets with $y = \pm 1$.

Electrically neutral states can mix with the H (even for $SU(2)$ singlets), leading to mass eigenstates h_i. In some cases, there are two or more h_i with scalar couplings to fermions, and

[36] We are using Higgs in the sense of any spin-0 color-singlet multiplets that are not forbidden by additional symmetries from mixing with the SM Higgs doublet.

one or more with pseudoscalar (γ^5) couplings, and in other cases (involving CP violation) the scalars and pseudoscalars mix. In the MSSM, for example, there are two Higgs doublets. In the decoupling limit, in which the second Higgs doublet is much heavier than the first, there is little mixing and the lighter scalar has couplings essentially the same as those of the SM Higgs. However, as will be discussed in Chapter 10, there can be considerable mixing between the two scalars (and with the pseudoscalar if CP is violated) in the non-decoupling region, in which the new supersymmetric particles and second Higgs are relatively light.

Mixing will modify the couplings of the mass eigenstates. For example, the overall strength of the Yukawa couplings of the up-type quarks are changed relative to the SM and to those of the down-type quarks and charged leptons in the MSSM and other two-doublet models (Section 10.2.5). Mixing can also lead to new invisible or exotic decay modes (increasing the width), or can decrease the width due to singlet mixing.

Many of these possibilities have been reexamined in light of the discovery of the Higgs-like particle, including models with additional singlets (Frank et al., 2013; Robens and Stefaniak, 2015) or doublets (Baglio et al., 2014; Craig et al., 2015). It has already been commented in Section 8.5.1 that there is a theoretical upper limit of ~ 135 GeV on the mass of the lightest Higgs scalar in the MSSM, due to the fact that the analog of λ in the SM is given by gauge couplings. Furthermore, even 125 GeV requires either very large stop masses in the multi-TeV range or large stop mixing. These issues, as well as some singlet-extended versions in which the upper limit is relaxed, are discussed in Section 10.2.5.

In addition to the implications for the Higgs spectrum and couplings, extended Higgs sectors can affect the origin of the fermion mass hierarchy and/or lead to flavor changing Higgs-fermion couplings (e.g., in non-supersymmetric two-doublet models in which both doublets couple to the same fermions). They can also modify the ρ_0 parameter (i.e., to the M_W/M_Z relation) at tree level due to the VEVs of Higgs triplets or higher-dimensional representations (Problem 8.1) unless a custodial symmetry is somehow imposed (e.g., Logan, 2014). There are also new mechanisms for explicit or spontaneous CP violation in the Higgs sector, and for a strongly first-order electroweak phase transition, both of which are relevant to the possibility of electroweak baryogenesis (e.g., Maniatis, 2010; Ellwanger et al., 2010; Huang et al., 2016)). There is even the danger of the spontaneous violation of electric charge conservation (Problems 8.3 and 8.4). Higgs singlets in extended models can also be dark matter candidates (Barger et al., 2009) or can serve as a *portal* connecting the standard model fields to a dark matter sector (Schabinger and Wells, 2005; Patt and Wilczek, 2006). Finally, Higgs singlets may offer a dynamical solution to the μ problem of the MSSM (Section 10.2.6).

It is also possible that there are no truly elementary Higgs particles. For example, electroweak symmetry breaking could be associated with the boundary conditions in a theory with extra dimensions of space-time [*Higgsless models* (Csáki et al., 2004)], for which there are no analogs of the Higgs field. The observation of the 125 GeV Higgs-like state implies that such models are no longer relevant to electroweak physics, but the ideas could play a role in, e.g., grand unified theories in extra dimensions. Similar statements apply to *dynamical symmetry breaking* mechanisms (Hill and Simmons, 2003). These are based on a new strong dynamics and usually do not have a light 0^+ scalar that could imitate the SM Higgs. In technicolor, for example, the SSB is associated with the expectation value of a fermion bilinear, analogous to the breaking of chiral symmetry in QCD (Section 5.7). Extended technicolor and top-color also fall into this class.

More promising are composite Higgs models. Typically, these involve a more fundamental strongly-coupled sector at the 1–10 TeV scale which (unlike technicolor) does not directly lead to electroweak symmetry breaking. However, the Higgs multiplet emerges as a composite state in a low-energy effective theory. A very attractive possibility is that, in the absence of electroweak or Yukawa couplings, all four Hermitian components of the Higgs doublet are the massless Goldstone bosons associated with a spontaneously broken global symmetry of the underlying theory. Turning on the electroweak couplings generates the Higgs potential, which in turn leads to $SU(2) \times U(1)$ breaking. Three of the Goldstone bosons are absorbed by the ordinary Higgs mechanism, while the Higgs-like scalar is a pseudo-Goldstone boson. The Higgs mass is generically suppressed by $\mathcal{O}(g/4\pi)$ compared to the scale Λ of the strong dynamics and of the other composite states. This suggests $\Lambda = \mathcal{O}(1 \text{ TeV})$, which is rather low for evading precision electroweak and FCNC constraints. A larger and safer Λ, e.g., 10 TeV, can be achieved by fine-tuning. Alternatively, Λ naturally increases by an order of magnitude (from an extra factor of $g/4\pi$) in Little Higgs models (Arkani-Hamed et al., 2002a), in which the Higgs mass is protected by two symmetries.[37] Agreement with precision constraints is further improved by the imposition of a discrete T-parity (Cheng and Low, 2003). Composite Higgs models can often be reinterpreted (i.e., are equivalent to) theories with extra space-time dimensions. In these dual descriptions the massless Higgs fields are identified as gauge fields in the extra dimensions. These ideas, related issues such as the generation of Yukawa couplings, and other dynamical mechanisms are reviewed in more detail in (Perelstein, 2007; Csáki et al., 2016; Panico and Wulzer, 2016) and in the articles on *Status of Higgs Boson Physics* and on *Dynamical Electroweak Symmetry Breaking* in (Patrignani, 2016). See also Section 10.1.

The various types of new physics mentioned above can affect the Higgs in a number of ways. As already mentioned, mixing in extended Higgs sectors can modify the couplings of the mass eigenstates. Mixing and other new physics can also change the width and allow new exotic or invisible decay modes (Chang et al., 2008; Curtin et al., 2014). New physics can lead to Higgs-mediated flavor changing effects (Buras et al., 2010a; Blankenburg et al., 2012), or can modify the Higgs self-interactions substantially, especially the induced H^3 coupling (Problem 8.22). Heavy particles such as superpartners, additional quarks and leptons, and heavy W' bosons are expected in many extended theories, including most of the composite Higgs models. These can enter into loops and significantly modify gluon fusion and such loop-induced decays as GG, $\gamma\gamma$, and $Z\gamma$, or perturb tree-allowed decays such as $H \to b\bar{b}$. Of course, there may be associated effects such as the direct production of the new heavy particles at the LHC, modification of the oblique parameters or other aspects of precision electroweak physics, the observation of FCNC, or the modification of high energy VV' scattering, where $V, V' = W$ or Z (see, e.g., Baak et al., 2013; Szleper, 2014; and Section 8.4).

Possible deviations from the SM predictions for the Higgs couplings have been studied quantitatively in specific classes of models and from a model independent or general effective operator framework, e.g., in (Giardino et al., 2014; Bélanger et al., 2013; Contino et al., 2013; Elias-Miro et al., 2013; Herrero, 2015; Ellis et al., 2015; Englert et al., 2014). Typical deviations are in the 1–10% range. Considerable improvement on the current observations shown in Figure 8.38 are expected from future running at the LHC, especially with a luminosity upgrade. These could reach a precision of $\sim$(5–10)%. Even more precise measurements would be possible at a future e^-e^+ collider, such as the International Linear Collider (ILC), proposed to be built in Japan. For example, the ILC running at 250 GeV would be able to determine the total H width by measuring the total $e^-e^+ \to ZH$ or $e^-e^+ \to ZH \to ZZZ^*$ rate and combining it with the $H \to ZZ^*$ branching ratio. Increasing the ILC energy to 500-1000 GeV would allow studies of WW fusion and measurements of

[37] Operationally, the Little Higgs models involve additional heavy vectors, fermions, and scalars that cancel the one-loop quadratic divergences.

the other couplings, including the $t\bar{t}H$ and H^3 couplings, many at the 1% level. Of course, the theoretical predictions of the SM would have to be computed to comparable precision (Denner et al., 2011; Almeida et al., 2014; Lepage et al., 2014). High intensity circular e^-e^+ colliders have also been proposed at CERN (the FCC-ee) and in China (the CEPC), and a very high energy (several TeV) linear collider (CLIC) at CERN. Future prospects for various facilities are surveyed in (Dawson et al., 2013; Asner et al., 2013; Bicer et al., 2014; Bechtle et al., 2014b; Arbey et al., 2015).

8.6 THE CKM MATRIX AND CP VIOLATION

We saw in Section 2.10 that CP violation can occur in a field theory whenever the Lagrangian density involves more complex parameters than can be removed by field redefinitions. Nevertheless, CP violation is observed to be very tiny, and it came as a surprise to most physicists when such a violation was first observed in 1964 (Christenson et al., 1964). It is straightforward to write down phenomenological models of new interactions that accomodate CP violation (for a review of early attempts, see Commins and Bucksbaum, 1983). However, when the SM was developed it was difficult to directly incorporate CP violation, because for one or two families the SM is sufficiently simple that CP emerges as an accidental symmetry. One possibility was to extend the Higgs sector, allowing CP-violating effects associated with the scalar exchanges (e.g., Weinberg, 1976). In this case the weakness of CP breaking would be attributed to the small Higgs Yukawa couplings. Another was to assume the existence of an entirely new very weak interaction to mediate the CP violation, such as in the superweak model (Wolfenstein, 1964). A third possibility, apparently realized in nature, is to introduce a third fermion family (Kobayashi and Maskawa, 1973). Observable CP violation would only occur when all three families are relevant to a process, accounting for the weakness.

On the other hand, most extensions of the SM have potential new sources of CP violation that are not naturally suppressed, so CP violation studies are important. Moreover, CP violation is necessary to explain baryogenesis. It is likely that the origin of the baryon asymmetry is related to CP phases beyond those in the CKM matrix, possibly those related to neutrino mixing or perhaps associated with BSM effects observable at the LHC.

This section will give a brief introduction to CP violation and mixing effects, mainly in the K and B meson systems, and the closely related status of the CKM matrix. Much more detailed treatments may be found in (Commins and Bucksbaum, 1983; Jarlskog, 1989; Branco et al., 1999; Kleinknecht, 2003; Ibrahim and Nath, 2008; Sozzi, 2008; Bigi and Sanda, 2009; Schubert, 2015; Patrignani, 2016). CP and time reversal (T) invariance are equivalent in a CPT-invariant theory in the sense that any CP-odd operator is also T-odd. Nevertheless, CP and T are different transformations with different consequences (e.g., Bernabéu and Martínez-Vidal, 2015). Recent direct observation of T-violation involving K and B oscillations, and constraints from electric dipole moments, will be considered in Section 8.6.5. Strong CP violation, associated with the parameter θ_{QCD} introduced in (5.2), and baryogenesis will be touched on in Section 10.1.

Most of the processes considered involve nontrivial complications from the strong interactions. Enormous effort involving perturbative techniques such as HQET and SCET, as well as non-perturbative lattice calculations and chiral perturbation theory have been required to overcome those obstacles.

8.6.1 The CKM Matrix

In Section 8.2.2 the quark and lepton mixing matrices V_q and V_ℓ were seen to arise from the mismatch between the fermion gauge and Yukawa interactions, i.e., between the weak and mass eigenstates. They also required a mismatch between the u and d (or e and ν) sectors, as is apparent in (8.60) on page 272. The empirical forms of V_q were briefly discussed for two and three quark families in (8.65) and (8.66), and in Chapter 7. Here, we consider V_q for three families, i.e., the CKM matrix, in more detail.

As already described, after imposing unitarity and removing unobservable phases by redefinitions of the chiral quark fields, V_q involves three mixing angles and one CP-violating phase. There are a number of possible parametrizations, but we will follow that of the Particle Data Group (see the article on the CKM matrix in Patrignani, 2016),

$$V_q \equiv \begin{pmatrix} V_{ud} & V_{us} & V_{ub} \\ V_{cd} & V_{cs} & V_{cb} \\ V_{td} & V_{ts} & V_{tb} \end{pmatrix} = \begin{pmatrix} 1 & 0 & 0 \\ 0 & c_{23} & s_{23} \\ 0 & -s_{23} & c_{23} \end{pmatrix} \begin{pmatrix} c_{13} & 0 & s_{13}e^{-i\delta} \\ 0 & 1 & 0 \\ -s_{13}e^{i\delta} & 0 & c_{13} \end{pmatrix} \begin{pmatrix} c_{12} & s_{12} & 0 \\ -s_{12} & c_{12} & 0 \\ 0 & 0 & 1 \end{pmatrix}$$
$$= \begin{pmatrix} c_{12}c_{13} & s_{12}c_{13} & s_{13}e^{-i\delta} \\ -s_{12}c_{23} - c_{12}s_{23}s_{13}e^{i\delta} & c_{12}c_{23} - s_{12}s_{23}s_{13}e^{i\delta} & s_{23}c_{13} \\ s_{12}s_{23} - c_{12}c_{23}s_{13}e^{i\delta} & -c_{12}s_{23} - s_{12}c_{23}s_{13}e^{i\delta} & c_{23}c_{13} \end{pmatrix}, \tag{8.214}$$

where $c_{ij} \equiv \cos\theta_{ij}$ and $s_{ij} \equiv \sin\theta_{ij}$. The mixing angles are θ_{12}, θ_{13}, and θ_{23}, and δ is the CP-violating phase. Assuming unitarity, the magnitudes of the elements of V_q are

$$|V_{ij}| \sim \begin{pmatrix} 0.9743 & 0.225 & 0.0036 \\ 0.225 & 0.974 & 0.041 \\ 0.0088 & 0.040 & 0.9992 \end{pmatrix}, \tag{8.215}$$

which implies $s_{12} \sim \sin\theta_c \sim 0.225$, while $s_{13} \ll s_{23} \ll s_{12}$. As we will see, the observed smallness of CP violation is not because δ is small, but rather because observable violations require that all three families contribute to the relevant transition amplitude, and therefore are suppressed by small mixing angles. It is often convenient to employ the approximate *Wolfenstein parametrization* (Wolfenstein, 1983)

$$V_q = \begin{pmatrix} 1-\lambda^2/2 & \lambda & A\lambda^3(\rho - i\eta) \\ -\lambda & 1-\lambda^2/2 & A\lambda^2 \\ A\lambda^3(1-\rho-i\eta) & -A\lambda^2 & 1 \end{pmatrix} + O(\lambda^4), \tag{8.216}$$

where $\lambda \sim \sin\theta_c$. The powers of λ incorporate the suggestive hierarchical pattern in (8.66) on page 273, while $A \sim 0.811$ (from V_{cb}), ρ, and η are real and of $\mathcal{O}(1)$. The displayed terms are unitary through $\mathcal{O}(\lambda^3)$. The higher-order corrections ensure unitarity to higher order, but are not important in practice. CP violation is associated with η, i.e., $\tan\delta = \eta/\rho$. There are various possible conventions for the CP phase in V_q. However, the quantity

$$\bar\rho + i\bar\eta \equiv -\frac{V_{ub}^* V_{ud}}{V_{cb}^* V_{cd}} \tag{8.217}$$

is independent of the convention. From (8.216)

$$\bar\rho = \rho(1-\lambda^2/2), \qquad \bar\eta = \eta(1-\lambda^2/2), \tag{8.218}$$

up to corrections of $\mathcal{O}(\lambda^4)$. V_q takes the same form as (8.216) when written in terms of $\bar\rho$ and $\bar\eta$ to that order. It is also convenient to define the *Jarlskog invariant*

$$J = \Im m\,(V_{us}V_{cb}V_{ub}^*V_{cs}^*) \sim c_{12}c_{23}c_{13}^2 s_{12}s_{23}s_{13}\sin\delta \sim A^2\lambda^6\bar\eta, \tag{8.219}$$

which is a convention-independent measure of CP violation.

Measurements of the elements of V_q are important to test the consistency of the SM and determine its parameters, which are relevant for other quantitites and processes such as electric dipole moments or the CKM contribution in models of baryogenesis. It is especially important to verify that V_q really is unitary, i.e., $V_q V_q^\dagger = V_q^\dagger V_q = I$. A violation or apparent violation would signal the presence of new physics. This could take the form of a fourth family or heavy quarks with exotic $SU(2)$ assignments, so that the 3×3 CKM submatrix would not by itself be unitary. Alternatively, it could involve new interactions like supersymmetry, leptoquarks, compositeness, or a heavy W' (e.g., coupling to right-handed currents) or Z' gauge boson that were not properly included in the analysis and therefore led to an incorrect determination of some of the elements. Unitarity studies include tests of weak universality (the diagonal elements), especially

$$\left(V_q V_q^\dagger\right)_{11} = |V_{ud}|^2 + |V_{us}|^2 + |V_{ub}|^2 = 1, \tag{8.220}$$

as well as unitarity triangle tests (off-diagonal elements), such as

$$\left(V_q^\dagger V_q\right)_{31} = V_{ub}^* V_{ud} + V_{cb}^* V_{cd} + V_{tb}^* V_{td} = 0. \tag{8.221}$$

Magnitudes of the CKM Matrix Elements and Weak Universality

We first briefly survey the magnitudes of the CKM matrix elements. For more detail see, e.g., (Antonelli et al., 2010b; Porter, 2016) and the articles on the CKM matrix, on $V_{ud,s}$, and on $V_{u,cb}$ in (Patrignani, 2016) (from which we take most of the numerical results). Lattice calculations of decay constants and other relevant form factor parameters, which are now very accurate, are reviewed in (Rosner et al., 2015; Aoki et al., 2017).

$|V_{ud}|$: As discussed in Section 7.2.5 the most precise determination is from superallowed $0^+ \to 0^+$ β decay, which only involves the vector current. The Ademollo-Gatto theorem ensures that the theoretical uncertainties are of second order in isospin breaking, e.g., $Z\alpha^2$, though these must be taken into account, allowing a clean extraction of $G_F|V_{ud}|$. Taking G_F from muon decay yields $|V_{ud}| = 0.97417(21)$. The neutron lifetime yields an independent determination of $G_F^2|V_{ud}|^2[1 + 3(g_A/g_V)^2]$, but the precision of the extracted $|V_{ud}|$ is limited by the measurements of the lifetime and of g_A/g_V from the decay asymmetries. Pion beta decay, $\pi^+ \to \pi^0 e^+ \nu_e$, also yields a theoretically clean measurement (Section 7.2.3), but with a larger uncertainty. The value extracted from the $\pi \to \mu\nu$ rate has a relatively large error dominated by the theoretical uncertainty in f_π.

$|V_{us}|$: The $K_{\ell 3}$ decays also involve only the vector current, implying by the Ademollo-Gatto theorem that corrections to the form factor $f_+(0)$ are second order in $SU(3)$-breaking. Nevertheless, there are still theoretical uncertainties in the determination of $|V_{us}|$, as described in the cited reviews and in Section 7.2.3. The experimental value $f_+(0)|V_{us}| = 0.2165(4)$ (with $f_+(0)$ normalized to unity in the $SU(3)$ limit), combined with the lattice value $f_+(0) = 0.9677(37)$, yields $|V_{us}| = 0.2237(9)$. A comparable result can be obtained from the ratio of $K_{\mu 2}$ and $\pi_{\mu 2}$ decay rates, using lattice estimates of the decay constants. $f_K/f_\pi = 1.1928(26)$ leads to $|V_{us}| = 0.2254(8)$, which averages with the $K_{\ell 3}$ result to give $|V_{us}| = 0.2248(6)$. Other determinations involving hyperon and hadronic τ decays each involve their own theoretical uncertainties, as detailed in Chapter 7, but yield reasonably consistent results.

$|V_{cd}|, |V_{cs}|$: $|V_{cd}|$ can be obtained from semi-leptonic ($D \to \pi \ell \nu$) and leptonic ($D \to \ell \nu$) decays, using lattice calculations of the form factor (at $Q^2 = 0$) and decay constant, leading to an average $|V_{cd}| = 0.218(5)$. $|V_{cd}| = 0.230(11)$ can also be obtained from the deep-inelastic processes $\nu(\bar{\nu})N \to \mu^{\mp}X + c(\bar{c})$ as described below (8.109), implying a combined value 0.220(5). Similarly, semi-leptonic ($D \to K\ell\nu$) and leptonic ($D_s \to \ell\nu$) decays imply $|V_{cs}| = 0.995(16)$. An indirect determination from the four LEP 2 experiments utilized the leptonic branching ratios $W \to \ell\nu$ measured in $e^-e^+ \to W^-W^+ \to f\bar{f}f\bar{f}$. From (8.155)–(8.157),

$$\left(1 + \frac{\alpha_s(M_W^2}{\pi} + \cdots\right) \sum_{\substack{m=u,c \\ n=d,s,b}} |V_{mn}|^2 \sim \frac{1}{3B_{\ell\nu}} - 1, \tag{8.222}$$

where $B_{\ell\nu}$ is the average of the $\ell = e, \mu$, and τ branching ratios. Assuming lepton universality, this implies (Patrignani, 2016, LEPEWWG website),

$$\sum |V_{mn}|^2 = 2.002(27), \tag{8.223}$$

consistent with the expectation of 2 from CKM universality. Using the measured values for the other elements, one obtains $|V_{cs}| = 0.976(14)$. Deep-inelastic neutrino scattering only determines $S|V_{cs}|^2$, where S is the s quark momentum fraction.

$|V_{ub}|, |V_{cb}|$: The values $|V_{ub}| = 0.00409(39)$ and $|V_{cb}| = 0.0405(15)$ are obtained from inclusive and exclusive semi-leptonic B decays. As described in Section 7.2.7 there is tension between the exclusive and inclusive determinations. There are also somewhat less precise measurements of $|V_{ub}|$ from $B \to \tau\nu$ and of $|V_{ub}|/|V_{cb}|$ from semi-leptonic $\Lambda_b = udb$ decays.

$|V_{td}|, |V_{ts}|, |V_{tb}|$: $|V_{td}|$ and $|V_{ts}|$ are mainly constrained by the mass differences $\Delta m_{d,s}$ obtained from $B^0 - \bar{B}^0$ and $B_s^0 - \bar{B}_s^0$ mixing, which are driven by box diagrams involving two W's, similar to Figure 7.3 (Section 8.6.4). Assuming $|V_{tb}| \sim 1$, one obtains $|V_{td}| = 0.0082(6)$ and $|V_{ts}| = 0.0400(27)$, with the uncertainties dominated by the unquenched lattice calculations used to obtain the matrix elements. These uncertainties are reduced in the ratio $\Delta m_d/\Delta m_s$, implying the more precise $|V_{td}/V_{ts}| = 0.216(11)$. Other constraints from radiative decays such as $B \to X_s\gamma$ or $B_s \to \mu^+\mu^-$ are described in the reviews. $|V_{tb}|$ is measured directly in single top production at the Tevatron and LHC, either through the exchange of a virtual W in the s or t channel in $q\bar{q}$ or qq scattering, or by the associated production $Gb \to tW$. The CDF, D0, ATLAS, and CMS results combine to give $|V_{tb}| = 1.009(31)$.

Combining these results one obtains

$$\underbrace{|V_{ud}|^2}_{\beta} + \underbrace{|V_{us}|^2}_{K_{\ell 3}} + \underbrace{|V_{ub}|^2}_{\text{negligible}} = 0.9996(5), \tag{8.224}$$

in impressive agreement with weak universality.[38] This is especially remarkable because it

[38]For many years there was an apparent 2-3σ deviation, which was often (wrongly) assumed to be due to some problem with the radiative corrections to superallowed β decays. The problem was resolved by a new generation of high precision $K_{\ell 3}$ measurements that yielded considerably higher values for $f_+(0)|V_{us}|$ than the earlier ones (which may not have treated the radiative corrections correctly). For discussions, see (Czarnecki et al., 2004; Antonelli et al., 2010a; Patrignani, 2016).

confirms the theory at the loop level: without radiative corrections applied to μ, β, and K decays the sum would have been ~ 1.04, and those corrections are only finite and meaningful in the full $SU(2) \times U(1)$ gauge theory. The agreement also strongly constrains such new physics as a heavy $W_R^\pm$ coupling to right-handed currents, $W - W_R$ mixing, and mixing between ordinary and heavy fermions.

Similarly, the universality sums for the second row and for the first two columns are

$$\sum_{n=d,s,b} |V_{cn}|^2 = 1.040(32) \quad [1.002(27)],$$
$$\sum_{m=u,c,t} |V_{md}|^2 = 0.9975(22), \qquad \sum_{m=u,c,t} |V_{ms}|^2 = 1.042(32), \tag{8.225}$$

where the first [second] value for $\sum_n |V_{cn}|^2$ utilizes D decays and the LEP 2 result in (8.223), respectively. Since all of these results are consistent with universality, one can fit to the CKM parameters (i.e., assume unitarity) to obtain the values in (8.215). In particular, this yields $|V_{tb}| = 0.99915(5)$, which is much more precise than the direct measurement from single top production.

8.6.2 CP Violation and the Unitarity Triangle

Universality tests whether the diagonal elements of $V_q^\dagger V_q$ and $V_q V_q^\dagger$ are equal to unity. The unitarity triangle tests involve whether the off-diagonal elements, such as $\left(V_q^\dagger V_q\right)_{31}$ in (8.221), vanish. The name derives from the fact that each term in the expression for $\left(V_q^\dagger V_q\right)_{31}$ is a complex number, which can be thought of as a vector in the complex plane. Unitarity implies that they should sum to zero, i.e., form a triangle.

There are six unitarity triangle tests, but we will concentrate on $\left(V_q^\dagger V_q\right)_{31}$, which is illustrated in Figure 8.41. The goal is to overconstrain the triangle by redundant tests, to check whether unitarity really is satisfied and to determine the parameters. In particular, the lengths of the sides can be determined from CP conserving rates, such as the $b \to u$ decay rate or $B^0 - \bar{B}^0$ mixing, as described above, while the angles can be determined independently by CP-violating effects. Any deviation would signal the presence of new physics, such as new fermions, new sources of CP-violation (e.g., loops involving supersymmetric particles), or new flavor changing neutral or charged current interactions (e.g., from extended Higgs sectors, a family-non-universal heavy Z', or a heavy $W_R^\pm$).

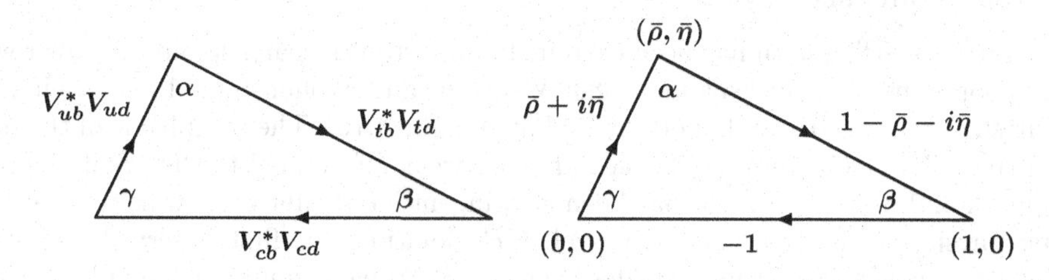

Figure 8.41 Left: the unitarity triangle for $\left(V_q^\dagger V_q\right)_{31} = 0$. Right: the triangle scaled by $V_{cb}^* V_{cd}$, so that the vertices are at $(0,0)$, $(1,0)$, and $(\bar\rho, \bar\eta)$.

In the Wolfenstein parametrization,

$$\begin{aligned}
\left(V_q^\dagger V_q\right)_{31} &= V_{ub}^* V_{ud} + V_{cb}^* V_{cd} + V_{tb}^* V_{td} \\
&\sim \left(1 - \frac{\lambda^2}{2}\right) A\lambda^3 \left(\rho + i\eta\right) - A\lambda^3 + A\lambda^3 \left(1 - \rho - i\eta\right) \\
&\sim A\lambda^3 \left(\bar{\rho} + i\bar{\eta}\right) - A\lambda^3 + A\lambda^3 \left(1 - \bar{\rho} - i\bar{\eta}\right) = 0.
\end{aligned} \tag{8.226}$$

$|V_{cb}^* V_{cd}| \sim A\lambda^3$ is known accurately, so it can be divided out,

$$\frac{\left(V^\dagger V\right)_{31}}{A\lambda^3} \sim \underbrace{\bar{\rho} + i\bar{\eta}}_{V_{ub}^* V_{ud}/A\lambda^3} \qquad \underbrace{-1}_{V_{cb}^* V_{cd}/A\lambda^3} \quad \underbrace{+1 - \bar{\rho} - i\bar{\eta}}_{V_{tb}^* V_{td}/A\lambda^3}. \tag{8.227}$$

The rescaled unitarity triangle is therefore the sum of the vectors $\bar{\rho} + i\bar{\eta}$, -1, and $1 - \bar{\rho} - i\bar{\eta}$. The sides are of length $\sqrt{\bar{\rho}^2 + \bar{\eta}^2}$, 1, and $\sqrt{(1-\bar{\rho})^2 + \bar{\eta}^2}$, respectively, while the angles[39] α, β, and γ defined in Figure 8.41 satisfy

$$\begin{aligned}
\beta &= \arg\left(-\frac{V_{cb}^* V_{cd}}{V_{tb}^* V_{td}}\right) \sim -\arg V_{td}, \qquad & \alpha &= \arg\left(-\frac{V_{tb}^* V_{td}}{V_{ub}^* V_{ud}}\right) \sim \arg(-V_{td} V_{ub}) \\
\gamma &= \arg\left(-\frac{V_{ub}^* V_{ud}}{V_{cb}^* V_{cd}}\right) \sim -\arg V_{ub} \sim \delta, \qquad & \beta_s &= \arg\left(-\frac{V_{tb}^* V_{ts}}{V_{cb}^* V_{cs}}\right) \sim \arg(-V_{ts})
\end{aligned} \tag{8.228}$$

γ is essentially the same as the CKM phase δ in the parametrization in (8.214), while β_s is the analog of β for $\left(V_q^\dagger V_q\right)_{32}$. The Jarlskog invariant J in (8.219) is twice the area of the original (unrescaled) version of the triangle. In fact, all six unitarity triangles can be shown to have the same area.

The consistency of this picture can be tested by determining the points $(\bar{\rho}, \bar{\eta})$ by redundant means to see whether they agree. One constraint follows from

$$\left|\frac{V_{ub}^* V_{ud}}{V_{cb}^* V_{cd}}\right| \sim \sqrt{\bar{\rho}^2 + \bar{\eta}^2} \sim 0.45(5), \tag{8.229}$$

which yields the circular annulus centered at the origin marked $|V_{ub}|$ in Figure 8.42. To proceed we need further inputs from the K and B systems.

8.6.3 The Neutral Kaon System

The neutral $K^0 - \bar{K}^0$ system has been extremely important in particle physics (for reviews, see, e.g., Commins and Bucksbaum, 1983; Winstein and Wolfenstein, 1993; Kleinknecht, 2003; Cirigliano et al., 2012; Schubert, 2015; Patrignani, 2016). The magnitude of the mixing between the K^0 and $\bar{K}^0$, induced by second-order weak effects, led to the prediction of the charm quark and of its mass, and has been a stringent constraint on new sources of flavor-violating physics at the tree and loop level. CP violation was first observed and studied in the neutral kaon system, and provides even more stringent limits on some kinds of new physics.

The states K^0 and $\bar{K}^0$ defined in Section 3.2.3 are the isospin partners of the K^+ and K^-, respectively, and carry strangeness $+1$ and -1. In the quark model,

$$|K^0\rangle = |d\bar{s}\rangle, \qquad |\bar{K}^0\rangle = |s\bar{d}\rangle. \tag{8.230}$$

[39] The alternative notation $\phi_1 = \beta$, $\phi_2 = \alpha$, and $\phi_3 = \gamma$ is used frequently, especially by the Belle collaboration.

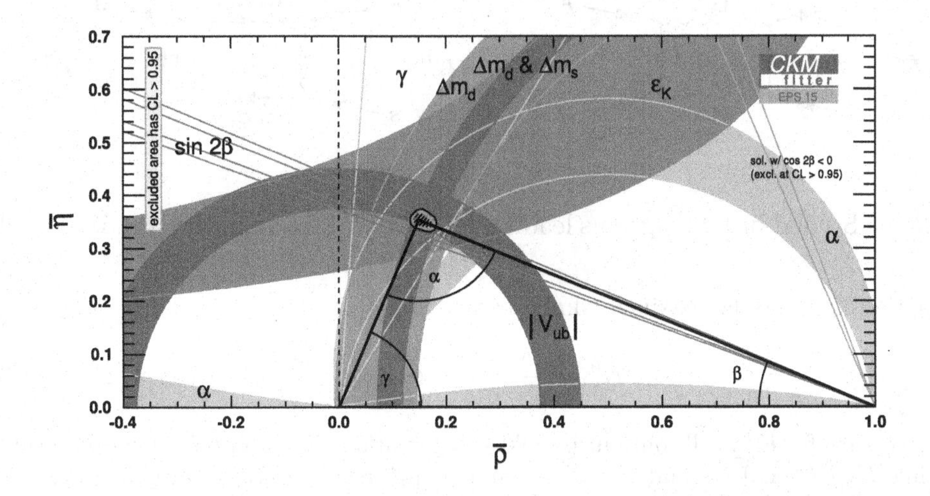

Figure 8.42 The unitarity triangle, showing the consistency of various CP-conserving and violating observables from the K and B systems. Plot courtesy of the CKMfitter group (Charles et al., 2005, `http://ckmfitter.in2p3.fr`), with kind permission of *The European Physical Journal* (EPJ).

Under CP

$$CP\,|K^0\rangle = \eta_K |\bar{K}^0\rangle, \qquad CP\,|\bar{K}^0\rangle = \eta_K^* |K^0\rangle, \tag{8.231}$$

where η_K is a phase. As discussed in Section 2.10 one can always perform a field redefinition on the kaons (or on the quarks) to change η_K; we use this freedom[40] to choose $\eta_K = -1$. The strong interactions conserve strangeness and are CP invariant, so K^0 and $\bar{K}^0$ are the relevant states for describing strong interaction transitions. For example, $\pi^- p \to K^0 \Lambda$ is allowed, where $|\Lambda\rangle = |sdu\rangle$ is the $S = -1$ isoscalar hyperon, while $\pi^- p \to \bar{K}^0 \Lambda$ is forbidden. The two states are degenerate by CPT with mass $m_{K^0} = 497.6$ MeV. This is 3.9 MeV larger than $m_{K^\pm}$. Similar to the $n - p$ mass difference, this is of opposite sign from the expected electromagnetic contribution, and is due to the quark mass difference $m_d > m_u$.

However, the weak charged current interactions can violate strangeness and lead to $K^0 - \bar{K}^0$ mixing at second order (in G_F). This implies that the $K^0 - \bar{K}^0$ system is described by a 2×2 mass matrix,

$$M \sim \begin{pmatrix} M_{K^0} & M_{K\bar{K}} \\ M_{K\bar{K}}^* & M_{\bar{K}^0} \end{pmatrix}, \tag{8.232}$$

where $M_{K^0} = M_{\bar{K}^0}$ ($\sim m_{K^0}$) by CPT invariance.[41] $M_{K\bar{K}}$ is the weak mixing term, generated by the diagrams in Figure 8.43. It is tiny, but its effects are important due to the degeneracy of the diagonal terms.

If we first ignore the third family then $M_{K\bar{K}}$ is real (i.e., CP is conserved), and the

[40]Other frequently used conventions are $\eta_K = +1$, and the Wu-Yang convention (Wu and Yang, 1964), which makes the amplitude A_0 defined in (8.283) real. The latter is somewhat awkward in the standard model. The $\eta_K = -1$ convention follows from (8.230) if we define the d and s quarks to have the same intrinsic C and P phases (Problem 8.26).

[41]See (Commins and Bucksbaum, 1983; Sozzi, 2008; Schubert, 2015) for a general discussion allowing for the possibility of CPT violation.

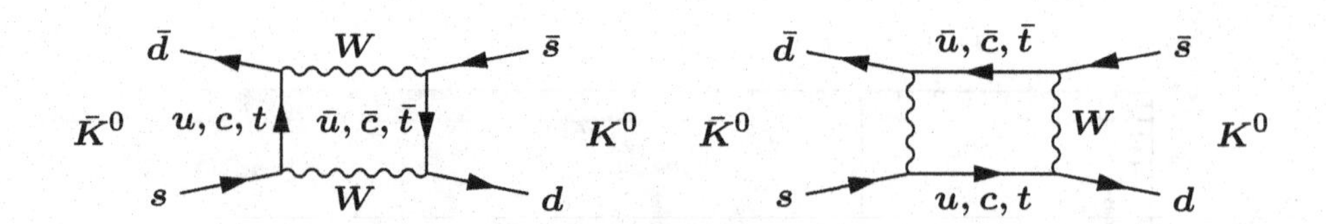

Figure 8.43 Second-order diagrams leading to $K^0-\bar{K}^0$ mixing in the standard model.

eigenstates of M are the maximal admixtures

$$K^0_{1,2} \equiv \frac{K^0 \mp \bar{K}^0}{\sqrt{2}} \simeq K_{S,L}. \tag{8.233}$$

The expression for $K^0_{1,2}$ will continue to hold as a definition even after CP-violation is turned on, while $K_{S,L}$ (S and L stand for "short" and "long," respectively) will represent the mass eigenstates in the more general case. K^0_1 and K^0_2 are CP eigenstates, with eigenvalues $+1$ and -1, respectively. The states $|\pi^+\pi^-\rangle$ and $|\pi^0\pi^0\rangle$ are also CP eigenstates,

$$CP|\pi^+\pi^-\rangle = +|\pi^+\pi^-\rangle, \qquad CP|\pi^0\pi^0\rangle = (-1)^L|\pi^0\pi^0\rangle. \tag{8.234}$$

But L is even for $|\pi^0\pi^0\rangle$ by Bose statistics, so CP conservation allows $K^0_1 = K_S \to 2\pi$ but forbids $K^0_2 = K_L \to 2\pi$. Similarly,

$$CP|\pi^+\pi^-\pi^0\rangle = (-1)^{I_{3\pi}}|\pi^+\pi^-\pi^0\rangle, \qquad CP|\pi^0\pi^0\pi^0\rangle = -|\pi^0\pi^0\pi^0\rangle, \tag{8.235}$$

where $I_{3\pi} = 0, 1, 2, 3$ is the total isospin of the $\pi^+\pi^-\pi^0$ system and we have assumed that the total angular momentum is zero. $I_{3\pi} = 0$ or 2 can only occur when there are internal orbital angular momenta for the 3π system, leading to a strong centrifugal suppression. Therefore, $K_L \to 3\pi$ is allowed, while $K_S \to 3\pi$ is strongly suppressed. Since the phase space for the 2π decay modes is much larger than for 3π, one expects a much shorter lifetime for K_S than for K_L, motivating the terminology. This picture is indeed approximately valid: The lifetimes τ_{K_S} and τ_{K_L} are, respectively, $0.8954(4) \times 10^{-10}$ s $\sim (2.7 \text{ cm })^{-1}$ and $5.12(2) \times 10^{-8}$ s $\sim (15.3 \text{ m })^{-1}$, with their non-leptonic decays almost exclusively 2π and 3π. The $K_{\ell 3}$ modes are competitive for K_L, but are only of $\mathcal{O}(10^{-3})$ for K_S, as can be seen in Table 8.7. However, the assumption that CP is absolutely conserved was invalidated by the observation of $K_L \to 2\pi$ decays at the 10^{-3} level (Christenson et al., 1964). We will return to the subject after discussing the calculation of $K^0 - \bar{K}^0$ mixing in the standard model.

Calculation of Δm_K

The mass difference between the eigenstates in (8.233) is

$$\Delta m_K = m_{K_2} - m_{K_1} = 2M_{K\bar{K}}. \tag{8.236}$$

We will see below that it is predicted to be

$$\Delta m_K \sim \frac{G_F^2}{6\pi^2} m_c^2 |V_{cd}|^2 |V_{cs}|^2 f_K^2 m_K B_K, \tag{8.237}$$

up to third family and QCD corrections, where $f_K \sim 1.2 f_\pi \sim 160$ MeV is the kaon decay constant defined in (7.78), and $B_K = \mathcal{O}(1)$ is associated with the matrix element of the

TABLE 8.7 Principal branching ratios for the neutral K decays.[a]

$K_S \to \pi^0\pi^0$	$(K_{2\pi})$	30.7%	
$\to \pi^+\pi^-$	$(K_{2\pi})$	69.2%	
$\to \pi^\mp e^\pm \nu_e(\bar{\nu}_e)$	(K_{e3})	7.0×10^{-4}	
$K_L \to \pi^\mp e^\pm \nu_e(\bar{\nu}_e)$	(K_{e3})	40.6%	
$\to \pi^\mp \mu^\pm \nu_\mu(\bar{\nu}_\mu)$	$(K_{\mu 3})$	27.0%	
$\to \pi^+\pi^-\pi^0$	$(K_{3\pi})$	12.5%	
$\to 3\pi^0$	$(K_{3\pi})$	19.5%	
$\to \pi^+\pi^-$		2.0×10^{-3}	CP-violating
$\to \pi^0\pi^0$		8.6×10^{-4}	CP-violating

[a]The charged K branching ratios are given in Table 7.2.

effective operator corresponding to Figure 8.43. Δm_K is much too small to be measured kinematically, but it can be determined indirectly by the regeneration technique described below to have the extremely small value $3.484(6) \times 10^{-6}$ eV. Using (8.237) with $B_K = 1$ one obtains the prediction $m_c \sim 1.5$ GeV (Gaillard and Lee, 1974b), in reasonable agreement with the value in (5.124) on page 194. In fact, the QCD corrections are non-negligible and there is still a non-trivial uncertainty from B_K, but it is clear that the SM prediction is qualitatively correct. Without the GIM mechanism (Glashow et al., 1970) the box diagram contribution would have been orders of magnitude larger, and there would also have been a tree contribution from the flavor changing $\bar{s}dZ$ vertices in (8.78). Assuming that no new physics contributions to Δm_K can be much larger than the experimental value also significantly constrains new tree-level physics leading to flavor changing couplings and certain types of new box diagram effects, as will be further considered in Section 8.6.6.

Now, let us derive (8.237). Box diagrams similar to Figure 8.43 will lead to an effective $|\Delta S| = 2$ operator

$$\mathcal{L}_{eff}^{|\Delta S|=2} = C^{|\Delta S|=2}\, \bar{d}\gamma^\mu(1-\gamma^5)s\, \bar{d}\gamma_\mu(1-\gamma^5)s + h.c. \tag{8.238}$$

We will work in the 't Hooft-Feynman gauge, which means that we must also include the box diagrams in which one or both of the W's are replaced by the Goldstone bosons $w^\pm$, as described in Section 4.4. We start with the schematic expression

$$i\mathcal{L}_{eff}^{|\Delta S|=2} = \frac{1}{4!}\mathcal{T}\left[(i\mathcal{L})^4\right], \tag{8.239}$$

where $\mathcal{L}$ includes both the WCC couplings in (8.55) and the Yukawa couplings in (8.91):

$$\begin{aligned} \mathcal{L} = &-\frac{g}{2\sqrt{2}}\Big\{\left(\bar{d}\gamma^\mu(1-\gamma^5)V_q^\dagger u\, W_\mu^- + \bar{u}\gamma^\mu(1-\gamma^5)V_q d\, W_\mu^+\right) \\ &+ \bar{d}\left[V_q^\dagger \frac{M_D^u}{M_W}(1+\gamma^5)\right] u\, w^- + \bar{u}\left[\frac{M_D^u}{M_W}V_q(1-\gamma^5)\right] d\, w^+\Big\}. \end{aligned} \tag{8.240}$$

We have neglected $m_{d,s}$ (the b terms are irrelevant) and have used $\sqrt{2}/\nu = g/\sqrt{2}M_W$. Expanding $\mathcal{L}^4$ and contracting the appropriate pairs of W^+W^-, w^+w^-, and $u_n\bar{u}_n$ fields, using the W and w propagators in (4.68) and (4.63), and neglecting external momenta,

$$\begin{aligned}\mathcal{L}_{eff}^{|\Delta S|=2} =& -2i\left(\frac{g}{2\sqrt{2}}\right)^4 \sum_{m,n=u,c,t} \lambda_m^*\lambda_n^* \int \frac{d^4k}{(2\pi)^4}\left(\frac{1}{k^2-M_W^2}\right)^2\left(\frac{1}{k^2-m_m^2}\right)\left(\frac{1}{k^2-m_n^2}\right)\\ &\times\left[\bar{d}\gamma^\mu \not{k}\gamma^\nu(1-\gamma^5)s\,\bar{d}\gamma_\nu\not{k}\gamma_\mu(1-\gamma^5)s - 2x_m x_n M_W^2 \bar{d}\gamma^\mu(1-\gamma^5)s\,\bar{d}\gamma_\mu(1-\gamma^5)s\right.\\ &\left.+x_m x_n \bar{d}\not{k}(1-\gamma^5)s\,\bar{d}\not{k}(1-\gamma^5)s\right] + h.c.,\end{aligned} \tag{8.241}$$

where

$$x_n \equiv \frac{m_n^2}{M_W^2},\quad n=u,c,t \tag{8.242}$$

and

$$\lambda_n \equiv V_{nd}V_{ns}^*, \qquad \sum_{n=u,c,t}\lambda_n = 0. \tag{8.243}$$

The three terms are associated with the WW, Ww, and ww diagrams, respectively. In the first and last terms the $\not{k}$ parts of the numerators of the fermion propagators survive, while in the Ww term the fermion mass term survives. The coefficient reflects the fact that there are $4!/2$ equivalent terms for the WW and ww diagrams, $4!$ for the Ww, and that $(1-\gamma^5)^2 = 2(1-\gamma^5)$. One can easily show that

$$\bar{d}\gamma^\mu\not{k}\gamma^\nu(1-\gamma^5)s\,\bar{d}\gamma_\nu\not{k}\gamma_\mu(1-\gamma^5)s = 4\bar{d}\not{k}(1-\gamma^5)s\,\bar{d}\not{k}(1-\gamma^5)s \tag{8.244}$$

by using the Fierz identities in (2.215) or by using the identity in Problem 2.11. Furthermore, one can replace $k_\rho k_\sigma$ by $g_{\rho\sigma}k^2/4$, since there is no other four-vector in the d^4k integral. Finally, one can employ the Wick rotation and angular factor as in (E.15) to obtain

$$C^{|\Delta S|=2} = -\frac{G_F^2 M_W^2}{16\pi^2}\sum_{m,n=u,c,t}\lambda_m^*\lambda_n^* F^{mn}, \tag{8.245}$$

where

$$\begin{aligned}F^{mn} &\equiv \left(1+\frac{x_m x_n}{4}\right)B_2^{mn} + 2x_m x_n B_1^{mn}\\ B_k^{mn} &\equiv \int_0^\infty \frac{z^k dz}{(z+1)^2(z+x_m)(z+x_n)}.\end{aligned} \tag{8.246}$$

The integrals are elementary, and yield

$$B_k^{mn} = \frac{A_k(x_m)-A_k(x_n)}{x_m - x_n}, \tag{8.247}$$

with

$$A_2(x) = \frac{1}{1-x} + \frac{x^2\ln x}{(1-x)^2}, \qquad A_1(x) = -\frac{1}{1-x} - \frac{x\ln x}{(1-x)^2}. \tag{8.248}$$

To verify (8.237), let us neglect the mixing with the third family, so that

$$\lambda_u = -\lambda_c = \cos\theta_c \sin\theta_c, \tag{8.249}$$

and

$$C^{|\Delta S|=2} = -\frac{G_F^2 M_W^2}{16\pi^2}\lambda_c^2\left[F^{uu} + F^{cc} - 2F^{uc}\right]. \tag{8.250}$$

x_c is very small ($\sim 3.5 \times 10^{-4}$ for $m_c \sim 1.5$ GeV), and x_u is negligible for either the current or dynamical mass. To leading order in x_c,

$$F^{uu} \sim 1, \qquad F^{cc} \sim 1 + 3x_c + 2x_c \ln x_c, \qquad F^{uc} \sim 1 + x_c + x_c \ln x_c, \tag{8.251}$$

so the quantity[42] in square brackets in (8.250) is x_c.

Now consider the matrix element

$$M_{K\bar{K}} = -\frac{\langle K^0 | \mathcal{L}_{eff}^{|\Delta S|=2} | \bar{K}^0 \rangle}{2m_K}. \tag{8.252}$$

The $1/(2m_K)$ follows from our covariant state normalization, noting that

$$\langle \vec{p}\,' | H | \vec{p} \rangle = E_p \langle \vec{p}\,' | \vec{p} \rangle = (2\pi)^3 \delta^3(\vec{p}\,' - \vec{p}) \langle \vec{p}\,' | \mathcal{H}(0) | \vec{p} \rangle, \tag{8.253}$$

and using (2.3) and translation invariance (1.15). To estimate the matrix element, we will use the *vacuum saturation approximation*, in which one inserts the vacuum state between the $\bar{d}s$ operators in all possible ways, i.e.,

$$\begin{aligned}
&\langle K^0 | \bar{d}\gamma^\mu (1-\gamma^5) s \, \bar{d}\gamma_\mu (1-\gamma^5) s | \bar{K}^0 \rangle_{vac} \\
=&2 \langle K^0 | \bar{d}^\alpha \gamma^\mu (1-\gamma^5) s_\alpha | 0 \rangle \, \langle 0 | \bar{d}^\beta \gamma^\mu (1-\gamma^5) s_\beta | \bar{K}^0 \rangle \\
&+ 2 \langle K^0 | \bar{d}^\alpha \gamma^\mu (1-\gamma^5) s_\beta | 0 \rangle \, \langle 0 | \bar{d}^\beta \gamma^\mu (1-\gamma^5) s_\alpha | \bar{K}^0 \rangle,
\end{aligned} \tag{8.254}$$

where we have reintroduced the color indices α and β to keep track of them: the second term, which corresponds to the second diagram in Figure 8.43, is obtained by performing a Fierz transformation before inserting the vacuum. The 2 comes from interchanging the two $\bar{d}s$ factors. The matrix elements of the vector currents vanish by reflection invariance, and the axial matrix elements are given in the $SU(3)$ limit by (7.76) on page 241, where $\bar{d}\gamma_\mu\gamma^5 s = A^6_\mu + iA^7_\mu = (\bar{s}\gamma_\mu\gamma^5 d)^\dagger$. Our phase choice $\eta_K = -1$ in (8.231) corresponds to

$$|K^0\rangle = \frac{1}{\sqrt{2}} |6 + i7\rangle, \qquad |\bar{K}^0\rangle = \frac{1}{\sqrt{2}} |6 - i7\rangle, \tag{8.255}$$

implying

$$\langle 0 | \bar{d}^\beta \gamma^\mu (1-\gamma^5) s_\beta | \bar{K}^0(q) \rangle = \langle 0 | \bar{s}^\beta \gamma^\mu (1-\gamma^5) d_\beta | K^0(q) \rangle = i f q^\mu, \tag{8.256}$$

with $f = f_\pi$. In fact, $SU(3)$ breaking is important, but from isospin one has that f should be identified with the f_K from $K^- \to \mu^- \bar{\nu}_\mu$ in (7.78). Similarly,

$$\langle 0 | \bar{d}^\alpha \gamma^\mu (1-\gamma^5) s_\beta | \bar{K}^0(q) \rangle = \frac{\delta^\alpha_\beta}{3} i f_K q^\mu, \tag{8.257}$$

so that[43]

$$\langle K^0 | \bar{d}\gamma^\mu (1-\gamma^5) s \, \bar{d}\gamma_\mu (1-\gamma^5) s | \bar{K}^0 \rangle_{vac} = \frac{8}{3} f_K^2 m_K^2. \tag{8.258}$$

Vacuum saturation is *not* expected to be a good approximation, but is useful as an order of magnitude estimate. It is therefore conventional to define a factor B_K to describe the departure from vacuum saturation, i.e,

$$\langle K^0 | \bar{d}\gamma^\mu (1-\gamma^5) s \, \bar{d}\gamma_\mu (1-\gamma^5) s | \bar{K}^0 \rangle \equiv \frac{8}{3} f_K^2 m_K^2 B_K. \tag{8.259}$$

[42]Without the c quark, the corresponding factor would be $F^{uu} \sim 1$, about 3000 times larger, as stated below (8.237).

[43]Older references often quote a value 3/2 larger. This was based on a pre-QCD calculation involving only one color of quark.

Combining (8.236), (8.250), (8.252), and (8.258), one reproduces (8.237).

QCD corrections (e.g., Buchalla et al., 1996) to $C^{|\Delta S|=2}$, which also involve additional penguin diagrams, are significant. Including the t quark,

$$C^{|\Delta S|=2} = -\frac{G_F^2 M_W^2}{16\pi^2}\left[\lambda_c^{*2}\eta_1 S_0(x_c) + \lambda_t^{*2}\eta_2 S_0(x_t) + 2\lambda_c^*\lambda_t^*\eta_3 S_0(x_c, x_t)\right]\mathcal{F}(\mu), \tag{8.260}$$

where $\eta_{1,2,3} = 1.87(76), 0.5765(65), 0.496(47)$ are short distance QCD corrections,[44] defined to be independent of the renormalization scale μ, and $\mathcal{F}(\mu)$ contains scale, scheme-dependent, and higher-order factors. B_K, defined in (8.259), also becomes μ-dependent, but

$$\hat{B}_K \equiv B_K(\mu)\mathcal{F}(\mu) \tag{8.261}$$

is an observable that is scale independent. $\hat{B}_K$ may involve both short distance and long distance contributions, where the latter are associated with virtual hadrons rather than quarks. Lattice calculations (Aoki et al., 2017) yield $\hat{B}_K = 0.763(10)$. The S_0 functions in (8.260) are obtained from (8.245) after eliminating $\lambda_u = -(\lambda_c + \lambda_t)$. After a somewhat tedious calculation, one finds

$$\begin{aligned}
S_0(x_c) &= 2.48 \times 10^{-4} \\
S_0(x_t) &= \frac{4x_t - 11x_t^2 + x_t^3}{4(1-x_t)^2} - \frac{3x_t^3 \ln x_t}{2(1-x_t)^3} \sim 2.33 \\
S_0(x_c, x_t) &= x_c\left[\ln\frac{x_t}{x_c} - \frac{3x_t}{4(1-x_t)} - \frac{3x_t^2 \ln x_t}{4(1-x_t)^2}\right] \sim 2.20 \times 10^{-3}
\end{aligned} \tag{8.262}$$

to leading order in x_c (Inami and Lim, 1981). The numerical values are for the $\overline{MS}$ masses $\hat{m}_c(\hat{m}_c) = 1.27$ GeV and $\hat{m}_t(\hat{m}_t) = 164.1$ GeV (Patrignani, 2016). Combining the various results one predicts $\Delta m_K \sim 3.08 \times 10^{-6}$ eV, close to the experimental value $3.484(6) \times 10^{-6}$ eV. The difference is presumably due to (difficult to estimate) long distance contributions not described by the box diagrams in Figure 8.43.

Complex phases in the λ_n's lead to CP violation. This requires not only the existence of three families, but that all three are relevant to the process. Otherwise, the factors could be made real by a redefinition of the fields.

$K_{S,L}$ *Decays, Oscillations, and Regeneration*

We saw following (8.235) that, assuming CP conservation, K_S decays rapidly to 2π, while $K_L \to 3\pi$ and to semi-leptonic modes with a lifetime nearly three orders of magnitude longer. However, in 1964 the decays $K_L \to 2\pi$ were observed, suggesting a tiny violation of CP. To see how this was done, let us consider an idealized experiment in which one starts with a pure K^0 beam, which could be produced for example by the strong interaction process $\pi^- p \to K^0\Lambda$ in a target. In particular, consider a single particle state $|\psi(\tau)\rangle$ that is initially a K^0 with definite velocity β,

$$|\psi(0)\rangle = |K^0\rangle = \frac{1}{\sqrt{2}}\left(|K_S\rangle + |K_L\rangle\right). \tag{8.263}$$

[44] The quoted values include NNLO or NLO corrections (Brod and Gorbahn, 2012, and references therein).

If one could ignore the decays, then at a later proper time $\tau = d/\beta\gamma$, where d is the distance travelled, the state would have evolved to

$$\begin{aligned}|\psi(\tau)\rangle &= \frac{1}{\sqrt{2}}\left[|K_S\rangle e^{-im_s\tau} + |K_L\rangle e^{-im_L\tau}\right] \\ &= \left[\cos\left(\frac{\Delta m_K \tau}{2}\right)|K^0\rangle - i\sin\left(\frac{\Delta m_K \tau}{2}\right)|\bar{K}^0\rangle\right]e^{-im_{K^0}\tau}.\end{aligned} \tag{8.264}$$

That is, there is a probability $\sin^2 \frac{\Delta m_K \tau}{2}$ that the K^0 will have *oscillated* into a $\bar{K}^0$. This is of course an example of the two-state problem familiar in quantum mechanics, which is itself reminiscent of the classical coupled oscillator system. We will encounter it again for $B^0 - \bar{B}^0$ and neutrino oscillations. However, in the neutral kaon system the oscillation time $2\pi/\Delta m_K \sim 1.2 \times 10^{-9}$ s is long compared to τ_{K_S}, so we must take the decays into account. This can be done by replacing (8.264) by

$$|\psi(\tau)\rangle = \frac{1}{\sqrt{2}}\left[|K_S\rangle e^{-im_s\tau} e^{\frac{-\Gamma_s\tau}{2}} + |K_L\rangle e^{-im_L\tau} e^{\frac{-\Gamma_L\tau}{2}}\right]. \tag{8.265}$$

The factors $\exp(-\Gamma_{S,L}\tau/2)$ represent the depletion of the beam into the 2π and 3π (or semi-leptonic) channels, respectively, with rates $\Gamma_{S,L} = 1/\tau_{K_{S,L}}$. (These factors could be incorporated in a more elegant manner by considering the multi-channel system including the decay states or by introducing a density matrix, but the simple exponential factors are adequate for our purposes.) Clearly, one observes 2π decays near the source from the K_S component. Far away, for $\tau_{K_S} \ll \tau \ll \tau_{K_L}$, one has an essentially pure K_L beam and one expects to observe only 3π and semi-leptonic decays.

Before turning to CP violation, let us consider the kaon regeneration technique that can be used to measure Δm_K. The basic idea is familiar from measurement theory in quantum mechanics, or from the use of polarizers in optics. Suppose one places a piece of matter, known as a regenerator, a distance d_1 downstream from the source. If the corresponding τ_1 is much larger than τ_{K_S} the beam entering the regenerator is a pure K_L, i.e.,

$$|\psi(\tau_1)\rangle = |K_L\rangle = \frac{|K^0\rangle + |\bar{K}^0\rangle}{\sqrt{2}}, \tag{8.266}$$

where we have renormalized the coefficient to unity. The two components can scatter or be absorbed in the regenerator with different amplitudes, so the state emerging on the other side at $\tau_2 = \tau_1 + \epsilon$ is

$$|\psi(\tau_2)\rangle = a|K^0\rangle + b|\bar{K}^0\rangle = \frac{a-b}{\sqrt{2}}|K_S\rangle + \frac{a+b}{\sqrt{2}}|K_L\rangle, \tag{8.267}$$

where we can again renormalize so that $|a|^2 + |b|^2 = 1$. For $a \neq b$ the K_S component has been *regenerated.* One typically expects $|b| \ll |a|$ since the $\bar{K}^0$ component can be strongly absorbed by $\bar{K}^0 p \to \pi^+\Lambda$ in the regenerator, while there is no analogous reaction for K^0. In the extreme case of $b = 0$, a pure K^0 emerges.

Now, suppose the regenerator is placed close to the original source, so that $|\psi(\tau_1)\rangle$ has a non-negligible K_S component. The $K_{S,L}$ components can then interfere with each other so that the (unnormalized) state emerging from the regenerator is

$$|\psi(\tau_2)\rangle = [(a+b)\lambda + (a-b)]\,|K_S\rangle + [(a-b)\lambda + (a+b)]\,|K_L\rangle, \tag{8.268}$$

where $\lambda \equiv \exp(i\Delta m_K \tau_2 - \Gamma_S \tau_2/2)$, we have removed a common phase, have approximated

$\exp(-\Gamma_L \tau_2/2) \sim 1$, and have neglected the thickness of the regenerator. The intensity of K_S in the regenerated beam can be measured by observing the 2π decays, with a rate proportional to $|\langle K_S|\psi(\tau_2)\rangle|^2$. For $b \sim 0$ this is

$$|\langle K_S|\psi(\tau_2)\rangle|^2 \propto \left[1 + e^{-\Gamma_S \tau_2} + 2e^{-\Gamma_S \tau_2/2} \cos\left(\Delta m_K\, \tau_2\right)\right]. \tag{8.269}$$

Δm_K can then be determined by varying $d_1 \sim d_2$. In practice, regeneration experiments usually involve two regenerators, allowing Δm_K and CP-violating parameters to be determined by varying the distance between them or utilizing the energy spread of the beam.

CP Violation in K decays

In 1964, Fitch, Cronin, and collaborators observed 2π decays in a neutral kaon beam produced at the Brookhaven AGS some 300 K_S decay lengths from the source, implying the CP-violating $K_L \to 2\pi$ decays at the 2×10^{-3} level. Subsequent observations utilized the regeneration technique, which allowed the measurement of both the magnitudes and phases of the CP-violating parameters (as well as Δm_K). Since regeneration measures interferences, it also eliminated possible alternative explanations of the events, such as the emission of an unobserved third exotic particle in the decay.

There are two possible sources of the CP violation:

CP violation in the $K^0 - \bar{K}^0$ mixing (*indirect violation*). This would be induced by an imaginary part in $M_{K\bar{K}}$ defined in (8.232), so the mass eigenstates $K_{S,L}$ no longer coincide with the CP eigenstates $K^0_{1,2}$ in (8.233).

CP-violation in the decay amplitude (*direct violation*).

We now understand that both effects are present, though the indirect CP-violation is much larger. They are both understandable in the SM as being due to the CP-violating phase in the CKM matrix for three families. This phase can be large, but the observed effects are strongly suppressed by small mixing angles.

Define the CP violating parameters

$$\begin{aligned} \eta_{+-} &\equiv \frac{A(K_L \to \pi^+\pi^-)}{A(K_S \to \pi^+\pi^-)} = |\eta_{+-}|e^{i\varphi_{+-}} \sim \epsilon + \epsilon' \\ \eta_{00} &\equiv \frac{A(K_L \to \pi^0\pi^0)}{A(K_S \to \pi^0\pi^0)} = |\eta_{00}|e^{i\varphi_{00}} \sim \epsilon - 2\epsilon', \end{aligned} \tag{8.270}$$

where $A(K_{S,L} \to \pi^+\pi^-, \pi^0\pi^0)$ are the decay amplitudes. The difference between them is due to the fact that the 2π state can have isospin 0 or 2. As commented briefly in Section 7.2, the $I = 2$ amplitude is much smaller than the $I = 0$ one (the $\Delta I = \frac{1}{2}$ rule). ϵ can be generated by mixing (indirect) or by direct breaking in the $I = 0$ amplitude, though the former is more important for our phase convention. ϵ' is due to a phase difference between the $I = 0$ and 2 amplitudes; it indicates direct violation but is suppressed by the $\Delta I = \frac{1}{2}$ rule.

$|\eta_{+-}|$ and $|\eta_{00}|$ can be measured from the rates for the various 2π decays, while the phases φ_{+-} and φ_{00} can be measured by interference effects in regeneration experiments. A fit to the results yields (Patrignani, 2016)

$$\begin{aligned} |\eta_{00}| &= 2.220(11) \times 10^{-3}, \qquad \varphi_{00} = 43.52(5)^\circ \\ |\eta_{+-}| &= 2.232(11) \times 10^{-3}, \qquad \varphi_{+-} = 43.51(5)^\circ, \end{aligned} \tag{8.271}$$

and the corresponding values

$$|\epsilon| = 2.228(11) \times 10^{-3}, \qquad \Re e(\epsilon'/\epsilon) \sim \frac{1}{3}\left(1 - \left|\frac{\eta_{00}}{\eta_{+-}}\right|\right) = 1.66(23) \times 10^{-3}. \tag{8.272}$$

The $\Re e(\epsilon'/\epsilon)$ measurement was especially difficult since it involves differences between two small effects, but was eventually determined precisely by the KTEV (Fermilab) and NA48 (CERN) collaborations (Sozzi and Mannelli, 2003). CP breaking is also observed in the difference between the $K_{\ell 3}$ decay rates,

$$\delta_L \equiv \frac{\Gamma\left(K_L \to \pi^- \ell^+ \nu\right) - \Gamma\left(K_L \to \pi^+ \ell^- \bar{\nu}\right)}{\Gamma\left(K_L \to \pi^- \ell^+ \nu\right) + \Gamma\left(K_L \to \pi^+ \ell^- \bar{\nu}\right)} = 3.32(6) \times 10^{-3}. \tag{8.273}$$

To interpret these results we need to find the eigenstates $|K_{S,L}\rangle$ (in the presence of CP violation) of the operator

$$H = M - i\frac{\Gamma}{2}, \qquad H|\psi\rangle = i\frac{\partial}{\partial t}|\psi\rangle, \tag{8.274}$$

which governs the time evolution of the system. $M = M^\dagger$ is the 2×2 mass matrix defined in (8.232), while $\Gamma = \Gamma^\dagger$ is an analogous decay matrix. It is defined as

$$\Gamma_{ab} = \sum_f \rho_f M(a \to f)^* M(b \to f), \tag{8.275}$$

where ρ_f are the coefficients and phase space factors for a decay into channel f defined in (D.7). In the absence of CP violation the eigenvalues of Γ are $\Gamma_{S,L}$ and (8.265) is the solution to (8.274). CPT conservation ensures that the diagonal entries are equal, i.e, $M_{K^0} = M_{\bar{K}^0}$ and $\Gamma_{K^0} = \Gamma_{\bar{K}^0}$. It is straightforward to find the eigenstates and eigenvalues (e.g., Commins and Bucksbaum, 1983),

$$\begin{aligned} |K_S\rangle &= p|K^0\rangle - q|\bar{K}^0\rangle = \frac{|K_1^0\rangle + \tilde{\epsilon}|K_2^0\rangle}{\sqrt{1+|\tilde{\epsilon}|^2}} \\ |K_L\rangle &= p|K^0\rangle + q|\bar{K}^0\rangle = \frac{\tilde{\epsilon}|K_1^0\rangle + |K_2^0\rangle}{\sqrt{1+|\tilde{\epsilon}|^2}}, \end{aligned} \tag{8.276}$$

where

$$\frac{q}{p} \equiv \frac{1-\tilde{\epsilon}}{1+\tilde{\epsilon}} = \left[\frac{M^*_{K\bar{K}} - i\frac{\Gamma^*_{K\bar{K}}}{2}}{M_{K\bar{K}} - i\frac{\Gamma_{K\bar{K}}}{2}}\right]^{1/2}, \tag{8.277}$$

so that nonzero $\tilde{\epsilon}$ requires imaginary parts for $M_{K\bar{K}}$ and/or $\Gamma_{K\bar{K}}$. The eigenvalues are

$$m_{S,L} - i\frac{\Gamma_{S,L}}{2} = \left(M_{K^0} - i\frac{\Gamma_{K^0}}{2}\right) \mp \sqrt{\left(M^*_{K\bar{K}} - i\frac{\Gamma^*_{K\bar{K}}}{2}\right)\left(M_{K\bar{K}} - i\frac{\Gamma_{K\bar{K}}}{2}\right)}, \tag{8.278}$$

so that

$$\Delta m_K = m_L - m_S \sim 2\,\Re e\, M_{K\bar{K}}, \qquad \Gamma_S - \Gamma_L \sim -2\,\Re e\, \Gamma_{K\bar{K}}, \tag{8.279}$$

neglecting corrections of $\mathcal{O}(\tilde{\epsilon})$. $|K_L\rangle$ and $|K_S\rangle$ are not orthogonal, but satisfy

$$\langle K_L | K_S \rangle = \frac{2\,\Re e\,\tilde{\epsilon}}{1+|\tilde{\epsilon}|^2} \sim 2\,\Re e\,\tilde{\epsilon}. \tag{8.280}$$

This is due to the non-Hermitian nature of H in (8.274). From (8.277) and (8.278) one finds to $\mathcal{O}(\tilde{\epsilon})$ that

$$\tilde{\epsilon} = \frac{2\,\Im m\, M_{K\bar{K}} - i\Im m\,\Gamma_{K\bar{K}}}{\Gamma_S - \Gamma_L - 2i\Delta m_K} = e^{i\theta_\epsilon} \sin\theta_\epsilon \left(\frac{\Im m\, M_{K\bar{K}}}{\Delta m_K} - i\frac{\Im m\,\Gamma_{K\bar{K}}}{2\Delta m_K} \right), \tag{8.281}$$

where

$$\theta_\epsilon = \tan^{-1} \frac{2\Delta m_K}{\Gamma_S - \Gamma_L} = 43.52(5)^\circ \tag{8.282}$$

is the *superweak phase.*[45]

Finally, let us define the amplitudes for K^0 and $\bar{K}^0$ to decay into 2π states of definite isospin $I = 0$ or 2 and $I_3 = 0$ by

$$\begin{aligned} \langle I\ I_3 = 0|\mathcal{H}_{nl}|K^0\rangle &= A_I e^{i\delta_I} \\ \langle I\ I_3 = 0|\mathcal{H}_{nl}|\bar{K}^0\rangle &= -A_I^* e^{i\delta_I}, \end{aligned} \tag{8.283}$$

where $\mathcal{H}_{nl}$ is the Hamiltonian density for non-leptonic transitions defined in (7.21). A_I is the weak part of the amplitude. It changes sign for $\bar{K}^0$ because of (8.231) (with $\eta_K = -1$) and (8.234), and the complex conjugation reflects the CP transformation, as described in Section 2.10. $\exp(i\delta_I)$ represents the S-wave phase shift induced by the strong interaction final state interactions of the two pions, and is the same for both K^0 and $\bar{K}^0$. Empirically, $\Re e\, A_2/\Re e\, A_0 \equiv \omega \sim 0.045 \ll 1$ (the $\Delta I = \frac{1}{2}$ rule), and

$$\theta'_\epsilon \equiv \delta_2 - \delta_0 + \frac{\pi}{2} = 42.3(1.5)^\circ. \tag{8.284}$$

The S-wave $\pi^+\pi^-$ and $\pi^0\pi^0$ states are related to the isospin states by

$$\begin{aligned} \frac{1}{\sqrt{2}} \left(|\pi^+\pi^-\rangle + |\pi^-\pi^+\rangle \right) &= -\sqrt{\frac{2}{3}} |I = 0\ I_3 = 0\rangle - \frac{1}{\sqrt{3}} |I = 2\ I_3 = 0\rangle \\ |\pi^0\pi^0\rangle &= -\sqrt{\frac{1}{3}} |I = 0\ I_3 = 0\rangle + \sqrt{\frac{2}{3}} |I = 2\ I_3 = 0\rangle, \end{aligned} \tag{8.285}$$

where we have used the conventions in (3.91).

Combining (8.270), (8.276), (8.283), and (8.285), one obtains

$$\begin{aligned} \eta_{+-} &\equiv \epsilon + \frac{\epsilon'}{1 + \omega e^{i(\delta_2 - \delta_0)}/\sqrt{2}} \sim \epsilon + \epsilon' \\ \eta_{00} &\equiv \epsilon - \frac{2\epsilon'}{1 - \sqrt{2}\omega e^{i(\delta_2 - \delta_0)}} \sim \epsilon - 2\epsilon', \end{aligned} \tag{8.286}$$

where

$$\epsilon = \tilde{\epsilon} + i\frac{\Im m\, A_0}{\Re e\, A_0}, \qquad \epsilon' = \frac{1}{\sqrt{2}} \omega e^{i\theta'_\epsilon} \left[\frac{\Im m\, A_2}{\Re e\, A_2} - \frac{\Im m\, A_0}{\Re e\, A_0} \right]. \tag{8.287}$$

Under the reasonable approximation that the 2π state with $I = 0$ dominates the decays the $i\Im m\,\Gamma_{K\bar{K}}$ term in (8.281) becomes

$$i\frac{\Im m\,\Gamma_{K\bar{K}}}{\Gamma_S} \sim i\frac{\Im m\, A_0}{\Re e\, A_0}, \tag{8.288}$$

[45] In the superweak model (Wolfenstein, 1964) all CP violation is due to a new superweak interaction that only contributes to the mixing. It was finally excluded by the observation of ϵ'.

so that

$$\epsilon \sim e^{i\theta_\epsilon} \sin\theta_\epsilon \left(\frac{\Im m\, M_{K\bar{K}}}{\Delta m_K} + \frac{\Im m\, A_0}{\Re e\, A_0} \right). \tag{8.289}$$

Both ϵ and ϵ' involve phase differences, which cannot be rotated away by field redefinitions. However, the relative importance of the two contributions to ϵ *can* be changed by choosing a different convention for η_K in (8.231). For the common conventions the $i\Im m\, A_0/\Re e\, A_0$ contribution is $\lesssim 5 - 10\%$ of the total and is usually neglected. If one neglects the second term in ϵ and approximates the superweak phase θ_ϵ by $\pi/4$ one obtains the commonly used approximate form

$$\epsilon \sim \frac{e^{i\pi/4}}{\sqrt{2}} \left(\frac{\Im m\, M_{K\bar{K}}}{\Delta m_K} \right). \tag{8.290}$$

ϵ' is predicted to be small because of the ω factor. For that reason and because of the accidental near equality of θ_ϵ and θ'_ϵ one expects

$$\varphi_{00} \sim \varphi_{+-} = \theta_\epsilon = 43.51(5)^\circ, \tag{8.291}$$

in agreement with (8.271). This is often viewed as a test of CPT invariance, which was critical in the derivation. When one uses the SM expressions for the WCC, the $K_{\ell 3}$ asymmetry parameter in (8.273) is predicted to be

$$\delta_L \sim 2\Re e\, \tilde{\epsilon} = 2\Re e\, \epsilon = 3.23 \times 10^{-3}, \tag{8.292}$$

consistent with the experimental value.

The SM prediction for ϵ is obtained by combining (8.260) and the related results with (8.290) (Buras, 2005),

$$\begin{aligned} \epsilon = &\underbrace{\frac{G_F^2 M_W^2 f_K^2 m_K}{6\sqrt{2}\pi^2 \Delta m_K}}_{3.63\times 10^4} \hat{B}_K k_\epsilon e^{i\pi/4} \\ &\times \Im m\, \lambda_t \Big(\Re e\, \lambda_c \left[\eta_1 S_0(x_c) - \eta_3 S_0(x_c, x_t) \right] - \Re e\, \lambda_t\, \eta_2 S_0(x_t) \Big), \end{aligned} \tag{8.293}$$

where we have used $\sum_n \lambda_n = 0$ and the conventions for V_q in (8.214), and have neglected $\Re e\, \lambda_t$ compared to $\Re e\, \lambda_c$. The factor $k_\epsilon \sim 0.94(2)$ estimates long-distance and other corrections (Buras et al., 2010b). The last term dominates. Using the experimental value for ϵ and expressing the result in terms of the Wolfenstein parametrization,

$$\bar{\eta}\big[(1-\bar{\rho}) + 0.27(9)\big] \sim 0.43(5), \tag{8.294}$$

which corresponds approximately to the hyperbola labeled ϵ_K in Figure 8.42. The uncertainties are dominated by the A parameter and the QCD factors.

ϵ' is due to direct CP violation in the decay amplitudes. It is believed to be dominated by the *penguin* diagrams (so-named for their appearance) shown in Figure 8.44. The gluon penguin contributes only to A_0 in the $SU(2)$ limit, and may be partially responsible for the $\Delta I = 1/2$ rule. However, because of the small value of ω, the electroweak (Z and γ) penguin is also important. There is a cancellation between the diagrams, increasing the theoretical uncertainty considerably. Furthermore, there are corrections for isospin breaking, and the matrix elements are sensitive to the s quark mass (which is rather uncertain—see (5.123)), which enters in some approaches using the PCAC equations in (5.116) and (5.119). Estimates have utilized large N_c, chiral perturbation theory, lattice, and other techniques for the long distance effects, with results varying considerably, from agreement with experiment to significantly smaller values (for recent reviews and discussions, see, e.g., Cirigliano et al., 2012; Buras et al., 2015b). For this reason, the uncertainties in the SM predictions for $\Re e\,(\epsilon'/\epsilon) \propto \Im m\,\lambda_t \propto \bar{\eta}$ are too large to usefully constrain the unitarity triangle or to draw definitive conclusions about the possible need for new physics.

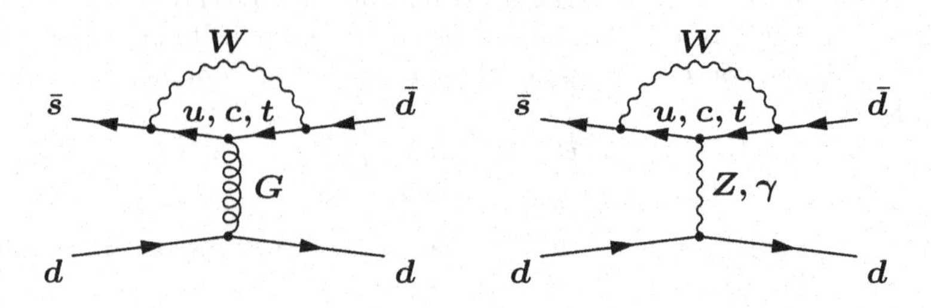

Figure 8.44 The penguin diagrams for $|\Delta S| = 1$ processes. Left: the gluon penguin, which only contributes to A_0. Right: electroweak penguin, which contributes to A_2.

8.6.4 Mixing and CP Violation in the B System

Mixing and CP violation in the neutral B system have been extensively studied, especially at LEP and the SLC; by CLEO at CESR (Cornell); at the asymmetric B factories BaBar at PEP-II (SLAC) and Belle at KEKB (KEK); by CDF and D0 at the Fermilab Tevatron; and by LHCb at the LHC (for reviews, see Artuso et al., 2009; Amhis et al., 2014; Patrignani, 2016).[46] The asymmetric B factories (Bevan et al., 2014) employed high intensity $e^\pm$ beams of unequal energy so that, e.g., $e^-e^+ \to \Upsilon(4S) \to B^0\bar{B}^0$ yields $B^0\bar{B}^0$ pairs that are boosted along the beam direction. This allows larger and more easily measured distances between the production and decay vertices, and allows the time between the decays to be measured by their spatial separation along the boost direction. The rates at the Tevatron and LHC are much higher, but are in the more difficult hadronic environment. The higher-energies at the hadron machines (and in some later Belle running) allow study of the B_s^0 system as well as the B^0.

The formalism for $B_i^0 - \bar{B}_i^0$ mixing and CP violation, with $i = d, s$ and $B_d^0 \equiv B^0$, is similar to that for the neutral kaons (see, e.g., Carter and Sanda, 1981; Buras, 2005; Nir, 2005; Patrignani, 2016). One important distinction is that in the kaon system there is a very large difference between the K_L and K_S lifetimes, because in the absence of CP breaking the 2π state is only accessible to the K_S. There is no analog for the neutral B decays because of the many decay channels, and one expects similar lifetimes for the two mass eigenstates, which are instead labeled B_{H_i} and B_{L_i}, where H and L denote "heavy" and "light." Defining $\Delta\Gamma_i \equiv \Gamma_{L_i} - \Gamma_{H_i} = 2|\Gamma_{B_i\bar{B}_i}|$, the SM prediction (Lenz and Nierste, 2011) is $\Delta\Gamma_d/\Gamma_d = 42(8) \times 10^{-4}$, which is negligible, while $\Delta\Gamma_s/\Gamma_s \sim 0.13(3)$ is small. Both are consistent with experiment.

[46]Mixing in the neutral D system has been observed by BaBar, Belle, CDF, and LHCb (e.g., Patrignani, 2016). The results are consistent with the SM expectations, though the latter have significant long distance uncertainties. CP violation is expected to be small in the SM since the mixing is dominated by the first two families. New physics implications are considered in (Golowich et al., 2007).

Similar to (8.230) and (8.231), we define the strong interaction eigenstates

$$|B_d^0\rangle \equiv |B^0\rangle = |d\bar{b}\rangle, \qquad |\bar{B}_d^0\rangle \equiv |\bar{B}^0\rangle = |b\bar{d}\rangle = -CP|B_d^0\rangle$$
$$|B_s^0\rangle = |s\bar{b}\rangle, \qquad |\bar{B}_s^0\rangle = |b\bar{s}\rangle = -CP|B_s^0\rangle, \tag{8.295}$$

where we have chosen an $\eta_{B_i} = -1$ phase convention. The B_i^0 and $\bar{B}_i^0$ are mixed[47] by box diagrams similar to Figure 8.43, but in this case only the t quark exchange is significant and the long-distance corrections are expected to be small. The eigenstates are[48]

$$|B_{H_i}\rangle = p_i|B_i^0\rangle + q_i|\bar{B}_i^0\rangle, \qquad |B_{L_i}\rangle = p_i|B_i^0\rangle - q_i|\bar{B}_i^0\rangle \qquad i = d, s, \tag{8.296}$$

where

$$\frac{q_i}{p_i} \equiv \frac{1-\tilde{\epsilon}_i}{1+\tilde{\epsilon}_i} = \left[\frac{M^*_{B_i\bar{B}_i} - i\frac{\Gamma^*_{B_i\bar{B}_i}}{2}}{M_{B_i\bar{B}_i} - i\frac{\Gamma_{B_i\bar{B}_i}}{2}}\right]^{1/2}. \tag{8.297}$$

Ignoring $\Gamma_{B_i\bar{B}_i}$, which is much smaller than $M_{B_i\bar{B}_i}$ for both B_d and B_s, one finds that q_i/p_i is a pure phase and that

$$\Delta m_i \equiv m_{H_i} - m_{L_i} = 2|M_{B_i\bar{B}_i}|, \qquad \tilde{\epsilon}_i = \frac{i\Im m\, M_{B_i\bar{B}_i}}{\Delta m_i \cos^2\varphi_M/2} \xrightarrow[\varphi_M \text{ small}]{} \frac{i\Im m\, M_{B_i\bar{B}_i}}{\Delta m_i}, \tag{8.298}$$

where φ_M is the phase of $M_{B_i\bar{B}_i}$. Δm_i can be measured by observing $B_i^0 - \bar{B}_i^0$ oscillations.[49] Suppose one starts with an initial state tagged as a B_i^0. The tagging is typically done by observing the decay of the other meson, e.g., from a $B^0 - \bar{B}^0$ or $B^0 - B^-$ pair. Same-side tags, utilizing other b-jet fragments, are also possible. In some cases, such as $\Upsilon(4S) \to B^0\bar{B}^0$ (or $\phi \to K^0\bar{K}^0$), the two are produced in an entangled coherent state, but we will ignore that complication. Similar to (8.263),

$$|\psi(0)\rangle = |B_i^0\rangle = \frac{1}{2p_i}\left(|B_{H_i}\rangle + |B_{L_i}\rangle\right). \tag{8.299}$$

At a later proper time τ,

$$\begin{aligned} |\psi(\tau)\rangle &\sim \frac{1}{2p_i}\left(|B_{H_i}\rangle e^{-im_{H_i}\tau} + |B_{L_i}\rangle e^{-im_{L_i}\tau}\right)e^{-\frac{\Gamma_i\tau}{2}} \\ &= \left[\cos\left(\frac{\Delta m_i\tau}{2}\right)|B_i^0\rangle - i\sin\left(\frac{\Delta m_i\tau}{2}\right)\frac{q_i}{p_i}|\bar{B}_i^0\rangle\right]e^{-im_i\tau}e^{-\frac{\Gamma_i\tau}{2}}, \end{aligned} \tag{8.300}$$

where $m_i = (m_{H_i} + m_{L_i})/2$ and we have ignored the lifetime difference. Δm_i can then be obtained by observing the proper time dependence for B_i^0 and/or $\bar{B}_i^0$ decays, e.g., by $b \to c\ell^-\bar{\nu}_\ell$ (from $\bar{B}^0$) or $\bar{b} \to \bar{c}\ell^+\nu_\ell$ (from B^0). The current values are $\Delta m_d = 0.510(3)$ ps^{-1} and $\Delta m_s = 17.757(21)$ ps^{-1}. The latter was especially difficult to measure accurately (initially by CDF and D0 and subsequently by LHCb) because of the much shorter oscillation time $2\pi/\Delta m_s$.

The standard model calculation from the box diagram yields

$$\Delta m_i = \frac{G_F^2 M_W^2}{6\pi^2}\eta_B m_{B_i^0}\, S_0(x_t)|V_{ti}^* V_{tb}|^2\, \hat{B}_{B_i} f_{B_i}^2, \tag{8.301}$$

[47] $B^0 - B_s^0$ and $B^0 - \bar{B}_s^0$ mixing are negligible because the diagonal terms are not degenerate.

[48] Some authors reverse the sign of q_i in (8.296).

[49] Early studies utilized time-integrated effects, such as the relative number of same-sign and opposite-sign lepton pairs resulting from associated $B^0 - \bar{B}^0$ production.

where $m_{B_d^0} \sim 5.28$ GeV, $m_{B_s^0} \sim 5.37$ GeV, $\eta_B \sim 0.55(1)$ is the short distance QCD correction (Buras, 2005), and $S_0(x_t)$ is defined in (8.262). $\hat{B}_{B_i}$ and f_{B_i} are analogous to $\hat{B}_K$ and f_K. From unquenched lattice QCD calculations (Aoki et al., 2017), $f_{B_s}\hat{B}_{B_s}^{1/2} \sim 270(16)$ MeV and $f_{B_d}\hat{B}_{B_d}^{1/2} \sim 219(14)$ MeV. Some of the theoretical uncertainties cancel in the ratio $\xi = 1.24(5)$. The values for $|V_{td}|, |V_{ts}|$, and the ratio quoted in Section 8.6.1 utilized (8.301) with $V_{tb} \sim 1$. Alternatively, one can use the Δm_i as a constraint on the unitarity triangle. In the Wolfenstein parametrization

$$|V_{td}^* V_{tb}|^2 = A^2\lambda^6\left[(1-\bar{\rho})^2 + \bar{\eta}^2\right], \qquad |V_{ts}^* V_{tb}|^2 = A^2\lambda^4, \tag{8.302}$$

so that Δm_d yields the circular annulus centered at $(\bar{\rho}, \bar{\eta}) = (1, 0)$ shown in Figure 8.42. The uncertainty is dominated by $\hat{B}_{B_d} f_{B_d}^2$, so one can obtain a tighter constraint from the ratio $\Delta m_d/\Delta m_s$, also shown in Figure 8.42. The $|V_{ub}|$ and $\Delta m_d/\Delta m_s$ constraints intersect at $(\bar{\rho}, \bar{\eta}) \sim (0.17, 0.36)$, with large uncertainties, and the ϵ_K hyperbola is consistent.

CP Asymmetries

Many CP-violating asymmetries have been measured or searched for in the neutral (and charged) B system, which have established indirect and direct CP violation and strongly supported the validity of the unitarity triangle predictions (for reviews, see Buras, 2005; Nir, 2005; Amhis et al., 2014; Porter, 2016; Gershon and Gligorov, 2017; Patrignani, 2016). We will illustrate with one especially clean and important case, i.e., the time-dependent asymmetry

$$a_f(\tau) \equiv \frac{\Gamma(B^0(\tau) \to f) - \Gamma(\bar{B}^0(\tau) \to f)}{\Gamma(B^0(\tau) \to f) + \Gamma(\bar{B}^0(\tau) \to f)}. \tag{8.303}$$

(A similar formalism applies to B_s^0 asymmetries.) $|B^0(\tau)\rangle$ and $|\bar{B}^0(\tau)\rangle$ are defined as the states that were initially tagged as B^0 or $\bar{B}^0$, and f is a CP-eigenstate, such as $J/\psi K_S$, $\pi^+\pi^-$, or ρK_S. One observes asymmetries associated with the interference between decays with and without mixing, and in some cases asymmetries associated with the decays themselves. Similar to (8.300),

$$\begin{aligned} |B^0(\tau)\rangle &= |B^0\rangle f_+(\tau) + |\bar{B}^0\rangle \frac{q}{p} f_-(\tau) \\ |\bar{B}^0(\tau)\rangle &= |B^0\rangle f_-(\tau)\frac{p}{q} + |\bar{B}^0\rangle f_+(\tau), \end{aligned} \tag{8.304}$$

where

$$\begin{aligned} f_+(\tau) &= \cos\left(\frac{\Delta m\,\tau}{2}\right) e^{-im\tau} e^{-\frac{\Gamma\tau}{2}} \\ f_-(\tau) &= -i\sin\left(\frac{\Delta m\,\tau}{2}\right) e^{-im\tau} e^{-\frac{\Gamma\tau}{2}}, \end{aligned} \tag{8.305}$$

and we have again ignored the lifetime difference. Also, define

$$A_f = \langle f|\mathcal{H}_{nl}|B^0\rangle, \qquad \bar{A}_f = \langle f|\mathcal{H}_{nl}|\bar{B}^0\rangle \tag{8.306}$$

in analogy with (8.283). A_f and $\bar{A}_f$ may be written

$$A_f = \sum_\alpha |A_\alpha| e^{i\delta_\alpha} e^{i\varphi_\alpha}, \qquad \bar{A}_f = -\eta_f \sum_\alpha |A_\alpha| e^{i\delta_\alpha} e^{-i\varphi_\alpha}, \tag{8.307}$$

where the sum is over the diagrams α that contribute to the decay, δ_α is the strong final state interaction phase associated with A_α, φ_α is the weak CP- phase generated (in the SM) by the phases in the CKM matrix, and $\eta_f = \pm 1$ is the CP parity of f, i.e., $CP|f\rangle = \eta_f|f\rangle$. The decay amplitudes at proper time τ are then

$$\begin{aligned} \langle f|\mathcal{H}_{nl}|B^0(\tau)\rangle &= A_f\left[f_+(\tau) - \lambda_f f_-(\tau)\right] \\ \langle f|\mathcal{H}_{nl}|\bar{B}^0(\tau)\rangle &= A_f\frac{p}{q}\left[f_-(\tau) - \lambda_f f_+(\tau)\right], \end{aligned} \tag{8.308}$$

where

$$\lambda_f \equiv -\frac{q}{p}\frac{\bar{A}_f}{A_f}. \tag{8.309}$$

The decay rates are

$$\begin{aligned} \Gamma\left(B^0(\tau)\to f\right) &= |A_f|^2\left[c_+ + c_-\cos(\Delta m\,\tau) - \Im m\,\lambda_f\,\sin(\Delta m\,\tau)\right]e^{-\Gamma\tau} \\ \Gamma\left(\bar{B}^0(\tau)\to f\right) &= |A_f|^2\left[c_+ - c_-\cos(\Delta m\,\tau) + \Im m\,\lambda_f\,\sin(\Delta m\,\tau)\right]e^{-\Gamma\tau}, \end{aligned} \tag{8.310}$$

where $c_\pm = (1 \pm |\lambda_f|^2)/2$ and we have used $|p/q| \sim 1$. (8.310) leads to the asymmetry

$$a_f(\tau) = C_f\cos(\Delta m\,\tau) - S_f\sin(\Delta m\,\tau), \tag{8.311}$$

where

$$C_f \equiv \frac{1-|\lambda_f|^2}{1+|\lambda_f|^2}, \qquad S_f \equiv \frac{2\Im m\,\lambda_f}{1+|\lambda_f|^2}. \tag{8.312}$$

(There are a number of conflicting notations for C_f and S_f and their signs in the literature.) Since $|p/q| \sim 1$, a nonzero C_f implies direct CP-breaking in the decay, i.e., $|\bar{A}_f/A_f| \neq 1$, while S_f is the mixing-induced term.

Especially simple to interpret are decays for which one diagram dominates (or two diagrams have the same weak phase). In that case the strong interaction effects cancel, and λ_f is just

$$\lambda_f = +\eta_f\frac{q}{p}e^{-2i\varphi} = \eta_f e^{-i\varphi_M}e^{-2i\varphi}, \tag{8.313}$$

where φ is the weak phase and $-\varphi_M$ is the phase of q/p, so that

$$C_f = 0, \qquad S_f = -\eta_f\sin(\varphi_M + 2\varphi). \tag{8.314}$$

In particular, for the golden modes $f = J/\psi K_{S,L}$ (with $\eta_f = \mp 1$, respectively), one has

$$\frac{q}{p} = \left(\frac{M^*_{B\bar{B}}}{M_{B\bar{B}}}\right)^{1/2} = \frac{V^*_{tb}V_{td}}{V_{tb}V^*_{td}} = e^{-2i\beta}, \tag{8.315}$$

as can be seen from Figure 8.41. From Figure 8.45 the contribution of the weak phase is

$$e^{-2i\varphi} = \frac{V^*_{cs}V_{cb}}{V_{cs}V^*_{cb}} \sim 1. \tag{8.316}$$

Neglecting small effects associated with $\Gamma_{B\bar{B}}$, CP breaking in the kaon system, and penguin diagrams[50] one obtains $S_{J/\psi K_{S,L}} = \pm\sin(2\beta)$. The $J/\psi K_{S,L}$ asymmetries have been measured by BaBar, Belle, and LHCb, leading to $\sin 2\beta = 0.691(17)$. As can be seen in Figure 8.42 this agrees well with the region in the $\bar{\rho}-\bar{\eta}$ plane determined by the other methods (other observations and the global analysis largely eliminate other branches for 2β).

[50]The gluon penguin diagrams for $B^0 \to J/\psi K_{S,L}$ are similar to those for $B^0 \to \phi K^0$ in Figure 8.45 except the $s\bar{s}$ pair is replaced by $c\bar{c}$. These are expected to be small, and in any case the dominant contributions have the same weak phase as the tree diagram.

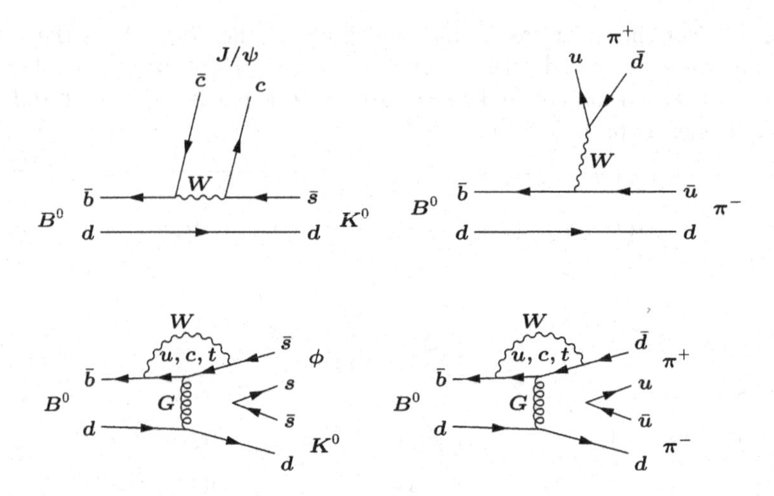

Figure 8.45 Top: tree-level diagram for $B^0 \to J/\psi K_{S,L}$ (left) and $B^0 \to \pi^+\pi^-$ (right). Bottom: Gluon penguin diagrams for $B^0 \to \phi K^0$ (left) and $B^0 \to \pi^+\pi^-$ (right). Electroweak penguins involving Z and γ are also possible.

Other decays, e.g., involving $\bar{b} \to \bar{c}u\bar{d}$, $\bar{b} \to c\bar{u}\bar{d}$, or the charmless $\bar{b} \to \bar{s}q\bar{q}$ decays such as $B^0 \to \phi K_S$ or $\pi^0 K_S$ can also be used to determine $\sin 2\beta$. All are consistent with the $J/\psi K_{S,L}$ results. In the SM the charmless decays are driven by the $\bar{b} \to \bar{s}q\bar{q}$ penguin diagrams shown in Figure 8.45. Since the SM amplitudes are loop-suppressed, they are a good place to search for the effects of new physics (Buras et al., 2004; Buchalla et al., 2005), which could enter in penguin diagrams, such as in supersymmetry (e.g., Artuso et al., 2008a), or at the tree-level, such as in models with an extended Higgs sector or a Z' with flavor changing couplings (e.g., Barger et al., 2004).

The direct measurement of the angles α and γ is more difficult. For example, the asymmetry for $B^0 \to \pi\pi$ would yield $\sin 2\alpha$ if it were due entirely to the tree-level diagram for $\bar{b} \to \bar{d}u\bar{u}$ shown in Figure 8.45. However, there is also a significant contribution from the $\bar{b} \to \bar{d}$ penguin diagram, which has a different phase. The tree and penguin effects can be sorted out with some effort, using different isospin channels (Gronau and London, 1990) for $B^0 \to \pi^+\pi^-, \pi^0\pi^0$ and $B^+ \to \pi^+\pi^0$ decays, as well as isospin, polarization, and Dalitz plot information on related $B \to \rho\pi$ and $\rho\rho$ decays, yielding $\alpha \sim 88(4)^\circ$. From (8.228) γ does not depend on V_{tn}. It can therefore be measured in the interference of tree-level decay amplitudes, e.g., between $B^- \to D^0 K^- \to fK^-$ and $B^- \to \bar{D}^0 K^- \to fK^-$, where f is a final state accessible to both D^0 and $\bar{D}^0$, yielding $\gamma = 73(7)^\circ$.

Studies of mixing and CP violation in the $B_s^0 - \bar{B}_s^0$ system have been carried out at the LHC and Tevatron, especially by LHCb, with generally good agreement[51] with the SM (for a general review, including both semi-leptonic and nonleptonic decays, see Artuso et al., 2016). One complication is that $\Delta\Gamma_s$ is no longer negligible. It is still a good approximation to ignore $\Gamma_{B_s\bar{B}_s}$ in (8.297), but (8.300) must be corrected for $\Gamma_{L_s} \neq \Gamma_{H_s}$. Similar to $B^0 \to J/\psi K_{S,L}$, the decays $B_s^0 \to J/\psi K^+K^-$ (or $B_s^0 \to J/\psi\phi$) are dominated in the SM by the

[51]There is a 3.6σ discrepancy from the SM prediction in a D0 measurement (Abazov et al., 2014) of the dimuon asymmetry $(N^{\mu^+\mu^+} - N^{\mu^-\mu^-})$/sum in the decays of a combination of $B^0\bar{B}^0$ and $B_s^0\bar{B}_s^0$ pairs. This has not yet been confirmed by other experiments.

$\bar{b} \to \bar{c}c\bar{s}$ tree-level diagram, with a very small penguin contribution. The time-dependent CP-asymmetry (including an angular analysis to separate the CP even and odd parts of the $J/\psi K^+K^-$ state) determines both $\Delta\Gamma_s$ and $\varphi_{M_s} \sim -2\beta_s$, where φ_{M_s} is the phase of $M_{B_s\bar{B}_s}$ and β_s is defined in (8.228). The result is (Amhis et al., 2014) $\Delta\Gamma_s = 0.084(7)$ ps^{-1} and $\varphi_{M_s} = -0.030(33)$ rad, consistent with the SM expectations 0.088(20) ps^{-1} and $-0.038(1)$, but with a much larger uncertainty.

The Wolfenstein Parameters

The constraints from the magnitudes of the CKM elements; K, D, and B mixing; ϵ, ϵ'; and CP violation in the B decays are generally consistent with the SM predictions and with the unitarity triangle, although there is still room for new physics and even hints of discrepancies. A global fit (see the CKM article in Patrignani, 2016) to the CKM parameters assuming unitarity yields

$$\lambda = 0.2251(5), \qquad A = 0.81(3), \qquad \bar{\rho} = 0.12(2), \qquad \bar{\eta} = 0.36(1) \tag{8.317}$$

as well as the results in (8.215). The J parameter defined in (8.219) is $J = 3.0(2) \times 10^{-5}$. For more information on the unitarity triangle, see, e.g., (Charles et al., 2005, 2015; Bona et al., 2006; Amhis et al., 2014; Porter, 2016), as well as the websites listed in the bibliography. Future prospects for the unitarity triangle and related physics studies, especially by LHCb and at the future Super B factory Belle II at KEK, are reviewed in (Harnew, 2016).

8.6.5 Time Reversal Violation and Electric Dipole Moments

Time reversal and CP violation (Section 2.10) are closely related, in the sense that any CPT invariant interaction that violates one must violate the other. We are assuming CPT here, so we will continue to use CP and T interchangeably for most theoretical discussions. Nevertheless, CP and T transformations are different, and possible T-violating observables open a new window on standard model tests and new physics searches (e.g., Bernabéu and Martínez-Vidal, 2015). In particular, the effects of true CP violation are essentially limited to flavor changing processes such as K and B decays, while some T-odd observables such as electric dipole moments are also relevant for flavor-diagonal channels.

The only direct observations of T violation to date are a difference in the $K^0 \to \bar{K}^0$ and $\bar{K}^0 \to K^0$ oscillation probabilities by the CPLEAR collaboration at CERN in 1998 (e.g., Angelopoulos et al., 2003), and in differences between, e.g., $B^0 \to B_+$ and $B_+ \to B^0$ oscillations, observed by BaBar in 2012 (Lees et al., 2012). The latter experiment involves pairs of quantum-entangled neutral B mesons, one of which decays to ℓ^+X (B^0) or ℓ^-X ($\bar{B}^0$), and the other to $J/\psi K^0_L$ (B_+) or $J/\psi K^0_S$ (B_-), where $B_\pm$ are appropriate linear combinations of B^0 and $\bar{B}^0$. The CPLEAR and BaBar results are consistent with the CP asymmetries and with CPT. In principle, T violation could be searched for in neutrino oscillations, e.g., by comparing $\nu_\mu \to \nu_e$ and $\nu_e \to \nu_\mu$, but this would be extremely difficult.

Another possibility is to directly observe T violation in scattering or decay processes. However, this is very difficult because T not only reverses spins and three-momenta, but also interchanges initial and final states. For example, consider the matrix element $\langle f|H_W|i\rangle$ of the weak Hamiltonian (or relevant effective operator) between an initial state i and final state f. Using (2.317) on page 59,

$$\langle f|H_W|i\rangle = \langle Tf|TH_WT^{-1}|Ti\rangle^* = \langle Ti|TH_WT^{-1}|Tf\rangle, \tag{8.318}$$

where Ti and Tf have their momenta and spins reversed compared to i and f and also have the roles of "in" and "out" states interchanged (which is relevant if there are initial or final state interactions), and we have used that H_W and TH_WT^{-1} are Hermitian. A direct test of whether $TH_WT^{-1} = H_W$ requires the comparison of two different reactions, which is very difficult (or impossible for decays). This argument is readily generalized to an arbitrary theory, using the transition matrix element defined by (B.1) on page 511. The interaction Hamiltonian in (B.2) is now understood to include *all* interactions, so that i and f are free particle states. If T commutes with the Hamiltonian, then the transition operator $U \equiv U(+\infty, -\infty)$ satisfies $TUT^{-1} = U^\dagger$, so that

$$\langle f|U|i\rangle = \langle Tf|U^\dagger|Ti\rangle^* = \langle Ti|U|Tf\rangle. \tag{8.319}$$

It is, however, possible to search for *pseudo* T violation by measuring such T odd quantities as the triple correlations $\vec{p}_1\cdot(\vec{p}_2\times\vec{p}_3)$ or $\vec{s}_1\cdot(\vec{p}_2\times\vec{p}_3)$, where $\vec{s}_j$ or $\vec{p}_j$ are spins or momenta in the process.[52] However, care must be taken with strong or electromagnetic final (or initial) state interactions, which can mimic the effects of T violation or at least complicate its determination. For definiteness, we will consider the triple product $\tau(\vec{p}) = \vec{p}_1 \cdot (\vec{p}_2 \times \vec{p}_3)$. Correlations involving spin and more detailed derivations are described in (Gasiorowicz, 1966; Sozzi, 2008). The expected value of τ is

$$\langle\tau(\vec{p})\rangle = \frac{\int D_f \tau(\vec{p})|\langle f|H_W|i\rangle|^2}{\int D_f|\langle f|H_W|i\rangle|^2}, \tag{8.320}$$

where D_f contains the appropriate phase space and flux factors. Any CP or T violating effect requires interference between contributions to the amplitude with different phases, analogous to (8.307). Since $\tau(\vec{p})$ is also odd under space reflection, one needs the interference of amplitudes with difference parities. Let us therefore assume that

$$\langle f|H_W|i\rangle = A(\vec{p})e^{i\delta_A}e^{i\varphi_A} + B(\vec{p})e^{i(\frac{\pi}{2}+\delta_B)}e^{i\varphi_B}, \tag{8.321}$$

where we have suppressed the spin indices. $\delta_{A,B}$ are the strong phases associated, e.g., with final state interactions; $\varphi_{A,B}$ are the weak phases associated with the T and CP-violating parts of the interaction; and the real amplitudes A and B have opposite parities, i.e., $A(-\vec{p}) = A(\vec{p})$ and $B(-\vec{p}) = -B(\vec{p})$. The phase $\pi/2$ could have been absorbed in δ_B, but is pulled out for convenience. Such a relative phase always occurs in the interference between even and odd parity amplitudes when summing over spins, as can be seen, for example, in (2.172) on page 36. It is clear that

$$\langle\tau(\vec{p})\rangle \propto \cos(\delta_A - \delta_B)\sin(\varphi_A - \varphi_B) + \sin(\delta_A - \delta_B)\cos(\varphi_A - \varphi_B), \tag{8.322}$$

so a non-zero value can be induced by final state interactions ($\delta_A - \delta_B \neq 0$), by T and CP violation ($\varphi_A - \varphi_B \neq 0$), or both. For this reason, it is difficult to isolate T violating correlations in processes involving more than one strongly interacting final or initial state particle. There is, however, some possibility of observing such effects in cascades in supersymmetric theories in which there are mainly leptons and neutralinos in the final state (e.g., Bartl et al., 2004; Langacker et al., 2007; Ellis et al., 2009).

More promising are electric dipole moments (EDMs) $\vec{d}_f$ for particle f (for recent reviews, see Ibrahim and Nath, 2008; Roberts et al., 2015; Bernabéu and Martínez-Vidal, 2015), defined by its interaction with an electric field $\vec{E}$,

$$H_{EDM} = -\vec{d}_f \cdot \vec{E}. \tag{8.323}$$

[52]Such correlations must be associated with Lorentz invariant quantities, such as $\epsilon_{\mu\nu\rho\sigma}p_0^\mu p_1^\nu p_2^\rho p_3^\sigma$, so the process must be sufficiently complicated to allow enough independent four-vectors.

For a spin-$\frac{1}{2}$ field this corresponds to an effective Langrangian density

$$\mathcal{L}_{EDM} = \frac{-i}{2} d_f \bar{\psi}_f \sigma^{\mu\nu} \gamma^5 \psi_f F_{\mu\nu}. \tag{8.324}$$

If we define the electromagnetic form factors F_i^f and g_i^f of $\bar{\psi}_f \gamma^\mu \psi_f$ analogous to those of the proton in (2.379) on page 77, then $\vec{d}_f$ is related to g_2^f by

$$\vec{d}_f = -i \frac{q_f e g_2^f(0)}{2m_f} \vec{\sigma} = -i \frac{q_f e g_2^f(0)}{m_f} \vec{S}, \tag{8.325}$$

where $q_f e$ and m_f are the electric charge and mass of f (Problem 2.30). For quarks one can also define analogous chromoelectric dipole moments (CEDMs) d_r^C by

$$\mathcal{L}_{CEDM} = \frac{-i}{2} d_r^C \bar{q}_r \sigma^{\mu\nu} \gamma^5 \frac{\lambda^i}{2} q_r G^i_{\mu\nu}. \tag{8.326}$$

It is obvious from (8.325) that an electric dipole moment (or chromoelectric dipole moment) violates time reversal and space reflection invariance, since by classical reasoning $\vec{d}_f$ must be even (odd) under T (P), while $\vec{S}$ is just the opposite. This extends to any non-degenerate system using the Wigner-Eckart theorem, and will be shown formally for spin-$\frac{1}{2}$ in Appendix G.

There are stringent experimental limits on possible EDMs for the neutron, paramagnetic atoms[53] (e.g., ^{205}Tl), diamagnetic atoms (e.g., ^{199}Hg), and molecules (e.g., ThO)

$$|d_n| < 3.0 \times 10^{-26}\,(90\%), \qquad |d_{Hg}| < 3.1 \times 10^{-29}\,(95\%), \qquad |d_{Tl}| < 9 \times 10^{-25}\,(90\%)$$
$$|d_e| < 9.8 \times 10^{-29}\,(90\%), \qquad |d_p| < 7.9 \times 10^{-25}\,(95\%), \tag{8.327}$$

all in units of e-cm. The limits on d_e and d_p are derived from those on ThO and d_{Hg}, respectively, assuming no other sources of T violation [the limits are weakened somewhat if one also allows for T and P-violating eN or πN interactions (e.g., Chupp and Ramsey-Musolf, 2015)]. Future experiments may improve the sensitivities significantly. There are also prospects for new or greatly improved sensitivities to other EDMs, such as the muon or deuteron.

The predicted EDM effects due to the CKM mixing in the standard model are extremely small due to approximate accidental symmetries. The quark EDMs are only generated at the three-loop level, and are expected to be of $\mathcal{O}(10^{-34}\, e-\text{cm})$, while d_e first enters in four-loop diagrams (at least for massless neutrinos) and should be smaller than $\sim 10^{-38}\, e$-cm. The contributions of the quark EDMs to d_n are expected to be negligible, with the largest CKM part of $d_n^{CKM} \sim 10^{-32}\, e$-cm from a two-loop diagram involving a gluon penguin. Other CKM contributions to atomic EDMs are also expected to be small.

One major complication, however, is from the strong CP violation (θ_{QCD}) term in the QCD Lagrangian density in (5.2), which contributes $d_n^\theta \sim 5 \times (10^{-16} - 10^{-15})\, \theta_{QCD}$ (Crewther et al., 1979; Kim and Carosi, 2010). One therefore needs $|\theta_{QCD}| \lesssim 10^{-10} - 10^{-11}$, while CKM effects are expected to shift θ_{QCD} by $\mathcal{O}(10^{-3})$. It is not known whether θ_{QCD} is small due to an accidental or other cancellation between the bare and shift values, or to some dynamical mechanism, and therefore whether or not $d_n^\theta = 0$. This strong CP problem will be elaborated in Section 10.1.

[53]The Schiff theorem (Schiff, 1963) states that atomic EDMs vanish in the non-relativistic limit for a pointlike nucleus because of electron screening effects. Fortunately, it is violated by relativistic and finite nuclear size effects.

Because the SM contributions are expected to be so small (except possibly for d_n^θ) EDMs are an excellent place to search for the effects of new physics (e.g., Pospelov and Ritz, 2005; Fukuyama, 2012; Engel et al., 2013). These typically have new CP-violating phases and allow EDMs at one-loop level, leading to values already excluded or within reach of future EDM experiments. For example, in the MSSM the e^- and quark EDMs and CEDMs (which feed into EDMs) can be generated by one-loop vertex diagrams involving neutralinos, charginos, gluinos and the fermion scalar partners (e.g., Ibrahim and Nath, 2008; Ellis et al., 2008), with new CP phases possible from the μ and $B\mu$ terms, gaugino masses, and A terms. There may also be significant effects from induced three-gluon operators (Weinberg, 1989a), and from Higgs exchange and two-loop diagrams involving Higgs fields (Barr and Zee, 1990). For large CP phases the existing EDM limits typically require SUSY masses $\gg \mathcal{O}(1 \text{ TeV})$ unless there are fine-tuned cancellations. Conversely, superpartner masses $\lesssim 1$ TeV would require small CP phases, creating some tension with models of electroweak baryogenesis (see Section 10.1). At any rate, future experiments have an excellent chance of observing EDMs if TeV scale supersymmetry, or many other SM extensions, exists, and would be a powerful diagnostic.

8.6.6 Flavor Changing Neutral Currents (FCNC)

Flavor changing neutral currents are similar to electric dipole moments and CP violation in that they are strongly suppressed by approximate accidental symmetries in the SM, but can be much larger in most extensions.

In the standard model the couplings of the Z to fermions are flavor diagonal at tree level because of the GIM mechanism (Glashow et al., 1970), i.e., because all fermions that have the same charge, color, and chirality and are therefore able to mix with each other are assigned to the same kind of $SU(2) \times U(1)$ representation. Off-diagonal Z-fermion couplings are induced at loop level, mainly by penguin diagrams such as in Figure 8.44, but these are small. The GIM mechanism also significantly suppresses the contributions of box diagrams such as in Figure 8.43. Similarly, the couplings of the Higgs to fermions are flavor diagonal in the SM because the same transformations that diagonalize the fermion mass matrices automatically diagonalize the H Yukawa couplings. This continues to hold in models with multiple Higgs doublets provided that only one doublet couples to $\bar{q}_L^0 u_R^0$, one to $\bar{q}_L^0 d_R^0$, and similarly for the leptons (Glashow and Weinberg, 1968).

Many types of new physics lead to FCNC, often including new sources of CP violation (see, e.g., Langacker, 1991; Artuso et al., 2008a; Altmannshofer et al., 2010; Isidori et al., 2010; Mihara et al., 2013; Buras and Girrbach, 2014). In many cases the new effects enter at tree level, where they are especially significant because they compete with SM loop effects. Other types enter at loop level, but may be enhanced compared with SM loop effects by stronger couplings or by not having the same cancellations. In some cases the new physics leads to processes that are forbidden or nearly so in the SM. For example, the SM exhibits an approximate *lepton flavor symmetry*: to the extent that the ν masses can be ignored there are separately conserved L_e, L_μ and L_τ numbers, with $L_e = +1$ for (e^-, ν_e), $L_e = 0$ for (μ^-, ν_μ) and (τ^-, ν_τ), and analogously for $L_{\mu,\tau}$. Processes such as $\mu \to e\gamma$ are therefore essentially forbidden in the SM up to negligible $(m_\nu/M_W)^4$ effects (Cheng and Li, 1977), but are allowed in many extensions.

One type of new physics that leads to FCNC effects at tree level involves heavy Z' gauge bosons with GIM-violating family non-universal couplings (Langacker and Plumacher, 2000), as frequently occur in string constructions (e.g., Blumenhagen et al., 2005). Closely related are Kaluza-Klein excitations of neutral gauge bosons in models in which the fermion families are located at different positions in large ($\mathcal{O}(\text{TeV}^{-1})$) or warped extra dimensions (Delgado et al., 2000), which therefore have family non-universal couplings. The mixing of ordinary fermions with heavy ones with exotic SM quantum numbers (e.g., left-chiral singlets or right-chiral doublets) can also lead to off-diagonal Z couplings (Langacker and London, 1988b). FCNC may also be mediated by heavy gauge boson exchange in alternatives to the Higgs mechanism, such as extended technicolor (Eichten and Lane, 1980; Hill and Simmons, 2003; Appelquist et al., 2004), or in models with gauged family symmetries (e.g., Cahn and Harari, 1980). Another possibility is multiple Higgs doublets that couple to the same types of fermion (Hall and Weinberg, 1993; Atwood et al., 1997; Branco et al., 2012). Leptoquarks are possible particles with couplings to quarks and leptons, such as $(\bar{e}_R^+ - \bar{\nu}_{eR}^c) q_L \tilde{d}^c$, while diquarks have couplings like $\bar{d}_R u_L^c \tilde{d}^c$. In this example, the antilepton fields are expressed in a conjugate form analogous to the conjugate Higgs doublet in (8.14), and $\tilde{d}^c$ is a heavy spin-0 color anti-triplet with electric charge $+\frac{1}{3}$. The tilde is suggestive that $\tilde{d}^c$ could be the scalar partner of the s^c (or the d^c for the leptoquark case) in R-parity violating versions of supersymmetry (Hewett and Rizzo, 1989). The two types of coupling violate lepton number and baryon number, respectively, and rapid proton decay would result if both were present simultaneously. Even if one is absent (stabilizing the proton) the other could mediate FCNC (Barbier et al., 2005; Doršner et al., 2016). We finally mention that some models involving composite quarks or leptons can lead to FCNC, e.g., by constituent interchange or configuration changes (Harari, 1984).

Loop effects in new physics models may also generate significant effective FCNC interactions, by box diagrams analogous to Figure 8.43 or by penguin-like vertex corrections that lead to off-diagonal G, Z, γ, or H vertices. These can be important in supersymmetry, especially in diagrams involving gluinos (which couple strongly) or in the limit of large $\tan\beta$ (for which the b Yukawa is large), as discussed in Section 10.2.5. Similarly, there can be important loop-induced effects in $SU(2)_L \times SU(2)_R \times U(1)$ models involving a heavy W_R coupling to $V + A$ (Section 10.3.2), or in extensions involving heavy neutrinos (Section 9.6).

There are stringent constraints on new sources of FCNC and CP violation from $K^0 - \bar{K}^0$, $D^0 - \bar{D}^0$, and $B^0 - \bar{B}^0$ mixing (Sections 8.6.3 and 8.6.4). There have also been extensive searches for and observations of rare K decays (e.g., Barker and Kettell, 2000; Cirigliano et al., 2012). Many focus on lepton flavor-violating decays such as $K_L \to \mu^\pm e^\mp$ or $K^+ \to \pi^+ \mu^\pm e^\mp$, which are completely negligible in the SM. Other decays are strongly suppressed in the SM, and serve as probes of the CKM matrix, new physics, CPT violation, or long-distance strong interaction physics. These include $K_L \to \pi^0 \ell^+ \ell^-$, which includes a significant direct CP violation component, and the decays $K^+ \to \pi^+ \nu\bar{\nu}$ and $K_L \to \pi^0 \nu\bar{\nu}$, which are, respectively, CP-conserving and violating and should be excellent probes of new physics (Buras et al., 2015a).

In addition to mixing and CP violation effects, rare decays of B mesons are especially promising for testing the SM and searching for new physics at the tree or loop level (Blake et al., 2015; Ali, 2016). This is in part because the charmless modes are strongly suppressed in the SM and in part because some types of new physics couple most strongly to heavier particles. For example, the inclusive radiative decays $B \to X_s \gamma$, where X_s carries strangeness, has a branching ratio $3.43(22) \times 10^{-4}$, in agreement with the SM expectation $3.36(23) \times 10^{-4}$, whereas the rate could have been strongly enhanced by the existence of a charged Higgs boson in some parameter regions of two-doublet models. Similarly, CMS and LHCb have recently measured $B(B_s^0 \to \mu^+ \mu^-) = 2.8(7) \times 10^{-9}$ (Khachatryan et al., 2015a), within about 1σ of the SM prediction $3.55(23) \times 10^{-9}$, excluding some parameter regions of supersymmetry. However, their constraint $B(B^0 \to \mu^+ \mu^-) = 3.9(1.5) \times 10^{-10}$ is higher than the expectation $1.01(9) \times 10^{-10}$ by about 2σ. Although this is not by itself significant, there are a number of other hints of discrepancies in B decays at the $(2-4)\sigma$ level. These include deviations at the $(2-3)\sigma$ level (depending on the theoretical modeling) in the angular analysis of the exclusive $B \to K^* \mu^+ \mu^-$ decays observed by LHCb and Belle. There is also a possibility of lepton-flavor nonuniversality in the ratio $\Gamma(B^\pm \to K^\pm \mu^+ \mu^-)/\Gamma(B^\pm \to K^\pm e^+ e^-) = 0.745(97)$ measured by LHCb, about 2.6σ below the SM value ~ 1, suggesting, e.g., a heavy Z'. (The absolute rate for $K^\pm \mu^+ \mu^-$ is also $\sim 1.8\sigma$ low.) Similarly, the charged current decays $B \to D^{(*)} \tau \nu$ as measured by BaBar, Belle, and LHCb appear to violate universality by $\sim 3.9\sigma$ when compared to those for $B \to D^{(*)} \ell \nu$, $\ell = e$, or μ, suggesting a leptoquark contribution. None of these hints of new physics are compelling, but together they hint at interesting things to come.

Similarly, there have been extensive searches for lepton flavor-violation, such as the leptonic processes $\mu \to 3e$, $\mu \to e\gamma$, $\mu \to e$ conversion in interactions with a nucleus, and analogous τ decays (Mihara et al., 2013; de Gouvêa and Vogel, 2013; Pich, 2014; Gorringe and Hertzog, 2015). For example, the MEG collaboration at PSI has recently obtained (Baldini et al., 2016) $B(\mu^+ \to e^+ \gamma) < 4.2 \times 10^{-13}$ at 90% c.l. There have also been stringent limits on lepton-flavor violation in B, D, and Z decays. Current limits on lepton-flavor violation in Higgs decay are weak (e.g., Khachatryan et al., 2015b; Aad et al., 2015b), but should become important in the future.

As will be described in Chapter 10 there are good reasons to suspect that new physics emerges at the TeV scale. As emphasized above, most types are likely to lead to observable FCNC, CP violation, and EDM effects. In fact, it is surprising that no such effects have yet been observed. The constraints on new physics relevant to FCNC can be parametrized by effective higher-dimensional operators, especially four-fermi operators (e.g., Buchmuller and Wyler, 1986; Grzadkowski et al., 2010) such as

$$\mathcal{L}_{4f}^{|\Delta S|=2} = \sum_{a,b=L,R} \frac{c_{ab}}{\Lambda_{ab}^2} (\bar{d}_a \gamma^\mu s_a)(\bar{d}_b \gamma_\mu s_b) + h.c. \tag{8.328}$$

For example, Δm_K and ϵ (defined in (8.270)) imply (Nir, 2015)

$$\frac{\Lambda_{LL}}{\sqrt{|c_{LL}|}} \gtrsim 10^6 \text{ GeV } [\Delta m_K], \qquad \frac{\Lambda_{LL}}{\sqrt{|\Im m\, c_{LL}|}} \gtrsim 2 \times 10^7 \text{ GeV } [\epsilon], \tag{8.329}$$

assuming only that the new physics part is not larger than the experimental values. There are even stronger limits on the LR operators. The analogous limits from B and D mixing are in the $10^5 - 10^6$ GeV range. The implication is that new physics at the $\Lambda \sim$ TeV scale should have been seen by now unless the coefficients like c_{ab} are very small. This could be due to very weak coupling, but the motivations for new TeV physics such as the Higgs/hierarchy problem or gauge unification suggest that it should have at least electroweak scale coupling. One therefore presumably requires that the coefficients of the flavor changing operators are strongly suppressed, similar to those from the SM (e.g., the factor of $m_c^2 |V_{cd}|^2 |V_{cs}|^2 / M_W^2$ in (8.237)). Since most new physics models do *not* have such strong suppressions, there is a strong tension between attempts to solve the Higgs/hierarchy problem and constraints from FCNC, CP, and EDMs. This tension has led to much recent discussion of *minimal flavor violation* (MFV), which is the hypothesis that all flavor violation, even that which is associated with new physics, is proportional to the standard model Yukawa matrices (D'Ambrosio et al., 2002; Nir, 2015), leading to a significant suppression of flavor changing effects similar to the SM. Examples include supersymmetry with anomaly or (simple forms of) gauge mediation of supersymmetry breaking (e.g., Chung et al., 2005).

We have already emphasized that for massless neutrinos the SM possesses an accidental lepton flavor symmetry. An extension of this idea, which motivates minimal flavor violation, is that there are no flavor changing or CP violation effects at all in the SM when one turns off the Yukawa interactions (θ_{QCD} becomes unobservable as well). In fact, in that limit there is a global $U(3)^5$ flavor symmetry associated with the three families each of q_L, ℓ_L, u_R, d_R, and e_R ($U(3)^6$ if we include the ν_R), with the diagonal generators corresponding to, e.g., conserved u_R, c_R, and t_R numbers. Of course, this symmetry is strongly broken by the third family Yukawas, but those of the first two families are small.

It is still uncertain whether MFV is employed by nature, or whether FCNC are suppressed by some other mechanism. It is clear, however, that FCNC, EDMs, CP violation, and rare decays have an enormous reach in searching for new physics. Effects involving the t quark and other third family members are especially interesting (Section 7.2.7), since it is so much heavier than the other two and may play a role in new physics, especially new physics associated with spontaneous symmetry breaking. Experiments at the LHC and lower-energy facilities, and at possible future colliders, will study and search for t quark decays; rare μ, τ, K, D, and B decays and processes; rare and flavor-changing Z and H decays; and EDMs. Lepton number and lepton flavor violation associated with neutrinos will be further discussed in Chapter 9, and baryon number violation, e.g., baryogenesis, proton decay, and neutron oscillations, in Chapter 10.

8.7 PROBLEMS

8.1 Consider a generalization of the $SU(2) \times U(1)$ model involving k multiplets $\phi_i, i = 1 \cdots k$, of complex scalars. The dimension of the ith multiplet is $2t_i + 1$, where t_i can be $0, 1/2, 1, 3/2 \cdots$, and the elements have T^3 eigenvalues $t_i^3 = -t_i, -t_i + 1 \cdots t_i$ (cf., the rotation group). Also, the ith multiplet has weak hypercharge y_i. Assume that each multiplet has one electrically neutral component ϕ_i^0, i.e., with $q_i = t_i^3 + y_i = 0$, and that that component acquires a vacuum expectation value $\langle \phi_i^0 \rangle = \nu_i/\sqrt{2}$.
(a) Show that the mass eigenstates $W^\pm$, Z, and A are the same as in the standard model.
(b) Calculate the W and Z masses in terms of g, g', t_i, t_i^3, and ν_i.
(c) The ρ_0 parameter, $\rho_0 \equiv M_W^2/(M_Z^2 \cos^2\theta_W)$ is predicted to be unity at the tree level in the standard model and in extensions involving additional Higgs doublets. Show that in the more general case

$$\rho_0 = \frac{M_W^2}{M_Z^2 \cos^2\theta_W} = \frac{\sum_{i=1}^k [t_i(t_i+1) - (t_i^3)^2]|\nu_i|^2}{2\sum_{i=1}^k (t_i^3)^2 |\nu_i|^2}.$$

(d) Specialize to the case of one doublet and two triplets

$$\phi = \begin{pmatrix} \phi^+ \\ \phi^0 \end{pmatrix}, \qquad \Phi = \begin{pmatrix} \Phi^{++} \\ \Phi^+ \\ \Phi^0 \end{pmatrix}, \qquad \Sigma = \begin{pmatrix} \Sigma^+ \\ \Sigma^0 \\ \Sigma^- \end{pmatrix},$$

where $\nu_\phi \equiv \sqrt{2}\langle \phi^0 \rangle \gg \nu_\Phi \equiv \sqrt{2}\langle \Phi^0 \rangle$ and $\nu_\phi \gg \nu_\Sigma \equiv \sqrt{2}\langle \Sigma^0 \rangle$. Calculate ρ_0 to leading nontrivial order in ν_Φ/ν_ϕ and ν_Σ/ν_ϕ.
(e) Now consider the case of multiple Higgs doublets but no higher-dimensional representations. Argue that the couplings of neutral physical Higgs bosons to fermions will no longer be flavor-diagonal. (Do not attempt to write the Higgs potential or find the exact Higgs mass eigenstates.)

8.2 Verify that there are no Y or Y^3 anomalies in the SM.

8.3 In the SM with a single Higgs doublet ϕ one can always perform an $SU(2) \times U(1)$ transformation so that $\langle\phi\rangle = \frac{1}{\sqrt{2}}\begin{pmatrix} 0 \\ \nu \end{pmatrix}$, with ν real. Therefore, $SU(2) \times U(1) \to U(1)_Q$ and there is a conserved electric charge, Q. However, for two or more doublets with the same hypercharge $y_n = +\frac{1}{2}$ one has

$$\langle\phi_n\rangle = \frac{\nu_n}{\sqrt{2}}\begin{pmatrix} \exp(i\rho_n)\sin\alpha_n \\ \exp(i\sigma_n)\cos\alpha_n \end{pmatrix},$$

where α_1, σ_1, and ρ_1 can be chosen to be zero by an $SU(2) \times U(1)$ transformation, but the other angles are determined by the potential. If any of the α_n are non-zero then $SU(2) \times U(1)$ is completely broken and there is no conserved electric charge. Nontrivial values of σ_n and ρ_n may be associated with CP violation, although in some cases they can be rotated away by field redefinitions. Analyze the two doublet case and show under what conditions it leads to the spontaneous breaking of $U(1)_Q$. To simplify the analysis: consider only renormalizable terms, impose CP invariance (i.e., assume that the coefficients in the potential are real), impose the Z_2 symmetry $\phi_1 \to -\phi_1$with $\phi_2 \to +\phi_2$, ignore terms involving the "tilde fields" analogous to (8.14), and assume that the parameters are such that $\nu_1 \neq 0$, $\nu_2 \neq 0$. Hint: it is not necessary to actually determine the values of the ν_i.

8.4 Suppose the scalar sector of the standard model is extended, so that it includes not only the ordinary Higgs doublet ϕ, but also a new complex field σ. Assume σ transforms as a singlet under the $SU(2)$ gauge group (and also under $SU(3)$ of color), but that it carries weak hypercharge $y_\sigma = 1$ and therefore electric charge $q_\sigma = 1$.
(a) What is the most general renormalizable gauge invariant potential $V(\phi, \sigma)$?
(b) Show that there are some choices of parameters for which not only ϕ but also σ will have non-zero vacuum expectation values.
(c) Suppose that the parameters are such that in unitary gauge,

$$\sigma = \frac{\nu_\sigma + \Sigma}{\sqrt{2}}, \quad \phi = \frac{1}{\sqrt{2}}\begin{pmatrix} 0 \\ \nu + H \end{pmatrix},$$

where $\nu_\sigma = \sqrt{2}\langle\sigma\rangle$ and $\nu = \sqrt{2}\langle\phi^0\rangle$ are the (real) vacuum expectation values, and Σ and H are physical real scalars of definite mass. Assume that $\nu_\sigma \ll \nu$. Show that the photon acquires a small mass, and calculate it to leading nontrivial order in ν_σ/ν.
(d) Show that Z can decay into $\gamma\Sigma$ if it is kinematically allowed, and calculate the rate to leading nonzero order in ν_σ/ν.

8.5 Add a non-chiral lepton doublet $\ell_{L,R}$ with $y = -\frac{1}{2}$, to the $SU(2) \times U(1)$ theory, i.e.,

$$\ell_L = \begin{pmatrix} N \\ E^- \end{pmatrix}_L, \quad \ell_R = \begin{pmatrix} N \\ E^- \end{pmatrix}_R.$$

(a) Write the Lagrangian density (before SSB) for $\ell = \ell_L + \ell_R$, including all allowed kinetic energy, gauge interaction, mass, and Yukawa terms.
(b) Now turn on SSB. Display the couplings of ℓ to $W^\pm$, A, and Z and to the Higgs scalar H.

8.6 The mass term for three chiral fermions $\psi^0_{L,R} = (\psi^0_{1L,R}\,\psi^0_{2L,R}\,\psi^0_{3L,R})^T$ is

$$-\mathcal{L} = \bar{\psi}^0_L M \psi^0_R + h.c., \text{ where } M = \begin{pmatrix} 1 & i & 0.5 \\ 0 & 3+\frac{i}{2} & 2 \\ 0.25 & \frac{i}{7} & 6 \end{pmatrix}.$$

Find the physical mass eigenvalues m_i and unitary matrices $A_{L,R}$ for which $A_L^\dagger M A_R = \text{diag}(m_1\, m_2\, m_3)$ (analogous to (8.47)). Make sure that the m_i are real and nonnegative. Hint: use any convenient numerical package.

8.7 Let ψ_1^0 and ψ_2^0 be two fermion fields, and define

$$\psi^0 \equiv \begin{pmatrix} \psi_1^0 \\ \psi_2^0 \end{pmatrix} \equiv \psi_L^0 + \psi_R^0.$$

Let

$$\mathcal{L} = \bar{\psi}^0 i \not\partial \psi^0 - \left(ic\bar{\psi}^0 \gamma^5 \Gamma \psi^0 + h.c.\right),$$

where c is real and positive and $\Gamma = \begin{pmatrix} 0 & 1 \\ 2 & 0 \end{pmatrix}$. Calculate the physical fermion masses, and express the mass eigenstate fields $\psi_{L,R}$ in terms of $\psi_{L,R}^0$.

8.8 Suppose one adds to the standard model an exotic non-chiral pair of charged leptons, $E_{L,R}^{-0}$, which are both $SU(2)$ singlets with $y_E = -1$. Then, ignoring the second and third families, the leptons are

$$\begin{pmatrix} \nu_L^0 \\ e_L^{-0} \end{pmatrix} \qquad e_R^{-0} \qquad E_L^{-0} \qquad E_R^{-0}.$$

The mass matrix for the charged leptons is (dropping the superscript $-$)

$$-\mathcal{L} = \begin{pmatrix} \bar{e}_L^0 & \bar{E}_L^0 \end{pmatrix} \begin{pmatrix} x & y \\ A & B \end{pmatrix} \begin{pmatrix} e_R^0 \\ E_R^0 \end{pmatrix} + h.c.,$$

where x and y are generated by the VEVs of the Higgs doublet, and A and B are bare masses (or can be generated by the VEVs of an $SU(2)$ singlet Higgs). e_R^0 and E_R^0 have the same quantum numbers, so w.l.o.g. we can take linear combinations such that $A = 0$. We also assume that the parameters are real with $B \gg x, y$.
(a) Find the mass eigenvalues and eigenstates to order x/B and y/B.
(b) Find the weak neutral current J_Z^μ to the lowest nontrivial order in x/B and y/B for each term, and show that it includes flavor changing components.

8.9 A hypothetical future accelerator allows the collisions of e^- with Higgs particles.
(a) Draw the tree-level diagrams for $e^- H \to e^- Z$, and show that only one of them is nonzero for $m_e = 0$.
(b) Calculate the spin-averaged center of mass differential cross section for $e^- H \to e^- Z$ as a function of the CM scattering angle θ, $s = E_{CM}^2$, M_Z, M_H, g, and θ_W. Take $m_e = 0$.

8.10 Suppose the $SU(2) \times U(1)$ model is extended by an additional global or gauge $U(1)$ symmetry that forbids an elementary down Yukawa, i.e., Γ^d in (8.13) must vanish. Introduce a vector pair of $SU(2)$ singlets $D_{L,R}$ and a complex scalar σ with charges that allow

$$-\mathcal{L} = \Gamma^D \bar{q}_L \phi D_R + M_D \bar{D}_L D_R + \kappa\sigma \bar{D}_L d_R + h.c.$$

(a) Display possible $U(1)$ charges consistent with $\mathcal{L}$.
(b) For large M_D one can integrate out the D. Show that this leads to a higher-dimension operator that generates an effective Yukawa coupling for d_R when $\langle\sigma\rangle \neq 0$.

8.11 Use the $C_{1u,d}$ part of the parity-violating e-hadron interaction in (8.118) to calculate the corresponding non-relativistic potential in (8.122). Use the formalism developed in Problem 2.33.

8.12 Verify the expression in (8.173) for the forward-backward asymmetry in $e^-e^+ \to f\bar{f}$ at the Z-pole directly using trace techniques. Neglect the fermion masses. Hint: it is not required to calculate σ_F and σ_B separately, only the combination A_{FB}.

8.13 (a) Find the spin and color-averaged $d\bar{\sigma}/dz$ for $q_r\bar{q}_r \to \mu^-\mu^+$ and $\bar{q}_r q_r \to \mu^-\mu^+$, due to s-channel γ and Z exchange, where z is the cosine of the scattering angle between q_r and μ^-. This can be obtained by appropriately modifying the results in (8.125)–(8.131).
(b) Now consider $pp \to \mu^-\mu^+$ at the LHC, where y is the rapidity of the $\mu^-\mu^+$ pair and $\sqrt{\hat{s}}$ and $\hat{z}$ are respectively the total energy and $\cos\hat{\theta}$ in the $\mu^-\mu^+$ CM. Since both $q_r\bar{q}_r$ and $\bar{q}_r q_r$ can contribute it is conventional to define $\hat{\theta}$ with respect to the direction of the rapidity, i.e., along $\vec{p}_A(\vec{p}_B)$ for $y > 0\,(< 0)$. Find $d\bar{\sigma}/dyd\hat{z}$ near the Z-pole in the narrow width approximation in terms of the luminosity functions, analogous to (6.28) and (6.29).
(c) Find expressions for the forward-backward asymmetry

$$A_{FB}(y) \equiv \frac{\int_0^1 d\hat{z}\frac{d\bar{\sigma}}{dyd\hat{z}} - \int_{-1}^0 d\hat{z}\frac{d\bar{\sigma}}{dyd\hat{z}}}{\int_0^1 d\hat{z}\frac{d\bar{\sigma}}{dyd\hat{z}} + \int_{-1}^0 d\hat{z}\frac{d\bar{\sigma}}{dyd\hat{z}}},$$

and for $A_{FB}(y_1, y_2)$, in which the numerator and denorminator are each integrated over y from $0 \le y_1 < y_2 \le \ln(\sqrt{s}/M_Z)$. Note that these are the same as the charge asymmetries

$$A_c \equiv \frac{\bar{\sigma}(|y_{\mu^-}| > |y_{\mu^+}|) - \bar{\sigma}(|y_{\mu^-}| < |y_{\mu^+}|)}{\bar{\sigma}(|y_{\mu^-}| > |y_{\mu^+}|) + \bar{\sigma}(|y_{\mu^-}| < |y_{\mu^+}|)}.$$

(d) Plot $A_{FB}(y)$ and calculate $A_{FB}(0, \ln(\sqrt{s}/M_Z))$ at the LHC for $\sqrt{s} = 14$ TeV.

8.14 (a) Write the tree-level amplitudes for $e^-(p_1)e^+(p_2) \to W^-(p_3)W^+(p_4)$ corresponding to the three diagrams in Figure 8.29. Neglect m_e and m_{ν_e}, but not M_W or M_Z.
(b) Use any convenient computer algebra program to calculate and plot the spin-averaged cross sections in pb as a function of $\sqrt{s}$ from $2M_W$ to 200 GeV, including just the ν_e, the ν_e and γ, and all three diagrams, and show that they roughly agree with Figure 8.30 (which includes higher-order corrections).

8.15 From (8.55), the interaction Lagrangian for the coupling of the top quark t to the bottom quark b and the W^+ is

$$\mathcal{L} = -\hat{g}W_\mu^- \bar{b}\gamma^\mu(1-\gamma^5)t + h.c.,$$

where $\hat{g} = gV_{tb}^*/2\sqrt{2}$.
(a) Calculate the differential decay rate $d\Gamma/d\cos\theta$ for $t \to bW^+$ in the t rest frame, where θ is the angle between the t spin direction and the b momentum. Sum over the b and W^+ spins. Neglect m_b but keep M_W.
(b) Calculate the numerical value (in s^{-1}) for the total t decay rate into bW^+, ignoring m_b.
(c) Calculate the fractions F_0, F_+, and F_- of all t decays that are into a W with helicity 0 (longitudinal), +1 (R) , or −1 (L), respectively.
(d) One of the $F_{0,\pm}$ should vanish. Interpret that fact in terms of angular momentum.

8.16 Show that the effective interaction in (8.192) reproduces the SM result in (8.190) for the couplings in (8.193).

8.17 Integrate the one-loop RGE in (5.33) and (8.199) numerically (using Mathematica, Maple, Fortran, C++, etc) and verify qualitatively the upper and lower limits on M_H in Figure 8.32.

8.18 The renormalization group equations sometimes have *infrared stable* or *ultraviolet stable fixed points*, which are constant values for couplings or their ratios that are approached asymptotically for $Q^2 \to 0$ or $Q^2 \to \infty$. As a simple example, show that the ratio $h_t(Q^2)/g_s(Q^2)$ has an infrared stable fixed point in the standard model if one neglects g and g', and find its value. The relevant RGE equations are given in (5.33) and (8.199). (The exact solution can be found in (Pendleton and Ross, 1981).) In practice the fixed point may not be reached until Q^2 is too small for the one-loop equations to be valid.

8.19 Derive the decay rate for $H \to GG$ in the standard model from the effective Lagrangian

$$\mathcal{L}_{eff} = \frac{\alpha_s}{12\pi\nu} H G^i_{\mu\nu} G^{i\mu\nu}$$

and show that it agrees with (8.212).

8.20 The discovery of the Higgs boson at the LHC involved pp scattering at 7 and 8 TeV, with ATLAS and CMS each accumulating ~ 5 fb^{-1} of luminosity at each energy. To model this, consider the production and decay of a 125 GeV Higgs at 8 TeV for $L = 10$ fb^{-1}.
(a) Calculate the cross sections at LO for the dominant production models ($GG \to H$, $W^+W^- \to H$, $ZZ \to H$, associated $W^\pm H$, and assocated ZH production). Compare these with the full (NNLO (QCD)+ NLO (EW)) results in Figure 8.35. Use the narrow width approximation for gluon-gluon fusion (GGF). For vector boson fusion (VBF) use the equivalent W approximation, which can be inferred from the Cross-Section article in (Patrignani, 2016) (although it tends to overestimate). The Higgstrahlung cross sections can be directly calculated or can be taken from the same article.
(b) Estimate the total number of signal events for the discovery channels $H \to \gamma\gamma$ and $H \to ZZ^* \to 4\ell$, i.e., $4e$, 4μ, or $2e + 2\mu$.
Hint: Use the theoretical SM values $B(H \to \gamma\gamma) \sim 0.0023$, $B(H \to ZZ^*) \sim 0.026$, $B(H \to GG) \sim 0.086$, $\Gamma_H \sim 4.1$ MeV, and $\Gamma(Z \to e^+e^-) \sim \Gamma(Z \to \mu^+\mu^-) \sim 0.034$.

8.21 Suppose that the standard model were extended by a fourth chiral family (t', b', e', ν') with masses (including that of the ν') $\gg m_t$. Estimate how the rates for $pp \to H \to ZZ^*$ and $pp \to H \to \gamma\gamma$ would be changed relative to the SM, assuming production by gluon fusion. Hint: assume that the widths for GG and $\gamma\gamma$ can be scaled from (8.212) and (8.213).

8.22 Modifications of the Higgs interactions due to new physics can often be described by higher-dimensional operators (HDOs). As a simple model, consider a *Hermitian* scalar field ϕ with

$$\mathcal{L}_\phi = \frac{1}{2}(\partial_\mu\phi)^2 + \frac{\sigma}{2}(\partial_\mu\phi^2)^2 - V(\phi), \qquad V(\phi) = \frac{\mu^2}{2}\phi^2 + \frac{\lambda}{4}\phi^4 + \frac{\rho}{6}\phi^6,$$

with $\mu^2 < 0$ and $\lambda > 0$. The renormalizable part of $\mathcal{L}_\phi$ models the Higgs component ϕ_3 in (8.18) in unitary gauge. The coefficients σ and ρ of the NROs have dimensions mass^{-2}.

They are assumed to be nonnegative and small. We will see that they not only induce new interaction vertices, but also modify the strength of the 3- and 4-point vertices when expressed in terms of the observable VEV and scalar mass.
(a) First take $\sigma = \rho = 0$. Show that $\nu^2 = -\mu^2/\lambda$ and $M_H^2 = 2\lambda\nu^2$, where $\nu \equiv \langle\phi\rangle$ and the "Higgs scalar" is $H = \phi - \nu$. Show that the coefficients of H^3 and H^4 in V are respectively $M_H^2/2\nu$ and $M_H^2/8\nu^2$. Hint: substitute $\mu^2 = -\lambda\nu^2$ in the expression for $V(H+\nu)$.
(b) Now take $\sigma = 0$, $\rho \neq 0$. Show that $\nu^2 = (-\lambda + \sqrt{\lambda^2 - 4\rho\,\mu^2})/2\rho$. Calculate M_H^2 and the coefficients of H^n, $n = 3,4,5,6$, in terms of M_H, ν, and ρ. It is again useful to substitute the expression for μ^2 in $V(H+\nu)$.
(c) For $\sigma \neq 0$, $\rho = 0$, one should perform a field redefinition $\phi = \kappa H + \nu$, where $\nu = \langle\phi\rangle = -\mu^2/\lambda$ and κ is chosen so that the kinetic energy term for H is canonical, i.e., $\frac{1}{2}(\partial_\mu H)^2$. Determine, κ, M_H^2, and the interactions in terms of M_H, ν, and σ to linear order in σ. Hint: there are terms $\propto H^2(\partial_\mu H)^2$ and $H(\partial_\mu H)^2$, as well as H^3 and H^4.

8.23 Calculate the amplitude and cross section for $Z_L Z_L \to W_L^+ W_L^-$ at high energy using the equivalence theorem.

8.24 Use the equivalence theorem to rederive the leading term in m_t/M_W for the polarized differential decay rate for $t \to bW^+$ considered in Problem 8.15a. Neglect m_b.

8.25 Derive (8.210) on page 331, and use the equivalence theorem to verify the leading term for large M_H/M_V.

8.26 Choose the phase conventions for the K^0 and $\bar{K}^0$ fields according to the $SU(3)$ convention in (3.91), and express the pseudoscalar octet fields in terms of the quark fields by $\phi_i = -i\bar{q}\frac{\lambda^i}{2}\gamma^5 q$, where $q = (u\,d\,s)^T$ (as in (5.102)). Calculate the CP phase η_K in (8.231) in terms of the CP phases for the d and s quarks defined in Section 2.10. Show that $\eta_K = -1$ under the additional convention that the d and s have the same CP phases.

8.27 Derive (8.234), (8.235), and the statement that the $I_{3\pi} = 0, 2$ states must have nonzero internal angular momenta.

8.28 Derive (8.244) using the Fierz identities in (2.215) on page 43.

CHAPTER 9

Neutrino Mass and Mixing

DOI: 10.1201/b22175-9

Neutrinos are a unique probe of many aspects of physics, geophysics, and astrophysics o
scales ranging from 10^{-33} to 10^{+28} cm. Neutrino scattering and decays involving neutrinc
have been essential in establishing the Fermi theory and parity violation, determining th
elements of the CKM matrix, and testing the weak neutral current predictions of the star
dard model, and therefore played a significant role in the precision electroweak program a
described in Section 8.3.6. Deep inelastic scattering involving neutrinos and charged lepton
has also been critical in establishing the existence and properties of quarks, the structur
of the nucleon, and the predictions of the short distance behavior of QCD. Similarly, neu
trinos are important for the physics and/or probes of the Sun, Earth, stars, core-collaps
supernovae, the origins of cosmic rays, the large scale structure of the universe, big ban
nucleosynthesis, and possibly baryogenesis.

Neutrinos are also interesting because their masses are so tiny and because, unlike th
quarks, some of the leptonic mixing angles are large. Small neutrino masses are sensitive t
new physics at scales ranging from a TeV up to the Planck scale, but because of their unusua
nature there is a good chance that they are somehow connected with the latter, possibl
shedding light on an underlying grand unification or superstring theory. The neutrinos ar
also unique in that they do not carry either color or electric charge. It is therefore possibl
(and many physicists think probable) that their masses are *Majorana* (lepton number vic
lating) rather than *Dirac* (lepton number conserving, analogous to the quark and charge
lepton masses). Establishing the nature of the neutrino masses, as well as understandin
the origin of the small masses and large mixings, is of fundamental interest.

The original version of the $SU(2) \times U(1)$ model did not have any mechanism to generat
nonzero masses at the renormalizable level. However, it is straightforward to extend th
original model by the addition of $SU(2)$-singlet right-chiral neutrinos,[1] allowing Dirac mas
terms. These could yield light Dirac neutrinos if the Yukawa couplings are extremely smal
as could occur, for example, if the Yukawa couplings are forbidden at tree-level by some nev
symmetry. Alternatively, $SU(2)$-singlet neutrinos could lead to light Majorana neutrinc
through the *seesaw* mechanism. One could also generate small Majorana masses withou
right-handed neutrinos via extended Higgs sectors or higher-dimension operators. Mos
extensions of the standard model (with the notable exception of the minimal $SU(5)$ gran
unified theory) involve either $SU(2)$-singlet neutrinos or extended Higgs sectors, thoug
they do not necessarily explain the smallness of the masses.

In this chapter we review the basic issues related to the neutrino masses and mixings
the major classes of models, and some of the experiments. More detailed discussions ma

[1]Also referred to as singlet, right-handed, or sterile neutrinos.

be found in a number of books (Bahcall, 1989; Langacker, 2000; Mohapatra and Pal, 2004; Giunti and Kim, 2007; Bilenky, 2010; Xing and Zhou, 2011; Barger et al., 2012; Zuber, 2012; Valle and Romao, 2015; Suekane, 2015) and review articles (Raffelt, 1999; Dolgov, 2002; Strumia and Vissani, 2006; Mohapatra et al., 2007; Gonzalez-Garcia and Maltoni, 2008; Camilleri et al., 2008; de Gouvêa et al., 2013; Patrignani, 2016).

9.1 BASIC CONCEPTS FOR NEUTRINO MASS

9.1.1 Active and Sterile Neutrinos

We saw in Chapter 2 that the minimal fermionic degree of freedom is a Weyl two-component field, as defined in (2.200) on page 41 and in Section 2.11. A Weyl field can be represented either in four-component notation, e.g., by $\psi_L = P_L\psi = \begin{pmatrix} \Psi_L \\ 0 \end{pmatrix}$, or in two-component notation as Ψ_L. We will use both in this section.

It is useful to first distinguish between *active* and *sterile* neutrinos. Active (a.k.a. ordinary or doublet) neutrinos are left-chiral Weyl neutrinos that transform as $SU(2)$ doublets with a charged lepton partner, and which therefore have normal weak interactions. The electron-type L doublet and its right-handed partner are

$$\ell_L \equiv \begin{pmatrix} \nu_{eL} \\ e^-_L \end{pmatrix} \xrightarrow[CP]{} \tilde{\ell}^c_R \equiv \begin{pmatrix} e^+_R \\ -\nu^c_{eR} \end{pmatrix}, \tag{9.1}$$

in four-component notation, where $\psi^c_R = C\bar{\psi}^T_L$ is the field related by CP to ψ_L up to γ matrices and a possible CP phase, as in (2.306) on page 57 (or (2.333) in two-component notation). We have also carried out a "tilde" transformation on the $SU(2)$ doublet indices, analogous to the one for the Higgs in (8.14) on page 260, so that $\tilde{\ell}^c$ transforms as a 2 rather than a 2*. We reemphasize that we define $\bar{\psi}_L \equiv (\psi_L)^\dagger \gamma^0 = (P_L\psi)^\dagger \gamma^0$, i.e., the Dirac adjoint acts on ψ_L and not on ψ.

Sterile (a.k.a. singlet or "right-handed") neutrinos, which are present in most extensions of the SM, are $SU(2)$ singlets. They do not interact except by mixing, Yukawa interactions, or beyond the SM (BSM) interactions. In four-component notation, a sterile right-chiral Weyl field will be written as ν_R and its conjugate as ν^c_L,

$$\nu_R \xrightarrow[CP]{} \nu^c_L. \tag{9.2}$$

In two-component notation, the L and R chiral fields will be written as $\mathcal{N}_L$ and $\mathcal{N}_R$, respectively, with their CP conjugates $\mathcal{N}^c_R$ and $\mathcal{N}^c_L$:

$$\mathcal{N}_L \xrightarrow[CP]{} \mathcal{N}^c_R \quad \text{(active)}, \qquad \mathcal{N}_R \xrightarrow[CP]{} \mathcal{N}^c_L \quad \text{(sterile)}. \tag{9.3}$$

9.1.2 Dirac and Majorana Masses

Dirac Masses

A fermion mass term converts a Weyl field of one chirality into one of the opposite chirality,

$$-\mathcal{L} = m\left(\bar{\psi}_{aL}\psi_{bR} + \bar{\psi}_{bR}\psi_{aL}\right) = m\left(\Psi^\dagger_{aL}\Psi_{bR} + \Psi^\dagger_{bR}\Psi_{aL}\right), \tag{9.4}$$

as in (2.347) on page 63. (We have taken m to be real.) A physical interpretation is that a massless fermion has the same helicity (chirality) in all frames of reference, while that of a massive particle depends on the reference frame and therefore can be flipped.

A *Dirac mass* connects two distinct Weyl fields, i.e., $\Psi_{bR} \neq \Psi^c_{aR}$. For a single type of neutrino, a Dirac mass connects an active neutrino with a sterile one,[2]

$$\begin{aligned} -\mathcal{L}_D &= m_D\left(\bar{\nu}_L\nu_R + \bar{\nu}_R\nu_L\right) = m_D\bar{\nu}_D\nu_D \\ &= m_D\left(\mathcal{N}_L^\dagger\mathcal{N}_R + \mathcal{N}_R^\dagger\mathcal{N}_L\right), \end{aligned} \tag{9.5}$$

where $\nu_D = \nu_L + \nu_R$ is a Dirac field. It has four distinct components, ν_L, ν^c_R, ν_R and ν^c_L, and there is a conserved fermion number (or lepton number L in this case), corresponding to the global phase symmetry $\nu_{L,R} \to e^{i\beta}\nu_{L,R}$. This L conservation ensures that there is no mixing between ν_L and ν^c_L, or between ν_R and ν^c_R. However, when embedded in the SM context the neutrinos are chiral, so m_D violates the third component T^3_L of weak isospin by $\Delta t^3_L = \frac{1}{2}$. It can be generated by the Higgs mechanism, as described in Section 8.2.1 and illustrated in Figure 9.1, and is in principle analogous to the quark and charged lepton masses. Dirac masses can be easily generalized to three or more families. However, the tiny values of the neutrino masses require that the Higgs Yukawa couplings $h_\nu = m_\nu/\nu$ defined in (8.46) on page 267 would have to be extremely small if they are due to a simple Dirac-type Higgs coupling: $m_\nu \sim 0.1$ eV would correspond to $h_\nu \sim 10^{-12}$, for example, to be compared with the t quark coupling $h_t = \mathcal{O}(1)$ or the electron coupling $h_e \sim 10^{-5}$. Of course, we do not understand the ratio h_e/h_t either, so some caution should be taken with such statements. In any case, most particle physicists believe that either an alternative mechanism or some explanation for a small h_ν is needed, as will be discussed below.

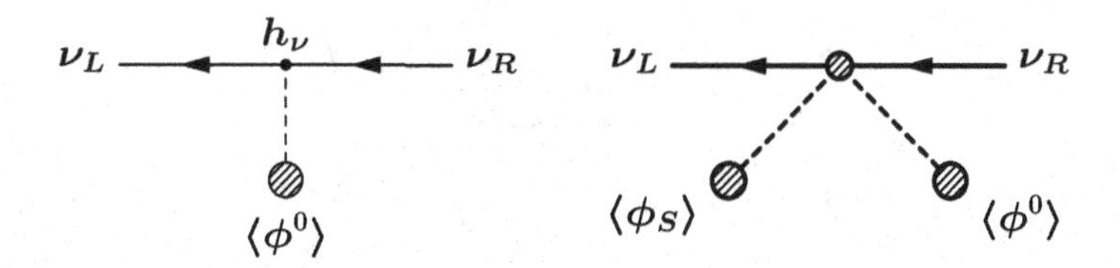

Figure 9.1 Mechanisms for generating a Dirac neutrino mass. Left: an elementary Yukawa coupling to the neutral Higgs doublet field ϕ^0. Right: a higher-dimensional operator leading to a suppressed Yukawa coupling.

Majorana Masses

Majorana mass terms are more economical in that they only require a single Weyl field, i.e., $\Psi_{bR} = \Psi^c_{aR}$ in (9.4). They are not as familiar as Dirac mass terms because they violate fermion number by two units. For the quarks and charged leptons such mass terms are forbidden because they would violate color and/or electric charge. However, the neutrinos do not carry any unbroken gauge quantum numbers, so Majorana masses are a possibility.

For an active neutrino, a Majorana mass term describes a transition between a left-handed neutrino and its conjugate right-handed antineutrino. In four-component language, it can be written

$$-\mathcal{L}_T = \frac{m_T}{2}\left(\bar{\nu}_L\nu^c_R + \bar{\nu}^c_R\nu_L\right) = \frac{m_T}{2}\left(\bar{\nu}_L C\bar{\nu}^T_L + \nu^T_L C\nu_L\right) = \frac{m_T}{2}\bar{\nu}_M\nu_M. \tag{9.6}$$

[2]There are variant forms of Dirac neutrino masses involving two distinct active or two distinct sterile neutrinos, as will be discussed below.

As is clear from the second form, $\mathcal{L}_T$ can be viewed as the annihilation or creation of two neutrinos, and therefore violates lepton number by two units, $\Delta L = 2$. In the last form in (9.6), $\nu_M \equiv \nu_L + \nu_R^c$ is a self-conjugate[3] two-component (Majorana) field satisfying[4] $\nu_M = \nu_M^c \equiv C\bar{\nu}_M^T$. A Majorana ν is therefore its own antiparticle and can mediate *neutrinoless double beta decay* ($\beta\beta_{0\nu}$), in which two neutrons turn into two protons and two electrons, violating lepton number by two units, as shown in Figure 9.2. A Majorana mass for an active neutrino also violates weak isospin by one unit, $\Delta t_L^3 = 1$ (hence the subscript T for triplet), and can be generated either by the VEV of a Higgs triplet or by a higher-dimensional operator involving two Higgs doublets (such as the minimal seesaw model), as in Figure 9.3. The $\frac{1}{2}$ in $\mathcal{L}_T$ is needed to yield the correct expression for the Hamiltonian. It is somewhat analogous to the extra $\frac{1}{2}$ in the free-field Lagrangian density for a Hermitian scalar. This is more obvious from the kinetic energy term

$$\mathcal{L}_{KE} = \frac{1}{2}(\bar{\nu}_M i \not{\partial} \nu_M) = \frac{1}{2}\left(\bar{\nu}_L i \not{\partial} \nu_L + \bar{\nu}_R^c i \not{\partial} \nu_R^c\right) = \bar{\nu}_L i \not{\partial} \nu_L, \tag{9.7}$$

where the two terms are equal because of (2.302) on page 56. The Majorana mass term can be rewritten in two-component language using (2.331) as

$$-\mathcal{L}_T = \frac{m_T}{2}\left(\mathcal{N}_L^\dagger \mathcal{N}_R^c + \mathcal{N}_R^{c\dagger} \mathcal{N}_L\right) = \frac{m_T}{2}\left(\mathcal{N}_L^\dagger i\sigma^2 \mathcal{N}_L^* - \mathcal{N}_L^T i\sigma^2 \mathcal{N}_L\right). \tag{9.8}$$

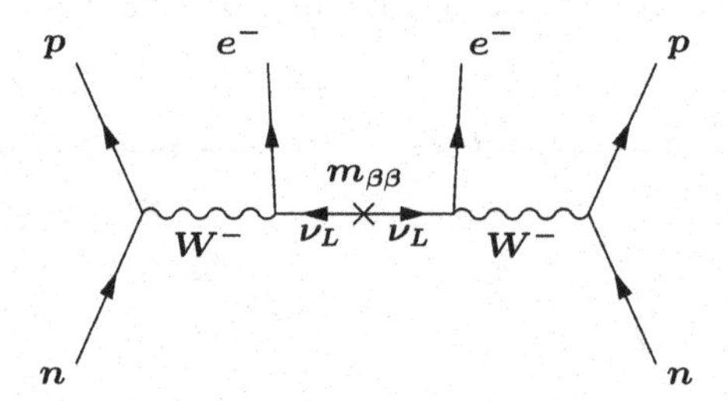

Figure 9.2 Diagram for neutrinoless double beta decay ($\beta\beta_{0\nu}$). For a single Majorana neutrino, $m_{\beta\beta}$ is just m_T in (9.6).

A sterile neutrino can also have a Majorana mass term of the form,

$$-\mathcal{L}_S = \frac{m_S}{2}\left(\bar{\nu}_L^c \nu_R + \bar{\nu}_R \nu_L^c\right) = \frac{m_S}{2}\left(\bar{\nu}_L^c C \bar{\nu}_L^{cT} + \nu_L^{cT} C \nu_L^c\right) = \frac{m_S}{2}\bar{\nu}_{M_S}\nu_{M_S}, \tag{9.9}$$

where $\nu_{M_S} \equiv \nu_L^c + \nu_R = \nu_{M_S}^c$. In this case, weak isospin is conserved, $\Delta t_L^3 = 0$, so m_S can be generated by the VEV of a Higgs singlet.[5] In two-component form

$$-\mathcal{L}_S = \frac{m_S}{2}\left(\mathcal{N}_L^{c\dagger} \mathcal{N}_R + \mathcal{N}_R^\dagger \mathcal{N}_L^c\right) = \frac{m_S}{2}\left(\mathcal{N}_L^{c\dagger} i\sigma^2 \mathcal{N}_L^{c*} - \mathcal{N}_L^{cT} i\sigma^2 \mathcal{N}_L^c\right). \tag{9.10}$$

[3]Unlike a Hermitian scalar, a Majorana state still has two helicities, corresponding to ν_L and ν_R^c. They only mix by the Majorana mass term, so there is still an approximately conserved lepton number to the extent that m_T is small. For example, there could be a cosmological asymmetry between ν_L and ν_R^c, even for Majorana masses, if the rate for transitions between them is sufficiently slow compared to the age of the universe (Barger et al., 2003a).

[4]We have taken a convention in which $\eta_\nu = 1$ for the CP phase in (2.306). More generally, $\nu_M \equiv \nu_L + \eta_\nu^* \nu_R^c = \eta_\nu^* C\bar{\nu}_M^T$, where ν_R^c is still defined as $C\bar{\nu}_L^T$.

[5]m_S could also be generated in principle by a bare mass, but this is usually forbidden by additional symmetries in extensions of the SM.

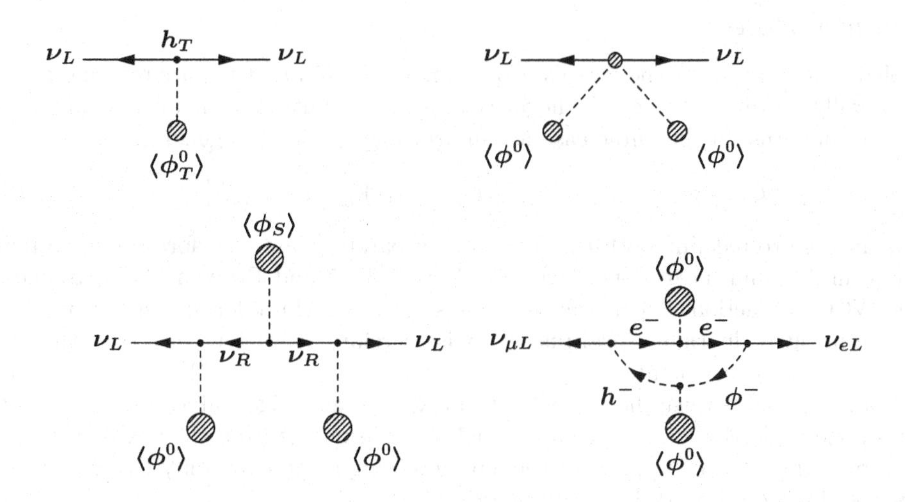

Figure 9.3 Mechanisms for a Majorana mass term. Top left: coupling to a neutral Higgs triplet field ϕ_T^0. Top right: a higher-dimensional operator coupling to two Higgs doublets. Botton left: the minimal seesaw mechanism (a specific implementation of the higher-dimensional operator), in which a light active neutrino mixes with a very heavy sterile Majorana neutrino. Bottom right: a loop diagram involving a charged scalar field h^-.

The four-component fields for Dirac and for active and sterile Majorana neutrinos can therefore be written

$$\begin{aligned} \nu_D &= \begin{pmatrix} \mathcal{N}_L \\ \mathcal{N}_R \end{pmatrix} = \begin{pmatrix} \mathcal{N}_L \\ i\sigma^2 \mathcal{N}_L^{c*} \end{pmatrix} \\ \nu_M &= \begin{pmatrix} \mathcal{N}_L \\ \mathcal{N}_R^c \end{pmatrix} = \begin{pmatrix} \mathcal{N}_L \\ i\sigma^2 \mathcal{N}_L^* \end{pmatrix}, \qquad \nu_{M_S} = \begin{pmatrix} \mathcal{N}_L^c \\ \mathcal{N}_R \end{pmatrix} = \begin{pmatrix} \mathcal{N}_L^c \\ i\sigma^2 \mathcal{N}_L^{c*} \end{pmatrix}, \end{aligned} \tag{9.11}$$

where two of the components are not independent in the Majorana cases.

The free-field equations of motion with a Majorana mass term obtained from the Euler-Lagrange equation (2.19) on page 9 are

$$i\,\partial\!\!\!/ \nu_L - m_T \mathcal{C}\bar{\nu}_L^T = 0, \qquad i\bar{\sigma}^\mu \partial_\mu \mathcal{N}_L - m_T i\sigma^2 \mathcal{N}_L^* = 0. \tag{9.12}$$

This leads to the free-field expression

$$\nu_M(x) = \int \frac{d^3\vec{p}}{(2\pi)^3 2E_p} \sum_{s=+,-} \left[u(\vec{p},s)\, a(\vec{p},s) e^{-ip\cdot x} + v(\vec{p},s) a^\dagger(\vec{p},s) e^{+ip\cdot x} \right], \tag{9.13}$$

which is of the same form as the free Dirac field in (2.159) on page 35 except that there is no distinction between a and b operators. Any spin basis can be used, but the helicity basis is usually most convenient.

A more compact notation for dealing with Majorana masses and fields will be developed in Section 10.2.2.

A Comment on Phases

We implicitly assumed that the masses m_D and m_T in (9.5) and (9.6) are real and positive. More generally, however, they can be negative or complex, but can be made real and positive by field redefinitions. In the Dirac case for an arbitrary ψ, one has generally

$$-\mathcal{L}_D = m_D\bar{\psi}_L\psi_R + m_D^*\bar{\psi}_R\psi_L = m_D\Psi_L^\dagger\Psi_R + m_D^*\Psi_R^\dagger\Psi_L. \tag{9.14}$$

There is freedom to redefine both ψ_L and ψ_R by separate phase transformations to remove any phase in m_D and make it positive (see Section 3.3.5 and Problem 3.32). In the SM, only the WCC interactions depend on the phases of the left-chiral fermion fields, and no SM interaction involves the right chiral phases. It is therefore convenient to choose the phases of the L-chiral mass eigenstates (ψ_L in the simple example in (9.14)) by any convenient convention, and then adjust the ψ_R fields to make the mass eigenvalues real and positive. This was done in Section 8.2.2 to remove unobservable phases from the CKM matrix, and a similar procedure can be applied to the lepton mixing if there are only Dirac masses.

However, for a general Majorana mass term,

$$-\mathcal{L}_M = \frac{1}{2}\left(m_M\bar{\psi}_L\mathcal{C}\bar{\psi}_L^T + m_M^*\psi_L^T\mathcal{C}\psi_L\right) = \frac{1}{2}\left(m_M\Psi_L^\dagger i\sigma^2\Psi_L^* - m_M^*\Psi_L^T i\sigma^2\Psi_L\right), \tag{9.15}$$

there is only one independent field ψ_L (or Ψ_L). One usually chooses to use the phase freedom in ψ_L to make m_M real and positive, in which case there is no remaining freedom. This will imply the existence of additional *Majorana phases* in the leptonic mixing matrix V_ℓ for the WCC for the case of Majorana neutrino masses. However, such phases are only observable in processes such as $\beta\beta_{0\nu}$ that involve the phases of the neutrino masses explicitly, as can be seen by working in an alternative convention of a simple V_ℓ but leaving m_M complex.

Even after a phase redefinition we will define the conjugate fields by the same convention as for the original ones. That is, for

$$\nu_L' \equiv e^{i\beta}\nu_L, \qquad \mathcal{N}_L' = e^{i\beta}\mathcal{N}_L, \tag{9.16}$$

we define

$$\nu_R^{c\prime} \equiv \mathcal{C}\bar{\nu}_L^{\prime T}, \qquad \mathcal{N}_R^{c\prime} = i\sigma^2\mathcal{N}_L^{\prime *}. \tag{9.17}$$

Mixed Models

When active and sterile neutrinos are both present, there can be Dirac and Majorana mass terms simultaneously. For one family, the Lagrangian density has the form

$$-\mathcal{L} = \frac{1}{2}\begin{pmatrix}\bar{\nu}_L^0 & \bar{\nu}_L^{0c}\end{pmatrix}\begin{pmatrix}m_T & m_D\\ m_D & m_S\end{pmatrix}\begin{pmatrix}\nu_R^{0c}\\ \nu_R^0\end{pmatrix} + h.c., \tag{9.18}$$

where 0 refers to weak eigenstates, and the masses are

$$\begin{aligned} m_T:&\quad |\Delta L| = 2, && \Delta t_L^3 = 1 && \text{(Majorana)}\\ m_D:&\quad |\Delta L| = 0, && \Delta t_L^3 = \frac{1}{2} && \text{(Dirac)}\\ m_S:&\quad |\Delta L| = 2, && \Delta t_L^3 = 0 && \text{(Majorana).}\end{aligned} \tag{9.19}$$

The two terms involving m_D are equal since

$$\begin{aligned} \bar{\psi}_{aL}^c\psi_{bR}^c &= \bar{\psi}_{bL}\psi_{aR}, & \bar{\psi}_{aR}^c\psi_{bL}^c &= \bar{\psi}_{bR}\psi_{aL}\\ \bar{\psi}_{aL}\psi_{bR}^c &= \bar{\psi}_{bL}\psi_{aR}^c, & \bar{\psi}_{aL}^c\psi_{bR} &= \bar{\psi}_{bL}^c\psi_{aR}\end{aligned} \tag{9.20}$$

for arbitrary $\psi_{a,b}$ by (2.296) on page 56. Diagonalizing the matrix in (9.18) yields two Majorana mass eigenvalues m_i and two Majorana mass eigenstates,[6] $\nu_{iM} = \nu_{iL} + \nu^c_{iR} = \nu^c_{iM}$, with $i = 1, 2$. The weak and mass bases are related by the unitary transformations

$$\begin{pmatrix} \nu_{1L} \\ \nu_{2L} \end{pmatrix} = A_L^{\nu\dagger} \begin{pmatrix} \nu_L^0 \\ \nu_L^{0c} \end{pmatrix}, \qquad \begin{pmatrix} \nu^c_{1R} \\ \nu^c_{2R} \end{pmatrix} = A_R^{\nu\dagger} \begin{pmatrix} \nu_R^{0c} \\ \nu_R^0 \end{pmatrix}, \tag{9.21}$$

similar to (8.47) on page 269. A_L and A_R are generally different for Dirac mass matrices, which need not be Hermitian. However, the general 2×2 neutrino mass matrix in (9.18) is symmetric because of (9.20). This implies that $A_L^\nu = A_R^{\nu *} K$, where K is a diagonal matrix of phases analogous to (8.52). (There is additional freedom in K in the presence of degeneracies.) K is in general arbitrary, but our phase convention, in which $\nu^c_{iR} = \mathcal{C}\bar{\nu}^T_{iL}$, implies $K = I$.

There are several important special cases of the mixed model in (9.18):

(a) Majorana: $m_D = 0$ is the pure Majorana case: the mass matrix is diagonal, with $m_1 = m_T$, $m_2 = m_S$, and

$$\begin{aligned} \nu_{1L} &= \nu_L^0, \qquad & \nu^c_{1R} &= \nu_R^{0c} \\ \nu_{2L} &= \nu_L^{0c}, \qquad & \nu^c_{2R} &= \nu_R^0. \end{aligned} \tag{9.22}$$

(b) Dirac: The Dirac limit is $m_T = m_S = 0$. There are formally two Majorana mass eigenstates, with eigenvalues $m_1 = m_D$ and $m_2 = -m_D$ and eigenstates

$$\begin{aligned} \nu_{1L} &= \frac{1}{\sqrt{2}}(\nu_L^0 + \nu_L^{0c}), \qquad & \nu^c_{1R} &= \frac{1}{\sqrt{2}}(\nu_R^{0c} + \nu_R^0) \\ \nu_{2L} &= \frac{1}{\sqrt{2}}(\nu_L^0 - \nu_L^{0c}), \qquad & \nu^c_{2R} &= \frac{1}{\sqrt{2}}(\nu_R^{0c} - \nu_R^0). \end{aligned} \tag{9.23}$$

Note that $\nu_{1,2}$ are degenerate in the sense that $|m_1| = |m_2|$, but the actual eigenvalues have opposite sign. To recover the Dirac limit, we will depart from our usual procedure of redefining the phase of ν_2 to make m_2 positive. Rather, let us expand the mass term

$$-\mathcal{L} = \frac{m_D}{2}(\bar{\nu}_{1L}\nu^c_{1R} - \bar{\nu}_{2L}\nu^c_{2R}) + h.c. = m_D(\bar{\nu}_L^0 \nu_R^0 + \bar{\nu}_R^0 \nu_L^0), \tag{9.24}$$

which clearly conserves lepton number (i.e., there is no $\nu_L^0 - \nu_L^{0c}$ or $\nu_R^{0c} - \nu_R^0$ mixing). Thus, a Dirac neutrino can be thought of as two Majorana neutrinos, with maximal (45°) mixing and with equal and opposite masses. This interpretation is useful in considering the Dirac limit of general models.

(c) Seesaw: The limit $m_S \gg m_{D,T}$ (e.g., $m_T = 0$, $m_D = \mathcal{O}(m_u, m_e, m_d)$, and $m_S = \mathcal{O}(M_X)$, where $M_X \sim 10^{14}$ GeV) is known as the *seesaw* (or *minimal seesaw*) (Minkowski, 1977; Gell-Mann et al., 1979; Yanagida, 1979; Schechter and Valle,

[6]For Majorana masses, and especially in mixed models, there is no conserved lepton number, and it is just a matter of definition to refer to the L states as particles, ν_{iL}, and the R states as antiparticles, ν^c_{iR}. Unfortunately, this becomes awkward in the Dirac limit where the Weyl states ν_R and ν^c_R are labeled as particle and antiparticle because they carry lepton number $+1$ and -1, respectively. There is no notation known to the author that is not awkward in some circumstances.

1980). The eigenstates and eigenvalues in the seesaw limit are

$$\begin{aligned} \nu_{1L} &\sim \nu_L^0 - \frac{m_D}{m_S}\nu_L^{0c} \sim \nu_L^0, & m_1 &\sim m_T - \frac{m_D^2}{m_S} \\ \nu_{2L} &\sim \frac{m_D}{m_S}\nu_L^0 + \nu_L^{0c} \sim \nu_L^{0c}, & m_2 &\sim m_S, \end{aligned} \tag{9.25}$$

with $|m_1| \ll m_D$ for $m_T = 0$. At energies low compared to m_S the ν_2 decouples and one obtains an effective theory involving a single active Majorana $\nu_{1M} \sim \nu_L^0 + \nu_R^{0c}$. The minimal seesaw mechanism is illustrated in Figure 9.3.

(d) **Pseudo-Dirac:** this is a perturbation on the Dirac case, with $m_T,\ m_S \ll m_D$. There is a small lepton number violation, and a small splitting between the magnitudes of the mass eigenvalues. As an example, $m_T = \epsilon,\ m_S = 0$ leads to $|m_{1,2}| = m_D \pm \epsilon/2$. The pseudo-Dirac case is also sometimes encountered for the variant Dirac forms involving two active or two sterile neutrinos.

(e) **Mixing:** The general case in which m_D and m_S (and/or m_T) are both small and comparable leads to non-degenerate Majorana mass eigenvalues and significant ordinary-sterile ($\nu_L^0 - \nu_L^{0c}$) mixing, such as were suggested by the LSND and some subsequent results to be described below. Only this and the pseudo-Dirac cases allow such mixings.

9.1.3 Extension to Two or More Families

These considerations can be generalized to two or more families, or even to the case of different numbers of active and sterile neutrinos. For $F = 3$ families, define the three-component weak eigenstate vectors

$$\nu_L^0 = \begin{pmatrix} \nu_{1L}^0 \\ \nu_{2L}^0 \\ \nu_{3L}^0 \end{pmatrix}, \qquad \nu_L^{0c} = \begin{pmatrix} \nu_{1L}^{0c} \\ \nu_{2L}^{0c} \\ \nu_{3L}^{0c} \end{pmatrix}, \tag{9.26}$$

and similarly for ν_R^{0c}, ν_R^0 and the two-component fields. In the Dirac case, (9.5) generalizes to

$$-\mathcal{L}_D = \left(\bar{\nu}_L^0 M_D \nu_R^0 + \bar{\nu}_R^0 M_D^\dagger \nu_L^0\right) = \left(\mathcal{N}_L^{0\dagger} M_D \mathcal{N}_R^0 + \mathcal{N}_R^{0\dagger} M_D^\dagger \mathcal{N}_L^0\right), \tag{9.27}$$

where M_D is a completely arbitrary 3×3 matrix. M_D can be diagonalized by separate left and right unitary matrices $A_{L,R}^\nu$, just as in (8.47) on page 269,

$$A_L^{\nu\dagger} M_D A_R^\nu = (M_D)_D = \begin{pmatrix} m_1 & 0 & 0 \\ 0 & m_2 & 0 \\ 0 & 0 & m_3 \end{pmatrix}, \tag{9.28}$$

with mass eigenstate fields

$$\nu_L = A_L^{\nu\dagger}\nu_L^0, \qquad \nu_R = A_R^{\nu\dagger}\nu_R^0. \tag{9.29}$$

The leptonic part of the weak charge raising current is then

$$J_W^{\ell\mu\dagger} = 2\bar{\nu}_L\gamma^\mu V_\ell e_L, \tag{9.30}$$

where

$$V_\ell = A_L^{\nu\dagger} A_L^e, \tag{9.31}$$

is the leptonic mixing matrix, analogous to the quark mixing matrix V_q in (8.61). $\mathcal{V} \equiv V_\ell^\dagger$ is also know as the PMNS matrix, after Maki, Nakagawa, and Sakata (Maki et al., 1962) and Pontecorvo (Pontecorvo, 1968). The counting of angles and phases in V_ℓ is the same as for the CKM matrix (see Equation 8.62). One can choose the arbitrary phases in A_L^ν (i.e., the K_L^ν matrix analogous to (8.51)) to remove unobservable phases from V_ℓ (leaving 3 angles and one phase for $F = 3$), and then choose those in A_R^ν (i.e., K_R^ν) to make the mass eigenvalues real and nonnegative.

For $F_A = 3$ active and $F_S > 3$ sterile neutrinos (and only Dirac masses), three linear combinations of the sterile fields will join with the active ones to form three massive Dirac fields, leaving $F_S - 3$ massless Weyl fields. The reverse situation occurs for $F_S < F_A$.

In the Majorana case the triplet mass term (9.6) generalizes to

$$-\mathcal{L}_T = \frac{1}{2}\left(\bar{\nu}_L^0 M_T \nu_R^{0c} + \bar{\nu}_R^{0c} M_T^\dagger \nu_L^0\right) = \frac{1}{2}\left(\bar{\nu}_L^0 M_T \mathcal{C} \bar{\nu}_L^{0T} + \nu_L^{0T} \mathcal{C} M_T^\dagger \nu_L^0\right), \tag{9.32}$$

where M_T is symmetric, $M_T = M_T^T$, by (9.20). This symmetric property holds for all Majorana mass matrices. The corresponding two-component expression is

$$-\mathcal{L}_T = \frac{1}{2}\left(\mathcal{N}_L^{0\dagger} M_T i\sigma^2 \mathcal{N}_L^{0*} - \mathcal{N}_L^{0T} i\sigma^2 M_T^\dagger \mathcal{N}_L^0\right). \tag{9.33}$$

Since M_T is symmetric, it can be diagonalized by a transformation analogous to (9.28), but with $A_L^\nu = A_R^{\nu *} K$, where as usual K is an undetermined diagonal phase matrix in the absence of degeneracies. We will choose $K = I$ in order to maintain $\nu_{iR}^c = \mathcal{C}\bar{\nu}_{iL}^T$ for the eigenstates (cf., Equation 9.17). In that case, the phases in A_L^ν are uniquely determined by the requirement that the mass eigenvalues m_i are real and positive.[7] Consequently, there is less freedom to remove phases from V_ℓ than in the Dirac case (or in the CKM matrix). The counting in (8.62) is modified in that one can only remove the F phases associated with the charged lepton fields from the F^2 parameters in a general unitary $F \times F$ matrix, so that there are $F(F-1)/2$ mixing angles and $F(F-1)/2$ observable CP-violating phases. The upshot is that V_ℓ can be written

$$V_\ell = K_M^\nu \hat{V}_\ell, \tag{9.34}$$

where $\hat{V}_\ell$ involves $\frac{1}{2}(F-1)(F-2)$ phases analogous to those in the CKM matrix, such as the single phase for $F = 3$ displayed in (8.214) on page 340. K_M^ν is a diagonal matrix of Majorana phases. Only the $F-1$ phase differences are observable, so one often takes it to be of the form $K_M^\nu = \text{diag}(e^{-i\alpha_1}, \cdots, e^{-i\alpha_{F-1}}, 1)$. Since the Majorana phases only multiply the mass eigenstate fields, they do not enter any amplitude involving only external neutrinos or those involving an ordinary (lepton-number conserving) internal neutrino line. They do affect amplitudes involving lepton number violation, such as the $\beta\beta_{0\nu}$ amplitude in Figure 9.2.

To summarize, for $F = 3$ families, the adjoint of the leptonic mixing matrix can be parametrized by

$$V_\ell^\dagger = \begin{pmatrix} 1 & 0 & 0 \\ 0 & c_{23} & s_{23} \\ 0 & -s_{23} & c_{23} \end{pmatrix} \begin{pmatrix} c_{13} & 0 & s_{13}e^{-i\delta} \\ 0 & 1 & 0 \\ -s_{13}e^{i\delta} & 0 & c_{13} \end{pmatrix} \begin{pmatrix} c_{12} & s_{12} & 0 \\ -s_{12} & c_{12} & 0 \\ 0 & 0 & 1 \end{pmatrix} \begin{pmatrix} e^{i\alpha_1} & 0 & 0 \\ 0 & e^{i\alpha_2} & 0 \\ 0 & 0 & 1 \end{pmatrix}, \tag{9.35}$$

[7]The phase would be undetermined but unobservable for a zero eigenvalue.

where $c_{ij} \equiv \cos\theta_{ij}$ and $s_{ij} \equiv \sin\theta_{ij}$ and δ are leptonic mixing angles and a CP-violating phase similar to (but numerically different from) the angles in the CKM matrix in (8.214), and $\alpha_{1,2}$ are Majorana phases. The same form holds for Dirac masses, except the last factor becomes the identity. Note that we are following the conventions in (Patrignani, 2016), and that (8.214) refers to V_q while (9.35) refers to $V_\ell^\dagger$. To conform to standard notations, we will also define the mixing matrix

$$\mathcal{V} \equiv V_\ell^\dagger. \tag{9.36}$$

It is often convenient to choose a basis for the lepton doublets in which the charged lepton mass matrix is already diagonalized, i.e., $A_L^e = I$. Then, $\mathcal{V} = A_L^\nu$.

In the general multi-family case, with both Dirac and Majorana masses,

$$-\mathcal{L} = \frac{1}{2} \begin{pmatrix} \bar{\nu}_L^0 & \bar{\nu}_L^{0c} \end{pmatrix} \begin{pmatrix} M_T & M_D \\ M_D^T & M_S \end{pmatrix} \begin{pmatrix} \nu_R^{0c} \\ \nu_R^0 \end{pmatrix} + h.c., \tag{9.37}$$

where $M_T = M_T^T$ and $M_S = M_S^T$. For $F = 3$ there are six Majorana mass eigenvalues and eigenvectors. The transformation to the mass eigenstate basis is

$$\nu_L = \mathcal{A}_L^{\nu\dagger} \begin{pmatrix} \nu_L^0 \\ \nu_L^{0c} \end{pmatrix}, \tag{9.38}$$

where $\mathcal{A}_L^\nu$ is a 6×6 unitary matrix and ν_L is a six-component vector. The analogous transformation for the R fields involves $\mathcal{A}_R^\nu = \mathcal{A}_L^{\nu*}\mathcal{K}$ because the 6×6 Majorana mass matrix is symmetric. Our phase convention $\nu_R^c = \mathcal{C}\bar{\nu}_L^T$ again implies the choice $\mathcal{K} = I$.

Analogous to (9.25), the seesaw limit of (9.37) occurs when the three eigenvalues of M_S are all large compared to the elements of M_D and M_T (the latter is usually assumed to be zero). Then, one has

$$\mathcal{A}_L^{\nu\dagger} = \mathcal{A}_R^{\nu T} = \begin{pmatrix} A_L^{\nu\dagger} & 0 \\ 0 & A_L^{\nu_S\dagger} \end{pmatrix} B_L^{\nu\dagger}, \tag{9.39}$$

where A_L^ν and $A_L^{\nu_S}$ are 3×3 unitary matrices, I and 0 are, respectively, the 3×3 identity and zero matrices, and

$$B_L^{\nu\dagger} \sim \begin{pmatrix} I & -M_D M_S^{-1} \\ M_S^{-1\dagger} M_D^\dagger & I \end{pmatrix}. \tag{9.40}$$

One finds

$$B_L^{\nu\dagger} \begin{pmatrix} M_T & M_D \\ M_D^T & M_S \end{pmatrix} B_L^{\nu*} \sim \begin{pmatrix} M_T - M_D M_S^{-1} M_D^T & 0 \\ 0 & M_S \end{pmatrix}. \tag{9.41}$$

That is, there are three light (approximately) active neutrinos with an effective Majorana mass matrix $M_T - M_D M_S^{-1} M_D^T$, diagonalized by $A_L^\nu = A_R^{\nu*}$. There are also three heavy (approximately) sterile neutrinos with mass matrix M_S, which is diagonalized by $A_L^{\nu_S} = A_R^{\nu_S*}$. The latter decouple at energies small compared with their masses.

In the more general case in which one or more of the eigenvalues of M_S are not large, some or all of the light states will include non-negligible sterile components. In particular, the active and sterile neutrinos of the same chirality will mix significantly (as was suggested by the LSND and some other experiments) if both Majorana and Dirac masses are of the same order of magnitude or in the pseudo-Dirac case. Constructing models with these features presents a special challenge compared to the Dirac, Majorana, and seesaw cases because one must find an explanation as to why two different types of mass terms are small. If all three sterile states remain light, the leptonic mixing matrix in the charge raising current in (9.30) becomes 6×3 dimensional, i.e.,

$$V_\ell = A_L^{\nu\dagger} P_A^\dagger A_L^e, \tag{9.42}$$

where $A_L^{\nu\dagger}$ is the 6×6 neutrino mixing matrix and $P_A = (I\ 0)$ is the 3×6 dimensional matrix that projects onto the active neutrino subspace. One can easily generalize to the case of $F_A = 3$ active neutrinos and F_S sterile neutrinos, in which case there are $3 + F_S$ Majorana mass eigenvalues and P_A becomes a $3 \times (3 + F_S)$ dimensional projection.

Our original definition of a Dirac mass term as one that couples two distinct Weyl fields and a Majorana mass as one that couples a Weyl field to itself is not very useful in the multi-family case. In the above discussion we implicitly referred to couplings between active and sterile neutrinos as Dirac, and active-active or sterile-sterile couplings as Majorana. However, an alternate definition, which we will now adopt, is to define Majorana or Dirac masses on the basis of the form of the mass eigenvalues and eigenvectors. One can then view Majorana mass terms as the generic case, and reserve the term Dirac for special or limiting cases in which there is a conserved lepton number. We already saw an example of the Dirac limit of the one family mixed model, which trivially generalizes to the F family case when $M_T = M_S = 0$, $M_D \neq 0$.

A less obvious example involves the two family Majorana mass matrix

$$M_T = \begin{pmatrix} 0 & m_{ZKM} \\ m_{ZKM} & 0 \end{pmatrix}. \tag{9.43}$$

This has the same form as the Dirac limit of (9.18), except that in this case the L and R components are both active. Let us be even more explicit, and assume that (9.43) holds in the basis in which the charged leptons (e and μ) are already diagonal. Then we can identify $\nu_{1L}^0 = \nu_{eL}$, $\nu_{1R}^{0c} = \nu_{eR}^c$, $\nu_{2L}^0 = \nu_{\mu L}$, and $\nu_{2R}^{0c} = \nu_{\mu R}^c$, so that

$$\begin{aligned} -\mathcal{L}_T &= \frac{m_{ZKM}}{2}\left(\bar{\nu}_{eL}\nu_{\mu R}^c + \bar{\nu}_{\mu L}\nu_{eR}^c + \bar{\nu}_{\mu R}^c\nu_{eL} + \bar{\nu}_{eR}^c\nu_{\mu L}\right) \\ &= m_{ZKM}\left(\bar{\nu}_{eL} + \bar{\nu}_{\mu R}^c\right)\left(\nu_{eL} + \nu_{\mu R}^c\right). \end{aligned} \tag{9.44}$$

This *Zeldovich-Konopinski-Mahmoud* model (Zeldovich, 1952; Konopinski and Mahmoud, 1953) involves a Dirac neutrino, in that there is a conserved quantum number ($L_e - L_\mu$, rather than $L = L_e + L_\mu$) and because two distinct Weyl neutrinos are involved. However, in many ways it is more closely related to the Majorana case, i.e., it violates weak isospin by one unit and is a limiting case of the general 2×2 Majorana matrix. Any perturbation involving nonzero diagonal elements of M_T would break the "degeneracy" $m_1 = -m_2$ of the two eigenvalues. A modern $F = 3$ version of the ZKM model involves the matrix

$$M_T = m_{ZKM}\begin{pmatrix} 0 & 1 & 1 \\ 1 & 0 & 0 \\ 1 & 0 & 0 \end{pmatrix}, \tag{9.45}$$

which leads to one massive Dirac neutrino $\nu_{eL} + \frac{1}{\sqrt{2}}(\nu_{\mu R}^c + \nu_{\tau R}^c)$, and one massless Weyl neutrino $\frac{1}{\sqrt{2}}(\nu_{\mu L} - \nu_{\tau L})$, with $L_e - L_\mu - L_\tau$ conserved. This actually yields a spectrum somewhat similar to the observed one, but would require nontrivial perturbations both in M_T and the charged lepton mixing to be fully realistic. An example of a perturbation on M_T leading to a generalization of the pseudo-Dirac case is considered in Problem 9.2.

An analogous situation sometimes occurs (especially in complicated models in which $F_S > F_A$) when two sterile neutrinos pair to form a Dirac neutrino, e.g., with M_S of a form analogous to (9.43).

Let us conclude this section by reemphasizing that there is no distinction between Dirac and Majorana neutrinos except by their masses (or by new BSM interactions). As the masses go to zero, the active components reduce to standard active Weyl neutrinos in both cases.

There are additional sterile Weyl neutrinos in the massless limit of the Dirac case, but these decouple from the other particles. We also repeat the comment from Section 8.2.2 that one can ignore V_ℓ in processes for which the neutrino masses are too small to be relevant. They are then effectively degenerate (with vanishing mass) and one can work in the weak basis.

9.1.4 The Propagators for Majorana Fermions

For free Dirac fields there is a conserved fermion number and therefore only a single type of propagator, given in (2.182) on page 38. For Majorana neutrinos, on the other hand, there is no conserved fermion number and there are three non-zero propagators,

$$\begin{aligned}
\langle 0|T[\nu_M(x), \bar{\nu}_M(x')]|0\rangle &= i\int \frac{d^4k}{(2\pi)^4} e^{-ik\cdot(x-x')} S_F(k) \\
\langle 0|T[\bar{\nu}_M(x), \bar{\nu}_M(x')]|0\rangle &= i\int \frac{d^4k}{(2\pi)^4} e^{-ik\cdot(x-x')} \mathcal{C}^\dagger S_F(k) \\
\langle 0|T[\nu_M(x), \nu_M(x')]|0\rangle &= i\int \frac{d^4k}{(2\pi)^4} e^{-ik\cdot(x-x')} S_F(k)(-\mathcal{C}),
\end{aligned} \tag{9.46}$$

where $S_F(k)$ is the usual $\frac{\not{k}+m}{k^2-m^2+i\epsilon}$. The first Majorana propagator is analogous to the Dirac case, while the second and third can be thought of as annihilating or creating two neutrinos, respectively. They can easily be derived from

$$\begin{aligned}
\sum_s u(\vec{p},s)\; v(\vec{p},s)^T &= (\not{p}+m)\,(-\mathcal{C}) \\
\sum_s \bar{u}(\vec{p},s)^T\; \bar{v}(\vec{p},s) &= \mathcal{C}^\dagger\,(\not{p}-m)
\end{aligned} \tag{9.47}$$

and two similar identities with $u \leftrightarrow v$ and $m \to -m$, which in turn follow immediately from (2.177), (2.290), and (2.291). Similar expressions apply to the Majorana fields that occur in supersymmetry.

Expressions for amplitudes involving the second and third terms in (9.46) take an unusual form, but they can usually be rendered more familiar by use of (2.290) and (2.291) on page 55, or by the use of (2.296) along with $\nu_M = \nu_M^c$. As a simple example, consider the process $W^-W^- \to e^-e^-$, assuming that the ν_e is Majorana and ignoring all family mixing effects. This proceeds via the diagrams in Figure 9.4, which form a critical part of those for $\beta\beta_{0\nu}$ in Figure 9.2. The relevant WCC interaction is

$$\mathcal{L} = -\frac{g}{2\sqrt{2}} J_W^\mu W_\mu^-, \tag{9.48}$$

where

$$J_W^\mu = \bar{e}\gamma^\mu(1-\gamma^5)\nu_L = \bar{e}\gamma^\mu(1-\gamma^5)\nu_M, \tag{9.49}$$

since $P_L\nu_R^c = 0$. The amplitude from the t-channel diagram is therefore

$$M_t = \left(\frac{-ig}{2\sqrt{2}}\right)^2 \epsilon_{1\mu}\epsilon_{2\nu}\left[\bar{u}_3\gamma^\mu(1-\gamma^5)\right]_\alpha \left[\bar{u}_4\gamma^\nu(1-\gamma^5)\right]_\beta \left[i\frac{\not{k}+m_T}{k^2}(-\mathcal{C})\right]_{\alpha\beta}, \tag{9.50}$$

where $k = p_3 - p_1$, m_T is the Majorana neutrino mass, the propagator follows from the last expression in (9.46), and we have assumed $|k^2| \gg m_T^2$. This expression can be simplified using (2.289), (2.291), and (2.292),

$$\left[\bar{u}_4\gamma^\nu(1-\gamma^5)\right]_\beta(-\mathcal{C})_{\delta\beta} = -\left[(1-\gamma^5)\gamma^\nu v_4\right]_\delta, \tag{9.51}$$

so that

$$M_t = \left(\frac{-ig}{2\sqrt{2}}\right)^2 \epsilon_{1\mu}\epsilon_{2\nu}\left(\frac{-2im_T}{k^2}\right)\bar{u}_3\gamma^\mu(1-\gamma^5)\gamma^\nu v_4, \tag{9.52}$$

which can be evaluated in the usual way. Note that the $\bar{u}_4$ spinor has been replaced by v_4. The $\not{k}$ part of the propagator has dropped out because of the pure $V-A$ interaction, but would be allowed if there were an admixture of $V+A$. The same result can be obtained more directly by rewriting

$$J_W^\nu = \bar{e}\gamma^\nu(1-\gamma^5)\nu_M = -\bar{\nu}_M\gamma^\nu(1+\gamma^5)e^c \tag{9.53}$$

obtained from (2.296) for *one* of the vertices. Then, the first propagator in (9.46) is the relevant one, and we must use a v spinor for e_4^- since e^c is the positron field.

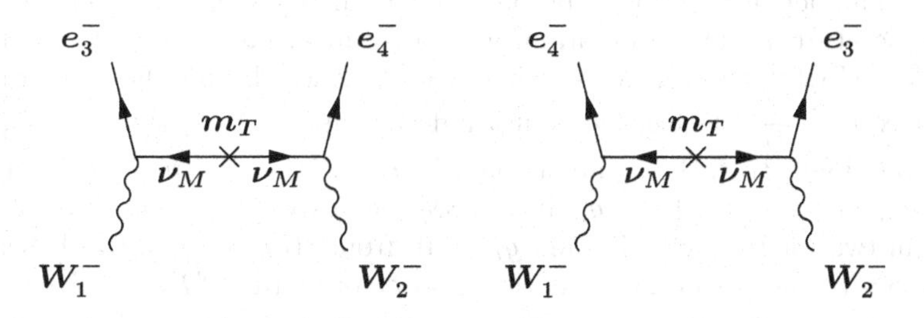

Figure 9.4 Diagrams for $W^-W^- \to e^-e^-$ assuming a Majorana neutrino. Left: t-channel. Right: u-channel. The cross represents a Majorana mass insertion.

9.2 EXPERIMENTS AND OBSERVATIONS

In this section we describe the principal laboratory and astrophysical constraints on the number of light active and sterile neutrinos, and their masses and mixings. The constraints on possible heavy Dirac or Majorana neutrinos and on neutrino decays[8] are reviewed in (Raffelt, 1999; Mohapatra and Pal, 2004; Atre et al., 2009; Giunti and Studenikin, 2015; Faessler et al., 2014; Drewes and Garbrecht, 2015; de Gouvêa and Kobach, 2016).

9.2.1 Neutrino Counting

As discussed on page 309, the width for Z to decay invisibly implies that there are only three active neutrinos with masses $\lesssim M_Z/2$. More precisely, $N_\nu^{inv} = 2.992 \pm 0.007$ from the global fit to precision data[9] (Patrignani, 2016), where N_ν^{inv} does not include sterile neutrinos or very heavy active neutrinos, but does include the effects of other possible invisible decay channels asssociated with new physics. Precision constraints also exclude additional heavy

[8] *Radiative decays*, such as $\nu_2 \to e^+e^-\nu_1$, if kinematically allowed, or $\nu_2 \to \nu_1\gamma$, are possible in the SM. The latter is loop suppressed and extremely slow. BSM physics could allow faster $\nu_2 \to \nu_1\gamma$ or invisible decays, such as $\nu_2 \to \nu_1\nu_1\bar{\nu}_1$, $\nu_2 \to \nu_1 F$, or $\nu_2 \to \bar{\nu}_1 M$, where F is a *familon* (a Goldstone boson associated with a hypothetical broken family symmetry), and M is a *Majoron* (a Goldstone boson associated with a spontaneously broken lepton number).

[9] There is also a more direct determination of the invisible width from $e^-e^+ \to \gamma+$ invisible, yielding $N_\nu^{inv} = 2.92 \pm 0.05$.

active neutrinos if they belong to a complete degenerate chiral family, but may be evaded, e.g., for non-chiral doublets or nondegenerate families.

Another major constraint comes from *big bang nucleosynthesis* (BBN) (e.g., Kolb and Turner, 1990; Dolgov, 2002; Steigman, 2012; Cyburt et al., 2016; Patrignani, 2016), which has also been a critical test of hot big bang cosmology and of many possible types of nonstandard particle physics. The basic point is that the reactions

$$n + \nu_e \leftrightarrow p + e^-, \qquad n + e^+ \leftrightarrow p + \bar{\nu}_e \tag{9.54}$$

kept the ratio of neutrons to protons in thermal equilibrium $\frac{n}{p} = \exp(-\frac{E_n - E_p}{T}) \sim \exp(-\frac{m_n - m_p}{T})$ in the early universe as long as the reaction rate $\Gamma \sim G_F^2 T^5$ was larger than the expansion rate (Hubble parameter) $H \sim 1.66\sqrt{g_*}\,T^2/M_P$, where M_P is the Planck mass. H^2 is proportional to the energy density $\rho = g_* \pi^2 T^4/30$, where $g_* \equiv g_B + \frac{7}{8} g_F$ and $g_{B,F}$ are the number of relativistic bosonic and fermionic degrees of freedom in equilibrium at temperature T. The equilibrium was maintained until the freezeout temperature $T_f \sim (\sqrt{g_*}/G_F^2 M_P)^{1/3} = \mathcal{O}(\text{few MeV})$ when $\Gamma \sim H$, at which time the n/p ratio was frozen at the value $\exp(-\frac{m_n - m_p}{T_f})$ except for neutron decay, and most of the neutrons were eventually incorporated into ^4He. By apparent coincidence T_f is close to $m_n - m_p$, so the expected abundance depends sensitively on g_*. In the SM, one expects $g_* = 43/4$ for $m_e < T < m_\mu$ ($g_B = 2$ from two photon helicities and $g_F = 10$ from $3(\nu_L + \nu_R^c) +$ two helicities each of $e^\pm$). This leads to the prediction that the ratio of primordial ^4He to H by mass should be ∼24%, in agreement with observations.[10] However, any additional contribution to the energy density for $T \gtrsim$ few MeV would increase H and T_f, and therefore the predicted helium abundance. This can be parametrized by writing

$$g_F = 4 + 2N_\nu^{BBN} \tag{9.55}$$

where any deviation of the effective N_ν^{BBN} from[11] 3.046 could indicate new light degrees of freedom in (partial) equilibrium, or such effects as neutrino masses of $\mathcal{O}$(MeV) or neutrino decay. There has long been some uncertainty and controversy in the observational primordial abundance, and therefore the limits on N_ν^{BBN}. Recent estimates include $N_\nu^{BBN} = 3.7 \pm 0.5$ (Steigman, 2012) and $N_\nu^{BBN} = 2.9 \pm 0.3$ (Cyburt et al., 2016). $\Delta N_\nu^{BBN} \equiv N_\nu^{BBN} - 3.046$ constrains not only additional active neutrinos with masses $\lesssim 1$ MeV, but also light sterile neutrinos of the type suggested by the LSND experiment, which could be produced by mixing with active neutrinos for a wide range of parameters (e.g., Dolgov, 2002; Cirelli et al., 2005; Hannestad et al., 2012). However, ΔN_ν^{BBN} does *not* include the sterile ν_R components of light Dirac neutrinos, which could not (for the currently relevant mass ranges) have been produced in equilibrium numbers unless they have new BSM interactions or properties (Barger et al., 2003b; Anchordoqui et al., 2013). Of course, other light BSM particles besides sterile neutrinos could contribute to ΔN_ν^{BBN}, such as (pseudo-)Goldstone bosons associated with some new symmetry (Weinberg, 2013).

Most new physics effects increase the predicted ^4He abundance, leading to a more stringent upper limit on ΔN_ν^{BBN}. One important exception is a possible large asymmetry between ν and $\bar{\nu}$, which would preferentially drive the reactions in (9.54) to the right, decreasing the n/p ratio and allowing a larger ΔN_ν^{BBN}. (Only the $\nu_e - \bar{\nu}_e$ asymmetry directly affects the reactions, but the observed neutrino mixing would probably have equilibrated

[10] There is also a weak dependence on the baryon density relative to photons, which is determined independently by the D abundance and by the cosmic microwave background anisotropies.

[11] The standard model value for N_ν^{BBN} differs slightly from 3 due to such effects as non-instantaneous neutrino decoupling.

the asymmetries between the families.) However, such an asymmetry would have to be enormous, $(n_\nu - n_{\bar{\nu}})/n_\gamma \sim \mathcal{O}(0.1)$, compared to the baryon or charged lepton asymmetries to have much effect. Even allowing it, the present constraints from BBN along with the cosmic microwave background (CMB) and large scale structure data imply that such asymmetries would not significantly perturb the constraints on ΔN_ν^{BBN} (Simha and Steigman, 2008). A large asymmetry could lead to important nonlinear effects in the case of active-sterile neutrino mixing, however (Foot et al., 1996).

There are also stringent constraints on the number of neutrinos and their masses from the CMB and from the distribution of galaxies (e.g., Wong, 2011; Lesgourgues and Pastor, 2012; Patrignani, 2016). For example, the CMB anisotropies depend on the number of relativistic degrees of freedom that were present at recombination, when the universe had cooled sufficiently (to $T \sim 0.26$ eV, or redshift $z \sim 1100$) for neutral atoms to form so that the photons decoupled from matter. The WMAP (Hinshaw et al., 2013) and Planck (Ade et al., 2016) collaborations have made very detailed studies of the CMB. When combined with galaxy distributions, the Planck analysis obtains $N_\nu^{CMB} = 3.15 \pm 0.23$ for the number of neutrinos (active and sterile, weighted by their abundance), as well as other BSM forms of dark radiation, that were relativistic at recombination.[12] This is consistent with 3.046, but also allows one or more additional species if they have a reduced abundance. The effects of and constraints on active or sterile neutrino masses are discussed in the the next subsection.

9.2.2 Neutrino Mass Constraints

A stringent kinematic limit on the effective ν_e mass-squared

$$m_{\nu_e}^2 \equiv \sum_i |\mathcal{V}_{ei}|^2 m_i^2 \tag{9.56}$$

can be obtained from the shape of the e^- spectrum near the endpoint in tritium β decay, $^3H \to {}^3He\, e^- \bar{\nu}_e$ (e.g., Otten and Weinheimer, 2008; Dragoun and Vénos, 2016). In (9.56) m_i is the ith mass eigenvalue, independent of whether it is Dirac or Majorana, and $\mathcal{V}$ is the leptonic mixing matrix as defined in (9.36). The current limits from experiments in Troitsk and Mainz are respectively $m_{\nu_e} < 2.05$ eV and < 2.3 eV at 95% c.l. The Karlsruhe KATRIN experiment should improve the sensitivity on m_{ν_e} down to around 0.2 eV. KATRIN and proposals for future experiments are reviewed in (Drexlin et al., 2013).

The kinematic limits on the ν_μ and ν_τ masses, defined analogously to (9.56), are much weaker: $m_{\nu_\mu} < 0.19$ MeV from $\pi^+ \to \mu^+\nu_\mu$ and $m_{\nu_\tau} < 18.2$ MeV from $\tau^- \to 3\pi^\pm\nu_\tau, 5\pi^\pm(\pi^0)\nu_\tau$. These bounds are now superseded by much more stringent ones from neutrino oscillations and cosmology. However, it is historically interesting that the combination of this bound on m_{ν_τ} from ALEPH with the BBN constraint (which becomes relevant because an $\sim (1-20)$ MeV neutrino contributes more than 1 to N_ν^{BBN}) excluded the possibility of a stable or long-lived ν_τ above 1 MeV (e.g., Fields et al., 1997).

We also mention the historically important observation of a burst of $\mathcal{O}(20)$ neutrinos (presumably mainly $\bar{\nu}_e$'s) from the core-collapse Supernova 1987A by the Kamiokande, IMB, and Baksan collaborations, which implied (amongst many other things[13]) that $m_{\nu_e} \lesssim 20$ eV,

[12] The various cosmological limits on the number and masses of neutrinos depend somewhat on the data set chosen and on possible correlations with other parameters.

[13] For example, the neutrinos arrived within a few hours of the supernova photons, whereas they would have arrived ~ 5 months sooner (after a 160,000 year journey) if they did not share the same gravitational interactions, testing the weak equivalence principle (Longo, 1988; Krauss and Tremaine, 1988).

because otherwise the arrival times of the detected neutrinos would have spread out more than was observed. It was hard to make the limit precise, however, because it depended on theoretical details of the neutrino emission (e.g., Bahcall, 1989; Raffelt, 1999). Core-collapse supernovae are expected to occur in our galaxy at the rate of several per century. Observation of the ν's from such a supernova in neutrino and other detectors would yield a wealth of information on neutrino properties as well as on the dynamics of the supernova explosion (e.g., Duan et al., 2010; Scholberg, 2012; Mirizzi et al., 2016). The diffuse background flux of neutrinos from core-collapse supernovae in other galaxies may also be observable (Beacom, 2010).

Light massive neutrinos would contribute to the cosmological energy density, and they would close the Universe, $\Omega_\nu = 1$, for $\Sigma \sim 35$ eV, where

$$\Sigma \equiv \sum_i |m_i| \tag{9.57}$$

is the sum of the masses of the light active neutrinos. Observationally, some 26-27% of the energy density is dark matter (Patrignani, 2016), but it is most likely cold dark matter (CDM), which was non-relativistic at decoupling, such as weakly interacting massive particles (WIMPs) or axions. Light neutrinos would be hot dark matter (relativistic at decoupling), which would free-stream away from density perturbations, preventing the formation of the observed smaller scale structures during the lifetime of the universe.[14] Smaller neutrino masses (close to the recombination temperature ~ 0.26 eV) would lead to subtle effects in the CMB and galaxy distributions (e.g., Wong, 2011; Abazajian et al., 2011; Lesgourgues and Pastor, 2012). For example, the Planck collaboration (Ade et al., 2016) finds

$$\Sigma < 0.23 \text{ eV at } 95\% \text{ c.l.} \tag{9.58}$$

for the sum of the active neutrino masses (assuming $N_\nu^{CMB} = 3.046$) from the combination of CMB and galaxy data, while some other recent data sets yield tighter or weaker limits (e.g., Abazajian and Kaplinghat, 2016). Future cosmological observations should be able to extend the sensitivity to Σ down to or below the minimum value 0.05 eV $\sim \sqrt{|\Delta m^2_{atm}|}$ allowed by the neutrino oscillation data (Abazajian et al., 2015).

Allowing for light sterile neutrinos as well one must take into account the extra contribution to the radiation. Furthermore, the cosmological observables depend on how the masses are distributed amongst the states, possible asymmetries, and possible non-thermal production. The Planck collaboration considered the example of one light ($m_S < 10$ eV) thermally-produced sterile neutrino. In that case, $\Delta N_\nu^{CMB} = (T_S/T_\nu)^4$, where T_S and T_ν are, respectively, the sterile and active neutrino temperatures (e.g., Kolb and Turner, 1990). (One expects $T_S \le T_\nu$ due to earlier decoupling.) Assuming also that $\Sigma = 0.06$ eV and no asymmetries, they obtained the correlated 95% c.l. limits

$$N_\nu^{CMB} < 3.7, \qquad m_S^{eff} < 0.52 \text{ eV}, \tag{9.59}$$

where the effective mass m_S^{eff} is m_S weighted by the sterile abundance, i.e., $m_S^{eff} = (T_S/T_\nu)^3\, m_S$.

[14]Warm dark matter, e.g., from keV mass sterile neutrinos, is an intermediate possibility. Theoretical, astrophysical, and cosmological implications of intermediate mass sterile neutrinos are reviewed in (Kusenko, 2009; Merle, 2013; Adhikari et al., 2017). An example, the *neutrino minimal standard model (νMSM)*, is reviewed in (Boyarsky et al., 2009).

9.2.3 Neutrinoless Double Beta Decay

Majorana masses can lead to $\beta\beta_{0\nu}$, i.e., $nn \to ppe^-e^-$, which violates lepton number by two units, by the diagrams in Figure 9.2 (for reviews, see Rodejohann, 2011; Vergados et al., 2012; Bilenky and Giunti, 2015; Päs and Rodejohann, 2015; Dell'Oro et al., 2016). Since there is no missing energy,[15] events should show up as a peak of known energy in the e^-e^- spectrum from a sample of $\beta\beta_{0\nu}$-unstable nuclei. However, the process would have an extremely long half-life, so problems of backgrounds are severe.

The amplitude for $\beta\beta_{0\nu}$ is $M \sim A_{nuc}\, m_{\beta\beta}$, where A_{nuc} contains the nuclear matrix element. A_{nuc} cannot be directly measured and therefore introduces considerable uncertainty into the interpretation of any upper limit or future observation (Vergados et al., 2012; Vogel, 2012; Šimkovic et al., 2013). $m_{\beta\beta}$ is the effective Majorana mass in the presence of mixing between light Majorana neutrinos,

$$m_{\beta\beta} \equiv \left|\sum_i (\mathcal{V}_{ei})^2 m_i\right|. \tag{9.60}$$

It is just the (e, e) element of m_T or of the effective Majorana mass matrix in a seesaw model (i.e., the $(1, 1)$ element in the family basis in which $A_L^e = I$). It involves the square of $\mathcal{V}_{ei}$ rather than the absolute square, allowing for the possibility of cancellations between terms. Such cancellations could occur even if the original mass matrix were real because some of the eigenvalues could be negative. (In our phase convention the m_i are taken to be positive, but the signs would appear because some of the Majorana phases in (9.35) would then be $\pm i$.) This also shows why the $\beta\beta_{0\nu}$ amplitude vanishes for a Dirac neutrino, which can be viewed as two Majorana neutrinos that give equal and opposite contributions. The cancellations could in principle allow the determination of (CP-violating) Majorana phases different from 0 or $\pm i$, though this is difficult in practice because the other parameters including the matrix elements would have to be known rather well (Barger et al., 2002; Pascoli et al., 2002).

There are several precise limits on $\beta\beta_{0\nu}$ in various nuclei,[16] including KamLAND-ZEN and EXO-200 [^{136}Xe], GERDA [^{76}Ge], and CUORE-0 [^{130}Te] (for a review, see Ostrovskiy and O'Sullivan, 2016). A combination of these yields $m_{\beta\beta} < (0.13-0.31)$ eV at 90% c.l. (Guzowski et al., 2015), with the range due to the nuclear matrix element uncertainties. The most recent KamLAND-ZEN result (Gando et al., 2016) (not included in the combination) obtains a 90% c.l. lower limit of 1.1×10^{26} yr on the ^{136}Xe $\beta\beta_{0\nu}$ half-life. This corresponds to $m_{\beta\beta} < (0.06 - 0.17)$ eV, very close to the range expected for the inverted hierarchy described in Section 9.4. Future experiments should be sensitive down to $\mathcal{O}(0.01 - 0.02 \text{ eV})$ or better (e.g., Ostrovskiy and O'Sullivan, 2016). They should be sufficient to observe $\beta\beta_{0\nu}$ if the neutrinos are Majorana with masses corresponding to the inverted or degenerate spectra, but no scheduled experiment would be sensitive to the normal hierarchy. (See Biller, 2013, however.)

A heavy Majorana neutrino could also contribute to $m_{\beta\beta}$, but its contribution would be

---

[15]Two-neutrino double beta decay, $\beta\beta_{2\nu}$, is the process $nn \to ppe^-e^-\bar{\nu}\bar{\nu}$, which can occur by ordinary second-order weak processes in some β-stable nuclei. It leads to a continuous e^-e^- spectrum and has been studied in a number of nuclei (Saakyan, 2013). It is helpful for testing calculations of the nuclear matrix elements entering $\beta\beta_{0\nu}$. A third possibiity is Majoron decay, $nn \to ppe^-e^-M$ or $\beta\beta_{0\nu M}$, where M is a Majoron (Goldstone boson). It would lead to a spectrum intermediate between $\beta\beta_{0\nu}$ and $\beta\beta_{2\nu}$.

[16]An observation of $\beta\beta_{0\nu}$ in ^{76}Ge with a half-life $\sim 2 \times 10^{25}$ yr has been claimed (Klapdor-Kleingrothaus and Krivosheina, 2006) by members of the Heidelberg-Moscow experiment. This would correspond to $m_{\beta\beta} \sim (0.25-0.60)$ eV. However, the result has not been confirmed by other experiments, and is apparently excluded by the ^{136}Xe results for plausible nuclear matrix elements.

suppressed by finite range effects (i.e., the m_i^2 in the denominator of the propagator would be important),

$$m_{\beta\beta} \to \left|\sum_i (\mathcal{V}_{ei})^2 m_i F(m_i, A)\right|, \qquad F(m_i, A) \equiv \frac{\langle e^{-m_i r}/r\rangle}{\langle 1/r\rangle}, \tag{9.61}$$

where A is the nucleon number. $F(m_i, A)$ is ~ 1 for $m_i \ll 10$ MeV, but falls rapidly for larger values (e.g., Vergados et al., 2012; Faessler et al., 2014). Lepton-number violating effects other than Majorana neutrino masses could also lead to $\beta\beta_{0\nu}$ (although the existence of a $\beta\beta_{0\nu}$ amplitude implies the existence of a Majorana mass at some level (Schechter and Valle, 1982)). For example, models involving both $V - A$ and $V + A$ interactions as well as lepton number violation can induce $\beta\beta_{0\nu}$ amplitudes not directly proportional to neutrino masses (Problem 9.3). There could also be effects from new interactions such as leptoquarks or R-parity violation in supersymmetry. If $\beta\beta_{0\nu}$ is observed, it would be useful to study it in several different nuclei, both to help control nuclear matrix element uncertainties and to shed some light on the underlying mechanism.

9.2.4 Relic Neutrinos

The BBN and CMB constraints on the cosmological neutrinos are indirect. Following their decoupling at $T \sim$ few MeV the neutrino wavelengths were redshifted so that their momentum distribution should at present have a thermal form, characterized by an effective temperature $T_\nu = (4/11)^{1/3}\, T_\gamma \sim 1.9\,\text{K}$, where $T_\gamma \sim 2.73\,\text{K}$ is the CMB temperature and the $(4/11)^{1/3}$ factor is because the γ's but not the neutrinos were reheated by e^-e^+ annihilation at $T \lesssim m_e$ (Steigman, 1979; Kolb and Turner, 1990; Weinberg, 2008). This corresponds to a number density of $\sim 50/\text{cm}^3$ for each neutrino degree of freedom, i.e., $\sim 300/\text{cm}^3$ for 3 flavors with two helicity states. Local clustering and modifications of the momentum distribution are not expected to be large unless the masses are $\gtrsim 0.1$ eV (e.g., Ringwald and Wong, 2004). Direct detection of these *relic neutrinos* appears extremely difficult. Effects involving macroscopic torques or forces (Stodolsky, 1975; Cabibbo and Maiani, 1982; Langacker et al., 1983) are tiny. Another possibility are *Z bursts*, in which ultra high energy cosmic ray neutrinos annihilate on relic neutrinos to produce Z's (Weiler, 1982; Eberle et al., 2004), which could be observed through their decay products or as absorption dips in the cosmic ray ν spectrum. However, this would only be feasible if there were some unexpected intense source of such high energy neutrinos. More promising are ν-induced $e^\pm$ emission by nuclei (e.g., Weinberg, 1962; Cocco et al., 2007; Lazauskas et al., 2008), which would show up as $e^\pm$ emission above the β decay endpoint. The expected rate for KATRIN is not encouraging (Faessler et al., 2017). However, the recent PTOLEMY proposal for a very large (100 g) tritium source deposited on a graphene substrate (Betts et al., 2013) might be sensitive to relic neutrinos for a degenerate spectrum with $m_i \gtrsim 0.1$ eV, with a larger rate for Majorana than Dirac neutrinos (Long et al., 2014). Such experiments would also have sensitivity to heavier sterile neutrinos and to $\nu - \bar{\nu}$ asymmetries. For reviews of relic neutrinos, see (Gelmini, 2005; Strumia and Vissani, 2006; Ringwald, 2009).

9.2.5 Electromagnetic Form Factors

Neutrinos have no electric charge, but they can acquire magnetic and electric dipole moments by diagrams analogous to the weak corrections to the muon magnetic moment in Figure 2.20 on page 72 or from new physics. These lead to effective electromagnetic interactions

$$\mathcal{H}_{\nu Q} = \frac{1}{2}\mu_{ij}\bar{n}_j\sigma^{\mu\nu}n_i F_{\mu\nu} + \frac{i}{2}d_{ij}\bar{n}_j\sigma^{\mu\nu}\gamma^5 n_i F_{\mu\nu}, \tag{9.62}$$

where n_i can represent either a Dirac (ν_{iD}) or Majorana (ν_{iM}) mass eigenstate field. The first (second) terms are magnetic (electric) dipole interactions, as can be seen from (2.358), (8.324), and Problem 2.30. The flavor-diagonal terms $i = j$ are known as *direct or intrinsic moments*, while those for $i \neq j$ are *transition moments.*

The magnetic and electric dipoles flip chirality. In the Majorana case, $\nu_{iM} = \nu_{iL} + \nu_{iR}^c$, so that

$$\bar{\nu}_{jM}\sigma^{\mu\nu}\nu_{iM} = \bar{\nu}_{jL}\sigma^{\mu\nu}\nu_{iR}^c + \bar{\nu}_{jR}^c\sigma^{\mu\nu}\nu_{iL}. \tag{9.63}$$

However, using $\nu_{iR}^c = \mathcal{C}\bar{\nu}_{iL}^T$,

$$\bar{\nu}_{jL}\sigma^{\mu\nu}\nu_{iR}^c = \nu_{jR}^{cT}\mathcal{C}\sigma^{\mu\nu}\mathcal{C}\bar{\nu}_{iL}^T = -\bar{\nu}_{iL}\sigma^{\mu\nu}\nu_{jR}^c, \tag{9.64}$$

so that $\mu_{ij} = -\mu_{ji}$. (This can also be seen from (2.303) on page 56.) Majorana neutrinos therefore cannot have direct magnetic moments, but can have transition moments $\mu_{ij} \neq 0$ for $i \neq j$, which can mediate decays such as $\nu_{iM} \to \nu_{jM}\gamma$. Similar statements apply to electric dipole moments.

Both direct and transition moments are possible for Dirac neutrinos $\nu_{iD} = \nu_{iL} + \nu_{iR}$. For a single flavor

$$\bar{\nu}_D\sigma^{\mu\nu}\nu_D = \bar{\nu}_L\sigma^{\mu\nu}\nu_R + \bar{\nu}_R\sigma^{\mu\nu}\nu_L. \tag{9.65}$$

One can write ν_L and ν_R in terms of two degenerate Majorana neutrinos using (9.23),

$$\nu_L = \frac{1}{\sqrt{2}}(\nu_{1L} + \nu_{2L}), \qquad \nu_R = \frac{1}{\sqrt{2}}(\nu_{1R}^c - \nu_{2R}^c), \tag{9.66}$$

where the unnecessary superscript 0 has been dropped. Then

$$\bar{\nu}_L\sigma^{\mu\nu}\nu_R = -\frac{1}{\sqrt{2}}\bar{\nu}_{1L}\sigma^{\mu\nu}\nu_{2R}^c + \frac{1}{\sqrt{2}}\bar{\nu}_{2L}\sigma^{\mu\nu}\nu_{1R}^c. \tag{9.67}$$

That is, a direct Dirac magnetic (or electric) moment is an antisymmetric combination of transition moments between degenerate Majorana states.

In the simplest extension of the SM with a small Dirac mass m_i, the direct neutrino magnetic moment is (Marciano and Sanda, 1977; Lee and Shrock, 1977)

$$\mu_i \sim \frac{3eG_F m_i}{8\sqrt{2}\pi^2} \sim 3.2 \times 10^{-19}\left(\frac{m_i}{1\text{ eV}}\right)\mu_B, \tag{9.68}$$

where μ_B is the Bohr magneton. This is negligibly small compared with laboratory limits $\lesssim$ few $\times 10^{-11}\mu_B$ (Vogel and Engel, 1989; Giunti and Studenikin, 2015; Giunti et al., 2016), and various astrophysical limits, e.g., from stellar cooling, $\mu_\nu \lesssim$ few $\times 10^{-12}$ (Raffelt, 1999; Giunti and Studenikin, 2015; Giunti et al., 2016). The latter often applies to electric dipole moments as well, and to both Dirac and Majorana transition moments.

One can construct models with magnetic moments that are much larger than (9.68), as was motivated by an alternative solution to the solar neutrino problem involving resonant spin-flavor precession in an assumed strong solar magnetic field (Akhmedov, 1988; Lim and Marciano, 1988). However, there is a limit as to how large they can be. For Dirac neutrinos, higher-dimensional operators that can generate a magnetic dipole moment μ_ν^D also contribute to m_ν (Bell et al., 2005). If the operators are generated by new physics at a scale $\gtrsim 1$ TeV then there is an upper limit on μ_ν^D of around $10^{-15}\mu_B$ for $m_\nu < 0.3$ eV. The corresponding limits in the Majorana case are much weaker because of Yukawa suppressions to the Majorana mass (Bell et al., 2006). Thus, observation of a dipole moment above $\sim 10^{-15}\mu_B$ would imply that the mass is Majorana (or that there are fine-tuned cancellations between contributions to m_ν).

9.3 NEUTRINO OSCILLATIONS

9.3.1 Oscillations in Vacuum

Neutrino oscillations are analogous to the neutral K and B meson oscillations described in Sections 8.6.3 and 8.6.4, and occur due to the mismatch between weak and mass eigenstates. They do not mix the neutrino helicities, and are therefore independent of whether the masses are Majorana or Dirac. First consider two neutrino flavors, ν_e and ν_μ, related to the mass eigenstates by

$$|\nu_e\rangle = |\nu_1\rangle \cos\theta + |\nu_2\rangle \sin\theta, \qquad |\nu_\mu\rangle = -|\nu_1\rangle \sin\theta + |\nu_2\rangle \cos\theta, \tag{9.69}$$

where θ, which corresponds to θ_{12} in (9.35), is the neutrino mixing angle. Suppose that one starts at time $t = 0$ with a pure state $|\nu(0)\rangle = |\nu_\mu\rangle$ of definite momentum[17] $|\vec{p}|$ from the decay $\pi^+ \to \mu^+\nu_\mu$. The two mass eigenstate components each develop with their own time dependence, so that

$$\begin{aligned} |\nu(t)\rangle &= -|\nu_1\rangle \sin\theta e^{-iE_1 t} + |\nu_2\rangle \cos\theta e^{-iE_2 t} \\ &\sim \left[-|\nu_1\rangle \sin\theta e^{-i\frac{m_1^2 t}{2E}} + |\nu_2\rangle \cos\theta e^{-i\frac{m_2^2 t}{2E}} \right] e^{-iEt}, \end{aligned} \tag{9.70}$$

at a later time t, where we have assumed that the neutrinos are extremely relativistic, so that $E_i = \sqrt{|\vec{p}|^2 + m_i^2} \sim E + m_i^2/2E$ where $E \sim |\vec{p}|$. After traveling a distance L there is a probability

$$\begin{aligned} P_{\nu_\mu \to \nu_e}(L) &= |\langle \nu_e | \nu(t) \rangle|^2 = \sin^2\theta \cos^2\theta \left| -e^{-i\frac{m_1^2 t}{2E}} + e^{-i\frac{m_2^2 t}{2E}} \right|^2 \\ &= \sin^2 2\theta \sin^2\left(\frac{\Delta m^2 L}{4E}\right) = \sin^2 2\theta \sin^2\left[\frac{1.27\Delta m^2(\text{eV}^2) L(\text{km})}{E(\text{GeV})}\right], \end{aligned} \tag{9.71}$$

for the neutrino to have oscillated into a ν_e, where $\Delta m^2 = m_2^2 - m_1^2$ and $L \sim t$. The ν_e could be detected, e.g., by the reaction $\nu_e n \to e^- p$. The *oscillation length* is defined as

$$L_{osc} = \frac{4\pi E}{\Delta m^2}. \tag{9.72}$$

Such *vacuum oscillations* depend only on $|\Delta m^2|$ and not on the absolute mass scale or on the hierarchy (which mass is larger). In *appearance experiments* one searches for the production of a different neutrino flavor than one started from, such as by the production of an e^- or τ^- in an initial ν_μ beam. In an idealized experiment with precisely known L and E for each event one observes not only the appearance of the new flavor, but the characteristic L/E

[17] This simple approximation yields the correct result. However, a complete treatment requires consideration of the coherence of the initial and final wave packets, the relation between t and the location L of the detector, entanglement with other particles, etc. For recent discussions, see, e.g., (Akhmedov and Smirnov, 2009; Cohen et al., 2009; Kayser et al., 2010).

dependence. However, for large $\Delta m^2 L/E$ the oscillations are averaged in practice by the spread and uncertainties in neutrino energy E and by finite detector/source sizes, so that

$$P_{\nu_\mu \to \nu_e}(L) \to \frac{1}{2}\sin^2 2\theta, \tag{9.73}$$

which is the same result one would have obtained from an incoherent superposition of ν_1 and ν_2. In a *disappearance experiment*, one searches for the reduction in the flux of the initial ν_μ (or other flavor) as a function of L and E, making use of the *survival probability* $P_{\nu_\mu \to \nu_\mu}(L) = 1 - P_{\nu_\mu \to \nu_e}(L)$ for the state to remain a ν_μ. For both types of experiment, careful attention has to be paid to the initial flux and spectrum (obtained from other measurements, theory, or by an initial calibration detector) and to backgrounds.

Even with more than two types of neutrino, it is sometimes a good approximation to use the two-neutrino formalism in the analysis of a given experiment, e.g., if some of the mixings are small or if some of the $\Delta_{ij} \equiv m_i^2 - m_j^2$ are small compared to E/L (see below), and most results are presented in terms of allowed or excluded regions in the $\sin^2 2\theta - \Delta m^2$ plane or the $\tan^2\theta - \Delta m^2$ plane,[18] whether or not that is really valid. However, a more precise or general analysis should take all three neutrinos into account. It is straightforward to show that the oscillation probability for $\nu_a \to \nu_b$ after a distance L is

$$\begin{aligned} P_{\nu_a \to \nu_b}(L) =& \delta_{ab} - 4\sum_{i<j} \Re e(\mathcal{V}_{ai}^* \mathcal{V}_{bi} \mathcal{V}_{aj} \mathcal{V}_{bj}^*) \sin^2\left(\frac{\Delta_{ij} L}{4E}\right) \\ &+ 2\sum_{i<j} \Im m(\mathcal{V}_{ai}^* \mathcal{V}_{bi} \mathcal{V}_{aj} \mathcal{V}_{bj}^*) \sin\left(\frac{\Delta_{ij} L}{2E}\right), \end{aligned} \tag{9.74}$$

where ν_a and ν_b are weak (flavor) eigenstates and $\mathcal{V}$ is the leptonic mixing matrix in (9.35). For antineutrinos, $P_{\bar\nu_a \to \bar\nu_b}(L)$ is given by the same formula, except the sign of the last term is reversed.[19] It is apparent from (9.74) that $P_{\nu_b \to \nu_a}(L)$ is the same as $P_{\nu_a \to \nu_b}(L)$ except that $\mathcal{V} \to \mathcal{V}^*$, and that any difference between them is due to CP-violating phases in $\mathcal{V}$. The combination $\mathcal{V}_{ai}^* \mathcal{V}_{bi} \mathcal{V}_{aj} \mathcal{V}_{bj}^*$ is a Jarlskog invariant, i.e., independent of phase conventions. The Majorana phases do not enter, so CP-violation in neutrino oscillations requires mixing between at least 3 families, just as in the CKM matrix. For $F = 3$, it is given by the phase δ in (9.35), and all CP-violating effects would vanish for $s_{13} = 0$. In practice, it is extremely difficult to compare $P_{\nu_a \to \nu_b}(L)$ and $P_{\nu_b \to \nu_a}(L)$ directly, because, e.g., $\bar\nu_e$ are mainly produced at reactors, and $\nu_\mu(\bar\nu_\mu)$ at accelerators. However, CPT, which is built into the expressions above, implies that $P_{\nu_b \to \nu_a}(L) = P_{\bar\nu_a \to \bar\nu_b}(L)$. Thus,

$$P_{\nu_a \to \nu_b}(L) \xrightarrow[\mathcal{V} \to \mathcal{V}^*]{} P_{\bar\nu_a \to \bar\nu_b}(L). \tag{9.75}$$

The oscillation rates for ν_a vs $\bar\nu_a$ can be compared, e.g., by using ν_μ ($\bar\nu_\mu$) beams from $\pi^+ \to \mu^+\nu_\mu$ ($\pi^- \to \mu^-\bar\nu_\mu$), and any difference must be due to leptonic CP violation. From (9.74) and (9.75) the survival probabilities in vacuum must be equal, $P_{\nu_a \to \nu_a}(L) = P_{\bar\nu_a \to \bar\nu_a}(L)$; CP violation in vacuum can therefore only occur in appearance experiments. The discussion above assumes the validity of CPT. Possible CPT violation in the neutrino sector is discussed in (Diaz and Kostelecky, 2012).

[18]One usually labels the mass eigenstates so that $\Delta m^2 \geq 0$. The cases $0 \leq \theta \leq \pi/4$ and $\pi/4 \leq \theta \leq \pi/2$ are physically different. However, they cannot be distinguished by vacuum oscillation experiments, so it was traditional to use the variable $\sin^2 2\theta$ to describe the results. Matter effects, however, *can* distinguish the two cases (de Gouvêa et al., 2000), so it is better to use $\tan^2\theta$ instead. The region $\tan^2\theta > 1$ is sometimes known as the "dark side". The cases $\theta <$ or $> \pi/4$ also differ by subleading effects for more than two flavors.

[19]ν_a and $\bar\nu_a$ are, respectively, the left- and right-chiral states annihilated by ν_{aL} and ν_{aR}^c.

Oscillations between active neutrinos of different flavors are known as *first class (flavor) oscillations*. The results in (9.74) and (9.75) generalize to *second class oscillations* (Barger et al., 1980a) between light active and sterile neutrinos of the same helicity, which can occur when there are both Majorana and Dirac mass terms. For example, mixing between 3 active and 3 sterile neutrinos would still be described by (9.74), except $\mathcal{V}$ is now a 6×6 unitary matrix. Other phenomena can sometimes mimic neutrino oscillations, including non-standard interactions in the source, detector, or matter (e.g., Gavela et al., 2009; Ohlsson, 2013); neutrino-antineutrino transitions (involving new interactions to flip helicity) (Langacker and Wang, 1998); and massless neutrinos that are non-orthogonal due to mixing with heavy states (Langacker and London, 1988a). The latter can be generalized to massive oscillating neutrinos with an effectively non-unitary mixing matrix, again due to neglecting the mixing with heavier neutrinos (Antusch et al., 2006; Antusch and Fischer, 2014).

An important special case of (9.74) occurs when the mass eigenstates can be divided into two sets, each of which is nearly degenerate compared to the E/L of the experiment. That is, consider F mass eigenstates, in which ν_i, $i = 1 \cdots n$, are close in mass, as are ν_j, $j = n+1 \cdots F$. If L/E is such that $\Delta_{kl} L/E$ can be neglected when k and l are in the same set, then it is straightforward to show that

$$P_{\nu_a \to \nu_b}(L) = P_{\bar{\nu}_a \to \bar{\nu}_b}(L) = \sin^2 2\theta_{ab} \sin^2 \left(\frac{\Delta m^2 L}{4E} \right), \tag{9.76}$$

for $a \neq b$, where

$$\sin^2 2\theta_{ab} \equiv 4 \left| \sum_{i=1}^{n} \mathcal{V}_{ai}^* \mathcal{V}_{bi} \right|^2 = 4 \left| \sum_{j=n+1}^{F} \mathcal{V}_{aj}^* \mathcal{V}_{bj} \right|^2, \tag{9.77}$$

and $\Delta \equiv m_j^2 - m_i^2$ for $i \leq n < j$. The last form in (9.77) follows from unitarity. Thus, if one can neglect all but one mass splitting the two-neutrino formula holds, although the effective mixing angle may be a complicated function of the elements of $\mathcal{V}$. In particular, there are no CP-violating effects in this limit. Similarly, the survival probabilities are given by

$$P_{\nu_a \to \nu_a}(L) = P_{\bar{\nu}_a \to \bar{\nu}_a}(L) = 1 - \sin^2 2\theta_{aa} \sin^2 \left(\frac{\Delta m^2 L}{4E} \right), \tag{9.78}$$

with

$$\sin^2 2\theta_{aa} = \sum_{b \neq a} \sin^2 2\theta_{ab} = 4 \sum_{i=1}^{n} |\mathcal{V}_{ai}|^2 \left(1 - \sum_{k=1}^{n} |\mathcal{V}_{ak}|^2 \right), \tag{9.79}$$

which implies

$$\sin^2 \theta_{aa} \equiv \sum_{j=n+1}^{F} |\mathcal{V}_{aj}|^2 = 1 - \sum_{i=1}^{n} |\mathcal{V}_{ai}|^2. \tag{9.80}$$

Examples will be given below for oscillations involving atmospheric and sterile neutrinos.

9.3.2 The Mikheyev-Smirnov-Wolfenstein (MSW) Effect

Equation (9.70) or its generalization to three or more flavors describes the time evolution of an (initial) weak eigenstate in vacuum. However, for propagation through matter, such as the Sun or Earth, one must take into account the phase changes associated with the coherent forward scattering of the neutrinos with the matter, very much like index of refraction effects in optics (Wolfenstein, 1978). Under appropriate conditions, the matter effects can combine with the neutrino masses to yield an effective degeneracy and therefore an enhanced transition probability, the MSW resonance (Mikheyev and Smirnov, 1985).

To see how this works, consider the time evolution of a neutrino state

$$|\nu(t)\rangle = \sum_a c_a(t)|\nu_a\rangle, \tag{9.81}$$

in vacuum. The weak eigenstates $|\nu_a\rangle$ are related to the mass eigenstates $|\nu_i\rangle$ by

$$|\nu_a\rangle = \sum_i |\nu_i\rangle\langle\nu_i|\nu_a\rangle = \sum_i \mathcal{V}_{ai}^*|\nu_i\rangle. \tag{9.82}$$

(In (9.82) and the following, we ignore the momentum integrals and delta functions, as well as the $(2\pi)^3 2E$ normalization factors, all of which cancel between the matrix elements and the sums over intermediate states.) Evolving the mass eigenstates analogously to (9.70), one finds that the coefficients satisfy the Schrödinger-like equation

$$i\frac{d}{dt}c_a(t) = \langle\nu_a|H_V|\nu_b\rangle\, c_b(t), \tag{9.83}$$

where the vacuum Hamiltonian (due to the masses) is

$$\langle\nu_a|H_V|\nu_b\rangle = \sum_i \mathcal{V}_{ai}E_i\mathcal{V}_{bi}^* \sim \sum_i \mathcal{V}_{ai}\frac{m_i^2}{2E}\mathcal{V}_{bi}^* + E\delta_{ab}. \tag{9.84}$$

The last term only affects the irrelevant overall phase of the state. This and similar multiples of the identity can be dropped in what follows. For two flavors, H_V becomes

$$H_V = \frac{\Delta m^2}{4E}\begin{pmatrix} -\cos 2\theta & \sin 2\theta \\ \sin 2\theta & \cos 2\theta \end{pmatrix} \tag{9.85}$$

using the conventions in (9.69), with $\Delta m^2 = m_2^2 - m_1^2$.

In the presence of matter, (9.83) must be modified by the addition of the effective (matter) Hamiltonian

$$H_M = \int d^3\vec{x} \sum_a \frac{2G_F}{\sqrt{2}}\bar{\nu}_{aL}\gamma^\mu\nu_{aL}\sum_r\langle\bar{\psi}_r\gamma_\mu(g_V^{ar} - g_A^{ar}\gamma^5)\psi_r\rangle, \tag{9.86}$$

which describes the scattering of ν_a from fermions $r = e, p, n$, where $g_{V,A}^{ar}$ are the effective vector and axial couplings, which receive contributions from Z exchange and (in the case of $r = e$) from W exchange, as in (8.93) on page 280. The brackets on the last term indicate an expectation value in the static medium. Assuming the medium is unpolarized,

$$\langle\bar{\psi}_r\gamma_\mu(g_V^{ar} - g_A^{ar}\gamma^5)\psi_r\rangle = g_V^{ar}n_r\delta_\mu^0, \tag{9.87}$$

where n_r is the number density of particle r (cf., Equation A.19 on page 509). Using also that $\int d^3\vec{x}\,\nu_{aL}^\dagger\nu_{aL}$ is the ν_a number operator, $N_{\nu_{aL}} - N_{\nu_{aR}^c}$, we obtain the propagation equation[20]

$$i\frac{d}{dt}c_a(t) = \langle\nu_a|H_V|\nu_b\rangle\, c_b(t) + \sqrt{2}G_F\left(\sum_{r=e,p,n} g_V^{ar}n_r\right)c_a(t). \tag{9.88}$$

[20]The correct numerical factor and sign for the matter term were derived in (Barger et al., 1980b) and (Langacker et al., 1983), respectively.

The vector coefficients are

$$\begin{aligned}
g_V^{ee} &= +\frac{1}{2} + 2\sin^2\theta_W, & g_V^{ep} &= +\frac{1}{2} - 2\sin^2\theta_W, & g_V^{en} &= -\frac{1}{2} \\
g_V^{\mu e} &= -\frac{1}{2} + 2\sin^2\theta_W, & g_V^{\mu p} &= +\frac{1}{2} - 2\sin^2\theta_W, & g_V^{\mu n} &= -\frac{1}{2} \\
g_V^{sr} &= 0,
\end{aligned} \tag{9.89}$$

for ν_{eL}, $\nu_{\mu,\tau L}$, and sterile ν_{sL}, respectively. The signs are reversed for ν_R^c, as can be seen from the number operator or from (2.302) on page 56. g_V^{ee} contains an extra $+1$ from the WCC, which makes the effect important for $\nu_e \leftrightarrow \nu_{\mu,\tau}$.

Specializing to two families,

$$i\frac{d}{dt}\begin{pmatrix} c_a(t) \\ c_b(t) \end{pmatrix} = \begin{pmatrix} -\frac{\Delta m^2}{4E}\cos 2\theta + \frac{G_F}{\sqrt{2}}n & \frac{\Delta m^2}{4E}\sin 2\theta \\ \frac{\Delta m^2}{4E}\sin 2\theta & \frac{\Delta m^2}{4E}\cos 2\theta - \frac{G_F}{\sqrt{2}}n \end{pmatrix} \begin{pmatrix} c_a(t) \\ c_b(t) \end{pmatrix}, \tag{9.90}$$

where

$$n \equiv \sum_{r=e,p,n} \left(g_V^{ar} - g_V^{br} \right) n_r, \tag{9.91}$$

and we have symmetrized the diagonal elements by subtracting the common term $\sqrt{2}G_F n/2$. For an electrically neutral medium, i.e., $n_e = n_p$, this yields

$$n = \begin{cases} n_e & \text{for} \quad \nu_{eL} \leftrightarrow \nu_{\mu L}, \nu_{\tau L} \\ n_e - \frac{1}{2}n_n & \text{for} \quad \nu_{eL} \leftrightarrow \nu_{sL} \\ -\frac{1}{2}n_n & \text{for} \quad \nu_{\mu L}, \nu_{\tau L} \leftrightarrow \nu_{sL} \end{cases}, \tag{9.92}$$

with the signs reversed for ν_R^c.

Under the right conditions, the matter effect can greatly enhance the transitions. In particular, if the Mikheyev-Smirnov-Wolfenstein (MSW) resonance condition $\frac{\Delta m^2}{2E}\cos 2\theta = \sqrt{2}G_F n$ is satisfied, the diagonal elements vanish and even small vacuum mixing angles lead to a maximal effective mixing angle. Because of the sign switch, an enhancement for ν_L corresponds to a suppression for ν_R^c and vice-versa (i.e., the presence of matter effectively breaks CPT). The matter effect breaks the sign degeneracy for vacuum oscillations, and allows a determination of the sign of Δm^2. Because of the E dependence it can lead to a distortion in the final neutrino spectra. Finally, if the matter density varies significantly along the neutrino path, as is the case for solar neutrinos produced near the solar core, one may encounter a level-crossing at the position for which the resonance condition is satisfied (for a given E), as illustrated in Figure 9.5. If the density varies sufficiently gradually, the transition is adiabatic, i.e., the neutrino remains on one level, with a maximal flavor transition probability. A more abrupt (non-adiabatic) transition will have a non-negligible probability to jump from one level to another, reducing the flavor transition probability (see, e.g., Kuo and Pantaleone, 1989; Strumia and Vissani, 2006; Gonzalez-Garcia and Maltoni, 2008, 2013; Blennow and Smirnov, 2013).

9.3.3 Oscillation Experiments

There have been many experimental searches for and observations of neutrino oscillations and transitions (for recent reviews, see Gonzalez-Garcia and Maltoni, 2008; Diwan et al., 2016; Patrignani, 2016), including experiments at accelerators and reactors, and those involving solar neutrinos and atmospheric neutrinos (from the decay products of particles

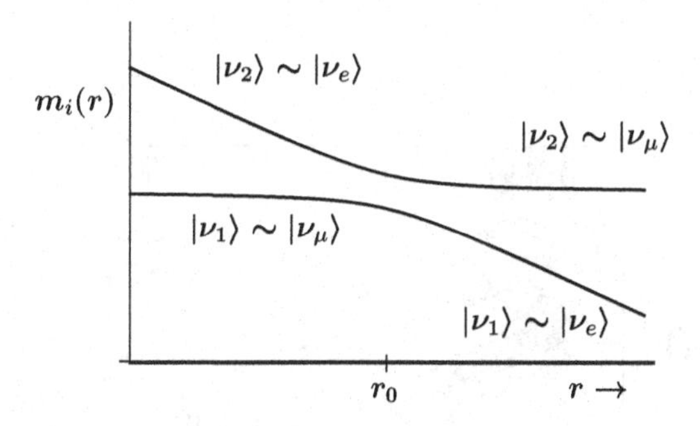

Figure 9.5 A level-crossing (resonance) at r_0 in the presence of matter density that decreases with distance r from the neutrino source. The $m_i(r)$ are the eigenvalues of the matrix in (9.90). (In the full three-neutrino case the state orthogonal to ν_e is actually a linear combination of ν_μ and ν_τ.)

produced by cosmic ray interactions in the atmosphere). Major observation and exclusion regions are plotted in the two-neutrino formalism in Figure 9.6.

Solar Neutrinos

Neutrinos (ν_e) are produced by fusion reactions in main sequence stars by the *pp* and *CNO* chains, which ultimately lead to $4p \rightarrow \alpha + 2e^+ + 2\nu_e$. The *standard solar model* (SSM) (Bahcall, 1989; Bahcall et al., 2006; Turck-Chieze and Couvidat, 2011; Haxton et al., 2013), which is well tested and constrained by helioseismology and other solar observations,[21] and by the properties of other stars, is dominated by the *pp* chain and leads to the predicted solar ν_e spectrum in Figure 9.7. The most important reactions are

$$p + p \rightarrow D + e^+ + \nu_e, \qquad D + p \rightarrow {}^3He + \gamma, \qquad 2\,{}^3He \rightarrow \alpha + 2p. \tag{9.93}$$

The first step leads to the most abundant (*pp*) neutrinos, which, however, have low energy and are hard to detect. About 15% of the time, however,

$${}^3He + \alpha \rightarrow {}^7Be + \gamma, \qquad {}^7Be + e^- \rightarrow {}^7Li + \nu_e, \tag{9.94}$$

for one of the ${}^3He$, leading to the two intermediate energy discrete ${}^7Be$ lines in the ν_e spectrum. Approximately 0.02% of the chains involve the sequence

$${}^7Be + p \rightarrow {}^8B + \gamma, \qquad {}^8B \rightarrow {}^8Be^* + e^+ + \nu_e, \tag{9.95}$$

which leads to the ${}^8B$ neutrinos. These are insignificant numerically, but because of their much higher energy are easiest to detect. The flux of *pp* neutrinos is well constrained by the observed solar luminosity, but the predicted ${}^7Be$ and (especially) ${}^8B$ fluxes are much more uncertain because of their strong dependence on the temperature of the solar core, low energy nuclear cross sections, and the solar composition.

The first solar neutrino experiment was the radiochemical ${}^{37}Cl$ experiment, which used

[21] Recent observations of heavy element abundances yield lower values than earlier (less precise) determinations, creating some tension with helioseismology (e.g., Haxton et al., 2013).

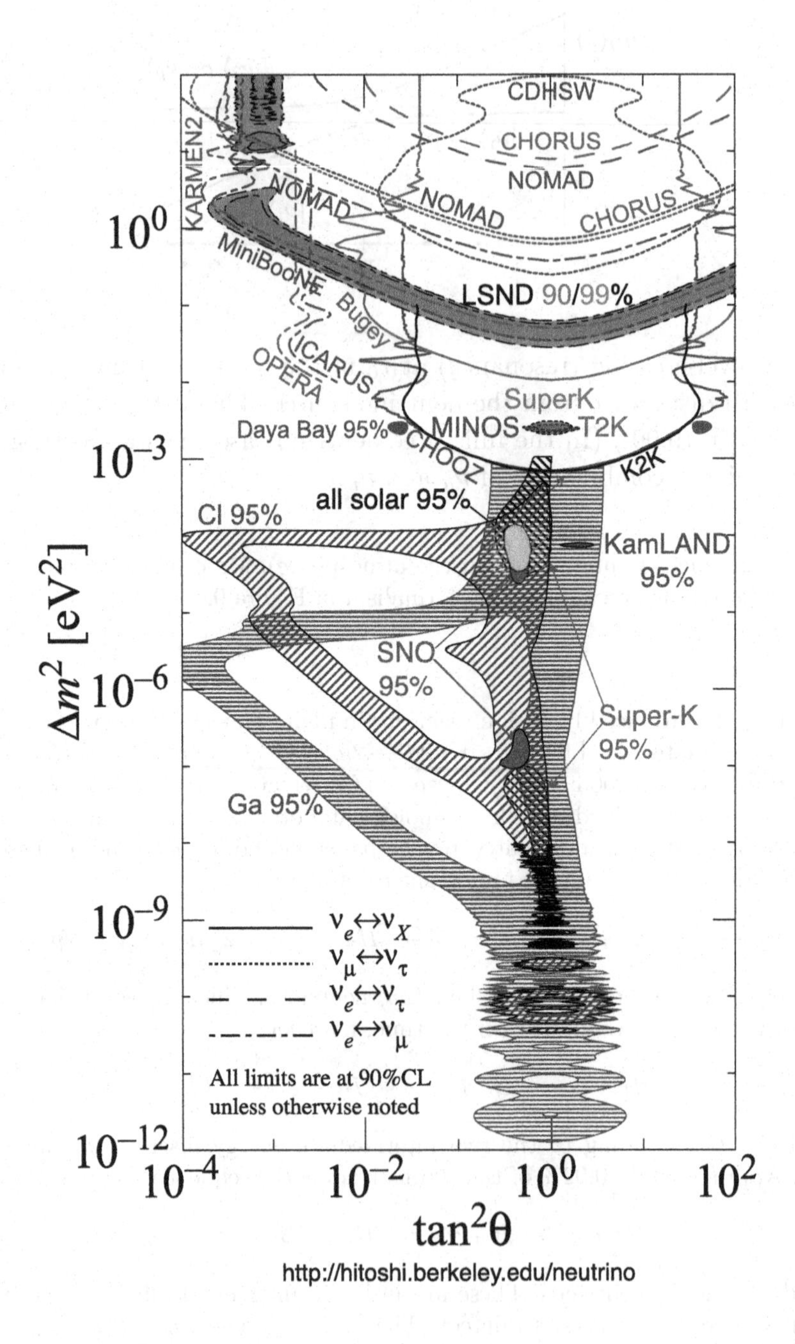

Figure 9.6 Neutrino oscillation results, showing the solar/KamLAND and atmospheric neutrino oscillation regions, the LSND and MiniBooNE regions, and various exclusion regions. Plot courtesy of H. Murayama.

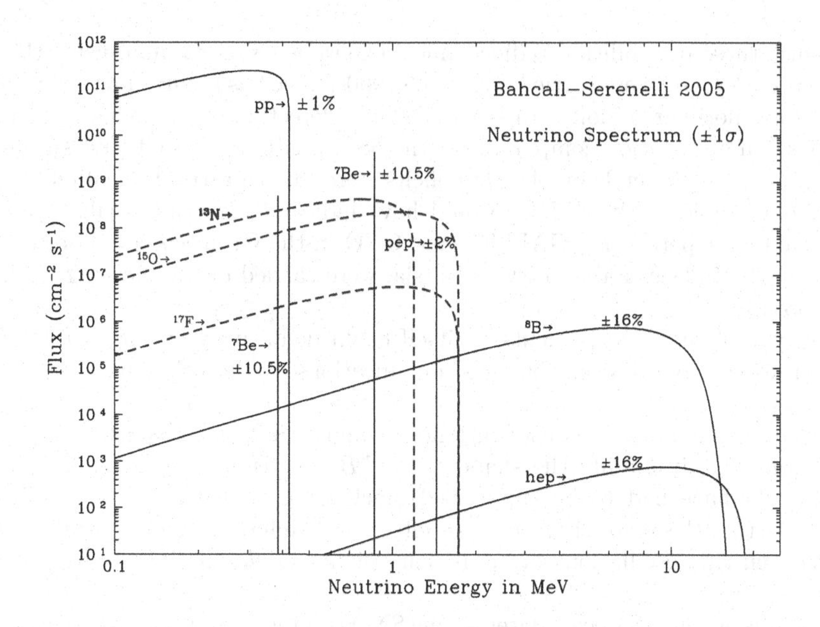

Figure 9.7 Spectrum of solar neutrinos predicted by the standard solar model. The gallium and liquid scintillator experiments are sensitive to the *pp* and higher energy neutrinos, the Homestake chlorine experiment to the higher energy ^7Be line and above, and the water Cherenkov experiments to the ^8B neutrinos. The other (minor) reactions, i.e., the *hep*, *pep*, and *CNO* neutrinos, are described in (Bahcall, 1989). Plot reproduced by permission of the AAS from (Bahcall et al., 2005).

a 10^5 gallon tank of cleaning fluid placed deep underground in the Homestake gold mine in South Dakota to shield from cosmic rays. Ray Davis and collaborators observed the decays of the ^{37}Ar atoms produced in the reaction $\nu_e + ^{37}Cl \rightarrow e^- + ^{37}Ar$ and separated chemically from the tank. The original goal was to probe the solar interior, but by the early 1970s it was apparent that they were only observing $\sim 1/3$ of the expected flux, beginning a 30-year enterprise that ultimately explored both solar and neutrino physics.

The Homestake results were confirmed in the late 1980s by the Kamiokande II water Cherenkov experiment in Japan, which also searched for proton decay and observed neutrinos from Supernova 1987A and atmospheric neutrinos. The reaction $\nu e^- \rightarrow \nu e^-$ was mainly sensitive to ν_e's, but because of WNC scattering had about 1/7 sensitivity to $\nu_{\mu,\tau}$, which could be produced if the ν_e's oscillated. They observed about 1/2 of the expected (without oscillations) solar flux, and also confirmed that the ν's actually came from the Sun because the e^- direction was correlated with that of the neutrino. These results were later confirmed and improved by the successor SuperKamiokande experiment (Abe et al., 2016), which extended the analysis to lower energies.

The water Cherenkov experiments were only sensitive to the upper part of the ^8B spectrum and the much rarer *hep* neutrinos. The Homestake experiment had a lower threshold and was sensitive to more of the ^8B spectrum and to some extent the ^7Be and *pep* neutrinos. The reduced fluxes that they observed could have been due to uncertainties in the standard solar model (e.g., by a 5% reduction in the temperature of the core) or other astrophysical effects, or to neutrino oscillations/transitions into ν_μ, ν_τ, or sterile neutrinos.

To distinguish these possibilities, radiochemical experiments on gallium, using the reaction $\nu_e + {}^{71}Ga \to e^- + {}^{71}Ge$, were carried out in the 1990s. This has a much lower threshold (233 keV) than the chlorine reaction (814 keV), allowing detection of the much more numerous *pp* neutrinos. An observation comparable to the SSM prediction would have suggested that the Homestake and (Super)Kamiokande deficits were due to astrophysical effects, while a reduction comparable to that for ${}^8B$ would have indicated neutrino oscillations. Eventually three gallium experiments, GALLEX and GNO in the Gran Sasso Laboratory in Italy and SAGE in the Baksan Laboratory in Russia, were carried out, indicating $\gtrsim 50\%$ of the predicted SSM flux.

No one type of these experiments by itself could definitively exclude an astrophysical explanation, especially allowing for large modifications of the SSM, but the three types together constituted a rough measurement of the distortion of the neutrino energy spectrum. Including the solar luminosity constraint and assuming that plausible astrophysical effects would not significantly modify the shape of the ${}^8B$ spectrum, it was found that the ${}^7Be$ neutrinos would have had to be suppressed much more than the ${}^8B$ ones (Hata et al., 1994). Because the ${}^8B$ is made from ${}^7Be$, this effectively excluded astrophysical explanations, but allowed neutrino oscillations or transitions (which *could* modify the shape of the ${}^8B$ spectrum).

The case was clinched a decade later by the SNO (Sudbury Neutrino Observatory) heavy water experiment in Ontario (Aharmim et al., 2013; Bellerive et al., 2016). SNO observed Cherenkov radiation from electrons and photons from neutron capture, allowing then to measure both WCC and WNC scattering from deuterium by

$$\nu_e + D \to e^- + p + p, \qquad \nu_x + D \to \nu_x + p + n, \tag{9.96}$$

where ν_x is any active neutrino. They also measured the electron scattering reaction $\nu e^- \to \nu e^-$, consistent with but less precisely than SuperKamiokande. The combination of these measurements allowed the SNO collaboration to separately determine the fluxes of ν_e and of $\nu_\mu + \nu_\tau$ arriving at the Earth, as shown in Figure 9.8. The result was that the sum of the three was consistent with the SSM flux prediction, and that about 2/3 had oscillated or been converted into $\nu_{\mu,\tau}$ (assuming that there are no sterile neutrinos involved), confirming both the SSM and neutrino oscillations.

More recently, the Borexino experiment (Bellini et al., 2014a,b) has studied solar neutrinos by observing $\nu e^- \to \nu e^-$ in a liquid scintillator detector in the Gran Sasso Laboratory. Borexino has measured the rates for the upper ${}^7Be$ neutrino line, *pp* neutrinos, and ${}^8B$ neutrinos at much lower energy than the water Cherenkov experiments. They also measured the rare *pep* neutrinos and set limits on neutrinos from the CNO cycle.

The survival probabilities $P_{\nu_e \to \nu_e}$ as measured by Borexino and the other experiments are shown as a function of neutrino energy in Figure 9.9.

At various stages the solar neutrino experiments allowed a number of solutions for neutrino oscillation parameters. These included the $\Delta m^2 \sim 10^{-10}$ eV^2 vacuum oscillation solutions, for which the Earth-Sun distance was of $\mathcal{O}(L_{osc})$. The matter effects were unimportant for these solutions and oscillations during the propagation between the Sun and Earth dominated. There were also several solutions with higher Δm^2, for which matter effects were significant. Eventually, the combination of the observed rates from the different reactions, as well as limits on the distortion of the ${}^8B$ spectrum and constraints (Bellini et al., 2014a) on or observation (Abe et al., 2016) of energy-dependent day-night asymmetries (due to reconversion to ν_e in the Earth), established the large mixing angle (LMA) solution characterized by (Bellerive et al., 2016)

$$\Delta m^2_\odot = 5.1^{+1.3}_{-1.0} \times 10^{-5} \text{ eV}^2, \qquad \tan^2\theta_\odot = 0.427 \pm 0.028, \tag{9.97}$$

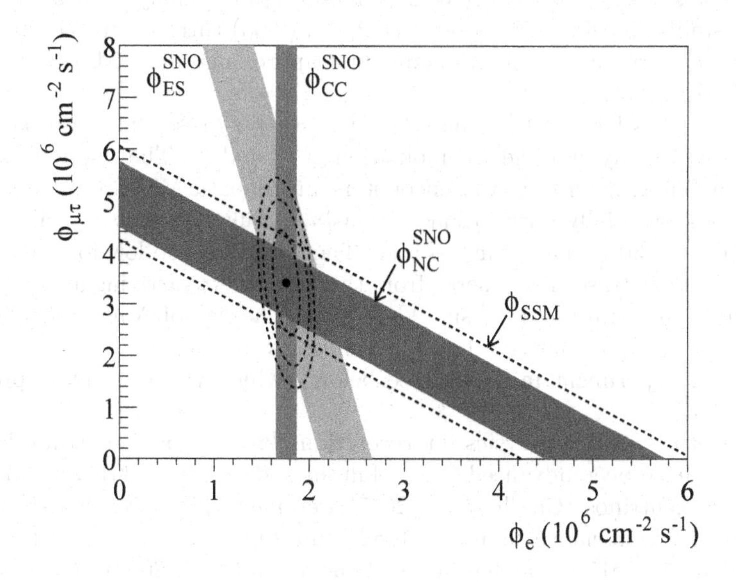

Figure 9.8 Fluxes from electron scattering (ES) and from the WCC and WNC reactions in the SNO heavy water experiment, compared with the standard solar model expectation. Plot reproduced with permission from (Bellerive et al., 2016).

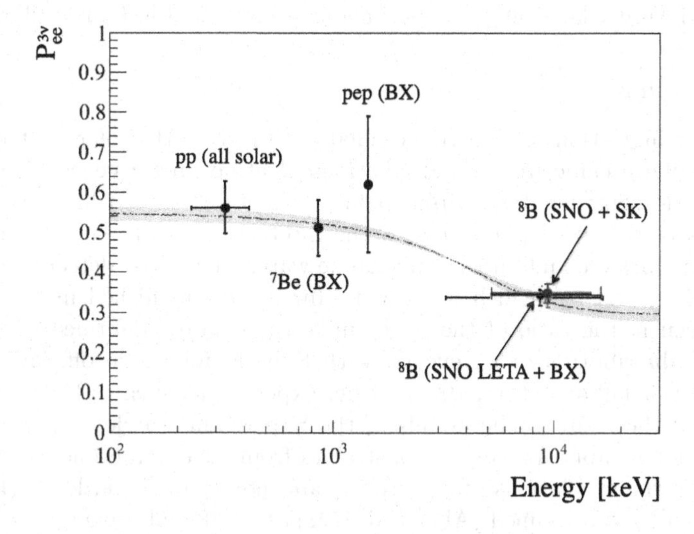

Figure 9.9 Survival probabilities $P_{\nu_e \to \nu_e}$ as a function of energy, compared with the predictions of the large mixing angle (LMA) solution. From (Bellini et al., 2014a). Not included is the recent direct measurement of the *pp* flux from the Phase 2 of Borexino (Bellini et al., 2014b).

corresponding[22] to $\sin^2\theta_\odot = 0.299 \pm 0.014$. A more detailed analysis involving all three neutrinos yields similar results. This is because θ_{13} in (9.35) turns out to be small, justifying the two-neutrino formalism as a reasonable first approximation, with the identification $\Delta m^2_\odot \sim \Delta m^2_{21}$ and $\theta_\odot \sim \theta_{12}$.

The solar angle $\theta_\odot$ is large but not maximal, i.e., $\theta_\odot \neq \pi/4$, so the ν_e is predominantly ν_1, while ν_2 consists mainly of a linear combination of ν_μ and ν_τ. The characteristics of the LMA solution are that a higher energy ν_e encounters an adiabatic MSW resonance, emerging from the Sun in an essentially pure ν_2 mass eigenstate. This does not oscillate, and has a probability $\sin^2\theta_\odot \sim 0.30$ of interacting as a ν_e. The density is too low for a resonance for the lower energy neutrinos, so they emerge from the Sun as a ν_e, arriving at the Earth with an average survival probability $\sim 1 - \frac{1}{2}\sin^2 2\theta_\odot \sim 0.58$. The sign of $\Delta m^2_\odot > 0$ is determined from the matter effect. Subsequently, the LMA solution was dramatically confirmed by the KamLAND reactor experiment in Japan (see below), which yielded a more precise (but slightly higher) $\Delta m^2_\odot$.

A number of other particle physics interpretations for the solar neutrino deficits and other observations have been advanced (e.g., Maltoni and Smirnov, 2016), including oscillations into sterile neutrinos (Cirelli et al., 2005); resonant spin-flavor transitions involving a large magnetic moment (Akhmedov, 1988; Lim and Marciano, 1988; Pulido, 1992; Giunti and Studenikin, 2015); neutrino decay (Beacom and Bell, 2002); new flavor changing or conserving interactions (Friedland et al., 2004; Miranda et al., 2006; Ohlsson, 2013); mass varying neutrinos due to interactions with the environment (Kaplan et al., 2004); Lorentz, CPT, or equivalence principle violations (Glashow et al., 1997; Diaz and Kostelecky, 2012); or CPT-violating decoherence between the quantum components of the wave function (Barenboim and Mavromatos, 2005). All of these are now excluded as the dominant effect for solar neutrinos, although they could still exist as perturbations on the basic picture.

Existing and future solar neutrino experiments are reviewed in (Antonelli et al., 2013).

Atmospheric Neutrinos

Although the first indications of neutrino oscillations involved the solar neutrinos, the first unambiguous evidence came from the oscillations of atmospheric neutrinos (e.g., Kajita, 2014). Atmospheric neutrinos result from pion and muon decays, which are produced in the upper layers of the atmosphere due to the interaction of primary cosmic rays. The data from the Kamiokande and SuperKamiokande water Cherenkov detectors indicated the disappearance of ν_μ and $\bar{\nu}_\mu$ (which will not be further distinguished in this paragraph). This was first seen in the ratio of the ν_μ/ν_e fluxes, and later confirmed dramatically by the zenith angle distribution of ν_μ events, with SuperKamiokande officially announcing their results in 1998 (Fukuda et al., 1998). Other experiments such as IMB, MACRO, and Soudan confirmed the results. The details of the SuperKamiokande ν_e and ν_μ events, as well as evidence for ν_τ appearance and constraints from other experiments show that the dominant effect is the oscillations of ν_μ into ν_τ, and not ν_e or a sterile ν_S (Kajita et al., 2016). SuperKamiokande obtained (Abe et al., 2011) the 90% c.l. ranges

$$|\Delta m^2_{atm}| \sim (1.7 - 3.0) \times 10^{-3}\ \text{eV}^2, \qquad \sin^2 2\theta_{atm} \sim (0.93 - 1.0), \tag{9.98}$$

from their 1998–2008 data, corresponding to $\sin^2\theta_{atm} \sim (0.37 - 0.63)$.

To a good approximation, $\Delta m^2_\odot$ can be neglected for the atmospheric neutrinos, justifying the two-neutrino formalism with $\Delta m^2_{atm} \sim \Delta m^2_{32}$. From (9.80) and (9.35) the effective

[22] Various authors quote the neutrino mixings as $\tan^2\theta, \sin^2\theta, \sin^2 2\theta$, or even as θ in degrees.

angle for ν_μ survival is then

$$\sin^2 \theta_{atm} = |\mathcal{V}_{\mu 3}|^2 = s_{23}^2 c_{13}^2, \tag{9.99}$$

so that $\theta_{atm} \sim \theta_{23}$ for small θ_{13}. The data is consistent with maximal mixing ($\theta_{23} = \pi/4$), and in fact the best fit is for that value. There are no Earth matter effects for $\nu_\mu \to \nu_\tau$, so the sign of Δm_{32}^2 is not determined in the approximation of neglecting both $\Delta m_\odot^2$ and θ_{13}.

The atmospheric neutrino fluxes were subsequently measured in the MINOS detector (Adamson et al., 2012), the steel-scintillator far detector for the NuMI-MINOS long-baseline facility described below. MINOS obtained results consistent with SuperKamiokande. The long-baseline experiments further confirm the atmospheric neutrino results, though there is some suggestion of non-maximal mixing.

Both MINOS and SuperKamiokande also showed that the ν_μ and $\bar{\nu}_\mu$ disappearance parameters are the same within uncertainties, consistent with CPT. This was done by MINOS on an event by event basis by exploiting the magnetization of the detector, while SuperKamiokande utilized the differences in the distortions of the zenith angle distributions.

Additional or future observations of atmospheric neutrinos include the high energy neutrino telescopes described below and the ICAL (Ahmed et al., 2015) detector at the Indian Neutrino Observatory (INO).

Comments similar to those for solar neutrinos concerning alternatives to neutrino oscillations apply to the atmospheric results. It is interesting that the atmospheric neutrino oscillations are a quantum mechanical coherence effect on the size scale of the Earth.

Accelerator Neutrinos

Early *short-baseline* accelerator experiments included ν_μ (or $\bar{\nu}_\mu$) disappearance experiments and searches for ν_e or ν_τ appearance at CERN, Brookhaven, and Fermilab (e.g., Gonzalez-Garcia and Maltoni, 2008). These typically involved energies in the $1 - 100$ GeV range and distances $100 - 1000$ m, with sensitivities to $\Delta m^2 \gtrsim 10^{-1} - 1$ eV^2 for large mixing. No evidence for oscillations was found (but see the discussion of possible sterile neutrinos below).

More recently, there have been *long-baseline* experiments involving beams from KEK (K2K) and J-PARC (T2K, for Tokai to Kamioka) (Abe et al., 2015a) to SuperKamiokande; the Fermilab NuMI beam to MINOS (Timmons, 2016) in the Soudan mine in Minnesota and (recently) to NOνA (Adamson et al., 2016a) in Ash River, Minnesota; and the CERN CNGS beam to the OPERA and ICARUS detectors in the Gran Sasso Laboratory (for reviews, see Feldman et al., 2013; Diwan et al., 2016; Nakaya and Plunkett, 2016). These experiments monitor the initial neutrino fluxes by their interactions in a near detector close to the accelerator (or by detailed simulations of the CNGS beam). The K2K and T2K experiments have baselines $L \lesssim 300$ km, those using the Fermilab-MINOS and CERN beams have $L \sim 735$ km, and Fermilab-NOνA has $L \sim 810$ km. These are sensitive to much lower Δm^2 than the traditional short-baseline accelerator experiments, down into the atmospheric neutrino region $\sim 10^{-3}$ eV^2. The long-baseline experiments have confirmed the SuperKamiokande and MINOS atmospheric oscillation results. They have significantly reduced the uncertainty in $|\Delta m_{32}^2|$ and $\sin^2 \theta_{23}$, as can be seen in Figure 9.10. For example, T2K obtained (Abe et al., 2015a)

$$|\Delta m_{32}^2| = 2.51(10) \times 10^{-3} \text{ eV}^2, \qquad \sin^2 \theta_{23} = 0.51(6) \tag{9.100}$$

at 1σ from their ν_μ disappearance data, assuming the normal hierarchy, with similar results for the inverted. The MINOS and NOνA data also suggest that θ_{23} may be non-maximal.

The results are consistent with dominantly $\nu_\mu \to \nu_\tau$ oscillations,[23] and that the ν_μ and $\bar{\nu}_\mu$ survival probabilities are the same. They have also observed sub-dominant oscillations into ν_e, which are associated with the (small) angle θ_{13}, and obtained strong constraints on possible sterile neutrinos.

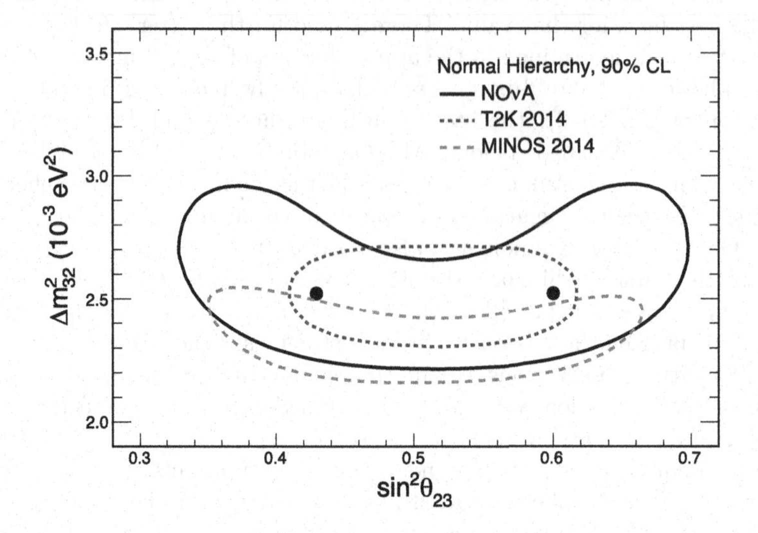

Figure 9.10 Allowed regions at 90% c.l. for the atmospheric neutrino oscillation parameters from atmospheric and long-baseline data, from (Adamson et al., 2016a).

The T2K and NOνA experiments both involve *off-axis* beams, taking advantage of the narrower energy spectrum away from the center, and both will benefit from planned upgrades to the intensity of the neutrino sources. These and other advantages should allow them to simultaneously search for leptonic CP violation (e.g., Nunokawa et al., 2008) and determine the type of neutrino hierarchy via matter effects in long-baseline ν_e and $\bar{\nu}_e$ appearance experiments. Both the appearance and disappearance channels will be sensitive to θ_{23}. There are also plans for two very large next generation experiments. One is the Deep Underground Neutrino Experiment (DUNE) (Acciarri et al., 2015), featuring a 40 kiloton liquid argon detector in the Sanford Laborary in Homestake, 1300 km from Fermilab. The other is Hyper-Kamiokande (Abe et al., 2015b), a megaton water Cherenkov detector near SuperKamiokande, utilizing a beam from J-PARC. These detectors will also be sensitive to proton decay, and to supernova, solar, and atmospheric neutrinos.

Reactor Neutrinos

There were a number of early reactor disappearance experiments, which compared the flux of $\bar{\nu}_e$ at short distances, $L \sim (10 - 100)$ m, with the theoretical expectations based on the known reactor energy output (see, e.g., Gonzalez-Garcia and Maltoni, 2008). These were sensitive down to $|\Delta m^2| \gtrsim 10^{-2}$ eV^2 for large mixing. They did not report any evidence for oscillations at the time, although a recent reanalysis of the expected fluxes (Mention et al., 2011) has suggested the possibilitiy of disappearance into sterile neutrinos, as described below.

[23]The OPERA hybrid emulsion experiment has directly observed five ν_τ appearance events (Agafonova et al., 2015).

The subsequent (around 2000) Palo Verde and Chooz experiments had longer baselines of $\mathcal{O}(1 \text{ km})$, and were able to exclude significant $\bar{\nu}_e$ mixing for $|\Delta m^2| \gtrsim 10^{-3}$ eV^2 (e.g., Qian and Wang, 2014). Since the atmospheric neutrino results already established $|\Delta m^2_{32}| >$ few $\times 10^{-3}$ eV^2, this implied that $\sin^2 2\theta_{13} < 0.12$ ($\tan^2\theta_{13} < 0.032$) at 90% c.l., which is small compared to the other leptonic mixing angles but comparable to quark mixings. The value of θ_{13} is critical: the small value motivated models of neutrino mass in which θ_{13} vanishes or is tiny (Section 9.5). Furthermore, the possibility of observing leptonic CP violation and significant matter effects in terrestrial experiments depends on having a sufficiently large θ_{13}.

For these reasons, an intensive effort was undertaken to observe or constrain θ_{13} for $|\Delta m^2|$ in the atmospheric neutrino range (Qian and Wang, 2014). Hints of a nonzero value were obtained from a global analysis of existing data (Fogli et al., 2008), but the first direct experimental evidence was obtained around 2011 by the MINOS and T2K long-baseline accelerator experiments, which observed $\nu_\mu \to \nu_e$ appearance at the several σ level. Subsequently, the Double Chooz[24] reactor experiment reported evidence for $\bar{\nu}_e$ disappearance. Finally, the Daya Bay[25] reactor experiment in China observed $\bar{\nu}_e$ disappearance at greater than 5σ, establishing that $\theta_{13} \neq 0$. RENO in South Korea also observed disappearance at nearly 5σ. Currently, The Daya Bay value (An et al., 2017) is

$$\sin^2 2\theta_{13} = 0.084(3) \quad \Rightarrow \quad \tan^2\theta_{13} = 0.022(1), \tag{9.101}$$

which is 25σ away from zero, and also close to the earlier upper limits. They also obtain $|\Delta m^2_{32}| = 2.45(8)[2.56(8)] \times 10^{-3}$ eV^2 for the normal [inverted] hierarchies, comparable to the long-baseline results. The RENO results (Choi et al., 2016) are consistent but less precise. Proposed future reactor experiments (e.g., Diwan et al., 2016) include JUNO in China and RENO-50 in Korea, each with baselines around 50 km. They would be able to distinguish between the normal and inverted hierarchies, independent of CP violation and matter complications, by exploiting the difference between $|\Delta m^2_{32}|$ and $|\Delta m^2_{31}|$, as well as constrain the oscillation parameters.

Because of the characteristic $\Delta m^2 L/E$ dependence of neutrino oscillations, long-baseline ($L \sim$ hundreds of km) experiments can probe to much lower Δm^2 than the traditional short-baseline ones. The KamLAND experiment was a liquid scintillator detector at the location of the original Kamiokande detector (Gando et al., 2013). It observed a $\bar{\nu}_e$ flux from a number of Japanese reactors, at a typical distance $L \sim 200$ km, which allows one to probe down into the region of the LMA solar neutrino solution. The KamLAND results dramatically confirmed the LMA interpretation of the solar neutrino deficit and gave a much more precise value for Δm^2_{21} around $7.5(2) \times 10^{-5}$ eV^2, as can be seen in Figure 9.11. KamLAND was also able to directly observe the L/E dependence expected from neutrino oscillations, and to observe ^8B and ^7Be solar neutrinos.

Geoneutrinos

Low-energy antineutrinos from radioactive decays in the Earth were initially an important background for the liquid scintillator experiments. Both KamLAND (Gando et al., 2011) and Borexino (Agostini et al., 2015b) were eventually able to measure the flux of these

[24]Double Chooz is a follow up to the Chooz experiment with near and far detectors. The initial result on $\bar{\nu}_e$ disappearance was obtained with only the far detector.

[25]The Daya Bay experiment consists of a number of near ($L \sim 0.5$ km) and far ($L \sim 1.5$ km) detectors relative to a reactor complex. Comparison of the fluxes in the near and far detectors essentially eliminates the uncertainties from the initial reactor flux. RENO is configured analogously.

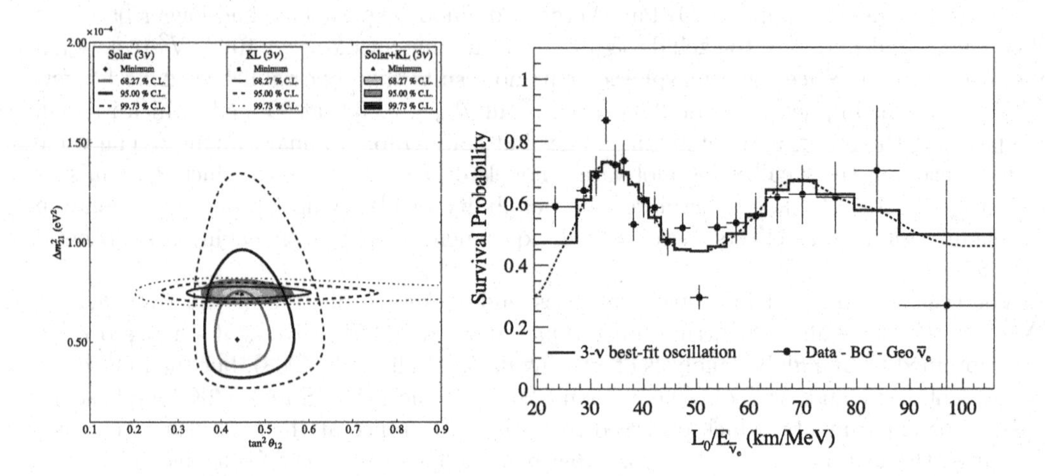

Figure 9.11 Left: oscillation parameters Δm^2_{21} (eV^2) vs $\tan^2\theta_{12}$ determined from solar neutrino data at 68, 95, and 99.7% c.l. (egg-shaped contours), compared with the KamLAND results (horizontal contours), and the combined fit. From (Bellerive et al., 2016). Right: L/E dependence of the $\bar{\nu}_e$ survival probability as determined by KamLAND. From (Gando et al., 2013).

geoneutrinos from the decay chains of ^{232}Th and ^{238}U (but not those from ^{40}K), constraining models of the element abundances and interior heating of the Earth. For a review, see (Ludhova and Zavatarelli, 2013).

High Energy Neutrinos

Ultra-high energy neutrinos, e.g., produced by pion decay, are potentially an extremely useful probe of violent astrophysical events. They are not significantly absorbed in interstellar/intergalactic media, and (unlike cosmic ray protons or nuclei) they are not deflected in magnetic fields and therefore point back to their sources. Neutrino telescopes are large detectors with good ability to measure the direction of the incoming neutrino. These typically involve strings of photomultiplier tubes deployed deep underwater (such as the ANTARES array in the Mediterranean) or under ice (such as the km^3 IceCube array at the South Pole). The expected cross sections for ultra-high energy neutrinos are reviewed in (Gandhi et al., 1998; Formaggio and Zeller, 2012), and the experiments in (Katz and Spiering, 2012; Gaisser and Halzen, 2014).

The IceCube collaboration has observed (Aartsen et al., 2013) several events with energies above 1 PeV (10^6 GeV), the highest energy neutrinos ever observed, as well as many additional events with energies above 30 TeV. These are most likely extra-terrestrial, and could be due to galactic or extra-galactic astrophysical events, or to exotic particle physics such as super-heavy dark matter decays (see Anchordoqui et al., 2014). IceCube has also shown (Aartsen et al., 2015) that the flavor ratio of high energy neutrinos arriving at the Earth is consistent with the value $\nu_e/\nu_\mu/\nu_\tau = 1/1/1$ expected for stable neutrinos produced by pion decay in distant sources when oscillations are taken into account. (Other sources, such as unstable neutrino decays, could lead to other ratios (e.g., Bustamante et al., 2015).) IceCube has also observed atmospheric oscillations at much lower energies (but still higher energy than the other atmospheric experiments), constrained oscillations into sterile states, and searched for neutrinos from dark matter annihilation in the Sun. It may also be sensitive to the *Glashow resonance* (Glashow, 1960), in which the cross section for $\bar{\nu}_e$ scattering is greatly enhanced at 6.3 PeV due to the resonant scattering $\bar{\nu}_e e^- \to W^- \to X$ from electrons in the atmosphere; to nonstandard interactions; and to neutrinos from galactic supernova. The proposed PINGU upgrade to the inner detector (Aartsen et al., 2017) would lower the energy threshold to around 5 GeV, allowing the determination of the mass hierarchy through resonant matter effects, and possibly allowing Earth tomography.

The proposed KM3NeT underwater telescope in the Mediterranean (Adrian-Martinez et al., 2016) and the Lake Baikal-GVD facility in Russia would have similar capabilities.

9.3.4 Possible Sterile Neutrinos

There are several indications of mixing between active and light (eV scale) sterile neutrinos, as well as many null experiments. For reviews, see, for example, (Abazajian et al., 2012; Palazzo, 2013; Kopp et al., 2013; Gariazzo et al., 2016).

The LSND experiment at Los Alamos observed a 3.8σ excess of $\bar{\nu}_e p \to e^+ n$ events in a $\bar{\nu}_\mu$ beam obtained from μ^+ decay at rest (Aguilar-Arevalo et al., 2001). This suggested $\bar{\nu}_\mu \to \bar{\nu}_e$ oscillations[26] with $L \sim 30$ m and $L/E \sim (0.4-1.5)$ m/MeV, corresponding to $|\Delta m^2_{LSND}| \sim (0.2-10)$ eV^2 and small mixing. This was not confirmed by the KARMEN2 experiment at Rutherford, but there was a small parameter region allowed by both. The LSND result, along with the solar and atmospheric oscillations, would imply three or more distinct Δm^2's, and therefore at least four light neutrinos that mix with each other. The extra neutrinos would have to be sterile because the invisible Z width result $N_\nu^{inv} = 2.992 \pm 0.007$ does not allow a fourth light active neutrino.

Subsequently, the Fermilab MiniBooNE experiment searched for ν_e appearance in a ν_μ beam with $L \sim 541$ m and $E_\nu \sim (20-1250)$ MeV, and later for $\bar{\nu}_e$ appearance in a $\bar{\nu}_\mu$ beam. The final MiniBooNE results[27] (Aguilar-Arevalo et al., 2013) showed excesses in both ν_e and $\bar{\nu}_e$-like events. The $\bar{\nu}_e$ excess was consistent with the LSND oscillation signal, though with a lower statistical significance. The energy dependence of the ν_e excess was only marginally consistent with the other results, however, at least within the framework of a single sterile neutrino.

$|\Delta m^2_{LSND}|$ and the associated E/L are large compared to $|\Delta m^2_{atm}|$ and $\Delta m^2_\odot$. Assuming the existence of a single sterile neutrino ν_S, one can therefore treat the three largely-active states as degenerate and use the effective two-neutrino formalism in Equations (9.76)–(9.80), with $|\Delta| = |m_4^2 - m_{1,2,3}^2| = |\Delta m^2_{LSND}| \sim (0.2-10)$ eV^2, where m_4 is the mass of the additional (largely sterile) mass eigenstate. The effective mixing for $\bar{\nu}_\mu \to \bar{\nu}_e$ and $\nu_\mu \to \nu_e$ appearance is therefore

$$\sin^2 2\theta_{\mu e} = 4 \sin^2 \theta_{\mu\mu} \sin^2 \theta_{ee} \sim \frac{1}{4} \sin^2 2\theta_{\mu\mu} \sin^2 2\theta_{ee}, \tag{9.102}$$

[26]Other possibilities to account for LSND, e.g., involving CPT violation, decoherence, new interactions, extra dimensions, mass-varying neutrinos, and hybrids, are surveyed in (Gonzalez-Garcia and Maltoni, 2008).

[27]The initial MiniBooNE $\nu_\mu \to \nu_e$ analysis was restricted to $E_\nu > 475$ MeV so as to coincide with the LSND L/E range. No excess was observed in this range, although there was an anomalous excess of ν_e events at lower energies. The excess observed in the subsequent $\bar{\nu}_\mu$ runs stimulated considerable discussion of possible CP violation to account for the $\nu - \bar{\nu}$ difference. However, the final MiniBooNE ν_e analysis included events with E_ν down to 200 MeV. The resulting excess was mainly due to (but did not fully account for) the low energy anomaly. The LSND and MiniBooNE experiments are reviewed in (Conrad et al., 2013b), while the experimental and theoretical status of low-energy ν scattering is reviewed generally in (Garvey et al., 2015).

where $\sin^2\theta_{\mu\mu} = |\mathcal{V}_{\mu 4}|^2$, $\sin^2\theta_{ee} = |\mathcal{V}_{e4}|^2$, and the last expression assumes small mixing. But $\sin^2\theta_{\mu\mu}$ and $\sin^2\theta_{ee}$, respectively, control the ν_μ and ν_e survival probabilities; e.g., the mixing for ν_e disappearance into ν_S is $\sin^2 2\theta_{eS} = 4|\mathcal{V}_{e4}^*\mathcal{V}_{S4}|^2 \sim 4\sin^2\theta_{ee}$. Thus, sterile-induced oscillations $\nu_\mu \to \nu_e$ necessarily imply disappearance of both ν_μ and ν_e into ν_S, and also the absence of any CP-violating difference between ν and $\bar{\nu}$, as long as the effective two-neutrino approximation is valid. For additional sterile neutrinos one can still usually treat the (mainly) active states as degenerate. The LSND-MiniBooNE signal still typically leads to ν_μ and ν_e disappearance, although the additional parameters would allow for cancellations and for observable differences between ν and $\bar{\nu}$.

Short-baseline ($L \lesssim 100$ m) reactor experiments should be sensitive to $\bar{\nu}_e$ disappearance into a sterile neutrino for Δm^2 in the LSND range. However, in existing experiments the measured fluxes must be compared with theoretical $\bar{\nu}_e$ spectra obtained from detailed modeling of the relevant fission decay chains in the reactors. A recent reanalysis (Mention et al., 2011) obtained a predicted flux about 3% higher than earlier estimates (see also Huber, 2011), leading to the *reactor anomaly*: whereas the previous measurements were consistent with no $\bar{\nu}_e$ disappearance ($P_{\bar{\nu}_e\to\bar{\nu}_e} = 0.976\pm 0.024$), the new theoretical spectrum suggested a 2.5σ deficit (0.943 ± 0.023), which could be due to oscillations into sterile neutrinos.[28] Some caution is required, however, since a later study (Hayes et al., 2014) argued that the systematic uncertainties in the flux are as large as the anomaly. The situation is further complicated by the observation by Daya Bay, RENO, and Double Chooz of a statistically significant enhancement ("bump") in the $\bar{\nu}_e$ flux in the 4-6 MeV region, which is not explained in the flux models. The entire situation is reviewed in (Hayes and Vogel, 2016).

Additional evidence for sterile neutrino oscillations comes from the *gallium anomaly*. The SAGE and GALLEX solar neutrino experiments each used intense radioactive ν_e sources (^{51}Cr, ^{37}Ar) of known intensity as a cross check on their efficiencies and on the theoretical estimate of the cross section for $\nu_e + ^{71}Ga \to e^- + ^{71}Ge$. The observed rates were around 2.8σ below the expectation (Gariazzo et al., 2016), suggesting $P_{\nu_e\to\nu_e} = 0.84\pm 0.05$ (or that the other uncertainties are underestimated).

There have also many negative searches for oscillations involving sterile neutrinos. These include searches for ν_e or $\bar{\nu}_e$ disappearance in accelerator and longer-distance reactor experiments, and global analyses of the solar neutrinos. Similarly, there are many null results on ν_μ and $\bar{\nu}_\mu$ disappearance[29] from short- and long-baseline accelerator experiments and atmospheric neutrinos, as well as searches for $\nu_\mu \to \nu_e$ or ν_τ (ν_e) appearance beyond the three flavor expectations by OPERA (ICARUS) in the CNGS beam. Especially strong limits on ν_μ disappearance by MINOS and $\bar{\nu}_e$ disappearance by Daya Bay constrain $\sin^2\theta_{\mu\mu}$ and $\sin^2\theta_{ee}$, respectively, and these can be combined in a joint analysis (Adamson et al., 2016b) to exclude a significant region in the $\sin^2 2\theta_{\mu e} - \Delta m^2_{41}$ plane. IceCube (Aartsen et al., 2016) has also exploited the nonobservation of resonant matter effects on atmospheric $\overset{(-)}{\nu}_\mu$ to strongly constrain $\sin^2\theta_{\mu\mu}$. These and earlier results are in considerable tension with the LSND-MiniBooNE data in the case of one sterile neutrino, and exclude most ot the low Δm^2_{41} (high $\sin^2 2\theta_{\mu e}$) region, as shown in Figure 9.12. A region around $\Delta m^2_{41} \sim 1.6$ eV^2 and $\sin^2 2\theta_{\mu e} \sim 0.0015$ appears to be the best compromise (e.g., Kopp et al., 2013; Gariazzo et al., 2016; Collin et al., 2016). Somewhat better agreement can be obtained if there are two or three sterile neutrinos (Conrad et al., 2013a), though this conclusion is weakened if one does not include the anomalous MiniBooNE low-energy data (Gariazzo et al., 2016).

As discussed in the sections on neutrino counting (page 385) and neutrino mass

[28]The value of θ_{13} is insensive to the sterile mixing.

[29]ν_μ oscillations into ν_τ can be distinguished from ν_S statistically by neutral current scattering, matter effects, and τ production.

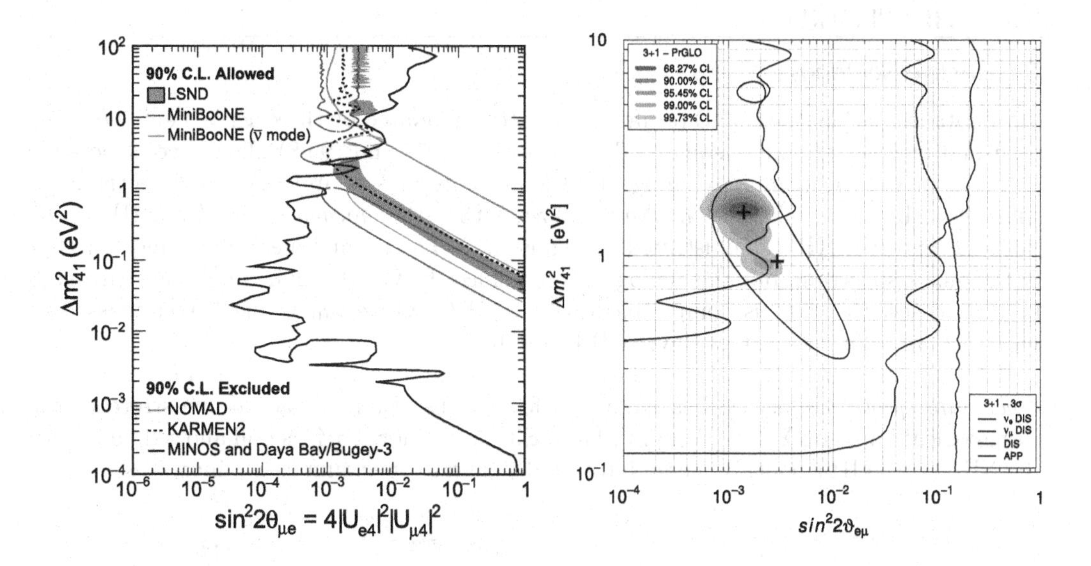

Figure 9.12 Left: 90% c.l. exclusion regions for sterile neutrino oscillations from the combined MINOS-Daya Bay analysis (to the right of the outer contour), contrasted with the allowed LSND and MiniBooNE regions (narrow, diagonal regions), from (Adamson et al., 2016b). Right: 3σ allowed region from short-baseline $\overset{(-)}{\nu}_e$ appearance experiments other than the anomalous low-energy MiniBooNE results (egg-shaped contour), and the 3σ exclusion regions from disappearance experiments (right of the leftmost line). The limits from $\overset{(-)}{\nu}_\mu$ (nearly vertical) and $\overset{(-)}{\nu}_e$ are also shown. The overall best fit is the cross around $\sin^2 2\theta_{\mu e} \sim 0.0015$, $\Delta m^2_{41} \sim 1.6$ eV^2. From (Gariazzo et al., 2016. Copyright IOP Publishing. Reproduced with permission. All rights reserved.)

(page 387) there are cosmological implications for both the numbers and masses of sterile neutrinos. In particular, in the standard cosmological scenario the sterile neutrinos should have thermalized efficiently by active-sterile mixing for the parameters suggested by Figure 9.12, i.e., $T_S \sim T_\nu$ so that $N_\nu^{CMB} \sim 4$ and $m_S^{eff} \sim m_4 \gtrsim 1$ eV, in tension with the Planck limits in (9.59). The situation would be worse if there are additional eV-scale sterile neutrinos with significant mixing. These constraints could be relaxed or evaded if the production is somehow suppressed, e.g., by large lepton asymmetries, late-time phase transitions, a low reheating temperature after inflation, or time-varying masses (for reviews, see, e.g., Langacker, 2005; Gariazzo et al., 2016). The possibility that one or more sterile states are lighter than the active ones is mentioned in the next section.

The situation concerning possible light sterile neutrinos and their mixing with active ones is confusing, and no compelling picture has emerged. Definitive experiments, e.g., involving intense radioactive sources, experiments very close to reactors, or specially-configured accelerator experiments, are needed to resolve the situation (for a review, see Gariazzo et al., 2016). It is also challenging to construct a theory in which active and sterile neutrinos mix, as will be discussed in Section 9.5.

9.4 THE SPECTRUM

Three Active Neutrinos

Most data other than the LSND results can be accommodated by three active neutrinos. The solar and atmospheric mass-squared differences in (9.97) and (9.98) allow several possible patterns for the spectrum. Assuming that the absolute masses are comparable to the mass splittings, the sign ambiguity in Δm^2_{32} allows either the *normal hierarchy* (NH) or the *inverted hierarchy* (IH), as illustrated in Figure 9.13. The normal hierarchy is most similar to the quark spectrum, but the analogy is poor since the CKM mixing angles are all small. In both cases, the data is compatible with $m_0 = 0$, where m_0 is the lowest mass, i.e., $m_0 = m_1$ (m_3) for the normal (inverted) hierarchy.

It is also possible that the absolute masses are larger than the mass differences (the *degenerate spectrum*), perhaps as large as a few tenths of an eV each if one stretches the cosmological limits on $\Sigma = \sum_i |m_i|$ in (9.58). Either sign for Δm^2_{32} is still allowed, and both the NH and the IH smoothly merge into the degenerate spectrum as m_0 increases.

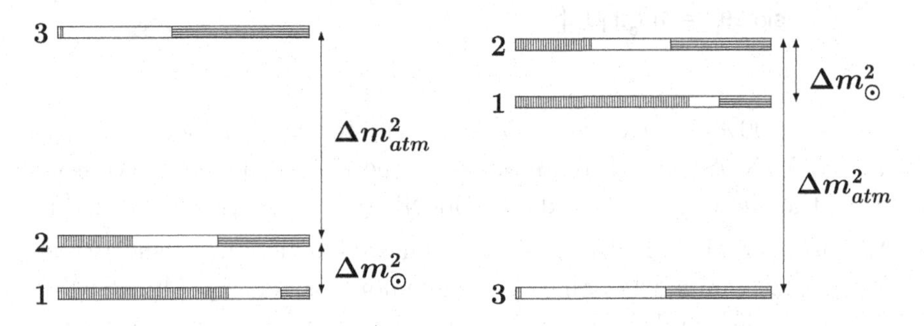

Figure 9.13 Left: the normal hierarchy for three neutrinos. Right: the inverted hierarchy. The vertical-dashed, open, and horizontal-ruled regions indicate the central values of $|\mathcal{V}_{ei}|^2$, $|\mathcal{V}_{\mu i}|^2$, and $|\mathcal{V}_{\tau i}|^2$, respectively, using the best fit parameters. The degenerate case corresponds to adding a large common mass to each state. In this context ν_3 is usually defined as the isolated state rather than heaviest.

A recent global three-neutrino analysis (Esteban et al., 2017) obtained

$$\begin{aligned} &\Delta m^2_{21} \sim 7.50(18) \times 10^{-5}\ \text{eV}^2, \qquad \Delta m^2_{31} \sim 2.52(4) \times 10^{-3}\ \text{eV}^2, \qquad \delta = 261(55)^\circ \\ &\sin^2\theta_{12} \sim 0.31(1), \qquad \sin^2\theta_{23} \sim 0.44(2), \qquad \sin^2\theta_{13} = 0.022(1), \end{aligned} \tag{9.103}$$

assuming the normal hierarchy ($\Delta m^2_{31} \gtrsim \Delta m^2_{32} > 0$). The results are similar for the inverted hierarchy except $\Delta m^2_{32} = -2.51(4) \times 10^{-3}\ \text{eV}^2 \lesssim \Delta m^2_{31}$, $\sin^2\theta_{23} \sim 0.59(2)$, and $\delta = 277(43)^\circ$. The χ^2 distributions for the parameters are shown in Figure 9.14. The analyses in (Forero et al., 2014; Capozzi et al., 2016) obtain similar results.

From these results it is apparent that

There is a slight ($\Delta\chi^2 \sim 0.8$) preference for the normal hierarchy.

θ_{13} is nonzero but small.

The solar angle θ_{12} is large but non-maximal.

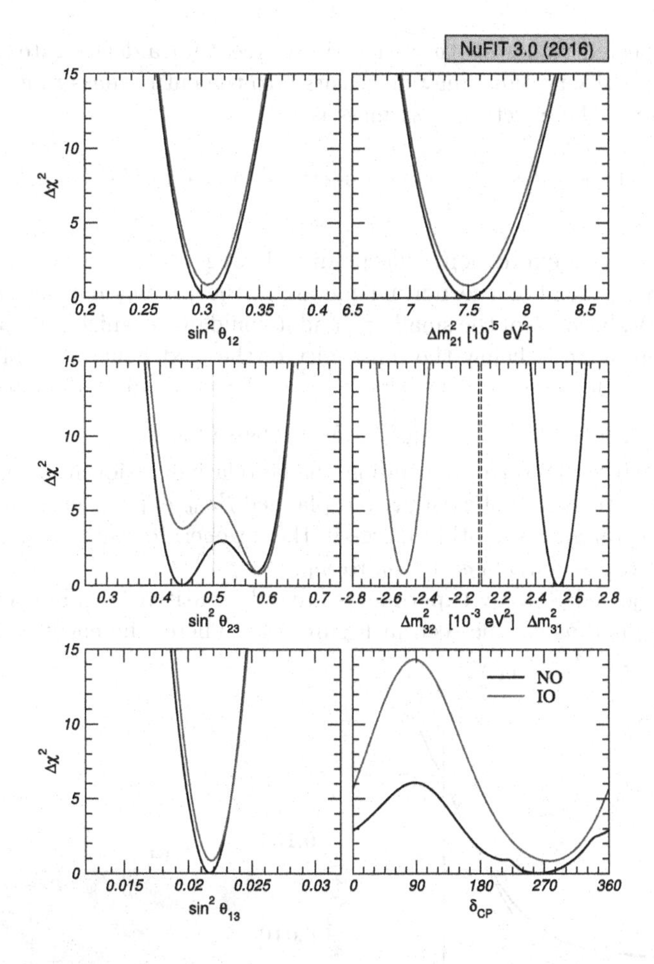

Figure 9.14 $\Delta\chi^2$ distributions (with repect to the best fit), from (Esteban et al., 2017), for the normal (lower $\Delta\chi^2$) and inverted hierarchies.

There is a preference for non-maximal θ_{23}, driven by NOνA and MINOS, but not T2K, as can be seen in Figure 9.10. However, maximal mixing ($\theta_{23} = \pi/4$) is still allowed at 2σ.

The first (second) octant for θ_{23} is preferred for the normal (inverted) hierarchy.

There is a preference for the Dirac CP phase δ to be nonzero and to lie between 1.1π and 1.8π. However, $\delta = \pi$ and 2π (i.e., no CP violation) are allowed at 2σ, and the entire range $0 - 2\pi$ at 3σ.

For $m_1 = 0$ in the normal hierarchy one finds $m_2 \sim 0.009$ eV and $m_3 \sim 0.050$ eV, with $m_2/m_3 \sim 0.17$, which is quite large compared to the analogous quark and charged lepton mass ratios. For the inverted hierarchy, $m_3 = 0$ implies $m_1 \sim 0.049$ eV and $m_2 \sim 0.050$ eV.

The predicted values for Σ are shown as functions of the lightest mass m_0 and compared with existing and projected experimental constraints in Figure 9.15. The existing cosmological limits on Σ (which, however, have a large theoretical uncertainty) are somewhat weaker than the expected values for three neutrinos except for a degenerate spectrum with $m_0 \gtrsim 0.05$ eV. Future observations may be sensitive to the entire mass range.

For three neutrinos, the effective $\beta\beta_{0\nu}$ mass is

$$m_{\beta\beta} = \left|\sum_i (\mathcal{V}_{ei})^2 m_i\right| \sim \left|e^{2i\alpha_1} c_{12}^2\, m_1 + e^{2i\alpha_2} s_{12}^2\, m_2 + e^{-2i\delta} s_{13}^2\, m_3\right|, \tag{9.104}$$

where $c_{13} \sim 1$ in the last expression. The predicted range is shown and compared with experiment in Figure 9.15. For the normal hierarchy the predicted range is very small, $m_{\beta\beta} \lesssim$ few $\times 10^{-3}$ eV, because of the small s_{13}^2, and it could even vanish due to cancellations for nonzero m_1. The range is below the sensitivity of the next generation of experiments. $m_{\beta\beta}$ is much larger for the inverted hierarchy, however. Up to small corrections from the last term, it is given by $\left|c_{12}^2\, m_1 + e^{2i(\alpha_2-\alpha_1)} s_{12}^2\, m_2\right|$, which can vary from $\sim (c_{12}^2 - s_{12}^2)\, |\Delta m_{32}^2|^{1/2}$ to $|\Delta m_{32}^2|^{1/2}$, i.e., $(0.02 - 0.05)$ eV, depending on the relative Majorana phases $(\alpha_2 - \alpha_1)$. This should be within the reach of existing and planned $\beta\beta_{0\nu}$ experiments, at least in their later phases. Both hierarchies smoothly merge in the degenerate region, where $m_{\beta\beta}$ ranges approximately from $(c_{12}^2 - s_{12}^2)\, m_0 \sim 0.4\, m_0$ to m_0.

Existing and projected β decay experiments are only sensitive to three-neutrino masses in the degenerate region, as can be seen in Figure 9.15. There, the effective β decay mass $m_{\nu_e} = \left(\sum_{i=1}^{3} |\mathcal{V}_{ei}|^2 m_i^2\right)^{1/2} \sim m_0$.

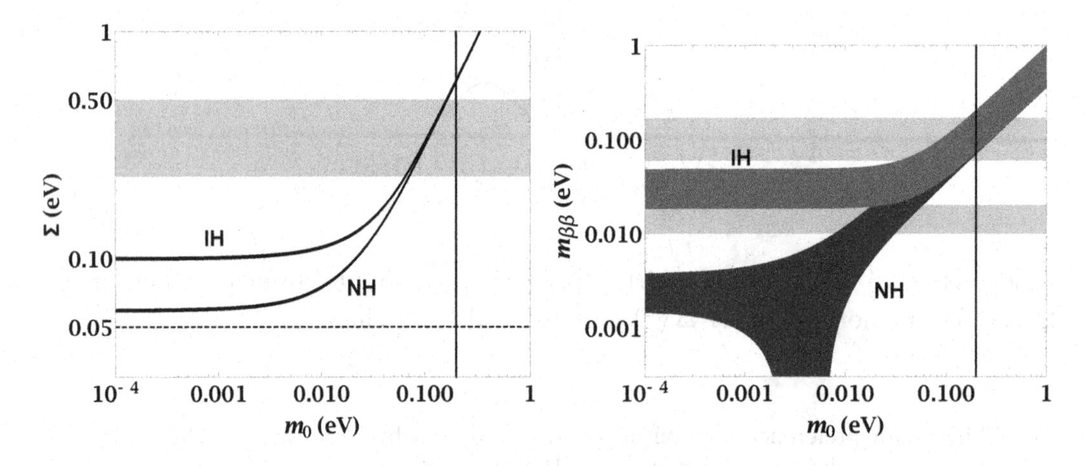

Figure 9.15 Predictions for the total mass Σ (left) and the effective $\beta\beta_{0\nu}$ mass $m_{\beta\beta}$ (right) for three neutrinos, as a function of the lightest mass m_0, for the normal hierarchy (NH) and the inverted hierarchy (IH). The angles and masses are allowed to vary over their 1σ ranges, while the phases $\alpha_{1,2}$ and δ vary from 0 to 2π. The light gray band in the Σ plot is an approximate spread of upper limits from cosmology, corresponding to different data sets and priors. The lower dashed line is an estimate of future sensitivity. The vertical line is the upper limit on $m_{\nu_e} \sim m_0$ anticipated from tritium β decay. The upper (lower) gray horizontal bands on the right represent the existing experimental upper limit (anticipated future sensitivity), with the widths due to uncertainties in the nuclear matrix elements.

Additional Sterile Neutrinos

The LSND, MiniBooNE, and reactor and gallium anomaly results, if interpreted in the context of neutrino oscillations, require the existence of one or more light sterile neutrinos that mix significantly with active neutrinos of the same helicity. One additional sterile neutrino could be accomodated in either the $2+2$ or the $3+1$ patterns, illustrated in Figure 9.16. The $2+2$ case would require that the solar and/or atmospheric results involve a significant admixture of oscillations into the sterile neutrino. However, it is well established that neither the solar nor the atmospheric oscillations are predominantly into sterile states, and this scheme is excluded. In the $3+1$ schemes a predominantly sterile state is separated from the others by $\sim |\Delta m^2_{LSND}|^{1/2}$. However, as mentioned in the Section 9.3.4 any scheme with just one relevant sterile neutrino is in strong tension with other experiments. One can generalize to $3+2$, $3+3$, or $3+n$ schemes with additional sterile states heavier than the three active ones, possibly obtaining better agreement with the other experiments, but at the expense of increased tension with the cosmological limits. One can also consider schemes in which the sterile neutrinos are lighter than the active ones, or in which some are lighter and some heavier. However, these are even more difficult to reconcile with the cosmological limit on Σ, since each of the three active states must be heavier than $|\Delta m^2_{LSND}|^{1/2}$.

Sterile neutrinos can give significant contributions to the effective masses for β decay (Esmaili and Peres, 2012) and $\beta\beta_{0\nu}$ in (9.56) and (9.60), especially if there are more than one (e.g., Barry et al., 2011). However, $m_{\beta\beta}$ can also vanish due to cancellations (i.e., if the (e, e) element of the Majorana mass matrix vanishes).

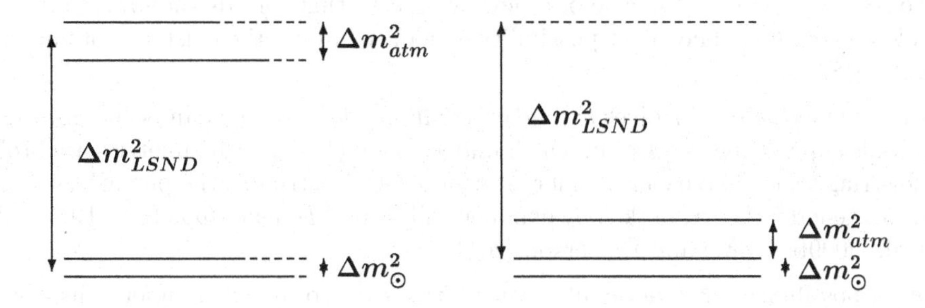

Figure 9.16 Left: an example of a $2+2$ pattern. Right: a $3+1$ pattern. The dashed (solid) lines indicate qualitatively the fractions of sterile (active) neutrinos prior to small mixings. Other patterns correspond to the inverted hierarchy for the atmospheric neutrinos or to placing the largely sterile state in the $3+1$ case on the bottom.

9.5 MODELS OF NEUTRINO MASS

There are an enormous number of models of neutrino mass and mixing, as reviewed in (Mohapatra and Smirnov, 2006; Mohapatra et al., 2007; Strumia and Vissani, 2006; Albright, 2009; Ma, 2009; Altarelli and Feruglio, 2004, 2010; Langacker, 2005, 2012; Barger et al., 2012; Antusch, 2013; King, 2015; de Gouvêa, 2016). Here we mention the major issues and possibilities.

9.5.1 General Considerations

There are a number of general issues concerning neutrino mass, including:

– Are the masses Dirac or Majorana?

The only distinction between Dirac or Majorana neutrinos in the standard model is the type of mass term and associated Yukawa interactions.[30] That is, both active and sterile neutrinos are described by Weyl two-component spinors, and in the massless limit any sterile spinors decouple.

Majorana masses have several advantages: they are not forbidden by any unbroken gauge symmetry, the active Majorana masses can be naturally small by the seesaw mechanism, and they may be connected to the leptogenesis mechanism for the baryon asymmetry. However, Dirac masses cannot be excluded: Majorana masses can be forbidden in a field theory by a conserved global lepton number symmetry,[31] there are a number of possible mechanisms for small Dirac masses, and there are alternative mechanisms for the baryon asymmetry (Chapter 10).

– Why are the masses so small compared with those of the quarks and charged leptons?

One possibility is that the neutrinos masses are associated with higher-dimensional operators in the fundamental or effective low-energy theory, i.e., they are suppressed by powers of S/M, where M is a large new-physics scale and $S \ll M$. Usually, S is associated with a symmetry breaking scale, such as the $SU(2)$-breaking scale for the seesaw model or the scale of a new symmetry that forbids the lowest-order Dirac Yukawa coupling. The latter possibility generally implies the existence of new physics around S.

Another mechanism involves exponential suppressions. These can occur, for example, in some superstring theories in which underlying $U(1)$ gauge symmetries are broken at the string scale. These can re-emerge as global symmetries at the perturbative level of the low-energy theory, broken by exponentially-small *D instantons* (e.g., Blumenhagen et al., 2009b; Cvetič and Halverson, 2011).

Other possibilities include small wave function overlaps in extra dimensions, couplings that only occur at loop level, and anthropic arguments (Tegmark et al., 2005).

– Why are the masses (or at least the mass-squared differences) in a hierarchical structure?

Mass hierarchies may arise from the *Froggatt-Nielsen mechanism* (Froggatt and Nielsen, 1979), i.e., different elements of a Dirac or Majorana mass matrix are suppressed by different powers of a discrete or continuous symmetry-breaking parameter. This parameter (or spurion) is often regarded as the vacuum expectation value of a scalar field known as a *flavon*.[32] An alternative possibility involves wave function overlaps or particle locations in extra dimensions. For example, three-point vertices in intersecting brane string constructions can be large for particles close to each other or exponentially suppressed if they are well separated. Both possibilities are similar

[30] There could also be a distinction if there are BSM interactions.

[31] It is generally believed that conserved global symmetries are not possible in string theory or other theories of quantum gravity (Banks and Dixon, 1988; Witten, 2001; Banks and Seiberg, 2011). However, models descended from string constructions may suppress Majorana masses because of underlying symmetries and selection rules. In particular, gravitationally-induced Majorana mass terms for the active neutrinos may be of $\mathcal{O}(\nu^2/M_P) \lesssim 10^{-5}$ eV, and perhaps very much smaller, i.e., they may be too small to be relevant.

[32] Another possibility is that the Yukawa couplings are actually the vacuum expectation values of a number of flavons, determined by the minimization of a potential (Alonso et al., 2013b).

in spirit to (and can occur simultaneously with) mechanisms for the small overall neutrino mass scale mentioned above, though they may differ in detail.

In fact, the neutrino mass hierarchies are not particularly large compared to those for the quarks and charged leptons ($\left|\Delta m^2_{31}/\Delta m^2_{21}\right|^{1/2} \sim 5.8$), and it is possible that their masses are essentially random numbers (*anarchy*) (Hall et al., 2000; de Gouvêa and Murayama, 2015; Lu and Murayama, 2014). An anarchical spectrum could be associated with superstring theory, which may have an enormous *landscape* of possible vacua, each with different values of the parameters.

- Why do the leptonic mixings have the observed pattern, and are they related to the (apparently different) pattern of quark mixings?

  The large mass hierarchies and small mixing angles in the quark sector suggest that the two are closely related, e.g., they may both be associated with different orders of family symmetry breaking, or with geometrical effects in extra dimensions.

  The situation is less clear for the neutrinos, for which there are two large and one small mixing angle, and only moderate hierarchies. Most theoretical work assumes that the structure of the leptonic mixing can be described by an underlying (probably broken) continuous or discrete symmetry, and in many but not all models the same symmetries account for the mass hierarchies. However, the observations are also consistent with random (anarchical) values for the parameters.

  The general issues of the fermion spectrum and mixings are further discussed in Chapter 10.

- Is there leptonic CP violation?

  CP violation in the quark sector was surprising when it was first observed, and it leads to the question of whether similar effects are present in the leptonic sector. CP violation is also necessary for leptogenesis, though not necessarily in the PMNS matrix. With the benefit of hindsight, however, CP breaking might have been anticipated: many parameters in field theory are complex numbers, and CP violation is expected if the system is sufficiently complicated that the complex phases cannot all be absorbed by field redefinitions. It is now understood that no such phases are possible in QCD,[33] QED, or in the weak neutral current. However, there *can be* observable phases in the quark and lepton mixings. Almost any model of neutrino masses and mixings can lead to CP violation unless CP is explicitly imposed.

We now turn to specific classes of models.

9.5.2 Dirac Masses

There are several possibilities for suppressing h_ν in the Yukawa interaction

$$-\mathcal{L}_{Yuk} = \sqrt{2}h_\nu \left(\bar{\ell}_L \tilde{\phi} \nu_R + \bar{\nu}_R \tilde{\phi}^\dagger \ell_L\right) \tag{9.105}$$

(cf., Equation 8.13 on page 260) to obtain a small Dirac mass. Usually one assumes that h_ν is forced to vanish at tree-level by some additional symmetry, such as an extra $U(1)'$ gauge symmetry and/or due to an underlying superstring construction. It is possible, however,

[33]However, the small value of the strong CP parameter discussed in Chapter 10 is not understood.

that higher-dimensional operators of the form

$$-\mathcal{L}_{HDO} = \sqrt{2}\,\frac{\Gamma_D}{M^p}\left(\phi_S^p \bar{\ell}_L \tilde{\phi} \nu_R + \phi_S^{\dagger p} \bar{\nu}_R \tilde{\phi}^\dagger \ell_L\right) \tag{9.106}$$

are allowed by the symmetries (e.g., Cleaver et al., 1998; Langacker, 2012), where ϕ_S is some SM-singlet spin-0 field and M is a new physics scale, such as the Planck mass. This is illustrated for $p = 1$ in Figure 9.1. If ϕ_S acquires a VEV $\langle \phi_S \rangle \equiv S$, an effective Yukawa coupling $h_{\nu_{eff}} = \Gamma_D (S/M)^p$ will be generated, which can be very small for $S \ll M$. For example, $\Gamma_D = 1$, $M = M_P$, $p = 1$, and S at an intermediate scale $\sim 10^7$ GeV yields $m_\nu \sim 0.1$ eV. Another possibility is that $h_{\nu_{eff}}$ is only generated at loop level, e.g., in a theory with an extended Higgs sector, or in a supersymmetric theory with heavy exotic quarks (Masiero et al., 1986) or with non-holomorphic soft scalar interactions (Demir et al., 2008). Small Dirac masses may also emerge in higher-dimensional theories, e.g., in which ν_L is confined to our 4-d brane, but ν_R is free to propagate in the extra dimensions, in which case h_ν is determined by the (possibly very small) overlap of their wave functions (Dienes et al., 1999b; Arkani-Hamed et al., 2002b). Related string-inspired possibilities involve large intersection areas in intersecting brane constructions (Blumenhagen et al., 2005, 2007b) or string instantons (Cvetič and Langacker, 2008), both of which yield exponential suppressions.[34]

It is possible that a dominant Dirac mass term is perturbed by a much smaller Majorana mass term for the active and/or sterile spinor. This is the pseudo-Dirac case mentioned on page 380. It implies that the Dirac neutrino splits into two Majorana neutrinos, each approximately half active and half sterile, with a small mass difference given by the perturbing Majorana mass. However, this would lead to an additional mass-squared difference that would have been observed by solar neutrinos oscillating into the sterile component unless the Majorana mass is smaller than $\mathcal{O}(10^{-9}$ eV) (de Gouvêa et al., 2009).

9.5.3 Majorana Masses

Majorana masses for active neutrinos violate weak isospin by one unit. A simple way to generate them at the renormalizable level is to extend the SM by the addition of a complex Higgs triplet $\vec{\phi}_T = (\phi_T^0, \phi_T^-, \phi_T^{--})^T$, with $t_T = 1$, and $y_T = -1$. The Yukawa couplings of ϕ_T to the leptons are

$$\begin{aligned} -\mathcal{L}_{\phi_T} &= \frac{1}{2} h_T\, \bar{\ell}_L \vec{\tau} \cdot \vec{\phi}_T\, \tilde{\ell}_R^c + h.c. \\ &= \frac{1}{2} h_T \begin{pmatrix} \bar{\nu}_L & \bar{e}_L \end{pmatrix} \begin{pmatrix} \phi_T^- & \sqrt{2}\phi_T^0 \\ \sqrt{2}\phi_T^{--} & -\phi_T^- \end{pmatrix} \begin{pmatrix} e_R^c \\ -\nu_R^c \end{pmatrix} + h.c., \end{aligned} \tag{9.107}$$

with $\phi_T^- = \phi_T^3$, $\phi_T^0 = \frac{1}{\sqrt{2}}(\phi_T^1 - i\phi_T^2)$, and $\phi_T^{--} = \frac{1}{\sqrt{2}}(\phi_T^1 + i\phi_T^2)$, and where $\tilde{\ell}_R^c$ is defined in (9.1) on page 374. This yields a Majorana mass $m_T = -h_T \nu_{\phi_T}$ if ϕ_T^0 acquires a VEV $\nu_{\phi_T}/\sqrt{2}$, as illustrated in Figure 9.3. The constraints on h_T are less stringent than those for a Dirac mass since one can have $|\nu_{\phi_T}| \ll \nu \sim 246$ GeV (the parameter $\rho_0 \equiv M_W^2/(M_Z^2 \cos^2\theta_W)$ requires $|\nu_{\phi_T}| \lesssim \mathcal{O}(10^{-2})\nu$ (Problem 8.1)). The original version of the model (Gelmini and Roncadelli, 1981; Georgi et al., 1981) involved a global lepton number symmetry (the coupling in (9.107) is invariant since ϕ_T can be defined to have $L = 2$). Therefore, the spontaneous breaking implies a massless Goldstone boson, known as the *triplet Majoron*. Astrophysical constraints (see, e.g., Raffelt, 1999) on stellar cooling

[34] For a review of superstring phenomenology, see (Ibáñez and Uranga, 2012). Implications of string theory for neutrino mass are reviewed in (Langacker, 2012).

n Majoron emission required $|\nu_{\phi_T}| < \mathcal{O}(10)$ keV, consistent with small neutrino mass. vever, there was no special reason for ν_{ϕ_T} to actually be that small. The triplet Majoron lel was eventually excluded by the invisible Z width, as already discussed below (8.170) page 309, because the decay into a Majoron and light scalar would have a partial width al to that of two additional active neutrinos.

The triplet model could survive, however, by adding a coupling

$$-\mathcal{L}_{\phi\phi_T} = V_{\phi\phi_T} = \kappa\, \tilde{\phi}^\dagger \vec{\tau} \cdot \vec{\phi}_T\, \phi + h.c. \tag{9.108}$$

ween ϕ_T and the Higgs doublet ϕ, where κ has dimensions of mass. The clash between 08) and (9.107) implies that the lepton number is explicitly violated, yielding a mass the Majoron that can be taken large enough to evade observational bounds. A currently ular version (the *type II seesaw* (Ma and Sarkar, 1998; Hambye et al., 2001)) includes ass term $\frac{1}{2}\mu_T^2 \vec{\phi}_T^\dagger \cdot \vec{\phi}_T$ for the triplet, where μ_T^2 is very large and positive and associated h a new physics scale. Substituting[35] $\phi^0 \to \nu/\sqrt{2}$ in (9.108), $V_{\phi\phi_T}$ becomes linear in ϕ_T^0, hat it is forced to acquire a VEV

$$\nu_{\phi_T} = -\frac{\kappa\nu^2}{\mu_T^2}, \qquad m_T = h_T \frac{\kappa\nu^2}{\mu_T^2}. \tag{9.109}$$

s is very small in the seesaw limit, $\mu_T^2/\kappa \gg \nu$.

Instead of a Higgs triplet, a Majorana mass for active neutrinos can be generated by the *inberg operator* (Weinberg, 1980), which is a higher-dimensional operator[36] in which two gs doublets are combined as an isospin triplet, as illustrated in Figure 9.3. The form of interactions can be obtained from (9.107) if we identify ϕ_T^i with $\phi^\dagger \tau^i \tilde{\phi}$, where the Higgs blet ϕ and its tilde form $\tilde{\phi}$ are $\phi = (\phi^+\ \phi^0)^T$ and $\tilde{\phi} = (\phi^{0\dagger}\ \ -\phi^-)^T$ as defined in (8.14) page 260. Then,

$$\begin{aligned} -\mathcal{L}_{\phi\phi} &= \frac{C}{2M}\left(\bar{\ell}_L \vec{\tau}\, \tilde{\ell}_R^c\right) \cdot \left(\phi^\dagger \vec{\tau} \tilde{\phi}\right) + h.c. = \frac{C}{M}\left(\bar{\ell}_L\, \tilde{\phi}\right)\left(\phi^\dagger\, \tilde{\ell}_R^c\right) + h.c. \\ &= \frac{C}{M}\bar{\ell}_L \begin{pmatrix} \phi^{0\dagger}\phi^- & \phi^{0\dagger}\phi^{0\dagger} \\ -\phi^-\phi^- & -\phi^-\phi^{0\dagger} \end{pmatrix} \tilde{\ell}_R^c + h.c., \end{aligned} \tag{9.110}$$

ere M is the relevant new physics scale and C is a coefficient that can be absorbed into M esired. The second form is obtained using the $SU(2)$ Fierz identity in Problem 1.1. (The ond term in the Fierz identity vanishes for a single Higgs doublet.) Thus, the Majorana ss is $m_T = -C\nu^2/M$. For example, $M \sim 10^{19}$ GeV (the Planck scale) and $C \sim 1$ implies $\sim 10^{-5}$ eV. $\mathcal{L}_{\phi\phi}$ describes an effective theory that can be generated by many different erlying models, including the type II seesaw mentioned above. It is easily generalized hree families, and is considered by many to be the favored description of small neutrino sses. The scale of M/C (around 10^{14} GeV for $m_T \sim 0.1$ eV) is suggestive of, though a orders of magnitude below, a grand unification or superstring scale.

The most familiar implementation of (9.110) is in the *minimal* or *type I see-* (Minkowski, 1977; Gell-Mann et al., 1979; Yanagida, 1979; Schechter and Valle, 1980),

---

One should in principle minimize V w.r.t ν and ν_{ϕ_T} simultaneously. In practice the modification to ν the coupling in (9.108) is usually negligible in the seesaw limit, and one can always adjust the other gs parameters to maintain the observed value of ν.

The Weinberg operator, with dimension $k = 5$ (i.e., the coefficient has mass dimension $4 - k$, as in ion 2.13), is the lowest-dimensional extension of the standard model (with no ν_R) that is consistent the gauge and Lorentz symmetries. Higher-dimensional SM operators relevant to neutrino mass are sified in (Babu and Leung, 2001; de Gouvêa and Jenkins, 2008).

in which the active neutrinos mix with heavy sterile Majorana neutrinos,[37] as in (9.25) or (9.41) and Figure 9.3, leading to Majorana masses of $\mathcal{O}(m_D^2/m_S)$. In some versions there is a spontaneously broken global lepton number, leading to a very weakly coupled Goldstone boson, the *singlet Majoron* (Chikashige et al., 1981). Type I seesaw models have been constructed with many different scales,[38] depending on m_D. For example, choosing m_D comparable to the electron mass and $m_\nu \sim 0.1$ eV implies $m_S \sim$ few TeV. However, most popular are those based on grand unified theories, where the masses in the Dirac matrix M_D are typically of $\mathcal{O}(m_{u,c,t})$. The sterile masses in M_S may be generated by large Higgs multiplets (e.g., the 126 of $SO(10)$) or by higher-dimensional operators, and must typically be several orders of magnitude below the grand unification scale M_X, which is $\gtrsim 10^{16}$ GeV in the supersymmetric case (Section 10.4). Such constructions are usually combined with family symmetries that restrict and relate the elements of M_D, M_S, and the quark and charged lepton mass matrices, and sometimes combine the type I and II seesaws (using the Higgs triplets in the 126). Generating the large observed leptonic mixings in the same GUT context that yields small quark mixings is a challenge that requires carefully chosen family symmetries and/or non-symmetric (*lopsided*) mass matrices (Albright et al., 1998). The latter leads to large mixing in the unobservable A_R^d and in the observable A_L^e mixing matrices. One especially attractive aspect of the minimal seesaw is that it opens the possibility of *leptogenesis* (Fukugita and Yanagida, 1986; Davidson et al., 2008; Hambye, 2012; Fong et al., 2012) for generating the observed baryon asymmetry of the Universe (Section 10.1).

Most of these GUT type models are hard to embed in known classes of superstring constructions, where any underlying grand unification is frequently broken in the higher-dimensional theory, and in any case it is difficult or impossible to obtain large representations like the 126. More promising is that the Majorana mass terms in M_S are generated by higher-dimensional operators in the superpotential[39] of the form $(\phi_S^{q+1}/M_P^q)\,\bar{\nu}_L^c \nu_R$, in which $q+1$ SM singlet fields ϕ_S (which need not be the same) acquire VEVs somewhat below the Planck scale M_P. This mechanism can lead to a sufficiently suppressed M_S, but is likely to lose the GUT and family symmetry structure of the $SO(10)$ models. String constraints sometimes forbid the simultaneous existence of such operators and those needed for M_D (Giedt et al., 2005). However, there are a few successful examples (Lebedev et al., 2008) in which both M_D and M_S emerge at high order and there are $\mathcal{O}(100)$ sterile neutrinos. Other possibilities for generating M_S involve higher-dimensional operators in the Kähler potential (Arkani-Hamed et al., 2001b) or string instanton effects (Blumenhagen et al., 2007a; Ibáñez and Uranga, 2007).

Seesaw models with lower scales have also been considered. For example, in some gauge extensions of the SM (e.g., with an extra $U(1)'$ (Kang et al., 2005)) the "sterile" neutrinos are charged under the extended group, so any Majorana masses for them cannot be much larger than the symmetry breaking scale. There are also various *extended seesaw* models, which involve additional $SU(2)$ singlets. For example, in the *double seesaw* (Mohapatra and Valle, 1986) one introduces a sterile n_L with a Majorana mass μ and a Dirac mass $\mathcal{M}_D$

[37]In the *type III seesaw* (e.g., Ma, 2009) the sterile neutrino is replaced by an $SU(2)$ triplet.

[38]The heavy sterile neutrino aggravates the Higgs/hierarchy problem (Section 10.1) unless $m_S \lesssim 10^7$ GeV (Vissani, 1998). It also tends to destabilize the electroweak vacuum (Elias-Miro et al., 2012). The lifetime was estimated to be sufficiently long for $m_S \lesssim 10^{13} - 10^{14}$ GeV, but see the caveats in Section 8.5.3.

[39]The Weinberg operator could also emerge directly in a superstring construction. However, C/M would typically be $\sim 1/M_P$, which is too small. This difficulty could be avoided if some of the extra dimensions are large compared to the inverse of the string scale (Conlon and Cremades, 2007; Cvetič et al., 2010).

coupling to ν_R:

$$-\mathcal{L} = \frac{1}{2} \begin{pmatrix} \bar{\nu}_L & \bar{\nu}_L^c & n_L \end{pmatrix} \begin{pmatrix} 0 & m_D & 0 \\ m_D & 0 & \mathcal{M}_D \\ 0 & \mathcal{M}_D & \mu \end{pmatrix} \begin{pmatrix} \nu_R^c \\ \nu_R \\ n_R^c \end{pmatrix} + h.c., \tag{9.111}$$

where we have taken the other masses to vanish. For $\mu = 0$ there is one massless Weyl neutrino and one Dirac neutrino, given approximately by ν_L and $n_L + \nu_R$ respectively for $\mathcal{M}_D \gg m_D$. For $\mu \ll m_D \ll \mathcal{M}_D$ the ν_L acquires a Majorana mass $m_\nu \sim \mu(m_D/\mathcal{M}_D)^2$. This *inverse seesaw* is actually driven by a small Majorana sterile mass μ rather than a large one, e.g., $\mu = 1$ keV, $m_D = 100$ GeV, and $\mathcal{M}_D = 10$ TeV corresponds to $m_\nu \sim 0.1$ eV. This model generalizes to three families. Variations of the double seesaw with larger μ and related models with small Dirac masses are also possible. Low-scale models may also have implications for LHC physics. They are reviewed in (Chen and Huang, 2011; Boucenna et al., 2014).

We briefly mention two other classes of models that lead to Majorana masses. One involves masses that are only induced at loop level, e.g., associated with an extended scalar sector. In one well-known example (Zee, 1980) an $SU(2)$-singlet charged scalar field h^- is introduced that couples to both leptons and Higgs doublets, leading to a loop-induced Majorana mass as shown in Figure 9.3 on page 377. This example actually leads to off-diagonal masses such as $\bar{\nu}_{eL}\nu_{\mu R}^c$, i.e, the ZKM model of (9.44) on page 383, and also requires a second Higgs doublet, because the h^- coupling is antisymmetric in lepton and Higgs family indices. Another possibility occurs in supersymmetric theories with R-parity violation (Section 10.2), which allows mixing and a type of seesaw between active neutrinos and neutralinos. The mixing can generate one small neutrino mass at tree level, with the other two masses entering at loop level (e.g., Grossman and Rakshit, 2004; Rakshit, 2004).

9.5.4 Mixed Mass Models

As discussed in Section 9.3.4 there are several experimental suggestions of possible mixing between active and light sterile neutrinos. Neutrino oscillations conserve helicity, so the mixing must be between states of the same chirality. Most extensions of the standard model involving neutrino mass introduce sterile neutrinos, which could in principle have mass at any scale. However, generating significant mixing is more difficult. For example, in the model with one active and one sterile neutrino in (9.18) on page 378 the appropriate $\nu_L^0 - \nu_L^{0c}$ or $\nu_R^{0c} - \nu_R^0$ mixing only occurs in the general case in which both Dirac and Majorana mass terms are present. Furthermore, for eV-scale sterile neutrinos with small mixing both the Dirac and Majorana mass terms must be tiny and not too different in magnitude. Similar statements apply to the multi-family case in (9.37). More generally, LSND-type active-sterile mixing requires two distinct types of mass terms to be simultaneously extremely small. These could be Majorana and Dirac, or, alternatively, two distinct types of Dirac masses, such as one connecting active and sterile states, and another connecting distinct steriles. Confirmation of such mixing would therefore require a major change in the paradigm, especially from the usual seesaw model.

There have been many models for such mixing, including higher-dimensional operators associated with an intermediate scale, sterile neutrinos from a *mirror world*, supersymmetry, extra dimensions, and dynamical symmetry breaking (see, e.g., Langacker, 2012, for a list of references). Especially promising is the (minimal) *mini-seesaw* model. This is just the ordinary seesaw model in (9.25) or in (9.41) (with m_T or $M_T = 0$), except that all of the Dirac and Majorana masses are assumed to be very small, e.g., $\lesssim \mathcal{O}(\text{eV})$. The minimal mini-seesaw has the advantage that the masses and mixings are related in a way that is roughly consistent with the LSND and other observations. For example, for one family one has

$$|m_1| \sim \frac{m_D^2}{m_S}, \qquad m_2 \sim m_S, \qquad |\theta| \sim \frac{m_D}{m_S} \tag{9.112}$$

from (9.25), e.g., $|m_1| \sim 0.04$ eV and $|\theta| \sim 0.2$ for $m_D = 0.2$ eV and $m_S = 1$ eV. There have been extensive studies of the minimal mini-seesaw for two or three families[40] (e.g., de Gouvêa and Huang, 2012; Donini et al., 2012), yielding qualitative agreement with the data. The minimal mini-seesaw predicts $m_{\beta\beta} = 0$ for the effective $\beta\beta_{0\nu}$ mass in (9.60), because for mass eigenvalues $\ll$ MeV it is just the (vanishing) (e, e) element of the mass matrix in (9.37).

The minimal mini-seesaw model parametrizes the mixing but does not explain the small values for m_D and m_S. Just as in the discussion of small Dirac masses, it suggests that both the Dirac Yukawa couplings and m_S are forbidden at the renormalizable level by some new discrete, global, or gauge symmetry, and only generated by symmetry-breaking (and possibly loop) corrections (e.g., Langacker, 1998, 2012; Sayre et al., 2005). For example, consider the case of one active and one sterile neutrino in which the lowest-order allowed mass terms are

$$-\mathcal{L}_{mini} = \sqrt{2}\Gamma_D \frac{\phi_S^p}{M^p} \bar{\ell}_L \tilde{\phi} \nu_R + \Gamma_S \frac{\phi_S^{q+1}}{2M^q} \bar{\nu}_L^c \nu_R + \Gamma_T \frac{\phi_S^{r-1}}{2M^r} \left(\bar{\ell}_L \vec{\tau} \tilde{\ell}_R^c\right) \cdot \left(\phi^\dagger \vec{\tau} \tilde{\phi}\right) + h.c., \tag{9.113}$$

rather than those in (9.105), (9.9) and (9.110). The new symmetry is broken by $\langle \phi_S \rangle \equiv S$, and M is a new physics scale.[41] Neglecting the last term, (9.113) yields a minimal mini-seesaw for $p, q \geq 1$. Taking $p = q = 1$ and $\Gamma_D = \Gamma_S = 1$, for example, one finds $m_D = S\nu/M$ and $m_S = S^2/M$, so that $|\theta| \sim \nu/S$ and $|m_1| \sim \nu^2/M$. Comparing with the numerical example above, one expects S in the TeV range and $S/M \sim 10^{-12}$. In general, one should not neglect the last term in (9.113). Any multiplicative symmetry of $\mathcal{L}_{mini}$ that allows the first two terms will also allow the third, with $r = 2p - q$. This corresponds to $r = 1$ in the $p = q = 1$ case, i.e, the new physics should be able to generate a contribution to m_1 directly that is of the same order as the mixing-induced term. This would modify the details of the minimal mini-seesaw model (and would allow $m_{\beta\beta} \neq 0$), but the general idea remains.

9.5.5 Textures and Family Symmetries

There are also many *texture models*, involving specific ansätze about the form of the 3×3 neutrino mass matrix, or of the Dirac and Majorana mass matrices entering seesaw models. These are often studied in connection with models also involving quark and charged-lepton mass matrices, such as family symmetries, left-right symmetry, or grand unification (Chapter 10). A major complication is that the form of a mass matrix depends on the basis chosen, e.g., whether the charged-lepton mass matrix is diagonal, and any underlying family symmetries may take a different form depending on the basis. Family symmetries may be continuous (global or gauged) or discrete. There has been considerable recent interest in discrete symmetries (for reviews, see Altarelli and Feruglio, 2010; Ishimori et al., 2010; King et al., 2014) because they more naturally lead to the large mixing angles observed

[40]From (9.40) the leading approximation to the active-sterile mixing for both the ordinary type I and mini seesaws is $-M_D M_S^{-1}$. M_D can be expressed in terms of the mass eigenvalues and PMNS matrix up to a complex orthogonal matrix (Casas and Ibarra, 2001).

[41]An explicit example of an *ultraviolet completion*, i.e., a renormalizable model of the underlying physics that leads to the effective operators in $\mathcal{L}_{mini}$, is given in (Heeck and Zhang, 2013).

in the neutrino sector. Also, they are free of the unwanted Goldstone bosons associated with spontaneously broken global continuous symmetries. Family symmetries are in principle symmetries of the underlying Lagrangian, but in practice the models are often very complicated, so studies are often restricted to the symmetry and its breaking pattern.

The observed values of $\theta_{23} \sim \pi/4$ and $\theta_{13} \sim 0$ suggest that the leptonic mixing matrix is close to

$$V_\ell^\dagger = \mathcal{V} = \begin{pmatrix} c_{12} & s_{12} & 0 \\ -\frac{s_{12}}{\sqrt{2}} & \frac{c_{12}}{\sqrt{2}} & \frac{1}{\sqrt{2}} \\ \frac{s_{12}}{\sqrt{2}} & -\frac{c_{12}}{\sqrt{2}} & \frac{1}{\sqrt{2}} \end{pmatrix} \tag{9.114}$$

up to field redefinitions and possible Majorana phases that are irrelevant for oscillations. The motivation for (9.114) is weakened by the observation of nonzero θ_{13} and likely deviation of θ_{23} from maximal mixing, but it may still be useful as a starting point in model building. The special case of $s_{12}^2 = 1/3$, known as *tri-bimaximal mixing* (Harrison et al., 2002; Ma, 2004), is consistent with but slightly above the current value in (9.103). Alternatively, the *golden ratio* prediction $s_{12}^2 = \frac{2}{5+\sqrt{5}} \sim 0.276$ (Kajiyama et al., 2007; Feruglio and Paris, 2011; Ding et al., 2012), is slightly below. *Bimaximal mixing*, $c_{12}^2 = s_{12}^2 = 1/2$ was also at one time a serious possibility. These and similar patterns may be associated with underlying permutation symmetries, such as S_4, or groups of even permutations, such as A_4 or (for the golden ratio) the icosahedral group A_5.

In practice, small deviations from any specific form such as tri-bimaximal mixing are possible. For example, they may be associated with broken discrete symmetries, and/or they could apply only to the neutrino mixing, with additional small mixing comparable to the CKM angles induced by the charged leptons [*Cabibbo haze* (Datta et al., 2005; Everett, 2006)]. It has been observed that θ_{12} is close to the quark-lepton complementarity (Raidal, 2004; Minakata and Smirnov, 2004) value $\theta_{12} + \theta_c = \pi/4$, where θ_c is the Cabibbo angle, reopening the possibility of bimaximal mixing in the neutrino sector. Similarly, θ_{13} is close to the value $\theta_c/\sqrt{2}$ that has been motivated in several schemes (Minakata and Smirnov, 2004; Antusch, 2013). Renormalization group evolution of couplings is another complication, since one might expect underlying symmetries to apply at a large GUT or string scale rather than at low energies, and mass degeneracies (in sign as well as magnitude) at a high scale may be destabilized by the running (e.g., Antusch et al., 2005).

In most quark texture models the small mixing angles are associated with the small ratios of mass eigenvalues, as in Problem 9.7. Two of the neutrino mixings are large, however, suggesting the possibility that the mixings and mass eigenvalues are decoupled. This can easily occur. In particular, any Dirac, triplet Majorana, or seesaw-induced 3×3 neutrino mass matrix of the form

$$M_\nu = \begin{pmatrix} A & B & -B \\ B & C & -D \\ -B & -D & C \end{pmatrix} \tag{9.115}$$

will lead (e.g., Altarelli and Feruglio, 2010) to the leptonic mixing matrix in (9.114), corresponding to $s_{13} = 0$ and $s_{23} = c_{23} = 1/\sqrt{2}$, with

$$\sin^2 2\theta_{12} = \frac{8B^2}{[A - (C+D)]^2 + 8B^2}. \tag{9.116}$$

The mass eigenvalues are related by

$$A = c_{12}^2 m_1 + s_{12}^2 m_2, \qquad C + D = s_{12}^2 m_1 + c_{12}^2 m_2, \qquad C - D = m_3, \tag{9.117}$$

which because of the phase conventions chosen in (9.114) may be negative or complex.

Clearly, the parameters can be chosen to yield any of the types of mass hierarchy. Tri-bimaximal mixing occurs whenever

$$A + B = C + D, \tag{9.118}$$

implying

$$\begin{aligned} s_{12}^2 &= \frac{1}{3}, \qquad c_{12}^2 = \frac{2}{3} \\ m_1 &= A - B, \qquad m_2 = A + 2B, \qquad m_3 = C - D. \end{aligned} \tag{9.119}$$

Bimaximal mixing corresponds to

$$A = C + D, \tag{9.120}$$

so that

$$s_{12}^2 = c_{12}^2 = \frac{1}{2}, \qquad m_{1,2} = A \mp \sqrt{2}B, \qquad m_3 = C - D. \tag{9.121}$$

The golden ratio occurs for

$$A = C + D - \sqrt{2}B. \tag{9.122}$$

The form of (9.114) and (9.115) exhibits an obvious $\nu_\mu \leftrightarrow \nu_\tau$ interchange symmetry, up to signs depending on our phase convention (Xing and Zhao, 2016).

9.6 IMPLICATIONS OF NEUTRINO MASS

Most extensions of the SM predict non-zero neutrino masses at some level, so it is difficult to determine their origin. Many of the promising mechanisms, such as the minimal seesaw, involve very short distance scales, e.g., associated with grand unification or string theories, and are therefore difficult to verify directly. Some models lead to other types of predictions. For example, lepton flavor violating processes[42] like $\mu \to e\gamma$, $\mu N \to eN$, or $\mu \to 3e$, by loop effects involving sneutrinos ($\tilde{\nu}$) and other superpartners in supersymmetry (e.g., Hisano et al., 1996; Casas and Ibarra, 2001; Masiero et al., 2004; de Gouvêa and Vogel, 2013) may be observable, but their connection to the neutrino mass generation mechanism is model dependent. Lepton-flavor violating effects or direct production of heavy Majorana neutrinos at hadron colliders (e.g., leading to same-sign dileptons) may also be observable in TeV-scale models for neutrino mass (Cheng and Li, 1977; Atre et al., 2009; Alonso et al., 2013a).

There are many unanswered questions. These include:

– Are the neutrinos Dirac or Majorana?

 Majorana masses, especially if associated with a type I or II seesaw, would allow the possibility of leptogenesis. The observation of $\beta\beta_{0\nu}$ would establish Majorana masses (or at least L violation), but foreseeable experiments will only be sensitive to the inverted or degenerate hierarchies. Dirac masses would suggest that additional TeV-scale symmetries or string symmetries/selection rules are forbidding Majorana mass terms and suppressing the Dirac Yukawa couplings.

– What is the absolute mass scale (with implications for cosmology)?

 This is very difficult, but ordinary and double beta decay experiments, as well as future cosmological observations, may be able to establish the scale.

[42]The nonzero neutrino masses and mixings themselves violate lepton flavor, but their effects are negligible except for neutrino oscillations.

- Is there leptonic CP violation? What is δ? Is the hierarchy normal or inverted? Is θ_{23} maximal?

  Leptonic CP violation is a necessary ingredient in leptogenesis. The CP phases in $\mathcal{V}$ are different from the ones relevant to leptogenesis, though they are often related in specific models. For three families, the phase δ in (9.35) may be observable, e.g., in differences between the $\nu_\mu \to \nu_e$ and $\bar{\nu}_\mu \to \bar{\nu}_e$ oscillation rates in long-baseline experiments such as T2K and NOνA, especially with their intensity upgrades, or in future experiments with even longer baselines. Such effects are proportional to θ_{13}, but fortunately the nonzero value measured in reactor experiments is sufficiently large. They also depend on the sign of Δm^2_{23} because of matter effects in the Earth. The nature of the hierarchy may be determined simultaneously with CP breaking in long-baseline experiments, in other matter effects involving atmospheric or supernova neutrinos, in future high precision reactor experiments, in the observation of $\beta\beta_{0\nu}$ if the neutrinos are Majorana, or by cosmological determinations of $\Sigma = \sum_i |m_i|$. The value of θ_{23} will also be measured more precisely in future long-baseline $\overset{(-)}{\nu}_\mu$ disappearance.

  On a longer time scale, CP violation, the hierarchy, and related issues may also be addressed at a dedicated neutrino factory (from a muon storage ring), or in *beta beams* involving $\overset{(-)}{\nu}_e$ emission from accelerated heavy ions.

  Ultra-high energy neutrinos from violent astrophysical events can be observed in large detectors in ice or water, such as IceCube and KM3NeT, and in fact IceCube has already detected a number of events in the 30 TeV-PeV range. These may possibly shed light on neutrino oscillations or decay, nonstandard properties, and on the astrophysical sources.

- Are there light sterile neutrinos?

  If the LSND, MiniBooNe, reactor, and gallium results are confirmed, it will suggest mixing between ordinary and sterile neutrinos, presenting a serious challenge both to particle physics and cosmology, or imply something even more bizarre, such as CPT violation.

- Are there any new ν interactions or anomalous properties such as large magnetic moments?

  Most such ideas are excluded as the dominant effect for the solar and atmospheric neutrinos, but could still appear as subleading effects.

Future experimental and observational prospects relevant to these issues are reviewed generally in (de Gouvêa et al., 2013), and the hierarchy determination in (Cahn et al., 2013; Qian and Vogel, 2015; Patterson, 2015).

9.7 PROBLEMS

9.1 Show that (9.13) is a solution to the free Majorana field equation (9.12).

9.2 Show that adding a small 23 element to the (symmetric) ZKM matrix in (9.45) (a generalization of the pseudo-Dirac case), will lead to three Majorana mass eigenvalues even though none of the diagonal elements is nonzero.

9.3 Suppose the WCC in the second form in (9.49) contained a small admixture of $V + A$,

$$J^\mu_W = \bar{e}\gamma^\mu \left[(1-\gamma^5) + \epsilon(1+\gamma^5)\right] \nu_M,$$

with $\epsilon \ll 1$. Calculate the leading correction to the t-channel amplitude in (9.52) for $W^-W^- \to e^-e^-$ and show that it survives even for $m_T = 0$. The interpretation is that the amplitude for the diagrams in Figures 9.2 or 9.4 requires a chirality flip on the internal neutrino line, which can be due either to a Majorana mass or to a new interaction. (This mechanism does *not* occur in normal $SU(2)_L \times SU(2)_R \times U(1)$ models, because the $V + A$ interactions involve ν_R rather than ν_R^c.)

9.4 Show that the neutral current coupling $\bar{\nu}_L\gamma^\mu\nu_L$ in (8.74) for an active massive neutrino can be written as $\frac{1}{2}\bar{\nu}_D\gamma^\mu(1-\gamma^5)\nu_D$ (i.e., $V - A$) if it is Dirac, and as $-\frac{1}{2}\bar{\nu}_M\gamma^\mu\gamma^5\nu_M$ (pure axial) if it is Majorana, where $\nu_D = \nu_L + \nu_R$ and $\nu_M = \nu_L + \nu_R^c$.

9.5 Suppose there existed a fourth stable active neutrino with mass m_ν in the GeV range but small compared with $M_Z/2$ (this is now excluded by the measurements of the invisible Z width, but was once considered an interesting possibility). Similar to the discussion in Problem 5.5, the relic density of such neutrinos depends on the thermal average of $\bar{\sigma}(\nu_D\bar{\nu}_D \to X)\beta_{rel}$ if the neutrino is Dirac, or $\bar{\sigma}(\nu_M\nu_M \to X)\beta_{rel}$ if it is Majorana, where β_{rel} is the relative velocity, X represents a sum over accessible final states, the energy is at or just above threshold, $s \gtrsim 4m_\nu^2$, and one averages over the (thermalized) initial spins. Calculate the leading contribution to $\bar{\sigma}\beta_{rel}$ in the CM via the s-channel Z annihilation diagram at threshold in each case, where $X = f\bar{f}$ is a light fermion with neutral current couplings given by (8.75). Show that it is S-wave (constant) for the Dirac case and P-wave ($\propto \beta^2 = \beta_{rel}^2/4$) in the Majorana case. Neglect m_f. (This calculation was used in the derivation of the Lee-Weinberg (and others) bound (e.g., Kolb and Turner, 1990; Dolgov, 2002), $m_\nu > 2$ GeV (Dirac) or $m_\nu > 8$ GeV (Majorana), to avoid overclosing the Universe, assuming that the neutrino is stable and heavier than $\mathcal{O}(100)$ eV and that $\Omega_\nu h^2 < 1$.)

9.6 Derive (9.74).

9.7 As a toy model of a quark texture, consider the empirical *Oakes relation* $\sin\theta_c \sim m_\pi/m_K$ (Oakes, 1969), where θ_c is the Cabibbo angle. Show how this might emerge from the three-quark Fermi theory described in Section 7.2 if one assumes that for some reason $m_u \sim M_{11}^d \sim 0$. Hint: recall the relation between the quark and pseudoscalar masses described in Section 5.8.

CHAPTER 10

Beyond the Standard Model

DOI: 10.1201/b22175-10

10.1 THE NEED FOR NEW PHYSICS

The structure of the electroweak part of the standard model was explored in detail in Chapter 8, but for convenience we summarize the Lagrangian density after spontaneous symmetry breaking:[1]

$$\begin{aligned}\mathcal{L} = \mathcal{L}_{gauge} + \mathcal{L}_\phi + \sum_r \bar{\psi}_r \left(i \not{\partial} - m_r - \frac{m_r H}{\nu} \right) \psi_r \\ - \frac{g}{2\sqrt{2}} \left(J_W^\mu W_\mu^- + J_W^{\mu\dagger} W_\mu^+ \right) - e J_Q^\mu A_\mu - \frac{g}{2\cos\theta_W} J_Z^\mu Z_\mu,\end{aligned} \tag{10.1}$$

where the self-interactions for the $W^\pm$, Z, and γ are given in (8.42) and (8.43), $\mathcal{L}_\phi$ is given in (8.39), and the fermion currents in (8.60), (8.70), and (8.74).

The standard electroweak model is a mathematically-consistent renormalizable field theory that predicts or is consistent with all experimental facts. It successfully predicted the existence and form of the weak neutral current; the existence and masses of the W and Z bosons; and those of the charm quark, as necessitated by the GIM mechanism. The charged current weak interactions, as described by the generalized Fermi theory, were successfully incorporated, as was quantum electrodynamics. The consistency between theory and experiment indirectly tested the radiative corrections and ideas of renormalization and allowed the successful approximate prediction of the top quark and Higgs boson masses. Although the original formulation did not provide for massive neutrinos, they are easily incorporated by the addition of right-handed states ν_R (Dirac) or as higher-dimensional operators perhaps generated by an underlying seesaw (Majorana). When combined with quantum chromodynamics for the strong interactions, the standard model is almost certainly the approximately correct description of the elementary particles and their interactions down to at least 10^{-16}cm, with the possible exception of new very weakly coupled particles. When combined with general relativity for classical gravity the SM accounts for most of the observed features of Nature (though not for the dark matter and energy).

However, the theory has far too much arbitrariness to be the final story. For example, the minimal version of the model has 20 free parameters for massless neutrinos and another (9) for massive Dirac (Majorana) neutrinos, not counting electric charge (i.e., hypercharge)

[1] $\mathcal{L}_{\nu_L}$ is given by $\sum_r \bar{\nu}_{rL} i \not{\partial} \nu_{rL} - \frac{1}{2} m_{\nu r} \left(\bar{\nu}_{rL} \nu^c_{rR} + h.c. \right) \left(1 + \frac{H}{\nu} \right)^2$ for Majorana ν_L masses generated by a higher-dimensional operator involving two factors of the Higgs doublet, as in the seesaw model.

assignments.[2] Most physicists believe that this is just too much for the fundamental theory. The complications of the standard model can also be described in terms of a number of problems.

10.1.1 Problems with the Standard Model

The Gauge Symmetry Problem

The standard model is a complicated direct product of three subgroups, $SU(3) \times SU(2) \times U(1)$, with separate gauge couplings. There is no explanation for why only the electroweak part is chiral and parity-violating. Similarly, the standard model incorporates but does not explain another fundamental fact of Nature: *charge quantization*, i.e., why all particles have charges that are multiples of $e/3$. This is important because it allows the electrical neutrality of atoms ($|q_p| = |q_e|$). The complicated gauge structure suggests the existence of some underlying unification of the interactions, such as one would expect in a superstring (e.g., Green et al., 1987; Polchinski, 1998; Becker et al., 2007; Ibáñez and Uranga, 2012) or grand unified theory (Section 10.3). Charge quantization can also be explained in such theories, though the "wrong" values of charge emerge in some constructions due to different hypercharge embeddings or non-canonical values of Y (e.g., some string constructions lead to exotic particles with charges of $\pm e/2$). Charge quantization may also be explained, at least in part, by the existence of magnetic monopoles (e.g., Preskill, 1984) or the absence of anomalies,[3] but either of these is likely to find its origin in some kind of underlying unification.

The Fermion Problem

All matter under ordinary terrestrial conditions can be constructed out of the fermions (ν_e, e^-, u, d) of the first family. Yet we know from laboratory studies that there are ≥ 3 families: (ν_μ, μ^-, c, s) and (ν_τ, τ^-, t, b) are heavier copies of the first family with no obvious role in Nature. The standard model gives no explanation for the existence of these heavier families and no prediction for their numbers. Furthermore, there is no explanation or prediction of the fermion masses, which are observed to occur in a hierarchical pattern that varies over 5 orders of magnitude between the t quark and the e^-, or of the quark and lepton mixings. Even more mysterious are the neutrinos, which are many orders of magnitude lighter still. It is not even certain whether the neutrino masses are Majorana or Dirac. A related difficulty is that while the CP violation observed in the laboratory is well accounted for by the phase in the CKM matrix, there is no SM source of CP breaking adequate to explain the baryon asymmetry of the universe.

There are many suggestions for new physics that might shed light on these questions. The existence of multiple families could be due to large representations of some grand unified theory. Alternatively, it could be associated with different possibilities for localizing particles in some higher-dimensional space, which could be associated with a superstring compactification, extra-dimensional grand unification, or by some effective brane world scenario involving large[4] (Arkani-Hamed et al., 1998; Dienes et al., 1999a) and/or warped (Randall

[2] 12 fermion masses (including the neutrinos), 6 mixing angles, 2 CP violation phases (+ 2 possible Majorana phases), 3 gauge couplings, M_H, ν, θ_{QCD}, M_P, Λ_{cosm}, minus one overall mass scale since only mass ratios are physical.

[3] The absence of anomalies is not sufficient to determine all of the Y assignments without additional assumptions, such as family universality.

[4] See (Witten, 1996; Lykken, 1996b) for early explorations of large dimensions associated with low superstring scales, and (Berenstein, 2014) for modern developments.

and Sundrum, 1999) extra dimensions. See, e.g., (Hewett and Spiropulu, 2002; Gherghetta, 2011; Ponton, 2013; Csáki and Tanedo, 2015; Patrignani, 2016) for general reviews and (Adelberger et al., 2009) for laboratory constraints. The hierarchies of masses and mixings could emerge from wave function overlap effects in such higher-dimensional spaces, or as exponential suppressions associated with intersection areas in the internal dimensions in intersecting brane constructions (e.g., Blumenhagen et al., 2005). Another interpretation, also possible in string theories, is that the hierarchies are because some of the mass terms are generated by higher-dimensional operators and therefore suppressed by powers of $\langle 0|S|0\rangle/M_X$, where S is some standard model singlet field and M_X is some large scale such as M_P. The allowed operators could perhaps be enforced by some family symmetry (Froggatt and Nielsen, 1979). Radiative hierarchies (e.g., Babu and Mohapatra, 1991), in which some of the masses are generated at the loop level, or some form of compositeness are other possibilities. Despite all of these ideas there is no compelling model and none of these yields detailed predictions. Grand unification by itself doesn't help very much, except for the prediction of m_b in terms of m_τ in the simplest versions.

As discussed in Chapter 9 the small values for the neutrino masses suggest that they are associated with Planck or grand unification physics, as in the seesaw model, but there are other possibilities.

Almost any type of new physics is likely to lead to new sources of CP violation.

The Higgs/Hierarchy Problem

In the standard model one introduces an elementary Higgs field to generate masses for the W, Z, and fermions. The observed mass $M_H \sim 125$ GeV of the physical Higgs boson is of the same order as the W mass. In fact, this is required by consistency for a weakly coupled theory: if M_H had been orders of magnitude larger than M_W the Higgs self-interactions would have been excessively strong. This is manifested by a theoretical upper limit of $\mathcal{O}(700$ GeV) on the mass of an elementary Higgs (see Section 8.5.1).

However, there is a complication. The tree-level (bare) Higgs mass receives quadratically-divergent corrections from loop diagrams such as those in Figure 10.1. One finds (Section 10.2.1)

$$M_H^2 = (M_H^2)_{bare} + \mathcal{O}(\lambda, g^2, h^2)\Lambda^2, \tag{10.2}$$

where Λ is the next higher scale in the theory. If there were no higher scale, one could simply interpret Λ as an ultraviolet cutoff and take the view that M_H is a measured parameter, with $(M_H)_{bare}$ not observable. However, the theory is presumably embedded in some larger theory that cuts off the momentum integral at the finite scale of the new physics.[5] For example, if the next scale is gravity, Λ is the Planck scale $M_P = G_N^{-1/2} \sim 10^{19}$ GeV. In a grand unified theory, one would expect Λ to be of order the unification scale $M_X \sim 10^{14}$ GeV. Hence, the natural scale for M_H is $\mathcal{O}(\Lambda)$, which is much larger than the expected value. There must be a fine-tuned and apparently highly contrived cancellation between the bare value and the correction, to more than 30 decimal places in the case of gravity. If the cutoff is provided by a grand unified theory there is a separate hierarchy problem at the tree-level. The tree-level couplings between the Higgs field and the superheavy fields lead to the expectation that M_H is close to the unification scale unless unnatural fine-tunings are done, i.e., one does not understand why $(M_W/M_X)^2$ is so small in the first place.

The *naturalness* paradigm (e.g., Giudice, 2008; Feng, 2013; Dine, 2015a), i.e., that the

[5] As discussed in Section 2.12.2, there is no analogous fine-tuning associated with logarithmic divergences. In QED, for example, corrections are of $\mathcal{O}(\alpha \ln(\Lambda/m_e))$, which is < 1 even for $\Lambda = M_P$. Similarly, from (5.38) the QCD coupling can become strong at $\mathcal{O}(1$ GeV) without excessive tuning, e.g., for $\alpha_s(M_P) \sim 0.02$.

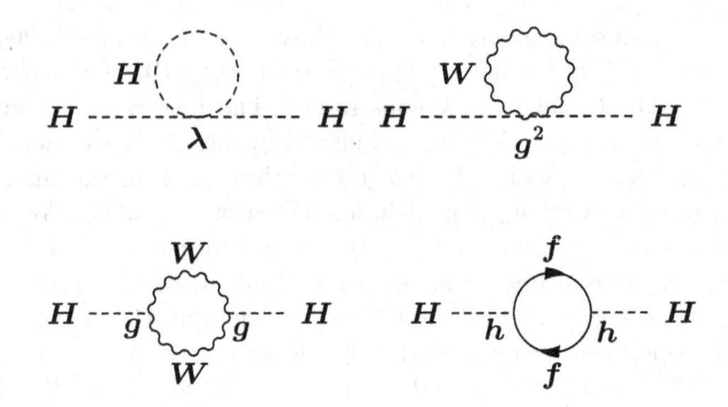

Figure 10.1 Radiative corrections to the Higgs mass, including self-interactions, interactions with gauge bosons, and interactions with fermions.

qualitative features of Nature should not depend on fine-tuned cancellations or on very large or small dimensionless ratios, has often served as a guide to possible new physics. For example, the observed value of the $K_L - K_S$ mass difference motivated the prediction of the charm quark (the GIM mechanism), which enforced the exact cancellation between much larger contributions and allowed an approximate prediction of m_c (Section 8.6.3).

Most of the solutions to the Higgs/hierarchy problem are similarly motivated by naturalness.[6] In TeV scale supersymmetry, for example, the quadratically-divergent contributions of fermion and boson loops cancel, leaving only much smaller effects of the order of supersymmetry-breaking. (However, supersymmetric grand unified theories still suffer from the tree-level hierarchy problem.) There are also (non-supersymmetric) extended models in which there are cancellations between bosons or between fermions, at least up to a Higgs compositeness scale (Section 8.5.3). This class includes Little Higgs models (Arkani-Hamed et al., 2002a), in which the (composite) Higgs is forced to be lighter than new TeV scale dynamics because it is a pseudo-Goldstone boson of an approximate underlying global symmetry, and Twin-Higgs models (Chacko et al., 2006). For a review, see (Csáki et al., 2016).

Large and/or warped extra dimensions can also resolve the difficulties by providing a cutoff at a fundamental scale M_F much lower than the Planck scale. The suppression of M_F/M_P may be associated with some (possibly fractional) inverse power of the size of a large extra dimension compared to M_F^{-1} (which, however, introduces its own naturalness problem), or may involve an exponential suppression in a gravitationally-warped dimension. Deconstruction models, in which no extra dimensions are explicity introduced (Arkani-Hamed et al., 2001a; Hill et al., 2001), are closely related.

Most of the models mentioned above have the potential to generate flavor changing neutral current effects and EDMs much larger than observational limits. Pushing the mass scales high enough to avoid these problems may conflict with a natural solution to the hierarchy problem, i.e., one may reintroduce a *little hierarchy problem.* Many are also strongly constrained by precision electroweak physics. In some cases the new physics does not satisfy the decoupling theorem (Appelquist and Carazzone, 1975), leading to large oblique corrections (Section 8.3.6). In others new tree-level effects may again force the scale to be too high. The most successful from the precision electroweak point of view are those which

[6]Of course, there was no Higgs/hierarchy problem in Higgsless models or models with dynamical symmetry breaking (Section 8.5.3), but the observation of the Higgs forces us to take it seriously.

have a discrete symmetry that prevents vertices involving just one heavy particle, such as R-parity in supersymmetry, T-parity in some little Higgs models (Cheng and Low, 2003), and KK-parity in universal extra dimension models (Appelquist et al., 2001).

A very different possibility is to accept the fine-tuning, i.e., to abandon the notion of naturalness[7] for the weak scale, perhaps motivated by *anthropic* considerations (Agrawal et al., 1998; Hall et al., 2014). (The anthropic idea will be considered below in the discussion of the gravity problem.) This could emerge, for example, in split supersymmetry (Arkani-Hamed and Dimopoulos, 2005).

Another *(relaxion)* mechanism that avoids the need for new physics at the TeV scale will be mentioned in the next subsection.

The Strong CP Problem

Another fine-tuning problem is the *strong CP problem* (e.g., Dine, 2000; Ramond, 2004; Peccei, 2008; Kim and Carosi, 2010). One can add an additional term[8] $\frac{\theta_{QCD}}{32\pi^2} g_s^2 G\tilde{G}$ to the QCD Lagrangian density that breaks P, T and CP symmetry, as in (5.2) on page 160. $\tilde{G}^i_{\mu\nu} = \epsilon_{\mu\nu\rho\sigma} G^{i\rho\sigma}/2$ is the dual field strength tensor introduced in (5.7). As mentioned in Section 8.6.5, such a term would induce an electric dipole moment d_n for the neutron. The stringent limits on d_n lead to the upper bound $|\theta_{QCD}| < 10^{-10} - 10^{-11}$. The question is, therefore, why is θ_{QCD} so small? It is not sufficient to just say that it is zero (i.e., to impose CP invariance on QCD) because of the observed violation of CP by the weak interactions, which is believed to be associated with phases in the quark mass matrices. The quark phase redefinitions that remove them lead to a shift in θ_{QCD} by $\mathcal{O}(10^{-3})$ because of the anomaly in the vertex coupling the associated global current to two gluons, as discussed following (5.130). An apparently contrived fine-tuning is therefore needed to cancel this correction against the bare value. Solutions include the possibility that CP violation is not induced directly by phases in the Yukawa couplings, as is usually assumed in the standard model, but is somehow violated spontaneously. θ_{QCD} then would be a calculable parameter induced at loop level, and it is possible to make θ_{QCD} sufficiently small. However, such models lead to difficult phenomenological and cosmological problems.[9] Alternatively, θ_{QCD} becomes unobservable (i.e., can be rotated away by the quark phase redefinition) if there is a massless u quark (Kaplan and Manohar, 1986). However, most phenomenological estimates (Aoki et al., 2017) are not consistent with $m_u = 0$.

Another possibility is the Peccei-Quinn mechanism (Peccei and Quinn, 1977), in which an extra global $U(1)$ symmetry is imposed on the theory in such a way that θ_{QCD} becomes a dynamical variable that is zero at the minimum of the potential. The spontaneous breaking of the symmetry, along with explicit breaking associated with the anomaly and instanton effects, leads to a very light pseudo-Goldstone boson known as an axion (Weinberg, 1978; Wilczek, 1978). It was initially assumed that the scale f_a at which the $U(1)$ symmetry is broken would be comparable to the electroweak scale ν, but this was soon excluded experimentally. Still allowed is the *invisible axion*, involving a very large $f_a \gg \nu$ and a very

[7]Naturalness ideas will be further tested at the LHC and other upcoming experiments, and are a major motivation (e.g., Arkani-Hamed et al., 2016) for even higher-energy facilities, such as $\mathcal{O}(100$ TeV$)$ pp colliders that have been proposed at CERN and in China.

[8]One could add an analogous term for the weak $SU(2)$ group, but it does not lead to observable consequences, at least within the SM (Anselm and Johansen, 1994; Dine, 2000).

[9]Models in which the CP breaks near the Planck scale may be viable (Nelson, 1984; Barr, 1984).

light axion, with a mass

$$m_a \sim \frac{\sqrt{m_u/m_d}}{1+m_u/m_d}\frac{f_\pi m_\pi}{\sqrt{2}f_a} \sim 6\text{ eV}\left(\frac{10^6\text{ GeV}}{f_a}\right), \tag{10.3}$$

estimated using chiral $SU(2)_L \times SU(2)_R$ considerations. Axion couplings also scale as f_a^{-1}. Laboratory, astrophysical, and cosmological constraints suggest that f_a is in the range $10^9 - 10^{12}$ GeV, corresponding to $m_a \sim (10^{-5} - 10^{-2})$ eV.

Axions and similar axion-like particles (ALPs), which are not necessarily associated with the strong CP problem and for which (10.3) need not hold, often occur in superstring theories (Svrcek and Witten, 2006). Experimental searches typically involve an electromagnetic coupling to $F_{\mu\nu}\tilde{F}^{\mu\nu}$ (i.e., to $\vec{E}\cdot\vec{B}$), analogous to the $G\tilde{G}$ coupling. This can induce axion-photon conversions in a strong magnetic field, which can lead to resonant excitations in a high-Q microwave cavity. Some ALP's might even allow "light shining through a wall", i.e., conversions of a photon to an axion and then a reconversion to a photon on the opposite side of a barrier. Coherent oscillations of a background axion field could also possibly generate oscillating electric dipole moments (Graham and Rajendran, 2013). Searches for axions and ALPs produced in the Sun, the laboratory, or from the early universe are reviewed in (Asztalos et al., 2006; Kim and Carosi, 2010; Jaeckel and Ringwald, 2010; Essig et al., 2013; Graham et al., 2015a; Patrignani, 2016). Their cosmological consequences, especially as viable dark matter candidates, are reviewed, e.g., in (Kawasaki and Nakayama, 2013; Marsh, 2016).

A novel solution to the Higgs/hierarchy problem involves coupling the Higgs to the QCD axion field, known in this context as the *relaxion* (Graham et al., 2015b). The coupling is such that the Higgs mass-squared changes from positive to negative as the relaxion evolves from an initial large value cosmologically, and then shuts the evolution off due to QCD effects soon after the Higgs acquires its VEV.

The Gravity Problem

Gravity is not fundamentally unified with the other interactions in the standard model, although it is possible to graft on classical general relativity[10] by hand. However, general relativity is not a quantum theory, and there is no obvious way to generate one within the standard model context. *Loop quantum gravity* (e.g., Rovelli, 2011) focuses on the properties of space-time itself and does not unify gravity with the other interactions. Kaluza-Klein (Chodos et al., 1987) and supergravity (e.g., Nilles, 1984; Wess and Bagger, 1992; Terning, 2006) theories connect gravity with the other interactions in a more natural way than the SM, but do not yield renormalizable theories of quantum gravity. More promising are superstring theories (which may incorporate Kaluza-Klein and supergravity), which unify gravity and may yield *finite* theories of quantum gravity and all the other interactions. String theories are perhaps the most likely possibility for the underlying theory of particle physics and gravity, but at present there appear to be a nearly unlimited number of possible string vacua (the landscape), with no obvious selection principle. As of this writing the particle physics community is still trying to come to grips with the landscape and its implications. Superstring theories naturally imply some form of supersymmetry, but the supersymmetry relevant to our sector of Nature might be manifest only at a high

[10]Tests of general relativity are reviewed in (Will, 2014). Particle physics implications of the direct observation of gravitational waves by the LIGO interferometers (Abbott et al., 2016) include a limit $m_g < 1.2 \times 10^{-22}$ eV on the graviton mass, and the possibility of future observations of signals from cosmological phase transitions.

scale and have nothing to do with the Higgs/hierarachy problem (split supersymmetry is a compromise, keeping some aspects at the TeV scale).

In addition to the fact that gravity is not unified and not quantized there is another difficulty, namely the cosmological constant. The cosmological constant can be thought of as the energy of the vacuum. However, we saw in Section 8.2.2 that the spontaneous breaking of $SU(2) \times U(1)$ generates a value $\langle 0|V(\nu)|0\rangle = -\mu^4/4\lambda$ for the expectation value of the Higgs potential at the minimum. This is a c-number which has no significance for the microscopic interactions. However, it assumes great importance when the theory is coupled to gravity, because it contributes to the cosmological constant (e.g., Bass, 2011), which becomes

$$\Lambda_{cosm} = \Lambda_{bare} + \Lambda_{SSB}. \tag{10.4}$$

$\Lambda_{bare} = 8\pi G_N V(0)$ is the primordial cosmological constant, which can be thought of as the value of the energy of the vacuum in the absence of spontaneous symmetry breaking (the definition of $V(\phi)$ in (8.12) implicitly assumed $\Lambda_{bare} = 0$) and Λ_{SSB} is the part generated by the Higgs mechanism:

$$|\Lambda_{SSB}| = 8\pi G_N |\langle 0|V|0\rangle| \sim 10^{56}\Lambda_{obs}. \tag{10.5}$$

It is some 10^{56} times larger in magnitude than the observed value $\Lambda_{obs}/8\pi G_N \sim (0.0022 \text{ eV})^4$ (assuming that the dark energy is due to a cosmological constant), and it is of the wrong sign.

This is clearly unacceptable. Technically, one can solve the problem by adding a constant $+\mu^4/4\lambda$ to V, so that V is equal to zero at the minimum (i.e., $\Lambda_{bare} = 2\pi G_N \mu^4/\lambda$). However, with our current understanding there is no reason for Λ_{bare} and Λ_{SSB} to be related. The need to invoke such an incredibly fine-tuned cancellation to 56 decimal places is probably the most unsatisfactory feature of the standard model.

The problem becomes even worse in superstring theories, where one expects a vacuum energy of $\mathcal{O}(M_P^4)$ for a generic point in the landscape, leading to $|\Lambda_{SSB}| \gtrsim 10^{123}\Lambda_{obs}$. The situation is almost as bad in grand unified theories. Finally, any solution must deal with other contributions to the vacuum energy, such as zero-point energies (which cancel in unbroken supersymmetry) and the smaller but still very large energy associated with the QCD vacuum condensates that break chiral symmetry. Contributions from strings, electroweak breaking, QCD, and from zero point energies are associated with very different physics scales, but somehow they must all add up to a negligible value.

So far no compelling solution to the cosmological constant problem has emerged. One intriguing possibility invokes the anthropic (environmental) principle (Barrow and Tipler, 1986; Hogan, 2000), i.e., that a much larger or smaller value of $|\Lambda_{cosm}|$ would not have allowed the possibility for life to have evolved because the Universe would have expanded or recollapsed too rapidly (Weinberg, 1989b). This would be a rather meaningless argument unless (a) Nature somehow allows a large variety of possibilities for $|\Lambda_{cosm}|$ (and perhaps other parameters or principles), such as might occur in different vacua, and (b) there is some mechanism to try all or many of them. In recent years it has been suggested that both of these needs may be met. There appears to be an enormous landscape of possible superstring vacua (Bousso and Polchinski, 2000; Kachru et al., 2003; Susskind, 2003; Denef and Douglas, 2004; Polchinski, 2015), with no obvious physical principle to choose one over the other. Something like eternal inflation (Linde, 1986, 2008) could provide the means to sample them, leading to an enormous *multiverse* of regions with different physical laws and parameters. Only the environmentally suitable vacua would lead to long-lived Universes suitable for life. Perhaps some of the other fine-tunings or arbitrary features of Nature, such as the Higgs/hierarchy problem or the relative values of the light fermion masses (and their implications for the stability of nuclei (Damour and Donoghue, 2008)), could similarly be associated with these ideas.[11] The landscape/multiverse is highly controversial and is currently being heatedly debated. My own view is that if any theory really has a landscape of vacua then one must take the multiverse as a serious possibility, even if it is hard to test. Recent reviews of the multiverse include (Schellekens, 2013, 2015; Linde, 2017; Donoghue, 2016). Such paradigms as uniqueness, naturalness, and minimality are further discussed in (Langacker, 2017).

10.1.2 New Ingredients for Cosmology and Particles

It is now clear that the standard model requires a number of new ingredients. These include:

A Consistent Incorporation of Quantum Gravity

Superstring theory is probably the most promising possibility.

The Initial Conditions on the Big Bang

The observed flatness, homogeneity, and isotropy of the Universe appear to require very fine-tuned initial conditions on the big bang. The fine-tuning could be avoided if there were an initial a period of exponentially rapid expansion known as *inflation* (Guth, 1981; Kolb and Turner, 1990; Lyth and Riotto, 1999; Linde, 2008), followed by a reheating. This inflation could have been driven by the energy density in a scalar field (the *inflaton*), most likely emerging from BSM physics.

A Mechanism for the Baryon Asymmetry

The observed excess of baryons with respect to antibaryons (e.g., Canetti et al., 2012), $n_B/n_\gamma \sim 6 \times 10^{-10}$, $n_{\bar{B}} \ll n_B$, is presumably due to a tiny asymmetry $(n_q - n_{\bar{q}})/n_q \sim 10^{-9}$ of quarks compared to antiquarks in the early Universe. This asymmetry could have been generated dynamically if the three *Sakharov conditions* (Sakharov, 1967; for general reviews, see Bernreuther, 2002; Dine and Kusenko, 2004; Cline, 2006), (a) baryon number violation (to allow B-violating transitions), (b) CP violation (to distinguish q and $\bar{q}$), and (c) nonequilibrium of the B-violating or other relevant processes (or CPT violation so that, e.g., $m_q \neq m_{\bar{q}}$).

Baryon number is conserved in the standard model at the perturbative level, but there are non-perturbative vacuum tunneling (instanton) effects ('t Hooft, 1976b), which violate B and L but preserve $B - L$. These are negligible at zero temperature where they are exponentially suppressed (of $\mathcal{O}(\exp[-2\pi \sin^2\theta_W/\alpha] \sim 10^{-80})$), but important at high temperatures due to thermal fluctuations (sphaleron configurations) (Klinkhamer and Manton, 1984; Kuzmin et al., 1985). This nonperturbative B violation would not have been enough to generate the asymmetry in the SM, however, because the only available candidate for non-equilibrium would be a first-order electroweak phase transition (i.e., the cosmological transition at critical temperature $T_c = \mathcal{O}(\nu)$ from a high temperature phase in which $SU(2) \times U(1)$ is unbroken to the broken phase at lower temperature), and this would only have been sufficiently strong for a Higgs mass $\lesssim 35$ GeV, far below the observed value.

[11] A good analogy is the relative radii of the planetary orbits, once thought by Kepler to have an elegant geometric explanation involving the nesting of Platonic solids, whereas we now understand that they are an accident of the initial conditions of the solar system. Similarly, the "lucky accident" that the Earth is in the Sun's habitable zone simply reflects that with many stars and many planets some will be "just right".

Furthermore, the CP violation from the CKM and PMNS matrices (or from θ_{QCD}) is too weak.

There have been a number of suggestions for explaining the baryon asymmetry in BSM scenarios, however. One involves the out of equilibrium decays of superheavy Majorana right-handed neutrinos [*leptogenesis* (Fukugita and Yanagida, 1986; Davidson et al., 2008; Hambye, 2012; Fong et al., 2012)], as found in the minimal seesaw model (Section 9.5). The decays are expected to occur at a temperature T that is small compared to the heavy Majorana mass (i.e., out of equilibrium), but still large compared to T_c, so that $SU(2)\times U(1)$ breaking can be ignored. If CP were exact, a heavy Majorana neutrino ν_M could decay to $\phi\,\ell_L$ (i.e., $\phi^0\nu_L$ or $\phi^+e^-_L$) or to $\phi^\dagger\,\ell^c_R$ (i.e., $\phi^{0\dagger}\nu^c_R$ or $\phi^-e^+_R$) with equal rates. Here, ϕ is the Higgs doublet and ℓ_L a lepton doublet, as in (8.13) on page 260, and we have suppressed flavor indices. However, there will in general be additional CP phases in the mass and Yukawa matrices that are not directly observable in the PMNS matrix. The interference between tree and loop diagrams could then lead to a rate difference $\Gamma(\nu_M \to \phi^\dagger\,\ell^c_R) > \Gamma(\nu_M \to \phi\,\ell_L)$ and a corresponding lepton asymmetry, $n_\ell < n_{\ell^c}$. The latter could then be partially converted to a baryon asymmetry by sphaleron processes. This leptogenesis scenario and more complicated versions are very attractive, but in practice a number of constraints must be satisfied to generate the observed asymmetry consistent with experiment and cosmology.

Another possibility assumes the existence of a strongly first-order electroweak phase transition [*electroweak baryogenesis* (e.g., Trodden, 1999; Morrissey and Ramsey-Musolf, 2012)]. A first-order transition would occur by the nucleation of bubbles of "true" ($SU(2)\times U(1)$ broken) vacuum inside the sea of "false" ($SU(2)\times U(1)$ unbroken) vacuum, which would eventually expand and fill the entire space. CP violation could occur by interactions of the high temperature plasma with the bubble wall, while $B+L$ violation would be due to sphalerons outside or near the wall. The needed first-order transition does not occur in the standard model for $M_H \sim 125$ GeV, but it could in extensions involving larger Higgs sectors.[12] For example, those involving SM singlet Higgs fields, such as extensions of the MSSM that can generate a dynamical μ term, have cubic Higgs interactions at tree-level that can easily lead to a strongly first-order transition (e.g., Maniatis, 2010; Ellwanger et al., 2010). Generic two-doublet models also have suitable parameter regions[13] in which the cubic interactions are generated by thermal loops (e.g., Cline, 2006; Dorsch et al., 2013). The extended Higgs sectors could also provide the needed new sources of CP violation, though some care is required to be consistent with EDM constraints (Section 8.6.5). They would likely yield signatures observable at the LHC.

Other possibilities for the baryon asymmetry include the decay of a coherent scalar field, such as a scalar quark or lepton in supersymmetry [the *Affleck-Dine mechanism* (Affleck and Dine, 1985)], or CPT violation (Cohen and Kaplan, 1987; Davoudiasl et al., 2004). Finally, one cannot totally dismiss the possibility that the asymmetry is simply due to an initial condition on the big bang. However, this possibility disappears if the Universe underwent a period of rapid inflation that diluted the asymmetry to essentially zero.

[12]This is a major motivation for probes of the Higgs cubic self-interactions at existing and future colliders (Huang et al., 2016; Arkani-Hamed et al., 2016).

[13]This could include the MSSM if one of the scalar top quarks is sufficiently light (Carena et al., 2013), but this possibility is in tension which LHC constraints.

What is the Dark Energy?

In recent years a remarkable concordance of cosmological observations involving the cosmic microwave background radiation (CMB), acceleration of the Universe as determined by Type Ia supernova observations, large scale distribution of galaxies and clusters, and big bang nucleosynthesis has allowed precise determinations of the cosmological parameters (Kolb and Turner, 1990; Peebles, 1993; Hinshaw et al., 2013; Ade et al., 2016; Patrignani, 2016): the Universe is close to flat, with some form of *dark energy* making up 68–69% of the energy density. *Dark matter* constitutes 26–27%, while ordinary matter (mainly baryons) represents only about 5%. The mysterious dark energy (Peebles and Ratra, 2003; Copeland et al., 2006; Frieman et al., 2008; Weinberg et al., 2013), which is the most important contribution to the energy density and leads to the acceleration of the expansion of the Universe, is not accounted for in the SM. It could be due to a cosmological constant that is incredibly tiny on the particle physics scale, to a slowly time varying field [*quintessence* (Zlatev et al., 1999)], or possibly to a modification of general relativity (Joyce et al., 2016). Is the acceleration somehow related to an earlier and much more dramatic period of inflation? If it is associated with a time-varying field, could it be connected with a possible time variation of coupling "constants" (e.g., Uzan, 2011)?

What is the Dark Matter?

There is abundant evidence from galactic rotation curves, the motion of galaxies in clusters, gravitational lensing, galactic mergers, and the CMB that there is much more matter in the Universe than can be accounted for in stars, gas, and other known forms. The missing matter does not seem to be ordinary baryonic matter that is somehow hidden, both because of direct astrophysical searches [e.g., for massive compact halo objects (*MACHOs*)] and more generally because the total baryonic density is determined independently by BBN and the CMB. It is therefore presumably some new (dark) form of matter that interacts at most weakly with electromagnetism.[14] For a historical introduction to dark matter, see (Bertone and Hooper, 2016). For general reviews, see, e.g., (Bertone et al., 2005; Hooper, 2010; Strigari, 2013; Klasen et al., 2015; Gelmini, 2015; Baudis, 2016; Lisanti, 2016).

One possibility is primordial black holes (*PBH*) that somehow formed prior to BBN. There are many astrophysical constraints on the density of PBHs (e.g., Green, 2015), but there is a window from around $10^{-13} - 10^{-8} M_\odot$ for which they could constitute the dark matter, and a second window around 10 $M_\odot$ (the range suggested by the LIGO gravitational wave signals from merging black holes (Abbott et al., 2016)) where they might account for at least some of it (e.g., Clesse and García-Bellido, 2016).

It is likely, though not certain, that the dark matter is associated with elementary particles. Candidate particles are usually classified as hot, cold, or warm, depending on whether they were relativistic, non-relativistic, or intermediate when they decoupled from equilibrium[15] in the early universe. It was briefly commented in Section 9.2.2 that hot dark matter (light massive neutrinos) is excluded because it does not cluster sufficiently on small scales, while keV-scale sterile neutrinos are a viable candidate for warm dark matter.

The most common cold dark matter (CDM) scenario involves a stable (at least on

---

[14]An alternative possibility is to modify Newtonian dynamics (*MOND*) or general relativity on large scales (Bekenstein and Milgrom, 1984; Famaey and McGaugh, 2012). This works reasonably well for galaxies but is less successful with larger scales, especially with colliding galaxies, which require some new component similar to dark matter.

[15]Some types of dark matter, such as PBH or axions, are considered cold if their gravitational clustering is similar even though their production mechanisms may be nonthermal.

cosmological time scales) neutral colorless particle χ, which usually has spin-0 or 1/2 and which may or may not be distinct from its antiparticle $\bar{\chi}$. In the *thermal* dark matter scenario χ and $\bar{\chi}$ are assumed to have equal number densities (if they are distinct), maintained in equilibrium number $\propto T^{3/2}e^{-M_\chi/T}$ by reactions such as $\chi\chi \leftrightarrow \sigma\bar{\sigma}$ or $\chi\bar{\chi} \leftrightarrow \sigma\bar{\sigma}$, where σ may be a lighter SM particle, until they freeze out at some temperature T_χ. A detailed analysis shows that

$$\Omega_\chi h^2 \sim \frac{3 \times 10^{-27}\ \text{cm}^3/\text{s}}{\langle \sigma v \rangle}, \tag{10.6}$$

where Ω_χ is the dark matter density in χ relative to the critical density, $h \sim 0.68$ is the Hubble parameter H at present in units of 100 km s^{-1} Mpc^{-1}, and $\langle \sigma v \rangle$ is the thermal average of the annihilation cross section times relative velocity. The form of (10.6) is easily understood: a larger annihilation cross section means that the freezeout occurs later (at a larger M_χ/T_χ) when the number density is lower. Assuming that $\Omega_\chi \sim 0.26$ (the observed dark matter density), one requires

$$\langle \sigma v \rangle \sim 3 \times 10^{-9}\ \text{GeV}^{-2}. \tag{10.7}$$

But for S-wave annihilation via a lighter mediator X (such as H or Z) one expects $\langle \sigma v \rangle \sim \alpha_X^2/M_\chi^2$, where $\alpha_X = g_X^2/4\pi$ and g_X is the relevant coupling. In particular, particles in the general range of 100 – 1000 GeV will automatically lead to dark matter densities via thermal production in the observed ballpark for weak interaction strength couplings, e.g., $\alpha_X \sim 10^{-2}$ for $M_\chi \sim 200$ GeV. This is referred to as the *WIMP miracle*, and such *weakly interacting massive particles* are known as WIMPs.

WIMPs candidates include the lightest supersymmetric partner (usually a neutralino) in supersymmetric models with R-parity conservation (Section 10.2.5), or analogous stable particles in Little Higgs or universal extra dimension models. As we will see, however, experimental constraints on these conventional candidates are significantly reducing the allowed parameter space. This has in part been the motivation for considerable attention to the possibility of much lighter dark matter particles, perhaps associated with some dark or hidden sector (e.g., Alexander et al., 2016) with very weak coupling to ordinary particles. Such sectors, which occur in some superstring constructions, may be connected to the ordinary sector by kinetic mixing of the Z with a new Z' gauge boson (Section 10.3.1), or by other types of portals involving Higgs bosons, heavy Z', neutrinos, axions, etc.

There are many direct, indirect, and collider searches for WIMPs and similar particles (see the general reviews, as well as (Jungman et al., 1996; Cushman et al., 2013) and the Dark Matter review in (Patrignani, 2016)). The direct experiments look for the nuclear recoil from an elastic (or possibly inelastic) scattering $\chi N \to \chi N$. The experiments are carried out on nuclear targets, and can search for spin-independent (SI) and spin-dependent (SD) scattering processes. SI scattering acts coherently on all of the nucleons and therefore scales as A^2. The SD case, which is proportional in amplitude to the net spin, is much less constrained. SI scattering is associated with interactions such as $\bar{\chi}\chi\bar{N}N$ or $\bar{\chi}\gamma^\mu\chi\bar{N}\gamma_\mu N$ (for spin-1/2), which could be, respectively, generated by t-channel H or Z exchange. The latter would require a Dirac χ. For a Majorana χ, such as neutralinos in the conventional MSSM, Higgs exchange would still lead to SI scattering, but Z exchange would involve the operator $\bar{\chi}\gamma^\mu\gamma^5\chi\bar{N}\gamma_\mu\gamma^5 N$, which only contributes to SD scattering. (These statements are easily seen by using the Pauli-Dirac representation for the γ matrices and spinors (Section 2.7.2) in the non-relativistic limit, and the fact that a Majorana spinor has no vector coupling to the Z (Problem 9.4.)) None of the experiments has observed the direct scattering of WIMPs. There have been some possible positive signals, especially from the DAMA/LIBRA experiment based on annual modulation (Bernabei et al., 2010), but these are difficult to reconcile with other exclusions within the canonical WIMP scenario. The limits on the SI cross sections per nucleon[16] are shown in Figure 10.2. Much but not all of the cross section region for heavy (100 – 1000 GeV) WIMPs in the MSSM is excluded. For lighter WIMPs the limits are approaching the *neutrino floor*, i.e., the cross sections for which the rates are comparable to backgrounds from solar, atmospheric, or diffuse supernova neutrino scattering. Going further would require some sort of directional sensitivity (e.g., Grothaus et al., 2014).

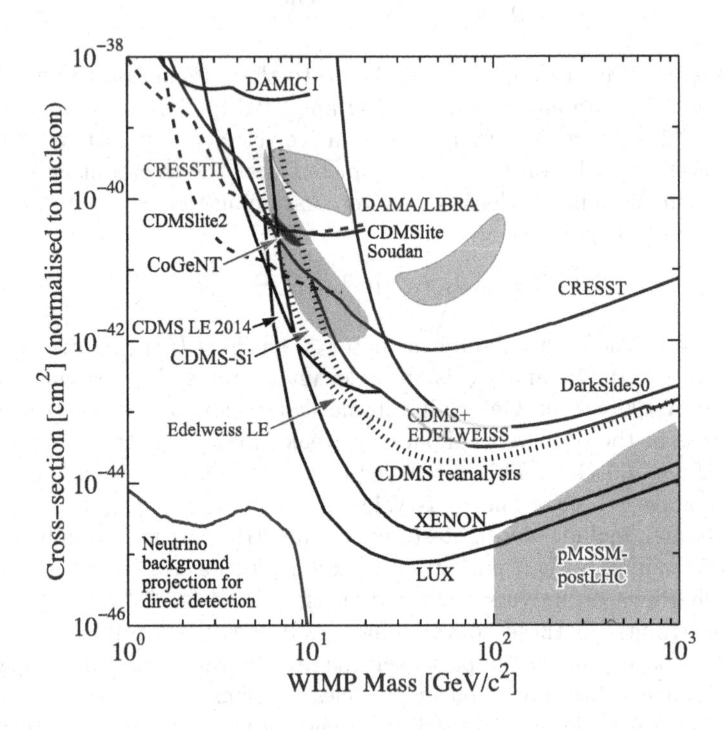

Figure 10.2 Limits on spin-independent cross sections for elastic WIMP-nucleon scattering vs. WIMP mass, assuming a theoretical model for the local density and velocity distribution consistent with $\Omega_\chi \sim 0.26$. The enclosed regions are reported positive signals. The low mass part of the neutrino floor and the part of the MSSM parameter space that is still allowed are also shown. Plot courtesy of the Particle Data Group.

Indirect detection refers to searches for γ's, ν's, e^+'s, $\bar{p}$'s, etc. that result directly from the annihilation or possibly decays of dark matter particles in regions of high concentration, or from the decays or secondary interactions of the directly produced particles. For example, neutrino signals could be due to WIMPs concentrated in the Sun or center of the Earth, while those near the galactic center or in other galaxies could lead to monoenergetic photons from direct annihilation, or to continuum photons or antiparticles from secondary processes. Indirect detection is sensitive to a wide mass range from small to multi-TeV. There have been various hints of signals involving photons or excesses of antiparticles (see Gaskins,

[16]The experiments on different nuclear targets can be directly compared assuming that the scattering is isospin-independent, i.e., the same for p and n. Isospin-dependent scattering is less constrained.

2016, for a recent review), but because of possible astrophysical backgrounds none have so far proved compelling.

ATLAS and CMS have searched for dark matter particles via $pp \to \chi\overset{(-)}{\chi} + X$, where X contains an energetic monojet, γ, Z, H, etc, typically radiated from an initial parton. The $\overset{(-)}{\chi}$'s are not directly observed, so the signature is X plus a large missing transverse momentum. The sensitivities are comparable to those of the direct searches for both SI and SD operators.[17] So far, no positive signals have been observed.

There are many variations on the thermal WIMP scenario. For example, equilibrium could be associated with more complicated processes, such as resonant annihilation (with $2M_\chi$ close to the mass of an s-channel resonance), coannihilation between χ and some other particle (presumably similar in mass), or in reactions other than $2 \to 2$. The WIMPs could also have strong interactions amongst themselves (SIMPs), there could be multiple dark matter components, or they could scatter inelastically to excited states. Increased numbers of WIMPs could be produced by non-thermal mechanisms, such as the decay of a coherent field, or in asymmetric scenarios, in which there is somehow an asymmetry between the numbers of χ and $\bar{\chi}$ analogous to the baryon asymmetry. Another possibility (SuperWIMPs) (e.g., Feng, 2010) is that the dark matter particles are extremely weakly interacting, and are produced by the late decays of WIMPs. An example are light gravitinos, expected in some versions of supersymmetry with a low breaking scale.

Invisible axions are a very different possibility for cold dark matter. They were originally introduced in connection with the strong CP problem, but axions and axion-like particles can occur more generallly. (See the discussion of the strong CP problem in Section 10.1.1.) For masses in the $10^{-5} - 10^{-4}$ eV range they would be viable CDM candidates, produced non-thermally by the decay of a coherent field. The details depend sensitively on the mass and scale, and on whether they were produced before or after a period of inflation (for reviews, see Kawasaki and Nakayama, 2013; Marsh, 2016).

The cold dark matter scenario is quite successful at describing large-scale structures, but is less so on scales smaller than our galaxy. Most CDM simulations suggest that there should be more dwarf and satellite galaxies and a steeper central density profile for dwarf galaxies than observed. These discrepancies may be due to inadequate modelling of the astrophysics, but they could also be hints that the CDM picture needs to be modified. Possible candidates include warm dark matter associated with keV-scale sterile neutrinos (Section 9.2.2). Another is self-interacting (SIMP) dark matter. We finally mention the interesting suggestion of *fuzzy dark matter* (Hu et al., 2000; Hui et al., 2017), which consists of ultralight, $\mathcal{O}(10^{-22}$ eV), scalars produced by the decay of a coherent field (similar to axions). These would have kpc-scale de Broglie wavelengths, with the small-scale dynamics controlled by its wave nature.

The Suppression of Flavor Changing Neutral Currents, Proton Decay, and Electric Dipole Moments

As discussed in Sections 8.6.5 and 8.6.6, the standard model has a number of (approximate) accidental symmetries and features that forbid proton decay at the perturbative level, preserve lepton number and lepton flavor (at least for vanishing neutrino masses), suppress transitions such as $K^+ \to \pi^+ \nu\bar{\nu}$ at tree-level, and lead to highly suppressed electric dipole

[17]Direct, indirect (for annihilations into quarks), and collider processes can be loosely thought of as crossed channel reactions, e.g., $\chi q \to \chi q$, $\chi\overset{(-)}{\chi} \to q\bar{q}$, and $q\bar{q} \to \chi\overset{(-)}{\chi}$, respectively. However, comparisons are complicated by the very different kinematic regimes.

moments for the e^-, n, atoms, etc. However, most extensions of the SM have new interactions that violate such symmetries, leading to potentially serious problems with FCNC and EDMs. There is a tradeoff/conflict between these constraints, which favor high mass scales for the new physics, and naturalness, which favors lower scales.

A Mechanism for Small Neutrino Masses

The most popular possibility is the minimal seesaw model, implying Majorana masses, but there are other plausible mechanisms for either small Dirac or Majorana masses, as described in Chapter 9.

Possible Types of New Physics

There is an enormous range of possibilities for new physics beyond the standard model (e.g., Eichten et al., 1984; Morrissey et al., 2012; Nath et al., 2010; Alves et al., 2012; Gershtein et al., 2013), many of which were mentioned above or in Chapter 8. Many types are "bottom-up," i.e., motivated by attempts to resolve problems of the standard model, to understand the dark matter, or to explain possible experimental anomalies. These include supersymmetry; extended Higgs sectors; family symmetries; extended TeV-scale gauge groups; new types of particles, such as leptoquarks or diquarks; large and/or warped extra dimensions; dark or hidden sectors; and strong coupling theories such as dynamical symmetry breaking, composite Higgs models, or composite and excited fermions[18] (Harari, 1984). Other "top-down" types, such as grand unification and superstring theories, attempt to achieve a more fundamental unified understanding of the microscopic interactions, to incorporate gravity, to describe the dark energy, or to make contact with ideas concerning the early universe. Bottom-up and top-down ideas are not mutually exclusive, and ideas such as supersymmetry can be motivated from either. Additional or exotic heavy fermions, additional Higgs particles, new types of interactions, hidden sector particles, etc., may also arise as *remnants* (e.g., Barger et al., 2007), i.e., particles emerging from an underlying theory characterized by a much larger scale that remain light by some accident in the dynamics or symmetry breaking. This often occurs, for example, in superstring vacua in which most but not all degrees of freedom achieve Planck scale masses. Such remnants are quite plausible from a top-down perspective, though they may appear superfluous and non-minimal from a bottom-up view.

In the following we describe three important examples of possible new physics: supersymmetry, extended (TeV-scale) gauge symmetries, and grand unification.

10.2 SUPERSYMMETRY

Supersymmetry is a hypothetical symmetry very different from those we have so far encountered, viz., between bosons and fermions. It is especially motivated by attempts to unify gravity with the other interactions, especially in supergravity and superstring theories, though that does not by itself imply that supersymmetry is relevant at the TeV scale. However, supersymmetry broken at the electroweak-TeV scale is one of the leading contenders for solving the Higgs/hierarchy problem discussed in Section 10.1. Other motivations include gauge coupling unification, which is much more successful in the supersymmetric

[18]Theories of composite quarks and leptons should involve new flavor-diagonal four-fermion interactions suppressed by the compositeness scale (Eichten et al., 1983), e.g., by the binding forces or by constituent interchange, as well as excited states. Experimental searches (Patrignani, 2016) indicate that the scales would have to be in the multi-TeV range, e.g., with the bound state masses protected by a chiral symmetry.

extension than in the SM, and the existence of natural candidates for cold dark matter in some versions. The possible anomaly in the anomalous magnetic moment of the $\mu^\pm$ could also be accounted for by relatively light supersymmetric partners, as mentioned in Section 2.12.3. No sign of supersymmetry has emerged from the LHC as of mid-2016, suggesting that if supersymmetry exists at all the mass scale for its breaking must be higher than early expectations. Nevertheless, the motivations are sufficiently strong to continue the search.

The technology and phenomenology of supersymmetry are extremely complicated, and here we can only give a brief introduction. These topics are thoroughly described in a number of articles (Fayet and Ferrara, 1977; Nilles, 1984; Haber and Kane, 1985; Lykken, 1996a; Martin, 1997; Peskin, 2008; Patrignani, 2016) and books (e.g., Wess and Bagger, 1992; Weinberg, 2000; Polonsky, 2001; Drees et al., 2004; Baer and Tata, 2006; Binétruy, 2006; Terning, 2006; Kane, 2010; Dine, 2015). The original idea of a symmetry between bosons and fermions involved the two-dimensional worldsheet of a string theory (Ramond, 1971; Neveu and Schwarz, 1971; Gervais and Sakita, 1971). Supersymmetry for a four-dimensional field theory was introduced in (Wess and Zumino, 1974a,b; Salam and Strathdee, 1974), and applied to the $SU(2) \times U(1)$ model in (Fayet, 1975). The history of the development of supersymmetry is described in detail in (Weinberg, 2000; Wess, 2009).

10.2.1 Implications of Supersymmetry

Corrections to the Higgs Mass

As described in Section 10.1 the SM leads to troubling quadratically-divergent contributions to M_H^2. For example, the top quark contribution (the first diagram in Figure 10.3), implies a radiative correction[19]

$$-i\left(\Delta M_H^2\right)_t = -N_c(-ih_t)^2 \int \frac{d^4k}{(2\pi)^4} \operatorname{Tr}\left(\frac{i}{\not{k} - m_t}\frac{i}{\not{k} - m_t}\right), \tag{10.8}$$

where -1 is due to the closed fermion loop, $N_c = 3$ counts the t-quark colors, and the vertices follow from (8.46) on page 267 with $h_t = m_t/\nu$. The quadratically divergent part can easily be calculated by applying a Wick rotation, as in (E.14), to yield

$$\left(\Delta M_H^2\right)_t = -4iN_c(h_t)^2 \int_0^\Lambda \frac{ik_E^3 dk_E\, d\Omega_4}{(2\pi)^4}\frac{-k_E^2 + m_t^2}{(k_E^2 + m_t^2)^2} \sim -\frac{N_c(h_t)^2}{4\pi^2}\Lambda^2, \tag{10.9}$$

since $\int d\Omega_4 = 2\pi^2$.

Now, suppose there are two complex color-triplet scalar fields $\phi_r, r = 1, 2$, with mass and couplings to H given by

$$\mathcal{L}_{\phi_r H} = -\lambda_r H^2 \phi_r^\dagger \phi_r - \kappa_r H \phi_r^\dagger \phi_r - m_r^2 \phi_r^\dagger \phi_r. \tag{10.10}$$

Each contributes to M_H^2 by the middle and right diagrams in Figure 10.3. The middle diagram is quadratically divergent, yielding

$$-i\left(\Delta M_H^2\right)_{\phi_r} = -2i\lambda_r N_c \int \frac{d^4k}{(2\pi)^4}\frac{i}{k^2 - m_r^2} \Rightarrow \left(\Delta M_H^2\right)_{\phi_r} \sim \frac{\lambda_r N_c}{8\pi^2}\Lambda^2. \tag{10.11}$$

The quadratic divergences cancel if we choose $\lambda_1 = \lambda_2 = (h_t)^2$. This would be a remarkable

---

[19]The diagram yields the t contribution to $\langle H|i\Delta\mathcal{L}|H\rangle$. (We are being careless about external momenta since we are considering the divergent part.) But $\mathcal{L} = -\frac{1}{2}M_H^2 H^2 + \cdots$, so the matrix element is $-i\left(\Delta M_H^2\right)_t$.

accident if not enforced by some symmetry. Fortunately, in the supersymmetric extension of the standard model it occurs naturally: in that case $\phi_{1,2}$ are the scalar partners of the t_L and t_L^c (i.e., t_R), respectively, and the necessary coupling constant relation is enforced by the supersymmetry. In fact, in the supersymmetric limit, one also has that $\kappa_r = 2\nu(h_t)^2$ and $m_r^2 = m_t^2$, which implies that the logarithmically divergent contributions to M_H^2 also cancel, as shown explicitly in, e.g., (Terning, 2006). This is an example of the *non-renormalization theorem* in the supersymmetric limit. In the presence of soft supersymmetry breaking the quadratic divergences continue to cancel, but there are finite contributions to M_H^2 related to the supersymmetry breaking scale. (See Equation 10.164.)

Supersymmetry is therefore successful at solving the part of the hierarchy problem associated with the loop corrections. However, as we will see below it introduces a new tree-level hierarchy problem, *the μ problem.*

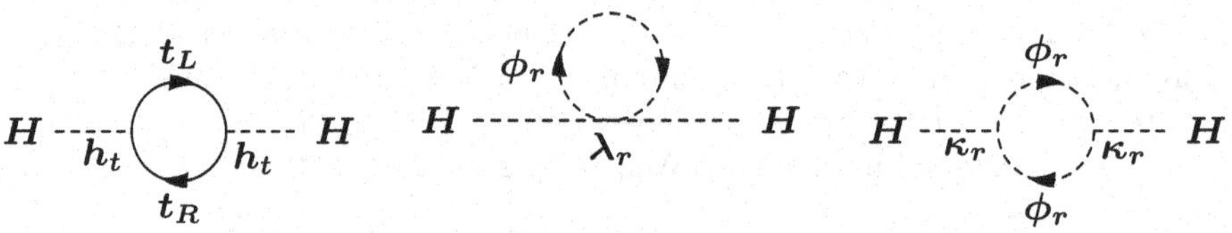

Figure 10.3 t quark and ϕ_r contributions to M_H^2.

The Supersymmetric Spectrum

Supersymmetry is elegant in its principles but not economical in its particle content: it requires a more than doubling of the SM spectrum. Each standard model particle must have a *superpartner* (or *sparticle*) differing in spin by 1/2 unit, which we will denote with a tilde. The sparticles have the same $SU(3) \times SU(2) \times U(1)$ assignments as their partners, and for phenomenological reasons none of the SM bosons or fermions can be each other's partners. In particular, each left- (right)-chiral quark q_L (q_R) is predicted to have a spin-0 *scalar quark* (*squark*) partner $\tilde{q}_L$ ($\tilde{q}_R$). Of course, the L and R labels for a spin-0 particle are not directly related to spin—they simply mean that the scalar is the partner of the corresponding quark. As discussed in Section 2.11, it is convenient to work in terms of the left-chiral particles and antiparticles, using the correspondence $q_L^c = C\bar{q}_R^T$ (Equation 2.301 on page 56) and $\tilde{q}_L^c = \tilde{q}_R^\dagger$. In the supersymmetric limit, the masses of the quarks and squarks would be the same and their interactions related in a definite way, as we saw in the example of the cancellation of the Higgs quadratic mass divergence. No light squarks or other superpartners have been observed, so the supersymmetry must be broken, with the sparticles all relatively heavy, e.g., in the TeV range.[20] Similarly, the leptons ℓ must have spin-0 *slepton* partners $\tilde{\ell}$, and the gauge bosons must have spin-1/2 *gaugino* partners, such as the *gluinos* ($\tilde{G}$), the *winos* ($\tilde{W}$), and the *bino* ($\tilde{B}$). Extensions to supergravity also predict a spin-3/2 partner of the graviton (g), the *gravitino* ($g_{3/2}$).

Supersymmetry also requires an extended Higgs sector. We saw in Section 8.1 that a single Higgs doublet ϕ could have the needed Yukawa couplings to generate masses for both the u and d quarks by making use of the conjugate $\tilde{\phi}$ defined in (8.14) on page 260 by

[20]It is fairly natural for the superpartners to be much heavier than the SM particles, because the former can all acquire large masses without the SSB of $SU(2) \times U(1)$, while the SM particle masses other than the Higgs all require SSB.

$\tilde{\phi} \equiv i\tau^2\phi^\dagger$ (this tilde does *not* represent a superpartner). However, supersymmetry does not allow the needed $\tilde{\phi}$ Yukawa couplings, so we must instead introduce two separate Higgs doublets

$$\phi_d = \begin{pmatrix} \phi_d^+ \\ \phi_d^0 \end{pmatrix}, \qquad \phi_u = \begin{pmatrix} \phi_u^0 \\ \phi_u^- \end{pmatrix}, \tag{10.12}$$

where ϕ_d is similar to the SM ϕ and has the couplings needed to generate masses for the d quarks and charged leptons, while ϕ_u plays the role of the SM $\tilde{\phi}$, and can lead to masses for the u quarks and Dirac neutrino masses. It is actually more convenient to introduce the conjugate doublets

$$h_u = \begin{pmatrix} h_u^+ \\ h_u^0 \end{pmatrix} \equiv -\tilde{\phi}_u, \qquad h_d = \begin{pmatrix} h_d^0 \\ h_d^- \end{pmatrix} \equiv \tilde{\phi}_d, \tag{10.13}$$

which will allow us to write the Higgs Yukawa interactions in a manifestly supersymmetric form. In any case, the minimal supersymmetric standard model (MSSM) involves 3 neutral and one conjugate pair of charged Higgs particles (not including the Goldstone bosons), as opposed to the single Higgs scalar of the SM.

Supersymmetry does not allow the quartic Higgs self-interaction $\lambda(\phi^\dagger\phi)^2$ of the SM. Rather, its role is played by (known) gauge couplings. This removes one of the most arbitrary aspects of the SM, and leads to a theoretical upper bound (at tree level) on the mass of the lightest neutral mass eigenstate h^0,

$$M_{h^0}^2 \leq \cos^2 2\beta\, M_Z^2, \tag{10.14}$$

where $\tan\beta \equiv |\langle 0|h_u^0|0\rangle|/|\langle 0|h_d^0|0\rangle|$ is the ratio of neutral Higgs VEVs. Equation 10.14 is inconsistent with $M_H = 125$ GeV (assuming, as is likely, that the observed H corresponds to h^0). However, there are large radiative corrections to the effective potential associated with top and scalar top loops that increase the upper bound to ~ 135 GeV.

The two Higgs doublets must also have spin-$\frac{1}{2}$ partners (*Higgsinos*). (If there were only one Higgs doublet the Higgsinos would introduce triangle anomalies, providing another rationale for the second doublet.) The Higgsinos can mix with the winos and bino to produce two mass eigenstate Dirac *charginos* ($\tilde{\chi}_r^\pm$, $r = 1, 2$), and 4 mass eigenstate Majorana *neutralinos* ($\tilde{\chi}_r^0$, $r = 1 \cdots 4$). (These are sometimes denoted $\tilde{C}_r^\pm$ and $\tilde{N}_r^0$ instead.)

The MSSM particles are listed in Table 10.1.

Other Implications and Difficulties

There are a number of other implications of supersymmetry, both good and bad.

We have not yet observed any superpartners, so supersymmetry must be broken, probably with sparticle masses in the TeV range or higher. To avoid the reintroduction of the hierarchy problem the breaking should be *soft*, i.e., appearing only in scalar and gaugino mass terms and in cubic scalar couplings. However, general soft breaking terms introduce an enormous number of free parameters.[21] Moreover, there are mass sum rule constraints that would be violated for spontaneous or dynamical breaking that occurs directly in the *ordinary sector* associated with the MSSM particles. Supersymmetry breaking therefore most likely occurs in some *hidden sector* which is only very weakly coupled to the ordinary sector. How this breaking occurs (e.g., Intriligator and Seiberg, 2007) and how the information is transmitted to our sector (the *mediation* mechanism) (e.g., Chung et al., 2005) introduce considerable uncertainties.

[21]The MSSM with R-parity conserved has 124 free parameters (Dimopoulos and Sutter, 1995; Patrignani, 2016), not including neutrino masses and mixings, right-handed scalar neutrinos, M_P, or Λ_{cosm}.

TABLE 10.1 Standard model particles and their supersymmetric partners.[a]

spin-0	$\begin{pmatrix}\tilde{u}_L\\ \tilde{d}_L\end{pmatrix}$	$\tilde{u}^c_L\ \tilde{d}^c_L$	$\begin{pmatrix}\tilde{\nu}_L\\ \tilde{e}_L\end{pmatrix}$	$\tilde{\nu}^c_L\ \tilde{e}^c_L$		$\begin{pmatrix}h^+_u\\ h^0_u\end{pmatrix}$	$\begin{pmatrix}h^0_d\\ h^-_d\end{pmatrix}$
spin-$\frac{1}{2}$	$\begin{pmatrix}u_L\\ d_L\end{pmatrix}$	$u^c_L\ d^c_L$	$\begin{pmatrix}\nu_L\\ e_L\end{pmatrix}$	$\nu^c_L\ e^c_L$	$\tilde{G}\ \tilde{W}^i\ \tilde{B}$	$\begin{pmatrix}\tilde{h}^+_u\\ \tilde{h}^0_u\end{pmatrix}$	$\begin{pmatrix}\tilde{h}^0_d\\ \tilde{h}^-_d\end{pmatrix}$
					$\underbrace{}_{\tilde{\chi}^0,\tilde{\chi}^\pm}$ (spanning $\tilde{W}^i\ \tilde{B}$ and Higgsinos)		
spin-1					$G\ \underbrace{W^i\ B}_{W^\pm,Z,\gamma}$		
spin-$\frac{3}{2}$	$g_{3/2}$						
spin-2	g						

[a]Family indices and mixing are ignored for the fermions and their scalar partners. The right-chiral fermions and their partners are related by $\psi_R = \mathcal{C}\bar{\psi}^{cT}$ and $\tilde{\psi}_R = \tilde{\psi}^{\dagger}_L$. The gauge and Higgs particles are listed in the weak basis. The graviton g and its partner the gravitino, $g_{3/2}$, are also listed.

Many supersymmetric models involve or impose a discrete R-parity symmetry (Farrar and Fayet, 1978), R_p, which requires that every allowed interaction vertex involves an even number of superpartners. This implies that the lightest superpartner (the LSP) is absolutely stable, and therefore a candidate to be dark matter. Neutralinos are the most promising possibility, although scalar neutrinos or the gravitino are a priori possible.

In the *decoupling limit*, in which the superpartners and extra Higgs fields are all much heavier than the electroweak scale, the contributions of the new particles to electroweak precision observables is small, consistent with the excellent agreement with the SM predictions. Also, the lightest neutral Higgs (h^0) acts very much like the SM Higgs in this limit.

On the other hand, the sparticles lead to possible new sources of FCNC and of CP violation, e.g., in the K and B systems and in EDMs. The non-observation of such effects suggests that sparticles should be very heavy and/or that sparticles of a given type are nearly degenerate and/or that there is some kind of alignment between the mixing effects in the quark and scalar quark sectors.

The combination of gauge invariance and supersymmetry does not allow masses for any of the MSSM particles (prior to the $SU(2) \times U(1)$ and supersymmetry breaking), with the exception of the Higgs scalars and their Higgsino partners, which are allowed to have a common arbitrary mass μ. If the supersymmetry derives from an underlying string theory one might expect μ to be comparable to the Planck or string scales, or possibly 0 if it is forbidden by some extra symmetry. However, neither 0 nor a very large value for μ is allowed phenomenologically. The μ problem (Kim and Nilles, 1984), which is a tree-level form of the Higgs/hierarchy problem, is to understand why μ should be nonzero and comparable to the soft supersymmetry breaking scale. Possible explanations are discussed in Section 10.2.6.

The Higgs soft mass parameters are typical soft supersymmetry breaking parameters. Assuming that the μ problem is somehow solved, the scale of electroweak symmetry breaking is therefore tied to the supersymmetry breaking scale, up to an order of magnitude or so. (A larger splitting reintroduces a more moderate version of the hierarchy problem, i.e., the little hierarchy problem.)

Many, but not all, supersymmetry breaking and mediation mechanisms imply that the Higgs and other scalar mass-squares are positive at some large scale, such as at M_P. However, electroweak symmetry breaking is most easily accomplished for a negative h_u mass-square $m^2_{h_u}$, though this is not absolutely essential. These constraints can be reconciled by the fact that the soft mass-squares are running quantities, just like coupling constants (e.g., Martin, 1997). A positive $m^2_{h_u}$ at a high scale can be driven negative at a low scale by a large top-Yukawa coupling h_t. This *radiative breaking* mechanism (e.g., Ibáñez and Ross, 2007) therefore requires a heavy top quark mass.

10.2.2 Formalism

In this section we summarize some of the formalism used in the construction of supersymmetric field theories.

The Lorentz and Poincare Groups

Let us briefly survey the Lorentz and Poincaré groups, which describe the classical spacetime symmetries. A Lorentz transformation can be defined by its action on four-vectors,

$$x^\mu \to x'^\mu = \Lambda^\mu{}_\nu x^\nu, \qquad x_\mu \to x'_\mu = x_\nu \left(\Lambda^{-1}\right)^\nu{}_\mu, \tag{10.15}$$

with $\left(\Lambda^{-1}\right)^\nu{}_\mu = \Lambda_\mu{}^\nu$. The group of such transformations is known as $SO(1,3)$, i.e., it leaves invariant $x^\mu y_\mu = g_{\mu\nu} x^\mu y^\nu$, where $g_{\mu\nu} = \text{diag}(1,-1,-1,-1)$. Λ can be expressed in terms of its generators[22] (i.e., elements of the Lie algebra) $M_{\rho\sigma}$ by

$$\Lambda = e^{-\frac{i}{2}\omega^{\rho\sigma} M_{\rho\sigma}}, \tag{10.16}$$

where the 6 group parameters $\omega^{\rho\sigma}$ are real and antisymmetric in the indices, as are the 6 generators $M_{\rho\sigma}$. The group action and commutation rules can be obtained using the classical representation

$$M_{\rho\sigma} = x_\rho p_\sigma - x_\sigma p_\rho, \tag{10.17}$$

where $p_\sigma = +i\partial_\sigma$ is the position space representation of the momentum operator P_σ. For small $\omega^{\rho\sigma}$ one finds

$$x'^\mu \sim (\delta^\mu{}_\nu - \omega^\mu{}_\nu) x^\nu. \tag{10.18}$$

It is useful to define the group parameters

$$\omega^i = \frac{1}{2}\epsilon^{ijk}\omega_{jk}, \qquad \zeta^i = \omega^{0i}, \qquad i,j,k = 1,2,3, \tag{10.19}$$

and generators

$$J^i = \frac{1}{2}\epsilon^{ijk} M_{jk} = (\vec{x} \times \vec{p})^i, \qquad K^i = M^{0i} = x^0 p^i - x^i p^0, \tag{10.20}$$

[22]These elements actually represent the proper orthochronous Lorentz group, with $\det \Lambda = +1$ and $\Lambda^0{}_0 \geq +1$. It must be supplemented with space reflection and time reversal to obtain the full Lorentz group.

so that

$$\frac{1}{2}\omega^{\rho\sigma}M_{\rho\sigma} = \vec{\omega}\cdot\vec{J} + \vec{\zeta}\cdot\vec{K}. \tag{10.21}$$

J^i are the rotation generators (angular momenta), and K^i generate Lorentz boosts in the i direction. The rapidity $\vec{\zeta}$, defined in Problem 1.4, is related to the velocity $\vec{\beta}$ of the Lorentz boost by $\vec{\zeta} = \hat{\beta}\tanh^{-1}\beta$, which is approximated by $\vec{\beta}$ for small β. Then

$$x'^0 \sim x^0 - \beta^i x^i, \qquad x'^i \sim x^i - (\vec{\omega}\times\vec{x})^i - \beta^i x^0, \tag{10.22}$$

which indeed represent[23] an infinitesimal rotation by $\vec{\omega}$ and boost by $\vec{\beta}$.

One can extend to the Poincaré group by including the translations

$$x' = Tx = e^{+ia^\rho P_\rho}\,x = x - a, \tag{10.23}$$

with a field $\Phi(x)$ transforming as in (1.14) on page 3, i.e.,

$$\Phi(x) \to \Phi'(x) \equiv e^{+ia^\rho P_\rho}\Phi(x)e^{-ia^\rho P_\rho} = \Phi(x+a) = \Phi(T^{-1}x). \tag{10.24}$$

The Lie algebra of the Poincaré group can be derived from $[x_\rho, p_\sigma] = -ig_{\rho\sigma}$,

$$\begin{aligned}
&\left[P_\mu, P_\nu\right] = 0, \qquad \left[M_{\mu\nu}, P_\rho\right] = -i(g_{\mu\rho}P_\nu - g_{\nu\rho}P_\mu) \\
&\left[M_{\mu\nu}, M_{\rho\sigma}\right] = i(g_{\nu\rho}M_{\mu\sigma} - g_{\nu\sigma}M_{\mu\rho} - g_{\mu\rho}M_{\nu\sigma} + g_{\mu\sigma}M_{\nu\rho}),
\end{aligned} \tag{10.25}$$

and correspondingly

$$\left[J^i, J^j\right] = i\epsilon_{ijk}J^k, \quad \left[K^i, J^j\right] = i\epsilon_{ijk}K^k, \quad \left[K^i, K^j\right] = -i\epsilon_{ijk}J^k. \tag{10.26}$$

This implies that $J^i_\pm \equiv \frac{1}{2}[J^i \pm iK^i]$ form a commuting $SU(2)_L \times SU(2)_R$ algebra,

$$\left[J^i_\pm, J^j_\pm\right] = i\epsilon_{ijk}J^k_\pm, \qquad \left[J^i_+, J^j_-\right] = 0, \tag{10.27}$$

which means that the compact representations of the Lorentz group can be labeled by (j_+, j_-), where $j_\pm = 0, \frac{1}{2}, 1, \cdots$ are defined by the Casimirs $\vec{J}^2_+ = j_+(j_+ + 1)$ and $\vec{J}^2_- = j_-(j_- + 1)$. A field $\Phi(x)$ transforming under the (j_+, j_-) representation goes into

$$\Phi(x) \to \Phi'(x) \equiv e^{-\frac{i}{2}\omega^{\rho\sigma}M_{\rho\sigma}}\Phi(x)e^{\frac{i}{2}\omega^{\rho\sigma}M_{\rho\sigma}} = \Lambda_{(j_+,j_-)}\Phi(\Lambda^{-1}x), \tag{10.28}$$

under a Lorentz transformation Λ, where $\Lambda_{(j_+,j_-)}$ is the representation matrix of $e^{\frac{i}{2}\omega^{\rho\sigma}M_{\rho\sigma}}$ (cf. Equations 3.28 and 3.30). For example, the four-dimensional $(\frac{1}{2}, \frac{1}{2})$ representation is just the defining four-vector representation, as will be shown in an example below. Similarly, the $(\frac{1}{2}, 0)$ and $(0, \frac{1}{2})$ representations[24] are two-dimensional, with $\vec{J} = \vec{\sigma}/2$ and $\vec{K} = \mp i\vec{\sigma}/2$, respectively,

$$\Lambda_{(\frac{1}{2},0)} = e^{i\vec{\omega}\cdot\frac{\vec{\sigma}}{2} + \vec{\zeta}\cdot\frac{\vec{\sigma}}{2}}, \qquad \Lambda_{(0,\frac{1}{2})} = e^{i\vec{\omega}\cdot\frac{\vec{\sigma}}{2} - \vec{\zeta}\cdot\frac{\vec{\sigma}}{2}} = \Lambda^{-1\dagger}_{(\frac{1}{2},0)}. \tag{10.29}$$

They describe the transformations of L- and R-chiral Weyl spinors Ψ_L and Ψ_R, respectively, as is shown in Problem 10.2. The conjugate representations

$$\Lambda_{(\frac{1}{2}^*,0)} = e^{-i\vec{\omega}\cdot\frac{\vec{\sigma}^*}{2} - \vec{\zeta}\cdot\frac{\vec{\sigma}^*}{2}} = \Lambda^{-1T}_{(\frac{1}{2},0)}, \quad \Lambda_{(0,\frac{1}{2}^*)} = e^{-i\vec{\omega}\cdot\frac{\vec{\sigma}^*}{2} + \vec{\zeta}\cdot\frac{\vec{\sigma}^*}{2}} = \Lambda^*_{(\frac{1}{2},0)} = \Lambda^{\dagger T}_{(\frac{1}{2},0)}, \tag{10.30}$$

are equivalent but useful for constructing Lorentz invariants.

---

[23] These are passive transformations, in which x' represents the coordinates of an event in a transformed coordinate system. For an active transformation, in which an event is rotated and boosted in a fixed reference system, the signs of $\vec{\omega}$ and $\vec{\zeta}$ must be reversed.

[24] These representations define the classical group $SL(2, C)$ of complex 2×2 matrices with unit determinant. It is obvious from this form that the Lorentz group is non-compact, i.e., ζ can take any real value, and $\mathrm{Tr}\,(K^iK^j) = -\frac{1}{2}\delta^{ij} \le 0$.

Spinor Notation for Two-Component Fields

The notations for L- and R-chiral fermions that we introduced in Chapter 2 can be confusing when discussing C and CP transformations, Hermitian conjugation, Majorana masses, or independent fields, especially in four-component notation. In supersymmetry one works mainly in terms of L-chiral particle and antiparticle fields. It becomes tedious to express Dirac or Majorana mass terms or Yukawa couplings using the notation developed so far (e.g., the Dirac mass terms in (2.347) on page 63 or the Majorana ones in (9.8) on page 376), so it is worthwhile here to introduce a more streamlined version of the two-component notation. The key points are that Hermitian conjugation reverses the chirality of a field, and that the $i\sigma^2$ that enters charge conjugation can be viewed as a raising/lowering operator.

We have used the notation that ψ_L and ψ_L^c (or Ψ_L and Ψ_L^c) are left-chiral fields, while Hermitian or Dirac conjugated fields, such as $\psi_R^c = \mathcal{C}\bar{\psi}_L^T \equiv \mathcal{C}\overline{(\psi_L)}^T$ (or $\Psi_R^c = i\sigma^2\Psi_L^* \equiv i\sigma^2(\Psi_L)^{\dagger T}$), and the corresponding conjugates of ψ_L^c, are right-chiral. A more compact notation is to write all left-chiral spinors by a symbol such as ξ (without a bar), and right chiral spinors by a symbol with a bar,[25] such as $\bar{\eta}$. (The bar has nothing to do with Dirac adjoint or with antiparticle). One can, if desired, refer to a left-chiral antiparticle spinor with a superscript[26] c. Thus, for example,

$$\Psi_L \to \xi, \qquad \Psi_L^c \to \xi^c, \qquad \Psi_R \to \bar{\eta}, \qquad \Psi_R^c \to \bar{\eta}^c. \tag{10.31}$$

It is often useful to display the components. It is conventional to use a lower undotted index for an L-spinor and an upper dotted one for an R-spinor, e.g.,

$$\Psi_{L\alpha} \to \xi_\alpha, \qquad \Psi_{R\alpha} \to \bar{\eta}^{\dot{\alpha}}, \tag{10.32}$$

where $\alpha = 1, 2$. The dotted index simply indicates that it is an R-spinor, and also serves as a warning since dotted and undotted indices are never contracted. The dot and bar are superfluous since they both indicate R, but both are useful depending on the context. The R-spinor is introduced with an upper index for convenience in the construction of Lorentz invariants and covariants. Under Lorentz transformations, ξ and $\bar{\eta}$ transform as $(\frac{1}{2}, 0)$ and $(0, \frac{1}{2})$, respectively, i.e.,

$$\xi_\alpha \to \mathcal{M}_\alpha{}^\beta \xi_\beta, \qquad \bar{\eta}^{\dot{\alpha}} \to (\mathcal{M}^{-1\dagger})^{\dot{\alpha}}{}_{\dot{\beta}} \bar{\eta}^{\dot{\beta}}, \tag{10.33}$$

where

$$\mathcal{M}_\alpha{}^\beta \equiv \left(\Lambda_{(\frac{1}{2},0)}\right)_{\alpha\beta}, \qquad (\mathcal{M}^{-1\dagger})^{\dot{\alpha}}{}_{\dot{\beta}} \equiv \left(\Lambda^{-1\dagger}_{(\frac{1}{2},0)}\right)_{\dot{\alpha}\dot{\beta}}, \tag{10.34}$$

with $\Lambda_{(\frac{1}{2},0)}$ defined in (10.29).

We next introduce raised and lowered indices, associated with the conjugate representations of the Lorentz group. Let

$$\epsilon^{\alpha\beta} = -\epsilon^{\beta\alpha} = -\epsilon_{\alpha\beta}, \qquad \alpha, \beta = 1, 2, \tag{10.35}$$

be an antisymmetric tensor with $\epsilon^{12} = -\epsilon_{12} = 1$, with a similar definition for $\epsilon^{\dot{\alpha}\dot{\beta}}$. Note that $\epsilon^{\alpha\beta}$ is just the $\alpha\beta$ component of $i\sigma^2$. Then

$$\epsilon^{\alpha\gamma}\epsilon_{\gamma\beta} = -\epsilon^{\alpha\gamma}\epsilon_{\beta\gamma} = \delta^\alpha_\beta. \tag{10.36}$$

[25] Some authors use a $\dagger$ or $*$.

[26] ξ^c and $\bar{\eta}^c$ are not independent of $\bar{\eta}^\dagger$ and $\xi^\dagger$.

We define the L and R spinors ξ^α and $\bar{\eta}_{\dot{\alpha}}$ with raised or lowered indices by

$$\xi^\alpha = \epsilon^{\alpha\beta}\xi_\beta, \qquad \bar{\eta}_{\dot{\alpha}} = \epsilon_{\dot{\alpha}\dot{\beta}}\bar{\eta}^{\dot{\beta}}. \tag{10.37}$$

These transform as $(\frac{1}{2}^*, 0)$ and $(0, \frac{1}{2}^*)$, respectively, i.e.,

$$\xi^\alpha \to \xi^\beta (\mathcal{M}^{-1})_\beta{}^\alpha, \qquad \bar{\eta}_{\dot{\alpha}} \to \bar{\eta}_{\dot{\beta}} (\mathcal{M}^\dagger)^{\dot{\beta}}{}_{\dot{\alpha}}. \tag{10.38}$$

Lorentz invariants can then be formed by contracting two L or two R spinors,

$$\xi_1\xi_2 \equiv \xi_1^\alpha \xi_{2\alpha} = -\xi_{2\alpha}\xi_1^\alpha = -\epsilon_{\alpha\beta}\epsilon^{\alpha\gamma}\xi_2^\beta\xi_{1\gamma} = \xi_2^\beta\xi_{1\beta} = \xi_2\xi_1, \tag{10.39}$$

which commute since the fields themselves are anticommuting fermions. The shorthand $\xi_1\xi_2$ always implies that the first index is upper and the second is lower. Similarly,

$$\bar{\eta}_1\bar{\eta}_2 \equiv \bar{\eta}_{1\dot{\alpha}}\bar{\eta}_2^{\dot{\alpha}} = \bar{\eta}_2\bar{\eta}_1, \tag{10.40}$$

where the convention for the barred spinors is that the first index is lower and the second is upper. Another way of saying this is that ξ_α and $\bar{\eta}^{\dot{\alpha}}$ are interpreted as column vectors, while ξ^α and $\bar{\eta}_{\dot{\alpha}}$ are row vectors. From (10.33) and (10.38) both $\xi_1\xi_2$ and $\bar{\eta}_1\bar{\eta}_2$ are Lorentz scalars.

Next, we recall that Hermitian conjugation converts L spinors into R spinors. Therefore, $(\xi_\alpha)^\dagger$ is a barred spinor with a lower index, and similarly for $(\bar{\eta}^{\dot{\alpha}})^\dagger$,

$$\bar{\xi}_{\dot{\alpha}} \equiv (\xi_\alpha)^\dagger, \qquad \eta^\alpha \equiv (\bar{\eta}^{\dot{\alpha}})^\dagger, \tag{10.41}$$

so that

$$\bar{\xi}\bar{\eta} = (\eta\xi)^\dagger. \tag{10.42}$$

We can use (2.200) on page 41 to express a four-component Dirac field as

$$\psi = \begin{pmatrix} \Psi_L \\ \Psi_R \end{pmatrix} = \begin{pmatrix} \xi \\ \bar{\eta} \end{pmatrix}, \qquad \begin{pmatrix} \Psi_{L\alpha} \\ \Psi_{R\beta} \end{pmatrix} = \begin{pmatrix} \xi_\alpha \\ \bar{\eta}^{\dot{\beta}} \end{pmatrix}. \tag{10.43}$$

The mass terms for ψ can therefore be written as

$$\begin{aligned} \bar{\psi}_L\psi_R &= \Psi_L^\dagger\Psi_R = \bar{\xi}\bar{\eta} = \bar{\xi}_{\dot{\alpha}}\bar{\eta}^{\dot{\alpha}}, \qquad \bar{\psi}_R\psi_L = \Psi_R^\dagger\Psi_L = \eta\xi = \eta^\alpha\xi_\alpha \\ \bar{\psi}\psi &= \bar{\psi}_L\psi_R + \bar{\psi}_R\psi_L = \Psi_L^\dagger\Psi_R + \Psi_R^\dagger\Psi_L = \bar{\xi}\bar{\eta} + \eta\xi, \end{aligned} \tag{10.44}$$

similar to the last form of (2.347) on page 63. For two Dirac fields $\psi_{1,2}$,

$$\bar{\psi}_{1L}\psi_{2R} = \bar{\xi}_1\bar{\eta}_2, \qquad \bar{\psi}_{1R}\psi_{2L} = \eta_1\xi_2. \tag{10.45}$$

We also introduce the vector bilinear forms

$$\bar{\eta}\,\bar{\sigma}^\mu\xi \equiv \bar{\eta}_{\dot{\alpha}}\,\bar{\sigma}^{\mu\dot{\alpha}\beta}\xi_\beta = (\bar{\xi}\bar{\sigma}^\mu\eta)^\dagger, \qquad \xi\sigma^\mu\bar{\eta} \equiv \xi^\alpha\sigma^\mu_{\alpha\dot{\beta}}\,\bar{\eta}^{\dot{\beta}} = (\eta\sigma^\mu\bar{\xi})^\dagger, \tag{10.46}$$

for arbitrary spinors ξ and $\bar{\eta}$, where we interpret the first (second) index of $\bar{\sigma}^\nu$ as a dotted (undotted) upper index, and the first (second) of σ^μ as undotted (dotted) lower. It is straightforward to show

$$\bar{\sigma}^{\mu\dot{\alpha}\beta} = \epsilon^{\dot{\alpha}\dot{\gamma}}\epsilon^{\beta\delta}\sigma^\mu_{\delta\dot{\gamma}}, \qquad \sigma^\mu_{\alpha\dot{\beta}} = \epsilon_{\alpha\gamma}\epsilon_{\dot{\beta}\dot{\delta}}\bar{\sigma}^{\dot{\delta}\gamma}, \tag{10.47}$$

which imply

$$\bar{\eta}\,\bar{\sigma}^\mu \xi = -\xi \sigma^\mu \bar{\eta}. \tag{10.48}$$

The bilinear forms in (10.46) transform as Lorentz four-vectors. For example,

$$\bar{\eta}\,\bar{\sigma}^\mu \xi \to \bar{\eta}\,\mathcal{M}^\dagger \bar{\sigma}^\mu \mathcal{M} \xi. \tag{10.49}$$

One can show (Problem 10.3) that

$$\mathcal{M}^\dagger \bar{\sigma}^\mu \mathcal{M} = \Lambda^\mu{}_\nu \bar{\sigma}^\nu, \tag{10.50}$$

establishing the result.

Vector currents for a Dirac field can be written

$$\begin{aligned} \bar{\psi}_{1L}\gamma^\mu \psi_{2L} &= \Psi^\dagger_{1L}\bar{\sigma}^\mu \Psi_{2L} = \bar{\xi}_1 \bar{\sigma}^\mu \xi_2 = -\xi_2 \sigma^\mu \bar{\xi}_1 \\ \bar{\psi}_{1R}\gamma^\mu \psi_{2R} &= \Psi^\dagger_{1R}\sigma^\mu \Psi_{2R} = \eta_1 \sigma^\mu \bar{\eta}_2 = -\bar{\eta}_2 \bar{\sigma}^\mu \eta_1. \end{aligned} \tag{10.51}$$

The QED Lagrangian density in (2.218) on page 43 in spinor notation is

$$\mathcal{L} = \bar{\xi}\bar{\sigma}^\mu \left[i\partial_\mu + eA_\mu \right] \xi + \eta \sigma^\mu \left[i\partial_\mu + eA_\mu \right] \bar{\eta} - m(\bar{\xi}\bar{\eta} + \eta\xi) - \frac{1}{4} F_{\mu\nu}F^{\mu\nu}, \tag{10.52}$$

where the electron field is $\psi = (\xi \;\; \bar{\eta})^T$. The second term in (10.52) can be rewritten as $\bar{\eta}\bar{\sigma}^\mu \left[i\partial_\mu - eA_\mu \right] \eta$, corresponding to the second form of the electromagnetic current in (2.346), i.e., in terms of Ψ_L and Ψ^c_L rather than Ψ_L and Ψ_R. Similarly, the tensor current is

$$\bar{\psi}_{1L}\sigma^{\mu\nu}\psi_{2R} = \Psi^\dagger_{aL}\, \bar{\mathfrak{s}}^{\mu\nu} \Psi_{bR} = 2\bar{\xi}_1\, \bar{\mathfrak{s}}^{\mu\nu} \bar{\eta}_2, \qquad \bar{\psi}_{1R}\sigma^{\mu\nu}\psi_{2L} = 2\eta_1\, \mathfrak{s}^{\mu\nu} \xi_2, \tag{10.53}$$

where $\bar{\mathfrak{s}}^{\mu\nu}$ and $\mathfrak{s}^{\mu\nu}$ are defined in (2.338) on page 61. Other identities can be found in, e.g., (Chung et al., 2005; Dreiner et al., 2010).

The expressions for the C, P, and T transformations can easily be translated into the spinor language

$$\begin{aligned} &\psi = \begin{pmatrix} \xi \\ \bar{\eta} \end{pmatrix} \underset{C}{\rightarrow} \psi^c = \begin{pmatrix} \eta \\ \bar{\xi} \end{pmatrix}, \qquad \psi \underset{P}{\rightarrow} \begin{pmatrix} \bar{\eta} \\ \xi \end{pmatrix}, \qquad \psi^c \underset{P}{\rightarrow} - \begin{pmatrix} \bar{\xi} \\ \eta \end{pmatrix} \\ &\psi \underset{CP}{\longrightarrow} \begin{pmatrix} \bar{\xi} \\ \eta \end{pmatrix}, \qquad \psi = \begin{pmatrix} \xi_\alpha \\ \bar{\eta}^{\dot{\beta}} \end{pmatrix} \underset{T}{\rightarrow} i \begin{pmatrix} \xi^\alpha \\ -\bar{\eta}_{\dot{\beta}} \end{pmatrix}, \qquad \psi \underset{CPT}{\longrightarrow} i \begin{pmatrix} -\bar{\xi}_{\dot{\alpha}} \\ \eta^\beta \end{pmatrix}, \end{aligned} \tag{10.54}$$

up to possible intrinsic phases.[27] It is understood that the appropriate transformations are made on the space-time variable x. We see from (10.54) that charge conjugation interchanges ξ and η, i.e., $\xi^c = \eta$ and $\eta^c = \xi$. As simple examples, the space reflection transformations expressed in 4 and 2 component language include

$$\begin{aligned} P\bar{\psi}_{1L}\psi_{2R}P^{-1} &= \bar{\psi}_{1R}\psi_{2L} &&\Longleftrightarrow \quad P\bar{\xi}_1\bar{\eta}_2P^{-1} = \eta_1\xi_2 \\ P\bar{\psi}_{1L}\gamma^\mu\psi_{2L}P^{-1} &= \bar{\psi}_{1R}\gamma_\mu\psi_{2R} &&\Longleftrightarrow \quad P\bar{\xi}_1\bar{\sigma}^\mu\xi_2P^{-1} = \eta_1\sigma_\mu\bar{\eta}_2, \end{aligned} \tag{10.55}$$

while charge conjugation leads to

$$\begin{aligned} C\bar{\psi}_{1L}\psi_{2R}C^{-1} &= \bar{\psi}^c_{1L}\psi^c_{2R} = \bar{\psi}_{2L}\psi_{1R} \Longleftrightarrow C\bar{\xi}_1\bar{\eta}_2C^{-1} = \bar{\eta}_1\bar{\xi}_2 = \bar{\xi}_2\bar{\eta}_1 \\ C\bar{\psi}_{1L}\gamma^\mu\psi_{2L}C^{-1} &= \bar{\psi}^c_{1L}\gamma^\mu\psi^c_{2L} = -\bar{\psi}_{2R}\gamma^\mu\psi_{1R} \\ &\Longleftrightarrow C\bar{\xi}_1\bar{\sigma}^\mu\xi_2C^{-1} = \bar{\eta}_1\bar{\sigma}^\mu\eta_2 = -\eta_2\sigma^\mu\bar{\eta}_1. \end{aligned} \tag{10.56}$$

---

[27]The transformations P, CP, and T raise or lower indices, implying an extra minus sign when acting on a spinor with the index in the "wrong" location. For example, $P\xi_\alpha P^{-1} = \bar{\eta}^{\dot{\alpha}}$, while $P\xi^\alpha P^{-1} = -\bar{\eta}_{\dot{\alpha}}$. There is also an extra sign in $P\psi^c P^{-1}$ because of the opposite intrinsic parity for an antifermion, as discussed below (2.332).

The Fierz identities in (2.340) and (2.341) become

$$\begin{aligned}(\bar{\xi}_1\bar{\sigma}^\mu\xi_2)\,(\bar{\xi}_3\bar{\sigma}_\mu\xi_4) &= (\bar{\xi}_1\bar{\sigma}^\mu\xi_4)\,(\bar{\xi}_3\bar{\sigma}_\mu\xi_2) = 2(\xi_2\xi_4)\,(\bar{\xi}_3\bar{\xi}_1)\\ (\eta_1\sigma^\mu\bar{\eta}_2)\,(\bar{\xi}_3\bar{\sigma}_\mu\xi_4) &= -2(\eta_1\xi_4)\,(\bar{\xi}_3\bar{\eta}_2).\end{aligned} \tag{10.57}$$

Finally, a Majorana fermion, such as the triplet Majorana neutrino introduced in (9.6) (page 375) and (9.8), can be written as

$$\psi_M = \begin{pmatrix} \Psi_L \\ \Psi_R^c \end{pmatrix} = \begin{pmatrix} \xi \\ \bar{\xi} \end{pmatrix}, \qquad \begin{pmatrix} \Psi_{L\alpha} \\ \Psi_{R\beta}^c \end{pmatrix} = \begin{pmatrix} \xi_\alpha \\ \bar{\xi}^{\dot{\beta}} \end{pmatrix}, \tag{10.58}$$

so the free field Lagrangian density in (9.7) and (9.8) becomes

$$\mathcal{L} = \bar{\xi}i\bar{\sigma}^\mu\partial_\mu\xi - \frac{m_T}{2}\left[\bar{\xi}\bar{\xi} + \xi\xi\right], \tag{10.59}$$

where $\nu_M \equiv (\xi\ \bar{\xi})^T$.

The Supersymmetry Algebra and Representations

Supersymmetry transformations connect fermions and bosons, and the associated generators Q must therefore be fermionic,[28] i.e., anticommuting. In analogy with the Weyl fields ζ and $\bar{\eta}$, we will introduce fermionic charges $Q_\alpha, \alpha = 1, 2$, and their conjugates $\bar{Q}_{\dot{\alpha}} \equiv (Q_\alpha)^\dagger$. The $N = 1$ supersymmetry algebra[29] is

$$\begin{aligned}&\{Q_\alpha, \bar{Q}_{\dot{\beta}}\} = 2\sigma^\mu_{\alpha\dot{\beta}}\,P_\mu, \qquad [Q_\alpha, P_\mu] = [\bar{Q}_\alpha, P_\mu] = 0\\ &\{Q_\alpha, Q_\beta\} = \{\bar{Q}_{\dot{\alpha}}, \bar{Q}_{\dot{\beta}}\} = 0\\ &[Q_\alpha, M_{\mu\nu}] = (\mathfrak{s}_{\mu\nu})_\alpha{}^\beta Q_\beta, \qquad [\bar{Q}^{\dot{\alpha}}, M_{\mu\nu}] = (\bar{\mathfrak{s}}_{\mu\nu})^{\dot{\alpha}}{}_{\dot{\beta}}\bar{Q}^{\dot{\beta}}.\end{aligned} \tag{10.60}$$

The non-vanishing commutators of the supersymmetry charges with $M_{\mu\nu}$ are an indication that supersymmetry connects states of different spin and statistics. Because of the antisymmetry, the product of three of more Q's must vanish

$$Q_\alpha Q_\beta Q_\gamma = 0, \tag{10.61}$$

and similarly for the $\bar{Q}$. The indices can be raised and lowered using $\epsilon^{\alpha\beta}$, just as for the Weyl spinors, e.g., $Q^\alpha = \epsilon^{\alpha\beta}Q_\beta$. The first relation in (10.60) implies

$$\langle 0|H|0\rangle = \frac{1}{4}\langle 0|Q_1(Q_1)^\dagger + (Q_1)^\dagger Q_1 + Q_2(Q_2)^\dagger + (Q_2)^\dagger Q_2|0\rangle \geq 0, \tag{10.62}$$

that is, the ground state energy of a supersymmetric field theory must have nonnegative energy. Furthermore, if the supersymmetry is not spontaneously broken, i.e, for $Q_\alpha|0\rangle = (Q_\alpha)^\dagger|0\rangle = 0$, the vacuum energy must vanish, $\langle 0|H|0\rangle = 0$. In the case of a scalar potential $V(\phi)$, for example, the condition for the spontaneous breaking of the supersymmetry is that $V(\nu) > 0$ at the minimum. This is to be contrasted with internal symmetries, for which the relevant issue is whether $\nu \neq 0$ at the minimum, as illustrated in Figure 10.4. The requirement $\langle 0|H|0\rangle \geq 0$ can be violated in supergravity (gauged supersymmetry) or in the presence of explicit supersymmety breaking terms.

[28] The Coleman-Mandula theorem (Coleman and Mandula, 1967) states that under reasonable assumptions the space-time symmetry of a field theory cannot be extended beyond the Poincaré algebra except for internal symmetry generators if the generators obey commutation rules. However, it *can* be extended by the inclusion of fermionic generators (Haag et al., 1975).

[29] The algebra can be extended to N fermionic charges $Q^i_\alpha, i = 1\cdots N$. The cases $N = 2, 4$, and 8 are of considerable interest for theoretical discussions, but do not appear to be directly applicable as extensions of the SM.

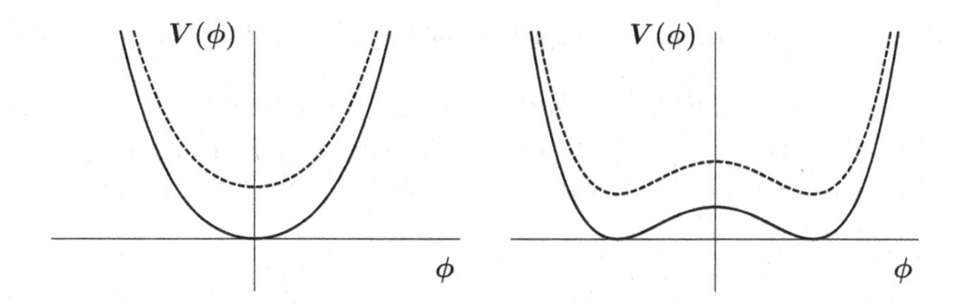

Figure 10.4 Left: potentials which preserve internal symmetries. Right: potentials which break internal symmetries. In both cases, supersymmetry is preserved for $V|_{min} = 0$ (solid) and spontaneously broken for $V|_{min} > 0$ (dashed).

The Q_α and $\bar{Q}_{\dot{\alpha}}$ acting on a single particle state either annihilate it or create a state with spin and helicity differing by $\frac{1}{2}$ unit. The most important massless irreducible multiplets for our purposes are: (a) the chiral supermultiplets, which consist of a Weyl fermion and a complex scalar.[30] Examples are the chiral quark or lepton fields and their partners, or the Higgs scalar and its partner. Chiral supermultiplets can be paired to form Dirac fermions. (b) The vector supermultiplets, consisting of a massless vector, e.g., with helicity $+1$, and a Weyl fermion partner. These are generally elements of the adjoint representation of a gauge group. A vector supermultiplet can be combined with its CP conjugate to form a massless vector with helicities ± 1 and a (massless) Majorana fermion.[31] (c) The gravity supermultiplet, consisting of the spin-2 graviton with helicity 2 and its spin-3/2 gravitino partner. This can again be combined with its CP conjugate involving a graviton with helicity $= -2$.

The massless chiral and vector supermultiplets are all that are needed for the MSSM. There are also massive representations, that are occasionally needed, e.g., for extensions of the SM involving an extended gauge symmetry broken at a scale large compared to supersymmetry breaking. We will illustrate the most important ones in Section 10.2.3. Before constructing the transformations of the supermultiplets and the rules for invariant Lagrangian densities, it is convenient to introduce Grassmann variables and superspace.[32]

Grassmann Variables and Superspace

Supersymmetry transformations mix particles with different spins, and therefore can be considered an extension of the rotation and Poincaré groups. It is convenient to extend the notation of four-dimensional spacetime with coordinates x^μ into a larger *superspace*, involving two additional anticommuting (*Grassmann*) coordinates θ_α, with $\alpha = 1, 2$, and their conjugates $\bar{\theta}_{\dot{\alpha}} \equiv (\theta_\alpha)^\dagger$. (Such Grassmann variables are similar to the ones introduced in field theory for dealing with fermions using functional methods.) The indices can be raised, lowered, or contracted using $\epsilon_{\alpha\beta}$, just as for chiral fields ξ and χ, or the supersymmetry generators Q and $\bar{Q}$. Because of the anticommuting nature, the product of three or more θ's vanishes, $\theta_\alpha\theta_\beta\theta_\gamma = 0$, and $\theta_\alpha\theta_\alpha = 0$ (no sum). This will enormously simplify the task of

[30]The supermultiplets also involve unphysical (non-propagating) auxiliary fields.
[31]It is convenient in this context to combine a Weyl spinor and its conjugate as in (10.58), even if the Majorana mass is zero.
[32]Our notations and conventions follow most closely those in (Wess and Bagger, 1992; Drees et al., 2004).

constructing general functions of superspace variables by expanding around $\theta = \bar{\theta} = 0$. Of course,

$$\theta^2 \equiv \theta^\alpha \theta_\alpha = -2\theta_1\theta_2, \qquad \bar{\theta}^2 \equiv \bar{\theta}_{\dot{\alpha}}\bar{\theta}^{\dot{\alpha}} = 2\bar{\theta}^{\dot{1}}\bar{\theta}^{\dot{2}}. \tag{10.63}$$

One can introduce the concepts of differentiation and integration in superspace, as are described in the more detailed books and articles. We will only utilize integration, defined by

$$\int d\theta_\alpha = 0, \qquad \int d\theta_\alpha \theta_\beta = \delta_{\alpha\beta}, \qquad \int d\theta_\alpha \bar{\theta}_{\dot{\beta}} = 0, \tag{10.64}$$

and similarly for $d\bar{\theta}$. Thus, for example

$$\int d\theta_\alpha \theta^2 = \epsilon^{\beta\gamma} \int d\theta_\alpha \theta_\gamma \theta_\beta = \epsilon^{\beta\alpha}\theta_\beta - \epsilon^{\alpha\gamma}\theta_\gamma = -2\theta^\alpha. \tag{10.65}$$

It is useful to define

$$d^2\theta \equiv -\frac{1}{4} d\theta^\alpha d\theta_\alpha, \qquad d^2\bar{\theta} \equiv -\frac{1}{4} d\bar{\theta}_{\dot{\alpha}} d\bar{\theta}^{\dot{\alpha}}, \qquad d^4\theta \equiv d^2\bar{\theta} d^2\theta, \tag{10.66}$$

from which one finds

$$\int d^2\theta \theta^2 = \int d^2\bar{\theta}\bar{\theta}^2 = 1, \qquad \int d^4\theta \theta^2 \bar{\theta}^2 = 1, \tag{10.67}$$

with all other integrals vanishing. These forms will be useful as projection operators onto the supersymmetric parts of operators.

The Grassmann variables can also be contracted with and anticommute with spinor fields and supersymmetry generators, and can be formed into vectors. Two useful identities are

$$\theta\xi\,\theta\chi \equiv \theta^\alpha \xi_\alpha\, \theta^\beta \chi_\beta = -\frac{1}{2}\theta^2\, \xi\chi, \qquad \theta\sigma^\mu\bar{\theta}\,\theta\sigma^\nu\bar{\theta} = +\frac{1}{2} g^{\mu\nu}\theta^2\bar{\theta}^2, \tag{10.68}$$

which can be derived by writing out the components and by using the Fierz identity in (10.57).

A Simple Example

Before we consider the general supersymmetry transformation rules, let us give a simple example. Consider the free field theory for a massless complex scalar ϕ, a massless left-chiral Weyl spinor ξ, and an auxiliary complex field F, with Lagrangian density

$$\mathcal{L} = (\partial_\mu \phi)^\dagger \partial^\mu \phi + i\bar{\xi}\bar{\sigma}^\mu \partial_\mu \xi + F^\dagger F. \tag{10.69}$$

F does not represent a physical particle (the Euler-Lagrange equation in this case is just $F = 0$), but it is needed to ensure a supersymmetric action. The supersymmetry transformations of the fields into each other turn out to be

$$\begin{aligned} \phi &\to \phi + \delta\phi, & \delta\phi &= \sqrt{2}\epsilon\xi \\ \xi &\to \xi + \delta\xi, & \delta\xi &= \sqrt{2}\epsilon F - i\sqrt{2}\sigma^\mu \bar{\epsilon}\,\partial_\mu\phi \\ F &\to F + \delta F, & \delta F &= -i\sqrt{2}\bar{\epsilon}\,\bar{\sigma}^\mu \partial_\mu \xi, \end{aligned} \tag{10.70}$$

where ϵ and $\bar{\epsilon}$ are Grassmann parameters that define the transformation, analogous to a^μ and $\omega^{\rho\sigma}$ for Poincaré transformations, or β^i for the internal global symmetries discussed in Chapter 3. It is straightforward to prove that $\mathcal{L}$ is invariant under this transformation (Problem 10.6). It can also be shown that these transformations indeed form a representation of the supersymmetry algebra in (10.72). We note that the shift in F is a total derivative (since $\bar{\epsilon}$ is a constant). Therefore $\int d^4x F(x)$ is invariant under the transformation. This will be very useful in more realistic examples.

Superfields

A *superfield* is an operator $\Phi(x, \theta, \bar{\theta})$ that is a function of the superspace coordinates x, θ, and $\bar{\theta}$. One can expand Φ in a power series in θ and $\bar{\theta}$, but no terms involving more than two factors of θ or $\bar{\theta}$ survive because of the antisymmetry. Consequently, the most general Lorentz invariant superfield is

$$\begin{aligned}\Phi(x, \theta, \bar{\theta}) &= \phi(x) + \sqrt{2}\theta\xi(x) + \sqrt{2}\bar{\theta}\bar{\chi}(x) + \theta\theta F(x) + \bar{\theta}\bar{\theta}G(x) \\ &\quad + \theta\sigma^\mu\bar{\theta}A_\mu(x) + \theta\theta\bar{\theta}\bar{\lambda}(x) + \bar{\theta}\bar{\theta}\theta\kappa(x) + \frac{1}{2}\theta\theta\bar{\theta}\bar{\theta}D(x),\end{aligned} \tag{10.71}$$

where ϕ, F, G, and D are Lorentz-scalar functions of x; ξ, $\bar{\chi}$, κ, and $\bar{\lambda}$ are Weyl spinors; and A_μ is a vector.

We are now ready to interpret the supersymmetry transformations as translations in superspace. The algebra in (10.60) can be rewritten in terms of commutators involving $\theta Q \equiv \theta^\alpha Q_\alpha$ and $\bar{\theta}\bar{Q} \equiv \bar{\theta}_{\dot{\alpha}}\bar{Q}^{\dot{\alpha}}$,

$$\begin{aligned}&\left[\theta Q, \bar{\theta}\bar{Q}\right] = 2\theta\sigma^\mu\bar{\theta}\, P_\mu, \qquad \left[\theta Q, P_\mu\right] = \left[\bar{\theta}\bar{Q}, P_\mu\right] = 0 \\ &\left[\theta Q, \theta Q\right] = \left[\bar{\theta}\bar{Q}, \bar{\theta}\bar{Q}\right] = 0 \\ &\left[\theta Q, M_{\mu\nu}\right] = \theta\mathfrak{s}_{\mu\nu}Q, \qquad \left[\bar{\theta}\bar{Q}, M_{\mu\nu}\right] = \bar{\theta}\bar{\mathfrak{s}}_{\mu\nu}\bar{Q}.\end{aligned} \tag{10.72}$$

Just as we defined the translation operator $\exp(+ia \cdot P)$ in ordinary space in (10.23), we can define a supersymmetry translation operator as

$$S(a, \epsilon, \bar{\epsilon}) \equiv e^{i(\epsilon Q + \bar{\epsilon}\bar{Q} + a\cdot P)}, \tag{10.73}$$

where ϵ and $\bar{\epsilon}$ are Grassmann variables like θ and $\bar{\theta}$. The multiplication of two group elements can be calculated exactly using the Baker-Campbell-Hausdorff construction in (3.6) because all of the commutators after the first vanish, with the result

$$S(a^\mu, \epsilon, \bar{\epsilon})S(b^\mu, \delta, \bar{\delta}) = S(a^\mu + b^\mu + i\epsilon\sigma^\mu\bar{\delta} - i\delta\sigma^\mu\bar{\epsilon}, \epsilon + \delta, \bar{\epsilon} + \bar{\delta}). \tag{10.74}$$

The transformation of a superfield $\Phi(x, \theta, \bar{\theta})$ can then be found in analogy to (10.28),

$$\Phi \to \Phi' = S\Phi S^{-1} \sim \Phi + i\left[\epsilon Q + \bar{\epsilon}\bar{Q} + a \cdot P, \Phi\right], \tag{10.75}$$

where the last form assumes infinitesimal translation parameters. The action of the generators can be found by casting them in the form of covariant derivatives with respect to the Grassmann variables. The derivation is carried out in the standard references, but here we will just give the result,

$$\Phi(x^\mu, \theta, \bar{\theta}) \to \Phi'(x^\mu, \theta, \bar{\theta}) = \Phi(x^\mu + a^\mu + i\epsilon\sigma^\mu\bar{\theta} - i\theta\sigma^\mu\bar{\epsilon}, \theta + \epsilon, \bar{\theta} + \bar{\epsilon}). \tag{10.76}$$

A general superfield such as that in (10.71) is highly reducible, but special cases are irreducible and close under the transformations. We are most interested in *left-chiral superfields* $\Phi_L(x, \theta, \bar{\theta})$, which only involve L-chiral spinors; right-chiral superfields ϕ_R (such as $(\Phi_L)^\dagger$); and *vector superfields*, which satisfy $V = V^\dagger$ (such as $\Phi_L^\dagger\Phi_L$ or $\Phi_L + \Phi_L^\dagger$). A left-chiral superfield and its associated *left-chiral supermultiplet* involve two complex scalars, $\phi(x)$ and an auxiliary field $F(x)$, and an L-spinor $\xi(x)$, with specific relations between the components. It can be written

$$\begin{aligned}\Phi_L(x, \theta, \bar{\theta}) &= \phi(x) + \sqrt{2}\theta\xi(x) + \theta\theta F(x) - i\theta\sigma^\mu\bar{\theta}\partial_\mu\phi(x) \\ &\quad - \frac{1}{4}\theta\theta\bar{\theta}\bar{\theta}\partial^\mu\partial_\mu\phi(x) + \frac{i}{\sqrt{2}}\theta\theta\left(\partial_\mu\xi(x)\right)\sigma^\mu\bar{\theta}.\end{aligned} \tag{10.77}$$

The supersymmetry transformation of Φ_L can be obtained from (10.76) (with $a = 0$), noting that, e.g.,

$$\phi(x^\mu + i\epsilon\sigma^\mu\bar{\theta} - i\theta\sigma^\mu\bar{\epsilon}) \sim \phi(x) + i(\epsilon\sigma^\mu\bar{\theta} - \theta\sigma^\mu\bar{\epsilon})\partial_\mu\phi(x), \tag{10.78}$$

to obtain

$$\begin{aligned}
\delta\phi &= \underbrace{\sqrt{2}\epsilon\xi}_{\theta\xi(x)} \\
\delta\xi &= \underbrace{\sqrt{2}\epsilon F}_{\theta\theta F(x)} \underbrace{- \frac{i}{\sqrt{2}}\sigma^\mu\bar{\epsilon}\,\partial_\mu\phi}_{\phi(x)} \underbrace{- \frac{i}{\sqrt{2}}\sigma^\mu\bar{\epsilon}\,\partial_\mu\phi}_{\theta\sigma^\mu\bar{\theta}\partial_\mu\phi(x)} = \sqrt{2}\epsilon F - i\sqrt{2}\sigma^\mu\bar{\epsilon}\,\partial_\mu\phi \\
\delta F &= \underbrace{- \frac{i}{\sqrt{2}}\bar{\epsilon}\bar{\sigma}^\mu\partial_\mu\xi}_{\theta\xi(x)} + \underbrace{\frac{i}{\sqrt{2}}(\partial_\mu\xi)\sigma^\mu\bar{\epsilon}}_{\theta\theta(\partial_\mu\xi(x))\sigma^\mu\bar{\theta}} = -i\sqrt{2}\bar{\epsilon}\bar{\sigma}^\mu\partial_\mu\xi,
\end{aligned} \tag{10.79}$$

where the origin of each term is indicated and we have used (10.68). This result reproduces the transformations in (10.70). Φ_L can be written in a more compact form by introducing the variable

$$y^\mu \equiv x^\mu - i\theta\sigma^\mu\bar{\theta}, \tag{10.80}$$

to obtain

$$\Phi_L(x,\theta,\bar{\theta}) = \Phi_L(y,\theta) = \phi(y) + \sqrt{2}\theta\xi(y) + \theta\theta F(y), \tag{10.81}$$

which depends explicitly only on θ. Equation (10.81) is easily verified by expanding the fields and using (10.68).

From now on we will drop the subscript L on a left-chiral superfield. Some important aspects to (re)emphasize are: (a) a chiral superfield transforms as an IRREP of the supersymmetry algebra. (b) The (auxiliary) F-component of a chiral superfield transforms as a total derivative, so that $\int d^4x F(x)$ is an invariant. (c) Superfields commute, $\Phi_1\Phi_2 = \Phi_2\Phi_1$, because each term has an even number of Grassmann variables and spinors. (d) The product $\Phi_1\Phi_2$ or sum $\Phi_1 + \Phi_2$ of two chiral superfields is also a chiral superfield. This can be seen for the product by multiplying out the expressions for $\Phi_{1,2}$,

$$\Phi_1\Phi_2 = \phi_1\phi_2 + \sqrt{2}\theta\,(\xi_1\phi_2 + \xi_2\phi_1) + \theta\theta\,(\phi_1 F_2 + \phi_2 F_1 - \xi_1\xi_2)\,, \tag{10.82}$$

where all of the fields are functions of y. This clearly has the form of a left-chiral superfield. The same is obviously true for the sum. We also record the product of three left chiral superfields for later use:

$$\begin{aligned}
\Phi_1\Phi_2\Phi_3 = {}& \phi_1\phi_2\phi_3 + \sqrt{2}\theta\,(\xi_1\phi_2\phi_3 + \xi_2\phi_3\phi_1 + \xi_3\phi_1\phi_2) \\
& + \theta\theta\,(\phi_1\phi_2F_3 + \phi_2\phi_3F_1 + \phi_3\phi_1F_2 - \xi_1\xi_2\phi_3 - \xi_2\xi_3\phi_1 - \xi_3\xi_1\phi_2)\,.
\end{aligned} \tag{10.83}$$

The *superpotential*[33] $W(\Phi_a)$ is a *holomorphic* function of left-chiral superfields Φ_a, i.e., it depends only on the Φ_a and not on their adjoints or on other right-chiral or vector superfields. We will mainly be concerned with superpotentials that are third-order polynomials

[33] An important consequence of supersymmetry is that the ultraviolet divergences are milder than in non-supersymmetric theories due to cancellations. One aspect of this is the non-renormalization theorems (Grisaru et al., 1979; Seiberg, 1993), which include the statement that the superpotential is not renormalized to any order in perturbation theory. This restricts the form of renormalization group equations for the coefficients, since couplings are only renormalized by wave function effects. It also means that if a term is absent from the superpotential or very small, it will not be generated by renormalization.

of the superfields, but in principle W could be any holomorphic function expressible as a power series. It follows from the above that W is itself a left-chiral superfield, and that its F component is a supersymmetric invariant of the action. The F component, denoted $[W]_F$, is given by

$$[W]_F = \int d^2\theta W(\Phi_a) = \sum_a F_a \frac{\partial W}{\partial \Phi_a}\bigg|_{\theta=\bar{\theta}=0} - \frac{1}{2}\sum_{a,b} \xi_a \xi_b \frac{\partial^2 W}{\partial \Phi_a \partial \Phi_b}\bigg|_{\theta=\bar{\theta}=0}, \tag{10.84}$$

where we have used (10.67). The second form is most easily seen from the examples in (10.82) and (10.83).

A vector superfield $V = V^\dagger$ is another important special case of (10.71), with the restrictions

$$\phi = \phi^\dagger, \quad \xi = \chi, \quad F = G^\dagger, \quad A_\mu = A_\mu^\dagger, \quad \lambda = \kappa, \quad D = D^\dagger. \tag{10.85}$$

One can use (10.76) to show that the change in the (auxiliary) D component of a vector superfield under a supersymmetry transformation is a total derivative, and therefore $D(x)$ in an invariant when integrated over x. From (10.67) and (10.71),

$$[V]_D \equiv \int d^4\theta \, V = \frac{1}{2} D(x). \tag{10.86}$$

The special case $V(x,\theta,\bar{\theta}) = \sum_a [\Phi_a(y,\theta)]^\dagger \, \Phi_a(y,\theta)$ leads (Problem 10.8) to

$$\int d^4\theta \sum_a [\Phi_a(y,\theta)]^\dagger \, \Phi_a(y,\theta) = \sum_a [(\partial_\mu \phi_a)^\dagger \partial^\mu \phi_a + i\bar{\xi}_a \bar{\sigma}^\mu \partial_\mu \xi_a + F_a^\dagger F_a], \tag{10.87}$$

which we recognize as the kinetic energy terms for ϕ_a and ξ_a and the auxiliary field term from (10.69). Vector superfields are also used to describe gauge bosons and their partners, as will be discussed below.

The supersymmetry transformations associated with $S(a,\epsilon,\bar{\epsilon})$ in (10.73) are global, i.e., ϵ, $\bar{\epsilon}$, and a are constants. They can be promoted to *local* (gauge) supersymmetry by allowing the parameters to be functions of spacetime, just as in an ordinary gauge symmetry. Local supersymmetry necessarily implies the existence of gravity, and is therefore known as *supergravity*. Supergravity is nonrenormalizable, and would presumably be an effective theory below some cutoff scale, e.g., the Planck scale, where it could emerge from an underlying superstring theory.

10.2.3 Supersymmetric Interactions

Yukawa and Scalar Interactions

The Lagrangian density for a supersymmetric theory of scalars and fermions (but no gauge interactions) can be written

$$\mathcal{L} = [K]_D + ([W]_F + h.c.) = \int d^4\theta \, K(\Phi,\Phi^\dagger) + \left(\int d^2\,\theta W(\Phi) + h.c.\right), \tag{10.88}$$

where $W(\Phi)$ is holomorphic and the *Kähler potential* $K(\Phi,\Phi^\dagger)$ is Hermitian. We will focus on the simplest case[34]

$$K(\Phi,\Phi^\dagger) = \sum_a [\Phi_a(y,\theta)]^\dagger \, \Phi_a(y,\theta), \tag{10.89}$$

[34]More general forms occur in supergravity and in models of supersymmetry breaking. Also, some authors reserve the term Kähler potential for related functions relevant to supergravity.

which yields the canonical kinetic energy terms in (10.87). The expression for $[W]_F$ is given in (10.84). The auxiliary fields F_a can be eliminated by using the Euler-Lagrange equations of motion,

$$F_a^\dagger = -\left.\frac{\partial W}{\partial \Phi_a}\right|_{\theta=\bar{\theta}=0} \equiv -W_a(\phi), \qquad F_a = -\left.\frac{\partial W}{\partial \Phi_a}\right|^\dagger_{\theta=\bar{\theta}=0} = -\left[W_a(\phi)\right]^\dagger. \tag{10.90}$$

The terms involving F_a therefore lead to the scalar potential $V(\phi)$, with

$$V(\phi) = -\sum_a \left(F_a^\dagger F_a + F_a W_a + F_a^\dagger W_a^\dagger\right) = \sum_a |F_a|^2 = \sum_a |W_a(\phi)|^2. \tag{10.91}$$

As expected, $V(\phi)$ is non-negative, and the condition for supersymmetry to be unbroken is for $W_a(\phi) = 0$ at the potential minimum.[35] We also define

$$W_{ab}(\phi) \equiv \left.\frac{\partial^2 W}{\partial \Phi_a \partial \Phi_b}\right|_{\theta=\bar{\theta}=0}, \tag{10.92}$$

so that

$$\mathcal{L} = \mathcal{L}_{KE} + \mathcal{L}_f - V(\phi), \tag{10.93}$$

where

$$\begin{aligned} \mathcal{L}_{KE} &= \sum_a \left[(\partial_\mu \phi_a)^\dagger \partial^\mu \phi_a + i\bar{\xi}_a \bar{\sigma}^\mu \partial_\mu \xi_a\right] \\ \mathcal{L}_f &= -\frac{1}{2}\sum_{a,b}\left(\xi_a \xi_b W_{ab}(\phi) + \bar{\xi}_a \bar{\xi}_b \left[W_{ab}(\phi)\right]^\dagger\right) \end{aligned} \tag{10.94}$$

are, respectively, the kinetic energy and fermion mass/interaction terms.

A Majorana Fermion

As a first example, consider the *Wess-Zumino model* (Wess and Zumino, 1974a), which contains a single left-chiral superfield Φ. We choose

$$W(\Phi) = a\Phi + \frac{m}{2}\Phi^2 + \frac{h}{3}\Phi^3, \tag{10.95}$$

with a, m, and h real. This implies

$$\mathcal{L} = (\partial_\mu \phi)^\dagger \partial^\mu \phi + i\bar{\xi}\bar{\sigma}^\mu \partial_\mu \xi - \frac{m}{2}(\xi\xi + \bar{\xi}\bar{\xi}) - h(\xi\xi\phi + \bar{\xi}\bar{\xi}\phi^\dagger) - V(\phi). \tag{10.96}$$

The potential is

$$V(\phi) = \left|a + m\phi + h\phi^2\right|^2, \tag{10.97}$$

which has supersymmetry preserving minima at the zeros of $a + m\phi + h\phi^2$. Let us take $a = 0$ and consider the minimum with $\langle\phi\rangle = 0$, so that

$$\begin{aligned} \mathcal{L} =& \mathcal{L}_{KE} - \frac{m}{2}(\xi\xi + \bar{\xi}\bar{\xi}) - h(\xi\xi\phi + \bar{\xi}\bar{\xi}\phi^\dagger) \\ &- m^2|\phi|^2 - hm|\phi|^2(\phi + \phi^\dagger) - h^2|\phi|^4. \end{aligned} \tag{10.98}$$

[35] This is generalized to $V = e^{(\phi_a^\dagger \phi_a/M_P^2)}\left[|\overline{W}_a|^2 - 3|W|^2/M_P^2\right]$ in supergravity with a minimal Kähler potential, where $\overline{W}_a \equiv W_a + \phi_a^\dagger W/M_P^2$ and M_P is the Planck mass. The condition for supersymmetry to remain unbroken turns out to be $\overline{W}_a = 0$ at the minimum, which allows $V < 0$.

We recognize the $\xi\xi + \bar{\xi}\bar{\xi}$ term as a Majorana mass term, as in (10.59), so the spectrum consists of a Majorana fermion $\psi_M = \begin{pmatrix} \xi \\ \bar{\xi} \end{pmatrix} = \psi_L + \psi_R^c$ and a complex scalar ϕ, both with mass m. The Majorana fermion mass and kinetic energy terms in four-component notation are

$$\bar{\psi}_L i \not\partial \psi_L - \frac{m}{2}\left(\bar{\psi}_L \psi_R^c + \bar{\psi}_R^c \psi_L\right) = \frac{1}{2}\bar{\psi}_M i \not\partial \psi_M - \frac{m}{2}\bar{\psi}_M \psi_M, \tag{10.99}$$

as in (9.7) and (9.9). There are Yukawa interactions

$$\mathcal{L}_{Yuk} = -h\left(\bar{\psi}_R^c \psi_L \phi + \bar{\psi}_L \psi_R^c \phi^\dagger\right) = -h\left(\bar{\psi}_M P_L \psi_M \phi + \bar{\psi}_M P_R \psi_M \phi^\dagger\right), \tag{10.100}$$

as well as cubic and quartic interaction terms for ϕ, with coefficients related by supersymmetry to the Yukawa coupling. It is sometimes useful to write ϕ in terms of Hermitian components, $\phi = (S + iP)/\sqrt{2}$, where S and P indicate scalar and pseudoscalar. Then,

$$\begin{aligned} V(\phi) &= \frac{m^2}{2}\left(S^2 + P^2\right) + \frac{hm}{\sqrt{2}} S\left(S^2 + P^2\right) + \frac{h^2}{4}\left(S^2 + P^2\right)^2 \\ \mathcal{L}_{Yuk} &= -\frac{h}{\sqrt{2}}\bar{\psi}_M\left(S - i\gamma^5 P\right)\psi_M. \end{aligned} \tag{10.101}$$

It is interesting to consider the special case $m = 0$. From (10.98) we see that $\mathcal{L}$ has a global $U(1)$ phase symmetry under $\xi \to e^{iq_\xi \beta}\xi$, $\phi \to e^{-2iq_\xi \beta}\phi$. However, the superfield Φ does not appear to have a simple transformation property. This is an example of an R-symmetry (e.g., Intriligator and Seiberg, 2007; Dine, 2015), which does not commute with supersymmetry because different components have different charges. One can formally express the R-symmetry by assigning a transformation to the Grassmann variable θ, i.e.,

$$\theta \to e^{i\beta}\theta, \qquad d^2\theta \to e^{-2i\beta} d^2\theta, \qquad W \to e^{2i\beta} W, \tag{10.102}$$

which implies

$$\Phi \to e^{\frac{2}{3}i\beta}\Phi, \qquad \phi \to e^{\frac{2}{3}i\beta}\phi, \qquad \xi \to e^{-\frac{1}{3}i\beta}\xi, \qquad F \to e^{-\frac{4}{3}i\beta}F. \tag{10.103}$$

A Dirac Fermion

As a second example, consider three chiral superfields U, U^c, and H. The symbols are chosen to be suggestive that the fermionic components of U and U^c will combine to form a Dirac field (i.e., the u quark), while H will play the role of the Higgs, but at this stage they are three independent superfields and there are no gauge interactions or chiral symmetries. We denote the scalar and spinor components as $(\tilde{u}, \xi_u)$, $(\tilde{u}^c, \eta_u)$, and $(h, \tilde{h})$, respectively, where the use of η_u rather than ξ_u^c is motivated by (10.54) and the subsequent discussion, and the notations $\tilde{u}, \tilde{u}^c$, and h for the scalars and $\tilde{h}$ for the H spinor are suggestive of the MSSM. We choose the superpotential

$$W = mUU^c + h_u UU^c H + \mathcal{W}(H), \tag{10.104}$$

where m and h_u are real and $\mathcal{W}$ depends only on H. We first take $h_u = 0$ and $\mathcal{W} = 0$, so that

$$\begin{aligned} \mathcal{L} &= \mathcal{L}_{KE} - m\left(\eta_u \xi_u + \bar{\xi}_u \bar{\eta}_u\right) - m^2\left(|\tilde{u}|^2 + |\tilde{u}^c|^2\right) \\ &= \mathcal{L}_{KE} - m\left(\bar{u}_R u_L + \bar{u}_L u_R\right) - m^2\left(|\tilde{u}|^2 + |\tilde{u}^c|^2\right), \end{aligned} \tag{10.105}$$

which corresponds to a Dirac fermion $u = \begin{pmatrix} \xi_u \\ \bar{\eta}_u \end{pmatrix} = u_L + u_R$, as well as two complex scalars, $\tilde{u}$ and $\tilde{u}^c$, all with mass m. Including h and $\mathcal{W}$,

$$\mathcal{L} = \mathcal{L}_{KE} + \mathcal{L}_u + \mathcal{L}_H - V(\tilde{u}, \tilde{u}^c, h), \tag{10.106}$$

with

$$\begin{aligned} \mathcal{L}_u =& - m\left(\eta_u \xi_u + \bar{\xi}_u \bar{\eta}_u\right) - h_u \left(\eta_u \xi_u h + \bar{\xi}_u \bar{\eta}_u h^\dagger\right) \\ & - h_u \left(\tilde{h} \xi_u \tilde{u}^c + \bar{\xi}_u \bar{\tilde{h}} \tilde{u}^{c\dagger}\right) - h_u \left(\eta_u \tilde{h} \tilde{u} + \bar{\tilde{h}} \bar{\eta}_u \tilde{u}^\dagger\right) \\ =& - m\left(\bar{u}_R u_L + \bar{u}_L u_R\right) - h_u \left(\bar{u}_R u_L h + \bar{u}_L u_R h^\dagger\right) \\ & - h_u \left(\bar{\tilde{h}}^c_R u_L \tilde{u}^c + \bar{u}_L \tilde{h}^c_R \tilde{u}^{c\dagger}\right) - h_u \left(\bar{u}_R \tilde{h}_L \tilde{u} + \bar{\tilde{h}}_L u_R \tilde{u}^\dagger\right), \end{aligned} \tag{10.107}$$

where $\tilde{h}_L$ is the L-chiral Higgsino field corresponding to $\tilde{h}$, and $\tilde{h}^c_R$ is its CP conjugate. Similarly,

$$\mathcal{L}_H = -\frac{1}{2}\left(\tilde{h}\tilde{h}\mathcal{W}_{HH}(h) + \bar{\tilde{h}}\bar{\tilde{h}}\mathcal{W}_{HH}(h)^\dagger\right), \tag{10.108}$$

and

$$V(\tilde{u}, \tilde{u}^c, h) = \left|m\tilde{u}^c + h_u \tilde{u}^c h\right|^2 + \left|m\tilde{u} + h_u \tilde{u} h\right|^2 + \left|h_u \tilde{u}\tilde{u}^c + \mathcal{W}_H(h)\right|^2. \tag{10.109}$$

Again, $u = \begin{pmatrix} \xi_u \\ \bar{\eta}_u \end{pmatrix}$ is a Dirac field with a bare mass m and/or a spontaneously generated mass $h_u \langle h \rangle$ if h acquires a VEV (which we assume is real). $\tilde{h}$ is either a massless Weyl spinor or acquires a Majorana mass, depending on $\mathcal{W}$ (we assume that $\langle \tilde{u} \rangle = \langle \tilde{u}^c \rangle = 0$). From (10.107) we see that the Yukawa couplings $\bar{u}_R u_L h$ and $\bar{u}_L u_R h^\dagger$ are accompanied by analogous couplings in which one fermion and one scalar are replaced by their superpartners, as illustrated in Figure 10.5.

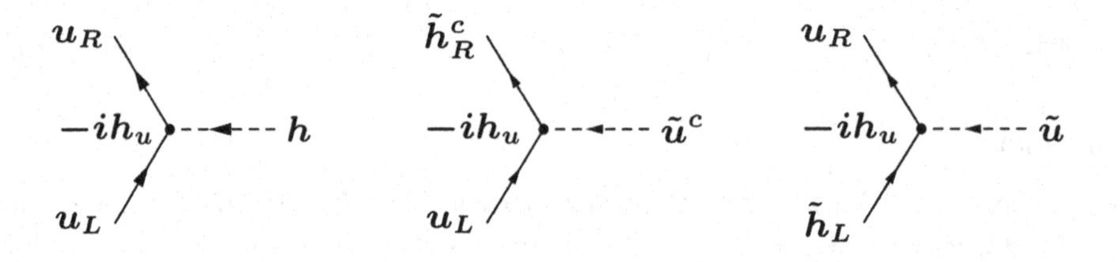

Figure 10.5 Yukawa vertices from (10.107). There are three more diagrams in which the incoming and outgoing particles are reversed.

Abelian Gauge Interactions

Let us first consider a $U(1)$ gauge symmetry. The gauge boson $A_\mu(x)$ is a component of a vector superfield V, which can be written

$$\begin{aligned} V(x,\theta,\bar{\theta}) =& \phi(x) + \sqrt{2}\theta\chi(x) + \sqrt{2}\bar{\theta}\bar{\chi}(x) + \theta\theta G(x)^\dagger + \bar{\theta}\bar{\theta}G(x) \\ & + \theta\sigma^\mu\bar{\theta}A_\mu(x) + \theta\theta\bar{\theta}\bar{\lambda}(x) + \bar{\theta}\bar{\theta}\theta\lambda(x) + \frac{1}{2}\theta\theta\bar{\theta}\bar{\theta}D(x), \end{aligned} \tag{10.110}$$

where ϕ, A_μ, and D are Hermitian. The form of V can be simplified by noting that $V + i\Lambda - i\Lambda^\dagger$ is also a vector superfield for any chiral superfield $i\Lambda$. In particular, if ϕ_Λ is the scalar component of Λ, then from (10.77) the *supergauge transformation*

$$V \to V' = V + i\Lambda - i\Lambda^\dagger \tag{10.111}$$

generates a new vector superfield with vector component

$$A'_\mu = A_\mu + \partial_\mu(\phi_\Lambda + \phi_\Lambda^\dagger), \tag{10.112}$$

which is just an ordinary $U(1)$ gauge transformation of the form in (4.9) on page 136. The D component $[V]_D = \frac{1}{2}D(x)$ is not only supersymmetry invariant (when integrated over x), but is invariant under supergauge transformations.

The other components of Λ can be used to put V in the form

$$V(x,\theta,\bar\theta) = \theta\sigma^\mu\bar\theta A_\mu(x) + \theta\theta\bar\theta\bar\lambda(x) + \bar\theta\bar\theta\theta\lambda(x) + \frac{1}{2}\theta\theta\bar\theta\bar\theta D(x). \tag{10.113}$$

This form, known as the *Wess-Zumino gauge* (WZ) is not manifestly supersymmetric, i.e., a supersymmetry transformation takes one back to the general form. However, it is extremely useful because it only involves the relevant physical degrees of freedom, i.e., the gauge boson A_μ, its superpartner the gaugino λ, and the auxiliary real D field, and is therefore analogous to the unitary gauge. One still has the freedom to perform ordinary gauge transformations while remaining in the WZ gauge. It is useful to note that in the WZ gauge

$$V^2(x,\theta,\bar\theta) = \frac{1}{2}\theta\theta\bar\theta\bar\theta A^\mu(x)A_\mu(x), \qquad V^3 = V^4 = \cdots = 0. \tag{10.114}$$

Under a supergauge transformation, a chiral superfield Φ_a and its conjugate transform as

$$\Phi_a \to e^{-2igq_a\Lambda}\,\Phi_a, \qquad \Phi_a^\dagger \to \Phi_a^\dagger\, e^{+2igq_a\Lambda^\dagger}, \tag{10.115}$$

where g is the gauge coupling and q_a is the charge of ϕ_a. It is easy to see that the special case of an ordinary gauge transformation reproduces (4.9), with $\beta(x) = -g[\phi_\Lambda(x) + \phi_\Lambda(x)^\dagger]$. Combining (10.111) and (10.115), we see that $\Phi_a^\dagger e^{2gq_aV}\Phi_a$ is supergauge invariant, and that

$$\mathcal{L}_g = \int d^4\theta \sum_a \Phi_a^\dagger e^{2gq_aV}\Phi_a = \Big[\sum_a \Phi_a^\dagger e^{2gq_aV}\Phi_a\Big]_D, \tag{10.116}$$

which generalizes the chiral kinetic energy term in (10.87), is supersymmetric and supergauge invariant. $\mathcal{L}_g$ can be written in terms of the component fields by expanding the exponential and using (10.114). One obtains

$$\begin{aligned}\mathcal{L}_g = \sum_a \Big[&(D_{a\mu}\phi_a)^\dagger D_a^\mu\phi_a + i\bar\xi_a\bar\sigma^\mu D_{a\mu}\xi_a - \sqrt{2}gq_a(\bar\xi_a\bar\lambda\phi_a + \phi_a^\dagger\xi_a\lambda_a)\\ &+ gq_a\phi_a^\dagger\phi_a D + F_a^\dagger F_a\Big],\end{aligned} \tag{10.117}$$

where $D_{a\mu}$ is the gauge covariant derivative, i.e.,

$$D_{a\mu}\phi_a = (\partial_\mu + igq_aA_\mu)\phi_a, \qquad D_{a\mu}\xi_a = (\partial_\mu + igq_aA_\mu)\xi_a. \tag{10.118}$$

We recognize the first two terms as the gauge covariant kinetic energies for ϕ_a and ξ_a, while the third is a related fermion-scalar-gaugino interaction required by the supersymmetry.

The construction of the kinetic energy terms for A_μ and λ is rather involved. It is possible to construct a supersymmetric derivative of V that contains the field strength $F_{\mu\nu}$ and a derivative of λ and that transforms as a chiral supermultiplet but with a spinor index. The F term of the contraction of this field with itself[36] yields the gauge kinetic terms

$$\mathcal{L}_A = -\frac{1}{4} F_{\mu\nu} F^{\mu\nu} + i\bar{\lambda}\bar{\sigma}^\mu \partial_\mu \lambda + \frac{1}{2} D^2. \tag{10.119}$$

One can also write a superpotential including any terms that are $U(1)$ invariant, such as $\Phi_a \Phi_b \Phi_c$ provided $q_a + q_b + q_c = 0$. Finally, for the special case of a $U(1)$ gauge symmetry, one can use the fact that $D(x)$ is gauge and supersymmetry invariant to add a *Fayet-Iliopoulos* (FI) term (Fayet and Iliopoulos, 1974)

$$\mathcal{L}_{FI} = \kappa D(x) = 2\kappa [V]_D, \tag{10.120}$$

where κ is a constant. There is no analog of the FI term for a non-abelian gauge symmetry or for a $U(1)$ embedded in a non-abelian group.

The auxiliary field D enters $\mathcal{L}_g$, $\mathcal{L}_A$, and $\mathcal{L}_{FI}$. The Euler-Lagrange equation implies

$$D(x) = -\left(\kappa + \sum_a g q_a \phi_a^\dagger \phi_a\right). \tag{10.121}$$

Putting everything together,

$$\begin{aligned} \mathcal{L} = &-\frac{1}{4} F_{\mu\nu} F^{\mu\nu} + i\bar{\lambda}\bar{\sigma}^\mu \partial_\mu \lambda - \frac{1}{2} \sum_{a,b} \left(\xi_a \xi_b W_{ab}(\phi) + \bar{\xi}_a \bar{\xi}_b W_{ab}(\phi)^\dagger \right) \\ &+ \sum_a \left[(D_{a\mu}\phi_a)^\dagger D_a^\mu \phi_a + i\bar{\xi}_a \bar{\sigma}^\mu D_{a\mu} \xi_a - \sqrt{2} g q_a (\bar{\xi}_a \bar{\lambda} \phi_a + \phi_a^\dagger \lambda \, \xi_a \right] - V(\phi), \end{aligned} \tag{10.122}$$

where

$$V(\phi) = \sum_a |F_a|^2 + \frac{1}{2}|D|^2 \equiv V_F + V_D. \tag{10.123}$$

F_a and D are given by (10.90) and (10.121), respectively.

As an example, consider a single chiral superfield Φ with $q = 1$. There is no gauge invariant holomorphic term in Φ, so the superpotential vanishes, and

$$\begin{aligned} \mathcal{L} = &-\frac{1}{4} F_{\mu\nu} F^{\mu\nu} + i\bar{\lambda}\bar{\sigma}^\mu \partial_\mu \lambda + \left| (\partial_\mu + igA_\mu)\phi \right|^2 \\ &+ i\bar{\xi}\bar{\sigma}^\mu (\partial_\mu + igA_\mu)\xi - \sqrt{2} g (\bar{\xi}\bar{\lambda}\phi + \phi^\dagger \lambda \, \xi) - V(\phi), \end{aligned} \tag{10.124}$$

where

$$V(\phi) = \frac{1}{2}|D|^2 = \frac{1}{2}\left|\kappa + g\phi^\dagger \phi\right|^2. \tag{10.125}$$

For $\kappa/g \geq 0$ the minimum of the potential is at $\phi = 0$, so the gauge symmetry is unbroken. However, $V = |D|^2/2 = \kappa^2/2 \neq 0$ at the minimum, so the supersymmetry is broken. This is manifested by the fact that the complex scalar ϕ acquires a mass

$$m_\phi^2 = gD\Big|_{min} = g\kappa, \tag{10.126}$$

[36] More general supersymmetric but non-renormalizable gauge kinetic terms, encountered in supergravity and string constructions, involve an additional *gauge kinetic function* that is holomorphic in the chiral superfields.

while the Weyl spinor ξ and the vector supermultiplet remain massless. The massless ξ, known as the *Goldstino*, is characteristic of a spontaneously broken supersymmetry. It is the analog of the Nambu-Goldstone boson of an internal symmetry.

For $\kappa/g < 0$ the minimum of the potential will be at $\langle\phi\rangle \neq 0$, so the gauge symmetry will be spontaneously broken. Writing $\kappa = -g\nu^2/2$, where ν is real and positive, one can choose $\langle\phi\rangle = \nu/\sqrt{2}$ and $\phi = (\nu + h)/\sqrt{2}$ (in unitary gauge), where h is Hermitian. Since $D = 0$ at the minimum, the supersymmetry is preserved. The spectrum consists of one massive scalar (h), a massive vector A_μ, and a massive Dirac fermion formed by combining the fermion from the chiral supermultiplet with the gaugino, $\psi = \begin{pmatrix} \xi \\ \bar{\lambda} \end{pmatrix}$. All of these have mass $g\nu$, i.e., they form a massive vector supermultiplet.

Non-abelian Gauge Interactions

Equation (10.122) generalizes in a fairly obvious way to the non-abelian case (which is given for a nonsupersymmetric theory in (4.16) on page 140). One finds

$$\begin{aligned}\mathcal{L} = &-\frac{1}{4}F^i_{\mu\nu}F^{i\mu\nu} + i\bar{\lambda}\bar{\sigma}^\mu D_{\lambda\mu}\lambda - \frac{1}{2}\sum_{a,b}\left(\xi_a\xi_b W_{ab}(\phi) + \bar{\xi}_a\bar{\xi}_b W_{ab}(\phi)^\dagger\right) \\ &+ \left[(D_{\Phi\mu}\phi)^\dagger D^\mu_\Phi\phi + i\bar{\xi}\bar{\sigma}^\mu D_{\Phi\mu}\xi - \sqrt{2}g(\bar{\xi}\bar{\lambda}^i L^i_\Phi\phi + \phi^\dagger\lambda^i L^i_\Phi\xi\right] - V(\phi),\end{aligned} \tag{10.127}$$

where

$$V(\phi) = V_F + D_D = \sum_a |F_a|^2 + \frac{1}{2}\sum_i |D^i|^2, \qquad D^i = -g\phi^\dagger L^i_\Phi\phi. \tag{10.128}$$

L^i_Φ is the group representation matrix for the matter fields (which is the same for ϕ and ξ but which may be reducible). The D term can be written more explicitly as

$$V_D \equiv \frac{1}{2}\sum_i |D^i|^2 = \frac{1}{2}\sum_i \left|-g\phi^\dagger_a \left(L^i_\Phi\right)_{ab}\phi_b\right|^2. \tag{10.129}$$

In (10.127) D_Φ and D_λ are, respectively, the covariant derivatives for Φ and λ.

$$D_{\Phi\mu} = \partial_\mu I + igA^i_\mu L^i_\Phi, \qquad D_{\lambda\mu} = \partial_\mu I + igA^i_\mu L^i_{adj}, \tag{10.130}$$

where $(L^i_{adj})_{jk} = -ic_{ijk}$ is the adjoint representation matrix. Thus,

$$(D_{\lambda\mu}\lambda)_i = \partial_\mu\lambda_i - gc_{ijk}A^j_\mu\lambda_k, \tag{10.131}$$

which should be compared with (4.27). The gaugino-gaugino-gauge interaction can be rewritten in four-component notation as

$$\begin{aligned}-igc_{ijk}\bar{\lambda}^i\bar{\sigma}^\mu A^j_\mu\lambda^k &= -igc_{ijk}\bar{\lambda}^i_L\gamma^\mu A^j_\mu\lambda^k_L = -igc_{ijk}\bar{\lambda}^{ic}_R\gamma^\mu A^j_\mu\lambda^{kc}_R \\ &= -\frac{i}{2}gc_{ijk}\bar{\lambda}^i_M\gamma^\mu A^j_\mu\lambda^k_M,\end{aligned} \tag{10.132}$$

with $\lambda^i \to \lambda^i_M = \begin{pmatrix} \lambda^i \\ \bar{\lambda}^i \end{pmatrix} = \lambda^i_L + \lambda^{ic}_R$. The vertices for the gauge interactions are shown in Figure 10.6.

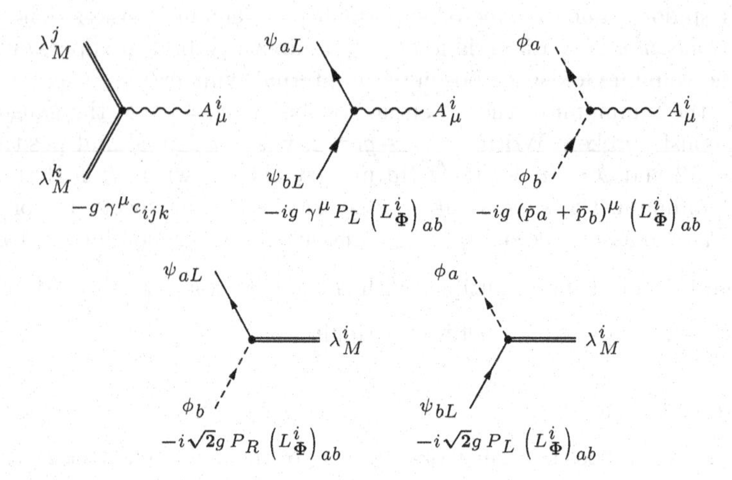

Figure 10.6 The three-point gauge and gaugino vertices corresponding to (10.132), (10.127), and (10.133). (There are two ways to contract the gauginos in the first diagram.) The notation $\bar{p}_{a,b}$ is defined in Figure 4.1. The double lines represent the Majorana gaugino $\lambda^i_M = \lambda^i_L + \lambda^{ic}_R$. There are additional scalar seagull and gauge self interactions, which are shown in Figure 4.1.

Similarly, the fermion-scalar-gaugino interactions are

$$\begin{aligned}
&-\sqrt{2}g\Big[\bar{\xi}_a\bar{\lambda}^i\left(L^i_\Phi\right)_{ab}\phi_b+\phi^\dagger_a\lambda^i\left(L^i_\Phi\right)_{ab}\xi_b\Big]\\
&\quad=-\sqrt{2}g\Big[\bar{\psi}_{aL}\lambda^i_M\left(L^i_\Phi\right)_{ab}\phi_b+\phi^\dagger_a\bar{\lambda}^i_M\left(L^i_\Phi\right)_{ab}\psi_{bL}\Big],
\end{aligned}\tag{10.133}$$

where the second form is in four-component notation, with $\xi_a \to \psi_{aL}$. Frequently, we consider a reducible representation involving pairs of left-chiral superfields Φ and Φ^c whose fermionic components will eventually pair to form Dirac fields. Similar to the example described above (10.105), we write their components as (ϕ,ξ) and (ϕ^c,η), respectively. Allowing for the possibility of a chiral gauge symmetry, we take $L_\Phi = L_L$ and $L_{\Phi^c} = -L_R^T$ for their representation matrices, where L_L and L_R are the representations for the L- and R-chiral fermions $\psi_L \leftrightarrow \xi$ and $\psi_R \leftrightarrow \bar{\eta}$. The fermion-scalar-gaugino interactions in this case become

$$\begin{aligned}
&-\sqrt{2}g\Big[\bar{\xi}_a\bar{\lambda}^i\left(L^i_L\right)_{ab}\phi_b+\phi^\dagger_a\lambda^i\left(L^i_L\right)_{ab}\xi_b-\phi^c_b\bar{\lambda}^i\left(L^i_R\right)_{ba}\bar{\eta}_a-\eta_b\lambda^i\left(L^i_R\right)_{ba}\phi^{c\dagger}_a\Big]\\
&\quad=-\sqrt{2}g\Big[\bar{\psi}_{aL}\lambda^i_M\left(L^i_L\right)_{ab}\phi_{bL}+\phi^\dagger_{aL}\bar{\lambda}^i_M\left(L^i_L\right)_{ab}\psi_{bL}\\
&\qquad-\phi^\dagger_{bR}\bar{\lambda}^i_M\left(L^i_R\right)_{ba}\psi_{aR}-\bar{\psi}_{bR}\lambda^i_M\left(L^i_R\right)_{ba}\phi_{aR}\Big],
\end{aligned}\tag{10.134}$$

where in the second expression we have introduced the suggestive notation $\phi_a \to \phi_{aL}$ and $\phi^c_a \to \phi^c_{aL} = \phi^\dagger_{aR}$.

10.2.4 Supersymmetry Breaking and Mediation

Since we do not observe degenerate supermultiplets, supersymmetry (if it is present at all) must be broken (for reviews, see Chung et al., 2005; Luty, 2005; Intriligator and Seiberg, 2007). Just as with an ordinary symmetry, the breaking can be explicit or spontaneous. There are a number of reasons that the breaking should be spontaneous, including the possibility of a naturally small supersymmetry breaking scale (compared to the Planck scale), limiting the number of parameters, suppressing flavor changing effects, and allowing a consistent extension to supergravity or superstring theory. Nevertheless, the effective theory relevant at low energies may well involve explicit symmetry breaking terms. These should be soft (i.e., mass terms and cubic scalar interactions, with dimension < 4), rather than hard to avoid the reintroduction of the Higgs hierarchy problem.

Spontaneous breaking occurs when the F and D terms cannot all vanish simultaneously. We already saw one example of *D-term breaking* in the abelian gauge model in (10.124) involving a single charged superfield. For $\kappa/g \geq 0$ the D term could not vanish, so the complex scalar acquired mass while the Weyl fermion (the Goldstino) remained massless. It is also possible to construct *F-term breaking (O'Raifeartaigh)* models, in which one or more of the $|F_a|$ is nonzero (O'Raifeartaigh, 1975). Consider three chiral superfields with superpotential

$$W = m\Phi_2\Phi_3 + h\Phi_1\left(\Phi_3^2 - \mu^2\right), \tag{10.135}$$

where for simplicity we will take m, h, and μ^2 to be real and positive. Then

$$F_1^* = -h\left(\phi_3^2 - \mu^2\right), \qquad F_2^* = -m\phi_3, \qquad F_3^* = -m\phi_2 - 2h\phi_1\phi_3, \tag{10.136}$$

which obviously cannot all vanish simultaneously, so $V_F > 0$ at the minimum. For example, for $m^2 > 2h^2\mu^2$ the minimum occurs for $\langle\phi_2\rangle = \langle\phi_3\rangle = 0$, with $\langle\phi_1\rangle$ undetermined (i.e., there is a flat direction at tree level), for which

$$V_F = |F_1|^2 = h^2\mu^4 > 0. \tag{10.137}$$

Quantizing around the minimum (taking $\langle\phi_1\rangle = 0$) yields scalar mass-squares

$$m_1^2 = 0, \qquad m_2^2 = m^2, \qquad m_{3R}^2 = m^2 - h^2\mu^2, \qquad m_{3I}^2 = m^2 + h^2\mu^2, \tag{10.138}$$

where $\phi_{3R,I}$ are the Hermitian components of ϕ_3, i.e., $\phi_3 = (\phi_{3R} + i\phi_{3I})/\sqrt{2}$. The fermion mass terms are from

$$\mathcal{L}_f = -m\xi_2\xi_3 + h.c. + \text{Yukawa terms}, \tag{10.139}$$

so that ξ_2 and ξ_3 combine to form a Dirac fermion with mass m, while ξ_1 remains massless. Similar to the D-term breaking example, ξ_1 is the massless Goldstino associated with the spontaneous supersymmetry breaking. As should be intuitive, it is associated with the supermultiplet with the nonzero F term.

The existence of the Goldstino is a generic feature of spontaneous breaking. In supergravity theories it is "eaten" to become the helicity $\pm\frac{1}{2}$ components of the massive gravitino (the *super-Higgs mechanism*). These issues are discussed in detail in the more extensive treatments listed in the bibliography.

Tree-level spontaneous supersymmetry breaking implies sum rules relating the mass-squares of the fermionic and bosonic degrees of freedom. For example, the sum of the scalar mass-squares for the O'Raifeartaigh model in (10.135),

$$m_B^2 = 2m_1^2 + 2m_2^2 + m_{3R}^2 + m_{3I}^2 = 4m^2, \tag{10.140}$$

is the same as that for the fermions, where in each case one must weight by the number of degrees of freedom (2 for a complex scalar, 4 for a Dirac fermion). There is a similar sum rule for the D term example in in (10.124), which however, involves a contribution from $|D|$. For this reason such tree-level breaking is not a phenomenologically viable option for the MSSM, because the sum rules would require that some of the superpartners are light, contrary to observational limits.

These constraints can be evaded if the breaking is radiative, i.e., associated with loop effects, or if it occurs in a *hidden sector* that is only weakly coupled to the SM particles. The hidden sector models are especially promising, e.g., because they may allow for suppressed flavor changing effects (although FCNC are still problematic in some cases). Breaking could occur in the hidden sector by tree level F or D mechanisms, or the breaking could be *dynamical*,[37] i.e., associated with some strong dynamics in the hidden sector (e.g., Witten, 1981, 1982; Affleck et al., 1985). The latter possibility could explain why the supersymmetry breaking scale is small compared to the Planck scale. For example, an asymptotically free gauge group that is weakly coupled at the Planck scale might become strong at some intermediate scale $\Lambda_S \ll M_P$, just as QCD becomes strong at Λ_{QCD}, leading to an effective F term and a gravitino mass (in supergravity) $m_{3/2} \sim F/M_P$. The breaking could be associated, for example, with a gaugino condensate,

$$\langle \lambda^i \lambda^j \rangle = \delta^{ij} \Lambda_S^3, \tag{10.141}$$

analogous to the dynamical breaking of the chiral $SU(3)_L \times SU(3)_R$ flavor symmetry of QCD or with some dynamical alternatives to the Higgs mechanism for electroweak breaking (Hill and Simmons, 2003). In this case, one expects $F \sim \Lambda_S^3/M_P$.

The supersymmetry breaking in the hidden sector must somehow be communicated to the MSSM particles, leading to effective soft masses for the gauginos, Higgsinos, squarks, and sleptons, as well as soft cubic scalar couplings, with a typical scale m_{soft}. This should be in the 100 GeV-few TeV range if supersymmetry is relevant to the Higgs/hierarchy problem. There are a number of possibilities for the *mediation* mechanism. One is *supergravity mediation*, in which the sectors are connected by higher-dimensional operators with coefficients suppressed by inverse powers of M_P. One usually finds $m_{soft} \sim F/M_P$ or $m_{soft} \sim \Lambda_S^3/M_P^2$, suggesting $\sqrt{F} \sim 10^{11}$ GeV. Simple versions of supergravity mediation may lead, e.g, to squark masses that are family universal at the Planck or GUT scale. However, RGE effects lead to splitting at lower energies, with possible difficulties for FCNC. The gravitino mass in supergravity theories is comparable to m_{soft}.

In the original *gauge mediation* models (e.g., Giudice and Rattazzi, 1999), the information about supersymmetry breaking is transmitted by *messenger fields* that interact with the hidden sector and also are charged under the SM gauge group. The effective soft breaking (due to loop effects) leads to

$$m_{soft} \sim \frac{\alpha}{4\pi}\frac{F}{M}, \tag{10.142}$$

where M is the mass of the messenger particles. For $\sqrt{F} \sim M$ one finds the much lower scale $\sqrt{F} \sim 10^4 - 10^5$ GeV, implying a very light gravitino. Since F is low and the SM gauge interactions are family universal, gauge mediation is much less problematic for flavor changing effects. Gauge mediation is treated generally in (Meade et al., 2009) and reviewed in (Kitano et al., 2010).

There are many other possibilities, including anomaly mediation, gaugino mediation,

[37] It is likely that such a supersymmetry breaking vacuum would be metastable, with a nearby supersymmetric minimum (Intriligator et al., 2006).

ion mediation, and D-term mediation, as reviewed in (Chung et al., 2005; Heinemeyer al., 2006; Patrignani, 2016). (Deflected) mirage mediation (e.g., Everett et al., 2008) bines some of these possibilities. Z' mediation is mentioned in Section 10.3.

2.5 The Minimal Supersymmetric Standard Model (MSSM)

s straightforward in principle though complicated in practice to write the Lagrangian sity for the MSSM. Here, we will show the main features, but for the details of the ark, slepton, neutralino, and chargino mixings, CP phases, etc., the reader is referred he reviews and books that have already been cited. We will follow a notation similar he example in (10.104), using upper case letters for left-chiral superfields and upper e letters with a superscript ^{c} for the left-chiral conjugates of right-chiral fields. Thus, the rk and lepton superfields will be written

$$Q \equiv \begin{pmatrix} U \\ D \end{pmatrix} \quad U^c \quad D^c; \qquad L \equiv \begin{pmatrix} N \\ E \end{pmatrix} \quad N^c \quad E^c, \tag{10.143}$$

ectively.[38] (Some authors use a hat notation, such as $\hat{Q}$, to indicate a chiral superfield.) ir scalar components will be written as, e.g.,

$$\tilde{u} \equiv \tilde{u}_L, \qquad \tilde{u}^c \equiv \tilde{u}^c_L = \tilde{u}^\dagger_R \tag{10.144}$$

U and U^c, and the fermion spinor components as

$$\xi_u \leftrightarrow u_L, \qquad \eta_u \leftrightarrow u^c_L \quad [\bar{\eta}_u \leftrightarrow u_R]. \tag{10.145}$$

will usually not display family indices, or superscripts ^{0} for weak eigenstates, as in (8.4) page 258, but these can easily be added. The Higgs chiral supermultiplets are

$$H_u = \begin{pmatrix} H^+_u \\ H^0_u \end{pmatrix}, \qquad H_d = \begin{pmatrix} H^0_d \\ H^-_d \end{pmatrix}, \tag{10.146}$$

n scalar and fermion components such as h^+_u and $\tilde{h}^+_u \leftrightarrow \tilde{h}^+_{uL}$ for H^+_u. The indices for (2) doublets are contracted using ϵ^{ab}, just as for the spinor indices of the Lorentz group. ike the spinor case, however, the order is significant since chiral superfields commute. example,

$$QH_u \equiv \epsilon^{ab} Q_a H_{ub} = UH^0_u - DH^+_u = -H_u Q. \tag{10.147}$$

(3) color indices are unambiguous, e.g., $QU^c \equiv Q_\alpha U^{c\alpha}$ or $U^c D^c S^c \equiv \epsilon_{\alpha\beta\gamma} U^{c\alpha} D^{c\beta} S^{c\gamma}$. The superpotential for the MSSM (assuming a conserved R-parity) is

$$\begin{aligned} W &= \mu H_u H_d - \Gamma^d Q H_d D^c + \Gamma^u Q H_u U^c - \Gamma^e L H_d E^c + \Gamma^\nu L H_u N^c \\ &= \mu \left(H^+_u H^-_d - H^0_u H^0_d \right) - \Gamma^d \left(U H^-_d - D H^0_d \right) D^c + \Gamma^u \left(U H^0_u - D H^+_u \right) U^c \\ &\quad - \Gamma^e \left(N H^-_d - E H^0_d \right) E^c + \Gamma^\nu \left(N H^0_u - E H^+_u \right) N^c, \end{aligned} \tag{10.148}$$

re the first term yields supersymmetric masses μ for the Higgs and Higgsino fields, and others yield the Higgs Yukawa vertices and their supersymmetric partners. The signs chosen so that the fermion mass terms have the "correct" signs for positive $\Gamma\langle h^0_{u,d}\rangle$.

We have included the right-handed neutrino fields N^c as an (optional) part of the MSSM, similar to definition of the SM in Section 8.1.

Family indices such as in (8.13) are suppressed. The kinetic energy and gauge interactions are derived in an obvious way from (10.127) and will not be displayed explicitly here.

The effective soft breaking terms for the MSSM include mass-squared terms for the scalars, Majorana mass terms for the gauginos, a scalar analog of the μ term (the *$B\mu$ term*), and scalar cubic interactions *(A terms)* similar to the superpotential Yukawa couplings. Thus,

$$\begin{aligned}\mathcal{L}_{soft} = &-\sum_r m_r^2 \phi_r^\dagger \phi_r - \frac{m_{\tilde{B}}}{2}\left(\tilde{B}\tilde{B} + \bar{\tilde{B}}\bar{\tilde{B}}\right) - \frac{m_{\tilde{W}}}{2}\sum_{i=1}^{3}\left(\tilde{W}^i\tilde{W}^i + \bar{\tilde{W}}^i\bar{\tilde{W}}^i\right) \\ &- \frac{m_{\tilde{G}}}{2}\sum_{i=1}^{8}\left(\tilde{G}^i\tilde{G}^i + \bar{\tilde{G}}^i\bar{\tilde{G}}^i\right) + \Big[-B\mu h_u h_d + A_d\Gamma^d \tilde{q} h_d \tilde{d}^c \\ &- A_u\Gamma^u \tilde{q} h_u \tilde{u}^c + A_e \Gamma^e \tilde{\ell} h_d \tilde{e}^c - A_\nu \Gamma^\nu \tilde{\ell} h_u \tilde{\nu}^c + h.c.\Big], \end{aligned} \tag{10.149}$$

where $\tilde{q} \equiv \begin{pmatrix} \tilde{u} \\ \tilde{d} \end{pmatrix}$, $\tilde{\ell} \equiv \begin{pmatrix} \tilde{\nu}_e \\ \tilde{e} \end{pmatrix}$, and the same conventions for the $SU(2)$ indices hold as in the superpotential. In the first term, $\sum_r$ sums over all of the scalars (Higgs, squarks, sleptons). The sum could be generalized to allow Hermitian mass-squared matrices for each scalar sector ($\tilde{q}$, $\tilde{u}^c$, $\tilde{d}^c$, etc), but without loss of generality one can choose family bases in which they are diagonal (this does, however, fix the family bases for the Yukawa matrices in (10.148).) A_u, A_d, A_e, and A_ν, which are matrices in family space, and B have dimensions of mass. We have extracted factors of μ and the Yukawa matrices Γ from the coefficients, because this occurs naturally in some (but not all) models of supersymmetry breaking/mediation.[39] The special case in which the A's are each multiples of the identity corresponds to minimal flavor violation (Section 8.6.6). Equation (10.149) is almost[40] the most general soft breaking allowed for the R_p-conserving MSSM.

In general (10.149) contains many free parameters after including family indices, but specific models, motivated by theoretical considerations, phenomenological constraints (e.g., from FCNC and CP violation), or simplicity typically have fewer. Most of the early phenomenological studies were motivated by supergravity mediation. For example, the *constrained MSSM* (CMSSM) or *minimal supergravity* (mSUGRA) models assume universal values for the soft parameters at the Planck or GUT scale,

$$m_{\tilde{B}} = m_{\tilde{W}} = m_{\tilde{G}} \equiv m_{1/2}, \qquad m_r \equiv m_0, \qquad A_u = A_d = A_e = A_\nu \equiv A, \tag{10.150}$$

where the universal A parameters multiply the full Yukawa matrices for each sector. It is usually further assumed that the soft parameters and μ are all real, so that altogether there are only 5 dimensionful parameters, $m_{1/2}$, m_0, A, B, and μ. Some specific models further relate the universal soft breaking parameters $m_{1/2}$, m_0, and A, as well as B, to each other and to the gravitino mass $m_{3/2}$ (see, e.g., Martin, 1997; Patrignani, 2016). (Some authors reserve the term "mSUGRA" for some of these more restricted versions.) In any case, these parameters must then be run down to the electroweak scale (e.g., Chung et al., 2005), which induces mass splittings and flavor off-diagonal couplings to the mass eigenstate fermions. Two of the parameters, μ and B, are then typically traded for M_Z and $\tan\beta = \langle h_u^0\rangle/\langle h_d^0\rangle$.

---

[39] There is no uniformity in the literature about the signs of the A terms or whether to extract μ and Γ.

[40] An explicit Dirac mass term for the Higgsinos can be absorbed in μ as long as $B\mu$ is free. One could add non-holomorphic cubic terms involving the "wrong" Higgs field, such as $\tilde{q}(i\sigma^2 h_u^\dagger)\tilde{d}^c$, but in most symmetry breaking/mediation schemes these are suppressed by an additional factor of m_{soft}/M and are therefore negligible. [A possible application for small Dirac neutrino masses is considered in (Demir et al., 2008)].

These are actually derived quantities, but M_Z is known and $\tan\beta$ is closely related to observables. The basic parameters are then $m_{1/2}$, m_0, A, M_Z, and $\tan\beta$, as well as the sign of μ, which is not determined, as well as the gauge couplings and fermion spectrum.

The CMSSM is extremely simple and is a useful benchmark, but it is likely that the real world is more complicated (assuming low energy supersymmetry exists). The nonobservation of evidence for supersymmetry in the early LHC running and also the heightened theoretical realization that there are really an enormous number of possibilities have led to considerable interest in more general possibilities, such as extending the CMSSM to allow Higgs masses at the Planck scale that differ from those of the squarks and sleptons. There have also been extensive studies of models motivated by mediation methods other than supergravity, such as gauge mediation and the other alternatives mentioned in Section 10.2.4, which can have very different spectra.

Even these more general frameworks leave open the possibility that some interesting cases might be missed. It is not really feasible to examine the entire $\geq$ 124-dimensional parameter space. However, the possibilities are reduced to a manageable level by making some reasonable simplifying assumptions in the *phenomenological MSSM* (pMSSM) (e.g., Djouadi et al., 2007; Berger et al., 2009), i.e. no R_p violation; no new CP violation or FCNC at tree level; degeneracy of the squarks and sleptons of each type between the first two genrations; and that A terms for the first two generations can be neglected. There remain 19 new observable parameters (ignoring N^c):

$$\tan\beta, \quad M_A, \quad \mu, \quad m_{\tilde{B},\tilde{W},\tilde{G}}, \quad m^2_{\tilde{q},\tilde{u}^c,\tilde{d}^c,\tilde{\ell},\tilde{e}^c}, \quad m^2_{\tilde{q}_3,\tilde{t}^c,\tilde{b}^c,\tilde{\ell}_3,\tilde{\tau}^c}, \quad A_{t,b,\tau}, \tag{10.151}$$

where M_A is the Higgs pseudoscalar mass in (10.157), $m^2_{\tilde{q},\tilde{u}^c,\tilde{d}^c,\tilde{\ell},\tilde{e}^c}$ are the scalar mass-squares for the first two families, and $m^2_{\tilde{q}_3,\tilde{\ell}_3}$ refer to the $(\tilde{t}\ \tilde{b})^T$ and $(\tilde{\nu}_\tau\ \tilde{\tau})^T$ doublets.

Another useful framework for analyzing experimental data (or for exploring the sensitivities of an experiment or analysis method) are *simplified models* (e.g., Alves et al., 2012). These are effective Lagrangians involving just a small number of particles and parameters that describe the essence of some process found in a more complicated model, such as the production of a gluino followed by its decay into jets and an unobserved LSP. Simplified models can also be parametrized in terms of quantities closely connected to what is actually measured, such as masses, cross sections, and branching ratios. They can be applied to any type of new physics, not just supersymmetry.

The Higgs Sector

The scalar potential $V(h_u, h_d, \tilde{q}, \tilde{\ell}, \tilde{u}^c, \tilde{d}^c, \tilde{\nu}^c, \tilde{e}^c)$ includes F-term, D-term, and soft contributions,[41]

$$V = V_F + V_D + V_{soft}. \tag{10.152}$$

We will assume that the minimum of V does not violate color, electric charge, B, or L, i.e., that the VEVs of the squark and slepton fields are zero.[42] Keeping only the Higgs doublets

[41] For more detailed discussions of the Higgs sector of the MSSM and extended supersymmetric models, see (Gunion and Haber, 1986; Gunion et al., 1990; Carena and Haber, 2003; Djouadi, 2008b; Accomando et al., 2006; Patrignani, 2016).

[42] This is not automatic and leads to restrictions on the soft parameters, especially the A parameter associated with the top quark Yukawa (e.g., Casas et al., 1996).

in V, the contributions are

$$\begin{aligned}
V_F =&|\mu|^2 \left(|h_u|^2 + |h_d|^2\right) \\
V_D =&\frac{g^2}{8} \left| h_u^\dagger \vec{\tau} h_u + h_d^\dagger \vec{\tau} h_d \right|^2 + \frac{g'^2}{8} \left| |h_u|^2 - |h_d|^2 \right|^2 \\
=&\frac{g^2 + g'^2}{8} \left| |h_u|^2 - |h_d|^2 \right|^2 + \frac{g^2}{2} \left| h_u^{+\dagger} h_d^0 - h_u^{0\dagger} h_d^- \right|^2 \\
V_{soft} =& m_{h_u}^2 |h_u|^2 + m_{h_d}^2 |h_d|^2 - \left[B\mu \left(h_u^0 h_d^0 - h_u^+ h_d^- \right) + h.c. \right],
\end{aligned} \tag{10.153}$$

where

$$|h_u|^2 = |h_u^+|^2 + |h_u^0|^2, \qquad |h_d|^2 = |h_d^0|^2 + |h_d^-|^2, \tag{10.154}$$

we have used the Fierz identity in Problem 1.1 to rearrange V_D, and have assumed the $U(1)_Y$ FI term is zero. One can make a field redefinition so that $B\mu$ is real and positive, and an $SU(2) \times U(1)$ transformation so that $\langle h_u^+ \rangle = 0$ and $\nu_u \equiv \sqrt{2} \langle h_u^0 \rangle$ is real and positive. V_D is then minimized for $\langle h_d^- \rangle = 0$, while the $B\mu$ term is minimized for $\nu_d \equiv \sqrt{2} \langle h_d^0 \rangle$ real and positive. That is, electric charge and CP conservation in the Higgs sector are automatic at tree-level in the MSSM, provided the squarks and sleptons do not have VEVs.[43] This is to be contrasted with general extensions of the SM involving additional Higgs multiplets which do allow electric charge violation (Problem 8.3 and 8.4).

The potential in terms of $\nu_{u,d}$ is

$$\begin{aligned}
V(\nu_u, \nu_d) &= \frac{1}{2} \mathfrak{m}_u^2 \nu_u^2 + \frac{1}{2} \mathfrak{m}_d^2 \nu_d^2 - B\mu \nu_u \nu_d + \frac{g^2 + g'^2}{32} \left(\nu_u^2 - \nu_d^2 \right)^2 \\
&= \frac{1}{2} (\nu_u \ \nu_d) \, M^2 \begin{pmatrix} \nu_u \\ \nu_d \end{pmatrix} + \frac{g^2 + g'^2}{32} \left(\nu_u^2 - \nu_d^2 \right)^2,
\end{aligned} \tag{10.155}$$

where

$$\mathfrak{m}_u^2 \equiv |\mu|^2 + m_{h_u}^2, \qquad \mathfrak{m}_d^2 \equiv |\mu|^2 + m_{h_d}^2, \qquad M^2 \equiv \begin{pmatrix} \mathfrak{m}_u^2 & -B\mu \\ -B\mu & \mathfrak{m}_d^2 \end{pmatrix}, \tag{10.156}$$

and it is understood that $\nu_{u,d}$ are real and positive (or at least nonnegative). We see that at tree level V depends only on three parameters, $\mathfrak{m}_u^2$, $\mathfrak{m}_d^2$, and $B\mu$ (as well as known gauge couplings). One combination of these will be fixed by the requirement that $\nu^2 \equiv \nu_u^2 + \nu_d^2 \sim (246 \text{ GeV})^2$ at the minimum. All aspects of the tree-level Higgs sector are therefore expressible in terms of the two remaining parameters, which we will take to be

$$M_A^2 \equiv \mathfrak{m}_u^2 + \mathfrak{m}_d^2 \ \text{ and } \ \tan\beta \equiv \frac{\nu_u}{\nu_d}. \tag{10.157}$$

The quartic term in V vanishes for $\nu_u = \nu_d$, so vacuum stability requires

$$\mathfrak{m}_u^2 + \mathfrak{m}_d^2 > 2B\mu. \tag{10.158}$$

Electroweak symmetry breaking requires that the origin is not a minimum, i.e., that one of the eigenvalues of M^2 is negative (the vacuum stability condition (10.158) does not allow both to be negative), so

$$\det M^2 < 0 \quad \Rightarrow \quad \mathfrak{m}_u^2 \mathfrak{m}_d^2 < (B\mu)^2. \tag{10.159}$$

[43] CP violating VEVs may be induced by loop effects in the effective potential associated with the Yukawa and soft couplings.

If (10.158) and (10.159) are satisfied, one can easily minimize V to obtain

$$\begin{aligned} -|\mu|^2 &= \frac{M_Z^2}{2} + \frac{m_{h_u}^2 \tan^2\beta - m_{h_d}^2}{\tan^2\beta - 1} \\ \sin 2\beta &= \frac{2B\mu}{\mathfrak{m}_u^2 + \mathfrak{m}_d^2} = \frac{2B\mu}{M_A^2}, \end{aligned} \tag{10.160}$$

where $M_Z^2 = (g^2 + g'^2)\nu^2/4$. (See Problem 3.35 for a very similar calculation.) The first equation relates ν^2 to the other parameters. We have separated $|\mu|^2$ from the soft masses to emphasize that electroweak breaking involves a relation between μ and the soft parameters that often requires a nontrivial fine-tuning. Note that $0 \le \sin 2\beta < 1$ $(0 \le \beta < \pi/2)$ by our phase conventions and vacuum stability. Most radiative breaking schemes lead to $m_{h_u}^2 < 0$ (or at least $m_{h_u}^2 < m_{h_d}^2$), because of the large top-Yukawa that drives $m_{h_u}^2$ to lower values at low energy, and $\tan\beta > 1$. The latter is also favored on phenomenological grounds. A non-zero $B\mu$ is required for both doublets to acquire VEVs, as is needed to generate all of the fermion masses.

It is straightforward though somewhat tedious to expand the potential around the minimum to find the Higgs mass eigenstates and eigenvalues (see Problem 3.35) and to write their Feynman rules, so we will only give the main results. We first write the physical neutral fields in terms of Hermitian components by

$$\begin{aligned} h_u &= \frac{1}{\sqrt{2}} \begin{pmatrix} \sqrt{2} h_u^+ \\ \nu_u + h_{uR} + i h_{uI} \end{pmatrix}, \qquad h_d = \frac{1}{\sqrt{2}} \begin{pmatrix} \nu_d + h_{dR} + i h_{dI} \\ \sqrt{2} h_d^- \end{pmatrix} \\ \tilde{h}_d &= i\tau^2 h_d^\dagger = \frac{1}{\sqrt{2}} \begin{pmatrix} \sqrt{2} h_d^+ \\ -\nu_d - h_{dR} + i h_{dI} \end{pmatrix} \end{aligned} \tag{10.161}$$

In the last form, the tilde refers to the $SU(2)$ conjugation and not a superpartner, and $h_d^+ = (h_d^-)^\dagger$. The mass eigenstates consist of two neutral (CP even) scalars h and H (with the convention $M_h < M_H$), one neutral (CP-odd) pseudoscalar A, and one charged pair $H^\pm$, as well as three Goldstone bosons z and $w^\pm$. The latter are analogous to those in the SM (Equation 8.85) and disappear in the unitary gauge. A does not mix with h and H at tree level, but CP-violating loop effects can induce mixing between all three. The mass eigenvalues are

$$\begin{aligned} M_A^2 &= \mathfrak{m}_u^2 + \mathfrak{m}_d^2, \qquad M_{H^\pm}^2 = M_A^2 + M_W^2 \\ M_{h,H}^2 &= \frac{1}{2}\left[M_A^2 + M_Z^2 \mp \sqrt{(M_A^2 + M_Z^2)^2 - 4 M_Z^2 M_A^2 \cos^2 2\beta} \right], \end{aligned} \tag{10.162}$$

so that $M_h^2 + M_H^2 = M_A^2 + M_Z^2$. One also finds the very important (tree-level) constraint

$$M_h^2 \le M_Z^2 \cos^2 2\beta \le M_Z^2, \tag{10.163}$$

with the first inequality saturated for $M_A \gg M_Z$. This is to be contrasted with the SM, where in principle there is no upper limit on the Higgs mass. For large stop masses there are important radiative corrections to this tree-level result, dominated by the top/stop loops in Figure 10.3. At one loop, the upper limit in (10.163) is replaced (under reasonable assumptions) by (Haber and Hempfling, 1991; Okada et al., 1991; Ellis et al., 1991)

$$M_h^2 < M_Z^2 \cos^2 2\beta + \frac{3 g^2 m_t^4}{8\pi^2 M_W^2} \left[\ln\left(\frac{M_S^2}{m_t^2}\right) + \frac{X_t^2}{M_S^2}\left(1 - \frac{X_t^2}{12 M_S^2}\right) \right], \tag{10.164}$$

where $M_S^2 = \frac{1}{2}(m_{\tilde{t}_1}^2 + m_{\tilde{t}_2}^2)$ is the average of the masses of the two stop mass eigenstates (mixtures of $\tilde{t}_L$ and $\tilde{t}_R$), and X_t is a stop mixing parameter that will be defined in the discussion of sparticles below. The largest value occurs for *maximal mixing*, $X_t = \sqrt{6}M_S$, while $X_t = 0$ is the *no mixing* scenario. Still higher-order corrections reduce the limit somewhat, yielding $M_h \lesssim 135$ GeV[44] for maximal mixing and M_S in the several TeV range, while $M_h \lesssim 125$ GeV for no mixing (see, e.g., Hahn et al., 2014; for a recent review, see Draper and Rzehak, 2016). In both cases, the upper limit is for large $\tan\beta \gtrsim 5 - 10$.

Such large masses are not typical, however, with most of the MSSM parameter space yielding M_h well below 125 GeV. One needs[45] either large stop masses in the multi-TeV range or (unexpectedly) large mixing (e.g., Hall et al., 2012; Patrignani, 2016). The smaller electroweak scale ν, which is closely associated with the supersymmetry breaking parameters, requires cancellations that are possible but appear rather unnatural (the little hierarchy problem). Of course, this scenario (*or* the absence of supersymmetry altogether) is also suggested by the nonobservation of superpartners during early LHC running.

The mass eigenstate fields are

$$\begin{pmatrix} h \\ H \end{pmatrix} = \begin{pmatrix} \cos\alpha & -\sin\alpha \\ \sin\alpha & \cos\alpha \end{pmatrix} \begin{pmatrix} h_{uR} \\ h_{dR} \end{pmatrix}, \tag{10.165}$$

where the mixing angle α is given by

$$\cos^2(\beta - \alpha) = \frac{M_h^2(M_Z^2 - M_h^2)}{M_A^2(M_H^2 - M_h^2)}, \tag{10.166}$$

with $-\pi/2 \leq \alpha \leq 0$. The massive pseudoscalar and charged Higgs fields are

$$A = \cos\beta h_{uI} + \sin\beta h_{dI}, \qquad H^+ = \cos\beta h_u^+ + \sin\beta h_d^+, \qquad H^- = \left(H^+\right)^\dagger, \tag{10.167}$$

while the Goldstone bosons are the orthogonal combinations. A simple way to derive (10.167) (and also (10.163)) is to introduce a new basis

$$\begin{pmatrix} h_I \\ h_{II} \end{pmatrix} = \begin{pmatrix} \cos\beta & -\sin\beta \\ \sin\beta & \cos\beta \end{pmatrix} \begin{pmatrix} h_u \\ -\tilde{h}_d \end{pmatrix}. \tag{10.168}$$

By construction, $\langle h_I^0 \rangle = 0$ and $\langle h_{II}^0 \rangle = \sqrt{2}\nu$, so the massive A and H^+ are associated with h_I, while the Goldstone bosons are in h_{II}. In unitary gauge

$$h_I = \frac{1}{\sqrt{2}} \begin{pmatrix} \sqrt{2}H^+ \\ h_{IR} + iA \end{pmatrix}, \qquad h_{II} = \frac{1}{\sqrt{2}} \begin{pmatrix} 0 \\ \nu + h_{IIR} \end{pmatrix}. \tag{10.169}$$

Combining (10.165) and (10.168), the CP-even components of $h_{I,II}$ are

$$\begin{pmatrix} h_{IR} \\ h_{IIR} \end{pmatrix} = \begin{pmatrix} \cos(\beta-\alpha) & -\sin(\beta-\alpha) \\ \sin(\beta-\alpha) & \cos(\beta-\alpha) \end{pmatrix} \begin{pmatrix} h \\ H \end{pmatrix}. \tag{10.170}$$

The $h_{I,II}$ basis is very convenient for deriving the Feynman rules for the gauge-Higgs

[44]The bound is increased to $M_h \lesssim 150 - 200$ GeV in extensions of the MSSM that remain perturbative up to the Planck or GUT scale, such as some of the singlet-extended models (Section 10.2.6) or other gauge extensions (Kane et al., 1993; Espinosa and Quiros, 1993; Batra et al., 2004; Accomando et al., 2006).

[45]We are assuming that the observed 125 GeV state corresponds to the h. However, the possibility that it corresponds to the heavier MSSM scalar is not entirely excluded (e.g., Bechtle et al., 2017).

interactions. h_{IIR} has the same gauge interactions as the SM H in (8.39) on page 265 and Table 8.1. For example, the induced SM HVV $(V = Z, W)$ vertices in (8.39) become

$$\mathcal{L}_{HVV} \to \left(2\frac{M_W^2}{\nu} W^{\mu +} W_\mu^- + \frac{M_Z^2}{\nu} Z^\mu Z_\mu\right) \Big[\sin(\beta - \alpha) h + \cos(\beta - \alpha) H\Big]. \tag{10.171}$$

There are no induced AVV or $H^\pm W^\mp Z$ vertices, since A and $H^\pm$ are associated with the other doublet. One can similarly read off the contributions of h_{II} to the four-point vertices. h_I is a second scalar doublet unrelated to SSB. It leads to new hAZ, HAZ, $hH^\pm W^\mp$, and $HH^\pm W^\mp$ interactions, as well as additional contributions to the four-point vertices. These can be read off from the SM R_ξ gauge interactions in (8.89), with the substitutions $z \to A$, $w^\pm \to H^\pm$, $H \to h_{IR}$, and $\nu \to 0$. (All of the rules are given in detail in Gunion et al., 1990). We give only one example here,

$$\mathcal{L}_{h_{IR}AZ} = \frac{g_Z}{2}\left[A \overleftrightarrow{\partial^\mu}\Big(\cos(\beta - \alpha) h - \sin(\beta - \alpha) H\Big)\right] Z_\mu, \tag{10.172}$$

with $g_Z = (g^2 + g'^2)^{1/2}$, which leads to the transitions $Z \to hA$ and $Z \to HA$, where the Z can be virtual. These have no SM analogs.[46]

The Higgs Yukawa couplings also differ from the SM by the presence of two doublets. The quark terms derived from (10.148) are (the leptons are similar)

$$-\mathcal{L}_{Yuk} = -\Gamma^d \left(\bar{d}_R u_L h_d^- - \bar{d}_R d_L h_d^0\right) + \Gamma^u \left(\bar{u}_R u_L h_u^0 - \bar{u}_R d_L h_u^+\right) + h.c., \tag{10.173}$$

which is to be contrasted with (8.44). Therefore, the Yukawa matrices defined in (8.46) are modified to

$$h^u = \frac{\Gamma^u}{\sqrt{2}} = \frac{M^u}{\nu_u} = \frac{M^u}{\nu \sin\beta}, \qquad h^d = \frac{\Gamma^d}{\sqrt{2}} = \frac{M^d}{\nu_d} = \frac{M^d}{\nu \cos\beta}. \tag{10.174}$$

The Yukawa couplings of t and b to h_{uR} and h_{dR} become

$$h_t = \frac{m_t}{\nu \sin\beta}, \qquad h_b = \frac{m_b}{\nu \cos\beta}, \tag{10.175}$$

respectively, rather than the SM values $m_{t,b}/\nu$. Since m_t is large, h_t would diverge (generate a Landau pole) below the Planck or GUT scale unless $\sin\beta$ is sufficiently large, suggesting $\tan\beta \gtrsim 1.7$ (Dedes et al., 2001). h_b is small in the SM, but can be considerably enhanced in the MSSM for large $\tan\beta$. In fact, it equals h_t for $\tan\beta \sim 45$. If we keep only the t and b Yukawa couplings and use (10.165) and (10.167), the Yukawa couplings to the mass eigenstate Higgs fields are

$$\begin{aligned} -\mathcal{L}_{Yuk} =& \frac{m_t}{\nu \sin\beta} \bar{t}t \,(\cos\alpha h + \sin\alpha H) + \frac{m_b}{\nu \cos\beta} \bar{b}b \,(-\sin\alpha h + \cos\alpha H) \\ & - i\frac{m_t}{\nu \tan\beta} \bar{t}\gamma^5 t\, A - i\frac{m_b}{\nu} \tan\beta\, \bar{b}\gamma^5 b\, A \\ & - \left[\frac{\sqrt{2}m_t}{\nu \tan\beta} \bar{t}_R b_L\, H^+ + \frac{\sqrt{2}m_b}{\nu} \tan\beta\, \bar{b}_R t_L\, H^- + h.c.\right]. \end{aligned} \tag{10.176}$$

[46] The $Z^* \to hA$ process was important prior to the discovery of the 125 GeV state. The LEP 2 lower limit on M_h from Higgstrahlung ($e^-e^+ \to Z^* \to Zh$) was weaker than the corresponding SM limit of 114.4 GeV because of the $\sin(\beta - \alpha)$ factor in (10.171). However, the complementary process (Problem 10.10) $e^-e^+ \to Z^* \to Ah$, with an amplitude proportional to $\cos(\beta - \alpha)$, was important for light enough h and A. By combining the channels, one obtained $M_{h,A} \gtrsim 90$ GeV assuming CP conservation (Schael et al., 2006b).

As expected, h and H couple as scalars and A as a pseudoscalar. (10.176) is easily extended to include the τ, the lighter fermions, and fermion mixing. For large $\tan\beta$ the couplings of A and of H and/or h to $\bar{b}b$ are enhanced, so that at high enough energies the associated production of $A, H,$ or h with a b, e.g., via $bG \to b \to (A, H, h)\, b$, becomes important at the LHC. The combination of the μ term and $\mathcal{L}_{soft}$ may lead to loop-induced flavor-changing neutral Higgs vertices (e.g., Hamzaoui et al., 1999; Gorbahn et al., 2011), which are especially dangerous for large $\tan\beta$.

A very important special case is the *decoupling limit*, $M_A \gg M_Z$, in which H, A, and $H^\pm$ form a degenerate heavy doublet and h acts like the SM Higgs. From (10.166), one has $\cos(\beta - \alpha) \to 0$, i.e, $\alpha \to \beta - \frac{\pi}{2}$. From (10.171), (10.172), and (10.176) h has the couplings of the SM Higgs, and the first tree-level inequality in (10.163) is saturated.

Squarks and Sleptons

The squarks and sleptons acquire masses from a number of sources. Consider the $\tilde{u}_{L,R}$ for a single family. The relevant mass terms in (10.152) are

$$\begin{aligned} V(\tilde{u}_L, \tilde{u}_R) = &\left|\Gamma^u h_u^0\right|^2 \left(|\tilde{u}_L|^2 + |\tilde{u}_R|^2\right) + \left|-\mu h_d^0 + \Gamma^u \tilde{u}_L \tilde{u}_R^\dagger\right|^2 \\ &+ \delta_{\tilde{u}_L}|\tilde{u}_L|^2 + \delta_{\tilde{u}_R}|\tilde{u}_R|^2 \\ &+ m_{\tilde{u}_L}^2|\tilde{u}_L|^2 + m_{\tilde{u}_R}^2|\tilde{u}_R|^2 + \left(A_u \Gamma^u h_u^0 \tilde{u}_L \tilde{u}_R^\dagger + h.c.\right), \end{aligned} \tag{10.177}$$

where the three lines represent V_F, V_D, and V_{soft}, respectively. The coefficients $\delta_{\tilde{u}_{L,R}}$ in V_D are

$$\frac{|h_d^0|^2 - |h_u^0|^2}{2} g_Z^2 \left[t_{\tilde{u}_{L,R}}^3 - \sin^2\theta_W\, q_{\tilde{u}_{L,R}}\right] \to \cos 2\beta M_Z^2 \left[t_{\tilde{u}_{L,R}}^3 - \sin^2\theta_W\, q_{\tilde{u}_{L,R}}\right], \tag{10.178}$$

where $t_{\tilde{u}_L}^3$ $(t_{\tilde{u}_R}^3) = \frac{1}{2}$ (0), and $q_{\tilde{u}_{L,R}} = \frac{2}{3}$ are the third component of weak isospin and charge. Replacing $|\Gamma^u h_u^0| \to m_u$, the 2×2 mass term is therefore

$$-\mathcal{L}_{\tilde{u}} = \begin{pmatrix} \tilde{u}_L^\dagger & \tilde{u}_R^\dagger \end{pmatrix} \begin{pmatrix} m_{\tilde{u}_L}^2 + m_u^2 + \delta_{\tilde{u}_L} & X_u^* m_u \\ X_u m_u & m_{\tilde{u}_R}^2 + m_u^2 + \delta_{\tilde{u}_R} \end{pmatrix} \begin{pmatrix} \tilde{u}_L \\ \tilde{u}_R \end{pmatrix}, \tag{10.179}$$

where

$$X_u \equiv A_u - \mu^* \cot\beta. \tag{10.180}$$

Similar expressions hold for $\tilde{d}$, $\tilde{\nu}$, and $\tilde{e}$, except the appropriate values of t^3 and q must be inserted, and $\cot\beta \to \tan\beta$ for $\tilde{d}$ and $\tilde{e}$. One can easily extend the discussion to allow mixing between the families, in which case the mass-squared matrix in (10.179) would be 6×6. Most consideration has been given to models in which such interfamily mixings are minimized (by choosing or generating family universal boundary conditions at the GUT or messenger scale) in order to suppress FCNC.

In most realistic scenarios the soft mass-squares $m_{\tilde{f}_{L,R}}^2$ dominate. The supersymmetric (but $SU(2) \times U(1)$ breaking) m_f^2 terms are negligible except for the stop, and to a lesser extent the sbottoms and staus; the $\delta_{\tilde{f}_{L,R}}$ contributions are also small but need to be included. The $X_f m_f$ terms lead to mixing between $\tilde{f}_L$ and $\tilde{f}_R$. These are also small except for the third family, especially the stops. The latter may be split significantly for large enough X_t, and could even lead to an unphysical unstable vacuum with a negative $m_{\tilde{t}_1}^2$ for large A_t. The significance of X_t for the lightest Higgs mass was discussed below (10.164).

In the simple minimal supergravity and gauge mediation scenarios, the squarks are usually heavier than the sleptons because of RGE running and/or because of their stronger coupling to messengers. One exception is the lighter stop eigenvalue, which can be the lightest sparticle for large enough X_t.

Squarks and sleptons can be produced efficiently in pp or $\bar{p}p$ scattering in the decay chains of gluinos, if they are heavier, or of other squarks/sleptons. Squarks can also also be produced directly by QCD processes (including t and u-channel squark or gluino exchange), such as $GG \to \tilde{q}\tilde{q}^\dagger$, $q\bar{q} \to \tilde{q}\tilde{q}^\dagger$, $qq \to \tilde{q}\tilde{q}$, or $Gq \to \tilde{q}\tilde{G}$ (the relevant cross sections for these and other processes are collected in the Cross Section article in Patrignani, 2016). Third family squarks have lower production rates because of the smaller number of corresponding quarks in the proton. For example, the production of $\tilde{u}_{L,R}$ by valence u quarks through t-channel gluino exchange is very efficient for pp, but negligible for $\tilde{t}$. The cross sections for the direct production of sleptons in hadronic processes are mainly due to s-channel Drell-Yan ($W^\pm$, Z, γ) exchange, and are much smaller.

Gluinos

The eight gluinos $\tilde{G}^i$ can acquire Majorana masses $m_{\tilde{G}}$ from $\mathcal{L}_{soft}$ in (10.149). Gluinos can be produced by ordinary QCD processes, as well as by processes such as t-channel squark exchange in $q\bar{q}$ scattering. The gluino can decay into a quark and squark, $\tilde{G} \to q\tilde{q}^\dagger$ or $\bar{q}\tilde{q}$, where the $\tilde{q}$ may be real or virtual. This initiates a cascade of decays, ultimately leading to a number of $q, \bar{q}, G, \ell$, and $\bar{\ell}$ as well as the LSP (e.g., a neutralino or the gravitino/Goldstino) in R-parity conserving models, with several jets + $\not{E}_T$ being the classic and most obvious signature (for an early study, see Hinchliffe et al., 1997). Since the $\tilde{G}$ is Majorana, conjugate pairs of decays such as $\tilde{G} \to q\bar{q}e^+\nu_e$+ LSP and $\tilde{G} \to \bar{q}qe^-\bar{\nu}_e$+ LSP occur with equal rates (up to CP violating effects), so pair-produced gluinos should lead to the same number of same-sign charged leptons as opposite sign. From (10.134) the gluino-quark-squark interactions (for a single family) are

$$\begin{aligned}\mathcal{L}_{\tilde{G}q\tilde{q}} = &-\sqrt{2}g_s\left[\bar{q}_L\frac{\lambda^i}{2}P_R\tilde{G}^i_M\tilde{q}_L - \bar{q}_R\frac{\lambda^i}{2}P_L\tilde{G}^i_M\tilde{q}_R\right.\\ &\left.+\tilde{q}^\dagger_L\frac{\lambda^i}{2}\bar{\tilde{G}}^i_M P_L q_L - \tilde{q}^\dagger_R\frac{\lambda^i}{2}\bar{\tilde{G}}^i_M P_R q_R\right],\end{aligned} \tag{10.181}$$

where $\tilde{G}^i_M = \tilde{G}^i_L + \tilde{G}^{ic}_R$, and we have inserted the $P_{L,R}$ to make clear the chirality.

Equation (10.181) is easily extended to three families, provided one interprets the quark and squark fields as weak eigenstates. When the interaction is rewritten in terms of mass eigenstate fields, the vertices will in general be flavor-changing due to a (probable) mismatch between the quark and squark unitary transformations, much like the generation of the CKM matrix due to a mismatch between the u and d type quarks. This can lead to potentially dangerous FCNC effects (Gabbiani et al., 1996; Misiak et al., 1998; Altmannshofer et al., 2010; Arana-Catania et al., 2014) due to diagrams such as those in Figure 10.7, unless the squark masses are nearly degenerate, the quark and squark mixing matrices are approximately aligned, or the squarks are very heavy (multi-TeV). The gluino vertices are especially problematic because of the strong QCD coupling, but there are analogous problems involving neutralino and chargino vertices and the sleptons (Section 9.6).

Neutralinos, and Charginos

From (10.148) and (10.149) we see that the neutral Higgsino pair $\tilde{h}^0_{u,d}$ have a Dirac mass $-\mu$, while the neutral gauginos $\tilde{B}$ and $\tilde{W}^3$ have Majorana masses $m_{\tilde{B}}$ and $m_{\tilde{W}}$, respectively.

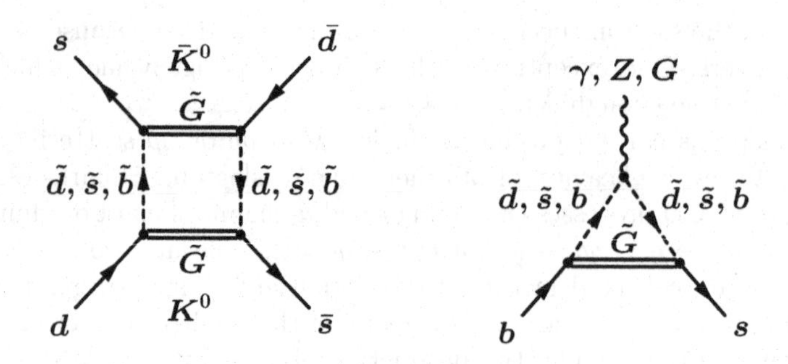

Figure 10.7 Typical gluino-quark-squark diagrams leading to FCNC effects. Left: a new contribution to the $K_L - K_S$ mass difference. Right: a contribution to $b \to s + (\gamma, Z, G)$. There are analogous diagrams involving neutralinos, charginos, and sleptons.

In addition, the gaugino-Higgsino-Higgs interactions from (10.133),

$$-\mathcal{L} = \sqrt{2}g\left[h_u^\dagger \tilde{W}^i \frac{\tau^i}{2}\tilde{h}_u + h_d^\dagger \tilde{W}^i \frac{\tau^i}{2}\tilde{h}_d\right] + \sqrt{2}g'\left[h_u^\dagger \frac{\tilde{B}}{2}\tilde{h}_u - h_d^\dagger \frac{\tilde{B}}{2}\tilde{h}_d\right] + h.c., \tag{10.182}$$

lead to gaugino-Higgsino mixing when the neutral Higgs fields acquire VEVs. The relevant terms are

$$-\mathcal{L} = \frac{1}{2}\left[-g\nu_u \tilde{W}^3 \tilde{h}_u^0 + g\nu_d \tilde{W}^3 \tilde{h}_d^0 + g'\nu_u \tilde{B}\tilde{h}_u^0 - g'\nu_d \tilde{B}\tilde{h}_d^0\right] + h.c., \tag{10.183}$$

leading to the 4×4 Majorana neutralino mass term

$$-\mathcal{L}_{\tilde{\chi}^0} = \frac{1}{2}\begin{pmatrix}\tilde{B} & \tilde{W}^3 & \tilde{h}_d^0 & \tilde{h}_u^0\end{pmatrix}\begin{pmatrix} m_{\tilde{B}} & 0 & -\frac{g'\nu_d}{2} & \frac{g'\nu_u}{2} \\ 0 & m_{\tilde{W}} & \frac{g\nu_d}{2} & -\frac{g\nu_u}{2} \\ -\frac{g'\nu_d}{2} & \frac{g\nu_d}{2} & 0 & -\mu \\ \frac{g'\nu_u}{2} & -\frac{g\nu_u}{2} & -\mu & 0 \end{pmatrix}\begin{pmatrix}\tilde{B} \\ \tilde{W}^3 \\ \tilde{h}_d^0 \\ \tilde{h}_u^0\end{pmatrix} + h.c. \tag{10.184}$$

Equation (10.184) implies four Majorana mass eigenstate neutralinos, $\tilde{\chi}_r^0, r = 1\cdots 4$, with masses $m_{\tilde{\chi}_r^0}$.

Similarly, the charged Higgsinos $\tilde{h}_u^+$ and $\tilde{h}_d^-$ have a Dirac mass $+\mu$, while $\tilde{W}^1$ and $\tilde{W}^2$ can be combined to form a Dirac pair $\tilde{W}^\pm = (\tilde{W}^1 \mp \tilde{W}^2)/\sqrt{2}$ (cf. Equation 8.28). The charged Higgsinos and gauginos are mixed by (10.182), leading to the 2×2 Dirac mass matrix

$$-\mathcal{L}_{\tilde{\chi}^\pm} = \begin{pmatrix}\tilde{W}^+ & \tilde{h}_u^+\end{pmatrix}\begin{pmatrix} m_{\tilde{W}} & \frac{g\nu_d}{\sqrt{2}} \\ \frac{g\nu_u}{\sqrt{2}} & \mu \end{pmatrix}\begin{pmatrix}\tilde{W}^- \\ \tilde{h}_d^-\end{pmatrix} + h.c. \tag{10.185}$$

The mass eigenstate Dirac charginos $\tilde{\chi}_{1,2}^\pm$ have mass eigenvalues

$$\begin{aligned} m^2_{\tilde{\chi}^\pm_{1,2}} =& \frac{1}{2}\left(|m_{\tilde{W}}|^2 + |\mu|^2 + 2M_W^2\right) \\ &\mp \frac{1}{2}\left(\left(|m_{\tilde{W}}|^2 + |\mu|^2 + 2M_W^2\right)^2 - 4|m_{\tilde{W}}\mu - M_W^2 \sin 2\beta|^2\right)^{1/2}. \end{aligned} \tag{10.186}$$

Some models predict gaugino unification, i.e., that the gaugino masses are in the ratio

$$m_{\tilde{G}} : m_{\tilde{W}} : m_{\tilde{B}} \sim \alpha_3 : \alpha_2 : \alpha_1 \sim 7 : 2 : 1, \tag{10.187}$$

where

$$\alpha_3 = \alpha_s, \qquad \alpha_2 = \frac{g^2}{4\pi}, \qquad \alpha_1 = \frac{5}{3}\frac{g'^2}{4\pi}. \tag{10.188}$$

α_1 (the GUT-normalized $U(1)$ coupling) is rescaled because $\sqrt{\frac{3}{5}}Y$ has the same Dynkin index for a fermion family as the $SU(2)$ and $SU(3)$ generators. This occurs, for example, in minimal supergravity schemes in which the gaugino masses are equal at the GUT scale, as are the GUT-normalized gauge couplings. The RGE equations preserve the ratios. Simple forms of gauge mediation lead to the same relation, while anomaly mediation predicts an entirely different pattern with the $m_{\tilde{W}}$ the smallest. Unfortunately, the actual measurement of the gaugino masses is complicated by mixing with the Higgsinos. We finally comment that gaugino mass terms break continuous R symmetries,[47] but allow a discrete R-parity.

The neutralinos and charginos are expected to be produced efficiently at hadron colliders through the decays of squarks and sleptons. Initial pairs of squarks and/or gluinos produced by QCD processes could cascade decay into multiple jets and leptons, e.g.,

$$\tilde{q} \to q\tilde{\chi}_2^0, \qquad \tilde{\chi}_2^0 \to \ell^\pm \tilde{\ell}^\mp, \qquad \tilde{\ell}^\mp \to \ell^\mp \tilde{\chi}_1^0. \tag{10.189}$$

If the $\tilde{\chi}_1^0$ is the LSP it will escape from the detector, implying large numbers of cascades with multiple jets and leptons and missing transverse energy. Detailed studies of the distributions of the number of jets, numbers of same and opposite sign leptons, multiple leptons, lepton flavors, etc., should yield information on the spectrum and test the Majorana character of the gluinos and neutralinos. If the intermediate particles in a cascade are on-shell, it should be possible to determine the masses by kinematic edge techniques (Problem 10.12). Some spin information may also be obtainable from the angular correlations of the decay products (e.g., Wang and Yavin, 2008). This would be important to distinguish supersymmetry from other scenarios with similar cascades, such as some versions of Little Higgs and extra dimensional models, as mentioned in the Higgs/hierarchy discussion in Section 10.1.

Neutralinos and charginos (and sleptons) may also be produced directly, but at a lower rate, by Drell-Yan processes, t-channel squark exchange, etc. For example, trileptons plus missing energy could result (e.g., Barger and Kao, 1999) from

$$W^{+*} \to \tilde{\chi}_1^+ \tilde{\chi}_2^0, \qquad \tilde{\chi}_1^+ \to \ell^+ \nu \tilde{\chi}_1^0, \qquad \tilde{\chi}_2^0 \to \ell^+ \ell^- \tilde{\chi}_1^0. \tag{10.190}$$

In supergravity and anomaly mediated scenarios with R-parity conservation the $\tilde{\chi}_1^0$ is usually the LSP, typically with a dominant $\tilde{B}$ composition in minimal supergravity or $\tilde{W}^0$ for anomaly mediation.[48] Direct, indirect, and collider dark matter searches have excluded much of the parameter space for neutralino dark matter,[49] but there is still an allowed region with a low spin-independent cross section (Figure 10.2). For a recent discussion, see (Baer et al., 2016). In models with a lower supersymmetry breaking scale, such as gauge mediation, the LSP is expected to be the gravitino/Goldstino, $g_{3/2}$. The next lightest

[47]The continuous R symmetries can be restored in models with *Dirac gauginos*, which are extensions of the MSSM involving additional gauge adjoint chiral supermultiplets (e.g., Patrignani, 2016).

[48]The dominantly $\tilde{W}^\pm$ chargino is expected to be somewhat heavier due to loop corrections (Pierce et al., 1997), allowing $\tilde{W}^\pm \to \tilde{W}^0 \pi^\pm$.

[49]The $\tilde{\nu}_L$ would be another cold dark matter candidate, but it is strongly excluded by direct CDM detection searches.

sparticle (NLSP) may be the $\tilde{\chi}_1^0$ or it could be a charged (or even colored) particle, such as the $\tilde{\tau}_R$. The NLSP may decay promptly, with a displaced vertex, or outside the detector, e.g., by $\tilde{\chi}_1^0 \to (\gamma g_{3/2}, Z g_{3/2},$ or $h g_{3/2})$ or $\tilde{\ell} \to \ell g_{3/2}$. Possibilities for dark matter and cosmological constraints from NLSP decays are discussed in (Giudice and Rattazzi, 1999).

Experimental Constraints on Supersymmetry

As of September 2016, there has not been any direct experimental evidence for supersymmetry, implying that if supersymmetry exists in the low-energy theory it is most likely in the decoupling limit, in which the sparticle and extra Higgs particle masses are large compared to M_Z. (Some possible loopholes are mentioned below.) This is also suggested by the relatively large value 125 GeV of the Higgs mass.

In the decoupling limit, the effects of the MSSM on the precision electroweak observables, such as the oblique parameters introduced in Section 8.3.6, are small (e.g., Erler and Pierce, 1998; Heinemeyer et al., 2006; Ramsey-Musolf and Su, 2008; Cho et al., 2011). Similarly, the predictions for FCNC processes such as $b \to s\gamma$ and $B_{s,d} \to \mu^+\mu^-$ (Section 8.6.6) are very similar to those of the SM (which roughly agree with experiment) for relevant regions of parameter space, at least in the CMSSM (Buchmueller et al., 2014). However, for heavy sparticles the contribution of the MSSM to the anomalous magnetic moment of the muon, $g_\mu - 2$, from diagrams such as in Figure 2.22 on page 73 is also expected to be small, and the condition (2.377) for the MSSM to explain the $g_\mu - 2$ discrepancy is most likely not satisfied.

The ATLAS and CMS collaborations have carried out a great many supersymmetry searches, including analyses motivated by the CMSSM, gauge mediation, R_p violation, simplified models, and the pMSSM.

There was no sign of the heavier Higgs states predicted by the MSSM or general two-doublet models in Run 1 at the LHC. Especially stringent were limits on H or A decaying to $\tau^+\tau^-$, which are strongly enhanced in the decoupling limit for large $\tan\beta$ (similar to the $\bar{b}b$ couplings in (10.176)), with $\tan\beta \gtrsim 10(40)$ excluded for $M_A \sim 300(800)$ GeV (Patrignani, 2016). The limits are weak for smaller $\tan\beta$, but it is then more difficult to achieve $M_h = 125$ GeV since the upper limit on the tree-level term in (10.163) becomes smaller.

The experimental limits and future prospects for sparticle masses depend on the mass of the LSP; the ordering of the masses, e.g., whether the gluino is lighter than all of the squarks; and other model dependent assumptions, so they will not be described in detail here. Searches at LEP and the Tevatron are surveyed in (Feng et al., 2010). The ATLAS and CMS Run 1 results,[50] analyzed in the pMSSM framework, are given in (Aad et al., 2015c; Khachatryan et al., 2016b), respectively, while general reviews may be found in (e.g., Melzer-Pellmann and Pralavorio, 2014; Autermann, 2016; Patrignani, 2016). Roughly, for supergravity models with a sufficiently light $\tilde{\chi}_1^0$ the LHC Run 1 was sensitive to masses of around 1.2 TeV for gluinos and for squarks of the first two families; 600 GeV for $\tilde{t}$ and $\tilde{b}$; 400 GeV for $\tilde{\chi}_1^\pm, \tilde{\chi}_2^0$; and 300 GeV for charged sleptons. There were also significant limits placed on the interaction strength of WIMP dark matter candidates (i.e., $\tilde{\chi}_1^0$) over a wide mass range from the nonobservation of an associated jet, Z, γ, etc. Similar sensitivities were achieved for gauge mediation, especially for decays of the NLSP into a photon. Preliminary Run 2 results through mid 2016 extended the exclusions by more than 50% in mass.

The various searches come with many caveats, coverage gaps, and loopholes. For example, limits are weaker for compressed spectra, in which parent and daughter particles are sufficiently close in mass to suppress a decay mode, or to reduce visible and missing energies

[50]See (Aad et al., 2009; Bayatian et al., 2007) for the respective capabilities of ATLAS and CMS.

below a trigger threshold. Thus, limits in supergravity are strongly dependent on the $\tilde{\chi}_1^0$ mass, and are usually presented as exclusions in the $m_{\tilde{p}} - m_{\tilde{\chi}_1^0}$ plane, where $\tilde{p}$ is a superpartner. Some limits are presented assuming a 100% branching ratio for a particular decay, and can be weakened by having multiple decay modes. (This is one of the motivations for a pMSSM analysis, which takes this possibility partially into account.) Most of the limits are not directly applicable to models with R-parity violation (below), in which there may be no missing E_T, and for which separate analyses are needed. Similar comments apply to models involving a long-lived particle that decays away from the main interaction vertex or outside the detector. Limits may also be weakened in variations on the MSSM. These include models with Dirac gluinos, for which the important production channel $uu \to \tilde{u}\tilde{u}$ is absent, and *stealth* models, in which a new light chiral supermultiplet drains off most of the energy in a decay chain before decaying back to SM particles and a light LSP.

10.2.6 Further Aspects of Supersymmetry

R-Parity Violation

The superpotential in (10.148) assumed a conserved R-parity, a discrete R symmetry under which each particle of spin S has $R_p = (-1)^{3(B-L)+2S}$. All of the ordinary particles (quarks, leptons, Higgs, and gauge bosons) have $R_p = +1$, while their superpartners have $R_p = -1$. R-parity therefore ensures that the LSP is stable and suppresses precision electroweak vertex corrections. However, it is possible add terms to W that violate R-parity,

$$W_{\not{R}_p} \sim \underbrace{H_u L}_{\not{L}\text{ mixing}}, \quad \underbrace{LLE^c}_{\not{L}\text{ dilepton}}, \quad \underbrace{LQD^c}_{\not{L}\text{ leptoquark}}, \quad \underbrace{U^c D^c D^c}_{\not{B}\text{ diquark}}, \tag{10.191}$$

where we have suppressed the coefficients (the LLE^c and $U^c D^c D^c$ terms must be antisymmetric in the LL or $D^c D^c$ family indices). The first three terms violate lepton number, but are otherwise allowed because L and H_d have the same SM gauge quantum numbers. The fourth term violates baryon number. The LQD^c and $U^c D^c D^c$ operators imply leptoquark and diquark couplings, respectively, for the squarks (Section 8.6.6), while the $\tilde{e}_R$ in LLE^c is a dilepton (Cuypers and Davidson, 1998). R-parity can also be violated spontaneously if $\langle \tilde{\nu}_{L,R} \rangle \neq 0$.

The simultaneous presence of the leptoquark and diquark couplings would lead to rapid proton decay via $\tilde{d}^c$ exchange, which would generate a $B - L$ conserving operator ℓqud. One of the two must therefore be absent or incredibly small, but the other would be allowed. The diquark operators could lead to *neutron oscillations*, i.e., $n \to \bar{n}$ transitions or nn annihilations in a nucleus[51] (e.g., Kuzmin, 1970; Calibbi et al., 2016). The various operators are also strongly constrained by precision electroweak measurements, FCNC, EDMs, and direct collider searches, especially for the first two families (Barbier et al., 2005; Allanach et al., 1999; Doršner et al., 2016). The $H_u L$ operator or sneutrino VEVs would lead to neutrino-neutralino (and charged lepton-chargino) mixing, which could give mass to one neutrino. The other neutrinos could obtain mass at loop level from the trilinear operators (e.g., Grossman and Rakshit, 2004; Rakshit, 2004).

In the presence of R-parity violation the LSP would no longer be stable, so other candidates (such as an axion) would be needed for the dark matter. For example, the couplings in (10.191) could lead to such decays as $\tilde{\chi}_1^0 \to \ell^+ \ell^- \overset{(-)}{\nu}, \ell^\pm q\bar{q}, \overset{(-)}{\nu} q\bar{q}$, or qqq. Such decays can be searched for if they decay inside the detector, or would resemble the R-parity conserving

[51]The possibility of neutron oscillations is more general than R-parity violation. For a general review, see (Phillips et al., 2016).

case if they occurred outside. There would also be the possibility of other types of LSP, such as a $\tilde{\tau}$ or even a gluino. Depending on the LSP and the flavor structure of the couplings many types of decays with unusual signatures are possible, including the $\tilde{\chi}_1^0$ decays and others such as $\tilde{\tau} \to \tau e^{\pm} \mu^{\mp} \overset{(-)}{\nu}$, $\tilde{\nu}_\tau \to \mu^{\mp} e^{\pm}$, $\tilde{t} \to \tau^+ b$, and $\tilde{G} \to tbs$.

The μ Problem

The μ problem was already introduced in Section 10.2.1. A nonzero μ is not actually required for the Higgs sector of the theory, since the needed terms in the tree-level potential can be generated by the soft terms $m^2_{h_{u,d}}$ and $B\mu$ in (10.153). However, a significant μ is required to generate large enough masses for the charginos, as can be seen from (10.186).

One possibility is to generate μ at the same time as the soft terms by non-renormalizable operators coupling the observable and hidden sectors (Giudice and Masiero, 1988; Casas and Munoz, 1993). To illustrate how this works, it is convenient to introduce a *spurion* superfield $X = \theta^2 F_X$, with a non-zero F component. A spurion is usually an effective or composite field (or superfield in this case) that is introduced with the right transformation properties to parametrize the effects of symmetry breaking.

One typically expects string or supergravity effects to generate allowed higher-dimensional operators involving X and the other superfields in the Kähler potential or superpotential, presumably associated with inverse powers of the Planck scale M_P. For example, a term $-c_r X^\dagger X \Phi_r^\dagger \Phi_r / M_P^2$ in the Kähler potential would lead to the soft mass-squared term

$$\mathcal{L}_{m_r^2} = -\int d^4\theta c_r \frac{X^\dagger X \Phi_r^\dagger \Phi_r}{M_P^2} = -c_r \frac{|F_X|^2}{M_P^2} \phi_r^\dagger \phi_r, \tag{10.192}$$

with $m_r^2 = c_r |F_X|^2 / M_P^2$. Gaugino masses and A terms can be similarly generated, e.g., a superpotential term $-c_{123} X \Phi_1 \Phi_2 \Phi_3 / M_P$ leads to a corresponding A term $c_{123} F_X \phi_1 \phi_2 \phi_3 / M_P$. In the *Giudice-Masiero mechanism* one also introduces a term $c_\mu X^\dagger H_u H_d / M_P + h.c.$ into the Kähler potential, which yields a μ term

$$W_\mu = \int d^2\bar{\theta} c_\mu \frac{X^\dagger H_u H_d}{M_P} = c_\mu \frac{F_X^\dagger}{M_P} H_u H_d, \tag{10.193}$$

with $\mu = c_\mu F_X^\dagger / M_P$ the same order of magnitude as the soft terms.

In this discussion, we have implicity assumed a supergravity mediation, so that $m_{soft} \sim F_X / M_P$. In schemes with a lower F_X, such as gauge mediation, such operators are still present but are negligibly small compared to m_{soft}, which is now $\lesssim F_X / M$, where $M \ll M_P$ is the messenger scale. It is also sometimes the case, even in supergravity mediation, that there are new discrete or continuous symmetries that allow the operators such as (10.192) but not the one in (10.193).

An alternative solution is to utilize renormalizable operators. In the singlet-extended versions of the MSSM[52] one replaces μ by a dynamical variable, by introducing a SM singlet superfield S with a superpotential coupling

$$W_{SH_u H_d} = \lambda_S S H_u H_d \tag{10.194}$$

to the doublets. If S acquires a VEV an effective μ parameter $\mu_{eff} = \lambda_S \langle S \rangle$ is generated. The singlet extensions usually impose additional symmetries on the theory to forbid an elementary μ term. There are a number of realizations of this mechanism (see Accomando

[52]For other possibilities, see (Komargodski and Seiberg, 2009).

et al., 2006; Barger et al., 2007; Maniatis, 2010; Ellwanger et al., 2010, for reviews). The best known is the *next to minimal model* (NMSSM), in which a discrete Z_3 symmetry forbids μ but allows the cubic terms $\lambda_S S H_u H_d$ and κS^3 in W (Ellis et al., 1989). The original form of the NMSSM suffers from cosmological domain wall problems because of the discrete symmetry. This can be remedied in more sophisticated forms involving a broken R symmetry (Panagiotakopoulos and Tamvakis, 1999b). A variation on that approach yields the *new minimal model* (nMSSM), in which the cubic S^3 term and its soft analog are replaced by tadpole terms linear in S with sufficiently small coefficients (Panagiotakopoulos and Tamvakis, 1999a). A $U(1)'$ symmetry, which is perhaps more likely to emerge from a string construction, is another possibility (Suematsu and Yamagishi, 1995; Cvetič and Langacker, 1996; Cvetič et al., 1997). This avoids the domain wall problem by embedding the discrete symmetry of the NMSSM into a continuous one. The singlet extended models involve an additional Higgs scalar and (except for the $U(1)'$ case) an additional pseudoscalar that may be light. The branching ratios (into SM particles) and couplings of the lightest Higgs may be reduced by decays into two pseudocalars (Chang et al., 2008) or by an admixture of singlet component (e.g., Barger et al., 2006). The theoretical *upper* limit in (10.164) is also increased (to around 150 GeV) by new F term and (for $U(1)'$) D-term contributions to the potential. As mentioned in Section 10.1, the singlet-extended models also make it relatively easy to achieve the strong first order electroweak phase transition needed for electroweak baryogenesis.

Gauge Unification

As will be discussed in Section 10.4 simple grand unified theories predict that the properly normalized gauge couplings in (10.188) should all be equal at and above the GUT (unification) scale M_X, above which the symmetry breaking can be ignored. Such unification may also occur in some superstring theories that compactify directly to the SM or the MSSM. Unification can be tested using the observed gauge couplings at M_Z and the known RGE equations for the gauge couplings. From (5.36) on page 171, at one loop, one has

$$\frac{1}{\alpha_i(Q^2)} = \frac{1}{\alpha_i(M_Z^2)} - 4\pi b_i \ln \frac{Q^2}{M_Z^2}, \tag{10.195}$$

where

$$\begin{pmatrix} b_1 \\ b_2 \\ b_3 \end{pmatrix} = \frac{1}{16\pi^2} \begin{pmatrix} \frac{41}{10} \\ -\frac{19}{6} \\ -7 \end{pmatrix}_{SM} \quad \text{or} \quad \frac{1}{16\pi^2} \begin{pmatrix} \frac{33}{5} \\ 1 \\ -3 \end{pmatrix}_{MSSM} \tag{10.196}$$

in the SM or the MSSM, respectively. There were early indications that unification was more successful in the MSSM than in the SM (Dimopoulos et al., 1981; Amaldi et al., 1987), but precise tests became possible (Ellis et al., 1990; Amaldi et al., 1991; Langacker and Luo, 1991; Giunti et al., 1991) after the gauge couplings at M_Z were determined accurately at LEP. (In practice, one needs to include two-loop effects; corrections or uncertainties associated with the t, Higgs, sparticle, and GUT particle thresholds; and a conversion of the $\overline{MS}$ gauge couplings to the dimensional reduction ($\overline{DR}$) scheme[53] appropriate for supersymmetry (Langacker and Polonsky, 1993, 1995; Carena et al., 1993; Bagger et al., 1995).) As can be seen in Figure 10.8, the couplings do not precisely unify when extrapolated assuming the SM, but do approximately meet at around $M_X \sim 5 \times 10^{16}$ GeV when the effects of the new particles in the MSSM are included in the β functions.

[53]The $\overline{DR}$ scheme is similar to dimensional regularization except that the Dirac algebra is evaluated in four dimensions.

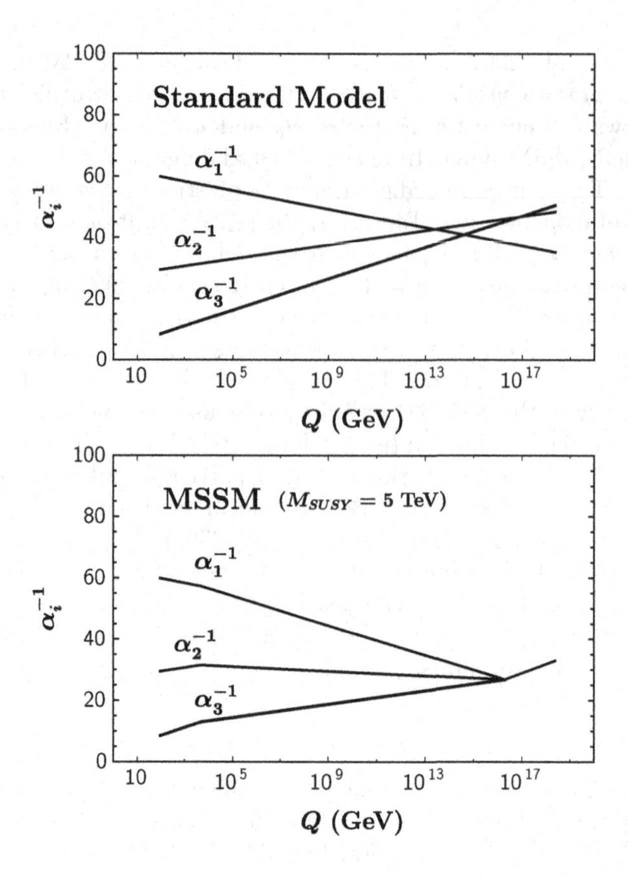

Figure 10.8 Extrapolation of the observed gauge couplings in the standard model and MSSM. The lower plot assumes that the new MSSM particles have mass around 5 TeV. The extension above M_X assumes supersymmetric $SU(5)$.

10.3 EXTENDED GAUGE GROUPS

There have been many proposed extensions of the gauge symmetry of the standard model or of the MSSM. One class, which is the focus of this section, involves extending the electroweak $SU(2) \times U(1)$ gauge symmetry. Such extensions are sometimes motivated by bottom-up considerations, e.g., solving the μ problem of the MSSM in $U(1)'$ extensions, or restoring P and C invariance in $SU(2)_L \times SU(2)_R \times U(1)$ models. Another, and perhaps stronger, motivation is top-down: such extra gauge symmetries often appear as accidental remnants of the breaking pattern of an underlying theory, such as a superstring theory or a theory in which new TeV scale physics is responsible for $SU(2) \times U(1)$ breaking.

Another class of extensions[54] involves the introduction of *horizontal* (or family or flavor) symmetries, which relate or distinguish between different fermion families and hopefully shed light on the spectrum. The horizontal symmetry could be discrete or continuous. If it is continuous and spontaneously broken it is presumably a gauge symmetry to avoid unwanted Goldstone bosons. FCNC, which can be mediated, e.g., by gauge bosons, are a serious con-

[54]For a general discussion, see (Ramond, 1979). Examples include (Altarelli et al., 2008; Luhn and Ramond, 2008). See Section 9.5.5 for applications to neutrinos.

straint on models with family symmetries broken at the electroweak or TeV scale. These are not necessarily present in the original weak eigenstate basis. Family transitions at one vertex may be compensated at the other, or there may be family diagonal but nonuniversal transitions. However, the interfamily mixing associated with the transformation to the mass eigenstate basis almost inevitably leads to FCNC. Unfortunately, there are no really compelling models of horizontal symmetries.

There have also been various extensions of QCD, such as a spontaneously-broken chiral or non-chiral $SU(3) \times SU(3)$ group, with the unbroken $SU(3)$ emerging as the diagonal subgroup. The massive color octet bosons are referred to as axigluons or colorons, respectively. See (e.g., Chivukula et al., 2015) and Problem 5.3. Various strongly-coupled gauge groups have also been suggested in connection with electroweak symmetry breaking (Section 8.5.3).

Grand unified theories, which unify the strong and electroweak interactions, will be discussed in Section 10.4.

10.3.1 $SU(2) \times U(1) \times U(1)'$ Models

Many extensions of the SM and MSSM involve additional $U(1)'$ factors and associated Z' gauge bosons (for reviews, see Hewett and Rizzo, 1989; Leike, 1999; Langacker, 2009; Han et al., 2013; Patrignani, 2016). These include superstring theories, grand unified theories, and many models involving new TeV-scale physics, such as dynamical symmetry breaking and little Higgs models. As a simple example of why $U(1)'$s so often survive, consider $SU(m)$ broken by a real adjoint Higgs, ϕ_i. Define the $m \times m$ matrix $\phi \equiv \phi_i L^i_m$, analogous to (3.31) on page 97. The VEV $\langle \phi \rangle$ can be diagonalized by an $SU(m)$ transformation, which makes it clear that the unbroken subgroup is $U(1)^m$ (or a larger group if some of the diagonal entries are equal).

The extra gauge bosons could be extremely heavy, massless or very light, or anywhere in between. Here we will mainly consider electroweak/TeV scale Z's. These are especially motivated in supersymmetric models, where the $SU(2) \times U(1)$ and $U(1)'$ breaking scales are usually both driven by m_{soft}. We will also comment briefly on very light "dark photons".

In addition, some models involving extra dimensions of space allow the Z and other SM gauge bosons to propagate in the bulk, implying Kaluza-Klein excitations with masses of order $R^{-1} \sim 2$ TeV $\times$ $(10^{-17}\text{cm}/R)$, where R is the scale of the extra dimension. These excitations would be similar to the bosons from new gauge symmetries, except that their couplings would be essentially the same as the SM ones.

$U(1)'$ *Gauge Interactions*

Consider the case of a single additional $U(1)'$ factor, with family universal couplings and (initially) no kinetic mixing (Section 2.14). The SM coupling of the three neutral gauge bosons to fermions in (8.55) on page 271 generalizes to[55]

$$-\mathcal{L}_{NC} = gJ_3^\mu W_\mu^3 + g' J_Y^\mu B_\mu + g_2 J_2^\mu Z_{2\mu}^0 = eJ_Q^\mu A_\mu + g_1 J_1^\mu Z_{1\mu}^0 + g_2 J_2^\mu Z_{2\mu}^0, \tag{10.197}$$

where $Z_{2\mu}^0$ is the new gauge boson, J_2^μ is the $U(1)'$ current, and g_2 the gauge coupling.[56] If we were to work in terms of W^3, B, and Z_2^0, then following SSB one would have to deal with a 3×3 gauge boson mass matrix. However, this can be avoided (so long as electric

[55]In some cases, the two $U(1)$ currents are linear combinations of J_Y^μ and a second current. However, if the model contains the SM as a subgroup, it can be written in this form by a rotation of the two abelian gauge bosons.

[56]The g_1 and g_2 used here differ from the $\sqrt{5/3}g'$ and g used in connection with gauge unification.

charge is not broken) by first transforming to A and $Z_1^0 \equiv Z$, which are related to W^3 and B by (8.30) and (8.33), i.e., they are the neutral SM gauge bosons. Similarly, $J_1^\mu \equiv J_Z^\mu/2$ where J_Z^μ is the SM current defined in (8.73), and $g_1 \equiv (g^2 + g'^2)^{1/2} = g_Z$. This second form is especially convenient when the mixing with the Z_2^0 is a perturbation. The currents $J_\alpha^\mu, \alpha = 1, 2$, are

$$J_\alpha^\mu = \sum_r \bar{\psi}_r \gamma^\mu [\epsilon_L^\alpha(r) P_L + \epsilon_R^\alpha(r) P_R] \psi_r = \frac{1}{2} \sum_r \bar{\psi}_r \gamma^\mu [g_V^\alpha(r) - g_A^\alpha(r)\gamma^5] \psi_r, \tag{10.198}$$

where the $\epsilon_{L,R}^1$ are the SM couplings in (8.76), and the $\epsilon_{L,R}^2$ depend on the $U(1)'$ model. The currents take the same form in the fermion weak and mass bases because of our assumption of family universality. When working in terms of left chiral fermion fields, it will be convenient to define the charges

$$Q_{\alpha f} \equiv \epsilon_L^\alpha(f), \qquad Q_{\alpha f^c} \equiv -\epsilon_R^\alpha(f). \tag{10.199}$$

for fermion f_L and its charge conjugate f_L^c. We similarly define $U(1)$ charges $Q_{\alpha i}$ for a complex scalar ϕ_i, with $Q_{1i} = t_i^3 - \sin^2\theta_W\, q_i$. The neutral current part of its gauge covariant derivative is therefore

$$D_\mu \phi_i = \left(\partial_\mu + ieq_i A_\mu + i \sum_{\alpha=1}^{2} g_\alpha Q_{\alpha i} Z_{\alpha\mu}^0 \right) \phi_i. \tag{10.200}$$

All members of an $SU(2)$ multiplet must have the same Q_2 since the two groups are assumed to commute.

Gauge Boson Masses and Mass Mixing

When some of the scalars acquire VEVs, they will generate masses for the neutral (and charged) gauge bosons. Assuming that electric charge is not broken, i.e., $q_i = 0$ for all scalars with $\langle\phi_i\rangle \neq 0$, one finds from (10.200) that the photon A_μ remains massless, while $Z_{1,2}^0$ develop a mass-squared matrix

$$M_{Z-Z'}^2 = \begin{pmatrix} 2g_1^2 \sum_i (t_i^3)^2 |\langle\phi_i\rangle|^2 & 2g_1 g_2 \sum_i t_i^3 Q_i |\langle\phi_i\rangle|^2 \\ 2g_1 g_2 \sum_i t_i^3 Q_i |\langle\phi_i\rangle|^2 & 2g_2^2 \sum_i Q_i^2 |\langle\phi_i\rangle|^2 \end{pmatrix} \equiv \begin{pmatrix} M_{Z^0}^2 & \Delta^2 \\ \Delta^2 & M_{Z'}^2 \end{pmatrix}, \tag{10.201}$$

where $Q_i \equiv Q_{2i}$. M_{Z^0} would be the Z mass in the absence of mixing. If the Higgs fields are all $SU(2)$ doublets and singlets, then

$$M_{Z^0}^2 = \frac{M_W^2}{\cos^2\theta_W} = \frac{g_Z^2}{4} \sum_{t_i = \frac{1}{2}} |\nu_i|^2 \equiv \frac{g_Z^2}{4} \nu^2, \tag{10.202}$$

where $\nu_i = \sqrt{2}\langle\phi_i\rangle$ and $\nu \sim 246$ GeV, as in the SM. The mass eigenstates corresponding to (10.201) are

$$\begin{pmatrix} Z_1 \\ Z_2 \end{pmatrix} = U \begin{pmatrix} Z_1^0 \\ Z_2^0 \end{pmatrix}, \qquad U = \begin{pmatrix} \cos\theta & \sin\theta \\ -\sin\theta & \cos\theta \end{pmatrix}, \tag{10.203}$$

with eigenvalues

$$M_{1,2}^2 = \frac{1}{2} \left[M_{Z^0}^2 + M_{Z'}^2 \mp \sqrt{(M_{Z^0}^2 - M_{Z'}^2)^2 + 4\Delta^4} \right]. \tag{10.204}$$

The mixing angle is given by

$$\theta = -\frac{1}{2}\arctan\left(\frac{2\Delta^2}{M_{Z'}^2 - M_{Z^0}^2}\right), \qquad \tan^2\theta = \frac{M_{Z^0}^2 - M_1^2}{M_2^2 - M_{Z^0}^2}. \tag{10.205}$$

An important limit is $M_{Z'} \gg (M_{Z^0}, |\Delta|)$, which typically occurs because an $SU(2)$ singlet field S has a large VEV $\gg \nu$ and contributes only to $M_{Z'}$. One then has

$$M_1^2 \sim M_{Z^0}^2 - \frac{\Delta^4}{M_{Z'}^2} \ll M_2^2, \qquad M_2^2 \sim M_{Z'}^2 \tag{10.206}$$

and

$$\theta \sim -\frac{\Delta^2}{M_{Z'}^2} \sim C\frac{g_2}{g_1}\frac{M_1^2}{M_2^2} \text{ with } C = -\frac{\sum_i t_i^3 Q_i |\langle\phi_i\rangle|^2}{\sum_i (t_i^3)^2 |\langle\phi_i\rangle|^2}. \tag{10.207}$$

C is model dependent, but typically $|C| \lesssim \mathcal{O}(1)$. From (10.205)–(10.207) one sees that both $|\theta|$ and the downward shift $(M_{Z^0} - M_1)/M_{Z^0}$ are of order M_1^2/M_2^2.

As an example, consider a complex $SU(2)$ singlet field S and two $SU(2)$ doublets $\phi_{u,d}$ or their conjugates $h_{u,d}$. The doublets are defined as in (10.12) and (10.13) for the MSSM, but we are not necessarily requiring supersymmetry. Then,

$$\begin{aligned} M_{Z^0}^2 &= \frac{1}{4}g_1^2(|\nu_u|^2 + |\nu_d|^2), \qquad \Delta^2 = \frac{1}{2}g_1 g_2 (Q_u|\nu_u|^2 - Q_d|\nu_d|^2) \\ M_{Z'}^2 &= g_2^2(Q_u^2|\nu_u|^2 + Q_d^2|\nu_d|^2 + Q_S^2|s|^2), \end{aligned} \tag{10.208}$$

with $Q_{u,d} \equiv Q_{h_u,h_d}$ and $s = \sqrt{2}\langle S\rangle$. In the supersymmetric version of this example (e.g., Cvetič et al., 1997) one usually assumes that the $U(1)'$ symmetry does not allow an elementary μ term $\mu H_u H_d$ in W, i.e., $Q_u + Q_d \neq 0$, but does allow the coupling $\lambda_S S H_u H_d$ in (10.194), i.e., $Q_S + Q_u + Q_d = 0$. The potential for S (which denotes the scalar component as well as the superfield), h_u^0, and h_d^0 is then $V = V_F + V_D + V_{soft}$, where the MSSM expressions in (10.153) are replaced by

$$\begin{aligned} V_F &= \lambda_S^2\left(|h_u^0|^2|h_d^0|^2 + |S|^2|h_u^0|^2 + |S|^2|h_d^0|^2\right) \\ V_D &= \frac{g_1^2}{8}\left(|h_u^0|^2 - |h_d^0|^2\right)^2 + \frac{g_2^2}{2}\left(Q_u|h_u^0|^2 + Q_d|h_d^0|^2 + Q_S|S|^2\right)^2 \\ V_{soft} &= m_{h_u}^2|h_u^0|^2 + m_{h_d}^2|h_d^0|^2 + m_S^2|S|^2 - \left(\lambda_S A_S S h_u^0 h_d^0 + h.c.\right). \end{aligned} \tag{10.209}$$

μ and $B\mu$ from the MSSM are replaced by $\mu_{eff} = \lambda_S\langle S\rangle$, and $(B\mu)_{eff} = \lambda_S A_S\langle S\rangle$. There is no MSSM analog of the first (second) term in V_F (V_D). One can make field redefinitions and gauge transformations so that $(B\mu)_{eff}$ and the VEVs are real and positive, and it is automatic (at least in the tree-level Higgs sector) that the minimum preserves $U(1)_Q$ invariance. (The contributions of h_u^+ and h_d^- to V are similar to the MSSM and are not displayed.) We see from (10.209) that all of the dimensional parameters in V are given by soft supersymmetry breaking terms, resolving the μ problem and suggesting that both the electroweak and $U(1)'$ breaking scales should be close to m_{soft}, up to an order of magnitude or so.[57] The new terms in (10.209) increase the tree-level limit in (10.163) for the lightest Higgs scalar to

$$M_h^2 \leq M_Z^2\cos^2 2\beta + \frac{1}{2}\lambda_S^2\nu^2\sin^2 2\beta + g_2^2[Q_d\cos^2\beta + Q_u\sin^2\beta]^2\nu^2, \tag{10.210}$$

allowing M_h as large as $\sim$ 150 GeV when higher-order corrections are included (Barger et al., 2006).

[57]More complicated models involving additional S fields with opposite sign charges may allow intermediate scale $U(1)'$ breaking along flat or nearly flat directions (Cleaver et al., 1998; Erler et al., 2002).

Kinetic Mixing

Kinetic mixing was discussed generally in Section 2.14. For a $U(1)_1 \times U(1)_2$ gauge theory, the most general gauge kinetic term[58] is (Holdom, 1986)

$$\mathcal{L}_{kin} \to -\frac{c_1}{4}\hat{F}_1^{0\mu\nu}\hat{F}_{1\mu\nu}^0 - \frac{c_2}{4}\hat{F}_2^{0\mu\nu}\hat{F}_{2\mu\nu}^0 - \frac{c_{12}}{2}\hat{F}_1^{0\mu\nu}\hat{F}_{2\mu\nu}^0, \tag{10.211}$$

where $\hat{F}^0_{\alpha\mu\nu}$ are the field strength tensors. Without loss of generality, we can rescale the $\hat{Z}^0_{1,2}$ so that $c_1 = c_2 = 1$ and (since the eigenvalues must be positive) $c_{12} = \sin\chi$. The kinetic mixing term is not expected initially if the $U(1)^2$ is embedded in a simple group, but it can be generated at the loop level due to multiplet mass splitting, e.g., it can arise from RGE effects if $\mathrm{Tr}_{\,m<\mu}(Q_1Q_2) \neq 0$, where the notation indicates that the trace is restricted to the particles lighter than the scale μ (e.g., Babu et al., 1998). It can also come about by superstring loop effects (Dienes et al., 1997) or by other constructions.

Kinetic mixing could provide a means to induce a weak coupling between the ordinary sector and a supersymmetric or dark matter hidden sector. For example, a massless Z' mixing with the photon would result in a small fractional electric charge for the hidden sector particles (Holdom, 1986). Here we restrict ourselves to the case in which kinetic mixing is a small (e.g., RGE induced) perturbation on the effects of a TeV scale $U(1)'$. Suppose one has

$$\mathcal{L} = -\frac{1}{4}\hat{F}_1^{0\mu\nu}\hat{F}_{1\mu\nu}^0 - \frac{1}{4}\hat{F}_2^{0\mu\nu}\hat{F}_{2\mu\nu}^0 - \frac{\sin\chi}{2}\hat{F}_1^{0\mu\nu}\hat{F}_{2\mu\nu}^0 + \frac{1}{2}\hat{Z}^{0\mu}\hat{M}^2_{Z-Z'}\hat{Z}^0_\mu, \tag{10.212}$$

where $\hat{M}^2_{Z-Z'}$ is induced by Higgs VEVs, and has the same form as (10.201). One can put $\mathcal{L}_{kin}$ into canonical form using the non-unitary transformation

$$\begin{pmatrix} \hat{Z}^0_{1\mu} \\ \hat{Z}^0_{2\mu} \end{pmatrix} = \begin{pmatrix} 1 & -\tan\chi \\ 0 & 1/\cos\chi \end{pmatrix} \begin{pmatrix} Z^0_{1\mu} \\ Z^0_{2\mu} \end{pmatrix} \equiv V \begin{pmatrix} Z^0_{1\mu} \\ Z^0_{2\mu} \end{pmatrix}, \tag{10.213}$$

in which case the mass-squared matrix becomes

$$\hat{M}^2_{Z-Z'} \to M^2_{Z-Z'} = V^T \hat{M}^2_{Z-Z'} V, \tag{10.214}$$

which must be diagonalized as in (10.203). It is straightforward to show that in the limit $(M_{Z^0}, |\Delta|) \ll M_{Z'}$ and $|\chi| \ll 1$ the effects of χ on M_1^2 and θ are second order and negligible (i.e., $\Delta^2 \to \Delta^2 - M^2_{Z^0}\chi$ in (10.206) and (10.207)). The only important change is then that the neutral current couplings of the Z_2^0 in (10.197) are shifted to include a first-order component proportional to J_1,

$$-\mathcal{L}_{NC} \to eJ^\mu_Q A_\mu + g_1 J^\mu_1 Z^0_{1\mu} + \left(g_2 J^\mu_2 - g_1\chi J^\mu_1\right) Z^0_{2\mu}. \tag{10.215}$$

The light boson couplings are not affected at this order, and one must still include the further effects of mass mixing (the θ rotation). This shift in the Z_2^0 couplings is especially important because in many models the fermion charges are orthogonal (i.e., $\mathrm{Tr}\,(Q_1Q_2) = 0$) prior to kinetic mixing; it can be used, e.g., to suppress the couplings of the Z_2 to leptons (*leptophobic* models).

---

[58]Analogous kinetic mixing terms are *not* allowed for non-abelian theories.

$U(1)'$ *Models*

There are an enormous number of possible $U(1)'$ models, distinguished by the Z' mass, the chiral couplings to the quarks and leptons and whether they are family universal, the extended Higgs sector, the possible *exotic* fields that may be necessary to avoid anomalies from the extended gauge sector, possible couplings to a hidden sector, possible kinetic mixing, etc. Even assuming electroweak/TeV scale masses, couplings comparable to electroweak, and family universal charges there are many models in the literature. Here we list a few of the major classes.

The *sequential* Z_{SM} boson is defined to have the same couplings to fermions as the SM Z boson. This is difficult to achieve in an ordinary gauge theory,[59] but is a useful benchmark.

The electric charge Q and the weak hypercharge $Y = Q - T_L^3$ in (8.5) can be written (at least for the SM particles) as

$$Q = T_L^3 + Y = T_L^3 + T_R^3 + T_{BL}, \tag{10.216}$$

where T_R^3 takes the values $t_R^3 = +\frac{1}{2}$ for ν_R and u_R; $t_R^3 = -\frac{1}{2}$ for e_R and d_R; and 0 for ψ_L. T_{BL} is defined as $(B - L)/2$, where B (L) are baryon (lepton) number. The *left-right* (LR) models, which can descend from the $SU(2)_L \times SU(2)_R \times U(1)$ models considered in the next section or from $SO(10)$ grand unified theories, involve the neutral current interaction

$$-\mathcal{L}_{NC} = gJ_{3L}^{\mu} W_{L\mu}^3 + g_R J_{3R}^{\mu} W_{R\mu}^3 + g_{BL} J_{BL}^{\mu} W_{BL\mu}, \tag{10.217}$$

in an obvious notation. Anticipating that $U(1)_{3R} \times U(1)_{BL}$ will be broken to $U(1)_Y$ at a scale $M_{Z'} \gg M_{Z^0}$, it is convenient to put (10.217) into the form (10.197) by rotating W_R^3 and W_{BL} to a new basis B and Z_2^0. This leaves invariant the kinetic energy terms (which we assume are canonical). One can take $B = \cos\gamma\, W_{3R} + \sin\gamma\, W_{BL}$ and choose γ so that B couples to $g'Y$. One then finds (Problem 10.16) that (8.69) becomes

$$\frac{1}{e^2} = \frac{1}{g^2} + \frac{1}{g'^2} = \frac{1}{g^2} + \frac{1}{g_R^2} + \frac{1}{g_{BL}^2}. \tag{10.218}$$

$Z_2^0 = \sin\gamma\, W_{3R} - \cos\gamma\, W_{BL}$ is associated with the charge

$$Q^{LR} = \sqrt{\frac{3}{5}} \left[\alpha T_{3R} - \frac{1}{\alpha} T_{BL} \right], \tag{10.219}$$

where $\alpha = \tan\gamma = g_R/g_{BL} = \sqrt{\kappa^2 \cot^2\theta_W - 1}$, with $\kappa \equiv g_R/g$. The coupling has been normalized to $g_2 = \sqrt{\frac{5}{3}} g \tan\theta_W \sim 0.46$. For example, the left-right symmetric version of the $SU(2)_L \times SU(2)_R \times U(1)$ model (Mohapatra, 2003), breaking at the TeV scale, implies $g_R = g$ so that $\alpha \sim 1.53$. Realistic breaking patterns for $SO(10)$ typically lead to $\alpha \sim 0.7 - 0.9$ (Robinett and Rosner, 1982b). The χ model in Table 10.2 coincides with the LR model with $\alpha = \sqrt{2/3} \sim 0.82$. Generalizations of the LR models, e.g., to $J_2 = J_{BL}$, are described in the reviews.

The LR models and their generalizations are attractive in that they are the unique

[59]It could occur as a diagonal subgroup in a complicated constructions involving new fermions, or as a Kaluza-Klein excitation in theories with extra TeV^{-1} scale dimensions.

nontrivial family-universal $U(1)'$ extensions of the SM that do not require the introduction of additional chiral fermions (other than ν_R) to cancel anomalies. However, they are perhaps less interesting in a supersymmetric context, because the Higgs superfields $H_{u,d}$ form a vector pair with $T^3_R = \pm\frac{1}{2}$ and $T_{BL} = 0$. Therefore, an elementary μ term in (10.148) is *not* forbidden by the extra $U(1)'$. One must also introduce SM singlet supermultiplets to break the $U(1)'$. They would most likely be introduced as non-chiral vector pairs to avoid anomalies, leading to their own "μ" problem. (One could instead give large VEVs to the scalar partners of the ν^c, but this would break R-parity and would be challenging for neutrino phenomenology.)

Many Z' studies focus on the two extra $U(1)$'s that occur in the decomposition of the E_6 GUT (Robinett and Rosner, 1982a; Langacker et al., 1984; Hewett and Rizzo, 1989), i.e., $E_6 \to SO(10) \times U(1)_\psi$ and $SO(10) \to SU(5) \times U(1)_\chi$. We consider them only as simple examples of anomaly-free $U(1)'$ charges and exotic fields, and do not assume a full underlying grand unified theory.[60] In E_6, each family of left-handed fermions is promoted to a fundamental 27-plet, which decomposes under $E_6 \to SO(10) \to SU(5)$ as

$$27 \to 16 + 10 + 1 \to (10 + 5^* + 1) + (5 + 5^*) + 1, \tag{10.220}$$

as shown in Table 10.2. In addition to the 15 fermions charged under the SM gauge group and the ν^c, each 27-plet contains a second SM singlet, the S. Both the ν^c and the S may be charged under the $U(1)'$. There is also an exotic color-triplet quark $\mathcal{D}$ with charge $-1/3$ and its conjugate $\mathcal{D}^c$, both of which are $SU(2)$ singlets, and a pair of color-singlet $SU(2)$-doublets, $h_u = \begin{pmatrix} h_u^+ \\ h_u^0 \end{pmatrix}$ and $h_d = \begin{pmatrix} h_d^0 \\ h_d^- \end{pmatrix}$ with $y_{h_{u,d}} = \pm 1/2$. The exotic fields, which are needed to cancel anomalies, are all singlets or non-chiral under the standard model, and therefore do not yield large corrections to the precision electroweak observables. However, they are usually chiral under the $U(1)'$, i.e., they are *quasi-chiral.*

TABLE 10.2 Couplings of the Z^0_χ, Z^0_ψ, Z^0_η, and Z^0_N to an E_6 27-plet of left-chiral fermions or superfields.

$SO(10)$		$SU(5)$	$2\sqrt{10}Q_\chi$	$\sqrt{24}Q_\psi$	$2\sqrt{15}Q_\eta$	$2\sqrt{10}Q_N$
	10	(u, d, u^c, e^+)	-1	1	-2	1
16	5^*	(d^c, ν, e^-)	3	1	1	2
	1	ν^c	-5	1	-5	0
	5	$(\mathcal{D}, h_u)$	2	-2	4	-2
10	5^*	$(\mathcal{D}^c, h_d)$	-2	-2	1	-3
1	1	S	0	4	-5	5

The E_6 *models* can be considered in both non-supersymmetric and supersymmetric versions. In the supersymmetric case, the scalar partners of the S (or the ν^c) can develop VEVs to break the $U(1)'$ symmetry. Similarly, the scalar partners of one $h_{u,d}$

[60] In a full supersymmetric grand unified theory with a TeV-scale $U(1)'$ the scalar partners of the $\mathcal{D}$ and $\mathcal{D}^c$ would have Yukawa couplings related to those of the Higgs, and could therefore mediate rapid proton decay.

pair can be interpreted as the two Higgs doublets of the MSSM. The two additional $h_{u,d}$ families may be interpreted either as additional Higgs pairs or as exotic-leptons (h_d has the same SM quantum numbers as an ordinary lepton doublet, while h_u would be conjugate to a right-handed exotic doublet).

$U(1)_\psi \times U(1)_\chi$ may survive to low energies, though most studies assume that only the $U(1)_\psi$, the $U(1)_\chi$, or a $U(1)$ associated with a linear combination of their charges does so. Examples of linear combinations include $Q_\eta = \sqrt{\frac{3}{8}}Q_\chi - \sqrt{\frac{5}{8}}Q_\psi$, which occurs in some heterotic string compactifications, or $Q_N = \frac{1}{4}Q_\chi + \frac{\sqrt{15}}{4}Q_\psi$, which allows a neutrino seesaw with a large ν^c mass (Ma, 1996) or can avoid cosmological and astrophysical constraints on Dirac neutrinos (Barger et al., 2003b).

Except for the χ model, all of these $U(1)'$s forbid an elementary μ term. They allow all of the other (Yukawa) interactions in the MSSM superpotential in (10.148), as well as the terms SH_uH_d and $S\mathcal{D}\mathcal{D}^c$, where $H_{u,d}$ are the Higgs/exotic lepton superfields, and $\mathcal{D}$ and $\mathcal{D}^c$ represent the superfields as well as the fermions. Therefore, the $U(1)'$ breaking induced by $\langle S\rangle$ can also generate masses for the Higgs and exotic fields. Leptoquark or diquark operators, such as $LQ\mathcal{D}^c$ or $U^cD^c\mathcal{D}^c$, which can allow $\mathcal{D}$ decay and also FCNC, are also allowed by the $U(1)'$. (They cannot both be present because they would then lead to rapid proton decay.) One problem with the E_6 models is that in its minimal form the particle content is not consistent with the minimal MSSM-type gauge coupling unification. This can be restored by adding a vector pair of Higgs-type doublets (e.g., an additional $H_u + H_u^c$) from an incomplete $27 + 27^*$, but only at the cost of introducing a μ-type problem for this new pair.

The supersymmetric LR and E_6 models all either involve a vector pair of chiral superfields, therefore leading to a version of the μ problem, or else are not consistent with the simple form of gauge unification found in the MSSM. The *minimal unification models* (Erler, 2000) remedy this by starting with the MSSM particle content and then choosing exotics in sets that preserve the MSSM unification at tree-level, with $U(1)'$ charges chosen to cancel anomalies. At least two SM singlets S_i with different $U(1)'$ charges are required to given mass to all of the exotics, however. It is possible but nontrivial to ensure that enough of them acquire VEVs, and to avoid unwanted accidental global symmetries and the associated Goldstone bosons (Langacker et al., 2009).

plications of a $U(1)'$

Z' at the electroweak-TeV scale would have important implications for WNC, Z-pole, and P 2 experiments, as well as for direct searches at colliders, e.g., in $\bar{p}p$ or $pp \to Z' \to \ell^+\ell^-$ the Drell-Yan (s-channel) process.

The low energy and LEP 2 experiments are influenced not only by $Z - Z'$ mixing, which difies the Z couplings, but by Z_2 exchange and by the shift in the Z_1 mass from its value. The effective four-fermi interaction relevant to low energy processes in (8.79) is laced (recalling that $J_1 = J_Z/2$) by

$$\mathcal{H}_{eff}^{NC} = -\mathcal{L}_{eff}^{NC} = \frac{4G_F}{\sqrt{2}}(\rho_{eff}J_1^2 + 2wJ_1J_2 + yJ_2^2), \tag{10.221}$$

where

$$\rho_{eff} = \rho_1 \cos^2\theta + \rho_2 \sin^2\theta \sim \rho_1, \qquad w = \frac{g_2}{g_1}\cos\theta\sin\theta(\rho_1 - \rho_2) \sim \frac{g_2}{g_1}\theta$$
$$y = \left(\frac{g_2}{g_1}\right)^2 (\rho_1 \sin^2\theta + \rho_2\cos^2\theta) \sim \left(\frac{g_2}{g_1}\right)^2 \rho_2. \tag{10.222}$$

In (10.222) $\rho_\alpha \equiv M_W^2/(M_\alpha^2 \cos^2\theta_W)$, and the second expressions are for small ρ_2 and θ. In the same limit, the Z_1 mass shift in (10.206) results in

$$\rho_1 \sim 1 + \theta^2/\rho_2 = 1 + \left(\frac{g_2}{g_1}\right)^2 C^2 \rho_2, \tag{10.223}$$

where C is defined in (10.207). The modification of (10.222) relevant to LEP 2 is straightforward. The Z-pole experiments are insensitive to Z_2 exchange, but are very sensitive to $Z - Z'$ mixing. This shifts the Z_1 mass downward according to (10.223), and therefore shifts the value of $\sin^2\theta_W$ obtained from M_Z relative to the values from the asymmetries and other observables. The mixing also modifies the tree-level vector and axial couplings of the Z_1 to fermion ψ_r. The couplings in (8.76) become

$$\begin{aligned} g_V^r &\to \cos\theta g_V^1(r) + \frac{g_2}{g_1}\sin\theta g_V^2(r) \sim g_V^1(r) + \frac{g_2}{g_1}\theta g_V^2(r) \\ g_A^r &\to \cos\theta g_A^1(r) + \frac{g_2}{g_1}\sin\theta g_A^2(r) \sim g_A^1(r) + \frac{g_2}{g_1}\theta g_A^2(r). \end{aligned} \tag{10.224}$$

These depend on the $U(1)'$ charges, so the oblique parameter formalism described in Section 8.3.6 is usually not a good parametrization of Z' effects. To the extent that the $U(1)'$ is a small tree-level perturbation it is usually sufficient to use the SM radiative corrections, though some care is needed in the definition of $\sin^2\theta_W$, e.g., by using the $\overline{MS}$ rather than the on-shell definition.

The limits set by the WNC and LEP 2 on a Z' mass M_2 were very model dependent, with typical sensitivities for the E_6 models in the several hundred GeV to $\gtrsim 1$ TeV range.[61] The Z-pole experiments limited the mixing to $|\theta| \lesssim$ few $\times 10^{-3}$.

The M_2 limits have been superseded by direct seaches at the Tevatron and LHC, searching for resonances in $\bar{p}p$ or $pp \to e^+e^-$ or $\mu^+\mu^-$ (and also $\tau^+\tau^-$, dijets, $\bar{t}t$, dibosons, etc) produced by the Drell-Yan process. Early Run 2 results by ATLAS (Aaboud et al., 2016) and CMS (Khachatryan et al., 2017) set lower limits of $\sim 3-4$ TeV on typical E_6-type bosons with electroweak coupling. These should be extended[62] to around 5-6 TeV later in Run 2. If a Z' were observed, various diagnostic studies of its couplings, by forward-backward asymmetries, rapidity distributions, heavy quark decays, associated productions, and rare decays, would in principle be possible (e.g., Cvetič and Godfrey, 1995; Han et al., 2013), but the statistical significance at Run 2 would be limited for masses much above the current exclusions. A later high luminosity phase at the LHC will extend the sensitivities somewhat, while a possible future 100 TeV pp collider would extend the typical reach to around 30 TeV (Arkani-Hamed et al., 2016). A possible future e^+e^- collider could also extend the LHC reach. Interference between s-channel Z and γ amplitudes with that of a heavy virtual Z' should be observable for masses well above the CM energy in cross sections,

[61] These and the other results assume an electroweak scale g_2, with the GUT-motivated value $g_2 \sim \sqrt{5/3}g' = \sqrt{5/3}g\tan\theta_W$ often assumed.

[62] These limits can be lower by as much as 1 TeV if the sparticle/exotic decay channels are open.

forward-backward and polarization asymmetries, etc. (e.g., Han et al., 2013; Arbey et al., 2015).

The observation of a TeV scale Z', especially in the context of supersymmetry, could have a significance far beyond the Z' itself. The possible physics implications include: (a) an extended Higgs/neutralino sector, complicating and modifying the collider physics signatures and expanding the range of possibilities for neutralino (or other kinds of) cold dark matter; (b) the possible existence of heavy (Z' scale) exotic particles, which may decay rapidly or be quasi-stable; (c) possible FCNC effects if the Z' couplings are family non-universal (which often occurs in string constructions); (d) an enhanced possibility for electroweak baryogenesis; (e) a number of possibilities for exponential or power law suppressed Dirac or Majorana neutrino masses; (f) a possible efficient source for sparticle production; and (g) a possible connection to an otherwise hidden sector, either by kinetic mixing or by direct $U(1)'$ couplings (which often occurs in string constructions). The latter possibility could lead to Z' mediation of supersymmetry breaking by a $Z' - \tilde{Z}'$ mass difference. All of these possibilities are reviewed in (Langacker, 2009).

Very Light Z's

We finally mention the possibility of very light Z' gauge bosons (often referred to as dark photons or as U bosons) with extremely small couplings to ordinary matter. A common scenario is that the new boson connects to ordinary matter only via kinetic mixing with the photon. Major motivations are that a dark photon could account for the discrepancy in the muon magnetic moment (Section 2.12.3) and/or that it could be a portal that mediates weak interactions between a dark sector (for the cold dark matter) and the SM particles. Typical masses considered are in the MeV-GeV range, with a kinetic mixing parameter χ of $\mathcal{O}(10^{-3} - 10^{-2})$, but much wider mass ranges and much smaller mixing are possible. Numerous constraints, e.g., from beam dump experiments, rare K and hyperon decays, fractional charges, QED, astrophysics, and cosmology, are reviewed in (Pospelov, 2009; Jaeckel and Ringwald, 2010; Essig et al., 2013; Alexander et al., 2016).

10.3.2 $SU(2)_L \times SU(2)_R \times U(1)$ Models

The simplest SM extension involving an additional charged gauge boson is the $SU(2)_L \times SU(2)_R \times U(1)$ model (Pati and Salam, 1974; Mohapatra and Pati, 1975; Senjanovic and Mohapatra, 1975), in which the first (second) $SU(2)$ couples to L (R)-chiral fermions. A left-right (LR) interchange symmetry between the factors is usually imposed, so that space reflection and charge conjugation invariance hold at the Lagrangian level even though the theory is chiral. The original version of the model, in which the extended gauge and LR symmetries are broken at the TeV scale, is disfavored by gauge unification and neutrino mass constraints. Most modern treatments assume that one or both symmetries are broken at a large scale, and they are usually considered in the context of a subgroup of the supersymmetric $SO(10)$ grand unified theory. The model has also been revived (Agashe et al., 2003) in the context of warped extra-dimensional models to provide a custodial symmetry to protect the ρ_0 parameter. Here, we will give a short overview of the original non-supersymmetric TeV scale model as useful background material. Modern developments are reviewed, e.g., in (Mohapatra, 2003; Maiezza et al., 2010; Cao et al., 2012; Patrignani, 2016).

The $SU(2)_L \times SU(2)_R \times U(1)_{BL}$ Fields and Gauge Interactions

We assume that the fermion fields

$$q^0_{mL} = \begin{pmatrix} u^0_m \\ d^0_m \end{pmatrix}_L, \quad q^0_{mR} = \begin{pmatrix} u^0_m \\ d^0_m \end{pmatrix}_R, \quad \ell^0_{mL} = \begin{pmatrix} \nu^0_m \\ e^0_m \end{pmatrix}_L, \quad \ell^0_{mR} = \begin{pmatrix} \nu^0_m \\ e^0_m \end{pmatrix}_R \tag{10.225}$$

transform under $SU(2)_L \times SU(2)_R \times U(1)_{BL}$ as

$$(t_L, t_R, t_{BL}) = (2, 1, \frac{1}{6}), \quad (1, 2, \frac{1}{6}), \quad (2, 1, -\frac{1}{2}), \quad (1, 2, -\frac{1}{2}), \tag{10.226}$$

generalizing (8.4). The $SU(2)_L$ and $SU(2)_R$ gauge bosons $W^{i\mu}_{L,R}$, $i = 1, 2, 3$, therefore couple to $V - A$ and $V + A$ currents, respectively. The $U(1)_{BL}$ gauge boson W^{μ}_{BL} couples to the charge $T_{BL} = (B - L)/2$ that was defined in (10.216). The gauge couplings $g_L \equiv g$, g_R, and g_{BL} are related to the g' of the SM and to e by (10.218). The fermion covariant kinetic energies are

$$\mathcal{L}_f = \bar{q}^0_L i \not{D} q^0_L + \bar{q}^0_R i \not{D} q^0_R + \bar{\ell}^0_L i \not{D} \ell^0_L + \bar{\ell}^0_R i \not{D} \ell^0_R, \tag{10.227}$$

where the sum over families is implied. $\mathcal{L}_f$ includes the fermion gauge interactions

$$\begin{aligned} \mathcal{L}_{fI} = &-\bar{q}^0_L \gamma_\mu \left[g_L \frac{\vec{\tau} \cdot \vec{W}^\mu_L}{2} + \frac{g_{BL}}{6} W^\mu_{BL} \right] q^0_L - \bar{q}^0_R \gamma_\mu \left[g_R \frac{\vec{\tau} \cdot \vec{W}^\mu_R}{2} + \frac{g_{BL}}{6} W^\mu_{BL} \right] q^0_R \\ &- \bar{l}^0_L \gamma_\mu \left[g_L \frac{\vec{\tau} \cdot \vec{W}^\mu_L}{2} - \frac{g_{BL}}{2} W^\mu_{BL} \right] l^0_L - \bar{l}^0_R \gamma_\mu \left[g_R \frac{\vec{\tau} \cdot \vec{W}^\mu_R}{2} - \frac{g_{BL}}{2} W^\mu_{BL} \right] l^0_R. \end{aligned} \tag{10.228}$$

A Higgs multiplet transforming as $(2, 2^*, 0)$ is needed to generate fermion masses and eventually break the $SU(2) \times U(1)$ symmetry. We introduce Φ and its tilde form $\tilde{\Phi}$,

$$\Phi = \begin{pmatrix} \phi^0_1 & \phi^+_1 \\ \phi^-_2 & \phi^0_2 \end{pmatrix}, \qquad \tilde{\Phi} = \tau_2 \Phi^* \tau_2 = \begin{pmatrix} \phi^{0*}_2 & -\phi^+_2 \\ -\phi^-_1 & \phi^{0*}_1 \end{pmatrix} \tag{10.229}$$

in analogy to (8.14). Like most extensions of the SM, there are two separate $SU(2)$ doublets. Unlike the MSSM, both neutral Higgs fields will have Yukawa couplings to both u and d when one includes the tilde couplings, implying Higgs mediated FCNC at some level. Φ transforms as

$$\Phi \to \Phi' = U_L(\vec{\beta}_L) \Phi U_R(\vec{\beta}_R)^\dagger = e^{i\vec{\beta}_L \cdot \frac{\vec{\tau}}{2}} \Phi e^{-i\vec{\beta}_R \cdot \frac{\vec{\tau}}{2}}, \tag{10.230}$$

and similarly for $\tilde{\Phi}$, and the corresponding gauge covariant derivative is

$$D^\mu \Phi = \partial^\mu \Phi + i \left(g_L \frac{\vec{\tau} \cdot \vec{W}^\mu_L}{2} \Phi - g_R \Phi \frac{\vec{\tau} \cdot \vec{W}^\mu_R}{2} \right). \tag{10.231}$$

The Φ kinetic energy is

$$\mathcal{L}_\Phi = \text{Tr} \left[(D_\mu \Phi)^\dagger (D^\mu \Phi) \right], \tag{10.232}$$

and the potential can be constructed in an obvious way from the invariants $\text{Tr}\,(\Phi^\dagger_i \Phi_j)$ and $\text{Tr}\,(\Phi^\dagger_i \Phi_j \Phi^\dagger_k \Phi_\ell)$, where $\Phi_{i,j,k,\ell}$ can be Φ or $\tilde{\Phi}$. We assume that the neutral components of Φ acquire VEVs,

$$\langle \Phi \rangle = \begin{pmatrix} \kappa & 0 \\ 0 & \kappa' \end{pmatrix}, \qquad \langle \tilde{\Phi} \rangle = \begin{pmatrix} \kappa'^* & 0 \\ 0 & \kappa^* \end{pmatrix}. \tag{10.233}$$

At least one additional Higgs field with $t_{BL} \neq 0$ is needed to break the electroweak symmetry to $U(1)_Q$, and it should be an $SU(2)$ singlet with a neutral component to ensure $M_{W_R} \gg M_{W_L}$. There are a number of possibilities, but two are usually considered.

In the doublet model, one introduces the pair of doublets

$$\delta_L = \begin{pmatrix} \delta_L^+ \\ \delta_L^0 \end{pmatrix}, \qquad \delta_R = \begin{pmatrix} \delta_R^+ \\ \delta_R^0 \end{pmatrix}, \tag{10.234}$$

which transform as $(2, 1, \frac{1}{2})$ and $(1, 2, \frac{1}{2})$, respectively. The δ_L is not really essential for $SU(2)_L \times SU(2)_R \times U(1)$, but is introduced to allow the possibility of a left-right interchange symmetry. We assume that the potential for $\Phi, \delta_{L,R}$, which we do not display, is such that $|v_{\delta R}| \equiv |\langle \delta_R^0 \rangle| \gg \nu/\sqrt{2}$, where the electroweak scale $\nu \sim 246$ GeV is given by $\nu^2 = 2(|v_{\delta L}|^2 + |\kappa|^2 + |\kappa'|^2)$, with $|v_{\delta L}| \equiv |\langle \delta_L^0 \rangle|$. In that case $SU(2)_L \times SU(2)_R \times U(1)$ is broken to $SU(2)_L \times U(1)_Y$ at scale $|v_{\delta R}|$. The $\delta_{L,R}$ do not contribute to fermion masses.

The other common model employs Higgs triplets transforming as $(3, 1, 1)$ and $(1, 3, 1)$,

$$\Delta_L = \begin{pmatrix} \Delta_L^{++} \\ \Delta_L^+ \\ \Delta_L^0 \end{pmatrix}, \qquad \Delta_R = \begin{pmatrix} \Delta_R^{++} \\ \Delta_R^+ \\ \Delta_R^0 \end{pmatrix}. \tag{10.235}$$

One assumes that $|v_{\Delta R}| \equiv |\langle \Delta_R^0 \rangle| \gg \nu/\sqrt{2} = (|\kappa|^2 + |\kappa'|^2)^{1/2}$ and that $|v_{\Delta L}| = |\langle \Delta_L^0 \rangle| \ll \nu/\sqrt{2}$. The second condition is required by the ρ_0 parameter in (8.182) on page 317 and Problem 8.1. The triplet model is popular because Δ_R can yield a $|v_{\Delta R}|$-scale Majorana mass term, $\sim h_{\Delta_R} v_{\Delta R}^* \bar{\nu}_L^c \nu_R/\sqrt{2} + h.c.$, for the right-handed neutrino (Mohapatra and Senjanovic, 1980), similar to (9.107) on page 416. For Dirac masses comparable to the charged leptons and $h_{\Delta_R} v_{\Delta R} = \mathcal{O}(1 \text{ TeV})$ this would yield $\nu_{e,\mu,\tau}$ seesaw masses of order eV, keV, and MeV, respectively. These values were quite acceptable when the model was proposed but are now excluded. However, the charged lepton and neutrino Dirac masses are in principle independent (see Equation 10.245 below), so one could choose the couplings so that $|v_{\Delta R}|$ and the heavy Majorana mass are at the TeV scale and the $\nu_{e,\mu,\tau}$ masses are in the experimentally-allowed range, i.e., the low-scale seesaw. Alternatively, the model can be reinterpreted as a GUT or high-scale model, probably embedded in $SO(10)$. We also mention that a VEV for Δ_L would yield a direct type II seesaw Majorana mass term $\sim h_{\Delta_L} v_{\Delta L}^* \bar{\nu}_L \nu_R^c/\sqrt{2} + h.c.$ The models also have the interesting feature of doubly charged Higgs fields $\Delta_{L,R}^{++}$ (Accomando et al., 2006).

The Gauge Bosons

The charged and neutral gauge boson mass matrices are obtained by a calculation similar to Problem 8.1. For the charged bosons,

$$\mathcal{L}_W = \left(M_W^2\right)_{ab} W_a^+ W_b^- = \sum_{i=1}^{2} M_i^2 W_i^+ W_i^-, \tag{10.236}$$

where $a, b = L, R$ and

$$M_W^2 = \begin{pmatrix} M_L^2 & M_{LR}^2 e^{+i\gamma} \\ M_{LR}^2 e^{-i\gamma} & M_R^2 \end{pmatrix}. \tag{10.237}$$

The elements are expressed in terms of the Higgs VEVs as

$$\begin{aligned} M_{L,R}^2 &= \frac{1}{2} g_{L,R}^2 \left(|\kappa|^2 + |\kappa'|^2 + |v_{\delta_{L,R}}|^2 + 2|v_{\Delta_{L,R}}|^2\right) \equiv \frac{1}{4} g_{L,R}^2 n_{L,R}^2 \\ M_{LR}^2 &= -g_L g_R |\kappa' \kappa^*|, \qquad e^{i\gamma} = \frac{\kappa' \kappa^*}{|\kappa' \kappa^*|}, \end{aligned} \tag{10.238}$$

where the mixing term is in general complex. The mass eigenstate fields are

$$\begin{pmatrix} W_1^+ \\ W_2^+ \end{pmatrix} = \begin{pmatrix} \cos\zeta & e^{-i\omega}\sin\zeta \\ -\sin\zeta & e^{-i\omega}\cos\zeta \end{pmatrix} \begin{pmatrix} W_L^+ \\ W_R^+ \end{pmatrix}, \tag{10.239}$$

where the mixing angle ζ and phase ω are

$$\tan 2\zeta = \mp \frac{M_{LR}^2}{M_R^2 - M_L}, \qquad e^{i\omega} = \pm e^{i\gamma}. \tag{10.240}$$

The $\pm$ signs represent alternative phase conventions for $W_2^{\pm}$. The mass eigenvalues $M_{1,2}^2$ are given by a formula analogous to (10.204). In the limit $n_R \gg n_L$,

$$\begin{gathered} M_1^2 \sim M_L^2 \sim \frac{1}{4} g_L^2 n_L^2, \qquad M_2^2 \sim M_R^2 \sim \frac{1}{4} g_R^2 n_R^2 \\ \zeta \sim \pm \frac{g_L}{g_R} \frac{4|\kappa\kappa'|}{|n_R|^2} \Longrightarrow \left| \frac{g_R}{g_L}\zeta \right| \leq \left(\frac{g_R}{g_L}\right)^2 \frac{M_1^2}{M_2^2}. \end{gathered} \tag{10.241}$$

The neutral gauge boson sector is the same as for the LR model in Section 10.3.1. The mass-squared matrix is given by (10.201), with

$$M_{Z^0}^2 = \frac{1}{4} g_Z^2 \bar{n}_L^2, \qquad M_{Z'}^2 = \frac{1}{4}\left(g_R^2 + g_{BL}^2\right) \bar{n}_R^2, \qquad \Delta^2 = -\frac{g g'}{2} \alpha \left(|\kappa|^2 + |\kappa'|^2\right), \tag{10.242}$$

where $\bar{n}_{L,R}^2$ are the same as $n_{L,R}^2$ defined in (10.238) except $2|v_{\Delta_{L,R}}|^2 \to 4|v_{\Delta_{L,R}}|^2$. The Z' charges and α are defined in (10.219). For $n_R \gg n_L$,

$$\frac{M_{Z_2}^2}{M_{W_2}^2} \to \left(1 + \frac{1}{\alpha^2}\right) \frac{|v_{\delta_R}|^2 + 4|v_{\Delta_R}|^2}{|v_{\delta_R}|^2 + 2|v_{\Delta_R}|^2}. \tag{10.243}$$

The Yukawa Couplings

The Φ and $\tilde{\Phi}$ Yukawa couplings to the quarks and leptons are

$$-\mathcal{L}_{Yuk} = \sum_{mn} \bar{q}_{mL}^0 \left(r_{mn}\Phi + s_{mn}\tilde{\Phi}\right) q_{nR}^0 + \sum_{m,n} \bar{\ell}_{mL}^0 \left(t_{mn}\Phi + u_{mn}\tilde{\Phi}\right) \ell_{nR}^0 + h.c. \tag{10.244}$$

There may be additional couplings of the $\Delta_{L,R}$ to the leptons, generating Majorana neutrino masses. The Φ and $\tilde{\Phi}$ Yukawa matrices are independent, allowing nontrivial quark and lepton mixings in the charged current interactions.[63] This leads to the quark mass matrices

$$M^u = r\kappa + s\kappa'^*, \qquad M^d = r\kappa' + s\kappa^*, \tag{10.245}$$

and similarly for the leptons. These can be diagonalized just as in the SM in Section 8.2.2, yielding a hadronic charged current interaction that generalizes (8.55),

$$-\mathcal{L}_W^h = \frac{1}{\sqrt{2}} \bar{u}\gamma_\mu \left[g_L V_q^L W_L^{\mu+} P_L + g_R V_q^R W_R^{\mu+} P_R\right] d + h.c. \tag{10.246}$$

in terms of the mass eigenstate quark fields. $V_q^L = A_L^{u\dagger} A_L^d$ is just the CKM matrix, while $V_q^R = A_R^{u\dagger} A_R^d$ is the analogous mixing matrix for the $SU(2)_R$ currents, which (unlike in

[63] Supersymmetric and $SO(10)$ extensions require a second independent Φ field.

the SM) is observable. A similar expression holds for the leptonic currents, except that the right-handed current is absent in the low energy effective theory if the ν_R acquires a large Majorana mass from Δ_R. In terms of the mass eigenstate gauge fields,

$$\begin{aligned}-\mathcal{L}_W^h = \frac{\cos\zeta}{\sqrt{2}} \{&\bar{u}\gamma_\mu \left[g_L V_q^L P_L + g_R \tan\zeta \, e^{i\omega} V_q^R P_R\right] d \; W_1^{+\mu} \\ &+ \bar{u}\gamma_\mu \left[-g_L \tan\zeta \, V_q^L P_L + g_R e^{i\omega} V_q^R P_R\right] d \; W_2^{+\mu}\} + h.c.\end{aligned} \tag{10.247}$$

The effective four-fermi interaction at low energies in (8.58) is replaced by

$$\mathcal{H}_{eff} = \frac{4\hat{G}_F}{\sqrt{2}} \left[J_{L\mu}^\dagger J_L^\mu + \frac{b}{a} J_{L\mu}^\dagger J_R^\mu + \frac{c}{a} J_{R\mu}^\dagger J_L^\mu + \frac{d}{a} J_{R\mu}^\dagger J_R^\mu \right], \tag{10.248}$$

where the apparent (measured) Fermi constant is

$$\frac{\hat{G}_F}{\sqrt{2}} = \frac{g_L^2 \cos^2\zeta \; a}{8M_1^2}. \tag{10.249}$$

The $V \mp A$ charge raising currents are

$$J_{L,R\mu}^\dagger = \bar{u}\gamma_\mu V_q^{L,R} P_{L,R} \, d + \bar{\nu}\gamma_\mu V_\ell^{L,R} P_{L,R} \, e, \tag{10.250}$$

with the usual caveat about ν_R. The SM current $J_{W\mu}^\dagger$ in (8.60) on page 272 is $2J_{L\mu}^\dagger$. The coefficients in $\mathcal{H}_{eff}$ are

$$\begin{aligned} a &= 1 + \beta \tan^2\zeta, \qquad b^* = c = e^{i\omega} \frac{g_R}{g_L} \tan\zeta \, (1 - \beta) \\ d &= \left(\frac{g_R}{g_L}\right)^2 (\tan^2\zeta + \beta), \text{ with } \beta \equiv \frac{M_1^2}{M_2^2}. \end{aligned} \tag{10.251}$$

Left-Right Symmetry

Our discussion of $SU(2)_L \times SU(2)_R \times U(1)$ has so far been quite general. However, most studies invoke a left-right interchange symmetry, under

$$\begin{aligned} &W_L^i \leftrightarrow W_R^i, \qquad W_{BL} \leftrightarrow W_{BL}, \qquad q_L^0 \leftrightarrow q_R^0, \qquad \ell_L^0 \leftrightarrow \ell_R^0 \\ &\Phi \leftrightarrow \Phi^\dagger, \qquad \delta_L \leftrightarrow \delta_R, \qquad \Delta_L \leftrightarrow \Delta_R. \end{aligned} \tag{10.252}$$

The Lagrangian density is invariant under LR[64] for

$$g_R = g_L, \quad r = r^\dagger, \quad s = s^\dagger, \quad t = t^\dagger, \quad u = u^\dagger, \quad h_{\Delta L} = h_{\Delta R}, \tag{10.253}$$

as well as appropriate constraints on the scalar potential. In this case, there is a formal P and C invariance at the Lagrangian level. However, the invariance is of a different character than for $SU(3)_c$ or $U(1)_Q$ because the gauge symmetries are still chiral under $SU(2)_L \times SU(2)_R$.

The gauge symmetry breaking also breaks the LR symmetry. In particular, even though r and s are assumed to be Hermitian, $M^u = r\kappa + s\kappa'^*$ is not Hermitian if κ and κ' are complex,

[64]A discrete LR symmetry broken at the TeV level would lead to problems with cosmological domain walls, similar to Section 3.3.2, would not allow the development of a baryon asymmetry while C is conserved, and is not consistent with gauge coupling unification, giving more reasons why current studies generally assume that the gauge symmetry and/or the LR symmetry is broken at a high scale.

implying $V_q^L \neq V_q^R$. However, many studies assume either: (a) *manifest LR symmetry*, i.e., r and s are complex (so that CP is explicitly broken) but κ and κ' are real (or at least their phases are small). If they are real, $V_q^L = V_q^R$. (b) *Pseudo-manifest LR symmetry*, i.e., r and s and the other Yukawa matrices are real and symmetric, but κ and κ' are complex (spontaneous CP breaking). The mass matrices are therefore complex-symmetric, so that $A_L^u = A_R^{u*} K^u$, where K^u is a diagonal phase matrix, and similarly for the other fermions. Unlike the somewhat similar Majorana neutrino case (Equation 9.33), it is more convenient to choose the phases in $A_L^{u,d}$ to put V_q^L in the canonical CKM form. Then $K^{u,d}$ must be chosen to make the mass eigenvalues real and positive. The result is that the elements of V_q^L and V_q^R may differ in phase, but they have the same magnitude, $|V_{qmn}^L| = |V_{qmn}^R|$.

Experimental Constraints

There are many modifications to the SM physics due to the W_R. The mass of the $W \sim W_1$ boson is lowered by the mixing effects (corrections to (10.241)). That and $Z - Z'$ mixing can modify the predicted M_W/M_Z ratio and the relation of the masses to other precision electroweak observables. Low energy observables are affected both by $W_L - W_R$ mixing and by W_R exchange, which lead to an admixture of $V + A$ in the hadronic interactions. A corresponding $V + A$ component to the leptonic current would only be present for a Dirac neutrino or a very light Majorana ν_R. Such effects have been searched for extensively in μ and β decay, in weak universality tests, in new box diagram contributions to the $K_L - K_S$ mass difference, as mentioned in Chapters 7 and 8, and in direct searches for a W' or (for the case of a TeV-scale seesaw) heavy Majorana ν_R at the Tevatron and LHC (e.g., Maiezza et al., 2010; Cao et al., 2012; Patrignani, 2016). No definitive evidence has been observed for such effects, but the extracted limits on M_2 and ζ are dependent on the assumptions concerning g_R/g_L, V_q^R, ω, and the nature of ν_R. Assuming manifest or pseudo-manifest left-right symmetry there is a stringent limit $M_2 \gtrsim 2.5$ TeV from the contribution of the box diagram with one W_L and one W_R to Δm_K defined in (8.236) (Beall et al., 1982; Zhang et al., 2007). However, somewhat lower M_2 would be possible more generally (even with $g_R = g_L$) (Langacker and Uma Sankar, 1989). The limits on $|\zeta|$ are stringent even for general V_q^R. For example, universality implies $|\zeta| \lesssim \mathcal{O}(0.0004)$ for $g_R = g_L$ and $\omega = 0$ (for heavy ν_R), but the constraint is weaker for $\omega \neq 0$ (Problem 10.17).

The low-scale seesaw versions of the model could also yield new contributions (Maiezza et al., 2010; Vergados et al., 2012; Faessler et al., 2014) to $\beta\beta_{0\nu}$ that could be comparable to the ordinary light-neutrino exchange diagram in Figure 9.2 on page 376. The most important are the analogous diagram in which ν_L is replaced by the heavy ν_R and W^- by W_R^- (the mixing effects are small), and one involving $(\Delta_R^\dagger)^{--}$.

The direct production limits on W' and ν_R depend on their relative masses, but generally extend into the several TeV range. W' could decay into $q\bar{q}$, $\ell\nu_R$, or possibly WZ. A Majorana ν_R could subsequently decay to $\ell q\bar{q}$ (i.e., $W' \to \ell\nu_R \to \ell\ell q\bar{q}$) via a virtual W', where the two leptons could be same sign or opposite sign with equal probability.

10.4 GRAND UNIFIED THEORIES (GUTS)

The gauge problem described in Section 10.1 suggests the possibility of *grand unification* (Pati and Salam, 1974; Georgi and Glashow, 1974), in which the SM gauge group $SU(3) \times SU(2) \times U(1)$ is embedded in a simple group G, with the quarks and leptons combined in the same multiplets. The Pati-Salam model achieved a partial unification linking quarks and leptons, based on the group $SU(4)_c \times SU(2)_L \times SU(2)_R$. The electroweak part incorporates the left-right symmetric $SU(2)_L \times SU(2)_R \times U(1)_{BL}$ model, with the T_{BL} generator embedded in $SU(4)_c$. The color $SU(4)_c$ group extends QCD to include a fourth "color," identified with lepton number, so that the fundamental representations for the first family are (u_R, u_G, u_B, ν_e) and (d_R, d_G, d_B, e). The $SU(4)_c$ symmetry would have to be spontaneously broken to $SU(3)_c \times U(1)_{BL}$ at a sufficiently high scale. Alternative versions of the model involved integer charged quarks and extended electroweak groups.

The first full unification of $SU(3) \times SU(2) \times U(1)$ into a simple group was the Georgi-Glashow $SU(5)$ model (extensively developed in Georgi et al., 1974; Buras et al., 1978), in which the strong, weak, and electromagnetic interactions are unified above the scale $M_X \gg M_Z$ of $SU(5)$ breaking. Thus, the properly normalized SM gauge couplings extrapolated from low energy should meet at M_X. At the time the model was written the gauge unification worked reasonably well, suggesting $M_X \sim 10^{14-15}$ GeV. As mentioned in Section 10.2.6 the unification (using the more precise couplings determined subsequently) works much better in the supersymmetric extension of the SM, yielding $M_X \sim 10^{16}$ GeV.

The left chiral q, q^c, ℓ and ℓ^c are unified in the same multiplets in $SU(5)$ (though each family is still reducible). This implies electric charge quantization since there are no $U(1)$ factors. It also predicts proton (and bound neutron) decay mediated by the new gauge bosons that connect quarks with leptons and quarks with antiquarks, by the diagrams in Figure 10.9. One expects decays into modes such as $p \to e^+\pi^0$, with a lifetime

$$\tau_p \sim \frac{M_X^4}{\alpha^2 m_p^5} \xrightarrow[M_X \sim 10^{14}\text{ GeV}]{} 10^{29} \text{ yr}. \tag{10.254}$$

More detailed estimates allowed a somewhat longer lifetime, but nevertheless the original model was eventually excluded by searches at SuperKamiokande and elsewhere.[65] The gauge mediated lifetime is much longer in the supersymmetric version, $\sim 10^{38}$ yr, but such models lead to faster decays into modes such as $\bar{\nu}K^+$ by processes involving sparticles. The simplest forms of $SU(5)$ also make predictions relating the quark and lepton Yukawa couplings, which are only partially successful.

In this section we will describe the structure and some of the implications of the non-supersymmetric $SU(5)$ model, and briefly comment on some extensions. Much more detailed discussions can be found in a number of reviews (Langacker, 1981; Ross, 1985; Hewett and Rizzo, 1989; Mohapatra, 2003; Patrignani, 2016).

10.4.1 The $SU(5)$ Model

The Fields

$SU(5)$ is a rank 4 group (like the SM), with fundamental 5×5 representation matrices $L^i = \lambda^i/2$, $i = 1 \cdots 24$, which generalize the $SU(3)$ matrices in Table 3.1 in an obvious way. The upper left 3×3 block corresponds to the $SU(3)$ subgroup (with indices $a, b = 1, 2, 3$ denoted by α and β), while the lower right 2×2 block (with $a, b = 4, 5$ denoted by r and s) corresponds to $SU(2)$. The normalized hypercharge generator $Y_1 \equiv \sqrt{\frac{3}{5}}Y$ is diagonal,

$$L_{Y_1} = \frac{1}{\sqrt{60}} \text{diag}(-2 \;\; -2 \;\; -2 \;\; 3 \;\; 3), \qquad \text{Tr}\, L_{Y_1}^2 = \frac{1}{2}, \tag{10.255}$$

and commutes with $SU(3)$ and $SU(2)$. $SU(5)$ contains 12 additional generators not contained in $SU(3) \times SU(2) \times U(1)$, with nonzero values for $L^i_{\alpha r}$ or $L^i_{r\alpha}$. We will use the tensor

[65] The current lower limit on the partial lifetime into $e^+\pi^0$ is 1.7×10^{34} yr (Takhistov, 2016). For reviews, see (Nath and Fileviez Perez, 2007; Babu et al., 2013).

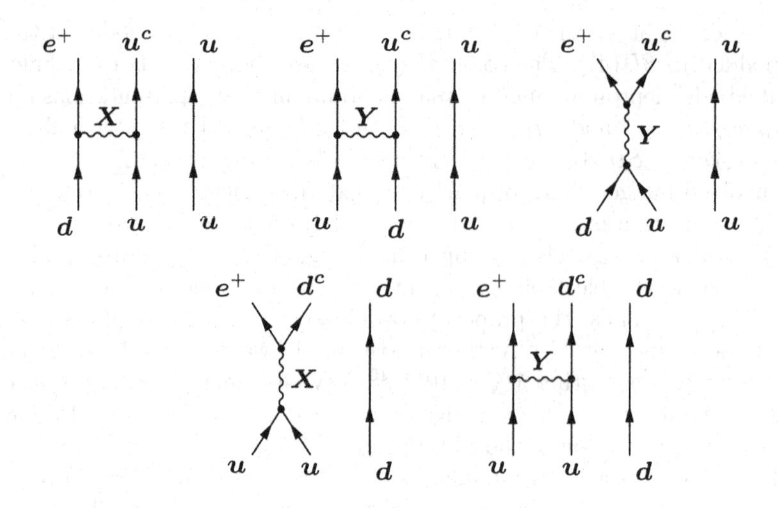

Figure 10.9 Diagrams leading to proton decay in the $SU(5)$ model. X and Y are color antitriplet gauge bosons with electric charges 4/3 and 1/3, respectively.

notation introduced in (3.115) on page 111 for the fields, with a fundamental 5 written with a lower index, such as Φ_a, and an anti-fundamental 5^* as Ψ^a.

The 24 adjoint gauge fields A^i decompose under $SU(3) \times SU(2) \times U(1)$ as

$$24 \to \underbrace{(8,1,0)}_{G_\alpha^\beta} + \underbrace{(1,3,0)}_{W^\pm, W^0} + \underbrace{(1,1,0)}_{B} + \underbrace{(3,2^*,-\frac{5}{6})}_{A_\alpha^r} + \underbrace{(3^*,2,+\frac{5}{6})}_{A_r^\alpha}, \tag{10.256}$$

where the third number is hypercharge, $y = q - t^3$. This can be written as a 5×5 matrix

$$A = \sum_{i=1}^{24} A^i \frac{\lambda^i}{\sqrt{2}}. \tag{10.257}$$

It is sometimes convenient to define the fields $A_a^b \equiv A_{ab}$, These are non-Hermitian for $a \neq b$, with $A_a^b = (A_b^a)^\dagger$. The diagonal elements are not independent and are constrained by $A_a^a = 0$. Displaying the entries explicitly,

$$A = \left(\begin{array}{ccc|cc} G_1^1 - \frac{2B}{\sqrt{30}} & G_1^2 & G_1^3 & \bar{X}_1 & \bar{Y}_1 \\ G_2^1 & G_2^2 - \frac{2B}{\sqrt{30}} & G_2^3 & \bar{X}_2 & \bar{Y}_2 \\ G_3^1 & G_3^2 & G_3^3 - \frac{2B}{\sqrt{30}} & \bar{X}_3 & \bar{Y}_3 \\ \hline X^1 & X^2 & X^3 & \frac{W^0}{\sqrt{2}} + \frac{3B}{\sqrt{30}} & W^+ \\ Y^1 & Y^2 & Y^3 & W^- & -\frac{W^0}{\sqrt{2}} + \frac{3B}{\sqrt{30}} \end{array} \right), \tag{10.258}$$

where $W^\pm = (W^1 \mp iW^2)/\sqrt{2}$, $W^0 = W^3$, and B are the $SU(2) \times U(1)$ bosons, and $G_\alpha^\beta \equiv \sum_{i=1}^8 G^i \lambda_{\alpha\beta}^i/\sqrt{2}$ are the gluon fields with $G_\alpha^\alpha = 0$. There are 12 additional gauge bosons A_r^α and A_α^r, which carry both $SU(2)$ and color indices. These can be written

$$\begin{array}{ll} A_4^\alpha \equiv X^\alpha \; [3^*, 2, q_X = \frac{4}{3}], & A_\alpha^4 \equiv \bar{X}_\alpha \; [3, 2^*, q_{\bar{X}} = -\frac{4}{3}] \\ A_5^\alpha \equiv Y^\alpha \; [3^*, 2, q_Y = \frac{1}{3}], & A_\alpha^5 \equiv \bar{Y}_\alpha \; [3, 2^*, q_{\bar{Y}} = -\frac{1}{3}], \end{array} \tag{10.259}$$

where the $SU(3)$ and $SU(2)$ representations and the electric charge are shown.

Each fermion family of left-chiral fields is assigned to a reducible $5^* + 10$ representation, where the 5^* is anti-fundamental and the 10 is the antisymmetric product of two 5's:

$$\begin{aligned}\underbrace{5^*}_{\chi^a} &\to \underbrace{(3^*, 1, \frac{1}{3})}_{\chi^\alpha} + \underbrace{(1, 2^*, -\frac{1}{2})}_{\chi^r} \\ \underbrace{10}_{\psi_{ab}=-\psi_{ba}} &\to \underbrace{(3^*, 1, -\frac{2}{3})}_{\psi_{\alpha\beta}} + \underbrace{(3, 2, \frac{1}{6})}_{\psi_{\alpha r}} + \underbrace{(1, 1, 1)}_{\psi_{45}},\end{aligned} \tag{10.260}$$

where $a, b = 1 \cdots 5$; $\alpha, \beta = 1, 2, 3$; and $r = 4, 5$. We emphasize that the χ and ψ are L-chiral, but we omit the subscript L to simplify the notation. The field content can be written explicitly as

$$\begin{aligned}\chi^a &= \left(d^{c1}\ d^{c2}\ d^{c3}\ e^-\ -\nu_e\right)_L^T \\ \psi_{ab} &= \frac{1}{\sqrt{2}} \left(\begin{array}{ccc|cc} 0 & u^{c3} & -u^{c2} & -u_1 & -d_1 \\ -u^{c3} & 0 & u^{c1} & -u_2 & -d_2 \\ u^{c2} & -u^{c1} & 0 & -u_3 & -d_3 \\ \hline u_1 & u_2 & u_3 & 0 & -e^+ \\ d_1 & d_2 & d_3 & e^+ & 0 \end{array}\right)_L,\end{aligned} \tag{10.261}$$

where the family indices and weak-eigenstate superscript ^{0} are suppressed. The lepton doublet $\begin{pmatrix} e^- \\ -\nu_e \end{pmatrix}_L = i\tau^2 \begin{pmatrix} \nu_e \\ e^- \end{pmatrix}_L$ in χ transforms as a 2^*, and the $1/\sqrt{2}$ in ψ is introduced to avoid double counting. The antisymmetric part of 3×3 in $SU(3)$ is a 3^*, so the upper 3×3 block of ψ is $\psi_{\alpha\beta} = \epsilon_{\alpha\beta\gamma} u_L^{c\gamma}/\sqrt{2}$. Similarly, the antisymmetric part of 2×2 in $SU(2)$ is a singlet, $\psi_{54} = -\psi_{45} = e_L^+/\sqrt{2}$. The fermion family multiplet can be extended to $5^* + 10 + 1$, where the singlet is the ν_L^c.

We also introduce a real adjoint Higgs multiplet Φ, and a fundamental complex 5-plet H that contains the SM Higgs doublet ($\phi^+ = H_4$ and $\phi^0 = H_5$),

$$\Phi = \sum_{i=1}^{24} \Phi^i \frac{\lambda^i}{\sqrt{2}}, \qquad H_a = \begin{pmatrix} \mathcal{H}_\alpha \\ \phi^+ \\ \phi^0 \end{pmatrix}. \tag{10.262}$$

The Φ fields have the same quantum numbers as the gauge fields in (10.258), and are introduced to break $SU(5)$ down to $SU(3) \times SU(2) \times U(1)$. The Higgs doublet $\begin{pmatrix} \phi^+ \\ \phi^0 \end{pmatrix}$ transforms as $(1, 2, \frac{1}{2})$, while its $SU(5)$ partner $\mathcal{H}_\alpha = (3, 1, -\frac{1}{3})$ is a color triplet with $q_\mathcal{H} = -\frac{1}{3}$.

The Fermion Gauge Interactions

The gauge covariant derivatives for the fermion fields are

$$\begin{aligned}(D_\mu \chi)^a &= \partial_\mu \chi^a - i\frac{g_5}{\sqrt{2}} (A_\mu)^a_b \chi^b \\ (D_\mu \psi)_{ab} &= \partial_\mu \psi_{ab} + i\frac{g_5}{\sqrt{2}} (A_\mu)^c_a \psi_{cb} + i\frac{g_5}{\sqrt{2}} (A_\mu)^d_b \psi_{ad},\end{aligned} \tag{10.263}$$

where g_5 is the $SU(5)$ gauge coupling and we have used $L^i_{5^*} = -L^{iT}_5$. The fermion gauge interactions are therefore

$$\mathcal{L}_f = \bar{\chi}_a i (\not{D}\chi)^a + \bar{\psi}^{ab} i (\not{D}\psi)_{ab} = \mathcal{L}_f^{SM} + \mathcal{L}_f^{XY}, \tag{10.264}$$

where the fields are weak eigenstates and a sum over families is implied. $\mathcal{L}_f^{SM}$ includes the kinetic energy terms and the SM fermion gauge interactions in (5.2) and (8.7), with the restriction that the gauge couplings are unified,

$$g_3 = g_2 = g_1 = g_5, \text{ where } g_3 \equiv g_s, \quad g_2 \equiv g, \quad g_1 \equiv \sqrt{\frac{5}{3}} g'. \tag{10.265}$$

$\mathcal{L}_f^{XY}$ contains the leptoquark and diquark interactions of the X and Y,

$$\begin{aligned}\mathcal{L}_f^{XY} = & \underbrace{-\frac{g_5}{\sqrt{2}}\left[\bar{e}_R^+ \not{X}^\alpha d_{R\alpha} + \bar{e}_L^+ \not{X}^\alpha d_{L\alpha} - \bar{\nu}_R^c \not{Y}^\alpha d_{R\alpha} - \bar{e}_L^+ \not{Y}^\alpha u_{L\alpha}\right]}_{\text{leptoquark vertices}} \\ & \underbrace{-\frac{g_5}{\sqrt{2}}\left[\epsilon^{\alpha\beta\gamma} \bar{u}_{L\gamma}^c \not{\bar{X}}_\alpha u_{L\beta} + \epsilon^{\alpha\beta\gamma} \bar{u}_{L\gamma}^c \not{\bar{Y}}_\alpha d_{L\beta}\right]}_{\text{diquark vertices}} + h.c.\end{aligned} \tag{10.266}$$

These expressions can be transformed using the identities in (2.301) on page 56 and (2.302). For example,

$$\bar{e}_R^+ \not{X}^\alpha d_{R\alpha} = -\bar{d}_{L\alpha}^c \not{X}^\alpha e_L^-, \qquad \epsilon^{\alpha\beta\gamma} \bar{u}_{L\gamma}^c \not{\bar{X}}_\alpha u_{L\beta} = -\epsilon^{\alpha\beta\gamma} u_{R\gamma}^T \mathcal{C}^{-1} \not{\bar{X}}_\alpha u_{L\beta}. \tag{10.267}$$

The transitions can be shown schematically (suppressing color) as

$$W^\pm \updownarrow \underbrace{\begin{pmatrix} \nu_L & \overset{X,Y}{\longleftrightarrow} & \\ e_L^- & & d_L^c \end{pmatrix}}_{5^*} \quad \underbrace{\begin{pmatrix} e_L^+ & \overset{X,Y}{\longleftrightarrow\ \ \longleftrightarrow} u_L & u_L^c \\ & d_L & \end{pmatrix}}_{10} \updownarrow W^\pm.$$

Since the X and Y act both as leptoquarks and diquarks, both baryon and lepton number are violated ($B - L$ is conserved), leading to proton and bound neutron decay by the diagrams in Figure 10.9. Prominent modes predicted by $SU(5)$ included $p \to e^+\pi^0, e^+\rho^0, e^+\omega, e^+\eta, e^+\pi^+\pi^-, \bar{\nu}\pi^+, \bar{\nu}\rho^+$, and $\bar{\nu}\pi^+\pi^0$. Bound neutrons can also decay into $\bar{\nu}\pi^0, \bar{\nu}\rho^0, \bar{\nu}\omega, \bar{\nu}\eta, e^+\pi^-, e^+\rho^-$, etc. One expects $\tau_{p,n} \sim \frac{M_{X,Y}^4}{\alpha_5^2 m_p^5}$, where $\alpha_5 \equiv \frac{g_5^2}{4\pi}$ (cf. the muon lifetime $\tau_\mu \sim M_W^4/g^4 m_\mu^5$). More detailed early estimates predicted $\tau_p \sim 10^{31\pm2}$ yr, with the lower end of the range favored by current input parameters. The $SU(5)$ prediction of proton decay motivated a number of sensitive searches involving large underground detectors. These eventually excluded most of the predicted range of lifetimes, i.e., they require $M_{X,Y} \gtrsim 10^{15}$ GeV, somewhat larger than expected from the gauge couplings. Proton decay continues to be a stringent constraint on more modern versions of grand unification, which usually predict longer lifetimes.

Spontaneous Symmetry Breaking

The adjoint Higgs Φ introduced in (10.262) has a potential (Buras et al., 1978)

$$V(\Phi) = \frac{\mu^2}{2}\text{Tr}\,(\Phi^2) + \frac{a}{4}\left[\text{Tr}\,(\Phi^2)\right]^2 + \frac{b}{2}\text{Tr}\,(\Phi^4) + \frac{c}{2}\text{Tr}\,(\Phi^3), \tag{10.268}$$

with $\text{Tr}\,(\Phi^2) \equiv \Phi^a_b \Phi^b_a$, etc. Invariance under $\Phi \to -\Phi$ is often imposed, implying $c = 0$. For $\mu^2 < 0$ one has $\langle 0|\Phi|0\rangle \neq 0$, and one can always take $\langle 0|\Phi|0\rangle$ to be diagonal by performing an $SU(5)$ transformation. One can show (Li, 1974) that for $b > 0, c = 0$ the minimum is at[66]

$$\langle \Phi \rangle = \text{diag}(\nu_\Phi \;\; \nu_\Phi \;\; \nu_\Phi \;\; -\frac{3}{2}\nu_\Phi \;\; -\frac{3}{2}\nu_\Phi) \quad \text{with} \quad \nu_\Phi^2 = \frac{-2\mu^2}{15a + 7b}. \tag{10.269}$$

(Vacuum stability requires $a > -7b/15$, so ν_Φ^2 is positive.) Therefore, $SU(5)$ is broken to $SU(3) \times SU(2) \times U(1)$ at the *GUT scale*

$$M_X^2 = M_Y^2 = \frac{25}{8} g_5^2 \nu_\Phi^2. \tag{10.270}$$

Twelve of the scalars in Φ are eaten by the X and Y. The others attain masses of $\mathcal{O}(|\mu|)$, but are of little phenomenological importance because they do not couple to fermions.

To further break $SU(3) \times SU(2) \times U(1)$ down to $SU(3) \times U(1)_Q$ one must utilize the 5-plet H_a in (10.262). The potential in (10.268) for Φ is extended to

$$\begin{aligned} V(\Phi, H) = {} & \frac{\mu^2}{2}\text{Tr}\,(\Phi^2) + \frac{a}{4}\left[\text{Tr}\,(\Phi^2)\right]^2 + \frac{b}{2}\text{Tr}\,(\Phi^4) + \frac{c}{2}\text{Tr}\,(\Phi^3) + \frac{\mu_5^2}{2} H^\dagger H \\ & + \frac{\lambda}{4}\left(H^\dagger H\right)^2 + \alpha H^\dagger H \, \text{Tr}\,(\Phi^2) + \beta H^\dagger \Phi^2 H + \delta H^\dagger \Phi H, \end{aligned} \tag{10.271}$$

where the $SU(5)$ indices are contracted in an obvious way, and $\Phi \to -\Phi$ invariance would imply $c = \delta = 0$. Two severe fine-tuning problems immediately arise: (a) One needs to have $\langle \phi^0 \rangle = \frac{\nu}{\sqrt{2}} \sim 10^{-13}\nu_\Phi$, where $\nu \sim 246$ GeV is the weak scale (the *GUT hierarchy problem*). (b) The color triplet mass $M_{\mathcal{H}}$ must be $\gtrsim 10^{14}$ GeV $\gtrsim 10^{12} M_\phi$ to avoid too rapid proton decay mediated by $\mathcal{H}_\alpha$ (the *doublet-triplet splitting problem*). (Note that the β and δ terms are the only ones in $V(\Phi, H)$ that can split the doublet mass from the triplet.)

Both of these enormous hierarchies must emerge from the parameters in $V(\Phi, H)$. If the dimensional parameters are all of $\mathcal{O}(M_X)$ and the dimensionless ones of $\mathcal{O}(1)$ then it is natural for $M_{\mathcal{H}}$ to be comparable to ν_Φ. One must then fine-tune the parameters to one part in $\nu_\Phi^2/\nu^2 \sim 10^{24}$ to obtain sufficiently small ν and M_ϕ. The fine-tuning problem is actually even worse, because it must apply to the entire effective potential. That is, one must "retune" the parameters at each order of perturbation theory to avoid destabilizing the hierarchy.[67]

Yukawa Couplings and Fermion Masses

The $SU(5)$ symmetry does not allow couplings of Φ to the fermions, but allows

$$\begin{aligned} \mathcal{L}_{Yuk} = {} & \gamma_{mn} \, \chi_m^{aT} C \, \psi_{ab} \, H^{\dagger b} \\ & + \Gamma_{mn} \, \epsilon^{abcde} \, \psi_{mab}^T \, C \, \psi_{ncd} \, H_e + \kappa_{mn} \nu_{mL}^{cT} C \chi_n^a H_a + h.c., \end{aligned} \tag{10.272}$$

where m and n are family indices, Γ is symmetric, ϵ is antisymmetric with $\epsilon^{12345} = 1$, and the weak eigenstate superscripts ^{0} have been suppressed. The fermion masses are generated

[66] The minimum is of the form $\text{diag}(\nu'_\Phi \;\; \nu'_\Phi \;\; \nu'_\Phi \;\; \nu'_\Phi \;\; -4\nu'_\Phi)$ for $b < 0, c = 0$, implying that $SU(5) \to SU(4) \times U(1)$.

[67] A small parameter is said to be *technically natural* if a new global symmetry emerges when it is set to zero, such as the chiral symmetry that is introduced when a fermion mass vanishes. That ensures that only one fine-tuning is needed, because renormalization effects are proportional to the original small parameter ('t Hooft, 1980). Unfortunately, there is no such new symmetry in the present case.

from $\langle H_a \rangle = \delta_a^5 \frac{\nu}{\sqrt{2}}$. The first term yields

$$\mathcal{L}_{de} = -\bar{d}_L M^d d_R - \bar{e}_L^+ M^{eT} e_R^+ + h.c. = -\bar{d}_L M^d d_R - \bar{e}_L M^e e_R + h.c., \tag{10.273}$$

where

$$M^d = M^{eT} = \frac{1}{2} \nu \gamma^\dagger. \tag{10.274}$$

That is, the $SU(5)$ symmetry predicts that the d quark and charged lepton mass matrices are the same up to a transpose,[68] because they derive from the same coupling. Equation (10.274) implies that the eigenvalues are equal at M_X, i.e.,

$$m_d = m_e, \qquad m_s = m_\mu, \qquad m_b = m_\tau. \tag{10.275}$$

This appears to be a disaster. However, the situation is partially remedied because the Yukawa couplings are running quantities, mainly from the gauge loops. The largest effects are from the gluons, which make the quark masses larger than the leptons' at low energies. To lowest order (Buras et al., 1978)

$$\ln \left[\frac{m_d(Q^2)}{m_e(Q^2)} \right] = \underbrace{\ln \left[\frac{m_d(M_X^2)}{m_e(M_X^2)} \right]}_{0} + \frac{4}{11 - 4F} \ln \left[\frac{\alpha_s(Q^2)}{\alpha_5(M_X^2)} \right] + \frac{3}{4F} \ln \left[\frac{\alpha_1(Q^2)}{\alpha_5(M_X^2)} \right], \tag{10.276}$$

where F is the number of families. This implies $m_b \sim 5$ GeV for $m_\tau \sim 1.7$ GeV and $F = 3$, consistent with the observed value. However, (10.276) fails for the light families. To avoid uncertainties in the overall scale of the light quark masses, one usually considers the predicted ratio,

$$\frac{m_\mu}{m_e} = \frac{m_s}{m_d}. \tag{10.277}$$

Equation (10.277) fails by an order of magnitude, with $m_\mu/m_e \sim 200$ and $m_s/m_d \sim 20$ from (5.122) on page 193. This suggests a more complicated Higgs structure, such as the introduction of an additional Higgs 45-plet with family symmetries to restrict the form of the Yukawa matrices (Georgi and Jarlskog, 1979). The second and third terms in (10.272) lead to the u quark and Dirac neutrino mass matrices

$$M^u = M^{uT} = \frac{4\nu}{\sqrt{2}} \Gamma, \qquad M_D^\nu = \frac{\nu}{\sqrt{2}} \kappa^\dagger. \tag{10.278}$$

Gauge Unification

We have seen in (10.265) that the properly normalized $SU(3) \times SU(2) \times U(1)$ gauge couplings should all be equal at M_X, implying that the value of the running weak angle is predicted at M_X,

$$\hat{s}_Z^2(M_X^2) = \frac{g'^2}{g^2 + g'^2} = \frac{3/5}{1 + 3/5} = \frac{3}{8}, \tag{10.279}$$

where we use the $\overline{MS}$ angle defined in (8.139) on page 299. To compare with experiment the couplings must be run to low energy (Georgi et al., 1974), using (10.195). The SM coefficients are given in (10.196), but it instructive to follow the historical approximation,

[68] The transpose is exploited in the lopsided neutrino mass models mentioned in Section 9.5 where a nonsymmetric γ induces large mixings in A_L^e and the unobservable A_R^d. The mechanism is more difficult to implement in $SO(10)$, because the simplest analog involving Higgs 10's is symmetric.

using the expressions in (8.198) for an arbitrary number F of families but ignoring the (small) contribution of the Higgs,

$$b_1 = \frac{F}{12\pi^2}, \quad b_2 = -\frac{1}{16\pi^2}\left[\frac{22}{3} - \frac{4F}{3}\right], \quad b_3 = -\frac{1}{16\pi^2}\left[11 - \frac{4F}{3}\right]. \tag{10.280}$$

These can be solved to yield

$$\frac{\alpha(M_Z^2)}{\alpha_s(M_Z^2)} = \frac{3}{8}\left[1 - \frac{11\alpha(M_Z^2)}{2\pi}\ln\frac{M_X^2}{M_Z^2}\right], \qquad \hat{s}_Z^2(M_Z^2) = \frac{1}{6} + \frac{5}{9}\frac{\alpha(M_Z^2)}{\alpha_s(M_Z^2)}, \tag{10.281}$$

independent of F. (Corrections do depend on F.) Plugging in $\alpha_s(M_Z^2) \sim 0.12$ and $\alpha(M_Z^2) \sim 1/128$ yields $\hat{s}_Z^2(M_Z^2) \sim 0.20$ and $M_X \sim 10^{15}$ GeV, which is a reasonable zeroth order prediction. However, as already discussed in Section 10.2.6, more precise two-loop calculations with the current inputs do not quite work, as can be seen in Figure 10.8. Starting with α and α_s, one finds $M_X \sim 10^{14}$ GeV and $\hat{s}_Z^2(M_Z^2) \sim 0.21$, which is too low. Taking the modern view of using α and $\hat{s}_Z^2(M_Z^2)$ as inputs yields a very low $\alpha_s(M_Z^2) \sim 0.07$ and $M_X \sim 10^{13}$ GeV, both of which are unacceptable.

Cosmology

The $SU(5)$ model apparently offered an elegant explanation for the observed baryon asymmetry of the Universe (Section 10.1), i.e., the out of equilibrium decays of the superheavy partners $\mathcal{H}_\alpha$ of the Higgs bosons (Equation 10.262) (Yoshimura, 1978). Although the lifetimes of the $\mathcal{H}_\alpha$ and their conjugates $\mathcal{H}_\alpha^c$ must be equal by CPT, the rates into specific corresponding modes could differ by C and CP violating effects (a *partial rate asymmetry* (Okubo, 1958)), e.g.,

$$\Gamma(\mathcal{H} \to q^c q^c) \neq \Gamma(\mathcal{H}^c \to qq), \qquad \Gamma(\mathcal{H} \to q\ell) \neq \Gamma(\mathcal{H}^c \to q^c \ell^c). \tag{10.282}$$

Combined with the intrinsic B violation and the nonequilibrium (i.e., the decays occurred at $T \ll M_{\mathcal{H}}$) all of the Sakharov ingredients to generate a baryon asymmetry were fulfilled and an adequate asymmetry (with $B - L = 0$) could be generated (though only in non-minimal models). Unfortunately, it was later recognized that the symmetry would be subsequently erased by nonperturbative sphaleron effects. The idea could be resurrected in more complicated models in which a nonzero $B - L$ was produced in the decays, the most plausible being leptogenesis.[69] Other possibilities for the baryon asymmetry, such as electroweak baryogenesis, do not involve out of equilibrium decays.

A serious difficulty for grand unification involved magnetic monopoles, which are topologically stable gauge/Higgs configurations present when a simple or semi-simple group is broken to a group including a $U(1)$ (Coleman, 1985). One expects a superheavy monopole mass $M_M \sim M_X/\alpha$, and most likely an efficient production during the phase transition resulting in the symmetry breaking. Although $M\bar{M}$ pairs would annihilate efficiently, enough would be left over to greatly overclose the universe (Preskill, 1979; Kolb and Turner, 1990). Suggestions for solving this monopole problem included the possibility that electric charge was not conserved at intermediate temperatures, leading to rapid annihilations (Langacker and Pi, 1980). However, much more compelling was the suggestion that a subsequent period of inflation would dilute the density to negligible values, as well as resolving other cosmological problems (Guth, 1981; Kolb and Turner, 1990; Lyth and Riotto, 1999; Linde, 2008). Magnetic monopoles and their experimental constraints are reviewed in a general context in (Patrizii and Spurio, 2015).

[69] Another difficulty with GUT baryogenesis was that any asymmetry would be erased by a subsequent period of inflation. This continues to be a constraint on leptogenesis models.

10.4.2 Beyond the Minimal $SU(5)$ Model

Supersymmetric Grand Unification

It is straightforward to extend the $SU(5)$ model to supersymmetry (Dimopoulos and Georgi, 1981; Raby, 2009). Just as in the MSSM, one must add an additional Higgs multiplet transforming as a 5^*. The Higgs part of the superpotential becomes

$$W_{\Phi,H} = \mu_\Phi \mathrm{Tr}\,(\Phi^2) + \lambda_\Phi \mathrm{Tr}\,(\Phi^3) + \mu_H H_u H_d + \lambda_{\Phi H} H_u \Phi H_d, \tag{10.283}$$

where

$$H_{ua} = \begin{pmatrix} \mathcal{H}_\alpha \\ h_u^+ \\ h_u^0 \end{pmatrix}, \qquad H_d^a = \begin{pmatrix} \mathcal{H}^{c\alpha} \\ h_d^0 \\ h_d^- \end{pmatrix}, \tag{10.284}$$

Φ is defined as in (10.262), and we are using the same symbols for the superfields and their scalar components. The Yukawa couplings to the matter supermultiplets are analogous to (10.272), with $H \to H_u$ and $H^\dagger \to H_d$.

As has already been emphasized in Section 10.2.6 and Figure 10.8, the gauge unification is much more successful when the couplings are extrapolated using the MSSM β functions (Dimopoulos et al., 1981), with the couplings approximately meeting at $M_X \sim 3\times 10^{16}$ GeV. Using the observed $\hat{\alpha}(M_Z^2)$, $\hat{s}_Z^2(M_Z^2)$, and $\alpha_s(M_Z^2)$ as inputs, the two-loop unification works precisely for a common scale $M_{SUSY} \sim 5$ TeV for all of the new sparticles and the second Higgs doublet. However, there could be additional threshold corrections at the TeV scale due to splitting between the masses of the new particles,[70] as well as ones due to multiplet splitting or higher-dimensional operators at the GUT scale (Langacker and Polonsky, 1993, 1995; Carena et al., 1993; Bagger et al., 1995). The $m_b(M_X) = m_\tau(M_X)$ prediction in the supersymmetric case works well for $\tan\beta = \nu_u/\nu_d \sim 1$ or $40 - 50$ (e.g., Barger et al., 1993; Langacker and Polonsky, 1994; Carena et al., 1994), with the larger value consistent with a Higgs mass of 125 GeV.

Because of the large value of M_X expected in the supersymmetric case, the contribution of *dimension-6* operators such as $uude/M_X^2$ (from X and Y exchange) are greatly suppressed, leading to proton lifetimes of $\mathcal{O}(10^{38})$ years. However, there are dangerous new *dimension-5* operators (Weinberg, 1982; Sakai and Yanagida, 1982; Dimopoulos et al., 1982; Ellis et al., 1982) such as $\tilde{u}\tilde{s}d\nu/M_X$ or $\tilde{c}\tilde{d}ue/M_X$. These are generated by the exchange of the fermionic partners of the $\mathcal{H}$ and $\mathcal{H}^c$ from the Higgs multiplets $H_{u,d}$, which have the same Yukawa couplings as those that generate the fermion masses. The operators must involve a second or third family because of the commuting nature of the superfields. They may be dressed by the exchange of a gluino, neutralino, or chargino, to produce a proton decay operator such as $usd\nu/M_X$. Since there is only one inverse power of the large mass scale, the proton lifetime $\tau_p \sim m_p^3/M_X^2$, which is dominantly into modes such as $p \to \bar{\nu}K^+$ or $n \to \bar{\nu}K^0$, is relatively short even after taking into account the small Yukawa couplings and additional couplings and loop factors from the dressing. The SuperKamiokande limit, $\tau(p \to \bar{\nu}K^+) > 6.6\times 10^{33}$ yr (Takhistov, 2016), excludes the minimal supersymmetric $SU(5)$ and $SO(10)$ models, while versions with non-minimal Higgs sectors predict lifetimes within a factor of a few of the current limits (see, e.g., the Grand Unification review in Patrignani, 2016). Of course, one must also exclude proton decay from squark exchange with R_p violating couplings, just as in the MSSM. These are known as *dimension-4* operators because they are not suppressed by any powers of the GUT scale.

[70] The low-scale threshold effects tended to be of the wrong sign for most breaking/mediation schemes (in which the colored sparticles are usually heavier) for the now excluded possibility that the masses were in the several hundred GeV range.

The $SU(5) \in SO(10) \in E_6$ Chain

Each left-chiral fermion family in $SU(5)$ is assigned to a reducible $5^* + 10$ representation, or $5^* + 10 + 1$ including a singlet ν_L^c. These are combined in an irreducible 16-plet when $SU(5)$ is embedded in the larger rank 5 $SO(10)$ group (Fritzsch and Minkowski, 1975; Georgi, 1975), which *requires* a ν_L^c. In addition to the breaking pattern $SO(10) \to SU(5) \times U(1)_\chi$ described in Section 10.3.1 and considered here, $SO(10)$ has alternative breaking patterns into (a) *flipped* $SU(5)$, in which the hypercharge generator Y includes a component of Q_χ, leading to different identifications of the particles in the 16-plet (Barr, 1982); (b) the Pati-Salam group.

In $SU(5)$ the mass matrices $M^d = M^{eT}$, M^u, and M_D^ν are all independent. In the simplest extension to $SO(10)$ the fermion masses are generated by a single Higgs field in the vector (10) representation, which decomposes into $5 + 5^*$ under $SU(5)$. This implies two Higgs $SU(5)$ multiplets H_u and H_d with two Higgs doublets, $h_{u,d}$, just as in (10.284), even in the non-supersymmetric version. (In $SU(5)$ a single Higgs can play both roles because the Yukawa couplings in (10.272) involve both H and $H^\dagger$, analogous to the SM tilde couplings in (8.13).) There is only a single $\psi_{16}\psi_{16}\phi_{10}$ Yukawa matrix, which is symmetric in family indices, with the immediate consequences

$$M^d = M^{dT} = M^e, \qquad M^u = \tan\beta M^d, \qquad M^u = M_D^\nu. \tag{10.285}$$

The first relation is similar to the $SU(5)$ case (except for being symmetric), while the second is a consequence of *Yukawa unification*, i.e., the b, τ, and t Yukawas are all equal at M_X. In the supersymmetric case the Yukawa unification for the third family works well for large $\tan\beta \sim 40 - 50$. However, both relations fail badly for the first two families. Not only do they give incorrect predictions such as $m_\mu/m_e = m_s/m_d = m_c/m_u$, but the second relation implies a trivial CKM matrix, $V_q = I$. The last relation equates the (Dirac) neutrino masses to the u quark masses, which is obviously a disaster.

A realistic spectrum therefore requires an extended Higgs sector. Introducing additional 10-plets resolves the problem with the CKM matrix, but not the other difficulties. Similar to the addition of a 45 in $SU(5)$ one can obtain a realistic spectrum by introducing very large 120 or 126 dimensional multiplets. The 126 also allows couplings that can generate Majorana masses for the ν_L^c, allowing a type I seesaw, and/or Majorana masses for ν_L (the type II seesaw), as described in Section 9.5. For that reason, $SO(10)$ models are frequently used, in connection with family symmetries, to generate models of neutrino mass, or more generally, models for fermion mass textures. Instead of large representations (which are very unlikely if there is an underlying superstring construction) one can supplement the $\psi_{16}\psi_{16}\phi_{10}$ Yukawa couplings with higher-dimensional operators (Babu et al., 2000; Pati, 2006).

The even larger E_6 (e.g., Slansky, 1981; Hewett and Rizzo, 1989), which has been suggested in some heterotic string compactifications, is rank 6 and includes the $SO(10) \times U(1)_\psi$ subgroup mentioned in Section 10.3.1. It has an alternative maximal $SU(3)_c \times SU(3)_L \times SU(3)_R$ subgroup, referred to as *trinification* when a permutation symmetry on the factors is imposed (Achiman and Stech, 1978). The $SO(10) \times U(1)_\psi$ embedding extends each family to a 27-plet, which breaks to $SO(10)$ and $SU(5)$ as in (10.220). The 16 contains a SM family, while the 10 are new exotic fermions,

$$10 = \begin{pmatrix} E^0 \\ E^- \end{pmatrix}_L + \begin{pmatrix} E^0 \\ E^- \end{pmatrix}_R + \mathcal{D}_L + \mathcal{D}_R, \tag{10.286}$$

and the 1 is a singlet S_L, similar to the ν_L^c. The 10 and 1 are quasi-chiral, i.e., vector or singlet under the SM but charged under the extra $U(1)'$ s (Table 10.2). The $\mathcal{D}_{L,R}$ are exotic charge $-\frac{1}{3}$ quarks, while the vector pair of doublets are usually considered to be exotic leptons in non-supersymmetric studies. However, in the supersymmetric case they are often considered to be Higgs doublet superfields instead, as mentioned in Section 10.3.1. (The interpretation depends on what global symmetries are imposed on the superpotential.) The E_6 model has an especially rich phenomenology because of the $U(1)'$ s and the exotic multiplets, and is often studied for that reason (e.g., King et al., 2006; Kang et al., 2008).

Extra Dimensions and Strings

In *orbifold GUTs* (Kawamura, 2001; Raby, 2009) the grand unification is present in a higher-dimensional space, but the gauge symmetry of the effective four-dimensional theory is (usually) that of the MSSM. Such constructions allow new symmetry breaking mechanisms associated with boundary conditions and background fields in the extra dimensions, and they may retain many of the desirable features of grand unification (such as gauge coupling unification, third family Yukawa relations, etc.) while avoiding some of the difficulties (e.g., the doublet-triplet problem, too rapid proton decay, and the need for large Higgs representations).

Grand unified theories do not incorporate gravity, and are therefore not as ambitious as superstring theories (e.g., Ibáñez and Uranga, 2012). Heterotic string theories include underlying grand unification symmetries. They may compactify into an effective four-dimensional GUT. However, it is difficult to generate the adjoint and other large Higgs multiplets introduced in most bottom-up constructions, so the more promising versions more closely resemble orbifold GUTs (e.g., Raby, 2011). The heterotic theories may also compactify directly to the SM or MSSM, or to an extended version, with limited memory of the underlying GUT. Constructions may retain simple MSSM-type gauge unification, or the unification may be modified (and complicated) by the effects of new matter multiplets that survive to low energy and/or by the string scale gauge coupling boundary conditions, especially for the $U(1)_Y$ (e.g., Dienes, 1997). The fermion families or the elements of the families may have different origins in the construction, breaking or modifying GUT Yukawa relations and possibly leading to family nonuniversal couplings to new $U(1)'$ s.

Type IIA intersecting D-brane constructions usually do not involve a full underlying GUT, but they often descend to four dimensions using a Pati-Salam group (e.g., Blumenhagen et al., 2005). Some versions with large enough stacks of branes do allow an $SU(5)$ (or flipped $SU(5)$) unification, as do some Type IIB theories (Blumenhagen et al., 2009a). In both cases, an underlying $U(5)$ gauge symmetry forbids a perturbative top Yukawa coupling, which could, however, be generated nonperturbatively by instantons. Especially promising are the related F-theory constructions, which allow more flexibility and can allow such extended groups as $SU(5)$, $SO(10)$, and E_6 (e.g., Heckman, 2010).

10.5 PROBLEMS

10.1 The two-loop generalization of (5.33) for a non-chiral $SU(m)$ gauge theory with n_f massless fermion multiplets (flavors) in the fundamental L_m representation is

$$\frac{dg^2}{d\ln Q^2} \equiv 4\pi\beta(g^2) = b_1(m,n_f)g^4 + b_2(m,n_f)g^6,$$

where (e.g., Dietrich and Sannino, 2007, and references therein)

$$b_1(m,n_f) = -\frac{1}{(4\pi)^2}\left[\frac{11}{3}C_2(G) - \frac{4}{3}T_F\right] \equiv -\frac{1}{(4\pi)^2}\hat{b}_1(m,n_f)$$

$$b_2(m,n_f) = -\frac{1}{(4\pi)^4}\left[\frac{34}{3}C_2(G)^2 - \frac{20}{3}C_2(G)T_F - 4C_2(L_m)T_F\right] \equiv -\frac{1}{(4\pi)^2}\hat{b}_2(m,n_f)$$

$$C_2(G) = m, \qquad T_F = \frac{n_f}{2}, \qquad C_2(L_m) = \frac{m^2-1}{2m}.$$

We will examine the running of $g(Q^2)$ from some initial large scale Q_0^2 down to lower scales as a function of m and n_f. It will be useful to think of $m \geq 2$ and n_f as continuous parameters, which can later be specialized to integer values.
(a) For small n_f both $\hat{b}_1$ and $\hat{b}_2$ are positive, but each reverses sign for larger n_f, i.e., at n_{AF} and n_{fp}, respectively. For $n_f < n_{AF}$ the theory is asymptotically free. If $n_{fp} < n_f$ as well, there is a value $g_* = (-b_1/b_2)^{1/2}$, known as the *conformal fixed point*, for which $\beta(g_*^2) = 0$. For some range of initial values $g(Q^2)$ will increase from $g(Q_0^2)$ until it reaches g_* and will then remain $\sim$ constant for smaller Q^2. (Such a behavior is invoked in some models of dynamical symmetry breaking.) Calculate n_{AF}, n_{fp}, and $\alpha_* = g_*^2/4\pi$.
(b) Plot $g(t)$ vs $t \equiv \ln(Q/M_Z)$ for $m = 3$ and some values of n_f and $g(t_0)$ that exhibit the fixed point behavior. Take $Q_0 = M_P$.
(c) For $\alpha_g = g^2/4\pi$ larger than some value α_c the $SU(n_f) \times SU(n_f)$ chiral flavor symmetry will be spontaneously broken (analogous to QCD), and the fermions will acquire mass and decouple. Thus, the fixed point behavior only occurs for $\alpha_* < \alpha_c$, which holds for n_f larger than some critical value n_c. The region $\max(n_c, n_{fp}) < n_f < n_{AF}$ is known as the *conformal window*. Calculate n_c using the approximate criterion $\alpha_c = \pi/[3C_2(L_m)]$ and plot n_{AF}, n_c, and n_{fp} vs m for $2 \leq m \leq 5$.

10.2 The left and right chiral projections $u_{L,R}$ of the u spinors in the chiral representation in (2.192) both reduce to the form $u_{L,R}(0,s) = \sqrt{m}\phi_s$ in the particle rest frame, $\vec{p} \to 0$. Show that the forms in (2.192) for arbitrary $\vec{p} = \vec{\beta}\gamma m$ can be obtained by acting on $u_{L,R}(0,s)$ with the (active) Lorentz boost $\Lambda(\vec{\zeta} = -\hat{\beta}\tanh^{-1}\beta)$, using the representation matrices $\Lambda_{(\frac{1}{2},0)}$ and $\Lambda_{(0,\frac{1}{2})}$, respectively.

10.3 Prove (10.50) using the explicit form for $\Lambda_{(\frac{1}{2},0)}$ in (10.29).

10.4 Let ψ_{1M} and ψ_{2M} be two Majorana fields written in 4-component notation, as in (10.58). Use the two-component spinor formalism developed in Section 10.2.2 to prove

$$\bar{\psi}_{1M}\Gamma\psi_{2M} = \sigma_\Gamma\bar{\psi}_{2M}\Gamma\psi_{1M},$$

with $\sigma_\Gamma = +1$ for $\Gamma = 1, \gamma^5$, and $\gamma^\mu\gamma^5$, and $\sigma_\Gamma = -1$ for $\Gamma = \gamma^\mu, \sigma^{\mu\nu}$. Note that this rederives the relations (9.20) (symmetric mass matrix), (9.64) (antisymmetric magnetic transition moments), and Problem 9.4 (no vector neutral current coupling) for Majorana fields, derived previously using four-component techniques.

10.5 Prove the Grassmann variable identities in (10.68).

10.6 Prove that the Lagrangian density in (10.69) is invariant under the supersymmetry transformations in (10.70). Hint: the easily derived identities

$$\sigma^\mu\bar{\sigma}^\nu + \sigma^\nu\bar{\sigma}^\mu = 2g^{\mu\nu}I, \qquad \bar{\sigma}^\mu\sigma^\nu + \bar{\sigma}^\nu\sigma^\mu = 2g^{\mu\nu}I$$

are useful.

10.7 Verify (10.81).

10.8 Derive (10.87) and (10.117).

10.9 Show using (8.199) that the logarithmic term in (10.164) can be interpreted as the running of the SM quartic Higgs coupling λ from a supersymmetry breaking scale M_S down to the electroweak scale $\nu = \mathcal{O}(m_t)$. Assume that all of the superpartners have masses M_S and that the running of λ is dominated by the h_t^4 term, with $h_t \sim$ constant.

10.10 (a) Calculate the cross sections for $e^-e^+ \to Zh$ and $e^-e^+ \to Ah$ via a virtual Z in the s-channel in the MSSM. Assume that one is well above the Z pole, and neglect the $e^\pm$ masses.
(b) You should find that

$$\frac{\sigma(Zh)}{\sigma(Ah)} = \tan^2(\beta - \alpha)\, \frac{k_{Zh}^3}{k_{Ah}^3} \left(1 + \frac{3M_Z^2}{k_{Zh}^2}\right),$$

where k_{Zh} and k_{Ah} are the magnitudes of the final CM momenta for each reaction. Interpret this result in terms of the equivalence theorem.

10.11 Derive (10.178).

10.12 Consider the cascade decay

$$\tilde{\chi}_2^0 \to e^- \tilde{e}_L^+, \qquad \tilde{e}_L^+ \to e^+ \tilde{\chi}_1^0$$

in the MSSM, where $\tilde{\chi}_{1,2}^0$ and $\tilde{e}_L^+$ are all on-shell, with (unknown) masses $m_{1,2}$ and $m_{\tilde{e}}$, respectively. Assume that the χ_1^0 is not observed and that the only kinematic information obtained for the event are the four-momenta $p_\pm$ of the e^+ and e^-. Show (neglecting the $e^\pm$ masses) that the maximum possible invariant mass of the e^+e^- pair is

$$M_{ee}^{max} = \frac{\sqrt{(m_2^2 - m_{\tilde{e}}^2)(m_{\tilde{e}}^2 - m_1^2)}}{m_{\tilde{e}}}.$$

Note that the actual M_{ee} distribution is expected to show a sharp cutoff (the *kinematic edge*) at M_{ee}^{max}, allowing a good measurement of that combination of superpartner masses.

10.13 Consider the reactions $pp \to \tilde{q}\tilde{G}$ or $\tilde{q}^c\tilde{G}$, where $\tilde{q}$ can be $\tilde{q}_L$ or $\tilde{q}_R$ with $q = u, d, c$, or s, and similarly for $\tilde{q}^c$. Assume that all of these squarks and the gluino have the same mass, m, and ignore mixing between families or between L and R.
(a) Draw, but do not calculate, the parton-level Feynman diagrams to lowest order. The parton-level cross section, which is the same for $\tilde{q}_L, \tilde{q}_R, \tilde{q}_L^c$, and $\tilde{q}_R^c$, is given in the Cross-Section article in (Patrignani, 2016).
(b) What is the largest m for which there would be at least 100 events at the 14 TeV LHC with an integrated luminosity of 100 fb^{-1}?

10.14 Suppose that the $SU(2) \times U(1)$ model with a single Higgs doublet ϕ is extended to $SU(2) \times U(1) \times U(1)'$, as in (10.197). Let Q_ϕ be the $U(1)'$ charge of ϕ, and assume that $M_{Z'}$ is much larger than M_Z and M_H due to the VEV of some SM singlet field.
(a) Calculate the $Z - Z'$ mixing angle θ.

(b) Calculate the decay widths for $Z' \to W^+W^-$ and $Z' \to ZH$ in the limit that M_Z, M_W, and M_H can be ignored. Hint: use the equivalence theorem with ϕ given in (8.85).
(c) The direct calculation of the $Z' \to W^+W^-$ rate (not using the equivalence theorem) is proportional to $|\theta|^2 \propto (M_Z/M_{Z'})^4$. Yet, the calculation in (b) indicates that it does not vanish for $M_Z/M_{Z'} \to 0$. Explain this apparent contradiction.

10.15 Consider the $U(1)'$ extension of the MSSM, with the neutral Higgs potential given in (10.209). There are now six neutralinos: the four MSSM ones, the singlino $\tilde{S}$, and the $U(1)'$ gaugino, $\tilde{Z}'$. (a) Write the 6×6 neutralino mass matrix corresponding to (10.184) in the weak basis, and analyze the spectrum in the $U(1)'$ *decoupling limit*, $s \equiv \sqrt{2}\langle S\rangle \gg (m_{soft}, \nu, \mu_{eff})$, with $g_2 Q_{S,u,d}$ fixed and comparable to g and g'.
(b) Show that the gauge/Higgs spectrum in the decoupling limit includes the the MSSM particles and a degenerate massive vector supermultiplet associated with Z', $\tilde{Z}'$, $\tilde{S}$, and S.

10.16 Derive (10.218) and (10.219).

10.17 Show how the universality test in (8.224) on page 342 is modified in the $SU(2)_L \times SU(2)_R \times U(1)$ model. Assume that ζ and β are small, $g_R = g_L$, $V_q^R = V_q^L$, and that the ν_R are very heavy. What are the approximate limits on ζ for $\omega = 0$ and for $\omega = \pi/2$?

10.18 (a) Assuming $\langle\Phi\rangle$ takes the form in (10.269), verify the expression for ν_Φ^2.
(b) Verify (10.270). Hint: use Problem 4.3.

APPENDIX A

Canonical Commutation Rules

In this appendix we briefly summarize the commutation rules and other aspects of standard field theories.

Spin-0 Fields

Consider a spin-0 field ϕ and Lagrangian density $\mathcal{L}$, as defined in Section 2.2. The *conjugate momentum* operator to ϕ is defined as

$$\pi(x) \equiv \frac{\delta \mathcal{L}}{\delta \dot{\phi}(x)}, \qquad \pi^\dagger(x) \equiv \frac{\delta \mathcal{L}}{\delta \dot{\phi}^\dagger(x)}, \tag{A.1}$$

where $\dot{\phi} \equiv \frac{\partial \phi}{\partial t}$ and the second definition is for the case in which ϕ is complex. For a spin-0 field, the *canonical equal time commutation rules* are

$$\begin{aligned} &[\phi(t,\vec{x}\,), \phi(t,\vec{x}\,')] = 0, \qquad [\pi(t,\vec{x}\,), \pi(t,\vec{x}\,')] = 0 \\ &[\pi(t,\vec{x}\,), \phi(t,\vec{x}\,')] = -i\delta^3(\vec{x} - \vec{x}\,'). \end{aligned} \tag{A.2}$$

If ϕ is complex, similar rules hold for $\phi^\dagger$ and $\pi^\dagger$, while $[\pi(t,\vec{x}\,), \phi^\dagger(t,\vec{x}\,')] = 0$. If there are distinct fields ϕ_i, then $[\pi_i(t,\vec{x}\,), \phi_j(t,\vec{x}\,')] = -i\delta^3(\vec{x} - \vec{x}\,')\delta_{ij}$.

The *Hamiltonian density* is

$$\mathcal{H}(\pi, \phi) = \pi\dot{\phi} + \pi^\dagger \dot{\phi}^\dagger - \mathcal{L}(\phi, \partial_\mu \phi). \tag{A.3}$$

The *Hamiltonian* $H(t)$ and *momentum* $\vec{P}(t)$ operators are

$$H(t) = \int d^3\vec{x}\; \mathcal{H}(\pi, \phi), \qquad \vec{P}(t) = \int d^3\vec{x}\; [\pi \vec{\nabla}\phi + \pi^\dagger \vec{\nabla}\phi^\dagger], \tag{A.4}$$

so the four-momentum operator is

$$P^\mu(t) = (H, \vec{P}\,) = \int d^3\vec{x}\; [\pi \partial^\mu \phi + \pi^\dagger \partial^\mu \phi^\dagger - g^{\mu 0}\mathcal{L}]. \tag{A.5}$$

(The upper case P is used to distinguish the momentum operators from the ordinary c-number momentum p^μ.) If ϕ is real then the second terms in (A.3)–(A.5) should be omitted. $H(t)$ and $\vec{P}(t)$ are actually independent of t unless there is explicit t dependence in $\mathcal{L}$.

The Hermitian scalar field is described in Section (2.3), with the Lagrangian density given in (2.20). The conjugate momentum is $\pi(x) = \delta\mathcal{L}/\delta\dot{\phi}(x) = \dot{\phi}(x)$, and the Hamiltonian density is

$$\mathcal{H}(\pi,\phi) = \frac{1}{2}\left[\dot{\phi}^2 + (\vec{\nabla}\phi)^2 + m^2\phi^2\right] + V_I(\phi). \tag{A.6}$$

Similarly, the Lagrangian density for the complex scalar field of Section 2.4 is given in (2.88), implying $\pi(x) = \dot{\phi}^\dagger(x)$, $\pi^\dagger(x) = \dot{\phi}(x)$, and

$$\mathcal{H}(\pi,\phi) = \left[\left(\frac{\partial\phi^\dagger}{\partial t}\right)\left(\frac{\partial\phi}{\partial t}\right) + (\vec{\nabla}\phi^\dagger)\cdot(\vec{\nabla}\phi) + m^2\phi^\dagger\phi\right] + V_I(\phi,\phi^\dagger). \tag{A.7}$$

For the free Hermitian field one can use the explicit form for $\phi_0(x)$ in (2.25) to obtain

$$\begin{aligned} P^\mu &= \frac{1}{2}\int \frac{d^3\vec{p}}{(2\pi)^3 2E_p}\, p^\mu \left[a(\vec{p})a^\dagger(\vec{p}) + a^\dagger(\vec{p})a(\vec{p})\right] \\ &= \int d^3\vec{p}\, p^\mu \left[\frac{\mathcal{N}(\vec{p})}{(2\pi)^3 2E_p} + \frac{\delta^3(0)}{2}\right], \end{aligned} \tag{A.8}$$

where

$$\mathcal{N}(\vec{p}) = a^\dagger(\vec{p})a(\vec{p}) \tag{A.9}$$

is the number operator, which counts the number of particles carrying momentum $\vec{p}$. The $\delta^3(0)$ term in (A.8) integrates to 0 for $\mu \neq 0$. For $\mu = 0$ it is the simple harmonic oscillator zero-point energy. It is infinite due to the integration over $\vec{p}$, e.g., it leads to an infinite vacuum energy

$$\langle 0|H|0\rangle = \frac{\delta^3(0)}{2}\int d^3\vec{p}\,\sqrt{\vec{p}^2 + m^2} = \infty. \tag{A.10}$$

(The $\delta^3(0)$ is an artifact of our use of continuum state normalization. For box normalization in a volume V it would be replaced by the density of states $V/(2\pi)^3$.) This constant of the energy is unobservable when considering the microscopic interactions,[1] so it is customary to ignore it. Such terms can be systematically removed by a technique known as *normal-ordering*, which means that creation operators are always written to the left of annihilation operators in a product (with an appropriate minus sign for fermions). The normal ordering of an operator $\mathcal{O}$ is sometimes indicated by the symbol $:\mathcal{O}:$, so that $:a_1^\dagger a_2: = :a_2 a_1^\dagger: = a_1^\dagger a_2$, while $:a_1 a_2: = a_1 a_2$. We will always assume normal-ordering for momentum and other related operators, such as the currents and charges associated with internal symmetries, but will not display the ::. Thus, (A.8) is replaced by

$$P^\mu \equiv :P^\mu: = \int \frac{d^3\vec{p}}{(2\pi)^3 2E_p}\, p^\mu\, a^\dagger(\vec{p})a(\vec{p}) = \int \frac{d^3\vec{p}}{(2\pi)^3 2E_p}\, p^\mu\, \mathcal{N}(\vec{p}). \tag{A.11}$$

Similarly, for the free complex scalar field in (2.93),

$$P^\mu = \int \frac{d^3\vec{p}}{(2\pi)^3 2E_p}\, p^\mu \left[\mathcal{N}_+(\vec{p}) + \mathcal{N}_-(\vec{p})\right], \tag{A.12}$$

where

$$\mathcal{N}_+(\vec{p}) = a^\dagger(\vec{p})a(\vec{p}), \qquad \mathcal{N}_-(\vec{p}) = b^\dagger(\vec{p})b(\vec{p}). \tag{A.13}$$

[1] It would become observable when gravity is included, so more care is needed in that case. The zero-point energies cancel between bosons and fermions in supersymmetric theories.

The Noether current in (2.100) corresponds, for the free field, to the charge in (2.104), i.e.,

$$Q = \int \frac{d^3\vec{p}}{(2\pi)^3 2E_p} \left[\mathcal{N}_+(\vec{p}) - \mathcal{N}_-(\vec{p})\right] \equiv N_{\pi^+} - N_{\pi^-}, \tag{A.14}$$

where $N_{\pi^\pm}$ are the overall number operators for $\pi^\pm$.

Spin-$\frac{1}{2}$ Fields

The Lagrangian density for a Dirac fermion in (2.152) implies that the conjugate field is $\pi_\alpha = \delta\mathcal{L}/\delta\dot{\psi}_\alpha = i\psi^\dagger_\alpha$. Fermion fields satisfy canonical equal time *anticommutation* rules

$$\begin{aligned} &\{\psi_\alpha(t,\vec{x}), \psi_\beta(t,\vec{x}\,')\} = 0, \qquad \{\psi^\dagger_\alpha(t,\vec{x}), \psi^\dagger_\beta(t,\vec{x}\,')\} = 0 \\ &\{\psi_\alpha(t,\vec{x}), \psi^\dagger_\beta(t,\vec{x}\,')\} = \delta^3(\vec{x}-\vec{x}\,')\delta_{\alpha\beta}. \end{aligned} \tag{A.15}$$

The expression for π_α implies

$$\mathcal{H} = \pi\dot{\psi} - \mathcal{L} = i\psi^\dagger\dot{\psi}, \qquad P^\mu = \int d^3\vec{x}\, \psi^\dagger i\partial^\mu\psi, \tag{A.16}$$

which hold even in the presence of gauge and Yukawa interactions. For the free field in (2.159), this implies

$$\begin{aligned} P^\mu &= \sum_{s=1}^{2} \int \frac{d^3\vec{p}}{(2\pi)^3 2E_p}\, p^\mu \left[a^\dagger(\vec{p},s)a(\vec{p},s) + b^\dagger(\vec{p},s)b(\vec{p},s)\right] \\ &= \sum_{s=1}^{2} \int \frac{d^3\vec{p}}{(2\pi)^3 2E_p}\, p^\mu \left[\mathcal{N}_+(\vec{p},s) + \mathcal{N}_-(\vec{p},s)\right]. \end{aligned} \tag{A.17}$$

The Noether current for fermion ψ is

$$J^\mu = \bar{\psi}\gamma^\mu\psi, \tag{A.18}$$

while the corresponding charge for the free Dirac field in (2.159) is

$$Q = \sum_{s=1}^{2} \int \frac{d^3\vec{p}}{(2\pi)^3 2E_p} \left[\mathcal{N}_+(\vec{p},s) - \mathcal{N}_-(\vec{p},s)\right] \equiv N_\psi - N_{\psi^c}. \tag{A.19}$$

Spin-1 Fields

The Lagrangian density for the free electromagnetic field is given in (2.111). Discussion of conjugate variables is complicated due to gauge issues, so we will only quote the Hamiltonian,

$$H = \frac{1}{2}\int d^3\vec{x}\left(\vec{E}^2 + \vec{B}^2\right). \tag{A.20}$$

For the free field in (2.113),

$$P^\mu = \sum_{\lambda=1,2} \int \frac{d^3\vec{p}}{(2\pi)^3 2E_p}\, p^\mu\, a^\dagger(\vec{p},\lambda)a(\vec{p},\lambda) = \sum_{\lambda=1,2} \int \frac{d^3\vec{p}}{(2\pi)^3 2E_p}\, p^\mu\, \mathcal{N}(\vec{p},\lambda). \tag{A.21}$$

APPENDIX B

Derivation of a Simple Feynman Diagram

Let us sketch the derivation of the first term in (2.30), i.e., the tree-level amplitude $M_{fi} = \langle \vec{p}_3\vec{p}_4|M|\vec{p}_1\vec{p}_2\rangle$ corresponding to the first diagram in Figure 2.3 for the Hermitian scalar field with $\kappa = 0$.

Our starting point is to consider a transition from an arbitrary initial state i to an arbitrary (but different) final state f with the same four-momentum. As shown in texts in quantum mechanics and field theory, the unitary *transition matrix element* U_{fi} and transition amplitude M_{fi} can be calculated in perturbation theory using the interaction picture, to yield[1]

$$U_{fi} = \langle f|U(+\infty, -\infty)|i\rangle \equiv (2\pi)^4\delta^4\left(\sum p_f - \sum p_i\right) M_{fi}, \tag{B.1}$$

up to subtleties concerning disconnected diagrams. In (B.1),

$$\langle f|U(t_2, t_1)|i\rangle = \langle f|\mathcal{T}\left[e^{-i\int_{t_1}^{t_2} H_I(t)\, dt}\right]|i\rangle \tag{B.2}$$

is the time evolution operator, $\mathcal{T}$ is the time-ordering operator, and

$$H_I(t) \equiv -L_I(t) = -\int d^3\vec{x}\; \mathcal{L}_I(\phi_0(x)) \tag{B.3}$$

is the interaction part of the Hamiltonian in (A.4). The fields occurring in H_I are free fields. For the non-gauge interactions of spin-0 particles, $\mathcal{L}_I = -V_I$. For our example, we can expand the exponential and keep just the linear term in H_I (the identity term doesn't contribute to the scattering amplitude),

$$U_{fi} \simeq \int d^4x\; \langle \vec{p}_3\vec{p}_4|i\mathcal{L}_I(\phi_0(x))|\vec{p}_1\vec{p}_2\rangle, \tag{B.4}$$

where

$$i\mathcal{L}_I(\phi_0(x)) = -iV_I(\phi_0(x)) = -i\frac{\lambda\phi_0^4(x)}{4!}. \tag{B.5}$$

[1] Some authors extract a factor of i from their definition of M.

Substituting the expression (2.25) for the free field $\phi_0(x)$, one finds

$$U_{fi} \simeq -i\frac{\lambda}{4!}\int \prod_{m=a}^{d} \frac{d^3\vec{p}_m}{(2\pi)^3 2E_m} \times \int d^4x \langle 0|a_3 a_4 \prod_{n=a}^{d} \left[a_n e^{-ip_n\cdot x} + a_n^\dagger e^{+ip_n\cdot x}\right] a_1^\dagger a_2^\dagger|0\rangle, \tag{B.6}$$

where $a_1 \equiv a(\vec{p}_1)$, etc. One can now move the a's to the right and the $a^\dagger$'s to the left. The nonvanishing terms in the matrix element in (B.6) are[2]

$$4!\; e^{i(p_a+p_b-p_c-p_d)\cdot x}\langle 0|a_3 a_a^\dagger\; a_4 a_b^\dagger\; a_c a_1^\dagger\; a_d a_2^\dagger|0\rangle, \tag{B.7}$$

where the 4! is due to the fact that there are 4! non-zero terms that differ only by the relabeling of the dummy indices $a \cdots d$. Finally, one has

$$a_d a_2^\dagger = (2\pi)^3 2E_d \delta^3(\vec{p}_d - \vec{p}_2) + a_2^\dagger a_d \;\rightarrow\; (2\pi)^3 2E_d\delta^3(\vec{p}_d - \vec{p}_2), \tag{B.8}$$

where the $a_2^\dagger a_d$ term vanishes since $a|0\rangle = 0$, and similarly for the other three terms. The momentum integrals can then be done using the delta functions, while the $\int d^4x$ integral yields $(2\pi)^4\delta^4(p_4 + p_3 - p_1 - p_2)$, so that

$$U_{fi} = -i\lambda(2\pi)^4\delta^4(p_4 + p_3 - p_1 - p_2), \qquad M_{fi} = -i\lambda. \tag{B.9}$$

The Feynman rules for more complicated diagrams involving internal lines can be derived using the *Wick ordering theorem.*

[2]The terms involving the commutators $[a_p, a_q^\dagger]$, $q = a \cdots d$, vanish for non-forward scattering. More generally, it can be shown that they should be ignored in such calculations.

Unitarity, the Partial Wave Expansion, and the Optical Theorem

The S matrix is a unitary operator describing the evolution of an initial state. It is given by $S = U(+\infty, -\infty)$, where the time evolution operator $U(t_2, t_1)$ is defined in (B.2), i.e., $S_{fi} = U_{fi}$. The transition matrix T is related by $S = I + iT$. It is convenient to extract a momentum-conserving δ function from a matrix element,

$$\langle f|S|i\rangle = \delta_{fi} + i\langle f|T|i\rangle \equiv \delta_{fi} + (2\pi)^4\delta^4(p_f - p_i)\, i\mathcal{T}_{fi}, \tag{C.1}$$

where the amplitude M defined in (B.1) is related by $M_{fi} = i\mathcal{T}_{fi}$. The unitarity of the S matrix,

$$(S^\dagger S)_{fi} = \sum_n S^\dagger_{fn} S_{ni} = \delta_{fi}, \tag{C.2}$$

implies the unitarity formula (for $p_f = p_i$)

$$-i\left(\mathcal{T}_{fi} - \mathcal{T}^\dagger_{fi}\right) = 2\Im m \mathcal{T}_{fi} = \sum_n (2\pi)^4\delta^4(p_n - p_i)\, \mathcal{T}^\dagger_{fn}\mathcal{T}_{ni}. \tag{C.3}$$

Applying this to the special case $f = i$ and using (2.51) yields the *optical theorem*,

$$2\,\Im m\, \mathcal{T}_{ii} = 4k\sqrt{s}\,\sigma_{tot}(k), \tag{C.4}$$

where $\sigma_{tot}(k)$ is the total (elastic plus inelastic) cross section at CM energy $\sqrt{s}$ and k is the center of mass three-momentum defined in (2.37). Thus, the optical theorem relates the imaginary part of the forward elastic scattering amplitude to the total cross section.

For elastic two-body scattering $\mathcal{T}$ is related to the traditional scattering amplitude $f(k, \theta)$ (especially familiar in quantum mechanics) by

$$\mathcal{T} = -iM = 8\pi\sqrt{s} f(k, \theta) \tag{C.5}$$

(see, e.g., the Kinematics article in Patrignani, 2016). Thus, the CM cross section is

$$\frac{d\sigma}{d\Omega} = |f(k, \theta)|^2 = \frac{1}{64\pi^2 s}|M|^2. \tag{C.6}$$

The partial wave expansion for spinless or spin-averaged particles (see Weinberg, 1995, for the general case) is

$$f(k,\theta) = \frac{1}{k}\sum_{\ell=0}^{\infty}(2\ell+1)a_\ell(k)P_\ell(\cos\theta), \tag{C.7}$$

where P_ℓ is the ℓth Legendre polynomial. The partial wave amplitude is

$$a_\ell(k) = \frac{k}{2}\int_{-1}^{1} d\cos\theta\, f(k,\theta)\, P_\ell(\cos\theta) = \frac{\left[\eta_\ell(k)e^{2i\delta_\ell(k)} - 1\right]}{2i}, \tag{C.8}$$

where δ_ℓ is the ℓth phase shift, and $0 \leq \eta_\ell \leq 1$ is the inelasticity parameter. From unitarity

$$|a_\ell(k)|^2 \leq \Im m\, a_\ell(k) \leq 1, \tag{C.9}$$

with the first inequality saturated for purely elastic scattering ($\eta_\ell = 1$). The partial wave amplitudes, and their graphical representation in terms of the *Argand plot*, are especially useful for parametrizing the behavior of low energy amplitudes (where few partial waves are important), searching for and studying the properties of resonances, etc., as described in introductory and quantum texts. They are also useful for unitarity bounds on the elastic cross section $\sigma(k)$ (integrated over angle), which are useful in connection with the breakdown of the Fermi theory and with theoretical constraints on the Higgs mass:

$$\sigma(k) \equiv \sum_{\ell=0}^{\infty}\sigma_\ell(k) = \frac{4\pi}{k^2}\sum_{\ell=0}^{\infty}(2\ell+1)|a_\ell(k)|^2, \tag{C.10}$$

so that

$$\sigma_\ell(k) \leq \frac{4\pi(2\ell+1)}{k^2}. \tag{C.11}$$

In terms of $f(k,\theta)$ the optical theorem is

$$\sigma_{tot}(k) = \frac{4\pi}{k}\Im m\, f(k,0) = \frac{4\pi}{k^2}\sum_{\ell=0}^{\infty}(2\ell+1)\Im m\, a_\ell(k) \geq \sigma(k). \tag{C.12}$$

The optical theorem, combined with some general considerations involving analyticity and crossing, can be used to derive the *Froissart bound* (Froissart, 1961; Pancheri and Srivastava, 2017) on the high energy behavior of the total cross section,

$$\sigma_{tot} \leq C\ln^2 s, \tag{C.13}$$

for $s \to \infty$, where C is a constant.

Two, Three, and n-Body Phase Space

Techniques for evaluating 2-body phase space were considered in Section 2.3.5. Here we formalize that discussion and consider some techniques for the 3-body case. More general discussions are given in (Barger and Phillips, 1997) and in the Kinematics article in (Patrignani, 2016).

Let us define the Lorentz invariant 2-body phase space factor

$$d_2\left(m_I^2, m_a^2, m_b^2\right) \equiv \delta^4\left(p_I - p_a - p_b\right)\frac{d^3\vec{p}_a}{E_a}\frac{d^3\vec{p}_b}{E_b}, \tag{D.1}$$

where p_I is the total four-momentum. For a decay process, $m_I^2 = p_I^2$ is the mass-squared of the decaying particle, while for a scattering process $m_I^2 = p_I^2 = s$. We saw in (2.55) that in the rest frame or CM (where $p_I = (m_I, 0)$),

$$d_2\left(m_I^2, m_a^2, m_b^2\right) = d\Omega\frac{p_f}{m_I} \to 2\pi d\cos\theta\frac{p_f}{m_I} \to 4\pi\frac{p_f}{m_I}, \tag{D.2}$$

where $d\Omega = d\varphi\, d\cos\theta$ is the solid angle element of $\vec{p}_a$, and

$$p_f = |\vec{p}_a| = |\vec{p}_b| = \frac{\lambda^{\frac{1}{2}}\left(m_I^2, m_a^2, m_b^2\right)}{2m_I}, \tag{D.3}$$

with $\lambda(x,y,z) \equiv x^2 + y^2 + z^2 - 2xy - 2xz - 2yz$. The differential cross section for 2-body scattering is

$$d\sigma = \frac{d_2}{64\pi^2 p_i\sqrt{s}}|M|^2 = \frac{1}{32\pi s}\frac{p_f}{p_i}|M|^2\, d\cos\theta, \tag{D.4}$$

where the initial momentum p_i is given in (2.38). Similarly, a 2-body differential decay rate is

$$d\Gamma = \frac{d_2}{32\pi^2 m_I}|M|^2 = \frac{p_f}{16\pi m_I^2}|M|^2\, d\cos\theta. \tag{D.5}$$

These results can be immediately generalized to n-body phase space, relevant to $I \to f_1 \cdots f_n$. Defining

$$d_n = \delta^4\Big(p_I - \sum_{i=1}^{n} p_{f_i}\Big)\prod_{i=1}^{n}\frac{d^3\vec{p}_{f_i}}{E_i}, \tag{D.6}$$

the differential cross section and decay rate are, respectively,

$$d\sigma = \frac{(2\pi)^4}{4p_i\sqrt{s}} c_n |M|^2 \, d_n, \qquad d\Gamma = \frac{(2\pi)^4}{2m_I} c_n |M|^2 \, d_n, \tag{D.7}$$

with $c_n = [2(2\pi)^3)]^{-n}$.

The case $n = 3$, with

$$d_3\left(m_I^2, m_a^2, m_b^2, m_c^2\right) = \delta^4\left(p_I - p_a - p_b - p_c\right) \frac{d^3\vec{p}_a}{E_a} \frac{d^3\vec{p}_b}{E_b} \frac{d^3\vec{p}_c}{E_c}, \tag{D.8}$$

is very important, especially for decay processes. We will sketch two techniques. The first method readily generalizes to n bodies. It involves the factorization of d_3 into a product of 2-body factors, one of which describes $b + c$ and the other describes $a + bc$, where bc represents the collective $b + c$ system. The technique is especially useful when b and c are not observed (a simplified version was used for μ decay in Section 7.2.1), when there is a resonance in the bc channel, or when a is the first particle emitted in a cascade. To begin, we introduce the factor

$$\begin{aligned} 1 = &\int_{(m_b+m_c)^2}^{(m_I-m_a)^2} dm_{bc}^2 \, \delta\left(m_{bc}^2 - (p_b + p_c)^2\right) \\ &\times \int d^4p_{bc} \, \delta^4\left(p_{bc} - p_b - p_c\right) \Theta\left(E_{bc}\right) \end{aligned} \tag{D.9}$$

in the expression for d_3. We can then rearrange the expression to yield

$$\begin{aligned} d_3 &= \frac{1}{2} \int dm_{bc}^2 \, \delta^4\left(p_I - p_a - p_{bc}\right) \frac{d^3\vec{p}_a}{E_a} \frac{d^3\vec{p}_{bc}}{E_{bc}} \delta^4\left(p_{bc} - p_b - p_c\right) \frac{d^3\vec{p}_b}{E_b} \frac{d^3\vec{p}_c}{E_c} \\ &= \frac{1}{2} \int dm_{bc}^2 \, d_2\left(m_I^2, m_a^2, m_{bc}^2\right) d_2\left(m_{bc}^2, m_b^2, m_c^2\right). \end{aligned} \tag{D.10}$$

Thus, d_3 is just the product of $b + c$ phase space, restricted to invariant mass m_{bc}, and phase space for $a + bc$, integrated over the allowed range shown in (D.9) for m_{bc}^2. Of course, the matrix element $|M|^2$ occurs under the integrals. The separation is especially useful if $|M|^2$ also factorizes. It should be remembered that the d_2's are Lorentz invariants, so it is often convenient to evaluate the factors in different Lorentz frames, such as the I and bc rest frames. As a simple example of factorization, suppose there is a narrow resonance in the bc channel, with, e.g.,

$$|M|^2 = \frac{f(m_I^2, m_{bc}^2)}{\left(m_{bc}^2 - M_R^2\right)^2 + M_R^2\Gamma_R^2} \to f(m_I^2, m_{bc}^2) \frac{\pi}{M_R\Gamma_R} \delta\left(m_{bc}^2 - M_R^2\right), \tag{D.11}$$

as shown in Figure D.1. Equation (D.11) is the Breit-Wigner resonance formula for an unstable particle R of mass M_R and width $\Gamma_R = 1/\tau_R$ propagating in the bc channel, as described in Appendix F. $f(m_I^2, m_{bc}^2)$ describes the splitting of the initial state into $a + R$ and the decay of R into bc. The final form is the narrow resonance approximation, valid for $\Gamma_R \ll M_R$. In this case

$$\begin{aligned} d_3|M|^2 &= \frac{1}{2} \int dm_{bc}^2 \, d_2\left(m_I^2, m_a^2, m_{bc}^2\right) f(m_I^2, m_{bc}^2) \times \frac{d_2\left(m_{bc}^2, m_b^2, m_c^2\right)}{\left(m_{bc}^2 - M_R^2\right)^2 + M_R^2\Gamma_R^2} \\ &\to d_2\left(m_I^2, m_a^2, M_R^2\right) f(m_I^2, M_R^2) \times \frac{\pi}{2M_R\Gamma_R} d_2\left(M_R^2, m_b^2, m_c^2\right). \end{aligned} \tag{D.12}$$

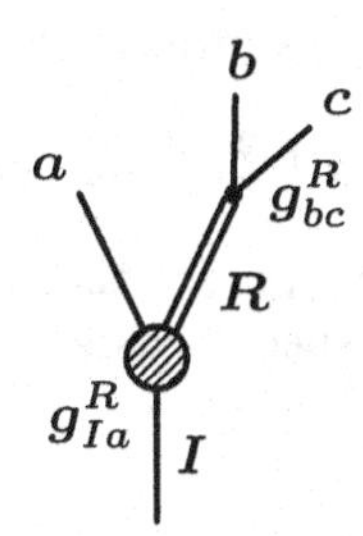

Figure D.1 Three-body final state with a resonance in the bc channel. I is the initial decay or scattering state, g^R_{Ia} is the amplitude to produce aR, and g^R_{bc} is the resonance decay amplitude. $f(m_I^2, m_{bc}^2)$ in (D.11) is given by $|g^R_{Ia}|^2|g^R_{bc}|^2$.

Now let us further suppose that f factorizes into terms involving the production and decay, which would be the case for a spin-0 resonance, i.e., $f(m_I^2, M_R^2) = |g^R_{Ia}|^2|g^R_{bc}|^2$. Then, incorporating the relevant factors of 2 and 2π from (D.7), one finds

$$\sigma(I \to abc) = \sigma(I \to aR)B(R \to bc), \quad \Gamma(I \to abc) = \Gamma(I \to aR)B(R \to bc), \tag{D.13}$$

where $B(R \to bc) = \Gamma(R \to bc)/\Gamma_R$ is the branching ratio for $R \to bc$. This reproduces the intuitive result that a cross section or decay rate involving an intermediate narrow (i.e., long-lived) state factorizes into the cross section or rate to produce the narrow state times the branching ratio into the final state. An interesting feature of this result is illustrated by assuming that R can only decay into bc, so that $B = 1$. In this case the overall rates in (D.13) are independent of the coupling strength g^R_{bc}. This is completely different from a virtual intermediate state, and is due to the fact that once R is produced it always decays. The same holds if there are other decay channels so long as $B(R \to bc)$ is held constant. Of course, the rates *do* depend on the production coupling $|g^R_{Ia}|$.

An alternative method is useful when $|M|^2$ only depends on $p_a \cdot p_b$, $p_a \cdot p_c$, and $p_b \cdot p_c$, such as a decay process summed and averaged over spins. In the I rest frame $\vec{p}_{a,b,c}$ lie in a plane. By assumption, $|M|^2$ does not depend on the orientation of the plane or the direction of $\vec{p}_a$ in the plane, so we can integrate over them. The spatial δ function can be used to eliminate $\vec{p}_c$. The angle between $\vec{p}_a$ and $\vec{p}_b$ can be expressed in terms of the energies of a and b using the remaining energy δ function, so that d_3 becomes an unconstrained integral over E_a and E_b within an allowed region. The result, derived in (Barger and Phillips, 1997), is

$$d_3\left(m_I^2, m_a^2, m_b^2, m_c^2\right) = 2\pi^2 m_I^2\, dX_a dX_b = \frac{2\pi^2}{m_I^2}\, dm_{bc}^2\, dm_{ac}^2, \tag{D.14}$$

where we define

$$X_i = \frac{2E_i}{m_I}, \qquad \mu_i = \frac{m_i^2}{m_I^2}, \qquad i = a, b, c, \tag{D.15}$$

and

$$m_{ij}^2 \equiv (p_i + p_j)^2 = (p_I - p_k)^2 = m_I^2 + m_k^2 - 2m_I E_k, \tag{D.16}$$

where i, j, and k are all different. The invariants can be expressed as

$$p_i \cdot p_j = \frac{1}{2} m_I^2 \left(1 + \mu_k - \mu_i - \mu_j - X_k\right), \tag{D.17}$$

which follows from (D.16). The energy or invariant mass variables satisfy

$$X_a + X_b + X_c = 2, \qquad m_{ab}^2 + m_{ac}^2 + m_{bc}^2 = m_I^2 + m_a^2 + m_b^2 + m_c^2. \tag{D.18}$$

However, the actual boundaries for $X_{a,b}$ are rather complicated (see Barger and Phillips, 1997), so we only give them for some special cases.

$$\begin{aligned}
&m_{a,b,c} = 0 : \\
&\qquad 0 \leq X_a \leq 1, \qquad 1 - X_a \leq X_b \leq 1 \\
\\
&m_a \neq 0, m_{b,c} = 0 : \\
&\qquad 2\mu_a^{1/2} \leq X_a \leq 1 + \mu_a, \qquad \alpha - \beta \leq X_b \leq \alpha + \beta \\
&\qquad \alpha \equiv \frac{1}{2}\left(2 - X_a\right), \qquad \beta \equiv \frac{1}{2}\left(X_a^2 - 4\mu_a\right)^{\frac{1}{2}} \\
\\
&m_{a,b} = 0, m_c \neq 0 : \\
&\qquad 0 \leq X_a \leq 1 - \mu_c, \qquad 1 - X_a - \mu_c \leq X_b \leq \frac{1 - X_a - \mu_c}{1 - X_a}.
\end{aligned} \tag{D.19}$$

An approximation to this technique was used for pion beta decay in Section 7.2.3. The method is applied to calculate the leading correction to that result in Problem 7.7.

This method is closely connected to the *Dalitz plot*, in which individual events observed in a 3-body process are plotted in the $X_a - X_b$ or the $m_{bc}^2 - m_{ac}^2$ plane. From (D.14) the phase space is uniform within the allowed region, so an excess of events would correspond to an enhancement of the matrix element. A resonance in a two-body channel would show up as an excess along a line parallel to the x or y axis, or along a line $x + y =$ constant.

APPENDIX E

Calculation of the Anomalous Magnetic Moment of the Electron

In this appendix we sketch the derivation of the one-loop (Schwinger) contribution of $\alpha/2\pi$ to $a_e = (g_e - 2)/2$, the anomalous magnetic moment of the electron, in part to illustrate some of the techniques for calculating Feynman integrals. The relevant contribution is from the diagram in Figure E.1, which will be of the form

$$ie_0 \bar{u}_2 \underbrace{[a\gamma^\mu + b(p_1 + p_2)^\mu]}_{\Gamma^\mu_{OL}} u_1, \tag{E.1}$$

where $\Gamma^\mu_{OL} \equiv Z_1^{-1}\Gamma^\mu - \gamma^\mu$ in the notation of Section 2.12.1. From the discussion following (2.359) and using the first Gordon identity in Problem 2.10, we identify $a_e = F_2(0) \sim -2mb$,

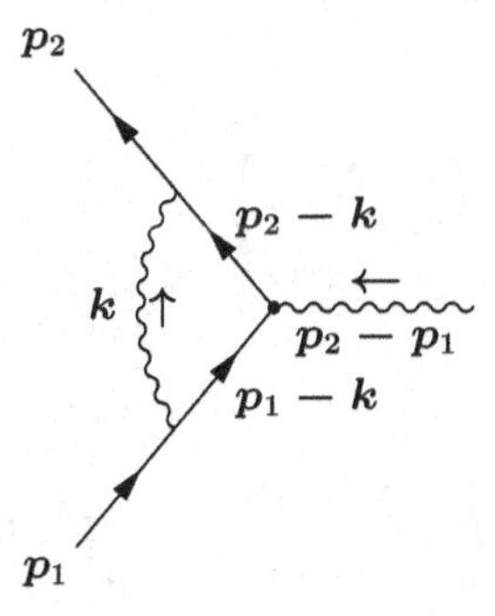

Figure E.1 One-loop vertex correction contributing to the anomalous magnetic moment of the electron. The photon momentum is $q = p_2 - p_1$.

which will turn out to be non-divergent. From the Feynman rules,

$$\begin{aligned}\Gamma^{\mu}_{OL} &= (ie)^2 \int \frac{d^4k}{(2\pi)^4} \frac{-i}{k^2+i\epsilon} \frac{i}{(p_2-k)^2-m^2+i\epsilon} \frac{i}{(p_1-k)^2-m^2+i\epsilon} \\ &\times \left[\gamma_\nu \left(\not{p}_2 - \not{k} + m\right)\gamma^\mu \left(\not{p}_1 - \not{k} + m\right)\gamma^\nu\right] \\ &= \frac{-ie^2}{(2\pi)^4} \int d^4k \frac{B^\mu(p_1,p_2,k)}{D},\end{aligned} \tag{E.2}$$

where

$$\begin{aligned}B^\mu(p_1,p_2,k) &= \gamma_\nu \left(\not{p}_2 - \not{k} + m\right)\gamma^\mu \left(\not{p}_1 - \not{k} + m\right)\gamma^\nu \\ D &= \left(k^2+i\epsilon\right)\left(k^2 - 2p_2\cdot k + i\epsilon\right)\left(k^2 - 2p_1\cdot k + i\epsilon\right).\end{aligned} \tag{E.3}$$

We have used $p_{1,2}^2 = m^2$ and have replaced e_0^2 by e^2 and m_0 by m since the differences are of higher order. (We set $Z_1 = 1$ in the identification of $F_2(0)$ for the same reason.) It is understood that the external electrons are on-shell and that Γ^μ_{OL} is sandwiched between u spinors. The d^4k integral appears formidable, but it can be handled by a series of tricks. First, one uses *Feynman parametrization* to combine the denominators. There are a number of ways to do this, but we will use the identity

$$\frac{1}{a_1 a_2 \cdots a_n} = (n-1)! \int_0^1 \frac{dz_1 \cdots dz_n \delta\left(1 - \sum_i z_i\right)}{\left[\sum_i a_i z_i\right]^n}, \tag{E.4}$$

for any $a_1 \cdots a_n$. For example,

$$\frac{1}{a_1 a_2} = \int_0^1 \frac{dz}{\left[a_1 z + a_2(1-z)\right]^2}. \tag{E.5}$$

Applying (E.4),

$$\frac{1}{D} = 2\int_0^1 \frac{dz_1 dz_2 dz_3 \delta\left(1-\sum z_i\right)}{\left[k^2 - 2p_2\cdot k z_2 - 2p_1 \cdot k z_1 + i\epsilon\right]^3} = 2\int_0^1 \frac{\mathcal{D}z}{\left[k'^2 - c + i\epsilon\right]^3}, \tag{E.6}$$

where

$$\begin{aligned}\mathcal{D}z &\equiv dz_1 dz_2 dz_3 \delta\left(1 - \sum z_i\right), \qquad k' = k - p_2 z_2 - p_1 z_1 \\ c &= \left(p_2 z_2 + p_1 z_1\right)^2 = m^2\left(1-z_3\right)^2 - q^2 z_1 z_2,\end{aligned} \tag{E.7}$$

with $c > 0$ for $q^2 \le 0$. Therefore,

$$\begin{aligned}\Gamma^\mu_{OL} &= \frac{-2ie^2}{(2\pi)^4} \int_0^1 \int \frac{d^4k B^\mu\left(p_1,p_2,k\right)}{\left[k'^2 - c + i\epsilon\right]^3} \mathcal{D}z \\ &= \frac{-2ie^2}{(2\pi)^4} \int_0^1 \int d^4k' \frac{B^\mu\left(p_1,p_2,k' + p_2 z_2 + p_1 z_1\right)}{\left[k'^2 - c + i\epsilon\right]^3} \mathcal{D}z,\end{aligned} \tag{E.8}$$

where we have shifted the integration variable $d^4k \to d^4k'$ in the second line. The terms linear in k' integrate to zero

$$\int \frac{d^4k'\, k'_\mu}{\left[k'^2 - c + i\epsilon\right]^3} = 0. \tag{E.9}$$

The quadratic terms involving $k'_\mu k'_\nu$ must be of the form,

$$\int \frac{d^4k'\; k'_\mu k'_\nu}{\left[k'^2 - c + i\epsilon\right]^3} = C g_{\mu\nu} \int \frac{d^4k'^2\; k'^2}{\left[k'^2 - c + i\epsilon\right]^3}, \tag{E.10}$$

since the l.h.s. of (E.10) is a Lorentz tensor, and $g_{\mu\nu}$ is the only tensor available. Contracting each side with $g^{\mu\nu}$ implies that $C = \frac{1}{4}$. Finally, using the γ matrix identities in (2.174) as well as $\bar{u}_2\, \not{p}_2 = m\bar{u}_2$ and $\not{p}_1 u_1 = m u_1$,

$$\begin{aligned} B^\mu \left(p_1, p_2, k' + p_2 z_2 + p_1 z_1\right) =& 2m \left(p_1 + p_2\right)^\mu \left[-2z_1 z_2 + z_2\left(1 - z_2\right) + z_1\left(1 - z_1\right)\right] \\ & + \gamma^\mu \text{ terms } + \text{ vanishing terms.} \end{aligned} \tag{E.11}$$

Therefore, dropping the prime on k,

$$F_2(0) = \frac{8im^2e^2}{(2\pi)^4} \int_0^1 \int d^4k \frac{\left[-2z_1z_2 + z_2\left(1 - z_2\right) + z_1\left(1 - z_1\right)\right] \mathcal{D}z}{\left[k^2 - c + i\epsilon\right]^3}, \tag{E.12}$$

where c in (E.7) is evaluated at $q^2 = 0$, i.e., $c = m^2\left(1 - z_3\right)^2$.

Now, consider the integral

$$I_n \equiv \int \frac{d^4k}{\left[k^2 - c + i\epsilon\right]^n} = \int_{-\infty}^{+\infty} dk_0 \int \frac{d^3\vec{k}}{\left[k_0^2 - \vec{k}^2 - c + i\epsilon\right]^n}, \tag{E.13}$$

where $c > 0$. I_n is convergent for $n > 2$. The k_0 integral can be viewed as a contour integral along the real axis in the complex k_0 plane. The integrand has poles at $k_0 = \pm(\sqrt{\vec{k}^2 + c} - i\epsilon)$, which lie in the second and fourth quadrants. One can therefore perform a *Wick rotation*, in which the contour is rotated to lie along the imaginary axis,

$$I_n = \int_{-i\infty}^{+i\infty} dk_0 \int \frac{d^3\vec{k}}{\left[k_0^2 - \vec{k}^2 - c\right]^n} = i \int_{-\infty}^{+\infty} dk_{0E} \int \frac{d^3\vec{k}}{\left[-k_{0E}^2 - \vec{k}^2 - c\right]^n}, \tag{E.14}$$

where $k_{0E} \equiv -ik_0$ is the Euclidean energy, and we have taken $\epsilon \to 0$ since the rotated contour is far from the poles. This can then be written

$$\begin{aligned} I_n &= (-1)^n i \int \frac{d^4k_E}{\left[k_E^2 + c\right]^n} = (-1)^n i \int d\Omega_4 \int_0^\infty \frac{k_E^3 dk_E}{\left[k_E^2 + c\right]^n} \\ &= (-1)^n i\pi^2 \frac{1}{(n-1)} \frac{1}{(n-2)} \frac{1}{c^{n-2}}, \end{aligned} \tag{E.15}$$

where k_E is a Euclidean four-vector with $k_E^2 = k_{0E}^2 + \vec{k}^2$, $k_E^3 dk_E = k_E^2 dk_E^2/2$, and $\int d\Omega_4 = 2\pi^2$ is the area of a unit 3-sphere in four dimensions (Problem 1.3). The integral has poles for $n = 1$ and $n = 2$, indicating the divergence of the integral, but is well behaved for $n > 2$. These techniques can be readily generalized to ones such as $\int d^4k k^{2p}/\left[k^2 - c + i\epsilon\right]^n$.

Putting everything together, the anomalous magnetic moment is

$$\begin{aligned} F_2(0) &= \frac{\alpha}{\pi} \int_0^1 \frac{\left[-2z_1z_2 + z_2\left(1 - z_2\right) + z_1\left(1 - z_1\right)\right)}{\left(1 - z_3\right)^2} dz_1 dz_2 dz_3 \delta\left(1 - \sum z_i\right) \\ &= \frac{\alpha}{\pi} \int_0^1 \left[-1 + \frac{1}{z_1 + z_2}\right] dz_1 dz_2 \Theta\left(1 - z_1 - z_2\right) = \frac{\alpha}{2\pi}. \end{aligned} \tag{E.16}$$

Breit-Wigner Resonances

An unstable particle or resonance of mass M_R and width $\Gamma = 1/\tau$ can be treated in Feynman diagrams by replacing the denominator of its propagator by the *Breit-Wigner* resonance form

$$\frac{1}{q^2 - M^2 + i\epsilon} \to D(q^2) \equiv \frac{1}{q^2 - M_R^2 + iM_R\Gamma}, \tag{F.1}$$

so that

$$|D(q^2)|^2 = \frac{1}{(q^2 - M_R^2)^2 + M_R^2\Gamma^2}. \tag{F.2}$$

If Γ is large, then additional energy dependent corrections may be important away from the peak (see, e.g., Gounaris and Sakurai, 1968). Away from the resonance peak a rate involving a resonance is suppressed compared to the on-shell rate by $M_R^2\Gamma^2/(q^2 - M_R^2)^2 \ll 1$, and far enough off shell the finite width effect is irrelevant. (There may be additional energy dependent factors associated with phase space, etc.)

For a narrow resonance, $\Gamma/M_R \ll 1$, it is sometimes useful to employ the *narrow width approximation*

$$|D(q^2)|^2 \to \frac{\pi}{M_R\Gamma}\delta(q^2 - M_R^2), \tag{F.3}$$

which effectively treats the production of the long-lived state as if it were stable. This was used in a 3-body phase space example in Appendix D, where it was shown that the production rate of a given final state through a narrow intermediate spin-0 resonance factorizes into the rate to produce the resonance, times its branching ratio into the final state. This result generalizes to more complicated production and decay processes and to differential cross sections and decay rates. One can still use (F.3) for narrow resonances with nonzero spin. However, the factorization into production rate times branching ratio only holds if one averages or sums over the external spins and integrates over all of phase space. Otherwise, there are spin and angular correlations that do not factorize.

Let us consider an example of an unstable Hermitian spin-0 field ϕ that can couple to fermions a or b, with

$$\mathcal{L} = \left(g_a\bar{\psi}_a\psi_a + g_b\bar{\psi}_b\psi_b\right)\phi, \qquad g_{a,b} = \text{real}. \tag{F.4}$$

The amplitude for $a\bar{a} \to b\bar{b}$ via the s-channel ϕ resonance is

$$M = i^3 g_a g_b \bar{u}_b v_{\bar{b}}\, \bar{v}_{\bar{a}} u_a\, D(s), \tag{F.5}$$

where we take $M_R = m_\phi$ and $\Gamma = \Gamma_\phi$ in $D(s)$. Neglecting m_a and m_b,

$$|\bar{M}|^2 = \frac{1}{4}g_a^2 g_b^2\, |D(s)|^2\, \mathrm{Tr}\,(\not{p}_{\bar{b}}\ \not{p}_b)\, \mathrm{Tr}\,(\not{p}_a\ \not{p}_{\bar{a}}) = g_a^2 g_b^2 s^2\, |D(s)|^2, \tag{F.6}$$

so that the cross section is

$$\bar{\sigma}(s) = \frac{1}{16\pi s}|\bar{M}|^2 = \frac{g_a^2 g_b^2}{16\pi} \frac{s}{(s - M_R^2)^2 + M_R^2 \Gamma^2}. \tag{F.7}$$

Note that it is usually not useful to use the narrow resonance approximation for an s channel process unless one is integrating over s. Equation (F.7) can be conveniently rewritten using

$$\bar{\Gamma}_{a\bar{a}} \equiv \sum_{s_a s_{\bar{a}}} \Gamma\left(\phi \to a\bar{a}\right) = \frac{1}{16\pi M_R}|\bar{M}_{a\bar{a}}|^2, \tag{F.8}$$

with

$$|\bar{M}_{a\bar{a}}|^2 = |ig_a|^2 \operatorname{Tr}\left(\not{p}_a \; \not{p}_{\bar{a}}\right) = 2M_R^2 g_a^2, \tag{F.9}$$

so that the spin-summed partial width into $a\bar{a}$ is

$$\bar{\Gamma}_{a\bar{a}} = \frac{g_a^2 M_R}{8\pi}, \tag{F.10}$$

with a similar formula for $\bar{\Gamma}_{b\bar{b}}$. If there are no other decay channels, then $\Gamma = \bar{\Gamma}_{a\bar{a}} + \bar{\Gamma}_{b\bar{b}}$. From (F.7) and (F.10),

$$\bar{\sigma}(s) = \frac{4\pi(s/M_R^2)\bar{\Gamma}_{a\bar{a}}\bar{\Gamma}_{b\bar{b}}}{(s - M_R^2)^2 + M_R^2\Gamma^2} \xrightarrow[s=M_R^2]{} \frac{4\pi}{M_R^2} B_{a\bar{a}} B_{b\bar{b}}, \tag{F.11}$$

where $B_{a\bar{a}} = \bar{\Gamma}_{a\bar{a}}/\Gamma$ is the branching ratio into $a\bar{a}$. This convenient form absorbs the couplings $g_{a,b}$ into the partial width or branching ratio factors. It also illustrates that the peak cross section at $s = m_R^2$ is independent of the coupling strengths for fixed branching ratios because they cancel between the partial and total widths. Similarly, for a massive vector resonance V_μ with interactions

$$\mathcal{L} = -\left[g_a \bar{\psi}_a \gamma^\mu \psi_a + g_b \bar{\psi}_b \gamma^\mu \psi_b\right] V_\mu, \tag{F.12}$$

one finds the same result as (F.11) except that $4\pi \to 12\pi$ (Problem 2.27).

By measuring $\bar{\sigma}(s)$ as a function of s one can determine M_R^2 from the position of the peak, Γ from the width of the peak, and a combination of branching ratios from the peak cross section $\bar{\sigma}_{peak} = \bar{\sigma}(M_R^2)$, as illustrated in Figure F.1. This was extremely powerful, for example, in the Z lineshape measurements at LEP. (In practice, one must apply additional energy dependent corrections.) In some cases, however, the detector energy resolution is not good enough to resolve the peak and determine Γ. One then effectively integrates over s to obtain (for the spin-0 case in (F.11))

$$\int \bar{\sigma}(s) ds \sim \frac{4\pi^2}{M_R} \bar{\Gamma}_{a\bar{a}} B_{b\bar{b}}. \tag{F.13}$$

The expression in (F.11) is typical for an s-channel resonance (see, e.g., Weinberg, 1995), provided that spins are averaged (summed) and the phase space is integrated over. The overall coefficient depends on the spins of the resonance and of the initial particles. Sufficiently near the peak, the cross section for $a_1 a_2 \to b_1 b_2$ is

$$\bar{\sigma}(s) = \frac{4(2S_R + 1)}{(2S_{a_1} + 1)(2S_{a_2} + 1)} \frac{4\pi(s/M_R^2)\bar{\Gamma}_{a_1 a_2}\bar{\Gamma}_{b_1 b_2}}{(s - M_R^2)^2 + M_R^2\Gamma^2}, \tag{F.14}$$

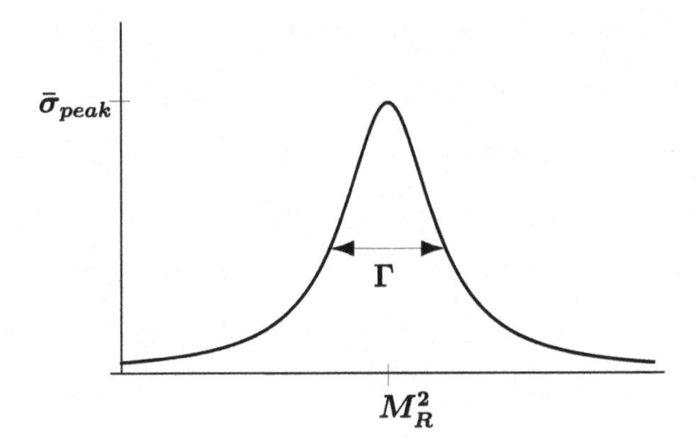

Figure F.1 The cross section vs. s for the Breit-Wigner resonance form in (F.11), illustrating the peak position, width, and peak cross section.

where S_R, S_{a_1}, and S_{a_2} are respectively the spins of R, a_1, and a_2 (with $2S+1$ replaced by 2 for a photon or gluon). The prefactor is because the $2S_R+1$ intermediate spin states contribute equally to (and do not interfere in) the total spin-averaged rate, while $(2S_{a_1}+1)^{-1}$ $(2S_{a_2}+1)^{-1}$ corrects for the fact that $\bar{\sigma}(s)$ is spin-averaged while $\bar{\Gamma}_{a_1 a_2}$ is spin-summed. There is an analogous correction if there is a color sum/average on a_1 and/or a_2, and a factor of 2 if a_1 and a_2 are identical (to correct for the statistical factor of 1/2 in the phase space integration in (2.81)). There may be additional s dependence away from the pole, due, e.g., to vertex factors or threshold effects.

APPENDIX G

Implications of P, C, T, and G-parity for Nucleon Matrix Elements

In this appendix we summarize the implications of various discrete symmetries for the matrix elements of vector and axial currents between on-shell nucleon or hyperon states. Recall that even though the weak interactions violate these symmetries, e.g., because of the $V - A$ structure of the charged current, the hadronic matrix elements are dominated by strong interaction effects, which are believed to be invariant under P, C, and T (Section 2.10), and approximately so under G-parity (Section 3.2.5). However, higher-order weak corrections could lead to small deviations from the symmetry predictions.

Crossing Symmetry

Consider an amplitude or matrix element involving an initial particle $b(p, s)$,

$$\langle \alpha | \mathcal{O} | \beta \, b(p, s) \rangle \equiv O(\bar{p}) u(\vec{p}, s), \tag{G.1}$$

where α and β are other external particles. The spinor-valued function $O(\bar{p})$ depends on $\bar{p} \equiv p$ and on the momenta and spins of the other particles. *Crossing symmetry* implies that the corresponding matrix element for a final b^c is described by the same function, i.e.,

$$\langle \alpha \, b^c(p, s) | \mathcal{O} | \beta \rangle = \pm O(\bar{p}) v(\vec{p}, s), \tag{G.2}$$

where in this case $\bar{p} \equiv -p$, and the $\pm$ sign depends on how many fermion interchanges are involved in the crossing. In both (G.1) and (G.2), p and s refer to the physical momentum and spin, while $\bar{p}$ refers to the momentum flowing in the direction of the arrow in a Feynman diagram. Although the same function $O(\bar{p})$ describes both processes, the physical values of $\bar{p}$ are in different regions, i.e., the two processes are related by analytic continuation. A similar relation applies to the interchange of a final particle and initial antiparticle. The crossing symmetry is obvious from Feynman diagrams but holds more generally, as can be proved using the LSZ reduction formalism. We will be concerned here with matrix elements of an operator $\mathcal{O}$ between single particle states,

$$\langle a(p_a) | \mathcal{O} | b(p_b) \rangle \equiv \bar{u}_a(p_a) O(p_a, p_b) u_b(p_b), \tag{G.3}$$

where the spin labels are not displayed. Applying crossing to both a and b,

$$\langle b^c(p_b)|O|a^c(p_a)\rangle = -\bar{v}_a(p_a)O(-p_a,-p_b)v_b(p_b) = +\bar{u}_b(p_b)O_c(-p_a,-p_b)u_a(p_a), \qquad (G.4)$$

where $O_c \equiv \mathcal{C}O^T\mathcal{C}^{-1}$ (Table 2.2) and the second form follows from (2.298).

Vector and Axial Form Factors

We will be considering matrix elements of the vector and axial vector currents

$$V^\mu_{ab} \equiv \bar{\psi}_a\gamma^\mu\psi_b = V^{\mu\dagger}_{ba}, \qquad A^\mu_{ab} \equiv \bar{\psi}_a\gamma^\mu\gamma^5\psi_b = A^{\mu\dagger}_{ba}. \qquad (G.5)$$

Analogous to (2.394), assuming only Poincaré invariance and the Gordon identities, the most general matrix element of V^μ between the corresponding single particle states is of the form

$$\begin{aligned}\langle a|V^\mu_{ab}(x)|b\rangle = \bar{u}_a &\left[\gamma^\mu F^V_{1ab}(q^2) + \frac{i\sigma^{\mu\nu}}{2m}q_\nu F^V_{2ab}(q^2) + q^\mu F^V_{3ab}(q^2)\right.\\ &\left. + \gamma^\mu\gamma^5 g^V_{1ab}(q^2) + \frac{i\sigma^{\mu\nu}\gamma^5}{2m}q_\nu g^V_{2ab}(q^2) + q^\mu\gamma^5 g^V_{3ab}(q^2)\right] u_b e^{iq\cdot x}\\ \equiv \bar{u}_a&\Gamma^{V\mu}_{ab}(q)u_b, \end{aligned} \qquad (G.6)$$

where $q \equiv p_a - p_b$ and the mass m can be taken to be $(m_a + m_b)/2$. The labels a and b on the states and spinors refer collectively to momentum, spin, and flavor. When we specify the special case $a = b$ we mean that the flavors are the same, but the initial and final momenta and spins may differ. We will henceforth take $x = 0$. A similar expression holds for $\langle a|A^\mu_{ab}|b\rangle$, with superscripts A on the form factors and matrix, F^A_{iab}, g^A_{iab}, $\Gamma^{A\mu}_{ab}(q)$.

Space Reflection

As was already discussed in Section 2.10, space reflection invariance requires

$$\Gamma^{V\mu}_{ab}(q) = \gamma^0\Gamma^V_{\mu ab}(q')\gamma^0, \qquad \Gamma^{A\mu}_{ab}(q) = -\gamma^0\Gamma^A_{\mu ab}(q')\gamma^0 \qquad (G.7)$$

where $q'_\mu = p'_{a\mu} - p'_{b\mu} = q^\mu$. This implies $g^V_{iab} = F^A_{iab} = 0$, up to weak interaction corrections (such as the anapole moment mentioned below (2.396) on page 78). All of the results are summarized in Table G.1.

Charge Conjugation

From (2.296) the currents transform under charge conjugation as

$$\mathcal{C}V^\mu_{ab}\mathcal{C}^{-1} = -V^\mu_{ba}, \qquad \mathcal{C}A^\mu_{ab}\mathcal{C}^{-1} = +A^\mu_{ba}. \qquad (G.8)$$

Therefore, using (G.5) and charge conjugation.

$$\langle a|V^\mu_{ab}(x)|b\rangle = \langle b|V^\mu_{ba}(x)|a\rangle^* = -\langle b^c|V^\mu_{ab}(x)|a^c\rangle^*. \qquad (G.9)$$

(G.6) and the crossing relation in (G.4) then yield

$$\bar{u}_a\Gamma^{V\mu}_{ab}(q)u_b = -\left[\bar{u}_b\Gamma^{V\mu}_{cab}(-q)u_a\right]^* = -\bar{u}_a\overline{\Gamma}^{V\mu}_{cab}(-q)u_b. \qquad (G.10)$$

TABLE G.1 Restrictions imposed by discrete symmetries on the form factors associated with vector and axial vector currents.

Space reflection:	$g^V_{iab} = 0$	$F^A_{iab} = 0$
Charge conjugation:	$F^V_{iab} = F^{V*}_{iab}$	$g^A_{iab} = g^{A*}_{iab}$
	$F^A_{iab} = -F^{A*}_{iab}$	$g^V_{iab} = -g^{V*}_{iab}$
Charge conjugation ($a = b$):	$F^V_{3aa} = g^V_{1,3aa} = 0$	$F^A_{1,2aa} = g^A_{2aa} = 0$
Hermiticity ($a = b$):	$F^{V,A}_{3aa} = -F^{V,A*}_{3aa}$	$g^{V,A}_{2aa} = -g^{V,A*}_{2aa}$
	others real	
Time reversal:	$F^{V,A}_{iab} = F^{V,A*}_{iab}$	$g^{V,A}_{iab} = g^{V,A*}_{iab}$
G-parity:	$F^V_{3pn} = g^A_{2pn} = 0$	$F^A_{1,2pn} = g^V_{1,3pn} = 0$

Applying similarly reasoning to A^μ, one obtains

$$\Gamma^{V\mu}_{ab}(q) = -\overline{\Gamma}^{V\mu}_{cab}(-q), \qquad \Gamma^{A\mu}_{ab}(q) = +\overline{\Gamma}^{A\mu}_{cab}(-q), \tag{G.11}$$

which are satisfied for real F^V_i and g^A_i, and for imaginary F^A_i and g^V_i. As we will see, time reversal invariance implies real form factors, in which case both P and C independently imply $F^A_i = g^V_i = 0$.

In the special case of $a = b$, one has $CV^\mu C^{-1} = -V^\mu$ and $CA^\mu C^{-1} = +A^\mu$, from which it follows that

$$\Gamma^{V\mu}_{aa}(q) = -\Gamma^{V\mu}_{caa}(q), \qquad \Gamma^{A\mu}_{aa}(q) = +\Gamma^{A\mu}_{caa}(q). \tag{G.12}$$

(This also follow from (G.11) and the Hermiticity of V^μ and A^μ, which imply that $\Gamma^{V,A\mu}_{aa}(q) = \overline{\Gamma}^{V,A\mu}_{aa}(-q)$.) Thus, the only C-allowed terms are $F^V_{1,2aa}$, g^V_{2aa}, $g^A_{1,3aa}$, and F^A_{3aa}. (g^V_{2aa} and F^A_{3aa} would have to be imaginary.)

Time Reversal Invariance

It follows from (2.317), (2.318), (2.325), and (2.326) that all of the form factors must be real (or at least relatively real for non-standard phase conventions) if time reversal holds. We already saw that for diagonal ($a = b$) transitions, Hermiticity requires g^V_{2aa} and F^A_{3aa} to be imaginary. Thus, observation of an intrinsic electric dipole moment ($g^V_{2aa} \neq 0$) would imply the violation of both space reflection and time reversal invariance.

G-Parity

The G-parity transformation $G = Ce^{i\pi T^2}$ in (3.131) is useful for the proton-neutron currents associated with the weak charged current. One has

$$\begin{gathered} G\bar{\psi}_p\gamma^\mu\psi_n G^{-1} = +\bar{\psi}_p\gamma^\mu\psi_n, \qquad G\bar{\psi}_p\gamma^\mu\gamma^5\psi_n G^{-1} = -\bar{\psi}_p\gamma^\mu\gamma^5\psi_n \\ G|p\rangle = |n^c\rangle, \qquad G|n\rangle = -|p^c\rangle. \end{gathered} \tag{G.13}$$

Combining with the crossing relation in (G.4), G-parity implies (Weinberg, 1958)

$$\Gamma^{V\mu}_{pn}(q) = -\Gamma^{V\mu}_{cpn}(q), \qquad \Gamma^{A\mu}_{pn}(q) = +\Gamma^{A\mu}_{cpn}(q) \tag{G.14}$$

for the weak charged current form factors. Equation (G.14) is satisfied by $F^V_{1,2pn}$ and $g^A_{1,3pn}$, which are known as *first class* form factors, and violated by the *second class* ones F^V_{3pn}

and g^A_{2pn}. (The situation is reversed for the parity-violating form factors.) There was once considerable activity in β decay searching for second class form factors (including some short-lived positive indications). Such effects, if significantly larger than those expected from isospin breaking, would presumably be due to *second class currents* with the opposite G-parities from those in (G.13). It is almost impossible to construct viable models involving second class currents within the general quark model framework, however (e.g., Langacker, 1977).

Quantum Mechanical Analogs of Symmetry Breaking

Many of the field theoretic possibilities for symmetry breaking and realization described in Section 3.3 have simple analogs in ordinary non-relativistic quantum mechanics. Consider a single particle of mass μ moving in the potential

$$V(x) = \frac{1}{2}\mu\omega^2 x^2 + \frac{\lambda x^4}{4}, \tag{H.1}$$

illustrated by the dashed curve in Figure H.1. (We take $\lambda > 0$ in all of the examples.) The minimum is at $x = 0$, and for sufficiently small λ the energy eigenvalues are approximately those of a simple harmonic oscillator, $E_n = (n + \frac{1}{2})\omega$. V exhibits a reflection symmetry, $V(x) = V(-x)$, so the energy eigenstates $\psi_n(x)$ will have either even or odd parity under $x \to -x$. This will be true even when the effects of the λx^4 term are non-negligible.

One can break the reflection symmetry explicitly, i.e., in the equation of motion, by adding a small linear term to the potential

$$V(x) \to \frac{1}{2}\mu\omega^2 x^2 - \epsilon x + \frac{\lambda x^4}{4}. \tag{H.2}$$

The minimum is now at $x_0 \sim \epsilon/\mu\omega^2 \neq 0$. Expanding around x_0,

$$V(x) \sim \frac{1}{2}\mu\omega^2 (x - x_0)^2 + \lambda x_0 (x - x_0)^3 + \frac{\lambda (x - x_0)^4}{4} + \mathcal{O}(\epsilon^2). \tag{H.3}$$

The energy eigenvalues (in this example) are unchanged to $\mathcal{O}(\epsilon)$, but the reflection symmetry (whether around the origin or around x_0) is broken. The energy eigenfunctions will no longer have a definite parity.

Another possibility is to maintain the exact reflection symmetry of the potential, but to break it spontaneously, i.e., in the solutions to the Schrödinger equation. For

$$V(x) = -\frac{1}{2}\mu\omega^2 x^2 + \frac{\lambda x^4}{4}, \tag{H.4}$$

(the solid curve in Figure H.1) the origin $x = 0$ is unstable, but there are degenerate minima at $\pm x_0$, where $x_0 = \left(\mu\omega^2/\lambda\right)^{1/2}$. One can expand around the minimum, e.g., at $+x_0$, to obtain

$$V(x) \sim +\mu\omega^2 (x - x_0)^2 + \lambda x_0 (x - x_0)^3 + \frac{\lambda(x - x_0)^4}{4}, \tag{H.5}$$

where we have dropped an irrelevant constant $V(x_0) = -\mu^2\omega^4/4\lambda$. Assuming that the barrier is sufficiently high that one can neglect tunneling, we can quantize around x_0. For small enough cubic and quartic terms, the energy eigenvalues are $\sim (n + \frac{1}{2})\sqrt{2}\omega$. Again, the reflection symmetry is lost. One can also combine explicit and spontaneous breaking by adding a perturbation $-\epsilon x$ to (H.4) (the dotted curve in Figure H.1). This breaks the degeneracy between the minima at $\pm x_0$, and also shifts their position slightly. For $\epsilon > 0$ the minimum near $+x_0$, known as the true or global minimum, is deeper. It is straightforward to calculate the $\mathcal{O}(\epsilon)$ shift in energy eigenvalues. For small enough ϵ there remains a shallower metastable minimum near $-x_0$, known as a local or false minimum.

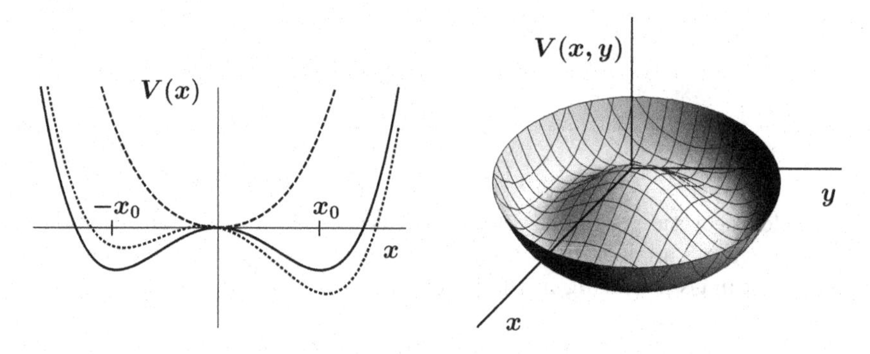

Figure H.1 Left: one-dimensional potential in (H.1) (dashed), (H.4) (solid), or (H.4) with an additional $-\epsilon x$ term (dotted). Right: two-dimensional Mexican hat potential in (H.7), which leads to spontaneous symmetry breaking.

Now consider a particle moving in the two-dimensional potential

$$V(x, y) = \frac{1}{2}\mu\omega^2(x^2 + y^2) + \frac{\lambda(x^2 + y^2)^2}{4} = \frac{1}{2}\mu\omega^2 r^2 + \frac{\lambda r^4}{4}, \tag{H.6}$$

where $r^2 = x^2 + y^2$. The potential is independent of the polar angle $\theta = \tan^{-1} y/x$, and is therefore invariant under the $SO(2)$ group of rotations on x and y, as well as under the reflections (sign changes) of x or y. The energy eigenstates can be classified by the IRREPs of $SO(2)$, i.e., as eigenstates of L_z with eigenvalue $m = 0, \pm 1, \pm 2, \cdots$. For small enough λ the energy eigenvalues are given by the harmonic oscillator formula $E_n = (n + 1)\,\omega$, where $n = 2k + |m|$ with $k = 0 \cdots \infty$ the radial quantum number. Equivalently, $n = n_x + n_y$ in a rectangular basis. The degeneracy of n is $n + 1$. Adding a perturbation $\epsilon y^2/2$ to $V(x, y)$ in (H.6) breaks the $O(2)$ symmetry down to discrete sign changes (i.e., rotations by π and sign changes in x or y). The degeneracy of the eigenvalues is broken, with $E_{n_x,n_y} = (n_x + \frac{1}{2})\,\omega + (n_y + \frac{1}{2})\,\omega'$, where $\omega' \sim \omega + \epsilon/(2\mu\omega)$.

The Mexican hat potential, illustrated in Figure H.1,

$$V(x, y) = -\frac{1}{2}\mu\omega^2(x^2 + y^2) + \frac{\lambda(x^2 + y^2)^2}{4} = -\frac{1}{2}\mu\omega^2 r^2 + \frac{\lambda r^4}{4}, \tag{H.7}$$

differs from (H.6) by the sign of the quadratic term. The origin is unstable, and there are rcle of degenerate minima at radius $r_0 = \left(\mu\omega^2/\lambda\right)^{1/2}$. Similar to (H.5) one can expand $,\theta)$ around r_0,

$$V(r,\theta) \sim +\mu\omega^2 (r - r_0)^2 + \lambda r_0 (r - r_0)^3 + \frac{\lambda(r - r_0)^4}{4}, \tag{H.8}$$

re we have again dropped a constant. For suitable parameter values, $V(r,\theta)$ can be roximated by a harmonic oscillator in the radial direction, centered at r_0 and with uency $\sqrt{2}\omega$, along with a flat potential in the θ direction, leading to non-degenerate nvalues

$$E_{n_r,m} \sim (n_r + \frac{1}{2})\sqrt{2}\omega + \frac{m^2}{2\mu r_0^2}. \tag{H.9}$$

second term represents the zero-frequency "rolling" of the particle along the flat θ ction, and corresponds to quantized momentum $k_m = m/r_0$. The $m = 0$ mode has no rgy in the angular variable and is analogous to the Nambu-Goldstone bosons encountered ection 3.3.3. For large r_0 the excitations are closely spaced.

One can add a small explicit breaking term $\epsilon y^2/2$ to (H.7). If the corresponding frequency $/\mu$ is large compared to the excitation energy $\sim (2\mu r_0^2)^{-1}$ of the rolling but still small ıpared to ω, the energy eigenvalues become

$$E_{n_r,n_\theta} \sim (n_r + \frac{1}{2})\sqrt{2}\omega + (n_\theta + \frac{1}{2})\sqrt{\frac{\epsilon}{\mu}}, \tag{H.10}$$

the Goldstone modes acquire a minimum vibrational energy.

References

R.1 FIELD THEORY

Baal, P. V. (2013). *A course in field theory.* New York: CRC Press.

Banks, T. (2008). *Modern quantum field theory: a concise introduction.* Cambridge: Cambridge Univ. Press.

Bjorken, J. D. and S. D. Drell (1964). *Relativistic quantum mechanics.* New York: McGraw-Hill.

Bjorken, J. D. and S. D. Drell (1965). *Relativistic quantum fields.* New York: McGraw-Hill.

Brown, L. S. (1992). *Quantum field theory.* Cambridge: Cambridge Univ. Press.

Casalbuoni, R. (2011). *Introduction to quantum field theory.* Hackensack, NJ: World Scientific.

Collins, J. C. (1986). *Renormalization.* Cambridge: Cambridge Univ. Press.

Fadeev, L. and A. A. Slavnov (1991). *Gauge fields: introduction to quantum theory; 2nd ed.* London: Benjamin.

Frampton, P. H. (2008). *Gauge field theories; 3rd ed.* Weinheim: Wiley.

Hübsch, Tristan (2015). *Advanced concepts in particle and field theory.* Cambridge: Cambridge Univ. Press.

Itzykson, C. and J.-B. Zuber (1980). *Quantum field theory.* New York: McGraw-Hill.

Mandl, F. and G. G. Shaw (2010). *Quantum field theory; 2nd ed.* New York: Wiley.

Peskin, M. E. and D. V. Schroeder (1995). *An introduction to quantum field theory.* Boulder, CO: Westview.

Pokorski, S. (2000). *Gauge Field Theories; 2nd ed.* Cambridge: Cambridge Univ. Press.

Ramond, P. (1997). *Field Theory: A Modern Primer.* New York: Westview Press.

Ryder, L. H. (1996). *Quantum field theory.* Cambridge: Cambridge Univ. Press.

Schwartz, M. D. (2014). *Quantum field theory and the standard model.* Cambridge: Cambridge Univ. Press.

Shifman, M. (2012). *Advanced topics in quantum field theory: a lecture course.* Cambridge: Cambridge Univ. Press.

Srednicki, M. (2007). *Quantum field theory.* Cambridge: Cambridge Univ. Press.

Sterman, G. F. (1993). *An introduction to quantum field theory.* Cambridge: Cambridge Univ. Press.

Streater, R. F. and A. S. Wightman (2000). *PCT, spin and statistics, and all that.* Princeton, NJ: Princeton Univ. Press.

't Hooft, G. (1994). *Under the spell of the gauge principle.* Singapore: World Scientific.

Weinberg, S. (1995). *The quantum theory of fields, vol I, II.* Cambridge: Cambridge Univ. Press.

Zee, A. (2010). *Quantum field theory in a nutshell.* Princeton, NJ: Princeton Univ. Press.

Zinn-Justin, J. (2002). *Quantum Field Theory and Critical Phenomena; 4th ed.* Oxford: Clarendon Press.

R.2 THE STANDARD MODEL AND PARTICLE PHYSICS

Aitchison, I. J. R. and A. J. G. Hey (2013). *Gauge theories in particle physics: a practical introduction; 4th ed.* Boca Raton, FL: CRC Press.

Anchordoqui, L. and F. Halzen (2009). Lessons in Particle Physics. `arXiv:0906.1271 [physics.ed-ph]`.

Bailin, D. and A. Love (1996). *Introduction to gauge field theory; Rev. ed.* Bristol: IOP.

Boyarkin, O. M. (2011). *Advanced particle physics.* Boca Raton, FL: CRC Press.

Burgess, C. P. and G. D. Moore (2007). *The standard model: a primer.* Cambridge: Cambridge Univ. Press.

Cheng, T. P. and L. F. Li (1984). *Gauge theory of elementary particle physics.* Oxford: Clarendon Press.

Cottingham, W. N. and D. A. Greenwood (2007). *An introduction to the standard model of particle physics; 2nd ed.* Cambridge: Cambridge Univ. Press.

Cvetič, M. and P. G. Langacker (1991). *Testing the standard model.* Singapore: World Scientific.

Donoghue, J. F., E. Golowich, and B. R. Holstein (2014). *Dynamics of the standard model; 2nd ed.* Cambridge: Cambridge Univ. Press.

Gasiorowicz, S. G. (1966). *Elementary particle physics.* New York: Wiley.

Hagedorn, R. (1964). *Relativistic kinematics.* New York: Benjamin.

Halzen, F. and A. D. Martin (1984). *Quarks and leptons.* New York: Wiley.

Langacker, P. (2017). *Can the Laws of Physics be Unified?* Princeton, NJ: Princeton Univ. Press.

Leader, E. and E. Predazzi (1996). *An introduction to gauge theories and modern particle physics.* Cambridge: Cambridge Univ. Press.

Mann, R. (2010). *An introduction to particle physics and the standard model.* Boca Raton, FL: CRC Press.

Morii, T., L. C. S, and S. N. Mukherjee (2004). *The physics of the Standard Model and beyond.* Singapore: World Scientific.

Pais, A. (1986). *Inward bound: of matter and forces in the physical world.* Oxford: Clarendon Press.

Pal, P. B. (2014). *An introductory course of particle physics.* Boca Raton, FL: Taylor & Francis.

Quigg, C. (2013). *Gauge theories of the strong, weak, and electromagnetic interactions; 2nd ed.* Princeton, NJ: Princeton Univ. Press.

Ramond, P. (2004). *Journeys beyond the standard model.* Boulder, CO: Westview Press.

Romao, J. C. and J. P. Silva (2012). A resource for signs and Feynman diagrams of the Standard Model. *Int. J. Mod. Phys. A27,* 1230025.

Tully, C. G. (2011). *Elementary particle physics in a nutshell.* Princeton, NJ: Princeton Univ. Press.

Walecka, J. D. (2010). *Advanced modern physics: theoretical foundations.* Singapore: World Scientific.

R.3 THE STRONG INTERACTIONS, QCD, AND COLLIDER PHYSICS

Adler, S. L. and R. F. Dashen (1968). *Current algebras and applications to particle physics.* New York: Benjamin. Collection of reprints.

Barger, V. D. and R. Phillips (1997). *Collider physics.* Redwood City, CA: Addison-Wesley.

Becher, T., A. Broggio, and A. Ferroglia (2015). *Introduction to soft-collinear effective theory.* Cham: Springer. `arXiv:1410.1892 [hep-ph]`.

Close, F. E., S. Donnachie, and G. Shaw (2007). *Electromagnetic interactions and hadronic structure.* Cambridge: Cambridge Univ. Press.

Collins, J. (2011). *Foundations of perturbative QCD.* New York: Cambridge Univ. Press.

Collins, P. D. B. (1977). *An introduction to Regge theory and high energy physics.* Cambridge: Cambridge Univ. Press.

DeGrand, T. and C. DeTar (2006). *Lattice methods for quantum chromodynamics.* Singapore: World Scientific.

Devenish, R. and A. Cooper-Sarkar (2004). *Deep inelastic scattering.* Oxford: Oxford Univ. Press.

Dissertori, G., I. G. Knowles, and M. Schmelling (2009). *Quantum Chromodynamics: High Energy Experiments and Theory* (1 ed.). Oxford: Oxford Univ. Press.

Dokshitzer, Yu. L., V. A. Khoze, A. H. Müller, and S. I. Troyan (1991). *Basics of perturbative QCD.* Gif-sur-Yvette: Ed. Frontires.

Donnachie, A., G. Dosch, P. V. Landshoff, and O. Nachtmann (2002). *Pomeron physics and QCD.* Cambridge: Cambridge Univ. Press.

Eden, R. J., P. V. Landshoff, D. I. Olive, and J. C. Polkinghorne (1966). *The analytic S-matrix.* Cambridge: Cambridge Univ. Press.

Ellis, R. K., W. J. Stirling, and B. R. Webber (2003). *QCD and collider physics.* Cambridge: Cambridge Univ. Press.

Elvang, H. and Y.-t. Huang (2015, Feb). *Scattering amplitudes in gauge theory and gravity.* Cambridge: Cambridge Univ. Press.

Field, R. D. (1989). *Applications of perturbative QCD.* Redwood City, CA: Addison-Wesley.

Fritzsch, H. and M. Gell-Mann (2015). *50 years of quarks.* Hackensack, NJ: World Scientific.

Gattringer, C. and C. B. Lang (2010). *Quantum chromodynamics on the lattice: an introductory presentation.* Berlin: Springer.

Green, D. (2005). *High p(T) physics at hadron colliders.* Cambridge: Cambridge Univ. Press.

Greiner, W., S. Schramm, and E. Stein (2007). *Quantum chromodynamics; 3rd ed.* Berlin: Springer.

Ioffe, B. L., V. S. Fadin, and L. N. Lipatov (2010). *Quantum chromodynamics: perturbative and nonperturbative aspects.* New York: Cambridge Univ. Press.

Kogut, J. B. and M. A. Stephanov (2004). *The phases of quantum chromodynamics: from confinement to extreme environments.* Cambridge: Cambridge Univ. Press.

Muta, T. (2010). *Foundations of quantum chromodynamics: an introduction to perturbative methods in gauge theories; 3rd ed.* Singapore: World Scientific.

Plehn, T. (2015). *Lectures on LHC physics; 2nd ed.* Cham: Springer.

Rothe, H. J. (2012). *Lattice gauge theories: an introduction; 4th ed.* Singapore: World Scientific.

Satz, H. (2012). *Extreme States of Matter in Strong Interaction Physics: An Introduction.* Berlin: Springer.

Schörner-Sadenius, T. (2015). *The Large Hadron Collider: harvest of run 1.* Cham: Springer.

Ynduráin, F. J. (2006). *The theory of quark and gluon interactions; 4th ed.* Berlin: Springer.

R.4 THE ELECTROWEAK INTERACTIONS

Barbieri, R. (2007). *Lectures on the electroweak interactions.* Berlin: Springer.

Bardin, D. Y. and G. Passarino (1999). *The standard model in the making: precision study of the electroweak interactions.* Oxford: Clarendon Press.

Bilenky, S. M. (1996). *Basics of introduction to Feynman diagrams and electroweak interaction physics.* Gif-sur-Yvette: Ed. Frontires.

Commins, E. D. and P. H. Bucksbaum (1983). *Weak interactions of leptons and quarks.* Cambridge: Cambridge Univ. Press.

D'Alfaro, V., S. P. Fubini, G. Furlan, and C. Rossetti (1973). *Currents in hadron physics.* Amsterdam: North-Holland.

Georgi, H. M. (1984). *Weak interactions and modern particle theory.* Menlo Park, CA: Benjamin-Cummings.

Greiner, W. and J. Reinhardt (2009). *Quantum electrodynamics; 4th ed.* Berlin: Springer.

Greiner, Walter and B. Müller (2009). *Gauge theory of weak interactions; 4th rev. version.* Berlin: Springer.

Gunion, J. F., S. Dawson, H. E. Haber, and G. L. Kane (1990). *The Higgs hunter's guide.* Boulder, CO: Westview.

Horejsi, J. (1994). *Introduction to electroweak unification: standard model from tree unitarity.* Singapore: World Scientific.

Horejsi, J. (2002). *Fundamentals of electroweak theory.* Prague: The Karolinum Press.

Kiesling, C. (1988). *Tests of the standard theory of electroweak interactions.* Berlin: Springer.

Kinoshita, T. (1990). *Quantum electrodynamics.* Singapore: World Scientific.

Langacker, P. G. (1995). *Precision tests of the standard electroweak model.* Singapore: World Scientific.

Maiani, L. (2016). *Electroweak interactions.* Boca Raton, FL: CRC Press.

Manohar, A. V. and M. B. Wise (2000). *Heavy quark physics.* Cambridge: Cambridge Univ. Press.

Marshak, R. E., M. Riazuddin, and C. P. Ryan (1969). *Theory of weak interactions in particle physics.* New York: Wiley.

Paschos, E. A. (2007). *Electroweak theory.* Cambridge: Cambridge Univ. Press.

Renton, P. B. (1990). *Electroweak interactions.* Cambridge: Cambridge Univ. Press.

Scheck, F. (2012). *Electroweak and Strong Interactions: Phenomenology, Concepts, Models.* Berlin: Springer.

Schwinger, J. S. (1958). *Selected papers on quantum electrodynamics.* New York: Dover.

Tenchini, R. and C. Verzegnassi (2007). *The physics of the Z and W bosons.* Singapore: World Scientific.

R.5 *CP* VIOLATION

Bigi, I. I. and A. I. Sanda (2009). *CP violation; 2nd ed.* Cambridge: Cambridge Univ. Press.

Branco, G. C., L. Lavoura, and J. P. Silva (1999). *CP violation.* Oxford: Clarendon Press.

Jarlskog, C. (1989). *CP violation.* Singapore: World Scientific.

Kleinknecht, K. (2003). *Uncovering CP violation: experimental clarification in the neutral K meson and B meson systems.* Berlin: Springer.

Sozzi, M. S. (2008). *Discrete symmetries and CP violation: from experiment to theory.* New York: Oxford Univ. Press.

R.6 NEUTRINOS

Bahcall, J. N. (1989). *Neutrino astrophysics.* Cambridge: Cambridge Univ. Press.

Barger, V., D. Marfatia, and K. Whisnant (2012). *The physics of neutrinos.* Princeton, NJ: Princeton Univ. Press.

Bilenky, S. (2010). *Introduction to the Physics of Massive and Mixed Neutrinos.* Berlin, Heidelberg: Springer.

Fukugita, M. and T. Yanagida (2003). *Physics of neutrinos: and applications to astrophysics.* Berlin: Springer.

Giunti, C. and C. W. Kim (2007). *Fundamentals of neutrino physics and astrophysics.* Oxford: Oxford Univ.

Kayser, B., F. Gibrat-Debu, and F. Perrier (1989). *The physics of massive neutrinos.* Singapore: World Scientific.

Langacker, P. (2000). *Neutrinos in physics and astrophysics from* $10^{-33} \to 10^{28}$ *cm.* Singapore: World Scientific.

Lesgourgues, J., G. Mangano, G. Miele, and S. Pastor (2013). *Neutrino cosmology.* Cambridge: Cambridge Univ. Press.

Mohapatra, R. N. and P. B. Pal (2004). *Massive neutrinos in physics and astrophysics; 3rd ed.* Singapore: World Scientific.

Suekane, F. (2015). *Neutrino oscillations: a practical guide to basics and applications.* Tokyo: Springer.

Valle, J. W. and J. C. Romao (2015). *Neutrinos in high energy and astroparticle physics.* Weinheim: Wiley-VCH.

Xing, Z.-Z. and S. Zhou (2011). *Neutrinos in particle physics, astronomy, and cosmology.* Hangzhou: Zhejiang Univ. Press.

Zuber, K. (2012). *Neutrino physics; 2nd ed.* Boca Raton, FL: CRC Press.

R.7 SUPERSYMMETRY, STRINGS, AND GRAND UNIFICATION

Acharya, B., G. Kane, and P. Kumar (2015). *Perspectives on string phenomenology.* Hackensack, NJ: World Scientific.

Aitchison, I. (2007). *Supersymmetry in particle physics: an elementary introduction.* Cambridge: Cambridge Univ. Press.

Baer, H. W. and X. Tata (2006). *Weak scale supersymmetry.* Cambridge: Cambridge Univ. Press.

Bailin, D. and A. Love (1994). *Supersymmetric gauge field theory and string theory.* Bristol: IOP.

Becker, K., M. Becker, and J. Schwarz (2007). *String theory and M-Theory: a modern introduction.* Cambridge: Cambridge Univ. Press.

Binétruy, P. (2006). *Supersymmetry: theory, experiment, and cosmology.* Oxford: Oxford Univ. Press.

Blumenhagen, R., D. Lüst, and S. Theisen (2013). *Basic Concepts of String Theory.* Berlin: Springer.

Chodos, A., P. G. O. Freund, and T. W. Appelquist (1987). *Modern Kaluza-Klein theories.* Merlo Park, CA: Addison-Wesley.

Dine, M. (2015). *Supersymmetry and string theory: beyond the standard model; 2nd ed.* Cambridge: Cambridge Univ. Press.

Drees, M., R. M. Godbole, and P. Roy (2004). *Theory and phenomenology of sparticles.* Singapore: World Scientific.

Gates, S. J., M. T. Grisaru, M. Rocek, and W. Siegel (1983). Superspace, or one thousand and one lessons in supersymmetry. *Front. Phys. 58*, 1–548.

Green, M. B., J. H. Schwarz, and E. Witten (1987). *Superstring theory.* New York: Cambridge Univ. Press.

Ibáñez, L. E. and A. M. Uranga (2012). *String theory and particle physics: an introduction to string phenomenology.* Cambridge: Cambridge Univ. Press.

Kane, G. L. (2010). *Perspectives on supersymmetry II.* Singapore: World Scientific.

Kiritsis, E. (2007). *String theory in a nutshell.* Princeton, NJ: Princeton Univ. Press.

Mohapatra, R. N. (2003). *Unification and supersymmetry: the frontiers of quark-lepton physics.* Berlin: Springer.

Müller-Kirsten, H. J. W. and A. Wiedemann (2010). *Introduction to supersymmetry; 2nd ed.* Singapore: World Scientific.

Nastase, H. (2015). *Introduction to AdS/CFT correspondence.* Cambridge: Cambridge Univ. Press.

Polchinski, J. (1998). *String theory.* Cambridge: Cambridge Univ. Press.

Polonsky, N. (2001). *Supersymmetry: structure and phenomena: extensions of the standard model.* Berlin: Springer. `arXiv:hep-ph/0108236`.

Rickles, D. (2014). *A brief history of string theory: from dual models to M-theory.* Berlin: Springer.

Ross, G. G. (1985). *Grand unified theories.* Reading, MA: Westview Press.

Srivastava, P. P. (1986). *Supersymmetry, superfields and supergravity: an introduction.* Bristol: Hilger.

Terning, J. (2006). *Modern supersymmetry: dynamics and duality.* Oxford: Clarendon Press.

Weinberg, S. (2000). *The quantum theory of fields, vol. III.* Cambridge: Cambridge Univ. Press.

Wess, J. and J. A. Bagger (1992). *Supersymmetry and supergravity.* Princeton, NJ: Princeton Univ. Press.

West, P. (2012). *Introduction to strings and branes.* Cambridge: Cambridge Univ. Press.

West, P. C. (1990). *Introduction to supersymmetry and supergravity; 2nd ed.* Singapore: World Scientific.

Zwiebach, B. (2009). *A first course in string theory; 2nd ed.* Cambridge: Cambridge Univ. Press.

R.8 ASTROPHYSICS AND COSMOLOGY

Bailin, D. and A. Love (2004). *Cosmology in gauge field theory and string theory.* Bristol: IOP.

Barrow, J. D. and F. J. Tipler (1986). *The anthropic cosmological principle.* Oxford: Clarendon Press.

Bertone, G. (2010). *Particle dark matter: observations, models and searches.* Cambridge: Cambridge Univ. Press.

Das, A. (1997). *Finite temperature field theory.* Singapore: World Scientific.

Dodelson, S. (2003). *Modern cosmology.* New York: Academic Press.

Gorbunov, D. S. and V. A. Rubakov (2011). *Introduction to the theory of the early universe, vol I, II.* Singapore: World Scientific.

Kolb, E. W. and M. S. Turner (1990). *The early universe.* Redwood City, CA: Addison-Wesley.

Liddle, A. (2015, Mar). *An introduction to modern cosmology; 3rd ed.* Chichester: Wiley.

Lyth, D. (2016). *Cosmology for physicists.* Boca Raton, FL: CRC Press.

Majumdar, D. (2015). *Dark matter: an introduction.* Boca Raton, FL: CRC Press.

Mukhanov, V. (2005). *Physical foundations of cosmology.* Cambridge: Cambridge Univ. Press.

Peebles, P. J. E. (1993). *Principles of physical cosmology.* Princeton, NJ: Princeton Univ. Press.

Peebles, P. J. E. and E. Peebles (2015). *Physical cosmology.* Princeton, NJ: Princeton Univ. Press.

Perkins, D. H. (2009). *Particle astrophysics; 2nd ed.* Oxford: Oxford Univ. Press.

Peter, P. and J.-P. Uzan (2009). *Primordial cosmology.* Oxford: Oxford Univ. Press.

Roos, M. (2015). *Introduction to cosmology; 4th ed.* Chichester: Wiley.

Weinberg, S. (1972). *Gravitation and cosmology: principles and applications of the general theory of relativity.* New York: Wiley.

Weinberg, S. (1983). *The first three minutes: a modern view of the origin of the universe.* London: Fontana.

Weinberg, S. (2008). *Cosmology.* Oxford: Oxford Univ. Press.

R.9 GROUPS AND SYMMETRIES

Barnes, K. J. (2010). *Group theory for the standard model of particle physics and beyond.* Boca Raton, FL: CRC Press.

Coleman, S. R. (1985). *Aspects of symmetry: selected Erice lectures.* Cambridge: Cambridge Univ. Press.

Das, A. and S. Okubo (2015). *Lie groups and Lie algebras for physicists.* Hackensack, NJ: World Scientific.

Georgi, H. M. (1999). *Lie algebras in particle physics.* Cambridge: Perseus.

Gilmore, R. (2005). *Lie groups, Lie algebras, and some of their applications.* New York: Dover.

Gilmore, R. (2008). *Lie groups, physics, and geometry: an introduction for physicists, engineers and chemists.* Cambridge: Cambridge Univ. Press.

Haywood, S. (2011). *Symmetries and conservation laws in particle physics: an introduction to group theory for particle physicists.* London: Imperial Coll.

Lee, B. W. (1972). *Chiral dynamics.* New York: Gordon and Breach.

Ma, Z.-Q. (2007). *Group Theory for Physicists.* New Jersey, NJ: World Scientific.

Mitra, P. (2014). *Symmetries and symmetry breaking in field theory.* Boca Raton, FL: CRC Press.

O'Raifeartaigh, L. (1986). *Group structure of gauge theories.* Cambridge: Cambridge Univ. Press.

Ramond, P. (2010). *Group theory: a physicist's survey.* Cambridge: Cambridge Univ. Press.

Tung, W.-K. (1985). *Group theory in physics.* Singapore: World Scientific.

Zelobenko, D. P. (1973). *Compact Lie groups and their representations.* Providence, RI: AMS. Trans. from the Russian.

R.10 ACCELERATORS, DETECTORS, AND DATA ANALYSIS

Behnke, Olaf and Kröninger, Kevin and Schott, Grégory and Schörner-Sadenius, Thomas (2013). *Data analysis in high energy physics: a practical guide to statistical methods.* Weinheim: Wiley-VCH.

Bevington, P. R. and D. K. Robinson (2003). *Data reduction and error analysis for the physical sciences; 3rd ed.* New York: McGraw-Hill.

Green, D. (2000). *The physics of particle detectors.* Cambridge: Cambridge Univ. Press.

Grupen, C. and B. Shwartz (2008). *Particle detectors; 2nd ed.* Cambridge: Cambridge Univ. Press.

James, F. E. (2006). *Statistical methods in experimental physics; 2nd ed.* Singapore: World Scientific.

Kleinknecht, K. (1998). *Detectors for particle radiation; 2nd ed.* Cambridge: Cambridge Univ. Press.

Narsky, I. and F. C. Porter (2013, Dec). *Statistical analysis techniques in particle physics: fits, density and supervised learning.* Weinheim: Wiley-VCH.

Roe, B. P. (2001). *Probability and statistics in experimental physics; 2nd ed.* Berlin: Springer.

Sessler, A. and E. Wilson (2014). *Engines of discovery: a century of particle accelerators; Rev. and expanded ed.* Singapore: World Scientific.

Wiedemann, H. (2015). *Particle accelerator physics; 4th ed.* Berlin: Springer.

R.11 ARTICLES

Aaboud, M. et al. (2016). Search for high-mass new phenomena in the dilepton final state using proton-proton collisions at $\sqrt{s} = 13$ TeV with the ATLAS detector. *Phys. Lett. B761*, 372–392.

Aad, G. et al. (2009). Expected performance of the ATLAS experiment - detector, trigger and physics. `arXiv:0901.0512 [hep-ex]`.

Aad, G. et al. (2012). Observation of a new particle in the search for the standard model Higgs boson with the ATLAS detector at the LHC. *Phys. Lett. B716*, 1–29.

Aad, G. et al. (2015a). Combined measurement of the Higgs boson mass in pp collisions at $\sqrt{s} = 7$ and 8 TeV with the ATLAS and CMS experiments. *Phys. Rev. Lett. 114*, 191803.

Aad, G. et al. (2015b). Search for lepton-flavour-violating $H \to \mu\tau$ decays of the Higgs boson with the ATLAS detector. *JHEP 11*, 211.

Aad, G. et al. (2015c). Summary of the ATLAS experiments sensitivity to supersymmetry after LHC Run 1 interpreted in the phenomenological MSSM. *JHEP 10*, 134.

Aad, G. et al. (2016). Measurements of the Higgs boson production and decay rates and constraints on its couplings from a combined ATLAS and CMS analysis of the LHC pp collision data at $\sqrt{s} = 7$ and 8 TeV. *JHEP 08*, 045.

Aaij, R. et al. (2015). Observation of $J/\psi p$ resonances consistent with pentaquark states in $\Lambda_b^0 \to J/\psi K^- p$ decays. *Phys. Rev. Lett. 115*, 072001.

Aaltonen, T. et al. (2013). Higgs Boson studies at the Tevatron. *Phys. Rev. D88*, 052014.

Aartsen, M. et al. (2013). Evidence for high-energy extraterrestrial neutrinos at the IceCube detector. *Science 342*, 1242856.

Aartsen, M. G. et al. (2015). Flavor ratio of astrophysical neutrinos above 35 TeV in IceCube. *Phys. Rev. Lett. 114*(17), 171102.

Aartsen, M. G. et al. (2016). Searches for sterile neutrinos with the IceCube detector. *Phys. Rev. Lett. 117*(7), 071801.

Aartsen, M. G. et al. (2017). PINGU: A vision for neutrino and particle physics at the South Pole. *J. Phys. G44*(5), 054006.

Abazajian, K. N. et al. (2011). Cosmological and astrophysical neutrino mass measurements. *Astropart. Phys. 35*, 177–184.

Abazajian, K. N. et al. (2012). Light sterile neutrinos: a white paper. `arXiv:1204.5379 [hep-ph]`.

Abazajian, K. N. et al. (2015). Neutrino physics from the cosmic microwave background and large scale structure. *Astropart. Phys. 63*, 66–80.

Abazajian, K. N. and M. Kaplinghat (2016). Neutrino physics from the cosmic microwave background and large-scale structure. *Ann. Rev. Nucl. Part. Sci. 66*(1).

Abazov, V. M. et al. (2014). Study of CP-violating charge asymmetries of single muons and like-sign dimuons in $p\bar{p}$ collisions. *Phys. Rev. D89*(1), 012002.

Abbott, B. P. et al. (2016). GW151226: Observation of gravitational waves from a 22-solar-mass binary black hole coalescence. *Phys. Rev. Lett. 116*(24), 241103.

Abdesselam, A. et al. (2011). Boosted objects: A probe of beyond the standard model physics. *Eur. Phys. J. C71*, 1661.

Abe, K. et al. (2011). Search for differences in oscillation parameters for atmospheric neutrinos and antineutrinos at Super-Kamiokande. *Phys. Rev. Lett. 107*, 241801.

Abe, K. et al. (2015a). Measurements of neutrino oscillation in appearance and disappearance channels by the T2K experiment with 6.6×10^{20} protons on target. *Phys. Rev. D91*(7), 072010.

Abe, K. et al. (2015b). Physics potential of a long-baseline neutrino oscillation experiment using a J-PARC neutrino beam and Hyper-Kamiokande. *PTEP 2015*, 053C02.

Abe, K. et al. (2016). Solar neutrino measurements in Super-Kamiokande-IV. *Phys. Rev. D94*(5), 052010.

Abele, H. (2008). The neutron: its properties and basic interactions. *Prog. Part. Nucl. Phys. 60*, 1–81.

Abers, E. S. and B. W. Lee (1973). Gauge theories. *Phys. Rept. 9*, 1–141.

Abouzaid, E. et al. (2008). Determination of the parity of the neutral pion via the four-electron decay. *Phys. Rev. Lett. 100*, 182001.

Abramowicz, H. et al. (2015). Combination of measurements of inclusive deep inelastic $e^{\pm}p$ scattering cross sections and QCD analysis of HERA data. *Eur. Phys. J. C75*(12), 580.

Abramowicz, H. et al. (2016). Combined QCD and electroweak analysis of HERA data. *Phys. Rev. D93*(9), 092002.

Acciarri, R. et al. (2015). Long-Baseline Neutrino Facility (LBNF) and Deep Underground Neutrino Experiment (DUNE). `arXiv:1512.06148 [physics.ins-det]`.

Accomando, E. et al. (2006). Workshop on CP studies and non-standard Higgs physics. `arXiv:hep-ph/0608079`.

Achiman, Y. and B. Stech (1978). Quark lepton symmetry and mass scales in an $E6$ unified gauge model. *Phys. Lett. B77*, 389–393.

Adamson, P. et al. (2012). Measurements of atmospheric neutrinos and antineutrinos in the MINOS far detector. *Phys. Rev. D86*, 052007.

Adamson, P. et al. (2016a). First measurement of muon-neutrino disappearance in NOνA. *Phys. Rev. D93*(5), 051104, copyright 2016 by the American Physical Society.

Adamson, P. et al. (2016b). Limits on active to sterile neutrino oscillations from disappearance searches in the MINOS, Daya Bay, and Bugey-3 experiments. *Phys. Rev. Lett. 117*, 151801, copyright 2016 by the American Physical Society.

Ade, P. A. R. et al. (2016). Planck 2015 results. XIII. Cosmological parameters. *Astron. Astrophys. 594*, A13.

Adelberger, E. G., J. H. Gundlach, B. R. Heckel, S. Hoedl, and S. Schlamminger (2009). Torsion balance experiments: A low-energy frontier of particle physics. *Prog. Part. Nucl. Phys. 62*, 102–134.

Ademollo, M. and R. Gatto (1964). Nonrenormalization theorem for the strangeness violating vector currents. *Phys. Rev. Lett. 13*, 264–265.

Adhikari, R. et al. (2017). A white paper on keV sterile neutrino dark matter. *JCAP 1701*(01), 025.

Adkins, G. S., C. R. Nappi, and E. Witten (1983). Static properties of nucleons in the Skyrme model. *Nucl. Phys. B228*, 552.

Adler, S. L. (1965a). Calculation of the axial vector coupling constant renormalization in beta decay. *Phys. Rev. Lett. 14*, 1051–1055.

Adler, S. L. (1965b). Consistency conditions on the strong interactions implied by a partially conserved axial vector current. *Phys. Rev. 137*, B1022–B1033.

Adler, S. L. (1969). Axial vector vertex in spinor electrodynamics. *Phys. Rev. 177*, 2426–2438.

Adler, S. L. (2004). Anomalies to all orders. `arXiv:hep-th/0405040`.

Adler, S. L. and W. A. Bardeen (1969). Absence of higher order corrections in the anomalous axial vector divergence equation. *Phys. Rev. 182*, 1517–1536.

Adrian-Martinez, S. et al. (2016). Letter of intent for KM3NeT 2.0. *J. Phys. G43*(8), 084001.

Affleck, I. and M. Dine (1985). A new mechanism for baryogenesis. *Nucl. Phys. B249*, 361.

Affleck, I., M. Dine, and N. Seiberg (1985). Dynamical supersymmetry breaking in four-dimensions and its phenomenological implications. *Nucl. Phys. B256*, 557.

Agafonova, N. et al. (2015). Discovery of τ neutrino appearance in the CNGS neutrino beam with the OPERA experiment. *Phys. Rev. Lett. 115*(12), 121802.

Agashe, K., A. Delgado, M. J. May, and R. Sundrum (2003). RS1, custodial isospin and precision tests. *JHEP 08*, 050.

Agostini, M. et al. (2015a). A test of electric charge conservation with Borexino. *Phys. Rev. Lett. 115*, 231802.

Agostini, M. et al. (2015b). Spectroscopy of geoneutrinos from 2056 days of Borexino data. *Phys. Rev. D92*(3), 031101.

Agrawal, V., S. M. Barr, J. F. Donoghue, and D. Seckel (1998). The anthropic principle and the mass scale of the standard model. *Phys. Rev. D57*, 5480–5492.

Aguilar-Arevalo, A. et al. (2001). Evidence for neutrino oscillations from the observation of $\bar{\nu}_e$ appearance in a $\bar{\nu}_\mu$ beam. *Phys. Rev. D64*, 112007.

Aguilar-Arevalo, A. et al. (2013). Improved search for $\bar{\nu}_\mu \to \bar{\nu}_e$ oscillations in the MiniBooNE experiment. *Phys. Rev. Lett. 110*, 161801.

Aguilar-Arevalo, A. et al. (2015). Improved measurement of the $\pi \to e\nu$ branching ratio. *Phys. Rev. Lett. 115*(7), 071801.

Aguilar-Saavedra, J. A. et al. (2006). Supersymmetry parameter analysis: SPA convention and project. *Eur. Phys. J. C46*, 43–60.

Aharmim, B. et al. (2013). Combined analysis of all three phases of solar neutrino data from the Sudbury Neutrino Observatory. *Phys. Rev. C88*, 025501.

Ahmed, S. et al. (2015). Physics potential of the ICAL detector at the India-based Neutrino Observatory (INO). `arXiv:1505.07380 [physics.ins-det]`.

Aidala, C. A., S. D. Bass, D. Hasch, and G. K. Mallot (2013). The spin structure of the nucleon. *Rev. Mod. Phys. 85*, 655–691.

Akhmedov, E. K. (1988). Resonant amplification of neutrino spin rotation in matter and the solar neutrino problem. *Phys. Lett. B213*, 64–68.

Akhmedov, E. K. and A. Y. Smirnov (2009). Paradoxes of neutrino oscillations. *Phys. Atom. Nucl. 72*, 1363–1381.

Albright, C. H. (2009). Overview of neutrino mixing models and ways to differentiate among them. `arXiv:0905.0146 [hep-ph]`.

Albright, C. H., K. S. Babu, and S. M. Barr (1998). A minimality condition and atmospheric neutrino oscillations. *Phys. Rev. Lett. 81*, 1167–1170.

Alekhin, S., J. Blumlein, and S. Moch (2014). The ABM parton distributions tuned to LHC data. *Phys. Rev. D89*(5), 054028.

Alexander, J. et al. (2016). Dark Sectors 2016 Workshop: Community report. `arXiv:1608.08632 [hep-ph]`.

Alford, M. G., A. Schmitt, K. Rajagopal, and T. Schafer (2008). Color superconductivity in dense quark matter. *Rev. Mod. Phys. 80*, 1455–1515.

Ali, A. (2016). Rare B-meson decays at the crossroads. *Int. J. Mod. Phys. A31*(23), 1630036.

Allanach, B. C. (2002). SOFTSUSY: a C++ program for calculating supersymmetric spectra. *Comput. Phys. Commun. 143*, 305–331.

Allanach, B. C., A. Dedes, and H. K. Dreiner (1999). Bounds on R-parity violating couplings at the weak scale and at the GUT scale. *Phys. Rev. D60*, 075014.

Alloul, A., N. D. Christensen, C. Degrande, C. Duhr, and B. Fuks (2014). FeynRules 2.0 - A complete toolbox for tree-level phenomenology. *Comput. Phys. Commun. 185*, 2250–2300.

Almeida, L. G., S. J. Lee, S. Pokorski, and J. D. Wells (2014). Study of the 125 GeV standard model Higgs boson partial widths and branching fractions. *Phys. Rev. D89*, 033006.

Alonso, R., M. Dhen, M. Gavela, and T. Hambye (2013a). Muon conversion to electron in nuclei in type-I seesaw models. *JHEP 1301*, 118.

Alonso, R., M. B. Gavela, D. Hernandez, L. Merlo, and S. Rigolin (2013). Leptonic dynamical yukawa couplings. *JHEP 08*, 069.

Altarelli, G. (2013). Collider physics within the standard model: a primer. `arXiv:1303.2842 [hep-ph]`.

Altarelli, G. and R. Barbieri (1991). Vacuum polarization effects of new physics on electroweak processes. *Phys. Lett. B253*, 161–167.

Altarelli, G. and F. Feruglio (2004). Models of neutrino masses and mixings. *New J. Phys. 6*, 106.

Altarelli, G. and F. Feruglio (2010). Discrete flavor symmetries and models of neutrino mixing. *Rev. Mod. Phys. 82*, 2701–2729.

Altarelli, G., F. Feruglio, and C. Hagedorn (2008). A SUSY $SU(5)$ grand unified model of tri-bimaximal mixing from $A4$. *JHEP 03*, 052–052.

Altarelli, G. and L. Maiani (1974). Octet enhancement of nonleptonic weak interactions in asymptotically free gauge theories. *Phys. Lett. B52*, 351–354.

Altarelli, G. and G. Parisi (1977). Asymptotic freedom in parton language. *Nucl. Phys. B126*, 298.

Altheimer, A. et al. (2012). Jet substructure at the Tevatron and LHC: New results, new tools, new benchmarks. *J. Phys. G39*, 063001.

Altmannshofer, W., A. J. Buras, S. Gori, P. Paradisi, and D. M. Straub (2010). Anatomy and phenomenology of FCNC and CPV effects in SUSY theories. *Nucl. Phys. B830*, 17–94.

Alves, D. et al. (2012). Simplified models for LHC new physics searches. *J. Phys. G39*, 105005.

Alwall, J. et al. (2007). A standard format for Les Houches event files. *Comput. Phys. Commun. 176*, 300–304.

Alwall, J. et al. (2014). The automated computation of tree-level and next-to-leading order differential cross sections, and their matching to parton shower simulations. *JHEP 07*, 079.

Amaldi, U. et al. (1987). A comprehensive analysis of data pertaining to the weak neutral current and the intermediate vector boson masses. *Phys. Rev. D36*, 1385, copyright 1987 by the American Physical Society.

Amaldi, U., W. de Boer, and H. Furstenau (1991). Comparison of grand unified theories with electroweak and strong coupling constants measured at LEP. *Phys. Lett. B260*, 447–455.

Ambrosino, F. et al. (2011). Measurement of $\sigma(e^+e^- \to \pi^+\pi^-)$ from threshold to 0.85 GeV^2 using initial state radiation with the KLOE detector. *Phys. Lett. B700*, 102–110.

Amhis, Y. et al. (2014). Averages of b-hadron, c-hadron, and τ-lepton properties as of summer 2014. `arXiv:1412.7515 [hep-ex]`, and updates at `www.slac.stanford.edu/xorg/hfag`.

An, F. P. et al. (2017). Measurement of electron antineutrino oscillation based on 1230 days of operation of the Daya Bay experiment. *Phys. Rev. D95*(7), 072006.

Anchordoqui, L. A. et al. (2014). Cosmic neutrino pevatrons: a brand new pathway to astronomy, astrophysics, and particle physics. *J. High Energy Astrophysics 1-2*, 1–30.

Anchordoqui, L. A., H. Goldberg, and G. Steigman (2013). Right-handed neutrinos as the dark radiation: status and forecasts for the LHC. *Phys. Lett. B718*, 1162–1165.

Anderson, P. W. (1963). Plasmons, gauge invariance, and mass. *Phys. Rev. 130*, 439–442.

Androic, D. et al. (2013). First determination of the weak charge of the proton. *Phys. Rev. Lett. 111*(14), 141803, copyright 2013 by the American Physical Society.

Angelopoulos, A. et al. (2003). Physics at CPLEAR. *Phys. Rept. 374*, 165.

Ansari, R. et al. (1987). Measurement of the standard model parameters from a study of W and Z bosons. *Phys. Lett. B186*, 440.

Anselm, A. A. and A. A. Johansen (1994). Can electroweak theta term be observable? *Nucl. Phys. B412*, 553–573.

Anthony, P. L. et al. (2005). Precision measurement of the weak mixing angle in Møller scattering. *Phys. Rev. Lett. 95*, 081601.

Antonelli, M. et al. (2010a). An evaluation of $|V_{us}|$ and precise tests of the standard model from world data on leptonic and semileptonic kaon decays. *Eur. Phys. J. C69*, 399–424.

Antonelli, M. et al. (2010b). Flavor physics in the quark sector. *Phys. Rept. 494*, 197–414.

Antonelli, V., L. Miramonti, C. Pena Garay, and A. Serenelli (2013). Solar neutrinos. *Adv. High Energy Phys. 2013*, 351926.

Antusch, S. (2013). Models for neutrino masses and mixings. *Nucl. Phys. Proc. Suppl. 235-236*, 303–309.

Antusch, S., C. Biggio, E. Fernandez-Martinez, M. Gavela, and J. Lopez-Pavon (2006). Unitarity of the leptonic mixing matrix. *JHEP 0610*, 084.

Antusch, S. and O. Fischer (2014). Non-unitarity of the leptonic mixing matrix: Present bounds and future sensitivities. *JHEP 10*, 094.

Antusch, S., J. Kersten, M. Lindner, M. Ratz, and M. A. Schmidt (2005). Running neutrino mass parameters in see-saw scenarios. *JHEP 03*, 024.

Aoki, S. et al. (2017). Review of lattice results concerning low-energy particle physics. *Eur. Phys. J. C77*(2), 112.

Aoyama, T., M. Hayakawa, T. Kinoshita, and M. Nio (2012). Complete tenth-order QED contribution to the muon $g-2$. *Phys. Rev. Lett. 109*, 111808.

Aoyama, T., M. Hayakawa, T. Kinoshita, and M. Nio (2015). Tenth-order electron anomalous magnetic moment — contribution of diagrams without closed lepton loops. *Phys. Rev. D91*(3), 033006.

Appelquist, T. and J. Carazzone (1975). Infrared singularities and massive fields. *Phys. Rev. D11*, 2856.

Appelquist, T., H.-C. Cheng, and B. A. Dobrescu (2001). Bounds on universal extra dimensions. *Phys. Rev. D64*, 035002.

Appelquist, T., N. D. Christensen, M. Piai, and R. Shrock (2004). Flavor-changing processes in extended technicolor. *Phys. Rev. D70*, 093010.

Appelquist, T. and H. D. Politzer (1975). Orthocharmonium and e^+e^- annihilation. *Phys. Rev. Lett. 34*, 43.

Arana-Catania, M., S. Heinemeyer, and M. J. Herrero (2014). Updated constraints on general squark flavor mixing. *Phys. Rev. D90*(7), 075003.

Arbey, A. et al. (2015). Physics at the e^+e^- linear collider. *Eur. Phys. J. C75*(8), 371.

Arbuzov, A. B. et al. (2006). ZFITTER: a semi-analytical program for fermion pair production in e^+e^- annihilation, from version 6.21 to version 6.42. *Comput. Phys. Commun. 174*, 728–758.

Arkani-Hamed, N., A. G. Cohen, and H. Georgi (2001a). Electroweak symmetry breaking from dimensional deconstruction. *Phys. Lett. B513*, 232–240.

Arkani-Hamed, N., A. G. Cohen, E. Katz, and A. E. Nelson (2002a). The littlest Higgs. *JHEP 07*, 034.

Arkani-Hamed, N. and S. Dimopoulos (2005). Supersymmetric unification without low energy supersymmetry and signatures for fine-tuning at the LHC. *JHEP 06*, 073.

Arkani-Hamed, N., S. Dimopoulos, and G. R. Dvali (1998). The hierarchy problem and new dimensions at a millimeter. *Phys. Lett. B429*, 263–272.

Arkani-Hamed, N., S. Dimopoulos, G. R. Dvali, and J. March-Russell (2002b). Neutrino masses from large extra dimensions. *Phys. Rev. D65*, 024032.

Arkani-Hamed, N., L. J. Hall, H. Murayama, D. Tucker-Smith, and N. Weiner (2001b). Small neutrino masses from supersymmetry breaking. *Phys. Rev. D64*, 115011.

Arkani-Hamed, N., T. Han, M. Mangano, and L.-T. Wang (2016). Physics opportunities of a 100 TeV proton-proton collider. *Phys. Rept. 652*, 1–49.

Arkani-Hamed, N. and N. Weiner (2008). LHC signals for a superunified theory of dark matter. *JHEP 12*, 104.

Armesto, N. (2006). Nuclear shadowing. *J. Phys. G32*, R367–R394.

nesto, N. et al. (2008). Heavy ion collisions at the LHC — last call for predictions. *J. Phys. G35*, 054001.

nstrong, D. S. and R. D. McKeown (2012). Parity-violating electron scattering and the lectric and magnetic strange form factors of the nucleon. *Ann. Rev. Nucl. Part. Sci. 62*, 37–359.

ison, G. et al. (1986). Recent results on intermediate vector boson properties at the ERN super proton synchrotron collider. *Phys. Lett. B166*, 484.

ington, J., C. D. Roberts, and J. M. Zanotti (2007). Nucleon electromagnetic form ictors. *J. Phys. G34*, S23–S52.

uso, M. et al. (2008a). B, D and K decays. *Eur. Phys. J. C57*, 309–492.

uso, M., E. Barberio, and S. Stone (2009). B meson decays. *PMC Phys. A3*, 3.

uso, M., G. Borissov, and A. Lenz (2016). CP violation in the B_s^0 system. *Rev. Mod. Phys. 88*(4), 045002.

uso, M., B. Meadows, and A. A. Petrov (2008b). Charm meson decays. *Ann. Rev. Nucl. Part. Sci. 58*, 249–291.

er, D. M. et al. (2013). Snowmass 2013: ILC Higgs white paper. `arXiv:1310.0763 [hep-ph]`.

talos, S. J., L. J. Rosenberg, K. van Bibber, P. Sikivie, and K. Zioutas (2006). Searches or astrophysical and cosmological axions. *Ann. Rev. Nucl. Part. Sci. 56*, 293–326.

e, A., T. Han, S. Pascoli, and B. Zhang (2009). The search for heavy Majorana neutrinos. *JHEP 05*, 030.

vood, D., S. Bar-Shalom, G. Eilam, and A. Soni (2001). CP violation in top physics. *Phys. Rept. 347*, 1–222.

vood, D., L. Reina, and A. Soni (1997). Phenomenology of two Higgs doublet models vith flavor changing neutral currents. *Phys. Rev. D55*, 3156–3176.

termann, C. (2016). Experimental status of supersymmetry after the LHC Run-I. *Prog. Part. Nucl. Phys. 90*, 125–155.

ik, M. et al. (2013). Working group report: Precision study of electroweak interactions. `arXiv:1310.6708 [hep-ph]`.

ou, K. and C. N. Leung (2001). Classification of effective neutrino mass operators. *Nucl. Phys. B619*, 667–689.

ou, K. S. et al. (2013). Snowmass 2013 working group report: Baryon number violation. `arXiv:1311.5285 [hep-ph]`.

ou, K. S., C. F. Kolda, and J. March-Russell (1998). Implications of generalized $Z - Z'$ mixing. *Phys. Rev. D57*, 6788–6792.

ou, K. S. and R. N. Mohapatra (1991). Top quark mass in a dynamical symmetry breaking scheme with radiative b quark and τ lepton masses. *Phys. Rev. Lett. 66*, 556–559.

Babu, K. S., J. C. Pati, and F. Wilczek (2000). Fermion masses, neutrino oscillations, and proton decay in the light of SuperKamiokande. *Nucl. Phys. B566*, 33–91.

Baer, H., V. Barger, and H. Serce (2016). SUSY under siege from direct and indirect WIMP detection experiments. *Phys. Rev. D94*(11), 115019.

Bagger, J., K. T. Matchev, and D. Pierce (1995). Precision corrections to supersymmetric unification. *Phys. Lett. B348*, 443–450.

Baglio, J., O. Eberhardt, U. Nierste, and M. Wiebusch (2014). Benchmarks for Higgs pair production and heavy Higgs searches in the two-Higgs-doublet model of type II. *Phys. Rev. D90*, 015008.

Bahcall, J. N., A. M. Serenelli, and S. Basu (2005). New solar opacities, abundances, helioseismology, and neutrino fluxes. *Astrophys. J. 621*, L85–L88.

Bahcall, J. N., A. M. Serenelli, and S. Basu (2006). 10,000 standard solar models: a Monte Carlo simulation. *Astrophys. J. Suppl. 165*, 400–431.

Baikov, P. A., K. G. Chetyrkin, J. H. Kuhn, and J. Rittinger (2012). Adler function, sum rules and Crewther relation of order $O(\alpha_s^4)$: the singlet case. *Phys. Lett. B714*, 62–65.

Baldini, A. M. et al. (2016). Search for the lepton flavour violating decay $\mu^+ \to e^+\gamma$ with the full dataset of the MEG experiment. *Eur. Phys. J. C76*(8), 434.

Ball, R. D. et al. (2015). Parton distributions for the LHC Run II. *JHEP 04*, 040.

Bandurin, D. et al. (2015). Review of physics results from the Tevatron. *Int. J. Mod. Phys. A30*(06), 1541001.

Banfi, A., G. P. Salam, and G. Zanderighi (2010). Phenomenology of event shapes at hadron colliders. *JHEP 06*, 038.

Banks, T. and L. J. Dixon (1988). Constraints on string vacua with space-time supersymmetry. *Nucl. Phys. B307*, 93–108.

Banks, T. and N. Seiberg (2011). Symmetries and strings in field theory and gravity. *Phys. Rev. D83*, 084019.

Barate, R. et al. (2003). Search for the standard model Higgs boson at LEP. *Phys. Lett. B565*, 61–75.

Barbier, R. et al. (2005). *R*-parity violating supersymmetry. *Phys. Rept. 420*, 1–202.

Barbieri, R., A. Pomarol, R. Rattazzi, and A. Strumia (2004). Electroweak symmetry breaking after LEP-1 and LEP-2. *Nucl. Phys. B703*, 127–146.

Bardeen, W. A. (1969). Anomalous Ward identities in spinor field theories. *Phys. Rev. 184*, 1848–1857.

Bardeen, W. A., A. J. Buras, D. W. Duke, and T. Muta (1978). Deep inelastic scattering beyond the leading order in asymptotically free gauge theories. *Phys. Rev. D18*, 3998.

Barenboim, G. and N. E. Mavromatos (2005). *CPT* violating decoherence and LSND: A possible window to Planck scale physics. *JHEP 01*, 034.

Barger, V., C.-W. Chiang, P. Langacker, and H.-S. Lee (2004). Z' mediated flavor changing neutral currents in B meson decays. *Phys. Lett. B580*, 186–196.

Barger, V., S. L. Glashow, P. Langacker, and D. Marfatia (2002). No-go for detecting CP violation via neutrinoless double beta decay. *Phys. Lett. B540*, 247–251.

Barger, V., J. P. Kneller, P. Langacker, D. Marfatia, and G. Steigman (2003a). Hiding relativistic degrees of freedom in the early universe. *Phys. Lett. B569*, 123–128.

Barger, V., P. Langacker, and H.-S. Lee (2003b). Primordial nucleosynthesis constraints on Z' properties. *Phys. Rev. D67*, 075009.

Barger, V., P. Langacker, H.-S. Lee, and G. Shaughnessy (2006). Higgs sector in extensions of the MSSM. *Phys. Rev. D73*, 115010.

Barger, V., P. Langacker, M. McCaskey, M. Ramsey-Musolf, and G. Shaughnessy (2009). Complex singlet extension of the standard model. *Phys. Rev. D79*, 015018.

Barger, V., P. Langacker, and G. Shaughnessy (2007). TeV physics and the Planck scale. *New J. Phys. 9*, 333.

Barger, V. D., M. S. Berger, and P. Ohmann (1993). Supersymmetric grand unified theories: two loop evolution of gauge and Yukawa couplings. *Phys. Rev. D47*, 1093–1113.

Barger, V. D. and C. Kao (1999). Trilepton signature of minimal supergravity at the upgraded Tevatron. *Phys. Rev. D60*, 115015.

Barger, V. D., P. Langacker, J. P. Leveille, and S. Pakvasa (1980a). Consequences of Majorana and Dirac mass mixing for neutrino oscillations. *Phys. Rev. Lett. 45*, 692.

Barger, V. D., K. Whisnant, S. Pakvasa, and R. J. N. Phillips (1980b). Matter effects on three-neutrino oscillations. *Phys. Rev. D22*, 2718.

Barker, A. R. and S. H. Kettell (2000). Developments in rare kaon decay physics. *Ann. Rev. Nucl. Part. Sci. 50*, 249–297.

Barr, S. M. (1982). A new symmetry breaking pattern for $SO(10)$ and proton decay. *Phys. Lett. B112*, 219–222.

Barr, S. M. (1984). A natural class of non Peccei-Quinn models. *Phys. Rev. D30*, 1805.

Barr, S. M. and A. Zee (1990). Electric dipole moment of the electron and of the neutron. *Phys. Rev. Lett. 65*, 21–24.

Barry, J., W. Rodejohann, and H. Zhang (2011). Light sterile neutrinos: models and phenomenology. *JHEP 1107*, 091.

Bartel, W. et al. (1984). A measurement of the electroweak induced charge asymmetry in $e^+e^- \to B\bar{B}$. *Phys. Lett. B146*, 437.

Bartl, A., E. Christova, K. Hohenwarter-Sodek, and T. Kernreiter (2004). Triple product correlations in top squark decays. *Phys. Rev. D70*, 095007.

Bass, S. D. (2011). The cosmological constant puzzle. *J. Phys. G38*, 043201.

Batra, P., A. Delgado, D. E. Kaplan, and T. M. P. Tait (2004). The Higgs mass bound in gauge extensions of the minimal supersymmetric standard model. *JHEP 02*, 043.

Baudis, L. (2016). Dark matter searches. *Annalen Phys. 528*, 74–83.

Bayatian, G. L. et al. (2007). CMS technical design report, volume II: physics performance. *J. Phys. G34*, 995–1579.

Bayes, R. et al. (2011). Experimental constraints on left-right symmetric models from muon decay. *Phys. Rev. Lett. 106*, 041804.

Beacom, J. F. (2010). The diffuse supernova neutrino background. *Ann. Rev. Nucl. Part. Sci. 60*, 439–462.

Beacom, J. F. and N. F. Bell (2002). Do solar neutrinos decay? *Phys. Rev. D65*, 113009.

Beall, G., M. Bander, and A. Soni (1982). Constraint on the mass scale of a left-right symmetric electroweak theory from the $K_L - K_S$ mass difference. *Phys. Rev. Lett. 48*, 848.

Beane, S. R. and N. Klco (2016). Chiral corrections to the Adler-Weisberger sum rule. *Phys. Rev. D94*, 116002.

Bechtle, P. et al. (2014c). HiggsBounds – 4: Improved tests of extended Higgs sectors against exclusion bounds from LEP, the Tevatron and the LHC. *Eur. Phys. J. C74*(3), 2693.

Bechtle, P. et al. (2017). The light and heavy Higgs interpretation of the MSSM. *Eur. Phys. J. C77*(2), 67.

Bechtle, P., S. Heinemeyer, O. Stal, T. Stefaniak, and G. Weiglein (2014a). *HiggsSignals*: Confronting arbitrary Higgs sectors with measurements at the Tevatron and the LHC. *Eur. Phys. J. C74*(2), 2711.

Bechtle, P., S. Heinemeyer, O. Stal, T. Stefaniak, and G. Weiglein (2014b). Probing the standard model with Higgs signal rates from the Tevatron, the LHC and a future ILC. *JHEP 1411*, 039.

Behrends, R. E. and A. Sirlin (1960). Effect of mass splittings on the conserved vector current. *Phys. Rev. Lett. 4*, 186–187.

Bekenstein, J. and M. Milgrom (1984). Does the missing mass problem signal the breakdown of Newtonian gravity? *Astrophys. J. 286*, 7–14.

Bélanger, G., F. Boudjema, A. Pukhov, and A. Semenov (2015). micrOMEGAs4.1: two dark matter candidates. *Comput. Phys. Commun. 192*, 322–329.

Bélanger, G., B. Dumont, U. Ellwanger, J. Gunion, and S. Kraml (2013). Global fit to Higgs signal strengths and couplings and implications for extended Higgs sectors. *Phys. Rev. D88*, 075008.

Belitsky, A. V. and A. V. Radyushkin (2005). Unraveling hadron structure with generalized parton distributions. *Phys. Rept. 418*, 1–387.

Bell, J. S. and R. Jackiw (1969). A PCAC puzzle: $\pi^0 \to \gamma\gamma$ in the sigma model. *Nuovo Cim. A60*, 47–61.

Bell, N. F., V. Cirigliano, M. J. Ramsey-Musolf, P. Vogel, and M. B. Wise (2005). How magnetic is the Dirac neutrino? *Phys. Rev. Lett. 95*, 151802.

Bell, N. F., M. Gorchtein, M. J. Ramsey-Musolf, P. Vogel, and P. Wang (2006). Model independent bounds on magnetic moments of Majorana neutrinos. *Phys. Lett. B642*, 377–383.

Bellerive, A., J. R. Klein, A. B. McDonald, A. J. Noble, and A. W. P. Poon (2016). The Sudbury Neutrino Observatory. *Nucl. Phys. B908*, 30–51.

Bellini, G. et al. (2014a). Final results of Borexino phase-I on low energy solar neutrino spectroscopy. *Phys. Rev. D89*, 112007, copyright 2014 by the American Physical Society.

Bellini, G. et al. (2014b). Neutrinos from the primary proton-proton fusion process in the Sun. *Nature 512*(7515), 383–386.

Bellm, J. et al. (2016). Herwig 7.0/Herwig++ 3.0 release note. *Eur. Phys. J. C76*(4), 196.

Belyaev, A., N. D. Christensen, and A. Pukhov (2013). CalcHEP 3.4 for collider physics within and beyond the standard model. *Comput. Phys. Commun. 184*, 1729–1769.

Benayoun, M. et al. (2014). Workshop proceedings: Hadronic contributions to the muon anomalous magnetic moment. $(g-2)_\mu$: Quo vadis? `arXiv:1407.4021 [hep-ph]`.

Bennett, G. W. et al. (2006). Final report of the muon E821 anomalous magnetic moment measurement at BNL. *Phys. Rev. D73*, 072003.

Benson, D., I. I. Bigi, T. Mannel, and N. Uraltsev (2003). Imprecated, yet impeccable: on the theoretical evaluation of $\Gamma(B \to X_c \ell \nu)$. *Nucl. Phys. B665*, 367–401.

Benvenuti, A. C. et al. (1974). Observation of muonless neutrino induced inelastic interactions. *Phys. Rev. Lett. 32*, 800–803.

Berenstein, D. (2014). TeV-scale strings. *Ann. Rev. Nucl. Part. Sci. 64*, 197–219.

Berger, C. F., J. S. Gainer, J. L. Hewett, and T. G. Rizzo (2009). Supersymmetry without prejudice. *JHEP 02*, 023.

Berman, S. M. and A. Sirlin (1962). Some considerations on the radiative corrections to muon and neutron decay. *Ann. Phys. 20*, 20–43.

Bernabei, R. et al. (2010). New results from DAMA/LIBRA. *Eur. Phys. J. C67*, 39–49.

Bernabéu, J. and F. Martínez-Vidal (2015). Time-reversal violation. *Ann. Rev. Nucl. Part. Sci. 65*(1), 403–427.

Bernard, V. (2008). Chiral perturbation theory and baryon properties. *Prog. Part. Nucl. Phys. 60*, 82–160.

Bernardi, G. and M. Herndon (2014). Searches for the standard model Higgs boson at hadron colliders at $\sqrt{s} = 2$, 7 and 8 TeV until the discovery of the 125 GeV boson. *Rev. Mod. Phys. 86*, 479.

Bernreuther, W. (2002). *CP* violation and baryogenesis. *Lect. Notes Phys. 591*, 237–293.

Bernstein, A. M. and B. R. Holstein (2013). Neutral pion lifetime measurements and the QCD chiral anomaly. *Rev. Mod. Phys. 85*, 49.

Bertone, G. and D. Hooper (2016). A history of dark matter. `arXiv:1605.04909 [astro-ph.CO]`.

Bertone, G., D. Hooper, and J. Silk (2005). Particle dark matter: evidence, candidates and constraints. *Phys. Rept. 405*, 279–390.

Bethke, S. (2007). Experimental tests of asymptotic freedom. *Prog. Part. Nucl. Phys. 58*, 351–386.

Betts, S. et al. (2013). Development of a relic neutrino detection experiment at PTOLEMY: Princeton tritium observatory for light, early-Universe, massive-neutrino yield. `arXiv:1307.4738 [astro-ph.IM]`.

Bevan, A. J. et al. (2014). The physics of the B factories. *Eur. Phys. J. C74*, 3026.

Bhattacharya, T. et al. (2012). Probing novel scalar and tensor interactions from (ultra)cold neutrons to the LHC. *Phys. Rev. D85*, 054512.

Bicer, M. et al. (2014). First look at the physics case of TLEP. *JHEP 01*, 164.

Bigi, I. I. Y., M. A. Shifman, and N. Uraltsev (1997). Aspects of heavy quark theory. *Ann. Rev. Nucl. Part. Sci. 47*, 591–661.

Bijnens, J. and G. Ecker (2014). Mesonic low-energy constants. *Ann. Rev. Nucl. Part. Sci. 64*, 149–174.

Bilenky, S. M. and C. Giunti (2015). Neutrinoless double-beta decay: a probe of physics beyond the standard model. *Int. J. Mod. Phys. A30*(04n05), 1530001.

Biller, S. D. (2013). Probing Majorana neutrinos in the regime of the normal mass hierarchy. *Phys. Rev. D87*(7), 071301.

Bjorken, J. D. (1969). Asymptotic sum rules at infinite momentum. *Phys. Rev. 179*, 1547–1553.

Blake, T., T. Gershon, and G. Hiller (2015). Rare b hadron decays at the LHC. *Ann. Rev. Nucl. Part. Sci. 65*, 113–143.

Blankenburg, G., J. Ellis, and G. Isidori (2012). Flavour-changing decays of a 125 GeV Higgs-like particle. *Phys. Lett. B712*, 386–390.

Blennow, M. and A. Y. Smirnov (2013). Neutrino propagation in matter. *Adv. High Energy Phys. 2013*, 972485.

Bloom, E. D. and F. J. Gilman (1971). Scaling and the behavior of nucleon resonances in inelastic electron-nucleon scattering. *Phys. Rev. D4*, 2901.

Blumenhagen, R., V. Braun, T. W. Grimm, and T. Weigand (2009a). GUTs in type IIB orientifold compactifications. *Nucl. Phys. B815*, 1–94.

Blumenhagen, R., M. Cvetič, S. Kachru, and T. Weigand (2009b). D-brane instantons in type II orientifolds. *Ann. Rev. Nucl. Part. Sci. 59*, 269–296.

Blumenhagen, R., M. Cvetič, P. Langacker, and G. Shiu (2005). Toward realistic intersecting D-brane models. *Ann. Rev. Nucl. Part. Sci. 55*, 71.

Blumenhagen, R., M. Cvetič, and T. Weigand (2007a). Spacetime instanton corrections in 4D string vacua — the seesaw mechanism for D-brane models. *Nucl. Phys. B771*, 113–142.

Blumenhagen, R., B. Kors, D. Lust, and S. Stieberger (2007b). Four-dimensional string compactifications with D-branes, orientifolds and fluxes. *Phys. Rept. 445*, 1–193.

Blumlein, J. (2013). The theory of deeply inelastic scattering. *Prog. Part. Nucl. Phys. 69*, 28–84.

Bodwin, G. T., E. Braaten, and G. P. Lepage (1995). Rigorous QCD analysis of inclusive annihilation and production of heavy quarkonium. *Phys. Rev. D51*, 1125–1171.

Bona, M. et al. (2006). The unitarity triangle fit in the standard model and hadronic parameters from lattice QCD: a reappraisal after the measurements of Δm_s and $BR(B \to \tau\nu_\tau)$. *JHEP 10*, 081.

Boos, E. et al. (2004). CompHEP 4.4: Automatic computations from Lagrangians to events. *Nucl. Instrum. Meth. A534*, 250–259.

Boos, E., O. Brandt, D. Denisov, S. Denisov, and P. Grannis (2015). The top quark (20 years after its discovery). *Phys. Usp. 58*(12), 1133–1158.

Boucenna, S. M., S. Morisi, and J. W. Valle (2014). The low-scale approach to neutrino masses. *Adv. High Energy Phys. 2014*, 831598.

Bouchendira, R., P. Clade, S. Guellati-Khelifa, F. Nez, and F. Biraben (2011). New determination of the fine structure constant and test of the quantum electrodynamics. *Phys. Rev. Lett. 106*, 080801.

Bousso, R. and J. Polchinski (2000). Quantization of four-form fluxes and dynamical neutralization of the cosmological constant. *JHEP 06*, 006.

Boyarsky, A., O. Ruchayskiy, and M. Shaposhnikov (2009). The role of sterile neutrinos in cosmology and astrophysics. *Ann. Rev. Nucl. Part. Sci. 59*, 191–214.

Boyle, P. A. et al. (2013). Emerging understanding of the $\Delta I = 1/2$ rule from lattice QCD. *Phys. Rev. Lett. 110*(15), 152001.

Brambilla, N. et al. (2011). Heavy quarkonium: progress, puzzles, and opportunities. *Eur. Phys. J. C71*, 1534.

Brambilla, N. et al. (2014). QCD and strongly coupled gauge theories: Challenges and perspectives. *Eur. Phys. J. C74*(10), 2981.

Branchina, V., E. Messina, and M. Sher (2015). Lifetime of the electroweak vacuum and sensitivity to Planck scale physics. *Phys. Rev. D91*, 013003.

Branco, G. et al. (2012). Theory and phenomenology of two-Higgs-doublet models. *Phys. Rept. 516*, 1–102.

Brandenberger, R. H. (2013). Probing particle physics from top down with cosmic strings. *The Universe 1*(4), 6–23.

Bressi, G. et al. (2011). Testing the neutrality of matter by acoustic means in a spherical resonator. *Phys. Rev. A83*(5), 052101.

Brock, R. et al. (1995). Handbook of perturbative QCD: Version 1.0. *Rev. Mod. Phys. 67*, 157–248.

Brod, J. and M. Gorbahn (2012). Next-to-next-to-leading-order charm-quark contribution to the CP violation parameter ϵ_K and ΔM_K. *Phys. Rev. Lett. 108*, 121801.

Brodsky, S. J., T. Kinoshita, and H. Terazawa (1971). Two photon mechanism of particle production by high-energy colliding beams. *Phys. Rev. D4*, 1532–1557.

Brooijmans, G. et al. (2012). Les Houches 2011: Physics at TeV colliders new physics working group report. `arXiv:1203.1488 [hep-ph]`.

Bryman, D., W. J. Marciano, R. Tschirhart, and T. Yamanaka (2011). Rare kaon and pion decays: Incisive probes for new physics beyond the standard model. *Ann. Rev. Nucl. Part. Sci. 61*, 331–354.

Buchalla, G., A. J. Buras, and M. E. Lautenbacher (1996). Weak decays beyond leading logarithms. *Rev. Mod. Phys. 68*, 1125–1144.

Buchalla, G., G. Hiller, Y. Nir, and G. Raz (2005). The pattern of CP asymmetries in $b \to s$ transitions. *JHEP 09*, 074.

Buchmueller, O. et al. (2014). The CMSSM and NUHM1 after LHC Run 1. *Eur. Phys. J. C74*(6), 2922.

Buchmuller, W. and D. Wyler (1986). Effective Lagrangian analysis of new interactions and flavor conservation. *Nucl. Phys. B268*, 621.

Buckley, A. et al. (2011). General-purpose event generators for LHC physics. *Phys. Rept. 504*, 145–233.

Buckley, A. et al. (2015). LHAPDF6: parton density access in the LHC precision era. *Eur. Phys. J. C75*, 132.

Buras, A. J. (1993). A 1993 look at the lower bound on the top quark mass from CP violation. *Phys. Lett. B317*, 449–453.

Buras, A. J. (2005). Flavour physics and CP violation. `arXiv:hep-ph/0505175`.

Buras, A. J. (2011). Climbing NLO and NNLO summits of weak decays. `arXiv:1102.5650 [hep-ph]`.

Buras, A. J., D. Buttazzo, J. Girrbach-Noe, and R. Knegjens (2015a). $K^+ \to \pi^+ \nu\bar{\nu}$ and $K_L \to \pi^0 \nu\bar{\nu}$ in the standard model: status and perspectives. *JHEP 11*, 033.

Buras, A. J., M. V. Carlucci, S. Gori, and G. Isidori (2010a). Higgs-mediated FCNCs: natural flavour conservation vs. minimal flavour violation. *JHEP 1010*, 009.

Buras, A. J., J. R. Ellis, M. K. Gaillard, and D. V. Nanopoulos (1978). Aspects of the grand unification of strong, weak and electromagnetic interactions. *Nucl. Phys. B135*, 66–92.

Buras, A. J., R. Fleischer, S. Recksiegel, and F. Schwab (2004). Anatomy of prominent B and K decays and signatures of CP-violating new physics in the electroweak penguin sector. *Nucl. Phys. B697*, 133–206.

Buras, A. J., J.-M. Gérard, and W. A. Bardeen (2014). Large N approach to kaon decays and mixing 28 years later: $\Delta I = 1/2$ rule, $\hat{B}_K$ and ΔM_K. *Eur. Phys. J. C74*, 2871.

Buras, A. J. and J. Girrbach (2014). Towards the identification of new physics through quark flavour violating processes. *Rept. Prog. Phys. 77*, 086201.

Buras, A. J., M. Gorbahn, S. Jäger, and M. Jamin (2015b). Improved anatomy of ϵ'/ϵ in the standard model. *JHEP 11*, 202.

Buras, A. J., D. Guadagnoli, and G. Isidori (2010b). On ϵ_K beyond lowest order in the operator product expansion. *Phys. Lett. B688*, 309–313.

Burda, P., R. Gregory, and I. Moss (2016). The fate of the Higgs vacuum. *JHEP 06*, 025.

Burgess, C. P. (2007). Introduction to effective field theory. *Ann. Rev. Nucl. Part. Sci. 57*, 329–362.

Burgess, C. P., S. Godfrey, H. Konig, D. London, and I. Maksymyk (1994). Bounding anomalous gauge boson couplings. *Phys. Rev. D50*, 7011–7024.

Burkardt, M., C. A. Miller, and W. D. Nowak (2010). Spin-polarized high-energy scattering of charged leptons on nucleons. *Rept. Prog. Phys. 73*, 016201.

Bustamante, M., J. F. Beacom, and W. Winter (2015). Theoretically palatable flavor combinations of astrophysical neutrinos. *Phys. Rev. Lett. 115*(16), 161302.

Butler, J. N. et al. (2013). Snowmass 2013 working group report: Quark flavor physics. `arXiv:1311.1076 [hep-ex]`.

Buttazzo, D. et al. (2013). Investigating the near-criticality of the Higgs boson. *JHEP 1312*, 089.

Butterworth, J. M., A. R. Davison, M. Rubin, and G. P. Salam (2008). Jet substructure as a new Higgs search channel at the LHC. *Phys. Rev. Lett. 100*, 242001.

Cabibbo, N. (1963). Unitary symmetry and leptonic decays. *Phys. Rev. Lett. 10*, 531–532.

Cabibbo, N. and L. Maiani (1982). The vanishing of order G mechanical effects of cosmic massive neutrinos on bulk matter. *Phys. Lett. B114*, 115.

Cabibbo, N., L. Maiani, G. Parisi, and R. Petronzio (1979). Bounds on the fermions and Higgs boson masses in grand unified theories. *Nucl. Phys. B158*, 295.

Cabibbo, N., E. C. Swallow, and R. Winston (2003). Semileptonic hyperon decays. *Ann. Rev. Nucl. Part. Sci. 53*, 39–75.

Cacciari, M., G. P. Salam, and G. Soyez (2012). FastJet user manual. *Eur. Phys. J. C72*, 1896.

Cahn, R. N. et al. (2013). White paper: measuring the neutrino mass hierarchy. `arXiv:1307.5487 [hep-ex]`.

Cahn, R. N. and H. Harari (1980). Bounds on the masses of neutral generation changing gauge bosons. *Nucl. Phys. B176*, 135–152.

Calibbi, L., G. Ferretti, D. Milstead, C. Petersson, and R. Pöttgen (2016). Baryon number violation in supersymmetry: n-$\bar{n}$ oscillations as a probe beyond the LHC. *JHEP 05*, 144.

Callan, C. G., J., S. R. Coleman, J. Wess, and B. Zumino (1969). Structure of phenomenological Lagrangians. 2. *Phys. Rev. 177*, 2247–2250.

Callan, C. G., J., R. F. Dashen, and D. J. Gross (1976). The structure of the gauge theory vacuum. *Phys. Lett. B63*, 334–340.

Callan, C. G., J. and D. J. Gross (1969). High-energy electroproduction and the constitution of the electric current. *Phys. Rev. Lett. 22*, 156–159.

Camargo-Molina, J. E., B. Garbrecht, B. O'Leary, W. Porod, and F. Staub (2014). Constraining the Natural MSSM through tunneling to color-breaking vacua at zero and non-zero temperature. *Phys. Lett. B737*, 156–161.

Camilleri, L., E. Lisi, and J. F. Wilkerson (2008). Neutrino masses and mixings: status and prospects. *Ann. Rev. Nucl. Part. Sci. 58*, 343–369.

Campbell, J. M., J. W. Huston, and W. J. Stirling (2007). Hard interactions of quarks and gluons: a primer for LHC physics. *Rept. Prog. Phys. 70*, 89.

Canetti, L., M. Drewes, and M. Shaposhnikov (2012). Matter and antimatter in the universe. *New J. Phys. 14*, 095012.

Cao, Q.-H., Z. Li, J.-H. Yu, and C. P. Yuan (2012). Discovery and identification of W' and Z' in $SU(2) \times SU(2) \times U(1)$ models at the LHC. *Phys. Rev. D86*, 095010.

Capozzi, F., E. Lisi, A. Marrone, D. Montanino, and A. Palazzo (2016). Neutrino masses and mixings: status of known and unknown 3ν parameters. *Nucl. Phys. B908*, 218–234.

Capstick, S. and W. Roberts (2000). Quark models of baryon masses and decays. *Prog. Part. Nucl. Phys. 45*, S241–S331.

Carena, M., G. Nardini, M. Quiros, and C. E. M. Wagner (2013). MSSM electroweak baryogenesis and LHC data. *JHEP 02*, 001.

Carena, M. S. and H. E. Haber (2003). Higgs boson theory and phenomenology. *Prog. Part. Nucl. Phys. 50*, 63–152.

Carena, M. S., M. Olechowski, S. Pokorski, and C. E. M. Wagner (1994). Electroweak symmetry breaking and bottom–top Yukawa unification. *Nucl. Phys. B426*, 269–300.

Carena, M. S., S. Pokorski, and C. E. M. Wagner (1993). On the unification of couplings in the minimal supersymmetric standard model. *Nucl. Phys. B406*, 59–89.

Carlson, C. E. (2015). The proton radius puzzle. *Prog. Part. Nucl. Phys. 82*, 59–77.

Carter, A. B. and A. I. Sanda (1981). CP violation in B meson decays. *Phys. Rev. D23*, 1567.

Casas, J. and A. Ibarra (2001). Oscillating neutrinos and $\mu \to e, \gamma$. *Nucl. Phys. B618*, 171–204.

Casas, J. A., A. Lleyda, and C. Munoz (1996). Strong constraints on the parameter space of the MSSM from charge and color breaking minima. *Nucl. Phys. B471*, 3–58.

Casas, J. A. and C. Munoz (1993). A natural solution to the μ problem. *Phys. Lett. B306*, 288–294.

Chacko, Z., H.-S. Goh, and R. Harnik (2006). The twin Higgs: natural electroweak breaking from mirror symmetry. *Phys. Rev. Lett. 96*, 231802.

Chang, S., R. Dermisek, J. F. Gunion, and N. Weiner (2008). Nonstandard Higgs boson decays. *Ann. Rev. Nucl. Part. Sci. 58*, 75–98.

Chanowitz, M. S. (2002). Electroweak data and the Higgs boson mass: a case for new physics. *Phys. Rev. D66*, 073002.

Chanowitz, M. S., M. A. Furman, and I. Hinchliffe (1978). Weak interactions of ultraheavy fermions. *Phys. Lett. B78*, 285.

Chanowitz, M. S. and M. K. Gaillard (1985). The TeV physics of strongly interacting W's and Z's. *Nucl. Phys. B261*, 379.

Chanowitz, M. S., M. Golden, and H. Georgi (1987). Low-energy theorems for strongly interacting W's and Z's. *Phys. Rev. D36*, 1490.

Charles, J. et al. (2005). CP violation and the CKM matrix: assessing the impact of the asymmetric B factories. *Eur. Phys. J. C41*, 1–131, and updates at `http://ckmfitter.in2p3.fr`.

Charles, J. et al. (2015). Current status of the standard model CKM fit and constraints on $\Delta F = 2$ new physics. *Phys. Rev. D91*(7), 073007.

Chatrchyan, S. et al. (2012). Observation of a new boson at a mass of 125 GeV with the CMS experiment at the LHC. *Phys. Lett. B716*, 30–61.

Chau, L.-L. (1983). Quark mixing in weak interactions. *Phys. Rept. 95*, 1.

Chen, C.-R., F. Larios, and C. P. Yuan (2005). General analysis of single top production and W helicity in top decay. *Phys. Lett. B631*, 126–132.

Chen, H.-X., W. Chen, X. Liu, and S.-L. Zhu (2016). The hidden-charm pentaquark and tetraquark states. *Phys. Rept. 639*, 1–121.

Chen, M.-C. and J. Huang (2011). TeV scale models of neutrino masses and their phenomenology. *Mod. Phys. Lett. A26*, 1147–1167.

Cheng, H.-C. and I. Low (2003). TeV symmetry and the little hierarchy problem. *JHEP 09*, 051.

Cheng, T. P. (1976). The Zweig rule and the πn sigma term. *Phys. Rev. D13*, 2161.

Cheng, T. P., E. Eichten, and L.-F. Li (1974). Higgs phenomena in asymptotically free gauge theories. *Phys. Rev. D9*, 2259.

Cheng, T.-P. and L.-F. Li (1977). Muon number nonconservation effects in a gauge theory with $V + A$ currents and heavy neutral leptons. *Phys. Rev. D16*, 1425.

Cheung, K.-M. (2001). Constraints on electron-quark contact interactions and implications to models of leptoquarks and extra Z bosons. *Phys. Lett. B517*, 167–176.

Chikashige, Y., R. N. Mohapatra, and R. D. Peccei (1981). Are there real Goldstone bosons associated with broken lepton number? *Phys. Lett. B98*, 265.

Chivukula, S. R., P. Ittisamai, and E. H. Simmons (2015). Distinguishing flavor nonuniversal colorons from Z' bosons at the LHC. *Phys. Rev. D91*(5), 055021.

Cho, G.-C., K. Hagiwara, and S. Matsumoto (1998). Constraints on four-Fermi contact interactions from low-energy electroweak experiments. *Eur. Phys. J. C5*, 155–165.

Cho, G.-C., K. Hagiwara, Y. Matsumoto, and D. Nomura (2011). The MSSM confronts the precision electroweak data and the muon $g-2$. *JHEP 11*, 068.

Chodos, A., R. L. Jaffe, K. Johnson, C. B. Thorn, and V. F. Weisskopf (1974). A new extended model of hadrons. *Phys. Rev. D9*, 3471–3495.

Choi, J. H. et al. (2016). Observation of energy and baseline dependent reactor antineutrino disappearance in the RENO experiment. *Phys. Rev. Lett. 116*(21), 211801.

Christenson, J. H., J. W. Cronin, V. L. Fitch, and R. Turlay (1964). Evidence for the 2π decay of the K_2^0 meson. *Phys. Rev. Lett. 13*, 138–140.

Chung, D. J. H. et al. (2005). The soft supersymmetry-breaking Lagrangian: theory and applications. *Phys. Rept. 407*, 1–203.

Chupp, T. and M. Ramsey-Musolf (2015). Electric dipole moments: a global analysis. *Phys. Rev. C91*(3), 035502.

Cianfrani, F. and O. M. Lecian (2007). E. C. G. Stueckelberg: a forerunner of modern physics. *Nuovo Cim. 122B*, 123–133.

Cirelli, M., G. Marandella, A. Strumia, and F. Vissani (2005). Probing oscillations into sterile neutrinos with cosmology, astrophysics and experiments. *Nucl. Phys. B708*, 215–267.

Cirigliano, V., G. Ecker, H. Neufeld, A. Pich, and J. Portoles (2012). Kaon decays in the standard model. *Rev. Mod. Phys. 84*, 399.

Cirigliano, V., S. Gardner, and B. Holstein (2013). Beta decays and non-standard interactions in the LHC era. *Prog. Part. Nucl. Phys. 71*, 93–118.

Cirigliano, V. and M. J. Ramsey-Musolf (2013). Low energy probes of physics beyond the standard model. *Prog. Part. Nucl. Phys. 71*, 2–20.

Cleaver, G., M. Cvetič, J. R. Espinosa, L. L. Everett, and P. Langacker (1998). Intermediate scales, μ parameter, and fermion masses from string models. *Phys. Rev. D57*, 2701–2715.

Clesse, S. and J. García-Bellido (2016). Detecting the gravitational wave background from primordial black hole dark matter. `arXiv:1610.08479 [astro-ph.CO]`.

Cline, J. M. (2006). Baryogenesis, In *Les Houches Summer School*. `arXiv:hep-ph/0609145 [hep-ph]`.

Cocco, A. G., G. Mangano, and M. Messina (2007). Probing low energy neutrino backgrounds with neutrino capture on beta decaying nuclei. *JCAP 0706*, 015.

Cohen, A. G., S. L. Glashow, and Z. Ligeti (2009). Disentangling neutrino oscillations. *Phys. Lett. B678*, 191–196.

Cohen, A. G. and D. B. Kaplan (1987). Thermodynamic generation of the baryon asymmetry. *Phys. Lett. B199*, 251.

Colangelo, P. and A. Khodjamirian (2000). QCD sum rules, a modern perspective. arXiv:hep-ph/0010175.

Coleman, S. R. (1966). The invariance of the vacuum is the invariance of the world. *J. Math. Phys. 7*, 787.

Coleman, S. R. and D. J. Gross (1973). Price of asymptotic freedom. *Phys. Rev. Lett. 31*, 851–854.

Coleman, S. R. and J. Mandula (1967). All possible symmetries of the S matrix. *Phys. Rev. 159*, 1251–1256.

Coleman, S. R. and E. Weinberg (1973). Radiative corrections as the origin of spontaneous symmetry breaking. *Phys. Rev. D7*, 1888–1910.

Coleman, S. R., J. Wess, and B. Zumino (1969). Structure of phenomenological Lagrangians. 1. *Phys. Rev. 177*, 2239–2247.

Collin, G. H., C. A. Argüelles, J. M. Conrad, and M. H. Shaevitz (2016). First constraints on the complete neutrino mixing matrix with a sterile neutrino. *Phys. Rev. Lett. 117*(22), 221801.

Collins, J. C. and D. E. Soper (1987). The theorems of perturbative QCD. *Ann. Rev. Nucl. Part. Sci. 37*, 383–409.

Combridge, B. L., J. Kripfganz, and J. Ranft (1977). Hadron production at large transverse momentum and QCD. *Phys. Lett. B70*, 234.

Conlon, J. P. and D. Cremades (2007). The neutrino suppression scale from large volumes. *Phys. Rev. Lett. 99*, 041803.

Conrad, J., C. Ignarra, G. Karagiorgi, M. Shaevitz, and J. Spitz (2013a). Sterile neutrino fits to short baseline neutrino oscillation measurements. *Adv. High Energy Phys. 2013*, 163897.

Conrad, J. M., W. C. Louis, and M. H. Shaevitz (2013b). The LSND and MiniBooNE oscillation searches at high Δm^2. *Ann. Rev. Nucl. Part. Sci. 63*, 45–67.

Conrad, J. M., M. H. Shaevitz, and T. Bolton (1998). Precision measurements with high energy neutrino beams. *Rev. Mod. Phys. 70*, 1341–1392.

Contino, R., M. Ghezzi, C. Grojean, M. Mühlleitner, and M. Spira (2013). Effective Lagrangian for a light Higgs-like scalar. *JHEP 1307*, 035.

Contino, R., M. Ghezzi, C. Grojean, M. Mühlleitner, and M. Spira (2014). eHDECAY: an implementation of the Higgs effective Lagrangian into HDECAY. *Comput. Phys. Commun. 185*, 3412–3423.

Copeland, E. J., M. Sami, and S. Tsujikawa (2006). Dynamics of dark energy. *Int. J. Mod. Phys. D15*, 1753–1936.

Costa, G., J. R. Ellis, G. L. Fogli, D. V. Nanopoulos, and F. Zwirner (1988). Neutral currents within and beyond the standard model. *Nucl. Phys. B297*, 244.

Cowan, C. L., F. Reines, F. B. Harrison, H. W. Kruse, and A. D. McGuire (1956). Detection of the free neutrino: A confirmation. *Science 124*, 103–104.

Craig, N., F. D'Eramo, P. Draper, S. Thomas, and H. Zhang (2015). The hunt for the rest of the Higgs bosons. *JHEP 06*, 137.

Crewther, R. J., P. Di Vecchia, G. Veneziano, and E. Witten (1979). Chiral estimate of the electric dipole moment of the neutron in quantum chromodynamics. *Phys. Lett. B88*, 123, [Erratum: *Phys. Lett. B91*, 487 (1980)].

Csáki, C., C. Grojean, L. Pilo, and J. Terning (2004). Towards a realistic model of Higgsless electroweak symmetry breaking. *Phys. Rev. Lett. 92*, 101802.

Csáki, C., C. Grojean, and J. Terning (2016). Alternatives to an elementary Higgs. *Rev. Mod. Phys. 88*(4), 045001.

Csáki, C. and P. Tanedo (2015). Beyond the standard model. In *2013 European School of High-Energy Physics*, Paradfurdo, Hungary. `arXiv:1602.04228 [hep-ph]`.

Curtin, D. et al. (2014). Exotic decays of the 125 GeV Higgs boson. *Phys. Rev. D90*(7), 075004.

Cushman, P. et al. (2013). Snowmass 2013: WIMP dark matter direct detection. `arXiv:1310.8327 [hep-ex]`.

Cuypers, F. and S. Davidson (1998). Bileptons: present limits and future prospects. *Eur. Phys. J. C2*, 503–528.

Cvetič, M., D. A. Demir, J. R. Espinosa, L. L. Everett, and P. Langacker (1997). Electroweak breaking and the μ problem in supergravity models with an additional $U(1)$. *Phys. Rev. D56*, 2861–2885.

Cvetič, M. and S. Godfrey (1995). Discovery and identification of extra gauge bosons. `arXiv:hep-ph/9504216`.

Cvetič, M. and J. Halverson (2011). TASI lectures: particle physics from perturbative and non-perturbative effects in D-braneworlds. `arXiv:1101.2907 [hep-th]`.

Cvetič, M., J. Halverson, P. Langacker, and R. Richter (2010). The Weinberg operator and a lower string scale in orientifold compactifications. *JHEP 1010*, 094.

Cvetič, M. and P. Langacker (1996). Implications of abelian extended gauge structures from string models. *Phys. Rev. D54*, 3570–3579.

Cvetič, M. and P. Langacker (2008). D-instanton generated Dirac neutrino masses. *Phys. Rev. D78*, 066012.

Cyburt, R. H., B. D. Fields, K. A. Olive, and T.-H. Yeh (2016). Big bang nucleosynthesis: 2015. *Rev. Mod. Phys. 88*, 015004.

Czarnecki, A., J. G. Korner, and J. H. Piclum (2010). Helicity fractions of W bosons from top quark decays at NNLO in QCD. *Phys. Rev. D81*, 111503.

Czarnecki, A. and W. J. Marciano (2000). Polarized Moeller scattering asymmetries. *Int. J. Mod. Phys. A15*, 2365–2376.

Czarnecki, A. and W. J. Marciano (2001). The muon anomalous magnetic moment: a harbinger for 'new physics'. *Phys. Rev. D64*, 013014.

Czarnecki, A., W. J. Marciano, and A. Sirlin (2004). Precision measurements and CKM unitarity. *Phys. Rev. D70*, 093006.

D'Ambrosio, G., G. F. Giudice, G. Isidori, and A. Strumia (2002). Minimal flavour violation: an effective field theory approach. *Nucl. Phys. B645*, 155–187.

Damour, T. and J. F. Donoghue (2008). Constraints on the variability of quark masses from nuclear binding. *Phys. Rev. D78*, 014014.

Danby, G. et al. (1962). Observation of high-energy neutrino reactions and the existence of two kinds of neutrinos. *Phys. Rev. Lett. 9*, 36–44.

Dashen, R. F. (1971). Some features of chiral symmetry breaking. *Phys. Rev. D3*, 1879–1889.

Datta, A., L. Everett, and P. Ramond (2005). Cabibbo haze in lepton mixing. *Phys. Lett. B620*, 42–51.

Davidson, S., S. Forte, P. Gambino, N. Rius, and A. Strumia (2002). Old and new physics interpretations of the NuTeV anomaly. *JHEP 02*, 037.

Davidson, S., E. Nardi, and Y. Nir (2008). Leptogenesis. *Phys. Rept. 466*, 105–177.

Davier, M., A. Hocker, and Z. Zhang (2006). The physics of hadronic tau decays. *Rev. Mod. Phys. 78*, 1043–1109.

Davier, M., A. Hoecker, B. Malaescu, and Z. Zhang (2011). Reevaluation of the hadronic contributions to the muon $g-2$ and to $\alpha(M_Z)$. *Eur. Phys. J. C71*, 1515, [Erratum: *Eur. Phys. J. C72,* 1874 (2012)].

Davies, C. T. H. et al. (2004). High-precision lattice QCD confronts experiment. *Phys. Rev. Lett. 92*, 022001.

Davoudiasl, H., R. Kitano, G. D. Kribs, H. Murayama, and P. J. Steinhardt (2004). Gravitational baryogenesis. *Phys. Rev. Lett. 93*, 201301.

Dawson, S. (1985). The effective W approximation. *Nucl. Phys. B249*, 42–60.

Dawson, S. et al. (2013). Snowmass 2013 working group report: Higgs. `arXiv:1310.8361 [hep-ex]`.

de Favereau, J. et al. (2014). DELPHES 3, A modular framework for fast simulation of a generic collider experiment. *JHEP 02*, 057.

de Florian, D. et al. (2016). Handbook of LHC Higgs cross sections: 4. Deciphering the nature of the Higgs sector. `arXiv:1610.07922 [hep-ph]`.

de Gouvêa, A. (2016). Neutrino mass models. *Ann. Rev. Nucl. Part. Sci. 66*(1), 197–217.

de Gouvêa, A. et al. (2013). Snowmass 2013 working group report: neutrinos. `arXiv:1310.4340 [hep-ex]`.

de Gouvêa, A., A. Friedland, and H. Murayama (2000). The dark side of the solar neutrino parameter space. *Phys. Lett. B490*, 125–130.

de Gouvêa, A. and W.-C. Huang (2012). Constraining the (low-energy) type-I seesaw. *Phys. Rev. D85*, 053006.

de Gouvêa, A., W.-C. Huang, and J. Jenkins (2009). Pseudo-Dirac neutrinos in the new standard model. *Phys. Rev. D80*, 073007.

de Gouvêa, A. and J. Jenkins (2008). A survey of lepton number violation via effective operators. *Phys. Rev. D77*, 013008.

de Gouvêa, A. and A. Kobach (2016). Global constraints on a heavy neutrino. *Phys. Rev. D93*(3), 033005.

de Gouvêa, A. and H. Murayama (2015). Neutrino mixing anarchy: alive and kicking. *Phys. Lett. B747*, 479–483.

de Gouvêa, A. and P. Vogel (2013). Lepton flavor and number conservation, and physics beyond the standard model. *Prog. Part. Nucl. Phys. 71*, 75–92.

De Roeck, A. and R. S. Thorne (2011). Structure functions. *Prog. Part. Nucl. Phys. 66*, 727–781.

De Rujula, A., M. B. Gavela, P. Hernandez, and E. Masso (1992). The selfcouplings of vector bosons: does LEP-1 obviate LEP-2? *Nucl. Phys. B384*, 3–58.

De Rujula, A., H. Georgi, and S. L. Glashow (1975). Hadron masses in a gauge theory. *Phys. Rev. D12*, 147–162.

de Swart, J. J. (1963). The octet model and its Clebsch-Gordan coefficients. *Rev. Mod. Phys. 35*, 916–939.

de Vries, J. et al. (2014). A study of the parity-odd nucleon-nucleon potential. *Eur. Phys. J. A50*, 108.

Dedes, A., C. Hugonie, S. Moretti, and K. Tamvakis (2001). Phenomenology of a new minimal supersymmetric extension of the standard model. *Phys. Rev. D63*, 055009.

Degrassi, G. et al. (2012). Higgs mass and vacuum stability in the standard model at NNLO. *JHEP 1208*, 098.

Delgado, A., A. Pomarol, and M. Quiros (2000). Electroweak and flavor physics in extensions of the standard model with large extra dimensions. *JHEP 01*, 030.

Déliot, F., N. Hadley, S. Parke, and T. Schwarz (2014). Properties of the top quark. *Ann. Rev. Nucl. Part. Sci. 64*, 363–381.

Dell'Oro, S., S. Marcocci, M. Viel, and F. Vissani (2016). Neutrinoless double beta decay: 2015 review. *Adv. High Energy Phys. 2016*, 2162659.

Delsart, P.-A., K. L. Geerlings, J. Huston, B. T. Martin, and C. K. Vermilion (2012). SpartyJet 4.0 user's manual. `arXiv:1201.3617 [hep-ex]`.

Demir, D. A., L. L. Everett, and P. Langacker (2008). Dirac neutrino masses from generalized supersymmetry breaking. *Phys. Rev. Lett. 100*, 091804.

Denef, F. and M. R. Douglas (2004). Distributions of flux vacua. *JHEP 05*, 072.

Deniz, M. et al. (2010). Measurement of $\bar{\nu}_e$-electron scattering cross-section with a CsI(Tl) scintillating crystal array at the Kuo-Sheng Nuclear Power Reactor. *Phys. Rev. D81*, 072001.

Denner, A., S. Heinemeyer, I. Puljak, D. Rebuzzi, and M. Spira (2011). Standard model Higgs-boson branching ratios with uncertainties. *Eur. Phys. J. C71*, 1753.

d'Enterria, D. and P. Z. Skands (2015). High-precision α_s measurements from LHC to FCC-ee. `arXiv:1512.05194 [hep-ph]`.

DeTar, C. and U. M. Heller (2009). QCD thermodynamics from the lattice. *Eur. Phys. J. A41*, 405–437.

Deur, A., S. J. Brodsky, and G. F. de Teramond (2016). The QCD running coupling. *Prog. Part. Nucl. Phys. 90*, 1–74.

Diaconu, C., T. Haas, M. Medinnis, K. Rith, and A. Wagner (2010). Physics accomplishments of HERA. *Ann. Rev. Nucl. Part. Sci. 60*, 101–128.

Diaz, J. S. and A. Kostelecky (2012). Lorentz- and CPT-violating models for neutrino oscillations. *Phys. Rev. D85*, 016013.

Diehl, M. (2003). Generalized parton distributions. *Phys. Rept. 388*, 41–277.

Dienes, K. R. (1997). String theory and the path to unification: a review of recent developments. *Phys. Rept. 287*, 447–525.

Dienes, K. R., E. Dudas, and T. Gherghetta (1999a). Grand unification at intermediate mass scales through extra dimensions. *Nucl. Phys. B537*, 47–108.

Dienes, K. R., E. Dudas, and T. Gherghetta (1999b). Light neutrinos without heavy mass scales: a higher-dimensional seesaw mechanism. *Nucl. Phys. B557*, 25.

Dienes, K. R., C. F. Kolda, and J. March-Russell (1997). Kinetic mixing and the supersymmetric gauge hierarchy. *Nucl. Phys. B492*, 104–118.

Dietrich, D. D. and F. Sannino (2007). Conformal window of $SU(N)$ gauge theories with fermions in higher dimensional representations. *Phys. Rev. D75*, 085018.

Dimopoulos, S. and H. Georgi (1981). Softly broken supersymmetry and $SU(5)$. *Nucl. Phys. B193*, 150.

Dimopoulos, S., S. Raby, and F. Wilczek (1981). Supersymmetry and the scale of unification. *Phys. Rev. D24*, 1681–1683.

Dimopoulos, S., S. Raby, and F. Wilczek (1982). Proton decay in supersymmetric models. *Phys. Lett. B112*, 133.

Dimopoulos, S. and D. W. Sutter (1995). The supersymmetric flavor problem. *Nucl. Phys. B452*, 496–512.

Dine, M. (2000). The strong CP problem. `arXiv:hep-ph/0011376`.

Dine, M. (2015a). Naturalness under stress. *Ann. Rev. Nucl. Part. Sci. 65*, 43–62.

Dine, M. and A. Kusenko (2004). The origin of the matter-antimatter asymmetry. *Rev. Mod. Phys. 76*, 1.

Ding, G.-J., L. L. Everett, and A. J. Stuart (2012). Golden ratio neutrino mixing and A_5 flavor symmetry. *Nucl. Phys. B857*, 219–253.

Dittmaier, S. and M. Schumacher (2013). The Higgs boson in the standard model — from LEP to LHC: expectations, searches, and discovery of a candidate. *Prog. Part. Nucl. Phys. 70*, 1–54.

Diwan, M. V., V. Galymov, X. Qian, and A. Rubbia (2016). Long-baseline neutrino experiments. *Ann. Rev. Nucl. Part. Sci. 66*, 47–71.

Dixon, L. J. (2014). European School of High-Energy Physics, 2012: A brief introduction to modern amplitude methods. `arXiv:1310.5353 [hep-ph]`.

Djouadi, A. (2008a). The anatomy of electro-weak symmetry breaking. I: The Higgs boson in the standard model. *Phys. Rept. 457*, 1–216.

Djouadi, A. (2008b). The anatomy of electro-weak symmetry breaking. II. The Higgs bosons in the minimal supersymmetric model. *Phys. Rept. 459*, 1–241.

Djouadi, A., J.-L. Kneur, and G. Moultaka (2007). SuSpect: A Fortran code for the supersymmetric and Higgs particle spectrum in the MSSM. *Comput. Phys. Commun. 176*, 426–455.

Dolan, L. and R. Jackiw (1974). Symmetry behavior at finite temperature. *Phys. Rev. D9*, 3320–3341.

Dolgov, A. D. (2002). Neutrinos in cosmology. *Phys. Rept. 370*, 333–535.

Domokos, S. K., J. A. Harvey, and N. Mann (2009). The Pomeron contribution to pp and $p\bar{p}$ scattering in AdS/QCD. *Phys. Rev. D80*, 126015.

Donini, A., P. Hernandez, J. Lopez-Pavon, M. Maltoni, and T. Schwetz (2012). The minimal 3+2 neutrino model versus oscillation anomalies. *JHEP 1207*, 161.

Donnachie, A. and P. V. Landshoff (2013). pp and $\bar{p}p$ total cross sections and elastic scattering. *Phys. Lett. B727*, 500–505, [Erratum: *Phys. Lett. B750*, 669 (2015)].

Donoghue, J. F. (2016). The multiverse and particle physics. *Ann. Rev. Nucl. Part. Sci. 66*(1), 1–21.

Dorsch, G. C., S. J. Huber, and J. M. No (2013). A strong electroweak phase transition in the 2HDM after LHC8. *JHEP 10*, 029.

Doršner, I., S. Fajfer, A. Greljo, J. F. Kamenik, and N. Košnik (2016). Physics of leptoquarks in precision experiments and at particle colliders. *Phys. Rept. 641*, 1–68.

Dragoun, O. and D. Vénos (2016). Constraints on the active and sterile neutrino masses from beta-ray spectra: past, present and future. *J. Phys. 3*, 77–113.

Draper, P. and H. Rzehak (2016). A review of Higgs mass calculations in supersymmetric models. *Phys. Rept. 619*, 1–24.

Drees, M., H. Dreiner, D. Schmeier, J. Tattersall, and J. S. Kim (2015). CheckMATE: Confronting your favourite new physics model with LHC data. *Comput. Phys. Commun. 187*, 227–265.

Dreiner, H. K., H. E. Haber, and S. P. Martin (2010). Two-component spinor techniques and Feynman rules for quantum field theory and supersymmetry. *Phys. Rept. 494*, 1–196.

Drell, S. D., D. J. Levy, and T.-M. Yan (1969). A theory of deep inelastic lepton-nucleon scattering and lepton pair annihilation processes. 1. *Phys. Rev. 187*, 2159–2171.

Drell, S. D. and T.-M. Yan (1971). Partons and their applications at high energies. *Ann. Phys. 66*, 578.

Drewes, M. and B. Garbrecht (2015). Experimental and cosmological constraints on heavy neutrinos. `arXiv:1502.00477 [hep-ph]`.

Drexlin, G., V. Hannen, S. Mertens, and C. Weinheimer (2013). Current direct neutrino mass experiments. *Adv. High Energy Phys. 2013*, 293986.

Duan, H., G. M. Fuller, and Y.-Z. Qian (2010). Collective neutrino oscillations. *Ann. Rev. Nucl. Part. Sci. 60*, 569–594.

Dubbers, D. and M. G. Schmidt (2011). The neutron and its role in cosmology and particle physics. *Rev. Mod. Phys. 83*, 1111–1171.

Dulat, S. et al. (2016). The CT14 global analysis of quantum chromodynamics. *Phys. Rev. D93*, 033006.

Dürr, S. et al. (2008). Ab initio determination of light hadron masses. *Science 322*, 1224–1227.

Eberhardt, O. et al. (2012). Impact of a Higgs boson at a mass of 126 GeV on the standard model with three and four fermion generations. *Phys. Rev. Lett. 109*, 241802.

Eberle, B., A. Ringwald, L. Song, and T. J. Weiler (2004). Relic neutrino absorption spectroscopy. *Phys. Rev. D70*, 023007.

Eichten, E., S. Godfrey, H. Mahlke, and J. L. Rosner (2008). Quarkonia and their transitions. *Rev. Mod. Phys. 80*, 1161–1193.

Eichten, E., I. Hinchliffe, K. D. Lane, and C. Quigg (1984). Super collider physics. *Rev. Mod. Phys. 56*, 579–707.

Eichten, E., K. Lane, and A. Martin (2012). A Higgs impostor in low-scale technicolor. `arXiv:1210.5462 [hep-ph]`.

Eichten, E. and K. D. Lane (1980). Dynamical breaking of weak interaction symmetries. *Phys. Lett. B90*, 125–130.

Eichten, E., K. D. Lane, and M. E. Peskin (1983). New tests for quark and lepton substructure. *Phys. Rev. Lett. 50*, 811–814.

Eidelman, S. and F. Jegerlehner (1995). Hadronic contributions to $g-2$ of the leptons and to the effective fine structure constant $\alpha(M_Z^2)$. *Z. Phys. C67*, 585–602.

Elias-Miro, J. et al. (2012). Higgs mass implications on the stability of the electroweak vacuum. *Phys. Lett. B709*, 222–228.

Elias-Miro, J., J. Espinosa, E. Masso, and A. Pomarol (2013). Higgs windows to new physics through $d = 6$ operators: constraints and one-loop anomalous dimensions. *JHEP 1311*, 066.

Ellis, J., F. Moortgat, G. Moortgat-Pick, J. M. Smillie, and J. Tattersall (2009). Measurement of CP violation in stop cascade decays at the LHC. *Eur. Phys. J. C60*, 633–651.

Ellis, J., V. Sanz, and T. You (2015). The effective standard model after LHC Run I. *JHEP 03*, 157.

Ellis, J. R., J. F. Gunion, H. E. Haber, L. Roszkowski, and F. Zwirner (1989). Higgs bosons in a nonminimal supersymmetric model. *Phys. Rev. D39*, 844.

Ellis, J. R., S. Kelley, and D. V. Nanopoulos (1990). Precision LEP data, supersymmetric GUTs and string unification. *Phys. Lett. B249*, 441–448.

Ellis, J. R., J. S. Lee, and A. Pilaftsis (2008). Electric dipole moments in the MSSM reloaded. *JHEP 10*, 049.

Ellis, J. R., D. V. Nanopoulos, and S. Rudaz (1982). GUTs 3: SUSY GUTs 2. *Nucl. Phys. B202*, 43.

Ellis, J. R., G. Ridolfi, and F. Zwirner (1991). Radiative corrections to the masses of supersymmetric Higgs bosons. *Phys. Lett. B257*, 83–91.

Ellis, S. D., J. Huston, K. Hatakeyama, P. Loch, and M. Tonnesmann (2008). Jets in hadron-hadron collisions. *Prog. Part. Nucl. Phys. 60*, 484–551.

Ellwanger, U., C. Hugonie, and A. M. Teixeira (2010). The next-to-minimal supersymmetric standard model. *Phys. Rept. 496*, 1–77.

Engel, J., M. J. Ramsey-Musolf, and U. van Kolck (2013). Electric dipole moments of nucleons, nuclei, and atoms: the standard model and beyond. *Prog. Part. Nucl. Phys. 71*, 21–74.

Englert, C. et al. (2014). Precision measurements of Higgs couplings: implications for new physics scales. *J. Phys. G41*, 113001.

Englert, F. and R. Brout (1964). Broken symmetry and the mass of gauge vector mesons. *Phys. Rev. Lett. 13*, 321–322.

Epelbaum, E., H.-W. Hammer, and U.-G. Meissner (2009). Modern theory of nuclear forces. *Rev. Mod. Phys. 81*, 1773–1825.

Erler, J. (1999). Global fits to electroweak data using GAPP. `arXiv:hep-ph/0005084`.

Erler, J. (2000). Chiral models of weak scale supersymmetry. *Nucl. Phys. B586*, 73–91.

Erler, J., C. J. Horowitz, S. Mantry, and P. A. Souder (2014). Weak polarized electron scattering. *Ann. Rev. Nucl. Part. Sci. 64*, 269–298.

Erler, J., A. Kurylov, and M. J. Ramsey-Musolf (2003). The weak charge of the proton and new physics. *Phys. Rev. D68*, 016006.

Erler, J., P. Langacker, and T. Li (2002). The Z - Z' mass hierarchy in a supersymmetric model with a secluded $U(1)'$ breaking sector. *Phys. Rev. D66*, 015002.

Erler, J. and D. M. Pierce (1998). Bounds on supersymmetry from electroweak precision analysis. *Nucl. Phys. B526*, 53–80.

Erler, J. and M. J. Ramsey-Musolf (2005). The weak mixing angle at low energies. *Phys. Rev. D72*, 073003.

Erler, J. and S. Su (2013). The weak neutral current. *Prog. Part. Nucl. Phys. 71*, 119–149.

Esmaili, A. and O. L. G. Peres (2012). KATRIN sensitivity to sterile neutrino mass in the shadow of lightest neutrino mass. *Phys. Rev. D85*, 117301.

Espinosa, J. R. (2016). Implications of the top (and Higgs) mass for vacuum stability. *PoS TOP2015*, 043.

Espinosa, J. R. and M. Quiros (1993). Upper bounds on the lightest Higgs boson mass in general supersymmetric standard models. *Phys. Lett. B302*, 51–58.

Espinosa, J. R. and M. Quiros (1995). Improved metastability bounds on the standard model Higgs mass. *Phys. Lett. B353*, 257–266.

Essig, R. et al. (2013). Snowmass 2013 working group report: New light weakly coupled particles. `arXiv:1311.0029 [hep-ph]`.

Esteban, I. et al. (2017). Updated fit to three neutrino mixing: exploring the accelerator-reactor complementarity. *JHEP 01*, 087, and updates at `www.nu-fit.org`. Updated from M. C. Gonzalez-Garcia et al., *JHEP, 1411, 052.*

Everett, L. L. (2006). Viewing lepton mixing through the Cabibbo haze. *Phys. Rev. D73*, 013011.

Everett, L. L., I.-W. Kim, P. Ouyang, and K. M. Zurek (2008). Deflected mirage mediation: a framework for generalized supersymmetry breaking. *Phys. Rev. Lett. 101*, 101803.

Faessler, A., M. Gonzalez, S. Kovalenko, and F. Simkovic (2014). Arbitrary mass Majorana neutrinos in neutrinoless double beta decay. *Phys. Rev. D90*, 096010.

Faessler, A., R. Hodak, S. Kovalenko, and F. Simkovic (2017). Can one measure the cosmic neutrino background? *Int. J. Mod. Phys. E26*, 1740008.

Famaey, B. and S. McGaugh (2012). Modified Newtonian dynamics (MOND): Observational phenomenology and relativistic extensions. *Living Rev. Rel. 15*, 10.

Farrar, G. R. and P. Fayet (1978). Phenomenology of the production, decay, and detection of new hadronic states associated with supersymmetry. *Phys. Lett. B76*, 575–579.

Fayet, P. (1975). Supergauge invariant extension of the Higgs mechanism and a model for the electron and its neutrino. *Nucl. Phys. B90*, 104–124.

Fayet, P. and S. Ferrara (1977). Supersymmetry. *Phys. Rept. 32*, 249–334.

Fayet, P. and J. Iliopoulos (1974). Spontaneously broken supergauge symmetries and Goldstone spinors. *Phys. Lett. B51*, 461–464.

Feger, R. and T. W. Kephart (2015). LieARTA Mathematica application for Lie algebras and representation theory. *Comput. Phys. Commun. 192*, 166–195.

Feldman, G., J. Hartnell, and T. Kobayashi (2013). Long-baseline neutrino oscillation experiments. *Adv. High Energy Phys. 2013*, 475749.

Feng, J. L. (2010). Dark matter candidates from particle physics and methods of detection. *Ann. Rev. Astron. Astrophys. 48*, 495–545.

Feng, J. L. (2013). Naturalness and the status of supersymmetry. *Ann. Rev. Nucl. Part. Sci. 63*, 351–382.

Feng, J. L., J.-F. Grivaz, and J. Nachtman (2010). Searches for supersymmetry at high-energy colliders. *Rev. Mod. Phys. 82*, 699–727.

Feruglio, F. and A. Paris (2011). The golden ratio prediction for the solar angle from a natural model with $A5$ flavour symmetry. *JHEP 1103*, 101.

Feynman, R. P. and M. Gell-Mann (1958). Theory of the Fermi interaction. *Phys. Rev. 109*, 193–198.

Field, R. D. and R. P. Feynman (1977). Quark elastic scattering as a source of high transverse momentum mesons. *Phys. Rev. D15*, 2590–2616.

Fields, B. D., K. Kainulainen, and K. A. Olive (1997). Nucleosynthesis limits on the mass of long lived tau and muon neutrinos. *Astropart. Phys. 6*, 169.

Flacher, H. et al. (2009). Gfitter-revisiting the global electroweak fit of the standard model and beyond. *Eur. Phys. J. C60*, 543–583, [Erratum: *Eur. Phys. J. C71*, 1718 (2011)].

Florkowski, W. (2014). Basic phenomenology for relativistic heavy-ion collisions. *Acta Phys. Polon. B45*(12), 2329–2354.

Fogli, G., E. Lisi, A. Marrone, A. Palazzo, and A. Rotunno (2008). Hints of $\theta_{13} > 0$ from global neutrino data analysis. *Phys. Rev. Lett. 101*, 141801.

Fong, C. S., E. Nardi, and A. Riotto (2012). Leptogenesis in the Universe. *Adv. High Energy Phys. 2012*, 158303.

Foot, R., M. J. Thomson, and R. Volkas (1996). Large neutrino asymmetries from neutrino oscillations. *Phys. Rev. D53*, 5349–5353.

Forero, D. V., M. Tortola, and J. W. F. Valle (2014). Neutrino oscillations refitted. *Phys. Rev. D90*(9), 093006.

Formaggio, J. A. and G. P. Zeller (2012). From eV to EeV: neutrino cross sections across energy scales. *Rev. Mod. Phys. 84*, 1307.

Frampton, P. H., P. Q. Hung, and M. Sher (2000). Quarks and leptons beyond the third generation. *Phys. Rept. 330*, 263.

Frank, M., L. Selbuz, L. Solmaz, and I. Turan (2013). Higgs bosons in supersymmetric $U(1)'$ models with CP violation. *Phys. Rev. D87*(7), 075007.

Friedland, A., C. Lunardini, and C. Pena-Garay (2004). Solar neutrinos as probes of neutrino-matter interactions. *Phys. Lett. B594*, 347.

Friedman, J. I. and H. W. Kendall (1972). Deep inelastic electron scattering. *Ann. Rev. Nucl. Part. Sci. 22*, 203–254.

Frieman, J., M. Turner, and D. Huterer (2008). Dark energy and the accelerating universe. *Ann. Rev. Astron. Astrophys. 46*, 385–432.

Fritzsch, H., M. Gell-Mann, and H. Leutwyler (1973). Advantages of the color octet gluon picture. *Phys. Lett. B47*, 365–368.

Fritzsch, H. and P. Minkowski (1975). Unified interactions of leptons and hadrons. *Ann. Phys. 93*, 193–266.

Froggatt, C. D. and H. B. Nielsen (1979). Hierarchy of quark masses, Cabibbo angles and CP violation. *Nucl. Phys. B147*, 277.

Froissart, M. (1961). Asymptotic behavior and subtractions in the Mandelstam representation. *Phys. Rev. 123*, 1053–1057.

Fubini, S. and G. Furlan (1965). Renormalization effects for partially conserved currents. *Physics 1*, 229–247.

Fujikawa, K., B. W. Lee, and A. I. Sanda (1972). Generalized renormalizable gauge formulation of spontaneously broken gauge theories. *Phys. Rev. D6*, 2923–2943.

Fukuda, Y. et al. (1998). Evidence for oscillation of atmospheric neutrinos. *Phys. Rev. Lett. 81*, 1562–1567.

Fukugita, M. and T. Yanagida (1986). Baryogenesis without grand unification. *Phys. Lett. B174*, 45.

Fukuyama, T. (2012). Searching for new physics beyond the standard model in electric dipole moment. *Int. J. Mod. Phys. A27*, 1230015.

Gabbiani, F., E. Gabrielli, A. Masiero, and L. Silvestrini (1996). A complete analysis of FCNC and CP constraints in general SUSY extensions of the standard model. *Nucl. Phys. B477*, 321–352.

Gagliardi, C. A., R. E. Tribble, and N. J. Williams (2005). Global analysis of muon decay measurements. *Phys. Rev. D72*, 073002.

Gaillard, J. M. and G. Sauvage (1984). Hyperon beta decays. *Ann. Rev. Nucl. Part. Sci. 34*, 351–402.

Gaillard, M. K. and B. W. Lee (1974a). $\Delta I = 1/2$ rule for nonleptonic decays in asymptotically free field theories. *Phys. Rev. Lett. 33*, 108.

Gaillard, M. K. and B. W. Lee (1974b). Rare decay modes of the K-Mesons in gauge theories. *Phys. Rev. D10*, 897.

Gaillard, M. K., B. W. Lee, and J. L. Rosner (1975). Search for charm. *Rev. Mod. Phys. 47*, 277.

Gaisser, T. and F. Halzen (2014). IceCube. *Ann. Rev. Nucl. Part. Sci. 64*, 101–123.

Gandhi, R., C. Quigg, M. H. Reno, and I. Sarcevic (1998). Neutrino interactions at ultrahigh-energies. *Phys. Rev. D58*, 093009.

Gando, A. et al. (2011). Partial radiogenic heat model for Earth revealed by geoneutrino measurements. *Nature Geo. 4*, 647–651.

Gando, A. et al. (2013). Reactor on-off antineutrino measurement with KamLAND. *Phys. Rev. D88*(3), 033001, copyright 2013 by the American Physical Society.

Gando, A. et al. (2016). Search for Majorana neutrinos near the inverted mass hierarchy region with KamLAND-Zen. *Phys. Rev. Lett. 117*(8), 082503.

Gariazzo, S., C. Giunti, M. Laveder, Y. F. Li, and E. M. Zavanin (2016). Light sterile neutrinos. *J. Phys. G43*, 033001.

Garvey, G. T., D. A. Harris, H. A. Tanaka, R. Tayloe, and G. P. Zeller (2015). Recent advances and open questions in neutrino-induced quasi-elastic scattering and single photon production. *Phys. Rept. 580*, 1–45.

Gaskins, J. M. (2016). A review of indirect searches for particle dark matter. *Contemp. Phys. 57*, 496–525.

Gasser, J. and H. Leutwyler (1982). Quark masses. *Phys. Rept. 87*, 77–169.

Gavela, M., D. Hernandez, T. Ota, and W. Winter (2009). Large gauge invariant nonstandard neutrino interactions. *Phys. Rev. D79*, 013007.

Gell-Mann, M. and M. Levy (1960). The axial vector current in beta decay. *Nuovo Cim. 16*, 705.

Gell-Mann, M., R. J. Oakes, and B. Renner (1968). Behavior of current divergences under $SU(3) \times SU(3)$. *Phys. Rev. 175*, 2195–2199.

Gell-Mann, M., P. Ramond, and R. Slansky (1979). Complex spinors and unified theories, in *Supergravity*, P. van Nieuwenhuizen and D. Z. Freedman (eds.). Amsterdam: North Holland. `arXiv:1306.4669 [hep-th]`.

Gelmini, G. B. (2005). Prospect for relic neutrino searches. *Phys. Scripta T121*, 131–136.

Gelmini, G. B. (2015). TASI 2014: The hunt for dark matter. `arXiv:1502.01320 [hep-ph]`.

Gelmini, G. B. and M. Roncadelli (1981). Left-handed neutrino mass scale and spontaneously broken lepton number. *Phys. Lett. B99*, 411.

Georgi, H. (1975). The state-of-the-art gauge theories. *AIP Conf. Proc. 23*, 575–582.

Georgi, H. and S. L. Glashow (1972). Gauge theories without anomalies. *Phys. Rev. D6*, 429.

Georgi, H. and S. L. Glashow (1974). Unity of all elementary particle forces. *Phys. Rev. Lett. 32*, 438–441.

Georgi, H. and C. Jarlskog (1979). A new lepton-quark mass relation in a unified theory. *Phys. Lett. B86*, 297–300.

Georgi, H., H. R. Quinn, and S. Weinberg (1974). Hierarchy of interactions in unified gauge theories. *Phys. Rev. Lett. 33*, 451–454.

Georgi, H. M., S. L. Glashow, and S. Nussinov (1981). Unconventional model of neutrino masses. *Nucl. Phys. B193*, 297.

Gershon, T. and V. V. Gligorov (2017). CP violation in the B system. *Rept. Prog. Phys. 80*, 046201.

Gershtein, Y. et al. (2013). Snowmass 2013 working group report: New particles, forces, and dimensions. `arXiv:1311.0299 [hep-ex]`.

Gervais, J.-L. and B. Sakita (1971). Field theory interpretation of supergauges in dual models. *Nucl. Phys. B34*, 632–639.

Gherghetta, T. (2011). TASI 09: A holographic view of beyond the standard model physics. `arXiv:1008.2570 [hep-ph]`.

Giardino, P. P., K. Kannike, I. Masina, M. Raidal, and A. Strumia (2014). The universal Higgs fit. *JHEP 1405*, 046.

Giedt, J., G. L. Kane, P. Langacker, and B. D. Nelson (2005). Massive neutrinos and (heterotic) string theory. *Phys. Rev. D71*, 115013.

Ginges, J. S. M. and V. V. Flambaum (2004). Violations of fundamental symmetries in atoms and tests of unification theories of elementary particles. *Phys. Rept. 397*, 63–154.

Giudice, G. F. (2008). Naturally speaking: the naturalness criterion and physics at the LHC. `arXiv:0801.2562 [hep-ph]`.

Giudice, G. F. and A. Masiero (1988). A natural solution to the μ problem in supergravity theories. *Phys. Lett. B206*, 480–484.

Giudice, G. F., P. Paradisi, and M. Passera (2012). Testing new physics with the electron $g-2$. *JHEP 11*, 113.

Giudice, G. F. and R. Rattazzi (1999). Theories with gauge-mediated supersymmetry breaking. *Phys. Rept. 322*, 419–499.

Giunti, C. et al. (2016). Electromagnetic neutrinos in laboratory experiments and astrophysics. *Annalen Phys. 528*, 198–215.

Giunti, C., C. W. Kim, and U. W. Lee (1991). Running coupling constants and grand unification models. *Mod. Phys. Lett. A6*, 1745–1755.

Giunti, C. and A. Studenikin (2015). Neutrino electromagnetic interactions: a window to new physics. *Rev. Mod. Phys. 87*, 531.

Glashow, S. L. (1960). Resonant scattering of antineutrinos. *Phys. Rev. 118*, 316–317.

Glashow, S. L. (1961). Partial symmetries of weak interactions. *Nucl. Phys. 22*, 579–588.

Glashow, S. L., A. Halprin, P. I. Krastev, C. N. Leung, and J. T. Pantaleone (1997). Remarks on neutrino tests of special relativity. *Phys. Rev. D56*, 2433–2434.

Glashow, S. L., J. Iliopoulos, and L. Maiani (1970). Weak interactions with lepton-hadron symmetry. *Phys. Rev. D2*, 1285–1292.

Glashow, S. L. and S. Weinberg (1968). Breaking chiral symmetry. *Phys. Rev. Lett. 20*, 224–227.

Glashow, S. L. and S. Weinberg (1977). Natural conservation laws for neutral currents. *Phys. Rev. D15*, 1958.

Gleisberg, T. et al. (2009). Event generation with SHERPA 1.1. *JHEP 02*, 007.

Godfrey, S. and N. Isgur (1985). Mesons in a relativized quark model with chromodynamics. *Phys. Rev. D32*, 189–231.

Golden, M. and L. Randall (1991). Radiative corrections to electroweak parameters in technicolor theories. *Nucl. Phys. B361*, 3–23.

Goldstone, J. (1961). Field theories with superconductor solutions. *Nuovo Cim. 19*, 154–164.

Goldstone, J., A. Salam, and S. Weinberg (1962). Broken symmetries. *Phys. Rev. 127*, 965–970.

Golowich, E., J. Hewett, S. Pakvasa, and A. A. Petrov (2007). Implications of D^0-$\bar{D}^0$ mixing for new physics. *Phys. Rev. D76*, 095009.

Gomez-Bock, M., M. Mondragon, M. Muhlleitner, M. Spira, and P. M. Zerwas (2007). Concepts of electroweak symmetry breaking and Higgs physics. `arXiv:0712.2419 [hep-ph]`.

Gondolo, P. et al. (2004). DarkSUSY: computing supersymmetric dark matter properties numerically. *JCAP 0407*, 008.

Gonzalez-Garcia, M. C. and M. Maltoni (2008). Phenomenology with massive neutrinos. *Phys. Rept. 460*, 1–129.

Gonzalez-Garcia, M. C. and M. Maltoni (2013). Determination of matter potential from global analysis of neutrino oscillation data. *JHEP 1309*, 152.

Gorbahn, M., S. Jager, U. Nierste, and S. Trine (2011). The supersymmetric Higgs sector and $B - \bar{B}$ mixing for large $\tan\beta$. *Phys. Rev. D84*, 034030.

Gorringe, T. and H. W. Fearing (2004). Induced pseudoscalar coupling of the proton weak interaction. *Rev. Mod. Phys. 76*, 31–91.

Gorringe, T. P. and D. W. Hertzog (2015). Precision muon physics. *Prog. Part. Nucl. Phys. 84*, 73–123.

Gounaris, G. et al. (1996). Triple gauge boson couplings. `arXiv:hep-ph/9601233`.

Gounaris, G. J. and J. J. Sakurai (1968). Finite width corrections to the vector meson dominance prediction for $\rho \to e^+e^-$. *Phys. Rev. Lett. 21*, 244.

Graham, P. W. et al. (2015a). Experimental searches for the axion and axion-like particles. *Ann. Rev. Nucl. Part. Sci. 65*, 485–514.

Graham, P. W., D. E. Kaplan, and S. Rajendran (2015b). Cosmological relaxation of the electroweak scale. *Phys. Rev. Lett. 115*(22), 221801.

Graham, P. W. and S. Rajendran (2013). New observables for direct detection of axion dark matter. *Phys. Rev. D88*, 035023.

Green, A. M. (2015). Primordial black holes: sirens of the early Universe. *Fundam. Theor. Phys. 178*, 129–149.

Greenberg, O. W. (2008). Discovery of the color degree of freedom in particle physics: a personal perspective. `arXiv:0803.0992 [physics.hist-ph]`.

Grisaru, M. T., W. Siegel, and M. Rocek (1979). Improved methods for supergraphs. *Nucl. Phys. B159*, 429.

Gronau, M. and D. London (1990). Isospin analysis of CP asymmetries in B decays. *Phys. Rev. Lett. 65*, 3381–3384.

Gross, D. J. (2005). The discovery of asymptotic freedom and the emergence of QCD. *Rev. Mod. Phys. 77*, 837–849.

Gross, D. J. and F. Wilczek (1973). Ultraviolet behavior of nonabelian gauge theories. *Phys. Rev. Lett. 30*, 1343–1346.

Grossman, Y. and S. Rakshit (2004). Neutrino masses in R-parity violating supersymmetric models. *Phys. Rev. D69*, 093002.

Grothaus, P., M. Fairbairn, and J. Monroe (2014). Directional dark matter detection beyond the neutrino bound. *Phys. Rev. D90*(5), 055018.

Grunewald, M. W. (1999). Experimental tests of the electroweak standard model at high energies. *Phys. Rept. 322*, 125–346.

Grzadkowski, B., M. Iskrzynski, M. Misiak, and J. Rosiek (2010). Dimension-six terms in the standard model Lagrangian. *JHEP 10*, 085.

Gunion, J. F. and H. E. Haber (1986). Higgs bosons in supersymmetric models. 1. *Nucl. Phys. B272*, 1.

Gupta, R. (1997). Introduction to lattice QCD. `arXiv:hep-lat/9807028`.

Guralnik, G. S., C. R. Hagen, and T. W. B. Kibble (1964). Global conservation laws and massless particles. *Phys. Rev. Lett. 13*, 585–587.

Guth, A. H. (1981). The inflationary universe: a possible solution to the horizon and flatness problems. *Phys. Rev. D23*, 347–356.

Guzowski, P. et al. (2015). Combined limit on the neutrino mass from neutrinoless double-β decay and constraints on sterile Majorana neutrinos. *Phys. Rev. D92*(1), 012002.

Haag, R. (1958). Quantum field theories with composite particles and asymptotic conditions. *Phys. Rev. 112*, 669–673.

Haag, R., J. T. Lopuszanski, and M. Sohnius (1975). All possible generators of supersymmetries of the S matrix. *Nucl. Phys. B88*, 257.

Haber, H. E. and R. Hempfling (1991). Can the mass of the lightest Higgs boson of the minimal supersymmetric model be larger than M_Z? *Phys. Rev. Lett. 66*, 1815–1818.

Haber, H. E. and G. L. Kane (1985). The search for supersymmetry: probing physics beyond the standard model. *Phys. Rept. 117*, 75–263.

Hagiwara, K., S. Ishihara, R. Szalapski, and D. Zeppenfeld (1993). Low-energy effects of new interactions in the electroweak boson sector. *Phys. Rev. D48*, 2182–2203.

Hagiwara, K., R. Liao, A. D. Martin, D. Nomura, and T. Teubner (2011). $(g-2)_\mu$ and $\alpha(M_Z^2)$ re-evaluated using new precise data. *J. Phys. G38*, 085003.

Hagiwara, K. and D. Zeppenfeld (1986). Helicity amplitudes for heavy lepton production in e^+e^- annihilation. *Nucl. Phys. B274*, 1.

Hahn, T. (2001). Generating Feynman diagrams and amplitudes with FeynArts 3. *Comput. Phys. Commun. 140*, 418–431.

Hahn, T., S. Heinemeyer, W. Hollik, H. Rzehak, and G. Weiglein (2014). High-precision predictions for the light CP-even Higgs boson mass of the minimal supersymmetric standard model. *Phys. Rev. Lett. 112*(14), 141801.

Hall, L. J., H. Murayama, and N. Weiner (2000). Neutrino mass anarchy. *Phys. Rev. Lett. 84*, 2572–2575.

Hall, L. J., D. Pinner, and J. T. Ruderman (2012). A natural SUSY Higgs near 126 GeV. *JHEP 1204*, 131.

Hall, L. J., D. Pinner, and J. T. Ruderman (2014). The weak scale from BBN. *JHEP 12*, 134.

Hall, L. J. and S. Weinberg (1993). Flavor changing scalar interactions. *Phys. Rev. D48*, 979–983.

Hambye, T. (2012). Leptogenesis: beyond the minimal type I seesaw scenario. *New J. Phys. 14*, 125014.

Hambye, T., B. Hassanain, J. March-Russell, and M. Schvellinger (2007). Four-point functions and kaon decays in a minimal AdS/QCD model. *Phys. Rev. D76*, 125017.

Hambye, T., E. Ma, and U. Sarkar (2001). Supersymmetric triplet Higgs model of neutrino masses and leptogenesis. *Nucl. Phys. B602*, 23–38.

Hambye, T. and K. Riesselmann (1997). SM Higgs mass bounds from theory. `arXiv:hep-ph/9708416`.

Hamzaoui, C., M. Pospelov, and M. Toharia (1999). Higgs-mediated FCNC in supersymmetric models with large $\tan\beta$. *Phys. Rev. D59*, 095005.

Han, T. (2005). Collider phenomenology: basic knowledge and techniques. `arXiv:hep-ph/0508097`.

Han, T., P. Langacker, Z. Liu, and L.-T. Wang (2013). Diagnosis of a new neutral gauge boson at the LHC and ILC for Snowmass 2013. `arXiv:1308.2738 [hep-ph]`.

Han, Z. (2008). Effective theories and electroweak precision constraints. *Int. J. Mod. Phys. A23*, 2653–2685.

Han, Z. and W. Skiba (2005). Effective theory analysis of precision electroweak data. *Phys. Rev. D71*, 075009.

Hanneke, D., S. Fogwell, and G. Gabrielse (2008). New measurement of the electron magnetic moment and the fine structure constant. *Phys. Rev. Lett. 100*, 120801.

Hannestad, S., I. Tamborra, and T. Tram (2012). Thermalisation of light sterile neutrinos in the early universe. *JCAP 1207*, 025.

Harari, H. (1984). Composite models for quarks and leptons. *Phys. Rept. 104*, 159.

Hardy, J. C. and I. S. Towner (2015). Superallowed $0^+ \to 0^+$ nuclear β decays: 2014 critical survey, with precise results for V_{ud} and CKM unitarity. *Phys. Rev. C91*(2), 025501.

Harland-Lang, L. A., A. D. Martin, P. Motylinski, and R. S. Thorne (2015). Parton distributions in the LHC era: MMHT 2014 PDFs. *Eur. Phys. J. C75*(5), 204.

Harnew, N. (2016). Future flavour physics experiments. *Annalen Phys. 528*, 102–107.

Harrison, P. F., D. H. Perkins, and W. G. Scott (2002). Tri-bimaximal mixing and the neutrino oscillation data. *Phys. Lett. B530*, 167.

Hasenfratz, A., K. Jansen, C. B. Lang, T. Neuhaus, and H. Yoneyama (1987). The triviality bound of the four component ϕ^4 model. *Phys. Lett. B199*, 531.

Hasert, F. J. et al. (1973). Observation of neutrino-like interactions without muon or electron in the Gargamelle neutrino experiment. *Phys. Lett. B46*, 138–140.

Hata, N., S. Bludman, and P. Langacker (1994). Astrophysical solutions are incompatible with the solar neutrino data. *Phys. Rev. D49*, 3622–3625.

Haxton, W., R. Robertson, and A. M. Serenelli (2013). Solar neutrinos: status and prospects. *Ann. Rev. Astron. Astrophys. 51*, 21–61.

Haxton, W. C. and B. R. Holstein (2013). Hadronic parity violation. *Prog. Part. Nucl. Phys. 71*, 185–203.

Haxton, W. C., C. P. Liu, and M. J. Ramsey-Musolf (2002). Nuclear anapole moments. *Phys. Rev. C65*, 045502.

Haxton, W. C. and C. E. Wieman (2001). Atomic parity nonconservation and nuclear anapole moments. *Ann. Rev. Nucl. Part. Sci. 51*, 261–293.

Hayes, A., J. Friar, G. Garvey, G. Jungman, and G. Jonkmans (2014). Systematic Uncertainties in the Analysis of the Reactor Neutrino Anomaly. *Phys. Rev. Lett. 112*, 202501.

Hayes, A. C. and P. Vogel (2016). Reactor neutrino spectra. *Ann. Rev. Nucl. Part. Sci. 66*, 219–244.

Heckman, J. J. (2010). Particle physics implications of F-theory. *Ann. Rev. Nucl. Part. Sci. 60*, 237–265.

Heeck, J. and H. Zhang (2013). Exotic charges, multicomponent dark matter and light sterile neutrinos. *JHEP 1305*, 164.

Heinemeyer, S. et al. (2013). Handbook of LHC Higgs cross sections: 3. Higgs properties. `arXiv:1307.1347 [hep-ph]`.

Heinemeyer, S., W. Hollik, and G. Weiglein (2000). FeynHiggs: A program for the calculation of the masses of the neutral CP even Higgs bosons in the MSSM. *Comput. Phys. Commun. 124*, 76–89.

Heinemeyer, S., W. Hollik, and G. Weiglein (2006). Electroweak precision observables in the minimal supersymmetric standard model. *Phys. Rept. 425*, 265–368.

Herczeg, P. (2001). Beta decay beyond the standard model. *Prog. Part. Nucl. Phys. 46*, 413–457.

Herrero, M. (2015). The Higgs system in and beyond the standard model. *Springer Proc. Phys. 161*, 188–252.

Hewett, J. L. and T. G. Rizzo (1989). Low-energy phenomenology of superstring inspired $E(6)$ models. *Phys. Rept. 183*, 193.

Hewett, J. L. and M. Spiropulu (2002). Particle physics probes of extra spacetime dimensions. *Ann. Rev. Nucl. Part. Sci. 52*, 397–424.

Hey, A. J. G. and R. L. Kelly (1983). Baryon spectroscopy. *Phys. Rept. 96*, 71.

Higgs, P. W. (1964). Broken symmetries, massless particles and gauge fields. *Phys. Lett. 12*, 132–133.

Higgs, P. W. (1966). Spontaneous symmetry breakdown without massless bosons. *Phys. Rev. 145*, 1156–1163.

Hill, C. T., S. Pokorski, and J. Wang (2001). Gauge invariant effective Lagrangian for Kaluza-Klein modes. *Phys. Rev. D64*, 105005.

Hill, C. T. and E. H. Simmons (2003). Strong dynamics and electroweak symmetry breaking. *Phys. Rept. 381*, 235–402.

Hillairet, A. et al. (2012). Precision muon decay measurements and improved constraints on the weak interaction. *Phys. Rev. D85*, 092013.

Hinchliffe, I., F. E. Paige, M. D. Shapiro, J. Soderqvist, and W. Yao (1997). Precision SUSY measurements at CERN LHC. *Phys. Rev. D55*, 5520–5540.

Hinshaw, G. et al. (2013). Nine-year Wilkinson microwave anisotropy probe (WMAP) observations: cosmological parameter results. *Astrophys. J. Suppl. 208*, 19.

Hisano, J., T. Moroi, K. Tobe, and M. Yamaguchi (1996). Lepton flavor violation via right-handed neutrino Yukawa couplings in supersymmetric standard model. *Phys. Rev. D53*, 2442–2459.

Hogan, C. J. (2000). Why the universe is just so. *Rev. Mod. Phys. 72*, 1149–1161.

Holdom, B. (1986). Two $U(1)'$s and ϵ charge shifts. *Phys. Lett. B166*, 196.

Hooper, D. (2010). TASI 2008: Particle dark matter. `arXiv:0901.4090 [hep-ph]`.

Hu, W., R. Barkana, and A. Gruzinov (2000). Cold and fuzzy dark matter. *Phys. Rev. Lett. 85*, 1158–1161.

Huang, P., A. Joglekar, B. Li, and C. E. M. Wagner (2016). Probing the electroweak phase transition at the LHC. *Phys. Rev. D93*(5), 055049.

Huber, P. (2011). On the determination of anti-neutrino spectra from nuclear reactors. *Phys. Rev. C84*, 024617.

Huber, P., J. Kopp, M. Lindner, M. Rolinec, and W. Winter (2007). New features in the simulation of neutrino oscillation experiments with GLoBES 3.0: general long baseline experiment simulator. *Comput. Phys. Commun. 177*, 432–438.

Hui, L., J. P. Ostriker, S. Tremaine, and E. Witten (2017). On the hypothesis that cosmological dark matter is composed of ultra-light bosons. *Phys. Rev. D95*(4), 043541.

Hyde-Wright, C. E. and K. de Jager (2004). Electromagnetic form factors of the nucleon and Compton scattering. *Ann. Rev. Nucl. Part. Sci. 54*, 217–267.

Ibáñez, L. E. and G. G. Ross (1992). Discrete gauge symmetries and the origin of baryon and lepton number conservation in supersymmetric versions of the standard model. *Nucl. Phys. B368*, 3–37.

Ibáñez, L. E. and G. G. Ross (2007). Supersymmetric Higgs and radiative electroweak breaking. *Comptes Rendus Physique 8*, 1013–1028.

Ibáñez, L. E. and A. M. Uranga (2007). Neutrino Majorana masses from string theory instanton effects. *JHEP 03*, 052.

Ibrahim, T. and P. Nath (2008). CP violation from standard model to strings. *Rev. Mod. Phys. 80*, 577–631.

Inami, T. and C. S. Lim (1981). Effects of superheavy quarks and leptons in low-energy weak processes $K_L \to \mu\bar{\mu}$, $K^+ \to \pi^+\nu\bar{\nu}$, and $K^0 \leftrightarrow \bar{K}^0$. *Prog. Theor. Phys. 65*, 297.

Intriligator, K. A. and N. Seiberg (2007). Lectures on supersymmetry breaking. *Class. Quant. Grav. 24*, S741–S772.

Intriligator, K. A., N. Seiberg, and D. Shih (2006). Dynamical SUSY breaking in meta-stable vacua. *JHEP 04*, 021.

Isgur, N. and M. B. Wise (1989). Weak decays of heavy mesons in the static quark approximation. *Phys. Lett. B232*, 113.

Ishimori, H. et al. (2010). Non-abelian discrete symmetries in particle physics. *Prog. Theor. Phys. Suppl. 183*, 1–163.

Isidori, G., Y. Nir, and G. Perez (2010). Flavor physics constraints for physics beyond the standard model. *Ann. Rev. Nucl. Part. Sci. 60*, 355.

Isidori, G., G. Ridolfi, and A. Strumia (2001). On the metastability of the standard model vacuum. *Nucl. Phys. B609*, 387–409.

Jackiw, R. and C. Rebbi (1976). Vacuum periodicity in a Yang-Mills quantum theory. *Phys. Rev. Lett. 37*, 172–175.

Jackson, J. D., S. B. Treiman, and H. W. Wyld (1957). Possible tests of time reversal invariance in beta decay. *Phys. Rev. 106*, 517–521.

Jaeckel, J. and A. Ringwald (2010). The low-energy frontier of particle physics. *Ann. Rev. Nucl. Part. Sci. 60*, 405–437.

Jaffe, R. L. (2005). Exotica. *Phys. Rept. 409*, 1–45.

Jegerlehner, F. and A. Nyffeler (2009). The muon $g - 2$. *Phys. Rept. 477*, 1–110.

Jimenez-Delgado, P. and E. Reya (2014). Delineating parton distributions and the strong coupling. *Phys. Rev. D89*(7), 074049.

Joyce, A., L. Lombriser, and F. Schmidt (2016). Dark energy vs. modified gravity. *Ann. Rev. Nucl. Part. Sci. 66*, 95–122.

Jungman, G., M. Kamionkowski, and K. Griest (1996). Supersymmetric dark matter. *Phys. Rept. 267*, 195–373.

Kachru, S., R. Kallosh, A. Linde, and S. P. Trivedi (2003). De Sitter vacua in string theory. *Phys. Rev. D68*, 046005.

Kajita, T. (2014). The measurement of neutrino properties with atmospheric neutrinos. *Ann. Rev. Nucl. Part. Sci. 64*, 343–362.

Kajita, T., E. Kearns, and M. Shiozawa (2016). Establishing atmospheric neutrino oscillations with Super-Kamiokande. *Nucl. Phys. B908*, 14–29.

Kajiyama, Y., M. Raidal, and A. Strumia (2007). The golden ratio prediction for the solar neutrino mixing. *Phys. Rev. D76*, 117301.

Kane, G. L., C. F. Kolda, and J. D. Wells (1993). Calculable upper limit on the mass of the lightest Higgs boson in any perturbatively valid supersymmetric theory. *Phys. Rev. Lett. 70*, 2686–2689.

Kane, G. L., G. A. Ladinsky, and C. P. Yuan (1992). Using the top quark for testing standard-model polarization and CP predictions. *Phys. Rev. D 45*(1), 124–141.

Kane, G. L. and M. E. Peskin (1982). A constraint from B decay on models with no t quark. *Nucl. Phys. B195*, 29.

Kane, G. L., W. W. Repko, and W. B. Rolnick (1984). The effective $W^{\pm}$, Z^0 approximation for high-energy collisions. *Phys. Lett. B148*, 367–372.

Kang, J., P. Langacker, and B. D. Nelson (2008). Theory and phenomenology of exotic isosinglet quarks and squarks. *Phys. Rev. D77*, 035003.

Kang, J.-H., P. Langacker, and T. Li (2005). Neutrino masses in supersymmetric $SU(3)_C \times SU(2)_L \times U(1)_Y \times U(1)'$ models. *Phys. Rev. D71*, 015012.

Kaplan, D. B. and A. Manohar (1988). Strange matrix elements in the proton from neutral current experiments. *Nucl. Phys. B310*, 527.

Kaplan, D. B. and A. V. Manohar (1986). Current mass ratios of the light quarks. *Phys. Rev. Lett. 56*, 2004.

Kaplan, D. B., A. E. Nelson, and N. Weiner (2004). Neutrino oscillations as a probe of dark energy. *Phys. Rev. Lett. 93*, 091801.

Karshenboim, S. G. (2005). Precision physics of simple atoms: QED tests, nuclear structure and fundamental constants. *Phys. Rept. 422*, 1–63.

Katz, U. and C. Spiering (2012). High-energy neutrino astrophysics: status and perspectives. *Prog. Part. Nucl. Phys. 67*, 651–704.

Kawamura, Y. (2001). Triplet-doublet splitting, proton stability and extra dimension. *Prog. Theor. Phys. 105*, 999–1006.

Kawasaki, M. and K. Nakayama (2013). Axions: theory and cosmological role. *Ann. Rev. Nucl. Part. Sci. 63*, 69–95.

Kayser, B., G. T. Garvey, E. Fischbach, and S. P. Rosen (1974). Are neutrinos always lefthanded? *Phys. Lett. B52*, 385.

Kayser, B., J. Kopp, R. Roberston, and P. Vogel (2010). On a theory of neutrino oscillations with entanglement. *Phys. Rev. D82*, 093003.

Kennedy, D. C. and P. Langacker (1990). Precision electroweak experiments and heavy physics: a global analysis. *Phys. Rev. Lett. 65*, 2967–2970.

Khachatryan, V. et al. (2015a). Observation of the rare $B_s^0 \to \mu^+\mu^-$ decay from the combined analysis of CMS and LHCb data. *Nature 522*, 68–72.

Khachatryan, V. et al. (2015b). Search for lepton-flavour-violating decays of the Higgs boson. *Phys. Lett. B749*, 337–362.

Khachatryan, V. et al. (2016a). Measurement of the W boson helicity fractions in the decays of top quark pairs to lepton+jets final states produced in pp collisions at $\sqrt{s} = 8$ TeV. *Phys. Lett. B762*, 512–534.

Khachatryan, V. et al. (2016b). Phenomenological MSSM interpretation of CMS searches in pp collisions at $\sqrt{s} = 7$ and 8 TeV. *JHEP 10*, 129.

Khachatryan, V. et al. (2017). Search for narrow resonances in dilepton mass spectra in proton-proton collisions at $\sqrt{s} = 13$ TeV and combination with 8 TeV data. *Phys. Lett. B768*, 57–80.

Kibble, T. W. B. (1967). Symmetry breaking in non-abelian gauge theories. *Phys. Rev. 155*, 1554–1561.

Kilian, W., T. Ohl, and J. Reuter (2011). WHIZARD: Simulating multi-particle processes at LHC and ILC. *Eur. Phys. J. C71*, 1742.

Kim, J. E. and G. Carosi (2010). Axions and the strong CP problem. *Rev. Mod. Phys. 82*, 557–602.

Kim, J. E., P. Langacker, M. Levine, and H. H. Williams (1981). A theoretical and experimental review of the weak neutral current: a determination of its structure and limits on deviations from the minimal $SU(2) \times U(1)$ electroweak theory. *Rev. Mod. Phys. 53*, 211.

Kim, J. E. and H. P. Nilles (1984). The μ problem and the strong CP problem. *Phys. Lett. B138*, 150.

King, S. F. (2015). Models of neutrino mass, mixing and CP violation. *J. Phys. G42*, 123001.

King, S. F., A. Merle, S. Morisi, Y. Shimizu, and M. Tanimoto (2014). Neutrino mass and mixing: from theory to experiment. *New J. Phys. 16*, 045018.

King, S. F., S. Moretti, and R. Nevzorov (2006). Theory and phenomenology of an exceptional supersymmetric standard model. *Phys. Rev. D73*, 035009.

Kinoshita, T. and A. Sirlin (1957). Muon decay with parity nonconserving interactions and radiative corrections in the two-component theory. *Phys. Rev. 107*, 593–599.

Kinoshita, T. and A. Sirlin (1959). Radiative corrections to Fermi interactions. *Phys. Rev. 113*, 1652–1660.

Kitano, R., H. Ooguri, and Y. Ookouchi (2010). Supersymmetry breaking and gauge mediation. *Ann. Rev. Nucl. Part. Sci. 60*, 491–511.

Klapdor-Kleingrothaus, H. V. and I. V. Krivosheina (2006). The evidence for the observation of $\beta\beta_{0\nu}$ decay: the identification of $\beta\beta_{0\nu}$ events from the full spectra. *Mod. Phys. Lett. A21*, 1547–1566.

Klasen, M., M. Pohl, and G. Sigl (2015). Indirect and direct search for dark matter. *Prog. Part. Nucl. Phys. 85*, 1–32.

Klebanov, I. R. (2000). TASI lectures: introduction to the AdS/CFT correspondence. arXiv:hep-th/0009139.

Klempt, E. and J.-M. Richard (2010). Baryon spectroscopy. *Rev. Mod. Phys. 82*, 1095–1153.

Klempt, E. and A. Zaitsev (2007). Glueballs, hybrids, multiquarks. experimental facts versus QCD inspired concepts. *Phys. Rept. 454*, 1–202.

Klinkhamer, F. R. and N. S. Manton (1984). A saddle point solution in the Weinberg-Salam theory. *Phys. Rev. D30*, 2212.

Kluth, S. (2006). Tests of quantum chromo dynamics at e^+e^- colliders. *Rept. Prog. Phys. 69*, 1771–1846.

Kobayashi, M. and T. Maskawa (1973). CP violation in the renormalizable theory of weak interaction. *Prog. Theor. Phys. 49*, 652–657.

Kodama, K. et al. (2001). Observation of tau-neutrino interactions. *Phys. Lett. B504*, 218–224.

Komargodski, Z. and N. Seiberg (2009). μ and general gauge mediation. *JHEP 03*, 072.

Konopinski, E. J. and H. M. Mahmoud (1953). The universal Fermi interaction. *Phys. Rev. 92*, 1045–1049.

Kopp, J., P. A. N. Machado, M. Maltoni, and T. Schwetz (2013). Sterile neutrino oscillations: the global picture. *JHEP 1305*, 050.

Kostelecky, V. A. and N. Russell (2011). Data tables for Lorentz and CPT violation. *Rev. Mod. Phys. 83*, 11–31.

Kraan, A. C., J. B. Hansen, and P. Nevski (2007). Discovery potential of R-hadrons with the ATLAS detector. *Eur. Phys. J. C49*, 623–640.

Krauss, L. M. and S. Tremaine (1988). Test of the weak equivalence principle for neutrinos and photons. *Phys. Rev. Lett. 60*, 176.

Kronfeld, A. S. (2012). Twenty-first century lattice gauge theory: results from the QCD Lagrangian. *Ann. Rev. Nucl. Part. Sci. 62*, 265–284.

Kronfeld, A. S. and C. Quigg (2010). Resource letter: quantum chromodynamics. *Am. J. Phys. 78*, 1081–1116.

Kröninger, K., A. B. Meyer, and P. Uwer (2015). Top-quark physics at the LHC. In T. Schörner-Sadenius (Ed.), *The Large Hadron Collider: Harvest of Run 1*, pp. 259–300. arXiv:1506.02800 [hep-ex].

Kuipers, J., T. Ueda, J. A. M. Vermaseren, and J. Vollinga (2013). FORM version 4.0. *Comput. Phys. Commun. 184*, 1453–1467.

mar, K. S., S. Mantry, W. J. Marciano, and P. A. Souder (2013). Low energy measure-aents of the weak mixing angle. *Ann. Rev. Nucl. Part. Sci. 63*, 237–267.

no, Y. and Y. Okada (2001). Muon decay and physics beyond the standard model. *Rev. Mod. Phys. 73*, 151–202.

o, T.-K. and J. T. Pantaleone (1989). Neutrino oscillations in matter. *Rev. Mod. Phys. 61*, 37.

senko, A. (2009). Sterile neutrinos: The dark side of the light fermions. *Phys. Rept. 481*, –28.

senko, A., P. Langacker, and G. Segre (1996). Phase transitions and vacuum tunneling nto charge and color breaking minima in the MSSM. *Phys. Rev. D54*, 5824–5834.

ti, J., L. Lin, and Y. Shen (1988). Upper bound on the Higgs mass in the standard nodel. *Phys. Rev. Lett. 61*, 678.

zmin, V. A. (1970). CP-noninvariance and baryon asymmetry of universe. *JETP Lett. 12*, 28–230.

zmin, V. A., V. A. Rubakov, and M. E. Shaposhnikov (1985). On the anomalous elec-roweak baryon number nonconservation in the early universe. *Phys. Lett. B155*, 36.

ak, Z., M. Lewicki, and P. Olszewski (2014). Higher-order scalar interactions and SM acuum stability. *JHEP 05*, 119.

ndau, L. (1948). On the angular momentum of a two-photon system. *Dokl. Akad. Nauk Ser. Fiz. 60*, 207–209.

ngacker, P. (1977). The general treatment of second class currents in field theory. *Phys. Rev. D15*, 2386.

ngacker, P. (1981). Grand unified theories and proton decay. *Phys. Rept. 72*, 185.

ngacker, P. (1989a). Implications of recent $M_{Z,W}$ and neutral current measurements for he top quark mass. *Phys. Rev. Lett. 63*, 1920.

ngacker, P. (1989b). Is the standard model unique? *Comments Nucl. Part. Phys. 19*, 1.

ngacker, P. (1991). W and Z physics. In *TeV Physics*, ed T. Huang et al., Gordon and Breach, New York, 1991.

ngacker, P. (1993). Five phases of weak neutral current experiments from the perspective of a theorist. `arXiv:hep-ph/9305255`.

ngacker, P. (1998). A mechanism for ordinary-sterile neutrino mixing. *Phys. Rev. D58*, 093017.

ngacker, P. (2005). Neutrino physics. *Int. J. Mod. Phys. A20*, 5254–5265.

ngacker, P. (2009). The physics of heavy Z' gauge bosons. *Rev. Mod. Phys. 81*, 1199–1228.

ngacker, P. (2012). Neutrino masses from the top down. *Ann. Rev. Nucl. Part. Sci. 62*, 215–235.

Langacker, P., J. P. Leveille, and J. Sheiman (1983). On the detection of cosmological neutrinos by coherent scattering. *Phys. Rev. D27*, 1228.

Langacker, P. and D. London (1988a). Lepton number violation and massless nonorthogonal neutrinos. *Phys. Rev. D38*, 907.

Langacker, P. and D. London (1988b). Mixing between ordinary and exotic fermions. *Phys. Rev. D38*, 886.

Langacker, P. and D. London (1989). Analysis of muon decay with lepton number nonconserving interactions. *Phys. Rev. D39*, 266.

Langacker, P. and M.-X. Luo (1991). Implications of precision electroweak experiments for m_t, ρ_0, $\sin^2\theta_W$ and grand unification. *Phys. Rev. D44*, 817–822.

Langacker, P., M.-X. Luo, and A. K. Mann (1992). High precision electroweak experiments: a global search for new physics beyond the standard model. *Rev. Mod. Phys. 64*, 87–192.

Langacker, P. and H. Pagels (1973). Nonrenormalization theorem in the chiral symmetry limit. *Phys. Rev. Lett. 30*, 630–633.

Langacker, P. and H. Pagels (1979). Light quark mass spectrum in quantum chromodynamics. *Phys. Rev. D19*, 2070.

Langacker, P., G. Paz, L.-T. Wang, and I. Yavin (2007). A T-odd observable sensitive to CP violating phases in squark decay. *JHEP 07*, 055.

Langacker, P., G. Paz, and I. Yavin (2009). Scalar potentials and accidental symmetries in supersymmetric $U(1)'$ models. *Phys. Lett. B671*, 245–249.

Langacker, P. and S.-Y. Pi (1980). Magnetic monopoles in grand unified theories. *Phys. Rev. Lett. 45*, 1.

Langacker, P. and M. Plumacher (2000). Flavor changing effects in theories with a heavy Z' boson with family non-universal couplings. *Phys. Rev. D62*, 013006.

Langacker, P. and N. Polonsky (1993). Uncertainties in coupling constant unification. *Phys. Rev. D47*, 4028–4045.

Langacker, P. and N. Polonsky (1994). The bottom mass prediction in supersymmetric grand unification: uncertainties and constraints. *Phys. Rev. D49*, 1454–1467.

Langacker, P. and N. Polonsky (1995). The strong coupling, unification, and recent data. *Phys. Rev. D52*, 3081–3086.

Langacker, P., R. W. Robinett, and J. L. Rosner (1984). New heavy gauge bosons in pp and $p\bar{p}$ collisions. *Phys. Rev. D30*, 1470.

Langacker, P. and S. Uma Sankar (1989). Bounds on the mass of W_R and the $W_L - W_R$ mixing angle ξ in general $SU(2)_L \times SU(2)_R \times U(1)$ models. *Phys. Rev. D40*, 1569–1585.

Langacker, P. and J. Wang (1998). Neutrino anti-neutrino transitions. *Phys. Rev. D58*, 093004.

Lazauskas, R., P. Vogel, and C. Volpe (2008). Charged current cross section for massive cosmological neutrinos impinging on radioactive nuclei. *J. Phys. G35*, 025001.

Leader, E. and C. Lorcé (2014). The angular momentum controversy: what's it all about and does it matter? *Phys. Rept. 541*(3), 163–248.

Lebedev, O. et al. (2008). The heterotic road to the MSSM with R parity. *Phys. Rev. D77*, 046013.

Lebrun, P. et al. (2012). The CLIC programme: towards a staged e^+e^- linear collider exploring the terascale: CLIC conceptual design report. `arXiv:1209.2543 [physics.ins-det]`.

Lee, B. W., C. Quigg, and H. B. Thacker (1977). Weak interactions at very high-energies: the role of the Higgs boson mass. *Phys. Rev. D16*, 1519.

Lee, B. W. and R. E. Shrock (1977). Natural suppression of symmetry violation in gauge theories: muon-lepton and electron lepton number nonconservation. *Phys. Rev. D16*, 1444.

Lee, B. W. and J. Zinn-Justin (1972). Spontaneously broken gauge symmetries. 1. Preliminaries. *Phys. Rev. D5*, 3121–3137.

Lee, B. W. and J. Zinn-Justin (1973). Spontaneously broken gauge symmetries. 4. General gauge formulation. *Phys. Rev. D7*, 1049–1056.

Lee, J. S., M. Carena, J. Ellis, A. Pilaftsis, and C. E. M. Wagner (2013). CPsuperH2.3: an updated tool for phenomenology in the MSSM with explicit CP violation. *Comput. Phys. Commun. 184*, 1220–1233.

Lee, T. D. and C.-N. Yang (1956). Question of parity conservation in weak interactions. *Phys. Rev. 104*, 254–258.

Lees, J. P. et al. (2012). Observation of time reversal violation in the B^0 meson system. *Phys. Rev. Lett. 109*, 211801.

Leike, A. (1999). The phenomenology of extra neutral gauge bosons. *Phys. Rept. 317*, 143–250.

Lenz, A. and U. Nierste (2011). Numerical updates of lifetimes and mixing parameters of B mesons. `arXiv:1102.4274 [hep-ph]`.

Lepage, G. P. and S. J. Brodsky (1980). Exclusive processes in perturbative quantum chromodynamics. *Phys. Rev. D22*, 2157.

Lepage, G. P., P. B. Mackenzie, and M. E. Peskin (2014). Expected precision of Higgs boson partial widths within the standard model. `arXiv:1404.0319 [hep-ph]`.

Lesgourgues, J. and S. Pastor (2012). Neutrino mass from cosmology. *Adv. High Energy Phys. 2012*, 608515.

Leutwyler, H. (2014). On the history of the strong interaction. *Mod. Phys. Lett. A29*, 1430023.

Leutwyler, H. and M. Roos (1984). Determination of the elements $V(us)$ and $V(ud)$ of the Kobayashi-Maskawa matrix. *Z. Phys. C25*, 91.

Li, L.-F. (1974). Group theory of the spontaneously broken gauge symmetries. *Phys. Rev. D9*, 1723–1739.

Liberati, S. (2013). Tests of Lorentz invariance: a 2013 update. *Class. Quant. Grav. 30*, 133001.

Lim, C.-S. and W. J. Marciano (1988). Resonant spin-flavor precession of solar and supernova neutrinos. *Phys. Rev. D37*, 1368.

Linde, A. (2017). A brief history of the multiverse. *Rept. Prog. Phys. 80*(2), 022001.

Linde, A. D. (1983). Decay of the false vacuum at finite temperature. *Nucl. Phys. B216*, 421.

Linde, A. D. (1986). Eternal chaotic inflation. *Mod. Phys. Lett. A1*, 81.

Linde, A. D. (2008). Inflationary cosmology. *Lect. Notes Phys. 738*, 1–54.

Lipkin, H. J. (1984). The theoretical basis and phenomenology of the OZI rule. *Nucl. Phys. B244*, 147.

Lisanti, M. (2016). TASI 2015: Lectures on dark matter physics. arXiv:1603.03797 [hep-ph].

Llewellyn Smith, C. H. (1983). On the determination of $\sin^2\theta_W$ in semileptonic neutrino interactions. *Nucl. Phys. B228*, 205.

Logan, H. E. (2014). TASI 2013 lectures on Higgs physics within and beyond the standard model. arXiv:1406.1786 [hep-ph].

Long, A. J., C. Lunardini, and E. Sabancilar (2014). Detecting non-relativistic cosmic neutrinos by capture on tritium: phenomenology and physics potential. *JCAP 1408*, 038.

Longo, M. J. (1988). New precision tests of the Einstein equivalence principle from SN1987A. *Phys. Rev. Lett. 60*, 173.

Lu, X. and H. Murayama (2014). Neutrino mass anarchy and the universe. *JHEP 1408*, 101.

Ludhova, L. and S. Zavatarelli (2013). Studying the Earth with geoneutrinos. arXiv:1310.3961 [hep-ex].

Luhn, C. and P. Ramond (2008). Anomaly conditions for non-abelian finite family symmetries. *JHEP 07*, 085.

Luo, M.-x., H.-w. Wang, and Y. Xiao (2003). Two loop renormalization group equations in general gauge field theories. *Phys. Rev. D67*, 065019.

Luscher, M. and P. Weisz (1989). Scaling laws and triviality bounds in the lattice ϕ^4 theory. 3. N component model. *Nucl. Phys. B318*, 705.

Luty, M. A. (2005). 2004 TASI lectures on supersymmetry breaking. arXiv:hep-th/0509029.

Lykken, J. D. (1996a). Introduction to supersymmetry. arXiv:hep-th/9612114.

Lykken, J. D. (1996b). Weak scale superstrings. *Phys. Rev. D54*, 3693–3697.

Lyth, D. H. and A. Riotto (1999). Particle physics models of inflation and the cosmological density perturbation. *Phys. Rept. 314*, 1–146.

Ma, E. (1996). Neutrino masses in an extended gauge model with $E(6)$ particle content. *Phys. Lett. B380*, 286–290.

Ma, E. (2004). $A(4)$ origin of the neutrino mass matrix. *Phys. Rev. D70*, 031901.

Ma, E. (2009). Neutrino mass: mechanisms and models. `arXiv:0905.0221 [hep-ph]`.

Ma, E. and U. Sarkar (1998). Neutrino masses and leptogenesis with heavy Higgs triplets. *Phys. Rev. Lett. 80*, 5716–5719.

Maalampi, J. and M. Roos (1990). Physics of mirror fermions. *Phys. Rept. 186*, 53.

Machleidt, R. and D. R. Entem (2011). Chiral effective field theory and nuclear forces. *Phys. Rept. 503*, 1–75.

Maiezza, A., M. Nemevsek, F. Nesti, and G. Senjanovic (2010). Left-right symmetry at LHC. *Phys. Rev. D82*, 055022.

Maki, Z., M. Nakagawa, and S. Sakata (1962). Remarks on the unified model of elementary particles. *Prog. Theor. Phys. 28*, 870.

Maksymyk, I., C. P. Burgess, and D. London (1994). Beyond S, T and U. *Phys. Rev. D50*, 529–535.

Maldacena, J. M. (1998). The large N limit of superconformal field theories and supergravity. *Adv. Theor. Math. Phys. 2*, 231–252.

Maltoni, M. and A. Yu. Smirnov (2016). Solar neutrinos and neutrino physics. *Eur. Phys. J. A52*(4), 87.

Mangano, M. L., M. Moretti, F. Piccinini, R. Pittau, and A. D. Polosa (2003). ALPGEN, a generator for hard multiparton processes in hadronic collisions. *JHEP 07*, 001.

Mangano, M. L. and T. J. Stelzer (2005). Tools for the simulation of hard hadronic collisions. *Ann. Rev. Nucl. Part. Sci. 55*, 555–588.

Maniatis, M. (2010). The next-to-minimal supersymmetric extension of the standard model reviewed. *Int. J. Mod. Phys. A25*, 3505–3602.

Marciano, W. J. and J. L. Rosner (1990). Atomic parity violation as a probe of new physics. *Phys. Rev. Lett. 65*, 2963–2966.

Marciano, W. J. and A. I. Sanda (1977). Exotic decays of the muon and heavy leptons in gauge theories. *Phys. Lett. B67*, 303.

Marciano, W. J. and A. Sirlin (1980). Radiative corrections to neutrino induced neutral current phenomena in the $SU(2) \times U(1)$ theory. *Phys. Rev. D22*, 2695.

Marciano, W. J. and A. Sirlin (1981). Precise $SU(5)$ predictions for $\sin^2\theta_W$, $m(W)$ and $m(Z)$. *Phys. Rev. Lett. 46*, 163.

Marsh, D. J. E. (2016). Axion cosmology. *Phys. Rept. 643*, 1–79.

Martin, S. P. (1997). A supersymmetry primer. `arXiv:hep-ph/9709356 [hep-ph]`, [Adv. Ser. Direct. High Energy Phys. 18, 1 (1998)].

Martin, S. P. and M. T. Vaughn (1994). Two loop renormalization group equations for soft supersymmetry breaking couplings. *Phys. Rev. D50*, 2282.

Masiero, A., D. V. Nanopoulos, and A. I. Sanda (1986). Observable physics from superstring exotic particles: small Dirac neutrino masses. *Phys. Rev. Lett. 57*, 663–666.

Masiero, A., S. K. Vempati, and O. Vives (2004). Massive neutrinos and flavour violation. *New J. Phys. 6*, 202.

Mason, D. et al. (2007). Measurement of the nucleon strange-antistrange asymmetry at next-to-leading order in QCD from NuTeV dimuon data. *Phys. Rev. Lett. 99*, 192001.

Mateu, V. and A. Pich (2005). $V(us)$ determination from hyperon semileptonic decays. *JHEP 10*, 041.

Meade, P., N. Seiberg, and D. Shih (2009). General gauge mediation. *Prog. Theor. Phys. Suppl. 177*, 143–158.

Mele, S. (2006). Measurements of the running of the electromagnetic coupling at LEP. arXiv:hep-ex/0610037.

Melnikov, K. and T. Ritbergen (2000). The three-loop relation between the MS-bar and the pole quark masses. *Phys. Lett. B482*, 99–108.

Melnitchouk, W., R. Ent, and C. Keppel (2005). Quark-hadron duality in electron scattering. *Phys. Rept. 406*, 127–301.

Melzer-Pellmann, I. and P. Pralavorio (2014). Lessons for SUSY from the LHC after the first run. *Eur. Phys. J. C74*, 2801.

Mention, G. et al. (2011). The reactor antineutrino anomaly. *Phys. Rev. D83*, 073006.

Merle, A. (2013). keV neutrino model building. *Int. J. Mod. Phys. D22*, 1330020.

Metz, A. and A. Vossen (2016). Parton fragmentation functions. *Prog. Part. Nucl. Phys. 91*, 136–202.

Michel, L. (1950). Interaction between four half spin particles and the decay of the μ meson. *Proc. Phys. Soc. A63*, 514–531.

Mihara, S., J. P. Miller, P. Paradisi, and G. Piredda (2013). Charged lepton flavor-violation experiments. *Ann. Rev. Nucl. Part. Sci. 63*, 531–552.

Mikheyev, S. P. and A. Y. Smirnov (1985). Resonance enhancement of oscillations in matter and solar neutrino spectroscopy. *Sov. J. Nucl. Phys. 42*, 913–917.

Miller, J. P., E. de Rafael, and B. L. Roberts (2007). Muon $g-2$: review of theory and experiment. *Rept. Prog. Phys. 70*, 795.

Minakata, H. and A. Y. Smirnov (2004). Neutrino mixing and quark-lepton complementarity. *Phys. Rev. D70*, 073009.

Minkowski, P. (1977). $\mu \to e\gamma$ at a rate of one out of 1-billion muon decays? *Phys. Lett. B67*, 421.

Miranda, O. G., M. A. Tortola, and J. W. F. Valle (2006). Are solar neutrino oscillations robust? *JHEP 10*, 008.

Mirizzi, A. et al. (2016). Supernova neutrinos: production, oscillations and detection. *Riv. Nuovo Cim. 39*(1-2), 1–112.

Mishra, S. R. and F. Sciulli (1989). Deep inelastic lepton-nucleon scattering. *Ann. Rev. Nucl. Part. Sci. 39*, 259–310.

Misiak, M., S. Pokorski, and J. Rosiek (1998). Supersymmetry and FCNC effects. *Adv. Ser. Direct. High Energy Phys. 15*, 795–828.

Mnich, J. (1996). Experimental tests of the standard model in $e^+e^- \to f\bar{f}$ at the Z resonance. *Phys. Rept. 271*, 181–266, copyright 1996, with permission from Elsevier.

Mohapatra, R. N. et al. (2007). Theory of neutrinos: a white paper. *Rept. Prog. Phys. 70*, 1757–1867.

Mohapatra, R. N. and J. C. Pati (1975). Left-right gauge symmetry and an isoconjugate model of CP violation. *Phys. Rev. D11*, 566–571.

Mohapatra, R. N. and G. Senjanovic (1980). Neutrino mass and spontaneous parity nonconservation. *Phys. Rev. Lett. 44*, 912.

Mohapatra, R. N. and A. Y. Smirnov (2006). Neutrino mass and new physics. *Ann. Rev. Nucl. Part. Sci. 56*, 569–628.

Mohapatra, R. N. and J. W. F. Valle (1986). Neutrino mass and baryon number nonconservation in superstring models. *Phys. Rev. D34*, 1642.

Mohr, P. J., B. N. Taylor, and D. B. Newell (2012). CODATA recommended values of the fundamental physical constants: 2010. *Rev. Mod. Phys. 84*, 1527–1605, and 2014 update in `arXiv:1507.07956 [physics.atom-ph]`.

Moretti, S., L. Lonnblad, and T. Sjöstrand (1998). New and old jet clustering algorithms for electron - positron events. *JHEP 08*, 001.

Mori, T. et al. (1989). Measurements of the e^+e^- total hadronic cross-section and a determination of m_Z and Λ_{MS}. *Phys. Lett. B218*, 499.

Morrissey, D. E., T. Plehn, and T. M. P. Tait (2012). Physics searches at the LHC. *Phys. Rept. 515*, 1–113.

Morrissey, D. E. and M. J. Ramsey-Musolf (2012). Electroweak baryogenesis. *New J. Phys. 14*, 125003.

Murray, W. and V. Sharma (2015). Properties of the Higgs boson discovered at the Large Hadron Collider. *Ann. Rev. Nucl. Part. Sci. 65*, 515–554.

Nakaya, T. and R. K. Plunkett (2016). Neutrino oscillations with the MINOS, MINOS+, T2K, and NOνA experiments. *New J. Phys. 18*(1), 015009.

Nambu, Y. (1960). Axial vector current conservation in weak interactions. *Phys. Rev. Lett. 4*, 380–382.

Nambu, Y. and M. Y. Han (1974). Three triplets, paraquarks, and colored quarks. *Phys. Rev. D10*, 674–683.

Nambu, Y. and G. Jona-Lasinio (1961). Dynamical model of elementary particles based on an analogy with superconductivity. I. *Phys. Rev. 122*, 345–358.

Nath, P. et al. (2010). The hunt for new physics at the Large Hadron Collider. *Nucl. Phys. Proc. Suppl. 200-202*, 185–417.

Nath, P. and P. Fileviez Perez (2007). Proton stability in grand unified theories, in strings, and in branes. *Phys. Rept. 441*, 191–317.

Nelson, A. E. (1984). Calculation of θ Barr. *Phys. Lett. B143*, 165.

Neubert, M. (1994). Heavy quark symmetry. *Phys. Rept. 245*, 259–396.

Neubert, M. (2005). Effective field theory and heavy quark physics. `arXiv:hep-ph/0512222`.

Neveu, A. and J. H. Schwarz (1971). Factorizable dual model of pions. *Nucl. Phys. B31*, 86–112.

Nielsen, H. B. and P. Olesen (1973). Vortex line models for dual strings. *Nucl. Phys. B61*, 45–61.

Nilles, H. P. (1984). Supersymmetry, supergravity and particle physics. *Phys. Rept. 110*, 1.

Nir, Y. (2005). CP violation in meson decays. `arXiv:hep-ph/0510413`.

Nir, Y. (2015). Flavour physics and CP violation. `arXiv:1605.00433 [hep-ph]`.

Nisius, R. (2000). The Photon structure from deep inelastic electron photon scattering. *Phys. Rept. 332*, 165–317.

Novikov, V. A., L. B. Okun, and M. I. Vysotsky (1993). On the electroweak one loop corrections. *Nucl. Phys. B397*, 35–83.

Nunokawa, H., S. J. Parke, and J. W. F. Valle (2008). CP violation and neutrino oscillations. *Prog. Part. Nucl. Phys. 60*, 338–402.

Oakes, R. J. (1969). $SU(2) \times SU(2)$ breaking and the Cabibbo angle. *Phys. Lett. B29*, 683–685.

Ochs, W. (2013). The status of glueballs. *J. Phys. G40*, 043001.

Ohlsson, T. (2013). Status of non-standard neutrino interactions. *Rept. Prog. Phys. 76*, 044201.

Ohnishi, A. (2012). Phase diagram and heavy-ion collisions: Overview. *Prog. Theor. Phys. Suppl. 193*, 1–10.

Okada, Y., M. Yamaguchi, and T. Yanagida (1991). Renormalization group analysis on the Higgs mass in the softly broken supersymmetric standard model. *Phys. Lett. B262*, 54–58.

Okubo, S. (1958). Decay of the Σ^+ hyperon and its antiparticle. *Phys. Rev. 109*, 984–985.

Okubo, S. (1977). Gauge groups without triangular anomaly. *Phys. Rev. D16*, 3528.

Olsen, S. L. (2015). XYZ meson spectroscopy. `arXiv:1511.01589 [hep-ex]`.

O'Raifeartaigh, L. (1975). Spontaneous symmetry breaking for chiral scalar superfields. *Nucl. Phys. B96*, 331.

Ostrovskiy, I. and K. O'Sullivan (2016). Search for neutrinoless double beta decay. *Mod. Phys. Lett. A31*(18), 1630017.

Otten, E. W. and C. Weinheimer (2008). Neutrino mass limit from tritium beta decay. *Rept. Prog. Phys. 71*, 086201.

Pacetti, S., R. Baldini Ferroli, and E. Tomasi-Gustafsson (2014). Proton electromagnetic form factors: basic notions, present achievements and future perspectives. *Phys. Rept. 550-551*, 1–103.

Pagels, H. (1975). Departures from chiral symmetry. *Phys. Rept. 16*, 219.

Paige, F. E., S. D. Protopopescu, H. Baer, and X. Tata (2003). ISAJET 7.69: a Monte Carlo event generator for pp, $\bar{p}p$, and e^+e^- reactions. arXiv:hep-ph/0312045.

Palazzo, A. (2013). Phenomenology of light sterile neutrinos: a brief review. *Mod. Phys. Lett. A28*, 1330004.

Panagiotakopoulos, C. and K. Tamvakis (1999a). New minimal extension of MSSM. *Phys. Lett. B469*, 145–148.

Panagiotakopoulos, C. and K. Tamvakis (1999b). Stabilized NMSSM without domain walls. *Phys. Lett. B446*, 224–227.

Pancheri, G. and Y. N. Srivastava (2017). Introduction to the physics of the total cross-section at LHC: a review of data and models. *Eur. Phys. J. C77*(3), 150.

Panico, G. and A. Wulzer (2016). The composite Nambu-Goldstone Higgs. *Lect. Notes Phys. 913*, pp.1–316.

Panofsky, W. K. H. and M. Breidenbach (1999). Accelerators and detectors. *Rev. Mod. Phys. 71*, S121–S132.

Parke, S. J. and T. R. Taylor (1986). An amplitude for n gluon scattering. *Phys. Rev. Lett. 56*, 2459.

Päs, H. and W. Rodejohann (2015). Neutrinoless double beta decay. *New J. Phys. 17*(11), 115010.

Paschos, E. A. and L. Wolfenstein (1973). Tests for neutral currents in neutrino reactions. *Phys. Rev. D7*, 91–95.

Pascoli, S., S. T. Petcov, and W. Rodejohann (2002). On the CP violation associated with Majorana neutrinos and neutrinoless double-beta decay. *Phys. Lett. B549*, 177–193.

Pati, J. C. (2006). Grand unification as a bridge between string theory and phenomenology. *Int. J. Mod. Phys. D15*, 1677–1698.

Pati, J. C. and A. Salam (1974). Lepton number as the fourth color. *Phys. Rev. D10*, 275–289.

Patrignani, C. (2016). Review of particle physics. *Chin. Phys. C40*(10), 100001, http://pdg.lbl.gov.

Patrignani, C., T. K. Pedlar, and J. L. Rosner (2013). Recent results in bottomonium. *Ann. Rev. Nucl. Part. Sci. 63*, 21–44.

Patrizii, L. and M. Spurio (2015). Status of searches for magnetic monopoles. *Ann. Rev. Nucl. Part. Sci. 65*, 279–302.

Patt, B. and F. Wilczek (2006). Higgs-field portal into hidden sectors. `arXiv:hep-ph/0605188 [hep-ph]`.

Patterson, R. B. (2015). Prospects for measurement of the neutrino mass hierarchy. *Ann. Rev. Nucl. Part. Sci. 65*, 177–192.

Peccei, R. D. (2008). The strong CP problem and axions. *Lect. Notes Phys. 741*, 3–17.

Peccei, R. D. and H. R. Quinn (1977). CP conservation in the presence of instantons. *Phys. Rev. Lett. 38*, 1440–1443.

Peebles, P. J. E. and B. Ratra (2003). The cosmological constant and dark energy. *Rev. Mod. Phys. 75*, 559–606.

Pendleton, B. and G. G. Ross (1981). Mass and mixing angle predictions from infrared fixed points. *Phys. Lett. B98*, 291.

Perelstein, M. (2007). Little Higgs models and their phenomenology. *Prog. Part. Nucl. Phys. 58*, 247–291.

Perez, E. and E. Rizvi (2013). The quark and gluon structure of the proton. *Rep. Prog. Phys. 76*, 046201.

Perl, M. L. et al. (1975). Evidence for anomalous lepton production in e^+e^- annihilation. *Phys. Rev. Lett. 35*, 1489–1492.

Peskin, M. E. (2008). Supersymmetry in elementary particle physics. `arXiv:0801.1928 [hep-ph]`.

Peskin, M. E. and T. Takeuchi (1990). A new constraint on a strongly interacting Higgs sector. *Phys. Rev. Lett. 65*, 964–967.

Peskin, M. E. and T. Takeuchi (1992). Estimation of oblique electroweak corrections. *Phys. Rev. D46*, 381–409.

Peskin, M. E. and J. D. Wells (2001). How can a heavy Higgs boson be consistent with the precision electroweak measurements? *Phys. Rev. D64*, 093003.

Phillips, II, D. G. et al. (2016). Neutron-antineutron oscillations: theoretical status and experimental prospects. *Phys. Rept. 612*, 1–45.

Pich, A. (1995). Chiral perturbation theory. *Rept. Prog. Phys. 58*, 563–610.

Pich, A. (1998). Effective field theory. `arXiv:hep-ph/9806303`.

Pich, A. (1999). Particle physics summer school, Triest: Aspects of quantum chromodynamics. `arXiv:hep-ph/0001118 [hep-ph]`.

Pich, A. (2014). Precision tau physics. *Prog. Part. Nucl. Phys. 75*, 41–85.

Pierce, D. M., J. A. Bagger, K. T. Matchev, and R.-J. Zhang (1997). Precision corrections in the minimal supersymmetric standard model. *Nucl. Phys. B491*, 3–67.

Pocanic, D. et al. (2004). Precise measurement of the $\pi^+ \to \pi^0 e^+ \nu$ branching ratio. *Phys. Rev. Lett. 93*, 181803.

Pohl, R. et al. (2010). The size of the proton. *Nature 466*, 213–216.

Pohl, R., R. Gilman, G. A. Miller, and K. Pachucki (2013). Muonic hydrogen and the proton radius puzzle. *Ann. Rev. Nucl. Part. Sci. 63*, 175–204.

Polchinski, J. (2015). Brane/antibrane dynamics and KKLT stability. `arXiv:1509.05710 [hep-th]`.

Politzer, H. D. (1973). Reliable perturbative results for strong interactions? *Phys. Rev. Lett. 30*, 1346–1349.

Pontecorvo, B. (1968). Neutrino experiments and the question of leptonic-charge conservation. *Sov. Phys. JETP 26*, 984–988.

Ponton, E. (2013). TASI 2011: Four lectures on TeV scale extra dimensions. `arXiv:1207.3827 [hep-ph]`.

Porod, W. and F. Staub (2012). SPheno 3.1: Extensions including flavour, CP-phases and models beyond the MSSM. *Comput. Phys. Commun. 183*, 2458–2469.

Porod, W., F. Staub, and A. Vicente (2014). A flavor kit for BSM models. *Eur. Phys. J. C74*(8), 2992.

Porter, F. C. (2016). Experimental status of the CKM matrix. *Prog. Part. Nucl. Phys. 91*, 101–135.

Pospelov, M. (2009). Secluded $U(1)$ below the weak scale. *Phys. Rev. D80*, 095002.

Pospelov, M. and A. Ritz (2005). Electric dipole moments as probes of new physics. *Annals Phys. 318*, 119–169.

Prades, J., E. de Rafael, and A. Vainshtein (2009). The hadronic light-by-light scattering contribution to the muon and electron anomalous magnetic moments. *Adv. Ser. Direct. High Energy Phys. 20*, 303–317.

Prescott, C. Y. et al. (1979). Further measurements of parity nonconservation in inelastic electron scattering. *Phys. Lett. B84*, 524.

Preskill, J. (1979). Cosmological production of superheavy magnetic monopoles. *Phys. Rev. Lett. 43*, 1365.

Preskill, J. (1984). Magnetic monopoles. *Ann. Rev. Nucl. Part. Sci. 34*, 461–530.

Pulido, J. (1992). The solar neutrino problem and the neutrino magnetic moment. *Phys. Rept. 211*, 167–199.

Punjabi, V. et al. (2015). The structure of the nucleon: elastic electromagnetic form factors. *Eur. Phys. J. A51*, 79.

Qian, X. and P. Vogel (2015). Neutrino mass hierarchy. *Prog. Part. Nucl. Phys. 83*, 1–30.

Qian, X. and W. Wang (2014). Reactor neutrino experiments: θ_{13} and beyond. *Mod. Phys. Lett. A29*, 1430016.

Quigg, C. (2011). LHC physics potential vs. energy: considerations for the 2011 run. arXiv:1101.3201 [hep-ph].

Quigg, C. (2015). Electroweak symmetry breaking in historical perspective. *Ann. Rev. Nucl. Part. Sci. 65*, 25–42.

Quiros, M. (1999). Finite temperature field theory and phase transitions. arXiv:hep-ph/9901312.

Raby, S. (2009). SUSY GUT model building. *Eur. Phys. J. C59*, 223–247.

Raby, S. (2011). Searching for the standard model in the string landscape: SUSY GUTs. *Rept. Prog. Phys. 74*, 036901.

Raffelt, G. G. (1999). Particle physics from stars. *Ann. Rev. Nucl. Part. Sci. 49*, 163–216.

Raidal, M. (2004). Relation between the neutrino and quark mixing angles and grand unification. *Phys. Rev. Lett. 93*, 161801.

Rakshit, S. (2004). Neutrino masses and R-parity violation. *Mod. Phys. Lett. A19*, 2239–2258.

Ramond, P. (1971). Dual theory for free fermions. *Phys. Rev. D3*, 2415–2418.

Ramond, P. (1979). The family group in grand unified theories. arXiv:hep-ph/9809459.

Ramsey-Musolf, M. J. and S. A. Page (2006). Hadronic parity violation: a new view through the looking glass. *Ann. Rev. Nucl. Part. Sci. 56*, 1–52.

Ramsey-Musolf, M. J. and S. Su (2008). Low energy precision test of supersymmetry. *Phys. Rept. 456*, 1–88.

Randall, L. and R. Sundrum (1999). A large mass hierarchy from a small extra dimension. *Phys. Rev. Lett. 83*, 3370–3373.

Ringwald, A. (2009). Prospects for the direct detection of the cosmic neutrino background. *Nucl. Phys. A827*, 501C–506C.

Ringwald, A. and Y. Y. Y. Wong (2004). Gravitational clustering of relic neutrinos and implications for their detection. *JCAP 0412*, 005.

Robens, T. and T. Stefaniak (2015). Status of the Higgs singlet extension of the standard model after LHC Run 1. *Eur. Phys. J. C75*, 104.

Roberts, B. M., V. A. Dzuba, and V. V. Flambaum (2015). Parity and time-reversal violation in atomic systems. *Ann. Rev. Nucl. Part. Sci. 65*, 63–86.

Robinett, R. W. and J. L. Rosner (1982a). Mass scales in grand unified theories. *Phys. Rev. D26*, 2396.

Robinett, R. W. and J. L. Rosner (1982b). Prospects for a second neutral vector boson at low mass in $SO(10)$. *Phys. Rev. D25*, 3036.

Rodejohann, W. (2011). Neutrino-less double beta decay and particle physics. *Int. J. Mod. Phys. E20*, 1833–1930.

Rosner, J. L. (1999). The arrival of charm. *AIP Conf. Proc. 459*, 9–27.

Rosner, J. L. (2007). Hadron spectroscopy: theory and experiment. *J. Phys. G34*, S127–S148.

Rosner, J. L., S. Stone, and R. S. Van de Water (2015). Leptonic decays of charged pseudoscalar mesons - 2015. `arXiv:1509.02220 [hep-ph]`.

Rovelli, C. (2011). Zakopane lectures on loop gravity. In *3rd Quantum Geometry and Quantum Gravity School*, Zakopane, Poland. *PoS QGQGS2011*, 003.

Ryd, A. and A. A. Petrov (2012). Hadronic D and D_s meson decays. *Rev. Mod. Phys. 84*, 65–117.

Saakyan, R. (2013). Two-neutrino double-beta decay. *Ann. Rev. Nucl. Part. Sci. 63*(1), 503–529.

Sakai, N. and T. Yanagida (1982). Proton decay in a class of supersymmetric grand unified models. *Nucl. Phys. B197*, 533.

Sakharov, A. D. (1967). Violation of CP invariance, C asymmetry, and baryon asymmetry of the universe. *JETP Lett. 5*, 24–27.

Salam, A. (1968). Weak and electromagnetic interactions. In *Elementary Particle Theory*, ed. N. Svartholm (Almquist and Wiksells, Stockholm 1968), 367-377.

Salam, A. and J. A. Strathdee (1974). Supergauge transformations. *Nucl. Phys. B76*, 477–482.

Salam, G. P. (2010a). Elements of QCD for hadron colliders. `arXiv:1011.5131 [hep-ph]`.

Salam, G. P. (2010b). Towards jetography. *Eur. Phys. J. C67*, 637–686.

Sapeta, S. (2016). QCD and jets at hadron colliders. *Prog. Part. Nucl. Phys. 89*, 1–55.

Sarantakos, S., A. Sirlin, and W. J. Marciano (1983). Radiative corrections to neutrino-lepton scattering in the $SU(2) \times U(1)$ theory. *Nucl. Phys. B217*, 84.

Sayre, J., S. Wiesenfeldt, and S. Willenbrock (2005). Sterile neutrinos and global symmetries. *Phys. Rev. D72*, 015001.

Schabinger, R. and J. D. Wells (2005). A minimal spontaneously broken hidden sector and its impact on Higgs boson physics at the large hadron collider. *Phys. Rev. D72*, 093007.

Schael, S. et al. (2006a). Precision electroweak measurements on the Z resonance. *Phys. Rept. 427*, 257, copyright 2006, with permission from Elsevier.

Schael, S. et al. (2006b). Search for neutral MSSM Higgs bosons at LEP. *Eur. Phys. J. C47*, 547–587.

Schael, S. et al. (2013). Electroweak measurements in electron-positron collisions at W-boson-pair energies at LEP. *Phys. Rept. 532*, 119–244, copyright 2013, with permission from Elsevier.

Schäfer, T. and E. V. Shuryak (1998). Instantons in QCD. *Rev. Mod. Phys. 70*, 323–426.

Schaile, D. and P. M. Zerwas (1992). Measuring the weak isospin of B quarks. *Phys. Rev. D45*, 3262–3265.

Schechter, J. and J. W. F. Valle (1980). Neutrino masses in $SU(2) \times U(1)$ theories. *Phys. Rev. D22*, 2227.

Schechter, J. and J. W. F. Valle (1982). Neutrinoless double-beta decay in $SU(2) \times U(1)$ theories. *Phys. Rev. D25*, 2951.

Schellekens, A. N. (2013). Life at the interface of particle physics and string theory. *Rev. Mod. Phys. 85*(4), 1491–1540.

Schellekens, A. N. (2015). The string theory landscape. *Int. J. Mod. Phys. A30*(03), 1530016.

Schiff, L. I. (1963). Measurability of nuclear electric dipole moments. *Phys. Rev. 132*(5), 2194–2200.

Schildknecht, D. (2006). Vector meson dominance. *Acta Phys. Polon. B37*, 595–608.

Scholberg, K. (2012). Supernova neutrino detection. *Ann. Rev. Nucl. Part. Sci. 62*, 81–103.

Schubert, K. R. (2015). T violation and CPT tests in neutral-meson systems. *Prog. Part. Nucl. Phys. 81*, 1–38.

Schwinger, J. S. (1957). A theory of the fundamental interactions. *Annals Phys. 2*, 407–434.

Seiberg, N. (1993). Naturalness versus supersymmetric non-renormalization theorems. *Phys. Lett. B318*, 469–475.

Semenov, A. (2016). LanHEP A package for automatic generation of Feynman rules from the Lagrangian. Version 3.2. *Comput. Phys. Commun. 201*, 167–170.

Senjanovic, G. and R. N. Mohapatra (1975). Exact left-right symmetry and spontaneous violation of parity. *Phys. Rev. D12*, 1502.

Severijns, N., M. Beck, and O. Naviliat-Cuncic (2006). Tests of the standard electroweak model in beta decay. *Rev. Mod. Phys. 78*, 991–1040.

Shelton, J. (2013). TASI 2012: Jet substructure. `arXiv:1302.0260 [hep-ph]`.

Shtabovenko, V., R. Mertig, and F. Orellana (2016). New developments in FeynCalc 9.0. *Comput. Phys. Commun. 207*, 432–444.

Simha, V. and G. Steigman (2008). Constraining the universal lepton asymmetry. *JCAP 0808*, 011.

Šimkovic, F., V. Rodin, A. Faessler, and P. Vogel (2013). $0\nu\beta\beta$ and $2\nu\beta\beta$ nuclear matrix elements, quasiparticle random-phase approximation, and isospin symmetry restoration. *Phys. Rev. C87*(4), 045501.

Sirlin, A. (1978). Current algebra formulation of radiative corrections in gauge theories and the universality of the weak interactions. *Rev. Mod. Phys. 50*, 573.

Sirlin, A. (1980). Radiative corrections in the $SU(2) \times U(1)$ theory: a simple renormalization framework. *Phys. Rev. D22*, 971–981.

Sirlin, A. and A. Ferroglia (2013). Radiative corrections in precision electroweak physics: a historical perspective. *Rev. Mod. Phys. 85*(1), 263–297.

Sirlin, A. and W. J. Marciano (1981). Radiative corrections to $\nu_\mu N \to \mu X$ and their effect on the determination of ρ^2 and $\sin^2\theta_W$. *Nucl. Phys. B189*, 442.

Sjöstrand, T. (2016). Status and developments of event generators. `arXiv:1608.06425 [hep-ph]`.

Sjöstrand, T. et al. (2015). An introduction to PYTHIA 8.2. *Comput. Phys. Commun. 191*, 159–177.

Skands, P. (2013). TASI 2012: Introduction to QCD. `arXiv:1207.2389 [hep-ph]`.

Skiba, W. (2011). TASI 2009: Effective field theory and precision electroweak measurements. `arXiv:1006.2142 [hep-ph]`.

Skyrme, T. H. R. (1962). A unified field theory of mesons and baryons. *Nucl. Phys. 31*, 556–569.

Slansky, R. (1981). Group theory for unified model building. *Phys. Rept. 79*, 1.

Sozzi, M. S. and I. Mannelli (2003). Measurements of direct CP violation. *Riv. Nuovo Cim. 26N3*, 1–110.

Spira, M., A. Djouadi, D. Graudenz, and P. M. Zerwas (1995). Higgs boson production at the LHC. *Nucl. Phys. B453*, 17–82.

Staub, F. (2014). SARAH 4 : A tool for (not only SUSY) model builders. *Comput. Phys. Commun. 185*, 1773–1790.

Steigman, G. (1979). Cosmology confronts particle physics. *Ann. Rev. Nucl. Part. Sci. 29*, 313–338.

Steigman, G. (2012). Neutrinos and big bang nucleosynthesis. *Adv. High Energy Phys. 2012*, 268321.

Sterman, G. (2004). QCD and jets. `arXiv:hep-ph/0412013`.

Sterman, G. F. and S. Weinberg (1977). Jets from quantum chromodynamics. *Phys. Rev. Lett. 39*, 1436.

Stodolsky, L. (1975). Speculations on detection of the neutrino sea. *Phys. Rev. Lett. 34*, 110.

Strassler, M. J. and K. M. Zurek (2007). Echoes of a hidden valley at hadron colliders. *Phys. Lett. B651*, 374–379.

Strigari, L. E. (2013). Galactic searches for dark matter. *Phys. Rept. 531*, 1–88.

Strumia, A. and F. Vissani (2006). Neutrino masses and mixings. `arXiv:hep-ph/0606054`.

Stueckelberg, E. C. G. (1938). Interaction energy in electrodynamics and in the field theory of nuclear forces. *Helv. Phys. Acta 11*, 225–244.

Sudarshan, E. C. G. and R. E. Marshak (1958). Chirality invariance and the universal Fermi interaction. *Phys. Rev. 109*, 1860–1862.

Suematsu, D. and Y. Yamagishi (1995). Radiative symmetry breaking in a supersymmetric model with an extra $U(1)$. *Int. J. Mod. Phys. A10*, 4521–4536.

Susskind, L. (1979). Dynamics of spontaneous symmetry breaking in the Weinberg-Salam theory. *Phys. Rev. D20*, 2619–2625.

Susskind, L. (2003). The anthropic landscape of string theory. `arXiv:hep-th/0302219`.

Svrcek, P. and E. Witten (2006). Axions in string theory. *JHEP 06*, 051.

Szleper, M. (2014). The Higgs boson and the physics of WW scattering before and after Higgs discovery. `arXiv:1412.8367 [hep-ph]`.

't Hooft, G. (1971a). Renormalizable Lagrangians for massive Yang-Mills fields. *Nucl. Phys. B35*, 167–188.

't Hooft, G. (1971b). Renormalization of massless Yang-Mills fields. *Nucl. Phys. B33*, 173–199.

't Hooft, G. (1973). Dimensional regularization and the renormalization group. *Nucl. Phys. B61*, 455–468.

't Hooft, G. (1974). A planar diagram theory for strong interactions. *Nucl. Phys. B72*, 461.

't Hooft, G. (1976a). Computation of the quantum effects due to a four- dimensional pseudoparticle. *Phys. Rev. D14*, 3432–3450.

't Hooft, G. (1976b). Symmetry breaking through Bell-Jackiw anomalies. *Phys. Rev. Lett. 37*, 8–11.

't Hooft, G. (1980). Naturalness, chiral symmetry, and spontaneous chiral symmetry breaking. *NATO Adv. Study Inst. Ser. B Phys. 59*, 135.

't Hooft, G. (1986). How instantons solve the $U(1)$ problem. *Phys. Rept. 142*, 357–387.

't Hooft, G. and M. J. G. Veltman (1972). Combinatorics of gauge fields. *Nucl. Phys. B50*, 318–353.

Takhistov, V. (2016). Review of nucleon decay searches at Super-Kamiokande. `arXiv:1605.03235 [hep-ex]`.

Tegmark, M., A. Vilenkin, and L. Pogosian (2005). Anthropic predictions for neutrino masses. *Phys. Rev. D71*, 103523.

Timmons, A. (2016). The results of MINOS and the future with MINOS+. *Adv. High Energy Phys. 2016*, 7064960.

Tishchenko, V. et al. (2013). Detailed report of the MuLan measurement of the positive muon lifetime and determination of the Fermi constant. *Phys. Rev. D87*(5), 052003.

Trócsányi, Z. (2015). QCD for collider experiments. `arXiv:1608.02381 [hep-ph]`.

Trodden, M. (1999). Electroweak baryogenesis. *Rev. Mod. Phys. 71*, 1463–1500.

Turck-Chieze, S. and S. Couvidat (2011). Solar neutrinos, helioseismology and the solar internal dynamics. *Rept. Prog. Phys. 74*, 086901.

Uzan, J.-P. (2011). Varying constants, gravitation and cosmology. *Living Rev. Rel. 14*, 2.

van Ritbergen, T., A. N. Schellekens, and J. A. M. Vermaseren (1999). Group theory factors for Feynman diagrams. *Int. J. Mod. Phys. A14*, 41–96.

van Ritbergen, T., J. A. M. Vermaseren, and S. A. Larin (1997). The four-loop beta function in quantum chromodynamics. *Phys. Lett. B400*, 379–384.

Veltman, M. J. G. (1977). Limit on mass differences in the Weinberg model. *Nucl. Phys. B123*, 89.

Vergados, J., H. Ejiri, and F. Simkovic (2012). Theory of neutrinoless double beta decay. *Rept. Prog. Phys. 75*, 106301.

Vicari, E. and H. Panagopoulos (2009). Theta dependence of $SU(N)$ gauge theories in the presence of a topological term. *Phys. Rept. 470*, 93–150.

Vilenkin, A. (1985). Cosmic strings and domain walls. *Phys. Rept. 121*, 263.

Vissani, F. (1998). Do experiments suggest a hierarchy problem? *Phys. Rev. D57*, 7027–7030.

Vogel, P. (2012). Nuclear structure and double beta decay. *J. Phys. G39*, 124002.

Vogel, P. and J. Engel (1989). Neutrino electromagnetic form-factors. *Phys. Rev. D39*, 3378.

Vos, K. K., H. W. Wilschut, and R. G. E. Timmermans (2015). Symmetry violations in nuclear and neutron β decay. *Rev. Mod. Phys. 87*, 1483.

Voutilainen, M. (2015). Heavy quark jets at the LHC. *Int. J. Mod. Phys. A30*(31), 1546008.

Wang, D. et al. (2014). Measurement of parity violation in electron-quark scattering. *Nature 506*(7486), 67–70.

Wang, L.-T. and I. Yavin (2008). A review of spin determination at the LHC. *Int. J. Mod. Phys. A23*, 4647–4668.

Weiler, T. J. (1982). Resonant absorption of cosmic ray neutrinos by the relic neutrino background. *Phys. Rev. Lett. 49*, 234.

Weinberg, D. H. et al. (2013). Observational probes of cosmic acceleration. *Phys. Rept. 530*, 87–255.

Weinberg, S. (1958). Charge symmetry of weak interactions. *Phys. Rev. 112*, 1375–1379.

Weinberg, S. (1962). Universal neutrino degeneracy. *Phys. Rev. 128*, 1457–1473.

Weinberg, S. (1967a). A model of leptons. *Phys. Rev. Lett. 19*, 1264–1266.

Weinberg, S. (1967b). Dynamical approach to current algebra. *Phys. Rev. Lett. 18*, 188–191.

Weinberg, S. (1973a). Current algebra and gauge theories. 1. *Phys. Rev. D8*, 605–625.

Weinberg, S. (1973b). Current algebra and gauge theories. 2. nonabelian gluons. *Phys. Rev. D8*, 4482–4498.

Weinberg, S. (1973c). General theory of broken local symmetries. *Phys. Rev. D7*, 1068–1082.

Weinberg, S. (1973d). Perturbative calculations of symmetry breaking. *Phys. Rev. D7*, 2887–2910.

Weinberg, S. (1975). The $U(1)$ problem. *Phys. Rev. D11*, 3583–3593.

Weinberg, S. (1976). Gauge theory of CP violation. *Phys. Rev. Lett. 37*, 657.

Weinberg, S. (1977). The problem of mass. *Trans. New York Acad. Sci. 38*, 185–201.

Weinberg, S. (1978). A new light boson? *Phys. Rev. Lett. 40*, 223–226.

Weinberg, S. (1979). Implications of dynamical symmetry breaking: an addendum. *Phys. Rev. D19*, 1277–1280.

Weinberg, S. (1980). Varieties of baryon and lepton nonconservation. *Phys. Rev. D22*, 1694.

Weinberg, S. (1982). Supersymmetry at ordinary energies. 1. Masses and conservation laws. *Phys. Rev. D26*, 287.

Weinberg, S. (1989a). Larger Higgs exchange terms in the neutron electric dipole moment. *Phys. Rev. Lett. 63*, 2333.

Weinberg, S. (1989b). The cosmological constant problem. *Rev. Mod. Phys. 61*, 1–23.

Weinberg, S. (2004). The making of the standard model. *Eur. Phys. J. C34*, 5–13.

Weinberg, S. (2009). Effective field theory, past and future. *PoS CD09*, 001.

Weinberg, S. (2013). Goldstone bosons as fractional cosmic neutrinos. *Phys. Rev. Lett. 110*(24), 241301.

Weisberger, W. I. (1965). Renormalization of the weak axial vector coupling constant. *Phys. Rev. Lett. 14*, 1047–1051.

Wess, J. (2009). From symmetry to supersymmetry. `arXiv:0902.2201 [hep-th]`.

Wess, J. and B. Zumino (1974a). A Lagrangian model invariant under supergauge transformations. *Phys. Lett. B49*, 52.

Wess, J. and B. Zumino (1974b). Supergauge transformations in four-dimensions. *Nucl. Phys. B70*, 39–50.

Weyl, H. (1929). Electron and gravitation. *Z. Phys. 56*, 330–352.

Wiebusch, M. (2015). HEPMath 1.4: A Mathematica package for semi-automatic computations in high energy physics. *Comput. Phys. Commun. 195*, 172–190.

Wietfeldt, F. E. and G. L. Greene (2011). Colloquium: The neutron lifetime. *Rev. Mod. Phys. 83*, 1173–1192.

Wilczek, F. (1978). Problem of strong P and T invariance in the presence of instantons. *Phys. Rev. Lett. 40*, 279–282.

Will, C. M. (2014). The confrontation between general relativity and experiment. *Living Rev. Rel. 17*, 4.

Willenbrock, S. and C. Zhang (2014). Effective field theory beyond the standard model. *Ann. Rev. Nucl. Part. Sci. 64*, 83–100.

Winstein, B. and L. Wolfenstein (1993). The search for direct CP violation. *Rev. Mod. Phys. 65*, 1113–1148.

Witten, E. (1981). Dynamical breaking of supersymmetry. *Nucl. Phys. B188*, 513.

Witten, E. (1982). Constraints on supersymmetry breaking. *Nucl. Phys. B202*, 253.

Witten, E. (1996). Strong coupling expansion of Calabi-Yau compactification. *Nucl. Phys. B471*, 135–158.

Witten, E. (2001). Lepton number and neutrino masses. *Nucl. Phys. Proc. Suppl. 91*, 3–8.

Witten, E. (2004). Perturbative gauge theory as a string theory in twistor space. *Commun. Math. Phys. 252*, 189–258.

Wobisch, M., D. Britzger, T. Kluge, K. Rabbertz, and F. Stober (2011). Theory-data comparisons for jet measurements in hadron-induced processes. `arXiv:1109.1310 [hep-ph]`.

Wolfenstein, L. (1964). Violation of CP Invariance and the possibility of very weak interactions. *Phys. Rev. Lett. 13*, 562–564.

Wolfenstein, L. (1978). Neutrino oscillations in matter. *Phys. Rev. D17*, 2369–2374.

Wolfenstein, L. (1983). Parametrization of the Kobayashi-Maskawa matrix. *Phys. Rev. Lett. 51*, 1945.

Wong, Y. Y. (2011). Neutrino mass in cosmology: status and prospects. *Ann. Rev. Nucl. Part. Sci. 61*, 69–98.

Wu, C. S., E. Ambler, R. W. Hayward, D. D. Hoppes, and R. P. Hudson (1957). Experimental test of parity conservation in beta decay. *Phys. Rev. 105*, 1413–1414.

Wu, S. L. (1984). e^+e^- physics at PETRA: the first 5-years. *Phys. Rept. 107*, 59–324.

Wu, T. T. and C.-N. Yang (1964). Phenomenological analysis of violation of CP invariance in decay of K^0 and $\bar{K}^0$. *Phys. Rev. Lett. 13*, 380–385.

Xing, Z.-z. and Z.-h. Zhao (2016). A review of $\mu - \tau$ flavor symmetry in neutrino physics. *Rept. Prog. Phys. 79*(7), 076201.

Yanagida, T. (1979). Horizontal gauge symmetry and masses of neutrinos. In *Workshop on the Baryon Number of the Universe and Unified Theories*, Tsukuba, Japan.

Yang, C.-N. (1950). Selection rules for the dematerialization of a particle into two photons. *Phys. Rev. 77*, 242–245.

Yang, C.-N. and R. L. Mills (1954). Conservation of isotopic spin and isotopic gauge invariance. *Phys. Rev. 96*, 191–195.

Yndurain, F. J. (2007). Elements of group theory. `arXiv:0710.0468 [hep-ph]`.

Yoshimura, M. (1978). Unified gauge theories and the baryon number of the universe. *Phys. Rev. Lett. 41*, 281–284.

Yukawa, H. (1935). On the interaction of elementary particles. *Proc. Phys. Math. Soc. Jap. 17*, 48–57.

Zahed, I. and G. E. Brown (1986). The Skyrme model. *Phys. Rept. 142*, 1–102.

Zee, A. (1973). Study of the renormalization group for small coupling constants. *Phys. Rev. D7*, 3630–3636.

Zee, A. (1980). A theory of lepton number violation, neutrino Majorana mass, and oscillation. *Phys. Lett. B93*, 389.

Zeldovich, Y. B. (1952). *Dok. Akad. Nauk. CCCP 86*, 505.

Zeller, G. P. et al. (2002). A precise determination of electroweak parameters in neutrino nucleon scattering. *Phys. Rev. Lett. 88*, 091802.

Zhang, C. and S. Willenbrock (2011). Effective-field-theory approach to top-quark production and decay. *Phys. Rev. D83*, 034006.

Zhang, Y., H. An, X. Ji, and R. N. Mohapatra (2007). Right-handed quark mixings in minimal left-right symmetric model with general CP violation. *Phys. Rev. D76*, 091301.

Zlatev, I., L.-M. Wang, and P. J. Steinhardt (1999). Quintessence, cosmic coincidence, and the cosmological constant. *Phys. Rev. Lett. 82*, 896–899.

Websites

- Particle Physics

PDG: The Review of Particle Physics, pdg.lbl.gov. Particle properties and many useful review articles in particle physics, astrophysics, and cosmology.

SPIRES: www.slac.stanford.edu/spires/hep.
Extensive data base of high energy physics publications.

ArXiv: arXiv.org. Archive of preprints in physics and related fields.

HepForge: www.hepforge.org. Collection of HEP software tools.

Cern Documents: cds.cern.ch. Books, proceedings, articles, preprints.

Online: library.web.cern.ch/particle_physics_information.
List of particle physics links.

SMB: www.sas.upenn.edu/~pgl/SMB2/. Supplements and corrections to this book.

- Tevatron and LHC Detectors/Results

CDF: www-cdf.fnal.gov/physics/physics.html.

D0: www-d0.fnal.gov/results.

ALICE: aliceinfo.cern.ch/Public/en/Chapter1/Chap1Physics-en.html.

ATLAS: atlas.ch, twiki.cern.ch/twiki/bin/view/AtlasPublic.

CMS: cms.cern.ch, cms.web.cern.ch/news/cms-physics-results.

LHC B: lhcb-public.web.cern.ch/lhcb-public.

LHC Physics Working Groups: twiki.cern.ch/twiki/bin/view/LHCPhysics.

- QCD, PDFs, and Cross Section Data

CTEQ: The Coordinated Theoretical-Experimental Project on QCD (Dulat et al., 2016), www.physics.smu.edu/scalise/cteq.

HepData: The Durham HepData Project, hepdata.cedar.ac.uk.

LHAPDF: Les Houches Accord PDFs (Buckley et al., 2015), lhapdf.hepforge.org.

MMHT: PDFs in various formats (Harland-Lang et al., 2015), www.hep.ucl.ac.uk/mmht.

NNPDF: Neural network PDFs (Ball et al., 2015), nnpdf.hepforge.org.

Cross Section/Luminosity Plots:
www.hep.ph.ic.ac.uk/~wstirlin/plots/plots.html.

- Event Generators, Analysis, Detectors (for general descriptions, see Mangano and Stelzer, 2005; Buckley et al., 2011; Sjöstrand, 2016; for the Les Houches Accord, see Alwall et al., 2007)

ALPGEN: `mlm.home.cern.ch/mlm/alpgen` (Mangano et al., 2003).

Herwig: `herwig.hepforge.org` (Bellm et al., 2016).

Isajet: `www.nhn.ou.edu/~isajet` (Paige et al., 2003).

Pythia: `home.thep.lu.se/Pythia` (Sjöstrand et al., 2015).

Sherpa: `sherpa.hepforge.org` (Gleisberg et al., 2009).

WHIZARD: `whizard.hepforge.org` (Kilian et al., 2011).

FastJet: Jet finding and analysis in pp and e^+e^- (Cacciari et al., 2012), `fastjet.fr`. See also, SpartyJet, `spartyjet.hepforge.org` (Delsart et al., 2012).

ROOT: Analysis package, `root.cern.ch`.

DELPHES: Detector simulator (de Favereau et al., 2014), `cp3.irmp.ucl.ac.be/projects/delphes`.

GEANT: Particle interactions in matter, `www.geant4.org`.

- Feynman Rules, Matrix Elements, Models

CalcHEP: `theory.sinp.msu.ru/~pukhov/calchep.html` (Belyaev et al., 2013).

CompHEP: `comphep.sinp.msu.ru` (Boos et al., 2004).

FeynArts: `www.feynarts.de`, `www.feynarts.de/formcalc`. Generation and calculation of Feynman diagrams (Hahn, 2001).

FeynCalc: `feyncalc.org`. Examples at `www.sns.ias.edu/~pgl/SMB`. Mathematica package for algebraic calculations (Shtabovenko et al., 2016).

FeynRules: `feynrules.phys.ucl.ac.be`. Calculation of Feynman rules (Alloul et al., 2014).

FORM: `www.nikhef.nl/~form`. Symbolic manipulation, e.g., for Dirac algebra (Kuipers et al., 2013).

HEPMath: `hepmath.hepforge.org` (Wiebusch, 2015).

LanHEP: `theory.sinp.msu.ru/~semenov/lanhep.html`. Feynman rules from Lagrangian density (Semenov, 2016).

LieART: `lieart.hepforge.org`. Lie algebras (Feger and Kephart, 2015).

MadGraph: `madgraph.hep.uiuc.edu` (Alwall et al., 2014).

HEPMDB: `hepmdb.soton.ac.uk`. High Energy Physics Model Database (Brooijmans et al., 2012).

SARAH: `sarah.hepforge.org`. SUSY and other models (Staub, 2014).

CheckMATE: `checkmate.hepforge.org`. LHC limits on model (Drees et al., 2015).

HEPfit: `hepfit.roma1.infn.it`. Direct and indirect limits.

- ## Electroweak Physics

GAPP: Global Analysis of Particle Properties (Erler, 1999), `www.fisica.unam.mx/erler/GAPPP.html`.

Gfitter: A Generic Fitter Project for HEP Model Testing (Flacher et al., 2009), `project-gfitter.web.cern.ch/project-gfitter`.

LEPEWWG: LEP Electroweak Working Group (LEP and SLC results), `lepewwg.web.cern.ch/LEPEWWG`.

ZFITTER: Electroweak radiative corrections and fits (Arbuzov et al., 2006), `zfitter.desy.de`.

- ## Higgs Physics

CPsuperH: MSSM Higgs with CP violation (Lee et al., 2013), `www.hep.man.ac.uk/u/jslee/CPsuperH.html`.

eHDECAY: Higgs boson decays in the SM, MSSM, and other extensions (Contino et al., 2014), `www.itp.kit.edu/~maggie/eHDECAY`.

FeynHiggs: MSSM Higgs (Heinemeyer et al., 2000), `wwwth.mppmu.mpg.de/members/heinemey/feynhiggs`.

HiggsBounds/HiggsSignals: LEP, Tevatron, and LHC results compared with arbitrary model (Bechtle et al., 2014a,c), `higgsbounds.hepforge.org`.

- ## Flavor, K and B Physics, CKM Matrix

CKMfitter: `ckmfitter.in2p3.fr` (Charles et al., 2005, 2015).

FLAG: Flavour Lattice Averaging Group (Aoki et al., 2017), `itpwiki.unibe.ch/flag`.

HFAG: Heavy Flavor Averaging Group, `www.slac.stanford.edu/xorg/hfag` (Amhis et al., 2014).

UTfit: `www.utfit.org` (Bona et al., 2006).

FlavorKit: Flavor extension of SARAH (Porod et al., 2014), `sarah.hepforge.org/FlavorKit.html`.

- ## Neutrinos

GLoBES: Global Long Baseline Experiment Simulator (Huber et al., 2007), `www.mpi-hd.mpg.de/personalhomes/globes`.

Neutrino oscillation industry: General neutrino links, `www.hep.anl.gov/ndk/hypertext`.

NuFIT: Global analysis (Esteban et al., 2017), `www.nu-fit.org`.

- ## Supersymmetry: spectrum, renormalization group evolution (see also Aguilar-Saavedra et al., 2006)

SoftSUSY: `softsusy.hepforge.org` (Allanach, 2002).

Spheno: `spheno.hepforge.org` (Porod and Staub, 2012).

SuSpect: `suspect.in2p3.fr/updates.html` (Djouadi et al., 2007).

- Dark Matter Abundance

DarkSUSY: `www.darksusy.org` (Gondolo et al., 2004).

MicrOMEGAs: `lapth.in2p3.fr/micromegas`
(Bélanger et al., 2015).

Index

C

D

G

I

J

T

T

Printed in the United States
by Baker & Taylor Publisher Services

Printed in the United States
by Baker & Taylor Publisher Services